"Theories, Techniques, Strategies"
For Spatial Planners & Designers

Murat Özyavuz (ed.)

"Theories, Techniques, Strategies" For Spatial Planners & Designers

Planning, Design, Applications

PETER LANG

Bibliographic Information published by the Deutsche Nationalbibliothek
The Deutsche Nationalbibliothek lists this publication in the Deutsche Nationalbibliografie; detailed bibliographic data is available online at http://dnb.d-nb.de.

Library of Congress Cataloging-in-Publication Data
A CIP catalog record for this book has been applied for at the Library of Congress.

ISBN 978-3-631-83922-5 (Print)
E-ISBN 978-3-631-85437-2 (E-PDF)
E-ISBN 978-3-631-85438-9 (EPUB)
E-ISBN 978-3-631-85439-6 (MOBI)
DOI 10.3726/b18405

Peter Lang – Berlin · Bern · Bruxelles · New York · Istanbul · Oxford · Warszawa · Wien

This publication has been peer reviewed.

www.peterlang.com

Contents

List of Contributors

Prof. Dr. Kağan GÜNÇE
Faculty of Architecture, Eastern Mediterranean University, Famagusta, Northern Cyprus (via Mersin 10 Turkey)

Assist. Prof. Dr. Damla MISIRLISOY
Faculty of Architecture, European University of Lefke, Lefke, Northern Cyprus (via Mersin 10 Turkey).

Assoc. Prof. Dr. Tuğba DÜZENLİ
Karadeniz Technical University, Faculty of Forestry, Department of Landscape Architecture, Trabzon, Turkey

Assoc. Prof. Dr. Serap YILMAZ
Karadeniz Technical University, Faculty of Forestry, Department of Landscape Architecture, Trabzon, Turkey

Assist. Prof. Dr. Emine TARAKÇI EREN
Afyon Kocatepe University, Faculty of Fine Arts, Interior Architecture Environmental Design, Trabzon, Turkey

Talia ÖZCAN AKTAN
Tekirdağ Namık Kemal University, Landscape Architecture PhD Programme, Tekirdağ, Turkey.

Assoc. Prof. Dr. Özge CORDAN
İstanbul Technical University, Faculty of Architecture, Department, of Interior Architecture, Istanbul, Turkey.

Assoc.Prof.Dr. Canan CENGİZ
Bartın University, Faculty of Engineering, Architecture and Design, Department of Landscape Architecture, Bartın, Turkey.

Res.Asst. Aybüke Özge BOZ
Bartın University, Faculty of Engineering, Architecture and Design, Department of Landscape Architecture, Bartın, Turkey.

Assoc. Prof. Dr. Elif BAYRAMOĞLU
Karadeniz Technical University, Faculty of Forestry, Landscape Architecture Department, Trabzon, Turkey

Res. Asst. Seyhan SEYHAN
Karadeniz Technical University, Faculty of Forestry, Landscape Architecture Department, Trabzon, Turkey

Lect. Dr. Ertan VARLI
Trakya University, Faculty of Architecture, Architecture Department, Edirne, Turkey

Assoc. Prof. Dr. Elif AKPINAR KULEKCI
Ataturk University, Faculty of Architecture-Design Department of Landscape Architecture, Erzurum, Turkey

Assoc. Prof. Dr. Feran ASUR
Van Yuzuncu Yil University, Faculty of Architecture-Design, Department of Landscape Architecture, Van, Turkey

Assoc. Prof. Dr. Sanem ÖZEN TURAN
Karadeniz Technical University, Faculty of Architecture, Urban and Regional Planning Department, Trabzon, Turkey

Assoc. Prof. Dr. Ömer K. ÖRÜCÜ
Suleyman Demirel University, Faculty of Architecture, Department of Landscape Architecture, Isparta, Turkey

Assist. Prof. Dr. Duygu DOĞAN
Pamukkale University, Faculty of Architecture and Design, Department of Landscape Architecture Denizli Turkey

Prof. Dr. Şükran ŞAHİN
Ankara University, Faculty of Agriculture, Department of Landscape Architecture Ankara, Turkey

Res. Ass. Duygu AKYOL
Karadeniz Technical University, Landscape Architecture Department, Trabzon, Turkey

Ass. Prof. Dr. Doruk Görken ÖZKAN
Karadeniz Technical University, Landscape Architecture Department, Trabzon, Turkey.

Assist. Prof. Dr. Volkan MÜFTÜOĞLU
Bursa Technical University, Faculty of Forestry, Landscape Architecture Department, Bursa, Turkey

Prof. Dr. Halim PERÇİN
Ankara University, Faculty of Agriculture, Landscape Architecture Department, Ankara, Turkey (retired faculty member)

Asssist. Prof. Dr. Ayşen ÇOBAN
Mehmet Akif Ersoy University, Faculty of Engineering Architecture, Department of Landscape Architecture, Burdur, Turkey

Assoc. Prof. Dr. Candan ŞAHİN
Suleyman Demirel University, Faculty of Architecture, Landscape Architecture Department, Isparta, Turkey

Res. Asst. Büşra ONAY
Suleyman Demirel University, Faculty of Architecture, Landscape Architecture Department, Isparta, Turkey

Esra MİRZA
Suleyman Demirel University, Faculty of Architecture, Landscape Architecture Department, Isparta, Turkey

Assoc. Prof. Dr. Gökçen BAYRAK
Trakya University, Faculty of Architecture, Department of Landscape Architecture, Edirne, Turkey

Res. Asst. M. Bihter BİNGÜL BULUT
Kirikkale University, Faculty of Fine Arts, Department of Landscape Architecture, Kirikkale, Turkey

Assoc. Prof. Dr. Gül YÜCEL
Istanbul Gelisim University, Engineering and Architecture Faculty, Department of Architecture, Istanbul, Turkey

Res. Asst. Seda ÖZLÜ
Karadeniz Technical University, Faculty of Architecture, Urban and Regional Planning Department, Trabzon Turkey

Res. Asst. Sinem DEDEOĞLU ÖZKAN,
Karadeniz Technical University, Faculty of Architecture, Urban and Regional Planning Department, Trabzon Turkey

Prof. Dr. Dilek BEYAZLI
Karadeniz Technical University, Faculty of Architecture, Urban and Regional Planning Department, Trabzon Turkey

Assoc. Prof. Dr. Serap YILMAZ
Karadeniz Technical University, Landscape Architecture Department, Trabzon, TURKEY

Assoc. Prof. Dr. Tuba DÜZENLİ
Karadeniz Technical University, Landscape Architecture Department, Trabzon, TURKEY

Assoc. Prof. Dr. Elif Merve ALPAK
Karadeniz Technical University, Landscape Architecture Department, Trabzon, TURKEY

Assist. Prof. Dr. Selin ARABULAN
Trakya University, Faculty of Architecture, Department of Architecture, Edirne, Turkey

Prof.Dr. H. Selma ÇELİKYAY
Bartın University, Faculty of Engineering, Architecture and Design, Bartın, Turkey

Assoc.Prof.Dr. Miray GÜR
Bursa Uludag University, Faculty of Architecture, Department of Architecture, Bursa, Turkey

Assist. Prof. Dr. Onur GÜNGÖR
Iskenderun Technical University, Faculty of Architecture, Department of Landscape Architecture, Iskenderun/Hatay, Turkey

Dr. Mehmet KARACA
Metroplan Consulting and Engineering, Department of Project Management, Ankara, Turkey.

Assist. Prof. Dr. Elif KARACA
Çankırı Karatekin University, Kızılırmak Vocational School, Department of Park and Garden Plants, Çankırı, Turkey.

Prof. Dr. Gülseven U. TÖNÜK
Gazi University, Faculty of Architecture, Department of Urban and Regional Planning, Ankara, Turkey.

Assoc. Prof. Dr. Serkan SINMAZ
Kırklareli University, Faculty of Architecture, Urban and Regional Planning Department, Kırklareli, Turkey

Assist. Prof. Dr. İlke CİRİTCİ
Istanbul Gelişim University, Faculty of Engineering and Architecture, Department of Architecture, Istanbul, Turkey.

MSc. Architect, PhD. Candidate Cenk CİHANGİR
Trakya University, Institute of Natural Science, Department of Architecture, Edirne, Turkey.

Assoc. Prof. Dr. Pınar KISA OVALI
Trakya University, Faculty of Architecture, Department of Architecture, Edirne, Turkey.

Ezgi ÜLKER BARIŞ
Architect, Guest Lecturer; Tekirdağ Namik Kemal University, Faculty of Fine Arts, Design and Architecture Department of Architecture, Tekirdag, Turkey.

Assoc. Prof. Dr. Alper SAĞLIK
Çanakkale Onsekiz Mart University, the Landscape Architecture Major of Architecture and Design Faculty, 17020 Çanakkale/Turkey

Busenur YAMAN
Çanakkale Onsekiz Mart University, School of Graduate Studies, Department of Landscape Architecture, 17020 Çanakkale/Turkey

Fatma AKBULUT
Çanakkale Onsekiz Mart University, School of Graduate Studies, Department of Landscape Architecture, 17020 Çanakkale/Turkey

Şura YILDIRIM
Çanakkale Onsekiz Mart University, School of Graduate Studies, Department of Landscape Architecture, 17020 Çanakkale/Turkey

Tuğçe ÖZDEMİR
Çanakkale Onsekiz Mart University, School of Graduate Studies, Department of Landscape Architecture, 17020 Çanakkale/Turkey

Assist. Prof. Dr. E. Seda ARSLAN
Süleyman Demirel University, Faculty of Architecture, Department of Landscape Architecture, Isparta, Turkey.

Assoc. Prof. Dr. Ayşe DURUKAN KOPUZ
Tekirdağ Namık Kemal University, Department of Architecture, Tekirdağ, Turkey.

Prof. Dr. Aybike Ayfer KARADAĞ
Düzce University, Faculty of Forest, Landscape Architecture Department, Düzce, Turkey

Assist. Prof. Dr. Demet A. DİNÇAY
Istanbul Technical University, Faculty of Architecture, Department of Interior Architecture, İstanbul, Turkey

Çağıl Y. TOKER
Istanbul Technical University, Faculty of Architecture, Department of Interior Architecture, İstanbul, Turkey

Elif B. ÖKSÜZ
Istanbul Technical University, Faculty of Architecture, Department of Interior Architecture, İstanbul, Turkey

Res. Asst. Sena SEMİZOĞLU
Istanbul Technical University, Faculty of Architecture, Department of Interior Architecture, İstanbul, Turkey

Reva ŞERMET
Tekirdağ Namık Kemal University Faculty of Fine Arts, Design and Architecture, Department of Landscape Architecture, Tekirdağ, Turkey, PhD. Student

Res. Asst. Nehar BÜYÜKBAYRAKTAR
Kırklareli University Faculty of Architecture, Department of Landscape Architecture, Kırklareli, Turkey

Assoc. Prof. Dr. Fürüzan ASLAN
Kırklareli University Faculty of Architecture, Department of Landscape Architecture, Kırklareli, Turkey

Lect. Banu GÖKMEN ERDOĞAN
Trakya University, Faculty of Architecture, Architecture Department, Edirne, Turkey

Assist. Prof. Dr. Gildis TACHIR
Trakya University, Faculty of Architecture, Architecture Department, Edirne, Turkey

Assist. Prof. Dr. Hatice KIRAN ÇAKIR
Trakya University, Faculty of Architecture, Architecture Department, Edirne, Turkey

Prof. Dr. Öner DEMİREL
Kırıkkale University, School of Fine Arts, Landscape Architecture Department, Yahşihan, Kırıkkale

Assist. Prof. Dr. Sultan Sevinç KURT KONAKOĞLU
Amasya University, Faculty of Architecture, Urban Design and Landscape Architecture Department, Amasya, Turkey

Assoc. Prof. Dr. Sema MUMCU
Karadeniz Technical University, Landscape Architecture, Trabzon, Turkey.

Assoc. Prof. Dr. Alev P. GÜRBEY
İstanbul University-Cerrahpaşa, Faculty of Forestry, Landscape Planning and Design Department, İstanbul, Turkey

Gülden KARABUDAK
Forest Bathing Guide, Certificated by the Forest Therapy Institute (FTI), İzmir, Turkey

Res. Ass. Demet Ülkü GÜLPINAR SEKBAN
Karadeniz Technical University, Faculty of Forestry, Landscape Architecture Department, Trabzon, Turkey

Assoc. Prof. Dr. A. Esra CENGİZ
Çanakkale Onsekiz Mart University, Faculty of Architecture and Design, Landscape Architecture Department, Çanakkale, Turkey

Zeynep Burke ÖZDEMİR[2]
Çanakkale Onsekiz Mart University, Graduate Education Institute, Landscape Architecture Department, Çanakkale, Turkey

Assoc. Prof. Dr. Müberra PULATKAN
Karadeniz Technical University, Faculty of Forestry, Department of Landscape Architecture, Trabzon, Turkey

Zeliha AKTÜRK
Karadeniz Technical University, Faculty of Architecture, Department of Urban and Regional Planning, Trabzon, Turkey

Assoc. Prof. Dr. Demet DEMİROĞLU
Kilis 7 Aralık University, Vocational School of Technical Sciences, Kilis, Turkey

Assoc. Prof. Dr. Ahmet BENLİAY
Akdeniz University, Faculty of Architecture, Department of Landscape Architecture, Antalya, Turkey

Assoc. Prof. Dr. Arzu ALTUNTAŞ
Siirt University, Faculty of Agriculture, Department of Landscape Architecture, Siirt, Turkey

Çağrı SAVAŞ
Canakkale Onsekiz Mart University, School of Graduate Studies, Çanakkale, Turkey

Assist. Prof. Dr. MEltem GÜNEŞ TİGEN
Tekirdağ Namık Kemal University, Faculty of Fine Arts, Design and Architecture, Department of Landscape Architectıre, Tekirdağ, Turkey

Prof. Dr. Murat ÖZYAVUZ
Tekirdağ Namık Kemal University, Faculty of Fine Arts, Design and Architecture, Department of Landscape Architectıre, Tekirdağ, Turkey

Assoc. Prof. Dr. Timur KAPROL
Kırklareli University, Faculty of Architecture, Department of Architecture, Kırklareli, Turkey

Assist. Prof. Dr. Oğuz ATEŞ
Kırklareli University, Faculty of Architecture, Department of Landscape Architecture, Kırklareli, Turkey

Assoc. Prof. Dr. Mehmet Kıvanç AK
Düzce University, Faculty of Forestry, Department of Landscape Architecture, Düzce, Turkey

Assoc. Prof. Dr. Tahsin YILMAZ
Akdeniz University, Faculty of Architecture, Department of Landscape Architecture, Antalya, Turkey

Assoc.Prof. Dr. Mehmet İNCEOĞLU
Eskisehir Technical University, Faculty of Architecture and Design, Department of Architecture, Eskisehir, Turkey

Bircan İNAN
PhD Student, Eskisehir Technical University, Faculty of Architecture and Design, Department of Architecture, Eskisehir, Turkey

Assist. Prof. Dr. Ayşe ÖZYETGİN ALTUN
Kırklareli University, Faculty of Architecture, Department of Urban and Regional Planning, Kırklareli, Turkey

Assist. Prof. Dr. Elif SAĞLIK
Çanakkale Onsekiz Mart University, Faculty of Architecture and Design, Landscape Architecture Department, Çanakkale, Turkey

Prof. Dr. Atilla GÜL
Süleyman Demirel University Faculty of Architecture, Landscape Architecture, Çünür/Isparta

Res. Asst. Beste KARAKAYA AYTİN
Trakya University, Faculty of Architecture, Department of Landscape Architecture, Edirne, Turkey

Assoc. Prof. Dr. Ömer Lütfü ÇORNACI
Recep Tayyip Erdoğan University, Department of Landscape Architecture, Rize, Turkey.

Assist. Prof. Dr. Simten SÜTÜNÇ
Siirt University, Faculty of Agriculture, Department of Landscape Architecture, Siirt, Turkey.

Assist. Prof. Dr. İsmet OSMANOĞLU
Trakya University, Faculty of Architecture, Department of Architecture, Edirne Turkey

Res. Asst. Makbulenur ONUR
Karadeniz Technical University, Faculty of Forest, Department of Landscape Architecture, Trabzon, Turkey

Res. Asst. Selver KOÇ ALTUNTAŞ
Karadeniz Technical University, Faculty of Architecture, Department of Interior Architecture, Trabzon, Turkey

Prof. Dr. Rüya YILMAZ
Tekirdağ Namık Kemal University, Faculty of Fine Arts, Design and Architecture, Department of Landscape Architecture, Tekirdağ, Turkey

Derya SERBEST ŞİŞEK
Tekirdağ Namık Kemal University, Faculty of Fine Arts, Design and Architecture, Department of Landscape Architecture, Tekirdağ, Turkey

Assist. Dr. Shirin IZADPANAH
Antalya Bilim University, Faculty of Fine Art and Architecture, Interior Architecture and Environmental Design Department, Antalya, Turkey

Kağan GÜNÇE and Damla MISIRLISOY

Developing 'Adaptive Reuse Strategies' for Heritage Buildings: A Case Study of Agios Panteleimon Monastery in Cyprus

Abstract: Heritage buildings play a significant role in transmitting ones culture to the generations to come. Adaptive reuse strategies can become a practical solution in preserving national heritages once their original roles diminish in time. When applying adaptive reuse strategies, the most crucial stage is to determine an appropriate function.

In order to decide the most appropriate adaptive reuse strategy for the heritage buildings all the factors must be taken into consideration holistically. The aim of the study is to propose a method that can be used in developing adaptive reuse strategies.

As the method, a proposed model, which is developed for adaptive reuse strategies, was investigated. Secondly, The Agios Panteleimon Monastery in Cyprus was observed through site survey as a case study and then, a matrix of various adaptive reuse strategies and alternatives have been formulated and discussed.

The aim of the proposed method is, therefore, of two-folds: to support decision makers in adaptive reuse projects and to inspire architects, designers, engineers, urban planners and restoration experts practicing strategies for adaptive reuse projects. Additionally, the method can be used as a guide for further researches in determining adaptive reuse strategies.

In adaptive reuse decision-making process, it is important to perform a detailed analysis of the heritage building. Incorporated onsite surveys are essential in understanding the behaviour of the heritage building, its background and the needs of the area. In-depth site surveys will ensure a full understanding of the area yielding a better choice of new use for each heritage building if their physical and original values are to be preserved. The continuity of the heritage buildings rely upon the correct choices made. In determining a new function, the factors that impact the adaptive-reuse-strategies need to be evaluated in terms of perspective and adaptability.

Keywords: Heritage buildings, adaptive reuse strategies, decision-making, holistic approach.

1 Introduction

Various internal and external factors induce adaptation of heritage buildings. Buildings become redundant for many reasons like changes in economic and industrial practices, demographic shifts or increasing cost of their maintenance. The main reason why they become redundant is because they are no longer suited for the function they were once built for [1]. Once they lost their original function, adaptive reuse is inevitable.

The decision-making process of reuse strategies requires a complex set of thinking over location of heritages, architectural assets and market trends [2]. The participants of the process should have a clear understanding on how to determine the most appropriate future use for buildings [3]. Conflicts often arise between professionals, public, government representatives, architects, architectural historians, developers and owners due to discrepancies in their perspectives [4].

Adaptive reuse of heritage buildings is a challenge for designers [5]. New use proposals should consider if the building is appropriate for this use or preserve the cultural significance of the heritage [1]. Deciding how to use a building is difficult since there are many actors in play. Finding the most appropriate function within a context is crucial in order to preserve and sustain its cultural significance, which would ensure its continuity. During adaptive reuse decision making, finding the most appropriate function for the building and considering the social, economic and physical benefits in different dimensions are important for a successful heritage adaptation.

The tendency is to focus on the technical issues of maintenance and to integrate the preservation activity with the planning of the general land use. However, little attention is paid to the role of management strategies and tactics during their adaptation. The conservation activity should be considered holistically and care should be given to their management for a sustainable heritage adaptation [6]. While assigning new functions, the existing fabric needs to be analyzed in depth. Surveys should be conducted during the decision-making process. The heritage values of the buildings should be considered, the needs of the district should be discovered and possible users of the buildings need to be identified [7].

The main issue in the process is the random decision of the functional changes without methodology, which may result in a short-term use thus ensuring no continuity. In order to find the most appropriate strategy for heritage buildings, the decisions taken for their new use should be based on analytic and scientific methods. Since conservation work is expensive, in the quest for funding economically, socially and physically sustainable buildings, the new uses should be compatible with the heritages. Unfortunately, there is a lack of clear and holistic methodology for an adaptive reuse decision-making of heritage buildings. Although there are some studies for adaptive reuse decision-making, there is a need of a holistic approach in adaptive reuse decision-making. There are existing models that propose methods for adaptive reuse decision-making just for specific functions or for a specific district. On the other hand, Burra Charter [8] suggests steps in planning for and managing the place of cultural significance under three main headings such as understanding significance, developing policies and managing in accordance with policy. However, the model proposes detailed guidance for the adaptive reuse of heritage buildings.

The study presents a holistic approach in identifying factors affecting the adaptive reuse process and proposes a model that would help decision-makers in developing strategies. The purpose of this research is to question the proposed method when developing adaptive reuse strategies of heritage buildings. In this context, a model proposal has been developed for heritage buildings that are abandoned, inappropriately functioned or disused.

As a method, the model for adaptive reuse strategies of heritage buildings was applied to a case study in Cyprus. New strategies and alternatives were developed for the Agios Panteleimon Monastery in Çamlıbel/Myrtou and a matrix was devised to discuss different proposed strategies. The model has been tested through the application of the case study and a matrix developed in order to propose a quantitative approach. The developed matrix compared three different adaptive reuse strategies. The selected case study was analysed in accordance with the proposed steps and factors. Adaptive reuse strategies were developed and new use alternatives and possibilities were discussed.

2 Literature Review on the Adaptive Reuse of Heritage Buildings

Adaptation of a building is defined as the process of transforming an existing building to accommodate new uses [9]. The term adaptation means the process of adjustment and alteration of a structure or building to fit new conditions. It can be defined as any work to a building to change its capacity, function or performance, which is beyond the maintenance [10]. It may include alterations, extensions, improvements and other works that modifies the building [11].

The Department of Environment and Heritage (2004) defines adaptive re-use as "a process that changes a disused or ineffective item into a new item that can be used for a different purpose" [12]. It is often described as a process by which structurally sound older buildings are developed for economically viable new uses [13]. Adaptive re-use is a process of change that requires creativeness from the architects involved in finding a way to fit a new function for the building and also all those involved in the process [14].

Adaptive reuse is not a recent phenomenon since reuse occurred in the past simply because demolition and the construction of new buildings would simply require more time, energy and money than reuse [14]. It has been started to discuss architecturally during the 1960s and 1970s due to the growing concern for the environment [15]. Much of the original inspiration came from examples of adaptive re-use documented by Sherban Cantacuzino in his two books, *New uses for old buildings* and *Re-architecture: old buildings/new uses* [16].

The building reuse and adaptation has become an increasing trend within the last decade [17] and there is a growing perception that it is cheaper to convert old buildings to new uses than demolish and rebuild [18]. Adaptive reuse of buildings is a viable alternative to demolition and replacement since it needs less energy and waste. Also it can offer social benefits by revitalizing familiar landmarks and giving them a new life [19]. New life into existing buildings has environmental and social benefits and helps to retain our national heritage; on the other hand, the focus on economic factors alone has contributed to destruction of buildings physical lives [20].

There are various reasons of adapting buildings such as conservation and sustainability. Firstly, reuse of an old building is more ecological than erecting a new building. Redevelopment activities spend more energy and expose more waste than adapting the existing building. Secondly, the historic and architectural significance of existing building can be satisfactory reasons why it should be sustained [11]. The main argument on the reasons of adaptive reuse is a very simple one: it is better to use what is there than making effort to build a new one [21].

At the end of the second millennium, in many European cities there is a clear sign that construction of new buildings is in decline; on the other hand, the reuse of existing buildings is becoming increasingly important. Today, society is more aware of ecological issues and the demolition of old buildings is perceived as an ecological waste and also as the eradication of local identity, of cultural heritage and of socio-economic values [22]. As a strategy to promote sustainability within the built environment, many buildings with cultural and historical values have been adapted and reused instead of demolition [2].

Adaptive reuse can transform heritage buildings into accessible and useable places by providing new places to be lived in a sustainable manner. The most successful adaptive reuse projects respects heritage significance of the building and also add a contemporary layer for its future [12].

3 Research Methodology

3.1 Model for Developing Strategies for Heritage Buildings

The model proposed steps to be followed in the decision-making process of adaptive reuse. It is applicable to any abandoned, inappropriately functioned or disused heritage building [23]. The adaptive reuse approaches may differ in any other context due to political and planning issues. However, the proposed model can be adapted and used for adaptive reuse strategies of any heritage buildings in any context while considering the policy and planning issues of the locations in concern.

The model can also be used in proposing new strategies for heritage buildings or evaluating the appropriateness of their new functions given that the adaptive reuse project was not successful and further identification of the problems required.

The model proposed a qualitative approach; never the less, the final vote depended on the decision variables of the actors in the adaptive reuse project, context of the heritage building, strategy issues of the related context, and so forth.

The final decision could only be interpreted and decided after considering pertinent variables.

According to a critical investigation, there are three options of adaptive reuse projects in the decision making process [23, 24]:

i. Possible new uses of abandoned or disused buildings.
ii. Sub-functions to support the existing function.
iii. Researching an appropriate building for the function in hand. For this category, an appropriate building should be chosen before the application of the model. Further research is required regarding the building and its listing by the authorities. For the listed ones, the regulations are the indicators of the decision-making.

3.2 Use of the Model

Deciding on a new use for heritage buildings requires a thorough consideration together with other factors as explained in previous steps. The model represents steps needed in the decision-making. However, the model was not a one-way approach as can be seen by the arrows in Fig. 1. If needed, before moving on it should be re-evaluated before making any decisions.

One of the biggest mistakes in adaptive reuse process of architectural heritages is the application of all conservation actions and finishing interventions then deciding on the new use. This may cause undesirable interventions and inappropriate additions to the heritage buildings by the users in order to adapt the buildings to their new use. User participation in decision-making is crucial. The original and contextual users are the potential users of the buildings but in most of the adaptive reuse projects, they are ignored although users' contributions, in decision-making, provide healthier decisions.

The objectives of the model were of two-folds: to evaluate the appropriateness of the re-functioned heritage building and to define problems faced in decision-making. The decision was interpreted according to the steps of the model. The second objective was to propose new strategies for the abandoned or disused heritage buildings requiring new lives. The model can be applied to different case studies in order to make appropriate decisions. The new proposals for the future use of the heritage buildings should not only focus on a single category of functions, but be the functional variations of the new proposals.

The decision of the new use alternatives is directly related to the decision-makers. These decision makers are stakeholders of the building. These actors discuss new strategies or as a second option an expert is hired to apply the decided model to buildings in concern and bring the strategy alternatives to the actors. The final decision is taken by the actors in the light of the proposals presented by the decision-making experts.

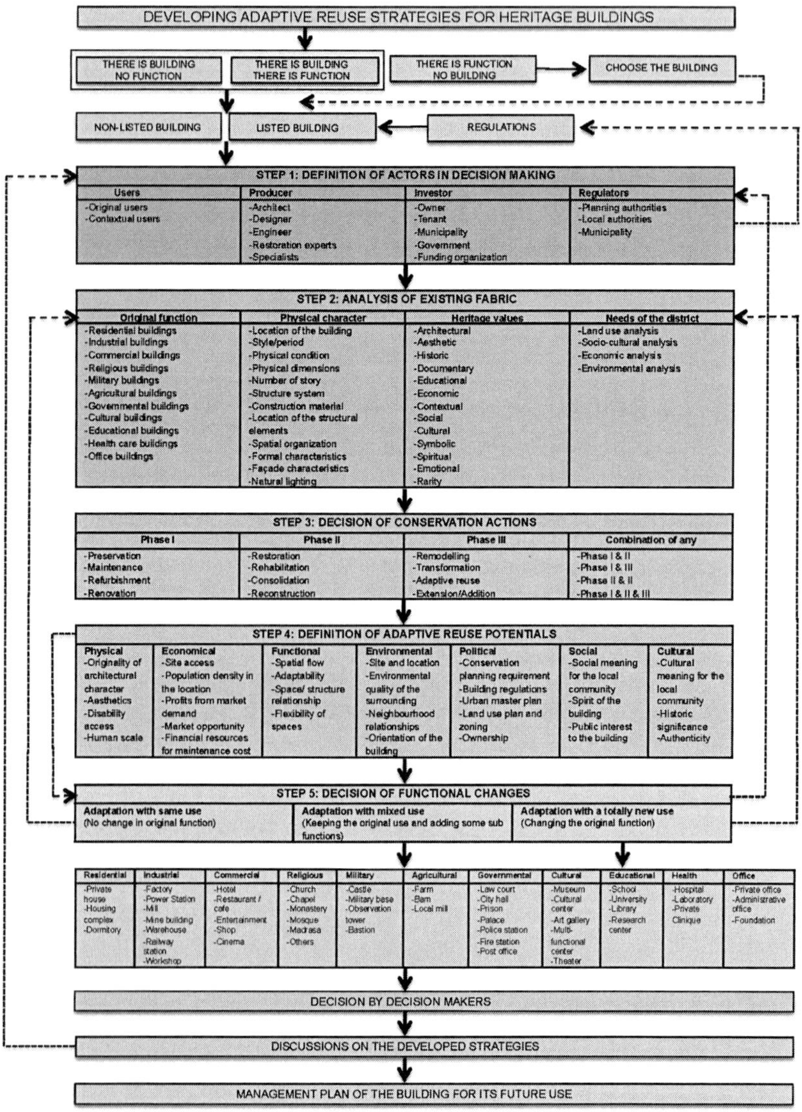

Fig. 1. A model for developing adaptive reuse strategies of heritage buildings [23, 24]

This model is suitable for all locations and listed or non-listed buildings. It can be used by architects, designers, engineers, urban planners and restoration experts in the professional field. Additionally, it can shed light onto future researches.

4 Political and Economic Issues in Conservation of Heritage Buildings in Cyprus

The island is divided into two territories as North and South separated by a UN Buffer Zone since 1974. The buffer zone cuts through the city of Nicosia and divides the whole island from northwest to southeast. A 180 km line of demarcation separates Greek and Turkish Cypriots. The division is the main reason for the island's on-going problems in terms of restricting development and imposes issues for future planning. In spite of the division, there are efforts on both sides for the conservation and revitalization of architectural heritages. The division limits development and creates diverse problems for planning [25].

Starting from 1986 to the present day, many projects have been conducted funded by *The United States Agency for International Development* (USAID) and the European Union through *United Nations Development Programme (*UNDP) [26]. Since 2001, the European Union funded Partnership for the Future Programme (UNDP-PFF) aims at contributing to the peace and confidence building process in Cyprus through different levels of interventions. UNDP-PFF started working in Cyprus under a bi-communal programme focusing on the rehabilitation of Nicosia; however, in 2004, the programme was extended to other cities namely Famagusta and Kyrenia. In recent years UNDP-PFF's focus returned to bi-communal projects to support the Technical Committee on Cultural Heritages (TCCH) for the preservation and promotion of the immovable cultural heritages of Cyprus [27]. Generally, TCCH gives more importance to the conservation of abandoned religious heritage buildings in both sectors since it is seen as a threat to the peace and reunion of the island.

Agios Panteleimon, which is selected as the case study of the research is a monastery located in the Northern territory. The different religious beliefs of the two communities living on the island adversely affected conservation of religious heritage buildings. Before the division of the island, Greek Cypriots living in villages were the stakeholders, however, the current stakeholders are Turkish Cypriots. Owing to the current stakeholders of the village and lack of budget to restore heritage buildings, the monasteries and churches are not used today.

5 Application of the Model on a Case Study

Çamlıbel/Myrtou is an ancient village, on the east outskirts, with the remains of the Byzantine village of Margi. It is located 28 km southeast of Kyrenia city next to Kyrenia – Morphou road (Fig. 2). The village is located on an intersection that connects roads from Nicosia, Kyrenia and Morphou. It has neighbourhood to other villages.

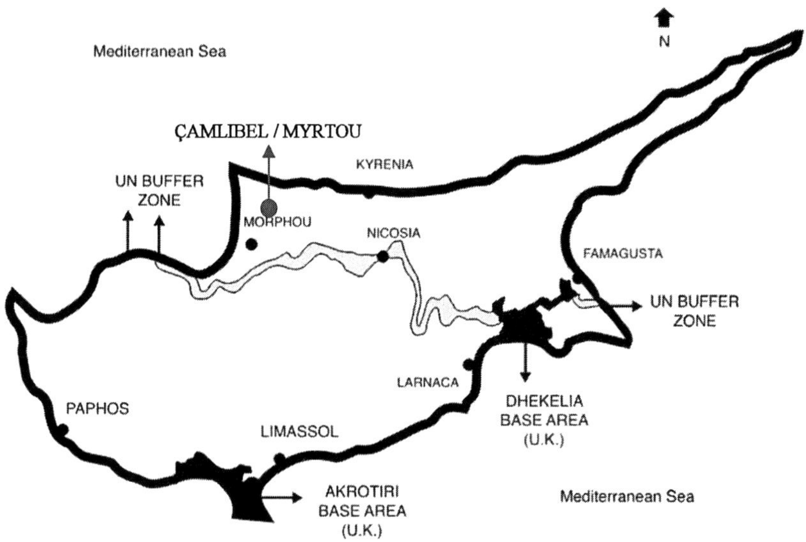

Fig. 2. Location of Çamlıbel/Myrtou on the map

Its principal monument is the Monastery of Agios Panteleimon, which was the residence of the Bishop of Kyrenia from 1571 to 1921 [28]. Agios Panteleimon Monastery in Çamlıbel/Myrtou is a disused and abandoned building (Fig. 3). Adaptive reuse strategies were developed and new use alternatives and possibilities were discussed.

The renovation works on the monastery started in October 2015 and completed in 2016. The conservation project was selected by the Technical Committee on Cultural Heritage; (TCCH) United Nations Development Program- Partnership for the Future (UNDP-PFF) supported the project financially [27].

The monastery is a complex consisting of 5 separate blocks of buildings, including a church, dorms, toilets, guest rooms and a fountain. Agios Panteleimon Monastery is unique in terms of its symbolic meaning.

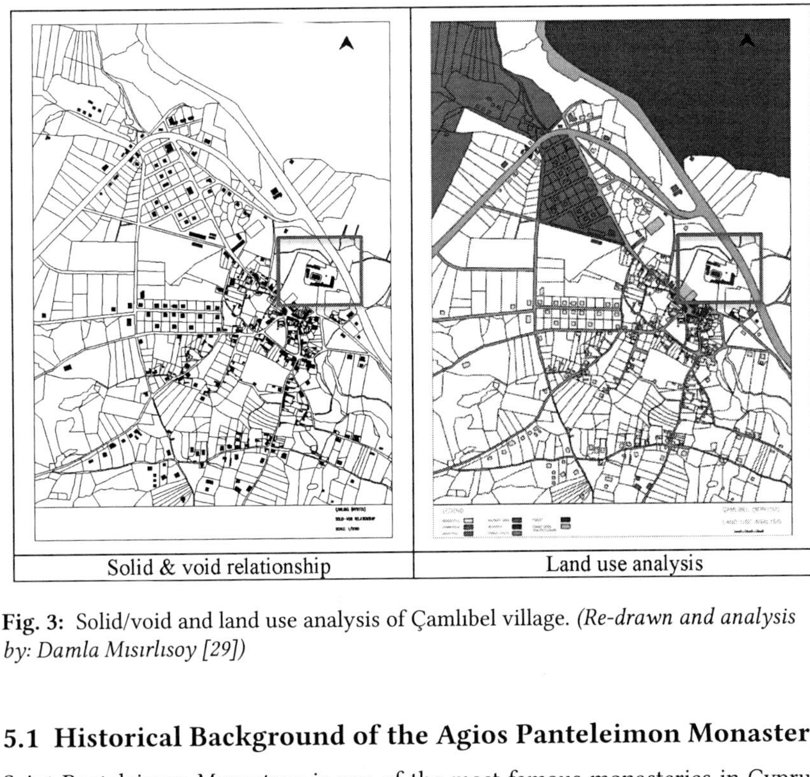

Fig. 3: Solid/void and land use analysis of Çamlıbel village. *(Re-drawn and analysis by: Damla Mısırlısoy [29])*

5.1 Historical Background of the Agios Panteleimon Monastery

Saint Panteleimon Monastery is one of the most famous monasteries in Cyprus. The monastery takes its name from Agios Panteleimon, a saint martyrized at Nicomedia by Maximian, A.D. 303. He was born in 275, and was the son of a rich Pagan father and a Christian mother. The monastery was used by the Bishop of Kyrenia since he had no hegemon of his own [30]. Agios Panteleimon was the patron of saint of doctors. In his pagan youth he studied medicine at Constantinople, and after his conversion to Christianity he cured the deaf, the blind and the lame by prayer alone. Following his martyrdom, his healing powers were said to have transferred into his silver gilt icon at the monastery [31]. The Monastery of St. Panteleimon is the principal monument of the village and its history of the church goes back to the 5th century AD. It is a listed building and its ownership belongs to a foundation (EVKAF) like other churches and monasteries located in the Northern part of Cyprus. Originally, it was a monastery made up of 3 detached buildings and a church in the middle of these three buildings. Block A is the Southern dorms of the monastery, and also the parliament building of the priests. The shorter leg of the L shape was the refectory. Block B in the north was also a dorm and Block C was the common toilets of the monastery. There is a fountain and aqueducts in the courtyard (Fig. 4).

AGIOS PANTELEIMONAS MONASTERY		
General Information		
Location: Çamlıbel, Cyprus, **Date of Construction:** 1600s		

Exterior view*	Exterior view*	Courtyard*

Toilets *	Church *	Church Interior*

Extension*	Southern Dorm*	Northern Dorm*
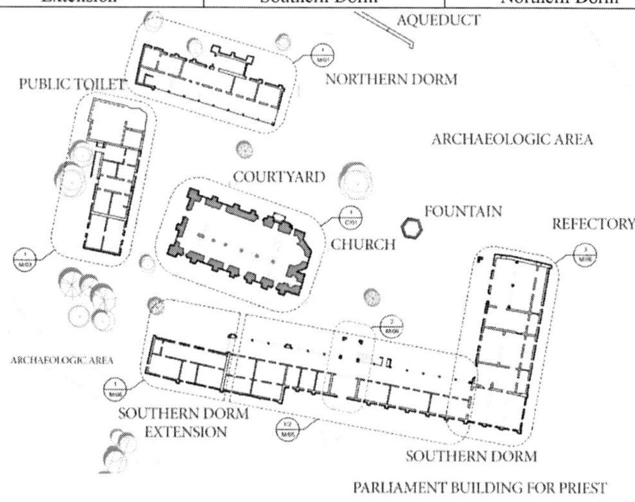		
Plan – Source: UNDP-PFF [27] / *Photos taken by Author, 2015		

Fig. 4: Visual materials of the Agios Panteleimon Monastery

5.2 Application of the Model on Agios Panteleimon Monastery

The proposed model was applied step by step to the monastery and new strategies were developed for its future use.

Strategy 1 – Adaptation with the same use:

The original users and investors of the heritage strongly believed in preserving the original use of the building. Original users being the Greek Cypriots living on the other side of the island apprehended the building pertinent to their cultural heritages in terms of spiritual and emotional values. The current stakeholders of the village, on the other hand, perceived the building physically as part of a cultural heritage of the district since religiously it has no value to them.

The building was used by the Bishop of Kyrenia from 1571 to 1921 as residence and if returned to this original function, some risks for the future of the heritage building were at stake. First of all, if the building was used as a monastery, lack of use was one of them since the current stakeholders of the village were Muslims. The building then may not be used as effectively and would cause lack of financial sustainability for future maintenance costs. Secondly, using the building as a monastery would engender privacy problems as the locals and tourists might request to visit the building as a preserved monument thus causing further issues for the stakeholders of the monastery. Thirdly, the building had a potential to be used with a function that can create facilities for the current stakeholders especially for the young people and women due to the exiguity of socio-cultural activities in the village. If it is used as a monastery, the building would have no contribution to the socio-cultural development of the area. Therefore, changing the use of the building and selecting the most appropriate new use played an important factor in the economic development of the district thus helping towards new investments.

Strategy 2 – Adaptation with mixed use:

The second strategy for the building was to keep the church with its original function but attribute new uses for the monastic buildings as supporting functions. The church continues to be open for rituals and prayers of the visitors yet its other buildings could accommodate socio-cultural activities of the district.

The monastic buildings could be used as cultural activity centres. Workshops could be organized for children and young people. Art and craft courses for the women living in the village could be administered. Their produce could be exhibited and sold to visitors. Moreover, collaborations could be made with nearby universities as some parts of the building could cater educational activities for their students.

Places like cafés or restaurants should also be provided for the visitors and users of the building. The former refectory of the monastery could be used for this purpose. A block of the monastic buildings could be converted into a boutique hotel and used for accommodation purposes.

Excavations needed to be done in the archaeological area around the site and the art works found exhibited in the building. Brief historical background information of the monastery should be given to the visitors since the building has distinct narrative. These kinds of activities would promote the economic sustainability of the building for its future maintenance cost. Additionally, they would contribute to the development of the district and its economic growth.

Strategy 3 – Adaptation with a totally new use:
A totally new use could be proposed for the future use of the complex, none the less, according to the factors discussed in the model; this was not a recommended approach in preserving the merit of the heritage building since a totally new use might harm the originality of the building. The physical conservation of the heritage building was not enough, the character and spirit of the building should also be preserved. Throughout the globe, there are many examples of the adaptive reuse of ecclesiastic buildings with obsolete new uses. The local authorities face problems regarding the conservation and rehabilitation of these religious buildings; as a result, they rent or sell them to private users on the basis that they stay the true controlling authority of their future uses. This of course is decided after analysing their heritage values. Every religious building cannot be attributed with a new use. However, if a building is disused, in order to sustain it, a new function is inevitable.

5.3 Findings of the Application of the Case Study

After a thorough discussion of the strategies, in order to attain a quantitative approach, a table which helped comparing the three proposed strategies and select the most appropriate one for Agios Panteleimon Monastery was made. All factors in adaptive reuse decision – making are listed and their effects on the abovementioned strategies are discussed. These effects are classified in three categories: negative, positive and neutral.

A factor may have a negative effect on the proposal of the new use alternatives in decision-making process thus forming a barrier in the adaptation process. On the other hand, a factor may have a positive effect on the relevant strategy and become the driving force in the adaptive reuse of the heritage building. What is more, a factor may not affect any related strategy. These three effects of the factors and their relationships between the developed strategies are represented in Tab. 1.

The table helps to evaluate the three proposed strategies in a quantitative and rational approach. Negative and positive effects of the factors should be taken into consideration and proposals should be re-interpreted holistically.

When all factors are evaluated holistically, it is evident that the building if used with the original use, is going to be under a financial threat in future. On the other hand, keeping the church's original function and proposing other appropriate functions for the monastic buildings may bring satisfaction to both communities while preserving the values of the monastery. Lastly, proposing a completely new use may damage the authenticity and heritage values of the monastery since it is an important religious building and demand in worshipping here remains.

As seen in the analysis, when the negative and positive effects of the proposed strategies are taken into consideration, the most appropriate function for the building is its multi- functional adaptation in compliance with cultural and educational uses for a sustainable heritage adaptation. However, in order to make a final decision, the contributions of all stakeholders and decision makers are essential.

Before the final decision, all actors should come together, discuss relevant issues and reach to a common decision as to which appropriate strategy for the future of the heritage building should be chosen. Community participation and stakeholders' collaboration is required in developing better adaptive reuse strategies for a sustainable heritage adaptation.

It is of great important to have an in-depth analysis on the needs of the district if an appropriate function is to be assigned for the heritage building. Probing heritage buildings independently of their setting is wrong. The conservation and promotion of the cultural significance of the region should also be taken into consideration. The successful adaptive reuse project of the monastery can act as a catalyst in the district and precipitate other conservation activities in the region. The final decision on the new use of the heritage buildings should be given prior to any interventions; the decision on the conservation actions is the last step in the model. Initially, the new use should be decided and then the building should be designed according to the space requirements of this new use. However, for the monastery, the conservation works started before determining any future use for the building, which might result in undesirable interventions by the users in the future (Tab. 1).

Tab. 1. Defining the factors as barriers or drivers for Agios Panteleimon Monastery

RELATIONSHIP BETWEEN FACTORS IN DECISION-MAKING AND STRATEGIES FOR AGIOS PANTELEIMON MONASTERY				
FACTORS AFFECTING THE STRATEGY		STRATEGY 1: Adaptation with Same Use	STRATEGY 2: Adaptation with Mixed Use	STRATEGY 3: Adaptation with Totally New Use
1. ACTORS IN DECISION MAKING	Users (Original Users / Current Users)	**-** / 0 / + -Current stakeholders of the village belong to another religion; in this respect, they will not use the monastery. -On the other hand, original users of the monastery wish to see the heritage building with its original use and they do not agree on changing its function.	- / 0 / **+** -Current stakeholders of the village want the building to be used with a function that will help socio-cultural and economic contribution of the village.	**-** / 0 / + -Neither original nor current stakeholders of the village agree on the use of the monastery with a totally new use.
	Producer	**-** / 0 / + -Producers of the project do not agree on the adaptation of the monastery with the same use since the project will not be sustainable.	- / 0 / **+** -Producers of the project agree on the adaptation with multi-functional use by keeping the church for the rituals and re-functioning of the monastic buildings.	**-** / 0 / + -Producers of the project do not agree on the adaptation of the monastery with a totally new use.
	Investor	**-** / 0 / + -Investor of the project does not agree on the adaptation of the monastery with the same use.	- / 0 / **+** -Investor of the project agrees on the adaptation with the mixed use by keeping the church for the rituals and re-functioning of the monastic.	**-** / 0 / + -Investor of the project does not agree on the adaptation of the monastery with a totally new use.
	Regulator	**-** / 0 / + -Regulators of the project do not agree on the adaptation of the monastery with the same use. -On the other hand, Bishop of Kyrenia wants to use the building as a monastery again.	- / 0 / **+** -Regulators of the project agree on the adaptation of the monastery with the mixed use by keeping the church for rituals and re-functioning the monastic buildings.	**-** / 0 / + -Investor of the project does not agree on the adaptation of the monastery with a totally new use.
2.ANALYSIS OF EXISTING FABRIC	Original Function	**-** / 0 / + -It is a religious building so due to its values adaptive reuse options is more limited than the other types of buildings.	**-** / 0 / + -It is a religious building so due to its values adaptive reuse options is more limited than the other types of buildings. The church might be kept as it is and monastic buildings can be re-functioned.	**-** / 0 / + -Reusing the monastery with a totally new use may harm the values and its originality.
	Physical Character	- / **0** / + -Using the building with its original use is more suitable in terms of the physical character of the building.	- / 0 / **+** -Keeping the church with its original use is an appropriate approach in terms of preservation of its physical character but suitable functions should be given for the monastic buildings.	- / 0 / **+** -If the appropriate functions are assigned to the buildings according to the physical character, the originality of the heritage will be preserved.
	Heritage Values	**-** / 0 / + -Original users (Greek Cypriots) that live in the other part of the island see the building as a part of their cultural heritage in terms of spiritual and emotional values.	- / 0 / **+** - By the adaptation of the monastery with multi-functional use like keeping the church for the rituals and re-functioning the monastic buildings may preserve the heritage values of the building.	**-** / 0 / + -There are important values of the heritage and a totally new use may harm its originality. The physical conservation of the heritage building is not enough; also the spirit of place should be preserved.
	Needs Of The District	**-** / 0 / + -Using the building as a monastery does not answer the needs of the district.	- / 0 / **+** - There is lack of socio-cultural activities for the current stakeholders of the village especially for the young people and the women.	**-** / 0 / + -The building has a potential to be used with a totally new function that may create facilities for the stakeholders of the village and answer the needs of the district.

Tab. 1. Continued

3.CONSERVATION ACTIONS		- **0** +	**-** 0 +	**-** 0 +
		-Conservation actions do not affect the decision since the function will not change in this strategy.	-Final decision of the new use for the building should be decided before starting to apply the necessary interventions. For the monastery, the restoration works started but the future use of the building is not determined yet. Firstly, the new use should be decided and then the building should be designed according to the space requirements of the new use. This may cause inappropriate interventions by the users.	- The new use should be decided before starting to apply necessary interventions.
4. ADAPTIVE REUSE POTENTIALS (+)	Physical Potential	- **0** +	- 0 **+**	- 0 **+**
		-Physical potential does not affect the decision since the function will not change in this strategy.	-The building is important in terms of being the only monastery in the district. Accordingly, the proposed function should be appropriate in order to preserve its originality.	-The building is important in terms of being the only monastery in the district. Accordingly, the proposed function should be appropriate in order to preserve its originality.
	Economic Potential	**-** 0 +	- 0 **+**	- 0 **+**
		-The use of the building as a monastery forms a risks for the future of the heritage. It may cause lack of economic sustainability of the building for the future maintenance cost.	-New activities may support economic sustainability of the building for the future maintenance costs. Also, it will contribute to the promotion of the district and its economic development.	-Changing the use of the building and selecting the most appropriate new use may contribute to economic development of the district and new investments can be made.
	Functional Potential	- **0** +	**-** 0 +	**-** 0 +
		-Functional potential do not affect the decision since the function will not change in this strategy.	-The spaces of the monastic buildings are small spaces attached to each other since they were designed for accommodation purposes. These space divisions should be kept as they are due to the structural issues and also preservation of architectural characteristics of the heritage buildings. However, these considerations create problems in terms of adaptability of the building for the new space requirements. Appropriate new uses for the monastic buildings should be proposed.	-The spaces of the monastic buildings are small spaces attached each other since they were designed for accommodation purposes. These space divisions should be kept as it is due to the structural issues and also preservation of architectural characteristics of the heritage buildings. However, these considerations create problems in terms of adaptability of the building for the new space requirements. Accordingly, appropriate new uses for each block should be proposed.
	Environmental Potential	**-** 0 +	- 0 **+**	- 0 **+**
		-The building has environmental potentials to be used with various functions as mentioned in other two strategies. Using the building as monastery is neglecting these potentials.	-The monastery has potentials due to its location. It is very close to the village settlement and also next to the main road. Also, there are important landmarks in the district, which attracts many visitors.	-The monastery has potentials due to its location. It is very close to the village settlement and also next to the main road. There are important landmarks in the district all of which attracts tourists.
	Politic Potential	**-** 0 +	**-** 0 +	**-** 0 +
		-The biggest issue in adaptive reuse of the monastery is the political issues since it is a religious building. The future use of the building is beyond the decision of the owner. It will be a political decision since it is belongs the Bishop of Kyrenia.	-The political issues are the barriers for multi-functional adaptation of the monastery.	-The political issues are the barriers for the adaptation of the monastery with a totally new use.

(continued)

Tab. 1. Continued

	Social Potential	- / 0 / **+**	- / 0 / **+**	- / 0 / **+**
		The monastery has a social meaning for the community, especially for the original users. There is a public interest towards the building even today; Greek Cypriots come to visit the monastery even it is in a demolished condition. The emotional and spiritual values of the building are important indicators of the decision-making as seen in the interview results. Coming to visit the church is a kind of social activity for the former stakeholders of the village.	-Today, the building lost its public interest since the current stakeholders are Muslims and they do not visit church. The spiritual and emotional values of the building do not make sense to them. They believe the monastery should be preserved as a cultural heritage of the community. They want the building to be used for social and cultural activities for the stakeholders of the village since there is not much today.	-There is lack of socio-cultural activities for the current stakeholders of the village.
	Cultural Potential	**-** / 0 / +	- / 0 / **+**	- / 0 / **+**
		-The cultural potentials of the building and its benefits for the current stakeholder of the village should be taken into consideration. There is no contribution to the cultural development of the district with the original function.	-If it is used as a monastery, the building will not be able to contribute to the cultural development of the district.	-The building has a historic significance and authenticity for the current users. It is the common cultural heritage of both communities and must be physically preserved by respecting to the original function its spiritual value.
5.DECISION OF FUNCTIONAL CHANGES		**-** / 0 / +	- / 0 / **+**	**-** / 0 / +
		- If the building is used as a monastery, there is a threat of lack of use since the current stakeholders of the village are Muslims. The building might not be used effectively and this may cause lack of economic sustainability for its future maintenance cost.	-Changing the use of the building and selecting the most appropriate new use may contribute to the economic development of the district and new investments can be made. The church might be kept open for the rituals and prayers of the visitors; on the other hand, other buildings can be used as socio-cultural activities. The building may be used for cultural tourism and ensure experiencing the history, folklore and culture of the district.	-Changing the function of the religious building completely is not an appropriate approach due to its heritage values.
FINAL COMMENTS		Using the building as monastery is not an appropriate decision in social and economic manners. Also, it means ignorance of the context and stakeholders of the village.	Keeping the church as it is and proposing appropriate functions to the monastic buildings may satisfy both communities and also preserve the values of the monastery.	Proposing a completely new use may damage the authenticity and heritage values of the monastery since it is a religious building.
CODES: (-), (0), (+)		(-): The mentioned factor has a negative effect on the related strategy and it may form a barrier against the adaptive reuse of the heritage building. (0): The mentioned factor does not affect the related strategy. (+): The mentioned factor has a positive effect on the related strategy and it might become a driving force in the adaptive reuse of the heritage building.		

It is important to appraise the potentials of the heritage buildings; especially the economic and social potentials should not be ignored for their sustainability. Management and sustainable renovation of heritage buildings, in northern part of the island, incorporate problems. For instance, buildings are physically preserved yet lack living functions due to political issues affecting final decisions. For the continuity of the buildings, international principles are applicable in the adaptive reuse of architectural heritages and the decisions should not be based solely on political matters.

Predominantly, the Department of Antiques and Museums or EVKAF is the property owner of heritage buildings in northern part of the island and all interventions regarding heritage buildings are under their supervision. There are guidelines and regulations in conservation strategies of heritage buildings, however, there are no policies proposing their adaptive reuse strategies. Therefore, new policies should be elaborated in order to develop adaptive reuse strategies while decision makers and community are educated on these matters.

It is important to fully comprehend the significance of the built heritages as living assets rather than just physical assets and preserve them accordingly. It is not possible to convert all heritage buildings into museums, art galleries, and cultural centres. None the less, successful adaptive reuse projects should also be economically sustainable. If there is no threat against their significance, functions like shops, department stores, boutique hotels, cafes or restaurants could be attributed to heritage buildings.

6 Conclusion

Heritage buildings are our identity as they transfer knowledge to new generations regarding lifestyles and cultures of people pertinent to that era. They are the reflections of their time and safeguard the history of communities. A comprehensive conservation and preservation of architectural heritages and giving them new functions according to their location, size, and potential can make heritage buildings sustainable. By the adaptive reuse of cultural heritages, a framework of cultural significance is created.

The life span of a physical structure is longer than its function; therefore, it is necessary to assign a new use for the building once its original function demeans in time. Desolation of heritages causes irrevocable havoc to structures. Nevertheless, any given new function must be appropriate in terms of preserving the cultural significance of their historic fabric. When buildings are reassigned disparate functions, the new use and the interventions should preserve the originality and architectural character of the buildings in order not to give wrong or missing information to future generations.

Before deciding on how to use an abandoned or disused heritage building, a clear definition of all factors affecting adaptive reuse decision-making is of essence. Adaptive reuse of heritage buildings is a challenging process since there are many factors involved in an integrated approach, for example, original function, physical characteristics, heritage values, needs of the district and potentials of the building should all be holistically contemplated on. In order to question the success of an adaptive reuse project, it is far from being enough to evaluate the project in terms of conservation principles only. An in-depth analysis of existing heritage buildings is fundamental prior to giving them new functions. In a decision-making process, a comprehensive understanding of the district and its context is important and should be secured by surveys conducted on site. It is not enough only to retain the building physically, but the originality of the building must also be preserved by giving an appropriate function and passing it on to appropriate users.

The research proposed a comprehensive methodology for the development of adaptive reuse strategies of heritage buildings and further provided a complete review on the factors affecting decision-making. A model for developing strategies for heritage buildings including industrial heritage buildings or buildings of the Modernism period was proposed all of which were abandoned, inappropriately functioned or disused.

In determining the most appropriate strategy of a heritage building, the adaptive reuse decision-making process should be placed under detailed scrutiny. All the factors of adaptive reuse should be considered from different perspectives. The significance of the heritage, its meaning to the local community and other physical aspects combined should all be taken into consideration. The success of an adaptive reuse project depends on details. It cannot be evaluated in terms of preservation since a successful adaptive reuse project should also be economically sustainable; thereof, bringing in the importance of a solid management plan.

The international principles of conservation of heritages should be harmonized with local needs, beliefs, practices and traditions. No heritage exists without communities as communities create heritages. Loss of unison between communities and their cultural heritages result in loosing identity and bond to locale. For the continuity of heritage buildings, the spirit of the place and local culture should be preserved as well as the physical characteristics of buildings.

The focal point of conservation is to sustain tangible and intangible values of the place. Adaptive reuse of heritage buildings can act as a catalyst and onset other projects in the proximity. As mentioned, the contributions of reuse to environment and community are important and cannot be overlooked since there is always an interaction between conversion projects and environment. In order to achieve a successful conversion, the history of heritages should be assessed before entitling appropriate functions.

The model proposed a qualitative approach; the final decision depended on such variables as decision makers, actors in the adaptive reuse project, context of the heritage buildings, policy issues of the related context, etc. The final decision could be interpreted and then decided considering all these variables. The final model can also be applied to several case studies in order to make comparisons and generate discussions as a testing method. In the research, the term 'sustainability' is discussed in combination with the term 'continuity of the cultural heritage'. The proposed model can also be adapted and reinterpreted by discussions on social, economic and environmental levels of sustainability.

References

1. Orbaşlı, A., *Architectural Conservation.* 2008, London, Blackwell Publishing.
2. Bullen, P.A. and Love, P.E.D., *Adaptive Reuse of Heritage Buildings.* Structural Survey, 2011-a, 29(5), 411–421.
3. Kincaid, D., *Adapting Buildings for Changing Uses: Guidelines for change of use refurbishment.* 2002, London, Taylor and Francis.
4. Yıldırım, M., *Assessment of the decision-making process for re-use of a historical asset: The example of Diyarbakir Hasan Pasha Khan, Turkey.* Journal of Cultural Heritage, 2012, 13(4), 379–388.
5. Langston, C. & Shen, L.Y., *Application of the Adaptive reuse application model in Hong Kong: A case study of Lui Seng Chun.* International Journal of Strategic Property Management, 2007, 11, 193–207.
6. Worthing, D and Bond, S., *Managing Built Heritage The Role of Cultural Significance.* 2008, Oxford, Blackwell Publishing.
7. Mısırlısoy, D. and Günçe, K., *A critical look to the adaptive reuse of traditional urban houses in the Walled City of Nicosia.* Journal of Architectural Conservation, 2016-a, 22 (2), pp. 149–166.
8. ICOMOS, Burra Charter, The *Charter for Places of Cultural Significance,* Online available from: http://australia.icomos.org, 1999.
9. Brooker, G. & Stone S., *Context and Environment.* 2008, Switzerland, Ava Publishing.
10. Chudley, R., *Maintenance and Adaptation of Buildings.* 1983, London, Longman.
11. Douglas, J., *Building Adaptation.* 2002, London, Butterworth-Heinemann Publishing.
12. DEH - Department of Environment and Heritage, *Adaptive reuse: Preserving our past, building our future.* ACT:, 2004, Commonwealth of Australia.
13. Austin, R. L., *Adaptive Reuse: Issues and Case Studies in Building Preservation.* David G. Woodcock, W. Cecil Steward, and R. Alan Forrester, editors. 1988, New York, Van Nostrand Reinhold Company.
14. Velthuis, K. and Spennemann, *D.H.R., The future of defunct religious buildings: Dutch approaches to their adaptive re-use.* 2007, Cultural Trends, 16 (1), 43–66.
15. Cantell, S. F., *The Adaptive Reuse of Historic Industrial Buildings: Regulation Barriers, Best Practices and Case Studies.* Unpublished Master Thesis, 2005, Virginia Polytechnic Institute and State University.

16. Falk, N., *New uses for old industrial buildings in Industrial Buildings: Conservation and Regeneration.* 2000, London, Taylor and Francis.
17. Bullen, P.A. and Love, P.E.D., *A new model for the past: a model for adaptive reuse decision-making.* Built Environment Project and Asset Management, 2011-b, 1(1), 32–44.
18. Ball, R., *Re-use potential and vacant industrial premises: revisiting the regeneration issue in stoke-on-trent.* Journal of Property Research, 2002, 19(2), 93–110.
19. Conejos, C., Langston, C., & Smith, J., *Improving the implementation of adaptive reuse strategies for historic buildings,* Aversa and Capri, Naples, 2011, Italy: La scuola di Pitagora s.r.l.
20. Shen, L.-Y. & Langston, C., *Adaptive reuse potential: An examination of differences between urban and non-urban projects.* Facilities, 2010, 28(1), 6 -16.
21. McCallum, D., *Regeneration and the historic environment in Understanding historic building conservation.* Edited by Forsyth, M., 2007, Oxford, Blackwell Publishing.
22. Cramer, J., and Breitling S. *Architecture in Existing Fabric.* 2007, London, Birkhauser.
23. Mısırlısoy, D., *A holistic model for adaptive reuse strategies of heritage buildings.* Unpublished PhD thesis, 2016, Eastern Mediterranean University, Famagusta
24. Mısırlısoy, D. and Günçe, K., *Adaptive reuse strategies for heritage buildings: A holistic approach,* Sustainable Cities and Society, 2016-b, 26, pp. 91–98.
25. Oktay, D. *An analysis and review of the Divided City of Nicosia, Cyprus and new perspectives.* Geography, 2007, 92, No. 3, pp. 231–247.
26. Vehbi, B. O. and Günçe, K., *Revitalization Studies for the Divided Historic Center: Walled City of Nicosia, Cyprus.* XXVI. International Building and Life Congress – Re-invention of City Center, Bursa, April 2014, pp. 313–323.
27. UNDP in Cyprus (2016) URL 1: https://www.cy.undp.org/
28. Robertson, I., *Blue Guide: Cyprus.* 1990, London, A&C Black.
29. Turkish Republic of Northern Cyprus - Land Registry and Cadastre Office/ Archive 2017.
30. Hackett, J., *A History of the Orthodox Church of Cyprus.* 1901, London, Methuen & Co.
31. Darke, D., *Guide to North Cyprus,* 1993, England: Bradt Publications.

Tuğba DÜZENLİ, Serap YILMAZ and Emine TARAKÇI EREN

Changes in Open Green Spaces in Post-Pandemic Era

Abstract: After the World Health Organization declared Covid-19 a pandemic, in a period where everything including economy, social policies, management styles, production relations and lifestyles were discussed and transformed, architectural and urbanization policies have also been addressed. Because it was observed that the Covid-19 contamination areas were mainly the cities and megacities. Due to easy and fast transportation between nations that increased due to urbanization, led to rapid spread of Covid-19 globally. It was observed that urban planners, landscape architects and designers took global action to develop urban spaces adequate for social distancing. In addition to measures such as widening the sidewalks, concepts such as "parks where you will not meet anyone" and "safe eating zones" were introduced. The present study aimed to discuss the changes in urban open green spaces observed during the pandemic.

Keywords: Pandemic, urban open green spaces, urban design.

1 Introduction

When measures against biological hazards could not be managed, epidemics could break and when these affect the world, they are called pandemics. Thus, it is important to analyze the effects of pandemics on urban policies and strategies from the perspective of urban sustainability and urban resilience. It was observed that the social distancing measures against COVID-19 began to change the cities. Furthermore, experts predict that the effects of these measures will be permanent. It is foreseen that the sidewalks, the parks, and the places where we eat a quick snack will transform due to social distancing rules.

The idea of keeping people apart during the pandemic contradicts with the designers' and planners' traditional emphasis on human interaction. Architects who design the parks or social housing create meeting points to improve cooperation, participation and socialization between the people. It is known that most questions urban designers will face after the pandemic will be about the management of green spaces that would be appreciated by the society. Greener cities could be more flexible and resilient during future pandemics in addition to proven physical and psychological benefits. The changes induced by the COVID-19 pandemic were predicted to be mostly in urban green area planning, urban policies and the content of urban activities and lead to significant global consequences [1,2].

Due to the curfews or limitations in cities around the world, economic activities slowed down. Thus, in cities, which are considered the foundation of the economy, discussions on the continuity of the socio-ecological system during and after the pandemic and the achievement of urban sustainability and urban resilience in these periods gradually increased [3,4]. Planners and designers, who have advocated for years that cities should be clean and free of motor vehicles to allow the people walk on the streets comfortably, had an unprecedented opportunity to test their ideas for a few months. COVID-19 quarantine measures drastically reduced the road and public transport system occupancy, while the authorities in several cities widened the sidewalks by closing the streets to motor vehicle traffic, while allowing bicycle traffic to help the urban residents maintain the 2-meter distance recommended by global health authorities. One of the common application areas of sustainability as a natural and anthropogenic environmentalism method is urban open green spaces. It is possible to approach sustainability through several existing urban structures and from other perspectives. However, the present study aimed to analyze the spatial restructuring of urban open green spaces, which became an important part of our lives, as spaced of socialization based on sustainability.

2 Pandemics and Covid-19 Pandemic

Cities have been the source of epidemics that led to mass deaths throughout history due to dense human population, activities, pollution and the movement of people between different regions. Another factor that accelerates epidemics is the destruction of natural resources as a result of urban sprawl. Urban development in previously uninhabited rural areas played a key role in the spread of new diseases [5,6].

Epidemics, which developed in cities for different reasons in history, led to tragic consequences for individuals, and were among the main factors that initiated changes in planning, design, infrastructure and urban policies for sustainable and resilient cities.

Throughout human history, epidemics have led to great events. It is known that pandemics such as the plague, cholera, typhus, smallpox, Ebola and flu had significant impact on human history among others. Humankind faced the coronavirus pandemic a decade after the SARS and MERS epidemics [7,8]. On December 31, 2019, the existence of a new epidemic was announced by the Wuhan Health Commission, the National Health Commission of the People's Republic of China and the Center for Disease Control and Prevention of the People's Republic of China. On January 8, 2020, a new coronavirus was officially declared by the Center for Disease Control and Prevention of the People's Republic of China as responsible for the new epidemic [9]. The epidemic that started in Wuhan, China, spread to all continents except Antarctica and was declared a pandemic by the World Health Organization (WHO) (2020) on March 12, 2020 [10]. The COVID-19 pandemic led to significant transformations in all areas in early 2020. According to the United Nations (UN) Habitat (2020) [11], 1430 cities in 210 countries were affected by COVID-19. Ninety-five percent of COVID-19 cases were diagnosed in urban areas. It was reported by the UN Development Program [12] (UNDP, 2020a) that the

COVID-19 pandemic defined our era and is much more significant than a global health crisis.

There is some inconclusive evidence that the virus is contagious even before disease symptoms are observed, where the incubation period could last between 2 and 27 days after the contraction of the virus and before the onset of the symptoms [10,13,14]. Disease symptoms include high fever, cough, and difficulty in breathing, and the virus can could lead to death [12,13]. The first death was reported on January 9, 2020. The virus is transmitted with the air droplets released through the mouth or node or through the surfaces [15]. There is evidence that the disease could be transmitted by individuals without any symptoms; the first transmission in Germany was such a transmission [16,17]. The contagious carrier spreads the virus by breathing, speaking, sneezing, coughing, and even when the infectious individual leaves the environment, the viruses could survive for a while based on the ambient conditions, when the environment is not ventilated, and an individual who arrives in the environment could be infected by the virus [18].

Close contact was effective on the transformation of COVID-19 epidemic into a pandemic. The possibility of future diseases such as coronavirus that are easily infect individuals require the urban areas to reduce population density. Due to COVID-19, people started to reconsider their consumption habits which have been accepted as normal despite the increases in extreme weather events and environmental disasters due to climate change, and furthermore, individuals started to widely utilize the technologies that they never used despite their availability in various activities such as distance education and job interviews, and the occupation of open has changed significantly [3].

The urbanization of the global population still continues. The population growth in cities introduce significant problems, especially in developing countries (Lima, Mexico, Nairobi, Mumbai, etc.). The lack of access to basic services in slums, where nearly one billion people live globally, combined with the population density, rendered these places potential areas for contamination by the diseases and led to negative consequences in urban sustainability and resilience [19].

Even though urban areas cover only 3 % of the terrestrial surface of the earth, the results of the interactions between urban areas are significant. It was predicted that COVID-19 would lead to significant economic, environmental and social consequences, especially in cities, affecting the lifestyles of current and future generations [20, 21, 22]. Provision of urban sustainability and urban resilience is important in urban planning and policy development during COVID-19.

3 Post-Pandemic Urban Green Spaces

Instead of accepting nature as an entity and living in harmony with it, it was observed that the distinction between rural and urban habitats disappeared as a significant wave of immigration to urban areas was observed due to exploitation of natural areas as resources for capital. Stream beds, valleys, forests, wetlands, basins, coasts, and pastures are urbanized in an uncontrolled way. Ecological thresholds and ecological integrity were broken, and the natural habitats and animal nests were destroyed. The drama experienced in global metropolitan areas included loss of life, calls to stay at home, curfews, measures taken to reduce the population density in urban spaces to reduce the risk of isolation contamination. The display of positive COVID cases in pandemic maps, the quarantine of the condominiums and residences that are as big as 10 villages were the indication of past errors of extreme urbanization policies.

However, nature and open spaces are very important both physiologically and psychologically for individuals to socialize with others, to experience different environments and to conduct different outdoor activities. Socialization of the individual is quite important to conduct a healthy life due to innate attributes. Being a part of the crowd and the community, to see and to be seen, resting, spending time, being not alone, and the getting rid of the troubles of daily life are among the most important reasons for visiting urban open green spaces. Thus, for contemporary individuals who need socialization, green spaces both meet their needs and allow them to participate in social life [23].

Urban open spaces that ensure urban integrity could be described as open spaces that individuals of different ages could freely access[24, 25, 26, 27]. Urban open green spaces connect and unite different urban quarters, as well as integrating individuals and raising social awareness [24,28,29].

Humans could occupy activity areas in open spaces

- to become aware of the social developments,
- to realize the psychological status of the society,
- to become a part of the society, and to communicate [30].

In short, literature review on the health benefits of urban open green spaces would demonstrate that one of the primary benefits of green spaces is social connections. People can see their neighbors and become part of the community. Designers and planners discuss the development of 'sticky streets' where people linger and remain longer periods. After the pandemic, this approach became a contradiction. Now the question is, will these efforts continue or how will they need to be changed? Can we both maintain social distance and ensure that people are socialized?

It was observed that low-density cities, reduced urban organization, more green spaces, breathing homes, green roads and a nature-oriented local government approach became mandatory. After the pandemic, it is expected that there will be migration from the big cities towards the countryside. Upper income groups have historically moved to the suburbs after each epidemic due to hygiene and health concerns. Local governments should ensure than everyone, not only the wealthy, should benefit from this trend and providing an egalitarian urban planning, spatial arrangement and urban services for all.

As the pandemic drives people away from one another due to isolation, digital communications, desolates cities and public spaces, forces us to utilize new multipurpose spaces, modular, easy-to-produce, transparent designs that communicate with open spaces, reveals the significance of windows and balconies, it also allows the development of new social means of communication that would overcome urban alienation.

Social distancing measures taken to combat COVID-19 are beginning to alter our cities. As the COVID-19 quarantine measures drastically reduced the use of roads and public transport systems, officials in cities including Liverpool and Lima closed streets to motor vehicle traffic, allowed bicycle traffic and widening sidewalks to help the residents to maintain the 2-meter distance recommended by global health authorities. Pedestrians and cyclists took the roads they never dared before, as jellyfish returned to the Venetian canals and flamingos to Mumbai.

Holland (2020) argued that solutions to reduce motor vehicle traffic in cities increased, but it will not be easy to change the habits [31]: There are several benefits of "reclaiming the streets" during an epidemic. Encouraging cycling could reduce the crowds in buses and subways, where people experience difficulties to maintain social distancing. Roads closed to motor vehicles also allow those who do not live close to the parks to exercise safely on the streets.

The Italian city of Milan plans to build 35 kilometers of new bike lanes and permanently widen sidewalks once curfew restrictions expire. In Budapest, the capital of Hungary, officials announced that if the measures prove to be "positive", new bike lanes could become permanent. Providence officials in the US state of Rhode Island announced that the traffic light buttons will not return.

However, it was determined that the number of such decisive cities was limited. It was also argued that it will not be easy to implement pedestrian and bicycle-friendly streets later, especially in countries that depend on cars such as the USA.

In fact, it was observed that the cities that are likely to adhere to pandemic measures were the resilient cities that were ready for change. For example, in Paris, it was announced that after May 11, when the curfew restrictions were lifted in France, more than 650 kilometers of bicycle paths named "Coronapist" will be opened. The Mayor of Paris announced that there will be no return to motor vehicle dominance in the city.

In other words, the pandemic may have only served as a catalyst. However, urban planning is a long road where change is patchy and the legacy of past decisions takes a long time to overcome. Public spaces and public spaces cannot be expanded or restructured at will.

Holland [31] also argued that, as in any major pandemic, urban structure, order, and behavior will need to change. Austrian design studio Precht developed a public park concept where 3 m high fences formed a labyrinth walkway network. The park allows a 20-minute walk where social distancing is maintained thanks to the gates that display whether the path is full.

Meanwhile, in the Czech Republic, it was suggested to draw geometric brightly colored marks on the ground to keep the pedestrians distant from the customers who eat outdoors; thus, creating "Safe Food Zones". Designer Antonio Lanzillo designed benches that separate the seating areas with "plexiglass shields" for the streets of Milan, one of the most affected cities.

According to Jensen Carr, the 2-meter social distance rule is the most important concept since it will soon redefine the plan and layout of public areas and spaces: "So when we think about the city and parks, the new unit we use could be 2 meters."

Furthermore, if a definite association is determined between pollution and the virus, green urban design and planning will also become prominent. Thus, it was

Tab. 1. The transformation in open space occupation during the pandemic

Certain post-pandemic open green space occupancy examples	
 Parc de la Distance [35]	 Safe eating zones [36]
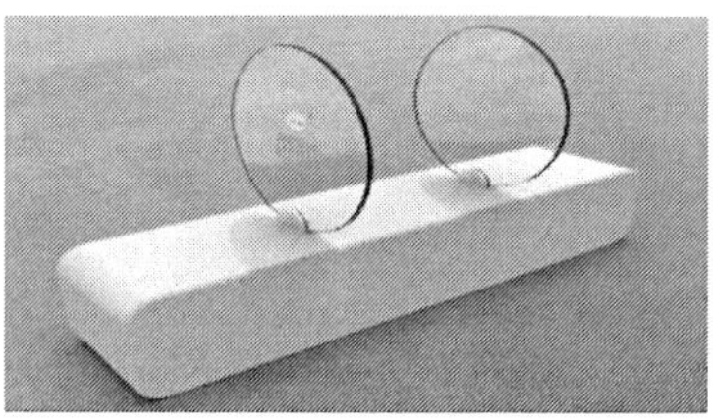 Antonio Lanzillo design public furniture [37]	 Urban green spaces designed for social distancing in Turkey [38]
 Umberto Menasci design plexiglass boxes [39]	 Pandemic Babylon Tower [40]

considered that our cities and streets will need to be redesigned not only because we need social and physical distance, but also to improve our chance of survival.

Italian designer Umberto Menasci designed a series of plexiglass boxes that will maintain a distance between the beachgoers The last eVolo skyscraper design competition was won by the prefabricated emergency health tower concept titled "Pandemic Babylon Tower" design, which the Chinese designers claimed that it could be built rapidly in a future outbreak.

Furthermore, in addition to all other benefits, it was also proposed that field hospitals, temporary shelters, or logistic support units could be built in outbreaks or other disasters in urban open green spaces. Thus, it is very important to ensure the connection between different urban areas and maintain social distancing by banning vehicle traffic in more quarters, building pedestrian and bicycle routes and organizing green spaces along these axes, and increasing sidewalk width [32,33,34]

4 Conclusion

In conclusion, urban open green spaces are important for natural ventilation, air quality, recreation facilities, human health and the urban environment; however, these could be ignored under the pressure of other occupancies. COVID-19 pandemic revealed the significance of flexible and mixed-occupancy open and green spaces in providing urban sustainability and resilience. In addition to other benefits, field hospitals, temporary shelters or logistic support units could be built rapidly during pandemics and other disasters in these areas. Thus, it is very important to ensure the connection between various urban areas and to maintain social distancing by closing further areas to vehicle traffic, constructing pedestrian and bicycle routes and organizing green areas along these axes, and increasing sidewalk width.

Urban green spaces should provide access, safety, need-activity-space diversity, and should appeal to users of all ages and their requirements, be self-sufficient, but should be connected to other areas. It should have an area adequate for the type of targeted activities and the number of occupants, a location adequate for the regional conditions (i.e., topography, human requirements) should be selected, and it should serve various uses.

References

1. UN Habitat. UN-Habitat COVID-19 Response Plan. https://unhabitat.org/sites/default/files/2020/04/final_unhabitat_covid-19_response_plan.pdf. 2020.
2. Tuğaç, Ç. Kentsel Sürdürülebilirlik ve Kentsel Dirençlilik Perspektifinden Tarihteki Pandemiler ve Covid-19 Pandemisi. *Yüzüncü Yıl Üniversitesi Sosyal Bilimler Enstitüsü Dergisi*, (Salgın Hastalıklar Özel Sayısı), 2020: 259–292.

3. UNDP (2020b). *Covid-19 is Testing Our Resilience.* https://www.tr.undp.org/content/turkey/en/home/presscenter/articles/2020/03/COVID-19-Eurasia Director.html
4. Srivastava,A.AfterCOVID.19.https://www.wri.org/blog/2020/04/after-covid-19-5-waysindia-can-pursue-sustainable-and-resilient recovery. 2020.
5. Jakob, T. Urbanization and Epidemic Disease, Joseph Byrne (Der.). *Encyclopedia of Pestilence, Pandemics, and Plagues,* 2008.
6. Özden, K. and Özmat, M. Salgın ve Kent: 1347 Veba Salgınının Avrupa'da Sosyal, Politik ve Ekonomik Sonuçları. *İdealkent,* 2014: 12, 60–87.
7. Perlman S. Another decade, another coronavirus. *The New England Journal of Medicine,* 2020.
8. İd, A. G. E. Dünyada ve Türkiye'de 2019 Koronavirüs Pandemisi. *FLORA*, 2020:*25*(1), 1–8.
9. Li Q, Guan X, Wu P, Wang X, Zhou L, Tong Y, et al. Early transmission dynamics in Wuhan, China, of novel coronavirus-infected pneumonia. *The New England Journal of Medicine,* 2020.
10. World Health Organization. *Novel Coronavirus (2019-nCoV) Situation Report-*51. 2020.
11. UN International Day for Biological Diversity, 22 May. https://www.un.org/en/observances/biologicaldiversityday/message 2020
12. UNDPCovid19.https://www.tr.undp.org/content/turkey/tr/home/prscenter/pressreleases/2020/04/COVID19de-duze-cikma.html.2020a.
13. Hessen, M.T. Novel Coronavirus Information Center: Expert guidance and commentary. *Elsevier Connect.* 2020.
14. Rothe C, Schunk M, Sothmann P, Bretzel G, Froeschl G, Wallrauch C, et al. Transmission of 2019-nCoV infection from an asymptomatic contact in Germany. N Engl J Med 2020.
15. NBC News. *How does coronavirus spread?.* 2020.
16. Chan, J. F. W., Yuan, S., Kok, K. H., To, K. K. W., Chu, H., Yang, J., ... & Tsoi, H. W. (2020). A familial cluster of pneumonia associated with the 2019 novel coronavirus indicating person-to-person transmission: a study of a family cluster. *The Lancet, 395*(10223), 514–523.
17. Söğüt, A. Covid-19 Pandemisi Sonrası Normalleşme Sürecinin Sürdürülebilirliğe Etkisi. *Mimarlık Bilimlerinde Güncel Araştırmalar,* 2020:55.
18. https://www.seyahatsagligi.gov.tr/site/HastalikDetay/Mers-CoV-Hastaligi
19. Vojnovic, I., Pearson, A. L., Asiki, G., DeVerteuil, G. and Allen, A. *Handbook of Global Urban Health.* New York: Routledge. 2019.
20. UNDP Goal 11: Sustainable Cities and Communities. Erişim: https://www.tr.undp.org/content/turkey/en/home/sustainabledevelpment-goals/goal-11-sustainable-cities andcommunities.html,2020c.

21. UN Habitat *The World's Cities in 2018 Data Booklet.* UNDESA. 2018a.
22. UN Habitat Urban Resilience Hub. http://urbanresiliencehub.org/wp content/uploads/2018/08/WCD-Concept-Note2.pdf. 2018b.
23. Aytar Sever, İ., *'Geçmişten Günümüze Alışveriş Merkezleri İç Mekân Kurgusu'* Kitabı, ISBN:978-625-7957-73-1, 2020:89-99
24. Madanipour, A., Why Are The Design and Development of Public Spaces Significant for Cities? *Environment and Planning B: Planning and Design,* 1999. 26, 879–891.
25. Nasution, A.D. and Zahrah, W. Public Open Space Privatization and Quality of Life, Case Study Merdeka Square Medan. *Procedia-Social and Behavioral Sciences.* 2012: 36, 466–475
26. Pekala, N. Conquering the generational divide. *Journal of Property Management.* November/December, 2001: 30–38.
27. Alpak, E.M., Düzenli, T., Kentsel Tasarımda Esneklik ve Adaptasyon: Kentsel Meydan Senaryoları Üretimi, *Journal Of Social & Humanities Sciences Research (Jshsr), 2018:16–22*
28. Ertekin M., Çorbacı Ö. L., Environmental Design Art in Seljuks *Journal of Scientific and Engineering Research,* 2018:122–126.
29. Bayramoğlu E. , Yurdakul N. M., The Effects of the Places as Urban Open Spaces on the Quality of Life: The Case of Trabzon, *Journal Of Hıstory Culture And Art Research,* 2019:425–435.
30. Project for Public Spaces. *How to Turn a Place Around: A Handbook for Creating Successful Public Places.* New York: Project for Public Spaces Inc. 2000.
31. Holland, O., CNN. *Our cities may never look the same again after the pandemic* https://edition.cnn.com/style/article/cities-design coronavirus/index.html.2020.
32. Lubell, S. *Commentary: Past Pandemics Changed the Design of Cities.* https://www.latimes.com/entertainment-arts/story/2020-04-22/coronavirus-pandemics-architecture urban-design. 2020.
33. Pérez, A. J., Y. Urban Planning in Times of COVID-19- Resilience and Inclusiveness. https://www.heriland.eu/2020/04/03/urban-planningin-times-of-covid-19-resilience-and inclusiveness/2020.
34. Snowhite, A. Pandemics: A New Piece of the City PlanningPuzzle. https://newcities.org/the-big-picture pandemics-a-new-piece-ofthe-city-planning-puzzle/.2020.
35. https://www.vbenzeri.com/tasarim/sosyal-mesafe-parki
36. https://www.archdaily.com/tag/hua-hua-architects

37. https://www.designboom.com/design/shield-bench-to-fight-covid-19-antonio-lanzillo-partners-04-17-2020/
38. https://www.internethaber.com/istanbulda-sosyal-mesafe-onlemleri-yesil-alanlar-cembere-aliniyor-foto-galerisi-2104247.htm
39. https://www.designboom.com/design/umberto-menasci-plexiglass-boxes-beach-summer-05-01-2020/
40. http://www.evolo.us/epidemic-babel-healthcare-emergency-skyscraper/

Talia ÖZCAN AKTAN and Özge CORDAN

Designing for/with Syrian Migrants: Sultanbeyli Case

Abstract: Syrians who have been displaced by the forced migration re-making home in the newly settled environment is to create a link between their previous and current living environment. This study aims to develop design proposals for residential interiors by approaching Syrians' remaking of home practices from an interior architecture/design perspective. Following the phases of *design research, design approaches* and *design proposals*, the final result of the design activity is reached in two stages. The first stage is that the initial ideas were developed and furniture design suggestions were proposed through the user centered design, with the *design for user* approach. The second stage includes the participation of a Syrian extended family to the design process of their home, which is designed according to cultural clues and living habits, by following *design with user* approach. According to the study results, the lifestyle, needs, behavioral and cultural habits of Syrian families, their income levels and tenant status are important factors for developing furniture design proposals, which are multi-functional, affordable, modular, portable and adaptable. This study contributes to all stakeholders related to the design discipline, who intent to work with a specific cultural group for developing design ideas appropriate to culture and for acquiring user friendly living environments.

Keywords: Design for user, design with user, forced migration, furniture design, re-making home.

1 Introduction

The civil war has started in Syria in 2011 and millions of Syrian forced migrants have been displaced and mostly fled to neighboring countries. Turkey applied to "open door policy" [1] to Syrians in the initial phase of the mass migration and hosted many Syrian migrants from the beginning of the crisis [2]. The number of Syrians living in Turkey is around 3.6 million according to current statistical data [1].

There is no doubt that the migration has a traumatic phenomenon affects people psychologically, economically, socially and politically. One of the main problems that Syrian migrants facing with is *housing*. In the initial phases of the migration, Syrians were accommodated in the temporary accommodation centers to meet their one of the basic human needs, which is sheltering [3]. Whereas those centers became as an immediate and temporal solution for sheltering, they were lack of giving the feeling of being at home for Syrian migrants [4].

Home is a key factor to attach to a new life after migration. Lose of home is not just a loss of economic or a loss of property, it also includes the lack of everyday life practices, culture, territory, solidarity networks, belonging and identity [2, 3]. According to Kissoon (2015), "Homelessness and refugeeness are major disruptive events in a person's life" [5]. Liu (2014) also stated that "migrants, never completely leave their original home behind, or totally arrive at a new one." [6]. This can be explained being here and there at the same time.

Therefore, finding a solution for housing problem of Syrian migrants becomes an important issue for adaptation of Syrians in relocation. In this sense, Syrians' past living experiences, ways of thinking and their attachment bonds and feelings are ultimately a matter of design, specifically in the process of remaking home. It is also important for interpreting Syrians' interior use and reflecting their resettlement experiences.

Home environments has always been crucial importance for architecture and interior disciplines [7]. The aim of the study is to focus on finding a solution for Syrian migrants housing problem through designing of residential interiors for/with Syrians in the scope of interior architecture/design discipline. More spesifically, focusing on the design process requiring some degree of user's involvement. Sultanbeyli district of Istanbul was chosen as the case of this study. The data collected from voluntarily participated Syrian families in their home environment and the design proposals developed in two stages.

2 Background

The design process includes the stages as research, analysis, programing, developing initial ideas and concepts, developing design proposals, detailing, visualization, and construction [8]. Rapoport (2005) defines the design as an action where the designer is a kind of "advocate" of the user [9]. Blurring roles between the user and designer results in the inevitable contribution of the user to the design process.

User/human centered design is the main concern of the interior design/architecture [10]. It requires to deeply understand people to be served, to dream up the ideas and to create innovative solutions rooted in actual needs [11] and *empathy* becomes important. User-centered design has been introduced by Donald Norman and Stephan Draper in 1986 with a book titled "*User-Centered System Design: New Perspectives on Human-Computer Interaction*" [12]. Another early references on user centered design principles was Gould and Lewis' (1985) article titled "*Design for Usability: Key principles and what designers think*", discussing the contacting with user [13, 14].

The Usability Professionals' Association (2008) defines user centered design as an approach to design that grounds the process in information about the people who will use the product or any other items [15]. Regarding user centered design discipline, Sanders (1992) groupes the research methodologies by mentioning; observation, classification, conversation, description and participation [16]. Then she stated that the "participatory experience is not simply a method or set of methodologies" [17], it offers a design strategy standing alone, a mindset and an attitude about people [17–19].

Therefore, it is worth to say that the role of designer has been changed while the participation of the user to design process is changing. Design practices, in which users participate in the design process are often defined by participatory design or co-designing [20]. Users can become part of the design team as "expert" of their experiences [19]. While the roles of the user and the designer are distantly related and the user is not completely as a part of the team in user centered design, "the roles of the designer and the researcher blur and the user becomes a critical component of the process" in the co-designing culture [17]. According to Hacıalibeyoğlu (2017), participatory design model is an extension of the traditional design process evolving communication [21].

Participatory processes need to evolve from a multidimensional texture that allows reversals, by discussing and evolving design alternatives, they are directly linked to the application area based on existing experiences and accumulations [21]. Therefore, it is a developed and synthesized design strategy that is built on the past experiences and created by evaluating the characteristics of these experiences.

In design and social science disciplines, the *user centered design* and *co-design/participatory design* are two confusing approaches generally used instead of one another. Mattelmäki (2005) states that the attention has shifted from improving usability to wider perspectives such as experiences in everyday life [22]. Also Sanders (2002) emphasizes the changing from user centered design process to that of participatory experiences as well as shifting from "*designing for* users" to "*designing with* users" [17].

The main difference between these two approaches might be explained by referring the targets. While the traditional design disciplines focus on the designing of "products" for the user, the emerging design disciplines priotarize designing with the user for a "purpose" [19]. Therefore, experiencing, experiences of things and situations, events and places etc. become remarkable in design process. Experiencing, a "constructive" [17] activity, structures the ways of interaction between the "communicator and the communicatee" and overlapping of these two sides brings the communication, so knowing the users' experiences becomes possible [17]. Thus, design for experiencing, emotion, interacting become important rather than just designing.

3 The Study: Designing for/with Syrian Migrants

Sultanbeyli, where the most Syrians live on the Anatolian side of Istanbul, has been determined as the study area of the research project[1]. The participants compose of the Syrian families living in the Battalgazi and Fatih neighborhoods in the district of Sultanbeyli. Syrian families with lower-middle income groups were contacted with a native translator. The database of the Syrian Coordination Center (SUKOM) [23] was used to define the families with the help of Sultanbeyli Municipality. The data were collected from the families through interviews via in person communication, observation, note taking, photo shooting and sketching methods and techniques in their domestic living environment. Since the user group is the focal point of the design, the field studies were conducted in their residences to get know the users well, to interact, to make direct observations and to define their needs. Thus, the data were collected regarding spatial use of home, activity types, living habits, privacy thresholds, behavioral patterns, tastes, preferences, family structure and, the relationships and roles between the household [2].

Within the scope of the research project, the field studies were carried out with five Syrian families in 2016 and the second one was conducted in 2019 with the participation of fifteen families. The design process, which consist of *design research, design approaches* and *design proposals* has carried out in two stages[2] as well. In the first stage of designning, furniture designs are developed according to the user's needs and requirements. The second stage of designning is based on placement of these furniture in the space developed in the first stage. The second stage was achieved with the participation of the choosen family. While the *design for user* approach was used in the first stage, the *design with user* approach was used in the second stage of the design process.

---

1 This study was prepared from Talia Özcan Aktan's M.A. thesis [4] as a part of TUBITAK (The Scientific and Technological Research and Council of Turkey) research project titled "Determination of the Housing and Interior Design Principles and Developing Design Proposals Suitable for the Modern Life and Cultural Conditions for the Adaptation of the Syrians Under Temporary Protection", run by Assoc.Prof.Dr.Özge Cordan with the grant number 118K441 in 2018. The thesis title is the same as the research project title.

2 The first design phase (2016) of the study was proposed under the International Masters in Interior Architectural Design (IMIAD) graduate program in Istanbul Technical University and in the Interior Architecture Project III course by Talia Özcan Aktan with supervision of Assoc.Prof.Dr. Özge Cordan. The second phase (2019) of the design study was developed as part of aforementioned thesis study.

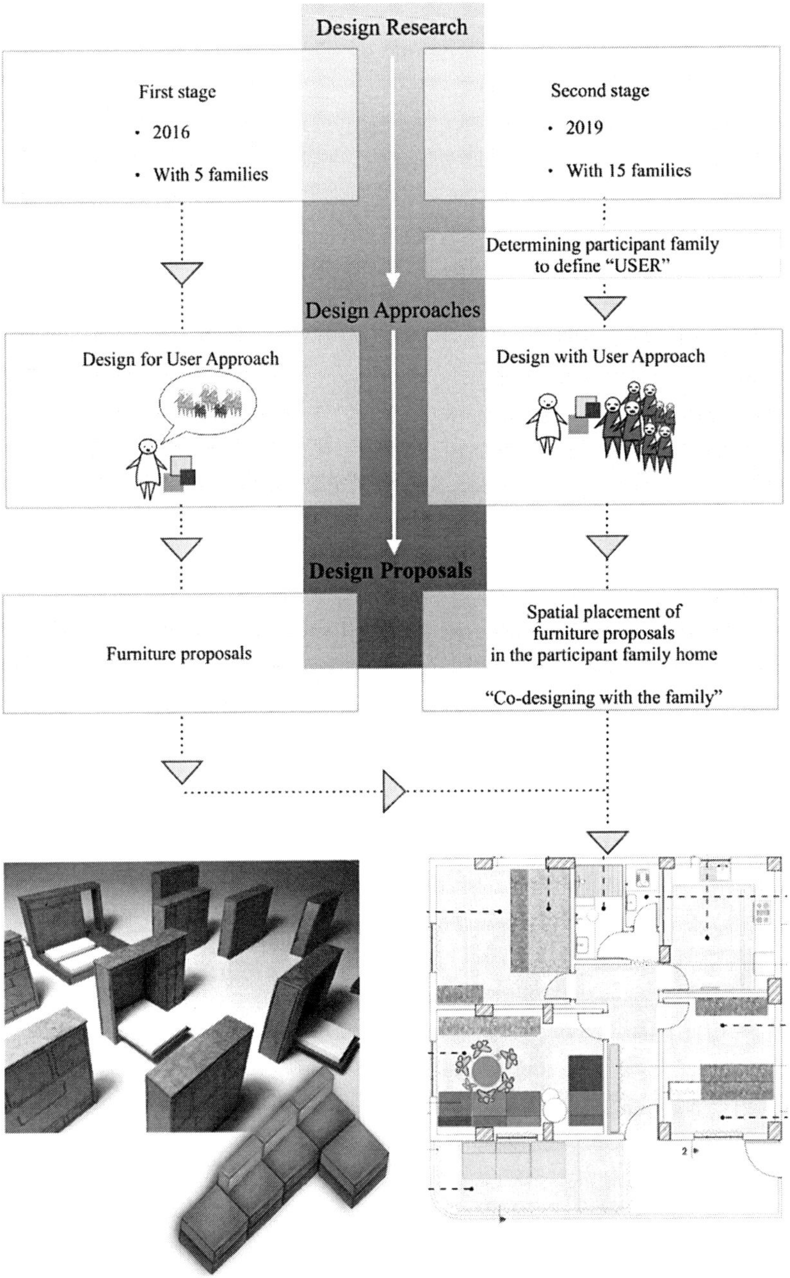

Fig. 1: Design process.

3.1 Designing for user

In the first stage of the designning, the data have been collected from Syrian families living in Battalgazi neighbourhood. Based on the initial obtained data, it becomes clear that a home environment should meet the needs and cultural habits of users in home making process. Despite facing some difficulties in finding a home, it is inferred that the migrants still find a way to feel at home, spesifically with the help of furniture. Rottmann (2020), who conducts studies with migrants, emphasizes the importance of furniture in home environments for the households and states that "the purchase of furniture signals the symbolic comfort of being at home" [24]. Thus the migrants create their own living environment through placement of furniture or some other decorating items such as accessories.

Considering the frequent moving of Syrians, focusing on furniture design was determined as a design decision regarding developing design proposals for Syrian's home environment. Because, the income level and being tenancies prioritize the use of semi-fixed and easily movable furniture and equipment for home making spesifically for personalization and appropriation of the home. Rapoport (2005) focusses on "fixed", "semi-fixed" and "non-fixed" elements in the process of home making [9]. Semi-fixed elements such as furniture and accessories are important regarding defining and expressing the personality, meaning, ethnic and other group identity, etc. [9].

Design proposals of the first stage are based on *modular furniture* design solutions appropriate to the user needs. Developed in the IMIAD Interior Architecture Project III course according to *design for user* approach in 2016, these modular furniture are designed to meet the basic needs of the users such as sleeping, sitting, resting, and storing. Thus, modular, multi-functional, affordable, portable and adaptable furniture proposals including *bed and storage units* (Fig. 2) and *storage units* (Fig. 3) were developed. A *modular ground cushion* was also designed for sitting, sleeping and resting activities of users (Fig. 4). *TV units* and *shelf units* have been proposed for enabling crowded families different needs and activities. The *bed and storage units*, acting as a spatial seperator and providing flexibility and privacy in spatial use, allow day and night use by folding bed units. In addition, the *shelf units* with polished and painted surfaces for hanging photographs, paintings etc. and for displaying objects have been integrated to the design proposals for meeting personalization and decoration needs of the users. Thus, modular design proposals have been developed with inexpensive and available materials such as aluminium prolife and OSB, which are appropriate for mass production to meet the various needs of the users.

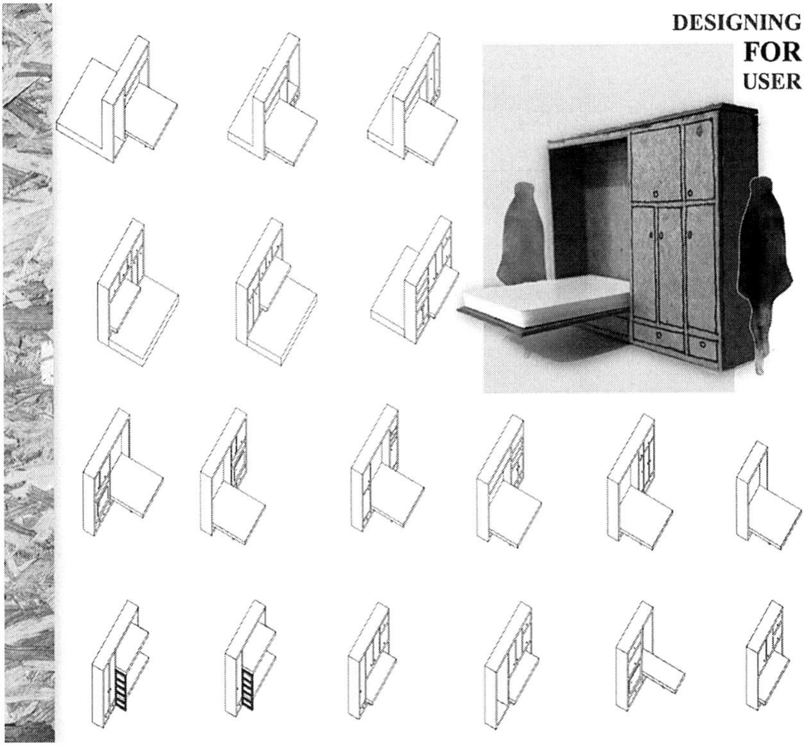

Fig. 2: Bed and storage units.

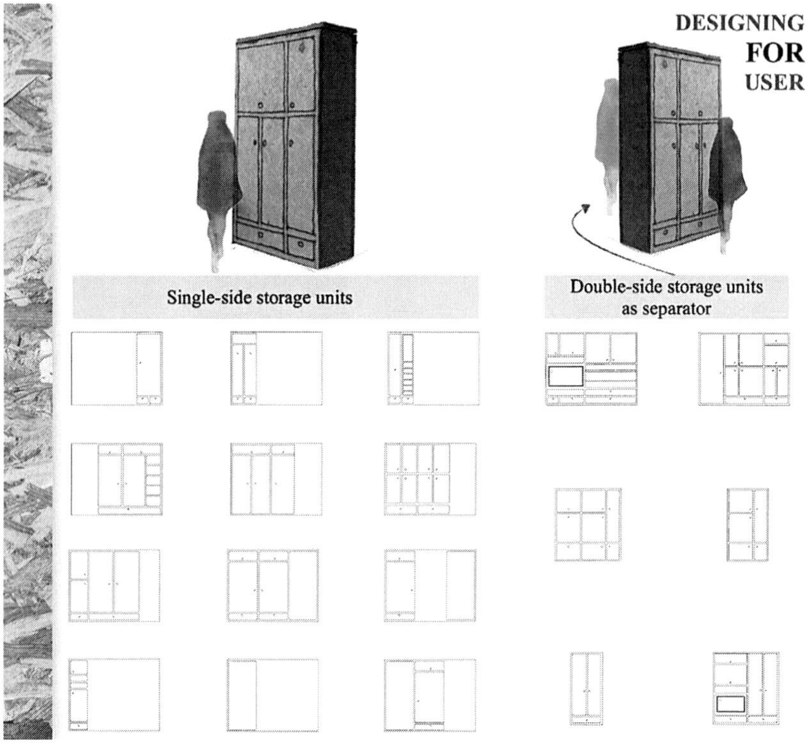

Fig. 3: Storage units.

According to the initial observations, the ground cushion is an important cultural trace taking place almost in all Syrian homes. The ground cushion was designed for meeting the cultural needs of the users. This design supports the sitting, sleeping, resting activities on the ground according to the lifestyle of the users. Additionally, the ground cushions that can be extented by additions are easy to carry, light, massively storable and mass producable. For instance, a modular ground cushion that is used as a sofa during day time, can be used as a bed at night with the modifications and displacements in the modules. Thanks to the cushion modules placed on top of each other, the opportunity to sit at different heights is provided (Fig. 4).

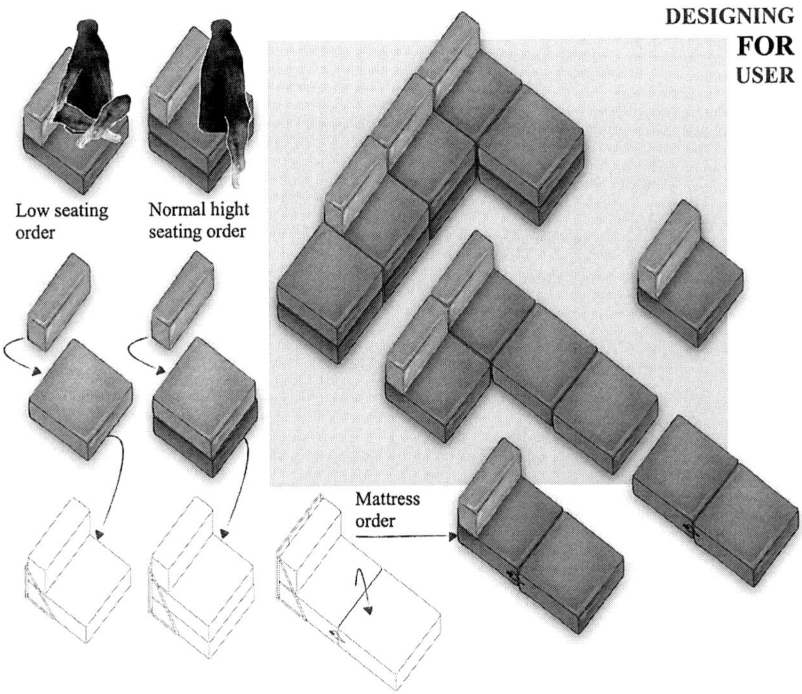

Fig. 4: Modular ground cushion.

3.2. Designing with User

The second stage of the designning was carried out with participation of an extended Syrian family living in the area between Seyhan and Şalgamlı Streets in Fatih neighborhood (Fig. 5). A design suggestion has been developed considering the cultural habits and the needs of this family. Besides the ethical approval from the Social and Anthropic Sciences Human Research Ethics Committee of Istanbul Technical University, a legal permit from the Republic of Turkey Ministry of Interior Directorate General of Migration Management and a letter of support from the Sultanbeyli Municipality, the family signed a content form. Additionally, a signed permission was obtained from the landlady. Thus, an "ethical" research [14] has been conducted.

The family lives on the ground floor of a three-storey apartment building. The house consists of two bedrooms, one living room, kitchen and wet areas. The family was visited two times in the designning process. In the first visit, the house plan was drawn schematically, technical measurements were taken and the family structure and the number of the households were determined (Fig. 6) and the spatial use of residential interiors have been analyzed.

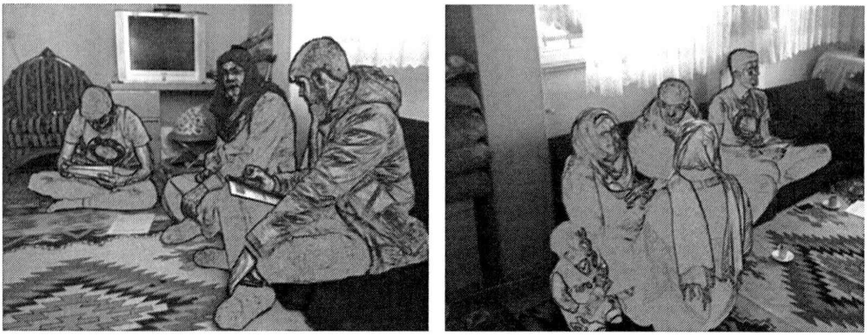

Fig. 5: The extended Syrian family participating the second stage of the design.

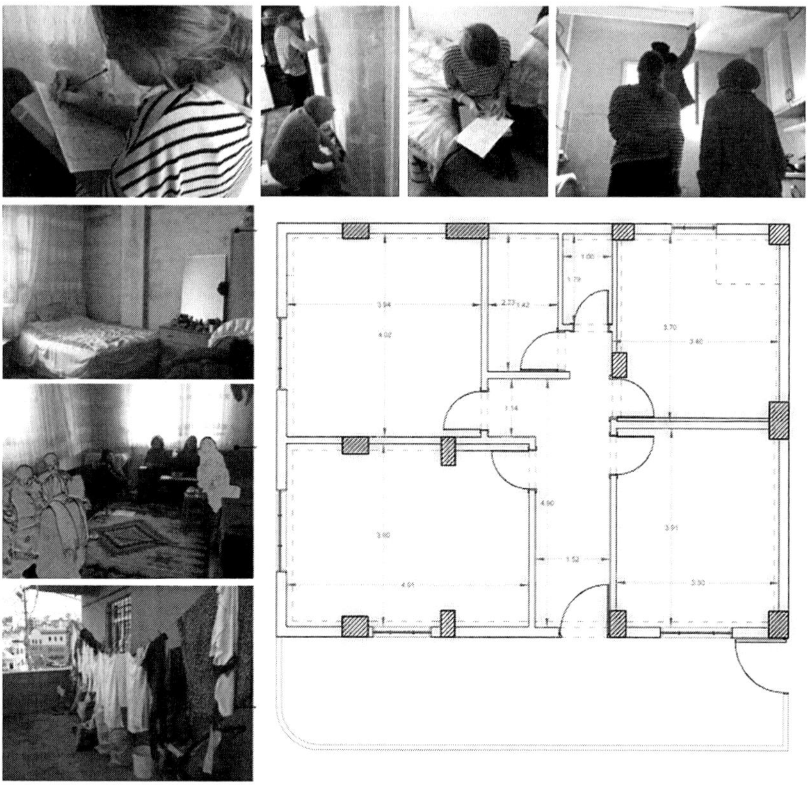

Fig. 6: Photos and technical drawing from the first visit.

The chosen family consists of households who are prone to participatory design. The family includes seven members, who came from the rurals of Aleppo in Syria. In this family, there is a talkative/communicative mother (age 54) and a father (age 64) wearing a traditional *gillabiya* dress. They have two sons, aged 22 and 19, and a 14 years old daughter. Their 22 years old son is married to a young woman (age 16) and the couple have a 10 months old baby. She was also pregnant to another baby. In this family, only two sons has worked temporal jobs periodically. While the women in the house do housework, the father does not work due to his age.

According to the interviews in 2019, the family defines the home as "privacy", "family" and "peace" [4]. This house is their third house in Sultanbeyli as tenant. They have changed their housing environment frequently due to the increase in rents. The furniture and other goods have been given by the neighbors and they are second hand items. In addition to the furniture purchased or donated by the neighbors, elderly family members sit on the ground cushions depending on their living habits. The 22 years old young man explains "This cushion that you are sitting on was covered by my mother's dress. We have used the fabric of the dress to cover it, she was wearing the dress during our migration to Turkey, so it is worth to keep" [4]. Memories are of great importance for family members who could not bring any special items or mementos with them. The relationship with the relatives living in Sultanbeyli are quite often for family members, who have difficulty in socializing with the host society due to the language barrier mostly. These home visits are quite crowded and hosting the guests is based on the sexes. For male, the most important offer is smoking hookah in this visits.

During the second visit to the family, a closer relationship was established with the family thanks to the sincerity of being acquainted and it was possible to get know the family members more closely. Thus, communal and individual needs of the family members, their past experiences regarding home environment both in Sultanbeyli and in Syria, cultural, behavioral and religion based characteristics of home, spatial use and appropriation of home environment, privacy thresholds and roles between the household and income status of the family were revealed (Fig. 7).

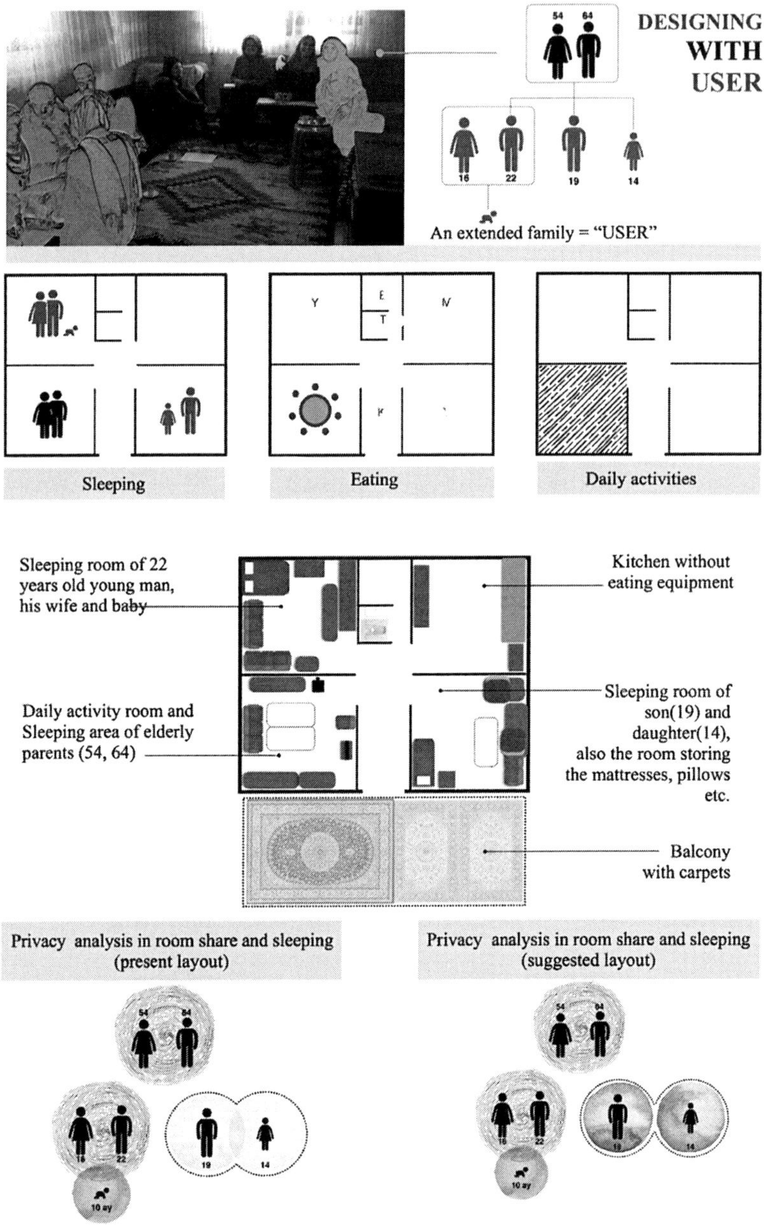

Fig. 7: Analysis from the second visit of the family.

The family structure, needs and cultural habits were influential in the development of the design proposal for the chosen family. Taking into account the family's financial insufficiency and status of being tenancies, affordable solutions have been developed to meet the needs of crowded family life. Providing comfort thresholds is also a crucial design decision to meet the need for privacy within the family and among the individuals. Due to the fact that the house is not large enough compared to the number of households, the space was organized flexible that supports the multi-functional use of the interior during day and night time. The living space and the bedrooms are arranged in accordance with the privacy thresholds. In addition, the *bed and storage units, storage units* and, *modular ground cushion* proposed in the previous stage of the designing activity are placed in the plan of the family's current residence (Fig. 8, 9).

In the design proposal, the living room used as the room where elderly parents sleep as suggested by the family due to the spatial deficiency. Beside, the room used by the young couple is divided into two by a *bed and storage unit.* Thus, it is aimed to prevent the visual relation between the parents and the baby who still sleep in the same room with his parents and to ensure the privacy need of the parents.

Two different *bed and storage units* standing back to back were used for the son (19) and the daughter (14) who still must share the same room because of the spatial deficiency. Thus, for both of them, the personal spaces and privacy zones have been achieved by dividing the room. One of the units has an opening as a door for transition between the spaces. Beside, the gap between the units and the ceiling was left open to let the sun shine in, however, there is a curtain proposal for the noise control on the users' initiative.

All the beds have folding feature in *bed and storage units.* In design proposal, the beds are supposed to be folded up during day time in the room where the elderly parents sleep, so the space became available to be used as a living room. Likewise, a *TV unit and shelf units* were integrated on the *bed and storage units* placed in the living room, to make it suitable for the daily activities and decoration. Therefore, the households are provided with watching TV, hosting guests and spending day time etc. in this room.

Modular ground cushions were used to accommodate the activities such as sitting, sleeping and resting in the living room. According to the living habits and needs of the family members, the modular ground cushion proposal is placed to be formed over each other (in sofa form, with or without reclining) or from one module (as cushion form). Since the living room is also used for the eating activities (breakfast, lunch, dining), an empty space was created in the middle of the room. Thus, the family is supposed to eat on the ground table and offer hookah to their guests etc.

In that family home, a shower and a sink were added to the bathroom. Additionally, a faucet for ablution (located lower) and a stool are suggested. Due to the fact that the family coming from the rural areas of the Aleppo, they are used to use an *arabi* (the name given to the squatting toilet in Syrian Arabic) toilet. Therefore, a squatting toilet was placed non oriented towards Qibla. Also the bathroom and the toilet area, opening to the corridor, was covered with a curtain as suggested by the family.

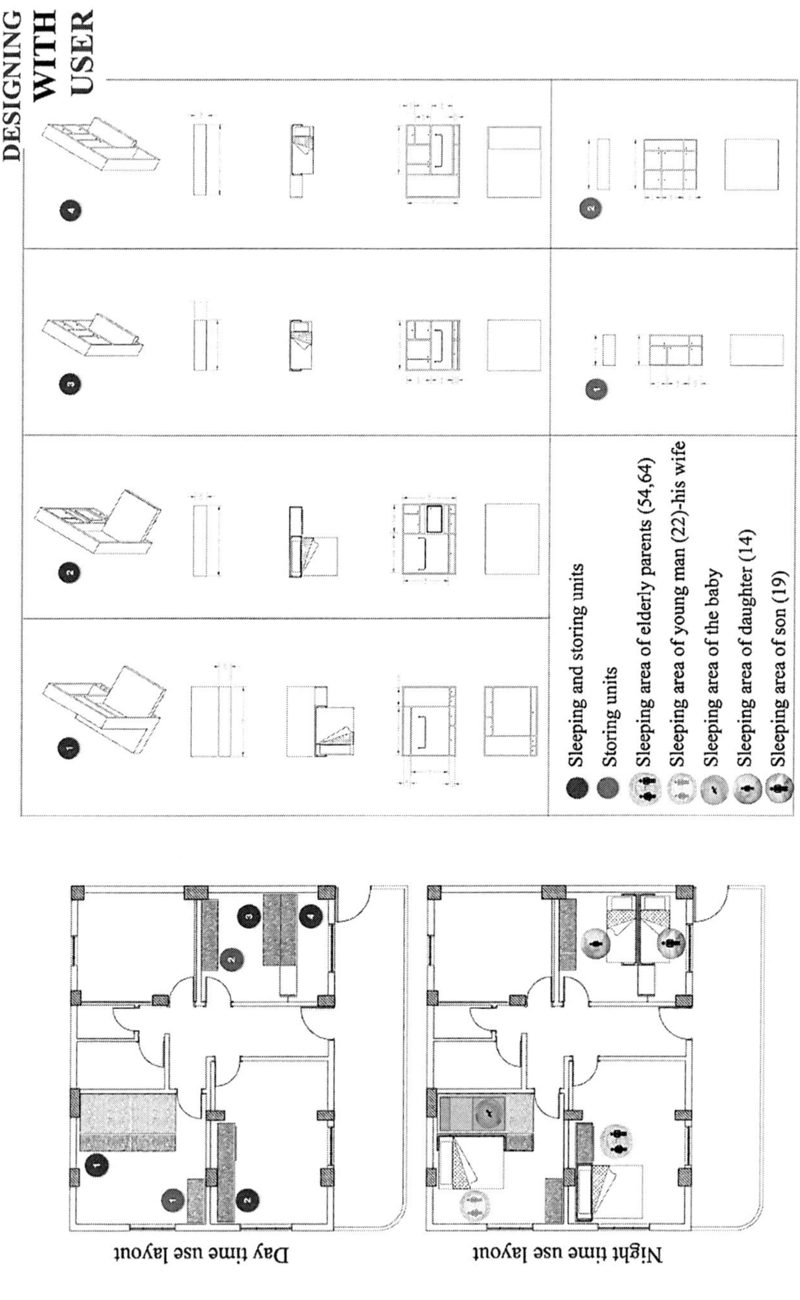

Fig. 8: Placement of furniture (folded up- folded down).

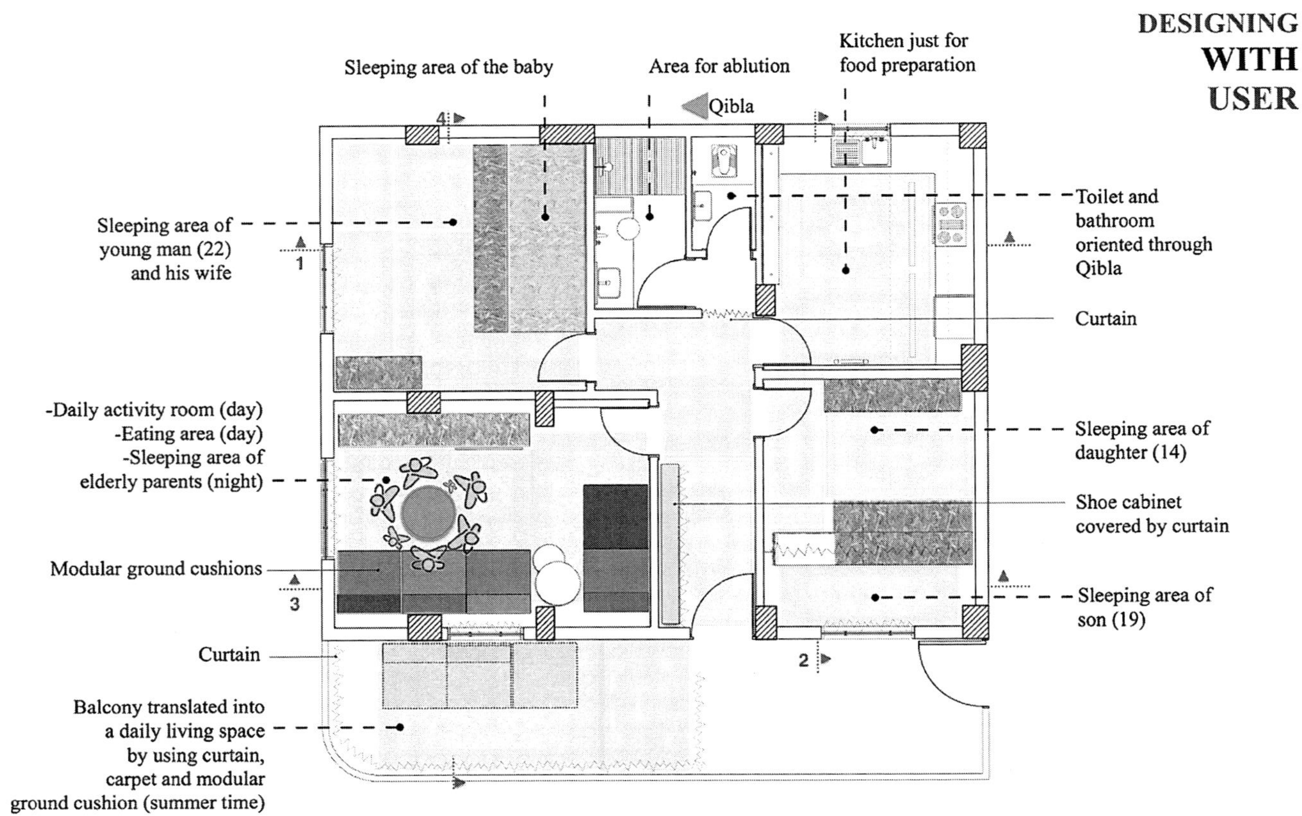

Fig. 9: Placement of the furniture, activities in spaces and solutions to the wet areas.

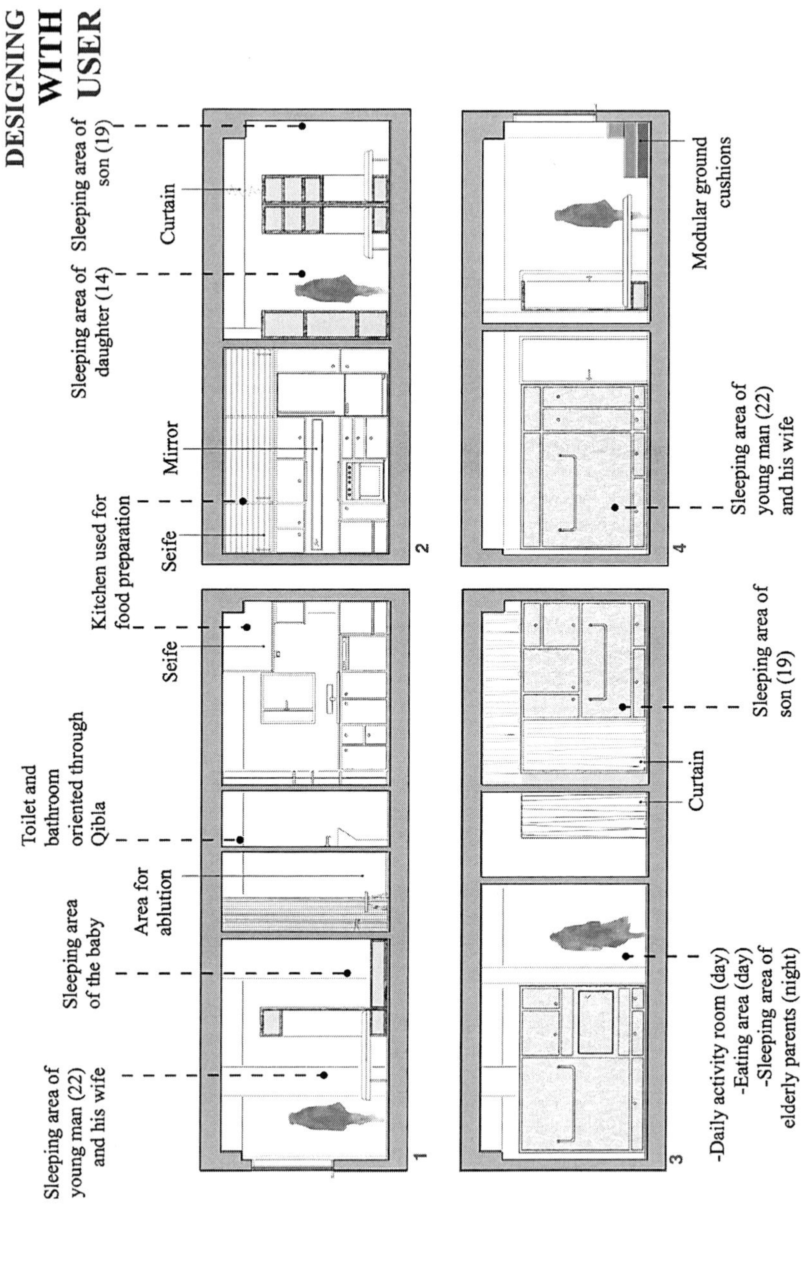

Fig. 10: Sections.

In spatial planning, the placement of the furniture, activities in spaces, solutions to the wet areas and other suggestions proposed are shown in Figure 8, 9 and 10. It is proposed that the floor is to be covered with traditionally patterned carpets reflecting the cultural taste of the users. The use of accessories such as carpets and curtains that complement the interior design can also be seen on the plan and the sections of the design proposal (Fig. 9, 10).

In addition, the balcony (also an entry) of the house on the ground floor is currently used by the family as storage, hosting and living area by laying carpet during summer time. For this reason, carpets and modular ground cushions were placed on the balcony, the required privacy was provided by closing the space with curtains and some solutions about storing were developed.

4 Discussion and Conclusion

Interior architecture/design discipline is an act for finding solutions to meet the ever changing needs of user's for creating functionally, aesthetically and psychologically supportive environments for humans. The residential areas which consists of the big part of the built environment is one of the main concern of this discipline.

Without no doubt, one of the basic need of Syrian forced migrants who have left their home and homeland massively due to ongoing civil war in Syria since in 2011 is sheltering. Forcibly migrated Syrians have an important issue for whole world and developing some solutions for their housing need is quite essential. This study was carried out to find solutions to the housing problem of Syrians through designning and in the light of interior architecture/design principle. The *design research* was done by the field studies in two steps in 2016 and in 2019.

The *design proposals* were based on the *design approaches* as *design for user* and *design with user.* The first stage of the designing, which was conducted with the *design for user* approach in 2016, for developing *design proposals* has been resulted with defining the initial design decisions, specifically for proposing modular furniture design proposals as *sleeping and storing units*, *storing units* and *modular ground cushion design.*

The second stage of the designing, which was conducted with the *design with user* approach in 2019, has been resulted with a *design proposal* to a chosen family home. In that stage, the user experience was emphasized with the participatory design and the co-designing was ensured. According to cultural habits and the needs of the family, the necessary analysis have been done and then the *affordable* and *flexible* design solutions were proposed for this family. The home environment was organized according to the privacy thresholds, religious beliefs, spatial use, roles between the family members, cultural and behavioral routines of the households and the furnishing/placement of furniture, which were designed in the first stage of the study. So, it was possible to test the functionality of the furniture developed in 2016, while designing with the users.

In the case of forced migration, changes in needs are inevitable. Facing with unfamiliar conditions in newly settled environment results in changes and transformations in lifestyle. The families participating to the study are from lower-middle income group who lost their homes in Syria due to the civil war. These people, who left their memories, homes, and homelands behind, also suffered from serious economic losses. Although the habits, lifestyle and needs are decisive in home making, the income level is also an important parameter. Additionally, crowded lifestyle of Syrian families and frequent moving are other arguments that should be taken into consideration in designing and remaking the home environments.

As Rapoport (2005) states there should be flexible and adaptible "loose" ties in housing design [9]. In this study, to achieve the flexibility in spatial organization, *multi-functional, affordable, modular, portable and adaptable* furniture design solutions were offered. In addition, the involvement of the users in the design process made the development of need-based design solutions by strengthening the interaction with the participants.

All stakeholders, intented to work with a specific cultural group might benefit from this study interms of developing design ideas appropriate to culture for increasing the living conditions of users. More spesifically, ethnographic based and in-depth studies can identify the ever changing needs of Syrians over time and some other furniture recommendations might be developed.

References

1. DIGM [Directorate General of Migration Management]. *Temporary Protection in Turkey.* (https://en.goc.gov.tr/general-principles-of-temporary-protection) last accessed: 31 December 2020.
2. Cordan, Ö., Aktan Özcan, T., *Home-Making Process of Syrians through Sustainable and Flexible Reuse of Existing Building Interiors: The Sultanbeyli Case, Istanbul, Turkey.* In Theory and Practice in Sustainable Planning and Design, ed: Murat Özyavuz, 2020, p. 445–460, Berlin: Peter Lang. ISBN 978-3-631-81792-6
3. Aktan, Ö.T., Cordan, Ö. *Re-establishing the "Home": A Case on Syrians in Turkey.* 5th International Conference on New Trends in Architecture and Interior Design, April 26–28, 2019, Istanbul, p.104–114.
4. Aktan, Ö.T., *Determination of the Housing Interior Design Principles and Developing Design Proposals Suitable for the Modern Life And Cultural Conditions for the Adaptation of the Syrians Under Temporary Protection.* Masters Thesis, International Masters of Interior Architectural Design Programme, Social Sciences Institute of Istanbul Technical University, 2019, İstanbul.
5. Kissoon P., *Intersections of Displacement: Refugees' Experience of Home and Homelessness.* 2010, Newcastle upon Tyne: Cambridge Scholars.

6. Liu L.S., *A Search for a Place to Call Home: Negotiation of Home, Identity and Sense of Belonging Among New Immigrants from the People's Republic of China (PRC) to New Zealand.* Emotion, Space and Society, 2014, **10**: p.18–26.
7. Brooker, G., Stone, S., *What is Interior Design?.* 2010. SA: Roto Vision. ISBN-13: 978-2888930174
8. Cordan, Ö., Dinçay, D., Teixeria Fialho, F., *Adaptive Reuse in Interior Architecture: A Case Study in Famagusta, Cyprus.* The International Journal of Architectonic, Spatial, and Environmental Design, 2017. **8** (1): p. 1–15.
9. Rapoport, A., *Culture, Architecture and Design.* 2005, Chicago: Locke Publishing.
10. Cordan, Ö., *Research-based Design: IMIAD Design Studios.* The International Journal of Design Education, 2017. **11** (3): p.15–27.
11. IDEO., *The Field Guide to Human-Centered Design.* 2015, Canada: by IDEO. ISBN 978-0-9914063-1-9
12. Norman, D., Draper S., *User-Centered System Design: New Perspectives on Human-Computer Interaction.* 1986, Florida: CRC Press.
13. Gould, J.D., Lewis, C., *Design for Usability: Key principles and What Designers Think.* Communications of the ACM, 1985, **28** (3): p.300–311.
14. Keinonen, T., *Protect and Appreciate – Notes on the Justification of User-Centered Design.* International Journal of Design, 2010, **4**(1): p: 17–27.
15. Usability Professionals' Association., *What is User-Centered Design?* (https://www.interaction-design.org/literature/topics/user-centered-design), last accessed: 28 December 2020.
16. Sanders, E.B., *Converging Perspectives: Product Development Research for the 1990s.* Design Management Journal, 1992, **3** (4): p.49–54.
17. Sanders, E.B., *From User-Centered to Participatory Design Approaches.* In Design and the Social Sciences, ed: J. Frascara, 2002, p. 1–9, Netherlands: Taylor & Francis Books Limited. ISBN 9780415273763
18. Sanders, E.B., *Design Research in 2006.* Design Research Society, 2006, **V.I** (1): p.1–9.
19. Sanders, E.B., Stappers, P.J., *Co-Creation and the New Landscapes of Design.* Co-Design, 2008, **4** (1): p.5–18.
20. Süner, S., Kaygan, H., *Tasarım Aktivizmi Olarak Katılımcı Tasarım ve Tasarımcının Katılım Deneyimi: ODTÜ Asistan Dayanışması Örneği* [Participatory Design and Participation Experience of the Designer as a Design Activism: METU Asistant Cooperation Example]. Ulusal Tasarım Araştırmaları Konferansı. Eğitim, Araştırma, Meslek ve Sosyal Sorumluluk, 10–12 Eylül 2014 Ankara, p.119–130.
21. Hacıalibeyoğlu, F., *Mimarlıkta Kullanıcı Katılımlı Tasarım Modeli. Ege Mimarlık,* 2017. **Ocak:** p.40–43.

22. Mattelmäki, T., *Design Probes.* Publication Series of the University of Art & Design Helsinki A69. 2006, Finland: Gummerus Printing.
23. SUKOM [Syrian Coordination Center]. (https://multeciler.org.tr/tag/sukom/), last accessed: 23 January 2021.
24. Rottmann, S.B., *RESPOND - Multilevel Governance of Mass Migration in Europe and Beyond.* (EU Horizon 2020 Research Project, Report No. 770564). 2020, European Union Funding for Research & Innovation.

Canan CENGİZ and Aybüke Özge BOZ

Nature-Based Solutions for Urban Resilience as an Innovative Ecosystem Approach in Sustainable Urbanization

Abstract: Nature-based solutions stand out for improving the quality of life, fighting against the effects of climate change and increasing urban resilience. Nature-based solutions encompass the smart use of natural resources and play an important role in attaining sustainability. In this scope, ecosystem services and urban green infrastructure are related with nature-based solutions and nature-based solutions act as an umbrella concept. Nature-based solutions have been taken into consideration in the present study during which the concept of urban resilience has been examined. The need for green infrastructure as a nature-based solution approach has been emphasized for improving urban resilience and ensuring resilience against the impacts of climate change.

Keywords: Nature-based solutions, urban resilience, green infrastructure

1 Introduction

Cities are faced with demographic, environmental, economic, social and spatial difficulties. In recent years, various nature-based practices are implemented for improving urban resilience and coping with climate change. These practices provide effective solutions for creating resilience in addition to providing environmental, social and economic benefits [1].

Nature-based solutions are defined by International Union for the Conservation of Nature (IUCN) as *actions to protect, sustainably manage, and restore natural or modified ecosystems, that address societal challenges effectively and adaptively, simultaneously providing human well-being and biodiversity benefits* [2].

Environmental difficulties can be overcome through the implementation of nature-based solutions via the smart utilization of the intrinsic resources of nature (clean air, water and soil). This is a nature-friendly approach operating in unison with nature to present sustainable and low-cost solutions that make efficient use of resources. New employment opportunities and economic growth are made possible through the use of nature-based solutions by way of new products and services that support natural resources [3].

Nature-based solutions are accepted as a concept encompassing different approaches. Nature-based solutions approaches that focus on ecosystem services with an objective to take into consideration social difficulties can be classified as below [4]:

(i) Ecosystem restoration approaches (e.g. ecologic restoration, ecology engineering and forest landscape restoration),
(ii) Ecosystem related approaches unique to problems (e.g. ecosystem based adaptation, ecosystem based reduction and ecosystem based disaster risk mitigation),
(iii) Infrastructure related approaches (e.g., natural infrastructure and green infrastructure approaches),
(iv) Ecosystem based management approaches (e.g. integrated coastal zone management and integrated water resources management) and
(v) Ecosystem protection approaches (e.g. space-based protection including protected area management).

European Commission [5] has identified four principal goals that can be addressed by nature-based solutions:

- *Enhancing sustainable urbanization nature-based solutions can stimulate economic growth as well as improving the environment, making cities more attractive, and enhancing human well-being.*
- *Restoring degraded ecosystems using nature-based solutions can improve the resilience of ecosystems, enabling them to deliver vital ecosystem services and also to meet other societal challenges.*
- *Developing climate change adaptation and mitigation using nature-based solutions can provide more resilient responses and enhance the storage of carbon.*
- *Improving risk management and resilience using nature-based solutions can lead to greater benefits than conventional methods and offer synergies in reducing multiple risks.*

2 Nature-Based Solutions

Cities are faced with environmental, social and economic hardships that threaten the resilience of urban areas as well as the public living there. The effects of climate change also reinforce these hardships. Nature-based solutions emerged as a concept for the integration of a series of ecosystem based approaches in order to overcome certain societal hardships. Studies are ongoing for the development of mechanisms and tools that will enable the implementation of this approach that has direct contributions to increasing urban resilience in cities. There is a potential to popularize nature-based solutions by way of integration to urban planning approaches, however these have been developed sufficiently neither in research nor in practice [6].

The scope of a “nature-based solutions” can be identified when different factors are taken into consideration with regard to operation [7];

- Nature-related with biological diversity, biological diversity elements (individual species, habitats, ecosystems) and/or ecosystem services.
- Nature-based-ecosystem approaches indicate the direct use of ecosystem based approaches, biomimicry or biodiversity elements.
- Solutions -denote the presence of a problem or hardship with certain identifiable solutions.

Nature-based solutions aim to support the society in reaching its development goals while preserving human welfare in ways that improve the resilience of ecosystems, their capacities for renewal and service presentation. Nature-based solutions have been designed to take into consideration certain significant societal hardships such as food safety, climate change, water safety, human health, disaster risk, social and economic development [4].

According to Brink et al. [8], nature-based solutions focus on returning the ecological flows in cities back to their original states in addition to new infrastructure solutions that improve the resilience of a city [9].

Frantzeskaki [9] developed seven comprehensive suggestions related with all stages of nature-based solutions implementation in cities. (a) Nature-based solutions need to be aesthetically appealing for citizens to appreciate and protect them (b) Nature-based solutions create new green urban commons, (c) Nature-based solutions experiments require and feed into trust between the city and its citizens both for the aim of the experiment and for the experimenting process itself, d) collaborative development of nature-based solutions require diversity and social innovation learning, (e) Nature-based solutions require a collaborative governance approach, (f) An inclusive narrative of mission for nature-based solutions can bridge knowledges and agendas across different departments of the city and (g) Nature-based solutions can be designed to learn and scale these in the long run.

According to Bommarco et al. [10] and Edwards et al. [11], opportunities for encouraging nature-based solutions are already present for many ecosystem service practices. Nature-based use in food production may be carried out by switching fossil fuel and fertilizer input with natural processes and functions (partially). Innovations in agroecology and ecological intensification may increase productivity while providing opportunities for qualified labor [12]. Ecosystems combine lower maintenance costs with longer lives since they are renewable. Ultimately, long term benefit is preserving food safety. Similar benefits in the short, medium and long terms can be evaluated for other ecosystem services especially in cities. As an example, relatively small increases in the nature-based use of urban systems through the implementation of green roofs, nature pockets or sustainable urban drainage systems have significant potential of generating significant socio-economic benefits [12].

The nature-based solutions taken into consideration in the report developed by IUCN are as presented below [4]:

- Ecological restoration
- Ecological engineering
- Forest landscape restoration
- Ecosystem-based adaptation
- Ecosystem-based mitigation
- Climate adaptation services
- Ecosystem-based disaster risk reduction
- Green infrastructure
- Natural infrastructure
- Ecosystem-based management
- Area-based conservation

In addition to other subjects, the idea of nature-based solutions are also used for the rearrangement of policy debates on the conservation of biological diversity, strategies for climate change adaptation and reduction as well as the sustainable use of natural resources. In the meantime, it is also used to define the approaches designed to increase resilience and reduce the risks for humans in large settlement areas as well as generating nature-based strategies in addition to learn biomimicry and industrial design from nature [7]. Nature-based solutions takes into consideration global topics such as reducing climate change and adaptation within the framework of the health-environment relationship thus expanding the understanding of mutual benefits while also making them more applicable in urban planning and implementation.

Increasing the priority of nature-based solutions in the urban design and planning agenda provides opportunities for preparing our cities for the future by providing an innovative ecosystem approach that may contribute to the resilience and economic growth of a city as well as human welfare. Sustainable urban planning has a positive impact through the use of nature-based solutions. It especially contributes to the process of adapting to climate change. It provides opportunities for reducing the impacts of small scale climate change through increasing carbon storage while also increasing urban resilience against risks such as drought, flood and hot air waves. In addition, it may also reduce the pressure on natural areas. As an example, water can be increased at locations closer to residential areas through the use of urban green infrastructure facilities within the scope of waste water management. Thus, recreation areas are obtained close to our homes while reducing the need for traveling to connect with nature [5].

Even though there is no global consensus on the classification of ecosystem services, Millennium Ecosystem Assessment (MEA) is generally accepted [13] (Tab. 1).

Tab. 1: Ecosystem services table according to MEA [5].

Provisioning Services	**Regulating Services**	**Cultural Services**	**Supporting Services**
Food	Air quality regulation	Cultural diversity	Soil formation
Fiber	Climate regulation	Spiritual and religious values	Photosynthesis
Fuel	Water regulation	Knowledge systems (traditional and formal)	Primary production
Genetic resources	Erosion regulation	Educational values	Nutrient cycling
Biochemicals, natural medicines, pharmaceuticals	Water purification and waste treatment	Inspiration	Water cycling
Ornamental resources	Disease regulation	Aesthetic values	
Fresh water	Pest regulation	Social relations	
	Pollination	Sense of place	
	Natural hazard regulation	Cultural heritage values	
		Recreation and ecotourism	

There is increasing proof that ecosystem restoration may play an important role in increasing resilience against the approaching risks and threats. For example, coastal restoration restarts the natural sedimentation processes thereby making coastal areas more resilient against storms and rising sea levels. Slopes can be stabilized through forest restoration practices thus protecting them against floods while reducing surface flow and erosion [5].

Nature-based solutions initiative started by the European Commission at the EU level combines various European policies and actions on adaptation to climate change, reducing disaster risk, energy efficiency, air quality, biological diversity and circular economy. Innovating with nature has been included in many global action plans including UN Sustainable Development Goals [14].

According to European Commission [15], Horizon 2020 large scale marketing projects provide opportunities for putting forth innovative solutions to the difficulties faced by European cities with regard to ensuring the use of nature-based solutions throughout Europe including the "Smart and Sustainable Cities with Nature-Based Solutions" focus area [14]. Projects such as CONNECTING, GROW

GREEN, UNALAB and URBAN GreenUP will employ nature-based solutions in cities for climate and water resilience. These projects will illustrate the benefits of re-naturalization of cities while providing a base of proof across Europe for the effectiveness, efficiency and comparative advantages of a series of tested, scalable and easy to market nature-based solutions. The findings of these projects support other projects such as Nature4Cities and NATURVATION that will examine the new governance, business and financing models and economic impact assessment tools [14].

Nature-based solutions are utilized for ensuring that cities and human settlements are more inclusive, resilient and sustainable. The benefits of implementing nature-based solutions in the urban context for addressing the goals of Sustainable Development Goals 11 are presented in the Fig. 1 [14].

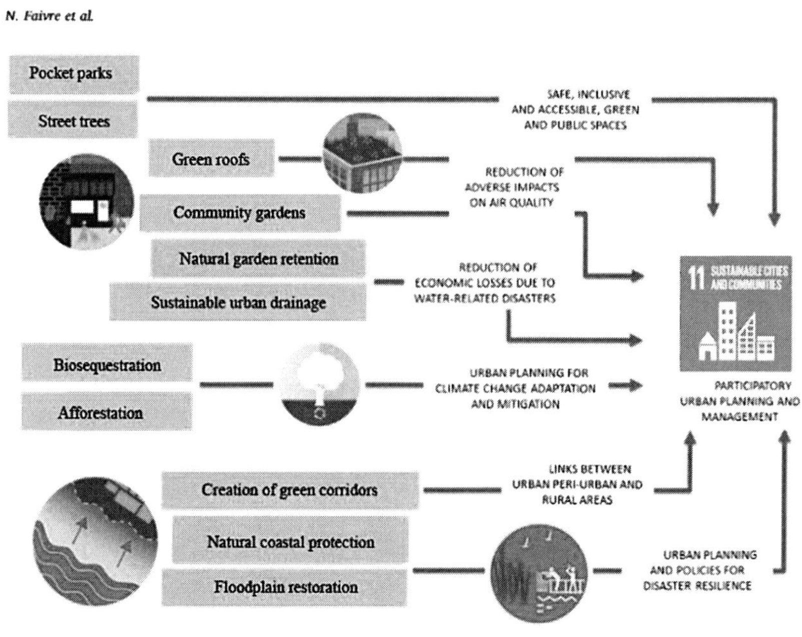

Fig. 1: Sustainable Development Goals Sustainable cities and communities [14].

3 Urban Resilience

Urban resilience can be measured by how well a city can simultaneously balance the ecosystem and human functions [16, 17]. According to Holling [18], the term "resilience" is generally used to define the sustainability related characteristics of a system in the field of ecosystem services and related fields of study. As a technical term, the idea of "resilience" has emerged in the field of ecology [19]. Demirel and Velibeyoğlu [20] identified that sustainability and resilience strategies are based on the protection of resources as well as the effective short and long term management of their use.

Urban resilience generally expresses the ability of a city or an urban system to resist various shocks and stresses. Urban resilience studies are based on a variety of literature that can broadly be classified in four different categories: (1) urban ecologic resilience; (2) urban threats and the mitigation of disaster risk; (3) resilience of urban and regional economies; and (4) the encouragement of resilience through urban governance and institutions [21].

Jennings and Bamkole [22] state that it is possible to increase urban resilience through the provision of nature-based solutions related ecosystem services at urban areas. In addition to supporting welfare through physical and mental well-being, nature-based solutions and urban green spaces also provide spatial opportunities in establishing recreation, social interaction, and social cohesion [6].

According to Foster [23], resilience is important for two reasons. The first is that it is not possible to have an exact knowledge of the weakness levels of technological and social systems. For this reason, resilience is of significant importance for ensuring that disasters do not lead to severe outcomes and the ability to adapt to such situations. Secondly, people and the built environment in resilient cities are able to overcome disaster situations more easily compared with less flexible cities. Risks of death and injury in cases of building collapse, electrical failure, shortcomings in communication and coordination activities are lower in resilient cities. Moreover, less homes and workplaces are under risk [24].

Godschalk [24] indicates that cities should have the eight characteristics listed in Tab. 2 in order to create cities that are more resilient against risks.

Tab. 2: Attributes of resilient cities [24].

Redundancy: Development of polycentric cities to build cities that are more resilient against risks and the prevention of the whole system taking damage in case of failure in one of its components;
Diversity: Protection against area specific hazards with multiple components or centers;
Efficiency: Lower energy supply that what suffices for the dynamic system;
Autonomy: Ability for independent management without external control;
Strength: Strength to resist against a threat or attack;
Interdependence: Integrated system components that support each other;
Adaptability: Learning from experiences and flexibility to changes;
Collaboration: Multiple opportunities and priorities for wider shareholder participation.

4 Green Infrastructure Planning

Studies put forth that traditional grey infrastructure systems are not sufficient against the impacts of globalization, urbanization and climate change and that nature-based solutions such as green infrastructure systems are necessary [25].

Green infrastructure may provide more than one function and benefit in the same spatial area. These functions can be environmental (preserving biological diversity, providing drainage or adapting to climate change etc.), social (food production and provision of green spaces etc.) and economic (providing employment opportunities and increasing real estate costs etc.) (Tab. 3). Grey infrastructure solutions that are typically used for single functions such as drainage or transportation make green infrastructure attractive because green infrastructure has the potential to solve several problems simultaneously. Traditional grey infrastructure is still required, however it can be reinforced through the use of nature-based solutions [26].

Tab. 3: Ecological, economic and social benefits of green infrastructure [27, 28, 29, 30].

Environmental Benefits	Economic Benefits	Social Benefits
Improved air quality	Increased property prices	Enhanced quality of life
Improved water quality	Improved tourist and recreation facilities	Cultivation of public education opportunities
Preservation of terrestrial and aquatic habitats	Increased land values	Encouraged physical activity
Increased groundwater recharge	Increased worker productivity among office employee	Improved community livability
Increased available water supply	Improved chances of gaining planning permission	Improved physical health
Better management stormwater runoff	Reduced use of energy (for cooling), salt (for icy roads), and water	Improved mental health
Protection to enhance geologically important sites, such as nature preserves and heritage sites	Reduced costs associated with infrastructure, development, flooding, water treatment, and healthcare	Improved landscape aesthetics
Reduced atmospheric CO_2	Urban agriculture/sustainable food production	
Reduced urban heat island		
Reduced grey infrastructure needs		
A reduced ecological footprint		
Resilient infrastructure and climate change adaptation/mitigation		
Reduced noise pollution		

Local administrations can use green infrastructure for helping the cities to be more resilient against excessive weather conditions, strengthening food production and supporting urban gardens for increasing green and blue spaces, improving the quality of life for dwellers in the region and creating popular recreation areas [31].

According to Kearns et al. [32], green infrastructure can be especially attractive for city authorities since it provides a tangible approach for improving the different aspects of urban resilience [33]. According to Rouse and Bunster-Ossa [34], short and long-term resilience can be supported through the utilization of green infrastructure subject to technology and scale of implementation thanks to its ability to negate the urban heat island effect, reduce the need for building cooling, reduce storm vulnerability through natural absorption of water, decrease runoff and overflows of untreated storm water into bodies of water, and even provide local sources of food [33].

According to Maes and Jacobs [12], nature-based solutions approach enables sustainable solutions that can respond to environmental changes and threats in the long run thus focusing on individuals and their benefits to their environment. For this reason, nature-based solutions may help us in staying inside a study area safe for people and improving local ecological and social sustainability in addition to ensuring long term productivity [35]. Pauleit et al. [36], nature-based solutions is in unison with various concepts such as ecosystem based adaptation, urban green infrastructure and ecosystem services [35]. Because these three are subsets that maximize the benefits and integration of nature-based solutions. Hence, nature-based solutions functions act as an umbrella for other concepts, however it focuses on the popularization of nature-based actions (e.g. city forests) especially compared with high technology based solutions [35].

Green infrastructure planning and design process utilizes natural and manmade structural elements to cover the scope of the preservation of natural resources in addition to ecological benefits and resolutions. Green infrastructure system enables the establishment of relations between open and green areas such as natural habitats, green corridors etc. in urban areas and their environs. Physical and functional connections are facilitated through the implementation of urban green infrastructure between the divided habitats within the scope of its components thus contributing to the formation of sustainable landscapes that support and improve the bonds between the society and nature [30].

5 Conclusion

There is a requirement for innovative solutions that will improve the ecosystem status and resilience in order to build sustainable, low carbon and climate resilient and adaptable cities. It is important for improving cities to design their green open spaces through the implementation of sustainable, restorative and regenerative approaches. In this context, the use of nature-based solutions stands out as a framework for improving public health in urban areas based on the importance of green spaces for health. Improving environmental quality of life through the provision and arrangement of ecosystem services varies subject to the impact area of the implemented nature-based solutions and spatial attributes.

Urban planning contributes to processes and mechanisms, participatory approaches for acquiring information and data and decision making processes for land use in order to determine the strategic goals and objectives. These urban planning based solutions may potentially support and encourage increasing number of applications. In this scope, more applications can be enabled through the integration of nature-based solutions in land use planning. Urban planning is human centric to a large extent; cities should be considered as habitats shared by all living beings. It is important to support the implementation of nature-based solutions for making a comprehensive and effective urban plan. Resilience, presents opportunities for questioning and developing the urban planning approach. Similarly, urban planning also provides mechanisms that enable the more effective and comprehensive implementation of nature-based solutions in urban areas [6].

References

1. Beceiro, P., Galvão, A., and Brito, R.S., Resilience Assessment Framework for Nature Based Solutions in Stormwater Management and Control: Application to Cities with Different Resilience Maturity. Sustainability, 2020. **12**: p. 1–19.
2. IUCN, Available from: https://www.iucn.org/commissions/commission-ecosystem-management/our-work/nature-based-solutions Accessed 27.12.2020.
3. European Commission, Available from: https://ec.europa.eu/info/research-and-innovation/research-area/environment/nature-based-solutions/research-policy_en Accessed 20.12.2020.
4. Cohen-Shacham, E., Walters, G., Janzen, C., and Maginnis, S. (Eds.), *Nature-Based Solutions to Address Global Societal Challenges.* 2016, Gland, Switzerland: IUCN.
5. European Commission, *Nature-Based Solutions & Re-Naturing Cities.* Final report of the Horizon 2020 Expert Group on "Nature-Based Solutions & Re-Naturing Cities", 2015.
6. Bush, J., and Doyon, A., *Building Urban Resilience with Nature-Based Solutions: How can Urban Planning Contribute?.* Cities, 2019. **95**: 102483.
7. Potschin, M., Kretsch, C., Haines-Young, R., Furman, E., Berry, P., and Baró, F. (2016). *Nature-Based Solutions.* OpenNESS ecosystem services reference book, ed: M. Potschin, and K. Jax, EC FP7 Grant Agreement.
8. Brink, E., Aalders, T., Adam, D., Feller, R., Henselek, Y., Hoffmann, A., ... and Wamsler, C., *Cascades of Green: A Review of Ecosystem-Based Adaptation in Urban Areas.* Global Environmental Change, 2016. **36**: p. 111–123.
9. Frantzeskaki, N., *Seven Lessons for Planning Nature-Based Solutions in Cities.* Environmental Science & Policy, 2019. **93:** p. 101–111.

10. Bommarco, R., Kleijn, D., and Potts, S.G., *Ecological Intensification: Harnessing Ecosystem Services for Food Security.* Trends in Ecology and Evolution, 2013. **28**(4): p. 230–238.
11. Edwards, P.E.T., Sutton-Grier, A.E., and Coyle, G.E., *Investing in Nature: Restoring Coastal Habitat Blue Infrastructure and Green Job Creation.* Marine Policy, 2013. **38**: p. 65–71.
12. Maes, J., and Jacobs, S., *Nature-Based Solutions for Europe's Sustainable Development.* Conservation Letters, 2017. **10**(1): p. 121–124.
13. Tolunay, D., İklim Değişikliğinin Ekolojik Sistemlerdeki Yeri. İklim Değişikliği Alanında Ortak Çabaların Desteklenmesi Projesi (iklimİN), İklim Değişikliği Eğitim Modülleri Serisi 5. 2019. Available from: http://www.iklimin.org/moduller/ekolojimodulu.pdf
14. Faivre, N., Fritz, M., Freitas, T., de Boissezon, B., and Vandewoestijne, S., *Nature-Based Solutions in the EU: Innovating with Nature to address Social, Economic and Environmental Challenges.* Environmental Research, 2017. **159**: p. 509–518.
15. European Commission, *Demonstrating Innovative Nature-Based Solutions in Cities.* 2015.
16. Alberti, M., Marzluff, J.M., Shulenberger, E., Gordon, B., Ryan, C., and Zumbrunnen, C., *Integrating Humans into Ecology: Opportunities and Challenges for Studying Urban Ecosystems.* Bioscience, 2003. **53**(12): p. 1169–1179.
17. Aygün, A., *Climate Change And Urban Resilience: Vulnerability and Risk Assessment for Istanbul.* M.Sc. Thesis, Istanbul Technical University Graduate School of Science Engineering and Technology, Department of Urban and Regional Planning, İstanbul. 2015, p. 199.
18. Holling, C.S., *Resilience and Stability of Ecological Systems.* Annual Review of Ecology Systematics, 1973. **4**: p. 1–23.
19. Carpenter, S., Walker, B., Anderies, J.M. and Abel, N., *From Metaphor to Measurement: Resilience of What To What?.* Ecosystems, 2001. **4**(8): p. 765–781.
20. Demirel, Ö., and Velibeyoğlu, K., *Yeni Kentsel Gündem: Çevresel Sürdürülebilirlik.* Türkiye Peyzajları II.Ulusal Konferansı, 2017, İstanbul.
21. Leichenko, R., *Climate Change and Urban Resilience.* Current Opinion in Environmental Sustainability, 2011. **3**(3): p. 164–168.
22. Jennings, V., and Bamkole, O., *The Relationship Between Social Cohesion and Urban Green Space: An Avenue for Health Promotion.* International Journal of Environmental Research and Public Health, 2019. 1**6**(3), 452.
23. Foster, H.D., *The Ozymandias Principles: Thirty-one Strategies for Surviving Change.* 1997, Victoria, B.C.: UBC Press.

24. Godschalk, D.R., *Urban Hazard Mitigation: Creating Resilient Cities.* Natural Hazards Review, 2003. **4**(3): p. 136–143.
25. Zuniga-Teran, A.A., Gerlak, A.K., Mayer, B., Evans, T.P., and Lansey, K.E., *Urban Resilience and Green İnfrastructure Systems: Towards a Multidimensional Evaluation.* Current Opinion in Environmental Sustainability, 2020. **44**: 42–47.
26. Diege, G., *Green Infrastructure: Better Living through Nature-Based Solutions.* European Environment Agency, 2015.
27. Kim, D., and Song, S.K., *The Multifunctional Benefits of Green Infrastructure in Community Development: an Analytical Review Based on 447 Cases.* Sustainability, 2019. **11**(14): p. 1–17.
28. ARUP, *Cities Alive: Rethinking Green Infrastructure.* 2014, London, UK.
29. Center for Neighborhood Technology, *The Value of Green Infrastructure. A Guide to Recognizing: Its Economic, Environmental and Social Benefits.* 2010. Available from: https://www.cnt.org/sites/default/files/publications/CNT_Value-of-Green-Infrastructure.pdf Accessed 30.12.2020.
30. Cengiz, C., and Boz, A.Ö., *Climate Compatible Green Infrastructure Applications for Sustainable Cities: Bartın Case Study.* Journal of Environmental Protection and Ecology, 2020. **21**(6): p. 2093–2099.
31. Global Communications Team, *How Cities are Using Nature for Sustainable Urban Development.* 2017. Available from: https://talkofthecities.iclei.org/how-cities-are-using-nature-for-sustainable-urban-development/ Accessed 25.12.2020.
32. Kearns, A., Saward, R., Houlston, A., Rayner, J., and Viraswamy, H., *Building Urban Resilience through Green Infrastructure Pathways.* Resilient sustainable cities: A future ed: J. Leonie, P. Peter, W. Newman, and P.Roberts, 2014, New York, NY: Routledge, p. 52–65.
33. Meerow, S., and Newell, J.P., *Urban Resilience for Whom, What, When, Where, and Why?.* Urban Geography, 2016. **40**(3): p. 309–329.
34. Rouse, D.C, and Bunster-Ossa, I.F., *Green Infrastructure: A Landscape Approach.* 2013, Chicago: American Planning Association.
35. Lafortezza, R., Chen, J., Van Den Bosch, C. K., and Randrup, T.B. (2018). *Nature-Based Solutions for Resilient Landscapes and Cities.* Environmental research, 2018. **165**: p. 431–441.
36. Pauleit, S., Zölch, T., Hansen, R., Randrup, T.B., and van den Bosch, C., 2017. *Nature-Based Solutions and Climate Change-Four Shades of Green.* Nature-Based Solutions to Climate Change Adaptation in Urban Areas ed: N. Kabisch, H. Korn, J. Stadler, and A. Bonn, 2017, Springer, Cham, p. 29–49.

Elif BAYRAMOĞLU and Seyhan SEYHAN

Drought Tolerant Landscape with Ecological Approach in Sustainable Cities

Abstract: In recent years, extreme temperatures and precipitation occurring in cities due to climate change, global warming and urban heat island effect have occurred on cities. However, warming cities negatively affect the quality of life of people, natural resources in cities and natural habitats are destroyed, ecosystems are intervened as a result of the destruction of natural resources, and ecosystems are disintegrated as a result of uses that destroy nature. In order to prevent this situation in cities; new concepts have emerged in the field of landscape architecture such as new urbanization, green architecture and sustainability. Landscape architects, who play an active role as urban open green space planners, have turned to researches for more effective use of existing water in order to be a solution to changes in cities. Drought tolerant designs in ecological approaches in cities are at the top of these studies. Drought tolerant landscaping also means saving water.

Urban landscape designs developed within the scope of rainwater management plans, one of the important components of landscape plans, are considered as components of the urban green space system.

Keywords: Drought tolerant landscaping, Sustainable Cities, Urban design, Saving on Water

1 Introduction

Humanity, which aims to develop and progress in the historical process, has constantly developed itself and its environment depending on technology in this process. But in addition to every positive, this process has also created negative effect. The pressures and pollution experienced in cities with uncontrolled have brought with it many environmental problems. Advanced technology, along with depleted and non-renewable natural resources, leads to pollution of nature and the environment, which can no longer defend itself. Global warming has come to the fore with the depletion of natural resources and the reduction of open green spaces in cities. Climate change, which is one of the most prominent effects of global warming, has brought the necessary measures to be taken regarding water shortage. It is impossible to completely eliminate the impact of urbanization on global warming and associated climate change, but it is possible to reduce it.

Sustainable cities have created new approaches in the cities that have recently taken action in this direction. In line with the concept of sustainability, existing resources must be protected and delivered without causing enough damage to future generations. Increasing the use of designs suitable for this purpose in architecture is one of the most important steps to achieve success in this regard. Under the general heading "Water-Efficient Landscaping", new landscape concepts have been developed that differ from classical landscape landscaping concepts such as "Water-Wise, Water-Smart", "Low-Water" and "Natural Landscaping" [1].

Especially in recent years, new urbanization, green architecture and sustainability have been adopted as new concepts in landscape architecture [2]. The increase in the water problem in recent years as a result of problems related to changing climatic conditions such as temperature increase, irregularity of rainfall has led designers and planners to research more efficient use of available water. Water is of vital importance especially for the sustainability of the use of plants, the most prominent material of landscape architects' designs. Plants that consume less water should be preferred instead of water-consuming plants and various applications in places such as parks and gardens. In this respect, it has become mandatory in landscape designs to introduce plants and practices that enable more efficient use of available water [3, 4].

In this respect, drought tolerant designs have started to be discussed in ecological approaches in sustainable cities. Drought tolerance does not only mean that vegetative elements are succulent. It also means that their designs will be able to address future drought problems.

2 Ecological Approach and Landscape

Towards the end of the 20th century, ensuring the sustainability of natural resources began to take place in the planning process. The inclusion of these natural resource values in the planning process has brought the ecological approach to the agenda [5]. The ecological approach in landscape design and management requires solutions that are compatible with natural processes, structural and ecological features of the area. The ecological approach in landscape design and management requires taking the nature itself as a model in the field design and bringing solutions that are compatible with natural processes and the structural and ecological characteristics of the area. Today's urban green spaces, which are organized by prioritizing human use, are an approach that brings solutions to the change of existing design understanding and ecological problems of the city and puts human-nature relations to the fore [6]. The main goal is to develop a self-sufficient sustainable system that can be part of the urban ecosystem [7, 8].

The aim of the ecological approach, which is taken on the basis of landscape planning, is to realize plans that preserve natural values, reveal the characteristics of natural and cultural values, landscape sensitivities and risks, and detect environmental pressures that may be caused by land use [9]. The basic principles of ecological design have been developed from McHarg (1969). The principles of McHarg are listed as preserving the current landscape character, using local resources in design and design in accordance with climatic data based on ecological conditions, aesthetics, sustainability, vegetative design, wildlife organization, management and maintenance with ecological approaches [10].

In this context, ecological design in landscape; it aims to protect the existing landscape character, to limit intervention to natural development, to protect the natural appearance, to protect and restore existing habitats. In addition, the design in accordance with ecological conditions and historical and cultural values, emphasizing the landscape aesthetics with ecological value increases the interaction between human and nature, which is compatible with the natural structure of the area. Including appropriate activities for social needs and using local and recycled materials with low maintenance costs and local plant species are important elements [11].

Landscape designs with an ecological approach mostly require the use of local resources in vegetative and structural design. In such designs, durable materials are used that can easily adapt to the region. Applications that minimize the consumption of natural resources and low cost are included in the designs [7]. The ecological texture should be in balance by protecting the water coasts and wetlands within the urban fabric. Natural formations and natural vegetation should be considered in the context of protection of natural resources. Depending on climate change in urban areas by taking measures to reduce the urban heat island effect is added to urban désign criteria. The drought tolerant landscape, which will be considered as planning criteria, is in today's world with a contemporary approach.

3 Drought Tolerant Designs

Climate has vital importance for all living things; it changes as a result of human activities and these changes affect both human and other living things negatively. The effects of climatic changes emerge slowly and for long years and begin to feel harmful signals [1]. Global warming impacts are drought, desertification, increased risk of flooding and a possible decrease in water resources over time. Considering this situation, water should be used rationally and recycled [12].

The primary effects of climate change in the city and the environment are the development of events following the increase in temperature and change in precipitation patterns. Secondary and tertiary effects are rising water, longer summer season, increase in endangered species and impacts on human health. According to the report to the IPCC, since the 1970s, global researches and data from continents and oceans prove that many natural systems are physically and biologically affected by regional climate change. The first results published by NASA show that global cooling has decreased since 1940 and that there was an increase of 0.4 ° C between the 1880s and 1970s [13, 14].

In landscape applications, improving the quality of the environment-space and repairing the deteriorating environmental conditions were considered as the main objectives. In recent years, due to global warming and climate change, plans for drought tolerance and rational use of water have come to the fore. As one of the products of urban landscape planning, open green spaces are important places to create the adaptation process by reducing the returns of climate change in urban spaces.

Increasing green areas by reducing hard surfaces: In today's world, where global warming has shown its effect more and more, green areas are needed more in cities. In order to prevent the increase in temperature in urban areas, it is necessary to reduce hard floors and increase green areas. An organic texture should be created by establishing vertical gardens on the surfaces and walls and roof gardens on the rooftops [11].

- **Creating roof gardens:** With the effect of global warming, industrialization and technology have reduced green areas and consequently hard floors have increased. Therefore, the lack of green space on the ground has led to the evaluation of building roofs. Green roofs prevent the increase of air and surface temperature by replacing the damaged vegetation in the urban area. Plants contribute to the prevention of temperature increase thanks to their evapotranspiration, shading and heat storage properties in the plant carrier layer [15].

Use of natural plant species: It is very important to use natural species specific to the region in planting designs by adopting "naturalistic" arrangements based on recreating nature rather than artificial arrangements in urban open green space designs. Because natural plants are more resistant than foreign native plants. In planting studies, using natural plant species specific to the region minimizes maintenance and repair work in the application area and it easily adapts to the environment.

In addition, since it is prone to the climate structure of the region, it meets the water it needs with natural rainfall without the need for additional water [16]. In addition, natural plants adapt best to local environmental conditions. They contribute to soil fertility, reduce erosion, and generally require less water, fertilizers and pesticides than other plant species [17].

Plants existing in their field are the most suitable plants to use. Therefore, while designing, it should be as faithful as possible to the existing plant tissue in the area. It will be successful in terms of efficient use of water by making rainwater and a small amount of additional irrigation with selected plant species [4]. The use of natural species in landscape arrangements has both ecological and economic benefits. It protects the rare or endemic species that are endangered ecologically by preserving the biological diversity in nature. Natural species improve soil, prevent erosion and improve environmental quality [18]. Economically, the use of natural plant species minimizes maintenance costs such as irrigation, spraying and fertilization. As it is resistant to plant diseases and pests, its negative effects on soil and water resources are also reduced. One of the most important issues in terms of aesthetics is ensuring visual quality.

Visual quality enhances urban aesthetics, breaking the mass effect of buildings, preparing people more livable environments. In addition, the harmony between the building masses and the natural environment is supported, increasing the visual quality of the environment as well as ensuring the environmental integrity [19].

Using water economically: Water is indispensable for living life. Today, water is used extensively, directly and indirectly, for aesthetic purposes, except for life concerns. This highlights the importance of using water-efficient designs for irrigating landscape areas [20]. Rainwater management, one of the important components of landscape plans, is included in urban planning. There are special application areas such as water harvesting, water meadows, water retention gardens, rain gardens in the developed urban landscape designs [21].

- **Restricted irrigation approach:** In landscape applications, restricted irrigation applications can be considered as an alternative in order to save water in places where irrigation water is limited. Especially in landscape architecture applications, the plants used in planting designs can be applied by determining the specific cut-off value for each plant in order not to lose their texture and form characteristics visually. In this way, the survival of plants, which are the living materials of landscape architecture applications, is ensured and water is saved at the same time [22]. The plant species to be irrigated should be evaluated to determine the amount or consumption of water they need to sustain their nutrition and vitality during their growth period. Ground covers need 12.7–25.4 mm/week of water, while grass areas require 38.1 - 50.8 mm/week water.
- **Rain water harvesting:** In the urban landscape, rainwater management and planning, water harvesting, water retention gardens, ground covers are preferred over large grass surfaces [23]. The preparation of irrigation programs by classifying the amount of water consumed by plant groups can be achieved with the help of special applications such as rain gardens [24]. Rain water harvesting applied within the scope of rain water management is also an application evaluated in this section. It has been developed for maximum benefit from precipitation waters. With urbanization, which rapidly increases with the population, impermeable surfaces are prevented from leaking into the underground in the natural cycle of water, and the surface flows of rain water increase dangerously. This situation leaves the residential areas insufficient for the management of heavy rains. The collection and accumulation of rainwater and runoff water is considered to be the provision of water required for plant and animal production and the provision of water for domestic consumption [25].
- **Rain garden:** Systems are used to manage, slow down and direct the rainwater by controlling it. There are many applications such as rain gardens; landscape channels, vertical gardens, artificial wetlands, plant water arches created for this purpose [26]. In this context, rain gardens increase the quality of water in nature by improving the gathering areas of flow water and create more beneficial water resources for the environment. It helps to reduce the amount of rainwater that occurs in cities mixing with wastewater, thereby reducing the velocity of runoff water on surfaces [27].
- **Landscape channels:** Landscape channels are used to collect, slow down and direct rainwater in urban areas. It is generally created at the edge of vehicle roads, bicycle and walking paths to direct the water instead of collecting it and to filter it while reducing the speed of the current [26].

- **Creating permeable surfaces:** It aims to reduce the amount of runoff water, remove pollutants, and temporarily store rainwater. Permeable surfaces are very effective in preventing small-scale puddles and large-scale floods. The permeable surfaces provide the feeding of groundwater, prevent the cleaning of the surface water and the accumulation of excess water on the surface [24].
- **Limiting the use of grass areas:** Using alternative ground covers instead of large grass surfaces in urban open green areas. Grass surfaces are highly effective materials in creating recreational spaces in terms of aesthetics and functionality in landscape architecture. However, when using grass, it is necessary to be more careful considering the drought and water scarcity. The use of drought-resistant grass type is also important for water conservation. These species go into a dormant period to save water during drought. Also; It is necessary to avoid creating grass areas in the form of narrow, long and separate parts independent of the main area. Because the maintenance of these areas is more costly, irrigation may be a problem and sometimes irrigation may be unnecessary [28, 29].

Xeriscape Landscape Practices: The high water consumption of urban designs in open green areas has made the applications that can be used as little as possible. Rational use of water has created a xerophytic landscape [30, 1]. The Xeriscape current, which was first presented as a solution to long-term drought in the US state of Colorado in 1981, came to the fore as an ecological approach [31, 32]. The method is applied according to 7 basic principles (appropriate planning and design, soil preparation, appropriate plant species selection, designing easy and practical grass areas, effective irrigation, use of mulch and proper care) [33, 34].

- **Planning and design:** When carefully and according to the characteristics of the area such as climate, topography, flora and fauna, erosion and water loss due to wind and rain are minimized.
- **Soil analysis and preparation:** At this stage, the soil is analyzed. The missing minerals and their structure are corrected and made ready for landscape application..
- **Proper plant selection:** Plants with limited need for water and fertilizers should be preferred. In direct proportion to this situation, re-vegetation will decrease with pruning of trees and shrubs. Therefore, a less expensive garden will be created.
- **Reduction of grass areas:** It is the adjustment of the amount of grass areas to be sufficient only for the necessary use. A large lawn covered with exotic grass will always require much more water, maintenance and expense than other landscape materials.
- **Effective irrigation:** Effective irrigation and water savings are made by using less and pure water. Efficient irrigation can also be done with sprinkler on grass areas, drip irrigation on trees and bushes and automatic irrigation with moisture sensors.

- **Mulch use:** Mulch application with materials such as tree bark, sawdust, paper/cardboard, dry leaves and stone, gravel has many benefits [32].

7 Conclusion

Global climate change and temperature differences experienced in recent years firstly affect cities negatively. With urbanization, the landscape is changing, and hard surfaces such as buildings and roads replace green areas that are permeable and humid. Thus, the surface of the urban area and the air mass above are warmer than the surrounding rural area. Climate has a vital importance on living things. Climate is changing as a result of human activities. These changes affect both humans and other living things negatively. The effects of climatic changes appear slowly and for long years, and their harmful signals are beginning to be felt.

Some regions are areas where the use of existing water resources is limited or inadequate to use existing water resources. It has become necessary to implement water saving models where technologies are not available in these regions. The most important stage in landscape design and application is the planting stage. The most important stage in landscape design and application is the planting stage.

Arrangements with plants that have minimum water consumption and need should come to the fore. In planted areas and rain gardens, water should be taken from cities to slow down, direct, collect and help protect. Another opportunity that provides ecological and natural rainwater management is rainwater storage and utilization systems. The green infrastructure system facilitates the discharge of water in the face of disasters such as floods that are expected to increase during the climate change process. On the other hand, it creates a resource for the water supply of the city dwellers during dry periods.

It is one of the water resources that do not have a water Alternative. There is a limited amount all over the world and available water resources are decreasing day by day. Therefore, every drop of water is very important. Water resources are limited in many parts of our country. For this reason, it is necessary to develop economical use methods, especially in open green areas, despite a potential water shortage problem. In this respect, the protection of water quality and the effective use of users has a big role. It will be a solid step forward for all other water conservation programs, policies and regulations, such as education and awareness.

References

1. Bayramoğlu, E., *Sürdürülebilir Peyzaj Düzenleme Yaklaşımı: KTÜ Kanuni Kampüsü'nün Xeriscape Açısından Değerlendirilmesi.* Artvin Çoruh Üniversitesi Orman Fakültesi Dergisi. 2016, **17(2),** 119–127.
2. Taylor, P. D., *Fragmentation and Cultural Landscapes: Tightening the Relationship between Human Beings and the Environment.* 2002, Landscape and Urban Planning. 58:93–99

3. McHarg, I., *Design with Nature.* 1969, Natural History Press, Garden city, New York.
4. Topkurulu, S.C., *Peyzaj Tasarımında Enerjinin ve Suyun Etkin Kullanımı Yaklaşımlarının Ev Bahçesi Örneğinde İrdelenmesi.* 2019, İstanbul Üniversitesi-Cerrahpaşa Lisansüstü Eğitim Enstitüsü. İstanbul.
5. Özügül, M.D., *Ekoljik Planlamada Kullanılabilecek Analitik Bir Model Önerisi: Ömerli İçme Suyu Havzası Örneği,* YTÜ Mimarlık Fakültesi E-Dergisi. 2006, **1(4),** 201–217.
6. Cranz, G., *Changing Roles Of Urban Park: From Pleasure Garden To Open Space.* 2000, San Francisco Planning and Urban Research Association, SPUR, USA.
7. Emery, M., *Promoting Nature in Cities and Towns: A Practical Guide.* 1986, Croom Helm, London.
8. Eugenio A., and Ortiz L., Questioning *Ecological Design: A Deep Ecology Perspective,* 2003.
9. Demir, S., *Tarihi Peyzaj ve Peyzaj karakter Değerlendirilmesi Yaklaşımları ile Doğa Koruma Turizm Odaklı Peyzaj Planlama: Meryemana Vadisi Örneği,* 2017. Trabzon, Doktora Tezi, Karadeniz Teknik Üniversitesi, Peyzaj Mimarlığı Bölümü, Trabzon, Türkiye.
10. Onur, E.B., *Peyzaj Tasarım ve Yönetiminde Ekolojik Yaklaşım ve Sürdürülebilir Kent Hedefine Katkıları,* İnönü Üniversitesi Sanat ve Tasarım Dergisi. 2012, ISSN: 1309-9876, **2(5):** 245–252.
11. Korkut, A., Kiper, T., Topal, T. Ü. *Kentsel Peyzaj Tasarımda Ekolojik Yaklaşımlar.* 2017, Artium, **5(1)**, 14–26.
12. Kayli, A., & Gölbey, A. G. G., *Yeşil Altyapı ve Yeşil Bina Bileşeni Olarak Kurakçıl Peyzaj Uygulamaları.* Ege Üniversitesi Ziraat Fakültesi, Dergisi. 2020, 140–150.
13. Hansen, J., Ruedy, R., Sato, M. and Lo, K., *Global Surface Temperature Change.* 2010, Rev. Geophys, 48(4)
14. Uygur, İ., *Peyzaj Ve Küresel İklim Değişikliği Etkileşiminin Görsel Anlatım Biçimleri Üzerinden Değerlendirilmesi.* 2015, Doktora Tezi, İstanbul Teknik Üniversitesi, Fen Bilimleri Enstitüsü, İstanbul.
15. Liu, K., *Sustainable Building Envelope – Garden Roof System Performanc*e. 2004, NRC – CNRC, RCI Building Symposium, Nov. 4–5: 1–14, New Orleans.
16. Ertop, G., *Küresel Isınma ve Kurakçıl Peyzaj Planlaması.* 2009, Yüksek Lisans Tezi, Ankara Üniversitesi, Ankara.
17. Barış, M.E., *Kurakçıl Peyzaj.* 2007, Bilim Teknik Dergisi, **478:** 22–27.
18. Çakıroğlu, G., *Peyzaj Tasarımında Su Tasarrufuna Yönelik Güncel Uygulamaların Irdelenmesi: İstanbul Örneği.* 2011, Yüksek Lisans Tezi, İstanbul Üniversitesi Fen Bilimleri Enstitüsü, İstanbul

19. Atik, M., ve Karagüzel, O., *Peyzaj Mimarlığı Uygulamalarında Su Tasarrufu Olanakları ve Süs Bitkisi Olarak Doğal Türlerin Kullanım Önceliği.* 2007, Tarımın Sesi TMMOB Ziraat Mühendisleri Odası Antalya Şubesi Yayını, Sayı **15**, s. 9–12.
20. Hilaire, R., Arnold, M. ve Wilkerson, D., Devitt, D., Hurd, B., Lesikar, B., Lohr, V., Martin, C., McDonald, G., Morris, R., Pittenger, D., Shaw, D., Zoldoske D., *Efficient Water Use in Residential Urban Landscapes Hortscience.* 2008, Vol. **43(7),** Arizona.
21. Çorbaci, Ö. L., Özyavuz, M., & Yazgan, M. E., *Peyzaj Mimarlığında Suyun Akıllı Kullanımı: Xeriscape.* 2011, International Journal of Agricultural and Natural Sciences, **4(1),** 25–31.
22. Bayramoğlu, E., Ertek, A., & Demirel, Ö. *Su Tasarrufu Amacıyla Peyzaj Mimarlığı Uygulamalarında Kısıntılı Sulama Yaklaşımı.* İnönü Üniversitesi Sanat ve Tasarım Dergisi. 2013,**3(7).**
23. Bayramoğlu, E., Demirel, Ö., *Xerophytic Landscape. In: Environment and Ecology at the Beginning of 21st Century.* 2015, ST. Kliment Ohridski University Press, Sofia, pp180–190
24. Büyükkurt, U., *Sürdürülebilir Kampüslerde Su Tasarrufuna Yönelik Çalışmalar Karadeniz Teknik Üniversitesi Kanuni Kampüs.* 2019, Yüksek Lisans Tezi. Trabzon
25. Mengü, G. P., ve Akkuzu, E.. *Küresel Su Krizi ve Su Hasadı Teknikleri.* Adnan Menderes Üniversitesi Ziraat Fakültesi Dergisi. 2008, **5(2),** 75–85.
26. Sert, E., *Enerji Etkin Kentsel Peyzaj Tasarımında Yağmur Suyu.* 2013, Doktora Tezi, İstanbul Teknik Üniversitesi
27. Jaber, F., Woodson, D., LaChance, C. and York, C., *Stormwater Management: Rain Gardens.* 2012, the Department of Soil and Crop Sciences and Texas A&M AgriLife Communications. the Texas A&M System, USA, p. 20. 11
28. Strong, R., Small Garden Designer's Handbook. 1987, Conran Octopus Limited, London.
29. Swindells, P., *The Complete Book of The Water Garden.* 1989, Ward Lock Ltd. London.
30. Vasishth, A. A., *Scale-Hierarchicecosystemapproachtointeg-Rative ecologicalplanning.* 2008, Progress in Planning, **70(3)**:99–132.
31. Perry, B., *Landscape Plants for Western Regions:An Illustrated Guide to Plants for Water Conservation.* 1992, Claremont: Land Design Publishing.
32. Wade, G., Midcap, J., Coder, K., Landry, G., Tyson, A., Weatherly Jr, N., *Xeriscape, a Guide to Developing a Water-Wise Landscape.* 2007, Georgia Environmental Landscape Design Department, Georgia.

33. Gary L. Wade, Midcap, J. T., Coder, K.D., Landry, G., Tyson, A.W., Weatherly, N. J., *A Guide to Developing a Water-Wise Landscape.* 2009, University of Georgia Environmental Landscape Design Department, Georgia 30602, pp 44.
34. Çetin, N., Mansuroğlu, S., *Akdeniz Koşullarında Kurakçıl Peyzaj Düzenlemelerinde Kullanılabilecek Bitki Türlerinin Belirlenmesi: Antalya/ Konyaaltı Örneği.* Ege Üniversitesi Ziraat Fakültesi Dergisi. 2018, **55(1),** 11–18

Ertan VARLI

Parametric Design in Architecture Education

Abstract: Advancements in digital technologies have affected architecture education as well. In this direction, in order to prepare the students for the occupational environment, alterations have been made in architecture education program as well and lectures comprising digital technologies have been included. Digital technologies are being used in architecture education not only in presentation and visualization fields, but also as a design supporting the creation and expression of creative design. Therefore, search for new methods for creative design and communication in architecture education has become a necessity. Parametric design, the leading method of the new methods in which digital technologies are commonly used in architecture education in these days, involves all designing movements that enable renewal of the design by changing a series of parameters at any stage of the design. The thing that changes in the parametric design is rather the parameters that constitute the form, not the form itself. Changes in the parameters that have impact on the design affect the final product and as a result, different design alternatives are introduced. In parametric design, design is based on specific parameters and is made with the help of parametric modelling.

In this study, effect of the parametric design on architecture education is examined. Within the content of the study, architecture students who have never used parametric design principles in architecture education were given 6 hours of parametric design training. Parametric design principles were explained to the students within the training program and they were asked to create a cover coat based on the parameters in design processes with the aid of a computer.

As a result, it is thought that with the usage of parametric design in architecture education, students will be provided advantages in producing different design combinations through the models they design, increasing their senses of three dimensional thought by using computer aided design tools and developing creative designs.

Keywords: Parametric Design, Architecture Education, Parametric Design Education, Digital Technologies, Parameter

1 Introduction

"Architectural design is interpretation, determination and documentation of conceptual, functional, formal, structural and performative properties and qualities of all of the elements involved by the construction of the entire structure that will fulfil the functions determined in order to meet the requirements" [1]. Nowadays, algorithm that underlies the computer and mathematics sciences as a part of architectural design and algorithmic thoughts that can be defined as the necessary stages to solve a problem compose the parametric design. In its most fundamental form, parameter can be expressed as qualitative and quantitative data that can be defined and changed in accordance with a specific circumstance. Number of parameters can vary depending on the situation. What matters is to make the connection between the parameters and to be able to change these parameters arbitrarily.

The objective of this study is to explore the effect of the parametric design on architecture education with the advancement of digital technologies. Within the training carried out within the content of the study, effects of the parametric designs created by the architecture students on architecture education were observed and analyzed.

2 Parametric Design

In the dictionary of Turkish Language Association, parameter is defined as "variable quantity" [2]. In its most fundamental form, parameter can be expressed as qualitative and quantitative data that can be defined and changed in accordance with a specific circumstance. Number of parameters can vary depending on the situation. What matters is to make the connection between the parameters and to be able to change these parameters arbitrarily.

Nowadays, algorithm that underlies the computer and mathematics sciences as a part of architectural design and algorithmic thoughts that can be defined as the necessary stages to solve a problem compose the parametric design. Despite the fact that history of algorithm and algorithmic thought is dated, application of this way of thinking in architecture is pretty recent. Terzidis, who has tried to adopt the algorithmic thinking to architecture field argues that "instead of considering the programming as an addition to architecture, programming should be syncretized with and internalized by architecture" [3]. Parametric design is used in design process, concept developing, formation of the space and analyze of the application details.

In the usage of parametric design in concept development stage, data of the environment to which the design belongs and other factors decided by the designer are determined as the parameters. With the change of these parameters, change of the form is ensured and by this means, design alternatives are determined. BMW Dynaform Pavilion that took place in 2001 Frankfurt International Auto Show is a project that can be an example of using parametric systems in concept developing stage (Fig. 1).

Fig. 1: BMW Pavilion-Dynaform

In analyze of application details, parametric design is used to bring various solutions based on detail solution by changing the values of the parameters through generating a formula in places where dimension, angle, thickness changes are required. Foster & Partners Architectural Firm has used parametric design approach for the design of cover coat that covers the gallery between the exhibition hall and the library of British Museum (Fig. 2).

Fig. 2: British Museum

In parametric design, design is based on specific parameters and is made with the help of parametric modelling. Parametric modelling is a modelling method that is established based on the parameters and that allows the changes in these values and as a result, in which end result can be changed consistently under the control of parameters [4]. Parametric modelling uses topological geometry as base. Because while Euclidean geometry functions in a metric space based on concepts like length, area and volume, in topological geometry space, distance is expressed with a dynamic length type that cannot be defined by proximity and that does not remain constant. In topological space, objects can be stretched without changing the characteristics of the identified points or the scales of the objects can be changed [5].

According to Mark Burry, in order to be able to model a structure or a space [6],

- First of all, constants and variables should be defined.
- Space or the structure should have a regular geometry composed of a number of sub-systems that are associated with each other.
- Space or the structure should have computable associations.

3 Method

Within this study, parametric design training was given to architecture students. Parametric design principles were explained to the students within the training and they were asked to create a cover coat based on the parameters in design processes with the aid of a computer. Within the scope of the study, parametric modelling techniques were used and cover coat design was made on Saraçlar Street. At the beginning of the study, parameters were designed and relations between the parameters were built. Execution of the design was made with the usage of "Rhinoceros" software.

For the cover coat that was considered as a parametric design element, parameters were determined by constituting a design concept that would be a combination of function, space, form and sustainability concepts. In parametric design, design elements that were studied within the context of continuity and mobility sub-contexts were also considered and attention was paid to the harmony of the design with Saraçlar Street. Parametric design approaches were studied within the contexts of scale, ratio, dimension and height and different design alternatives were generated.

Alternative 1

Cover coat that was generated by starting out with the human brain section was considered in a way that it would not cut the pedestrian traffic. In the meantime, cover coat that covers the book sales function aimed to create a small culture corridor on Saraçlar Street (Tab. 1).

City Location	Sketch
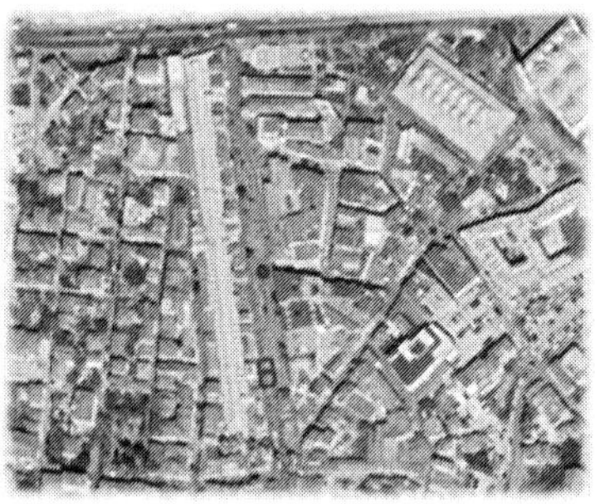	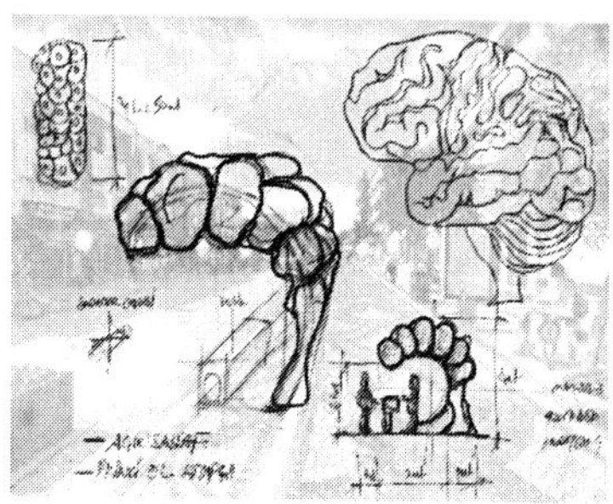
Rhinoceros Model	**End Result**
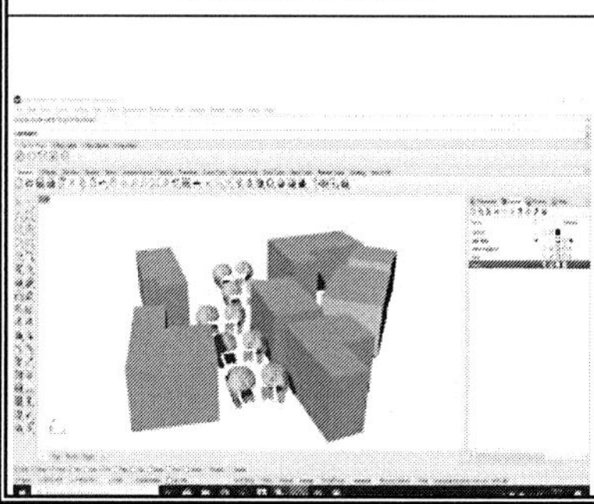	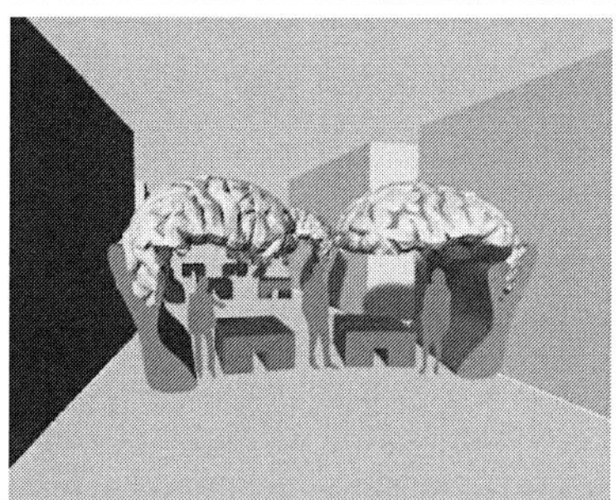

Tab. 1: Alternative 1

Alternative 2

Cover coat that was designed to cover the furniture placed in parallel with Saraçlar Street was considered in a way that it would pay attention to the sense of circulation of the street. Cover coat was formed by also considering the sun's angle of incidence depending on its location and the form was deformed with drainages and/or pargetting on the modules separated with the parameters (Tab. 2).

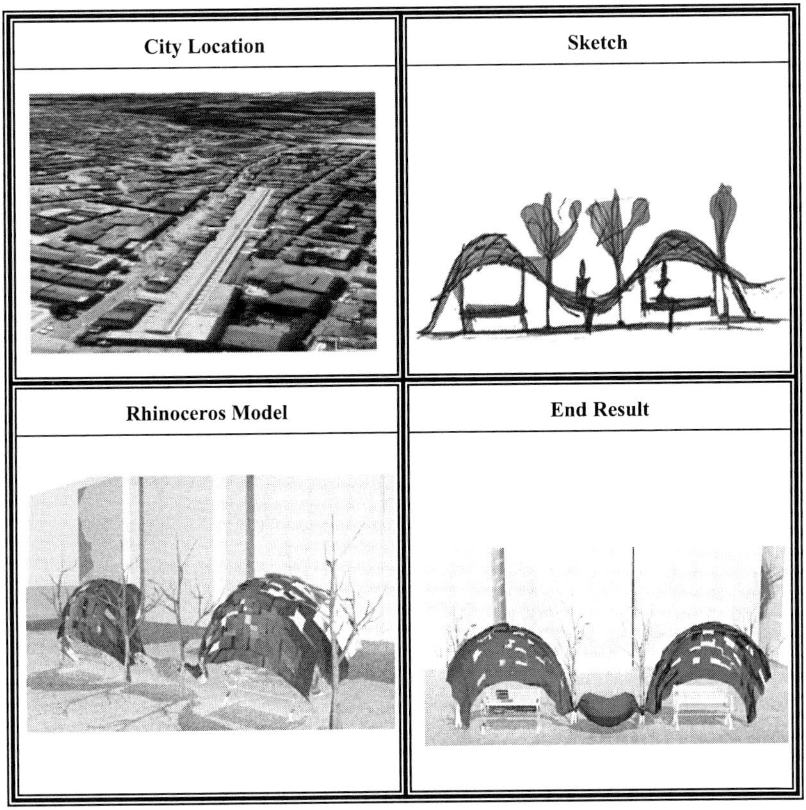

Tab. 2: Alternative 2

Alternative 3

Parasite coat which was designed to cover the bazaar function of Saraçlar Street at the location where the pedestrian traffic density is the highest was tied to the masses on the eastern and southern sides of the street and was composed of pentagon parameters (Tab. 3).

City Location	Sketch
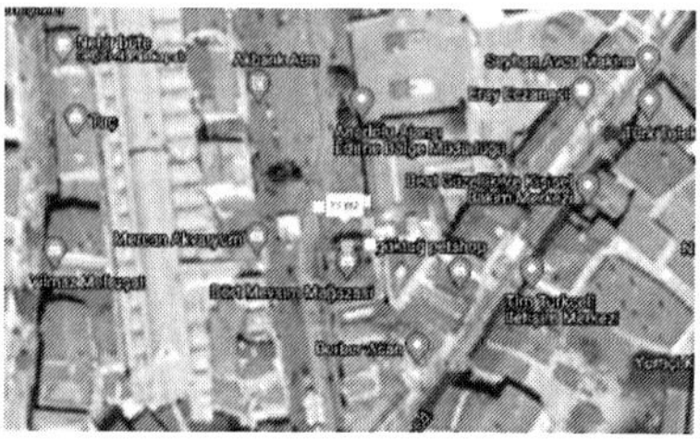	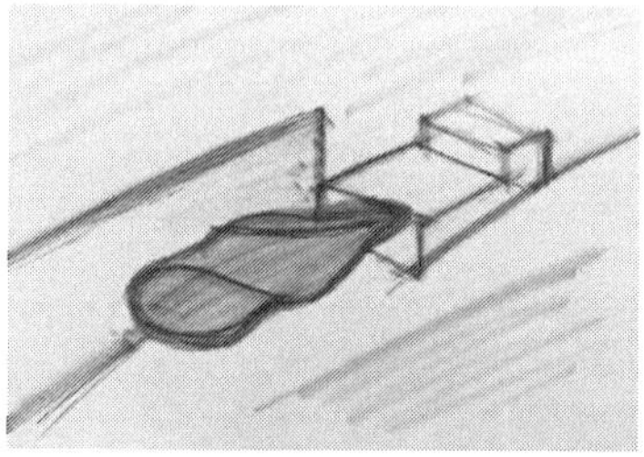
Rhinoceros Model	**End Result**

Tab. 3: Alternative 3

Alternative 4

With its corrugated lines, cover coat which was designed with the purpose of covering the exhibition function of Saraçlar Street without disturbing the human density on the street aimed to refer to the dynamism and fluidity of the street (Tab. 4).

City Location	Sketch
	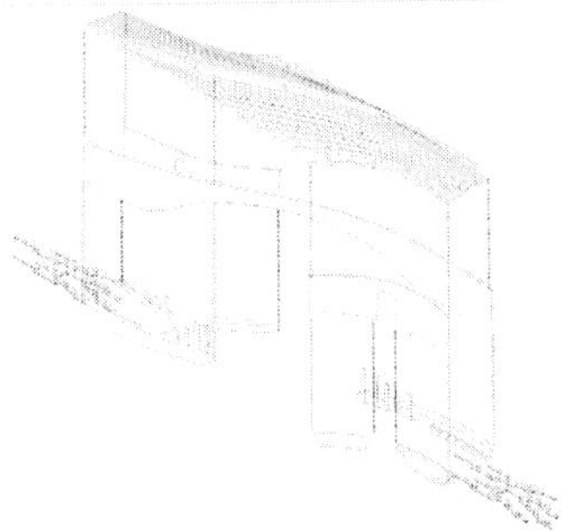
Rhinoceros Model	**End Result**

Tab. 4: Alternative 4

3.1 Sample Group

Study was conducted for 6 hours with 38 students who took New Approaches in Architectural Design in Trakya University, Faculty of Architecture, Architecture Department in 2020–2021 fall term.

3.2 Data Collection Tool

In this study, goal-directed data collection tool is the "Student View Survey" that was conducted over Google Form.

3.3 Student View Survey

A parametric design survey consisted of 6 questions was developed in order to learn at which stages, do the students who participated in the survey thought of using parametric design approach throughout their educations (design process, concept development, forming the space, analyzing the application details), if the duration of the parametric design training within the scope of the lecture was enough or not and whether the course content and the parametric design training made contribution or not in terms of professional development. With the help of this survey, personal views of the students regarding the course content and the education given to them were analyzed and assessed.

4 Findings

When the parametric cover coat designs that were designed on Saraçlar Street within the scope of the study was considered within the context of continuity and mobility concepts, it was concluded that the cover coats that were designed conformed with Saraçlar Street and that they were generated based on a flexible design process.

In the 6 questions "Parametric Design Education" survey that was prepared with the purpose of finding out the positive and negative personal views of the students regarding the course content and the attainments they had throughout their education, positive views regarding the outcomes of the lecture weighed.

94,7 % of the students stated that they were planning to use parametric design approach in the design process while 89,5 % of the students stated that they were planning to use it in concept development process, 89,5 % while forming the space and 86,8 % of the students stated that they were planning to use it in analyzing the details (Tab. 5).

All of the students who answered the survey had the opinion that the course content and the parametric design education made contributions to them in terms of professional development. In the 6th question of the survey, sufficiency of the duration of the parametric design education was asked as an open ended question. 63,2 % of the students stated that the parametric design education received within the scope of the lecture was insufficient, that the practice implementations were not adequately repeated due to lack of time and the course hours should be at least 10 to 12 hours (Tab. 6).

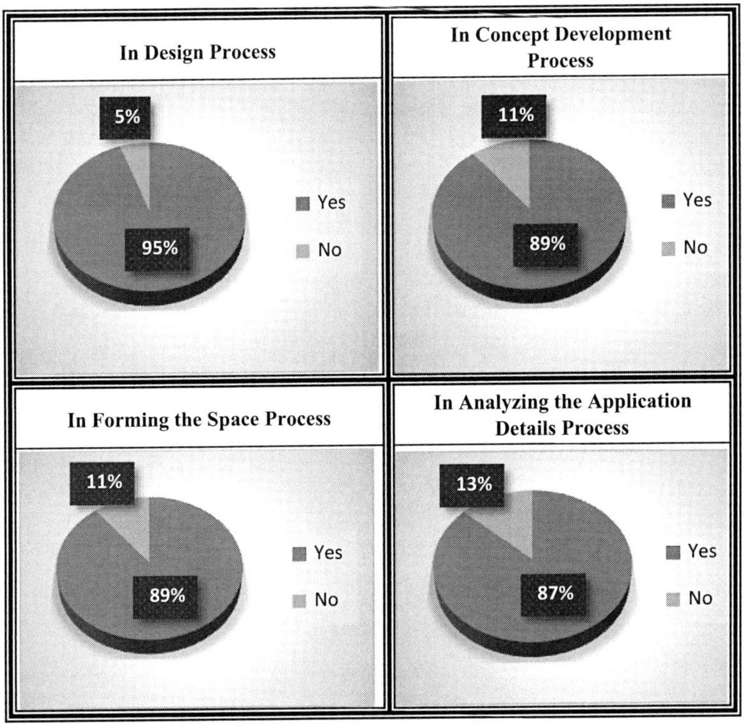

Tab. 5: Usage of Parametric Design Approach throughout the Architecture Education

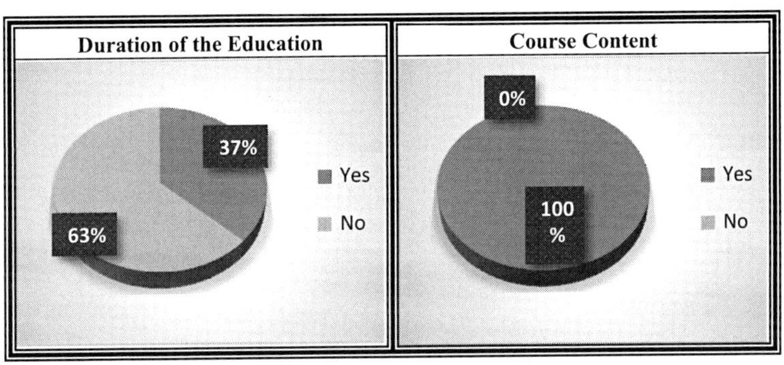

Tab. 6: Contributions of Parametric Design Education Duration and Course Content in terms of Professional Development

5 Conclusions

By using the parametric design created by the algorithmic thoughts in architecture education, students will be provided advantages in creating different design combinations through the models they will design, in increasing their senses of three dimensional thinking using computer aided design tools and in generating creative designs.

As a result of the findings that were obtained, it is weighed in the student view survey that the parametric design will make contributions in terms of their professional development. When parametric design was considered within the limits of the practical works carried out throughout the education process, it was stated that it was insufficient, however, it was also stated that a programming education to be given by blending it with occupational information within the education life would be more effective with regard to adopting parametric design approaches.

In conclusion, it is suggested that parametric design should be thought together with the traditional design methods in universities giving architectural design educations and by extension, software that supports parametric design should be featured at the early stages of the architecture education.

References

1. İzgi, U., *Mimarlıkta Süreç, Kavramsal İlişkiler*, Yapı Endüstri Merkezi Yayınları, 1999, İstanbul.
2. *www.tdk.gov.tr./tdksozluk.*
3. Erdoğan, E., Sorguç A. G., *Hesaplamalı Modeller Aracılığıyla Mimari ve Doğal Biçim Türetim İlkelerini İlişkilendirmek*, JFA, Cilt:28, Sayı.2, Syf:269–281, 2011.
4. Schmal, P., *Digital-Real, Blobmeister: First Built Projects*, Birkhauser, 2001, Berlin.
5. Hensel, M., Menges, A., *Morpho-Ecologies, Towards an Inclusive Discourse in Heterogenous Architectures*, Architectural Association, AA Agendas, No:4, 2008, Londra.
6. Burry, M., *From Real to Virtual, Towards New Design Conventions*, Faculty of Architecture University of Bialystok, Bialystok, 1997, Poland.

Elif AKPINAR KULEKCI and Feran ASUR

Autism Spectrum Disorder (ASD) and Garden

Abstract: Garden culture, which emerges in different ways in different cultures and is shaped according to the social, cultural and ecological conditions of societies, is one of the living spaces of human beings from past to present. Especially in landscape architecture studies, the garden has a great importance in bringing autistic individuals to life within the framework of designs for human health, psychology and in their physical, spiritual and cognitive development. In this study, it is aimed to determine the garden design approach for autistic individuals who have difficulties in social interaction and communication. In this context, the concept of garden and autism was discussed in detail, and the points to be considered in garden designs that appeal to the senses and development of autistic individuals were tried to be determined. As a result of the study, the importance to design gardens that support social skills and environmental education, which are integrative with green spaces that have high sensory stimulation, in a way that encourages their mental development, where autistic individuals can act with their own free will without restricting their movement and behavior was revealed once again. In addition, the necessity of creating spacious spaces with natural vegetation and large green areas, which provide various sensory opportunities to these individuals, especially in bringing autistic individuals to life, arises. For this purpose, it was emphasized that joint studies should be carried out with different professional disciplines, so that such studies will play an important role in the social development of individuals with autism by contributing to their physical, psychological and cognitive development as well as their social development.

Keywords: Autism Spectrum Disorder (OSB), Garden, Landscape Design

1 Introduction

Garden culture; in the historical process, has developed with the idea of watching the view in the western in general terms and living in the eastern societies. Regardless of civilization, it is known that all of them have developed a unique garden idea and these gardens have changed in line with the topographical, ecological, social and cultural characteristics that dominate the geography of the societies [1]. The garden, which is defined as an enclosed area, has developed over time as a space that has certain boundaries and has been planned and arranged by people according to the purposes, and has become a living space created by people to integrate with nature and to be closer to nature [2].

The word *"garden"* is actually Persian and means "little vineyard". In general; it is also expressed as a piece of land where flowers, ornamental plants, fruits, vegetables and herbs are grown, as well as the beauty, greenery, and resting of nature are controlled by human beings [3]. Gardens allow people to experience colors and textures visually, perceive the sound of the wind on trees, and experience the environment with their senses [4]. Gardens are complementary elements that serve human beings for various purposes by combining different plant, flower, animal and insect species. Gardens have existed since the world came into existence. Traces of different geographies and cultures can be found in gardens from ancient cultures to modern cultures.

Also, gardens come in many types. Peace, tranquility and wildlife are also a kind of revitalization for the cultivation of the land or the places where agricultural resources are developed. Globally, gardens have helped all living people survive and relax. Additionally, they have inspired important artists, poets, musicians and writers [5]. Gardens have therapeutic benefits for humans, such as reducing depression, anxiety and stress. A wide variety of choices in the world of autistic individuals give them a sense of control over themselves and their environment, and consequently improve their self-confidence.

All over the world, there is increasing interest in research results showing the impact of the physical environment on people's mental and body health. It has been shown in many studies that the designs made indoors and outdoors not only provide functional efficiency but also strengthen and improve people's health processes. All these studies led to the emergence of a new architectural field called "Design and Health" [6,7].

Recently, it has been observed that knowledge and awareness of how good design can affect people's well-being has increased among architects, interior designers and landscape architects. Especially in Landscape Architecture studies, designs for human health and psychology have started to show up. In this study, it was tried to determine the points that should be taken into consideration when designing a garden in the landscape design process for psychological, physical and cognitive development, especially for autistic individuals.

2 What is Autism Spectrum Disorder (ASD)?

ASD is a neurological problem that manifests itself in the first 3 years of an individual's vital journey and affects the functioning of the brain. It can cause an obstacle to social life, behaviors and even communication, which causes lagging behind from the peers in various stages of development and sometimes deviations from the goal. Therefore, these individuals, who show a different course of development, have different prognoses in the disease [8,9.] Due to this disease, which spreads rapidly throughout the world, the awareness of society is increased by carrying out various research activities in the world in order to increase the knowledge of individuals living in the society about autism.

Today, the causes of autism spectrum disorder is not clearly known yet, but there is evidence that it is genetically based. However, the responsible genes are not known yet. There are opinions that environmental factors can also lead to autism. There is a lot of research on the effects of both genetic bases and environmental factors.

Autism has nothing to do with the child-rearing characteristics or the economic conditions of the family; Therefore, autism spectrum disorder is encountered in all kinds of societies, different geographies, races and families.

3 What Are the Symptoms of ASD?

Children with Autism (ASD);

- Do not make eye contact with others,
- Do not look when you say his name,
- Act like they do not hear what is being said,
- Do not point to what he wants with his finger,
- Do not know how to play with toys,
- Not interested in the games their peers play,
- Say certain words repeatedly and in unrelated settings,
- They lag behind their peers in speaking,
- Have strange movements such as swinging or struggling,
- Overly mobile, always behaving in their own way,
- Their eyes are stuck on something,
- They do unusual movements such as rotating some items or lining up,
- They overreact to changes in daily life.

If these symptoms are observed, the individuals should be evaluated in terms of autism. Approximately fifty percent of children who receive intensive education with early diagnosis and a correct education method can control the symptoms of autism, develop, make great progress, and even some children with autism may not be different from their peers when they reach adolescence [8, 9, 10, 11, 12, 13].

The main cases associated with the concept of autism spectrum disorder can be listed as follows;

- Autism spectrum disorder is thought to be caused by neurological causes. In a significant proportion (approximately 35 %) of individuals with autism spectrum disorder, due to abnormal electrical movements in the brain; seizure, involuntary movements, loss of consciousness, etc. neurological problems can also be seen.
- Autism spectrum disorder is not a mental illness; however, its symptoms may recall some mental illnesses.
- Scientific research has shown that autism spectrum disorder is not related to child-rearing characteristics or family's socio-economic characteristics.
- There is evidence that autism spectrum disorder may be inherited; however, the gene or genes that cause this have not yet been explored.

- While the incidence of autism spectrum disorder was accepted to be one in 500 in previous years, it is thought that autism spectrum disorder affects approximately one in every 59 children according to the latest data. In addition, its prevalence in males is 4.3 times higher than females.
- Contrary to popular belief, most individuals with autism spectrum disorder have different levels of mental retardation. Also, in intelligence tests, certain areas can be much more backward than other areas.
- Very few (about 10 %) of individuals diagnosed with autism spectrum disorder have superior characteristics like strong memory, good musical abilities, etc.

According to the guide (DSM-V) published by the American Psychiatric Association, Autism is called "Autism Disorder" and manifests itself in two areas of inadequacy [14].

a) Difficulties in Social Communication and Interaction
Difficulty in building and maintaining relationships
Inability to make eye contact
Inability to express feelings
Difficulty initiating and maintaining interactions
b) Restricted-Repetitive Behavioral Patterns (Repetitive Behaviors)
Stereotypes and repetitive motor movements
Insistence on same routine, strict adherence to routine
Limited and intense interest
Being stimulated less or more than normal

4 Autism and Garden

A wide variety of options in the world of autistic individuals give them a sense of control over themselves and their environment, and consequently improve their self-confidence.

Physical effects that may cause distress for individuals with autism in an open space, such as; lightings containing excessively bright light, a perfume smell that may cause nausea, certain foods or textures that may reveal gas reflexes. Individuals with autism who experience some or all of these reactions are easily upset by the discomfort they feel. At this point, sensory gardens help them cope with these troubles [4]. The garden has three main functions for individuals with autism [15, 12, 13, 16]:

a) Therapeutic (healing) function: It is defined as a series of interactions to provide the desired changes in the emotions, attitudes and behaviors of the autistic individuals.
b) Educational function: It can be defined as the set of activities that involve the person with autism in activities and games and increase their verbal or non-verbal interaction cognitively (Fig. 1).

Fig. 1: An example of a garden designed for individuals with autism [16].

c) Restorative (allowing individuals with autism to have control over their overstimulated senses) function: Gardens made for autistic individuals, which are increasing in popularity due to their positive effects on users, attract more and more countries around the world. These gardens, which appeal to the senses of individuals, have to be considered in terms of design. In order to provide the expected effects and benefit from the sensory gardens, the user population must first be determined and it is essential to design according to the user. Sensory arousal occurs when the five senses come together. Our five senses can be counted as sight, hearing, touch, taste and smell;

- Sight: Using materials which will trigger the stimulation of the sense of sight is important for sense gardens. Although color, texture, light, shape have a strong effect on people, the perception they arouse is different for each person (Fig. 2).

Fig. 2: A design that improves the eyesight of individuals with autism [17].

- Hearing: To attract living things that can be sound elements or natural sound elements to the garden by adapting to the auditory senses, using water elements, wind and leaf sounds of tree bushes as when the wind blows represent calmness and peace for many people.
- Touch: Gardens designed for autistic individuals are defined as theme gardens that provide a sense of touch, smell, hearing, taste and sight which benefit young children, elderly and people with mental disabilities as a self-sufficient area where sensory experiences can develop in general.
- Taste: The sense of taste can be loaded into the sensory gardens with different activities such as the use of edible plants and the use of information boards as well as their maintenance.
- Smelling: Scent is often preferred by people, and many plants create attractive spaces with their unique fragrances. Visually impaired users can find directions or plants by using their sense of smell.

5 Considerations in Garden Designs that Appeal to the Senses and Development of Autistic Individuals

Considering today's living conditions, these gardens help reduce stress, encourage exercise and relax in the open green areas needed; by offering places where users can enjoy nature. Such gardens can maintain their existence for many years by establishing their own cycle in a healthy way. It is possible to list the design principles and features of these gardens that contribute to the senses and development of individuals with autism as follows [18, 19, 15, 20, 21, 22, 23, 24, 25, 4, 26, 27].

- When designing a garden for autistic individuals, it should be ensured that the five senses are stimulated by using vegetation such as trees, bushes, water elements and other garden elements, and evaluating their color, texture, smell, sound and taste (Fig. 3).

Fig. 3: An example of a garden design with different landscape design elements designed for individuals with autism [28].

- A comfortable environment should be created by developing a sense of curiosity and secret. Gardens covered with comfortable, peaceful and soft textured trees, flowers and vegetation can offer a relaxing and tranquil environment.
- Balanced spaces should be created in gardens, thanks to the regular and rhythmic placement of plants and other elements. In this case, balance may be an important criterion for creating calming and relaxing spaces.
- Proportional spaces should be created in terms of harmony of color, form, texture, smell and sound. The fact that the sensory arousal is at a controllable level should be considered an important factor in preventing the users from getting bored with the sensory confusion in the space.
- In lice plants or plant groups that can have a large top diameter and have a shadow effect should be used. In this way, psychological protection, comfort and confidence should be created.
- The use of plant groups in different colors, textures and sizes should try to provide a sense of visual attraction and depth.
- The water element should create soft and naive sounds and movements, and should be used in a way to help visually relax. Otherwise, too much noise, use of motion or visual discomfort can cause concern and anxiety, which can lead to stress.
- Rhythmic sounds should be created with water elements. These sounds can be created by cascades, ripples, and fountains of various forms. The sounds created by the water element enable the senses to be stimulated and benefit from the healing effect of water. Another feature of water is that it can absorb sound. These factors should be considered in order to create a calm and relaxing effect.
- Separate garden sections should be created for different senses. Spaces should be designed where autistic individuals can use their preferences for any sensory department they want to experience.
- In addition to flowering plants, spaces and facilities such as birdhouses and shelters, should be designed in order to attract animals such as birds and butterflies that give peace to people.
- An area should be reserved in the garden for users who want to participate actively in plant care. This area is important in terms of allowing them to establish a psychological, emotional and physical relationship between the users and the garden.
- Smooth turns and curvilinear lines should be in the foreground on the roads and should be designed to reach the place directly. If the roads are to be made with steeper angles, the road edges and turning points should be softened with plants.
- Paving materials for walkways should have a soft texture and make individuals feel natural. The roads are designed to be smooth and easy to pass, allowing them to be maneuvered for wheelchair users (Fig. 4).

Fig. 4: An example of a garden designed for individuals with autism [29].

6 Conclusion and Recommendations

From past to present, gardens have been places where people can relax, perform recreational and sports activities, and have physical, social and physical relaxing effects. These healing and positive effects of gardens can turn into an advantage with the right landscape design approaches for the healing and development processes of individuals with Autism Spectrum Disorder (ASD). Thus, it is ensured that these individuals are reintegrated into the society and become less dependent individuals who are self-sufficient. Especially in the designs to be made, attention should be paid to components such as floor preferences, plant choices, colors, textures and forms, and designs that emphasize sensory components, that is, provide maximum stimulation.

When the literature studies conducted are examined, it is seen how important it is to design gardens that support social skills and environmental education with green spaces with high sensory stimulation, in a way that especially autistic individuals can act with their own free will without restricting their movements and behaviors and encourage their mental development. Rapid increase in concrete consolidation and the decrease in the amount of open green space per person is a big problem for people living in cities as well as for individuals with mental and physical disabilities or developmental disabilities. Garden designs, which include active recreational designs, are often a breathing space for autistic individuals who cannot cope with the world they live in and cannot show the same development as their peers. In this study, it is recommended to pay attention to the following criteria in garden designs for autistic individuals:

- For people with autism, the world they live in is often full of difficulties. Therefore, regularity, calmness and consistency are important for individuals with autism. Seeing more than one item at the same time can overstimulate or tire an autistic person. It is recommended to pay attention to all these factors in the designs.
- The time a person with autism will spend outdoors in a garden can offer him a valuable and therapeutic experience. It is thought that designs with aromatic and evergreen oak species that are set up in a garden, will be beneficial as they influence sensory stimuli.
- Garden designs that can be useful in reducing stress, creating a balanced sensory environment, increasing motivation and improving one's skills should be included.
- Creating lines that provide clear vision for individuals to take part in the environment is very important in the development of autistic individuals who are disturbed by excessive light. For this, a garden design should be included in order to create shadows in places if necessary.
- In a landscape design to be designed for individuals with autism, design elements such as large grass areas, movement, direction, rhythm, texture, color, shape, light, shadow should be included and have different textures that activate the senses of vision and hearing, feeling and taste. Surfaces, pools with effective safety precautions, plants without strong scent, direction and planting boards, gravel roads, fun water fountains should be included.
- For autistic individuals who are sensitive to sounds, the gardens in areas with heavy traffic, a noise curtain can be created with plants. In addition, music corridors can be created by taking into account the relaxing effect of music.
- Flooring materials and sandboxes with different texture can be designed.
- Aromatic, slightly textured, non-coniferous plant species (tree, shrub, shrub and groundcover) plants such as: *Acacia farnesiana* (acacia), *Buddleia davidii* (butterfly bush), *Lavendula* spp. (lavender), *Rosmarinus officinalis* (rosemary), *Salix babylonica* (weeping willow) [30] can be used in road routes and recreational areas.
- Plants such as: *Solanum lycopersicum* (tomato), *Capsicum frutescens* (ornamental pepper), *Lactura sativa* (lettuce), *Daucus corata* (carrot), *Mentha spicata* (curly mint), *Mentha suaveolens* (mint), *Salvia* spp. (sage) which are among the plant species grown in edible gardens [30] should be preferred.
- Colorful plants should be used for the sense of sight. Color provides a visual stimulation, create balance, unity rhythm and focal points. In this context, *Hibiscus rosasinensis* (Japanese rose), *Justicia brandegeeana* (shrimp flower), *Lagerstroemia indica* (needle tree), *Gazania rigins* (sheep's eye flower) [31, 32, 33] plants can be used.
- Plants should be touchable by autistic individuals in gardens designed for individuals with autism. Soft textured flowers, fluffy leaves, juicy leaves, prickly seed pods, fruit and sweet plant juices can give an educational excitement to

individuals with autism. *Salvia officinalis* (sage), *Salix caprea* (goat willow), *Geranium sanguineum* (geranium), *Petroselinum crispum* (parsley), *Rumex acetosella* (sheep's sorrel), *Cyclamen* sp. [35]. plants can be used.

- You can sit under a tree to listen to the sound of the wind. Bamboo stalks rustle, palm leaves shake. Leaves can be left to step on by the feet. The sounds of animals can stimulate feelings. Areas can be created to attract birds and butterflies. The water element can be used to bring sound to the garden.
- *Bambuseae* spp. (bamboo), *Briza maxima* (Big quakinggrass), *Miscanthus giganteus* (chinese cane), *Quercus* spp. (oak), *Phoenix dactylifera* (palm) [35, 36] plants can be preferred.

As a result, we can say that gardens designed for autistic individuals are very important to meet the physical, mental and psychological needs of these individuals. In order to meet these needs, designers need to design the characteristics of the garden in a sustainable way, presenting ideas to the users in a way that indicates how to behave and perceive the gardens. Especially in bringing autistic individuals to life, it is the duty of a landscape designer to create spacious spaces that provide various sensory opportunities to them, contain natural vegetation and large green areas. While doing this, multidisciplinary collaborative studies should be carried out by different areas of profession. Thus, it is thought that such studies will play an important role in the social development of individuals with Autism Spectrum Disorder by contributing to their physical, psychological and cognitive development as well as their social development.

References

1. Çınar, H.S. and Kırca, S. Türk Kültüründe Bahçeyi Algılamak. İ.Ü. Orman Fakültesi Dergisi, 2010, vol.**60**, pp.59–68.
2. Kuş Şahin, C. K. and Erol, U. E., Türk Bahçelerinin Tasarım Özellikleri. Süleyman Demirel Üniversitesi Orman Fakültesi Dergisi, 2009. **2,** 107–181.
3. Demiröz, Z., Tarihsel Süreç İçinde İslam Bahçe Sanatı Hint-Moğol Bahçeleri Örneği ve İslam Bahçeleri'nin Türk Bahçe Sanatı'na Etkileri. İstanbul Teknik Üniversitesi, Fen Bilimleri Enstitüsü, Y. Lisans Tezi, 2003. **129s**, İstanbul.
4. Şensoy, N. Sensory Garden Design for Individuals with Autism Spectrum Disorder. Inonu University Journal of Art and Design. 2017. **Vol. 7,** No.15: 115–128.
5. Çetin, U., İngiliz Bahçe Sanatı. Meriç Uluslararası Sosyal ve Stratejik Araştırmalar Dergisi, 2013. **2**(4), 1–19.
6. Marcus, C. C., & Barnes, M. (Eds.). Healing gardens: Therapeutic benefits and design recommendations, 1999. Vol. **4.** John Wiley & Sons.
7. Dilani, A., Psychosocially Supportive Design- Scandinavian Health Care Design, World hospitals and health services: the official journal of the International Hospital Federation, 20013. **7(1)**:20–4, 33, 35.

8. Borazancı, Person, S., AQ Otistik Zekâ ve Seviyeleri Otizm. İstanbul: 2000. Sistem Yayıncılık.
9. Aydın, A., Otizmde İlk Adım. 2. Baskı. Epsilon Yayıncılık. 2003. pp. 144. ISBN: 9789753314374. İstanbul.
10. Tokuç, F., Otistik Çocuk ve Aile Özelliklerinin Aile İşlevlerine Etkisi. T.C. Marmara Üniversitesi Sağlık Bilimleri Enstitüsü Yüksek Lisans Tezi, 2009. İstanbul.
11. Üstüner Top, F., Otistik çocuğa sahip ailelerin yaşadıkları sorunlar ile ruhsal durumlarının değerlendirilmesi: Niteliksel araştırma. Çocuk dergisi 2009. **9** (1), 34–42.
12. Bashir, A., Bashir, U., Lone, A. & Ahmad, Z., Challenges faced by families of autistic children. İnternational journal of interdisciplinary research and innovations. 2014. **2** (1), 64–68.
13. Gona, J. K., Newton, C. R., Rimba, K. K., Mapenzi, R., Kihara, M., Vijver, F. V. & Abubakar, A., Challenges and coping strategies of parents of children with autism on the Kenyan coast. Rural Remote Health, 2016. **16**(2), s 3517.
14. APA., American Psychiatric Association (DSM-V). Diagnostic and Statistical Manual of Mental Disorders Individuals with Autism Spectrum Text Revision, 4th ed. Washington D.C., 2013. APA.
15. Hussein, H., Sensory garden in special schools: The issues, design and use. Journal of Design and Built Environment, 2009. **5**, 77–95.
16. URL-1, 2021. https://tr.pinterest.com/pin/324329610666260112/ (Accessed: January 2021).
17. URL-2, 2021. https://jyldyzai.files.wordpress.com/2012/10/800px-bca_sensory_garden.jpg ((Accessed: January 2021).
18. Baron-Cohen, S., The eye direction detector (EDD) and the shared attention mechanism (SAM): Two cases for evolutionary psychology. In C. Moore & P. J. Dunham (Eds.). Joint attention: Its originsand role in development 1995. (pp. 41–60) Hillsdale, NJ: Erlbaum.
19. Tufan, Otistik Çocuklar dahi mi, engel mi? İLETİŞİM Yayınları, 2003. pp.244. ISBN: 9789754708981.
20. Adevi, A. and Martensson, F., Stress rehabilitation through garden therapy: The garden as a place in the recovery from stress. Urban Forestry and Urban Greening, 2013.**12**, 230–237.
21. Lee, C.M. and Bo, J., Motor Functioning in Children with Autism Spectrum Disorder. Med Pub Journals, 2015. **1**(2:13), 1–3. Acta Psychopathologica ISSN 2469-6676.
22. Gonzalez, M. T., & Kirkevold, M., Clinical use of sensory gardens and outdoor environments in Norwegian nursing homes: A cross-sectional e-mail survey. Issues in Mental Health Nursing, 2015. **36**(1), 35–43.

23. Akfidan, A., Otizm'de Görülen Duyusal Bozuklukların Gelişim Basamaklarına Olan Etkisi. Yüksek Lisans Tezi, Üsküdar Üniversitesi, 2016. İstanbul.

24. Wit, S., Sensory landscape experience: Stepping outside the visual landscape of the motorway in the garden of birds. Journal of Landscape Architecture, 2016. **11**(3), 20–31.

25. Pouya, S., Bayramoğlu, E., and Demirel, Ö., Bir engelli okulu bahçesi tasarım ana kararları: Doğan Çağlar ortopedik engelliler okulu. Kastamonu Üniv., Orman Fakültesi Dergisi, 2017. **17**(4), 682–690.

26. Akpınar Külekçi, E., Sezen, I., Peyzaj Tasarım Sürecinde İyileştirme Bahçeleri. Bursa Uludağ Üniversitesi Ziraat Fakültesi Dergisi, 2020. **34**(Özel Sayı), 337–350.

27. URL-3, 2021. https://www.tohumotizm.org.tr/otizm/otizm-spektrum-bozuklugu/ (Accessed: January 2021).

28. URL-4, 2021. https://tr.pinterest.com/pin/478437160408619426/ (Accessed: January 2021).

29. URL-5, 2021. https://dg2design.com/healing-gardens-for-mental-health-behavioral-health/ (Accessed: January 2021).

30. Flick, K. M., The application of a horticultural therapy program for pre-school children with autism spectrum disorder. Journal of Therapeutic Horti-culture, 2012. **22**(1), 39–45.

31. Bentley, A., Sensory gardenshealing mind, body, and soul. The Journal of Horticultural Science and Biotechnol-ogy, 2007. **82**, 45–345.

32. Philips, C., Butler, P., and Brooks, M., The healing circle: A sensory garden for all abilities. Barwon Commu-nity Leadership Program, Community Project, 2011. **1**–17.

33. Lipscomb, M., and Stewart, A., Analysıs of therapeutıc gardens for chıldren wıth autısm spectrum dısor-ders. Perkins Will Research Journal, 2014. **6**(2), 42–56.

34. Balode, L., & Skujeniece, S., The design guidelines for therapeutic sen-sory gardens. Annual 19th International Scientific Conference Proceedings, Re-search for Rural Development Latvia University of Agriculture, 2013. **2,** 114–119.

34. Aşur F. Sağlıkta Etkili Olan İyileştirme Bahçelerinin Tasarım İlkeleri, II. Uluslararası Türk Dünyası Eğitim Bilimleri ve Sosyal Bilimler Kongresinin, Antalya, Türkiye, 7 - 08 Aralık 2018, cilt.2, ss.261–270

36. Raveendra, Y., Sensory gardens for disabled: A review. HortFlora Research Spectrum, 2014. **3**(3), 288–291.

Sanem ÖZEN TURAN

Urban Transplantation: An Analogical Approach on Urban Conservation Interventions

Abstract: Historical environments/cultural built heritage areas are the accumulations and products of the civilizations of the human communities who lived in the settlement areas from prehistoric to the present. Cities gain their own identity with these values. For these reasons, historical environments/cultural built heritage areas are the mirrors of social identity, and they must be carefully protected, guarded and handed over to future generations under better conditions than they are handed over to us, by taking precautions to keep the verbal, written and especially architectural culture within them alive. Based on the theory that cities are living organisms and that historical environments/cultural built heritage areas are an important part of this system, a proposal process that includes different stages, methods and interventions has been developed for the conservation of historical environments/cultural built heritage areas with different characteristics through the treatment process of a living organism. The aim of this study is to discuss "the concept of conservation qualitatively" with the questions of what, why, how and for whom we should conserve in the proposal process. This discussion fictionalizes the intervention types in urban conservation, especially the demolition and reconstruction intervention and the intervention process through the organ transplant analogy.

Keywords: Cultural built heritage areas, urban conservation, city as a living organism, analogical approach

1 Introduction

Cities are dynamic, living organisms that constantly change in a complex structure. Their similarity to organisms stems from the life cycle. For this reason, as in every organism, it is frequently mentioned in the literature [1, 2, 3, 4, 5, 6, 7] that cities are born, grow and live, and even die.

The necessity of the life cycle, the length of its life is related to how long the living organism maintains its healthy existence. Any deterioration in the system of the organism may occur in disorder or loss of the function in the organism, and may even result in the death of the organism. For this, first of all, before the death/inability of the organism come to the stage, different interventions with varying methods may be required to the living organism. In order for the living organism to heal, it is necessary to focus on the following questions:

1) Taking medical complains and history of the patient/prediagnosis and diagnosis:
   - What are the factors leading to morbidity and mortality of the patient?
   - What are the signs for the disease? How does this dysfunction or loss of function emerge?

2) Treatment procedure:
   - Which form and dose of the medication should be given and how should be the management of the disease?
   - Does the treatment differ for different patients with different presentation?
   - What are the steps of treatment according to stages of the illness?
   - What are the possible reasons for rejection or acceptance of therapy by the patient?

3) Posttreatment phase:
   - How should be the post-recovery care of the patient?
   - Could the cured patient be rediagnosed with the same illness?
   - If so, would the treatment of the patient be different than the previous attack?

In line with these questions, the objective of this study is to discuss the interventions that can be applied to cities defined as living organisms/systems and the cultural built heritage that are important parts of these organisms/systems through the interventions made during the life cycle of living organisms.

2 Proposal Approach for Urban Conservation Interventions to Cultural Built Heritage

With an analogical approach, a proposal process (Tab. 1) has been developed that includes different stages, methods and interventions to historical environments with different characteristics, which face different threats of different severities and therefore have different needs:

Tab. 1. An analogical proposal process and its questions for urban conservation interventions to cultural built heritage

Problem Definition and Current Situation Analysis Process:	←	Taking medical complains and history of the patient/prediagnosis and diagnosis:
What will cause the deterioration of cultural built heritage/What are the threats that cultural built heritage faces? What can be considered as signs of the deterioration of cultural built heritage?		What are the factors leading to morbidity and mortality of the patient? What are the signs for the disease? How does this dysfunction or loss of function emerge?
Planning and Intervention Process:	←	Treatment procedure:
What methods/interventions to be applied to cultural built heritage with different qualities/characteristics should be sought at what level? How is the preparation process of cultural built heritage for interventions? What are the criteria and stages of the intervention?		Which form and dose of the medication should be given and how should be the management of the disease? Does the treatment differ for different patients with different presentation? What are the steps of treatment according to stages of the illness? What are the possible reasons for rejection or acceptance of therapy by the patient?
Monitoring and Evaluation Process After the Intervention:	←	Posttreatment phase:
What are the elements of post-implementation monitoring and evaluation process of inventions to cultural built heritage as a response to the maintenance process?		How should be the post-recovery care of the patient? Could the cured patient be rediagnosed with the same illness? If so, would the treatment of the patient be different than the previous attack?

2.1 Problem Definition and Current Situation Analysis Process = Prediagnosis and Diagnosis

Today, historical environments and cultural built heritage can be under pressure for various reasons. These pressures can be grouped into economic, social, physical and legal-administrative (Tab. 2) [8, 9, 10]:

Tab. 2. Pressures on the cultural built heritage

Economic Pressures	**Social Pressures**	**Physical Pressures**	**Legal-Administrative Pressures**
– Rapid urbanization process – Inability to measure effectively and accurately the limits of development potential – Rising land values/income – Tourism pressure, insufficient protection-tourism balance and restrictions being imposed – Functional obsolescence – Ownership – Expropriation (new regulation and barter) – Economic obsolescence/ Depreciation – Lack of resources/ funds allocated for conservation inventions – Not informing different layers of the society about the available possibilities (incentives and exemptions)	– Immigration phenomenon – Population concentration – Historical environments gradually becoming areas of depression and urban poor settling in these areas – Weak groups (elderly, people in need and low income groups with no economic power recently immigrated to the city) – The relocation or abandonment of property owners – Loss of social coherence – Not informing different layers of the society about the available possibilities (incentives and exemptions) – Ignoring social values by fragmentary conservation approaches	– Spatial clutter as a result of rapid urbanization process – Inability to measure effectively and accurately the limits of development potential – Transportation and parking problems caused by overstructure and population concentration – Zoning interventions that do not fit the purpose and buildings contrary to the traditional urban texture – Visual obsolescence – Physical obsolescence and dilapidated	– Development plan applications that do not fit the purpose – Tourism pressure, insufficient protection-tourism balance and restrictions being imposed – Ownership – Expropriation (new regulation and barter) – Confusion of authority – Very messy and complex legislation – Following a prohibitive approach in the legislation – Fragmented and incoordination management structure – Gradually erosion of the integrated planning approach

– Economic Pressures:

Issues such as inability to measure effectively and accurately the limits of development potential and tourism pressure, are among the leading economic problems faced by historical environments. In addition, functional obsolescence can also be a matter because over time, the demands for change develop depending on the needs of the society, and the usage of the space in question is outdated. It can be transformed into other functional uses according to the living conditions of the day. In addition to functional obsolescence, the change of supply-demand balances depending on today's conditions also requires functional change. The fact that the urban texture in historical environments is not economical against increasing rent causes economic obsolescence/loss of value [11].

Also, issues such as implementation challenges of multi-part ownership, expropriation, insufficient resources/funds allocated for protection, insufficient information on incentives are also economic problems faced by historical environments.

– Social Pressures:

Historical environments have begun to become areas of social collapse due to the inability of the conservation phenomenon to develop in relation to urbanization [12]. The settlement of the urban poor (weak groups and low income groups that have recently migrated to the city without economic power) in these urban depression areas creates social problems. One of the reasons for this is that the original owners were replaced by reasons such as death or inheritance. In addition, many areas of the cities have undergone changes with the problems such as social transformation, slums and social collapse that occurred with the abandonment of the property owners in these areas and this change affected the cities badly [10]. However, the right thing is to establish a proper connection between the old and the new. Thus, historical environments turn into livable places [12].

In addition to such social problems, the fragmented conservation approach also triggers problems. This threat, in general terms, in conservation projects, addressing a certain part of the protected area (such as a limited zone, street rehabilitation), ignoring social life and problems, originality, traditional values, and applying conservation practices aimed at improving the physical appearance with purely aesthetic concern [9].

– Physical Pressures:

In addition to the economic threats experienced by historical environments, there are also physical threats. There are uncontrolled urban sprawls as a result of rapid urbanization. Urbanization necessitates change. If change is not controlled, the problem of unplanned urbanization comes to the fore. This change can not be experienced moderate and controlled in Turkey, most of the city was faced with important questions [12].

In our cities green areas have decreased and most of the valuable examples of civil architecture reflecting the architectural character of a certain period with a qualified identity have been demolished and destroyed. Instead of these, unqualified concrete piles, which do not conform to any architectural style, aesthetic understanding, or even the life style of the Turkish people, were erected without a plan [13].

Apart from these threats, one of the most important threats faced by historical environments is the deterioration process. Historical environments become unresponsive to today's conditions after the physical obsolescence caused by the aging of the materials, visual obsolescence caused by unconscious additions and arrangements, and eventually they experience the process of deterioration [11].

– Legal-Administrative Pressure:

Development plan applications and settlement decisions that do not fit the purpose are among the main threats against historical environments. Making plans that do not aim to protect the traditional urban texture, increase the development rights, etc. under the name of zoning, even aiming to destroy it completely can be given as an example [10].

Issues such as the fragmented and incoordination management structure, confusion of authority, the absence of integrated planning approach are legal-administrative threats against historical environments and cultural built heritage.

In the problem definition and current situation analysis process, which can be considered as disease diagnosis, identifying the problems specific to the field clearly with determining the deterioration level and size correctly by analyzing these pressures faced by the historical environments that are a part of the living organism is very important about the rehabilitation process and format. Of course, in urban spaces where time-space-human relations are experienced, these threats can remain open to interpretation because they are values that cannot be measured clearly in numerically.

With the implementation of urban transformation approaches, which are defined in the literature [14] as the whole of strategies and actions implemented to improve the economic, social, physical and environmental conditions of the urban space where collapse and degradation are experienced with comprehensive and integrated approaches, historical environments can resist these pressures and continue to exist within the system.

The results of the Sulukule and Tarlabaşı urban transformation projects, which are among the first examples of the "renewal area" concept that came with the Law No. 5366 (in 2005) on the "Renewal and Preservation of Deteriorated Historical and Cultural Immovable Property" in the name of conservation and renewal of the historical environments in our country, have been widely discussed since the beginning of the planning process.

The intervention method of these projects is to demolish all tissue (both in physical and social terms) other than the registered buildings and create a brand new and completely different space. So, are demolition and rebuilding interventions absolutely wrong practices, do their results always create such problems? Is this destruction and rebuilding intervention similar to kidney transplantation for failed kidneys or placement of prosthesis for amputated hand due to gangrane? Then, in order to prevent the damage of an organ that has lost its function, for example kidney, to the body, can the transplantation, which is the process of replacing a new one, be successful in cities? These questions prompted the intervention types in urban transformation, especially the demolition and reconstruction intervention and the intervention process, to discuss the organ transplant analogy.

2.2 Planning and Intervention Process = Treatment Procedure

The process for intervention to cultural built heritage areas that are determined to be transformed with problem definition and current situation assessment includes implementation with the evaluation of:

- at what level/at what dosage and with what methods/interventions a solution should be sought for cultural built heritage of different qualities/characteristics,
- the preparation process of cultural built heritage areas for interventions,
- the criteria of the intervention.

2.2.1 The Types of Intervention

The types of interventions to be made, of course specific to each field, for the conservation, improvement and development of historical environments/cultural built heritage areas should change in relation to the location of the area in the city, which described as the living organism itself, to its quality, to its function in the context of its cultural, emotional and use value, and to how much it can perform this function.

The types of intervention that can be applied when necessary was grouped by Kuban (2000) as consolidation, reintegration, rehabilitation/active re-use, reconstruction [15]. Ahunbay (2004) lists it as consolidation, reintegration, renewal, contemporary addition, reconstruction, cleaning and transportation [16]. Hasol (1997) offers two approaches for historical environments; not to touch and keep alive [17].

ICOMOS (1993) offers a wide range of interventions [18]:

- Prevention of deterioration: Indirect protection such as robust maintenance program and control of environmental pollution
- Preservation: Keeping the cultural built heritage in place and preserving it from further decay
- Consolidation: Adding or applying supportive materials to the actual tissue to ensure permanent strength and structural integrity
- Restoration: Revitalizing the original concept for texture or use, or both

- Rehabilitation: Adapting the building to contemporary use that will make it sustainable
- Reproduction: Copying an existing human artefact to replace some lost and decaying parts, or, under extreme conditions, to move objects to more suitable environments
- Reconstruction: Making new by imitating the old.

Günay (1994) lists all these forms of intervention as follows [19]:

- Urban Reconstruction,
- Urban Improvement and Urban Renewal,
- Urban Redevelopment,
- Urban Rehabilitation,
- Urban Revitalisation,
- Urban Preservation/Urban Conservation,
- Infill Development,
- Urban Refurbishment,
- Re-Urbanisation
- Urban Strengthening and
- Urban Relocation.

In diagnosis of kidney insufficiency the level of decrease in glomerular filtration rate (GFR) which is a parameter for renal functions are important. Drugs and diets to decrease the progression of kidney disease are prescribed to the patients. Once end stage kidney disease is diagnosed, renal replacement treatments such as kidney transplantation and/or dialysis are performed.

Again, with the analogical description of historical environments as kidney, an evaluation can be made on the level of deterioration and obsolescence in the historical urban tissue. According to this, It is necessary to consider the types of intervention as follows:

- Phase 1 (in stages where obsolescence is not so much): *only conservation-oriented intervention types* such as urban preservation/urban conservation), infill development, urban refurbishment, urban strengthening, urban rehabilitation and urban improvement
- Phase 2 (for dilapidated historical environments): *open to change intervention types* such as urban revitalisation, urban renewal and urban redevelopment
- Phase 3 (when historical environments/cultural built heritage completely lose their function in the city): *regenerative intervention types* such as re-urbanisation, urban relocation and urban reconstruction.

Otherwise, no one already will think of replacing an organ that can be healed with a healthy or simple treatment.

2.2.2 *The Preparation Process and Stages of Cultural Built Heritage Areas for Interventions*

The prepatory process that should be done before the invervention in the historical built environments where regenerative interventions are decided includes the stages of users' adoption of the intervention and the evaluation of whether the area is suitable for the intervention (patients' decision, contra-indications), last decision, pretransplant evaluation of the patient and addition of the patient to national organ transplant waiting list.

The acceptance of the intervention of the patient, that is, the users of the historical environment, is the first step of recovery, as in the treatment of every disease. For this reason, the effectiveness of the *"doctor-patient relationship"*, the active involvement of the users by decision-makers and practitioners have a key role in the success of the result.

After the patient accepts the treatment, it is evaluated whether the patient and the organ are suitable for treatment. At this stage, criteria such as whether the patient has another disease that is not possible to treat or is very difficult to treat, and whether the patient has bad habits such as substance addiction stand out. Here, the aim can be described as not wasting the investment with the interventions to be made in the city, which is already struggling with completely different and complex problems. It should not be overlooked that in applications to be made without solving other problems in the immediate vicinity of the area and in the city that can be improved, the area will be affected by these problems and the obsolescence and deterioration process will accelerate.

The success of the intervention to be made will affect whether the field has been intervened before, to what extent it has been found and what the results are. The area, which has been intervened repeatedly, will be exhausted and tired, its ties to its essence and its value of originality will be weakened. Even though it is simple, the system will react to the effect, since there will be an external effect on the operation of the system with every intervention. Therefore, when to intervene is also important.

Practices considered to be carried out in historical environments require that they be handled in an integrated context with all kinds of contexts such as physical, social, economic, etc. Therefore, the *teamwork of experts* from many disciplines like urban planning, architecture, sociology, restoration, economy, landscape architecture, etc. is essential. According to the ICOMOS (2003) "Principles for the Analysis, Conservation and Structural Restoration of the Architectural Heritage", it is stated that it is necessary to work with different disciplines in order to achieve success in planning interventions for the conservation of architectural heritage [20]. The final decision must also be made by all members of this interdisciplinary team.

When a decision involving a heavy intervention in the form of destroying and rebuilding the historical texture is made, *tissue compatibility (matching)* in the project to be implemented, that is, considering it compatible with other areas of the city and the functioning of the urban system, is one of the most important criteria in the intervention. A healthy harmony can be achieved by ensuring that the area where the project will be implemented is physically compatible with the city and its immediate surroundings, ensuring social integration, and maintaining the continuity of local economic activities and lifestyle.

2.3 Monitoring and Evaluation Process after the Intervention = Posttreatment Phase

It cannot be denied that change is continuous. The goal is not only to overcome the problems of the day, but also to make the city as the system itself and the historical environment, which is a part of the system, a healthy and livable place in the future. Therefore, the process will never be completed with the intervention. After the intervention, along with the monitoring-evaluation process, control mechanisms will be activated; the continuity of the cycle will be ensured by reviewing, maintenance and improvements.

Cultural heritage management stages are explained by UNESCO (2005) as determining the management structure, planning, preventive measures and monitoring – control [21]. Basically, a successful management mechanism should provide the planning, implementation, monitoring, evaluation, feedback loop [22]. In the "the Convention on the Conservation of the World Cultural and Natural Heritage", which was submitted for signature by UNESCO in 1972 and our country was a party in 1983, it is said that "every asset is required to have a management plan or management system whose purpose is to protect and maintain the asset for today and for the future" [23].

The concepts of "management area" and "management plan" gained a legal basis only with the law numbered 5226 of 2004 on amending "Law on the Conservation of Cultural and Natural Property" numbered 2863 of 1983, and in 2005 with the publication of the "Regulation on Foundation and Duties of Site Management and Monument Councils and Condition and Principles related to Establishment of Management Areas", the work has actually started.

Management plans are not a "physical" plan like conservation plans, but a plan that includes all the components of the strategic planning approach, in which how to protect the area with a rational, sustainable and holistic approach with which resources, experts and program, that is, it is a political process cycle. Since the management plan, which is defined as a continuous process, is not static, it should be arranged in accordance with the changing national and international conditions and goals, together with cultural value judgements [24].

This monitoring-evaluation phase is one of the most important but not productive stages in this process. Failure to define the evaluation criteria and performance criteria clearly due to the unmeasurable values in conservation practices reduces the efficiency at this stage of the prepared management plans.

Activities and practices that should be done in order to prolong the period in which the new tissue will remain healthy, which is the result of the intervention, should be regularly encouraged. Even in the preparation phase, informing and training should begin for the citizens and users regarding the maintenance process. Not only spatial, but also non-spatial and awareness-raising strategies should be given great importance. Thus, the adaptation of the citizen and the user to the process after the intervention will be easier. Continuous maintenance and simple interventions will be required in order for the city to internalize and not reject the new tissue. The size and duration of these simple interventions should be specific to the area. At certain intervals, action decisions within the management plan should be evaluated with the performance criteria to be determined before waiting for the completion of the management plan process.

3 Instead of the Conclusion

In the whole living organism system, the role/importance of historical environments and cultural built heritage areas, which are one of the important elements/organs of cities considered as living organisms, must be correctly described. Thus, it should not be overlooked what the essential quality is to be conserved and not lost.

It must be admitted that change is an inescapable and undeniable truth. In the Declaration of Amsterdam (1975) it is stated that "the future cannot be created at the expense of the past" [25]. With a similar expression, Kuban (2000) said, "in a dynamic interpretation of history, it is necessary to say that the values created by human beings in the past will contribute to the creation of a creative environment and bring a healthy attitude to the contemporary urban chaos" [15]. With this, he emphasized that protection will not prevent change. The answer to the question of what we will protect in order to carry out its basic function, which is essentially ensuring cultural continuity, is through the transmission of traces from the past to the present and future generations.

The producer and consumer actors of the cultural heritage defined by Litchfield (2009) are also at the basis of the questions "what should be protected for", "protected for whom" and "protected by whom" [26]. What is mentioned here is that what stakeholders gain or lose from the conservation response affects the stakeholder's perspective on the response. Conservation is a universal phenomenon that is now considered as an international area of responsibility. As a result, the universal effect of the intervention performed/to be made should be taken into account, not the individual benefit-cost results. In order for the intervention to be internalized and accepted, it is important to build trust as in the patient-doctor relationship, and to inform the stakeholders correctly in every process of the implementation.

While searching for the answer to the question of "how to protect", wrong treatments can be applied by acting in a hasty or negligent way about diagnosis. The framework of the theoretical and practical foundations of the phenomenon of conservation, which is a selection task, should be established and the limits of these values should be determined clearly through the legal system, and the criteria for these limits should be set accurately [27]. In this sense, the character analysis of the historical urban tissue, what this area means for all stakeholders and what kind of value it has should be read correctly. It should not be forgotten that it is unique to the place. It should be evaluated what the cultural heritage area means for that city and the world, how much it has been damaged, whether it prevents the social, cultural, economic and physical development of the city. In other words, the balance should be observed by paying attention to whether the cultural built heritage area has come to a situation that damages the city. In this sense, ways to find objective measurement criteria such as GFR (glomerular filtration rate) value that will enable to determine the intervention dose to be applied in cities should be sought.

It is not possible to say that the concepts for cultural heritage areas are addressed by creating a common language and terminology. This can be clearly seen from the terminology confusion in the forms of intervention and from the texts of legal regulations. Comprehensive studies are needed to simplify the complexity of the terminology of the concept of urban conservation, which includes definitions similar to each other.

The legislations, with analogy to *drugs*, used in Turkey until today for interventions to be applied to cultural built heritage are listed below:

- The Law on the "Old Assets" numbered 1710 (in 1973)
- Law on the "Conservation of Cultural and Natural Property" numbered 2863 (in 1983)
- The Law No. 3386 (in 1987) on Amending the Law on the "Conservation of Cultural and Natural Property" No. 2863 (in 1983)
- The Law No. 5226 (in 2004) on Amending the Law on the "Conservation of Cultural and Natural Property" No. 2863 (in 1983)
- The Law No. 5366 (in 2005) on the "Renewal and Preservation of Deteriorated Historical and Cultural Immovable Property"
- The Law No. 5663 (in 2007) on Amending the Law on the "Conservation of Cultural and Natural Property" No. 2863 (in 1983)
- The Law No. 5835 (in 2009) on Amending the Law on the "Conservation of Cultural and Natural Property" No. 2863 (in 1983)
- "Decree Law Regarding the Organization and Duties of the Ministry of Environment and Urbanization and the Decree Law Amending Some Laws and Decree Laws" numbered 648 (in 2011)
- The Law on the "Transforming Areas Under Disaster Risk" numbered 6306 (in 2012)
- The Law No. 6498 (in 2013) on Amending the Law on the "Conservation of Cultural and Natural Property" No. 2863 (in 1983)

The most radical change in Turkey's conservation legislation was experienced by The Law No. 5226 (in 2004) on Amending the Law on the "Conservation of Cultural and Natural Property" No. 2863 (in 1983). It can be said that this law has shown improvements in parallel with the concept of "Integrated Conservation", which started to be discussed with the Amsterdam Declaration of 1975 published internationally. The Law on which the most discussions and studies are carried out is the

Law No. 5366 (in 2005) on the "Renewal and Preservation of Deteriorated Historical and Cultural Immovable Property". This Law, which is based on applications for "urban renewal" areas, ignores the whole of the city with the "renewal area" logic that it considers in line with the urban site areas. This situation continued with the regulations of the "Decree Law Regarding the Organization and Duties of the Ministry of Environment and Urbanization and the Decree Law Amending Some Laws and Decree Laws" numbered 648 (in 2011) and The Law on the "Transforming Areas Under Disaster Risk" numbered 6306 (in 2012) [27].

Beyond conserving individual buildings and site areas, our legislation system, which is our treatment tool, should be improved in accordance with the integrated protection approach and, in a wider context and sense, to see cities and cultural built heritage as the focus of culture rather.

Interventions directed at a single organ in a living organism cannot be successful without considering its relationship with other organs/systems in the organism and looking at the whole. The social and economic dimension of protection is as important as its physical and aesthetic dimension. While implementing cultural built heritage policies, the social structure of the citizen should not be changed. It should not be forgotten who the intervention is for, who will use it. It is necessary to integrate the people with their buildings, to make them love their buildings and their surroundings, and thus to make the places livable. One way to achieve this is "integrated urban conservation" [12].

Every individual embraces the values that she/he believes to belong to her/him and strives for her/his protection and development. In this sense, one of the ways for the city dweller to embrace the city in which she/he lives social, cultural and economic life is to feel that the dweller belongs to the city she/he lives in. Conservation starts with belonging and continues with sustainability efforts. Belonging creates emotional bonds between the individual and the space, and is a driving force factor in conservation.

Cities shape urban dwellers and urban dwellers shape the city. One of the ways to have an urbanity identity is that users can perceive the environments they live in as familiar and familiar spaces. For this reason, it is necessary to conserve the historical environments that build a bridge between the past and the present and thus create familiar and familiar places, and to convey the awareness of conservation to the user [28].

Whether they are conservation plans as physical plans or cultural built heritage management plans as strategic plans, the most important key to ensuring sustainability in practice is the development of conservation awareness of stakeholders and ensuring their active participation in the process. It will be easier for the stakeholder who is informed, from whom information is obtained and the information exchange environment is presented, to adopt the intervention.

Urban sustainability requires continuous improvement and continuous improvement requires integrated urban conservation. The process based on participation and coordination, which is the most important key to urban sustainability, is important in the management of cultural built heritage areas. In the process, as well as the implementation, the monitoring and evaluation stages of the implementation should be carried out successfully. Otherwise, recovery cannot be achieved in its real form. In this sense, clearly determining the performance value criteria, which is the control mechanism in area management plans, will ensure the effective execution of feedback processes.

As a result of successful interventions developed with monitoring-evaluation process, social, economic and physical development in cultural built heritage areas will directly mobilize actor groups and trigger conscious individuals and institutions that form the impulse mechanisms to repeat the cycle [28], the self-regulation cycle will begin to function. Just as the patient has to make changes in lifestyle with individual precautions to protect her/his health after a severe intervention such as transplantation and tries to stay away from factors that threaten her/his health, the urban citizen will do this with self-control after successful interventions in the city and cultural heritage areas. This will bring with it continuous development and improvement and accordingly urban sustainability.

This analogical approach designed can create a basis for discussion for the answers of many questions regarding the intervention process towards cultural built heritage. However, in ensuring the sustainability of cultural heritage, there are questions such as;

- How threats that cause/will deteriorate the historical tissue can be measured,
- How to measure/determine the benefits and costs of conserving cultural heritage sites according to stakeholders,
- How to determine the breaking points that will affect the differentiation in intervention styles,
- How fundamentally unmeasurable values, such as when to intervene, can be embodied and
- What should be the essence that needs to be preserved when intervening, which include the main problem of conservation but still have no clear answers.

References

1. Le Corbusier, *The Radiant City,* 1933, Faber And Faber: London.
2. Le Corbusier, *Concerning Town Planning*, 1947, The Architectural Press.: London.
3. Mumford, L. *The Culture Of Cities*, 1938, George Routledge And Sons.: London.
4. Sert, J. L., *Biology Of Cities*, 1942, Time Magazine, http://www.time.com/time/magazine/article/0,9171,766697,00.html.
5. Jacobs, J., *The Death And Life Of Great American Cities*, 1961, Random House: New York, Ny.
6. Lynch, K., *A Theory Of Good City Form*, 1981, The Mit Press: Cambridge, Ma.

7. Newman, P.W.G., *Sustainability and Cities: Extending the Metabolism Model*, 1999, Landscape and Urban Planning, **44**, 219–226.
8. Huan, H., Smolders, B., Verweij, J., *Cultural Heritage Conservation in Historic Wuhan Urban Renewal And The Creative Industry*, 2008, http://www.isocarp.net/data/case_studies/1177.pdf
9. T.C. Bayındırlık Ve İskan Bakanlığı, *Kentsel Miras, Mekan Kalitesi Ve Kentsel Tasarım Komisyonu Raporu*, 2009, https://webdosya.csb.gov.tr/db/kentges/editordosya/kitap5.pdf
10. Özsoydan, G., *Kentsel Korumaya Stratejik Yaklaşım (Beşiktaş Köyiçi Kentsel Sit Alanı Örneği)*, Yüksek Lisans Tezi, 2007, Mimar Sinan Güzel Sanatlar Üniversitesi Fen Bilimleri Enstitüsü: İstanbul.
11. Özyaba, M., *Koruma ve Planlama Teknikleri*, Kentsel Alanların Planlanması ve Tasarımı, Aydemir, Ş., et al., 2004, Akademi Kitapevi Yayınları: Trabzon.
12. Özen, S.L., *Koruma Yaklaşımlarına Yeni Bir Bakış Açısı Olarak "Bütünleşik Koruma" – Trabzon Örneği*, Yüksek Lisans Tezi, 2005, KTÜ Fen Bilimleri Enstitüsü: Trabzon.
13. Avcı, N., *Türkiye'de Tarihi Kent Dokularının Korunması Ve Geleceğe Taşınması Sempozyumu – Önsöz*, Kültür Bakanlığı KTVKGM, 2002, Nurol Matbaacılık: Ankara.
14. Akkar, M., *Kentsel Dönüşüm Üzerine: Batıdaki Kavramlar, Tanımlar, Süreçler ve Türkiye*, Planlama, 2006, **2**: p.29–38 .
15. Kuban, D., *Tarihi Çevre Korumanın Mimarlık Boyutu: Kuram ve Uygulama*, 2000, Yapı Endüstri Merkezi Yayınları: İstanbul.
16. Ahunbay Z., *Tarihi Çevre Koruma ve Restorasyon*, 2004. Yapı Yayınları: İstanbul.
17. Hasol, D., *Yağma Var*, 1997, Yapı Endüstri Merkezi Yayınları: İstanbul.
18. ICOMOS, *Conservation Economics, Cost Benefit Analysis for the Cultural Built Heritage: Principles and Practices*, 1993, https://www.icomos.org/en/116-english-categories/resources/publications/295-isc-conservation-economics
19. Günay B., *Kentlerin yeniden üretilmesi süreçleri*, 1994, Mimarlık, **249**:111–114.
20. ICOMOS, *Principles for the Analysis, Conservation and Structural Restoration of the Architectural Heritage*, 1994, http://www.icomos.org.tr/Dosyalar/ICOMOSTR_tr0033791001536913477.pdf
21. UNESCO, *Convention on the Protection and Promotion of the Diversity of Cultural Expressions*, 2005, https://en.unesco.org/creativity/sites/creativity/files/passeport-convention2005-web2.pdf
22. Vicente, G.J., Smirnow, L., *Management Plans and The World Heritage Convention: A Bibliography*, 2010, ICOMOS Documentation Centre: Paris.

23. UNESCO, *Convention on the Conservation of the World Cultural and Natural Heritage*, 1972, https://whc.unesco.org/archive/convention-en.pdf
24. Thomas, L., Middleton, J., *Guidelines for Management Planning of Protected Areas. The World Conservation Union*, 2003, IUCN Gland: Switzerland and Cambridge
25. ICOMOS, *The Declaration of Amsterdam, Congress on the European Architectural Heritage,* 1975, https://www.icomos.org/en/and/169-the-declaration-of-amsterdam.
26. Litchfield N., *Economics in Urban Conservation*, 2009, Cambridge Books, Cambridge University Press: Cambridge.
27. Altuntaş, B., *Türkiye'de Kentsel Koruma Kavramının Değişim Sürecinin Lisansüstü Çalışmalar ve Yasal Düzenlemeler Üzerinden Değerlendirilmesi,* Yüksek Lisans Tezi, 2019, KTÜ Fen Bilimleri Enstitüsü: Trabzon.
28. Özen Turan, S., *Bütünleşik Kentsel Koruma ve Ekonomisi: Trabzon Kenti Modeli*, Doktora Tezi, 2009, KTÜ Fen Bilimleri Enstitüsü: Trabzon.

Ömer K. ÖRÜCÜ

Habitat Distribution Shifts of the Oriental Spruce (*Picea Orientalis* L.) under Climate Change Scenarios

Abstract This study seeks to identify the change that will occur in future in the current habitat of the *Picea orientalis* L. (oriental spruce) under climate change scenarios. To this end, a model was developed in the MaxEnt 3.4.1 software programme using 19 bioclimatic variables obtained from the Worldclim database. The study made use of the Community Climate System Model (CSSM) version 4 and the Representative Concentration Pathway (RCP) 4.5 and RCP 8.5 scenarios for 2050–2070 developed in accordance with the fifth report of the Intergovernmental Panel on Climate Change (IPCC). The results of the Jackknife test which was used to identify the bioclimatic variables that affect the geographical distribution of the species the most showed that isothermality was the variable that generated the greatest gain. In conclusion, the total extent of the "suitable" and "highly suitable" areas within the study area, constituting the current distribution area of the oriental spruce, is found to be 45,323 km^2. By 2070, according to the model, this figure rises to 65,042 km^2 in the RCP 4.5 scenario and 90,861 km^2 in the RCP 8.5 scenario. In view of the results of the study, it is anticipated that the oriental spruce will start to occur in those areas that will in future come to display climatic characteristics similar to those in its natural distribution area.

Keywords: *Picea orientalis* L., Ecological niche, Climate change, CCSM4, MaxEnt

1 Introduction

Climate change poses a growing global threat to ecosystems and biodiversity [1–4]. One of its effects is to alter the structures and functions of ecosystems through its impact on the species, communities and habitats of plants, resulting in changes or losses in the goods and services which the natural ecosystem supplies to society [5–7]. The ways in which species respond in the face of global climate change has been acknowledged to be the most important ecological factor in determining the basic characteristics of plant communities and their distribution areas [8, 9]. Understanding the direction and magnitude of the responses of species to climate change is therefore vital for protecting the species and ensuring their sustainability [10, 11].

The exposure of the plant species in natural forest landscapes to the negative impacts of climate change is a matter of concern not only for the sake of the ecosystems themselves but also for the ecological services they provide, such as the supply of raw materials (timber, medicinal plants, etc.), carbon storage and the preservation of biodiversity [12–17]. Recent research anticipates that loss of habitats linked to changes in climatic conditions will threaten global biodiversity in the years ahead [18–20]. It follows that mapping the spatial distributions of species under current and future climate conditions is of great importance for the development of nature conservation-based strategies for reducing the threat to biodiversity and for ensuring the continuity of species in the context of landscape management [21–23]. As climate change alters the bioclimatic conditions that species normally require, it also leads to changes in their geographical distributions [24–27]. These changes are related largely to increases in temperature and declines in rainfall in the growing season [28]. Numerous studies referencing the impact of climate change on species have investigated the effects of global warming and confirmed that the species will migrate towards the poles (higher latitudes) and higher altitudes [29–32].

Species Distribution Models (SDMs) are one of the main instruments used in ecologically-based studies that seek to predict the effects of climate change on the distributions of plant species [33, 34]. While there are many ecological models for modelling the distributions of plant species, the MaxEnt model, based on the maximum entropy algorithm, yields more successful results than the others even when the samples are small [35, 36]. MaxEnt has been used widely to model the distributions of various plant species including endangered and endemic species and invasive species under existing conditions and climate change scenarios for purposes of species conservation and management [37–41]. The main objective of the present study is to model the spatial distribution of *Picea orientalis* L. within the study area using the MaxEnt 3.4.1 and QGIS 3.10.5 software. *P. orientalis* L. is naturally distributed in the Eastern Black Sea region of Turkey and in the Caucasus. The aim is to identify the potential distribution areas of the species today and in the future under different climate change scenarios. The fact that *P. orientalis* L. is on the IUCN Red List of Threatened Species (2013) only adds to the importance of identifying the changes in its distribution areas [42].

The study seeks answers to the following three research questions:

- Which bioclimatic variable is the key environmental determinant of the distribution of *P. orientalis* L.?
- What are the potential current and future distributions of this species?

In conclusion, the potential distribution areas of the species *P. orientalis* L. have been mapped, the difference between its present and future distributions has been investigated, and recommendations have been made for landscape management and adaptation to climate change.

2 Materials and Methods

Picea orientalis L., commonly known as the Caucasian spruce or oriental spruce, constitutes the main material of the study. The appearance of the species and its cones can be seen in Fig. 1.

Fig. 1. Appearance (a) and cones (b) of the Picea orientalis L. (BitkiVT, 2020)

P. orientalis L. has a localised geographical distribution bounded by the Greater and Lesser Caucasian mountains in the Caucasus and the coastal mountains of Northeast Anatolia [43]. It is generally found at an altitude of between 1,200 m and 2,000–2,400m on humid slopes that overlook the Black Sea. It forms either pure forests or mixed forests alongside the oriental beech, the Black Sea spruce and the Scots pine. It grows well on lightly acidic, loamy soils with high relative humidity [43]. It is a significant forest tree which in favourable conditions can attain a height of 40–50m (A 68m specimen has been recorded in Torul, in Gümüşhane, Turkey) and a diameter of 1.5–2m [44, 45]. Being soft and light, the wood of the oriental spruce is used for the production of cellulose and paper, for woodwork and veneering in the furniture and construction sectors, and for packing cases [46, 47]. It is also used in the manufacture of the soundboards of bowed string instruments. Dwarf varieties in particular are grown as ornamental plants in urban parks and are popular as Christmas trees [43, 48]. With its very short, shiny dark green needles, the oriental spruce is a highly attractive tree for landscape architecture work [49].

In the first stage of the study, the coordinates of points representative of the geographical distribution of the oriental spruce were obtained and located on the map (Fig. 2). For the presence data used in the study, 50 sample points were identified with the aid of the existing literature [50, 51] in the regions where the species lives. The coordinates of these points were then indicated in the WGS84 coordinate system in the QGIS 3.10.5 software using Google Satellite Hybrid high spatial resolution satellite data as base data (Fig. 2).

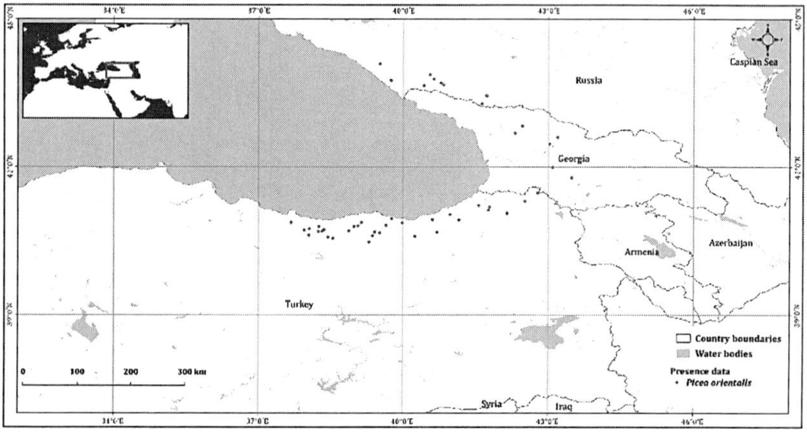

Fig. 2. Study area and Picea orientalis L. sample points

In order to model the current potential distribution area, use was made of the bioclimatic variables in the Worldclim database. Worldclim version 1 contains monthly minimum, maximum and average temperatures and monthly average rainfall for the years 1960–1990. The bioclimatic variables (in 30-second, or approximately 1,000m, spatial resolution) used to determine the current distribution were derived from the data observed in Worldclim version 1 [52]. For the climate model, the Community Climate System Model (CCSM) version 4 was used. This version enables researchers to carry out basic studies on the past, present and future climatic situations in the World. It combines the world's atmosphere, the oceans and the land surface and simulates them simultaneously. It is made up of four separate models. For the climate scenarios linked to this model, Representative Concentration Pathways (RCP) 4.5 and 8.5 were selected from among the scenarios given in the Fifth Assessment Report of the Intergovernmental Panel on Climate Change (IPCC5). Pearson correlation analysis of 19 bioclimatic variables (Tab. 1) was conducted using SPSS version 26 to remove the effect of multicollinearity on the modelling process and for selection of the most fitting variables that show high contribution degrees to the model, and the highly correlated variables with Pearson's correlation coefficients higher than 0.7 were removed to further improve the precision of the model simulation. In modelling the potential current habitat of the oriental spruce and the future potential shift in its habitat under climate change scenarios, 25 % of the presence data (12 points) were set aside as test data, then the MaxEnt modelling procedure was carried out using the 'auto features' setting. To determine the performance of the model, use was made of the 'Area under the ROC curve' (AUC) value obtained by means of Receiver Operating Characteristics (ROC) analysis [53, 54]. A Jackknife test was also conducted in the model in order to determine the importance of the contribution made by each environmental variable [55, 56]. This test is used to determine the importance of each independent variable in forming the model.

Five threshold values were used in creating the potential distribution maps. The models generated make use of five different degrees of suitability – namely, not suitable (0), not very suitable (0–0.25), moderately suitable (0.25–0.50), suitable (0.50–0.75) and highly suitable (0.75–1.0) [57]. They were mapped using the QGIS 3.10.5 software. The current and future spatial distribution of *P. orientalis* L. can be observed from the resulting maps. In this way, suitability classes were obtained by classifying the geographical distributions modelled using the climate change scenarios according to the threshold values mentioned.

3 Findings

3.1 Statistical Assessment and Performance of the Model

As a result of Pearson correlation analysis, eight variables were excluded and only eleven variables were kept to establish the potential distribution model of *Picea orientalis* L. The selected variables include BIO1 (Annual Mean Temperature), BIO3 (Isothermality ((BIO2/BIO7) ×100), BIO6 (Min. Temperature of Coldest Month), BIO10 (Mean Temperature of Warmest Quarter), BIO11 (Mean Temperature of Coldest Quarter), BIO12 (Annual Precipitation), BIO13 (Precipitation of Wettest Month), BIO14 (Precipitation of Driest Month), BIO16 (Precipitation of Wettest Quarter), BIO17 (Precipitation of Driest Quarter) and BIO18 (Precipitation of Warmest Quarter).

The AUC values of the training data and test data were examined in order to test the accuracy of the model (Fig. 3). The AUC values were found to be 0.987 for the training data and 0.971 for the test data. Both were therefore "very good". The model was taken to be sensitive and definitive. When the Jackknife test was used to determine the levels of the contributions of the bioclimatic variables (Fig. 4), the variable that generated the greatest gain was found to be BIO3 (Isothermality). This indicates that the variable BIO18 incorporates the most useful information. This variable was followed by BIO11 (Mean Temperature of Coldest Quarter) and BIO6 (Min. Temperature of Coldest Month). BIO3 (Isothermality) was also found to be the variable that reduces the gain most when ignored.

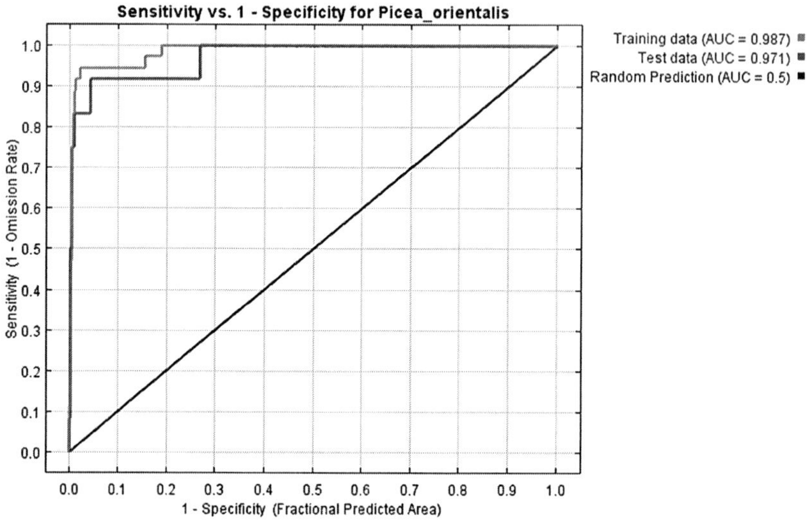

Fig. 3. AUC curve for Picea orientalis L.

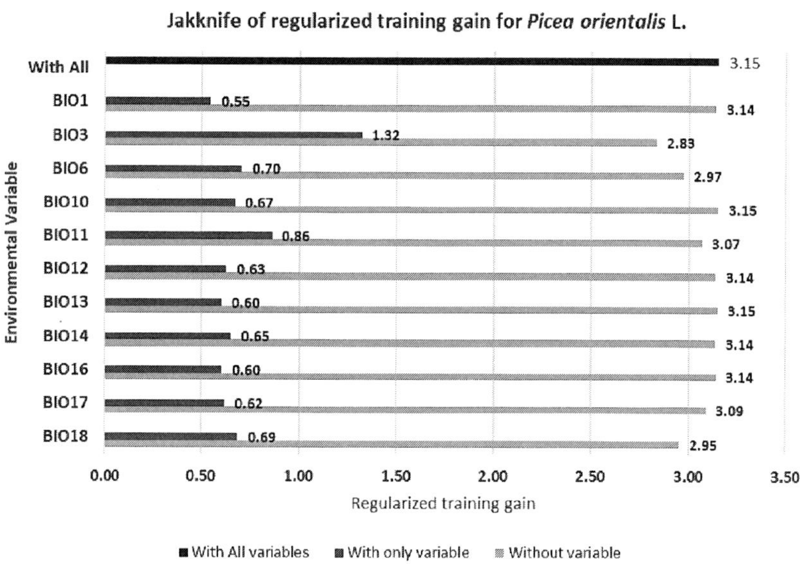

Fig. 4. Jackknife test results

3.2 Current and Future Potential Geographical Distribution of *P. Orientalis* L.

The predicted models of the current and future potential geographical distributions of *P. orientalis* L. are given in Figures 5, 6 and 7. A glance at the map of the current distribution of *P. orientalis* L. obtained from the model shows a strong correlation between the points representing its current dispersion and the zones that extend parallel to the shores of the Black Sea in the Eastern Black Sea region of Turkey, the western parts of Georgia and the southern coast of the Russian Federation (Fig. 5). The natural distribution area of the species in Turkey begins from the River Melet in the province of Ordu to the west and spreads eastwards along the Black Sea Mountains, the Kaçkar Mountains, the Mescit Mountains and the Yalnızçam Mountains as far as the Posof Basin, the Çoruh Valley and Yusufeli, in the province of Artvin [44, 58]. The tree is known to grow most intensively in the Black Sea coastal provinces of Ordu, Giresun, Trabzon, Rize, and Artvin and less intensively in the inland provinces of Gümüşhane, Bayburt, Erzincan, Erzurum, Kars, and Ardahan [44, 50, 59].

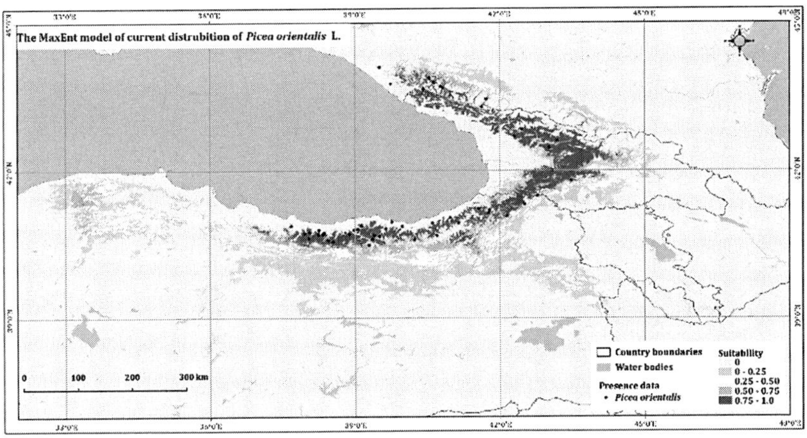

Fig. 5. Current potential distribution of Picea orientalis L.

With respect to the predicted future potential geographical distribution of the species, the maps generated by the model for 2050 (Fig. 6) and 2070 (Fig. 7) point clearly to an increasing trend in the locations where it is likely to occur. The extent of the areas in question (Tab. 1) suggest that the changes in the climate that will result in relatively moderate and high temperature increases may create a positive environment for the potential dispersion of the species.

Tab. 1. Predicted spatial distribution of Picea orientalis L. under current conditions and under the 2050–2070 RCP 4.5 and RCP 8.5 climate change scenarios

Suitability	**Current**	**RCP 4.5**				**RCP 8.5**			
		2050		**2070**		**2050**		**2070**	
		km²	%	**km²**	%	**km²**	%	**km²**	%
Not suitable	1034864	975572	-5.7	997329	-3.6	973295	-5.9	986586	-4.7
Not very suitable	68763	90647	31.8	72943	6.1	83021	20.7	62317	-9.4
Moderately suitable	27936	45867	64.2	41574	48.8	47865	71.3	37124	32.9
Suitable	22265	29348	31.8	32702	46.9	31091	39.6	38102	71.1
Highly suitable	23058	35453	53.8	32340	40.3	41615	80.5	52759	128.8
Total	1176886 km²								

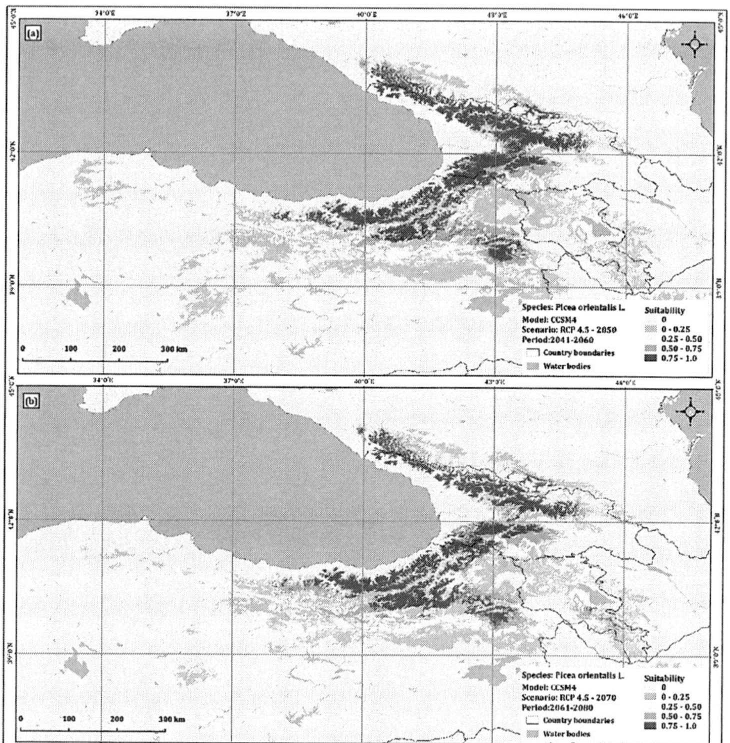

Fig. 6. Prediction of the future potential distribution under RCP 4.5 climate change scenarios for 2050 (a) and 2070 (b)

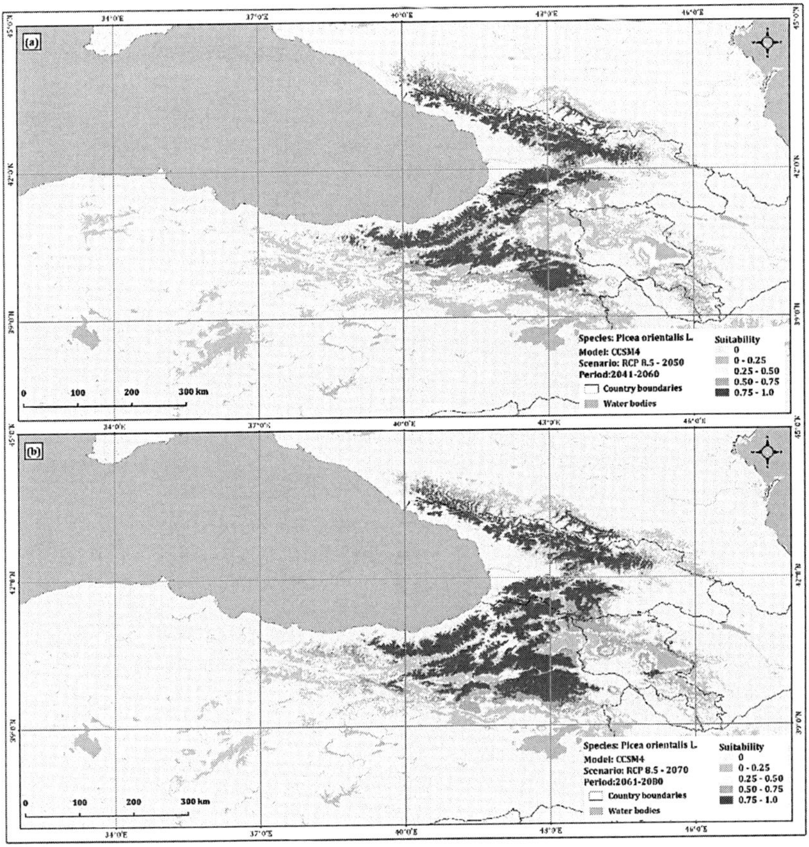

Fig. 7. Prediction of the future potential distribution under RCP 8.5 climate change scenarios for 2050 (a) and 2070 (b)

4 Discussion

At the global level, the responses of plant species to climate change, and the resulting shifts in their geographical reach, are expected to play an important role in the shaping of future forest ecosystems [60–62]. One of the preconditions for understanding the sustainability of a species, particularly in the context of ecosystem restoration, is to understand its geographical distribution in detail [60, 63–65]. The factors that influence geographical distribution are very much on the agenda as they are among the key phenomena in landscape ecology. These factors are used as variables in modelling ecological niches and are used widely and productively in developing strategies for managing forest areas in the face of climate change [66–68]. Determining priorities and taking effective conservation actions under climate change scenarios entails high-quality presence data [60, 64].

The present study has modelled the geographical range of *Picea orientalis* L. under future climate change scenarios with the maximum entropy algorithm. While the total areas of suitable habitats and the total areas of unsuitable habitats increase by 2050 and 2070 under the RCP 4.5 and RCP 8.5 scenarios, the extent of stable areas declines. Although the increase in the distribution area of the species increases over time, this situation raises the question of what the increase in the distribution area of the species means for the quality of the landscape. In other words, how does an increase in the area providing a suitable habitat for a species affect the quality of the landscape? Or will the increase in the area where the habitat is suitable cause the species to become dominant and to exert pressure on other species? The current study does not focus directly on the answers to these questions, but future research might seek answers using various habitat models, since previous studies have established that stable areas ensure a static balance that provides for ecological connectivity in the landscape [69–71]. Any disturbance to the static balance could result in the degradation of habitat connectivity, potentially leading to fragmentation of the landscape, even though the overall area where the habitat is suitable for the species has increased. Increases in suitable areas may appear to be beneficial for the continuity of the species itself, but they may also pose dangers to the other habitats in its surroundings, stemming from its dominance. In addition, significant declines might occur in the degree of sociability of the habitats in question. The changes that might occur in plant associations could force some species to shift location; they could also cause degradation in the short term and evolutionary adaptation in the long term [72, 73].

Looking at the areas to which the species might potentially spread, it is observed to tend to prefer areas of high temperatures. Moreover, although the results of the Jackknife test do not directly show BIO1 (Annual Mean Temperature) to be the most important variable, the climatic variables BIO3 (Isothermality), BIO11 (Mean temperature of Coldest Quarter) and BIO6 (Min. Temperature of Coldest Month) are all related to changes in temperature, which may indicate that it has significant impacts on the potential geographical distribution of the species. Most studies of the dispersion of plant species in the face of climate change according to the maximum entropy algorithm stress that the geographical ranges of the species in question may shift towards warmer regions [74–79]. Consequently, the research results presented in this paper show similarities or parallelisms with the results of previous studies.

It is recommended that the number of variables to be incorporated at the modelling stage should be considered carefully, since the inclusion of certain variables in the model, or their exclusion from it, could cause the model to perform poorly or lead to erroneous predictions. For this reason, the knowledge and views of experts on the ecological niche of the species should be taken into account whenever possible when determining and selecting environmental variables. It is also recommended that future studies should be conducted in the light of knowledge of the edaphic requirements of the species in question. It will also be beneficial to use strong algorithms that produce realistic results in climate estimate modelling. In the current study, the AOC curve results generated for *P.orientalis* L. demonstrate that the model used is reasonable and highly accurate.

The objectives of this study do not include the modelling of the distribution of the species based on potential future changes in land use. Yet in addition to climate change, changes in the land cover can also be influential on the distribution of a species. Hence different future prediction models (such as the Markov chain model, or the business-as-usual model) and modelling methods that reflect potential changes in the land cover could beneficially be integrated into climate-based species distribution studies [80, 81].

There are still major gaps in our knowledge of the effects of climate change on natural ecosystems, the related wildlife, and human beings, and of how adaptation measures can best be tailored to local conditions [82]. The development of species distribution models and improvement of spatial databases can contribute to the development of scientific conservation strategies beneficial to both individual species and ecosystems in the face of future climatic conditions. In this way, the linkages among the different ecosystems in the landscape can be protected [83]. By restoring and maintaining the linkages between habitat patches, functional connectivity can also be ensured among smaller landscape patches [84]. The relative significances of habitats from the points of view of environmental diversity, connectivity and isolation could also be incorporated into studies of the effects of climate change on the distribution of species. Related to this, the sustainability of the ecosystem services which people need could be ensured in the context of conservation-based approaches. In this way, the outputs of the theory can be put into practice to achieve the goals of landscape planning.

5 Conclusion

The modelling of ecological niches is acknowledged to be a fruitful and widely applicable approach for the development of guidelines for nature conservation in the context of global climate change. Moreover, high-quality geographical distribution data that provides inputs for climate change scenarios play an important role in the determination of habitat suitability and the implementation of effective conservations actions. In the present study, the current and future geographical distribution of *Picea orientalis* L. has been modelled successfully making use of the maximum entropy algorithm, a machine learning technique. The high spatial resolution assisted the model in making an accurate prediction of the suitable habitats. The findings of the research have yielded important information for landscape management and the shaping of the strategies that need to adopted for the conservation of the stands of the oriental spruce in the face of climate change. Predictions based on two climate scenarios (RCP 4.5 and RCP 8.5) show that the habitats currently suitable for P. *orientalis* L. will expand significantly by 2050–2070 as a result of climate change. Consequently, it is anticipated that the oriental spruce will start to occur in those areas that will in future come to display climatic characteristics similar to those in its natural distribution area. It is considered that this may increase the resilience of the species to climate change, and that the resulting situation will be to the benefit of the species.

References

1. Díaz, S., J. Settele, E. Brondízio, H. Ngo, M. Guèze, J. Agard, A. Arneth, P. Balvanera, K. Brauman, S. Butchart, K. Chan, L. Garibaldi, K. Ichii, J. Liu, S. Subrmanian, G. Midgley, P. Miloslavich, Z. Molnár, D. Obura, A. Pfaff, S. Polasky, A. Purvis, J. Razzaque, B. Reyers, R. Chowdhury, Y. Shin, I. Visseren-Hamakers, K. Wilis, and C. Zayas, *Summary for policymakers of the global assessment report on biodiversity and ecosystem services of the Intergovernmental Science-Policy Platform on Biodiversity and Ecosystem Services.* 2019.
2. Malhi, Y., J. Franklin, N. Seddon, M. Solan, M.G. Turner, C.B. Field, and N. Knowlton, *Climate change and ecosystems: threats, opportunities and solutions.* Philosophical Transactions of the Royal Society B: Biological Sciences, 2020. **375**(1794): p. 20190104.
3. Palomo, I., *Climate Change Impacts on Ecosystem Services in High Mountain Areas: A Literature Review.* Mountain Research and Development, 2017. **37**(2): p. 179–187.
4. Weiskopf, S.R., M.A. Rubenstein, L.G. Crozier, S. Gaichas, R. Griffis, J.E. Halofsky, K.J.W. Hyde, T.L. Morelli, J.T. Morisette, R.C. Muñoz, A.J. Pershing, D.L. Peterson, R. Poudel, M.D. Staudinger, A.E. Sutton-Grier, L. Thompson, J. Vose, J.F. Weltzin, and K.P. Whyte, *Climate change effects on biodiversity, ecosystems, ecosystem services, and natural resource management in the United States.* Science of The Total Environment, 2020. **733**: p. 137782.
5. Montoya, J.M. and D. Raffaelli, *Climate change, biotic interactions and ecosystem services.* Philosophical Transactions of the Royal Society B: Biological Sciences, 2010. **365**(1549): p. 2013–2018.
6. Runting, R.K., B.A. Bryan, L.E. Dee, F.J.F. Maseyk, L. Mandle, P. Hamel, K.A. Wilson, K. Yetka, H.P. Possingham, and J.R. Rhodes, *Incorporating climate change into ecosystem service assessments and decisions: a review.* Global Change Biology, 2017. **23**(1): p. 28–41.
7. Zarrineh, N., K.C. Abbaspour, and A. Holzkämper, *Integrated assessment of climate change impacts on multiple ecosystem services in Western Switzerland.* Science of The Total Environment, 2020. **708**: p. 135212.
8. Günal, N., *Türkiye'de iklimin doğal bitki örtüsü üzerindeki etkileri.* Acta Turcıca Çevrimiçi Tematik Türkoloji Dergisi, 2013. **1**: p. 1–22.

9. Pecl, G.T., M.B. Araújo, J.D. Bell, J. Blanchard, T.C. Bonebrake, I.C. Chen, T.D. Clark, R.K. Colwell, F. Danielsen, B. Evengård, L. Falconi, S. Ferrier, S. Frusher, R.A. Garcia, R.B. Griffis, A.J. Hobday, C. Janion-Scheepers, M.A. Jarzyna, S. Jennings, J. Lenoir, H.I. Linnetved, V.Y. Martin, P.C. McCormack, J. McDonald, N.J. Mitchell, T. Mustonen, J.M. Pandolfi, N. Pettorelli, E. Popova, S.A. Robinson, B.R. Scheffers, J.D. Shaw, C.J.B. Sorte, J.M. Strugnell, J.M. Sunday, M.-N. Tuanmu, A. Vergés, C. Villanueva, T. Wernberg, E. Wapstra, and S.E. Williams, *Biodiversity redistribution under climate change: Impacts on ecosystems and human well-being.* Science, 2017. **355**(6332).
10. Elsen, P.R., W.B. Monahan, E.R. Dougherty, and A.M. Merenlender, *Keeping pace with climate change in global terrestrial protected areas.* Science Advances, 2020. **6**(25): p. eaay0814.
11. Wan, J.Z., C.J. Wang, and F.H. Yu, *Spatial conservation prioritization for dominant tree species of Chinese forest communities under climate change.* Climatic Change, 2017. **144**(2): p. 303–316.
12. Belote, R.T., M.S. Dietz, P.S. McKinley, A.A. Carlson, C. Carroll, C.N. Jenkins, D.L. Urban, T.J. Fullman, J.C. Leppi, and G.H. Aplet, *Mapping Conservation Strategies under a Changing Climate.* BioScience, 2017. **67**(6): p. 494–497.
13. Bouchard, M., N. Aquilué, C. Périé, and M.-C. Lambert, *Tree species persistence under warming conditions: A key driver of forest response to climate change.* Forest Ecology and Management, 2019. **442**: p. 96–104.
14. Gülçin, D. and B. Deniz, *Remote sensing and GIS-based forest fire risk zone mapping: The case of Manisa, Turkey.* Türkiye Ormancılık Dergisi, 2020. **21**(1): p. 15–24.
15. Gülçin, D. and T. Yılmaz, *Evaluation of forest fragmentation with particular reference to landscape-based ecological assessment and wildlife conservation.* Türkiye Ormancılık Dergisi, 2020. **21**(1): p. 84–93.
16. Meineke, E., E. Youngsteadt, R.R. Dunn, and S.D. Frank, *Urban warming reduces aboveground carbon storage.* Proceedings of the Royal Society B: Biological Sciences, 2016. **283**(1840): p. 20161574.
17. Morelli, T.L., C.W. Barrows, A.R. Ramirez, J.M. Cartwright, D.D. Ackerly, T.D. Eaves, J.L. Ebersole, M.A. Krawchuk, B.H. Letcher, M.F. Mahalovich, G.W. Meigs, J.L. Michalak, C.I. Millar, R.M. Quiñones, D. Stralberg, and J.H. Thorne, *Climate-change refugia: biodiversity in the slow lane.* Frontiers in Ecology and the Environment, 2020. **18**(5): p. 228–234.
18. Fuentes-Castillo, T., R.A. Scherson, P.A. Marquet, J. Fajardo, D. Corcoran, M.J. Román, and P. Pliscoff, *Modelling the current and future biodiversity distribution in the Chilean Mediterranean hotspot. The role of protected areas network in a warmer future.* Diversity and Distributions, 2019. **25**(12): p. 1897–1909.

19. Segan, D.B., K.A. Murray, and J.E.M. Watson, *A global assessment of current and future biodiversity vulnerability to habitat loss–climate change interactions.* Global Ecology and Conservation, 2016. **5**: p. 12–21.
20. Urban, M.C., G. Bocedi, A.P. Hendry, J.B. Mihoub, G. Pe'er, A. Singer, J.R. Bridle, L.G. Crozier, L.D. Meester, W. Godsoe, A. Gonzalez, J.J. Hellmann, R.D. Holt, A. Huth, K. Johst, C.B. Krug, P.W. Leadley, S.C.F. Palmer, J.H. Pantel, A. Schmitz, P.A. Zollner, and J.M.J. Travis, *Improving the forecast for biodiversity under climate change.* Science, 2016. **353**(6304).
21. Bellard, C., C. Bertelsmeier, P. Leadley, W. Thuiller, and F. Courchamp, *Impacts of climate change on the future of biodiversity.* Ecology Letters, 2012. **15**(4): p. 365–377.
22. Synes, N.W., A. Ponchon, S.C.F. Palmer, P.E. Osborne, G. Bocedi, J.M.J. Travis, and K. Watts, *Prioritising conservation actions for biodiversity: Lessening the impact from habitat fragmentation and climate change.* Biological Conservation, 2020. **252**: p. 108819.
23. Çoban, H.O. and Ş. Gündoğdu, *Orman alanlarındaki değişimlerin CBS tabanlı belirlenmesi: Çamsu Orman İşletme Şefliği örneği.* Türkiye Ormancılık Dergisi, 2020. **21**(1): p. 60–69.
24. Arslan, E.S., A. Akyol, Ö.K. Örücü, and A.G. Sarıkaya, *Distribution of rose hip (Rosa canina L.) under current and future climate conditions.* Regional Environmental Change, 2020. **20**(3): p. 107.
25. Bellard, C., J.M. Jeschke, B. Leroy, and G.M. Mace, *Insights from modeling studies on how climate change affects invasive alien species geography.* Ecology and Evolution, 2018. **8**(11): p. 5688–5700.
26. Li, X., F. Mao, H. Du, G. Zhou, L. Xing, T. Liu, N. Han, Y. Liu, D. Zhu, J. Zheng, L. Dong, and M. Zhang, *Spatiotemporal evolution and impacts of climate change on bamboo distribution in China.* Journal of Environmental Management, 2019. **248**: p. 109265.
27. Zhang, K., L. Yao, J. Meng, and J. Tao, *Maxent modeling for predicting the potential geographical distribution of two peony species under climate change.* Science of The Total Environment, 2018. **634**: p. 1326–1334.
28. Ipcc, *Climate Change 2014: Synthesis Report. Contribution of Working Groups I, II and III to the Fifth Assessment Report of the Intergovernmental Panel on Climate Change*, R.K. Pachauri and A.L. Meyer, Editors. 2014: Geneva, Switzerland. p. 151.
29. Akyol, A., Ö.K. Örücü, and E.S. Arslan, *Habitat suitability mapping of stone pine (Pinus pinea L.) under the effects of climate change.* Biologia, 2020.
30. Fei, S., J.M. Desprez, K.M. Potter, I. Jo, J.A. Knott, and C.M. Oswalt, *Divergence of species responses to climate change.* Science Advances, 2017. **3**(5): p. e1603055.

31. Prasad, A., J. Pedlar, M. Peters, D. McKenney, L. Iverson, S. Matthews, and B. Adams, *Combining US and Canadian forest inventories to assess habitat suitability and migration potential of 25 tree species under climate change.* Diversity and Distributions, 2020. **26**(9): p. 1142–1159.

32. Zolkos, S.G., P. Jantz, T. Cormier, L.R. Iverson, D.W. McKenney, and S.J. Goetz, *Projected Tree Species Redistribution Under Climate Change: Implications for Ecosystem Vulnerability Across Protected Areas in the Eastern United States.* Ecosystems, 2015. **18**(2): p. 202–220.

33. Booth, T.H., *Species distribution modelling tools and databases to assist managing forests under climate change.* Forest Ecology and Management, 2018. **430**: p. 196–203.

34. Ghanbarian, G., M.R. Raoufat, H.R. Pourghasemi, and R. Safaeian, *9 - Habitat Suitability Mapping of Artemisia aucheri Boiss Based on the GLM Model in R*, in *Spatial Modeling in GIS and R for Earth and Environmental Sciences*, H.R. Pourghasemi and C. Gokceoglu, Editors. 2019, Elsevier. p. 213–227.

35. Phillips, S.J., R.P. Anderson, and R.E. Schapire, *Maximum entropy modeling of species geographic distributions.* Ecological modelling, 2006. **190**(3–4): p. 231–259.

36. Saatchi, S., W. Buermann, H. ter Steege, S. Mori, and T.B. Smith, *Modeling distribution of Amazonian tree species and diversity using remote sensing measurements.* Remote Sensing of Environment, 2008. **112**(5): p. 2000–2017.

37. Dagnino, D., M. Guerrina, L. Minuto, M.G. Mariotti, F. Médail, and G. Casazza, *Climate change and the future of endemic flora in the South Western Alps: relationships between niche properties and extinction risk.* Regional Environmental Change, 2020. **20**(4): p. 121.

38. Ferrarini, A., M.H.S.A. Alsafran, J. Dai, and J.M. Alatalo, *Improving niche projections of plant species under climate change: Silene acaulis on the British Isles as a case study.* Climate Dynamics, 2019. **52**(3): p. 1413–1423.

39. Koch, R., J.S. Almeida-Cortez, and B. Kleinschmit, *Revealing areas of high nature conservation importance in a seasonally dry tropical forest in Brazil: Combination of modelled plant diversity hot spots and threat patterns.* Journal for Nature Conservation, 2017. **35**: p. 24–39.

40. Lu, N., C.-X. Jia, H. Lloyd, and Y.-H. Sun, *Species-specific habitat fragmentation assessment, considering the ecological niche requirements and dispersal capability.* Biological Conservation, 2012. **152**: p. 102–109.

41. Rojas Briceño, N.B., D.A. Cotrina Sánchez, E. Barboza Castillo, M.Á. Barrena Gurbillón, F.O. Sarmiento, D.A. Sotomayor, M. Oliva, and R. Salas López, *Current and Future Distribution of Five Timber Forest Species in Amazonas, Northeast Peru: Contributions towards a Restoration Strategy.* Diversity, 2020. **12**(8): p. 305.

42. Farjon, A., *Picea orientalis (Oriental Spruce).* 2013.
43. Yaltırık, F., *Dendroloji Ders Kitabı.* 1988: Istanbul Universitesi Orman Fakultesi.
44. Erkuloğlu, Ö.S., *Doğu Ladini.* 1989, Ankara: Ormancılık Araştırma Enstitüsü.
45. Yaltırık, F. and A. Efe, *Dendrology Handbook, Gymnospermae-Angiospermea.* 2000, Istanbul: University of Istanbul Publication.
46. Akgül, E., *Doğu Ladininin Önemi.* Ormancılık Araştırma Enstitüsü Dergisi, 1970. **16**(2): p. 59–62.
47. Bostancı, Ş., *Türkiye Doğu Ladini (Picea orientalis (L.) Carr.) Odununun Kimyasal Bileşimi ve Doğu Ladini ile Çoğu Karadeniz Köknarı (Abies nordmanniana (sach.) Yongalarından Elde Edilen Mekanik Kağıt Hamurunun Endüstride Kullanma Olanakları.* 12. 1979: KTÜ Orman Fakültesi.
48. Euforgen, *European Forest Genetic Resources Programme.* 2019.
49. American Conifer, S., *Picea orientalis/Caucasian Spruce | Conifer Species.* Picea orientalis/Caucasian spruce | Conifer Species, 2020.
50. Davis, P.H., *Flora of Turkey and The East Aegean Islands - I.* Vol. One. 1965, Edinburgh: Edinburgh University Press.
51. Gbif, *Picea orientalis (L.) Peterm. in GBIF Secretariat. GBIF Backbone Taxonomy. Checklist dataset https://doi.org/10.15468/39omei accessed via GBIF.org on 2020-10-25.* 2019.
52. WorldClim, *WorldClim - Global Climate Data.* 2019.
53. Fielding, A.H. and J.F. Bell, *A review of methods for the assessment of prediction errors in conservation presence/absence models.* Environmental conservation, 1997. **24**(1): p. 38–49.
54. Merow, C., M.J. Smith, and J.A. Silander, *A practical guide to MaxEnt for modeling species' distributions: what it does, and why inputs and settings matter.* Ecography, 2013. **36**(10): p. 1058–1069.
55. Pearson, R.G., C.J. Raxworthy, M. Nakamura, and A.J.J.o.b. Townsend Peterson, *Predicting species distributions from small numbers of occurrence records: a test case using cryptic geckos in Madagascar.* 2007. **34**(1): p. 102–117.
56. Shcheglovitova, M. and R.P. Anderson, *Estimating optimal complexity for ecological niche models: A jackknife approach for species with small sample sizes.* Ecological Modelling, 2013. **269**: p. 9–17.
57. Srivastava, V., V.C. Griess, and M.A. Keena, *Assessing the Potential Distribution of Asian Gypsy Moth in Canada: A Comparison of Two Methodological Approaches.* Scientific Reports, 2020. **10**(1): p. 22.
58. Kayacık, H., *Doğu Ladininin Türkiyedeki Coğrafi Yayılışı ve Silvikültür Esasları ve Tabii Sınırlarının Genişletilmesi İmkanlarını Araştırma.* 1952: Orman Genel Müdürlüğü Yayınları.

59. Atalay, İ., *Türkiye Vejetasyon Coğrafyasına Giriş*. 1983, İzmir: Ege Üniversitesi Edebiyat Fakültesi Yayınları.

60. Arslan, E.S., D. Gülçin, A.G. Sarıkaya, Z. Ölmez, S. Gülcü, İ. Şen, and Ö.K. Örücü, *Kokulu Ardıç'ın (Juniperus foetidissima Willd.) Günümüz ve Gelecekteki Potansiyel Yayılışının Makine Öğrenmesi ile Modellenmesi*. Avrupa Bilim ve Teknoloji Dergisi, 2021(22): p. 1–12.

61. Gebrewahid, Y., S. Abrehe, E. Meresa, G. Eyasu, K. Abay, G. Gebreab, K. Kidanemariam, G. Adissu, G. Abreha, and G. Darcha, *Current and future predicting potential areas of Oxytenanthera abyssinica (A. Richard) using MaxEnt model under climate change in Northern Ethiopia*. Ecological Processes, 2020. **9**(1): p. 6.

62. Pecchi, M., M. Marchi, V. Burton, F. Giannetti, M. Moriondo, I. Bernetti, M. Bindi, and G. Chirici, *Species distribution modelling to support forest management. A literature review*. Ecological Modelling, 2019. **411**: p. 108817.

63. Qin, A.L., B. Liu, Q.S. Guo, R.W. Bussmann, F.Q. Ma, Z.J. Jian, G.X. Xu, and S.X. Pei, *Maxent modeling for predicting impacts of climate change on the potential distribution of Thuja sutchuenensis Franch., an extremely endangered conifer from southwestern China*. Global Ecology and Conservation, 2017. **10**: p. 139–146.

64. Zhang, K., H. Liu, H. Pan, W. Shi, Y. Zhao, S. Li, J. Liu, and J. Tao, *Shifts in potential geographical distribution of Pterocarya stenoptera under climate change scenarios in China*. Ecology and Evolution, 2020. **10**(11): p. 4828–4837.

65. Coban, H.O., A. Koc, and M. Eker, *Investigation on changes in complex vegetation coverage using multi-temporal landsat data of Western Black sea region-A case study*. Journal of environmental Biology, 2010. **31**(1/2): p. 169–178.

66. Kolanowska, M. and A. Jakubska-Busse, *Is the lady's-slipper orchid (Cypripedium calceolus) likely to shortly become extinct in Europe?—Insights based on ecological niche modelling*. PLOS ONE, 2020. **15**(1): p. e0228420.

67. Mavárez, J., S. Bézy, T. Goeury, A. Fernández, and S. Aubert, *Current and future distributions of Espeletiinae (Asteraceae) in the Venezuelan Andes based on statistical downscaling of climatic variables and niche modelling*. Plant Ecology & Diversity, 2019. **12**(6): p. 633–647.

68. Tsiftsis, S. and V. Djordjević, *Modelling sexually deceptive orchid species distributions under future climates: the importance of plant–pollinator interactions*. Scientific Reports, 2020. **10**(1): p. 10623.

69. Ewers, R.M., R.K. Didham, L. Fahrig, G. Ferraz, A. Hector, R.D. Holt, V. Kapos, G. Reynolds, W. Sinun, J.L. Snaddon, and E.C. Turner, *A large-scale forest fragmentation experiment: the Stability of Altered Forest Ecosystems Project*. Philosophical Transactions of the Royal Society B: Biological Sciences, 2011. **366**(1582): p. 3292–3302.

70. Holling, C.S., *Resilience and Stability of Ecological Systems.* Annual Review of Ecology and Systematics, 1973. **4**(1): p. 1–23.
71. Larsen, J., *Ecological stability of forests and sustainable silviculture.* Forest Ecology and Management, 1995. **73**: p. 85–96.
72. Alberto, F.J., S.N. Aitken, R. Alía, S.C. González-Martínez, H. Hänninen, A. Kremer, F. Lefèvre, T. Lenormand, S. Yeaman, R. Whetten, and O. Savolainen, *Potential for evolutionary responses to climate change – evidence from tree populations.* Global Change Biology, 2013. **19**(6): p. 1645–1661.
73. Frank, A., G.T. Howe, C. Sperisen, P. Brang, J.B.S. Clair, D.R. Schmatz, and C. Heiri, *Risk of genetic maladaptation due to climate change in three major European tree species.* Global Change Biology, 2017. **23**(12): p. 5358–5371.
74. Akyol, A. and O.K. Orucu, *Investigation and Evaluation of Stone Pine (Pinus Pinea L.) Current and Future Potential Distribution Under Climate Change in Turkey.* Cerne, 2019. **25**(4): p. 415–423.
75. Al-Qaddi, N., F. Vessella, J. Stephan, D. Al-Eisawi, and B. Schirone, *Current and future suitability areas of kermes oak (Quercus coccifera L.) in the Levant under climate change.* Regional Environmental Change, 2017. **17**(1): p. 143–156.
76. Arslan, E. and K. Örücü, *Present and future potential distribution of the Pinus nigra Arnold. and Pinus sylvestris l. usıng Maxent Model.* International Journal of Ecosystems and Ecology Science, 2019. **9**(4): p. 787–798.
77. Çoban, H.O., Ö.K. Örücü, and E.S. Arslan, *MaxEnt Modeling for Predicting the Current and Future Potential Geographical Distribution of Quercus libani Olivier.* Sustainability, 2020. **12**(7): p. 2671.
78. Sarıkaya, A. and Ö. Örücü, *Prediction of potential and future distribution areas of Anatolian Chesnut (Castanea sativa Mill.) by using maximum entropy (Maxent) modeling depending on climate change in Turkey.* International Journal of Ecosystems and Ecology Science (IJEES), 2019. **9**(4): p. 699–708.
79. Yi, Y.J., X. Cheng, Z.F. Yang, and S.H. Zhang, *Maxent modeling for predicting the potential distribution of endangered medicinal plant (H. riparia Lour) in Yunnan, China.* Ecological Engineering, 2016. **92**: p. 260–269.
80. Rathore, P., A. Roy, and H. Karnatak, *Assessing the vulnerability of Oak (Quercus) forest ecosystems under projected climate and land use land cover changes in Western Himalaya.* Biodiversity and Conservation, 2019. **28**(8): p. 2275–2294.
81. Schleuning, M., E.L. Neuschulz, J. Albrecht, I.M.A. Bender, D.E. Bowler, D.M. Dehling, S.A. Fritz, C. Hof, T. Mueller, L. Nowak, M.C. Sorensen, K. Böhning-Gaese, and W.D. Kissling, *Trait-Based Assessments of Climate-Change Impacts on Interacting Species.* Trends in Ecology & Evolution, 2020. **35**(4): p. 319–328.

82. Reside, A.E., K. Critchell, D.M. Crayn, M. Goosem, S. Goosem, C.J. Hoskin, T. Sydes, E.P. Vanderduys, and R.L. Pressey, *Beyond the model: expert knowledge improves predictions of species' fates under climate change.* Ecological Applications, 2019. **29**(1): p. e01824.

83. Littlefield, C.E., M. Krosby, J.L. Michalak, and J.J. Lawler, *Connectivity for species on the move: supporting climate-driven range shifts.* Frontiers in Ecology and the Environment, 2019. **17**(5): p. 270–278.

84. Mestre, F., B.B. Risk, A. Mira, P. Beja, and R. Pita, *A metapopulation approach to predict species range shifts under different climate change and landscape connectivity scenarios.* Ecological Modelling, 2017. **359**: p. 406–414.

Duygu DOĞAN and Şükran ŞAHİN

Landscape Corridors: The Case of Malatya Province

Abstract: The corridors, as linear landscape elements, defined by many different names starting with Frederick Law Olmsted's "Emerald Necklace" park system concept in 1867, then from the term of "Environmental Corridor" by Philip Lewis in the 1960s to the present day, are developed as a response to the ecological, cultural and recreational needs in terms of spatial planning and design purposes.

The corridors are mainly being considered in planning and design processes as spatial units providing connectivity and linear spaces. However, the functions of the corridors are of a character that cannot be limited to these mentioned ones. In this Chapter, an innovative method based on not only the mentioned functions but also the key processes that shape natural and cultural landscapes is proposed in determining the linear areas that will form corridors. Linear landscape elements mapped with such a method are named as "Landscape Corridor". The landscape corridors to be determined by applying the method described in the example of Malatya province are, on the one hand, primarily the linear areas that reflect and shape the original character of the landscape and therefore have high landscape protection value. On the other hand, these landscape corridors are both access lines and potential spaces for outdoor activities suggested to be determined in line with the user profile and demand as well as spatial requirements.

Anahtar Kelimeler: Landscape corridor, greenway, wildlife corridor, landscape connectivity, river corridor

1 Introduction

Frederick Law Olmsted introduced the park system notion by connecting parks and green areas together for the first time in 1867, in Boston USA [1]. In the 1960s, Philip Lewis defined the major and minor/additional recreational resources of the US state of Wisconsin, different from Olmsted's concept of the park system, and stated that these resources constitute the "Environmental Corridor"[2, 1]. Since 1867, corridor concept as a response to the environmental, cultural, and recreational needs are being studied in all the world. This chapter is also on the landscape corridors of Malatya province created with an innovative approach. It is claimed that the implementation of these landscape corridors will contribute to the protection and sustainable use of the ecological and cultural values of Malatya. The methodological ground and content of the terminology of "Landscape Corridor" is aimed to contribute to scientific knowledge within the scope of spatial planning and design studies.

Malatya Province is located in the Euphrates basin of East Anatolia region lies between 35°54' and 39° 03' Northern latitude and 38°45' and 39°08' Eastern longitude [3]. (Fig. 1).

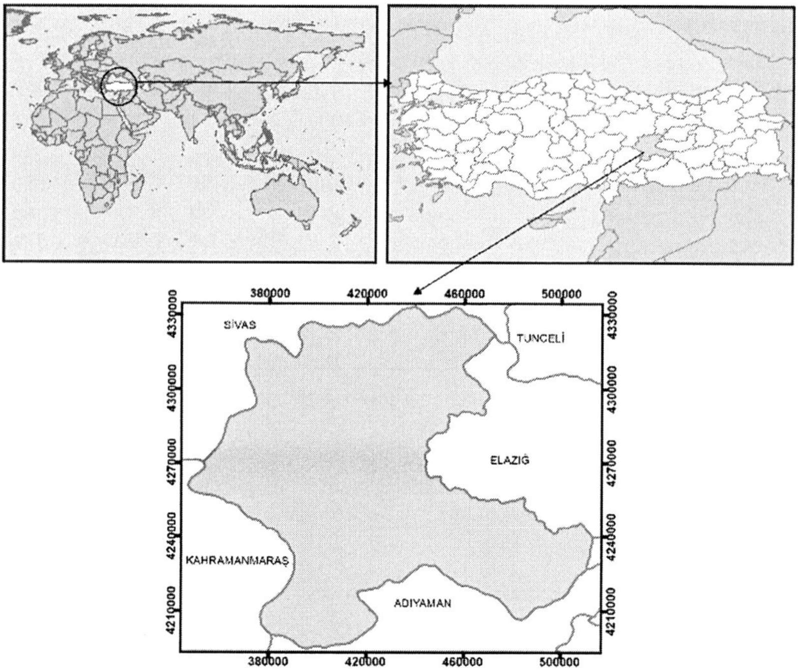

Fig. 1: Location map

In this study, different from the current approaches to define landscape patches, not only perceptible landscape elements were considered in fauna mobility and habitat analyzes, but also areas that have important functions in the formation of the landscape, in other words, landscape function analysis, were also included in the analysis and evaluation process. This is because the sustainability of appropriate habitats of the fauna depends on the healthy functioning of the ecological processes. Forman and Godron (1986) explain natural processes under the functions of the landscape while explaining the patch-corridor-matrix model under landscape structure. On the other hand, structural characteristics of the landscape are a result of functional characteristics and the changes in the functional characteristics affect the structural characteristics. In that sense, as a first stage, the areas at higher landscape (protection) value are identified in terms of structural and functional landscape characteristics based on this pattern-process concept within the study area extend, then each of those areas is defined as a patch. Connectivity is created by the use of ecologically and culturally linear elements in the designated areas. The stages of the methodology are summarized into three headlines. They are (1) Identification of high landscape (protection) value in terms of structure and function, (2) Identification of perceptible linear landscape structures which might be potential connectivity elements, and (3) Preparation of connectivity plan by the use of ecologically and culturally linear elements.

River corridors as linear spatial units and plant cover as that ecologically form the landscape structures and steering geomorphological processes as landscape functions (erosion risk, groundwater recharging potential, and rainfall surface run-off potential) were used within the scope of the study. Firstly, areas with a high degree of erosion risk, groundwater recharging potential, and surface run-off potential are identified, and secondly, those areas are overlapped with habitat patches and INA's (Important Nature Areas) to be able to identify ecologically the areas of high landscape (protection) value. Thirdly, potential ecological corridors providing ecological connectivity are created by overlapping the areas with high landscape (protection) value with river corridors as natural linear elements, as well as with roads and energy transmission lines as cultural linear elements. Both barrier function [5] and the connectivity function [6] of the linear landscape elements (rivers, roads, and energy transmission lines) were considered while identifying ecological connectivity. Buffer zones are set on this line when considering the barrier function.

Data obtained and produced within the scope of the Landscape-44 project [7] are used as inputs in this study (Tab. 1).

While determining cultural connectivity, the topographical, hydrological, historical, and archaeological features of the province are considered under two headings: sports routes and historical/archaeological tourism routes. Historical archeological routes and elements and the tourism/recreation points identified within the scope of the land survey conducted in the Project are used when both of the routes were specified.

Tab. 1: Data Table

Data source	***Data***
Abrogated Turkish Republic Ministry of Environment and Forestry data processıng	Province boundary
	Important areas accepted as fisheries/ living environments in Malatya provincial boundaries
The General Directorate of State Hydraulic Works	Rivers
	Lakes
	Dam
	Streambed
The General Directorate of Forestry	Forest
Turkish Electricity Works Incorporated	Powerline
General Directorate of Highways	Railway
	Road
The General Command of Cartography	Topographic Maps (1/25000 Scale)
Nature Associatıon	Important Natural Areas in Malatya
Landscape - 44: Şahin et al. 2013	CORINE 2010
Landscape - 44: Şahin et al. 2013	Landscape Survey Inventory and Landscape Functions

Observations carried out during land surveys as well as the topographic and hydrological data obtained were utilized in defining the routes of sports tourism. Nature sports and their standards were investigated, then suitable regions were identified in the process. Historical artifacts and archeological resources detected as a result of the land survey carried out under the scope of the Project Landscape-44 in the phase of identifying historical/archeological tourism were used for the specification of tourism routes. As a methodology, the closest settlement area was determined according to the density of the distribution of these points in both sites of the routes, and routes were established by using the existing roads nearby these settlement areas (Fig. 3).

2 Ecological Landscape Corridors

2.1 Identification of Areas with High Landscape (Protection) Value

Erosion risk, groundwater recharging potential, and surface water run-off potential maps were utilized to identify the areas with high landscape (protection) value.

Areas with high and very high potential of erosion risk, groundwater recharging potential, and surface water runoff potential are identified and they are combined with habitat patches and INA's (Important Nature Areas) while identifying high and very high landscape value areas (irrespective of which function is included). Assessment of the area in terms of mentioned components. Additionally, if there is vegetation cover on the areas with no high or very high value in terms of any function representing a geomorphological process, these areas are also considered as valuable from the perspective of ecological connectivity.

2.2 Identification of River Corridors

Rivers are areas chosen for settlement by humans in very early periods along with coastal areas [1]. Hence, river corridors are linear elements important both for ecological and cultural connectivity. Hence, river corridors were firstly defined while establishing the connectivity under the scope of this study.

To begin with, the bed width of the river was detected in the identification of river corridor under the scope of the study. This width is calculated with $B=3.7W^{1.12}$ formula (B= river belt, W= channel width). This is 6 times of the river channel width (slope<%2) in meandering rivers [8]. Riverbed width, slope, and river sensitivity (aquaculture, important habitat) are taken into consideration in the identification of river corridors in the study.

There are two ways of specifying the river width in identifying river corridors: (1) 1 to 4 times of the width of the riverbed is suggested for areas that are steep, lowly erodible and restricted to narrow valleys, where there is basic river type and special valley character. (2) Very sensitive river types or landslide regions require 6 times of the riverbed width [8]. According to this method, the aquaculture habitat of Malatya province is considered as sensitive river type, and 6 times of the bed width is used. 4 times of the riverbed width is used for nine important rivers in Malatya province (Söğütlü Brook, Morhamam Brook, Kuruçay, Tohma Brook, Sultansuyu, Sürgü suyu, Beylerderesi, Mamihan Brook ve Şiro Brook).

2.3 Roads from the Perspective of Ecological Connectivity

Roads lead to fragmentation acting like a barrier in the habitats they pass from. These effect distances that lead to fragmentation are used in the designation of buffer zones. Girvetz et al. (2008) assigned the effect range of highways as 20 m in California USA, main roads as 10 m, and rural roads as 6 m. Li et. al. (2010) assigned the effect range of national roads and railway as 100 m, regional roads as 60 m, rural roads as 30 m in China [9]. Buffer zones are designated within the scope of the study by taking into account the observations made during fieldwork. They are 30 m for state highways and railways, 20 m for provincial roads, 10 m for intercity roads, and 6m for rural roads.

2.4 Energy Transmission Lines from the Perspective of Ecological Connectivity

Human made roads, railways, energy transmission lines, and paths act as corridors in the landscape [6]. Energy transmission lines are assessed in terms of connectivity from this perspective.

As energy transmission lines carry high electricity energy, constructions are allowed only after certain distances in the regions where there are transmission lines. Tree plantation activities can be conducted in areas where conductive materials are present with lowest vertical distances concepts per the Electricity High Current Power Plants Regulation [10] to create a plant corridor. The buffer zones created considering this situation are given in Tab. 2.

Tab. 2: Distances used in identifying buffer zones (m)

Energy transmission line kV value	Distances used in identifying buffer zones
34.5	7 m from both sides of the transmission line
154	8 m from both sides of the transmission line
380	12 m from both sides of the transmission line

2.5 Ecological Corridors

Areas with high and very high landscape (protection) value, river corridors, state highways, and energy transmission lines are combined to create ecological corridors (Fig. 2). The data used in the identification of ecological corridors within the study are as follows: (1) Areas with very high and high landscape (protection) value, (2) Areas that do not have high or very high value in terms of landscape functions but have vegetation cover, (3) Areas with no high or very high functional value, (4) Important natural areas, (5) River corridors, highway routes, (6) Energy transmission lines. The ecological corridor route established using these are given in Fig. 2.

3 Cultural Landscape Corridors

The Landscape-44 project produced data as a result of a land survey on Malatya province regarding historical and cultural sources, potential nature sports areas, and canyons. Data obtained from this survey contains information on historical structures, archeological settlements, caves, burial mounds, tumulus, caravanserais, inns, domes, historical water fountains, visits, mosques, thermal resources, museums, shrines, bathhouses, and old graveyards as well as types of sports that could be done besides their locations [11].

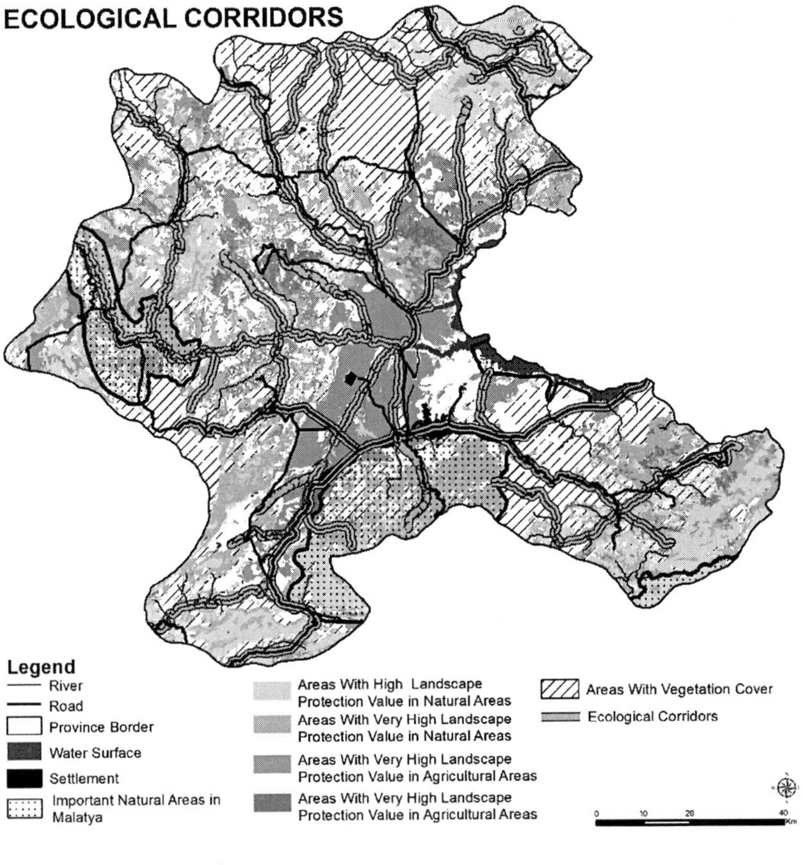

Fig. 2: Ecological corridors

3.1 Identification of the Sports Tourism Route

Sports that could be performed in the province were identified as a result of observations made during field works, considering the definitions and spatial requirements of nature sports of bouldering [12], rock climbing [12], hiking [13], trekking [14], paragliding [15], watercourse slalom [16], free flow river canoe [16], canoe marathon [16], and rafting [17].

Mountains, rocks, and canyons in Malatya provide ideal places for climbing and watercourses and dam lakes present suitable areas for water sports. Point datas gathered during fieldwork, mountainous areas were identified by using topographical maps, watercourses were identified from hydrological maps, canyons were identified by using topographical maps of General Command of Mapping with the scale of 1:25000 and rocky areas produced by using CORINE 2010 data which all were overlapped in order to determine suitable areas for aforementioned nature sports. In accordance with the scale of the study, the closest settlement place in areas suitable for nature sports were identified and access to these places were provided. Definition of routes by providing access to the place where the sport is performed must be done by carrying out studies on lower scales. The identified settlement places were connected to motorized roads and sports tourism route was developed.

3.2 Identification of Historical/Archaeological Tourism Routes

Malatya with its history dating back to Hittites, Seljukian, and Ottomans has never lost its historical texture and it is one of the most important cities in Turkey [18]. There are many historical artworks in the province. These artworks and archaeological resources identified as a result of fieldwork done within the framework of the project were used while defining tourism routes. Nearest settlement places to these historical artifacts and archaeological resources were identified while defining historical/archaeological tourism routes. In accordance with the scale of the study, nearest places to historical/archaeological areas were identified and access to these points were provided. Definition of routes by providing access to historical/archaeological areas must be done by carrying out studies on lower scales. These identified places were connected to motorized roads.

3.3 Cultural Corridors

Cultural corridors were created by combining the sports routes and historical/archaeological tourism routes, which were created taking into account the topographical, hydrological, historical, and archaeological features of the province of Malatya. Although these two routes mostly support each other, different routes also arise according to the locations of historical and archaeological resources in the province (Fig. 3).

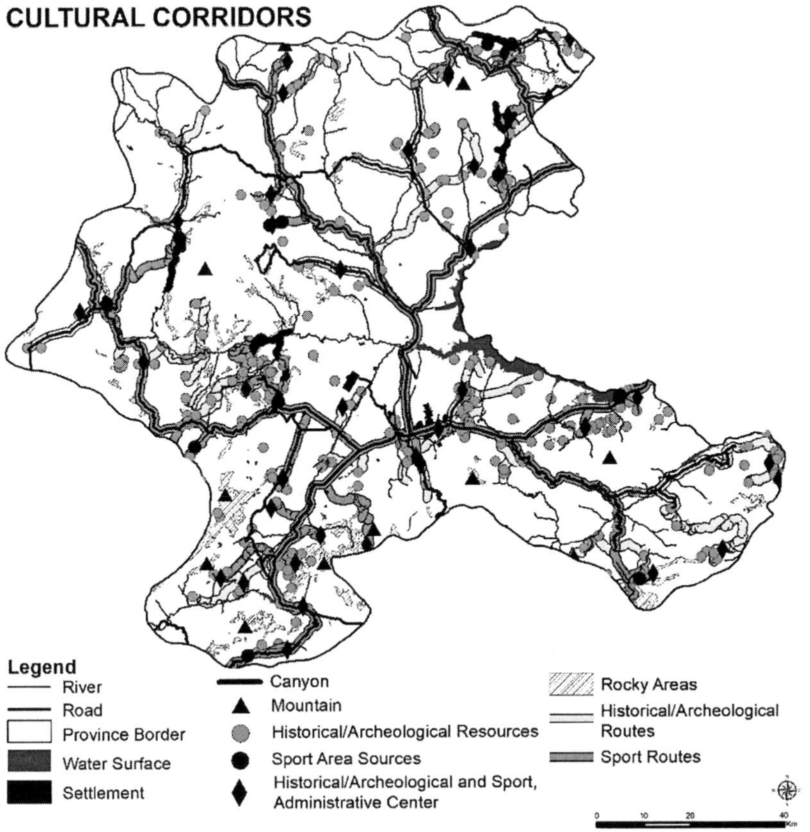

Fig. 3: Cultural corridors

4 Landscape Corridors

The map of landscape corridors was created by the integration of the ecological and cultural corridors in one map, and is given in Fig. 4. The map produced contains the following information. These two corridors may be lines that support each other or there may be where intersection will not be desirable. The result of the biodiversity analyses can be used to decide as to whether the determined corridors support each other or not. Biodiversity analyses should be executed at more detailed lower scales, considering the landscape fragmentation and important species.

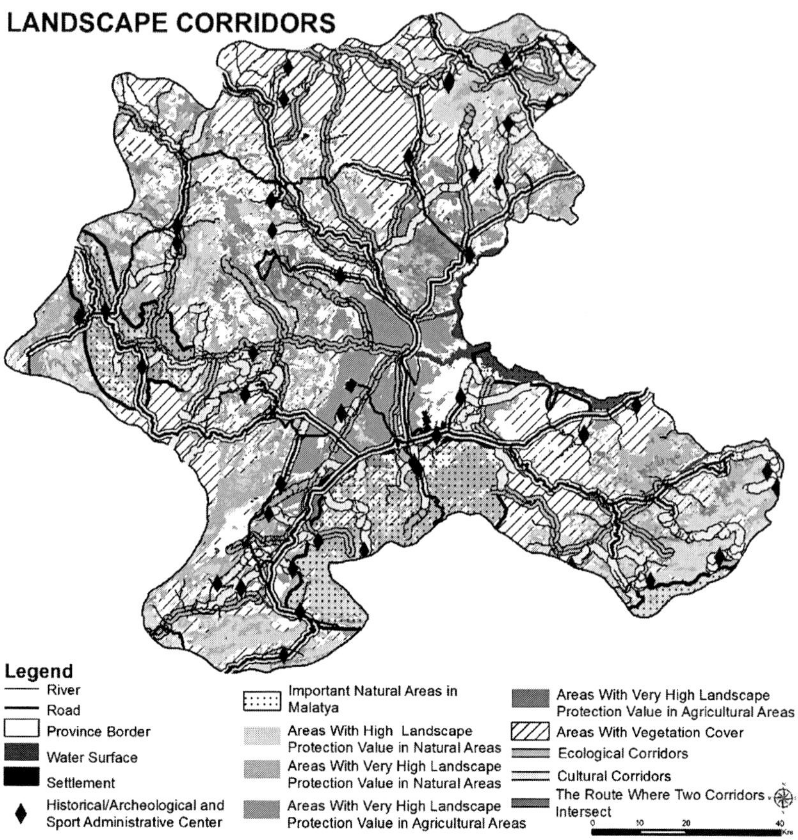

Fig. 4: Landscape corridors

5 Discussion and Conclusions

A model that takes landscape corridor concept and mechanisms that shape the landscape while ensuring the sustainability is introduced under the scope of this study. Studies carried out by Philip H. Lewis in 1960s coining the concept of "environmental corridor" and by Fabos (1995) in mid 1990s on greenways have guided the formation of this system.

Lewis defined 220 natural and cultural resources in the 1960s under the term of Environmental Corridor. Resources are scored according to a scale on recreational potentials [2]. Lewis provided the connection between these sources, which he defined, with the linear line he created along the streams and main drainage areas.

Lewis focused on sustainability in this study. This work by Lewis is the first to focus on protecting sensitive environmental areas and stream corridors [1].

Fabos (1995) defines green roads as corridors of different widths and divides greenway systems into 3: (1) Greenways created with important ecological systems and natural systems: mostly along streams, coastal areas and ridges, to conserve biodiversity, to support wildlife migrations and to be suitable for nature studies. (2) Recreational greenways: areas where road networks, water connections and water-based recreational activities are performed; roads and routes are roads passing through landscapes with high visual quality. Its scale can be local, regional or country. (3) Greenways with historical heritage and cultural values: attractive green roads for tourists, providing recreational, educational, natural and economic benefits. Fabos (1995) defined the details for these 3 types of greenways and used the details he defined in creating greenways.

The concept introduced within the scope of this study, different from the concept used by the mentioned researchers, includes the structure and functions of the landscape as a whole. In order to understand the processes that make up the concept of landscape, it is necessary to understand the mechanisms that shape a landscape (erosion risk, surface water runoff potential and groundwater recharge potential in this study) and their interactions [19]. At different scales, the landscape corridor will contain different details. It is aimed that the landscape corridors put forward within the scope of this study will guide the upper scale (regional and sub-regional scale) decisions to be taken within the scope of landscape planning and management.

Landscape corridors demonstrated with this Chapter could not include analysis for biodiversity as there is no available data for the pilot area. Nevertheless, biodiversity must be considered when the routes that create the landscape corridors and distance of buffer zones are identified. On the other hand, visual assessments should also be taken into account within the scope of cultural connectivity. Visual corridor point of visual resource, areas with a view to the visual resource and scenery from the visual resource (distance visible to human eye).

Areas to be designated for sports routes should be ascertained after more detailed lower scale landscape analyses and iterated expert opinion panels. For instance, routes for trekking should be classified according to difficulty levels and suitability analyses for camping locations should be carried out. Another element to consider for the identification of cultural corridors is bioclimatic comfort studies. Detection of the most appropriate periods in terms of thermal comfort is of great importance for the activities, sportive activities in particular.

Hedge grows in agricultural fields should also be taken into account within the scope of landscape corridor planning. Especially in agricultural areas where human intervention is high, plant hedge grows support the diversity of flora and fauna, especially creating feeding and hiding areas for fauna [20]. With this feature, hedge grows help to ensure the mobility of fauna by showing the feature of corridors. On the other hand, wire fences surrounding agricultural production areas create barriers for many species. In other words, it causes habitat fragmentation.

Habitat fragmentation refers to decreases in biodiversity and a reduction in the self-healing power of ecosystems [21]. In accordance with the working scale of this research, the issues mentioned above were not examined. However, they must be evaluated in lower scale studies.

According to Ahern (2005) "sustainable planning aspires to link knowledge about sustainability with actions to achieve it" [22]. Connectivity is an indispensable element of conservation actions in the management of natural areas. An important objective of EECONET, NATURA 2000 and the European Community Habitats Directive is largely based on the concept of connectivity [23, 24]. This concept ensures the continuity of the fragmented and isolated habitat spots and the protection of biological diversity, especially in areas with high human influence. In the study conducted by Eşbah et al. (2009), it was found that the probability of ensuring the continuity of the ecological quality of natural areas/protected areas surrounded by land uses that are less fragmented, with strong connections and compatible with their surroundings is higher [25].

Connectivity is one of the most important landscape characteristics that creates a link between landscape structure and function [26, 27]. And yet, when the scientific publications related to the topic are reviewed from the literature [For example 28,29,30,31;32], we see that most of them attribute landscape structure and biologic processes (fauna mobility, biodiversity) that shape the structure. This study focuses on that designation of corridors only by those means will not be adequate in creating the expected effect, but also geomorphological processes (erosion risk, surface flow potential, etc.) should be taken into consideration as determinants. Forman and Godron (1986) express the mechanisms that shape the landscape in three headlines: structure, function and change. It is for certain that these three mechanisms shape the landscape interactively and holistically.

Acknowledgements

This Chapter was produced from the Master's Thesis entitled "A Research on Landscape Corridor Concept within the Province of Malatya" which had been conducted by Duygu Doğan. The Thesis was constructed within the context of sixth sub-work package (WP6 Corridors Design and Mapping for Connectivity) of the main stage entitled IDA2 Landscape Information System of Tourism and Recreation Resources of a Project which was financially supported by TÜBİTAK (The Scientific and Technological Research Council of Turkey) Public Institutions Research Funding Group (KAMAG) Program 1007 with the number of 109G074 and with the title of "Landscape Character Analysis and Its Assessment for Tourism/Recreation (Landscape-44) at Province Level". The result maps of the Thesis was included with a citation in the Chapter 7.2.3 Landscape Corridors of the "National Technical Guideline for Landscape Character Analysis and Assessment at the Regional and Sub-Regional (Provincial) Levels [33] which is one of the outputs of the mentioned TUBITAK project coordinated by Prof. Dr. Şükran Şahin.

References

1. Fabos, J.G. Greenway Planning In the United States: Its Origins and Recent Case Studies, 2004. Journal of Landscape and Urban Planning. Vol. 68. 321–342
2. Belknap, R.K. And Furtado, J.G. Three Approaches to Environmental Resource Analysis, 1967. USA, Washington D.C. The Conservation Foundation.
3. T.C. Malatya Valiliği. Coğrafi Konum, 2021. http://www.malatya.gov.tr, Erişim Tarihi: 05.02.2021.
4. Deniz, B., Küçükerbaş, E.V. and Tunçay, H.E. Peyzaj Ekolojisine Genel Bakış, 2006. Ziraat Fakültesi Dergisi 3(2): 5 – 18.
5. Forman, R.T.T. Land Mosaics, 1995. The Ecology of Landscape and Region. Cambridge University Press. Cambridge, Uk. P 145–176
6. Dramstad, W.E, Olson, J.D. & Forman, R.T.T. Landscape Ecology Principles in Landscape Architecture and Land – Use Planning, 1996. Harvard University, Graduate School of Design, Island Pres, American Society of Landscape Architects. P 35–46
7. Şahin, Ş, Perçin, H., Kurum, E., Uzun, O., Bilgili, B.E., Tezcan, L., Çiçek, İ., Müftüoğlu, V., Çorbacı, Ö.L., Sütünç, S., Doğan, D., Koç, Ö., Ateş, E., Tarım, B., Kurdoğlu, G. ve Kaşko, Y. PEYZAJ-44: İl Ölçeğinde Peyzaj Karakter Analizi ve Turizm/Rekreasyon Açısından Değerlendirilmesi, 2011a. 109G074 Nolu TÜBİTAK KAMAG Projesi 2. Ara Raporu.
8. Anonymous. Vermont Dec River Management Programme Defining River Corridors Fact Sheet 2, 2004. River Management Section Catalog No:27
9. Çağlayan, D.S. Karayollarından Kaynaklanan Peyzaj Parçalanmasının Korunan Alanlar ve Önemli Doğa Alanları Üzerine Etkilerinin İrdelenmesi, 2011. Yüksek Lisans Tezi Ankara Üniversitesi Peyzaj Mimarlığı Anabilim Dalı
10. Resmi Gazete. Elektrik Kuvvetli Akım Tesisleri Yönetmeliği, 2000. http://www.mevzuat.adalet.gov.tr, Resmî Gazete Tarihi: 30.11.2000 Resmî Gazete Sayısı: 24246. Erişim Tarihi:05.02.2021.
11. Şahin, Ş., Perçin, H., Kurum, E., Uzun, O., Bilgili, B.E., Tezcan, L., Çiçek, İ., Müftüoğlu, V., Çorbacı, Ö.L., Sütünç, S., Doğan, D., Koç, Ö., Ateş, E., Tarım, B., Kurdoğlu, G. ve Kaşko, Y. PEYZAJ-44: İl Ölçeğinde Peyzaj Karakter Analizi ve Turizm/Rekreasyon Açısından Değerlendirilmesi, 2011b. 109G074 Nolu TÜBİTAK KAMAG Projesi 3. Ara Raporu.
12. UIAA. Recommendations on the Preservation of Natural Rock for Adventure Climbing, 2021. https://www.theuiaa.org, Erişim Tarihi:05.02.2021
13. İstanbul Teknik Üniversitesi Dağcılık Kulübü, 2021. http://www.itudak.itu.edu.tr, Erişim Tarihi:05.02.2021
14. Erdoğan, N. Çevre ve Ekoturizm, 2003. Ankara: Erk Yayıncılık.

15. Türk Hava Kurumu Türkkuşu Genel Müdürlüğü, Çok Hafif Hava Araçları Okulu Yamaç Paraşüt Başlangıç Pilot Eğitim El Kitabı, 2009. https://www.fenomenairsports.com, Erişim Tarihi:05.02.2021
16. Türkiye Kano Federasyonu. Akarsu Slalom, 2011. http://www.turkaf.org.tr, Erişim Tarihi:10.12.2011
17. Alkan, U. Rekreasyonel Amaçlı Doğa Sporlarının Turizm'de Kullanılması: Antalya Köprülü Kanyon Rafting Uygulaması, 2005. Akdeniz Üniversitesi Sosyal Bilimler Enstitüsü Spor Yöneticiliği Anabilim Dalı.
18. Karagülle, E. Seyahatnamelerde Malatya, 2007. Yüksek Lisans Tezi. Niğde Üniversitesi Sosyal Bilimler Enstitüsü Tarih Anabilim Dalı
19. Şahin, Ş.. Peyzaj Ekolojisi, 2009. Peyzaj Yönetimi, Editörler: Aslı Akay ve Münevver Demirbaş Özen, Ankara: TODAİE Yayın No: 354, Yerel Yönetimler Merkezi Yayın No: 27.
20. Forman, R.T.T and Godron, M. Landscape Ecology, 1986. New York: John Wiley & Sons. 619 P.
21. Flather, C.H, Brady, S.J, and Inkley, D.B. Regional Habitat Appraisals of Wildlife Communities: A Landscape-Level Evaluation of A Resource Planning Model Using Avian Distribution Data, 1992. Landscape Ecology 7: 137–147.
22. Pena, S.B., Abreu, M.M. Teles, R.E and Santo, M.D. A methodology for creating greenways through multidisciplinary sustainable landscape planning, 2010. Journal of Environmental Management (91): 970–983
23. Bennet, G. (Ed.). In: The European-North American Workshop Eeconet And The Wildlands Project, 1995. Luso-American Foundation for Development (Flad) And The Institute For European Environmental Policy (Ieep), Lisbon.
24. Leitao, A.B. and Ahern, J. Applying Landscape Ecological Concepts and Metrics in Sustainable Landscape Planning, 2002. Landscape and Urban Planning, 59, 65–93.
25. Görmüş, S. Korunan Alanlarda Peyzaj Karakter Analizi: Kastamonu-Bartın Küre Dağları Milli Parkı Örneği, 2012. Doktora Tezi. Ankara Üniversitesi Fen Bilimleri Enstitüsü. 278 S
26. Leitão, B.A, Miller, J., Ahern, J. and Mcgarigal, K. Measuring Landscapes: A Planner's Handbook, 2006. Island Press, Washington, P.118
27. Hepcan, Ç.C., Özeren, M., Hepcan, Ş. ve Özkan, M. B. İzmir İli Metropol Kıyı İlçelerinin Peyzaj Yapı Analizi, 2015. Ege Üniversitesi. Ziraat Fak. Dergisi. 52 (3):353–362
28. Bennett, A. F. Linkages in The Landscape: The Role of Corridors and Connectivity İn Wildlife Conservation, 2003. IUCN, Gland, Switzerland And Cambridge, Uk. Xiv + 254 Pp.
29. Kettunen, M., Terry, A., Tucker, G. and Jones, A. Guidance on The Maintenance of Landscape Features of Major Importance for Wild Flora and

Fauna - Guidance on The Implementation of Article 3 of the Birds Directive (79/409/Eec) and Article 10 Of The Habitats Directive (92/43/Eec), 2007 Institute For European Environmental Policy (Ieep), Brussels, 114 Pp. & Annexes.

30. Pascual-Hortal, L. and Saura, S. Impact of Spatial Scale on The Identification of Critical Habitat Patches for The Maintenance of Landscape Connectivity, 2007. Landscape and Urban Planning 83 176–186
31. Galpern, P., Manseau, M. and Fall, A. Patch-Based Graphs of Landscape Connectivity: A Guide to Construction, Analysis and Application for Conservation, 2001. Biol Conserv 144:44–55
32. Nuñez, T. A. Connectivity Planning to Facilitate Species Movements in Response to Climate Change, 2011. (Master of Science Master), University of Washington, California
33. Şahin, Ş., Perçin, H., Kurum, E., Uzun, O. and Bilgili, B.C. National Technical Guideline for Landscape Character Analysis and Assessment at the Regional and Sub-Regional (Provincial) Levels, 2014. TÜBİTAK KAMAG 1007 Program PEYZAJ-44 Project No 109G074 outcome, where Ministry of Interior, Ministry of Environment and Urbanization, and Ministry of Forest and Water Affairs are the Beneficiary Institutions and Ankara University is the Coordinating Institution, 146 Pages, Ankara.
34. Ahern, J., 2005. Theories, methods and strategies for sustainable landscape planning. Tress, B., Tress, G., Fry, G., Opdam, P. (eds.) 2005. From Landscape Research to Landscape Planning: Aspects of Integration, Education and Application, Springer, Frankfurt, pp. 119–131.
35. Doğan, D. Malatya İli Kapsamında Peyzaj Koridoru Kavramının İrdelenmesi, 2012. Yüksek Lisans Tezi, Ankara Üniversitesi Fen Bilimleri Enstitüsü (Basılmamış), Ankara
36. Eşbah, H., Deniz, B, Kara, B ve Keskin, B. Peyzaj Strüktür İndeksleri ile Koruma Alanları ve Çevresindeki Peyzajın Geçirdiği Değişimin Tespiti ve Alan Kullanım Planlaması Önerilerinin Geliştirilmesi: Dilek Yarımadası-Menderes Deltası Milli Parkı ve Bafa Gölü Koruma Alanı Örneği, 2009 Proje No: 106Y015, TÜBITAK Kesin Rapor, Mart '009, Ankara.
37. Girtevz E.H., Thorne J.H., Berry A.M. ve Jaeger J.A.G. Integration of Landscape Fragmentation Analysis into Regional Planning: A Statewide Multi Scale Study from California, USA, 2008. Landscape and Urban Planning, Vol. 86, Pp. 205–218
38. http://www.dogadernegi.org/userfiles/pagefiles/yayinlarimiz/09_Dogu_Anadolu.Pdf, (Erişim Tarihi: 04.12.2011)
39. Li, T., Shilling F., Thorne J., Li F., Schott H., Boynton R., and Berry A.M. Fragmentation of China's Landscape by Roads and Urban Areas, 2010. Landscape Ecology Vol.25 P.839–853.

40. Şahin Ş., Perçin H., Kurum E., Uzun O., Bilgili B.E., Tezcan L., Çiçek İ., Müftüoğlu V., Çorbacı Ö.L., Sütünç S., Doğan D., Koç Ö., Ateş E., Tarım B., Kurdoğlu G. ve Kaşko Y. Peyzaj-44: İl Ölçeğinde Peyzaj Karakter Analizi ve Turizm/Rekreasyon Açısından Değerlendirilmesi, 2013. 109G074 Nolu TÜBİTAK KAMAG Projesi Sonuç Raporu.

Duygu Akyol, Doruk Görkem Özkan and Sinem Dedeoğlu Özkan

Urban Development in the Perspective of Ecological Sustainability on a Metropolitan Scale

Introduction

In order to understand sustainable urbanization, it is necessary to first clarify the concepts of sustainability and sustainable development.

The concept of sustainability is focused on preventing environmental problems emerging in parallel with economic and technological developments and protecting the ecosystem. It is a very new phenomenon that the concept of sustainability is dealt with the concept of city. However, environmental, economic, political, social, demographic, institutional and cultural objectives, which are elements of sustainable development, can be addressed as part of the phenomenon of sustainable urbanization [1]. Considering that sustainability has three dimensions: economic, environmental and social sustainability [2]; the concept of sustainable development also found its most basic expression in the balance between economy, environment and society [3].

Today, urbanization policies for sustainable urban development and management, sustainable city and planning often confront economy and ecology, and since the economy is prioritized in this context, cities are formed for economic returns and environmental sustainability remains in the background, and the relationship with sustainability is only considered on the condition that a visible and profitable investment is planned. Considering the experiences of different countries; it is seen that the policies and practices applied within the framework of the philosophy of sustainable development differ in accordance with each country's own economic, environmental and social structure. At this point, especially the rapidly urbanizing underdeveloped and developing countries and the phenomenon of urbanization in these countries are notable.

Especially in developing countries, cities have been places where modernization has been experienced, technology has developed, rapid production has increased and uniform cities and human beings have been hardened as uniform individuals. These developments and changes have brought many problems in the cities that make up the framework of people's lives. Cities that have grown and changed over time have caused ecological balance and life-threatening effects as the source of various problems. The environmental and social problems created by today's urban life have resulted in the development of new urban planning approaches in order to provide healthier urban developments. The approach, which is often discussed in today's urban planning literature, is to plan and develop cities in a way that considers the principles of sustainability. In order to sustain urban life, it is required to create living environments in healthy environments in social and physical respects. For this, the development and management of cities must be healthy, natural and environmentally oriented.

Within the scope of the study, the concept of urban development will be examined, especially through izmir, Istanbul and Ankara, which are three important metropolitan areas, and proposals will be made both on a managerial and planning scale within the scope of sustainable urban development.

Sustainable Urban Development

Sustainable urban development is a concept with a wide range of scopes. The concept that its scope is so wide and related to so many different areas is abstract. Accordingly, many scientists are also lying that sustainable urban development is not an open concept. In this context, it is needed to make the concept more concrete.

First of all, the focus should be on urban sustainability indicators that reveal what should be maintained for the sustainable city. Then, the place of urban planning within the framework of sustainable urban development, which is necessary for urban development to be healthy, should be discussed and the concept of urban forms should be considered within the physical space of sustainable urban development, especially by identifying them with urban sizes that have a significant impact on the birth of environmental problems. Finally, the concrete relations of the concept of sustainable urban development with the urban environment and social structure should be revealed [4,5]. There is no agreed definition of sustainable city or sustainable human settlements. However, environmental, economic, political, social, demographic, institutional and cultural objectives, which are elements of sustainable development, can also be considered as part of the phenomenon of sustainable urbanization [1].

Although the concept of sustainable urban development cannot be fully reconciled in the light of many debates, a few of the accepted definitions are as follows; "the city that responds better to human requirements than today's cities and ensures that city systems are developed in a way that prevents future generations from meeting their needs [4] or "the city where socio-economic interests are aligned with environmental and energy concerns in order to ensure change in continuity [6]. When a synthesis of definitions about sustainable urbanization is made, three elements come to the forefront. These are as follows: The first is to overcome the problem of improving the quality of life of people living in cities in their relations with the city, in the use of public areas of the city and in the reception of public services. The second is the strengthening of the city's ability to maintain its own existence as a settlement. Finally, it is the need to question the environmental values of the city on the basis of the production and consumption patterns that exist in the transformation of resources by using them above their carrying capacity [7].

The European Environment Agency also describes the goals to be achieved to ensure urbansustainability as follows [8]:

- Minimizing the consumption of space and natural resources
- Effectively managing urban flows
- Protecting the health of the urban population
- Ensure equal access to resources and services
- Maintaining cultural and social diversity

The approach to sustainable urbanization, all environmental (built environment/natural environment) affected by urban development and affecting urban development, contains social and economic elements in relation to each other; it prevents the combining of economic and social development with the aim of environmental protection and improvement; it requires that the form of development be decided by participating processes. It provides the basic characteristics of sustainable communities that member states of the European Union agree on. These characteristics are derived from the environmental, social and economic development dimensions described above [9].

In accordance with the general framework drawn by the principle of sustainability, the approach to "sustainable urbanization" should include the following key issues [10,11]:

- Growth control and land use planning
- Urban design
- Housing
- Transportation
- Environmental protection and restoration
- Energy and material use
- Green architecture and construction
- Equality and environmental justice
- Economic development
- Population

Principles on which sustainable urbanization is based; environmental, economic, social and administrative principles [7,12]. In line with these principles, the objectives of sustainable urban development are as follows [13, 7];

- Improving the quality of life
- Finding options in development
- Countering poverty
- Solving employment and nutrition problems
- Meeting basic health requirements
- Protection and development of biological variety
- Technology restructuring
- Control of population growth
- Use of renewable energy sources and finding clean, reliable water
- Elimination of risks.

For sustainable urbanization, the following conditions must be provided. These: [14]

- Space-saving developments should be implemented.
- Natural habitat must be preserved.
- Urbanization should be planned.
- Large green space must be provided.
- Water resources must be protected.
- Motor vehicle use should not be encouraged.
- Recycling programs must be started.

What should be maintained and maintained in the process of sustainable urban development is important for urban planning. In this context, "sustainability is an ambiguth but important concept in terms of planning. The permeable nature of sustainability makes it attractive in terms of planning. Because the concept of sustainable development include concepts such as environment, development, social relations. In other words, the concept is intended to continue mutual dialogue between some kind of opposing interests." Therefore, the relationship of the concept with urban planning is actually due to the effort to insment the contrasts inherent in both. In this direction, one of the basic principles of urban planning is to create sustainable cities. That is, in line with sustainability, urban planning, all institutions involved in urban life; identifying, evaluating problems, producing alternative solutions; it can be carried out by taking responsibility for taking decisions, implementing them and monitoring the whole process as a whole. Accordingly, the city planning process should aim to reduce costs with minimal resource usage and thus provide the most benefits [15].

Urban planning is one of the most important tools for ensuring many dimensions of sustainability. As a matter of fact, urban planning ensures that all physical elements of the city are under control, both while guiding urban development with its high-scale decisions.

Sustainable Urban Development and Metropolitan in Turkey

The concept of sustainable urban development actually takes place on a spatial scale on an urban scale. This is especially important for developing countries. The concept is based on environmental social and managerial relations, the basis of which is economics. These relationships make sense on the scale of space and reveal the concept of self-governing cities that are self-capable for cities.

The focus of natural resource consumption and the first place in environmental pollution causes them to take part in sustainability discussions.

The rapidly increasing population in metropolitan areas triggers spatial growth, and after a while the cities spread to agricultural areas and natural resource limits.

In developing countries, this situation is very inverse to the axis of sustainable development, so sustainable development for these cities is a question mark.

Global relations, local relations feed each other today, which causes the cities to form a competitive spirit. In order to participate in the competition, a city must ensure its sustainability in both an environmental and economic and social context [16]. In this context, it is the social surface and social surface, which is the other dimension of the concept of sustainable development in cities [17].

There are many definitions of the concept of sustainable urban development, but it is very difficult to make a general definition for each city or country, but many administrations have expected that the concept of sustainable urban development should be defined as projection of human settlements in space with attention to environmental impacts.

In growing countries, cities that are pregnant with ecological problems triggered by structural collapse and such as non-humane living conditions, unemployment, poverty, social and economic, and the extreme destruction of the environment have started to form.

The hybridization of the sustainability principle with the subject of "urban planning" was carried out at the UN Conference on Environment and Development in 1992 with the search for "sustainable urban-rural development strategies" within the framework of the global partnership monitor known as Local Agenda 21 [18]. It can be stated that the policies created in the light of these strategies and the theme of sustainability are institutionalized as a Sustainable Cities Program by including the dimension of governance based on environmental, economic and social dimensions, participation and cooperation [18]. Accordingly, it is stated that sustainable urban development develops within the framework of space accompanied by nature, society, economy and political-management components.

The problems ad resulting from urban development are in line with the difficulties in the process of performing sustainable development. Accordingly, it is necessary to define the causes of urban development well in order to understand the difficulties of creating environmental policy at the local level and setting goals. As with sustainable development, the most effective dimension of sustainable urban development is the economy. The labor market, which forms the intersection of economy and environmental dimension, plays an important role at this point. Addressing environmental issues at an urban level will positively impact the business force and the formation of new jobs. While addressing the reflection of sustainable development on an urban scale in accordance with the dimensions acquired in the historical development of the concept; it is important to look not only at environmental protection and development efforts, but also to socio-economic development indicators, examine participation mechanisms, examine urban visions and address all these dynamics in integrity [18].

Urban Development Types and the Impact on Ecology Sustainability

It is possible to evaluate the concept of sustainable city within the scope of its environmental, economic, social, political, demographic, institutional and cultural purposes, which are considered as elements of sustainable development. As of the last point of the world, this concept necessitates its interpretation in sustainable development. The concepts of sustainable development and urbanization relationship include sustainable cities, sustainable urban development, sustainable human settlements, eco-cities, livable cities, green cities, slow or calm cities, energy efficient cities and smart cities. These expressions, which are differentiated from each other by small nuances, serve the same purpose that can be expressed as the harmony of the city with the environment.

Two types of views on the concept of sustainable city are put forward. The first is the singuistic view that 'the definition of an ideal sustainable city should and should be made in advance'. The aforseed approach focuses on environmental elements and deals with urban dynamics at the level of physical criteria.

On the other hand, there is no assumption based on the definition, since in multi-opinion, the lack of agreement on whether a definition of sustainable city can bemade [1].

Today, the concept of "sustainable urbanization" or "sustainable urban development" has become a fundamental concept that is often expressed in the policy producing bodies of the European Union. It works for the goal of sustainable urban development by insuring the main objectives such as strengthening economic prosperity and employment in cities and towns, promoting equality, social participation and urban renewal in urban areas, protecting and improving the urban environment, contributing to urban governance and local capacity building.

It is possible to diversify the deeds made in sustainable urban development. Definitions are made that are separated from each other by small nuances. However, if a whole of these things are to be made, it is seen that there are three main purposes in the phenomenon of sustainable urban development: the first is to solve the problem of improving the quality of life of individuals in cities in their relations with the city, in the benefit of the city's common areas and in the acquisition of public services. In this sense, the access of urban residents to public services such as education, health, housing and transportation, which will improve the quality of life and strengthen their relationship with the city, can be provided on a regularly. It is the establishment of a system in which the regular provision of public services minimizes the destruction it will cause on the environment. The existence of the city can be maintained by keeping an ecological perspective at the forefront on issues such as energy efficiency, waste and water management, air quality and urban climate change. The third is: it is imperative that production and consumption habits are questioned fundamentally with the use of the city above its natural transportation capacities. Strengthening the economic infrastructure of cities with concepts such as technology, innovation, investment, renewable energy, which enables the realization of the opportunities it offers to the people living on it [19].

Ecology Sustainable Urban Development on Istanbul Ankara Izmir City

One of the prominent factors in the concept of sustainability in urban management is the increase in living standards and public services in the city where the city lives.

In the 21st century, local governments began to have a significant say on behalf of the city administration with the importance and strengthening of the concept of governance and the start of its adopting. In this regard, the central administration began to leave some of its duties and powers to the municipalities, and the provincial and district municipalities began to move towards making decisions with the residents about choosing and making decisions about what was best for them.

Within the scope of these evaluations, it was observed that there were many district municipalities that were inadequate in infrastructure, although all three carried out metropolitan areas. Especially in the regions that we can define as wall districts, the new drinking water lines and the new addition of dirty water channels to the district are the main problems that stand out. It is observed with the current situation that all three metropolitan lyn ies are doing well in planning and management, but the decisions taken in this regard cannot go further than the targeting dimension. Many district municipalities have made planning decisions for planning and district management and targets have not been set for sustainability. However, it was not seen that these decisions were applied in the design, education and social activities section, which is the application part. In many district municipalities, whether the center or the districts, the only target in design activities has been recreation and green space increase. It is not concretely explained how to make these increases, what kind of activities to carry out. In addition, municipalities aiming to increase the amount of green space from 10 m^2 to 12–14 m^2 have not announced which criteria and action plans they will make according to these targets. Within the scope of education and social activities, it was determined that many district municipalities practiced and these practices were more in the central districts. As a plus aspect, some district municipalities have given environmental education projects as courses in schools and set targets for the number of people who will participate in the events to be held every year. This can be seen as an important and big step in raising public awareness and creating awareness of urbanity [20].

All planning in the hierarchy of urban development in metropolitan municipalities in our country requires an environmental perspective within the framework of sustainable urban development. On the other hand, applications for sustainable urban development should be considered in line with the unique dynamics of our country by considering global criteria. Because the content of sustainable urban development, which applies in developed countries, is very likely to change in developing countries. Local governments are developing local solutions for the sustainability of cities.

Recommendations on Results and Urban Management

The results obtained according to the projects implemented on ecological city management throughout Turkey and the strategy plan evaluations on the three metropolitan areas were classified as positive developments and inadequate services within the scope of ecological sustainability, and then proposals were developed. We can state that positive developments in services for ecological sustainability include innovative approaches, especially within the scope of green infrastructure creation. In other projects aimed at creating city forests and protecting water basins, it is positive for the protection of natural resources. In addition, pedestrianization projects or projects promoting the use of bicycles contribute to the ecological sustainability of cities. However, there are still inadequaies in the studies within this scope. Due to the expansion of the administrative boundaries of metropolitan areas with the law no. 6360, the service areas of metropolitan administrations have expanded and urban infrastructure units have not been developed and environmental policy has not been developed for many of these areas due to the previous status of villages of many settlements included in the metropolitan municipality boundaries.

Activities that can be carried out within the three metropolitan cities examined;

- Determination of environmental quality limits,
- Strategic environmental evaluation in all districts,
- Continue to develop policies focused on sustainable development by the district municipalities,
- Promoting environmental policies that contribute to income and employment formation, especially in rural areas and poor areas of metropolitan areas,
- Adopting integrated coastal management in seashore districts,
- Depending on the central administration, each district establishes its own environmental action plan,
- To declare and protect the areas of nature and biodiversity character that exist within the boundaries of the district,
- The creation and use of smart city applications,
- To prepare zoning plans for the creation of a healthy and safe environment with high quality of life while preserving natural, historical and cultural values and to ensure their sustainability,
- Preparing a report revealing the health-environment relationship,
- It is to reduce the number of people who do not have access to environmental services.

Recommendations for Design Activities; In all the district municipalities examined, there are projects that include the construction of parks of different scales throughout the district under the name of design activities in general. Some district municipalities suggest that these projects will increase or increase the amount of green space per capita in the district. The important issue here is what these areas add to the ecological data of the districts. Many park projects gradually lose their importance from the day they are first built and cannot perform their function within the district boundaries. In this context, the cities;

- It would not be wrong to install eco parking. Sustainable work within the eco park will be an example for the whole city and meet the city's need for recreation in the green area.
- However, it is important to focus on public transport systems that will work effectively in cities (route frequency, transfer, etc.), as well as to encourage alternative forms of transportation such as bicycle use.
- In order to prevent urban developments from growing towards natural areas such as agricultural areas and forest areas, development should not be allowed in these directions and compact development of cities should be preferred. Recommendations for education and social activities;
- In the three cities examined, it is seen that social and educational activities are carried out in 80 % district municipalities. In particular, it is ensured that campaigns and events are organized. Despite such intense awareness-raising activities, the environmental problem is still seen as the main problem in many cities. As can be seen from here, trainings and activities carried out by local governments cannot fully have an impact. For this;
- In order for activities other than campaigns to be effective, activities that the residents of the district can participate more actively (tax exemption, legal sanctions, etc.) should be carried out within the scope of incentives including rewards and penalties.
- Solid waste can be separated and collected with the waste collection system that can be created at the neighborhood level. In addition, studies of the type "Neighborhood Environmental Award" can be carried out to encourage the public in this regard.
- It is very important to educate the people of the city in this regard in the development of the concept of an ecological city. Through education, the people of the city can be provided with a very strong ecological structure awareness and the people of the city can participate in environmental activities in this way and work to keep the city healthy.

References

1. Satterthwaite D. Sustainable Cities or Cities that Contribute to Sustainable Development? *Urban Studies.* 1997;34(10).
2. Uysal, Y., (2002), "Environment on the International Platform, File: Sustainability and Architecture", Mimar.ist Magazine, Year 2, S.6, Istanbul, p.44–46.

3. Guler, M., Turan, A., (2013). Development Strategies for Sustainable Urbanization in Turkey: KENTGES Action Plan (2010–2023) Case. International Conference On Euraan Economies.
4. Yazar, K. H. (2006). "Urban Planning Method Proposal for Medium-Sized Cities within the Framework of Sustainable Urban Development". Ankara University Institute of Social Sciences Unpublished Doctoral Thesis.
5. Gülpinar Sekban, D. U., Bekar, M., & Acar, C., (2019). Evaluation of Sustainability Potential According to Cittaslow Criteria in Turkey/Trabzon. Fresenius Environmental Bulletin,vol.28, 5435–5446.
6. Geenhuisan, M. V., & Nijkamp, P., (1994), "How to Plan a Sustainable City?", Journal of Society and Science, p.129–133.
7. Bayram, F., (2001), "Sustainable Urban Development: Tools, Approaches and Turkey", Gift to Cevat Geray, Association of Property Owners Publications: 25, Ankara, p.254–258.
8. URL 1: European Environment Agency 1995, http://www1.wspgroup.fi/lt/propolis/PROPOLIS_Final_100204.pdf, Accessed November 2007
9. URL2: Bristol Consensus 2005; http://www.mimarlarodasi.org.tr/UIKDocs%5Cbristolmutabakati.pdf, Accessed October 2007
10. Tosun, K.E. (2009). Journal of Economics, Sociology and Politics Paradox, ISSN 1305-7979, Year:5 Issue:2.
11. Akyol, D., Özkan, D. G., & Dedeoglu Özkan, S., (2019). Measurıng The Qualıty Of Urban Lıfe In The Trabzon Provınce. *Fresenıus Envıronmental Bulletin*, vol.28, 7991–8000.
12. Gülpinar Sekban, D. U., Akyol, D., & Özkan, D. G., (2019). Sustainability of the Urban Landscape: An Investigation of theEffects of Ecology on Urban Parks within the Scope of Integrated Landscape Approach. New Approaches to Spatial Planning and Design Planning, Design, Applications (pp.199–210), Berlin: Peter Lang.
13. Bar, M.(2000) "21. Thoughts for Humanized and Sustainable Urbanization in the 19th century", Urban Environment and Sustainability Panel, Eastern Mediterranean University, Faculty of Architecture, Urban Research and Development Center.
14. Ulusoy A., Vural T. (2001) "Socio-Economic Effects of Urbanization", Municipal Journal, volume 7, issue 12, June.
15. Yalkı, T.(1998). "Sustainable City and Bursa 2020 Environmental Order Strategy Plan", Discussion Meeting on The Implementation of Sustainable Development, 11–12 December 1997, Turkish Environmental Foundation Publication, Publication No.126, Önder Printing Press, Ankara, 1998.
16. Düzenli, T., (2020). Occupant Views on Environmental Effects of Trabzon Square UrbanTransformation Project. Online Journal of Art and Design, vol.8.

17. Mumcu, S., Yilmaz, S., & Akyol, D., (2019). With nature in mind: 'Green metaphors' as an approach to reflect environmental concerns and awareness in landscape design. A/Z: ITU journal of Faculty of Architecture, vol.16, 131–144.
18. Varol, T. (2019). Participation in the Context of Sustainable Urban Development: Gaziantep Metropolitan Municipality Example, Ankara University Institute of Social Sciences Unpublished Master's Thesis.
19. Mega V., (1996), "Our City, Our Future: Towards Sustainable Development In European Cities", Environment and Urbanization, C.8, P.1, p.141.
20. Akyol,D. (2015). Ecological Sustainability in Urban Management,Dokuz EylülUniversity, Institute of Science, Unpublished Master's Thesis.

Volkan MÜFTÜOĞLU and Halim PERÇİN

An Implementation with Low Impact Development (LID) Techniques within the Scope of Sustainable Stormwater Management: An Example of Mass Housing Project

Abstract: Stormwater, which is tried to be rapidly collected from urban lands, is transferred to receiving waters in an uncontrolled manner, and this situation causes the receiving waters, and thus the entire water cycle to be negatively affected in terms of pollution, especially with the sediment and heavy metals carried by runoffs. Traditional stormwater management solutions in our cities are often inadequate in terms of size and capacity, and even fail to fully accomplish their purpose because of accumulated sediment and pollutants, as a result of excessive rainfall in a short amount of time due to global warming in recent years. On the other hand, sustainable stormwater management approaches, in which natural solutions are prioritized, aim to solve the stormwater problem at its source as much as possible and to store the water that will pass through the runoff or infiltrate it to groundwater. The widespread use of sustainable stormwater management solutions in our cities will provide rain harvesting, which can be a solution to water shortages due to drought, and will support traditional stormwater management solutions during excessive rainfalls. In this context, this study aims to propose appropriate stormwater solutions on a mass housing project with the lid approach one of sustainable stormwater management approaches, and to reveal the contribution of these solutions in stormwater management. It was concluded that the continuation of the sample area, which is managed with integrated management practices within the scope of lid approach, by preserving the pre-development hydrological conditions, is very important in terms of ensuring safer habitats and the sustainability of the water-stormwater cycle in the sample area.

Keywords: Water cycle, runoff, ecological stormwater management, low impact development, integrated management practices, Ankara Gölbaşı Yaylabağ Yaylaşehir mass housing project

1 Introduction

Water, one of the important elements of urban ecology, is the most valuable resource that is constantly in a cycle motion on earth and ensures the continuity of life [1]. In recent years, with climate change, there are changes in water cycles, and there are consequences such as the more frequent occurrence of overflows and floods due to severe weather activities and their increased effects, or on the contrary, as drought and desertification with these negative consequences, residential areas are damaged, freshwater resources are polluted, and water quality is also gradually decreasing. In addition to climate change, urban settlements mostly occupied with hard lands also negatively affect the water cycle. Runoff and groundwater recharge, which are among the basic components of the water cycle, is disrupted by the unplanned development of residential areas. Incorrect and/or insufficient location selection of urban open-green areas, as a result of unplanned development, cause stormwater to runoff, and infiltration to be very low. This situation causes flood risk on the valley bottoms and prevents recharge of groundwater due to the increasing volume of runoff. On the other hand, an increase in water consumption due to overpopulation also decreases the freshwater resources day by day [2].

Due to these problems, the importance of sustainable management of water resources has emerged in urban areas. The intense and sudden occurrence of irregular rainfalls due to climate change necessitates the discharge of stormwater falling on ground and roofs in residential areas much faster and more efficiently without causing floods [3]. It is extremely important in terms of sustainable stormwater management that the above mentioned discharge is carried out with practices that will slow runoff and increase infiltration, instead of taking stormwater away from the area as soon as possible through traditional methods such as trench drains. With the new stormwater management models (sustainable drainage systems, water sensitive urban design, lid, etc.) developed in addition to the traditional methods, runoff can be reduced and groundwater recharge can be increased in accordance with the natural structure in balance, and residential areas can be protected from flood risk, as well as excess stormwater can be stored and used as an alternative water source [4].

The aim of this study is the exemplification of environmentally friendly ecological practices aiming at sustainable management of stormwater in urban areas, as well as lid tools, which are one of the sustainable stormwater management models, on a mass housing project. There is a traditional stormwater infrastructure project in the study area. However, it is thought that the integrated management practices that will be proposed for the area within the scope of the lid development model, in addition to the traditional management approach, will contribute much more to the urban water cycle by increasing the rain infiltration. On the other hand, the proposed environmentally-friendly integrated management practices will also help traditional stormwater management practices in preventing floods and overflows that occur due to excess rainfalls which increased with climate change.

2 Sustainable Stormwater Management Approaches and LID

The management of the quantity and quality of runoff that occur after rainfalls and snowmelts, together with various structural and non-structural control systems, is called "stormwater management" [5,6]. In natural areas, most of the runoff that occurs after rainfalls and/or snowmelts is infiltrated by the soil, and the remaining water joins large watery areas via natural drainage. Infiltrated runoff feeds groundwater sources and thus recharges aquifer areas. On the other hand, runoff and groundwater recharge processes, which are among the basic components of the water cycle, are disrupted by unplanned and hard-surfaced urban areas. Incorrect and/or insufficient location selection of urban open-green areas, as a result of unplanned development, cause stormwater to runoff and infiltration to be very low. On the other hand, traditional stormwater harvesting systems in cities that are separated or combined with sewage systems are often inadequate and this may result in floods and overflows [7].

In this context, sustainable management is considered to be important to make rain useful in urban areas. Regarding sustainable stormwater management within the scope of ecological urban design, new approaches have emerged. "Water Sensitive Urban Design" from Australia, "Sustainable Drainage Systems" from the UK, and "Green Infrastructure" and "Low Impact Development" from the USA and Canada have emerged. There are common concepts in the basis of these approaches such as being respectful to nature, prioritizing natural characteristics, ensuring sustainability by considering the balance of protection-use, and these concepts are similar to each other in many ways [8].

Within the scope of this study, lid, one of the ecological stormwater management approaches commonly used in the USA and Canada, was evaluated on a sample mass housing project. Lid is an approach to land development that focuses on how water enters an area, how it is stored, and how it is taken away from the area. In this context, land development includes lid practices such as reducing impermeable surfaces, conserving natural vegetation and soils, and managing flood water at its source. Urbanization significantly changes the natural hydrological cycle. With the implementation of lid in urban areas, the hydrological cycle of cities improves, as well as their aesthetic values increase. Lid uses stormwater as a resource and reduces overflows in combined flood trenches [9].

The main goal of lid methods is to simulate post-developmental field hydrology to pre-developmental field hydrology using field design techniques such as storage, infiltration, evaporation, and runoff delay. The use of these techniques helps to reduce the runoff water going out of the area and provide adequate groundwater recharge. Since the development of the field affects the hydrological properties of the area, with all of its aspects, low-impact growth control techniques mainly focus on field hydrology [10].

Lid is a technology-based, comprehensive approach to managing urban stormwater. It provides stormwater solutions for each parcel by using low-effective development and cost-effective landscaping features, compared to expansive harvesting units located under drainage areas. Resource control in lid is quite different from traditional stormwater management (pipe and harvesting basin, stormwater management area design). Hydrological functions such as infiltration, recurrence, discharge volume, and groundwater recharge can be protected by using lands with reduced impermeability, separating functional excavation-filling activities, open sections of trenches, hydrological runoff paths, and using bioretention/filter landscape areas. Lid also includes multifunctional area design elements within the stormwater management plan. Alternative stormwater management practices such as plot-scale micro-storage, functional landscaping, open drainage ditches, reducing impermeability, reducing slopes, increasing runoff travel time, and pit storage can be integrated into a multi-functional area design [10,11].

Special low impact developmental control techniques called "integrated management practices" reduce runoff by integrating stormwater management controls across the area in many small separate units [12]. With the use of integrated management practices in a small part of each parcel close to the source of impact, the need for structural stormwater management techniques will decrease. Areas, developing thanks to this process, can be designed as an integral part of an environment, where pre-developmental hydrological functions are preserved, with careful use of low impact growth control measures [10].

Low impact developmental technology uses micro-scale and distributed management techniques called integrated management practices (IMP), in order to obtain the desired post-development hydrological conditions. The lid area planning process significantly reduces the hydrological effects of the development. After the area planning techniques are implemented, some additional practices are required in order to obtain the pre-development hydrological situation. In this context, the use of integrated management practices introduces additional hydrological controls (such as peak discharge and runoff volume) and ensures that pre-development runoff volume is maintained. Integrated management practices, suitable for lid approach, are bioretention units (rain garden, etc.), dry wells, filter/buffer strips, and other multifunctional landscaping areas, grassed swales, dry and wet swales, rain barrels, cisterns, and infiltration trenches [10,11,12].

3 Method

Within the scope of this study, the process of lid approach has been detailed on the sample mass housing project area, and appropriate integrated management practices are proposed for this area. In this context, the workflow of the study is as follows: data collection related to the sample area, analysis and mapping of these data, determination of the current and post-development runoff potentials of the sample area, comparison of the current and post-development runoff potentials within the scope of stormwater management of the sample area, and integrated management practices of the conflict points identified in the sample area within the scope of the lid approach.

Within the scope of analysis and mapping of the natural and cultural data taken from the study sample area as Ankara Gölbaşı Yaylabağ Yaylaşehir Mass Housing Area, annual rainfall is calculated using the climatic data of the sample area and its immediate surroundings, elevation group analysis, slope analysis, and aspect analysis were conducted using the topographic structure data of the area, hydrological soil groups were determined from soil data of the area (large soil groups and other features), a permeability analysis was conducted by overlapping the geological structure data, hydrological soil groups data, and the slope analysis, by using Corine 2006 data, existing plan, and satellite images, and up-to-date land cover analysis was conducted, and this analysis was supported with field trips to determine the vegetation status of the area and finally, the land uses (residential areas, commercial areas, green areas, etc.) in the implementation zoning plan of the study area were determined, and the traditional stormwater harvesting network system of the area was also analyzed.

On the other hand, the current situation of the sample area of the study - Ankara Gölbaşı Yaylabağ Yaylaşehir mass housing area - and the runoff potentials of the post-development area has been determined. In this context, the sample area was first divided into main basins (water harvesting areas), and then micro-basins in order to facilitate the calculations of runoff potential. The "design storm" concept has been presented after the geostatistical analysis of the climatic data of the study area to be used in the runoff calculation. After the design storm has been determined, it was deemed appropriate to use the TR-55 method [13] to estimate the runoff volume. SCS Curve Number Method, developed by the American Soil Conservation Service, is widely used because it is a flexible and simple method for estimating the runoff volume. In this method, basin parameters and climatic factors are combined into a single value called "curve number". However, since the cultivated areas in the United States generally have a slope level of less than 5 % and this slope range does not affect the curve number value to a great extent, the effect of the slope is not taken into account when estimating runoff volume [14,15].

The most important factor determining the movement of water in a basin is the land slope. According to Zhan et al. (2004), all runoff-related variables (such as the number of runoffs, runoff depth, and average curve number), aside from the minimum depth of runoff, increase with the slope factor [16]. On the other hand, according to Sharma (1986), studies on experimental runoff plots showed that areas with steep slopes have much higher runoff efficiency than areas with lower slope (near flat) [15,17]. The increase in runoff due to steeper slopes can be explained by decreasing the initial retention (I_a), decreasing the infiltration rate, and decreasing the stagnation time of the runoff [14]. In this context, the curve numbers determined on the sample area have been adapted to the slope due to the high slope of the sample area.

According to the workflow of the study, the runoff potentials in the current state of the sample area and the post-development state was compared in the next stage. In this context, in the first stage, the differences between the estimated runoff volumes for both the current situation and the post-development situation of each micro-basin in the sample area were determined. After the differences between runoffs are determined, the permeability analysis - conducted with the geological structure, soil structure, and slope situation - was overlapped with the implementation zoning plan of the study area. Thus, it has been determined whether the current permeable areas have turned into impermeable areas after development. As a result, according to the difference between the runoff volumes of the current situation and the post-development situation, the conflict areas that may cause problems, according to the permeable/impermeable area transformations, within the scope of stormwater management have been determined.

In the last phase of the study, lid integrated management practices were used within the scope of sustainable stormwater management, on samples selected from the determined conflict areas. Integrated management practices (such as biological retention areas, rain gardens, rain barrels, cisterns, etc.) have been proposed within the scope of the lid approach to solving conflict areas. In this context, due to the difference in runoff volumes between the current state and the post-development state of micro-basin, the most appropriate management practices for the area are selected with the utmost care, without disturbing the balance of groundwater recharge.

4 Sample Area (Ankara Gölbaşı Yaylabağ Yaylaşehir Mass Housing Project Area)

The sample area of the study is an area of approximately 833.40 ha planned as a mass housing area in Yaylabağ (Bursal) neighborhood, Gölbaşı district, Ankara province (seen Fig. 1). The total of residential islands covers 38.61 % of the entire area. The total number of residences will be 9,376. 3 different housing typologies have been determined as multistoried residential areas, commercial + residential areas, and independent housing areas provided that they do not exceed the determined total number of dwellings and total construction area [18].

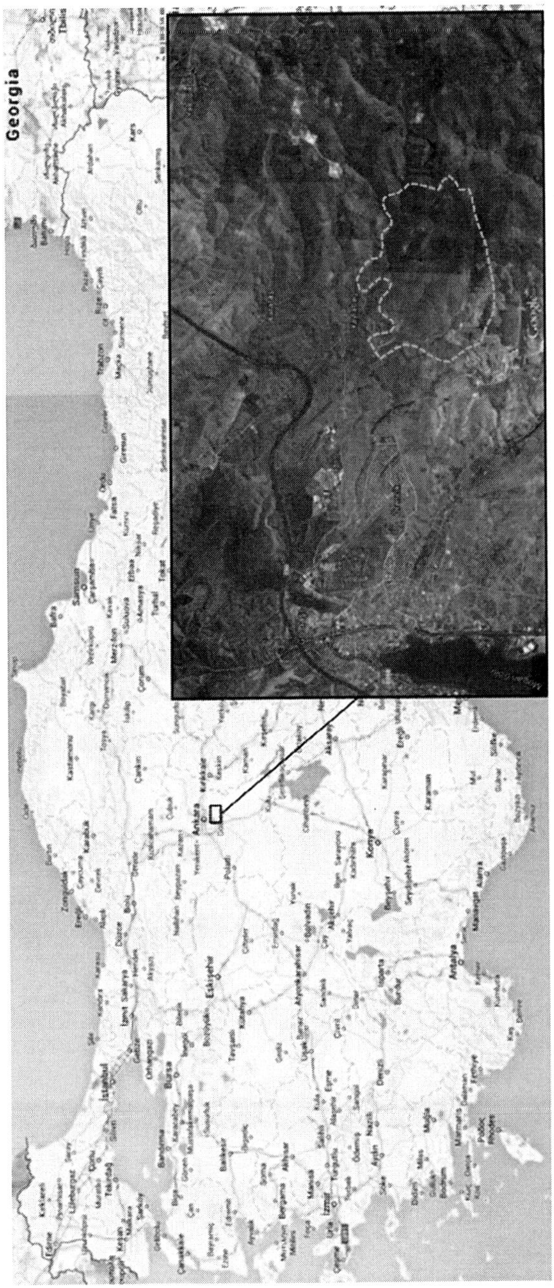

Fig. 1: Its location of the study area in Turkey.

According to the calculations, it has been determined that the average areal rainfall for the 51.17 km^2 drainage basin of the study area was 513 mm according to the isohyet method. The result was that rainfall in the city of Ankara has been higher than average rainfall ((1954–2013) (402 mm)) for many years. This difference also showed the necessity of stormwater management for the study area to prevent overflows and floods that may occur with the increased runoff after rainfalls.

When the slope analysis of the current situation of the area was conducted, it was seen that the study area was full of steep regions with great slopes [18]. On the other hand, the current land use map has been updated to use the curve number method together with satellite images, land trips, and land cover units. Plants in the form of shrubs and trees, that make up the existing massive greenery of the study area, generally follow the bottomlands and the valleys in a thin strip [19].

On the other hand, the valleys within the general geomorphology of the study area played an important role in the new green area planning. The fact that the creeks and valleys extend especially in the north-south direction in the sample area, provided suitable conditions for the use of open-green areas between residential areas. In this context, the areas and valley floors on a certain slope were not allowed for residents and it was decided to use them as open-green areas. This decision made it possible to determine the main backbone of the general open-green area system for the study area. Parks, playgrounds, and sports areas planned for neighborhood units, and residential gardens within the construction areas were also included in the general open-green area system. The two most important streams in the study area that form the main basins of the area are Kaplum (Kuzkuyu) Creek in the north and Büvet Creek in the south. Of these large creeks, Büvet Creek is generally dry and becomes fluid in rainy seasons. There are many dry rills in south-north and southeast-northeast directions that discharge towards both streams [18].

According to the determined hydrological soil groups, 55.06 % of the study area is in group D, 28.82 % is in group C, and 16.12 % is in group B. There was no hydrological soil group, which should be in group A (high permeability). In this context, the hydrological soil groups of the study area were evaluated and the existing soil permeability levels were classified. As a result of overlapping soil permeability and slope maps, the soil permeability zone was determined by evaluating the overlapping areas. When the soil permeability was examined, it was observed that 3.90 % of the study area had very high, 8.51 % had high, 32.52 % had medium, and 55.07 % had low permeability. The permeability zones of the rocks have been classified by interpreting the current geological maps and reports of the study area. In this case, 0,66 % of the study area was high-permeable, 34,52 % was permeable and 64,82 % was low-permeable. The overlapping areas were evaluated and geological permeability was obtained after overlapping rock permeability and slope maps. When the geological permeability was examined it was seen that 2,29 % of the area had very high, 8,77 % had high, 24,09 % had medium, 1,09 % had low and 63,74 % had very low permeability.

In order to determine the total permeability zones in the study area, overlapping areas were evaluated and the total permeability zone was obtained after overlapping the geology and soil permeability zone data. As a result, when the total permeability zone was examined it was detected that 2,55 % of the area had very high permeability, 4,72 % had high, 2,94 % had medium, 12,98 % had low and 76,78 % had very low permeability.

In addition, the traditional stormwater lines of Gölbaşı Yaylabağ Yaylaşehir mass housing area were planned to discharge into Kaplum (Kuzkuyu) Creek. Stormwater lines will primarily be planned as open trenches in order to ensure the gravity flow and transition from parcel to parcel. On the other hand, in villa parcels, the stormwater will be discharged to the road from suitable points. On the islands with multi-storied buildings, stormwater lines had been created within islands according to the settlement plans [20].

5 Determination and Comparison of Pre and Post Development Runoff Potentials of the Study Area

The Sample area was primarily divided into micro-basins in terms of water harvesting, design storm was determined and finally, the runoff potentials pre and post-development were revealed by the TR-55 method. In the hydrological evaluation of the rainfall-runoff potential of the sample area, it is necessary to divide it into water harvesting areas, simply basins. The size and slope of each determined basin are effective in harvesting rainfall and determining runoff directions. For this purpose, Gölbaşı Yaylabağ Yaylaşehir Mass Housing Area, which is the study sample area, was divided into 2 main basins (Kaplum Creek basin and Büvet Creek basin) and then into 29 sub-basins within the boundaries of the area.

On the other hand, for the lid approach, which is applied within the scope of stormwater management in the study area and which includes much more ecological solutions than traditional stormwater harvesting systems (trench drains, etc.), the goal of preserving the hydrological conditions of the study area pre-development was taken into account in selecting the recurrence interval of the design storm. The recommended recurrence interval for the low impact developmental approach is between 25 and 50 years [21]. While determining the design storm for the research area, 25 years was selected as the appropriate recurrence (repetition) interval in line with the goal of maintaining hydrological conditions and recommended recurrence interval. In the calculation of the probable design storm according to the 25-year recurrence interval of the study area, 5 stations were used due to the proximity of the rainfall stations included in Anlı and Öztürk's (2011) study [22].

The values obtained according to the daily rainfall averages of these 5 stations for many years were multiplied by the 25-year recurrence interval values and possible design storm for each station was calculated. Kriging method, one of the interpolation methods was used, with the help of GIS, for the design storm calculations made on the point scale (station-based), and the design storm calculations to be made at the study area scale.

With the implementation of the kriging method, an isohyet map was obtained for the study area. As a result, in order to find the average areal value, a connection between the isohyet and the areal size was established and 37.46 mm possible design storm was determined for the study area at the 25-year recurrence interval. However, in order to obtain more reliable rainfall data and consequently increase the accuracy of possible design storms, it is also a very important issue to establish rainfall observation stations in the regions where needed [23].

After dividing the study sample area into micro-basins and determining the design storm, the TR-55 method was used to estimate the pre-development and post-development runoff volumes. Within the scope of this method, curve numbers were assigned and average curve numbers were calculated by considering the land cover (updated with satellite images), hydrological soil groups, and implementation zoning plan decisions for both current (pre-development) status and post-development status of each micro-basin. In addition, considering the effect of the high slope factor of the research area, the weighted slopes were calculated for both pre-developmental and post-development conditions, and the average of the curve numbers adapted to the slope was also determined.

The differences between runoff volumes calculated for the current state (pre-development) and post-development state of the 29 micro-basins of the study area are determined. These values also show the minimum runoff volume that must be dissolved or retained for the study area. It can be seen that the greatest difference between the pre-development and post-development runoff volumes of the study area was in the micro-basin No. 19 (9,060,825 m^3) located in the main basin of Kaplum Creek. The total average difference value across 29 micro-basins had been found as 1,932,915 m^3. The big difference in micro-basin No. 19 reveals that the conditions in the current situation (such as land use and landcover) had changed quite negatively in the post-development situation.

6 Identifying Conflict Areas of the Study Area within the Scope of Sustainable Stormwater Management

After the comparison of the runoff potentials of a total of 29 micro-basins in the study area pre and post-development, the current permeability situation and the post-development impermeable area use (1/1000 Implementation Zoning Plan) were overlapped. The main purpose of the overlap is to reveal the transformation of the existing permeable areas in the study area into impermeable lands after development. Within the scope of stormwater management sensitive to ecological processes, acquiring this transformation is very important.

The areas where the permeable lands turn into impermeable lands as a result of overlap were named "conflict areas" seen in Fig. 2. The determined conflict areas were ranked from 1 to 4 in order of importance. Accordingly; 1st-degree conflict

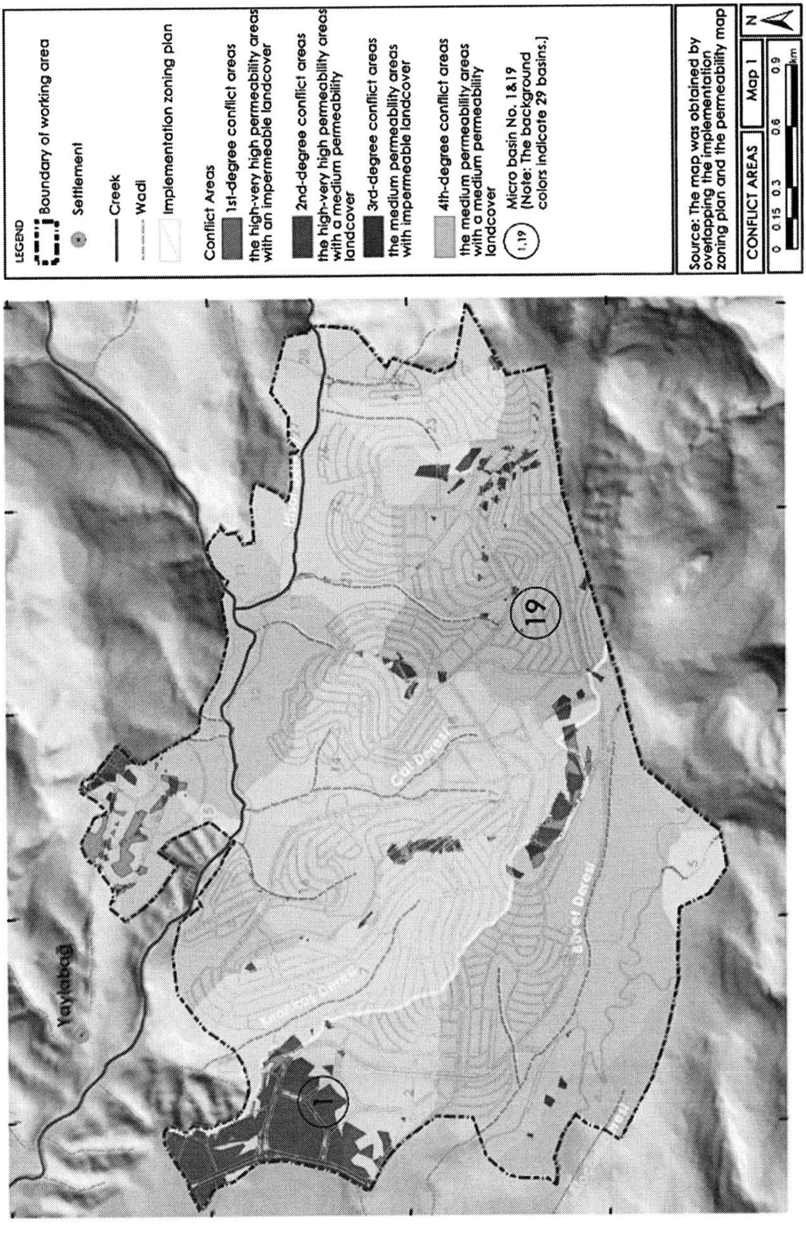

Fig. 2: The conflict areas identified in the sample area.

areas (the most important conflict areas) represent the high-very high permeability areas with an impermeable landcover. 2nd-degree conflict areas represent the high-very high permeability areas with a medium permeability landcover. 3rd-degree conflict areas represent the medium permeability areas with impermeable landcover, and finally 4th-degree conflict areas represent the medium permeability areas with a medium permeability landcover. The conflict areas analysis was parallel with the result of the current state and the post-development state runoff potential comparison.

7 Handling the Study Area Sample Conflict Areas within the Scope of LID Approach

Compared current/post-development runoff volumes and conflict areas classified according to four degrees of importance constitute important findings within the scope of stormwater management considered in the ecological framework in line with the objectives of the study. With the correct evaluation of both findings, solution(s) can be created for areas that pose a problem in terms of stormwater management for the study area. The created solutions will be within the scope of the lid approach that prioritizes environmentally-friendly integrated management practices.

In this context, it was considered to review sample micro-basins within the scope of the lid approach for stormwater management from both basins. The first micro-basin reviewed as an example is No. 1 micro-basin; it had the highest percentage of increase in Büvet Creek basin (approximately 415 %) in comparison of current/post-development state runoff volumes, and it was placed in 1st degree in the analysis of conflict areas. The second micro-basin reviewed as an example is the micro-basin No. 19, which has the largest difference (9,060,825 m^3) within the Kaplum Creek basin in comparison to the current/post-development state runoff volumes. In the case of pre and post-development of the micro-basins No.1 and 19, the runoff volumes, concentration times, and the peak runoff rates are determined. It can be determined that the concentration-time of the micro-basin No. 19 increased in post-development compared to the pre-development state. This situation was caused by the lengthening of the runoff path, as the grade level of the area determined in post-development for the micro-basin No. 19, changed the runoff direction.

On the other hand, within the scope of preserving the hydrological characteristics of the pre-development current state, revealing the drainage plan of the study area, that is, the direction of rainfall was very important. Considering the post-development rainfall-runoff directions and increasing and modifying new drainage runoff tracks that will preserve and/or increase the current concentration-time will protect against risks such as floods.

The micro-basin No. 1, in the pre-development state, is a micro-basin mostly occupied with hydrological soil groups B and C in general, covered with flat and nearly flat slopes in all other micro-basins, has a high-very high water permeability and non-irrigated arable lands throughout the entire research area. On the other hand, in the post-development state, the micro-basin No.1 is faced with a dense settlement texture.

The two uses that cover the most area percentage were multi-storied housing areas with about 60 % and asphalt road with 19 %, in the micro-basin No. 1. In addition, there are 90 villa-type houses in low-rise housing areas that cover approximately 10 % of the area. In this context, pre-development hydrological characteristics (runoff volume, concentration-time, peak runoff rate, drainage directions, etc.) are affected negatively.

Within the framework of one of the main objectives of the lid approach, the preservation, and improvement of pre-development hydrological characteristics by imitating them, an alternative combination of integrated management practices was chosen for the micro-basin No. 1. On the other hand, while choosing integrated management practices, criteria such as permeability, the slope of the area, runoff directions, and proximity to building foundations were also considered.

In this context, (a)Since the commerce space (shopping center) located in the north of the micro-basin No. 1 is on the runoff direction and is located on a high-permeability area, "porous pavement" was proposed as one of the integrated management practices. (b)For multi-storied housing areas that cover a large part of the micro-basin No. 1, an integrated management practice "bioretention area" had been proposed. (c)For 90 villas (low-rise housing areas) located in the south of the micro-basin No. 1, a "rain garden" had been proposed as an integrated management practice. (d)Finally, "grassed swale" had been proposed for vehicle roads as one of the management practices integrated into the medians.

The micro-basin No. 19 in the main basin of Kaplum Creek in the pre-development state, is a micro-basin mostly occupied with hydrological soil groups B and D, had areas with slopes varying between 1 and 39 %, low-very low water permeability, and covered with natural meadows. In the post-development state, the micro-basin No. 19 had encountered a dense settlement of low-rise housing areas.

In the post-development state of micro-basin No. 19, there are approximately 81 % low-rise housing areas (586 villas), 7 % asphalt road, 6 % medians, 2 % multi-storied housing areas, 2 % park and recreation areas, and the remaining 2 % are education, trade, religious and social facility areas.

As in the micro-basin No. 1, in the selection of integrated management practices for micro-basin No. 19, the area limitations of the integrated management practices were taken into account, and the slope, permeability, runoff-drainage directions, and proximity to building foundations of the area were also considered. In this context, (a)For low-rise housing areas that cover a large part of the micro-basin

No. 19, "cistern" as an integrated management practice had been proposed. (b) For the multi-storied housing areas located in the southwest of the micro-basin No. 19, "bioretention area (with underground drainage system)" as an integrated management practice had been proposed. (c)Grassed swale, one of the integrated management practices, for the medians had been proposed. (d)Finally, a "porous pavement", one of the integrated management practices for the sports area located at the beginning of the recreation area, in the middle of the micro-basin No. 19, had been proposed.

Integrated management practices proposed for both micro-basins within the scope of the lid approach were determined to improve pre-development hydrological characteristics by preserving them in post-development. With the use of the right combinations of integrated management practices in sample micro-basin, preserving or decreasing the runoff volume, preserving or increasing the concentration-time of pre-development, and also decreasing the peak runoff rates in post-development compared to the pre-development is the ideal state in hydrological terms.

8 Conclusion

In the lid approach, it is argued that there should be urban developments that will affect the rainfall-runoff cycle as little as possible and simple ecological solutions that will support the rainfall-runoff cycle. In this context, integrating hydrological characteristics into the site planning process, paying attention to details, controlling the runoff at its source, using simple solutions, and creating a multi-functional landscape and infrastructure are the basic concepts of the lid approach [10].

Within the scope of sustainable stormwater management, in addition to comparing the differences between the runoff volumes in pre- and post-development states of the study area, the permeability analysis of the study area and the 1/1000 implementation zoning plan are overlapped, and the "conflict areas" - which shows the permeable areas in pre-development that are covered with impermeable lands in post-development - were determined. The differences between the determined conflict areas and the pre-development and post-development runoff volumes were parallel. Conflict areas were graded from 1 to 4. Again, 1st-degree conflict areas are the areas that require an immediate solution in terms of stormwater management compared to other conflict areas. Accordingly, 1st-degree conflict areas, represents areas where high-very high permeability with an impermeable landcover, densely located in the northwest and north of the study area, according to the implementation zoning plan, considering the stormwater management and permeability state of the commercial space and the fuel and service station located in the northwest of the study area and at the entrance of the mass housing areas, it was found that the location choices were wrong. Necessary precautions should be taken within the scope of stormwater management for both areas. Especially, since the fuel service station opposite the commercial space is located in the micro-basin with the highest permeability, it would be more appropriate to move this station to an impermeable area due to the possibility of high contamination of groundwater with oil and derivatives. In addition, in the conflict area in the north of the study area, there were the urban technical infrastructure area and the social and cultural facility area. For this area, necessary measures should be taken in terms of stormwater management or the area in question should be left to open-green area regulations. Due to the fact that the study area is generally impermeable, other 1st-degree conflict areas within the study area are located on asphalt coated roads in smaller dimensions.

On the other hand, 2nd (high-very high permeability areas with medium permeability landcover), 3rd (where medium permeability areas with an impermeable landcover), and 4th degree (medium permeability areas with medium permeability landcover) conflict areas are gradually the areas where priority action is not required in terms of stormwater management.

Within the framework of the calculation and comparison of pre-development and post-development runoff potentials for the study area, and the analysis of conflict areas; within the scope of lid approach, solution proposals have been prepared for samples of micro-basins No.1 and 19, which have the greatest difference between the runoff volumes from the two main micro-basins, and have priority in terms of scope of stormwater management by being included in the 1st-degree conflict area in the analysis.

A lid approach, which ensures a healthy water-stormwater cycle, is a flexible stormwater management approach that can be used in the planning phase of an urban area and/or the implementation phase. Environmentally friendly, ecological, and simple solutions within the scope of the approach can be easily preferred for all types of urban areas. Within the scope of the lid approach, care has been taken to preserve and maintain pre-development hydrological features in sample micro-basins. In this context, in addition to calculating the pre-development and post-development runoff volumes for micro-basins No.1 and 19, more comprehensively, the concentration times (min) and the highest runoff rates (m^3/s) are also calculated and compared. In addition, drainage runoff directions have been determined for the whole study area.

While choosing solutions (integrated management practices) developed within the scope of lid approach for micro-basins No.1 and 19, the area limitations of integrated management practices and the physical characteristics of both micro-basins (topography, slope, permeability such as) have been evaluated, aside from considering the above-mentioned hydrological features (see Tab. 1). In this context, an alternative design has been proposed for both micro-basins. In addition, different alternatives can be developed for both basins.

Even though designs have been made within the boundaries of sample micro-basins based on the principle of managing stormwater at its source, within the scope of the lid approach, it should not be overlooked that the stormwater management planned for the study area should be a whole.

Within the scope of the study, considering that traditional stormwater management systems are inadequate in terms of ensuring the sustainability of the water-stormwater cycle of the sample area, a lid approach, which includes environmentally friendly, ecological, and simple solutions, has been adopted. Sustainable stormwater management solution proposals, offered to the sample area within the scope of the lid approach, will also support and help traditional stormwater management solutions. With the solution suggestions offered to the sample area, the state of the study area was returned to the pre-development hydrological properties (such as the runoff volume), and more reliable, comfortable, and healthy living spaces were provided to the sample area users, in terms of stormwater management problems.

Acknowledgements

This study was constituted from the PhD thesis titled "A Methodological Study on Urban Design Guidelines in terms of Landscape Architecture on Ecological Framework" at Ankara University, Turkey.

Tab. 1: Evaluation of integrated management practices applied within the scope of low impact development approach in micro-basin No.1 and 19

<table>
<tr><th colspan="6">Before applying integrated management practices within the scope of lid</th><th colspan="6">After applying integrated management practices within the scope of lid</th></tr>
<tr><th colspan="6">Micro Basin Runoff Potential (m³)</th><th colspan="6"></th></tr>
<tr><th colspan="6"></th><th colspan="2" rowspan="3">Integrated Management Practices</th><th colspan="2" rowspan="2">Micro Basins</th><th colspan="2" rowspan="2">Runoff Retention Potential of Integrated Management Practices (m³)</th></tr>
<tr><th colspan="3">No. 1</th><th colspan="3">No. 19</th></tr>
<tr><th>Pre-development</th><th>Post-development</th><th>Difference</th><th>Pre-development</th><th>Post-development</th><th>Difference</th><th>No.1</th><th>No.19</th><th>No.1</th><th>No.19</th></tr>
<tr><td rowspan="15">1,521.369</td><td rowspan="15">7,837.302</td><td rowspan="15">6,315.933</td><td rowspan="15">1,746.512</td><td rowspan="15">10,807.337</td><td rowspan="15">9,060.825</td><td rowspan="3">Bioretention Areas</td><td>With underground drainage</td><td></td><td></td><td></td><td>2,765.000</td></tr>
<tr><td>Without underground drainage</td><td></td><td></td><td>1,606.000</td><td></td></tr>
<tr><td>Rain Garden</td><td></td><td></td><td>527.000</td><td></td></tr>
<tr><td colspan="2">Dry Wells</td><td></td><td></td><td></td><td></td></tr>
<tr><td colspan="2">Filter Strips</td><td></td><td></td><td></td><td></td></tr>
<tr><td colspan="2">Vegetated Buffers</td><td></td><td></td><td></td><td></td></tr>
<tr><td colspan="2">Level Spreaders</td><td></td><td></td><td></td><td></td></tr>
<tr><td rowspan="2">Grassed Swales</td><td>Wet Grassed Swales</td><td></td><td></td><td></td><td></td></tr>
<tr><td>Dry Grassed Swales</td><td></td><td></td><td>985.000</td><td>1,088.000</td></tr>
<tr><td colspan="2">Rain Barrels</td><td></td><td></td><td></td><td></td></tr>
<tr><td colspan="2">Cisterns</td><td></td><td></td><td></td><td>3,019.000</td></tr>
<tr><td colspan="2">Infiltration Trenchs</td><td></td><td></td><td></td><td></td></tr>
<tr><td colspan="2">Green Roofs</td><td></td><td></td><td></td><td></td></tr>
<tr><td colspan="2">Porous Pavements</td><td></td><td></td><td>3,200.000</td><td>2,189.000</td></tr>
<tr><td colspan="4">Total Runoff Retention Potential (m³)</td><td>6,318.000</td><td>9,061.000</td></tr>
</table>

References

1. Endlicher, W., Langner, M., Hesse, M., Mieg, H.A., Kowarik, I., Hostert, P., Kulke, E., Nützmann, G., Schulz, M., van der Meer, E., Wessolek, G. and Wiegand, C., *Urban Ecology - Definitions and Concepts.* 2007, In Shrinking Cities: Effects on Urban Ecology and Challenges on Urban Development; Peter Lang: Frankfurt, Germany, p. 1–13.
2. Marsalek, J., Jimenez-Cisneros, B., Karamouz, M., Malmqvist, P.-A., Goldenfum, J. and Chocat, B., *Urban Water Cycle Processes and Interactions.* 2008, Taylor and Francis: Leiden, The Netherlands.
3. Fidani, D., Gesovska, V., *Sustainable Management of Water Resources in Urban Areas,* Int J Environ Sci Nat Res., 2019. **20**(5): p. 141–147.
4. Pyke, C., Warren, M.P., Johnson, T., LaGro Jr, J., Scharfenberg, J., Groth, P., Main, E., *Assessment of low impact development for managing stormwater with changing precipitation due to climate change.* Landsc. Urban Plan 2011, **103**: p. 166–173.
5. MA Department of Environmental Protection, MA Office of Coastal Zone Management, *Stormwater Management Volume One: Stormwater Policy Handbook.* 1997, U.S. Environmental Protection Agency.
6. MA Department of Environmental Protection, MA Office of Coastal Zone Management, *Stormwater Management Volume Two: Stormwater Technical Handbook.* 1997, U.S. Environmental Protection Agency.
7. Zhang, D., et al. *Conventional and decentralized urban stormwater management: A comparison through case studies of Singapore and Berlin Germany.* Urban Water Journal, 2017. **14**(2): p. 113–124.
8. Bogunovich, D., *Eco tech urbanism: Merging urban design with clean technology.* International conference on ecological and technological cities, 2008. Gazi Universiy, Ankara.
9. Nisqually River Council, *Low Impact Development and Architectural Guidelines for the Nisqually Watershed,* 2003.
10. Prince George's County Department of Environmental Resources Programs and Planning Division, *Low-Impact Development Design Strategies An Integrated Design Approach.,* 1999. 140 p., Maryland, USA.
11. Wang, X., Shuster, W., Pal, C., Buchberger, S., Bonta, J., Avadhanula, K., *Low impact development design-integrating suitability analysis and site planning for reduction of post-development stormwater quantity,* Sustainability **2010**. 2, p. 2467–2482.
12. Chang, N. B., Lu, J. W., Chui, T. F. M., Hartshorn, N., *Global policy analysis of low impact development for stormwater management in urban regions.* Land Use Policy, 2018. **70**, p. 368–383.

13. United States Department of Agriculture Natural Resources Conservation Service, Conservation Engineering Division Technical Release 55, *Urban Hydrology for Small Watersheds TR-55*, 1986. USA.
14. Huang, M., Gallichand, J., Wang, Z., Goulet, M., *A modification to the soil conservation service curve number method for steep slopes in the Loess Plateau of China.* Hydrol. Process. 2006, **20**, p. 579–589.
15. Ebrahimian, M., Nuruddin, A. A. B., Soom, M. A. B. M., Sood, A. M. and Neng, L. J., *Runoff Estimation in Steep Slope Watershed with Standard and Slope-Adjusted Curve Number Methods. Polish Journal of Environmental Studies,* 2012. **21**(5); p. 1191–1202.
16. Zhan, X., Huang, M.L., *ArcCN-Runoff: An ArcGIS Tool for generating curve number and runoff maps.* Environ. Modell. Softw. 2004. **19**; p.875.
17. Sharma, D., *Sustainable Drainage System (SuDs) for Stormwater Management: A Technological and Policy Intervention to Combat Diffuse Pollution.* 11th International Conference on Urban Drainage, 2008. Edinburgh, Scotland, UK.
18. MİA Planlama, *Ankara İli, Gölbaşı İlçesi Gölbaşı-Yaylabağ Toplu Konut Alanı Gölbaşı Özel Çevre Koruma Bölgesi İmar Planı Plan Açıklama Raporu,* 2010. 37 s., Ankara.
19. Çelebi, C. *Gölbaşı (Ankara) ve Çevresindeki Yeraltı Sularında Bor Kirliliğinin İncelenmesi.* Msc thesis, 2010. Gazi Üniversitesi, Fen Bilimleri Enstitüsü, Ankara.
20. Sudes Mühendislik Müşavirlik İnş. Tic. Ltd. Şti., *Ankara İli, Gölbaşı İlçesi Gölbaşı-Yaylabağ Toplu Konut Alanı Atıksu-Yağmursuyu Projeleri Açıklama Raporu,* 2013. 31 s., Ankara.
21. U.S. Army Corps of Engineers Baltimore District & U.S. Army Corps of Engineers Engineer Research and Development Center, *Army Low Impact Development Technical User Guide,* 364 p., USA.
22. Anlı, A. S., Öztürk, F., *Ankara'da Ölçülen Yıllık Maksimum Yağışların Bölgesel Frekans Analizi.* Gaziosmanpaşa Üniversitesi Ziraat Fakültesi Dergisi, 2011. **28**(1); p. 61–71.
23. Anlı, A. S., *Ankara'da Meydana Gelen Yağmurların L Moment Yöntemleri ile Bölgesel Frekans Analizi,* 2009. PhD Thesis, Ankara Üniversitesi, Fen Bilimleri Enstitüsü, Ankara.

Ayşen ÇOBAN

Urban Agriculture within the Scope of Sustainable and Productive Urban Strategies

Abstract: Today, cities have transformed into unsustainable areas where rural and urban balance is disturbed. Cities have destroyed the agricultural areas, which are the production sources of these areas, by putting the rural areas in the peri-urban under pressure. In addition, the neglect of agricultural areas in urban areas and the transformation of these areas into residential areas cause the disconnection of urban area with agricultural areas. Thus, cities have transformed from productive to consuming areas. The aim of this study is to reveal the contributions of urban agriculture to the sustainability and productivity of cities. The need for urban agriculture, one of the productive urban landscapes that contributes to the productivity, sustainability and achievement of the Sustainable Development Goals (SDG), has increased. In addition, the ecosystem services provided by urban agriculture emphasized that these areas are an integral part of cities.

Keywords: Sustainable development goals, sustainable cities, productive landscape, urban agriculture

1 Introduction

The world population reached 7.7 billion in mid-2019. The size of the global population is estimated to be between 8.5 and 8.6 billion in 2030, between 9.4 and 10.1 billion in 2050 and between 9.4 and 12.7 billion in 2100 [1]. The majority of the population that is constantly increasing all over the world lives in cities. While 51.1 % of the world's population lived in cities ten years ago, this rate increased to 55.7 % in 2019 [2].

The increase in the rate of population growth, the rapid development of technology, the decrease of the population in rural areas, the rapid urbanization that comes with it and the growth of urban areas lead to both the emergence of large urban masses and increased pressure on natural resources.

In the 1970s, environmental and development problems began to be high on the agenda all over the world, especially in urban areas [3]. Due to the increasing environmental problems and developing environmental awareness, the First World Environment Conference was organized by the United Nations (UN) between June 5–16, 1972 in Stockholm, the capital city of Sweden. The concept of sustainable development was used for the first time officially in 1987 in the report named "Our Common Future" published by the World Environment and Development Commission (WCED) under the chairmanship of the Norwegian Prime Minister

Gro Harlem Brundtland [4]. According to the report, sustainable development is defined as "Development that meets today's requirements without compromising the ability of future generations to meet their own needs" [5]. At the same time, this report predicts that environmental problems can be resolved through environmental and economic development integration [6]. The relationship of sustainable development should be established between the economy and the environment, as well as between the environment and other sectors and the city. Because sustainable development also provides an important turning point in the field of urban planning [7].

Sustainability of cities is not possible due to insufficient attention paid to the effects of policies, decisions and practices that directly affect the development of cities on the presence natural resources. This attitude causes the change of ecosystems from the landscape to the global scale, the depletion of natural resources and the deterioration of the relations between the natural environment and human beings. The most important reason for these problems is that the balance between protection and use cannot be established sufficiently and natural resources are not taken into account in land use decisions [8, 9].

In order to ensure sustainable development, 17 goals have been set by the UN in Agenda 2030. Among these, the Sustainable Development Goal (SDG) 11 aims at sustainable cities and communities. Food supply is one of the most basic needs in order to ensure the sustainability of cities. It aims at ending hunger, achieving food security which is the second item of the Sustainable Development Goals, and to promote sustainable agriculture. These goals should be achieved through many sub-targets, i.e. to increase the productivity while at the same time maintaining ecosystems, adapting to climate change, improving the quality of soil and maintaining the genetic diversity of plants. Achieving these goals require better access to food and the widespread promotion of sustainable agriculture [10].

Today, cities have been considered not as a sustainable habitat where rural and urban areas are balanced, but as an aggressive habitat that develops by suppressing the rural area towards its periphery. This acceptance has caused the destruction of agricultural areas, which are the source of existence of humanity. Besides, agricultural areas were ignored in urban areas and these areas were made available to different sectors [11]. Thus, the connection between rural and urban area is cut off. In this context, the connection of the agricultural landscape with the city, which contributes to the sustainability and productivity of the cities, must be firmly established.

Joint Programming Initiative Urban Europe considers urban and peri-urban agriculture to be strategic to guide urban transition to a sustainable future. Urban and peri-urban agriculture is interesting due to the promotion of local food procurement, biodiversity, carbon sequestration, reduction of wastes produced, participated agriculture and social cohesion [10].

Urban agriculture (UA), defined as the food growing in the city, is increasing in importance worldwide. UA can become a fundamental element of the urban green infrastructure, which improve the urban structure, the image that the city transmits and, in short, its sustainability and resilience [12].

In the light of this information, the aim of this study is to reveal the contributions of urban agriculture to the sustainability and productivity of cities.

2 Conceptual Framework

2.1 Sustainable Cities

One of the most important aspects of sustainable development is sustainable urban development. Sustainable urban development arguments started to form after 1990 and opened new horizons for engineers, scholars, and architects. This theory addresses issues such as preventing urban pollution, not supporting destructive developments and trying to reduce poverty in the society. Sustainable cities can be defined as compact, eco-friendly and appealing cities that provide equal opportunities for people from different levels of society to reach prosperity [13].

What is clear is that sustainable development has environmental, economic, and social dimensions, but how exactly the "triple bottom line" is to be understood, and how it applies to urban contexts needs to be analyzed [14].

The main challenge for today's cities is to manage the heavy dependence on ecosystem services, which results in the depletion of natural resources and biodiversity and the efforts to mitigate and adapt to climate change, while prioritising public health and quality of life [15]. So the sustainable city concept has assumed prominence. This is evident in the Sustainable Development Goal [SDG] 11, which is "*to make cities and human settlements inclusive, safe, resilient and sustainable*". In this regard, the primary goal of urban development should be to make cities and human settlements inclusive, safe, resilient and sustainable [16].

A sustainable city highlights the balance between economic development, environmental protection and equality in income, employment, housing, basic services, social infrastructure and equality in transport [16]. However, food production and supply, flora and fauna are not included or considered to the extent necessary. These are basic requirements for sustainable cities. Indicators determined to measure the sustainability of a city provide a framework that combines the economic, social and environmental dimensions of the city. The unified framework in Fig. 1 presents the indicators used to assess the relationship between urban agriculture and sustainable cities. The results of the review indicate that the urban agriculture sustainable city nexus manifests in economic, social and ecological terms [16].

UA seems to be one of the vital solutions for decreasing urban footprint and promoting urban sustainability [17]. The easy review of agricultural land for non-purpose uses, the evaluation of agriculture as an out-of-city activity, therefore the disconnection between city and nature are closely related to the way modern urban fiction and planning approach. This issue needs to be addressed. It is known that agricultural activities have existed in cities in every period of history except the modern period, and in recent years, UA has come to the fore in the world in the context of food security and there are important experiences in Asia, Europe and North America [18].

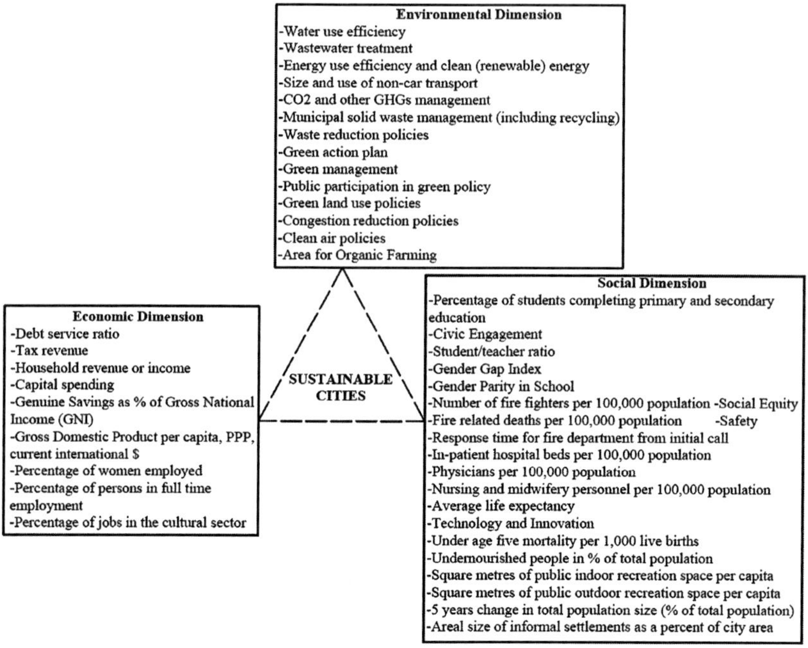

Fig. 1: Sustaianble City Framework [16].

UA is one of the key nature-based solutions to reduce urban footprint and promote urban sustainability. Researchers and practitioners focus on the link between urban and rural areas to improve the sustainability of food in urban areas without compromising the quality of life. One of the main aims of urban agriculture, food security has also an important role for cities future [18].

2.2 Productive Urban Landscape

Productive is important in ensuring the sustainability of cities. Sustainable concept of Productive Urban Landscapes with the spatial concept of Continuous Landscape, the proposes a new urban design strategy by integrating the appearance of contemporary cities with nature [19]. Thus, Bohn and Viljoen (2005) revealed a design concept for a new type of landscape layer for cities. This concept is a "continuous productive urban landscape" or CPUL. The CPUL is an active, productive, consistently planned and designed open urban landscape network that is integrated with a metropolitan scale sustainable landscape strategy. CPUL is considered as an uninterrupted city-wide open area network that includes agricultural and ecologically functional landscapes [20]. CPULs require land, but in return they will enrich cities by reducing their environmental impact and bringing in spatial qualities until then only associated with rural or natural conditions [19]. CPUL affects the city qualitatively through citizens, experiences and reduced negative environmental impact. However CPUL balances ecosystem services [21].

CPUL aims to economically produce food or crops, improve socio-cultural quality of life and environmentally reduce carbon dioxide, emission and urban heat island effect, improve biodiversity, air quality, capture and reuse of rainwater [20].

One of the main components of the concept of a productive urban landscape and/or CPULs is UA [20, 22, 23]. It provides not only access to food but also the opportunity to improve public participation and community management.

According to Ian Mcharg, the cities were not 'natural' but he believed that they could be better aligned with nature [22]. Therefore, the connection between reflections on the need of redefining the modern city by changing relations with its "humanized" nature, understood not only as UA, but also as the presence of nature and urban wildlife seems to be important [23]. It is also thought that wildlife will occur more frequently due to UA, greater security and much higher chance of obtaining food, which are food production areas that should be considered as part of green infrastructure systems.

In this context, productive urban landscape systems provide the additional benefit of reminding city inhabitants of the waste-nutrient cycles that exist within cities. Too frequently the systems of infrastructure which provide our drinking water, food, and energy as well as cart off our waste, mask the complex processes required to handle these overly engineered solutions. By utilizing landscape conditions to recreate resource cycles, there is an opportunity to expose city dwellers to the important links between natural processes and resources, cities and landscapes [24].

3 Urban Agriculture

Rapid urbanisation, mainly in terms of high population growth, continues to pose substantial threats to the sustainability of cities. This phenomenon also threatens urban ecological services, social stability and health systems [25].

Concerns about the impact of the food system - globally focused - along with the reality of today's fast-growing cities, raises a global awareness of how to feed cities sustainably. In this context, it has been determined that agriculture within a city-region is generally ignored and disconnected from the urban area. The urgency of reconnecting agriculture with the urban areas become evident that the global food system is under pressure due to uncertainties that affect local food supply such as geopolitical crises, natural disasters, climate change, water scarcity, hunger and Covid-19 pandemic [26].

In early times urban agriculture was an integral part of the urban domain, but with the industrial revolution agriculture lost that position. UA regained interest with Ebezener Howard's garden city design which revalued and integrated it in the urban and peri-urban planning [26]

UA refers to food production in cities [27]. According to Nugent (1997), UA means *"fresher and cheaper food, an action which utilizes and cleans domestic wastes, and more green areas".* Another concept of UA came to the agenda in the United Nations Habitat Program held in Istanbul in 1996 and was defined as *"the cultivation, growth and development of plant cultural assets and animal products in the city"* [11, 22]. The FAO (2014) defines UA as follows: "Urban and peri-urban agriculture refers to growing of plants and animals in and around a city". More general definitions of UA are based on such determinants as "types of economic activities, food/non-food categories of products and subcategories, the urban and peri-urban characters borne by the location of the practice, the type of lands where the practices are conducted, types of production systems, production scale, and production objective" [11].

In some parts of the world, UA's main function is to feed the rapidly growing population while in others it is more associated with lifestyle [28].

3.1 From Urban Agriculture to the Sustainability of Cities

In order to stop the unsustainability of the modern city, governments at national, regional and local levels are taking initiatives all over the world. UA provides opportunities for the sustainability development of cities around the world. It supports sustainability of cities ecologically, economically and socially [28]. In addition, urban agriculture helps achieve the Sustainable Development Goals (SDG 11) by providing positive effects on the health of urban and urban residents.

The rebalancing of urban systems based on a green space network of rings and vegetable corridors around urban centers, thereby linking agricultural lands with urban residents and neighborhoods. Within this network or green matrix, urban and peri-urban agriculture plays an important role as one of the most important strategies for moving towards sustainability of cities (Fig. 2) [28].

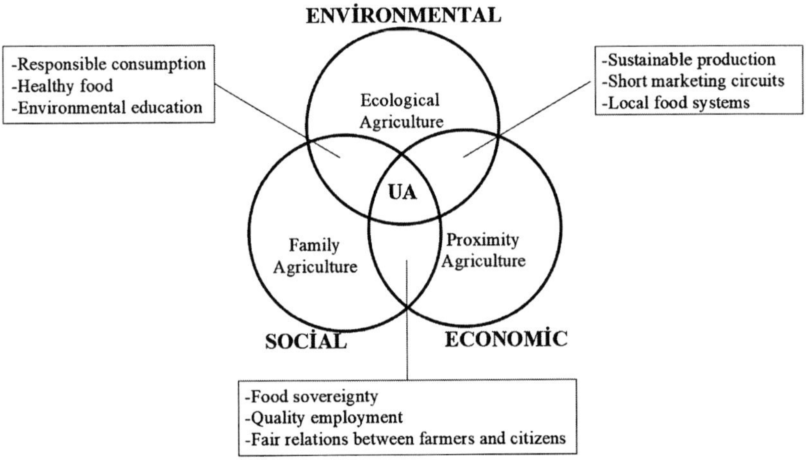

Fig. 2: The significance of urban agriculture in the context of sustainability of the cites [28].

It is particularly important to consider the social, economic and environmental development contribution of UA to the sustainability of cities. Hence, UA can positively affect all these (Fig. 3) [16, 28, 29].

UA helps social development by providing urban food security, healthy nutrition (access to fresh and healthy food), personal well-being (UA reduces psychological and biological health problems). Also, direct participation in UA provides physical, intellectual and psychological benefits), community building (promoting integration and participation), personal skills (e.g. gardening and communication skills), sense of place (environmental protection, environmental values associated with one's experience of place) [28, 30].

UA offers the potential to use land efficiently, self-procure, create employment and income, save time and money (reduce food miles), promote marketing initiatives within the scope of economic development [28,30].

From the environmental point of view, UA allows the protection of urban soil, improvement of microclimate, protection from urban island effects, reduction of pollution (reduction of the effects of fossil fuel combustion through carbon capture, waste and nutrient recycling), increasing biodiversity and an increase in environmental awareness [28, 30].

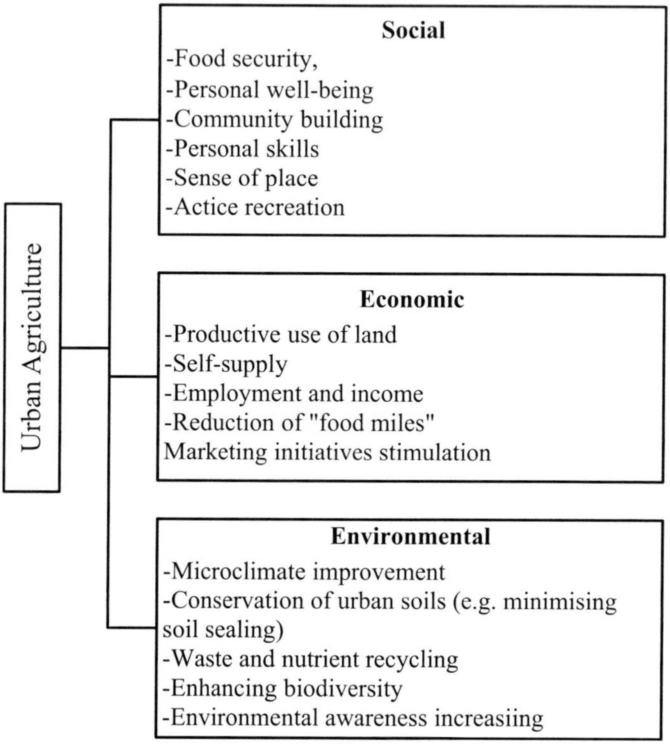

Fig. 3: Positive aspects of urban agriculture to sustainability of cities [28]

3.2 Contribution of Urban Agriculture to Sustainable Development Goals

The literature covered in the present study reveals the multi-functional roles of urban agriculture in the sustainability of cities. In addition, urban agriculture contributes directly or indirectly to the SDG (Tab. 1).

Tab. 1: Evaluation of UA within the scope of SDG (Developed from [31])

NO	SDG	DIRECT	INDIRECT
1	End poverty in all its forms everywhere	X	
2	End hunger, achieve food security and improved nutrition and promote sustainable agriculture	X	
3	Ensure healthy lives and promote well-being for all at all ages	X	
4	Ensure inclusive and equitable quality education and promote lifelong learning opportunities for all		
5	Achieve gender equality and empower all women and girls		X
6	Ensure availability and sustainable management of water and sanitation for all		
7	Ensure access to affordable, reliable, sustainable and modern energy for all		
8	Promote sustained, inclusive and sustainable economic growth, full and productive employment and decent work for all	X	
9	Build resilient infrastructure, promote inclusive and sustainable industrialization and foster innovation		
10	Reduce inequality within and among countries		X
11	Make cities and human settlements inclusive, safe, resilient and sustainable	X	
12	Ensure sustainable consumption and production patterns	X	
13	Take urgent action to combat climate change and its impacts	X	
14	Conserve and sustainably use the oceans, seas and marine resources for sustainable development		
15	Protect, restore and promote sustainable use of terrestrial ecosystems, sustainably manage forests, combat desertification, and halt and reverse land degradation and halt biodiversity loss	X	
16	Promote peaceful and inclusive societies for sustainable development, provide access to justice for all and build effective, accountable and inclusive institutions at all levels		
17	Strengthen the means of implementation and revitalize the Global Partnership for Sustainable Development		X

In this context, city authorities should strive to promote and sustain agriculture in the urban landscape to make cities safe, resilient and sustainable (SDG11). The ecological functions of UA can reduce the negative environmental impacts of cities (SDG 11.6) and increase people's access to green spaces (SDG 11.7). Similarly, UA can improve the links between urban, peri-urban and rural areas (SDG 11a) and improve resource use efficiency and adaptation to climate change (SDG 11b). Furthermore, the employment and income generation potentials of UA can accelerate poverty reduction (SDG 1) [25]. In the light of this information, UA plays an important role in achieving SDG (Tab. 2).

Tab. 2: Urban Sustainable Development Goals [32]

SDG 11	Make cities and human settlements inclusive, safe, resilient and sustainable
SDG 11.6	By 2030, reduce the adverse per capita environmental impact of cities, including by paying special attention to air quality and municipal and other waste management
SDG 11.7	By 2030, provide universal access to safe, inclusive and accessible, green and public spaces, in particular for women and children, older persons and persons with disabilities
SDG 11a	Support positive economic, social and environmental links between urban, peri-urban and rural areas by strengthening national and regional development planning
SDG 11b	By 2020, substantially increase the number of cities and human settlements adopting and implementing integrated policies and plans toward inclusion, resource efficiency, mitigation and adaptation to climate change, resilience to disasters, and develop and implement, in line with the Sendai Framework for Disaster Risk Reduction 2015–2030, holistic disaster risk management at all levels.

3.3 Ecosystem Service and UA

Although the scientific background of the concept of "Ecosystem Services (ES)" dates back to 1970s; the use of the concept in the literature gained momentum in the 1990s [33]. The concept of "Ecosystem Services" has been defined for the first time by Daily (1997) as "the state and processes realized by natural ecosystems and species in order to sustain human life" [34, 35].

Costanza et al. (1997), benefits from ecological functions; MEA (2005) benefits from both processes and functions; Boyd and Banzaf (2007) state that ecosystem services are only directly utilized products.

Although the number of ecosystem services we benefit from globally remains uncertain, in the "Millennium Ecosystem Assessment Report-MEA", one of the most important publications on this subject, the services provided by nature to people are discussed in four groups. These [34, 35, 36, 38, 39, 40];

1. Provisioning Services
2. Regulating Services
3. Cultural Services
4. Supporting Services

Considering the high population living in cities, it is the cities that benefit the most from ecosystem services. In cities parks, gardens, green areas in cemeteries, urban forests, green roofs, wetlands, stream corridors, lakes, ponds, natural areas on the urban periphery and rural areas are the main service providers of the urban ecosystem. Demirbaş Çağlayan et. al., (2020) revealed the most important ecosystem services provided in cities. These are included in Tab. 3.

Ecosystem services provided by the UA as direct or indirect benefit derived from this urban activity, which results in an improvement in the citizen's quality of life and the sustainability of cities. Considering its role in increasing the resilience and quality of life in cities and the sustainability of cities, UA's contribution to the achievement of ecosystem services is very important [12, 41].

Tab. 3. The most important ecosystem services provided in cities [40]

Ecosystem Services	Ecosystem Services Provided In Cities
Provisioning	Food Supply
	Clean Water
	Wood Products
	Medicines
Regulating	Regulation of the local climate
	Cleaning the air
	Cleaning the water
	Pollination
	Noise
	Carbon capture
Cultural	Ecotourism
	Recreation
	Inspirational values
	Moral and spiritual values
Supporting	Providing habitat
	Ensuring genetic diversity

As the UA is multifunctional [42] the concept of ecosystem services is useful to explicitly identify and evaluate its multiple functions [12].

Within the scope of ecosystem services provided by UA, MEA (2005), Barker et al. (2019) and Demirbaş Çağlayan et al. (2020) has been used. First of all, determining the forms of UA is important for analyzing the ecosystem services provided by UA.

Agriculture can take many forms as a result of moving away from common practices in rural areas towards urban areas with more intensive land use. The forms of UA described in Tab. 4 address very different food demands. It has radically different consequences for social and environmental costs and benefits, with different commercial or nonprofit models showing [31]

Tab. 4. The forms of UA (Developed from [31])

Form of agriculture	Characteristics
Home Gardens	Cultivation of food on privately owned land adjoining a residence
Allotment Garden	Small plots of cultivation are often made available in nature by the local authority, especially to hobbyist (non-commercial) people.
Communal Growing/ Community Gardens	Including community, confidence, welfare and skills. These spaces are typically open to the public, managed by groups of local people rather than by local authorities
Rooftop Gardening	Cultivation of food crops on the roofs of domestic or commercial or municipal buildings
Edible Landscape	Incorporating edible fruiting trees, shrubs or herbaceous plants into the design of the residential landscape
Community Supported Agriculture (CSA)	More commonly peri-urban and rural, food has more strongly defined roles among growers and community members. It is produced on a larger scale than community gardens. Characterized by an exchange relationship that allows growth risks to be shared more equally than in a typical producer-consumer relationship.
Vertical Agriculture	Integrating intensive agriculture into multi-use, multi-level urban structures, usually on a commercial basis
Informal Urban Agriculture	Opportunistic cultivation of food crops for sale or use on land that is not subject to strict land tenure enforcement, more prevalent
Commercial Urban Agriculture	Food production for profit. Sale via a typical producer–consumer relationship (although produce is often sold into a supply chain rather than directly to end consumers)

In evaluating the ecosystem services provided by UA forms, Barker et al. (2019) and Demirbaş Çağlayan et al. (2020)'s studies are based (Tab. 5).

Tab. 5. Evaluation of urban agriculture forms and ecosystem services [40, 43].

	Home Garden	**Allotment Gardens**	**Community Garden**	**Edible Landscape**	**Rooftop gardening**	**Community Supported Agriculture**	**Vertical Agricuture**	**Informal Urban Agriculture**	**Commercial Urban Agriculture**	**TOPLAM**
Provisioning	10	10	10	10	10	10	10	10	10	**90**
Food Supply	4	4	4	4	4	4	4	4	4	**36**
Clean Water	1	1	1	1	1	1	1	1	1	**9**
Wood Products	1	1	1	1	1	1	1	1	1	**9**
Medicines	4	4	4	4	4	4	4	4	4	**36**
Regulating	11	12	12	16	16	16	18	17	17	**135**
Regulation of the local climate	2	2	2	2	2	3	3	3	3	**22**
Cleaning the air	2	2	2	3	3	2	3	3	3	**23**
Cleaning the water	1	1	1	3	3	2	2	2	2	**17**
Pollination	4	4	4	4	4	4	4	4	4	**36**
Noise	1	2	2	2	2	2	3	2	2	**18**
Carbon capture	1	1	1	2	2	3	3	3	3	**19**
Cultural	9	13	13	12	11	14	10	14	9	**105**
Ecotourism/ Agrotourism	1	3	3	3	2	4	2	4	2	**24**
Recreation	1	3	3	3	2	3	2	3	2	**22**
Inspirational values	3	3	3	3	3	3	3	3	2	**26**
Moral and spiritual values	4	4	4	3	4	4	3	4	3	**33**
Supporting	6	6	6	7	6	7	6	7	7	**58**
Providing habitat	4	4	4	4	4	4	4	4	4	**36**
Ensuring genetic diversity	2	2	2	3	2	3	2	3	3	**22**

Tab. 6. Evaluation key

Point	Likelihood Of Benefit Being Achieved-Point Scale
4	Very Likely
3	Likely
2	Possible
1	Unlikely

Urban agriculture forms and ecosystem services provided by Barker et. al., (2019) was examined on the basis of the evaluation scale (Tab. 6).

In this context, 4 point scale was used in the evaluation of ecosystem services (4: Very Likely, 3: Likely, 2: Possible, 1: Unlikley).

When 9 urban agriculture forms are evaluated in terms of "Provisioning Services", it is determined that they received 90 points out of 144 points in total.

When they are evaluated in terms of other services, it is concluded that "Regulatory Services" scores 135 points out of 216 in total; "Cultural Services" scores 105 points out of 144 points in total and "Supporting Services" scores 58 points out of 72 points in total.

The 4 point scale used in this study shows the mathematical result of the ecosystem services provided by UA for cities. The results determined reveal the contributions of UA to urban areas in micro, meso and macro scales. In the light of this information, UA areas should not be considered as uses that are not suitable for urban form. These areas should be integrated into green infrastructure systems and it should be accepted that they are the inseparable components of these systems.

4 Conclusion

With the rapid population growth and urbanization problems, most of the agricultural areas have turned into residential area and housing areas in time. In addition, agricultural areas are considered independent from cities because they are not seen as suitable for urban forms. Thus, cities have turned into consuming areas separated from productive areas.

Productive urban landscapes are needed to ensure the sustainability of urban areas and to achieve the SDGs. Considering the benefits of UA (providing healthy, fresh and cheap food production, providing food security for the urban poor, creating a better living environment and green cities, providing a positive effect on human physical and mental health, creating new employment opportunities, increasing social communication, waste management, urban energy reducing footprint, contributing to climate change adaptation and mitigation etc), which is one of the productive urban landscapes, for the citizens, re-doing agriculture in the cities will make a significant contribution to increase food security by expanding the urban/local food system [18].

Agricultural spaces must be considered part of the permanent urban landscape, such as a park or an outdoor space in which people engage in civic activities, and therefore must be deliberately designed for such purposes [12]. Because when the "Provisioning Services, Regulating Services, Culturel Services and Supporting Services" provided by UA are evaluated, it reveals the requirements of these areas in urban areas. In this context, UA areas should be integrated into urban planning and the use of forms of these areas should be strengthened by urban design.

As stated by Gökçe and Çarıkçı (2019), agriculture should be defined as an urban activity, and "UA" should be included in the spatial plan and land use class of all scales. In this context, the Spatial Plans Construction Regulation should be rearranged and the concepts related to UA such as roof gardens, hobby gardens, community gardens, city farms should be defined in the regulation. An urban agriculture master plan should be prepared for each city, recognizing the nature of agriculture in relation to food, health and ecology at the local level.

Also, concepts related to UA should include forms of UA. These forms are home gardens, allotment gardens, communal growing/community gardens, rooftop gardening, edible landscape, Community Supported Agriculture (CSA), vertical agriculture, informal urban agriculture, commercial urban agriculture.

As a result, the inclusion of UA in urban areas will move these cities from consumer position to producer position. Thus, it will promote cities environmental, social and economic resilience.

References

1. United Nations (UN). The World Population Prospects 2019. 2019 [cited 2021 24 January]; Available from: https://population.un.org/wpp/Publications/Files/WPP2019_Highlights.pdf. Accessed 24.01.2021
2. United Nations Conference on Trade and Development (UNCTAD). Urbanization continues. 2019 [cited 2021 24 January]; Available from: https://stats.unctad.org/handbook/Population/Total.html#:~:text=Ten%20years%20ago%2C%2051.1%20per,two%20(65.4%20per%20cent). Accessed 24.01.2021
3. Deniz, T. *Kentleşme, Yerel Gündem 21 ve Kent Konseyleri.* International Periodical for the Languages, Literature and History of Turkish or Turkic, 2013. **8(12):** p. 391–402.
4. Aksu, C. Sürdürülebilir Kalkınma ve Çevre. Güney Ege Kalkınma Ajansı. 2011. [cited 2021 20 January]; Available from: http://geka.org.tr/yukleme/dosya/f6574f6e6b0a8d70a27bfbde52c53a47.pdf. Accessed:13.11.2015

5. Anonymous. Report of the World Commission on Environment and Development: Our Common Future. 2015. [cited 2021 22 January]; Available from: http://www.un-documents.net/our-common-future.pdf. Accessed13.11.2015
6. Güzelmansur, A. *Amik Ovası ve Yakın Çevresinin Sürdürülebilir Alan Kullanım Planlaması.* 2012, Doktora Tezi, Çukurova Üniversitesi, Fen Bilimleri Enstitüsü, Adana, p. 259.
7. Yazar, K.H. *Sürdürülebilir Kentsel Gelişme Çerçevesinde Orta Ölçekli Kentlere Dönük Kent Planlama Yöntem Önerisi.* 2006. Doktora Tezi, Ankara Üniversitesi, Sosyal Bilimler Enstitüsü, Ankara, p. 363.
8. Özyurt, S.S. *Kentsel Gelişme Planlarının Doğal Yapı Üzerindeki Etkisinin Adana Kenti İmar Planı Örneğinde İrdelenmesi.* 2009. Yüksek Lisans Tezi, Çukurova Üniversitesi, Fen Bilimleri Enstitüsü, Adana, p. 143.
9. Steiner, F. F*rontiers in urban ecological design and planning research.* Landscape and Urban Planning, 2014. **125:** p. 304–311.
10. Caputo, P., Zagarella, F., Cusenza, M.A., Mistretta, M. and Cellura, M. *Energy-environmental assessment of the UIA-OpenAgri case study as urban regeneration project through agriculture.* Science of the Total Environment, 2020. 729: 138819. https://doi.org/10.1016/j.scitotenv.2020.138819.
11. Cengiz A.E., PEKİN TİMUR, U, KARADAĞ, A.A. and DEMİROĞLU, D. A Contribution to Sustainable Urban Development: Urban Agricultural Lands: Urban and Urbanization, R. Efe, T.T., Onay, I. Sharuho, E., Atasoy Editors. 2014. ISBN: 798-954-07-3772-0, St. Kliment Ohridski University Press, Sofia. pp. 85–104.
12. G´omez-Villarino, M.T. and Ruiz-Garcia, L. *Adaptive design model for the integration of urban agriculture in the sustainable development of cities. A case study in northern Spain.* Sustainable Cities and Society, 2020. **65:** p. 1–8. https://doi.org/10.1016/j.scs.2020.102595
13. Mohammadi, M., Lau, S.K., Kim, Y., Ahmed Ghoneim, Dwivedi, S., Mastalerz, C.J. and Klose M. Urban Planning: Sustainable Cities III, E. Pollock, A., Reinbold Editors. 2020. ISBN 978-952-328-223-0, ISSN 2669-8021, Helsinki Metropolia University of Applied Sciences, Helsinki. pp. 137–244.
14. Del-Aguila-Arcentales, S., Alvarez-Risco, A. and Rosen M.A. Chapter 5 Sustainable Cities: Building Sustainable Cities, A. Alvarez-Risco, Rosen, M., S, Del-Aguila-Arcentales., D., Marinova Editors .2020. ISBN 978-3-030-45533-0, Springer, Cham, Switzerland, pp. 51–63. https://doi.org/10.1007/978-3-030-45533-0_5
15. European Commission. Indicators for Sustainable Cities, 2018. [cited 2021 25 January]; Available from: https://ec.europa.eu/environment/integration/research/newsalert/pdf/indicators_for_sustainable_cities_IR12_en.pdf

16. Azunre, G.A., O. Amponsah, C. Peprah and S.A. Takyi. *A review of the role of urban agriculture in the sustainable city discourse.* Cities, 2019. **93:** p. 104–119.
17. Akyol, M. *Evolution Of Urban Agriculture Concept and Determination Of Design Criteria.* 2011. M.Sc. Thesis, Istanbul Technical University Institute Of Science and Technology, İstanbul, p. 99.
18. Gökçe, D. and Çarıkçı, A., *Ortak Geleceğimizin Sağlıklı İnşası İçin Tarım Odaklı Kentler.* Journal of Social and Humanities Sciences Research. 2019, **6(32):** p.53–63.
19. Bohn, K. and Viljoen, A. More Space Wıth Less Space: An Urban Desıgn Strategy: Continuous Productive Urban Landscapes: Designing Urban Agriculture For Sustainable Cities, A. Viljoen Editor. 2005. ISBN 0 7506 55437, Architectural Press, Oxford. pp. 10–16.
20. Davey, C.A. *Productive urban Landscapes: The Relationship Between Urban Agriculture and Property Values in Minnepolis, Minnesota.* 2010. Doctor of Philosophy, Kansas State Unıversıty, Environmental Design and Planning, College of Architecture, Planning and Design, Manhattan, Kansas, p. 227.
21. Upadhyay, P. and Chaudhary, M., *Continuous Productive Urban Landscapes.* Journal of Civil Engineering and Environmental Technology, 2016. **3(7):** p. 665–666.
22. Akyol, M. and Eşbah Tunçay, H. *Productive landscapes and resilient cities.* ITU A|Z, 2013. **10(2):** p. 133–147.
23. Kleszcz, J. *Urban Farm As A System Of Productive Urban Green-Challenges and Risks.* Architecture Civil Engineering Environment, 2018. **1:** p. 23–36.
24. Napawan, N.C. Multi-Productive Landscapes of the Sustainable City: Opportunities for Managing Resource Needs through Urban Landscapes. Journal of Environmental Design and Planning, 2010. **6:** p. 1–18.
25. Ayambire, R.A., Amponsah, o., Perah, C. and Takyi, S.A. *A review of practices for sustaining urban and peri-urban agriculture: Implications for land use planning in rapidly urbanising Ghanaian cities.* Land Use Policy, 2019. **84:** p. 260–277
26. Jansma, J.E. and Werheim-Heck, S.C.O. Thoughts for urban food: A social practice perspective on urban planning for agriculture in Almere, the Netherlands. Landscape and Urban Planning, 2021. **206:** 103976. https://doi.org/10.1016/j.landurbplan.2020.103976
27. Davies, J., Hannah, C., Guido, Z., Zimmer, A., McCann, L., Battersbdy, J. and Evans, T. *Barriers to urban agriculture in Sub-Saharan Africa.* Food Policy, 2020. 101999. https://doi.org/10.1016/j.foodpol.2020.101999
28. Maćkiewicz, B. Asuero, R.T. and Almonacid, A.G. *Urban Agriculture As The Path To Sustainable City Development. Insights Into Allotment Gardens In Andalusıa.* Quaestiones Geographicae, 2019. **38(2):** p. 121–136.

29. Kontothanasis, G. *Social practices of urban agriculture in the metopolitan region of Thessaloniki.* Procedia Environmental Sciences, 2017. **38:** p. 666–673.
30. Haywood, K. Urban Agriculture A Guide for Municipalities. 2017. [cited 2020 7 January]; Available from: http://d-scholarship.pitt.edu/33657/1/Urban%20 Agriculture%20-%20A%20Guide%20for%20Municipalities.pdf
31. Nicholls, E., Ely, A., Birkin, L., Basu, P. and Goulson, D. The contribution of small-scale food production in urban areas to the sustainable development goals: a review and case study. Sustainability Science, 2020. **15:** p. 1585–1599.
32. Klopp, J.M. and Petretta, D.L. *The urban sustainable development goal: Indicators, complexity and the politics of measuring cities.* Cities, 2017. **63:** p. 92–97
33. Costanza, R., Darge, R., Groot, R., Farber, S., Grasso, M., Hannon, B., Limburg, K., Naeem, S., O'neil, R.V., Paruelo, J., Raskin, G.R., Sutton P. and Belt, M. *The value of the world's ecosystem services and natural capital.* Nature, 1997. **387:** p. 253–260.
34. Demiroğlu, D. and Karadağ, A.A. Ecosystem Services Approach To Spatial Planning In Turkey, I. Uluslararası Kent Araştırmaları Kongresi, 16–17 April 2015. 2015: Eskişehir, Turkey. pp. 153–168.
35. Cengiz, A.E. and Demiroğlu, D. Integrating Ecosystem Services into Spatial Planing: New Approaches to Spatial Planning and Design - Planning, Design, Applications, M. Özyavuz Editors. 2019. ISBN:ISBN 978-3-631-78274-3, Peter Lang, Berlin, pp.221–241.
36. MEA (Millennium Ecosystem Assessment). Ecosystems and Human Well-being: Synthesis. 2005, ISBN 1-59726-040-1, Millennium Ecosystem Assessment, Island Press: Washington DC. 137 p. [cited 2021, 24 January]; Available from: https://www.millenniumassessment.org/documents/ document.356.aspx.pdf
37. Boyd, J. and Banzhaf, S. *What are ecosystem services? The need for standardized environmental accounting units.* Ecological Economics, 2007. **63 (2–3):** pp. 616–626.
38. Hassan, R., Scholes, R. And Ash, N. *Ecosystems and Human Well-being: Current State and Trends.* Island Press, London, 2005. **1:** pp. 25–36.
39. Çoban, A. and Yücel, M. *Kent Planlamasında Ekosistem Hizmetlerinin Rolü.* Düzce Üniversitesi Bilim ve Teknoloji Dergisi, 2018. **6:** pp.444–454.
40. Demirbaş Çağlayan, S. Balkız, Ö., Arslantaş, F., Ceviz Sanalan, K., Lise, Y. and Zeydanlı, U. *Şehir Planlama Aracı Olarak Ekosistem Hizmetleri: Çankaya İlçesi Örneği.* Doğa Koruma Merkezi, Ankara. 2020. pp. 236.
41. G´omez-Baggethun, E., Gren A., Barton, D.N., Langemeyer, J. McPhearson, T., O'Farrell, P., Andersson, E., Hamstead, Z. And Kremer, Peleg. Urban Ecosystem Services: Urbanization, Biodiversity and Ecosystem Services: Challenges and Oppotunities, T. Elmqvist, M. Fragkias, J. Goodness,

B. Güneralp, P.J. Marcotullio, R.I. McDonald, S. Parnell, M. Schewenius, M. Senstad, K.C. Seto, C. Wilkinson Editors. 2013, ISBN 978-94-007-7088-1, Springer Dordrecht Heidelberg New York London. pp. 175–251.

42. Lovell, S.T. *Multifunctional Urban Agriculture for Sustainable Land Use Planning in the United States.* Sustainability, 2010. **2:** p. 2499–2522.
43. Barker, A., Clay, G., Morrison, R., Payne, S., Gilchrist, A., Rothwell, J. and Tantanasi, I. Understanding Green Infrastructure at Different Scales: A Signposting Guide, University of Manchester, Manchester, UK. 2019.

Candan ŞAHİN, Büşra ONAY and Esra MİRZA

A Natural Sustainable Material for Urban Design: Wood

Abstract: Wood material has been used with different forms and shapes in urban design practices for centuries. Nowadays, it is widely utilized material as environmentally friendly material to use in many architectural plans. Since it is a natural and sustainable material, the demand and expectations should be considered in accordance with the applications. Selecting correct material and protective agent can be augment the beauty of wood pieces requires considering the dominant tones in end uses. However, choosing species that are resistant to outdoor use is important in terms of ensuring for a longer time. The technological and economic necessities have made wood to select and eliminate some negative properties. Landscape architects should be considered environmental product that measuring and verifying the life cycle of products and convince the people to do it. Therefore, natural infrastructure, including sustainable materials have widely used, especially in urban environments. In order for wooden material to be considered as an urban design element, it must first be properties considered. In this study, the properties of the wood material in general are explained, and the advantages of using wood in outdoor and urban design principles are discussed with some examples.

Keywords: Urban design, wood material, urban furniture, weathering, wood strength,

1 Introduction

Wood has been benefited since human beings. Because of versality and easy of workability, it has been utilized for many industrial and domestic applications including as; construction element, energy source, raw material for engineering technologies (pulp and paper, furniture, composite) and for some household purposes [1,2]. It is often used a general term that '*people need wood from born (as cradle) to death (as coffin)*'.

Wood used in the landscape practices could be found in nearly all cultures throughout history. Even in very early times, it was used in the construction of some garden structures (i.e., pergolas, arbors, boxes, fences) [3]. It was also used extensively in the places of worship to reflect naturalness. The oldest known wooden worship structure is the five-story *Horyu-Ji temple*, which was built in 670 A.D. Although this building had been destroyed many times throughout history, it has been rebuilt in accordance with the original and has survived in its original place up to date [3].

Its wide range of specialty with aesthetic feature, wood is well suited to apply in many landscape practice implementations [3–10]. In many cultures, the wooden materials are widely used in open space arrangements in villages, cities and metropolitan areas. Therefore, the use of wooden materials in architectural applications has an important issue that societies have placed the features with reflecting their own cultural heritage [3,4].

However, wood is a unique material that takes people back to the nature. Because it is a lightweight and easy to joint, which is one of the most versatile design materials in modern architectural applications [2, 4, 7–10]. Hence, wood in constructional approaches could be a wonderful source of ideas that will inspire people to explore the many exciting possibilities to enhance landscape designs. In this context, wood in the urban design is a valuable resource for architects, landscape designers, and engineers who provide site-planning services [3–5].

Landscape architects could be combined wood-based materials with technical response and creative natural environment to develop considered planning process and delight end user. However, as designers within an environmental and engineering consultancy, it is able to draw on a wealth of environment within knowledge to develop holistic, cross-disciplinary solutions for complex sites [3–7]. But some basic considerations should be done for selection and use of wood-based material for landscape practices [4]. Those are typically:

- whether or not the wood-based material will make soil-contact
- whether or not it exposed to continuously cycles of wet and dry conditions,
- whether or not it is wished to preserve the natural appearance,
- whether or not it is wished to change the appearance

As a result of technological developments and engineering improvements, many of new wood-based materials produced and has become to utilize in construction and design applications [2,11]. Therefore, engineeringly design, wood-based materials (i.e., composites), could be manufactured by adding different shapes, strengths and forms to nature and landscape environment [2].

The objective of this study is to understand the suitability of wood-based materials for urban design applications (e.g., urban furniture). Hence the general properties and specifications of wood material has been presented. The use of wood-based material in landscape applications for general properties are discussed herein.

2 General Properties of Wood

Wood is light and flexible material when compared to steel, concrete or other common construction materials [2]. However, it has a very interesting feature, exhibits different properties in different directions (anisotropic). It is a natural biological material whose properties vary not only from one species to another but within the same species [1–3,12]. Therefore, the properties are influences on end use and durability. This is especially important when wood-based elements are used decoratively open spaces and natural environment (outdoor). In this respect, some of the important factors affecting the useability of wood can be given as;

density, hygroscopicity/moisture content, natural defects, strength, durability and insulation [9, 12].

Hygroscopicity/moisture content: wood is naturally a hygroscopic material. However, it evaporates or absorb moisture until its moisture content is in equilibrium with the surrounding air. This is called the *equilibrium moisture content* (EMC) while it is a function of temperature and relative humidity [12]. It is a general rule that the higher the moisture content, the lower the strength of the wood. The reason for this is the lack of load resistance capacity of water [2,9,12]. Moreover, the moisture ultimately effects volume differences, which causes the wood to creep, shrink and swell. Therefore, the excess water in the cell wall reduces the volume of the structure. But many mechanical properties are affected by changes in moisture content below the fiber saturation point [2]. In addition, the excessive water in the pores causes the growth of fungi and insects that cause the destruction of the wood texture, which indirectly reduces the strength and quality [12,13].

Density: the density of a material is the amount of mass per unit volume under certain conditions. In the case of wood as hygroscopic material, density primarily depends on two factors: the weight of the cell wall constituents and the moisture retained in it. However, the higher the density value, the higher its strength may be [9,12]. The reason for this is that the pore walls with high intrinsic mass value are thicker, which provides more constituents per unit volume.

Natural defects: because wood is a biological and natural material, it could be carrying some natural growing defects such as; reaction wood, spiral fiber orientations, knots, crevices, so on. These could be caused a reduction in the strength as well as variations in many end use properties [2,12,13].

Strength: wood obtained from various type of trees is extremely resistant to high pressure load. However, wood strength expresses the resistance against the forces acting on the structure and causing to break. Due to anisotropy, wood is usually show higher resistant to forces parallel to the fiber directions than forces perpendicular to the fiber directions [2]. Hence, the tensile strength of some wood species against the forces parallel to the fibers varies between 500 and 2000 kg/cm^2, while this value is measured between 10 and 100 kg/cm^2 for perpendicular forces when considering the same type of wood. Moreover, the bending strength is very high (300 to 900 kg/cm^2) and it is frequently used in beams for construction applications [2,9].

Durability: one of the most important features of wood is that it can last for a long time if suitable protected. Archaeologists have found many wooden objects or the remains of some large wooden buildings, hundreds of years old [2,12,13]. However, like other natural materials, it is subject to natural degradation processes known as decay, which refers to fungi, termites or insects destroying it over time by eating cell wall constituents (carbohydrates and lignin). Moreover, as a result of atmospheric conditions (i.e., rain, snow, sun light, heat, etc.), it caused the formation of crevices, color changes and cracks resulting destroying of surfaces [9,13].

Insulation: wood is a relatively good heat and sound insulator because of the tiny air pockets within its cellular structure. However, dry wood catches heat easily and converts sound energy into heat by frictional and viscous resistance in the cells to make a good acoustical material. But if the wood is exposed to high temperature for a long time, the strength value decreases due to the deterioration of the cell wall and the decrease in weight [2,12,13].

3 Factors Effects on Wood Properties

There are many factors could be effects on wood durability. But two of the most important factors that are effective on wood and wood-based materials in external used conditions without any treatment are; moisture changes and sun rays (UV radiation) [12–16]. Since wood is a hygroscopic material, it is inevitable for dimensional changes. This effects simply by increasing fine cracks, roughness color changes and make better target for microorganisms. However, especially short wavelength solar rays (<340 nm) can react with the chemical compounds (cellulose, hemicellulose, lignin) [14–17]. In this regard, photochemical reactions start within cell wall constituents. Especially, damage of the lignin, which adheres/holds the cells together in the matrix structure, causes the structure of the carbohydrate enriched cells to loosen, resulting in abrasion or degradation on the surfaces [14–16]. Moreover, in photochemical reactions, it can be easily understood by the occurrence of natural color changes or color irregularities [12,13]. In order to prevent these important changes that occur as simple but considerably impact on its performance, wooden materials should be prevented from contact of the sun's rays and air humidity. This is achieved simply by treating with light and moisture resistant pigment and/or hydrophobic binder additives [1–2,12–13].

3.1 Outdoor Conditions Effects on Wood Material

When wood used in the natural environmental conditions without any protective actions, significant changes in dimensions and structure may occur because of humidity, significant deformations (i.e. distortion, bending, warping) may occur along with cracks on the surfaces [2,9]. In general, these defects may occur more in wood material with high density than in low density species. Because the higher the density of the wood, the higher the size change rate [2,12]. However, one of the most stable condition for the wooden material is the moisture condition of 12 % and below. This is because wood material does not change significantly below this humidity level [2–4, 12–17].

However, wood is also regarded as a source of food and attacked by living microorganisms [13]. As a result of the microorganism's attack, biological aspects may be weakened and differentiation caused by living beings settled on the wood structure (rot, mold, fungi, insects etc.) [2, 13].

Fungi and insects typically consume chemical nutrients incidentally to unprotected wood materials that have favorable conditions (nutrient, air, moisture and temperature). Microscopically, changes in the structure were first identified as the growth of edged passages and subsequent microcracks in the cell walls [2,13]. The degraded cells easily separate from the middle lamella and undergo significant degradation (collapse) in the secondary wall. Moreover, the microcracks occurring in the structure, grow with the shrinkage of the cell walls, and the development of these cracks is facilitated by the plasticizing and washing effects of water [2,13]. Fig. 1 show decay development on wood materials with fungi [2,3,16,18].

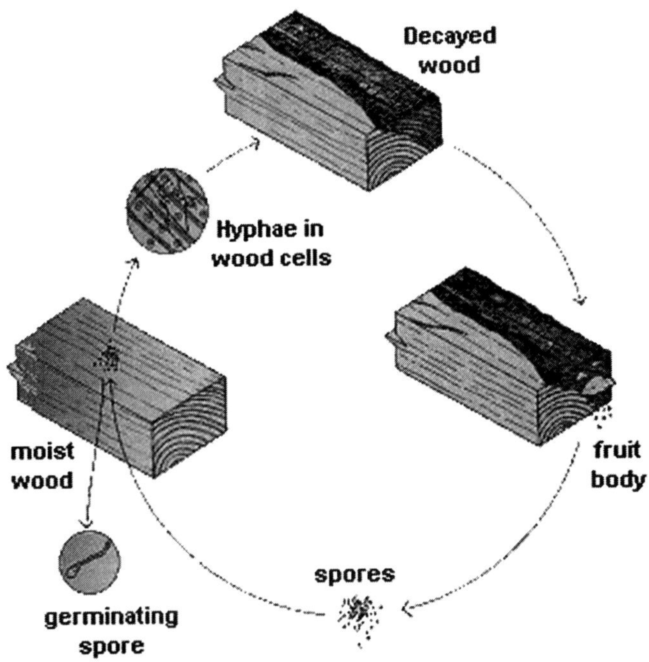

Fig. 1. Decay development on wood materials

As a result of the weakening in the cell structure, the degree of intercellular spaces and openings increase and loss of adhesion between different layers occurs. It has explained that full surface degradation of the wood material which used in outdoor conditions without any protective treatment, loss of some tissues could be completed in the first 6 months [2,12–13]. However, it has also summarized that lignin and followed by cellulose microfibrils has the highest resistance while the hemicellulose is the lowest in terms of fungi attack [2, 11–17]. Tab. 1 show micro-organism effects on wood material and their specifications.

Tab. 1. Microorganism effect on wood material

Bacteria	• It is usually called parasite if in a living tree or saprophyte if in a death tree (wood). • Bacteria first destroy cell storage substances and core rays and then parenchymal cells. • Bacterial destruction may occur as a result of use in wet and humid environment of wooden materials. • They produce neoplasm and lump on tree trunks and dark spots on wooden surfaces. • Their effects are not as high as insects and fungus • They are generally show abrasive effect (erosion) on surfaces.
Fungus	• Since they are simple creatures that cannot perform photosynthesis, produce their own nutrients and feed. • Fungi multiply by spore germination. The spores form molds (small spinner/spindle structure). • Hypeas provides nutrients by creating a chemical reaction with the enzymes. • Fungus need proper nutrients (cell wall material), adequate temperature (20–50 °C), appropriate humidity (12–36 %) and air (oxygen) to cause decay. • They are divided into two general classes, which cause color change and impart decay. • Those that cause color change create aesthetic problems. • Decayers change the physical and chemical properties as they destroy the cell components. • Since low nutrient level and some toxic substances are stored in heartwood, sapwood is less resistant to fungus damage than heartwood. • As a result of fungal destruction, physical (weight, strength, permeability, hardness) and chemical (chemical constituents) properties changed.
Insects	• Insects are malfunctioning both trees and wood to shelter, feed or lay eggs. • Insects make their lives in two different stages, larvae and adolescence. • In order for insects to develop and to destroy wood material; they need nutrients, suitable humidity (20–40 %) and temperature (<55 ° C). • Since low nutrient level and some toxic substances are stored in heartwood, sapwood is less resistant to insect damage than heartwood.
Termites	• Termites are also a kind of insects that live in colonies and form long galleries/cavities in the internal structures of wood material. • They can affect for both living trees and wood material. • They are mostly consumed cellulose. • Since their creations are not noticeable at the beginning, they make galleries and gaps in the interior part of wood-based materials.

4 Protection of Wood

In urban design with wooden materials, it could be provided relevant technical aspects of working, including the physicochemical properties as well as finishes, fasteners, and adhesives [2,3,7]. In this sense, the design methodologies used to build a range of common structures, from fences and decking to gazebos and bridges, especially important when it comes to outdoor settings can adversely affect durability.

Because of easy and abundant supply in the early days, wood was used directly in its natural state and without any protection process. But in these days, it has become a valuable and expensive material that needs to be preserved due to the continuous increase in its price and to maintain its aesthetic and technical value for a long time. Many tree species were last assessed more than a decade ago, so it allows many woods that are no longer plentiful.

Paint and varnish applications, which form a film layer on the surface, are commonly applied in order to prevent complex changes caused by outdoor conditions. The dyes provide a more effective protection against sunlight. However, they could crack, swell, and peel over time. In this case, the surface should be cleaned and painted again, otherwise, suitable environments for fungi will occur [1–2,7, 12–13].

Varnish applications have a shorter life span than paints. Oils, water repellents, varnishes and covering paints, do not show the effective preservations against cracking and peeling rather than deep penetrating techniques (impregnations) [1–3,12]. However, there are many technical approaches for the protection of wood material against abiotic and biotic factors. These operations are carried out according to a certain order and purpose. In general, the application properties of the materials used for the protection can be examined under two main groups. These are [1,2];

a) **Form a film, a layer or coating on the wood surface:** It includes paints and varnishes.
b) **Penetrating chemicals into wood structure:** Oils, preservatives, and water repellents.

Tab. 2 show a general classification of woods, considering the natural durability. But these is not for lasting and only be durable for less than decade. Detailed information on the protection of wood material and its natural durability properties can be obtained from other sources [1–2, 18].

Tab. 2. Natural durability characteristics of some tree species

High durability	Medium durability	Low durability
Cypress (Cupressus spp.)	Douglas fir (Pseudotsuga menziesii)	Redwood (Alder spp.)
Catalpa (Catalpa spp.)	Larch (Larix spp.)	Aspen (Populus spp.)
Chestnut (Castanea spp.)	Pine (Pinus spp.)	Linden (Tilia spp.)
Walnut (Juglans spp.)	Wattles (Acacia spp.)	Maple (Acer spp.)
White oak (Quercus alba)	Juniper (Juniperus spp.)	Spruce (Picea spp.)
Black cherry (Prunus avium)	Birch (Betula spp.)	Plane (Platanus spp.)

Solid wood material can only be used for a long time at outdoor atmospheric conditions with an effective impregnation process. However, the effect of the impregnation process cannot be the same in every wood [2,13]. For this reason, the type of wood to be impregnated and the place of use of the impregnated material are effective factors for choosing the preservative type. The preservatives make enable wood material to be used in external atmospheric conditions for a long time. These impregnation agents could be classified in three main groups. These are [2,12–13].;

1. Oily impregnating agents
2. Organic solvent-based preservatives
3. Water soluble preservatives

Tab. 3 summarized wood preservatives and their specifications.

Tab. 3. Wood impregnation agents

Oily impregnating agents	These agents are recommended when there is little chance of skin contact and the wood must last a long time. Creosote, Carbolineum, tar and tar oil, petroleum products are some of these agents.
Organic solvent-based agents	They are obtained from petroleum distillation, dissolved in organic solvents. These preservatives consist of active chemicals, an insecticide and/or a fungicide, dissolved in an organic solvent. Since the organic solvent impregnants are insoluble in water, they can maintain their protection for a long time. Chlorinated hydrocarbons, Pentachlorphenol, tributyl-tin-oxide (TBTO), Naphthenates, Copper 8-quinolinolate, Benzalkoniumchlorid compounds, chlorinated hydrocarbons are some of these agents.
Water soluble impregnation agents	Since these agents are water soluble, the wood impregnated with these agents can be washed in contact with precipitation and soil. They are frequently preferred because of their well properties such as, cheap, easily accessible, usually odorless and do not affect the performance of the wood. CCA (Copper-Chrome-Arsenic) type impregnants, CCB (Copper-Chrome-Boron) type impregnants, Boron compounds, ACA (Ammonia copper arsenic), PAS (Pentachlorphenol/Ammonia/Solvent) type impregnants are some of these agents.

5 Use of Wood in Open Spaces and Residential Gardens

Wood in open spaces and urban areas usually makes a strong statement that it lends itself to its own appearance and design. However, families residing in houses create their gardens primarily for their own living comfort. A typical residential house gardens have usually uniform vegetative formation with herbal materials [4]. But people make use of wood materials together with garden herbal elements in a various type arrangement suitable for their eye tastes [4,7]. It is understood from many studies in the literature that people reflect their psychological situations to their gardens, and in many cases, it is possible to see the expression of their inner worlds in vegetal material and garden arrangement [3,7]. Wood has a special feeling and a compatible material in people due to its naturalness and aesthetic appearance. Even a dead, overturned tree can be used as an aesthetic and support element in outdoor spaces, parks and gardens, creating positive effects on aesthetics and comfort. Fig. 2 shows an example of a dead tree trunk as a sitting element in an urban park area, right next to the growing trees. In this sense, an old, dead tree trunk could be placed in harmony with nature in an open area that has adapted to a green environment, has its naturalness and aesthetics in the foreground.

Fig. 2. A sitting element from dead tree trunk placed in an urban park (Picture; Şahin, 2014)

Some examples of the ways of using wood material in some open areas and residential gardens are shown in Fig. 3. As can be realized from these examples that wood can be used as a shelter for plants at the edges of the gardens, a support for wrap around plants, a swing for vertical arrangement of plants, as a boundary-defining element at the garden edges, and as an integrator with nature [4]. Therefore, wood is a material that can be hot, easily shaped, processed and used in many ways. It can provide an aesthetic appearance and functionality that cannot be achieved with any other natural or synthetic material.

Wood in the landscape should be given information need to exploit wood's full potential in design work. However, the comprehensive guide to the properties, use, selection, and installation of wood should be carefully considered from the landscape architect's perspective. Fig. 4 show some common wood based constructional applications [2,12].

Fig. 3. A utilization approaches from wood material in garden and parks [4].

Fig. 4. Some common wood based constructional applications (A: Circular arch; B: Parabolic arch; C: Gothic arch; D: A-frame; E: Bowstring trust)

6 Conclusions

Architects are getting a handle on how to employ infrastructure together in a functional and artful way. Wood material can be obtained from self-renewable natural resources and can be easily used, processed, shaped and versatile in harmony with nature. Because wood is a special aesthetic material and has technological properties, it is utilized since ancient times. However, although many engineering materials have been developed, its usage areas and the need are constantly increasing. In this sense, the knowledge on abiotic and biotic effects on wood substrate are given in above sections. Thus, could help architectures to urban designers to gain the understanding and to make meaningful contributions to this field.

However, landscape architects are design sites with collaboration architectures and civil engineers in the green area management. These resulting sustainable site design has been really successful. Over the past few decades, extensive progress has been made towards achieving sustainable landscapes. There are many wood lumber alternatives now on the market. These are also very durable product, as; thermally modified-, acetylated-, or polymerized wood products. These are based on materials to modify the sugars of the wood, so that decay organisms no longer recognize them as food.

References

1. Bozkurt, Y., Göker, Y., and Erdin, N. *Emprenye Tekniği*, (Turkish language), 1993. İstanbul Üniversitesi Orman Fakültesi, Yayın No:3779/425, İstanbul.
2. Bowyer J.L., Shmulsky, R and Haygreen J.G. *Forest products and wood science-An introduction.* Fourth edition, Iowa State University, Ames, IA. 2003: 553p.
3. Winterbottom, D.M. *Wood in the Landscape*, A Practical Guide to Specification and Design,. John & Wiley Sons, 2000. New York, USA.
4. De Cristoforo, R.J. *Wood Projects for the Garden*, Ortho Books, Chevron Chemical Company, 1976. pp.96. San Francisco, CA,USA.
5. Sahin C.K and Onay, B. *Alternative wood species for playgrounds: wood from fruit trees*, Wood Research, 2020. 65(1):149–160.
6. Sahin C.K., Topay, M and Var, A.A. *A study on suitability of some wood species for landscape applications: surface color, hardness and roughness changes at outdoor conditions*, Wood Research, 2020: 65(3):395–404.
7. Topay M., and Şahin C. Dış *mekânın etkin kullanımında ahşap malzeme seçimi* (Turkish language), Plant Peyzaj ve Süs Bitkiciliği Dergisi. 2016: 18:154–158.
8. Aslan, M. *Wood Use for Urban Furniture From Point of Ecological Design*, MSc. Thesis (Turkish, Abstract in English), Istanbul University-Cerrahpasa, Institute of Graduate Studies, Department of Landscape Architecture, (Turkish, Abstract in English), 2018. 341p. İstanbul
9. Aykut, F. *Wooden Material Consumption in Outdoor Urban Fitting Elements: Bartın Manucipality Park Case Study*, Zonguldak Karaelmas University, Graduate School of Natural and Applied Sciences, Department of Landscape Architecture, (Turkish, Abstract in English), 1997. 117p. Zonguldak.

10. Çetinkata Karafaki, F. *Usage of Wooden Materials in Urban Landscape Design*, Ankara University, Graduate School of Natural and Applied Sciences, Department of Landscape Architecture (Turkish, Abstract in English), 2009. 108p.Ankara.
11. Sahin, H.T. *Athmospheric Conditions Effects on Wood*, In: Recent Research Advances in Biology Vol. 3, Ch. 13; pp.132–139. M. Ramalho de Lima, (Editor), Book Publisher International, 2020. London, UK.
12. FPL. *Wood Handbook Wood as an Engineering Material*, Department of Agriculture, Forest Service, Forest Products Laboratory, General Technical Report, FPL–GTR–190, 508p.Madison,
13. Zabel R. A and Morrell J. J. *Wood microbiology: decay and its prevention.* Academic press.2020: 576p.
14. Hon D. N. S., Chang S. T and Feist W. C. *Protection of wood surfaces against photooxidation.* Journal of Applied Polymer Science, 1985. *30*(4): 1429–1448.
15. Hon N-S. D. *Photochemistry of Wood*, in: Wood and Cellulosic Chemistry, D. Hon and N. Shiraishi (Eds), 1991. Markel Dekker. Inc. NY,
16. Feist W. C and Hon D. N. S. *Chemistry of weathering and protection*, Ch 11. in: The chemistry of solid wood, Roger M. Rowell (ed), ACS Publ. Washington D.C. 1984.
17. Cassens D. L. and Feist W. C. *Exterior Wood in the South: Selection.* USDA FPL-GTR-69. Madison, WI, 1991: 60p.
18. Kubler, H. *Wood as Building and Hobby Material*, John wiley & Sons, 1980. New York, USA

Gökçen BAYRAK

How to Integrate Green and Grey Infrustructure

Abstract: Sustainable stormwater management approaches, which have a significant place in integrated water management, are getting widespread throughout different geographies and countries in the world. As an application method of these approaches, the integration of green infrastructures with the grey infrastructure has certain processes to follow. This study examines various sustainable urban water management approaches adopted around the world. In addition, it puts forward the ways to be followed in the integration of green infrastructure to the existing and future grey infrastructure, the difficulties encountered during its implementation and the issues that need to be attended. Although there are applications of different scales, in order to perform all applications efficiently, it is necessary to adopt green infrastructure-based approaches, to establish legislative provisions and standards for implementation, to draw up guidelines for the dissemination of practices, and to finance the practices through various support programs. It is anticipated that the widespread use of green infrastructures and the integration of green infrastructure with grey infrastructure will only be achieved with the integration of engineering, planning and design disciplines and the participation of decision makers, financial providers and citizens.

Keywords: Sustainable urban water management, stormwater management, green infrastructure.

1 Introduction

55 % of the population in the world live in cities. The United Nations predict that this rate will be 68 % by 2050 [1, 2]. "Water", which is the main subject that connects the Sustainable Development Goals of the United Nations, will be a critical determinant in achieving these goals [3]. However, rapid urban growth, climate change and economic development pose a challenge for water resources to deliver clean water into cities and to provide a safe environment. Traditional water management approaches prove to be no longer sufficient to meet the diverse and complex needs of cities.

Grey infrastructure is a traditional municipal infrastructure dominated by single-function municipal engineering consisting of roads, bridges, railways, sewer systems, pipelines and other networks that keep the industrial economy running smoothly. In cities adopting traditional stormwater management, stormwater is rapidly removed from the environment with grey infrastructure components (separated or combined sewage system, sidewalks, gutters, grids, ponds and vaults, etc.). The main function of these highly standardized infrastructures is the transfer, treatment and discharge of pollutants, but they cannot solve the basic pollution problem, and the construction cost is quite high [4–7]. Neglecting the combination of infrastructure and urban open green spaces together with the social, aesthetic and ecological functions which they should have seriously affect the overall contribution of the infrastructure to the city [7, 8]. It is necessary to eliminate this problem and to integrate grey and green infrastructures for many environmental, economic and social benefits, namely sustainable urban water management [5].

Unlike grey infrastructure, green infrastructure (GI) provides some vital services such as clean air, drinking water, and food. GI is an evolving ecological planning and design concept based on the idea of establishing a link between green spaces and built infrastructure, and it can be defined as an approach relying on particularly hydrological networks that are dwindling in number but important to fulfill ecological functions. It is an integrated system that covers the parts of nature that are valuable in terms of ecosystem features (ecological corridors, core areas, etc.) and that protects, improves and supports plant and animal habitats [4, 9, 10].

Over the past three decades, urban stormwater management has focused on drainage and flood control as well as non-point pollution reduction. However, due to the high costs of providing water and energy, many studies are carried out to develop GI systems and to spread the applications. In the future, it is anticipated that water supply, stormwater and sewage systems will be managed together in a closed cycle rather than separately by increasing the hydrological and ecological functionality of urban landscapes [9].

2 Approaches in Sustainable Stormwater Management

Alternative approaches to traditional stormwater management are developed in order to improve water quality and quantity, reduce flood effects, protect public health and cope with water scarcity in the world. Such approaches have been conceptualized as hybrid socio-technical engineering systems where grey infrastructure and green infrastructure are integrated. Drawing upon solely technical approaches is not sufficient; on the contrary, sustainable solutions should be developed technically, institutionally, financially, socially and environmentally sensitive concurrently with an integrated approach. With this approach, it is argued that calls for integrated system approaches involving large engineering projects should include the integration of more transparent and open design, alternative designations and decision processes besides the integration of social factors and effects [2, 11].

Examples to such approaches in the world are Low Impact Development (LID) in the United States of America (USA), Sustainable Urban Drainage Systems (SuDS) in the United Kingdom (UK), Water Sensitive Urban Design (WSUD) in Australia, Sponge Cities in China, Alternative Techniques in France and LID and Vancouver Rain City Strategy in Canada. The common feature of these approaches is to ensure the sustainable management of stormwater with less energy and low costs by using green areas.

2.1 Best Management Practices (BMPs)

Best Management Practices (BMPs) are a common approach, especially in the USA, and are based on control means for the quantity and quality of stormwater runoff. In the USA, each state has its own guidebook containing information and observations regarding the environment, especially on climate, soil type and urban development characteristics. Minimum standards are set to help local governments comply with legal requirements imposed by national and federal programs [12].

2.2 Low Impact Development (LID)

Since the 1990s, LID has been implemented with BMPs in the USA. Water quality standards were determined with the 1972 Clean Water Act, which was updated in 2001. While BMPs were designed to control contaminants from surface runoff, the LID approach was developed to reduce the flow volume. The basic principles of the LID approach are to preserve current natural hydrological functions. LID is also used in Canada. This stormwater management strategy aims to reduce the impacts of increased runoff and stormwater pollution by managing surface runoff as close to its source as possible [12–16].

2.3 Water Sensitive Urban Design (WSUD)

WSUD is an approach developed in Australia in the 1990s and later adopted in the UK and New Zealand to conserve water resources. In this approach, it is aimed to minimize the hydrological effects of urban development on the environment by means of urban planning and design [17]. With the WSUD, environmental protection strategies have been determined to ensure water supply, sanitation, flood protection along with long-term sustainability, habitability, durability and prosperity [2]. In addition, long-term planning and localized strategies require continuous monitoring, evaluation and flexible institutional arrangements [12].

2.4 Sustainable Drainage System (SuDS)

Developments in stormwater management approach began in the UK in the late 1980s, and, in 1992, the "Scope for Control of Urban Runoff" guidelines were published, providing guidance on a range of technical control options. In the 1990s, the adoption of storm water management in Scotland was realized faster than in England and Wales. In 2000, a significant set of guidelines was published, and the term "Sustainable Urban Drainage Systems (SuDS)" was formalized. In UK practice, SuDS consist of a set of technologies and techniques used to drain stormwater more sustainably than traditional solutions. SuDS focus on water quality, water quantity and amenity/biodiversity issues, called the SuDS triangle [12, 17]. The Sewerage Sector Guidance contains new rules to be applied to all water and sewage companies in the UK as of April 1, 2020. Non- statutory technical standards for the UK have been produced by DEFRA (Department for Environment Food & Rural Affairs) and are currently under review (2020). The Flood and Water Management Act 2010, which came into force on 7 January 2019 in Wales, requires developments in Welsh national standards for SuDS features [18].

2.5 European Union (EU) Approach

GI makes a significant contribution to the implementation of most of the EU's basic policy objectives, especially regional and rural development, climate change, disaster risk management, agriculture/forestry and environment. The EU Strategy as to GI determines how to deploy a network with consistency and connectivity, and encourages action at all levels. The development of GI is regarded as an important step for the success of the "EU 2020 Biodiversity Strategy" [19]. On May 6, 2013, the European Commission adopted the "Green Infrastructure (GI)-Enhancing Europe's Natural Capital strategy", which encourages GI investments, improves the health of ecosystems, ensures that natural areas remain connected and allows species to thrive in all natural habitats, thus ensuring the continuity of nature. The Strategy promotes the deployment of GI throughout Europe as well as the development of a "Trans-European Green Infrastructure Network" in Europe, called TEN-G, equivalent to existing networks for transport, energy and communication. Similarly, the reforms of "Common Agricultural Policy" identify a number of important elements that will help improve GI in the wider rural area [20]. Member States support GI through programs integrated into their development strategies. In addition, the "Natura 2000" network makes up the central part of the European green infrastructure and provides a legal and organizational framework that can contribute to the long-term safety, efficiency and cost effectiveness of GI investments [21].

2.6 Sponge Cities

Frequent floods and spreading pollutant sources caused by the rapid progress of global climate change and urbanization threaten the sustainable development of modern cities seriously. Since 2014, sponge city practices, which were initiated in China to alleviate these problems, have been getting widespread. Sponge city is a city that can absorb stormwater like a sponge, and can filter, store, treat and discharge it for later use. Compared to traditional urban stormwater management, sponge city relies on natural-based solutions and increases the longevity of stormwater systems. Sponge city is a typical green-grey-blue system with GI at the source, grey infrastructure in the middle and receiving water bodies (blue system) downstream [7, 22, 23].

3 Green Infrastructures

Green infrastructure (GI) is a concept combined with settlements, ecological conservation and green technologies. The first prototypes of the GIs date back to the advent of urban parks in the 1950s. Then it was practiced in the early stages of the environmental protection movement from the 1960s to the 1990s [10]. Since the 1990s, a rapid development phase has been experienced in coordination with development and expansion in many areas. Land conservation, smart growth, green road, LID and river restoration have led to a clear conceptual consensus on the GI, and related research and applications have been developed rapidly and comprehensively. The development of GI is a result of the joint promotion of parks, park systems, open spaces, greenways, ecological networks, biological corridors, and stormwater management [4, 7, 10].

GI is a term used by the United States Environmental Protection Agency (USEPA) since 1990s. Section 502 of the Clean Water Act defines green infrastructure as "... the range of measures that use plant or soil systems, permeable pavement or other permeable surfaces or substrates, stormwater harvest and reuse, or landscaping to store, infiltrate, or evapotranspirate stormwater and reduce flows to sewer systems or to surface waters" [10]. GI is described in a variety of ways in the US stormwater management literature as "a network of decentralized stormwater management practices, such as green roofs, trees, rain gardens and permeable pavement, that can capture and infiltrate rain where it falls, thus reducing stormwater runoff and improving the health of surrounding waterways" [17].

GI have become a significant tool in ensuring the sustainability of planning especially in urban areas [24–26] and have focused on the integrated management of natural and cultural vegetation and surface waters. The concept of GI is defined at different scales and deals with disciplines related to urban design, nature conservation and planning. It is defined as a "multi-functional open area network" at the city and regional scale, and as "a stormwater management approach" at the local and field scale. GI mimics natural hydrological processes [27–29]. GI is a network of natural and semi-natural areas, rural and urban open green spaces and terrestrial, freshwater, coastal and marine areas which improve ecosystem health and resilience, contribute to the conservation of biodiversity and benefit human communities [6]. It can cover various application areas such as large parks, green spaces, agricultural lands, green roofs, trees, rain gardens, flower beds, artificial wetlands as well as floodplains, wetlands and forests, which are a significant component of stormwater infrastructure [10, 28].

GI techniques for sustainable stormwater management include the following designs [6, 13, 30–34]: Rain gardens and bio-retention areas, green roofs and green walls, disconnecting of rain gutter from wastewater channel, rainwater harvest and reuse systems, natural rainwater drainage systems (biological swales), flower beds, rainwater retention basins, ponds, xeriscaping, permeable pavements (permeable asphalt, concrete and pavement systems), green streets, roads and parking lots.

Furthermore, city parks, cemeteries, public parks and gardens, river and stream corridors, bike paths, main transportation roads (roads, railways and tramlines), golf courses, school/corporate playgrounds, sports and entertainment facilities, private/semi-private gardens, block of flats, backyards, balconies, national parks and nature reserves, wetlands and coasts, quarries, airports, vineyards, vegetable gardens, orchards and farms contribute to a city's green infrastructure network [5, 6].

Green infrastructures create mutually reinforcing effects between water, food, transport, energy, health and integrated social-ecological-technical systems, and provide many benefits [4–6, 11, 13, 15, 30, 31, 35–40]:

1. *Water quality and quantity control*
   - Stormwater management,
   - Reducing flood risks and effects,
   - Reducing grey infrastructure construction, operation and maintenance costs,
   - Improving river water quality by reducing the load on the combined sewage system,
   - Reducing the amount of water and pollution load in the sewage system,
   - Reducing treatment costs,
   - Water efficiency through rainwater harvesting,
2. *Reducing air pollution*
   - Improving air quality by removing pollution,
   - Carbon capture/storage and reduced greenhouse gas generation,
   - Odor and noise control,
3. *Ensuring efficient use of energy*
   - Reducing the urban heat island effect,
   - Reducing the load on traffic and using public transportation vehicles,
   - Energy saving and energy efficiency in buildings,
4. *Providing ecosystem services*
   - Increase in habitats for different species,
   - Increase in biological diversity,
5. *Establishing social-ecological link*
   - Urban aesthetics and neighborhood quality perception,
   - Recreation opportunities and social cohesion,
   - Urban agriculture,
   - Environmental justice (reduction of problems compared to grey infrastructure projects),
   - Education opportunities,
   - Stress management,
   - Increase in socio-economic benefit-cost ratio (compared to grey infrastructure projects).

4 Grey and Green Infrastructure Integration Process

The positive results of GI applications that have been in use for the last thirty years in the world have now been proven. The reasons for the integration of green infrastructure with grey infrastructure can be listed as economic outputs, the reduction of negative environmental impacts, the multi-functionality of green infrastructure and its contribution to the fulfillment of policies that try to improve some negative effects caused by increasing population, resource consumption and economic growth [5, 41].

In different countries of the world, some difficulties and issues worth noting arise during its applications, which are carried out with different names but with similar purposes. These are [11];

- Design standards,
- Regulatory ways,
- Socio-economic factors,
- Fundability,
- Determination of tangible and intangible economic benefits and evaluation of benefit/cost,
- Public/private sector participation,
- Innovation,
- Integrating water system planning with urban planning.

The goal of urban development plans and strategies should be the protection and management of existing GI, improvement of its functionality, quality and connectivity and creation of new GI. The issues that need to be addressed in GI planning, which is not a strict land use plan but a flexible arrangement, are as follows [2, 5, 7, 23, 26, 41];

- Planning scale and stages of GI,
- Local contexts,
- Connections between human and nature, not just green items,
- Interdisciplinary perspective towards urban water management,
- Determination of partnerships at an appropriate scale in line with relevant environmental and socio-economic targets,
- Maintenance and improvement of biodiversity,
- Integrity of areas of international importance,
- Access of disadvantaged groups and communities to open spaces,
- Mitigation of the effects of climate change and adaptation,
- The necessity of applying GI to urban systems such as energy, transportation, waste, food production and built environment with integrated approaches.

Planning the integration reduces investment and implementation costs and grey infrastructure applications [2, 26]. Depending on the land use, landowners and other variables, some sewer basins may be more suitable for GI applications. Therefore, integration should be considered in the whole sewer basin. Estimating the maximum or optimum amount of GI that can be implemented in a sewer basin requires analysis of institutional support to facilitate implementation as well as technical/environmental factors such as land use and soil types and topography. The determined implementation target may require many decentralized GI implementations [26]. In all scenarios, not only flow rate and quality control but also local conditions for groundwater recharge and mitigation of urban heat island effects should be considered. Based on regional data, it is necessary to standardize the regional design parameters, building material characteristics, and maintenance plan and schedule, which will significantly affect the infiltration and filtration processes of the facility [23, 26].

Municipal properties in cities, parcels that may be suitable for a right of way and GI applications (institutional campuses, school campuses and empty parcels without near-term demand, etc.) should be determined. It is an advantage that the city or sewerage administration is in charge of supervision for proper construction, conservation and maintenance of applications. Monitoring the performance of the GI application is important to ensure that prescribed storage and control levels are achieved [23, 26].

The methodology to be followed in the integration of GI into grey infrastructure has been arranged considering various application examples and is presented below [2, 5, 23, 26, 42–45];

1. A sample sewage basin is chosen that represents the service area as a whole in terms of land use, land ownership, soil and topography.
2. Considering the connectivity of green areas in sub-basins, existing ecological quality, pollutant levels, planning constraints, access characteristics, cultural heritage characteristics, settlement characteristics, property and flooding possibilities, the land use/land cover is determined by use of aerial photographs and/or geographic information system (GIS).
3. Templates are created for various land uses in sewer basins (e.g. typical detached residential area, typical commercial/office space), and permeable and impermeable areas are estimated for the templates.
4. GI options are determined for different templates, considering area needs, soil types and slopes.
5. From the templates, the total GI area is estimated that can be applied to the sewer basin as a whole. In this estimation, the attitude of private owners towards GI practices, current and future zoning, institutional issues, public property and parks should also be taken into consideration.
6. A number of GI scenarios are developed.
7. Hydrological and hydraulic modeling is performed with various green and grey infrastructure combinations in accordance with the scenarios. The hydrological component of the model estimates what time and how stormwater will flow to the soil and groundwater, and will diffuse into the atmosphere, and how much it will flow into the sewer network. The hydraulic component of the model is used to simulate how flows will move through the sewer network.

8. The investment/operating costs and environmental benefits of each scenario are compared.
9. Inferring from the sample set of the sewer sub-basins examined, GI options are determined as a whole.
10. Needs and a regular maintenance program are determined to ensure proper operation and management of GI.
11. Methods are determined so as to monitor and report the performance/effects for 12 months immediately after the GI implementation.
12. Adaptive management/corrective actions are determined if GI performance does not meet the expectations.

5 Conclusion

Sustainable stormwater management approaches, which have an important place in integrated water management, are becoming widespread throughout the world. The common feature of these approaches is to ensure the sustainable management of rainwater with less energy and low costs by using green areas. In many countries, GI is an implementation tool of these approaches to achieve their targets. This study puts forward the ways to be followed in the integration of GI with the current and future grey infrastructure, the difficulties encountered in implementation and the issues that need to be considered.

Although applications can be implemented at different scales, there are common necessities to be considered in order to perform all applications efficiently as follows:

- Creating an approach that is not only technical, but also corporate, financial, social and environmentally sensitive,
- Establishing the legislative provisions and standards for implementation,
- Integration of engineering, planning and design disciplines,
- Establishing guidelines for the dissemination of practices,
- Funding applications with various support programs,
- Understanding the current situation of local stakeholders and setting clear targets,
- Identifying, understanding and estimating complex relationships between technological, ecological and social factors,
- Including decision makers, financial providers and citizens in the process.

As a result, regarded as one of the most important tools in tackling the obstacles in adapting to the problems brought about by climate change and in ensuring sustainability, green infrastructures can only be made widespread by adhering to a common vision by stakeholders, developing a common change model in line with long-term cooperation and strategic priorities. These stakeholders are central and local governments, water services, legislators, planners, engineers, designers, R&D organizations, academia and citizens.

References

1. UN, https://www.un.org/development/desa/en/news/population/2018-revision-of-world-urbanization-prospects.html, 09.12.2020, 2020b.
2. Wong, T.H.F., Rogers, B.C. and Brown, R.R., *Transforming Cities through Water-Sensitive Principles and Practices*, https://doi.org/10.1016/j.oneear.2020.09.012, 2020.
3. UN, *Global indicator framework for the sustainable development goals and targets of the 2030 agenda for sustainable development*, A/RES/71/313. https://unstats.un.org/sdgs/indicators/indicators-list/, 09.12.2020, 2020a.
4. Benedict, M.A. and McMahon, E.T., *Green infrastructure: smart conservation for the 21st century*, Renewable Resources Journal, 20(3): 2002, p.12–17.
5. Natural Economy Northwest (NENW), *Putting the green in the grey" Creating sustainable grey infrastucture A guide for developers, planners and Project managers*, 2009.
6. Ely, M., Pitman, S., *Green Infrastructure Life support for human habitats The compelling evidence for incorporating nature into urban environments*, Infrastructure Project Botanic Gardens of South Australia, Department of Environment, Water and Natural Resources, 2014.
7. Sun, Y., Deng, L., Pan, S.Y., Chiang, P.C., Sable, S.S. and Shah, K.J., *Integration of green and grey infrasturctures for sponge city: Water and energy nexus*, Water- Energy Nexus 3 (2020) 29- 40, 2020, p. 29–40.
8. Kim, E., Fiore, A. M. and Kim, H., *Fashion Trends Analysis and Forecasting*, London: Bloomsbury Academic, 2017.
9. Heaney, J.P. and Sansalone, J.J., *A Vision For Urban Stormwater Management in 2050*, 2012.
10. USEPA, www.epa.gov, 07.12.2020.
11. Staddon, C., Ward, S., De Vito, L., Zuniga-Teran, A., Gerlak, A.K., Schoeman, Y., Hart, A. and Booth, G., *Contributions of green infrastructure to enhancing urban resilience*, Environment Systems and Decisions (2018) 38:330–338, https://doi.org/10.1007/s10669-018-9702-92018, 2018, p.330–338.
12. Perini, K. and Sabbion, P., *Urban Sustainability and River Restoration: Green and Blue Infrastructure*, First Edition by John Wiley & Sons Ltd., 2017.
13. USEPA, *Low Impact Development (LID) A Literature Review*, United States Office of Water (4203) EPA-841-B-00-005, Washington, 2000, p. 2–30.
14. Econorthwest, *The Economics of Low-Impact Development: A Literature Review*, 2007, p.1–2.
15. USEPA, *Reducing Stormwater Costs through Low Impact Development (LID) Strategies and Practices*, December 2007, EPA 841-F-07-006, 2007, p.2–8.

16. Econorthwest, *Low Impact Development At The Local Level: Developers' Experiences and City and County Support,* February 25, 2009.
17. Fletcher, T.D., Shuster, W., Hunt, W.F., Ashley, R., Butler, D., Arthur, S., Trowsdale, S., Barraud, S., Semadeni-Davies, A., Bertrand-Krajewski, J.L., Mikkelsen, P.S., Rivard, G., Uhl, M., Dagenais, D. and Viklander, M., *SUDS, LID, BMPs, WSUD and more – The evolution and application of terminology surrounding urban drainage,* Urban Water Journal, 12:7, 525–542, DOI:10.1080/1573062X.2014.916314, 2015, p.525–542.
18. https://www.gov.uk/government/publications/sustainable-drainage-systems-non-statutory-technical-standards, 30.01.2021.
19-. EC (European Commission), Our Life Insurance, Our Natural Capital: An Eu Biodiversity Strategy To 2020, 2011.
20. EC, *Building a green infrastructure for Europe,* ISBN 978-92-79-33428-3 doi: 10.2779/54125 c European Union, 2013.
21. EC, https://ec.europa.eu/environment/nature/ecosystems/strategy/index_en.htm, 07.12.2020.
22. Leng, L., Mao, X., Jia, H., Xua, T., Chen, A.S., Yin, D. and Fu, g., *Performance assessment of coupled green-grey-blue systems for Sponge City construction,* Science of the Total Environment 728 (2020) 138608, https://doi.org/10.1016/j.scitotenv.2020.138608, 2020.
23. Yin, D., Chen, Y., Jia, H., Wang, Q., Chen, Z., Xu, C., Li, Q., Wang, W., Yang,. Y., Fu, G., Chen, A.S., *Sponge city practice in China: A review of construction, assessment, operational and maintenance,* Journal of Cleaner Production, Available online 5 November 2020, 124963, 2020.
24. Foster, J., Lowe, A. and Winkelman, S., *The Value of Green Infrastructure for Urban Climate Adaptation,* The Center for Clean Air Policy, February 2011, 2011, p. 4–28.
25. Melboune Water, http://www.melbournewater.com.au, 06.12.2020.
26. USEPA, *Combined sewer overflows,* Green Infrastructure Permitting and Enforcement series: Factsheet 2, 20.09.2020.
27. Scott, M., Lennon, M., Collier, M., Foley, K., Integrating Ecosystem Approaches, Green and Spatial Planning, EPA Research Programme 2014–2020, (2012-B-DS-5), EPA Research Report, Prepared for the Environmental Protection Agency by,School of Architecture, Planning and Environmental Policy, Earth Institute, University College Dublin, Ireland, 2012.
28. Fisch, J., *Green Infrastructure and the Sustainability Concept: A Case Study of the Greater New Orleans Urban Water Plan,* University of New Orleans, University of Florida, United States, (2014), p. 3.

29. Bayrak, G., Keleş, E., Ölmez, Z., *Investigation of Green Infrastructure Applicationability in Edirne City Center*, XIth International Sinan Symposium Proceedings Book, 11–12 April 2019, Trakya University Faculty of Architecture, Edirne, 2019. (In Turkish)
30. Kloss, C. and Calarusse, C., *Rooftops to Rivers Green Strategies for Controlling Stormwater and Combined Sewer Overflows*, Natural Resources Defense Council, 2006, p. 2–11.
31. Maryland Department of Environment (MDE), 2007, *Bioretention Manual*, Environmental Services Division Department of Environmental Resources, The Prince George's County, Maryland, Revised December 2007, p. 2–44,
32. Ecojustice, *Green Cıtıes, Great Lakes Using Green Infrastructure to Reduce Combined Sewer Overflows*, August 2008, p. 47–48
33. CNT (Center for Neighborhood Technology), *The Value of Green Infrastructure A Guide to Recognizing Its Economic, Environmental and Social Benefits*, 2010, p. 3.
34. Ertin, D.G., Bayrak Yılmaz, G., Zülfikar, C., *Rain Gardens of Green Infrastructure Applications in Sustainable Landscape Design: The Case of Edirne*, GreenAge Symposium, Mimar Sinan Fine Arts University, Faculty of Architecture 26–27 April 2012, Istanbul, Türkiye, 2012.
35. USEPA, *Storm Water Technology Fact Sheet Bioretention*, United States Environmental Protection Agency Office of Water Washington, D.C. EPA 832-F-99–012 September 1999, 1999, p. 2–6.
36. Dietz, M.E., *Low Impact Development Practices: A Review of Current Research and Recommendations for Future Directions*, Water Air Soil Pollut (2007) 186:351–363. 2007, p. 351–363.
37. Jacobs, S., *Green Infrastructure and Storm water management*, United States: US Environmental Protection Agency (EPA), 2014, p. 2–5.
38. Kim, S.Y. and Kim, B.H.S., *The Effect of Urban Green Infrastructure on Disaster Mitigation in Korea*, Sustainability, 1026; doi:10.3390/su9061026, www.mdpi.com/journal/sustainability, 2017.
39. Low Impact Development (LID) Center, www.lid-stormwater.net/bio_costs.htm, 11.12.2020.
40. Liu, B-W., Wang, M-H., Chen, T-L., Tseng, P-C., Sun, Y., Chiang, A. and Chiang, P-C., *Establishment and Implementation of Green Infrastructure Practice for Healthy Watershed Management: Challenges and Perspectives*, Water-Energy Nexus (2020), doi: https://doi.org/10.1016/j.wen.2020.05.003, 2020.
41. Browder, G., Ozment, S., Bescos, I.R., Gartner, T. and Lange, G.M., *Integratıng Green And Gray Creating Next Generation Infrastructure*, WORLDBANK, 2019, p.19–23.

42. USEPA, *Greening CSO Plans: Planning and Modeling Green Infrastructure for Combined Sewer Overflow (CSO) Control,* March 2014 Publication # 832-R-14-001, 2014.
43. Jersey Water Works, *Balancing Green and Gray Solutions to CSO Management Report on How to Balance Green and Gray Infrastructure in CSO Long Term Control Plans,* December 2018, p.9–12.
44. Skinner, R. and Satur, P., *Integrated Water Management: Principles and best practice for water utilities,* prepared for the Water Services Association of Australia, Monash Sustainable Development Institute, Monash University, Melbourne, 2020, p.39–40.
45. Tawfik, S. and Chesterfield, C. *Facilitating water sensitive urban development through planning integration — A discussion paper,* Melbourne, Australia: Cooperative Research Centre for Water Sensitive Cities., 2020, p.8–12.

M. Bihter BİNGÜL BULUT

Human Dimensions of Landscape Planning and Design

Abstract: Human dimensions research in landscape planning and design is a complex scientific knowledge that requires multidisciplinary perspectives and a wide range of applications and implications. This chapter examines the term human dimensions why is needed in landscape discipline and how social science approaches seek to fill this need due to landscape planning and design deals with the management of human and landscape interactions. Planners and designers need to evaluate this relationship and its consequences.

Keywords: human dimensions, social sciences, social aspects, landscape planning and design

1 Introduction

Landscape defined as 'an area, perceived by people, whose character is the result of the action and interaction of natural and/or human factors by The European Landscape Convention (ELC) (Council of Europe, 2000). All kinds of landscapes, natural, rural, urban etc. determines the quality of people's living environment. Landscape is a subject of interest not only for landscape architects, urban designers and planners but also other different disciplines such as natural sciences, social sciences, humanities, and arts (Tress, Tress, Decamps & Hauteserre, 2001). Each of these different disciplines share common ground 'human-landscape interactions'. People are inseparably linked with landscape and the actions and behaviors of people directly or indirectly effect the quantity and quality of the landscape. Landscape quality and availability likewise effects humans. Sustainability of these landscapes involves a deeper understanding of human and sharing the knowledge about it.

Understanding landscape contexts and audiences are essential in the landscape planning and design. Yet many researchers and practitioners do not have the necessary resources to effectively address the human dimensions of landscape planning and design. The term human dimensions in landscape planning and design refer to how and why humans value landscapes, how humans want landscapes managed, and how humans affect or are affected by landscape management decisions. Human dimensions research helps to understand effects of human actions and how to incorporate that knowledge into planning and design. It focuses on various issues and practices including physical, biological, cultural, social, and economic values; individual and social behavior; demographics; legal and institutional frameworks; communication and education; and decision-making processes of management (Decker, Brown & Siemer, 2001).

2 The Term 'Human Dimensions'

In the literature, there are several studies discuss directly the term 'human dimensions' in landscape planning and design. In the study of exploring research priorities in landscape architecture (Meijeringa, Tobi, Brink, Morris & Bruns, 2015), 'human dimensions of planning and design' (place attachment, landscape perception, human–environment interaction) was chosen as the most important domain for landscape architecture as a research discipline and the most useful domain in professional practice. In the study, some experts who evaluated the given domains from the academia perspective explained that this domain is needed to understand how people perceive and respond to landscapes and they highlighted "*planning and design are for people. So, changing attitudes and perceptions of the people towards the landscape in a changing world should be the subject of research.*" Some others stated that successful landscape planning and design meets people's needs therefore, "*social studies in landscape architecture are integral to planning and designing most constructed places that succeed in meeting user needs.*" From the professional perspective, some experts mentioned '*human dimensions provide landscape architects with knowledge about how people perceive, value, use, and are affected by landscapes'*.

Tress and her colleagues (2001) give recommendations for future landscape research in their study to bridge the gap between the human dimensions and natural sciences in landscape research. One of the recommendations was landscapes should be seen as holistic and dynamic systems. In this system, each aspect has equal importance and the relationship between human and landscape is mutual. Not only people influence the landscape, but landscape also influences people. This mutual relationship is the most significant factor between natural- and human-oriented sciences in landscape research.

Another study conducted by Matsuaka and Kaplan (2008) examines human dimensions and issues or human needs in urban settings with the argument that meeting with human needs in the urban settings is important because the world is becoming more urbanized. They categorized human needs into two groups; nature needs, and human interaction needs. Nature needs includes contacting with nature, aesthetic preference, and recreation and play. Human interaction needs includes the issues of social interaction, citizen participation in the design process, and community identity. They found that a very few studies dealt with topics related to human dimensions and outdoor environments and concluded a wide variety of spaces such as green open spaces, greenways, hospital grounds, parks and residential neighborhoods can meet the same needs and a particular setting can meet multiples needs. For instance, natural settings on the one hand promote social interactions on the other hand nature elements are also provide privacy in a space.

Gobster and Westphal (2004) examined human dimensions for an urban greenway. Six items related human dimensions were identified cleanliness, naturalness, aesthetics, safety, access, and appropriateness of development to understand how people perceive and use the greenway for recreation and other experiences. Their findings shows that cleanliness was the most ascendent dimension for Chicago Greenway and all the six dimensions can be used to inform greenway planners and managers.

Soini (2001) examines human dimensions through mapping techniques mental and concept mapping of different landscapes. She argues mental maps are associated with the behavioral approach and represents the human mind. In other words, it's a way of how people read the landscape. She claims that knowledge about human perception of landscape provides a better understanding for current and future land uses.

Ryan (2011) uses the term social landscape (social relations, perceptions, values and attitudes) to refer human dimensions and highlight the importance of social, perceptual and spatial information for landscape planning and needs mapping of this kind of information through GIS and suggests that social scientist and landscape planners work together for creating a spatial integration of social information.

Hunziker, Buchecker and Hartig (2007) analyze human-landscape relationship on two theories landscape as a space or as a place. As space, people use the landscape for their physical and biological needs, but as place, they see the landscape from the view of personal experiences, values, norms, social meanings. It is important to understand what kind of experience occurring with the landscape, recreational meanings or representation of social meanings.

3 Why Human Dimensions Research?

Social aspects of landscape planning and design has become increasingly important and it will get even more attention in the future.

Comprehensive understanding of landscapes requires to understand human beings because people play a crucial user role in this dynamic system and causes the changes of landscape (Nassauer, 1997). From the beginning of the history people have modified landscape through agricultural and industrial activities, urban development etc. (Harden et al. 2014). For sustainable development, knowledge of people's needs is a necessity for planning and designing landscapes.

Human Dimensions Research is defined *'the scientific investigation of the physical, biological, sociological, psychological, cultural and economic aspect of resource utilization at the individual and community level'* (Evert, 1996). In order to clearly articulate how human dimensions information can be used and integrated into landscape planning and design, there is a need to develop human dimensions framework that enhance the ability of landscape researchers and practitioners to understand and address the social aspects of landscape architecture. The framework should include not only social aspects, but also economic, cultural, political and governance aspects that need to be taken into consideration in landscape planning and design processes (Sowman et al., 2014). The social dimension involves consideration of gender, ethnicity, religion, social cohesion, values, attitudes and perceptions regarding landscape, as well as people's goals and aspirations. The economic dimension focuses on poverty, food insecurity, employment, income and assets, livelihoods. It also includes consideration of ecosystem services. The cultural dimension is concerned with examining cultural practices and rights, including local and indigenous knowledge. This theme also requires consideration of people's way of life, sense of place, culture and cultural heritage, and considers how communities have adapted to management decisions and regulations over time. The political dimension relates to issues of equity and access rights, representation and legitimacy, how benefits and losses are distributed among stakeholders, and issues of politics. The governance dimension involves issues and elements such as community organization and stakeholder participation, and how information sharing. Consideration of the enabling laws and policies for landscape planning and design, and the institutional arrangements governing landscapes are also important issues to take into account. In this context, the human dimensions framework will provide an understanding of the human dimensions of landscape planning and design as well as guidance on how human dimensions can be understood and integrated into planning and design processes.

4 Social Sciences

As human beings continue move on the landscape, planners and designers need to better approaches understanding reasons of changes and need to innovative responds to those changes. Social science research can make major contribution providing answers to address these challenged issues that occurs by humans, because social science disciplines focus on different social aspects (e.g., culture, economics, politics, law, values, beliefs, knowledge, motivations, preferences, perceptions, and behaviors).

Social science research can be conducted on planning and design issues at different scales from individual to local or regional. While individual scale studies focus on perceptions, values, norms, attitudes of users, local or regional scale examines culture, economics, law and policy effects (Bennett et al., 2017). The social science studies are conducted by various research approaches, tools and methods. Several methods to collect data: qualitative, quantitative, participatory and spatial. Qualitative methods include interviews, focus groups, observations. Quantitative methods are surveys, field experiments. Participatory methods are group facilitation methods such as Delphi technique, community-based research and spatial methods are mapping techniques: Geographic Information Systems (GIS), mind maps.

The most distinct characteristics of social science research reliability, validity, representativeness, and generalizability. Reliability addresses the repeatability of the results, validity deals with the accuracy of generalizations, representativeness is a sampling issue, generalizability addresses the broadness of results (Manfredo, Vaske and Sikorowski, 1996).

Landscape planners and designers is better to collaborate with social and behavioral scientists to advance understanding of human-landscape relationships and need to focus on the long-term effects of human activities on landscapes.

5 Human Dimensions and Landscape Planning and Design Process

Landscape planning and design process, in general, consists of defining goals, identifying current situations (inventory), identifying problems and opportunities, analysis, developing alternatives, evaluation and decision (preferred plan), detailed designs for the chosen option, implementation and monitoring (Hersperger, et al., 2021). At all stages of the process, human dimensions information is needed.

Defining goals is a key step in the process and represents human and landscape values toward desired conditions. This knowledge facilitates planners/designers defining goals. Identifying current situations, problems and opportunities, analysis and developing stages include social (gender, ethnicity, religion, social cohesion, values, attitudes and perceptions etc.), cultural (people's way of life, sense of place, culture and cultural heritage etc.) and economic (poverty, food insecurity, employment, income etc.) dimensions. Evaluation alternatives and choosing the best option requires public participation to determine public preferences for the alternatives. Monitoring as the last step touches the satisfaction that shows fulfillment level of expectations and needs.

In this context, human dimensions research provides a comprehensive and dynamic understanding of human behavior, interactions between humans and their environment; shapes a framework for a more holistic and comprehensive way of incorporating human landscape information into the planning and design process; helps to communicate with complex inter-related issues in human-landscape interaction to the decision making and results with a comprehensive framework for human monitoring, understanding human use patterns, and as a decision support system for defining and testing alternative management responses to changing landscapes.

6 Conclusion

This chapter focused on the human dimensions of landscape planning and design although human dimensions are certainly not new to the landscape architecture profession. Human dimensions research is a scientific exploration to help understand the cultural, sociological, psychological, economic, biological, and physical aspects of landscape planning and design. Social science data must be linked to biological and physical science information if a planner and designer are to recognize the complex, interactive role that humans play in natural systems (Tress, et al., 2006).

The European Landscape Convention highlights the need for an integrative approach, where ecological, social and cultural dimensions are included in any analysis or assessment concerning the landscape (Mikusinski, et al., 2013). However, unlike the ecological dimension, social and cultural values are difficult to measure and to communicate to planners and designers (English & Lee, 2004). Human dimensions contribute to an analysis of social and cultural aspects to an overall understanding of landscape planning and design.

References

1. Bennett, N. J., Roth, R., Klain, S. C., Chan, K., Christie, P., Clark, D. A., Cullman, G., Curran, D., Durbin, T. J., Epstein, G., Greenberg, A., Nelson, M. P., Sandlos, J., Stedman, R., Teel, T. L., Thomas, R., Veríssimo, D., & Wyborn, C., *Conservation Social Science: Understanding and Integrating Human Dimensions to Improve Conservation.* Biological Conservation, 2017. 205 93–108.
2. Council of Europe, *European Landscape Convention. European Treaty,* 2000. Series No. 176. Florence.
3. Decker, D. J., Brown, T. L. and Siemer, W. F., *Human Dimensions of Wildlife Management in North America.* 2001, Bethesda, MD: The Wildlife Society.
4. Downes, M., & Lange, E. *What you see is not always what you get: a qualitative, comparative analysis of ex ante visualizations with ex post photography of landscape and architectural projects.* Landscape Urban Plan, 2015. 142:136–146.
5. English, A. J., & Lee, E. *Managing the Intangible,* The George Wright Forum, 2004. 21, pp. 23–33.
6. Francis, M., *Behavioral approaches and issues in landscape architectural education and practice.* Landscape Journal, 1982. **1**, (2), pp. 92–95.
7. Harden, C. P., Chin, A., English, M. R., Fu, R., Galvin, K. A., Gerlak, A. K., McDowell, P. F., McNamara, D. E., Peterson, J. M., Poff, N. L., Rosa, E. A., Solecki, W. D. & Wohl, E. E., *Understanding Human–Landscape Interactions in the "Anthropocene".* Environmental Management, 2014. 53:4–13.
8. Hersperger, A. M., Gradinaru, S. R., Pierri Daunt, A. B., Imhof, C. S. & Fan, P., *Landscape ecological concepts in planning: review of recent developments.* Landscape Ecol. online version https://doi.org/10.1007/s10980-021-01193-y
9. Hunziker M., Buchecker M., & Hartig T., *Space and Place – Two Aspects of the Human-landscape Relationship.* In: Kienast F., Wildi O., Ghosh S. (eds) A Changing World. Landscape Series, 2007 vol 8. Springer, Dordrecht.
10. Matsuoka, R. H. & Kaplan, R., *People needs in the urban landscape: Analysis of Landscape and Urban Planning contributions.* Landscape and Urban Planning, 2008. 84, 7–19.

11. Meijeringa, J.V., Tobia, H., van den Brinkb, A., Morrisb, F. and Bruns, D., *Exploring research priorities in landscape architecture: An international Delphi study.* Landscape and Urban Planning, 2015. 137, 85–94.
12. Mikusinski, G., Blicharska, M., Antonson, H., Henningsson, M., Göransson, G., Angelstam, P. & Seiler, A., *Integrating Ecological, Social and Cultural Dimensions in the Implementation of the Landscape Convention,* 2013. Landscape Research, Vol. 38, No. 3, 384–393.
13. Nassauer J.I., *Placing Nature – Culture and Landscape Ecology.* (ed.), 1997. Washington, DC: Island Press.
14. Sowman, M., Raemaekers, S., and Sunde, J., *Guidelines for integrating human dimensions into MPA planning and management.* Environmental Evaluation Unit, 2014. University of Cape Town.
15. Tress, B., Tress, G., De´camps, H., d'Hauteserre, A. *Bridging human and natural sciences in landscape research.* Landscape and Urban Planning, 2001. 57, 137–141.
16. Tress, B., Tress, G., Fry, G. & Opdam, P. *From Landscape Research to Landscape Planning. Aspects of Integration, Education and Application,* 2006. Dordrecht, The Netherlands: Springer.

Gül YÜCEL

The Role of Disaster Museums for Disaster Awareness

Abstract: Disaster museums are founded in the aftermath of a great disaster having occurred in its native region. As opposed to disaster prevention centres, disaster museums also function as a tool for collective memory besides its practical function as a museum. Museums are among the salient tools in life-long learning as well. In the study, disaster museums founded within the context of earthquake, tsunami and flood disasters have been examined. Within the context of natural disasters, the most recent disaster museum samples from all over the world have been evaluated in detail. In the analyzed museums features of located environment, exhibition main themes and exhibition methods have been investigated. A disaster museum constitutes a major component of disaster tourism for the region hit by disaster by virtue of its museum function that bears the traces of history and is also highly valued as a memoir. It can be said that exhibition design and contents within the scope of museum undertake a major function in terms of disaster awareness.

Keywords: Disaster, disaster museum, public awareness, dark tourism.

1 Introduction

Many settlements in the world face disasters due to natural or technological hazards and they then experience enormous losses. Global Disaster Statistics (1970–2013) ranks flood disaster first in terms of number of cases but in terms of lost lives, earthquake and tsunami disasters come first in the list [1]. In relation to structural damage and destruction, first rank belongs to flood and earthquake. Moreover, disasters require evacuating millions of people while a great number of people are forced to relocate [1]. On the other hand, disaster statistics indicate that during the 21st century number of disasters relatively increased compared to the past. In an aim to reduce disaster risks, at first, it is essential to take required measures in physical-built environment. In addition, to achieve disaster resilience in the physical environment, increasing disaster awareness of the society also offers great contribution to disaster resilience. In Sendai Risk Reduction Frame 2015–30, disaster risk reduction and resilience against disaster are the featured approaches. To create a resilient society in the face of disasters, awareness trainings must decidedly be conducted by employing different tools and methods [2]. As it is yet an open field for further exploration, it is also feasible to develop and employ novel tools aside from available tools so as to elevate disaster awareness among society. In addition to boosting disaster awareness starting with the school age, voluntary actions of nongovernmental organizations and media can also be utilized as effective mediums in accessing large-size groups. In certain periods disaster-related campaigns were projected to grab people's attention. Within the context of disaster risk reduction; organizing campaigns to heighten public disaster awareness, participatory learning, informal and school-based formal educational approaches are presented [3]. Disaster awareness-raising activities are also provided directly by disaster education centers (It is named as "Life Safety Learning Center" in Japan [4]) in a way to entail different types of disasters. It can be argued that museums are also, in some way, a form of informal education. Artistic activity, exhibition or studies associated with disaster in varied periods could also be included within that context.

As it was in the past effects caused by disaster are narrated in the works of a great number of artists today. In Farnsworth Art Museum an exhibition named as "Disaster Art" (2016) was opened in which disaster-related works were presented [5]. In the Museum of International Folk Art, "The Arts of Survival: Folk Expression in the Face of Natural Disaster" exhibition some of the works that reflect the gravity of support offered by folk art and artists in the specific settlements having faced a natural disaster can be seen [6]. Likewise artist Takashi Murakami depicts earthquake and tsunami inspired works of art [7]. Global Facility for Disaster Reduction and Recovery (GFDRR) pointed out that art can be included within the scope of disaster resilience and it called for artists' contribution for "The Art of Resilience" exhibition [8]. It has been widely attested that via its educational role too, art can contribute to the development of public disaster awareness. Dakka Art Summit with its broadly detailed "Seismic Movements" title opened 2020-exhibition [9]. In "Designing for Disasters" exhibition, National Building Museum (May 11, 2014–September 13, 2015) exhibited the kind of approach that is focused on where and how to build structures in the face of a disaster [10]. The said exhibition, in some sense, urged museum visitors to reflect more about disasters and realize their presence. In its exclusive exhibition named as "Quake City", Canterbury Museum (A Special Exhibition from Canterbury Museum) geared at publicizing the earthquake and its effects that hit Christchurch and Canterbury cities [11]. Natural sciences and natural history museums, though not directly related with disaster theme, can also entail exhibitions that offer information on disaster and display experimental exhibitions having natural events.

Museums are an important part of lifelong learning [12]. Museums are vital components of life-long learning and museums- by their ties with history, nature, science, art and a wide range of branches- are the foundations empowering societal link with national history. Currently it is argued that globewide the total number of museums has reached to nearly hundred thousand [13]. Based on different themes museums can also act as a tool to transfer knowledge from a wide perspective. Disaster museums in particular can play a vital role by virtue of its educational aspect on disaster. With this role it is safe to claim that disaster museums can also play a part in raising disaster awareness within a society. In many parts of the world there are museums that focus on a specific disaster such as earthquake, flood or tsunami. Disaster museums are also components of dark tourism that includes a wide range of subthemes from war zones to archeological settlements hit by a disaster in the antiquity. In that sense it can also be included within global-scale tourism mobility. Aside from regional visitors disaster museums could also attract international visitors. As seen in Chernobyl Nuclear Accident zone, thousands of visitors across the entire world pay a visit to this disaster-hit region. Dating back to over two thousand years, Pompeii-impacted by Vesuvius Volcano- is one example to such disaster-hit settlements having attracted millions of visitors now from all over the world [14].

Despite the devastating power of disasters, remnants of a disaster can offer an opportunity to boost post-disaster resilience among younger generations. Damaged structures displaying impacts of a disaster or its aftermath effects in the nature could be useful in grasping disaster mechanism better [15]. Thus, a complete removal of disaster impacts could alter collective memory. Taiwan 921 Earthquake Museum is the one that was erected around a school building damaged in 1999 Earthquake [16]. Designing the museum centered around a damaged school campus and examining many topics related to earthquake in practice leave a great effect on visitors and play active role in raising public awareness [17]. On the other hand aside from main function of museums as information and education centers, it is suggested to simplify and make them more accessible [16]. Although designing disaster-hit zones and buildings as museums is costly, their educational role indeed matters strongly [15].

Led by Japan, there are museums all over the world focusing on different types of disasters such as earthquake, tsunami or flood etc. In the Eastern Japan regions that were hit by earthquake and tsunami social and cultural life has been significantly altered and many things were lost during flood. Yet it is also known that around damaged buildings in the impacted region, there was a boost in disaster museum works in a way to encompass affected zones [18]. In museums that are, in general, built in the aftermath of a disaster having occurred in their region, effects of disaster in the region are displayed and in specific sections there exist exhibitions allowing to visualize features and effects of a hazard causing the disaster via employing different simulation techniques. Also viewed as one extension of life-long learning, museums can play role in informing public about a disaster and heighten social awareness. A research that examined motives in visiting a disaster museum through a sample revealed that visitors mostly opted for spending time in there rather than learning [19]. Disaster prevention and education centres have important role for public awareness and disaster education. Nevertheless disaster museums can also offer art and memory through education and by means of its rich content, it can -as a touristic factor- attract visitors outside of the region. In disaster-related education, museum can expand to an even wider base with its spaces. Since they are mostly built in the aftermath of a great disaster and integrate memoirs, documents etc. from the region it allows the disaster museum to act as a memorial for visitors to receive education and know about earlier disasters specific to the region. Disaster history of a settlement can thus be transferred via museums.

In addition to museums of which main theme is disaster, there are museums in which some part of collection is reserved to disasters or periodical exhibitions on disaster are held. Such exhibitions that correspond to anniversaries of great disasters causing massive losses play a huge role in remembering disaster's impacts, reviving urban memory on the history of disaster and boosting preparedness against a disaster. Next to independent virtual museums, it is feasible to watch exhibitions of a museum as virtual museum. Exhibitions of virtual museums can roam in the internet domain. During pandemic, most museums were temporarily closed off and only after implementing various restrictive measures could they be reopened under strict supervision. Depending on the course of disaster, number of visitors and the length they remained open were also impacted. Besides, in disaster museums too, experimental sites that require contact and frequently used in an exhibition got affected as a result of pandemic situation. Online exhibition options of disaster museums could likewise be helpful in reaching larger masses to improve disaster awareness.

As is the case in all other museums, in disaster museums too, use of venue and exhibition is of utmost importance in terms of viewer interaction. In relation to disaster museums' role in boosting public disaster awareness it is also essential to examine the effects of museum architecture and exhibition style. Venue setup of the museum, its relationship with used material and the city, exhibition techniques and other themes are the headings that can be analyzed within the context of disaster museums. By moving beyond traditional museum approach; it is vital to analyze exhibition methods, exhibited collection and effects of architectural venue where it meets with viewers and investigate in relation to its physical structure sheltering all its constituents. That being the case, next to the disasters that led to building the museum, world samples that focused on earthquake, flood and tsunami disaster have also been examined. The study covers architecture of disaster museums and a detailed analysis of four different recent museums of which exhibition model focus on different disasters has been provided. Analysis was based on the information retrieved from internet websites of the museums.

2 Disaster Museums

Museums that refer to the cultural dimension of tourism are formed under a great number of sub-headings. Disaster museums, in general, are regarded as one part of dark tourism. Disaster tourism basically refers to paying a visit to a disaster region in the aftermath of a natural or technological disaster. A vast portion of disaster museums is built in a settlement hit by a great disaster so as to educate about disaster and also to commemorate the event. Mainly founded in relation to earthquake, flood and tsunami museums are expanded in a way to hold the memory of experienced disaster in its located city. Being a disaster site, Chernobyl Nuclear Accident zone (26 April 1986) is also now a global attraction center for visitors [20]. Great Eastern Japan Earthquake and Nuclear Disaster Memorial Museum having covered tsunami in the aftermath of Eastern Japan Earthquake and Fukushima

Nuclear plant accident (11 March 2011) was opened in Futaba in 2020 [21]. Tsunami Museum opened in Iwate Region hit significantly hard by the tsunami after the Great Eastern Japan Earthquake (22 September 2019) is also among the samples of recently opened museums [22].

Tab. 1 Earthquake museums in the world

;No	Museum Name, City/Country	Related disaster
1	Great Kanto Earthquake Memorial Museum (1930) Tokyo, Japan	September 01, 1923, 7.9 Mw Great Kanto Earthquake
2	Kita-ku, Tokyo Disaster Prevention Center and Earthquake Science Museum (1984), Tokyo, Japan	-
3	Great Hanshin-Awaji Earthquake Memorial Museum (2002), Kobe, Japan	January 17, 1995,6.9Mw Great Hanshin-Awaji Earthquake
4	Ojiya Earthquake Museum Sonaekan(2011), Niigata, Japan	October 23, 2004, 6.6Mw Chuetsu Earthquake
5	Hanshin Expressway Earthquake Museum (1996, re-opened 2009) Kobe, Japan	January 17, 1995,6.9Mw Great Hanshin-Awaji Earthquake
6	The Ruins Great East Japan Earthquake Kesennuma City Memorial Museum (2019) Kesennuma, Japan	March 11, 2011, 9.0–9.1 Mw, The Great East Japan Earthquake and Tsunami
7	Hokudan Earthquake Memorial Park & Nojima Fault Preservation Museum (1998), Hyogo, Japan	January 17, 1995, 6.9Mw Great Hanshin-Awaji Earthquake
8	The Great East Japan Earthquake and Nuclear Disaster Memorial Museum(2020), Futaba, Japan	March 11, 2011, 9.0–9.1 Mw, The Great East Japan Earthquake and Tsunami
9	Ningxia Haiyuan Earthquake Museum (2010), China	December 16, 1920, 7.8 M_L 1920 Haiyuan earthquake
10	The Memory of Earth: Epicenter Memorial Hall in Yingxiu (2012), Wenchuan, China	May 12, 2008, 7.9 Mw Sichuan Earthquake
11	5.12 Wenchuan Earhquake Memorial Museum (2013), Beichuan, China	May 12, 2008, 7.9 Mw Sichuan Earthquake
12	Earthquake Memorial Hall and Relics Park in Longtou mountain Town (2018),Yunnan, China	3 August 2014, 6.1 Mw The Ludian Earthquake
13	Muter Earthquake Museum (2014), Soriano Calabro, Italy	The 1783 Calabrian earthquake
14	Earthquake Museum and Tourism Burou, Tulbagh, South Africa	September 29, 1969 6.3Mw Tulbagh Earthquake
15	Tayvan 921 Earhquake Museum (2001), Taichung, Taiwan	September 21, 1999 7.7 Mw 1999 Jiji Earthquake
16	17 August Izmit Earthquake Memorial and Awareness Museum (2018), İzmit, Turkey	August 17, 1999, 7.6 Mw Marmara Earthquake

In Miyagi, Japan, Rias Ark Museum of Art which is a contemporary art and local history museum received damage due to 2011 Eastern Japan Earthquake and stayed close for a while during when it served as a center for those in the region. In the same period of time, museum staff documented the disaster and next, in order to tell the story of this disaster, they added a collection named as "Records of the Great Eastern Japan Earthquake and the History of Tsunami Disaster" to the permanent exhibition of the museum [23]. Disaster museums centered on earthquake theme were founded in earthquake-hit cities most of which belonged to Japan (Tab. 1). It can also be stated that many other museums were also founded around the main themes of tsunami and flood raids. Flood disaster centered museums are mostly located in the USA (Tab. 2, Tab. 3). As for the disaster museums in general, it is apparent that disaster museums centered on tsunami theme in particular were largely established in places hit by tsunami disaster after 2000. Third World Tsunami Museum Virtual Conference which was held 5 November 2020 (5 November as World Tsunami Awareness Day), focused preserving memories past disasters and discuss for achieving disaster awareness, particularly under Covid-19 conditions [24]. It is safe to argue that many of the disaster museums were designed in memory of the disaster having occurred in a region.

In disaster museums, exhibition model relate to; a section in which the disaster and natural hazards are discussed, impacts of emergent disaster, fight against the disaster and reconstruction process are elaborated. The section related to disaster impacts includes losses due to the disaster having hit the region where the museum is located and there are other sections where certain items, papers etc. from the disaster region are exhibited. In the section where hazards that may cause a disaster are explained interactive exhibition methods are employed to offer visitors a close experimentation. Impacts of the disaster provide the most dramatic section. Visitors can personally experience impact caused by the danger via exhibition methods such as damage corridor and as seen in specific museum samples, edifices impacted by the disaster can also constitute a part of the exhibition.

Tab. 2 Tsunami museums in the world

No	Museum Name, City/Country	Related disaster
1	Tsunami and Storm Surge Disaster Prevention Station, Osaka/Japan	For Future Tonankai/Nankai Earthquakes and tsunamis risk
2	IWATE Tsunami Memorial Museum (2019) Iwate/Japan	March 11, 2011, 9.0–9.1 Mw, The Great East Japan Earthquake and Tsunami
3	Inamura-no-Hi no, Yakata, (2007), Wakayama, Japan	December 24, 1854 8.4 ML 1854 Nankai Earthquake and Tsunami
4	Okushiri Isle Tsunami Museum (2001), Hokaido, Japan	Hokkaido Nansei-Oki Earthquake July 12, 1993, 7.7 Mw
5	(Hilo Tsunami Museum) Pasific Tsunami Museum, (1994), Hilo, Hawai,USA	May 23, 1960 Hawaii's 1960 Tsunami
6	Banda Aceh Tsunami Museum, 2009, Banda Ace, Indonesia	2004 Indian Ocean earthquake and tsunami 26 December 2004
7	International Tsunami Museum, (2006) Khao Lak,Tailand	2004 Indian Ocean earthquake and tsunami
8	Community Tsunami Education Centre and Tsunami Museum (2007) Telwatta, Sri Lanka	2004 Indian Ocean earthquake and tsunami
9	Kerala Tsunami Museum (2008) Kerala, India	2004 Indian Ocean earthquake and tsunami

Edifices damaged by the disaster or ground impact caused by the earthquake fault-line can also be a component of disaster museums. Similar to museum sample in Burdur Turkey they can be fully open- air museums. Samples in which impacted site is open for direct visits should be treated like archeological sites and be supported with additional measures for safe use and long-lasting protection. Samples collected from damaged buildings or from the city can also be an exhibition item in a museum. From a general perspective, remaining items in a destructed site are regarded as ruins that must be removed but despite that, they constitute a part of life and some form of daily life archaeology; hence they too can be exhibited in disaster museums. The process of turning remnants of a disaster into an exhibition item was emphasized during the stage of forming a permanent exhibition in the aftermath of disaster in Kesennuma Museum.

Tab. 3 Flood museums in the world

No	Museum Name, City/Country	Related disaster
1	The Houstion Flood Museum US	Hurricane Harvey August 17-September 2, 2017
2	Johnstown Flood Museum (1989) Johnstown,US	The Johnstown Flood May 31, 1889
3	LaCrosse Ice Age Flood Museum (2017) Washington, US	-
4	Teton Dam Flood Museum (Rexburg Museum), (1983) Rexburg, Idaho,US	Teton Dam Flood June 5, 1976
5	1927 Flood Museum Greenwille, US	Great Mississippi Flood, April 15, 1927
6	DSI Flood Museum (2020) Trabzon, Turkey	-
7	Watersnood Museum Ouwerkerk, Holland	North Sea flood of 1953 February 1, 1953
8	Laingsburg Flood Museum South Africa	the Laingsburg Flood January 25, 1981
9	1931 Floods Museum at Wenyou Tai, Gaoyou City, China	The Great Floods of 1931 July–November 1931

Museum structures can not only be re-planned but should, as seen in "MUTER Earthquake Museum" sample, also be constructed within a historical complex [25]. Similar to Taiwan 921 Earthquake Museum, samples in which the site is directly opened for visits can form a museum structure with a top layer covering a massive open site. As particularly apparent in earthquake museums in terms of form there are specific models such as 5.12 Wenchuan Earthquake Memorial Museum, Haiyuan Earthquake Museum in which earthquake is a symbolic manifestation in the plan and front face.

3 Architecture and Exhibition Features in Disaster Museums

Within the context of losses caused by a disaster and significance of disaster in regions particularly hit hard by a great disaster, there exist a good number of disaster museums and memorials. There are different museum samples related to a wide range of disasters the foremost of which is earthquake and the rest belongs to flood, tsunami etc. In modern age a wide range of trainings to raise disaster awareness are offered by varied disaster education centers but in addition to them, it can be claimed that museums founded within the context of disaster can also offer service in raising disaster awareness among people. In Japan, following the Great Eastern Japan Earthquake and subsequent tsunami disaster that resulted in massive damages, two important museums were founded. In the aftermath of China Earthquake (The 2008 Sichuan Earthquake, May 12, 2008, 8.0 Ms, 7.9 Mw) disaster museums designed accordingly have also been examined at length within that context.

3.1 The Great East Japan Earthquake and Nuclear Disaster Memorial Museum

Museum was initially designed as an archive that collected all the records related to Eastern Japan Earthquake (11.03.2011, 9.0–9.1 Mw) disaster (Fig. 1). Opened on 20 September 2020 the museum aimed to share the information and lessons learnt from Fukushima Nuclear Disaster, achieve disaster prevention reduction and speed up reconstruction attempts [21]. Great Eastern Japan Earthquake and Nuclear Disaster Memorial Museums are located in Futaba-gun, Fukushima Prefecture and four kilometers away from Fukushima Daiichi. It is situated on a vast land and close to the seaside. During the first twenty days the museum attracted 10 thousand visitors daily.

In the museum, exhibition venue consists of six sections (Fig. 1). In the exhibition venue the first section is reserved to introduction and demonstration. This is an area with a cylindrical form and seven screens and in here it is possible to view different scenes from the evacuation process after the earthquake and tsunami to the reconstruction stage. Viewers can receive information on these different stages from screens. Second section tells the entire chronological list of stages during nuclear disaster. Third section depicts recovery interventions after the nuclear plant accident and in the fourth section the way daily life was before the disaster and its change after the disaster can be visualized. Fifth section depicts impacts caused in the process of ongoing nuclear disaster and coping with such impacts. Final section demonstrated works from the reconstruction period. Exhibition model is on the same level and linear form in the exhibition sections having varied spatial designs. Aside from exhibition the museum also offers earthquake and disaster prevention education, shares the memories of people having experienced the disaster and organizes trips to see the plants damaged after tsunami and nuclear disaster and to witness reconstruction works in place.

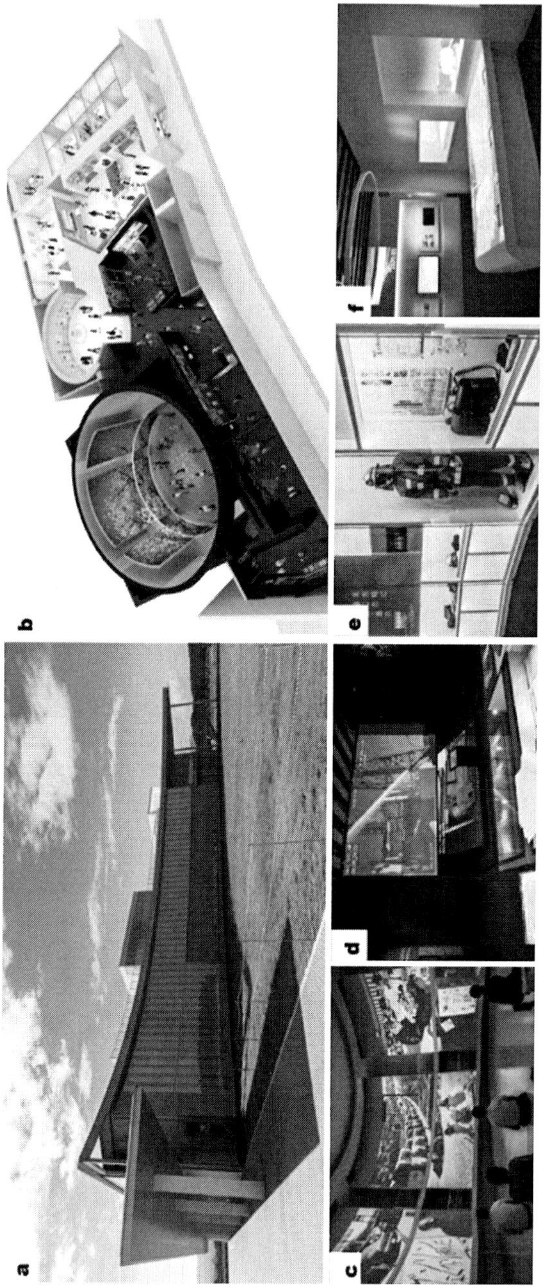

Fig. 1: The Great East Japan Earthquake and Nuclear Disaster Memorial Museum a) general view of museum, b) axonometric plan of museum with exhibition zones, c- d- e- f) museum exhibition hall views [21]

Museum is situated onto 5200 m^2 of land and has three floors. It is a concrete and semi-steel complex. This museum with a linear plan is built on a vast land and the region it is located is near the seashore but it has no barrier walls on the shore to protect against the impacts of tsunami. This structure that was initially designed as an archive having collected all the records and documents related to disaster region eventually turned into a museum. Museum plays a vital role as it functions like an archive in which all the stages of nuclear disaster occurring in the region were preserved. Its educational aspect related to planning site analysis surveys is considered to be significantly effective in raising disaster awareness. Within the scope of exhibition visits during Covid-19 pandemic; observing the physical distancing rules, using touch-screen exhibition panels or wearing single-use gloves are some of the measures posted on the museum website [21].

3.2 Iwate Tsunami Memorial Museum

Iwate Tsunami Memorial Museum is located in Takatamatsubara Memorial Park for Tsunami Disaster in Japan [22]. The museum was opened on 22 September 2019 which is after eight years after big east Japan Earthquake and Tsunami Disaster. The museum building has linear form and consisting two parts as one floor (Fig. 2). The passage and museum entrance is settled middle of the building. Museum exhibition is consisting four parts as tracing history, learning facts, and learning lesson and moving towards restoration together. The museum is located on a large land close to the sea.

Fig. 2: Iwate Tsunami Memorial Museum, a) general view of museum, b) axonometric plan of museum with exhibition zones, c- d- e- f) museum exhibition hall views [22]

The Miracle Lone Pine Tree, which has become a symbol of the disaster, is also located near the museum in the area significantly affected by the tsunami. Tsunami precaution sets have been set up along the shore (Massive sea wall) against tsunami risks.

3.3 The Ruins Great East Japan Earthquake Kesennuma City Memorial Museum, 2019

The goal of Great Eastern Japan Earthquake Ruins Kesennuma City Memorial Museum is, by utilizing disaster memoirs and remnants as "visible proof", to function as a warning sign so as to ensure that in the future Kessennuma city receives no damage from potential tsunami disasters [26]. Museum is located on a vast plain in which school building damaged by the disaster is settled. Damaged school is also a major component of the museum. First section in the museum is reserved to a documentary show consisting of the videos shot at the moment of disaster and at its immediate aftermath and an exhibition section displaying threat and damage caused by the earthquake and tsunami. Through a connection passage it is feasible to enter the second section having disaster ruins (former Kesennuma Koyo High-school Building) (Fig. 3).

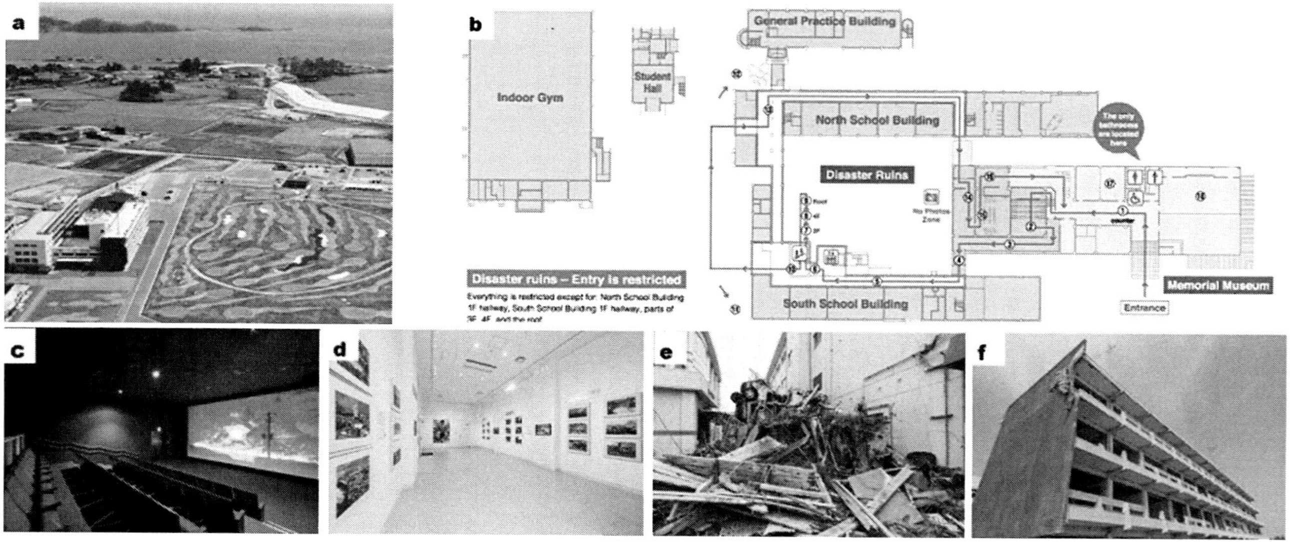

Fig. 3: The Ruins Great East Japan Earthquake Kesennuma City Memorial Museum, a) general site view, b) museum layout, c)video theather d) general view of exhibition hall, e) piles of cars, f) damaged school building [26].

In the school building it is seen that due to tsunami a car penetrated into the school structure and in here, highest peak reached by tsunami and evacuation area can also be observed. In the open area, demolished wall and piled cars after the tsunami are included into the exhibition. In the third section of the museum earthquake and tsunami threats and damages, search and rescue of lost people, scenes from evacuation center and living quarters in temporary residences are exhibited. Some sections display video shows of those rescued from the disaster and recovery process after the disaster. In the museum there are also training and workshop halls. Museum consists of a section in which condition of a disaster-hit high school is included in the exhibition and exhibition sections supported by a video and visuals. In this museum memorial and exhibition section is mainly designed to principally reflect ruined buildings impacted by the disaster and forming a vital component of the exhibition.

3.4 Earthquake Memorial Hall and Relics Park in Longtou mountain Town, China, 2018

The Museum was erected in Longtoushan city of China in memory of losses due to 6.5-magnitude earthquake in 2014 [27]. The starting point of museum design also including earthquake-park is formed by eucalyptus tree in the area. Museum and park are set on vast land which is situated on a sloping hillside. Buildings, park and memorial on the land were projected by taking into account symbolic lines referring to earthquake crack (Fig. 4).

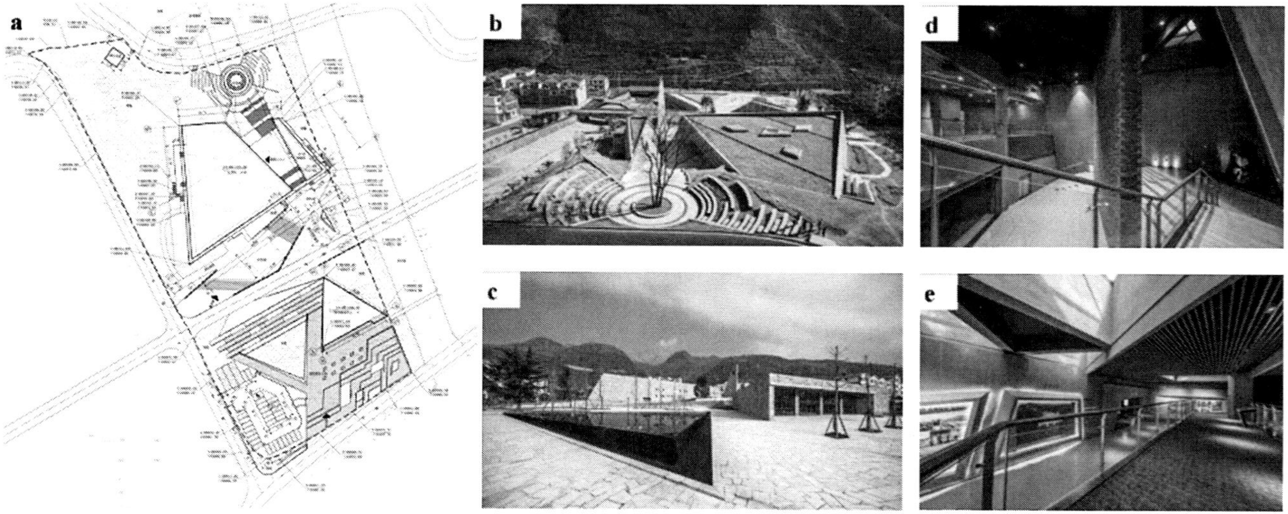

Fig. 4: Earthquake Memorial Hall and Relics Park in Longtou mountain Town, China a) site plan, b-c)exterior view, d-e) interior view of museum [27]

Spaces depicted below four themes such as peace, fracture, rebirth and remembrance integrate a park site, memorial venue and earthquake sites. Designed by HUAXIA Architects Studio the museum covers 5300 m^2 of area [27]. Situated on a sloping land, memorial park and museum are coordinated with the land slope and its front faces are ornamented with natural stone. Structure is harmonious with the land. Earthquake crack was utilized as the principal symbol in designing the complex having the earthquake park, memorial and museum and through this symbolic use it was feasible to obtain triangle forms in planning. This notion is more explicit in the plan setting.

4 Conclusion

Disaster museums belong to group of thematic museums as one group comparatively fewer in quantity among thousands of museums globewide. Also being a part of dark tourism too it is expected to witness these museums to create widespread effects outside of their native region. It can reasonably be argued that disaster museums having focused on varied types of disasters contribute to raise public disaster awareness. On the other hand, one of the other themes is that their native region is a disaster-risk open site.

Museums that relate to tsunami caused by the quake at sea and emergent disaster were founded particularly after the 2000-dated disaster having claimed great losses. A simulation to better recognize tsunami and museums displaying its impacts and featuring relevant memoirs were designed. Discovering the way tsunami occurs and effects it causes is vital to determine which measures to implement within that context in a risky situation.

In the aftermath of the Great Eastern Japan Earthquake (2011) and China Shichuan Earthquake (2008) many museums and disaster memorials were established to share the impacts of disaster directly with the public. Museums where memorial section constitutes a significant portion is an important tool in allowing the locals to form connection with their past. In a few of the museums set within the theme of earthquake disaster, direct impacts of earthquake on the ground can also be offered as an exhibition item. In disaster museums that also portray damaged structures as one component of exhibition, the size of damaged structure is a serious matter to manage so as to ensure that the structure can be protected for many long years ahead.

Effects of Covid-19 pandemic resulted in major shifts in visiting museums in place. Just like other public sites museums have also been temporarily closed off due to pandemic but they are still being operational on condition that personal or corporate measures are taken. Those museums examined in detail have been founded on a vast area. The Great Eastern Japan Earthquake and Nuclear Disaster Memorial Museum spaces provide sufficient means to pay a visit by observing physical distancing. As they are founded on a vast area, that can be seen as another advantage in pandemic situation. Despite being located within a landscape its connection with the city is possible via public transportation. For the museums that host an abundance of video exhibitions it is significant to popularize online visits in a distance. With the online exhibitions organized it is viable to reach greater numbers of people. It is true that disasters can cause devastating effects for the region it directly hit and yet in order to gain disaster experience and know the measures to take, disaster knowledge could be continuously updated. In the same vein disaster museums should, aside from being just a center for disaster education, play an active role as an art venue.

References

1. Desinventar, "GAR Global Assesment Report on Disaster Risk Reduction. Desinventar Disaster Information Management System," 2015. [Online]. Available: http://www.desinventar.net/DesInventar/profiletab.jsp?countrycode=g15&continue=y. [Accessed 20 December 2020].
2. UN United Nations, "Sendai Framework for Disaster Risk Reduction 2015–2030. United Nations UN.," 2015. [Online]. Available: https://www.unisdr.org/files/43291_sendaiframeworkfordrren.pdf. [Accessed 20 October 2020].
3. IFRC, "Public awareness and public education for disaster risk reduction: a guide," 7 June 2011. [Online]. Available: https://www.ifrc.org/Global/Publications/disasters/reducing_risks/302200-Public-awareness-DDR-guide-EN.pdf. [Accessed 01 January 2021].
4. Tokyo Fire Department, "Life Safety Learning Center," Tokyo Fire Department, [Online]. Available: https://tokyo-bskan.jp/en/. [Accessed 10 01 2021].
5. Farnsworth Art Museum, "Art of Disaster, October 8, 2016 – April 23, 2017," [Online]. Available: https://www.farnsworthmuseum.org/exhibition/art-of-disaster/. [Accessed 10 January 2021].
6. Museum of International Folk Art, "The Arts of Survival: Folk Expression in the Face of Natural Disaster, July 3, 2011- April 29, 2012," [Online]. Available: http://www.internationalfolkart.org/events/details/. [Accessed 10 January 2021].
7. J. Jones, "Takashi Murakami's Art From Disaster," 14 April 2017. [Online]. Available: https://www.thedailybeast.com/takashi-murakamis-art-from-disaster. [Accessed 15 January 2021].
8. GFDRR, "The Art of Resilience," 16 April 2019. [Online]. Available: https://www.gfdrr.org/en/art-of-resilience/details. [Accessed 10 January 2021].
9. Dhaka Art Summit, "Seismic Movements," 7–15 February 2020. [Online]. Available: https://www.dhakaartsummit.org/. [Accessed 10 January 2021].
10. The National Building Museum, "Designing for Disasters," May 11, 2014–September 13, 2015. [Online]. Available: https://www.nbm.org/exhibition/designing-for-disaster/. [Accessed 10 January 2021].

11. Canterbury Museum, “Quake City,” 2013–2017. [Online]. Available: https://quakecity.co.nz/. [Accessed 15 January 2021].
12. B. Feyzioğlu, “Üç Nesil Bir Arada,” *UNIMUSEUM,* vol. 2, no. 1, pp. 12–14, 2019.
13. UNESCO, “Museums,” [Online]. Available: https://en.unesco.org/themes/museums. [Accessed 15 January 2021].
14. Archaeological Park of Pompeii, “Visitor Data,” [Online]. Available: http://pompeiisites.org/en/archaeological-park-of-pompeii/visitor-data/). [Accessed 15 January 2021].
15. P. Migoń ve E. Pijet-Migoń, «Natural Disasters, Geotourism, and Geo-interpretation,» *Geoheritage,* no. 11, pp. 629–640. https://doi.org/10.1007/s12371-018-0316-x, 2019.
16. C. Ryan ve S.-Y. Hsu, «Why do Visitors Go to Museums? The Case of 921 Earthquake Museum, Wufong, Taichung,» *Asia Pacific Journal of Tourism Research,* cilt 16, no. 2, pp. 209–228, DOI:10.1080/10941665.2011.556342, 2011.
17. Ö. Avşar, “Importance of Earthquake Museums to Increase Public Awareness: The Taiwan Example,” in *2ND International Symposium on Natural Hazards and Disaster Management 04–06 MAY 2018 (ISHAD2018 Sakarya Turkey),* Sakarya, 2018.
18. A. Littlejohn, “Museums of themselves: disaster, heritage, and disaster heritage in Tohoku,” *Japan Forum,* p. DOI: 10.1080/09555803.2020.1758751, 2020.
19. B.-J. Yan, J. Zhang, H.-L. Zhang, S.-J. Lu and Y.-R. Guo, “Investigating the motivation–experience relationship in a dark tourism space: A case study of the Beichuan earthquake relics, China,” *Tourism Management, 53,* pp. 108–121, 2016.
20. “CHERNOBYL POWER PLANT AND PRIPYAT TOUR,” [Online]. Available: https://www.chernobylwel.com/chernobyl-history. [Accessed 10 January 2021].
21. The Great East Japan Earthquake and Nuclear Disaster Memorial Museum, 5 January 2021. [Online]. Available: https://www.fipo.or.jp/lore/about. [Accessed 10 January 2021].
22. Iwate Tsunami Memorial Museum, [Online]. Available: https://iwate-tsunami-memorial.jp/en/. [Accessed 15 January 2021].
23. Rias Ark Museum of Art, “1st floor Permanent Exhibition Documentary of East Japan Earthquake and Tsunami and History of Tsunami Disaster,” [Online]. Available: http://rias-ark.sakura.ne.jp/2/en/sinsai/. [Accessed 10 January 2021].
24. United Nations, “Third World Tsunami Museum Virtual Conference,” United Nations Office for Disaster Risk Reduction, 05 November 2020. [Online].

Available: https://www.undrr.org/event/third-world-tsunami-museum-virtual-conference. [Accessed 25 January 2021].

25. SAO, "MUTER - Earthquake Museum," SAO Schiavello Architects Office, [Online]. Available: http://www.schiavello.it/Projects/MUTER+-+Earthquake+Museum. [Accessed 15 January 2021].
26. The Ruins Great East Japan Earthquake Kesennuma City Memorial Museum, [Online]. Available: https://www.kesennuma-memorial.jp/english/. [Accessed 15 December 2020].
27. Archdaily, "Earthquake Memorial Hall and Relics Park," 24 September 2019. [Online]. Available: https://www.archdaily.com/924959/earthquake-memorial-hall-and-relics-park-in-longtou-mountain-town-huaxia-architects-studio. [Accessed 10 December 2020].

Seda ÖZLÜ Sinem DEDEOĞLU ÖZKAN and Dilek BEYAZLI

Spatial Distribution of Human Capital: The Case of Turkey

Abstract: Although there is no clear definition of the popular smart city concept, which is constantly transformed around the recent contemporary technological and economic trends, the concept is tried to be defined by considering it in technology, human, and institution dimensions. The smart city concept, which tries to connect the physical infrastructure, information technology infrastructure, business infrastructure and social infrastructure of the city, has social, economic and political sub-components. It is critical to address these components, especially smart people and communities that generate and benefit from social capital as part of smart cities. One of the hallmarks of the concept of smart cities is that people play essentially a central role and the well-being of their inhabitants. Özellikle son dönemlerde; it has become an important issue to reveal the domestic spatial distribution of urban human capital in smart people dimension, which is one of the components of smart cities, and the differences observed in these distributions. In this study, it was aimed to define the human capital structure of Turkey provinces through the selected variables and to discuss their spatial distribution. Hierarchical (ward) and two-step clustering models were used in the study, which approached the smart person dimension, which is the component of the smart city, from a human capital perspective. With the detection of similarities and differences at the provincial level, the smart city approach is not only the information and communication technology applications at the urban scale; it is aimed that it should be a people-oriented approach that targets human development. It is thought that the study will guide the determination of interventions that will ensure the human capital-based development of people living in cities.

Keywords: Human Capital, Information and Communication Technologies, Cluster Analysis, NUTS-3

1 Background

In the last two decades, the concept of smart cities has become a key element in international politics and future strategic plans and has been becoming increasingly popular. In the most general sense, the concept of smart cities which is considered as urban innovation and transformation initiatives, "reflects the idea of creating smart infrastructures, smart spaces and communities by adopting developing information and communication technologies as the most important tool, and thus, providing high efficiency for human and nature" [1]. This concept aims to use physical infrastructures, Information and Communication Technologies (ICT), information resources and social infrastructure that help cities to better use resources in different urban areas for economic regeneration, social cohesion, and a better city and infrastructure management [2; 3]. Komninos (2006) considers smart cities as "... the regions with high learning and innovation capacity based on the creativity of their populations, knowledge creation institutions and digital infrastructures for communication and information management" [4]. Caragliu et al., (2012) also indicated factors such as "a strong focus on the purpose of ensuring the social participation of different urban residents in public services, an emphasis on the vital role of high technology and creative industries in long-term urban growth, and a deep interest in the role of social and relational capital in urban development" as the characteristics of smart cities [5]. Angelidou (2016) indicates two distinctive characteristics of smart cities as a) improving urban functions using technology as a means to accumulate and organize a large amount of information and make it accessible to a growing number of people and using it to save resources, and b) the advancement of human and social capital through knowledge creation and dissemination, advanced participation, and the creation of new forms of innovation. As technologies and societies change dynamically, the smart city will begin to redefine itself and develop new ways of thinking about technology [6]. Therefore, although it is not possible to define this process, which is open to continuous change and transformation, there are six components that define the methodology according to the "Smart Cities Wheel (SCW)," which was featured by Cohen and adopted by the European Union (EU). They are indicated below (Tab. 1);

Tab. 1. Dimensions and factors of a smart city

Smart Economy	**Smart Government**	**Smart Environment**	**Smart Mobility/ Transportation**	**Smart Living**	**Smart People**
Innovative spirit	Participation in decision-making	Attractive of natural conditions	Local accessibility	Culturel facilities	Level of qualification
Entrepreneurship	Public and social services	Pollution	(Inter-)national accessibility	Health conditions	Affinity to life long learning
Economic image&trademarks	Transparent governance	Environmental protection	Availability of ICT-infrastructure	Individual safety	Flexibility
Productivity	Plitical strategies & perspectives	Sustainable resource management	Sustainable, innovative and safe transport systems	Housing Quality	Creativity
Flexibility of labour market				Education Facilities	Cosmopolitanism/ Openmindedness
International embeddedness				Touristic attractivity	Participation in public life
Ability to transform				Social cohesion	

Source: [7]

1. *Smart Economy*: It includes all factors related to economic competitiveness such as innovation, entrepreneurship, trademarks, productivity/flexibility of the labor market/integration into the national market [8].
2. *Smart Government*: Efficient and effective communication, political participation, transparency in public administration and participatory decision-making mechanisms are established between stakeholders at different levels with information and communication technology solutions, and it includes services for citizens and the functioning of the administration [9].
3. *Smart Environment*: This component, which includes solutions such as renewable energy, smart grids and meters, microgrids, advanced air pollution monitoring systems, environmentally friendly buildings and urban planning, energy-efficient smart street lighting, solid waste management, smart water management and drainage systems, is defined by the efforts for the protection of the environment [8; 9]
4. *Smart Mobility/Transportation*: Local and international accessibility is an important aspect of the component, and it includes modern, sustainable and integrated transportation systems supported by information and communication [8;9]. Furthermore, this component primarily aims to provide inclusive transportation solutions for environmentally friendly and disadvantaged groups [9].
5. *Smart Living*: It includes various aspects of quality of life such as culture, health, safety, housing and tourism [8]. With information and communication technologies, people's lives are made easier, and urban residents are provided with healthier and safer environments [9]
6. *Smart People*: It is aimed to create an inclusive society that encourages creativity and innovation by improving people's ability to use and produce information and communication technologies [9]. Smart people are defined not only by the qualification or educational levels of citizens but also by the quality of their social interactions related to integration and public life and openness to the "outside" world [8]. Therefore, the success of smart cities is directly proportional to the adoption by the people in the city, and it is also accepted that the precondition for creating smart cities is to create smart people [10; 11].

As can be seen from these six components that nurture and develop each other, smart cities are related to a mixed mix of social enterprise, cultural enterprise and economic enterprise that encompasses education/training, culture/arts, business/commerce [11]. As a result, when the smart city indicators defined as smart economy, smart government, smart environment, smart mobility/transportation, smart living and smart people achieve success together-holistically, the existence of the smart city is in question. Especially, it is not possible to talk about sustainable urban systems without human capital [12].

Intelligent human awareness in the human-oriented definition that has been gaining more acceptance in recent years; it is defined as an individual who has a high participation and creativity, lifelong learner, is included in the life of information technologies, and is the main element of human and social capital and the focal point of city life. [11; 13]. According to this people-oriented approach that emphasizes the role of human capital and education; rather than blindly believing that ICT can transform and improve cities automatically, it should start by focusing on the human capital, the human capital [5; 11]. In addition to this; smart city projects, which have an impact on the quality of life of citizens, also aim to raise more informed, educated and participatory citizens [14]. At this point, it is important to consider the smart people factor, which is the component of smart cities, from the perspective of human capital and to identify spatial differences.

Human capital concept; it is a concept that includes strategic elements intertwined with many disciplines or subjects such as economy, politics, industry, sociology, demography, law, biology, psychology, health, education, geography, technological development, R&D, innovation [15]. The first studies on the concept of human capital, which is generally defined as the sum of the knowledge, skills and experiences of the workforce, date back to the 1960s [16; 17; 18; 19; 20]. According to Schultz (1961), while human capital reflects the value of human capabilities [18]; Husz (1998) expresses the time, experience, knowledge and skill that a household or a generation can use in the production process [21]. The creation and development of human capital is a phenomenon studied in the context of issues such as life expectancy, literacy rate, nutrition, child mortality and various health indicators [22]. According to Yumuşak; today, the concept of human capital, defined as qualified manpower in the most general sense, requires addressing all aspects of the population [23]. Human capital, which is among the most important investments in the success of countries, is questioned with various indicators in the socio-cultural and economic context while today's researches are conducted [23; 24].

In summary; although there is no clear definition of the popular smart city concept, which is constantly being transformed around the recent technological and economic trends, the concept is tried to be defined by considering it in technology, human and institutional dimensions [11; 25; 26]. Based on these points, the smart people, one of the components of smart cities, are handled from the human capital perspective. The spatial distribution of human capital within the country and revealing the differences observed in these distributions become an important issue. With this study, it is aimed to discuss the spatial distribution of human capital structure in Turkey provinces by defining it with the help of selected variables. In the study, primarily the concepts of smart city, smart city components and smart people are dealt with theoretically. While giving information about the methods and variables in the methodological process of the study; The results of the cluster analysis made with the human capital variables selected in the findings were discussed. The clusters revealed in the cluster analysis were evaluated at the provincial level.

2 Method

Cluster analysis is a statistical method to "partition a set of observations into a distinct number of unknown groups or clusters in such a manner that all observations within a group are similar, while observations in different groups are not similar" [27]. According to Shih, et al. (2010), the purpose of clustering is to divide the data that can show high similarity to each other into several groups in the same group [28]. Copus (1996) identified one of the advantages of the methodologie as their capacity to tackle large numbers of variables easily and their adequacy for explorative data analysis [29]. The degree of similarity in one group is defined by the distance between the observations within a multidimensional co-ordinate system where each axis represents one feature. According to its characteristics, each region is definitely positioned in this multidimensional space. The closer to each other regions are, the more likely they are to be grouped into the same cluster [30; 31]. Furthermore, aggregative approaches to cluster analysis lead to useful and sometimes unexpected information about data patterns [29]. The following could be suggested when hierarchical and non-hierarchical methods are detailed:

- ***Agglomerative or hierarchical methods***: In these methods, the number of clusters is equal to the number of observations, which are then merged into larger clusters. The method aims to combine the observations into increasing sizes of clusters, using a measure of similar distances [32]. The main approach is that the number of clusters is reduced one by one by merging two existing clusters. In the first step, each region represents a single cluster. After the last step, all regions are included in one cluster. A dendrogram visualises the steps in a hierarchical clustering procedure [33]. According to Baum et al., 2004 the elbow criterion that creates a sudden jump upwards in the agglomeration coefficients, the dendrogram, various statistical values of the clusters, and the plausibility of the grouping are means of deciding on the number of clusters [30].
- ***Two Step Cluster analysis***: This method was developed to cope with the problems in K-means algorithm. This algorithm produces solutions with a mixture of continuous and categorical variables. It is a single-pass data approach that allows to use quantitative and qualitative variables simultaneously, determining pre-sets in the first step, then performing hierarchical clustering [34; 35]. In the pre-clustering phase of the method, the data is scanned one by one and it is decided whether the current situation will be combined with previously formed clusters and whether it will start a new cluster according to the distance criteria [35]. In the cluster step; Using the standard hierarchical clustering algorithm, the data are grouped into any number of clusters according to the subsets obtained from the pre-set step [36]. The SPSS algorithm leads to an optimal number of automatically determined clusters; however, since cluster analysis does not test the hypothesis, the researcher needs to check the accuracy of the solution. The most important features of this algorithm; to be applied in large data sets, to process categorical and continuous variables, to automatically

determine the most appropriate number of clusters, and to extract observations that do not comply with the obtained clusters from the data when desired [37]. Cluster quality bar represents the silhouette coefficient, which is a measure of both cohesion (i.e., the similarity of the elements in a cluster), separation (i.e., the difference of the clusters) and ranges between −1 and 1 [38; 39].

In principle, the cluster analysis as an explorative method can be considered as suitable for classifying regions. Revealing so far unknown regional structures and coherences and, thus, contributing to new insights it can motivate argumentations of regional policy and contribute to initiate political effects [30].

The variables used in the evaluation of the 'human capital' component obtained from domestic/foreign literature to make cluster analysis were determined by taking into account the existing data. Turkey NUTS 3 level in the study of human capital variables describing the structure of the provinces; Turkey Statistical Institute (TUIK), the Turkish Patent Institute and the Insurance Information and Monitoring Center at the provincial level consists of 16 variables related to 2017, 2018 and 2019 (Tab. 2).

Tab. 2. Data sources with selected human capital variables

Variables	**Source**	**Year**
Urbanization rate	TUIK	2019
Gross domestic product per capita	TUIK	2018
Gross value added per capita	TUIK	2018
Number of registered patent applications	Turkish Patent Institute	2019
Literacy rate (age 6 and over)	TUIK	2019
Number of vocational school students	TUIK	2018
Higher education graduate rate (age 15 and over)	TUIK	2019
Number of higher education students	TUIK	2019
Number of hospital beds per 100000 person	TUIK	2018
Life expectancy at birth	TUIK	2017
Health staff ratio	TUIK	2018
Sickness ınsurance policy rate	Insurance Information-Monitoring Center	2019
Number of cinema seats	TUIK	2019
Number of theater seats	TUIK	2019
Number of materials borrowed from the library	TUIK	2019
Number of cars per 1000 people	TUIK	2019

Source: [40; 41; 42]

3 Finding

After determining the variables, in order to determine the spatial distributions, similarities/differences of human capital, which is the intelligent human component, at the provincial level, firstly, "Two-Step Clustering Analysis" was used as the basic method. 2017, 2018 and 2019 of the "human capital" as a result of clustering performed with human capital variables Turkey the status of the province (adequacy), analyzing results and spatial distribution were mapped and reviewed. Since there are differences between the variable values used in the study, clustering analyzes were conducted with standardized data. During the process, "Two-Step Clustering-Ward-Two-Step Clustering" analyzes were used

Determining the optimum number of clusters: Two-Step Clustering Analysis was performed with the help of standardized 16 human capital variables and SPSS 25.0 package program. At this stage, where the number of clusters is automatically determined by the program, 2 clusters have been formed. The first of the clusters formed as a result of the analysis are from the provinces of Istanbul, Izmir, Eskişehir, Ankara and Antalya; the second cluster consists of the other 76 provinces (Fig. 1). Since the number of clusters (2) formed as a result of the Two-Step Clustering Analysis does not reflect the diversity in the human capital structure of the country; Ward method dendograms were used to determine the optimum number of clusters.

As a result of Ward cluster analysis made with standardized human capital data, there are "human capital variables of provinces" and "province names" in the axes of the dendograms, and the lines show the relationships. Considering the clusters formed at unit distance and the number of provinces in the clusters, 11 clusters at 1 unit distance, 7 at 2 units distance, 6 at 3 units distance and 4 clusters at 4 units distance are formed (Fig. 2).

Two-Step Cluster Analysis According to the Number of Clusters Determined: Two-step clustering analyzes were performed separately with (11–7-6–4) with all the cluster numbers formed according to the Ward analysis distances. The most appropriate number of clusters was decided according to the provinces that vary according to the number of clusters and the causality of the change. Considering the diversity of 81 provinces, the distribution of human capital within the country and the causality of the distribution, the cluster number was determined as 7. As a result of the analysis, it is seen that the variables of "Number of higher education students, number of cinema seats and GDP" are the most effective in forming clusters, and the variable "life expectancy at birth" is the least effective. While 39.50 % of the provinces are in cluster 2; there is only one province each in cluster 5 and cluster 7 (Tab. 3).

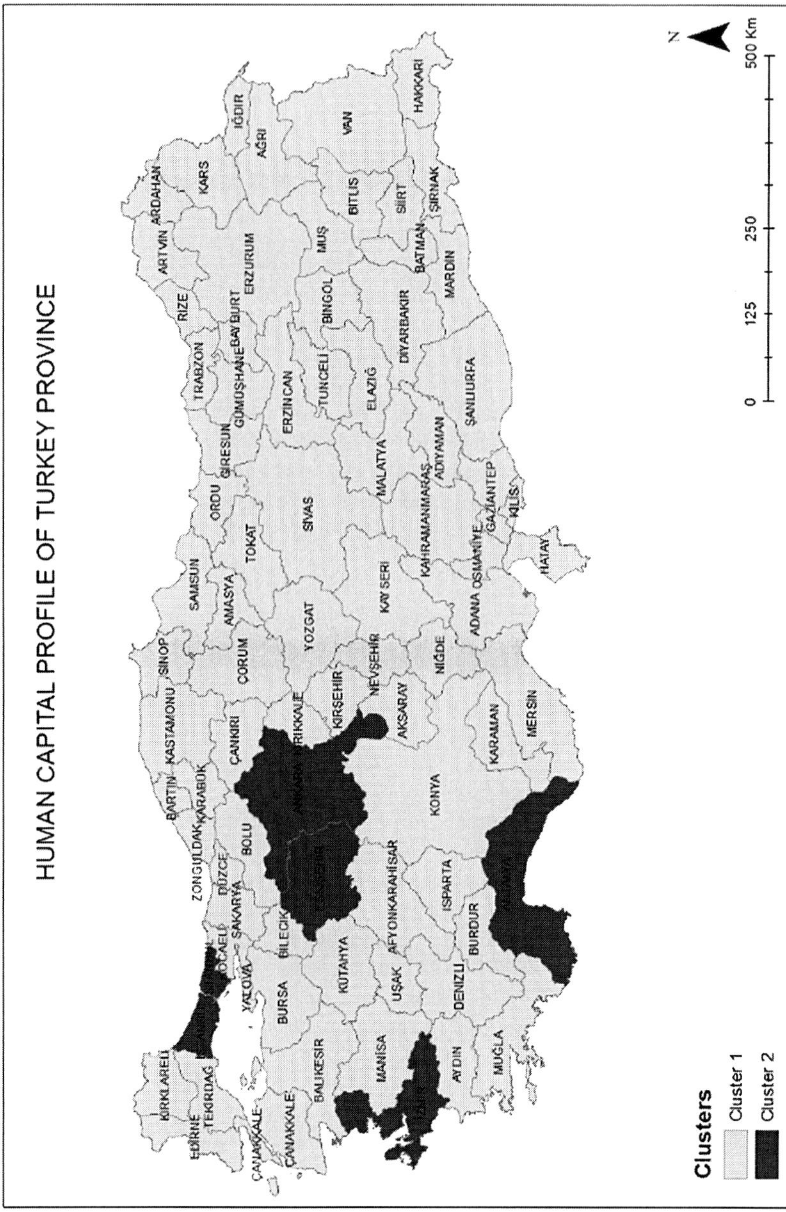

Fig. 1. Distribution of clusters formed according to two-step cluster analysis

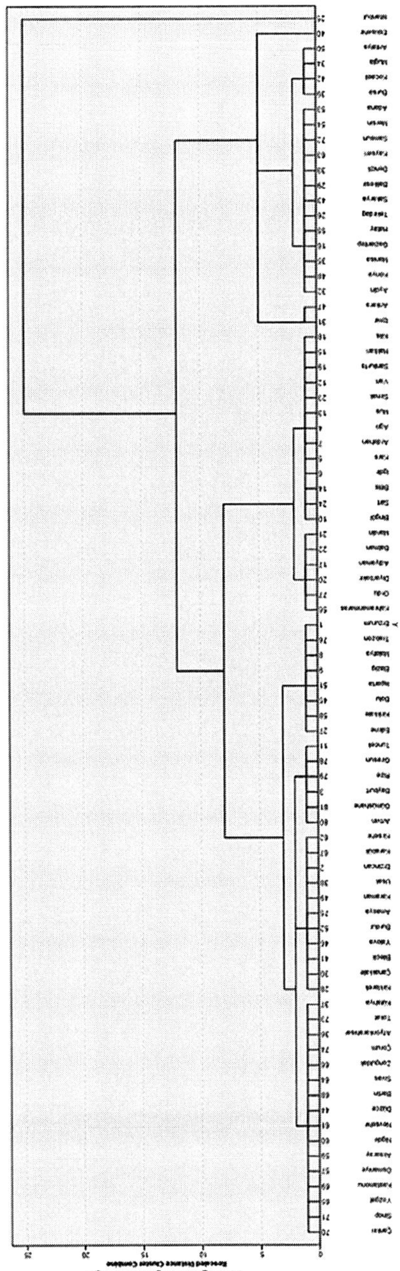

Fig. 2. Distribution of human capital according to Ward analysis

Tab. 3. Distribution of provinces according to two-step clustering analysis

Cluster1	Cluster2	Cluster3	Cluster4	Cluster5	Cluster6	Cluster7
Adıyaman	Afyonkarahisar	Bolu	Adana	Eskişehir	Ankara	İstanbul
Ağrı	Aksaray	Edirne	Antalya		İzmir	
Ardahan	Amasya	Elazığ	Aydin			
Batman	Artvin	Erzurum	Balıkesir			
Bingöl	Bartın	Isparta	Bursa			
Bitlis	Bayburt	Kırıkkale	Denizli			
Diyarbakır	Bilecik	Malatya	Gaziantep			
Hakkâri	Burdur	Sivas	Hatay			
Iğdır	Çanakkale	Trabzon	Kayseri			
K.maraş	Çankırı		Kocaeli			
Kars	Çorum		Konya			
Kilis	Düzce		Manisa			
Mardin	Erzincan		Mersin			
Muş	Giresun		Muğla			
Ordu	Gümüşhane		Sakarya			
Şanlıurfa	Karabük		Samsun			
Siirt	Karaman		Tekirdağ			
Şırnak	Kastamonu					
Van	Kırklareli					
	Kırşehir					
	Kütahya					
	Nevşehir					
	Niğde					
	Osmaniye					
	Rize					
	Sinop					
	Tokat					
	Tunceli					
	Uşak					
	Yalova					

Tab. 3. Continued

Cluster1	**Cluster2**	**Cluster3**	**Cluster4**	**Cluster5**	**Cluster6**	**Cluster7**
	Yozgat					
	Zonguldak					
19	**32**	**9**	**17**	**2**	**1**	**1**
(%23,5)	**(%39,50)**	**(%11,1)**	**(%21)**	**(%2,5)**	**(%1,2)**	**(%1,2)**

Provinces with metropolitan status are shown underlined.

While the human capital structure has the highest values in Cluster 7, variables such as 'number of cinema and theater seats', 'Gross value added per capita', 'number of registered patent applications', 'number of health personnel', 'number of policies' have the highest values among all clusters. These variables, which are the most effective in determining the clusters, show their highest values in Cluster 6 after Cluster 7. Cluster 1 and Cluster 2, covering 43 % of the country, consist of provinces with the lowest human capital level. Cluster 1 consists of 'number of higher education students', 'number of cinema and theater seats' and 'number of registered patent applications' are the lowest. Cluster 2 consists of the settlements with the lowest health variables such as ' Gross value added per capita', 'number of personnel and policies'. Provinces in Cluster 3 and Cluster 4 have average values in terms of human capital (Tab. 4).

As a result of the two-step clustering analysis, it is noteworthy that in general, provinces form a homogeneous structure with neighboring provinces with spatial proximity, and many of the sub-regions throughout the country have diversity in terms of human capital structure. In this context; being diverse in social, economic and physical etc. aspects, having an important population share in the country and being one of the most important metropolises in the country, Istanbul alone constitutes the cluster with the highest human capital level. İzmir, which has metropolitan status, and Ankara, which has capital and metropolitan status, are important focal points in the country and are located in the same-high human capital cluster with similar human capital structure. Eskişehir province, which is in the neighborhood of the capital, has formed a cluster alone, as it is an important point in the country in terms of the number of higher education students. It is also observed that most of the provinces located in the east of the country are in a low cluster in terms of human capital level. Cluster 4 consists entirely of provinces with metropolitan status; It can be traced spatially that it formed an arc that is continuous from south to north. A clustering based on spatial proximity also occurs, especially in the north-inner parts (Fig. 3).

Tab. 4. Importance of variables that are effective in the formation of clusters - mean values

Variables	Cluster 1	Cluster 2	Cluster 3	Cluster 4	Cluster 5	Cluster 6	Cluster 7
Number of higher education students	-0,2*	-0,2	-0,02	-0,1	8,48**	0,27	2,25
Number of cinema seats	-0,25*	-0,24	-0,16	0,12	-0,03	1,82	8,28**
Gross value added per capita	-0,24	-0,26*	-0,21	0,17	-0,03	1,75	8,25**
Number of registered patent applications	-0,2*	-0,19	-0,18	0,06	-0,09	0,92	8,59**
Number of theater seats	-0,29*	-0,28	-0,16	0,22	0,15	2,1	7,96**
Health staff ratio	-0,27	-0,35*	-0,16	0,29	0,05	2,72	7,4**
Sickness insurance policy rate	-0,34	-0,36*	-0,31	0,52	0,15	2,33	7,08**
Number of cars per 1000 people	-1,42*	0,27	0,3	0,59	1,15	1,71**	1,08
Number of materials borrowed from the library	-0,35	-0,39*	-0,11	0,53	0,06	3,85**	3,43
Literacy rate (age 6 and over)	-1,29*	0,15	-0,07	0,92	1,13**	1,21	1,02
Number of hospital beds per 100000 person	-0,75*	-0,22	2,02	0,06	1,59**	0,34	-0,18
Gross domestic product per capita	-1,03*	0	-0,04	0,68	1,13	1,9	3,51**
Higher education graduate rate (age 15 and over)	-0,98*	-0,04	0,35	0,42	1,89	2,82**	2,11
Urbanization rate	-0,34	-0,65*	0,28	1,17	1,17**	1,17**	1,17**
Number of vocational school students	0,5	-0,77*	-0,35	0,77	0,25	0,98	2,85**
Life expectancy at birth	-0,13	0,08	0,21	-0,24	-0,28	0,83**	0,5

*Not: The lowest values of the variables in the clusters are indicated with *, and the highest values with **.*

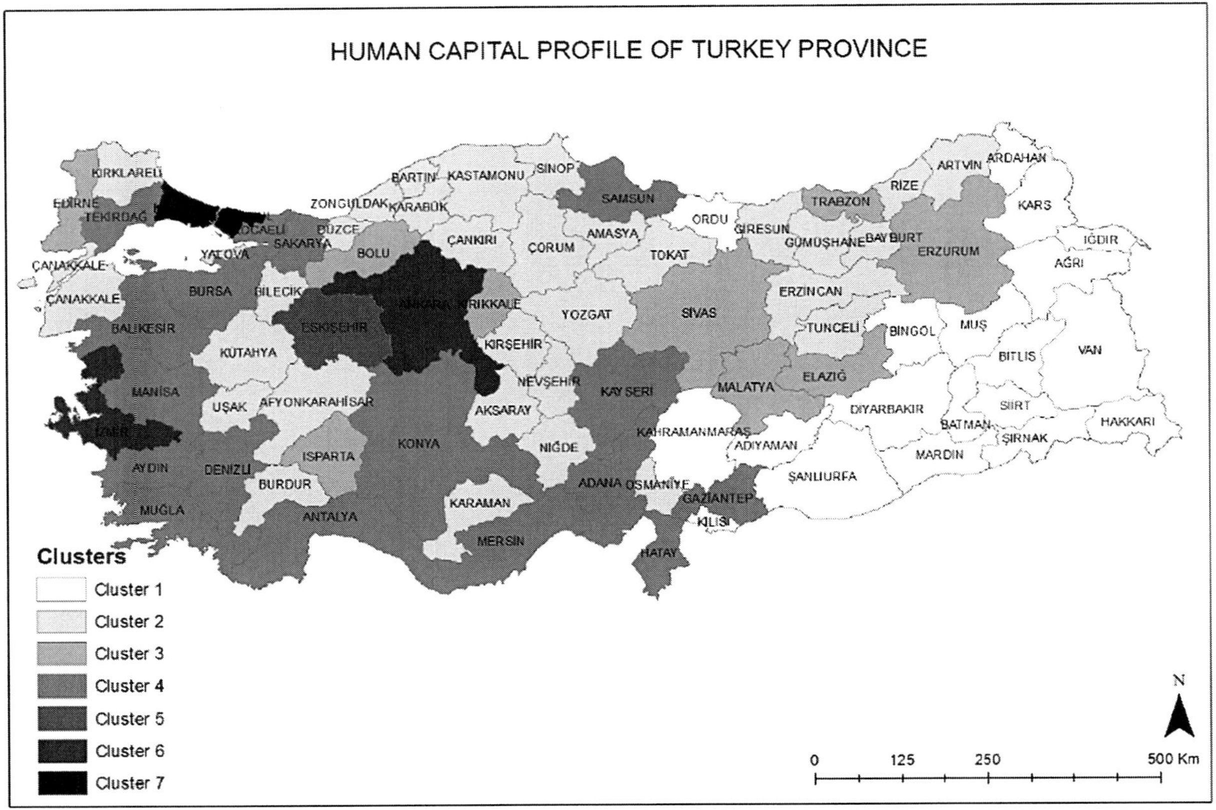

Fig. 3. Distribution of human capital according to Two-Step Clustering Analysis

The human-focused study, which reflects the distribution of human capital, was compared with the 6-level grouping of the 2017 "SEGE-2017 City and Regions Socio-Economic Development Ranking Research", which is also a human-oriented study. When the comparison is made in terms of both studies; in the SEGE-2017 study, it is seen that all provinces included in the Level 6 and a few of the provinces in the Level 5 center are in Cluster-1 with a low human capital level. When an evaluation is made between the two studies, it is seen that the most similarity is between Level 6 provinces and Cluster-1. It is observed that Ankara, Izmir, Istanbul and Eskişehir, among the provinces in Level 1, form three clusters with high human capital levels. Also; Cluster 2, Cluster 3 and Cluster 4 consist of provinces at different levels in SEGE rankings; it is seen that Cluster 2 consists mainly of provinces in the Level 4 and Level 5, and Cluster 4 consists of provinces in Level 2 (Tab. 5).

Tab. 5. Socio-Economic Development Ranking of Provinces and Regions (Ministry of Development, 2017)

Level 1	**Level 2**	**Level 3**	**Level 4**	**Level 5**	**Level 6**
Ankara	Aydın	Adana	Amasya	Bayburt	Adıyaman
Antalya	Balıkesir	Burdur	Afyonkarahisar	Çankırı	Ağrı
Bursa	Bilecik	Düzce	Aksaray	Erzurum	Ardahan
Eskişehir	Bolu	Gaziantep	Artvin	Giresun	Batman
İstanbul	Çanakkale	Karaman	Bartın	Gümüşhane	Bingöl
İzmir	Denizli	Kırıkkale	Çorum	K.maraş	Bitlis
Kocaeli	Edirne	Kütahya	Elazığ	Kilis	Diyarbakır
Muğla	Isparta	Mersin	Erzincan	Niğde	Hakkâri
Tekirdağ	Karabük	Rize	Hatay	Ordu	Iğdır
	Kayseri	Samsun	Kastamonu	Osmaniye	Kars
	Kırklareli	Trabzon	Kırşehir	Sinop	Mardin
	Konya	Uşak	Malatya	Tokat	Muş
	Manisa	Zonguldak	Nevşehir	Tunceli	Siirt
	Sakarya		Sivas	Yozgat	Şanlıurfa
	Yalova				Şırnak
					Van
					Küme

As a result; when examined clusters occurring on human capital structure in Turkey, it is necessary to increase the values of the variables with state policies, local policies or planning decisions for the variables with the lowest values of the provinces with low human capital levels. Thus, one of the most important dynamics will be successfully achieved with the investments to be made in smart people in order to speak of the existence of smart cities. In addition, with these strategies to be produced; Turkey can be reduced at times sharp distinction between the eastern and western provinces that has existed from the past, there will be more sustainable and livable city.

4 Conclusion

Smart cities are models that aim to achieve urban sustainability by using information communication technologies to create cities based on people, collaboration and technology. All solutions to be produced in these smart systems that produce location-specific models instead of today's uniform city models should be planned and organized in an integrated manner. It is also aimed to improve the quality of urban life and to save on natural resources and energy with technology-based solutions for cities that are handled in a holistic manner. While the development of smart cities and the sustainability of cities is possible with the development of smart people, a healthy and happy urban environment will be achieved as long as people transfer their needs and expectations to the system well. Human-centered smart city approaches should make use of participatory technologies to create collective intelligence by connecting all communities in the city. The human component of the smart city is directly related to learning/education and should eliminate digital and technological information differences. For this reason, in the process of being smart, it is essential that all stakeholders in the city gain awareness and education and increase human capital.

Smart city approaches, which emerged as a solution proposal as a result of urban problems and technological developments in today's world, are considered as a strategy generation or development tool. In the Tenth and Eleventh Development Plan and Information Society Strategy (2015–2018) [43;44] documents are recommended developing cities with smart systems. Lack of smart human resources, lack of accessible data, lack of using technology-based learning methods, and implementation differences between provinces are the deficiencies in front of smart city creation. With this study, a sharp distinction was observed in terms of human capital clusters between eastern and western provinces, which have differences in terms of socio-economic development levels within the country. In addition to clusters determined based on human capital, the values of the variables and the variables that are effective in the formation of the clusters will guide urban policies and strategies to be proposed in the formation of smart people and therefore smart cities. As a result, in the technology-based developing world, applications that address smart city components should be developed in order to improve urban sustainability and increase the quality of urban life, and it should be addressed in integrated ways at different levels.

References

1. Ateş, M., & Önder, D. E. (2019). 'Akıllı Şehir' Kavramı ve Dönüşen Anlamı Bağlamında Eleştiriler. Megaron, 14(1).
2. Neirotti, P., De Marco, A., Cagliano, A. C., Mangano, G., & Scorrano, F. (2014). Current trends in Smart City initiatives: Some stylised facts. Cities, 38, 25–36.
3. Ojo, A., Dzhusupova, Z., & Curry, E. (2016). Exploring the nature of the smart cities research landscape. In Smarter as the New Urban Agenda (pp. 23–47). Springer, Cham.
4. Komninos, N. (2006). The architecture of intelligent clities: Integrating human, collective and artificial intelligence to enhance knowledge and innovation. In *2006 2nd IET International Conference on Intelligent Environments-IE 06* (Vol. 1, pp. 13–20). IET.
5. Caragliu, A., Del Bo, C., & Nijkamp, P. (2011). Smart cities in Europe. *Journal of urban technology, 18*(2), 65–82.
6. Angelidou, M. (2016). Four European smart city strategies. Int'l J. Soc. Sci. Stud., 4, 18.
7. Giffinger, R., & Gudrun, H. (2010). Smart cities ranking: an effective instrument for the positioning of the cities?. ACE: architecture, city and environment, 4(12), 7–26.
8. Giffinger, R., Fertner, C., Kramar, H., Kalasek, R., Pichler-Milanović, N., & Meijers, E. (2007). Smart cities: Ranking of european medium-sized cities. vienna, austria: Centre of regional science (srf), vienna university of technology. www. smart-cities. eu/download/smart cities final report. pdf.
9. Elvan, L. (2017). Akıllı şehirler: lüks değil ihtiyaç. İTÜ Vakfı Dergisi, 77, 7–8.
10. Öktem, M. K., & Mutdoğan, A. S. (2020). Yeşil Kampüs: Kapsam| Uygulama| Yönetim.
11. Nam, T., & Pardo, T. A. (2011). Conceptualizing smart city with dimensions of technology, people, and institutions. In Proceedings of the 12th annual international digital government research conference: digital government innovation in challenging times (pp. 282–291).
12. Erdoğan, G. (2019). Akıllı Kent Göstergeleri Ve Stratejileri. *Yönetim Bilişim Sistemleri Dergisi, 4*(2), 1–23.
13. https://www.akillisehirler.gov.tr/akilli-insan/
14. Chourabi, H., Nam, T., Walker, S., Gil-Garcia, J. R., Mellouli, S., Nahon, K., & Scholl, H. J. (2012). Understanding smart cities: An integrative framework.

In *2012 45th Hawaii international conference on system sciences* (pp. 2289–2297).

15. Aksu, L. (2016). Türkiye'de beşeri sermayenin önemi: iktisadi büyüme ile ilişkisi, sosyal ve stratejik analizi. *İktisat Politikası Araştırmaları Dergisi, 3*(2), 68–129.
16. Denison, E.W., (1962)."Education,Economic Growth and Gaps in Information", The Journal of Political Economy, Vol:LXX,No:5,Part:2,s.124–128.
17. Becker, G. S. (1962). Investment in human capital: A theoretical analysis. *Journal of political economy, 70*(5, Part 2), 9–49.
18. Schultz, T. W. (1961). "Beşeri Sermayeye Yapılan Yatırım". (Çev.) F. GÜRKAN. İktisadi Kalkınma Seçme Yazılar, ODTÜ, B.F. Yayın No:6:406–424, Ankara, 1966
19. Özyakışır, Deniz. (2011). Beşeri Sermayenin Ekonomik Kalkınma Sürecindeki Rolü: Teorik Bir İnceleme. Girişimcilik ve Kalkınma Dergisi, 1(6), ss.46–71.
20. Berber, M., YILDIZ, E. B., & DİNDAROĞLU, Y. A. (2013). Bölgesel beşeri sermaye yeterliliğinin ölçülmesi: piramit model. *Kocaeli Üniversitesi Sosyal Bilimler Dergisi*, (25), 1–18.
21. Husz, M. (1998); Human Capital, Endogenous Growth, and Government Policy, Peter Lang Pub. Inc., Frankfurt am Main, New York.
22. Kızılkaya, O., Alver, A., Sofuoğlu, E. (2015). Sosyal Sermayenin Beşeri Sermaye Ve Bilgi Ekonomisi Oluşumundaki Rolü. 1. Uluslararası Uygulamalı Bilimler Kongresi: "Sosyal Sermaye".
23. Yumuşak, İ. (2008). Beşeri sermayenin iktisadi önemi ve Türkiye'nin beşeri sermaye potansiyeli. *Sosyal Siyaset Konferansları Dergisi*, (55).
24. Sezgin, Y., Bozdağlıoğlu, E.Y. (2017). Türkiye'de beşeri sermayenin durumu (2005–2015). *Osmaniye Korkut Ata Üniversitesi İktisadi ve İdari Bilimler Fakültesi Dergisi, 1*(1), 48–67.
25. Neirotti, P., De Marco, A., Cagliano, A. C., Mangano, G., & Scorrano, F. (2014). Current trends in Smart City initiatives: Some stylised facts. *Cities, 38*, 25–36.
26. Angelidou, M. (2014). Smart city policies: A spatial approach. *Cities,41*(Supplement), S3-S11.http://dx.doi.org/10.1016/j.cities.2014.06.007
27. Timm, N.H. (2002): Applied Multivariate Analysis. New York: Springer.
28. Shih, M. Y., Jheng, J. W., & Lai, L. F. (2010). A Two-Step Method For Clustering Mixed Categroical and Numeric Data. Tamkang Journal o Science and Engineering, 13(1), 11–19.
29. Copus A (1996) A rural development typology of European NUTS III regions, working paper 14, agricultural and rural economics department, Scotish Agricultural College, Aberdeen

30. Baum, S., Trapp, C., & Weingarten, P. (2004). Typology of rural areas in the CEE new Member States. Assessing rural development of the CAP.
31. Vincze, M., & Mezei, E. (2011). The increase of rural development measures efficiency at the micro-regions level by cluster analysis. A Romanian case study. Eastern Journal of European Studies, 2(1), 13.
32. Ballas, D., Kalogeresis, T., & Labrianidis, L. (2003). A comparative study of typologies for rural areas in Europe, 43rd Congress of the European Regional Science Association: "Peripheries, Centres, and Spatial Development in the New Europe", 27th - 30th August 2003, Jyväskylä, Finland
33. Haır, J.F., Anderson, R.E., Tatham, R.L. And W.C. Black: Multivariate Data Analysis. Fifth Edition, Upper Saddle River, New Jersey 1998.
34. Wu, X., Zhan, F. B., Zhang, K., & Deng, Q. (2016). Application of a two-step cluster analysis and the Apriori algorithm to classify the deformation states of two typical colluvial landslides in the Three Gorges, China. Environmental Earth Sciences, 75(2), 146.
35. Michailidou, C., Maheras, P., Arseni-Papadimititriou, A., Kolyva-Machera, F., & Anagnostopoulou, C. (2009). A study of weather types at Athens and Thessaloniki and their relationship to circulation types for the cold-wet period, part I: two-step cluster analysis. Theoretical and applied climatology, 97(1–2), 163–177.
36. Satish ve Bharadhwaj, 2010. Information Search Behaviour among New Car Buyers: A Two-Step Cluster Analysis. IIMB Management Review, 22(1–2), 5–15.
37. Ceylan, Z., Gürsev, S. ve Bulkan, S. (2017). İki Aşamalı Kümeleme Analizi ile Bireysel Emeklilik Sektöründe Müşteri Profilinin Değerlendirilmesi. Bilişim Teknolojileri Dergisi, 10(4), 475–485
38. Raggi, M., Mary, S., Santini, F., & Paloma, S. G. Y. (2013). A classification of European NUTS3 regions. Publications Office of the European Union.
39. Bacher, J., Wenzig, K., & Vogler, M. (2004). SPSS TwoStep Cluster-a first evaluation.
40. http://www.turkstat.gov.tr/Start.do;jsessionid=mWB1fLkPkvc6dJqv0sCn8Mhn8cd3h9W5KrlLpPXq2ryD2gbQTTl4!-1533787143
41. https://www.turkpatent.gov.tr/TURKPATENT/?lang=en
42. https://www.sbm.org.tr/en
43. http://www.sbb.gov.tr/wp-content/uploads/2018/11/Onuncu-Kalk%C4%B1nma-Plan%C4%B1-2014-2018.pdf
44. http://www.sbb.gov.tr/wp-content/uploads/2019/07/OnbirinciKalkinmaPlani.pdf

Serap YILMAZ, Tuğba DÜZENLİ, Elif Merve ALPAK

Aesthetics in Landscape Design

Abstract: Aesthetics is described as the science of beauty or learning natural harmony, and aesthetics are derived from the natural order and the sense of satisfaction that an individual feels due to this order. Landscape aesthetics, on the other hand, originates not from one of its elements, but the whole landscape and the sensory data created in the observer. Thus, the aesthetics in landscape is defined as the satisfaction of an individual with the landscape. It was observed that there was a strong correlation between landscape aesthetics and the visual quality created by all the components of a landscape. Thus, landscape architects should understand two landscape components; aesthetic objects that trigger an aesthetic experience and the aesthetic experience itself. The present study was structured in two stages: in the first stage, the study focused on the analysis of aesthetic experience and landscape aesthetics, and the second stage discussed how the designer could guide the aesthetic experience of the landscape user based on landscape silhouettes. Thus, the features of aesthetically successful silhouettes were determined.

Keywords: Landscape aesthetics, landscape design, visual quality.

1 Introduction

Landscape architects often use the term aesthetics as a synonym for design style or form, yet aesthetics have the power to alter the way people react to the environment. Therefore, the basic philosophy in landscape aesthetics and its association with design are important. Because the analysis of visual satisfaction in outdoor spaces is an important topic in landscape research and has been generally associated with landscape aesthetics. Landscape aesthetics do not originate in its components, but from the whole landscape and the sensory data formed in the observer [1]. Richard E. et al. (1990), on the other hand, described landscape aesthetics as the satisfaction of an individual with the landscape [2]. Thus, it could be observed that there was a strong correlation between landscape aesthetics and the visual quality created by all landscape components. Daniel (2001) supported this idea and described visual quality in a landscape as associated with the aesthetic perfection of a landscape. Aesthetic perfection is associated with the pleasure of the viewer with the landscape [3]. Thus, the individual assessment of a landscape depends on the visual data collected from the landscape, as well as the personality and culture of the viewer [4]. Thus, the features of the landscape components directly affect the aesthetic quality of the space. To determine the design criteria for these components, their effects on the visual quality of the landscape; thus, on landscape aesthetics should be scrutinized. The designer could not affect the personal traits of the user but could guide the aesthetic satisfaction levels with the landscape based on these criteria.

Lang's (1994) statement that aesthetics are hidden in the clues provided by the designer, not in the eyes of the beholder, confirm this approach. Thus, how to combine plants to provide aesthetic satisfaction and how the aesthetics of a designed landscape can be analyzed are important issues. Landscape aesthetics are analyzed based on symbolic and formal aesthetics:

1. Symbolic aesthetics investigate meaningful relationships in the environment
2. Formal aesthetics investigate the order, form, rhythm, and complexity of the visual world [5].

Thus, the designer seeks answers to the following questions in the scope of the present study:

1. How can the order, form and rhythm of the plants in a silhouette be organized to provide aesthetic satisfaction for the viewers in a landscape?
2. Can the plant silhouettes be analyzed based on formal aesthetics to obtain satisfaction?

Thus, the conceptual framework of the study was constructed based on the formal aesthetic approach and Gestalt principles. Gestalt theory, which was particularly based on the perception of form, is an important theory that could be employed to achieve formal aesthetics in plant silhouettes.

2 Formal Aesthetics

Formal aesthetics analyze the actual geometric qualities of environmental forms and the measurable physical properties and visual structure of the designed-natural world with a holistic approach [5, 6, 7]. Formal aesthetics focus on the existence of formal features such as order, rhythm, repetition, and harmony in the whole, and assume that the forms have an effect on the neurological processes that make the environment legible and pleasing. The features that should include a visual composition were analyzed based on the following criteria in various studies on formal aesthetics (Tab. 1):

Tab. 1. The properties of visual composition in formal aesthetics

Visual composition in formal aesthetics	
Nasar (1994)	Form, proportion, rhythm, scale, complexity, color, light, shadow, order, hierarchy
Lang (1987)	Order, rhythm, harmony and complexity
Daniel & Vining (1983)	Order, rhythm, harmony and complexity
Ylmaz et al. (2018)	Order, rhythm, harmony and complexity

Formal aesthetics is associated with design principles that allow decisions - on the geometric structure of the environment. Therefore, researchers, who determined the features of visual experiences, based their analyses on attitudes established by Gestalt psychology and cognitive science [7, 8, 9].

3 The Gestalt Theory

The Gestalt theory influenced architects and environmental designers more than other perception theories and allowed them to understand the aesthetic perceptions about the environment [6]. Gestalt theory discussed how we perceive objects in the environment and addresses the relationships between the entire composition and its parts. According to the Gestalt theory, the whole means more than its parts. Because an individual does not perceive the elements but the whole composition. Thus, the Gestalt Theory is focused on grouping and form-background perception [9].

According to the Gestalt theory, a well-organized form is perceived as a form with respect to the background formed by all natural and artificial elements in the environment. When the "form" is strengthened, aesthetic appreciation and pleasure start [6]. Thus, Gestalt psychologists discussed the factors that affect the perception of the form [10, 11] and how to group visual elements to express harmony with the principles listed in Tab. 2 [12].

Tab. 2. Gestalt principles

Proximity: Closer elements are perceived as a group.	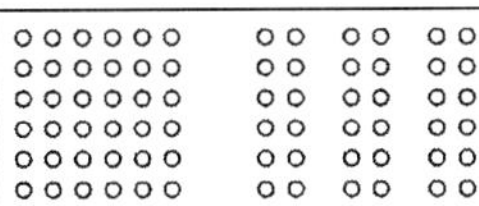
Similarity: If the elements of an organization have similar size, form, color and texture, they create a group.	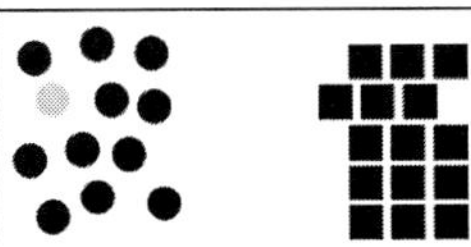
Closure: When certain parts of previously known objects are missing, the mind completes the object and it is perceived as a whole.	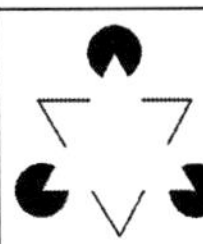
Common region: In a composition where elements move in different directions, forms that tend to move towards the same direction are perceived as a whole.	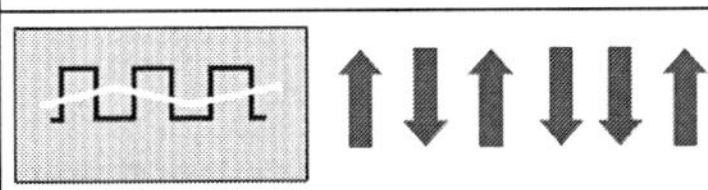
Continuity: Similar elements that move to the same direction are perceived as a group.	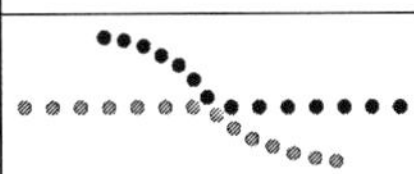

The psychological organization of a visual composition that includes these principles is defined as "good" (not an assessed state, but a quality defined by form). Good forms include features such as harmony, order, similarity, closure, and simplicity. Environments with these features are considered "orderly", and "easily perceived". Thus, these environments have "unity" [13] and this is very important for both formal aesthetics and visual perception. Good environment and good form determined by the Gestalt Theory are accepted by most designers [6].

4 Planting Design

The combination of the form, size, texture and color of the landscape space and the combination of the plants that are the most effective spatial components are important to achieve visual quality [14]. Thus, plant compositions are created based on certain criteria (Tab. 3):

Tab. 3. Planting design principles

Planting design principles	
Booth (1990) [15].	Harmony, contrast, balance, emphasis, order and diversity
Robinson (1994) [16].	Harmony, contrast, balance, emphasis, hierarchy, scale, movement and perspectives
Bell (1999) [17].	Harmony, contrast, rhythm, unity, balance, emphasis, measure and diversity

Planting design aims to contribute to the spatial functions and aesthetics; thus, respond to the desires of the occupants, a sense of satisfaction [16], facilitate the recognition of natural plant communities, and the restoration of the human/nature balance by creating a suitable wildlife habitat [18]. Planting design directly affects the beauty of the landscape which recently became an important component of landscape planning and management strategies [3, 19]. The aesthetic satisfaction with the landscapes with good aesthetic qualities and scenic beauty directly affects the user preference, occupancy and positive perception [2]. Thus "aesthetic satisfaction" is an important goal in planting design.

5 The Correlations Between Formal Aesthetics, the Gestalt Theory, and Planting Design

In order to speak of "unity" and therefore "order" in a design, the features of the elements of that design such as size, form, color and texture should be combined based on the Gestalt principles. This fact provides the association between formal aesthetics, which originates in the perception of form, gestalt theory and planting design [9]. The correlations between formal aesthetics, the Gestalt theory, and planting design are summarized below:

The aesthetic value of the landscape stems from the whole landscape, not from a component [1].

Thus, landscape aesthetics coincides with the Gestalt theory that the whole reflects more than the total of the parts.

The correlation between landscape aesthetics and Gestalt theory

According to Gestalt principles, the meaning of the whole is not based on the parts but the principles that combined the parts.

When the plants are organized based on design principles (harmony, contrast, rhythm, unity, etc.), the whole, not the parts, is perceived [9].

The correlation between Gestalt theory and planting design

Plant silouettes with unity

6 How to Acquire Formal Aesthetics in Landscape Silhouette?

The individual intuitively understands the rules in the environment, analyzes these rules based on the transmissions between the eye and the brain, and transforms this information into an intuitive response [20, 21, 22, 23, 24]. Thus, the designer could guide the individual viewing the landscape the way the designer desires with clues that would lead to aesthetic satisfaction. How could this be accomplished in a landscape silhouette? The answer could be determined with the analysis of the plant silhouettes based on formal aesthetics. Thus, the correlations between planting design, formal aesthetics, gestalt theory and aesthetic satisfaction could be summarized as follows (Adapted from Yılmaz et al., 2018) [9] (Fig. 1):

The designer creates groups of plants with similar properties (form, texture, color) based on Gestalt principles, especially the similarity and proximity principles.

Plant silhouettes include a formal aesthetic property when designed as follows (Fig. 2):

- The rhythm is enforced by continuity between the visual properties of the plants and by repeating the plant groups at certain intervals. Thus, the plant composition acquires a distinct "order".
- The "complexity" should be improved by the principles of harmony and contrast between the visual plant features.

- The harmony in form, color and texture allows the plants in the plant silhouette to create a holistic visual effect, and the contrasts created in the whole silhouette prevents monotony. Thus, achievement of the harmony and prevention of monotony leads to order and similarity.
- Groups are formed with the visual features of the plants (form, texture, color) to achieve a he rhythmic change in plant composition. Because this amplifies complexity.

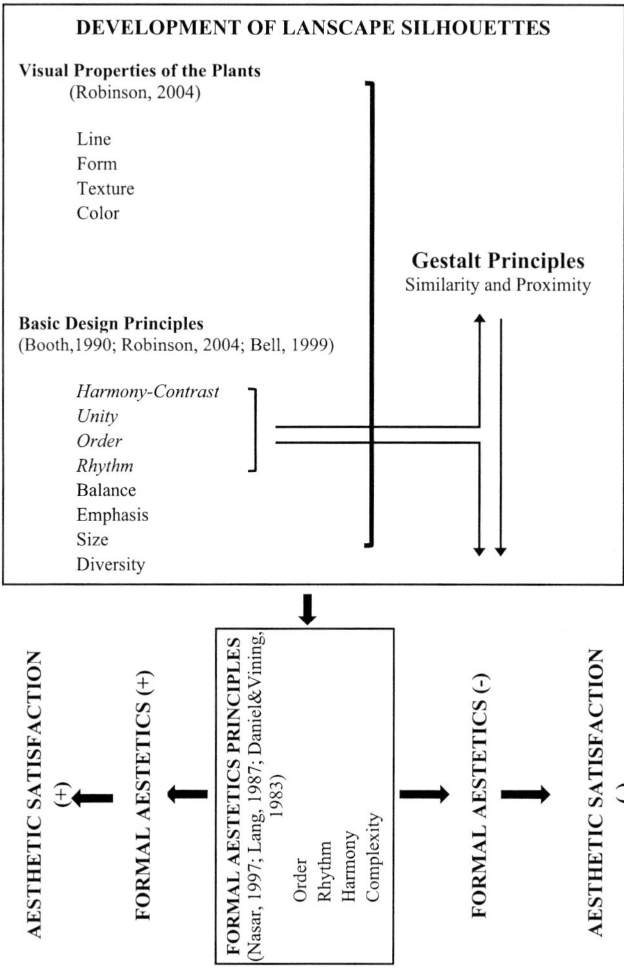

Fig. 1. The correlations between planting design, formal

The "rhythm" was enforced with continuity between the texture and form of the plants and repetition of the plant groups at certain intervals..

The "complexity" was improved with the employment of the principles of harmony and contrast between forms and textures.

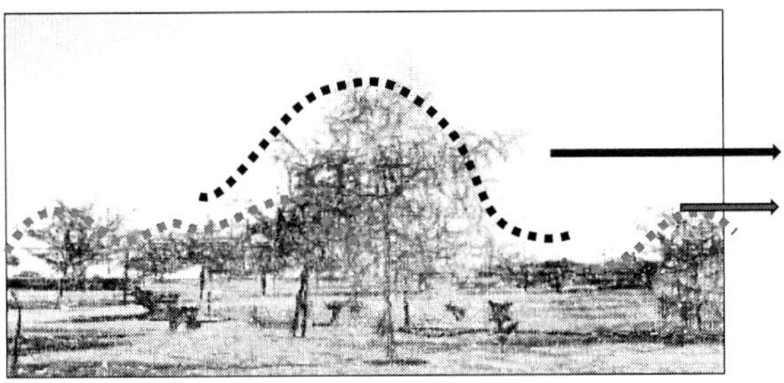

To ensure harmony between plant forms and textures;

- The repetition of plant forms and textures should be diversified with contrast components to improve the complexity of silhouettes and reduce the feeling of monotony caused by repetition.
- There should be hierarchical variation and repetition between the sizes of plants of similar form and texture.

Fig. 2. The formal aesthetics analysis of plant silhouette

7 Conclusion

One of the important functions of plants is the aesthetic value they add to the space. This aesthetic value is provided due to the dendrological features of plants as well as visual features such as form, color and texture. Thus, the value created by the formal and textural juxtaposition of plants in a silhouette is very important. This value is measured with "formal aesthetics", which investigates the order, form, rhythm and complexity of that silhouette. The formal aesthetics of a landscape silhouette facilitate the environmental perception of individuals. Because formal aesthetics entails order, rhythm, harmony and complexity. All these principles are provided by the designer's approach to grouping the elements in a silhouette. Grouping could be achieved by Gestalt organization principles in a silhouette. Because the Gestalt theory provides perception rules that organize the factors which affect the perception of form. These rules, which include proximity and similarity, allow a holistic perception and provide a planting design with high aesthetics due to improved readability and consistency. Thus, the relationship between formal aesthetics and the Gestalt theory, which focuses on the perception of the form, should not be neglected in planting design. When plant silhouette lacks repetitive texture, repetitive form and rhythmic variations among these, neither formal aesthetics nor aesthetic satisfaction are possible. The present study investigated the principles required for a successful plant silhouette based on formal aesthetics, and it was observed that the silhouettes with similarity-closeness Gestalt principles included order and harmony relations. Complexity is achieved by the textural and formal diversity in addition to these relationships. Well-adopted diversity encorces unity in the silhouette and leads to a more successful formal effect of the silhouette.

Environments designed with these principles are important for the perception of the environment by individuals, reading the environment, and aesthetic satisfaction. Thus, these principles guide the designers to create strong silhouettes with formal aesthetics.

References

1. Beza, B.B. (2010). The aesthetic value of a mountain landscape: A study of the Mt. Everest Trek. *Landscape and Urban Planning*, 97, 306–317.
2. Richard, E.C., Gobster, P.H. (1990). The Nature and Ecology of Aesthetic Experiences in the Landscape. Landscape Journal, 9, 1–8. doi- 10.3368/lj.9.1.1
3. Daniel, T. (2001). Whither scenic beauty? visual landscape quality assessment in the 21st century. Landscape and Urban Planning, 54, 267–281.
4. Bishop, I.D., Ye, W.S., Karadaglis, C. (2001). Experiential approaches to perception response in virtual worlds. Landscape and Urban Planning, 54, 115–123.

5. Lang, J. (1988). Symbolic aesthetic in architecture: toward a research agenda. In: Nasar J. L. (Eds), Enviromental aesthetics theory, research, and applications (pp.11–26).New York: Cambridge University Press.
6. Lang, J. (1987). Creating Architectural Theory: The Role of the Behavioural Sciences in Environmental Design. New York: Van Nostrand Reinhold.
7. Lang, J. (1994). Urban Design: The American Experience. New York: John Wiley&Sons. Inc.
8. Habibi, A. (2017). New Approaches to the Landscape Aesthetics Research. BAGH-E NAZAR, 14(49), 69–76.
9. Yılmaz, S., Özgüner, H., & Mumcu, S. (2018). An aesthetic approach to planting design in urban parks and greenspaces. Landscape research, 43(7), 965–983.
10. Zusne, L. (1970). Visual Perception of Form. New York: Academic Press.
11. Koffka, K. (1999). Principles of Gestalt Psychology. (1st ed.). London: Routledge Taylor and Francis Group.
12. Reardon, A.P. (2004). Beyond Intuition: A Perceptual Basis for The Elements and Principles of Design. Master of Science in Interior Design, Michigan University.
13. Reis, A.T.L., Lay, M.C.D. (2010). Internal and external aesthetics of housing estates. Environment and Behavior, 42, 271–294. doi-10.1177/0013916509334134
14. De La Fuente De Val, G., Mühlhauser S. H. (2014). Visual quality: An examination of a South American Mediterranean landscape, Andean foothills east of Santiago (Chile). Urban Forestry and Urban Greening, 13(2), 261–271.
15. Booth, N. (1990). Basic Elements of Landscape Architectural Design. USA: Waveland Press.
16. Robinson, N., 2004. The Planting Design Handbook. (2nd ed.). England: Ashgate Publishing Limited.
17. Bell, S. (2004). Elements of visual design in the landscape. (2nd ed.) London: Spon Press.
18. Dee, C. (2001). Form and Fabric in Landscape Architecture. (1st ed.). London: Spon Press.
19. Scott, A. (2002). Assessing public perception of landscape: The LANDMAP experience. Landscape Research, 27, 271–295.
20. Laurie, I. C. (1975). Aesthetic Factors in visual evaluation. In Ervin H. Zube, Robert O. Brush, & Julius Gy. Fabos (Eds.) Landscape assessment: Values, perceptions, and resources (pp. 102–117). Stroudsburg, PA: Dowden, Hutchinson & Ross.

21. Tarakci-Eren, E. Duzenli, T. (2017). Determınatıon Of The Vısual Preference Levels and Perceptual Dıfferences in The Appearance of Certain Taxa in Dıfferent Seasons. Feb-Fresenius Environmental Bulletin, 26/12A, 8304–8315.
22. Düzenli, T., Alpak, E. M., Tarakci Eren, E., (2017). Artistic Plant Representation Techniques in Landscape Archıtecture. NWSA Fine Arts, 12, 177–184.
23. M. Ertekin, Ö. L. Çorbacı, (2018). Environmental Design Art in Seljuks, Journal of Scientific and Engineering Research, 5, 122–126.
24. Tarakci Eren E., Düzenli T. (2017). Determinatıon of The Visual Preference Levels And Perceptual Dıfferences in The Appearance of Certain Taxa in Dıfferent Seasons. Fresenius Environmental Bullettin, 26(12A), 8304–8315.

Selin ARABULAN

Human Effect on Urban Identity

Abstract: Identity is the way of defining an entity with different features such as visual perception, past information, and behavioral patterns. Urban Identity, on the other hand, is the integrity formed by the determinants that differs the city from others.

Designs, to be made by investigating the identity of city or part of the city, will be designs that belong to the place, exist for the place. According to this view, first the concept of identity was explained in the study, the use and importance of identity in design was specified, identity components, which are the main subject of the study, were defined item by item and evaluated through examples. After the environmental and social components are cited in this way, the human factor, the cornerstone of society, was detailed. The study differentiates from other similar studies in that it focuses on the human influence on urban identity. It is thought that the study will be beneficial on identity-based studies in terms of both providing a theoretical background and presenting current topic recommendations.

Keywords: Design, urban identity, environmental identity, social identity, identity component, identity actors.

1 Introduction

The city is getting more and more into social life every day. All the patterns, behavioral styles, problems and expectations of human life are becoming urbanized day by day. According to World Bank 2019 data, 55 % of the world's population lives in cities [1]. This ratio was determined as 92.78 % according to Turkstat data in Turkey [2].

Cities and parts of cities, that have experienced a continuous improvement due to demographic changes caused by migration from rural to urban and also due to economic and social developments, were re-planned, designed and built in this process. Due to political developments, globalization and the opposite developed ideas of localization, the cities and their parts have entered the process of competing, highlighting of unique qualities, differentiation and branding. In the design and production process between this racing and living, it is of great importance that the city dwellers and cities maintain their identity by preserving.

The main purpose of this study is to understand and convey the city and its unique identity better and more accurately.

The questions of the study are as follows:

What are the main features of urban identity? What are the elements that generates the urban identity? Who are the actors effective in the formation of urban identity? According to which headings can identity analysis be done?

By finding the answers to these questions to understand the cities and their inhabitants, to find the original values of the city and to highlight it, identifying structures and areas that damage identity and most importantly, to contribute to the city with appropriate urban designs is targeted. The unique aspects of the research are that all factors that form the urban identity are dealt within all dimensions and the human factor is emphasized.

The research methods as, concept analysis based on the detailed explanation of a concept and its features, evaluation methods in which the operability of a process is examined and evaluated, were being used in the study. Studies that the identity of a city or its part was examined require extensive and different investigations. Ethnographic, Historical Research, Review-Meta Analysis/Syntesis, Hermeneutics methods are also used in the studies conducted within this scope. Document analysis was used as a data collection tool. By searching under the keyword of urban identity; newspaper articles, papers, articles, books and theses published in written and electronic media were examined.

identity and urban identity being the basic concepts are defined and identity component in design and its importance are explained. In the next step, the components that form the identity were explained and briefly sampled. As a result, using identity value determined with the help of these components as data in designs and keeping cities up-to-date and livable in this way is aimed.

2 Identity and Urban Identity

The word identity is defined by TDK (2020) as "the totality of signs, qualities and characteristics that show what kind of person a human is as a social entity" and "The whole of the features that are used to determine any object"[3].

Identity, according to Kevin Lynch, is the discernible uniqueness of an object with other objects and he emphasizes that it will be formed by differences, not by similarities. Identity is used in the sense of individual or uniqueness, not equality with other entities. Identity elements are the main determinants of urban perception and urban image [4].

Cities are organized places where diverse human communities are located and concentrated, where various goods, services, and ideas are produced and exchanged [5]. They differ from other settlements with social and environmental concepts such as socio-cultural level, economic structure and size [6].

Kevin Lynch explains the concept of city, that has many functions together, makes sense according to people's perception characteristics and as a constantly developing and changing structure. According to Lynch, cities are places where certain circulation axes, large usage - facilities, focal - landmarks and residents create a common memory [4]. Urban spaces, which are both the founders and products of urban life, are places where the citizens create a common production by coming together and the perceptions of the space are being lived [7]. Urban spaces are formed by the multiple identities, lifestyles and cultural diversity of the people living in them. Beyond being a place/space, it corresponds to social belonging, identity and cultural pluralism. Urban spaces emerge spontaneously as those who live on it continue their vital activities. Through these spaces, the citizens of the city establish a relationship of belonging, identity and attachment [8].

Urban identity in its broadest sense can be explained as a set of determinative components that define a city and distinguish it from others. Urban identity is the result of a collective process based on interpretation and expression rather than purely design features [9,10,11].

3 Identity Factor in Design

It can be possible to intervene in the identity of a city, to change it, only by understanding it and being a part of it. Without thinking much about the search for identity, either the arrangements made by exaggerating a past symbol of the city, or a project implemented in a hurry, often, it results, not the enrichment of the city's identity, but rather the degeneration and becoming commonplace. [12].

People design, direct and use the space as they perceive it. Urban identity arises from the interaction between human and space. The way the space is perceived by city users is important for the formation, definition, determination and analysis of urban identity.

4 Urban Identity Elements

Urban identity is formed by the combination of many layers in a historical process, as a product of the interactions between human and space. Urban identity, which includes multiple parameters such as natural environmental features, structured environmental features, social environment features, behavioral styles, traditions, and cultural level, has a dynamic structure due to the change and development of these characteristics over time.

Urban identity has social, economic and physical aspects that affect the city image. It is formed by considering the meaning, society attributes to the space, together with the physical environment. In addition to these factors, the political, economic, cultural and historical features of the society also constitute the social identity of the period and affect the urban identity and also its urban space [13]. While natural and artificial environmental data are also highly effective in the formation of urban identity, the most important share belongs to people. For this reason, when all data are examined in detail, all actors involved in the formation of identity under the heading of social identity are discussed in more detail.

4.1 Environmental Identity

It is the identity of the city's physical structure. This physical structure can be examined in two ways: Natural Environmental Features, Artificial Environmental Features.

Natural environmental features, geographical data and climate and vegetation characteristics are studied under the subtitles. Geographical data is the component of the settlement that gives the land features and topographic formations. The city has a character and identity based on these features. The texture characteristics of the region where the city was founded and expanded affect the identity both in settlement and building scale. The settlement type may differ depending on the condition of the land on which it is located. Apart from the topographical features, some geographical formations also contribute to the urban identity. This can be a lake, river or valley. Serengeti National Park in Tanzania is on the migration route of many wild animal species due to its geographical conditions. This has caused the region to develop in terms of safari tourism and to be identified in this way.

It is seen that climatic features have an important role in the formation of urban identity by affecting the characteristics of settlements and even the characters of people. According to climatic conditions, buildings and therefore settlements can have very different characteristics. It has an effect from the choice of material to be used in buildings to the roof shape. An example for this feature can be the given name as "city of rain" to Seattle because of its cool and rainy weather.

Artificial environment is created by human interaction with the environment and also defines the area where the structures for protection are predominant. We can examine this environment in three groups in terms of settlement, reinforcement and symbolic elements.

Urban settlements consist of many components that come together for different purposes, in different conditions and in different forms. There are many different interpretations in defining and classifying these components.

Lynch revealed the sensory and mental processes of space perception. He emphasized that there are some components that affect and support the perception of space. He defined five main components that determine the perception level of cities. These components are defined as image elements; they are grouped as paths, districts, edges, nodes, landmarks. The perceived effect of an area or a point where the relationship established between the individual and his environment becomes concrete, creates the urban image. The determinants of urban image and urban perception are also identity elements existing in the city (Lynch, 1960).

Identification of New York City can be made through its regions. Manhattan is one of the five regions of city with Brooklyn, Queens, Bronx and Staten Island and is the district of the city identified as the financial center.

The equipment scale consists of urban furniture. This group consist from the elements that establish a connection between the city and the people and try to attract people to urban life. Seating elements, street lamps, post boxes, telephone booths, waste bins etc. are effective components in urban life. London's most well-known identity items are red phone boxes.

Symbolized elements, some characteristic features of the city and are elements that add a "symbolic" value to the city's identity. These may be a television antenna, colorful linens hanging from a rope stretched from house to house on the streets, or advertising signs covering a street. An element on this scale is a component with symbolic value and can be enough to describe a city alone, like the glow of Las Vegas...

4.2 Social Identity

The main factor of identity formation is "society". Man creates the city and integrates with the city. Social identity can be examined under three sub-headings as Socio-Cultural Identity, Socio-Economic Identity and Psychological-Perceptual Identity.

Socio-cultural identity includes the social and cultural structure of the society and the characteristics of the society as a result of its "existence" and "relations".

Historical characteristics express the urbanization process that societies have created throughout history. This process is the most important factor of urban civilization and the existence in urban of the "life experience" accumulated over the centuries is only possible when these historical traces take a permanent place in the urban identity. Natural disasters faced by the city, social events and wars experienced leave deep marks in the memory of the society. The Berlin City Germany of is known for its wall built after World War II.

The most important characteristic of the population is the population density of that city. Apart from the numerical characteristics of the population, the qualitative characteristics of the population also play an important role. Demographic factors such as, distribution of population density to city, its cultural and economic structure, features of the family, the characteristics and distribution of the "urbanized" or "non-urbanized" groups of the immigrant population are effective factors in the identity formation [14]. Demographic structure (population size, structure, density, age group), institutional structure (political, administrative, legal, economic) and the knowledge, skills, traditions, customs, talents etc. gained by the individual in the social structure are the elements of socio-cultural structure. Identification of China and India with their large populations or England with the royal family can be given as an example.

Culture is the whole of tradition, life, thought and art entities that come from the past and still continue in a community. In general, all the values of society become part of the culture. Within these values, there is an important component that gives the life of society. It is the folkloric characteristics of the society. All kinds of traditions, customs, eating and drinking habits, oral and written literature, speaking style, in short all the basic features of the "daily life" of the society constitute the folkloric structure.

Globalization is a concept used to describe dissolution of national identities, economies and borders, the integration of the people of the world in a global society, the increasing process of single field perception of the world. While the world was uniting, everything that was national lost its meaning in this process. However, while this political trend was on the agenda, being distinguished and defined within the community became more important and difficult. Cultural and historical heritages have been strategically researched in order to be regional and local identity data. It can be said that the cultural elements that make up Italy's identity are the textile industry and the illegal families in history.

Socio-economic identity is the definition that forms from the employment structure of the society. The distribution of the labor force in the society to various branches (worker, civil servant, tradesman, etc.) and their place in the society give the "wealth" and therefore the economic identity of the society. The labor force type also provides the region to be defined as rural or urban. Germany's automotive technology is a part of its urban identity as it hosts important automotive brands.

Psychological-perceptual identity is the way in which people's expectations from the environment and their different perceptions of the environment are defined. This different perception leads to differences in behavior. As the psychological structure of the society manifests itself with its behavior, the way of behavior and its effects on the city affect the urban identity.

Perception is the acquisition of physical data with sensory organs, and the evaluation of these data by brain filters and it consists of two processes. These two processes, which are departed as emotional and informational, have made the space meaningful by following each other. Sensory process is the process in which data from the environment is interpreted through senses. The mental process is the process of conceptualizing environmental information, acquired with the senses or unnoticed, in the mind based on experience [15]. Tekeli explains the perception of the environment at three different levels. The first of these is cognitive perception. Cognitive perception is knowledge of the properties of objects being in the environment such as function and form. The other level of perception is perception at symbolic level. It is the attribution of associative meanings that are not related to their cognitive meaning to objects in the environment through symbols. The identity and image attributed to the city are related to this kind of perception. Another level of perception is relational perception. This level of perception is related to the aesthetic dimension, namely visual values [16].

While the person who accumulated good memories during her time there described Venice as entertaining, the same person might have described the city with more pessimistic expressions if she had unfortunate accidents on these dates.

4.3 Actors of City

City actors should be evaluated separately under the heading of social identity, because these actors are differently identity definers depending on their roles in the city.

Cities and therefore urban spaces are the subject of many legal regulations. Government bodies, local administrations, private sector, non-governmental organizations and universities are highly influential on the production of policies related to the use of urban space and thus the shaping of urban identity [17].

As far as the Central Government is evaluated, the head of the executive body in parliamentary systems is the Council of Ministers. TGNA, the Council of Ministers and the deputies in the parliament are the main actors in the change and transformation of the urban identity with the policies of migration, slum, development, economy, industry, environment, urbanization.

Governors and district governors, who are elements of local government, work for their cities on concepts and strategies such as competition around the world, investment or branding. Cities are shaped in parallel with these studies and are underwent a spatial and identity transformation. There are duties that affect the urban identity assigned to municipalities with the Municipality Law No. 5393. These duties can be listed as; urban infrastructures such as zoning, transport, geographical and city information systems, environment, cleanliness and solid waste, urban traffic, cemeteries, afforestation, parks and green areas, housing, culture and art, tourism and promotion, youth and sports; protection of cultural and natural assets, historical texture and places and functions that are important in terms of city history; reconstruction of those who cannot be preserved in accordance with their originals [18].

Non-governmental organizations and professional associations are the most important social tools in the development and updating of urban identity. Associations and foundations are being established with the aim of protecting social structure and fabric, preserving ethnic identities, protecting different groups, historical/cultural structures and identity. And the role they play on the transformation of urban identity is one of the reasons for the existence of non-governmental organizations.

Universities are seen as institutions carrying the responsibility of transforming the cities in which they are located, interacting with the social dynamics in the city and in this context, they have urban development and improvement missions.

Rousseau's statement as "houses constitute towns in; whereas citizens make up the city" in "The Social Contract" named work, is of great importance in terms of showing that it is not the buildings but the people who make the cities [19]. Recently, inhabitants or visitors of the city reveal their views about the cities by sharing their experiences on the city and through the media [20]. This shows that not only the inhabitants but also the visitors are very important for the identity of the city.

The lifestyle of the inhabitants, the way of earning income, the lives of their ancestors, their traditions and structures-buildings have determine the culture of the city. However, another very important identity data is the people who have left mark on the society and who were born or lived in this city. The house where these people were born, the workspace they used for a part of their life or the room where they made an important decision are extremely valuable for the city they are located. Netherlands - Van Gogh Museum, Mexico - Frida Kahlo Museum can be given as an example in this sense.

Fig. 1: Van Gogh Museum Netherlands [21]

Fig. 2: Frida Kahlo's Blue House, Mexico City [22]

5 Conclusion

The identity of a city; identity elements originating from the natural and artificial environment, spatial units of these elements, history, culture, geography, vegetation, climate, geopolitical position, architecture, economic level, social, political, ethnic, religious structure, lifestyle, worldview, the civilizations that they cradle, whether it was the capital of these civilizations, its relations with other cultures, the occupations, wars, disasters and so on can change according to many factors and thus gains unique features. [4,23,24,25,26,27]. In short; the information about how the city developed, history of the city and all current data reflect urban identity.

In the study, the elements that form the city's identity are explained. While the definition of identity components can be encountered a lot in the literature, there are very few studies explaining the main element, human and its role. In this context, the city actors were explained in detail in the study. The titles to be evaluated while examining identity of the cities were given respectively in the study.

Recently, the number of studies for identity inquiries on social platforms has increased. Studies can be done to examine the identity data given in the study in specified cities by these platforms. Apart from photo-based studies, identification studies can also be done through interpreting the more personal comment-oriented blogs where travel experiences are shared. Considering that the human is the most important factor in determining the identity, the methods to obtain the most accurate and numerous data, surveys and interviews should be evaluated for this studies.

Considering the direct relations of identity with the city, society and people, it is necessary to develop solutions for the correct determination and protection of urban identity. In this framework, participatory mechanisms of central and local governments should be effective. In cooperation with non-governmental organizations and universities, training, seminars or social activities can be organized in order to increase the awareness of urbanity and to bring the city together with the citizens.

In this way, to be more informed about our cities and its user and to make designs to create more livable environments is aimed.

References

1. https://data.worldbank.org/indicator/SP.URB.TOTL.IN.ZS
2. https://www.tuik.gov.tr/
3. Turkish Language Association Dictionaries, 2020, http://www.tdk.gov.tr/ date of access: 10.12.2020.
4. Lynch, K., The Image Of The City, 1960, Boston: Harvard University Pres.
5. Lozabo, E. E., Community Design and the Culture of Cities: The Crossroad and the Wall, 1990,Cambridge University Press, New York.
6. Doxiadis, C. A., Existics: An Introduction to the Science of Human Settlements, 1968, Hutchinson&co Ltd., Londra.
7. Al, M., Kentte Bellek Yıkımı ve Kimlik İnşası – Palimpsest: Ankara Atatürk Bulvarı Bağlamında Bir İnceleme. İdealkent Dergisi, 2011, Sayı:4, 22–36.
8. Aytaç, Ö., Kent Mekânlarının Sosyo-kültürel Coğrafyası. Fırat Üniversitesi Sosyal Bilimler Dergisi, 2007, Cilt:17, Sayı:2, 199–226.

9. C. Hague Planning and place identity, C. Hague, P. Jenkins (Eds.), Place Identity, Participation and Planning, Routledge, London and New York, 2005, pp. 3–17
10. Weichhart, P, Weiske, C and Werlen, B., 2006, *Place Identity und Images: Das Beispiel Eisenhüttenstadt [Place Identity and Images: The Case of Eisenhüttenstadt]*. Institut für Geographie und Regionalforschung der Universität Wien.
11. Tölle A.,Urban identity policies in Berlin: From critical reconstruction to reconstructing the Wall, Cities, Volume 27, Issue 5,2010,Pages 348-357,ISSN 0264-2751,https://doi.org/10.1016/j.cities.2010.04.005. (http://www.sciencedirect.com/science/article/pii/S0264275110000697)
12. Tekeli, İ., Post Modernizm Tartışmaları Üzerine Düşünceler, O.D.T.Ü Şehir ve Bölge Planlama Bölümü, 1991, Sonbahar Yarıyılı Seminer Programı, Ankara.
13. Özer, M. N., Kentlerimizde Kimlik Arayışlarında Bir Yaklaşım Aracı Olarak Kimlik Çözümleme Rehberi, Kent ve Bölge Üzerine Çalışmalar. 2007, Gazi Üniversitesi, Mühendislik ve Mimarlık Fakültesi, Ankara.
14. Fersan, K., "Tek Derste Uygulamalı Şehir Kimliği", Arredamento Dekorasyon, 1990, s.128
15. Lang, J., Creating Architectural Theory, The Role of Behavioral Sciences In Environmental Design, 1987, Van Nostrand Reinhold, New York.
16. Tekeli, İ., Bir Kentsel Tasarım Kuramının Geliştirilmesi Üzerinde Düşünceler. Kent Planlama, Politika, Sanat, 1994, ODTÜ Mimarlık Fakültesi Yayını, Ankara.
17. Beyazıt, E., Kent Kimliği Bağlamında Hatay, Kent Aktörleri Gözüyle Kentsel Kimlikte Kırılma Noktaları, Süleyman Demirel Üni., Sos.B.E. Yayınlanmamış Doktora Tezi, 2016, Isparta, s: 74
18. https://www.mevzuat.gov.tr/MevzuatMetin/1.5.5393.pdf erişim tarihi: 20.12.2020
19. Bookchin, M., Kentsiz Kentleşme, (Cev.B.Ozyalcın), 1992, İstanbul: Ayrıntı Yayınları, s:59
20. Kaypak, Ş., Küreselleşme Sürecinde Kültürel Kimlik Açılımları ve Kentsel Çevreye Yansıması. Uluslararası Davraz Kongresi Küresel Diyalog Bildiriler Kitabı, 2009, Süleyman Demirel Üniversitesi, 2373–2389.
21. https://conscioustravelguide.com/amsterdam/things-to-do-see/van-gogh-museum date of access: 29.12.2020
22. https://cocobetty.com/cocobettytravels-photography-travel-blog/frida-kahlos-blue-house-mexico-city/ date of access: 29.12.2020
23. Ocakcı, M., "Kimlik Elemanlarının Sehirsel Tasarıma Yonlendirici Etkisi", Kentsel Tasarıma Ekolojik Yaklasım-5.Kentsel Tasarım ve Uygulamalar Sempozyumu, 1994, İstanbul: Mimar Sinan Universitesi, ss.239–245.

24. Ocakcı, M.-Southworth, M., "Elements of Urban Identity: The Case of the Beykoz District of Istanbul", IURD, Working Paper 650, 1995, Berkeley: University of California.
25. Suher, H.-Ocakcı, M.-Karabay, H., "İstanbul Metropoliten Sehir Planlama Surecinde Kent Kimliği ve Kent İmgesi", İstanbul 2020 Sempozyumu Bildiriler, İTU Şehir Bölge Planlama Bolumu,1996, ss.57–72.
26. Onem, A.B.-Kılıncaslan, İ., "Halic Bolgesinde Cevre Algılama ve Kentsel Kimlik", İTU Dergisi/Mimarlık, Planlama ve Tasarım, 2005, 4(1), ss.115–125.
27. Koyuncu, A., "Kimliğin İnsasında Kent: Konya Orneği", Akademik İncelemeler Dergisi (Journal of Academic Inquiries), 2013, 8(2), ss.155–179.

H. Selma ÇELİKYAY

Planning and Design Approaches for Liveable, Self-Sufficient and Smart Neighbourhoods

Abstract: In the pandemic process, which captures most cities of the world and we are still in, there is a need for paradigm changes in the field of planning in terms of urban health and public health. For a resilient and prepared urban approach at the local level, a planning approach that requires an individual-oriented, public benefit-oriented, preserving ecological values, providing health care at all scale and creating a qualified public space hierarchy is required. Could compact settlement models be a development model that will increase the resistance of cities? Concentrated urban pattern is envisaged as a sustainable city model. Small-scale settlements with environmental growth control, defined border relations, and controlled access may perform more resistively. In this context, in the urban space hierarchy, the neighbourhood emerges as units that need to be emphasized and developed in a way that is self-sufficient and that will ensure the food and water security of its residents, in addition to the energy demand. In the face of the unavoidable increase of urbanization, crowding and urban chaos, it is now the time of the inevitable rise of the neighbourhood, which promises quality of life, for the future of human settlements.

Keywords: Sustainable neighbourhood, self-sufficiency, smart neighbourhood, urban innovation.

1 Introduction

In the twenty-first century in which we are still living, in the new world order created by the Covid-19 outbreak, which has affected most of the human settlements in the world, starting from China, the lifestyle of individuals and society and therefore the way they use urban space are changing. Today, cities need a change in the planning perspective for sustainable development after the Covid-19 outbreak [1]. It is important to develop new perspectives and approaches in planning science, which aims to produce a liveable urban space for the society.

On the one hand, while the problems caused by rapid population growth and crowding brought about by urbanization are experienced in most of the cities in the world, on the other hand, global climate change has started to cause bigger and vital problems such as the decrease in water resources. It is emphasized by scientists that the negative effects of climate change on agricultural production are increasing, that problems such as food supply and security will be on the agenda of the whole world in the coming years. Perhaps in the near future, human beings may face difficulties in providing basic vital needs such as water and food, along with other possible outbreaks such as Covid-19.

Therefore, in the context of sustainable, durable and resilient urban development, it is necessary to develop methods and policies that ensure that the needs of the urban population are met in an efficient, effective and sustainable manner and their reflection on spatial planning and design processes. In this context, the neighbourhood, which is the basic unit that forms the city, gains importance in terms of relationship between urban sustainability, planning and urban design [2].

In addition, the production of human and environment-oriented urban policies, the development of smart systems that offer urban innovations that will respond to the changing needs of the society are the factors that will ensure the construction of liveable and self-sufficient neighbourhoods and cities that offer living comfort.

For the design of smart cities, the major principle is the integration of ecology and technology. Together with technological development, settlement should be re-evaluated within the framework of ecological sensitivity, sustainable resource management and making sustainable living spaces should be ensured [3].

2 Basic Approaches in Neighbourhood Planning and Design

The neighbourhood is one of the settlements that make up the city in terms of urban area hierarchy and is an important part of the city [2]. The neighbourhood unit concept is a model for residential environments including community facilities, parks, local shopping and housing for the population needed to support an elementary school [4].

Creating sustainable neighbourhoods is one of the first steps towards sustainable urban development [5, 6]. The carrying capacity of the neighbourhoods must also be determined in order to be manageable and to create urban and social reinforcement areas that the population living within their borders will need. In terms of planning, a number of criteria are needed to distribute the reinforcement rationally. Population is the main determining factor for planning studies that determine the land use at the neighbourhood scale as in the urban scale. Clarence Perry's neighbourhood plan applied to a 100-acre was designed for single-family, row, and multifamily housing to accommodate 8,000 residents [7]. Clarence Perry's neighbourhood unit plan of the 1920s saw the urban neighbourhood as a means to organize space and socialize residents [8]. It is necessary to make up-to-date additions to these basic objectives, which we can say are still valid today, in the context of improving the chaos of urban environments and meeting the social needs of the society today.

The neighbourhood-scale organization of urban space primarily involves the planning and design of a self-sufficient small-scale settlement. Urban Planning is the process of preparing a spatial scenario for all members of the society. Therefore, the basic principle of planning processes aiming at the welfare of the society should be to construct a spatial scenario focused on human and environment.

Highlights of sustainable neighbourhood unit planning and design are the topics following:

- Human and environment oriented planning and design
- Self-sufficiency for sustainability
- Smart urban innovations

3 Human and Environment Oriented Planning and Design

Humans interact with their physical environments and how it affects people in negative and positive ways is the main subject of environmental psychology [9]. The human's interaction with both social and spatial environments is important determiner his behavioral patterns. For example, research on green spaces has shown that the visual effect of green spaces on people can change their physiology and these effects are reflected in their nervous systems [10],[11],[12]. Scientific evidences on human-environment interaction provide a key data for planners and urban designers involved in the spatial organization in the neighbourhood unit to balance the natural environment and built environment.

Planning decisions, urban design and landscape design projects are the most important tools for creating a built environment that people can relate to at the neighbourhood scale. In planning and design processes, it is possible to create a liveable urban environment by taking into account the balance between human-built environment and natural environment. In the spatial organization process of a liveable and sustainable neighbourhood, following subjects should be considered and be reflected to the planning process.

- Pedestrian accessibility and walkability
- A primary school, a religious facility, a shopping unit and a clinic with primary care services in the center, bank etc.
- Residential areas surrounding of the center
- Open green spaces
- Urban agricultural areas on the periphery of the neighbourhood

Pedestrian Accessibility, Facilities and Residential Buildings Location

Pedestrian accessibility and human scale in the built environments are of great importance for making liveable and safety urban environments. A liveable neighbourhood structured based on walkability includes the mixture of land-uses, and the interconnecting of street patterns to facilitate all modes of movement [13].

Walkability is becoming a necessity in cities. Demand for walkable neighbourhoods is increasing and this is also changing real estate values. People of all ages are aware of the benefits of living in a place where they can walk to nearby shops, restaurants, recreation, and public transportation [14]. Walkable neighbourhoods are significant for health, happiness and economic growth [15]. In order to make a decision about which service buildings will be placed in the center of neighbourhood, a weighted index of amenities depending on how crucial they are for daily life (eg. access to healthcare is more important than access to nightlife) could be built or an empirical data about what amenities people visit most often could be used [15]. For example, dwellers of the neighbourhoods must have access to a clinic with primary care services.

Placement of a primary school in the center is the basis in neighbourhood planning. The size of a neighbourhood is 5-minute (400m) walking distance radius is proposed by Perry (1929) and then by Plater-Zyberk (1994). On the other hand, Farr (2007) proposed a 10-minute (800m) walking distance radius for a neighbourhood size (Fig. 1). The basic approach in these propositions is to take the walking distance scale into account primarily [16], [2].

Recently, in some cities in the world, governments advocate the 15-minute (15-min) walkable neighbourhoods to give citizens with 15-min walkable access opportunity to basic public services, and ultimately to improve walking behavior and overall health [17]. There are a lot of scientific researches about 10-minute, 15-minute and 20-minute neighbourhoods accessibility. One of them is seen in Fig. 2 [18].

Residential areas should be located in an equally walkable distance to the amenities in the center. Detached houses with garden, row houses, multi-storey housing blocks are involved in the planning of residential areas where the foreseen population will settle. Residential buildings in a neighbourhood contain single-family, row, and multifamily housing. The amount of open space and open green areas per capita in the neighbourhood should be correlated with the amount of open space created by the proposed building layouts on a parcel basis. Connection between residential buildings and open spaces and green areas is an important indicator for wellbeing of dwellers living in the neighbourhood. Today, the pandemic process has shown that home gardens, open spaces, public spaces, green spaces are of vital importance. However, due to the constraints created by the pandemic process, the need for open space of individuals who have to live more in their homes and who are overwhelmed in closed spaces have increased. Those living in detached houses with gardens meet the need for open space and green space thanks to their private gardens (Fig. 3). However, for those living in row houses and apartments without gardens, the existence of public open green spaces and the pedestrian accessibility of these areas are important.

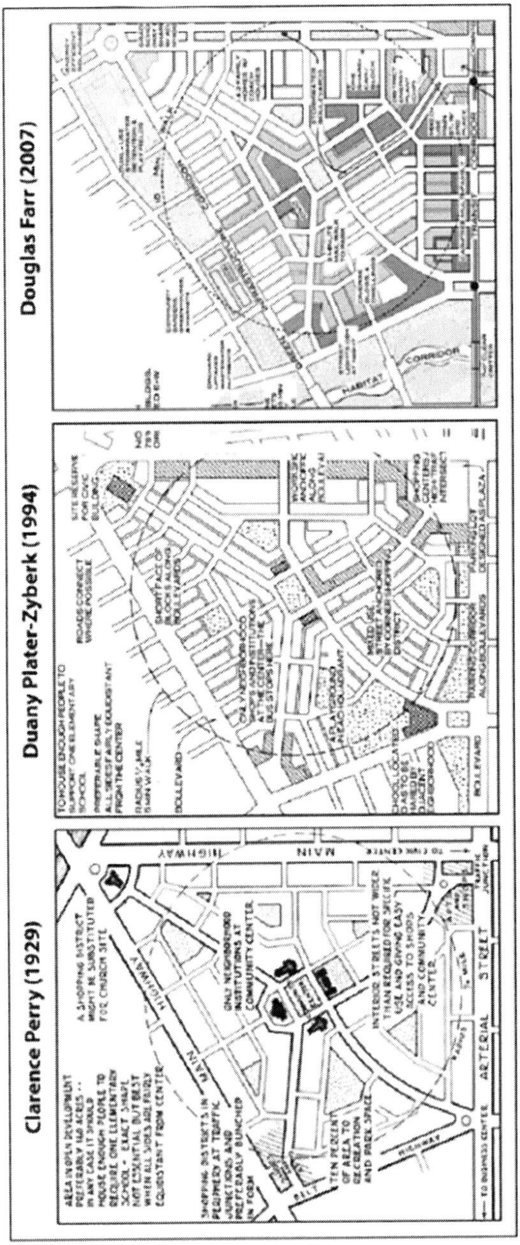

Fig. 1: The evolution of the neighbourhood unit concept (different scale). Source: Farr (2007) [16].

Fig. 2: Features of a 20-minute neighbourgood [18].

Fig. 3: Pasadena, California: Best Neighbourhood [19]

Open/Green Spaces

Campbell (2001), "Open spaces in neighbourhoods; "Any land that provides environmental, social and economic benefits to communities directly or indirectly" [20],[4],[2]. Open space includes green space, schoolyards, playgrounds, public seating areas, public plazas, vacant lots. Open space provides recreational areas for residents and helps to enhance the beauty and environmental quality of neighbourhoods. Green space (land that is partly or completely covered with grass, trees, shrubs, or other vegetation) includes parks, community gardens, and cemeteries [21]. Urban green spaces are ecosystems of vital significance in enhancing the quality of life in an urban environment and supplying ecosystem services, such as biodiversity, climate regulation [22].

Green spaces, which are important for both the physical sustainability of the neighbourhood and the social sustainability and psychological comfort of the living people, should be integrated into the urban green network system in the planning process. Parks, playgrounds and sports areas, children's gardens, botanical parks, zoos, promenade areas, recreation areas, green corridors and green belts are important components of the urban green system that establishes a balance between the built environment and the natural environment. Green spaces should be located within a distance of up to 300 m (five-minute walk) from the place of residence [23].

In a neighbourhood, it is not possible to plan all residential buildings in a garden layout, as in larger-scale settlements. On the one hand, it is necessary to consider the possibility of providing spatial options to the members of the society while meeting the housing needs of the population. In the spatial organization in the planning process, the backyards of row houses and multi-family houses can be planned to be used by those living in these houses as common gardens. Thus, backyards, which are private properties, can be designed as semi-public spaces in terms of use. On the other hand, thanks to this spatial organization, the amount of green space in the neighbourhood and the possibility of oxygen production will also increase. In some cities around the world, the backyards of the apartment buildings with high population density, which do not have front and side gardens, have been used as common areas in this way. As in shown in Fig. 4 and Fig. 5, the backyards in Amsterdam, which is the capital and most populous city of the Netherlands, are used as common open green spaces by residents living multifamily adjacent houses.

Fig. 4: Adjacent buildings with common backyards in Amsterdam (Google Earth, 2020)

Fig. 5: Backyards in Amsterdam as semi-private and semi-public spaces (Google Earth, 2020)

Urban Agricultural Areas

Urban agriculture has been shortly defined by Wagstaff and Wortman (2013) as "all forms of agricultural production (food and non-food products) occurring within or around cities." [24]. In terms of social sustainability and food supply, it is important to encourage local food networks, and in this context, planning urban agricultural areas and providing urban agriculture services at the neighbourhood level are also important. Urban agricultural areas should be located on the lands having agricultural potential and could be planned on the periphery of the neighbourhood. This issue will be mentioned again under the heading of self-sufficiency for sustainability.

4 Self-Sufficiency for Sustainability

The prediction that the conservation of resources will be a reality of life in the new millennium, emphasized by Storey and Baird [25] is now fully realized. As a result, today we have to make our habitats in which we live more self-sufficient than they are now. Self-sufficiency means the state of being able to provide everything needed, especially food, without the help of other people or countries [26]. Self-sufficiency refers to meeting basic needs from within, rather than outsourcing. Sufficiency-based development requires a set of policies aiming to induce the development of human settlements in such a way that they are self-sufficient [27]. If the population is too large, housing in the city is insufficient to serve the population. However, since the system includes a loop that maintains the proper housing rate [27], it creates an unavoidable and uncontrollable increase in regulating the amount of housing. Therefore, instead of dealing with a large-scale city, it is a more rational action to determine development, sustainability and self-sufficiency strategies for the neighbourhood, which is a small-scale settlement. The self-sufficient neighbourhood concept comes from the idea that cities should be able to produce the resources they need locally [28].

The main indicator of self-sufficiency is to meet basic vital needs without being externally dependent. These basic needs are housing, energy, water and food. As the housing issue is included in the texts above, energy, water and food supply will be focused here in the context of meeting other needs.

Just as sustainable cities require sustainable neighbourhoods and sustainable buildings, self-sufficient cities need to design and build self-sufficient neighbourhoods and self-sufficient buildings. But, the design of most of the current buildings in the cities is unsustainable and a different approach is necessary. In terms of energy it is possible, using existing and new technologies, to supply self-sufficiency of life-cycle by minimising the total energy load of the building, fully using available, renewable energy sources [25]. In the last years, the technologies associated with generation, storage, measurement and control of electricity have improved increasingly [29]. Thanks to these advances, it is possible to design a new energy model, either for residential users or cities, which target to achieve sustainable and efficient growth in future scenarios [29].

Water is vital for humans because water is essential for the survival of life. Planning process making spatial decisions for the future of society should also take account water need and consumption of the population. Water management is of great significance for rational resource use because non-renewable resources are declining. Thus, it is necessary to reduce water usage per person and to use water efficiently [30]. Water management deals with the problem of optimal allocation of reusable water among different water-using operations [31]. Development of sustainable water management strategies includes determination of vulnerability and adaptation possibilities, followed by an impact analysis of these strategies regarding different possible futures [32]. Both domestic and public water consumption and water footprint should be determined to save water and increase useful water usage [30]. It should be taken into account not only reduction of water consumption in buildings but also rainwater reclamation from water supply and water management aspects.

Food supply is the other important issue of self-sufficiency. A country's food production for its own domestic consumption may provide both economic and political benefits [33]. According to the economic criteria, in cities defined as "an economic mechanism that emerges to meet the constantly changing needs of the society in the process of production, distribution and consumption of goods and services" [34], at the neighbourhood-scale urban agriculture opportunities will facilitate the provision of agricultural production to consumers. In addition, it will save time and provide economic benefit by minimizing the cost required by the distribution mechanism (equipment, labor, etc.). Urban agricultural area as a neighbourhood garden, which is a big piece of ground used to grow vegetables, fruit, herbs, or flowers, are necessary to supply essential food for dwellers living in the neighbourhood.

In according to United Nations Environment Programme (UNEP), one of the basic elements for achieving sustainable development is the transition towards Sustainable Consumption and Production (SCP). It is essential to develop policies for not only improvement of production, but also supporting consumers to move towards sustainable consumption choices [35]. Everyone in society has to take responsibility in this transition including governments, educators, the private sector and each and every consumer. Thanks to the participation of all to this process, SCP can lead to an improved quality of life.

For sustainability, metabolic processes should be monitored with input-output analysis regarding urban metabolism at neighbourhood scale and building metabolism at building scale [36], [37], [2]. By means of information and communication technologies (ICT), Smart Urban Metabolism model can provide real-time feedback on energy and material flows, from the level of the household to the urban district [38].

5 Smart Urban Innovations

Urban public spaces are areas that are outside of the buildings in a city and have the potential to provide environmental, social and economic benefits to the society [39]. Public spaces are areas that belong to everyone and all citizens have the right to use equally [40]. In smart cities that use information and communication technologies to ensure the livability, workability and sustainability of the city, "smart city solutions/applications" that offer access, comfort and security to urban public space users have been developed [41].

The main purpose of smart cities, which is defined as the technology-supported future of sustainable, durable and liveable cities, is to increase the quality of life with creative solutions. Smart cities where ICTs are used to provide better urban services are related to urban innovation. However, ICTs improving people's quality of life can increase inequalities and promote a digital divide [42].

Smart cities, which are structures created to solve the problems of cities with technology, activate smart urban systems to improve the quality of life [43]. The same approach is applied for the neighbourhoods, which are small-scale settlements as the main component that make up cities. Intelligent systems using information and communication technologies in order to increase the quality of life of the residents in the neighbourhood and to meet all their basic needs add comfort to life by providing service throughout the day.

On the one hand, the pandemic process has affected daily life negatively by restricting the use of urban space, on the other hand, it has revealed the importance and necessity of ICTs that enable people to continue their work, shopping and education process in the home environment [1]. The use of ICTs changes and transforms the urban life, culture and habits of the citizens very quickly. With this transformation and development, the usage patterns of urban areas and the demands and expectations of individuals are completely different [44]. This also provides an opportunity to integrate smart urban innovations into everyday life.

As an example; Southern Company's research and development team built a compound comprising solar panels, battery storage, and a natural gas-fed engine to create a resilient mix of generation resources – nearly adjacent to the neighbourhood itself. These resources are able to communicate in real time with the homeowners' water heaters and air conditioners, ensuring that power is both generated and used in the most efficient manner [45]. These enhanced efficiencies give opportunities to Smart Neighbourhood homeowners who enjoy maximum comfort and convenience regardless of the weather or the time of day. For another example, since its inception in 2012, Neighbourhood Services unit in Texas has developed programs and services for neighbourhoods to revitalize themselves using community resources [46].

ICTs, which are used in every moment of daily life, have gained a spatial dimension in the public sphere by making urban furniture smart. Self-powered smart urban street furnitures can help increase the attractivity of public spaces by providing smart services, information, and connectivity, enabling the collection of valuable data for optimizing processes and reducing costs [47]. Fig. 6 shows the smart designed solar bench where people can charge their mobile devices. Everyone can enjoy fast 4G internet and at the same time charge their smartphone or tablet. This solar bench to work only on clean solar energy, which is stored and is available throughout day, is a great innovative solution [48].

Fig. 6: Smart solar bench [48]

In addition to the few examples of smart urban innovation presented here, there are a greater number of both smart and eco-technological tools and spatial elements in public spaces to serve people living in a neighbourhood or city.

6 Conclusion

This chapter emphasizes the necessity of planning sustainable neighbourhoods and designing sustainable buildings for sustainable cities. In this context, it refers to the principles of spatial organization as well as self-sufficiency and the presentation of smart urban innovations in the context of meeting the changing tendencies and needs of the society in a general approach.

The last year of the world has been recorded with the fear created by the pandemic process and the increasing mortality rates in almost all countries. Outbreaks are a type of disaster that deeply affects human life enough to change the social order. While this process has kept most societies from the social rituals of daily life that they are used to, people's perception of space has also changed. In addition to being private spaces of their own, people's homes have also turned into a business environment, school and shopping environment. This permeability and intimacy has turned houses, which are areas of privacy, into prisons that have lost physical connection with the outside world but continue to communicate. This overwhelming process, which is not known when it will end, has made people need more outdoor spaces, public spaces, and socialization areas.

Therefore, all urban actors working to increase the welfare of the society and create more liveable settlements need to develop new approaches in their fields of expertise in the context of society's changing urban and spatial sociology. Planners and designers, who play a decisive role in urban space organization, are the main urban actors.

Sustainable development goal has been on the world's agenda since it was emphasized with the heading of "Our Common Future" in Brundtland Report of the World Commission on Environment and Development in 1987 [49]. Sustainable conservation and sustainable growth goals accompanying sustainable development have turned into goals of building smart cities, intelligent cities, digital cities and eco-technological cities today due to technological advances to date and the use of ICTs. With all this change, it has become even more important to create human and environment-oriented settlements as an unchanging necessity.

Planning and designing small-scale and compact neighbourhoods as smart, high-quality and safe urban environments that offer comfort of life also provides convenience for units serving the neighbourhood. There are innovative vehicles to serve the inhabitants both indoors and outdoors. Smart and eco-technological elements are important tools in urban public spaces for the creation of new living spaces that are comfortable, healthy, non-polluting, minimizing carbon consumption, and are self-sufficient. Smart and eco-tech urban furnitures are innovative tools for development of public spaces in neighbourhoods.

Finally, the approach for neighbourhood planning and design presented in this chapter could be a sustainable development model that will increase the resistance of cities. Small-scale settlements with environmental growth control, defined border relations, and controlled access may perform more resistively. Planning approaches individual-oriented, public benefit-oriented, preserving ecological values, providing health care at all scale and create qualified public spaces.

References

1. Öztaş Karlı, R.G., Çelikyay. S. *Covid-19 Effect on the Cities Becoming Smart.* Van Yüzüncü Yıl University the Journal of Social Sciences Institute, 2020. Outbreak Diseases Special Issue: p. 321–338.
2. Çelikyay, S., Öztaş, R.G. *Sürdürülebilir Kentsel Gelişmede Mahalle Ölçeği.* Mimarlık, Planlama ve Tasarım Alanında Araştırma ve Değerlendirmeler, 2019. p.227–238.
3. Çelikyay, S., Öztaş, R.G. *Eco-Design Principles For Sustainable And Smart Cities.* V. International Multidisciplinary Congress of Eurasia, Barcelona, Spain, 2018. p.314–318.
4. Brody, J. *The neighbourhood unit concept and the shaping of land planning in the United States 1912–1968.* Journal of Urban Design, 2013. **18**(3): p. 340–362.

5. Al-Hagla, K. *Towards A Sustainable Neighborhood: The Role of Open Spaces.* International Journal of Architectural Research, 2008. **2**(2): p.162–177.
6. Dehghanmongabadi, A., Hoşkara Ş., Shirkhanloo N. *Introduction to achieve sustainable neighborhoods.* International Journal of Arts and Commerce, 2014. **3**(9): p. 16–26.
7. Silver, C. *Neighbourhood Planning in Historical Perspective.* Journal of the American Planning Association, 1985. **51**(2): p.161–174, DOI: 10.1080/01944368508976207
8. Keating, W. D., Krumholz, N. *Neighbourhood Planning.* Journal of Planning Education and Research. 2000. 20: p.111–114.
9. Gifford, R., *Introduction Environmental Psychology and its Methods, Research Methods for Environmental Psychology.* ed: R. Gifford, 2016. John Wiley & Sons.
10. Ellard, C., *Cities and Their Psychology: How Neuroscience Affects.* Urban Planning. 2014. The Guardian.
11. Ulrich, R.S., *View Through a Window May Influence Recovery from Surgery.* Science, 1984. **224**(4647): p. 420–421.
12. Thompson, C.W., Roe, J., Aspinall, P., Mitchell, R., Clow, A., and Miller, D. *More Green Space is Linked to Less Stress in Deprived Communities.* Landscape and Urban Planning, 2012. **105**(3): p. 221–229.
13. Jones, E. *Walkable towns: the Liveable Neighbourhoods strategy.* Sustainable Transport, 2003. p. 314–325.
14. https://www.nar.realtor/on-common-ground/walkability-is-becoming-a-necessity
15. Jones, N. *Measuring pedestrian accessibility.* Towards Data Science, 2018. https://towardsdatascience.com/measuring-pedestrian-accessibility-97900f9e4d56 .
16. Park, Y., Rogers, G.O. *Neighborhood planning theory, guidelines, and research: can area, population, and boundary guide conceptual framing.* Journal of Planning Literature, 2015. **30**(1): p. 18–36.
17. Weng, M., Ding, N., Li, J., Jin, X., Xiao, H., He, Z., Su, S. *The 15-minute walkable neighborhoods: Measurement, social inequalities and implications for building healthy communities in urban China.* Journal of Transport & Health, 2019. Vol. 13: p.259–273
18. Gould, E. *The 15-minute city.* https://transform.iema.net/article/15-minute-city, 2020.
19. https://www.sunset.com/travel/california/best-neighbourhood-pasadena-california

20. Campbell, K. *Rethinking open space, open space provision and management: a way forward.* Report presented by Scottish Executive Central Research Unit, Edinburgh, Scotland, UK. 2001.
21. https://www3.epa.gov/region1/eco/uep/openspace.html
22. Vargas-Hernández J.G., Pallagst K., Zdunek-Wielgołaska J. *Urban Green Spaces as a Component of an Ecosystem.* In: Dhiman S., Marques J. (eds) Handbook of Engaged Sustainability. Springer, 2018. p 885–916.
23. Senetra, A., Krzywnicka, I., Mielke, M. *An analysis of the spatial distribution, influence and quality of urban green space – a case study of the Polish city of Tczew.* Bulletin of Geography. Socio–economic Series, 2018. 42: p. 129–149.
24. Neith Little, N., Lynch, K.R., Johnson, D., Cook, N., Myers, G. *From Surviving to Thriving: Strategies for Urban Farm Success.* 1st edition, Ed: Neith Little, 2019. Published by University of Maryland Extension.
25. Storey, J.B., Baird, G. *Towards the Self-sufficient City Building.* Transactions of the Institution of Professional Engineers New Zealand: General Section, 1999. Vol. **26** (1): p.1–8.
26. https://dictionary.cambridge.org/tr/s%C3%B6zl%C3%BCk/ingilizce/self-sufficiency
27. Park, M., Kim, Y., Lee, H., Han, S., Hwang, S., Choi, M.J. *Modeling the dynamics of urban development project: Focusing on self-sufficient city development.* Mathematical and Computer Modelling. 2013. **57**: p. 2082–2093.
28. https://transform.iema.net/article/15-minute-city
29. Arcos-Vargas, A., Gomez-Exposito, A., Gutierrez-Garcia, F. *Self-Sufficient Renewable Energy Supply in Urban Areas: Application to The City of Seville.* Sustainable City and Society, 2019. Vol. **46**, 101450.
30. Öztaş Karlı, R.G., Çelikyay. S. *An Alternative Indicator on Water Management in Planning: Water Footprint.* V.International Multidisciplinary Congress of Eurasia, 2018. 319–323.
31. Bandyopadhyay, S., Ghanekar, M. D., Pillai, H. K. *Process Water Management.* Industrial & Engineering Chemistry Research, 2006. **45**(15): p. 5287–5297.
32. Haasnoot, M., Middelkoop, H., E. van Beek, E., van Deursen, W.P.A. *A method to develop sustainable water management strategies for an uncertain future.* Sustainable Developmet, 2011. **19**(6): p. 369–381.
33. Clapp, J. *Food self-sufficiency: Making sense of it, and when it makes sense.* Food Policy, 2017. (66): p.88–96.
34. Keleş, R. *Kentleşme politikası* (8.bs.). Ankara: İmge Kitabevi Yayınları. 2014.
35. UNEP. *Sustainable Consumption and Production A Handbook for Policymakers.* Ed: Emily Briggs, 2015.

36. Çelikyay, S., Öztaş, R .G. *Kentsel metabolizma yaklaşımının bina ölçeğine indirgenmesi: Bina metabolizması.* International Eurasian Conference onScience, Engineering and Technology, Ankara, Turkey, 2018. p.2025–2030.
37. Çelikyay, S., Öztaş, R .G. *A framework based on urban metabolism approach for sustainable urban planning* (Abstract). International Symposium Ecology Kastamonu, Türkiye, 2018. p. 454.
38. Shahrokni, H., Lazarevic, D., Brandt, N. *Smart Urban Metabolism: Towards a Real-Time Understanding of the Energy and Material Flows of a City and Its Citizens.* Journal of Urban Technology, 2015. Vol. **22**(1): p.65–86.
39. Uzgören, G., Erdönmez, M.E. *Kamusal açık alanlarda mekan kalitesi ve kentsel mekan aktiviteleri ilişkisi üzerine karşılaştırmalı bir inceleme.* Megaron, 2017. **12**(1): p. 41–56
40. Çelikyay, S., *The Role of Urban Image Elements on Definition of Public Space.* In: Çelikyay, H.S. (ed) Spatial Organizations of Public Spaces. 2017. p.19–41.
41. Çelikyay, S., Özbek, M.Ö. *Innovative Tools For Development Of Public Spaces: Smart And Eco-Tech Urban Furnitures.* 7th International Symposium on Academic Studies in Science, Engineering and Architecture Sciences, 2019. Ankara, Turkey, p.887–903.
42. Chourabi, H., Nam, T., Shawn, W., Gil-García, J. R., Mellouli, S., Nahon, K., et al. *Understanding Smart Cities: An Integrative Framework.* 45th Hawaii international conference on system sciences, 2012.
43. Terzioğlu, M.K. *İnovasyon ve Kentsel Verimlilik.* TOKİ Haber e-Dergi, 2021, **131**: p.42–44. https://www.tokihaber.com.tr/e-dergi/
44. Örselli, E., Akbay, C. (2019). *Teknoloji ve Kent Yaşamında Dönüşüm: Akıllı Kentler.* Uluslararası Yönetim Akademisi Dergisi, 2019. **2**(1): p.228–241.
45. Knakmuhs, H. *The Neighbourhood of the Future.* Global Energy Institute. 2018. https://www.globalenergyinstitute.org/neighbourhood-future.
46. https://www.roundrocktexas.gov/departments/planning-and-development-services/community-development/neighbourhood-services/
47. Mueller, T., *Smart Urban Street Furniture Solutions In Smart Cities.* 2017. https://hub.beesmart.city/en/solutions/smart-urban-street-furniture-solutions.
48. https://solar-bench.com/solar-bench/
49. Report of the World Commission on Environment and Development: Our Common Future, 1987.

Miray GÜR

The Analysis of Urban Health in Bursa Based on Resident Perceptions

Abstract: The Healthy Cities Movement initiated by the WHO, urban health, the significant of which increased over time, and the associated higher quality of life in cities, support long-term sustainability. Attributes such as green spaces, healthy habitats, safe and inclusive urban facilities, and governance, awareness and education assistance that constitute healthy cities and support the health of urban citizens lead to democratic, healthy and high quality of life, safe and habitable cities with social interaction, and as it is transformed into built environment, transportation and governance decisions in urban planning, it supports sustainability. In this context, in Bursa, an urban area with significant historical and cultural heritage and the first capital of the Ottoman Empire, both the metropolitan municipality and the district municipalities have prioritized urban health. Turkish Healthy Cities Association is located in Bursa, leading to the organization of numerous seminars, meetings and conferences where academicians, municipalities, self-employed architects and planners could meet with stakeholders such as NGOs and share and implement ideas. One of the studies on urban health, high participation consultation meetings were organized before the by Bursa Metropolitan Municipality. The published Consultation Meetings Analysis Results with the participation of Bursa Uludag University, NGOs, municipalities and Academic Chambers, and workgroups provides a significant data source for the analysis of urban health. Regional data discussed in the meetings, HDI-D results, and the findings of various studies were analyzed and recommendations to support healthy cities were proposed in the publication based on various studies conducted in different cities in the world.

Keywords: Healthy Cities, Urban Health, Public Health, Bursa, Resident Perceptions, Livability, Sustainable Development

1 Urban Health and Related Parameters

The Healthy Cities Project, which was initiated as a small European project the World Health Organization (WHO) in 1986, turned into a significant movement and an approach that included global policies and spatial applications within a short time. The Healthy Cities Project is a long-term development project [1] that aims to improve the physical, psychological and environmental well-being of individuals who live and work in a city. In the project, which focuses on the physical, social, psychological and environmental well-being of individuals who live and work in a city based on the "health for all" approach, aimed multidimensional development.

The Healthy Cities Network, founded by the WHO, is an active and vibrant platform that aims to contribute to the health, well-being and equality in health and well-being since 1988. City is important spatial scale since they correspond to a geography defined by a multidimensional organism. The city, which is both an economic and institutional process and at the centre of social actions, is also a symbol of the collective harmony among various occupant groups [2]. All the individuals who live in the city or just pass through are at the focus of both national and global spaces and processes since all these individuals benefit from the healthy urban life and form the city and health. In the Healthy Cities Project, healthy cities were not only described as cities with a certain health level, but also cities are aware about the urban health and strive to improve it. Thus, regardless of the health status, all cities that strive for health could be healthy cities. The important component is the availability of structure, functions and firm determination to succeed. Thus, it is possible to describe a baseline for all cities based on the associated dynamics. Furthermore, administrative dimensions and policies and actors should mobilize stakeholders, especially the public, to achieve this goal.

A review of the strategic targets of the Healthy Cities Network would underline concepts such as healthy and sustainable development, equality, support of local policies and applications, participatory governance, solidarity and cooperation between networks established by local governments and organizations in urban problems. The main themes of the current 7th Phase effective between 2019 and 2024 include investments in urban population, improvement of the health and well-being of urban 'places', building wider participation and partnerships, promotion of social well-being, protection of the planet, and promotion of peace and safety through inclusive societies [3]. Thus, the individual is at the center of the urban construct targets that prioritize the improvements of the partnership, participation, governance and welfare of the individuals.

To achieve national and global success in strategic targets of healthy cities, not only the health industry, but other industries, the inhabitants and employees in the region should be included in the process. Not the passive population, but health initiatives conducted with this population are important. To develop innovative ideas for a healthy environment, Burris et al. (2007) suggested the adaptation of innovative strategies by the administration for a healthy and strong city [4]. These strategies emphasized governance and creativeness and assertiveness of urban administrators to improve participation and effectiveness. In fact, in the 48th European Regional Committee Meeting organized by the WHO, the main ethics included joint participation of all stakeholders for the acceptance of health as a basic human right and the improvement of health in the "21 goals in the 21st century" slogan [5].

The importance of addressing health inequalities in studies on healthy cities has increased [6]. The theme of the Human Development Quality of Life 2019 report also supported this approach [7]. In the final 2019 report published by UNDP, inequalities in human development in the 21st century were addressed globally. The main theme of the 2019 HDI was the inequalities. Due to the developments in the 21st century, new generation competencies and new generation inequalities emerged among individuals. Universal policies were developed with a comprehensive analysis of the inequalities since birth based on income, well-being, climate change, gender, and structural transformation in the report. Based on these developments, it could stressed that the most important current problem is various spatial and social inequalities.

Urban administrations are responsible for the welfare and health of various social groups, including residents, immigrants, tourists and visitors. Governments have the potential to influence administrative, regulatory, planning, commercial and social components through policies. The built or natural environment policies are one of the most important determinants of health. This is even more important for disenfranchised groups such as the poor, unemployed, ethnic minorities and disabled. Inequalities prevent the access of these communities to healthy life and work environments [6] (WHO, 2012). This is even more challenging for individuals during the COVID 19 pandemic who had to live and work and could not hunker down. "Environmental justice" is a catalyst in addressing this problem, defining environmental inequalities, and converting them into policies in various countries [8, 9].

2 Successful Healthy City Applications

For a healthy city with high quality of life, interesting experiences were observed in various European cities where environments were developed for different age groups and to ensure mass participation. For example, Sandnes Municipality in Norway developed strategies for children to grow up in a healthy environment. The growth of children in a safe and stable environment, priority of children in urban planning and inter-sectoral cooperation were prioritized, and the young generation was encouraged to participate in political processes and decision-making mechanisms in a city council for children [10]. The child-friendly cities approach that encouraged the next generation to grow and live in a healthy environment is significant. Similarly, Portland became one of the cities where sustainability and habitability were ensured by child-friendly urban plans in public policies, planning, action plans and area strategies to improve the quality of life [11]. Urban walkability and learning opportunities are also important. The political dynamics, various decision-making mechanisms and planning decisions for urban growth and sustainability were significant in Curitiba, another city where active participation was significant in the development of habitability along with Portland [12] (Irazábal, 2016). Curitiba was a process and an outcome where the necessity of a holistic system that includes transportation systems, decision-making processes, local community structures that promote habitability was pronounced and experienced.

Environmental health, as a field defined by specific criteria in the literature, constantly evolves due to changes in lifestyle, complicating the field [13]. The National Environmental Health Association (NEHA), which aims to improve specialization in this field and provide healthy environment for all, determined various environmental health dimensions. These could be summarized as the areas that should be managed and improved based on the human factor, control of health hazards, development and implementation of policy guidelines, providing communication on health and educational materials, development of recommendations for building and land use, and conducting research on the correlation between health and environment. Thus, concrete improvements should be implemented initially via integrated and participatory policies, and the development of communication and awareness via the educational dimension, and the analysis of the resulting data as feedback in periodical research on physical and social environment would improve healthy environments.

3 Healthy City Indicators

WHO defined an innovative and participatory vision for healthy cities as spaces that serve individuals and the planet, including the participation of all communities, good governance, and combating the inequalities. Main topics in healthy cities vision include people, participation, prosperity, planet, place and peace and the following could be stated about spatial strategies:

- Prioritizing investments in individuals, investing in human and social capital as a strategic approach to access to common goods and services by all,
- Leadership in the participation of the public in healthy city decisions,
- Governance of common goods and services based on values (support of social welfare),
- All internal or external urban policies centered on human health with a universal inclusive approach.

An in-depth analysis of health based on "Place" would reveal dimensions such as a human-oriented urban development and planning for an accessible social, physical and cultural environment, integration of equality and sustainability into urban development and planning, and enhanced inclusiveness in the use and management of common spaces. It should be noted that the prominent dimensions should support and trigger each other with an interactional approach. This interaction is associated with the fact that the health indicators in habitable, quality and sustainable cities already feeds these identities. The requirements for healthy settlements coincide with the habitability indicators determined by the AARP, such as a safe and healthy environment for all, opportunities for different income levels, options in meeting the needs, affordability, accessible services for different age and education groups, and social integration. Similarly, factors that contribute to sustainable urban settlements include social education, local democracy, health and quality of life, safety, social interaction, sense of community, cultural traditions, while physical factors include attractive public realm, decent housing, environmental quality and amenity, accessibility, sustainable urban design and walkable neighborhoods [14]. Continuous improvement of urban health is important for habitable cities for quality of life in the present and sustainable cities in the future.

Several studies were conducted to determine health indicators in Europe and Turkey for the analysis, comparison and association of urban health. These included institutional large-scale studies and academic research. Webster and Sanderson (2013) developed a scale that allowed the collection of qualitative and quantitative data, including health, health services, environmental and socioeconomic indicators, Ritsatakis (2009) discussed the issue as social determinants of health based on the space, process and individual [15, 16]. In the study, which emphasized that spatial differences triggered the disintegration in healthy life, the main caused were determined as transportation, air pollution, road safety, neighborhood and facilities, urban planning, green spaces, crime and fear of crime, urban environment and climate. The scale clarified spatial studies. Another index that further established the correlation based on spatial scale and planning was the International Healthy Cities Index, developed with the coordination of Building Research Establishment. The index included air quality, food access, green infrastructure, housing and buildings, leisure and recreation, noise pollution, resilience, safety and security, transportation, and utilities and services factors [17]. Health disparities, socioeconomic status across the life cycle, the physical environment, the social environment, access to health and social services were included in most health indicator approaches for physical and social spaces [18, 19]. A different approach was proposed by Dye (2008), who argued that the main obstacles to the improvement of urban health were associated with governance and the organization of the civil society, beyond technical and financial parameters [20]. The analysis of the policy and management dimension was obvious due to the interaction between the inequalities in the life cycle and the properties of the habitat as determined in the above-mentioned studies.

To determine an accurate road map that would lead to a progress in the health of urban stakeholders, Turkish Urban Health Indicators were developed in 2013 by academicians and government organizations under the leadership of Healthy Cities Association. These health indicators developed for provincial and municipal planning included demographical, educational, socio-economical, health, environmental, transportation, cultural and art data [21]. The urban health index (UHI) is a newer and more comprehensive scale developed by the WHO. The index, which is a combined index to measure and mapp the differences in health determinants in urban areas, allows working with local and selected indicators in a specific geographical area. UHI provides a framework and method to determine the health index in urban areas. An important and positive aspect of the index was its ability to measure local dynamics based on the time, conditions and stakeholders during the implementation. It was reported that the scale could vary based on the location, accessibility and aim, and that each dimension was equal. The main and subdimensions were as follows [22]:

- Health: disease, conditions, disability, injury, healthcare distribution, healthcare access, healthcare costs, immunization
- Environment: infrastructure, built environment, transportation, water quality, air quality, contaminants, noise
- Geography: land use, natural barriers, population density, neighborhood configuration, boundaries, spatial patterns
- Economics: occupation, employment patterns, unemployment, income and wages, services, economic growth, natural resources
- Socio-demographics: population, growth, migration, natality, mortality, education.

The analysis of the conceptual approaches, determinants and indicators associated with urban health, habitability and life quality should address the built environment, human (and socioeconomics) and governance dimensions in architecture and urban planning disciplines. Environmental and urban facilities that affect the health of the individuals and cities, created by the interaction between the architecture and urban planning disciplines, affect the quality of society, personal activities, satisfaction, and physical health. These environmental facilities include natural resources, recreational areas, lakes, rivers, forests, parks, water and air quality, noise levels, solid waste, hazardous waste, and the occupation of these areas directly affects human health. Urban facilities, on the other hand, could affect urban and human health and urban quality through androgenous recreation and cultural areas. Thus, urban natural and built environment facilities include the determinants of urban health and the physical and psychological health of humans.

4 The Analysis of Urban Health in Bursa Based on Occupant Perceptions

As mentioned in the section on urban health, the well-being of the urban population, the health of urban spaces, the promotion of a safe environment, governance, participation and partnership environment are significant in the current Phase 7. Concurrently, inequalities in human development were addressed and new generation inequalities were listed in the last 2019 Human Development Index report, which was one of the most valid indexes with global acceptance in the assessment of the urban and national quality of life [7]. Thus, in the analysis of urban health in Bursa with the urban health parameters and indicators was also approached with the inequality perspective addressed in Phase 7 and HDI. Based on the physical environment, the concepts that determine urban and personal health and quality of life were categorized in sub-components based on the main dimensions of environmental factors, human factors, policy and services, and the urban health in Bursa was analyzed.

The method employed in the analysis of urban health in Bursa was based on the urban health parameters matrix presented in Tab. 1. The matrix was developed based on the three main components, namely policy and services dimensions that affect humans, the environment, and the whole system in environmental planning that leads to a healthy city. The main and sub-components determined in the study with an inclusive approach for a sustainable environment, society, economy and development included physical environment, safe environment and healthy housing, including the natural environment based on the environmental factors. The human factors that determine the social environment included awareness, responsibility and equality and inclusivity since inequalities were important, commitment and will of individuals, solidarity and cooperation between various stakeholders (individuals, organizations), healthy development and lifestyle, and physical activity habits. And the policies and services that affected the whole system included participatory governance and education and services determinants (i.e., health services, social services, transportation).

Tab. 1. Healthy city parameters included in the current analysis

HEALTHYCITIES SUSTAINABLE ENVIRONMENT - SOCIETY - ECONOMY - DEVELOPMENT	
ENVIRONMENT DIMENSION	Health of the physical environment Environment that supports healthy development, social interaction, and physical activities (green and recreational areas)
	Safe environment
	Environment suitable for various social and age groups
	Healthy housing Housing that meets different needs of the families
HUMAN/SOCIAL DIMENSION	Awareness Responsibility
	Equality and inclusivity
	Commitment of individuals, solidarity and cooperation between various stakeholders (individuals, organizations)
	Healthy development and lifestyle, physical activity
POLICIES and SERVICES	Participatory governance
	Improvement in determinants such as education, health services, social services, and transportation
INCREASE IN HABITABILITY AND QUALITY OF LIFE	

Information about Bursa

Bursa, which became the first capital of the Ottoman Empire in 1335, is a metropolitan city with significant historical and cultural heritage. Inclusion of the 700-year-old Cumalıkızık village and Sultan Complexes (Emir Sultan/Yeşil Kulliye/Murat Hüdavendigar) among UNESCO World Heritage sites since 2014 allowed Bursa to acquire global significance and recognition. In addition to the historical and cultural background, industry plays an important role in Bursa. There are 17 active industrial zones in the city with negative effects on urban pollution and environmental health. The 3 largest districts in the city center include Osmangazi, Nilüfer, and Yıldırım districts (Fig. 1), while Bursa has a total of 17 districts.

Fig. 1. Locations of Osmangazi, Nilüfer and Yıldırım districts on Bursa Googe Earth Image

Located in the center of the city, Osmangazi district includes an important number of historical and cultural locations in Bursa. Historical and religious monuments such as historical inns, the Grand Bazaar, Salt Market, Muradiye social complex, Bursa Castle, the tombs of Osman and Orhan, Ulu Mosque, Tophane Clock Tower are among the important places in the district. Osmangazi District is the most economically and culturally developed district in the city [23]. The analysis of the education level in the district would reveal that most residents are

(75.6 %) primary, middle and high school graduates, and 13.1 % are with undergraduate or graduatc degrees. The socio-economic analysis would demonstrate the majority of the population is in Group B (33.4 %), followed by Group C (24.7 %) and group D (20.1 %) [24]. In other words, the district residents are middle, middle-upper, and low-income group individuals. In the district, where the old urban texture is prominent in built environment, the topography includes sloping narrow streets and a dense mixed typology, and planned and unplanned construction is next to each other (Fig. 2).

Fig. 2. An image of Osmangazi District

Nilüfer district, which is named after the Nilüfer Stream, is the most preferred residential district in Bursa with high urban planning levels. In Nilüfer, which contributes significantly both to the urban and national economy, the industrial zones provide employment to the majority of the urban population. Bursa Uludağ University is also located in this district, and 30.6 % of the district residents are college and higher education graduates, 60.6 % are primary, middle and high school graduates [25]. 35.3 % of the district population is in Group B, followed by Group C (21.4 %) and white-collar Group A (18.7 %). In other words, the population is predominantly middle and upper income. It is the most recently developed district in the urban center and dominated by planned neighborhoods (Fig. 3).

Fig. 3. An image of Nilüfer District [26]

There are several historical heritage buildings in the Yıldırım District, established on the skirts of Uludağ, dating back to the Ottoman period, including the Yeşil Complex and the Emir Sultan Mosque. 75 % of the population are primary, middle and high school graduates, and 12 % are college or higher education graduates. The socio-economic level of most district residents is Group B (32.4 %), 26 % are in Group C, followed by 21.2 % in Group D [27]. In other words, the districy is mostly preferred by middle and low income groups and blue collar workers. There are large slum areas in the Yıldırım district, which hosts the most unplanned and illegal buildings in Bursa, receiving blue-collar immigration from various provinces (Figure 4, 5).

Fig. 4 and 5. Images of Yıldırım District [28]

The most preferred residential districts in Bursa are Nilüfer (32.8 %), Osmangazi (22.9 %) and Yıldırım (10.5 %). Based on residential and healthy housing, the planned areas in the western section of the city where Nilüfer district is located and the unplanned areas in the eastern section where Yıldırım is located reflect spatial differences based on socioeconomic segregation. The preferred residential areas include the planned development neighborhoods in the west, while the regions of unplanned developed in the east are less preferred [29]. Physical and social inequalities between the districts were analyzed and presented in the following paragraphs.

Numerous studies have been and are conducted on urban health in Bursa. Bursa Metropolitan Municipality, a member of the Turkish Healthy Cities Association, organized several urban planning and process management meetings, and invited different stakeholders at different stages. Bursa Metropolitan Municipality that conducted detailed studies on urban health was awarded the WHO European Healthy Cities Network 3rd Phase membership in 2000. The municipality that played an active role in the Healthy Cities Project with City Council projects such as the improvement of environmental-friendly transportation alternatives, traffic regulations, preservation of historical and cultural heritage, climate change control, integrated waste management, development of industrial inventory and participatory democratic governance and reported that it prioritized equality in health in urban administration [30]. In this context, a consultation meeting was held by Bursa Metropolitan Municipality in December 2019, with the collaboration of Bursa Uludağ University and Academic Chambers, on urban center planning, emphasizing the urban health. The urban health in Bursa was analyzed based on the Consultation Meeting Analysis Results, where various stakeholders produced ideas about the city. The results of the aforementioned analysis was conducted based on the developmental perception about Bursa and the future expectations of the residents, and the data were analyzed based on the environmental, socioeconomical and urban facilities in environmental, residential, policies and services dimensions in Tab. 1.

The analysis of the satisfaction of the urban residents in Bursa as the social determinants of urban health demonstrated that satisfied individuals were the majority (80 %), and about 60 % of the occupants stated that they did not want to live in another city unless they had to (Tab. 2). Based on the views of the residents on urban health and facilities, satisfaction with basic services such as housing, education and health was about 65 %. On raising children and especially education, the satisfaction level was also 65 %. Based on the analysis of satisfaction with urban employment opportunities in Bursa, 50 % stated that they could find a job. Thus, it could be suggested that the economic opportunities provided by the city were lower when compared to other facilities. The level of belonging and basic services in the city generally satisfied more than half of the population, while job opportunities remained unsatisfactory. It could be stated that these facilities have decreased even more during the COVID 19 pandemic.

Tab. 2. Resident views on facilities and belongingness

		Strongly disagree (%)	Disagree (%)	I have no idea(%)	Agree(%)	Strongly agree (%)	TOTAL AGREEMENT (%)
FACILITIES	The facilities available in Bursa (housing, education and other services) are adequate for a comfortable life.	6,7	19	8,3	39,2	26,9	66,1
	There are adequate opportunities for Bursa to become a better city in the future.	6,8	17,2	8,6	44,8	22,6	67,4
	It is easy to find a job for those who want to work in Bursa	10,3	23,8	11,5	35,7	18,8	54,5
	Bursa provides good opportunities for raising children (education, health etc.)	7,2	15	11,5	44,5	21,7	66,2
BELONGINGNESS	I am satisfied with living in Bursa	4,1	9,5	3,8	58,6	23,9	82,5
	I do not want to live in a city other than Bursa unless I have to.	9,2	18,7	10,6	35,7	25,9	61,6

The analysis of the future concerns of the urban residents revealed the following ranking, from most to least common:

- Migration/population growth
- Economic factors (inflation, irregular investments, etc.)
- Administrative problems
- Inadequacy of services
- Extreme concretion.

The association of the reasons for the concerns for the future among the residents and the sense of belonging and opportunities demonstrated that the data were consistent and explained the causality. Satisfaction with housing, education and health services, which was above average, directly correlated with migration and population growth as the leading causes of concern. According to TURKSTAT data, 22

660 individuals migrated to Bursa in 2018 and about 20280 individuals migrated in 2019 [31, 32]. In addition to the migration to Bursa from various provinces in recent years, significant number of Syrian citizens also migrated to the city. Certain healthcare and education services were required by the immigrants, especially the Syrian citizens. In addition to primary, middle and high school education, certain other advantages were provided for immigrant citizens such as university education. Both Turkish and foreign citizens lived in certain neighborhoods, and various services were provided in these areas. Certain urban citizens moved to other neighborhoods, especially from Osmangazi to Nilüfer over time. The number of urban residents who moved to Osmangazi and Yıldırım districts was higher than Nilüfer. Fifty percent of the urban residents stated that both services and employment opportunities were not satisfactory, explaining the concerns about decreasing job opportunities due to migration and population growth. Thus, migration and population growth, economic factors and inadequacy of the services constituted a mechanism that triggered one another, and administrative problems were another completely associated cause of concern. Furthermore, the increase in construction and excessive concretion fueled by the population growth led to a different cause of anxiety for the residents and affected urban health negatively based on the physical environment.

Tab. 3. The perception of development about the last 5 years

Development areas	Agreement rates
Public services (health, education, security, etc.) were good	% 55,6 (about half)
The opportunity to access products and services that would improve comfortable living was good	% 55,5 (about half)
Urban economy (employment, livelihood, etc.) has been improving	% 54,1 (about half)
The urban architecture is good/the urban architectur and aesthetics improve	% 50,4 (half)
The city is clean and green	% 42,3 (less than half)
Ethical values (social ethics, business ethics, etc.) improve	% 47,9 (less than half)
Social relationships (neighborhood, family relationships, etc.) are good/improve	% 45 (less than half)

The analysis of the physical environmental health revealed the satisfaction with the architectural and aesthetic urban development was about 50 %, and the percentage of the residents who considered Bursa clean and green was quite low (40 %). Urban green spaces close to residential areas support personal health, healthy development, social interaction and physical activities. A healthy and active urban life recognizes the value of physical activity and provide a habitat that allows physical activities and sports facilities for all. In this approach that focuses on the physical and social environment land use patterns, transportation systems, urban design, green spaces and the built environment (schools, homes, workplaces and recreational areas) are included. In the social environment, participation in physical activities was associated with income level, equality, culture and social support [33]. Thus, the relatively low percentage of green spaces in Bursa was negative for urban health.

As seen in Tab. 3, the fact that development in public services and urban economy led to the satisfaction of half of the residents during the last five years could be associated with inadequate job opportunities and anxiety factors. The rate of development perception in employment and livelihood opportunities accompanied the public concern due to economic reasons. This, within the context of policies and services, reduced the rate of the people with a high sense of belonging and desired to continue living in Bursa. On the other hand, the reduction in ethical values such as social and business ethics, the deterioration of neighborhood and social relations negatively affected the human factor (50 %), in other words, the social determinants of urban health. In the human dimension of urban health, social solidarity, equality and perception of justice are among the most important social determinants. The findings obtained in Bursa indicated a reduction in social ties and solidarity. This led to weakening in the social determinants of urban health and also negatively affected the spirit of participation.

An important factor that affected Bursa in both physical and social determinants was the environmental and socioeconomic inequalities among the districts. As the green spaces gradually decrease from Nilüfer district towards the east and Yıldırım district, the building density and unplanned housing increase. The lack of infrastructure in unplanned urban areas, the relatively limited access to public services, socio-economic differences, and environmental and human inequalities had a significant negative impact on urban health. The Human Development Index - Districts research conducted by INGEV in 2017 revealed these inequalities in Bursa more clearly based on statistical data. The findings of the report that included social, economic and environmental factors and 50 indicators in governance, social inclusion, income level, education, health, social life, environment and transportation dimensions revealed that Nilüfer district had the highest development score. In the report that compared 150 districts with the highest population in 30 metropolitan municipalities in Turkey, Nilüfer and Osmangazi districts in Bursa were among the group with the highest human development. In a system that was scored between 0 and 1, similar to the HDI scores, Nilüfer district was the leading district with 0.511 points and Osmangazi was the second with 0.460 points. Yıldırım, on the other hand, was among medium human development districts with 0.346 points [34]. District scores in various indices are presented in Tab. 4.

Tab. 4. The main and sub-index scores of the districts [35]

HDI-D Index	Nilüfer	Osmangazi	Yıldırım
Main index	5	35	50
Governance and transparency	25	55	93
Social inclusion	20	21	6
Economic conditions	19	36	48
Education	8	91	112
Health	11	50	60
Social life	17	20	45
Environmental performance of municipality and transportation	11	42	22

As seen in Tab. 4, Nilüfer was the leader in the main index, but there were differences in sub-indices. Nilüfer ranked first and Yıldırım ranked last in governance and municipal performance dimensions in politics. In the new residential areas in the west and Mudanya-Izmir axis preferred by the upper-middle income individuals, the education and income levels and participation in social life were higher, while these attributes were relatively lower in the east [29]. Osmangazi ranked 2nd in overall sub-indices after Nilüfer but ranked last in social inclusion and municipal environmental performance and transportation indices. It was interesting that Yıldırım, with the most negative socio-physical environment score, led in social inclusion index, and based on INGEV data, which demonstrated that the Yıldırım district provided good services for various age and social groups, suggested that Nilüfer and Osmangazi could not provide equal access to these services. The data reported in a study by Mutlu and Varol (2017) were consistent with the INGEV data. In the study, individuals who stated that public services were adequate lived predominantly in Yıldırım and Osmangazi districts and in neighborhoods that could not meet the Spatial Plans Construction Regulation requirements, the residents of the western axis with planned development and relatively better technical and social facilities stated that the services were not sufficient. In the study, it was suggested that the residents were not aware of inadequate level of services due to their low expectations [29]. Holistic analysis of the study data clearly demonstrated that there were physical and social injustices and inequalities in Bursa and this fact was reflected in the space.

The analysis of the policy and services data demonstrated that the satisfaction with urban facilities was 65 %, and those who agreed with the statement "Everyone could contribute to social and administrative decisions" were 49.5 %. In urban health that requires a democratic and transparent governance, this ratio in collaborative decision-making and strengthening stakeholders, which was close to

50 %, was insufficient based on both political and social determinants. Participatory mechanisms should be improved, and the governance environment should be supported to establish solidarity and cooperation between various stakeholders (individuals, organizations), to determine the responsibilities and to raise commitment. In fact, one of the basic underlying ideas behind the human development index was the improvement of life preferences and options of individuals. Thus, the choices and ability to select improve the quality of life and social determinants of urban health through participation mechanisms and democracy.

General satisfaction with the environmental and social determinants of urban health and the governance dimension varied between 50 % and 60 %; however, it would be a good approach to examine the parameters preferred by the public. The order of significance attributed to various factors that affected urban health by the residents is presented in Tab. 5.

Tab. 5. The significance attributed to various factors that affected urban health by the residents

1. A clean and green city	% 50,4
2. A safe and peaceful city	% 49,6
3.A fun and modern city	% 36,5
4. A city that preserves the historical values and traditions	% 33,6
5. A city with easy transportation	% 30,2

Based on the data presented in Tab. 5, it could be suggested that the people of Bursa attributed the most significance to the environmental dimension of urban health. Half of the urban residents prioritized keeping the city clean, green and safe, followed by fun and modern facilities, preservation of historical values and urban traditions and ease of transportation. Thus, the findings demonstrated that urban cleanliness and green spaces were important, but the residents were not satisfied with the current status. Also, the urban areas reserved for physical activity were insufficient and there was inequality between the districts. In preserving historical values and traditions, although there was deterioration in social and ethical values, the residents valued historical and cultural heritage.

When the residents were asked to describe "the Bursa of their dreams", their responses were as follows (in the descending order of frequency):

- Plenty of green spaces, parks and trees,
- Peaceful/Calm/Quiet
- Clean (air, water, environment)
- Successful/Advanced/Modern
- No traffic/transportation problems

- Safe
- High employment
- Happy
- No immigration (Syrians, etc.)/Low population
- Protected natural resources/In touch with the nature.

It was observed that the correlation between the perceptions about the urban facilities in Bursa and about the areas of development and concern was also determined in the Bursa of dreams. However, similar to the order of importance, environmental factors were associated first, followed by social determinants (and administrative decisions) with the imagined factors. Environmental qualities that could be defined with concrete and abstract expressions such as "a greener, cleaner, more peaceful city where natural resources are preserved and in touch with nature association with the environmental factor" were consistent and could positively affect happiness. Employment and low population, and no immigration would affect each other positively and reduce the resident concerns about the future.

5 Conclusion and Recommendations

Certain present study findings on Bursa demonstrated that employment opportunities were not sufficient, and as a result, migration, population growth and economic problems were among the concerns about the future. Also, inadequacy of services and administrative problems and views that negatively affected the social determinants of urban health were the causes of concern. The findings also demonstrated that there was an interaction between all parameters that affected urban health. To support the habitability and long-term sustainability in Bursa, it was necessary for different stakeholders to produce ideas in collaboration, especially about social integration, improve inequalities, and balance migration and population growth. Similar to most metropolitan cities, there is social and spatial injustice in Bursa, and various determinants should be addressed to balance the differences between district facilities. Policies that aim to improve the inequality could positively affect the environmental and social dimensions of urban health and reduce the concerns about inadequate services and administrative problems. Awareness about various determinants of urban health should be raised through education in disadvantaged urban districts and groups. Policies that support the disadvantaged groups would promote the principle of equality, which is important in the 7th stage of urban health.

The improvement of educational opportunities in social groups starting from early ages and to construct lifelong awareness tools would promote elimination of inequalities. The popularization of distance education methods is an advantage to achieve this goal. Knowledge about the basic determinants of health at an early age, utilization of technological facilities for this purpose, and finding solutions in a participatory environment would support healthy habitats. The improvement of social welfare in healthy cities is also closely associated with social awareness.

Suggestions about the physical environment may not fulfill their potential or fade out within a short time in the lack of occupant awareness. A society without awareness could easily compromise the health of physical environment for economic benefits and rent without being aware of the deprivation. Since the urban consciousness develops based on the environment of growth since childhood, this awareness should be supported with various tools.

It is possible to define a baseline for all cities based on the specific dynamics. Both the responsibility of the urban authorities and the responsibility of the residents are important in the development of policies associated with a healthy environment and various age groups and the sense of responsibility and cooperative approach should be ensured among all stakeholders in the participative and governance environments. High participation and partnership levels, emphasized in Phase 7, is a requirement for an improvement in environmental and social health and the reflection of public expectations and preferences on the environment. Since a society that is not included in the decision-making mechanisms does not have the right to choose, satisfaction levels decrease, inequalities are triggered, and urban health cannot be supported due to the lack of governance. Regulations that ensure fair and easy access to urban facilities such as green spaces, healthy housing, affordable health and education services, and supporting the participation of various age groups in the decision-making process would support a habitable, sustainable, healthy habitat that improves the quality of life of families which are the building blocks of the society.

References

1. https://www.who.int/healthpromotion/healthy-cities/en, Access Date: 18.11.2020
2. Mumford, L., *What Is a City?, The City Reader,* ed: R. T. LeGates & F. Stout, 2000, London and New York: Routledge.
3. WHO, *Implementation framework for Phase VII (2019–2024) of the WHO European Healthy Cities Network: Goals, Requirements and Strategic Approaches,* 2019. WHO Regional Office for Europe UN City Marmorvej 51, DK-2100 Copenhagen, Denmark
4. Burris, S., Hancock, T., Lin, V., Herzog, A., *Emerging Strategies for Healthy Urban Governance.* Journal of Urban Health: Bulletin of the New York Academy of Medicine, 2007. **84** (1): p. 54–63
5. Akın, S., Arı, H.O., Demirok, A. vd., *21 Hedefte Türkiye: Sağlıkta Gelecek,* ed: N. Yardım, S. Mollahaliloğlu, Ü. Hülür Ü, S. Aydın S, N. Ünüvar, 2013. Ankara: T.C. Sağlık Bakanlığı Refik Saydam Hıfzıssıhha Merkezi Başkanlığı Hıfzıssıhha Mektebi Müdürlüğü; 2017. p. 1.
6. WHO, *Healthy cities: Promoting health and equity – evidence for local policy and practice Summary evaluation of Phase V of the WHO European Healthy*

Cities Network, 2014. WHO Regional Office for Europe UN City Marmorvej 51, DK-2100 Copenhagen, Denmark.

7. Conceição, P., *Human development report. 2019: Beyond income, beyond averages, beyond today: inequalities in human development in the 21st century, Human Development Report, 2019,* New York: United Nations Development Programme,
8. Mitchell G, Dorling D., *An Environmental Justice Analysis of British Air Quality*. Environment and Planning A, 2003, **35**: p. 909–929
9. Pye, S., Skinner, I., Meyer-Ohlendorf, N., Leipprandi A., Lucas, K., Salmons, R. 2008. *Addressing the social dimensions of environmental policy: A study on the linkages between environmental and social sustainability in Europe Report*, European Commission, Directorate-General for Employment, Social Affairs and Equal Opportunities, Unit for Social and Demographic Analysis. 2008, European Commission.
10. Zanussi M., *Health-integrated spatial planning: the experience of Sandnes, Norway,* Healthy urban planning in practice: experience of European cities, ed: H. Barton, C. Mitcham, C. Tsourou,. Copenhagen. 2003, WHO Regional Office for Europe.
11. Han X., H., Chen, K., *Child-friendly City Research: A Case Study of Portland Pearl District*. Urban Development Studies, 2016, **59** (4): p. 443–468.
12. Irazábal, C. *City Making and Urban Governance in the Americas: Curitiba and P*ortland. 2016, 1st ed. Routledge Taylor & Francis Group: New York:
13. Harrison, M., Coussens, C., *Global Environmental Health in the 21st Century: From Governmental Regulation to Corporate Social Responsibility: a Workshop Summary*. Institute of Medicine (U.S.). 2007, Washington, DC.
14. Dempsey, N., Bramley, G., Power, S., Brown, C., *The Social Dimension of Sustainable Development: Defining Urban Social Sustainability*. Sustainable Development, 2011. **19:** p. 289–300
15. Webster, P., Sanderson, D. *Healthy Cities Indicators—A Suitable Instrument to Measure Health?*. J Urban Health, 2013.**Oct 90 Suppl** 1(Suppl 1): p. 52–61.
16. Ritsatakis, A., *Equity and Social Determinants of Health at a City Level*. Health Promot Int, 2009. **24**: p. 81–90
17. Pineo H, Zimmermann N, Cosgrave E, Aldridge R, Acuto M, Rutter H., *Promoting a Healthy Cities Agenda through İndicators: Development of a Global Urban Environment and Health Index*. Cities & Health, 2018. **2** (1): p. 27–45
18. Fiscella K, Williams DR., 2004. *Health Disparities based on Socioeconomic Inequities: Implications for Urban Health Care*. Acad Med, 2004. **79**(12): p.1139–47.
19. Galea, S., Vlahov, V., *Urban Health: Evidence, Challenges, and Directions*. Annual Review of Public Health, 2005. **26** (1): p. 341–365

20. Dye, C., *Health and Urban Living*. Science, 2008. **319** (5864): p. 766–769
21. TSKB (Türkiye Sağlıklı Kentler Birliği), *Türkiye Kent Sağlık Göstergeleri (Tanımlama, Sınıflama, Değerlendirme, Öneriler),* Türkiye Sağlıklı Kentler Birliği. *2013,* Bursa.
22. Weaver, S. R., Dai, D. Luo, R., Rothenberg,R. *The Urban Health Index: A handbook for its calculation and use.* World Health Organization. 2014.
23. https://www.bursa.com.tr/osmangazi-3-1496.html, Access Date: 19.11.2020
24. https://www.endeksa.com/tr/analiz/bursa/osmangazi/demografi, Access Date: 7.12.2020
25. https://www.endeksa.com/tr/analiz/bursa/nilufer/demografi, Access Date: 7.12.2020
26. https://www.olay.com.tr/bursaya-bir-de-boyle-bakin-fsm-caddesi-558754), Access Date: 8.1.2021
27. https://www.endeksa.com/tr/analiz/bursa/yildirim/demografi, Access Date: 7.12.2020
28. https://www.yildirim.bel.tr/yildirim-detay/YILDIRIM-HAKKINDA, Access Date: 8.1.2021
29. Mutlu, E. Varol, Ç. *Sosyo-Ekonomik Farklılaşma ve Mekânsal Ayrışma: Bursa Metropoliten Alanı Analizi.* Megaron, 2017. **12** (1): p. 87–105.
30. Fidan, N. *Şehir sağlık profili hazırlamak=şehir sağlığını ölçmek.* Kentli, 2017. **Ocak-Şubat-Mart 2017**: p. 25.
31. https://data.tuik.gov.tr/Bulten/Index?p=Uluslararasi-Goc-Istatistikleri-2019-33709
32. https://data.tuik.gov.tr/Bulten/Index?p=Uluslararasi-Goc-Istatistikleri-2018-30711
33. Edwards, P., Tsouros, A. D., 2008. *A Healthy City is an Active City: A Physical Activity Planning Guide*, WHO Regional Office for Europe. 2008, Copenhagen, Denmark
34. İNGEV (İnsani Gelişme Vakfı), *İnsani Gelişme Endeksi –İlçeler (İGE-İ) Raporu*, Basın Bülteni, 2017. (http://ingev.org/wp-content/uploads/2017/10/IGE-I-ozet.pdf)
35. http://www.bursaport.com/yazarlar/kayihan-pala-31/bursa-da-ilcelerin-insani-gelisme-endeksi-sonuclari-2212.html, Access Date: 20.12.2020

Onur GÜNGÖR

The Examination of the Open and Green Areas of Iskenderun (Hatay) in Terms of Urban Green Infrastructure Planning Principles

Abstract: In the study, the urban open-green areas of Iskenderun were examined via multiple-criteria analysis in line with the green infrastructure planning principles. In this context, the presence of open-green areas of the district was assessed in terms of compliance with the four principles of basic green infrastructure planning principles, namely green-gray integration, connectivity, multifunctionality and social inclusion. Principal evaluations were carried out based on a) analyses of compliance with the size of urban active green areas, their amount per capita and accessibility, and b) SWOT analysis including determinations on the status of available and planned open-green areas. In the light of the findings, objectives of green infrastructure development were presented for Iskenderun. The results of the research revealed that there was not a system complying with the green infrastructure planning principles in the district. It was found out that the active green areas in the district were not distributed in an organized way, did not meet the accessibility criteria, and did not comply with the national and international standards with the availability of 4,52 m^2 of active green areas per capita. However, the potentials of Iskenderun's natural and physical characteristics to form an urban green infrastructure system were identified. It was concluded that the development of Iskenderun's open-green areas in line with the urban green infrastructure planning principles and handling them with a systematic approach would provide the district with many benefits in ecological, economic, social and cultural terms.

Keywords: Green infrastructure, open-green areas, urban, urban landscape, landscape planning

1 Introduction

The population of our world is rapidly increasing, and more than 50 % of the population began to live in cities in the 21st century for the first time in human history. This is predicted to rise to 70 % by 2050 [1]. This causes the natural environment to be affected severely by the built environment. To survive, humans consume our natural capital by using environmental resources which are considered free products, the real value of which is not fully met, and the sustainability of the environment is imperiled. The European Union defines investing in green infrastructure as a significant step towards the protection of the natural capital as a part of orientation to smart, sustainable and inclusive growth. For this purpose, the

European Union Commission published a notification numbered 0249 and titled "Green Infrastructure - Development of Europe's Natural Capital" in 2013. In this document, green infrastructure was defined as "a strategically planned network of natural and semi-natural areas with other environmental features designed and managed to deliver a wide range of ecosystem services. It incorporates green spaces (or blue if aquatic ecosystems are concerned) and other physical features in terrestrial (including coastal) and marine areas. On land, GI is present in rural and urban settings" [2]. However, the roots of the concept of green infrastructure date back to distant past. Benedict and McMahon (2012) [3] refer to the activities which began in the USA in the 1900s and were named as green structure, green road plans etc. and the green area plans produced for the city of Boston by F.L. Olmsted as the pioneering activities of the idea of green infrastructure. In the more recent scientific literature, it is highlighted that the concept of green infrastructure combines the green areas, water ecosystems, and different physical characteristics in lands, coasts and sea regions in its own body, and, with this approach, the principle of consciously integrating the benefits humans get from the nature with the protection and improvement of the nature and natural processes into the spatial planning is emphasized [4–7]. It is observed that the approach of green infrastructure planning focuses on an urban context as an alternative to the engineering infrastructure services which fulfill the constantly increasing needs of cities for water, food, transportation and energy, and the basic elements of the urban green infrastructure planning appear under the titles of the green-gray integration, connectivity, multifunctionality and social inclusion principles. ***The principle of green-gray integration*** stresses the role of the integration of built/gray infrastructure and green infrastructure in the reduction of urban water and energy consumption. ***The principle of connectivity*** emphasizes the characteristic of green areas and water surfaces to set a spatial and ecologic relation and the ability of habitat protection and development, climatic regulation, and building social connection by providing recreational opportunities. ***The principle of multifunctionality*** refers to making use of ecosystem services by protecting biodiversity. In this respect, it embodies the protection of social welfare and human health, sustainable water and land management, and reinforcement of the development of green economy. ***The principle of social inclusion*** underlines the topics of fair and free access of the groups with different social, cultural and economic conditions to environmental services, balancing the interests of different stakeholders and ensuring social consensus [2], [5], [8–12].

Green infrastructure plans can be made at different scales such as the region, sub-region, district and neighborhood scale [4], [7], [13]. There are many scientific studies on measuring whether the open-green areas and the green infrastructure system of a city fulfill the principles of green-gray integration, connectivity, multifunctionality and social inclusion by using quantitative and qualitative methods. Spatial location, size and distance analyses are crucial to measure the spatial integration and connectivity of green infrastructure [14–17]. In national regulations [18], international reports [19], [20], decision guides [21] and scientific literature [22–26], open-green areas were defined, classified and analyzed according to land size, per capita size and accessibility criteria. In the assessment of multifunctionality and social inclusion principles, qualitative and semi-quantitative measurement methods such as survey and SWOT analysis come into prominence [27].

This study aimed to identify the current situation of open-green areas and the potential of creating a green infrastructure system for Hatay's district of Iskenderun, and to develop the objectives of urban green infrastructure planning in the light of the findings. In this sense, the open-green areas that constitute the urban landscape of Iskenderun were examined with multiple-criteria analyses.

2 Materials and Methods

Research area. Iskenderun district of Hatay is located on the coast of Iskenderun Bay, in the region where 36° 34' 54" north latitude and 36° 09' 54" east longitude intersect. The district is bordered by the Mediterranean Sea in the west and the Amanos Mountains with rich vegetation in the east. The district is founded on a narrow waterside plain, which has a width of approximately 5 kilometers, and develops in the north-south direction in the coastal, plain and threshold lands. Iskenderun has turned into a significant industrial and port town today with the effect of the iron and steel factory established in the 1970s. The surface area of the district is 247 km^2 [28], and its urban population of 2018 is 248.335 people [29]. The urban area consists of 45 neighborhoods.

Research material. The main materials of the study are as follows:

- Publications on the concepts of urban open and green areas, urban green infrastructure systems and planning, legal legislation and institutional reports,
- 1/5000 and 1/1000 scaled Iskenderun city development plans
- 1/100.000 scaled Hatay Environmental Plan,
- Unpublished reports of Municipal Department of Park and Gardens
- Google Earth images of the study area,
- CORINE Land Cover Data published in 2018 [30]
- Image processing and analysis software.

Method. The method of the study consists of three stages. In the first stage, previous studies on the subject were examined and data related to the research area were collected. In the second stage, multiple-criteria analyses were carried out. It was aimed to produce the data that would enable setting the green infrastructure planning objectives of the district via three basic analyses.

The three basic analyses that constitute the second stage of the study are as follows:

- Analysis of existing land uses and presence of open-green areas
- Urban green infrastructure analysis
- SWOT analysis

The tools and techniques used in the analyses are explained below:

With the analysis of existing land uses and presence of open-green areas, it was researched whether the existing open-green areas within the borders of the district had a systematic plan or not. For this purpose, land use/land cover classification was done for Iskenderun according to CORINE level 3. The potential of open-green areas to form a green infrastructure system was examined by identifying the types and sizes of land use. Analyses on the patch-corridor relationship were conducted by using ArcGIS software, and citywide current land use/land cover map was produced.

With the urban green infrastructure analysis, the current status of the green areas in Iskenderun urban area was defined. The green areas of the 45 neighborhoods that constituted the urban area of Iskenderun were examined on a neighborhood scale. In this scope, analyses were carried out based on size, number, and accessibility criteria. The data of actively used urban green areas were obtained through the information provided by Iskenderun Municipality, 1/1000- and 1/5000-scale development plans and Google Earth images, and confirmed with area observations. Firstly, urban green areas were digitized in the ArcGIS Software, and the spatial distribution and accessibility maps of these areas were obtained. In the accessibility analysis, the distance of accessibility to children's parks and playgrounds from residential areas was considered as 250 m [21], and the distance of accessibility to neighborhood parks as 400 m [20]. Secondly, the population, size of green areas, and the amount of active green areas per person were determined on a neighborhood scale.

With the SWOT analysis, the opportunities, threats, strengths and weaknesses of the research area were determined by paying attention to their territorial and regional location, natural structure and resources, current situation, problems, potentials and area-related demands.

In the third stage of the study, the analysis results were synthesized, and the objectives of green infrastructure planning were developed for the district of Iskenderun.

3 Results and Discussion

3.1 General Characteristics of the Research Area

The research area, which is located in the Mediterranean climate zone, has a micro-climate characteristic due to its topography and the presence of the Amanos Mountains. It exhibits characteristics similar to those of the oceanic climate, which is arid-humid, mesothermal in summer, and has very strong excess water in winter. The range of the Amanos Mountains, which is a branch of the Taurus Mountains along the Mediterranean Region, lies in the northeast-southwest direction, and has an altitude of 2240 m, is a natural border for Iskenderun, and has specific characteristics within the whole Mediterranean Region. These mountains rise quite steeply starting from the sea, and here, there is a forest cover, which has been formed since the tertiary (glacial) period and is similar to the Black Sea and Marmara Regions. The air mass hits these mountains with a high altitude and turns into moist winds and heavy rain in the canyons and corridors created by the dynamic topographic structure. The Amanos Mountains, which have rich vegetation in terms of flora and lie very close to the shore, have become the symbol of Iskenderun by integrating into the district together with the coastal parks along the shore and beaches around. The forests of the region, which resemble research laboratories with their rich vegetation, and are the insurance of the sustainability of the district ecosystem, should be protected with appropriate protection statuses. Many ridges and valleys have been formed in the Amanos due to the dynamic topographic structure. Accordingly, there are numerous water parting and water flow lines and seasonal streams in the region. In millions of years, beaches have been formed in places where the streams created by the water lines meet the sea. This system creates drinking and utility water resources by feeding underground and surface water resources. Protection of the topographic structure without destruction is necessary for the sustainability of the water existence.

The total built-up area of the district, which develops on a waterside plain and threshold areas and is mainly characterized by residential, industrial and port areas, is 75,92 km^2. This area constitutes 30 % of the surface area of the district. With the law numbered 6360 and enacted in 2012, the villages gained the status of neighborhoods, and the number of neighborhoods of the district increased from 28 to 45 [31]. The buildings in the center and surroundings of the district were built in a multistory and adjacent order. However, there are multistory buildings surrounded by small gardens or detached houses with 2–3 stories and gardens in the neighborhoods between D817 and E91 highways. Settlements transformed from villages into neighborhoods are the neighborhoods located inside the urban area and forest area on the eastern side of E91 highway.

There are seasonal streams, which are significant elements of a green area system and function as natural corridors, within the boundaries of the district. Feyezan Stream is an important water element flowing through the urban area and pouring into the sea.

3.2 Multiple-Criteria Analysis of the Research Area

As mentioned before, this stage of the study consists of three steps.

3.2.1 Analysis of Existing Land Uses and Presence of Open-Green Areas

Existing land uses of Iskenderun were investigated with the level 3 land use/land cover classification according to the 2018 data of CORINE (Fig. 1). In the district with a surface area of 247 km^{2,} large, coniferous and mixed forests cover an area of 191 km^2 and constitute the largest land cover. On the slopes of the mountains facing the sea, garrigue formation is seen, including *Olea europaea, Ceratonia siliqua, Quercus coccifera* macquis and bulbed tuberous geophytes up to 800 meters from the shore. *Pinus brutia* forests begin from a height of 300 meters. These *Pinus brutia* forests are rather mixed with garrigue and maquis, agricultural and residential areas and are exposed to intense anthropogenic effects. In the forests between 800–1200 meters, where the species of *P.brutia* and *Quercus cerris* are dominant, the species of *Fagus sp, Cornus sp, Platanus sp.* are available. Above 1200 meters, large forest areas with *P. brutia, P. nigra, Cedrus libani and Juniperus foetidissima* were formed in general.

It was noticed that the forest cover was fragmented by mining and material extraction sites at five spots covering an area of 3 km^2 in total. This causes a negative effect on the characteristics of the green-gray integration, connectivity, and multifunctionality in the area. Arable lands, agricultural lands, orchards, olive groves and pastures have a significant surface area covering 51 km^2 in total. These areas are interrupted by continuous and intermittent urban structure, industrial and commercial units and highways covering an area of 48 km^2. The entrance to the city in the north is enabled by D817 highway passing through the organized industrial zone.

The industrial zone starts in the district of Payas and extends along the coast to the urban residential area of Iskenderun. The small industrial zone remained in residential areas in the center. Iron-steel, cement and fertilizer factories in the organized industrial zone and port activities lead to serious environmental problems such as air, soil, water and sea pollution [32]. Industrial activities are among the important anthropogenic effects on the vegetation of the region. There are neighborhoods containing workers' suburbs around the industrial areas. The Sarıseki neighborhood is one of them and has a marsh with a calculated size of approximately 29,03 hectares. Ulus (2008) reported that the marsh area was 85 hectares with the data of 2008 [33]. This means that 56 hectares of marsh area had been lost in ten years. Other marsh areas in the center were dried up and turned into residential areas. However, these areas are significant biotope areas enriching the landscape with the presence of flora and fauna that are unique to them, and they need to be protected and improved within the scope of wetlands. Residential, industrial and transport networks are the land uses causing marshes and seasonal streams that feed marshes to disappear.

3.2.2 Urban Green Infrastructure Analysis

Within the framework of urban green infrastructure analysis, 45 neighborhoods that constitute the urban residential area were examined. It was revealed that there were no neighborhood parks, children's parks, playgrounds and sports areas in 19 of the 45 neighborhoods. The open-green areas along the Iskenderun coast are district parks and serve all the residents of the district. It has children's parks, playgrounds and sports areas, social facilities, walking and cycling paths, water show pools, ceremony and festival areas. The size of the green areas on the coastline, which is about 2.9 km long, is 600,000 m^2, and the amount of green area per capita is 2.5 m^2 on average when proportioned to the urban population. The green

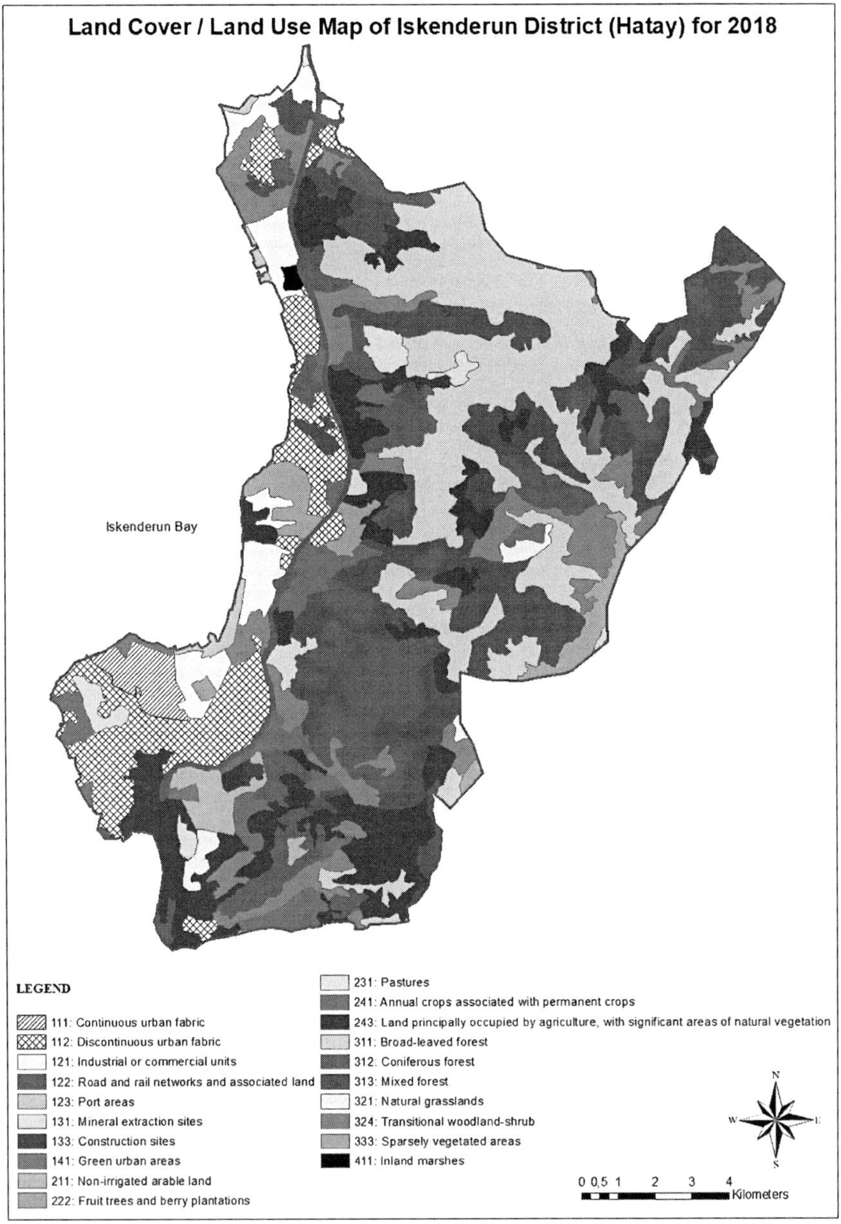

Fig. 1: The level 3 Land Use/Land Cover Classification of Iskenderun according to the 2018 Data of CORINE (Original, 2021)

areas along the coast, with their linear structure, and Atatürk Boulevard form the single green corridor of the district but cannot connect with the inner regions of the district. The role of this largest open-green area of the district in creating the green infrastructure is crucial. The area should be examined with detailed researches in terms of plant species diversity, plant composition, richness of visual and audio presentation, functionality and aesthetics of urban furniture, diversity of social, cultural, sports areas and activities, and the adequacy of engineering infrastructure such as irrigation, drainage and lighting systems. It is necessary to establish landscape design and implementation strategies which would integrate the gray-green infrastructure in this area. Many buildings and landscaping arrangements along the Iskenderun coast are situated on the embankment area. This makes it even more obligatory to establish the green infrastructure to support the engineering infrastructure against natural disasters such as earthquakes and floods.

Feyezan Stream, which reaches the sea at the western border of the district by traveling for approximately 6 km, pours into the concrete canal in the district, and it is under the pressure of the intense urbanization around it. This stream has a high potential of transforming into a green corridor. When the development plans of the district are reviewed, it is seen that intensive green space planning is made along the stream; however, it is not implemented. Except for the Feyezan canal throughout the district, 7 seasonal streams meet the sea. These mostly pass through the industrial areas and neighborhoods developing around the industrial areas. For the Sarıseki marsh, which is fed by three different stream branches in the Sarıseki neighborhood, to fulfill its ecosystem services, it needs to be re-approached and evaluated as a part of the system within the scope of green infrastructure planning.

The size and accessibility characteristics of the green areas (children's park, neighborhood park, playground and sports areas) at the neighborhood level were examined; the results of the analysis are summarized and given in Tab. 1. Accordingly, the distances of access to green areas on the coastline from 7 neighborhoods along the coast comply with international criteria [20,21]. The coastal green areas, which contain different types of urban open-green areas in different quantities and qualities, meet the connectivity and multifunctionality principles of the green infrastructure approach for the region of the district including these neighborhoods, with the benefits such as establishing ecological connections, enabling various types of recreational activities and contributing to the economy of the district. However, accessibility (travel time and distance on foot and in a vehicle) to these areas on the coastline from the remaining 38 neighborhoods does not comply with the international standards. These neighborhoods either do not have active open-green areas at all or have green areas far below the standards. For this reason, the size and accessibility analyses performed on a neighborhood scale show us that the presence of active green areas is inadequate to create an urban green infrastructure within the district. Besides, the principle of green-gray integration is not provided since there are no systems such as bioswales, biofiltration swales, which support the gray infrastructure within the green areas in the district. The principle of social inclusion emphasizes collaborative and participatory planning, and for the urban green areas of Iskenderun, this principle can be interpreted as follows. Even though the participation of the stakeholders is encouraged in the legislation on spatial planning in Turkey, it is not clearly put forward how the procedure will Iskenderun take place. Views of the stakeholders such as land owners, entrepreneurs, investors, local administrators, relevant non-governmental organizations and urban planning and urban design experts should be included in the planning in an objective and measurable manner. More than 50.000 objections to the additional and revised 1/5000-scale urban development plan made in Hatay in 2019 significantly indicate that relevant stakeholders could not be included in plan-making processes properly.

Tab. 1: The quantity of active green areas of Iskenderun at the neighborhood level and their compliance with green infrastructure planning principles.

Quantity of Active Green Areas at Neighborhood Unit Level								**Compliance with Green Infrastructure Planning Principles**			
Name of the Neighborhood		**The population of the neighborhoods in 2018**	**Number of active green areas**	**Size of active green areas (m²)**	**Area of green spaces per capita (m²/per capita)**	**Percentage size of the active green areas in relation to the area of the neighborhood (%)**	**Distance of access (meter)**	**Green-gray integration**	**Connectivity**	**Multifunctionality**	**Social inclusion**
1	Akarca	1.400	*	*	*	*	-	-	-	-	-
2	Akçay	2.152	*	*	*	*	-	-	-	-	-
3	Aşkarbeyli	697	*	*	*	*	-	-	-	-	-
4	Azganlık	2.632	*	*	*	*	-	-	-	-	-
5	Barbaros	4.254	1	2900	0,68	1,32	+	-	-	-	-
6	Barıştepe	2.673	1	2000	0,75	1,62	+	-	-	-	-
7	Bekbele	7.754	1	1650	0,21	0,03	+	-	-	-	-
8	Bitişik	1.035	*	*	*	*	-	-	-	-	-
9	Buluttepe	5.302	1	1200	0,23	0,15	+	-	-	-	-
10	Büyükdere	1.333	*	*	*	*	-	-	-	-	-
11	Cırtıman	2.908	*	*	*	*	-	-	-	-	-
12	Cumhuriyet	6.992	*	*	*	*	+	-	-	-	-

(continued)

Tab. 1: Continued

Quantity of Active Green Areas at Neighborhood Unit Level								Compliance with Green Infrastructure Planning Principles			
Name of the Neighborhood		The population of the neighborhoods in 2018	Number of active green areas	Size of active green areas (m^2)	Area of green spaces per capita (m^2/per capita)	Percentage size of the active green areas in relation to the area of the neighborhood (%)	Distance of access (meter)	Green-gray integration	Connectivity	Multifunctionality	Social inclusion
13	Çay	5.305	4	9785	1,84	0,63	+	-	+	+	-
14	Çınarlı	815	*	*	*	*	-	-	-	-	-
15	Denizciler	23.461	9	60.476	2,58	1,50	-	-	-	-	-
16	Dumlupınar	7.918	*	*	*	*	-	-	-	-	-
17	Düğünyurdu	751	*	*	*	*	-	-	-	-	-
18	Esentepe	6.482	1	500	0,08	0,03	-	-	-	-	-
19	Fatih S. Mehmet	3.560	1	7630	2,14	1,30	+	-	-	-	-
20	Gültepe	6.410	2	3500	0,55	0,20	-	-	-	-	-
21	Gürsel	2.236	1	1000	0,45	0,32	+	-	-	-	-
22	Güzelçay	1.018	3	9145	8,98	0,18	+	-	-	-	-
23	Hürriyet	3.623	2	1270	0,35	0,79	+	-	+	+	-
24	İsmet İnönü	16.370	5	17500	1,07	0,86	-	-	-	-	-
25	Kaledibi	349	*	*	*	*	-	-	-	-	-
26	Karayılan	9.700	2	85330	8,80	2,26	+	-	+	+	-
27	Kavaklıoluk	325	*	*	*	*	-	-	-	-	-
28	Kocatepe	5.405	*	*	*	*	-	-	-	-	-
29	Kurtuluş	6.414	1	650	0,10	0,32	+	-	-	-	-
30	Meydan	4.757	1	1500	0,32	0,12	+	-	-	-	-
31	Modernevler	12.108	5	11038	0,91	0,56	+	-	-	-	-
32	Muradiye	6.503	*	*	*	*	-	-	-	-	-
33	Mustafa Kemal	18.794	4	12450	0,66	0,84	+	-	-	-	-

Tab. 1: Continued

Quantity of Active Green Areas at Neighborhood Unit Level								**Compliance with Green Infrastructure Planning Principles**			
Name of the Neighborhood		**The population of the neighborhoods in 2018**	**Number of active green areas**	**Size of active green areas (m^2)**	**Area of green spaces per capita (m^2/per capita)**	**Percentage size of the active green areas in relation to the area of the neighborhood (%)**	**Distance of access (meter)**	**Green-gray integration**	**Connectivity**	**Multifunctionality**	**Social inclusion**
34	Numune	9.504	4	13495	1,42	1,13	-	-	-	-	-
35	Orhangazi	723	*	*	*	*	-	-	-	-	-
36	Pınarbaşı	2.697	*	*	*	*	+	-	+	+	-
37	Pirireis	8.611	4	5250	0,61	0,86	+	-	+	+	-
38	Sakarya	15.601	4	12609	0,81	1,01	+	-	+	+	-
39	Sarıseki	3.550	6	169860	47,85	1,63	+	-	+	+	-
40	Savaş	1.024	4	7093	6,93	7,13	+	-	+	+	-
41	Suçıkağı	1.779	*	*	*	*	-	-	-	-	-
42	Süleymaniye	3.911	1	500	0,13	0,21	+	-	+	+	-
43	Yenişehir	4.518	3	2888	0,64	1,35	+	-	+	+	-
44	Yıldırımtepe	4.269	*	*	*	*	-	-	-	-	-
45	Yunus Emre	10.712	5	30800	2,88	3,44	+	-	+	+	-

**: none in the neighborhood unit. +: provides the relevant feature. -: does not provide the related feature.*

3.2.3 SWOT Analysis

SWOT analysis is frequently used in green infrastructure research [27]. In this study, the present strengths and weaknesses of the open-green areas of Iskenderun city to create green infrastructure and the opportunities and threats for the future were examined by SWOT analysis (Tab. 2).

Tab. 2: SWOT Analysis

Strengths:	**Weaknesses:**
- The presence of forest and agricultural areas surrounding the district, seasonal streams, sea, seashore, beaches, marshes and active green areas in the district, - The scientific knowledge of the university, - University-Local Government collaborations *(for instance, a bicycle path master plan is being prepared on a district scale)* - University-Industry Collaborations *(for instance, development of environmentally friendly technology, electric, hydrogen powered cars, R&D studies)*	- The presence of numerous mining and material extraction sites in the district. - Air, water, soil and sea pollution caused by the industrial zone. - Distortion of vegetation, topography and water flow lines caused by various land uses. - The division of the forest cover by E91 Highway crossing the district in the east and approximately defining the borders of the urban area. - The strength of Iskenderun's open and green areas to establish a spatial and ecologic connection is weak.
Opportunities: - Green corridors can be created in the district by planting E91 and D817 highways, which cross the district in the north-south direction, in accordance with the principles of landscape planning for highways. - The presence of cooperation of university-industry-local administrations to improve the gray infrastructure weakening the urban ecosystem with environmentally friendly new technologies and engineering studies and to develop projects that integrate the gray-green infrastructure. - The fact that a significant part of the EXPO 2021 Garden Exhibition will be held in İskenderun and the projects to be implemented within this context will turn into investments supporting the urban green infrastructure system.	**Threats:** - The pressure of the decisions made in the development plans on increasing the density of buildings and urban section on agriculture, forest areas and urban green areas.

3.3 Green Infrastructure Planning Objectives Proposed for the City of Iskenderun

As stressed in many articles, developing an urban green infrastructure system requires relevant stakeholders to collaborate [26, 34]. For Iskenderun, the stakeholders of such a collaboration should be Hatay Metropolitan Municipality,

Iskenderun Municipality, Iskenderun Technical University, representatives of the rural organization of the relevant ministries, landscape architects as private sector representatives, city planners and architects, non-governmental organizations, land owners and contractors (implementers). Stakeholder groups should be expanded and diversified in line with the objectives. The first step in planning the urban green infrastructure system in Iskenderun should be the definition of priority objectives by the stakeholders, and the second step should be determining the strategies and actions to accomplish the objectives. Within the context of this study, it was attempted to set forth basic guiding objectives in the light of the analyses (Tab. 3).

Tab. 3: Basic Objectives in Developing the Urban Green Infrastructure System of Iskenderun.

Objectives:	
1.	Proposing appropriate conservation statuses for the protection and development of forest areas surrounding the city
2.	Protection of stream beds taking their sources from forests and pouring into the sea in the district and its peripheries.
3.	Conservation and improvement of marshes in the district and its immediate surroundings within the framework of wetland protection rules.
4.	Controlling and monitoring the consumption of natural resources by industries and bringing restrictions and sanctions for them.
5.	Conservation and development of agricultural lands under the urbanization pressure in the immediate vicinity of the district.
6.	Improvement of the size, per capita amount and accessibility of urban active green areas to a level complying with national and international norms.
7.	Organization of the urban pedestrian and vehicle transportation network to create green corridors.
8.	Establishing a habitat connection by connecting urban active green areas with green corridors.
9.	Increasing the diversity of recreational activities of the residents by integrating the green areas which are separated and in the form of corridors, and strengthening social communication.
10.	Creating smart systems and innovative engineering solutions as a part of the green infrastructure system to reduce the district's need for materials, energy and water.

4 Conclusion

In this research, to shed light on the current status of Iskenderun district of Hatay, the presence of open-green areas and urban active green areas were analyzed through quantitative and qualitative criteria with an approach based on compliance with the principles of urban green infrastructure planning. The results demonstrated that open-green areas in Iskenderun are fragmented and distributed irregularly, and they do not form an open-green area network with ecological and recreational functions or green infrastructure. This problem is common for many cities; for example, Wang (2009) reported that the pattern created by green areas in China could not create an urban green area system, and the size of the fragments and their per capita amount were inadequate [35]. Similarly, in their study carried on Antwerp, Ghent, Aalst and Kortrijk cities of Belgium, Van Herzele and Wiedemann (2003) expressed that green areas were divided and fragmented by other land uses, and therefore, their size and accessibility were reduced [36].

The population of Iskenderun in 2018 was 248.335, and its annual population increase rate was 2.4 % [29]. The total size of active green areas in 45 neighborhoods within the district is 1.124.259 m^2, and the amount of active green area per capita is 4,52 m^2, which has remained below the active open-green areas standard of 10 m^2 announced by Spatial Plans Production Regulation [18].

Even though they do not exactly meet the functions of district parks in Iskenderun, the green area arrangements along the coastline constitute a significant open-green area with the biggest size in the whole district. The National Garden, which was opened in 2018 and located in the center, is inadequate in terms of size, plant composition, and diversity of social and recreational activities; however, it is the single large park in an adjacent and constricted order within the district. It will assume a significant role in the preservation of biological diversity particularly in the urban landscape, where housing is intense, over time. Most of the parks in Iskenderun are small, and their impact areas are limited. For this reason, the coastal green areas and the national garden are among the most important components of the potential open-green area system as the largest urban parks of the district. It is extremely difficult to create new open-green areas in the existing urban structure. It is necessary to evaluate the urban transformation activities in Iskenderun as an opportunity to increase the number of open-green areas, revisit the issue of the Feyezan canal as a stream corridor and establish an open-green area system. It is necessary to prepare the urban green infrastructure plan of Iskenderun, which suggests the establishment of the connections of the Amanos forests surrounding the district in the east with active green areas, the fertile agricultural areas that form the Arsuz plain by extending along the coastline in the southern and southwestern directions of the Mediterranean coast in the west, and seasonal streams.

Apart from production, agricultural lands in urban landscapes bring people together with the land and provide the residents with recreational opportunities, support wildlife by creating habitats, and contribute to the development of sustainable cities [13]. In this sense, it is crucial to protect the agricultural areas in the periphery, especially in the Karaağaç region as a part of the Arsuz plain, even if no lands are left in the built area in Iskenderun. However, these areas are under the pressure of intense construction in the form of summer housing.

The largest public open area in the urban district is 785.942 m^2, which is covered by the campus of Iskenderun Technical University. This is a new campus area, and intensive planting has been done within the scope of landscape arrangement activities for the last few years, and attention has been paid to using species belonging to the Mediterranean natural vegetation. The open-green areas of the campus, where there are mainly *Cupressus sempervirens, Pinus brutia, Pinus pinea* and *Olea europaea*, will be a very valuable component of a constructed urban green infrastructure system. Moreover, the pond, which was transformed from a marsh area in the campus, and its surrounding region are home to various bird species and other animals since the region is on the route of migratory birds. All these benefits should be considered and evaluated in the context of the ecological services offered by the area.

Similarly, graveyards also have a significant potential in terms of the open-green area system. The graveyard in Karaağaç neighborhood at the urban area border of Iskenderun especially provides a feeding and shelter area for many wild species since it has old trees and contributes to the biological diversity of the district.

Green area corridors are not adequate in the district. Therefore, there is a need for forming new green corridors with sufficient widths and lengths. The physical association of the corridors to be formed with independent fragments is significant. In this sense, 47 km corridors can be created through the afforestation of E91 and D817 highways, 3 km corridors through the afforestation of the railway periphery, and 27 km corridors through the improvement of the stream beds; hence, physical connection can be established with some of the available green fragments. As a part of the urban green infrastructure system, streets, roads and boulevards with a width of 10 meters or above in Iskenderun should be afforested and transformed into corridors extending to the forest areas surrounding the city.

When the research findings are interpreted, it is revealed that there are no green infrastructure systems in the district, the condition of the open-green areas in the urban landscape is not adequate in respect of quality and quantity; however, there are green areas not inside the district, but in its periphery, which can become a part of the green infrastructure system. Furthermore, there are also steps that can be taken towards increasing the quality of the exiting green areas in the urban residential area, and the neighborhoods for which urban transformation is planned have new opportunities. Here, it is necessary to create open-green areas that come close to international norms in respect of quantity and quality and to connect them to the peripheral areas by means of natural and/or man-made corridors. Open-green areas to be produced by planning will be functional in terms of many ecosystem services such as habitat protection and development, supporting biodiversity, rainwater management, and fulfillment of recreational needs, and serve as the backbone of the city.

The objectives proposed within the context of the study are the key points during the preparation of an urban green infrastructure plan for Iskenderun and its reflection on the physical development plans of the district. Preparing plans and developing objectives and strategies involve multiple stakeholders. It is necessary to ensure the participation of specialized scientists, local and central administrations, universities, non-governmental organizations and the public, to evaluate the objectives and strategies for creating a green infrastructure in the light of national and international criteria, and to make legislative arrangements in the urban planning system.

In the light of the law numbered 6360 enacted in 2012, the boundaries of Hatay Metropolitan Municipality have been determined as the provincial administrative boundaries [31]. Therefore, it is appropriate to evaluate the analyses conducted within the scope of the study, results and recommendations together with the regional development plans and decisions on land use in rural settlements. In conclusion, there is a need for urban green infrastructure planning that integrates the open-green areas in Iskenderun and around it by paying attention to the spatial planning scale hierarchy. A solution for this need can be provided with the introduction of the concept of the green area planning into the spatial planning system of our country and their integration into urban plans.

References

1. UN - United Nations Population Division World, *Population prospects: the 2010 revision.* 2011. New York: United Nations.
2. European Commission (EC), *Green Infrastructure (GI) - Enhancing Europe's Natural Capital.* Brussels, 6.5.2013. COM(2013) 249 final, SWD(2013) 155 final, 2013, 11 pages. [cited 2020 23 November]; Available from: https://www.europarl.europa.eu/RegData/docs_autres_institutions/commission_europeenne/com/2013/0249/COM_COM(2013)0249_EN.pdf
3. Benedict, M.A. and E.T. McMahon *Green Infrastructure: Linking Landscapes and Communities.* 2012. Washington, D.C., USA: Island Press.
4. Davies, C., Hansen, R., Rall, E., Pauleit, S., Lafortezza, R., De Bellis, Y., A. Santos, and I. Tosics, *Green infrastructure planning and implementation - the status of European green space planning and implementation based on an analysis of selected European city-regions.* GREEN SURGE Deliverable 5.1 Technical Report of the Green Surge Project. 2015.
5. Hansen, R., Rolf, W., Rall, E., Pauleit, S., Erlwein, S., Fohlmeister, S., Santos, A., Luz, A.C., Branquinho, C., Santos-Reis, M., Gerőházi, E., Száraz, L., Tosics, I., Davies, C., De Bellis, Y., Lafortezza, R., Vierikko, K., van der Jagt, A., Cvejić, R., Železnikar, S., Nastran, M., Pintar, M., Hjorth, Caspersen, O., Stahl, Olafsson, A., Gentin, S., Andersson, E., Kronenberg, J., Delshammar, T., T. Mattijssen and R. Otten, *Advanced Urban Green Infrastructure Planning and Implementation - Innovative Approaches and Strategies from European*

Cities. GREEN SURGE Deliverable 5.2 Technical Report of the Green Surge Project. 2016.

6. Pauleit, S., Hansen, R., Rall, E.L., Zolch, T., Andersson, E., A. Luz, and K. Vierikko, *Urban landscapes and green infrastructure*, in *Research Encyclopedia of Environmental Science*, H. Shugart, Editor. 2017, USA: Oxford.
7. Artmann, M., O. Bastian, and K. Grunewald, *Using the Concepts of Green Infrastructure and Ecosystem Services to Specify Leitbilder for Compact and Green Cities - The Example of the Landscape Plan of Dresden (Germany).* Sustainability, 2017. **9**(2): 198.
8. European Environment Agency (EEA), *Green Infrastructure and Territorial Cohesion. The Concept of Green Infrastructure and Its Integration into Policies Using Monitoring Systems.* 2011, Publications Office of the European Union: Luxembourg.
9. Ignatieva, M., G.H. Stewart, and C. Meurk, *Planning and design of ecological networks in urban areas.* Landsc. Ecol. Eng., 2011. **7**: p. 17–25.
10. European Commission (EC), *The Multifunctionality of Green Infrastructure. Science for Environment Policy.* In depth Reports, 2012.
11. European Commission (EC), *Green Infrastructure (GI) - Enhancing Europe's Natural Capital.* Technical information on Green Infrastructure (GI), Commission Staff Working Document, Brussels, 6.5.2013. COM(2013) 249 final, SWD(2013) 155 final, 2013, 24 pages. [cited 2020 23 November]; Available from: https://ec.europa.eu/environment/nature/ecosystems/docs/green_infrastructures/1_EN_autre_document_travail_service_part1_v2.pdf
12. Girma, Y., Terefe, H., S. Pauleit, and M. Kindu, *Urban green infrastructure planning in Ethiopia: The case of emerging towns of Oromia special zone surrounding Finfinne.* Journal of Urban Management, 2019. **8**: p. 75–88.
13. Hepcan, Coşkun, Ç. and Ş. Hepcan, *Kentsel yeşil altyapı analizi: Bornova örneği.* Mediterranean Agricultural Sciences, 2018. **31** (1): p. 37–43.
14. Weber, T. and J. Wolf, *Maryland's green infrastructure - using landscape assessment tools to identify a regional conservation strategy.* Environmental Monitoring and Assessment, 2000. 63(**1**): p. 265–277.
15. Schäffler, A. and A. Swilling, *Valuing green infrastructure in an urban environment under pressure - The Johannesburg case.* Ecological Economics, 2013. volume: **86**, p. 246–257.
16. Liquete, C., et al., *Mapping green infrastructure based on ecosystem services and ecological networks: A Pan-European case study.* Environmental Science and Policy, 2015. **54**. p. 268–280.
17. Geberemariam, T.K., *Post Construction Green Infrastructure Performance Monitoring Parameters and Their Functional Components.* Environments, 2017. **4** (1): 2.

18. T.C. Resmi Gazete, *Mekânsal Planlar Yapım Yönetmeliği, EK-2: Kentsel, Sosyal ve Teknik Altyapı*, Tarih: 14.06.2014, Sayı: 29030, Ankara, 2014.
19. WHO - World Health Organization, *Urban Planning, Environment and Health: From Evidence to Policy Action* – Meeting Report, 2010.
20. National Recreation and Park Association, *2016 NRPA Field Report, Park and Recreation Agency Performance Benchmarks*, 2016, Ashburn, VA 20148–5401.
21. NPFA - National Playing Fields Association (Fields in Trust), *Guidance for Outdoor Sport and Play: Beyond the Six Acre Standard-London*. 2008.
22. Mertes, J.D. and J.R. Hall, *Park, Recreation, Open Space and Greenway Guidelines*, National Recreation and Park Association (NRPA) Published Report, 1995, Arlington, VA, USA.
23. Williams, S., *Recreation in the Urban Environment*, 1995, London: Routledge.
24. Willis, K.G., R.K. Turner, and I.J. Bateman (eds.), *Urban Planning and Management*, 2001, Cheltenham, Edward Elgar Publishing, UK.
25. Cvejić, R., Eler, K., Pintar, M., Železnikar, Š., Haase, D., N. Kabisch, and M. Strohbach, *A Typology of Urban Green Spaces, Eco-System Services Provisioning Services and Demands*. GREEN SURGE Project Technical Report, Report D5.1, EU FP7 Project. 2015.
26. Manavoğlu, E. and V. Ortaçeşme, *Antalya kenti yeşil alanlarının çok ölçütlü analizi ve planlama stratejilerinin geliştirilmesi*. Akdeniz Univ. Ziraat Fak. Derg., 2015. **28**(1): p. 11–19.
27. Yılmaz, R. and O. Yılmaz, *Evaluation of the Concept of Green Infrastructure Systems, Planning and Management by Using Swot Analysis (Strengths, Weaknesses, Opportunities and Threats)*, in *New Approaches to Spatial Planning and Design*, 31th Chapter, p 425–434. M. Özyavuz, Editor. 2019, Peter Lang AG: Bern.
28. GDM - General Directorate of Mapping, *Province and District Areas* (in Turkish), 2014. [cited 2021 10 January]; Available from: http://www.harita.gov.tr/images/urun/il_ilce_alanlari.pdf
29. TUIK - Turkish Statistical Institute. *Address based population registration system results – İskenderun, 2018*, [cited 2021 5 January]; Available from: https://biruni.tuik.gov.tr/medas/?kn=95&locale=tr
30. Copernicus Land Monitoring Service, *CORINE Land Cover, CLC 2018*. [cited 2020 5 February]; Available from: https://land.copernicus.eu/pan-european/corine-land-cover/clc2018
31. T.C. Resmi Gazete, *On Dört İlde Büyükşehir Belediyesi ve Yirmi Yedi İlçe Kurulması ile Bazı Kanun ve Kanun Hükmünde Kararnamelerde Değişiklik Yapılmasına Dair Kanun*. Kanun Numarası: 6360, Tarih: 06.12.2012, Sayı: 28489. 2012. Ankara.

32. T.R. Hatay Governorship - Provincial Directorate of Environment and Urbanization, *Hatay Province 2018 Environmental Status Report (in Turkish).* 2018. [cited 2021 5 January]; Available from: https://webdosya.csb.gov.tr/db/ced/icerikler/hatay_2018_-cdr_son-20191031135340.pdf
33. Ulus, A., *İskenderun Kenti ve Yakın Çevresinin Peyzaj Ekolojisi Açısından İncelenmesi.* İstanbul Üniversitesi Orman Fakültesi Dergisi, 2008. **58**: p. 53–73.
34. UK-GBC, *Practical how-to guide: Developing and implementing a green infrastructure strategy.* The UK Green Building Council. 2015.
35. Wang, X. J., *Analysis of problems in urban green space system planning in China.* Journal of Forestry Research, 2009. **20**(1): p. 79–82.
36. Van Herzele, A. and T. Wiedemann, *A Monitoring Tool for the Provision of Accessible and Attractive Urban Green Spaces.* Landscape and Urban Planning, 2003. **63**: p. 109–126.

Mehmet KARACA, Elif KARACA and Gülseven U. TÖNÜK

Rainwater Sensitive Urban Planning – Hydrological Foundations and Planning Principles

Abstract: It is emerging as a necessity to introduce a rainwater sensitive urban planning perspective, focused on ensuring the harmony with the hydrological structure of the basin for improving the durability and adaptation capacity of the urban areas under the negative effects that will be caused by climate change. In this study, a general framework is introduced to be realised a rainwater sensitive urban planning. Within this framework; 3 basic precepts based on analysis, protection and use put forward. The precept based on analysis introduces 11 agents, which are effective on the rainfall-runoff process in the basin, and 3 hydrological responses (depression, infiltration and flow), which define the hydrological characteristic of the basin. Also, the precept based on use introduces 7 urban planning criteria to ensure the balance of use and preservation.

Keywords: Rainwater management, Urban Planning, Urban Hydrology, Hydrological characteristic, Hydrological response

Introduction

As a result of climate change, the effects of which are now noticeable, the frequency, severity and duration of hydro-meteorological disasters such as heavy precipitation events, drought and extreme temperature differences increase. Therefore, environmental problems such as floods, heat and water stress are increasingly experienced. And these changes effect all social, economic, environmental and institutional systems from the global scale to the urban scale [1–5].

Cities, since containing large human populations and hence a socio-economically important system are susceptible to those extreme hydro-meteorological events that will develop under the influence of climate change. Therefore, to develop the plan decisions, that will increase the durability and adaptability is of great importance in terms of ensuring the sustainability of the cities.

In this context, the rainwater subject is accepted as a key issue in the literature.

Therefore, to increase the degree of durability and adaptation capacity of the cities for coping with the effects of the climate change, it is a necessity to develop a rainwater sensitive urban planning perspective that focuses on ensuring harmony with the hydrological structure of the basin; beyond conventional urban planning approaches based on just rainwater drainage.

Rainwater Sensitive Urban Planning

Land-use changes caused by urbanization, on the one hand, have negative effects on the hydrological processes, and on the other hand, deteriorate the hydrological structure, this situation causes urban areas to become more vulnerable to the negative effects of runoff.

Considering especially the extreme hydro-meteorological events that will be caused by the effect of the climate change; it is of great importance to transfer hydrological information to urban planning studies, for both, to protect the natural hydrological structures of the basins and to increase the degree of durability and adaptation of the cities.

The basic paradigm put forward in the literature focuses on the necessity of urban planning to be compatible with the hydrological structure of the basin and it is aimed to ensure the harmony of the hydrological structure of the urban areas and the basin.

Within this scope, a rainwater sensitive urban planning approach based on the hydrological structure of the basin in terms of rainfall-runoff processes should be introduced.

To ensure the harmony of the hydrological structure of the basin and the urban areas, namely rainwater sensitive urban planning approach, depends on the definition of the rainfall-runoff processes, the correct understanding of the interactions of these processes with the urban areas and thus the development of compatible urban planning decisions with the hydrological structure of the basin.

In this context, to realize the rainwater sensitive urban planning, three basic precepts, depends on analysis, preservations and use, should be applied. These are, respectively;

- *The first precept – Analysis:* Each basin has a unique hydrological structure. Thus, to develop rainwater sensitive urban planning approaches, the hydrological character and behaviours of the basin should be accurately defined.
- *The second precept- Preservation:* Hydrologically sensitive areas should be protected and of the hydrological functions should be improved.
- *The third precept- Use:* To ensure sustaining natural hydrological conditions and minimizing the negative effects caused by urbanisation, the urban planning decisions should be developed correspondingly with hydrological characteristic and behaviours of the basin.

1.1 The First Precept- Analysis

The first of the 3 precepts, to realize rainwater sensitive urban planning, consists of analysis phase and is based on to define hydrological characteristic and behaviours of the basin.

If to assess the precepts, it is seen that the second and third depend on the realization of the first. Therefore, in the initial stage, it is of importance to realise the first precept, accurately.

While the literature on hydrology is being reviewed; it is seen that the rainfall-runoff processes of a basin are examined over a complex structure that consists of different temporal and spatial scales of many interacting processes, such as rainfall, infiltration, percolation, runoff, interflow, groundwater flow, evapotranspiration, etc., and is formed according to highly heterogeneous environmental factors, such as climate, topography, soil properties, geological structure, vegetation, hydrological and hydrogeological situation etc. [6–17].

Therefore, to define the hydrological characteristic and behaviours of the basin;

i. It should be defined, the agents effective on the rainfall-runoff processes
ii. Depending on those agents, it should be defined, rainfall-runoff responses and their spatial variations, in the basin.

Under the natural circumstances, both the rainfall and the basin features have unlimited physiological variations that make it practically impossible to understand in all details the rainfall-runoff processes [18]. In this case, the common approach accepted in the literature in defining the hydrological systems with unlimited heterogeneity is to define and simplify the basic and the most important components of the system [7, 8, 13, 14, 17, 19].

With this aim, a simplified hydrological concept defining the relationships among the rainfall and spatial agents is schematized as in Fig. 1.

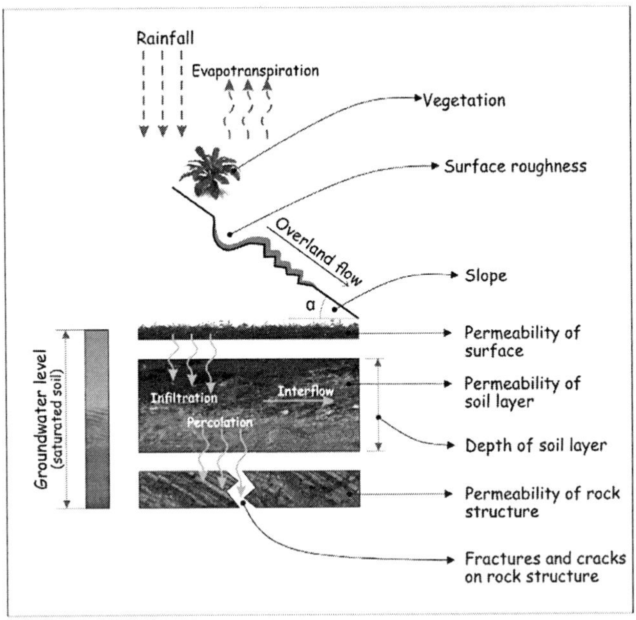

Fig. 1: Hydrological concept

The hydrological concept introduces 11 spatial agents that determine the formation of rainfall-runoff processes (Tab. 1).

Tab. 1: The spatial agents

1. Runoff amount
2. Surface slope
3. Topographic effect
4. Surface roughness
5. Permeability of the soil layer
6. Depth of the soil layer
7. Permeability of the rock structure
8. Fractures and cracks on the rock structure
9. Water table at shallow level
10. Proximity to water bodies
11. Vegetation density

Each of these agents has quantitative variations that are determinative on the pattern of the hydrological responses in the basin.

The hydrological response is the behaviour of the surface against runoff, and the pattern of the hydrological responses within the basin defines the hydrological character of that basin.

Therefore, a correct definition of the hydrological character of the basin depends on the correct definition of the variability of these factors within the basin.

In the context of rainfall-runoff processes, 3 basic hydrological responses determine the hydrological character of a basin. These are; (1) depression, (2) infiltration and (3) flow (Fig. 2).

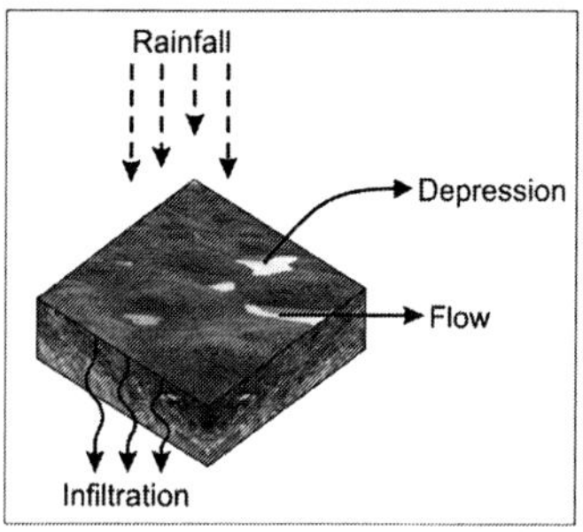

Fig. 2: Hydrological responses

Depression; in other words, ponding is the accumulation of the water on the surface or a shallow level. So, the depression fields can be easily observed with the water accumulating on the surface or a shallow level during or immediately after the rainfall events.

In plain or concave fields with large drainage areas, the presence of low permeable surface layer or impermeable rock layer or high groundwater level under the shallow soil layer encourages the formation of depression.

Infiltration; it is the entry of water into a porous environment by passing through the soil surface [20]. While some of the water infiltrating into the soil structure remains suspended in the aeration zone (unsaturated zone), some may seep into a water table located below and deeper in the root zone. This situation is called deep infiltration or percolation.

The amount of infiltration varies depending on the permeability of the surface and sub-surface layers, the level of the groundwater, the saturation of the soil, the factors effecting the application time of the water to the soil such as slope, cavities and vegetation, as well as the severity and duration of the precipitation.

In addition, another factor effecting the amount of infiltration is the presence of water that will infiltrate. For this reason, the two factors effecting the water availability, the amount of precipitation and the size of the associated drainage area (contributing area) are also determinants on the amount of infiltration in the area.

Flow; it includes overland flows and shallow subsoil flows in which the rainwater joins a stream in about a day or little more [21]. According to this, flow is the rapid movement of the water accumulating on the surface or just below the surface at a shallow level, as a result of heavy or high-intensity rainfall and also with the effect of the slope.

When the intensity of the rainfall exceeds the infiltration rate of the soil, water accumulates on the surface as a thin layer. This accumulated water begins to fill the cavities on the surface. Filling of the cavities, the waters move on the surface with the effect of the slope, creating the flow. Also, the flow can occur when the soil is saturated. In this case, all cavities, cracks and gaps of the soil profile are filled with water and consequently, excess water starts to accumulate on the soil surface and flows with the effect of the slope.

Factors such as the intensity and volume of rainfall, the size and shape of the drainage area, the surface slope, the infiltration rate of the surface, the permeability of the surface and sub-surface layers, the level of saturation of the soil, amount of the cavities and the vegetation density effect the formation, volume and velocity of the flow.

1.2 The Second Precept- Preservation

The second precept of rainwater sensitive urban planning involves the preservation of hydrologically sensitive areas.

As mentioned, each factor effecting rainfall-runoff processes varies quantitatively within the basin, and this variability determines the intensity and frequency of the hydrological responses. In this context, hydrologically sensitive areas refer to areas where the intensity and frequency of any hydrological response (depression, infiltration and flow) have a high value.

These areas have key roles to ensure the continuity of natural hydrological processes in the basin. Therefore, to sustain the natural hydrological structure in the basin, the hydrologically sensitive areas should be preserved, and their hydrological functions should be improved if necessary.

Depression sensitive areas; these areas are very sensitive to meteorological events. With the rainfall, the soils in these areas quickly reach the saturation level, and even in high volumes of rainfall, the water level rises on the surface. The ascent rate of the water depends on the severity of the rainfall and the concentration-time of the drainage area.

Since the stored water is on the surface or in a shallow soil layer, it is in direct interaction with the meteorological conditions and the plant presence.

One of the most important functions of the depression areas is to keep rainfall waters in the basin and so contribute positively to the water budget of the basin. They create an aquatic environment by allowing rainwater to accumulate on the surface or at a shallow level, thereby supporting the development of an amphibious ecosystem.

In addition, these areas, where the flowing waters are stored, extend the concentration period of the basin and create a lowering effect on the peak flows and flow amounts at the basin outlet hydrograph.

As the flowing rainwaters move along the surface, they carry the particles and even urban pollutants accumulated on the surface. Depression areas also act as a filter, as they allow these substances carried to settle by resting the waters.

Infiltration sensitive areas; in short, the infiltration areas are the natural regulators where the water is kept at a depth that is not affected by climatic variables and is given back to nature only sparingly in dry periods, and also where large amounts of water are stored in underground reservoirs. Because of all these features, infiltration is widely accepted as an indicator of a healthy rainfall-runoff process in the literature and the infiltration areas are seen as the important resource areas for the sustainability of biological diversity.

The infiltration process is the basic water-saving system of the basin, as the losses in the water budget through flow and evaporation are significantly slowed down as a result of the transfer of rainwater from the surface to the soil structure and groundwater. The infiltration is responsible for recharging and regenerating underground water reserves that constitute an important part of the available freshwater resources in the world. Also, groundwater flows and interflows originating from infiltration form the base flows of streams.

Taking the rainfall waters into the soil structure by infiltration creates a large-scale lowering effect on flow rates. This lowering effect enables the concentration and recession times of the basin to be longer at the outlet discharge hydrograph and consequently, the peak flow to decrease.

Another important effect of the infiltration process is the filtration of the particles and substances that the rainfall waters add to their bodies while flowing over the surface, thereby cleaning the water.

Flow sensitive areas; flow is an issue that should be carefully addressed as it causes rapid and dramatic changes both in the water budget and on the morphological structure of the basin.

Rapid and high-volume flows created by rainfall cause rapid changes in the water budget of the basin, and consequently, shorten the concentration and recession times and increase the peak flows. Mostly this rapid moving water with high volume results in sudden inundations in the depression areas in the lower basins. Also, flowing water has an abrasive effect on the surface. The extent of the abrasion is related to the intensity, volume and duration of the water flowing and can have dramatic effects on the surface. On the other hand, the hydro-ecological balance in the basin is the result of a long-term adaptation process, and the basins have developed natural control mechanisms against these effects such as drainage networks, floodplains, vegetation patterns.

Also, the flows are the fast water transmission processes in the basin hydrology. The flows rapidly transport the rainwaters to depressions or infiltration zones in the lower basin to feed these systems. So, the flow processes involved in the collection and transport of rainwaters dispersed over the surface have a key role in ensuring the healthy functioning and continuity of the hydrological processes in the basin and consequently the ecosystem.

Although the flow is basically a water movement, this process as well as involves the transportation of many organic and inorganic substances that the water washes over the surface. This substance transport, which includes nutrients, seeds, larvae and even living species, is very important in terms of ensuring and sustaining a dynamic ecosystem structure and biodiversity.

1.3 The Third Precept- Use

The third principle includes approaches for developing urban use decisions that focus on ensuring the balance of preservation and use within the scope of rainwater sensitive urban planning. To achieve the balance, 7 basic rainwater sensitive urban planning strategies must be implemented. These are, respectively;

i. <u>Developing decisions of construction corresponded with the hydrological characteristics of the area:</u>

As mentioned, the hydrological character of an area is related to the hydrological responses to the rainfall. In other words, the hydrological character is determined by the realization potential of each of the depression, infiltration or flow responses of the area.

Accordingly, an area with a high depression potential is defined as an area with depression character, an area with high infiltration potential, an area with infiltration character and an area with a high flow potential, an area with flow character.

In this context; for a rainwater sensitive urban planning, the construction decisions should be corresponded with the hydrological characteristic of each area.

For example, considering that in the areas with depression character, rainwater will collect on the surface or at a shallow level and the soil moisture level will be high for long periods depending on the seasons; open space uses in these areas should be planned in a way that allows water to be collected during rainy periods, open water channels should be established instead of closed drainage systems, constructions which are underground level should not be allowed or special measures should be defined for obligatory situations.

In areas having an infiltration character, measures should be developed to protect the natural structure of the soil layer, and impermeable surface applications that will prevent infiltration should be limited. In the limitation of impermeable surfaces; reducing the density of housing by keeping the population density constant, that is, developing multi-storey housing for the same population density are effective solutions that reduce the building floor areas and the needs for transportation infrastructure.

ii. Developing planning approaches to prevent water pollution caused by urbanization:

Land-use changes such as urbanization have the significant impacts on hydrological processes and therefore, during urbanization, those of the basin are changed natural energy and material flow processes, land cover, topographic structure, soil texture and natural drainage character. And the effects of all those show themselves on the rainfall characteristics, flow processes and water quality.

The rainwater sensitive urban planning should put planning decisions forward to prevent this impact caused by urbanization. However, while doing this, it should be evaluated, considering the hydrological character of each area. Because the urbanization will cause effects in different ways for each hydrological response (depression, infiltration, runoff). Therefore, the measures put forward should also be different depending on the hydrological responses.

iii. Developing planning approaches to preserve natural water systems:

Since the visible processes of the rainfall-runoff phenomenon, in general, are the rainfall and the overland flow, it is a very common misconception that the flow is a phenomenon that occurs only on the surface. In reality, the flow event includes the processes that in all three states of water (gas, liquid and solid) both at the surface (overland flow), in the soil (infiltration, interflow, percolation and groundwater flow) and the atmosphere (evaporation and transpiration).

However, the chain of events taking place in the hydrological cycle under natural conditions is extremely complex and therefore it is almost impossible to reveal it in all details. Therefore, to understand whether the hydrological processes in a basin function properly or not, the water bodies in that basin are taken as a reference.

The state of the water bodies within the basin is also an indicator of the state of hydrological processes.

Therefore; Putting planning decisions forward for the preservation and development of water systems within the basin will also ensure that the natural hydrological processes are protected and sustained.

iv. <u>Developing rainwater management solutions corresponded with the hydrological characteristic of the area:</u>

The hydrological responses of the areas within the basin are also determinant on the rainwater management approaches. Because each hydrological response requires a different management solution.

For example, infiltration-based stormwater management techniques are the most referenced and suggested alternative techniques in the literature. However, when these techniques are applied in areas with a depressive character, the desired results will not be achieved, and even negative results could arise by increasing the depression formation.

Therefore, to achieve the desired results, the hydrological character of the basin should be analysed correctly, and rainwater management solutions tailored to the hydrological response of each area should be developed (Tab. 2).

Tab. 2: Instances for rainwater management solutions

Hydrological Responses	**Instances for Solutions**
Depression	• Solutions for rainwater harvesting • Multifunctional urban areas allowing the collecting rainwater • Linking depression areas and water bodies by the open channel systems • Increasing evapotranspiration amount by the vegetation
Infiltration	• Using infiltration-based rainwater management systems • Preservation and rehabilitation of soil structure and vegetation
Flow	• Preservation and rehabilitation of natural channels and floodplains • Preventing sedimentation and erosion • Grading, if necessary

v. Developing solutions to enhance landscape values and biological diversity:

In order to ensure that both hydrological systems within the basin and urban life can be sustained healthily, targets should be developed to increase landscape values and support urban biodiversity.

In this context, initially, the rainwaters should be considered as an important resource for the ecosystem, not as a problem. And planning decisions should be put forward to allow the development of natural vegetation and biological diversity.

vi. Developing solutions for multifunctional uses:

Planning the areas to be used for rainwater management, to serve as urban open spaces at the same time is important in terms of both increasing the quality and satisfaction of environment and transforming these areas into a part of urban life and so gaining acceptance.

In addition, planning these areas in a way that can serve multiple urban usage needs will also contribute to preventing urban sprawl.

In these areas, uses such as recreational activities, sports activities, training, walking or gathering areas can be planned for dry periods. Yet, the uses in these areas should be planned in a way that will not cause compression, deterioration or pollution of the soil structure.

vii. Creating public awareness of rainwater efficient urban life to support sustainability.

Raising the awareness of the society is the basic condition for the creation and sustaining of an urban life suitable for the natural hydro-ecological structure.

The modern pragmatic perspective considers the cities as systems organized to serve human needs at the highest level by isolating them from the constraints of the natural environment. However, it is necessary a paradigm shift about that the cities are a part of the natural system and it is the necessity of protecting, developing and most importantly living in harmony to ensure the establishment and continuity of a healthy urban environment.

The protection and sustainability of the environment is possible with this awareness.

Conclusion

Rainwater issue plays a key role in efforts to increase the durability and adaptability capacities of the cities in the face of extreme hydro-meteorological events that will develop under the effect of climate change.

In this context, adapting a perspective that takes the rainfall-runoff processes as a reference to urban planning will provide significant gains in efforts to ensure urban sustainability.

References

1. EEA, *Climate change, impacts and vulnerability in Europe 2012.* 2012: Copenhagen, Denmark.
2. Izrael, Y.A., et al., *The fourth assessment report of the intergovernmental panel on climate change: Working group II contribution.* Russian Meteorology and Hydrology, 2007. **32**(9): p. 551–556.
3. Lehner, B., et al., *Estimating the Impact of Global Change on Flood and Drought Risks in Europe: A Continental, Integrated Analysis.* 2006. **75**: p. 273–299.
4. Munich_RE, *Topics Geo natural catastrophes 2010: Analyses, assessments, positions,* A. Wirtz and M. Re, Editors. 2011, Munich Reinsurance Company Rep: München. p. 58.
5. Voskamp, I.M. and Van de Ven, F.H.M., *Planning support system for climate adaptation: Composing effective sets of blue-green measures to reduce urban vulnerability to extreme weather events.* Building and Environment, 2015. **83**: p. 159–167.
6. Betson, R.P., *What is watershed runoff?* Journal of Geophysical Research, 1964. **69**(8): p. 1541–1552.
7. Flügel, W.A., *Delineating hydrological response units by geographical information system analyses for regional hydrological modelling using PRMS/MMS in the drainage basin of the River Bröl, Germany.* Hydrological Processes, 1995. **9**(3-4): p. 423–436.
8. Flügel, W.A., *Application of GIS to derive Hydrological Response Units for hydrological modelling in the Brol catchment, Germany,* in *HydroGIS'96: Application of Geographic Information Systems in hydrology and water resources management.* 1996: Vienne, Autriche. p. 413–420.
9. Hewlett, J.D. and Hibbert, A.R., *Factors Affecting the Response of Small Watersheds to Precipitation in Humid Areas,* in *Proceedings of the International Symposium on Forest Hydrology,* W.E. Sopper and H.W. Lull, Editors. 1967, Pergamon: New York. p. 275–290.
10. Horton, J.H. and Hawkins, R.H., *Flow path of rain from the soil surface to the water table.* Soil Science, 1965. **100**(6): p. 377–383.
11. Horton, R.E., *The Rôle of infiltration in the hydrologic cycle.* Transactions, American Geophysical Union, 1933. **14**(1): p. 446.
12. Hursh, C.R., *Appendix B—Report of sub-committee on subsurface-flow.* Transactions, American Geophysical Union, 1944. **25**(5): p. 743.
13. Khan, U., Tuteja, N.K., and Sharma, A., *Delineating hydrologic response units in large upland catchments and its evaluation using soil moisture simulations.* Environmental Modelling & Software, 2013. **46**: p. 142–154.
14. Kirkby, M.J., *Hillslope hydrology.* 1978: Wiley.

15. Leavesly, G.H., et al., *Precipitation-runoff modeling system; user's manual*, in *Water-Resources Investigations Report*. 1983.
16. Sklash, M.G. and Farvolden, R.N., *The role of groundwater in storm runoff*. Journal of Hydrology, 1979. **43**(1–4): p. 45–65.
17. Walter, M.T., et al., *Hydrologically sensitive areas: Variable source area hydrology implications for water quality risk assessment*. Journal of Soil and Water Conservation, 2000. **55**(3): p. 277–284.
18. Chow, V.T., Maidment, D.R., and Mays, L.W., *Applied Hydrology*. 1988: McGraw-Hill.
19. Khan, U., et al., *Applicability of Hydrologic Response Units in low topographic relief catchments and evaluation using high resolution aerial photograph analysis*. Environmental Modelling & Software, 2016. **81**: p. 56–71.
20. DSİ. *Devlet Su İşleri Genel Müdürlüğü Hidroloji Sözlüğü*. 2019 [22.11.2019]; Available from: http://dsi.gov.tr/.
21. Tarboton, D.G., *Rainfall Runoff Processes*, in *Civil and Environmental Engineering Faculty Publications*. 2003, Utah State University.
22. Karaca, M., *Yağmur suyu etkin kentsel planlama için hidrolojik tepki potansiyeli model önerisi Çankırı örneği (In Eng.: Hydrologic response potential model proposal for rainwater effective urban planning- the case of Çankırı) in Graduate School of Natural and Applied Sciences*. 2020, Gazi University: Ankara.

Serkan SINMAZ

Evaluation of Renewable Energy Sources in Spatial Decision Process in Urban Planning, Case Study of Balikesir

Abstract: Renewable energy technologies and energy management systems are rapidly developing topics of the urbanism agenda. Ensuring the integration of renewable energy sources with the city and managing them on a common plane with conventional sources is an important strategy that has started to be implemented in recent years. Maximum benefit from renewable energy sources at the optimum application field is an inevitable requirement for effective energy management in cities. The aim of this study is to create the spatially optimal evaluation base in a particular region in order to meet this requirement. Balıkesir has been designated as an evaluation area due to its renewable energy potentials. As a result of the analysis, the most efficient application zones are identified and evaluated within the framework of urban planning principles and practices.

Keywords: Renewable Energy, Sustainability, Urban Planning, Geographic Information Systems, Zoning Plans.

1 Introduction

In the last half-century, with the increase in energy demand in the face of technological innovations and the increase in urbanization, the relationship between energy and city has begun to be questioned in the field of urban planning and design. In addition, the reality of global warming and climate change, which are developing in the face of increasing consumption factors, have brought this issue to the world agenda. In this process, technological development diversifies and facilitates the way of intervention in the city, ecological threat integrates the process of energy production and consumption with the concept of the city, and social transformation reshapes the urban land and fabric. Accordingly, less energy consumption is an one of the main strategies according to the innovative planning approaches developed since the 90s within the framework of the theme "sustainability". Urban reconstruction is carried out through the theme of energy. İn this process, the main topics of plans and applications are

- environmental and energy action plans
- energy efficient urban transformation
- smart infrastructure
- renewable energy
- energy efficient innovative point interventions.

In particular, the steps to be taken to integrate renewable energy sources into urban space offer great importance for the future of cities. Thus, it is necessary to establish legal, managerial and design bases of renewable energy sources in the field of urban planning in Turkey immediately.

Material and Method

Numerical, verbal and spatial data obtained from various public institutions (Ministry of Environment and Urbanisation – General Directorate of Spatial Planning, Ministry of Agriculture and Forestry, Ministry of Energy and Natural Resources, South Marmara Development Agency, Balıkesir Metropolitan Municipality) for four kind of renewable energy sources related to Balıkesir, which were identified as an area of case study, were transferred to the GIS environment and conformity analysis was performed in the specific thresholds filter. In this process, different analysis tools (density analysis, spatial deciduous estimation method (IDW), buffer analysis and map algebra, overlay analysis) were used in GIS platform.

In this context, spatial maps were produced by identifying the regions that have potential for biomass energy, wind energy, solar energy and geothermal energy production and installation in Balıkesir province. According to produced base map, it is revealed at what level Balıkesir settlements can benefit from which type of renewable energy source. At the next stage, the content of physical plans and the adequacy of legal regulations are discussed so that this potential can be integrated into cities.

2 Renewable Energy and the City

Fossil resource depletion, global warming prevention policies and energy import issues are focusing attention on renewable energy efficiency and renewable energy sources. The operability of these potentials requires an effective energy management policy.

- Photovoltaic panel technology for solar energy,
- Wind turbine technology,
- City grids and heat pump technologies for geothermal energy,
- Combined heat systems technologies for biomass energy and biofuel production,
- Urban infrastructure systems for hydrological energy,
- Wave energy technologies are the main systems used for power generation .

While these systems present significant energy recovery potentials, they also present some threats to be considered in the field of Environment and urbanism. For example; biomass energy production process can cause deforestation, air pollution due to incineration action, loss of biodiversity, liquid wastes damage water life, increase the amount of nitrogen in the water due to the use of fertilizers. Wind turbines can cause environmental disturbance due to noise, aesthetic pollution with system installation and bird deaths. Solar panels cause toxic waste accumulation due to the use of copper in collectors. Micro hydroelectric power plants damage local flora and fauna and negatively affect the agricultural and tourism sectors. During the production of geothermal energy systems can pollute the environment. If the system is not close to residential areas there may be a risk of soil pollution and thermal pollution, energy transport may be inefficient. But the dimensions of these effects vary according to the forms of application. For example, biomass energy is considered an efficient energy source if re-planting is performed [1].

Renewable energy sources establish and will establish strong links with cities as providers of electricity and fuel. Over time, with technological development, cost efficiency, expansion and increased public support, new innovations will be able to developed in this sector [2]. The integration of renewable energy sources into the urban system is an issue that should be addressed in the physical plans of settlements in the perspective of the future. Building applications for solar energy, building and open space qualities for wind energy, soil conservation and transportation fuel technology for biomass energy, stream regimes for hydrological energy, integration of geothermal energy into urban networks should be discussed in the urban planning process [3].

Wind energy; low-sloping continuous wind valleys, hills, peaks and coastlines are recommended for large-scale power plants [4][5]. Ease of connection to the main electric grid, proximity to settlements and transport areas, ownership, nature of the land (land site, Natural Protection, National park), visuality and landscape, aircraft safety, local economy impact requires a sensitive approach in the field of urban planning [6][7]. In addition, the potential for integration of small-scale wind plants into buildings and urban pattern is an issue that needs to be addressed in the field of urban design. At this point, developing nonpropelled wind energy systems will be able to be developed as an effective urban furniture.

Solar energy; plants is constructed according to low slope, shadow and aspect qualities. Land quality (protected areas, geographical character) is the determinant element. Plants are required certain offset distances. Transportation and connection to the energy network are important and ease of integration into urban spaces is an element that needs to be evaluated in the field of urban planning.

Geothermal energy; is the subject of evaluation in urban planning as a source for heating and electricity generation systems. Design of district heating networks and heat regulators depends on target temperature ratings [8]. Applications that can be evaluated in urban areas are heat pumps, geothermal energy storage, direct use in agriculture and industry, and district heating and combined heat power systems [9].

Biomass energy; is considered an important energy source because it helps socioeconomic development for rural areas. Biomass energy is produced in the form of electricity, fuel and heat [10]. The use of energy plants, energy forests, wood industry waste, animal waste, urban waste, agricultural industrial waste for energy production and installation in rural areas should be examined in urban planning processes.

Apart form these, smart grid systems to ensure optimum utilization of conventional and renewable energy sources, combined heat and power systems that reduce energy losses, district heating systems as an application field in public housing, zero energy building systems are energy management systems that need to be developed in urban areas. In addition, the integration of green building applications with a waste water conversion system, natural ventilation system, active and passive solar energy systems, heat pump system becomes important in the urban system [11].

3 Balıkesir Renewable Energy Potentials

Balıkesir province has been chosen because it contains a wide range of renewable energy potential and is already active in energy investments. Our aim is to prepare a base for the development of renewable energy sources in the most suitable regions.

Wind Power Plant Installation Potential in Balıkesir

According to the wind energy potential map, Balıkesir's data on the annual average wind speeds (m/s) and the spatial distribution of wind capacity factors (%) at 50 m (Fig. 1) were examined. Accordingly, most of the areas outside the east and southeast of Balıkesir have the quality of speed (7 m/s and above) and capacity factor (35 % and above) required for economic wind power plant investment [12].

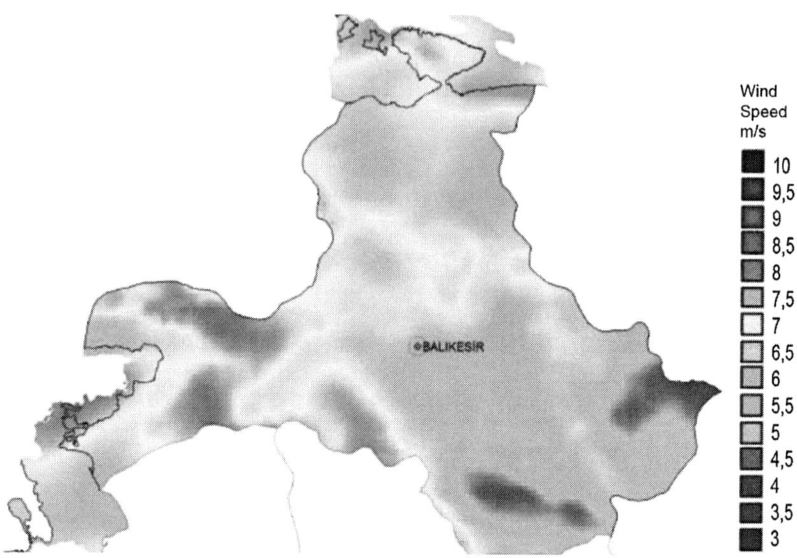

Fig. 1. Annual average wind speeds (m/s) [13].

According to REPA (Wind Energy Potantial Map), the power potentials of the wind power plants that can be installed in Balıkesir province at an altitude of 50 meters are 13,827, 36 MW for the medium – extraordinary wind class and 6,270,24 MW for the good - extraordinary wind class (Tab. 1).

Tab. 1. Power capacities of wind power plants (50 m) that can be ınstalled in Balıkesir province [14]

Wind Level	**Wind Class**	**Wind Power (w m^{-2})**	**Wind Speed (M S^{-1})**	**Total Area (Km^2)**	**Potential To Be Installed (Mw)**
Medium	3	300–400	6,8–7,5	1.511,42	7.557,12
Good	4	400–500	7,5–8,1	8.50,96	4.254,80
Very Good	5	500–600	8,1–8,6	2.84,51	1.422,56
Excellent	6	600–800	8,6–9,5	1.15,23	576,16
Extraordinary	7	>800	>9,5	3,34	16,72
			Total	2.765,47	13.827,36

21 wind farms established by 18 different companies within Balıkesir province have installed wind power of 925.5 MW (covering 18.8 % of Turkey's total installed wind power). This capacity makes Balıkesir the first in Turkey in terms of installed wind power.

Solar Power Plant Installation Potential in Balıkesir

According to the data of the Turkish solar energy potential Atlas (GEPA) published by the General Directorate of renewable energy of the Ministry, the annual global radiation value is 1527 kWH/m2. (Tab. 2).

Tab. 2. Solar Energy Potential [15]

	Global Radiation Value (kWh/ m^2-year)	**Global Radiation Value (kWh/ m^2-day)**	**hours of insolation (hour/year)**	**hours of insolation (hour/day)**
TÜRKİYE	1527	4.18	2741	7.51
BALIKESİR	1422	3.90	2690	7.37

Balıkesir annual global radiation value is less than the Turkish average (1.527 kWh/m2), but for Germany, which holds 35.6 % of the World's photovoltaic capacity at the end of 2011, this value is 1.007 kWh/m2 on average. In addition, an average of 762 kWh of electricity can be produced per year in Germany from a 1 kW phovoltaic system, while Balıkesir has a potential of 1,064 kWh of electricity generation.

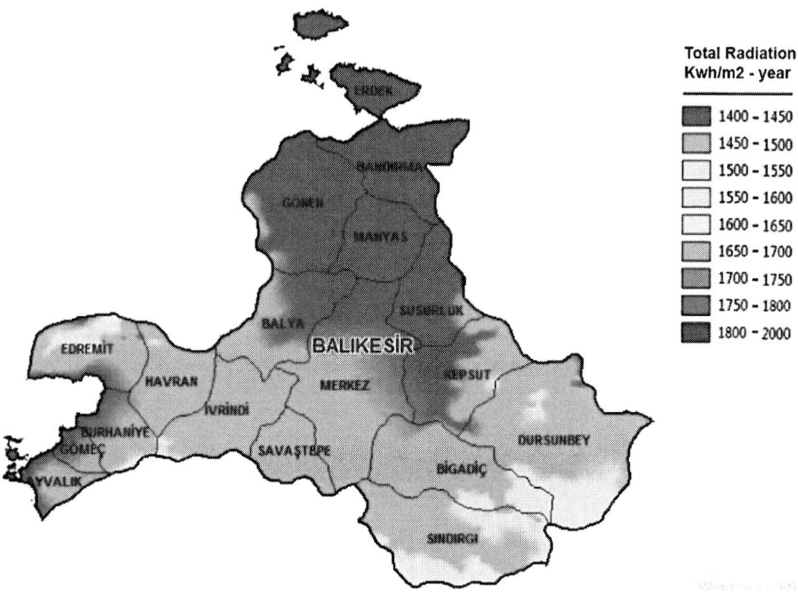

Fig. 2. Balıkesir Global Radiation Values (kWh/m²-year) [15]

The highest global radiation value during the year occurs in June for Balıkesir, and the average daily energy value obtained for this month is 6.29 kWh/m2-day. The highest insulation time by month occurs in July (11.04 hours) (Fig. 2).

Biomass Energy Potential in Balıkesir

31.1 % of the Turkey land is agricultural area, 27.6 % is forest area, 18.6 % is pasture area. According to the 2015 data of the General Directorate of Renewable Energy, the biomass potential of Turkey is 20.307,069.02 TEP/year (Fig. 3). This value is 1746.24 GWh when converted to electrical energy (1 TEP=11.629 kWh).

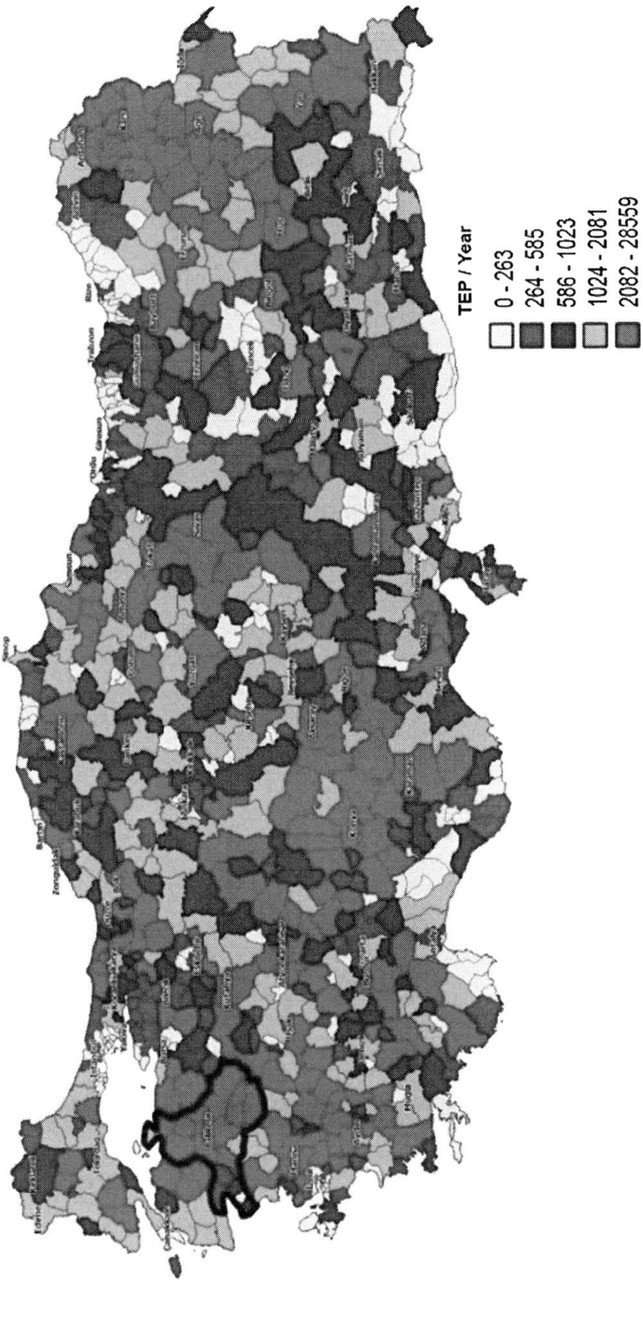

Fig. 3. Energy value of animal wastes by districts (TEP/year) [16]

In Balıkesir, the average presence of cattle is 490,643 units, the average presence of small cattle is 813,395 units, the presence of poultry is 24,057,823 units. Biogas can be produced from the feces of animals.

Tab. 3. Biogas potential and thermal capacity that can be obtained from animal waste in Balıkesir [14]

	Number of animals (2008–2012)	Wet fertilizer (ton/year)	Dry fertilizer (ton/year)	Useable dry fertilizer (ton/year)	Amount of biogas (m3-year)	Calorific capacity (TJ/year)
Cattle	49643	3223680	409407	266115	53223000	1208
Small Cattle	813395	594220	148555	19312	3862400	88
Poultry	24057823	702625	175656	175480	35096000	797
Total	25361861	4520525	733618	460097	92181400	2093

Balıkesir province has an average useable dry fertilizer capacity of 460.097 tons/year. Against this capacity, an average biogas value of 92.181.400 m3/year and an average thermal value of 10,462.589 MJ/year can be obtained (Tab. 3). The biogas potential, which can be obtained especially from cattle and poultry, is remarkable. Although there is a very high capacity in Balıkesir province, there is no working biogas plant to evaluate this capacity.

On the other hand, solid waste management in Balıkesir for the utilization of urban waste is carried out by the activities of Balıkesir Sustainable Environmental Management Association established through EU funds. The total amount of waste is 512719 tons/year [17].

Tab. 4. Waste and thermal capacity in Balıkesir City [14]

Waste	**Amount of waste Ton/year**	**Unit Calorific value MJ/year**	**Calorific capacity GJ/year**
Wastewater Sludge	6.077	2,09	12.701
Vegetable Waste Oil	26,799	37,62	1.008
Waste Paper	1.153	9,15	10.545
Waste Plastic	408	19,90	8.119
Household Garbage	85.580	8,99	792.364
Total	93.244,799		824.737

The total calorific capacity that can be obtained from waste in Balıkesir provincial center is 824.737 GJ/year (Tab. 4).

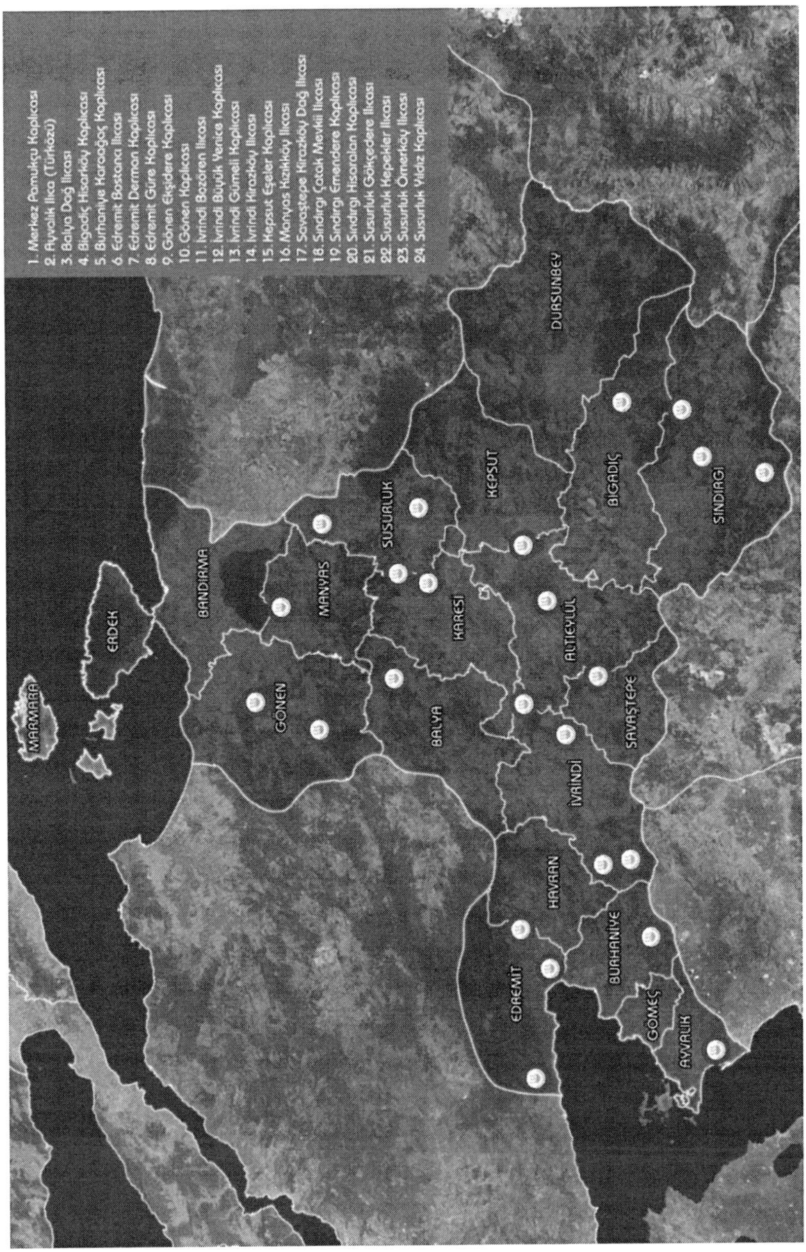

Fig. 4. Balıkesir Province Hot Water Resources (Produced by MTA data)

Balıkesir Geothermal Energy Potential

Balıkesir ranks fourth in Turkey in terms of geothermal energy potential. As a result of the work carried out by the MTA General Directorate in previous years, 7 geothermal resources with a temperature of 30°C and higher were identified in Balıkesir [18].

These resources are mainly used for tourism facilities, but are used for residential heating purposes at a limited level.

4 Analysis of Potential Regions According to Renewable Energy Sources in Balıkesir

"Conformity analysis" was applied by using geographic information systems according to designated criteria for the determination of potential regions for wind energy, solar energy, biomass energy and geothermal energy sources. First, areas defined as thresholds under the Environmental Impact Assessment regulation have been excluded from the assessment. Potential areas were considered within the scope of marginal agricultural areas, taking into account the sustainability of agricultural lands.

Analysis of Potential Regions for Wind Energy Production

Criteria for identifying potential areas for wind power plant:

- Proximity to substations and power transmission lines
- Transportation-Accessibility
- Wind speed (50 m height)
- Slope and morphological structure of the land.

Threshold areas where wind power plant cannot be built:

- Forest areas
- Olive Groves
- Pasture Areas
- Drinking water basins
- Sites
- Residential areas
- Military areas and security zones
- Res protection zones
- Obstacle plans
- Irrigation Areas
- Water surfaces (lake, pond, stream, dam)
- Special environmental protection zones, national parks, nature reserves
- Areas with a slope of more than 20 %
- Land elevation is above 1500 meters.

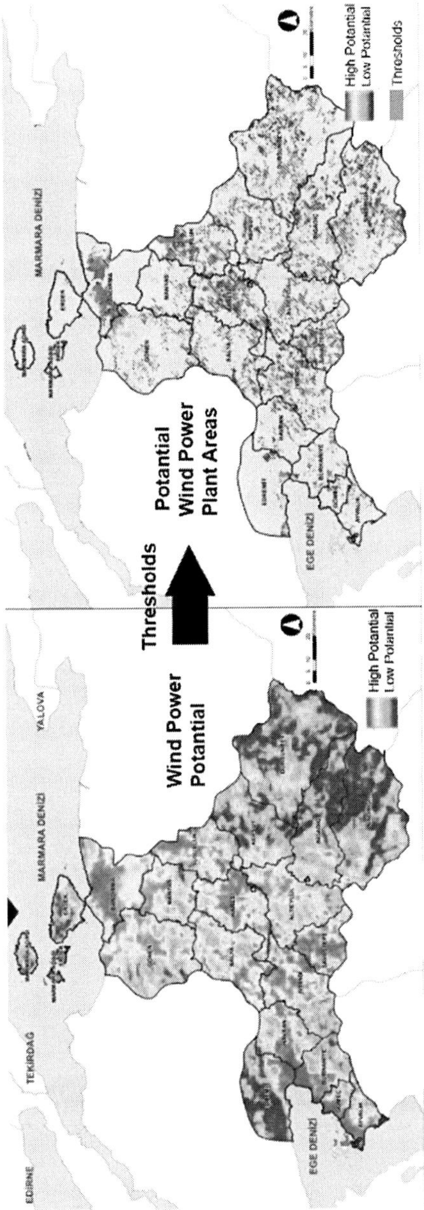

Fig. 5. Spatial analysis of wind power plant potential areas of Balıkesir

Firstly, a surface map of the wind speed distribution was produced (IDW analysis). In accordance with the criteria determined, the parameters that are important in choosing the location of the wind power plant areas were determined and the potential location value map for the wind power plant areas was produced in Balıkesir with spatial overlay analysis. After each layer is classified in the generated location value map, it is collected and the score map is produced. (Map algebra, reclassify). The scores of the parameters used are of equal value. Depending on the generated location score map, areas that cannot be wind power plant established were identified and cleared from the location score map. As a result, potential areas for wind power plant facility have been identified.

As a result of the analysis, Bandırma district has the most potential in terms of spatial size among the regions where wind power plant can be installed. Although the score of Edremit Bay and the areas west of the Bay are high in terms of wind speed and position score, there is no potential for installation due to the fact that most of this region remains in threshold areas. (Fig. 5)

Analysis of Potential Regions for Solar Energy Production

for identifying potential solar power plant areas:

- Average slope not higher than 15 %
- South axis terrain orientation
- Absence of elevations with shading effect from sunrise to sunset
- Convenient location in terms of operation and connection to the network (Max distance: 3km)

Threshold areas where GES cannot be established:

- Forest areas
- Pasture areas
- Olive groves
- Areas with an average slope above 15 %
- Residential areas
- Drinking water basins
- Sites
- Water surfaces
- Valley Areas
- Airports
- Special environmental protection zones, natural parks, natural monument, National Parks
- Military areas and security zones
- Tourism Areas, Cultural Tourism Protection Development Zones
- RES protection zones

Spatial global radiation values can be evaluated as basic data to detect potential solar power plant fields in Balıkesir. Valleys, ridges, sloping areas, Hills were determined

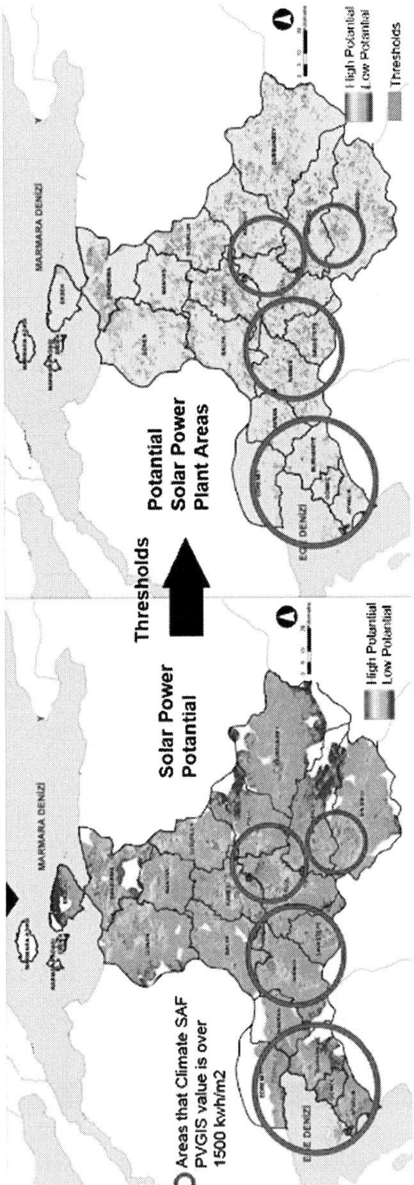

Fig. 6. Spatial analysis of Solar Powe Plant potential areas of Balıkesir province

by the analysis method used (Topographic position index). In this direction, valleys and sloping regions are areas that cannot be evaluated within the scope of location selection for solar power plant and are cleared from the data set. Another important criterion is the identification of areas that are overshadowed. Areas that remain under shadow throughout the year were determined by shadowing analysis using a digital terrain model (Hillshade analysis). Other important criteria are proximity to the main transport arteries and electric grid proximity were analyzed in spatial level by buffer analysis. Slope and orientation analysis and other produced layers are classified within themselves (reclassify tool) and associated with each other by map mathematics method (mapalgebra tool).

Thus, potential areas for solar power plants have been identified throughout the province. As a result, it was determined that the average annual global radiation values in the southern and southwestern areas of Balıkesir were above 1500 kwh/m2-year. (Fig. 6)

Analysis of Potential Regions for Biomass Energy

Criteria used to identify potential areas for biomass power plant:

- Amount of animal waste based on settlement
- Ease of connecting to the National Electric Grid
- Proximity to water sources
- Slope and slope direction of the land
- Proximity to the main transport arteries

Threshold areas that cannot be installed for biomass power plant:

- Forest areas
- Pasture areas
- Olive groves
- Areas with an average slope above 10 %
- Residential areas
- Drinking water basins
- Sites
- Water surfaces
- Airports
- Special environmental protection zones, natural parks, natural monument, National Parks
- Military areas and security zones
- Tourism Areas, Cultural Tourism Protection Development Zones

In order to analyze the biomass energy potential associated with animal waste in Balıkesir, regions where the presence of animals is spatially concentrated were identified in the first place (density Analysis tool). According to literature research, the most suitable zone where waste can be collected in terms of location selection of biomass power plants ranges from 10 km to 20 km.

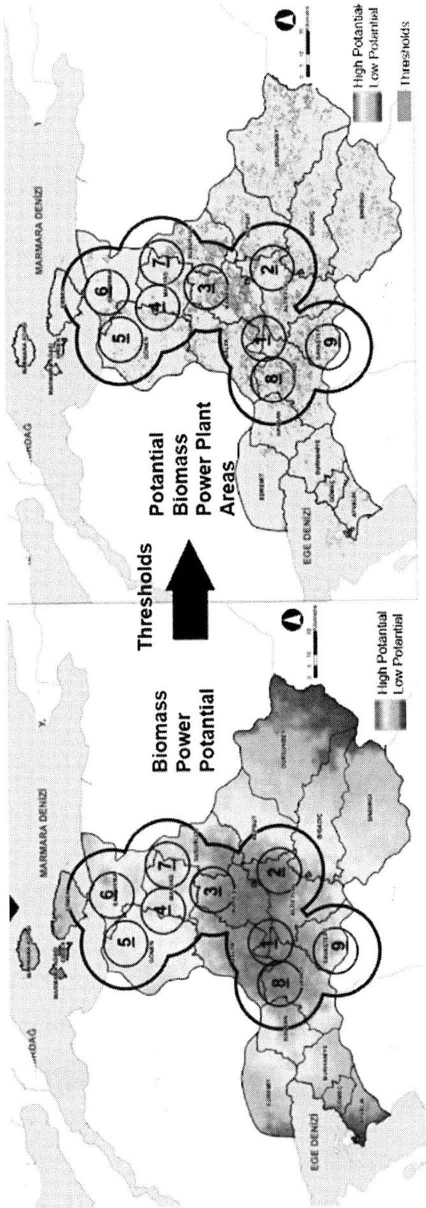

Fig. 7. Potential areas where biomass energy can be produced based on the presence of cattle, small cattle and poultry in Balıkesir

In order to determine the suitable areas for plants based on biomass energy, the numbers of cattle, small cattle and poultry that exist in each settlement in Balıkesir were obtained. Data on the presence of animals within a radius of 10 km and 20 km of settlements were transferred to 1 km^2 cells and regions with high density were identified. Based on the method used by the southern Marmara Development Agency, the potential amount of biogas and electricity that will be produced from animal waste on the basis of settlement was calculated. Potential areas for biomass energy plants based on animal waste have been identified by matching the criteria stated above with spatial density analysis (using tools of overlay analysis, buffer analysis, map algebra in ARCGIS).

Potential installation zones for biomass energy from animal waste in Balıkesir point to the inner parts of the province. The nine identified subregions can be evaluated based on cattle and small cattle waste, and Regions 6 and 3 can be evaluated based on poultry waste (Fig. 7).

Analysis of Potential Regions for Geotermal Energy

Criteria used to identify potential areas for geotermal energy:

- Proximity to production wells,
- Proximity to substations and power transmission lines
- Accessibility

Threshold areas that cannot be evaluated for facilities based on geothermal energy:

- Forest areas
- Pasture areas
- Olive groves
- Residential areas
- Drinking water basins
- Sites
- Water surfaces
- Airports
- Special environmental protection zones, natural parks, natural monument, National Parks
- Military areas and security zones
- Wind energy Plant protection zones

The entire area is divided into 1 km^2 cells to identify potential areas that can be utilized from geothermal energy sources. These cells were encoded by the values of distance to production wells, distance to transformer power transmission lines, and distance to transport arteries. (using tools of reclassify, map algebra in ARCGIS). By removing threshold zones from the designated areas, potential areas of facilities based on geothermal energy were identified.

As a result of the analysis, Bandırma, Susurluk and Karesi districts offer potential for geothermal energy plants (Fig. 8).

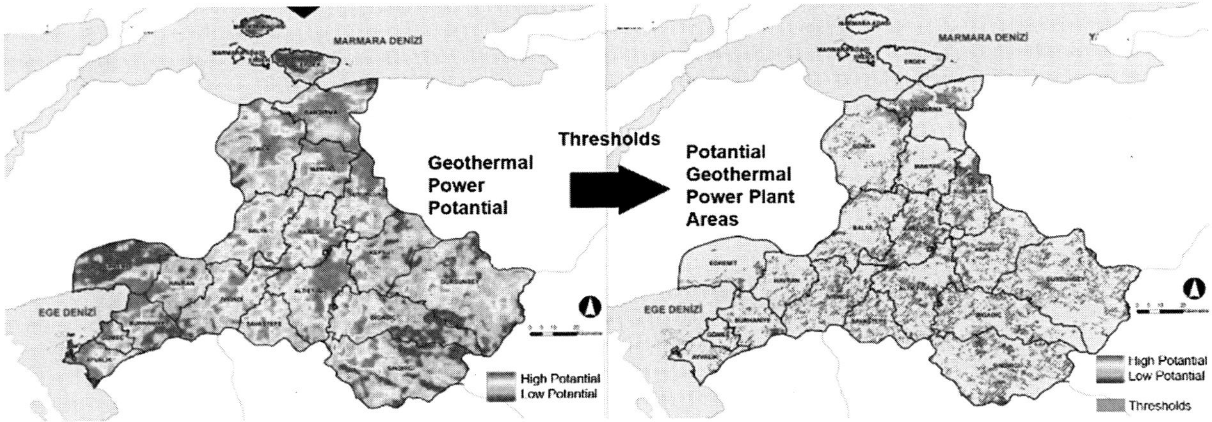

Fig. 8. Analysis of potential areas that can be utilized for geothermal energy in Balıkesir

5. Conclusion

Important studies and applications related to renewable energy sources are being carried out in the World and Turkey. Considering the laws and regulations of the planning legislation, there is no guiding decision on how to select a location for renewable energy power plants and how to build relations with settlements. In the spatial planning regulation, which describes the construction of city plans, there are no definitions and regulations in terms of the integration of renewable energy sources into the city. It is defined only as an "energy production area" in the upper scale plan legends[1]. But at this point, the most general definition for renewable energy production areas is the 1/1000 scale implementary zoning plan legend, which is the lowest scale planning stage in Turkey.[2] Therefore, there are legant definitions that are used independently of the planning hierarchy. Renewable energy production areas are generally determined in accordance with investor demand, depending on the investment programs of the relevant institutions, regardless of a holistic plan.

Apart from urban planning legislation, the importance of renewable energy sources in the legal plane has started to increase since 2010 in Turkey. The law on renewable energy sources of 2011 aims to provide incentives in terms of price determination and domestic production. However, the law does not provide an approach on the need for site selection, compliance with field criteria, and the status of settlements being affected. According to the regulations, the distances of wind turbines to each other are defined, the maximum power density for solar energy is taken into account and license applications are not accepted for agricultural land that needs to be protected. On the official site of the General Directorate of renewable energy, the selection criteria have been defined in more detail, but not yet sufficient legal regulations have been implemented. In addition, applications in areas that need to be protected are left to the ministry's opinion. An objective scientific assessment of naturally sensitive areas is required [19].

On the physical planning scale, the only and important aspect is unlicensed energy production (maximum 1 MW)[3] in Turkey. This encourages energy production in settlements, and provides potential for public housing, agricultural and industrial facilities, and public facilities. Lower installed power systems offer the opportunity for individual applications. However, the legislation does not contain more detailed implementation provisions. Thus, the risk of developing haphazard practices arises. On the other hand, the General Directorate of Renewable Energy's

1 Energy production area: these are facilities that will be built after a license issued by the Energy Market Regulatory Authority and/or permits obtained from the relevant institutions.
2 It is defined as "the area of a production plant based on renewable energy sources" at the implementary development plans
3 https://www.resmigazete.gov.tr/eskiler/2016/03/20160323-5.htm

analysis of potential regions related to "renewable energy resource areas" around the country is general and not sufficiently guided.

As a result, in order to benefit from renewable energy sources in a way that is sensitive to planning principles, it is necessary to re-evaluate the city planning legislation and include more directive regulations. This study presents an exemplary field study in terms of creating a base that can be addressed in the planning process. In this way, a method is put forward to control the mentioned legal deficiencies and approaches related to investor demand. This method can be developed as a binding preliminary work within the legal framework and presented as a standard institutional document in the planning process of each city.

References

1. Goldemberg, J., Lucon O., *Energy, Environment and Development.* 2013, Earthscan, UK. p. 194
2. Girardet H., Mendonça, M., A Renewable World Energy. 2009, Ecology, Equality, World Future Council, London, p.77
3. Droege, P., Renewable Energy and the City, Encyclopedia of Energy, 2004, Academic Press, p.307
4. Özkaya, G.M., Variyenli H.İ., Uçar S., Electricity Production With Wind Energy and The Evaluation of Environmental Effects for Kayseri, Cumhuriyet Science Journal, 2008, 29 (1): p. 1–18.
5. Çetin, A.C., Wind Energy And The Selection Of A Facility Location For A Wind Plant In Isparta, Proceedings Of The International Davraz Congress, Isparta, 2009, p.370
6. Özön, N.N., ve Kılıçaslan, İ., Kocaeli İlinde Rüzgar Çiftliği İçin Yer Seçimi, IV. Renewable Energy Resources Symposium, 2007, p. 3
7. Stankovic, S., Campbell N., Harries A., Urban Wind Energy, 2009, BDSP Partnership Ltd., London.
8. Quaschning,V., Renewable Energy and Climate Change, 2010, John Wiley & Sons, United Kingdom, p. 221
9. EREC, Revewable Energy in Europe: Markets, Trends and Technologies, 2010, Washington DC., p. 62–64
10. Topal M. ve Arslan E. I., Biyokütle Enerjisi ve Türkiye, VII. National Clean Energy Symposium, 2008, İstanbul, p.243
11. Maczulak, A., Renewable Energy: Sources and Methods, 2010, Infobase Publishing, p.137
12. T.C. Ministry of Energy and Natural Resources, 2015, https://enerji.gov.tr/bilgi-merkezi-enerji-ruzgaren

13. T.C. Ministry of Energy and Natural Resources, 2020, https://repa.enerji.gov.tr/REPA/
14. GMKA, Renewable Energy Research Results Report, 2012, https://www.gmka.gov.tr/dokumanlar/yayinlar/Yenilenebilir-Enerji-Arastirmasi.pdf
15. T.C. Ministry of Energy and Natural Resources, 2015, https://gepa.enerji.gov.tr/MyCalculator/
16. T.C. Ministry of Energy and Natural Resources, 2020, https://bepa.enerji.gov.tr/
17. General Directorate Of Mineral Research And Exploration, 2011
18. ÇŞB, Balikesir ili 2018 yılı Çevre durum Raporu, 2018, https://webdosya.csb.gov.tr/db/ced/icerikler/bal-kes-r_2018_-cdr_rev-ze-20190807131228.pdf
19. Sınmaz, S., Akıllı Yerleşme Kurgusu ve Küçük Ölçekli Yerleşmelerin Enerji Verimli Gelişimi: Lapseki Üzerine Bir Değerlendirme, 2014, PhD Thesis, Yıldız Technical University, İstanbul, p.247
20. Seifried, D. ve Witzel, W., Renewable Energy - The Facts, 2010, Earthscan/Energieagentur Regio, Freiburg.
21. Lund, H., Renewable Energy Systems, A Smart Energy Systems Approach To The Choice And Modelling Of %100 Renewable Solutions, 2010, Elsevier, USA

İlke CİRİTCİ

Understanding Olivetti Factory as a Sustainable Industrial Heritage

Abstract Although they were designed to survive for centuries during the period they were built, some buildings have become unusable with the loss of importance of business lines and have been left to their fate in cities for many years. The continuous change of production with today's technological speed, and the loss of use of many industrial structures, which were used especially in the 19th century, brought the reuse of these structures and usage suggestions for different functions have been introduced.

Industry heritage takes an important part in the sustainable development of the city. Olivetti's sustainable project is mostly based on the link between an ethical concept and a development-oriented idea (Castano, Gallo, 2013). The spreading of industrial structures in the Olivetti campuses to the region, ensuring production and sales, attracted attention from service optimization to the quality of life and the environment. Understanding the philosophy of Olivetti Settlements, which took its place in the UNESCO World Heritage List in 2018, will contribute to the concept of Industrial Heritage and the sustainable development of cities.

Keywords: Industrial Heritage, Sustainability, Reuse, Olivetti, UNESCO

1 Introduction

Industrial heritage buildings are described as structures that have gained functionality with the technology developed during the Industrial Revolution and witnessed that process and remained from that process. Initially the industrial structures established on the waterfront, have become available in places where the raw materials and stock are obtained and also where the produced materials will be marketed with the increase of transmission and transportation possibilities, thus the rate of spread of industrial structures has increased (Köksal, 2012). These abandoned buildings, which are important today as industrial heritage, have been considered as part of the cultural heritage of the world for approximately fifty years. Historical, architectural, social, technological and scientific heritage of industrial culture are included in these industrial structures.

The preservation of cultural heritage is mainly based on the Venice Charter, but within the scope of the new conditions and requirements developing today, some ideas and protection measures are reconsidered and modern understandings for the preservation and restoration of cultural heritage have been redefined in the meetings held by UNESCO, ICOMOS and the COUNCIL OF EUROPE. In these meetings, significant emphasis was made on the intangible cultural heritage, reuse decisions of the existing industrial heritage and sustainability, and new approaches were introduced. Organizations involved in the conservation of Industrial Heritage include TICCIH (The International Committee for Conservation of the Industrial Heritage), ERIH (European Route of Industrial Heritage), DOCOMOMO, ICOHTEC, SHOT, E-FAITH, CILAC, SIA, AIA (Saner, 2012).

Although modern technological development brought significant improvements in all areas of human life, it also contained difficulties in terms of the ecological balance of the natural environment, weakened by the overuse of natural resources. These difficulties have raised the question of the role and importance of local-historical forms of cultural and architectural heritage in modern technological society without borders (Blagojevic, Tufegdzic, 2016). Although it can be said that new technological developments are of great importance in the precautions and interventions regarding the protection and reuse of buildings in historical residential areas, it would not be wrong to say that the environmental pollution caused by the uncontrolled consequences of the same technological development has negative effects on cities. The abandonment of production methods in today's cities due to the negative effects they have on the environment, has brought forward the suggestions to bring these buildings to the city.

In addition to reflecting the production style and technology of the period in which industrial structures were built, they are the values that are the bearers of social memory that enable them to define the social, economic and political structure of the society (Yavaşoğlu, Özgül, 2020). In this respect, before the re-functioning decision, the structure to be transformed should be handled with a multi-layered evaluation. Accurate evaluation of the spatial and structural qualities of industrial buildings, simple and reversible changes to be made carefully, presenting technical equipment on site, researching the needs of the area where the buildings are located by conducting survey and market analysis will enable successful applications. (Köksal, 2012).

Büyükarslan and Güney (2013) gathered these criteria under the titles of architectural transformation, spatial transformation, functional transformation, structural, contextual transformation evaluations and building-place relationship in their article on the Re-functioning of Industrial Heritage Buildings. They emphasized the importance of detailed examination of the building to be transformed under these titles and explained the advantages of re-functioning in being sustainable both architecturally, socio-culturally and economically. Yıldız and Güçhan (2018), in their study on the Ayvalık Ertem Olive Oil Factory as part of the Transformation of Industrial Heritage, mentioned that industrial structures represent a symbolic and memorable value for collective memory as a witness of industrial-technological history. They stated that industrial buildings are generally transformed into multi-functional uses and/or museums, arising from the technical values of the production equipment in the buildings, which also add "aesthetic value" to them. Although there is a general tendency to be converted into museums, exhibition halls or concert halls, some industrial heritage buildings can also be used as movie sets. Alpler et al (2020) mentioned in their study on the

Re-use of the industrial structures as film sets in Turkey, that the transformation of industrial buildings into a movie set, which is a different re-functioning alternative due to their spatial suitability, the historical importance and values of these spaces can reach more people by meeting the audience.

2 Industrial Heritage and Sustainability

The advancement of technology has led to some changes in the application areas of our daily work and professions, and our lifestyles have adapted to this situation in a short time, so some business lines and even professions have lost their importance. For this reason, although they were designed to survive for centuries in the period they were built, some buildings have been left to their fate in cities for many years, which have become unusable due to the loss of importance of business lines. The continuous change of production with today's technological speed, especially the loss of use of many industrial structures that were used in the 19th century, brought the reuse of these structures in different functions were made. Historical buildings, which are a component of the human environment and a part of cultural heritage, are nowadays a sustainable producer for cities. The industrial heritage plays an important role in the sustainable development of the city. In recent years, it has been seen that cultural heritage has been attributed a more important role within the framework of development models based on local characteristics and increasing the internal resources of the region, in international researches and discussions (Spina et al 2020). In the development of countries, in order to ensure collective interest in the development of cultural heritage, it plays an important role to determine sustainable strategies that focus on conservation as a priority and can also trigger regional and local development models.

Community expectations are decisive in defining reuse strategies and can contribute to improving the quality of life, enhancing functional equipment, infrastructures and services. Users' needs and behavior must be ethically justified in order to preserve the historical values, originality and integrity of the heritage. Sustainable renovation is the proper management of use and change in and around historic sites and places to respect and increase their value to society (Blagojevic, Tufegdzic, 2016). Adaptive reuse describes the re-functioning of buildings abandoned in the city or not used within the framework of their potential, while preserving the historical/archaeological, visual/cultural, economic, functional and psychological values of the existing building. Blagojevic and Tufegdzic (2016) examined Sustainability under three titles as Environmental, Economic and Social and made a table under which subjects should be addressed for Industrial structures for Adaptive Reuse. In this case, for a sustainable reuse, the energy efficiency value of the industrial structure, which is located at the intersection of environmental and economic sub-titles of sustainability, should be examined with the inputs of the embedded energy in the building, construction-structural limitations and energy systems. The ethical value of sustainability, located at the intersection of economic and social subtitles, should be examined with the inputs of evaluation, authenticity and integrity.

In the case of industrial buildings, the principle that unites all elements as a whole is technological functionality, since production technology is an important factor affecting the development of architectural features (except for stylistic ones). Considering that technological functionality defines the architectural integrity of the building and facility, it can be understood as a principle of the aesthetic integrity of the industrial heritage. To understand the technology of the manufacturing process, machines and buildings representing their physical frameworks are equally important for industrial archeology. Specifically, based on the belief that the formal features of an industrial building reflect the technological process, technological functionality is a limiting factor in reuse. According to Blagojevic, Tufegdzic (2016), choosing any other function for the existing industrial facility or structure, other than refunctioning it into an industrial museum is paradoxical in terms of its archaeological value. Mısırlıoğlu and Günçe (2016) identified the factors affecting reuse decisions in their study and created a holistic model to determine the most appropriate function for adaptive reuse of historical buildings. As a result of their studies, they emphasized that it is not enough to evaluate the success of the reuse of historical buildings that are cultural heritage not only in terms of conservation principles, but also the importance of preparing the management plan for sustainable heritage adaptations.

In terms of environmental performance, facing an existing building can be limiting due to the fixed site layout. Limitations will be imposed on the scope of energy efficiency improvements when changes need to be made that be irreversible and may affect the originality and integrity of the building, especially to minimize energy use in reuse.

3 Olivetti

Numerous industrial buildings built in the early 20th century have become an important part of the historiography of contemporary architecture. These include the Fagus Factory and AEG Workshops in Germany, the Van Nelle Factories in the Netherlands, the Ford Factories in the United States, and the Olivetti Factories built in the mid-thirties to the end of the fifties in Ivrea, Italy.

Adriano Olivetti took over the management of the typewriter factory from his father Camillo at the end of 1932 and started a development process with positive results in the economic field. (Astaria, 2012). Adriano Olivetti has created an 'industrial strategy' that values the entire company system by using creativity and technological culture, which are two elements required for sustainable development today. Olivetti's sustainable project is mostly based on the link between an ethical concept and a development-oriented idea (Castano, Gallo, 2013). Olivetti;

Ethical sensitivity, defined as a fundamental aspect of sustainable development by Luca Davico (2012), and the ability to bring modern demand for continuous and increasing progress can be cited as an example of "political and cultural mediation". The spreading of industrial structures in the region of Olivetti's buildings and ensuring production and sales has gathered attention from service optimization to the quality of life and to the environment.

Olivetti's modern approach is also reflected in the LeCorbusian glass façade of his new factories built in 1934. Inspired by a strong urban context, architect Figini and Pollini emphasized the 'transparent wall' of local culture with the glass facade, which is the symbol of industrial modernity, thus enabling workers to work on typewriter production under a stronger and more natural lighting and to feel the sun rays more in the space. (Manfioletti, 2012).

Adriano Olivetti followed the principles of corporate social responsibility, ensuring that the company respected its production plans and also contributed to improving the well-being of its employees and the quality of life of local communities (Sciarelli and Tani, 2015). He invested his earnings particularly in urban landscape design and construction. Their initiatives have been supported by the active collaboration of intellectuals, architects, urban planners, industrial designers, sociologists, psychologists and writers from around the world (Bonfati et al, 2015). This Italian practice has also enabled industrial companies to create social value through their investments in architecture.

4 Establishment, History and Philosophy of Olivetti Company

Camillio Olivetti was born in 1868 to a wealthy family in Ivrea. He was graduated from the Polytechnic of Turin University in 1891 as an electrical engineer (Conte, 2017). After graduation, he worked in London for a while and gained work experience in a factory. He met his mentor Ferraris in the USA at an electrical congress in Chicago in 1893, and during this time he attended physics courses at Stanford University and was offered a job as an electrical engineering assistant. This American experience gave Camillo the opportunity to visit the country, the most important cities, laboratories and factories, and to internalize the foundation of American industrial economic dynamism (Labò 1957). The rapid development of the mechanical industry and its products observed in America drove Camillo's attention in 1896 to the CGS building, a small tools factory for electrical measurements he designed, in Ivrea (Conte, 2017). The exterior is typical of the proto-industrial buildings of the late 19th century, with two stories high and light red bricks on the facade, but the building was actually one of the first buildings in Italy to use a reinforced concrete system (Fig. 1). As an architectural language, it is very restrained and at the same time conveys the idea of stability and simplicity.

Fig. 1. The original red brick building built by Camillo Olivetti for CGS in 1895 'Eng. C. Olivetti & C. First National typewriter factory '. (AASO, Conte 2017)

In the search for new partners and new commercial outlets in 1904, Camillo Olivetti moved the company to Monza but left management decisively to return to Ivrea in 1907. Olivetti refused to use American models when designing his products and designed a new machine and a new writing system with all its parts. In 1911, Camillo Olivetti presented his first Italian typewriter at the Universal Exhibition in Turin with the words: *'A typewriter should not be an ornamented living room trinket, but should also have a serious and elegant appearance'* (Conte, 2017). The years between 1911 and 1914 were the richest years for the Italian economy and very important years for Olivetti. The 6000 parts needed to assemble and the machines needed to make them, were all produced in the laboratory in Ivrea, and Olivetti had the opportunity to prove to be a reliable product. Increased orders with the introduction of the machine to the market; it required new staff, better organization to increase production, new financing, and above all a successful distribution and sales network. In 1912, the first branch was opened in Scala square in Milan, followed by Genova, Rome, Naples branches in 1913 and the first store in the Vittorio Emanuele Gallery in Milan in 1914.

Camillo's son, Adriano Olivetti, worked as an apprentice at the factory, his father's training allowed him to get to know the working class. His innovative 'human scale' factory vision will be shaped by the memory of his experience at the age of 13.

Italy's entry into the First World War in 1915 led to the minimization of typewriter production in the following years, significantly limiting the development of new products due to the war-based transformation of the company. Despite this, the company needed a new area for production in a short time and in 1918 the first additional building was built along Castellamonte Street (Conte,2017). With its new products, the Olivetti brand started to spread in Europe, especially in Spain and the Netherlands and South America, especially Argentina, while sales branches increased in Italy: Trieste in 1920 and Turin in 1923 were added to the number of four buildings. The slow and continuous expansion of the buildings accompanied the development of the company without significant style changes until the 30s. (Boltri et al, 1998).

Adriano Olivetti graduated from Torino Politechnico in 1924 with a degree in industrial chemistry and then lived in America for six months, encountering a very different country than his father visited. During his stay abroad, he had the opportunity to visit 105 factories, including his competitors, accompanied by Domenico Buzio, who worked as the technical manager of the company, in addition to patenting his father's inventions. After returning from America between 1926 and 1927, with the support of his friend Gino Levi Martinoli and Eng, he began to test what he had learned. He worked on early changes in work rhythms and intervention programs to modernize Olivetti production. Between 1927 and 1928, it marked a defining turning point in production organization. Adriano, with his father's approval, introduced a personal form of Taylorism that would start from the press atelier and reach the typewriter assembly line, and reduced production time from 12 hours to 4 and a half hours, thus moving the company from a semi-artisan to mass production.

For Olivetti, the 1930s were a period of renewal of the organizational and company's image. Servizio Pubblicità (Advertising Service) began operating as early as 1928, by order of Adriano, who had been in management for some time, and soon became an indispensable tool for the systematic promotion of commercial development. At the invitation of distinguished figures such as Xanti Schawinsky from the Bauhaus school, designers Edoardo Persico and Marcello Nizzoli, architects Luigi Figini and Gino Pollini, Bruno Munari and Luigi Veronesi, advertising communication has changed style: the drawings used in the early years to coordinate the corporate image are replaced by more refined messages, photographic images and more detailed graphic representations. Showcases are rearranged every 15 days; Intense colors, geometric shapes, avant-garde references were always included not only for purchasing, but also as products where you can find different images in the store you entered (Conte, 2017).

According to Adriano, a company and its management should not only aim at a high efficiency, an innovative production that anticipates the future demands of society, and a profit generation, but also set an ethical goal. The idea of a new factory, moving in a global design and supporting the economic and social progress of the community, must follow innovation in the product space. In this period, a series of social rights and operations to protect these rights were organized against the employees, their families and the working woman figure. The factories are equipped with the first medical care service, a company transport service, company canteen, library and kindergartens, which will later find space in detached buildings. These are rights acquired as employees of the company. The company of which the worker is an integral part has produced not only material goods, but also wealth and culture for the context in which he is found. Adriano began to give concrete form to the idea of an ethical and cultural industry by creating new buildings for industrial production to recognize human dignity and increase individual protection at work. Instead of relying on a single designer to achieve this goal, Adriano Olivetti invites sociologists, artists, philosophers, young architects, designers, and urban planners to Ivrea, exchanging ideas about, 'Architectural structures, the organization of the environment and the region that can live together with formal beauty and functionality, the working conditions in the company and improving the quality of life outside the company' (Olivetti, 2015).

During his life in Milan between 1931–1934, he assigned two young architects who prioritized rational environmental design in building design; Luigi Figini and Gino Pollini. In 1934, they designed the first factory in relation to nature and its environment and defined their structures with the words *"in human scale and size, in a happy relationship with nature, because the factory is for human"* (Caizzi 1962).

Adriano, who established new factories in Paris in 1939 and in Spain (1939–1942) with the end of the civil war, had to leave the country in 1944 due to suspicion of spying against the fascist regime in Italy. During his exile, Adriano matured the ideas of factory and district development and completed writing his manifesto, in which he described his proposal to reform the society with a definitive constitutional Project.

The 1950s are of fundamental importance for Olivetti: it is a period of major expansion in the national and European market. In 1950, it took its place in the American market along the Fifth Avenue in New York and thus established the Olivetti Corporation of America with all its technological and commercial facilities (Conte, 2017). It was not enough to produce a beautiful product, it was necessary to present the innovative features of the exhibited product in a showroom that could impress the customers. Customers had to have an emotional experience, get to know, see, touch and especially try the product. This was Adriano's innovative idea based on the American shops opened between 1953 and 1954.

The different architectural solutions applied for the showroom were an innovation, especially in terms of the interface between customers and products. The display window behind the sidewalk created a portico that was almost an entrance hall for the shop; his new portable product, Lettera 22, came out of the gallery and is positioned on a pedestal like a work of art (Fig. 2). It was a revolution in marketing, a new way of seeing a brand connect values that go beyond the product itself and its use.

Fig. 2: The showcase of Leo Lionni's Chicago store from 1953; Store on Fifth Avenue in New York in 1954 (Conte, 2017)

Those years were also years when production was reorganized for Olivetti, which allowed achieving better standards and production levels by focusing on technological progress. Considering the huge demand for products, in 1950, on the idea of Adriano, Ufficio Tecnico began reorganizing the machine assembly line to a continuous line system with the use of conveyors (Conte, 2017). Reducing working time required the introduction of new special machine tools and the realization of innovative factories.

For ten years, it is planned to add functional new buildings to the factory along Castellamonte Street in Ivrea and Canavese region simultaneously and to build new spaces and services necessary for the development of the Community. The first true example of functional architecture began in 1952 and changed the linear expansion used for historic buildings. The Bernando industrial area is strategically located near Ivrea and the Turin-Milan highway. For construction, a city planning where green spaces and the user are linked with each other has been implemented; buildings are designed by applying industrial production to the modular architecture. The same logic was applied by Vittoria in collaboration with Zanuso for the realization of the Olivetti Scarmagno region, the latest project born with Adriano's will and designed to increase production through the transition to electronic technology. These and other productive practices have been adopted as an economic and social regeneration plan tool for the Canavese community, promoted by Adriano to combat unemployment and population decline. The company has acted both by taking over the companies in the mechanical sector by turning them into new functions and by building buildings in new production sectors.

In 1955, in collaboration with Pietro Porcinai for green design and Marcello Nizzoli for interior design, he designed another factory in a dominant location in Pozzuoli bay. The factory had an architecture that adapted to the slopes of the ground, allowing a perfect integration with the surrounding landscape. At the same time, the factory has enabled the development of the relationship with the outside of the building, with the maximum level of light transmission inside. A few years later, following the company's social policy, a residential area was built as in Ivrea (Conte, 2017).

Employee needs and social assistance system also play an important role in Olivetti company policy. In 1908, Camillio took a series of protective precautions against workers and set up a joint fund to provide support assistance. The construction of the first residences for company employees was also added to these aids as a solution to the housing problem as a result of the increase in the number of workers and the insufficiency of existing buildings. In 1926, on Camillo's initiative, six two-storey detached houses were built in an area called Borgo Olivetti, close to the factory. The stylistic model is a traditional hut and the houses also have their own garden for food livelihoods for families who remain connected to the area while working at the factory.

Adriano believes that workplace and housing conditions affect social quality of life and productivity. In this context, a systematic social service planning action took place consecutively in 1932. The Adriano factory has, above all, the moral mission of being a pioneer of economic and social community progress towards the region in which it is located.

Since this period, protective practices have been implemented for employees, their families and working women. By the late 1930s, Olivetti workers were able to use discounted transportation to work and free canteen service within the factory, and with free home calls from the Olivetti family's doctor, they were able to receive medical care both at the infirmary and outside the factory. In addition, the first library service was opened in these years. In 1941, the regulation on insuring female employees during pregnancy and breastfeeding entered into force. According to the regulation, it was stated that workers could be on leave for nine and a half months for almost full wage, and a range of services from pediatric clinics to medical care to the possibility of using the kindergarten. The company also offered its employees vacation opportunities at sea and mountain summer camps or daily summer camps in the city. One of the most important activities of social services is for housing, first in Ivrea region where housing needs are obtained from the growing population with the development of the company, and then in all these existing campuses.

The first residential and social buildings started to develop from 1939; In Figini and Pollini's project, the first social housing buildings for 24 large families were built on the opposite side of Castellamonte Street compared to the factories in Borgo Olivetti. Luigi Piccinato, Luigi Pollini and Egisippo Devoti carried out the first comprehensive regional development between 1938–1942 with the Ivrea city plan. The construction of the region took place at different times and with different actors, and it became an experimental laboratory where research on the life imagined would be applied. Between 1940 and 1942, five rows of buildings with 4 rows of houses were built by Figini and Pollini. Later, between 1948 and 1952, Marcello Nizzoli and Gian Mario Olivieri built six detached villas for managers, other villas for management staff and several apartment blocks for families. Designed by Nizzoli and Olivieri in 1955, the house with eighteen flats is placed in a large green area and presented by combining three elements at different heights with different formal solutions for facades.

The company's design activity has taken a pioneering place in many different areas, from the design and production of typewriters to the establishment of publishing houses, from the design of city plans to the development of welfare policies.

A lot has been done in this direction, starting in the 80s, thanks to the establishment of the AASO certified advisory center by the Andriano Olivetti Foundation and the municipality of Ivrea. Benjamino de 'Liguori Carino, who participated in the Italian Architects Seminar held in Istanbul on 22 October 2018, gave information about life in factories from Olivetti's approach to architecture. Carino represented at the seminar that Camillo Olivetti answered the question *"What does architecture mean to you?"* as *"It is a society's expression of itself"*, and introduced photographs from the collection belonging to the Foundation on the contact between the industry and the city.

Fig. 3: I. Kindergarten, II. Library, Archieve of Fondoziane Adriano Olivetti (Carino, 2018)

In Fig. 3, in the left photo, the kindergarten structure, which was built for the children of the workers working in the factories and whose architecture is integrated with nature, and in the next photo, the library where the workers spend time in their spare time, shows two of the social services of Olivetti Company.

5 Olivetti Buildings and UNESCO

The industrial city of Ivrea is located in the Piedmont region, Italy, and was developed as a testing ground for Olivetti, a manufacturer of typewriters, mechanical calculators and office computers. It includes a large factory and buildings designed to serve administrative and social services as well as housing units. Ivrea, a model social project designed by leading Italian urban planners and architects, mostly between the 1930s and 1960s, expresses a modern vision of the relationship between industrial production and architecture.

The Olivetti campus in Ivrea, the evaluation of which started in 2017, was designated as a World Heritage Site in UNESCO's 42nd meeting held in Bahrain in 2018. Ivrea is also the 54th area of Italy that has been accepted to the UNESCO World Heritage List. UNESCO World Heritage buildings include 27 modern buildings that characterize the industrial city founded by famous city planners and architects of the period. Since the architectural features and environmental features of the original projects of the buildings are preserved, they have been evaluated in an original and holistic manner [1]. The industrial city of Ivrea represents an important example of 20th century urban development and architectural theories in response to industrial and social transformations, including the transition from the mechanical sector to the digital industry. The facility, which is stated to exhibit Outstanding Universal Value in the evaluations available from UNESCO's website, is mentioned that Ivrea is able to show a response and a contribution to the 20th century theories of urbanism and industrialization, as it represents a model of the modern industrial city and a response to the challenges of rapid industrial change [2].

Among the buildings in Ivrea Olivetti campus are 1.Social Services Center, 2.Kindergarten, 3.Borgo Olivetti House, 4.Thermal Power Plant, 5.Office Building, 6.New Office Building, 7.Old Carpentry Offices (now ARPA offices), 8.Ex-Sertec Building, 9.18 accommodation units, 10.House for employees, 11.House for managers, 12.4 houses, 13.Western settlement (Talponia), 14.Villa Capellaro, 15.ICO Workshop, 16.Canteen and Rest Club of the company, 17.Olivetti Education and Experience Center, 18.UCCD- Villa Rossi is located (Fig. 4).

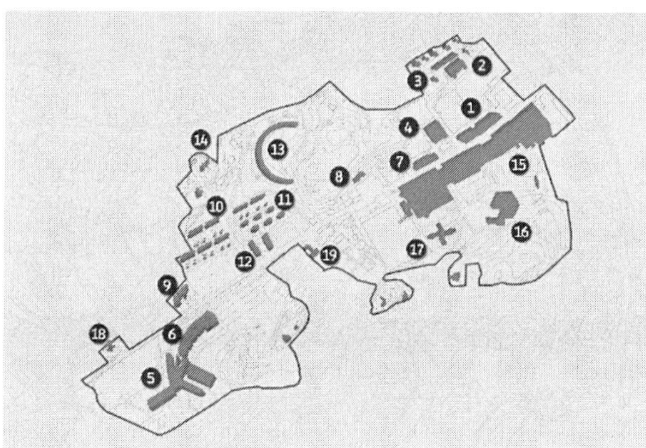

Fig. 4. Ivrea Olivetti Industrial Area Map [1]

The movement that Adriano proposes with this campus is not only architectural but also a political and social movement related to architecture. All buildings follow a structural and moral line determined by the Olivetti family, which facilitates vertical and horizontal communication between users and has been developed to leave the least impact on the environment (Suarez, 2018). Fig. 5 shows the aerial view of Ivrea Olivetti Campus on the left and the photo on the right is of the Western Housing Unit-Talponia, built between 1968–1971 by Architect Roberto Gabetti and Aimaro Oreglia d'Isola belonging to the Olivetti complex in Ivrea; Accommodates 13 duplexes and 72 single flats (Suarez, 2018).

Fig. 5. I. Ivrea Olivetti Campus, UNESCO World Heritage Site; II. Western Housing Unit-Talponia (Suarez, 2018)

Olivetti had his law enforced in the architectural language of his buildings; easy connection and communication. This effect is felt in the grand staircase, which is one of the best preserved parts of the complex, which provides fluid communication between different floors at the entrance with its hexagonal, central radial shape of the Office Building (Fig. 6).

Fig. 6. The stairs of the Office Building (Suarez, 2018).

According to the evaluations of UNESCO; Ivrea represents one of the first and highest expressions of a modern vision of manufacturing, architectural design and social aspects on a global scale in connection with the transition to digitized industrial technologies [2]. Although most non-residential buildings have ceased to function, the Community Movement's impact on residential and social buildings is an important intangible element. One of the most important details in the Olivetti campus is the coexistence of architecture with nature and landscape. By integrating the buildings with nature, not only a massive harmony was achieved, but also the comfort conditions of the users (workers, their families and managers) were improved in the psycho-social sense.

In the UNESCO's World Heritage Site report, the Integrity criterion is evaluated with the explanation of; *"The integrity of this urban space is based on the inclusion of the buildings, spaces and urban form needed to express the importance of Ivrea's 20th century development. The preservation status of the city's components is variable. Most of the residential buildings have a good/adequate protection status".* The criterion of Authenticity/Originality is evaluated with the explanation of; *"Ivrea's originality is based on the number and quality of urban and architectural projects that date back to the first period of its development as an industrial city. A detailed analysis of individual components was carried out in terms of their forms, designs and materials, as well as their location and immediate surroundings. Despite the production changes that have affected the city in recent years, many elements have preserved their original characteristics. Many residential, administrative and service buildings are intact, some have been renovated, and many buildings are now empty and have an uncertain future. There is a risk of gradual loss of the authenticity of the buildings due to the decay of the exterior of the buildings and the deterioration of interior decoration and details. Typically similar to its original uses - such as telecommunications, production or cultural activities - efforts have been observed to develop new uses"* [2]. In UNESCO's Assessment of the Protection and Management Requirements for the Campus, it was pointed out that the national protection laws are applied only for some buildings, and it was stated that the legal protection system is complex and multi-level, relying heavily on the commitment, resources and expertise of both national and municipal authorities. It was recommended to pay attention to the need for coordination between local, regional and national institutions, and it was emphasized by the Council of Ivrea that the regional landscape setting should be adopted, the directives directly related to the conservation and improvement of structures should be integrated and strengthened. It was also noted that plans must be made to increase tourism capacity and interaction with residents and other users is a priority [2].

6 Conclusion

Olivetti's strength stems from its ability to renew the entire factory concept by giving a new life to its processes, services, products and architecture that point to a healthy innovation.

In Olivetti Campuses, the company has become a representation mechanism by itself, by implementing an architectural set of representation systems. What distinguishes Olivetti campuses from others is this unique representation approach, the use of land, the importance of urban planning, the rationality of architectural forms, the functionality of the spaces, the comfort it offers to its users and the integrated design with the environment.

Thanks to the planning idea in Olivetti Campuses, it is possible to continue to follow a functional path for the modern production system that supports economic, social and cultural development. Consciousness determined by the understanding of the changes going on today may make it more possible to design new formal, cultural and technological languages that can educate the society. The need to identify key points for a sustainable planning of the factory system; It requires the development of a long-term growth strategy that triggers changes in work production along with the changing culture. As in Olivetti Campuses, architectural planning, which covers issues such as the creation and organization of the workspace as a tool of social and economic growth, should stimulate and export creativity by expressing the quality of the environment, the work done and the people involved in this process. Thanks to the creation of open, flexible and versatile systems in architecture, it is possible to make factories human-oriented with a design approach that can encourage progress through communication, knowledge and experience sharing.

References

1. [1] https://blog-italia.com/actualidad/unesco-incluye-ivrea-como-patrimonio-de-la-humanidad-de-italia.html Accessed on January 10, 2021.
2. [2] UNESCO, https://whc.unesco.org/en/list/1538/, Accessed on January 10, 2021.
3. AASO, Associazione Archivio Storico Olivetti, https://www.archiviostoricolivetti.it/
4. Alpler Z. B., Şahin N.P., Dağlı U.U., 2020, A Critical Discussion of Industrial Heritage Buildings Adaptive Re-Use As Film Spaces, Case Study: Industrial Heritage Buildings at Istanbul, *Journal of Architectural Conservation.* https://doi.org/10.1080/13556207.2020.1782105
5. Astaria R.,2012. Gli Architetti di Olivetti; Una Storia di Committenza Industriale, Storia Dell'Architettura e Della Citta, FrancoAngeli, Milano, Italy.
6. Bonfanti A., BATTİSTİ E., PASQUALİNO L., 2015, Social Entrepreneurship and Corporate Architecture: Evidence From Italy, *Management Decision Vol.54, No.2; p390–417.* DOI 10.1108/MD-08-2014-0532
7. Blagojevic M.R., Tufegdzic A., 2016. The new technology era requirements and sustainable approach to industrial heritage renewal. *Energy and Buildings 115 (2016), 148–153.* http://dx.doi.org/10.1016/j.enbuild.2015.07.062

8. Boltrı D., Maggıa G., Papa E., Vıdarı P. P., *Architetture olivettiane a Ivrea - I luoghi del lavoro e i servizi socio-assistenziali di fabbrica*, Fondazione Adriano Olivetti, Gangemi, Roma 1998.
9. Büyükarslan B., Güney E.D., 2013. Endüstriyel Miras Yapılarının Yeniden İşlevlendirilmesi Süreci ve İstanbul Tuz Ambarı Örneği, *Beykent Üniversitesi Fen ve Mühendislik Bilimleri Dergisi Cilt 6(2) 2013, 31–58.*
10. Caizzi B., 1962, *Camillo e Adriano Olivetti*, Utet, Torino, 1962
11. Carıno, B.L., 2018, L'idioma di Olivetti Eredita 'e Futuro Della Legacy Olivettiana, Fondoziane Adriano Olivetti, Sala Teatro Casa d'Italia, Italian Cultural Center, Unpublished Seminar Notes 22 October 2018, The Event: Italian Architects #3, İstanbul.
12. Castano F., Gallo A., 2013. The Olivetti Factory as a Paradigm of Sustainable Growth, *Proceeding of the 2nd International Conference On Sustainable Intelligent Manufacturing, Lisbon Portugal, June 26–29 2013, Green Design, Materials and Manufacturing Process, pp67–70, ISBN:978-1-138-00046-9*
13. Conte S.,2017. Dal Progetto Di Adriano Olivetti: Eduardo Vittoria. Ricerca, Disegno e Design; Nuovi Sistemi di Rappresentazione Per La Valorizzazione Dell'Architettura Moderna. Area Di Ricerca: Disegno e Rappresentazione-ICAR/17, Politecnico di Milano, Dipartimento di Design, Dottorato di Ricerca in Design. Coordinatore: Prof.ssa. Paola Bertola, Ciclo XXXI.
14. Davıco, L. 2012. Etica e Sostenibilita. *Lo Sguardo – Rivista di Filosofia* 8 (1): 75–76.
15. Köksal G., 2012. Endüstri Mirasını Koruma ve Yeniden Kullanım Yaklaşımı, *Güney Mimarlık TMMOB Mimarlar Odası Adana Şubesi Haziran 2012 Sayı 8 s18–23.*
16. Labò M.*, 1957. L'aspetto estetico dell'opera sociale di Adriano Olivetti,* Görlich, Milano,1957
17. Maffioletti M., 2012. L'Olivetti d'Adriano. Une image industrielle du personnalisme et du communautarisme, *COnTEXTES [En ligne], 12 | 2012, mis en ligne le 21 août 2012, consulté le 04 janvier 2021.* DOI : https://doi.org/10.4000/contextes.5545
18. Mısırlıoglu D., Günçe K., 2016. Adaptive Re-Use Strategies For Heritage Buildigs: A Holistic Approach, *Sustainable Cities and Society 26(2016) 91–98.* http://dx.doi.org/10.1016/j.scs.2016.05.017
19. Olıvettı A.,2015. *Città dell'uomo*, Edizioni di Comunità, Ivrea/Roma, 2015
20. Saner M., 2012. Endüstri Mirası: Kavramlar, Kurumlar ve Türkiye'deki Yaklaşımlar, *Planlama 2012/1–2, s53–66*
21. Sciarelli M., Tanı M. 2015. Sustainability and stakeholder approach in Olivetti from 1943 to 1960: a lesson from the past, *Sinergie Italian Journal of Management, Vol. 96, January- April, pp. 19–36.*

22. Spına L D., Gıorno C., Casmıro R.G. 2020. An Integrated Decision Support System to Define the Best Scenario Fort he Adaptive Sustainable Re-Use of Cultural Heritage in Southern Italy, *C. Bevilacqua et al (Eds.) NMP 2020, SIST 177, pp.251–267.* https://doi.org/10.1007/978-3-030-52869-0_22

23. Suarez, M. 2018, Ocak 2021 tarihinden https://elpais.com/elpais/2018/10/04/icon_design/1538641860_640062.html sitesinden alınmıştır.

24. Yavaşoğlu F., Özgül C. G., 2020. Endüstri Mirasının Korunması ve Dönüşümü:Malatya Şeker Fabrikası Yerleşkesi Örneği, *İdealkent Kent Araştırmaları Dergisi 30, Cilt11, 2020--2, 676–700.* Doi: 10.31198/idealkent. 680959

25. Yıldız G., Güçhan N.S., 2018. An Industrial Heritage Case Study in Ayvalık: Ertem Olive Oil Factory, *Journal of Contemporary Urban Affairs, 2018, Volume 2, Number 3, p20–30.*https://doi.org/10.25034/ijcua.2018.4715

Cenk CİHANGİR and Pınar KISA OVALI

Architects' Opinions on Ecological Building Design and Application Practice (Edirne/Turkey)*

Abstract: Developing formal and informal processes, which have an important place in the holistic structure of architectural education, within the scope of "sustainable architecture and ecological design" principles and strategies are necessary for the organization of habitable environments. The main purpose of this study is to question the level of application of the theoretical knowledge acquired by architects within the scope of ecological energy-efficient design, climate-compatible architecture, solar architecture or environmental protection approaches in their formal education processes and the reasons for not being able to apply them. Supporting the professional development of the architect in the work (lifelong learning) informal processes and their contribution to professional development towards ecological building design are also discussed. The study questionnaires were applied to 165 of 184 architects registered to TMMOB Chamber of Architects Edirne and Keşan Representatives (2017 year data). Survey questions; In addition to determining the theoretical knowledge acquired by architects in their formal and informal processes, it was created within the scope of environmentally friendly energy-efficient building designs of ecological architecture. The questionnaires were processed using the SPSS program and the data were visualized with graphic explanations. In addition to the general views of the architects on ecological building design and application practices, beside this the problems they encounter were also determined.

Keywords: Architect, Architectural education, Ecological design

* This study was produced from the first author's master thesis.

1 Introduction

Technological developments that started with the industrial revolution caused a rapid increase in population, urbanization and construction and the abandonment of the traditional building design concept. The understanding of adaptation to nature, environmental and climate conditions (which are the essence of today's ecological architecture), which constitutes the essence of traditional building design, has been replaced by rent-based buildings and unqualified buildings that ignore human health. This understanding of structuring, which ignores the context, geographical and climatic characteristics, and causes too much energy consumption during the construction, use and re-evaluation stages, also harms the global ecosystem. Climate change, species losses, environmental pollution and natural disasters have brought the concept of ecological architecture to the agenda with the concept of sustainability

After the "dependency for sustainable future" decisions taken in the General Assembly of the World Association of Architects in 1993, the Architects Community took important steps towards "Sustainable/Ecological Architecture". In the declaration addressing addiction decisions; "The buildings and the built environment play an active role in the natural environment and quality of life of people, sustainable/ecological design indicates the efficient use of resources and energy, includes healthy buildings and building materials, and land use with high ecological and social sensitivity, so it is emphasized that the negative impact of humans on nature can be reduced"[1]. Architects have entered the 21st century's sustainable architecture and ecological design era within the scope of this declaration. The ecological design approach, which started in the first quarter of the 1990s, took its place in the formal processes of architectural education [2] (Courses on sustainability/ecology/energy efficient design/solar architecture/environmental protection/renewable energy use/climate adaptation were added to undergraduate and graduate programs)

Another way of learning that contributes to the professional development of architects is the organizations that are handled in informal processes (congress, symposium, workshop, seminar, publications and other educational activities). In order to question the effect of informal processes The publications and organizations of the professional association dealing with "ecological architecture" have been examined. In this context, the formal and informal processes that form the views of architects on ecological architecture were defined within the scope of the manual research and were used in the preparation and interpretation of the questionnaire questions.

Edirne selected as an example; In addition to being a cultural city that served as the capital of the Ottoman Empire as the face of our country opening to Europe, it is a city of high ecological importance, surrounded by first-class agricultural lands, wetlands and river ecosystems. In this context, it is important for the architects who carry out reconstruction activities in the city to organize ecological/sustainable buildings and living spaces.

2 Ecological Architecture and Design Criteria

It is seen that the architectural profession is evaluated in three different periods from the perspective of environment and ecology. First one; The "Traditional Architecture" period before the Industrial Revolution, the second "Modern Architecture" period after the Industrial Revolution, and the third "Ecological Architecture" period specific to the 21st century. The aims of the ecological architecture period, which is also the main subject of the study; To construct a harmonious relationship with nature in a continuous manner, to organize the built environment within the scope of principles of ecological design and to create designs that take artistic and spiritual dimensions into account [3].

Ecological design; It means designing the built environment and life forms to integrate the life forms on earth in harmony and flawlessly with the natural environment in which the biosphere is located [4]. It is seen that today's conventional (conventional) architecture and ecological architecture have different characteristics, today's architectural design exhibits the same attitude in all conditions, and ecological design leaves its place to bio-regionalism, focusing on design with local data. [5, 6, 7, 8]. Simos and Yannas stated that knowledge, performance and urban context are important elements of environmentally friendly architecture as 3 basic criteria, and that thermal and visual comfort emphasizes the necessity of meeting with non-renewable energy [9]. Ecological design according to Edwards (2007); problems in ecology-environment-energy (3E) issues; It seeks solutions with the concepts of "reduce", "reuse", "recycle" and "recover" (4R) [10].

The design criteria of ecological architecture can be discussed under two headings, namely physical and built environment. Physical environmental factors include the basic approaches of bioclimatic building design and consist of land data, topography, climate data, orientation and organization of green tissue. Design criteria for the built environment; It includes building form, space organization, building envelope, material selection, sanitary installation and the use of renewable energy sources [11,12]. Survey questions were created within the scope of this classification.

3 Forming Opinions on Ecological Architecture: Formal and Informal Processes

Education is "the activity of helping new generations acquire the necessary knowledge, skills and understanding and develop their personalities while preparing to take their place in the life of society" [13]. It is known that architectural education is not limited to occupational acquisition, but informal processes as well as formal processes are of great importance within the scope of lifelong learning.

3.1 Formal Process

Architectural education programs have a triple fiction that supports and develops each other based on theoretical/applied courses, studio courses and professional practice, and each university's distinctive institutional identity [2]. However, the study does not focus on these weights. The aim is to reveal the level, number and content information of the courses in the informational process of the architect candidate through the data. A formal examination has been made through the education programs of universities. architectural education in Turkey in this context that 82 universities in the 2016 academic year of undergraduate and graduate programs in environmental/ecological/energy/sustainability on compulsory and elective courses have been investigated through the internet. By comparing the data with the research of Ciravoğlu (2011), the course increase rate in the programs over a 5-year period was determined.

Referring to compulsory courses in architecture department of the undergraduate program in Turkey, the information reached 80 universities in 66 (83 %) compulsory course for students can earn environmental protection in respect of vocational curriculum is seen that there [14]. In 2011, it was determined that 33 of the architecture departments of 42 universities had compulsory courses [2]. It was determined that the rate of 78 % in 2011 reached 83 % in 2016 and there was a 5 % increase in the compulsory courses on ecology/sustainability in education programs in a short period of five years. In many departments, it has been observed that there are physical environmental control/control courses in the second or third year.

When we look at the elective courses in the language programs, it is seen that in 65 (81 %) of 80 universities, there are also courses in the elective course curriculum that students can gain awareness of environmental protection [14]. It was determined that there were elective courses in 30 of the architecture departments of 40 universities in 2011 [2]. It was determined that the ratio of 75 % in 2011 reached 81 % in 2016 and increased 6 %. When the contents of the elective courses were examined, it was seen that certain principles and strategies (ecology, environment, energy, sustainability) were mostly addressed at the point of protecting natural resources.

In addition, some universities give a wider place to compulsory and elective courses on environment, ecology, energy and sustainability in architectural education compared to other universities. According to 2016 data, Mimar Sinan Fine Arts University 26, Yıldız Technical University 11, Okan University 10, Istanbul Commerce University 9, Dokuz Eylül University 9, Dicle University 9, Istanbul Technical University 8, Middle East Technical University 8, Trakya University 8, Kırklareli University has established its program with 8 courses [14].

Architecture graduate program in Turkey 45 (96 %)of 47 universities, environment, ecology, energy, courses was determined to be addressing the sustainability issues [14]. It has been observed that universities have extensive study and research areas in postgraduate education programs on issues that evaluate the relationship between architecture and environment. In 2011, it was stated that there were courses focusing on environmental issues in 28 of the graduate programs of 30 universities [2]. It has been determined that the rate of 93 % in 2011 increased 3 % in 2016, reaching 96 %. In addition to this increase, it has been observed that some universities give a wide place to courses on environment, ecology, energy and sustainability in their graduate programs. YTU stands out in the environmental perspective with a total of 23 courses in its master's and doctoral programs, but it has a total of 18 in ITU master's and doctoral programs, 9 in MSGSU master's and doctoral programs, 9 in Kocaeli University master's and doctoral programs, Beykent University master's and doctoral programs 8 courses in doctorate programs and 8 courses in Dicle University master and doctorate courses [14].

In this section, the education programs of the universities' courses on environment, ecology, energy, sustainability in architectural education were questioned, and it was observed that there were program updates and course increases in parallel with the importance of the subject. However, it should not be forgotten that the main problem in architectural education, besides the way the course is given, is what knowledge or acquisition a student can receive and how much he/she can transfer this to his designs. In this context, the research is essentially a self-evaluation.

3.2 Informal Process

According to education-oriented researches, even with the most successful education programs, 30–40 % of the learning takes place through formal ways and 60–70 % of it is realized through informal ways [15]. When we look at the subject of learning in terms of architectural education; The discipline of architecture, which is a multidisciplinary, interdisciplinary perception-interpretation and production process, must be supported at different levels, through different channels and with different methods. In university education, it is assumed that the student acquires information about the profession structure, application of the profession, and the culture of the profession and that the necessary guidance is provided [15].

For this reason, it is important for the architect to participate in trainings with various events (congresses, symposiums, panels, conferences, exhibitions, workshops, competitions, etc.), especially with ecology and sustainability in the professional development process. It has been observed that there are many activities organized or supported by various branches of TMMOB Chamber of Architects, dealing with ecological, energy, environment and sustainability issues. In addition to these, the continuous professional development branch of the professional association organizes trainings and publishes in many provinces with the support of experts at universities. Mimarlık, Mimarlıkta Malzeme, Mimar.ist, Ege Mimarlık etc. It contributes to the professional development of architects by many publications

4 Opinions on Ecological Building Design and Application Practice

165 of 184 architects registered in Edirne and Keşan Representatives of TMMOB Chamber of Architects participated in this research aimed at determining an opinion on ecological architecture. The survey was conducted with one-to-one interviews by reaching 90 % of the study universe. The first part of the questionnaire is about determining the profiles of the participants and the second part of their views on ecological architecture. The first part consists of category and open-ended question types, and the second part consists of 5-point likert scale questions. Likert questions cover environmental problems, general acceptances of ecological architecture, ecological design criteria, factors related to physical and

built environment, renewable energy use titles and the impact of regulations on ecological design and individual competencies.

4.1 Evaluation of Findings

53.30 % of the architects participating in the survey are between the ages of 26–45 and 69.70 % have received undergraduate education. 19,40 % of the participants have a master's degree and 10,92 % have a doctorate education. The high rate of doctorate can be explained by the participation of Trakya University academicians in the survey. Considering the duration of the profession of architects, it was determined that the majority of architects working for over 21 years (38.20 %), (Tab. 1).

Participants have graduated from 26 different universities and Trakya University graduates are in the majority (63.60 %). This value is followed by the graduates of Mimar Sinan Fine Arts University with 6.70 %, Yıldız Technical University with 4.20 % and Istanbul Technical University with 3.60 %. 38.20 % of the participants who graduated from Trakya University graduated in 1996 and before.

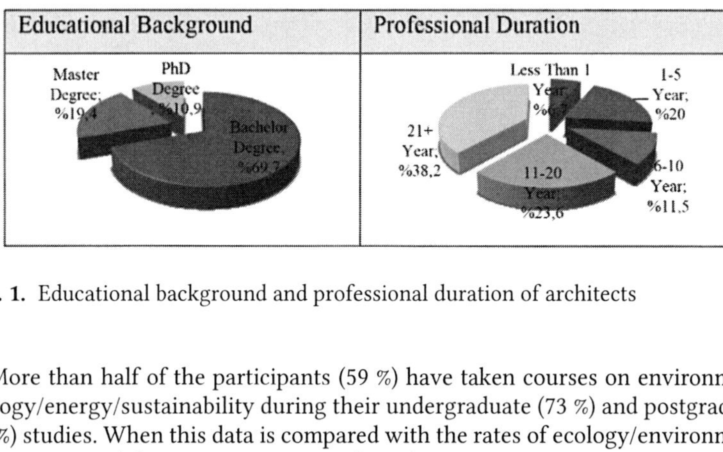

Tab. 1. Educational background and professional duration of architects

More than half of the participants (59 %) have taken courses on environment/ecology/energy/sustainability during their undergraduate (73 %) and postgraduate (27 %) studies. When this data is compared with the rates of ecology/environment/energy/sustainability courses given in formal processes in universities, it is seen that the theoretical courses in all four universities are sufficient and help architects gain basic knowledge and opinion in designing ecological buildings.

However, 79.4 % of the participants stated that they did not design an ecological building. Of the architects (20.6 %) who designed ecological buildings, 64.7 % answered "I designed it myself", 29.4 % "I designed with the group" and 5.9 % "designed in the competition" (Tab. 2) . It was learned that the participants did not design ecological buildings within the provincial borders (97.6 %) and that 1 bioclimatic building was applied within the provincial borders. These results show that the majority of the participants could not transfer their theoretical knowledge to practice and could not realize ecological design. It is thought that the reason for this may be the lack of practice within the training programs, the need for specialization and the inability to form a team.

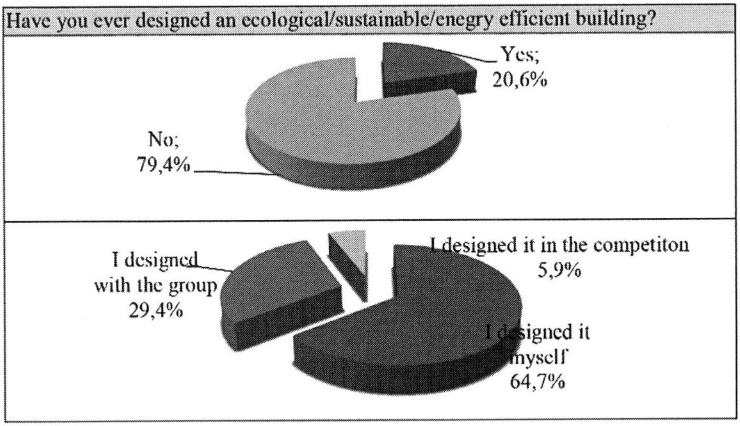

Tab. 2. Rate of architects to design ecological buildings

Within the scope of the study, it was aimed to learn how architects benefit from informal activities among the ways of informing about ecological architecture. In this context, the participants' congress, symposium, etc. It was determined that they participated in activities (61,8 %). This result is thought to be influenced by the International Sinan Symposium, which has been organized regularly every two years since 2005 by the Faculty of Architecture of Trakya University. However, it was observed that 51.5 % of the participants did not follow a specific publication. The fact that publication follow-ups are mostly done by academics, the scarcity of architects who design ecological buildings, the low following of the literature, and the inability to transfer theoretical knowledge to practice are unfortunately a finding that negatively supports the rate of ecological building in the province.

Within the scope of the first part of the survey study, besides the findings evaluations of the category and open-ended question types, significance inquiries were made between the participant profiles (age/gender) and the information acquisition questions to determine the consistency between the data. In this context, it is stated that in the relationship between age range and formal education, all of the participants took any courses on environment/ecology/energy/sustainability within their formal education, but the rate of taking lessons decreases as the age increases; Therefore, it was determined that there is an inverse relationship between age range/taking lessons. This determination is in line with the increase in the course rates of universities on environment/ecology/energy/sustainability between 2011–2016. In the relationship between age range and informal activity participation, any congress, symposium, etc. on environment/ecology/energy/sustainability issues in all age groups. The participation of young architects to the organization is high in line with the lessons they have taken in their education process; however, it was determined that the rate of participation decreases with increasing age. This determination shows that there is a direct proportion between the course increases in the undergraduate or graduate course programs of universities and the young population after 2011 (Tab. 3).

When the participants are questioned about taking any courses on environment/ecology/energy/sustainability within gender and formal education, it is seen that women are in the majority. The same situation continues with the question of gender and participation in informal activities. Despite the low number of women architects throughout the province, the high participation in informal activities can be attributed to the fact that the academicians of the Faculty of Architecture of Trakya University are predominantly women (Tab. 3.)

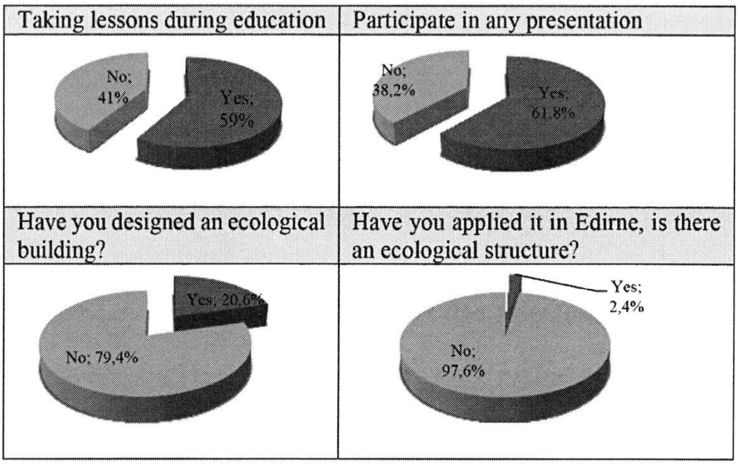

Tab. 3. Architects' taking courses, participating in presentations, designing and applying ecological buildings

The second part of the survey study consists of 5-point Likert scale questions to determine the views of architects on ecological architecture and building design. 48 questions were prepared within the conceptual framework of ecological architecture, reducing from general to specific. In this context, it is understood that the participants have a pluralist view of the environmental problem and share the common view that ecological architecture can create a solution to reduce these problems. At the same time, regarding ecological architecture; They have a common view when it is necessary to adapt to the environment, to create self-sufficient buildings as well as useful living spaces. Designers agreeing that climate-compatible/bioclimatic designs are a prerequisite for ecological architecture think that the number of energy-efficient buildings should increase (Tab. 4).

<table>
<tr><td>There are energy and environmental problems in your century.</td><td>Ecological building designs are important for sustainability.</td></tr>
<tr><td>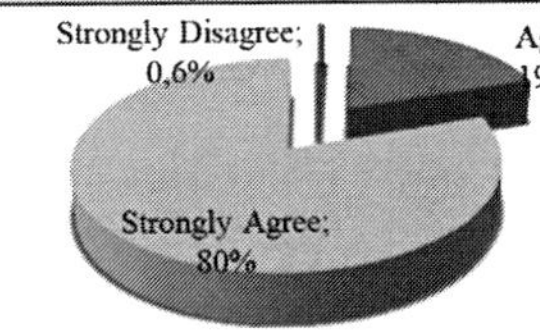
</td><td>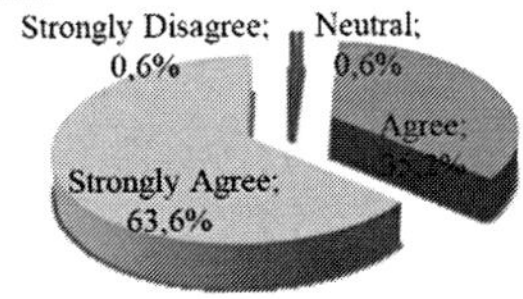
</td></tr>
<tr><td>Ecological approaches can solve energy and environmental problems.</td><td>Ecological buildings should be designed to be self-sufficient.</td></tr>
<tr><td>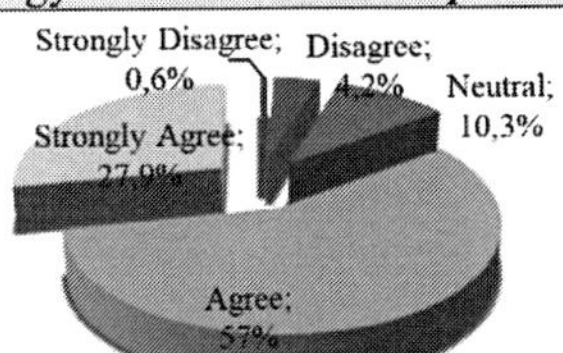
</td><td>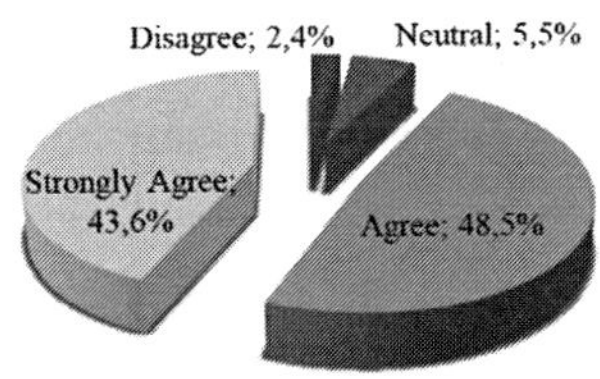
</td></tr>
<tr><td>Climate-compatible building designs are pioneers of ecological designs.</td><td>Nowadays environment-friendly energy efficient building designs need to increase.</td></tr>
<tr><td>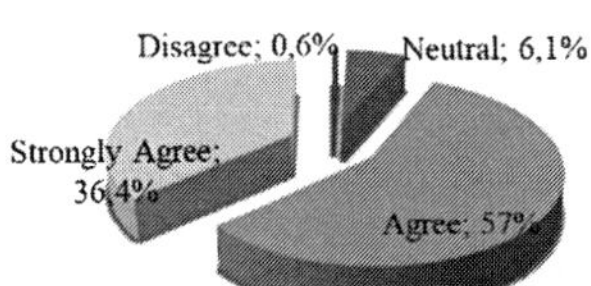
</td><td>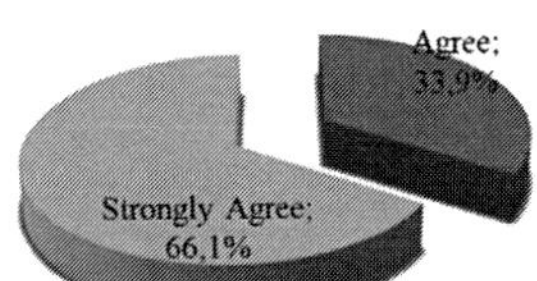
</td></tr>
</table>

Tab. 4. Architects' responses to likert questions about ecological architecture.

Participants included physical environmental factors (land conservation, integration with the place, harmony with the topographic structure, importance of determination of land data, location, orientation, knowledge of basic climate data, use of landscape, etc.) among the ecological architecture criteria. They have a good opinion. They also agree that the criteria of form, surface area, crust formation, material selection and recyclability are important in creating indoor comfort conditions after the control of external climatic conditions (Tab. 5).

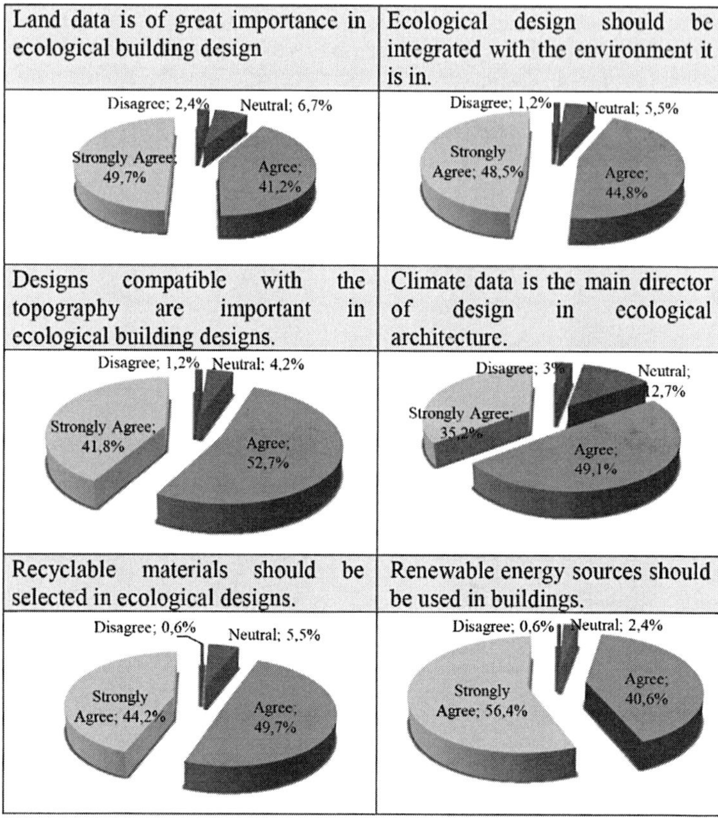

Tab. 5. Architects' responses to likert questions on ecological architectural criteria

Participants stated that the knowledge they gained on "ecological building design" during the undergraduate period of formal education processes was not sufficient (70.3 % agree and strongly agree), (Tab. 6). This finding does not coincide with the number of ecology comprehensive courses in educational processes. The primary reason for this contradiction is, as stated before, qualitative content rather than quantitative numbers, methods of transferring knowledge and individual differences in learning outcomes. The secondary reason may be that courses on environment/ecology/sustainability have been actively involved in the undergraduate programs of universities after the 2000s. Because the findings show that informal processes became prominent in informing the graduate architects about the subject before 1996. In addition, it is seen that the architects with 11–21 years of professional experience completed their undergraduate education between

1997–2006 and some of the architects were partially informed about ecological architecture. Although it is seen that architects, who have been working for 6–10 years in 2007 and after, effectively took courses on environment/ecology/sustainability/energy at undergraduate and graduate levels, the inadequacy of transferring the theoretical knowledge acquired in the courses into practice;

- Lessons are theoretically weighted,
- Limitation of application opportunity at the undergraduate level,
- The way the person acquires and processes information,
- It is estimated that the multi-component structure of ecological architecture is due to reasons such as difficulties in transferring to design (knowledge, interdisciplinary work, regulation, customer, etc.).

Architects agree that it is necessary to collaborate with other professional groups in terms of the multi-component ecological building designs. At this point, the architects did not find the criteria for designing ecological buildings as restrictive in terms of their designs. On the contrary, they accepted the necessity of these and stated that environmental problems may decrease especially with ecological designs (Tab. 6).

In addition, the architects, who evaluated their personal knowledge by making a self-assessment, stated that they needed more information and expert support in order to design ecological buildings within the scope of ecological architecture (Tab. 6). At this point, it is seen that the information acquired in formal and informal processes is insufficient at the point of designing ecological buildings that require expertise. This actually reveals the need for today's interdisciplinary cooperation of the architectural profession.

It is seen that architects question the energy-efficient approaches in their design methods and pay attention to the direction criteria in the relations between functions, especially in the organization of space. However, it is understood that the architects do not have sufficient application knowledge especially on renewable energy applications and accordingly they cannot apply active, passive or mixed methods in their designs. Therefore, architects do not find themselves competent in ecological architecture and ecological building applications (Tab. 6). This result overlaps with the rate of architects not applying ecological buildings (97,6 %) within the provincial borders. In order to eliminate this negative cycle, "Sustainable/Ecological Design" internship can also be suggested in addition to increasing the applied courses during undergraduate education.

Although the participants found ecological buildings applicable, they stated that laws and regulations do not support ecological architectural designs (Tab. 6). Because cliché zoning plans and construction provisions (towing distances, roof slopes, etc.) that do not address the values of the settlements within the scope of ecological integrity, direct the designers to the mathematicized mass organizations. Especially, withdrawal distance, fixed roof slope, etc., which prevent using the sun through passive, active and mixed methods. provisions need to be reconsidered.

In undergraduate education, sufficient information is not given on ecological building design.	The ecological building concept is not really applicable.
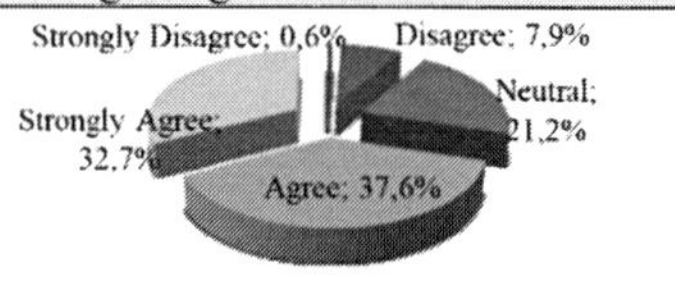	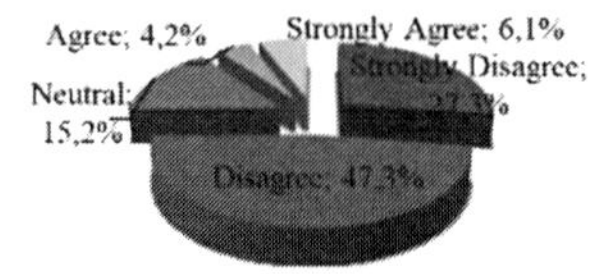
Laws and regulations adequately support ecological building design.	The criteria for ecological building design restrict the designer.
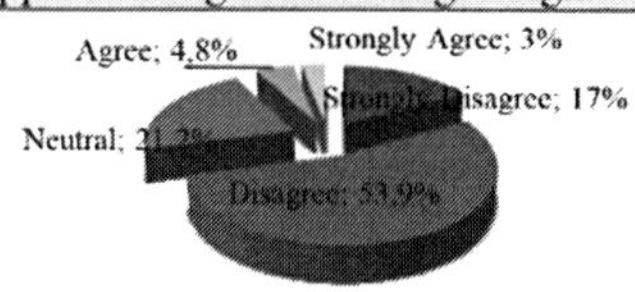	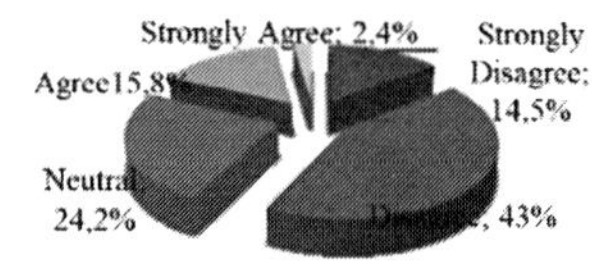
I need more information to be able to design ecological buildings.	Expert support is required to applying an ecological building.
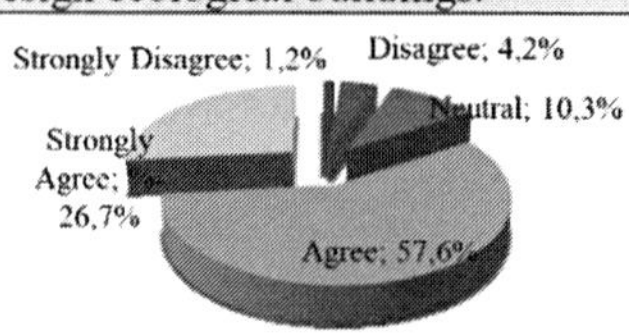	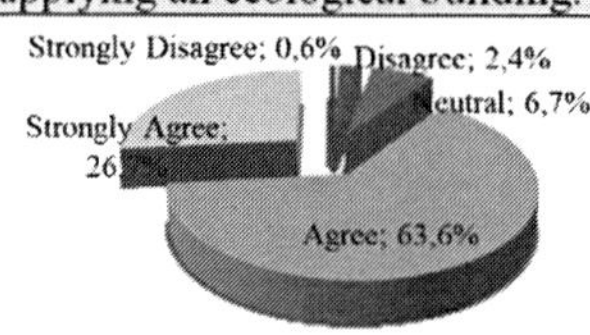
I have enough knowledge about renewable energies and their applications.	The lessons I took in my educational background helps me design ecological buildings.
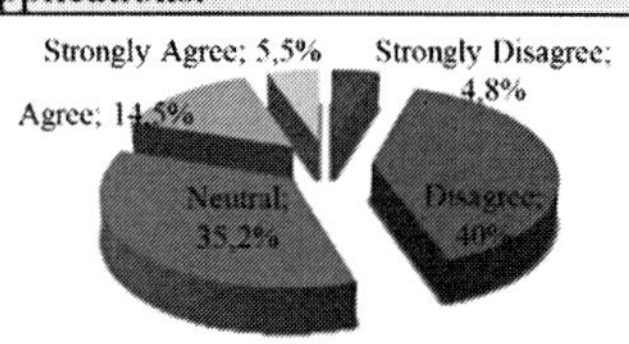	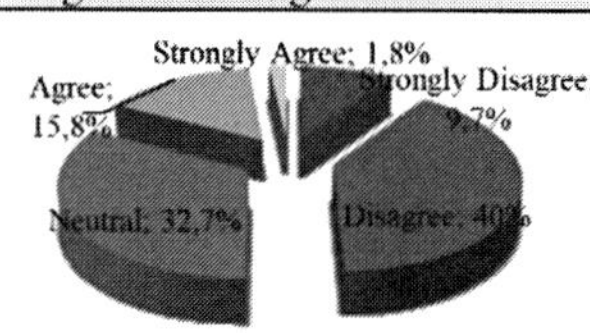
I have enough knowledge about ecological building design.	
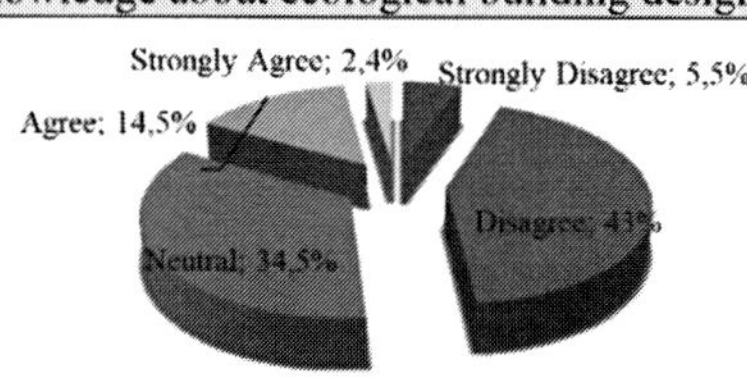	

Tab. 6. Architects' responses to likert questions on ecological architectural competencies

5 Results and Recommendations

The architectural profession, which entered the 21st century with the acceptance of sustainability and ecological architecture, re-examines its formal and informal processes within this scope. First of all, undergraduate and graduate level courses covering environment/ecology/sustainability/energy were added to the educational programs of universities. Afterwards, the awareness and knowledge of individuals and consequently architects were increased through global, regional and local-scale organizations and activities covering environmental awareness, recycling, energy saving and ecological construction.

This study aims to determine the views of architects on ecological design as decision makers who organize the built environment; As a factor that creates and improves the design and application abilities of architects, it has been the subject of formal and informal education processes. In formal education programs, it is understood that candidate architects are given compulsory or elective courses on environment/ecology/energy/sustainability at undergraduate and graduate levels. Lack of application has been identified in the training programs that have been determined to have a theoretical infrastructure. It was concluded that the lectures with theoretical content helped the architect gain basic knowledge in terms of ecological building design, but it was not sufficient. Because architects did not consider themselves competent in applying ecological buildings. This indicates that applied courses on ecological building design at undergraduate or graduate level should be included in formal processes. As a suggestion to overcome this problem, a special project course that can take place in the third or fourth grade can be proposed in addition to the "Sustainable/Ecological Design" internship.

It is important for architects to participate in various organizations on a global, regional or local scale in order to increase their personal knowledge and application experience, especially after undergraduate education. The acceptance that vocational education should continue lifelong shows the importance of informal processes. In the near future, it is foreseen that the applications related to sustainability/energy-efficient design/ecological building design will not be made by architects who do not have a certificate or have not specialized in the field. At this point, it is imperative that architects make their informal processes effective. The survey study also showed that architects should increase their informal efficiency.

Healthy cities, healthy buildings and all kinds of positive approaches that will protect and develop human existence with their physical and socio-cultural structure are important for the common future. One of these approaches is the zoning plan provisions that will direct and support the designers to design ecological buildings. The architects who participated in the study were at the point of designing ecological buildings in Edirne-specific zoning plan conditions (the constant distance between buildings prevents the use of the sun, zoning plan construction conditions incompatible with the topography reveal type block formations, opening agricultural lands, inconvenience of the existing roof slope for active systems, etc.) They stated that they did not support them. The understanding of construction of a city like Edirne that developed on 1st class agricultural lands should be questioned again in the name of ecological architecture. This issue is a macro-scale problem discussed throughout the country; Construction provisions that ignore topography, insist on fixed distances between buildings, or support peer increases for rent purposes will never be related to ecological architecture.

The effective realization of the proposals developed within the scope of the study depends on the ability of the architects who organize the built environment to transfer their knowledge of ecological architecture to their designs competently and to apply them. This study reveals that although the architects have a positive opinion on ecological architecture and ecological building design, they have serious difficulties in applying their designs.

References

1. Eryıldız, D. I., *Sürdürülebilirlik ve Mimarlık Dosyasında Ekolojik Mimarlık*, 2003, Arredamento Mimarlık Dergisi, **154**, p. 71–75, İstanbul.
2. Ciravoğlu, A., *Türkiye'de Mimarlık Eğitiminin Çevresel Bağlamı: İyimser Tabloya Kuşkucu Sorgulamalar*, 2011, Mimar.ist Dergisi, **41**, p. 52–60.
3. Yılmaz, M., *Mimarlık ve Çevre*, Çevre ve Politika: Başka Bir Dünya Özlemi, Editör: Ayşegül Mengi, 2007, İmge Kitabevi Yayınları, p. 75–92.
4. Yeang, K., *Ekotasarım: Ekolojik Tasarım Rehberi*, Çev: Semih Eryıldız, Demet Eryıldız, 2012, Yem Kitabevi, ISBN 9789944-757-64-5, 1. Baskı, İstanbul,).
5. Schittich, C., *In Detail, Solar Architecture, Strategies, Visions, Concepts*, 2003, Kösel GmbH&Co. KG, Kempten.
6. Van der Ryn, S., *Ecological Design*, 1996, Island Press, Washington D.C., p. 26–28.
7. Orr, D., *Architecture, Ecological Design and Human Ecology*, The Green Braid, Ed. By Kim Tanzer and Rafael Longoria, 2007, Routledge, Great Britain, p. 15–33.
8. Olgyay, V., *Design with Climate*, 1963, Bioclimatic Approach to Architectural Regionalism, Princeton University Press, USA, p. 1–13.
9. Yannas, S., *Towards Environmentally Responsive Architecture*, 2003, PLEA 2003, Chile, p. 68.
10. Edwards, B., *Sürdürülebilirlik Kültürü ve Mimari Tasarımın Önündeki Güçler*, 2007, Ekolojik Mimarlık ve Planlama Ulusal Sempozyumu, Antalya, p. 22–34.
11. Tönük, S., *Bina Tasarımında Ekoloji*, 2001, YTÜ Yayınları, Yayın No Mf. Mim-01.005, YTÜ Basım-Yayın Merkezi, İstanbul, p. 4–105.
12. Kısa Ovalı, P., *Türkiye İklim Bölgeleri Bağlamında Ekolojik Tasarım Ölçütleri Sistematiğinin Oluşturulması Kayaköy Yerlesmesinde Örneklenmesi*, 2009, Yayınlanmamış Doktora Tezi, Trakya Üniversitesi Fen Bilimleri Enstitüsü, Edirne.

13. Türk Dil Kurumu, *Eğitim*, 1974, BSTS/Eğitim Terimleri Sözlüğü, (http://www.tdk.gov.tr/index.php?option=com_bts&view=bts&kategori1=veritbn&kelimesec=107568, Erişim Tarihi: 11.05.2017)
14. Motor, C., *Mimarların Ekolojik Mimariye Ilişkin Görüşlerini Belirlemeye Yönelik Ampirik Bir Çalışma: Edirne Örneği*, 2017, Yayınlanmamış Yüksek Lisans Tezi, Trakya Üniversitesi Fen Bilimleri Enstitüsü, Edirne.
15. Barkul, Ö., Tuncer, E., *Mimarlık Eğitiminde İnformel Çalışmaların Önemi: Bir Atölye Çalıması Deneyimi Ve Öğrencilerle Paylaştıklarımız,* 2007, Tasarım Eğitimi Semineri '07 Kitabı, Mimarlık Fakültesi, Çevre Düzenleme Bilim Dalı, Yıldız Teknik Üniversitesi, Sayı No: MF.MİM.2007.002, İstanbul, p. 99–108.

Ezgi ÜLKER BARIŞ

Negative Reflections of Political Decisions on Design in Public Spaces

Abstract: Squares and parks are among the primary public spaces which establish a city's identity and determine the life quality in that city. Public spaces are places where citizens communicate and interact with other socio-economic groups, take part in activities; therefore, they are places in which citizens' life quality is affected. Public spaces' qualities are as significant as their quantities. Local governments have power over the planning, designing and building public spaces. The projects local governments get designed without determining their qualities and the revision decisions they take during the construction stages cause public spaces to be insufficient and of poor quality. Designing and construction of the square, which is a public space, is examined considering the local government's and contractor firm's decisions with the example of "Süleymanpaşa Multi-Storey Underground Car Park and Square Project", located in Tekirdağ Province. The differences between the project offered to the public and the project construction of which is completed have been compared, and how the revision decisions have affected the project has been analyzed. The aim of this study is to find out the importance of the decisions taken by local governments in the planning, designing and building stages of public spaces.

Key Words: Local Government, Public Spaces, Construction of Public Space, Spatial Quality, City Square

1 Introduction

Public spaces not only allow citizens life quality to improve but also define the city. Although many philosophers, sociologists and urban planners have different opinions on public spaces, what they have in common is that public spaces are meeting places which are open to everyone and in which we can be seen [3]. When we classify public spaces as open spaces and indoor spaces, places such as parks and squares are public open spaces while public institutions such as municipal buildings and tax offices are indoor spaces. These public spaces form an important part of settled life, providing spaces for settlement, entertainment and recreation [2, 19].

Squares which are a type of public open spaces are the most important components of a city. A square of high quality improves the life quality of the citizens and is important in terms of the legibility of a city.

The only authority in terms of planning of public open spaces is local governments. Local governments are responsible for city planning considering the future periods, designing public spaces which the public in the city can benefit from and the construction of the spaces they design. However, in local governments, "quality" projects cannot be created as a result of the issues during the tender process, changes in the authorities of the people in the liable units for the project, or lack of experience. Moreover, the mechanism of authority failing to function properly and the interference during the construction make this issue worse. This situation has been analyzed considering "Süleymanpaşa Multi-Storey Underground Car Park and Square Project", located in Tekirdağ Province. The differences between the designed project and the constructed project have been compared. The aim of this study is to emphasize the significance of the authorized local governments' decisions and responsibilities during the designing and construction of these public spaces belonging to the citizens.

2 The Square as a Public Space

The discussions about the concept of public space started after the second half of the twentieth century in Europe, and many researchers have handled this topic with different opinions. While Jurgen Habermas[1] defines it as the space of freedom to which everyone can access and where common interests are discussed and the truth is found, Hannah Arendt[2] says that public space is an area which can be seen and heard by everyone and is open to everyone. In Turkey, according to Turkish Language Association, the concept of public space is defined as "the place which belongs to the public and where the work regarding the public is done" [12]. And, İlhan Tekeli refers to it as an appearance and meeting place [3, 11].

Public space is necessary for the public world to be continuous, understandable, accessible and meaningful, and for the protection and the maintenance of the common heritage [9]. Public spaces are individuals' fundamental means of sharing their wishes, expectations and defending and sharing their rights.

Christian Norberg-schulz[3] defines city as a greeting place consisting of public spaces where there are different people [3]. In the view of this definition, the public spaces in a city have to be of good quality for the city to exist.

Squares, a type of public open spaces, are the most important part of a city. Square meaning an open place or a broadened street is derived from the words "Platea" in Latin, "Place" in English and French, "Plaza" in Spanish and "Piazza" in Italian [7, 10].

1 Jurgen Habermas: Philosophy Professor, from his work titled "The Structural Transformation of the Public Sphere" (Strukturwandel der Öffentlichkeit), published in 1962 [4,5].

2 Hannah Arendt: Political Scientist, from her work titled "The Human Condition", published in 1958 [1, 4].

3 Christian Norberg-schulz: Architect, from his work titled "Genius Loci: Towards a Phenomenology of Architecture", published in 1980 [8].

In the history of Turkey, even though constructions of squares are not common, squares have a great significance in defining the city [6]. When these spaces are looked at in an upper scale, definition of the space can be made, and in human scale, they allow people to meet and find other places. The correct planning of squares allows building blocks to be constructed properly; thus, the city gains architectural and aesthetic value. The quality of a square is evaluated according to its own physical structure as well as what it adds into the lives of the citizens. Like all public places, squares are spaces where citizens find activities, communicate and interact with different socio-economic groups; consequently, they improve citizens' life quality.

Even though designs of squares can be different from each other according to the climate, nature and culture, which are specific to the city, there are certain main criteria for squares in terms of quality. Although whether squares are important in societies which are modern and have extremely high communication skills or not is discussed today, urban areas still have importance for citizens.

3 Local Governments' Responsibilities in Public Spaces

The authority in planning, design and construction of public open spaces is local governments. According to the articles in the Municipal Law No. 9469[4] and the Metropolitan Municipality Law No. 8901[5], local governments have authority to create and change master plans, and to implement and supervise these master plans. They also have authority and responsibility in the building of infrastructure, dwelling, and cultural and social spaces.

When the design processes of public spaces are analyzed, it is seen that some local governments create projects with design groups working in their bodies whereas some other local governments get support from architectural design offices. This support can be found by announcing a tender and accepting the winning firm's design as well as by organizing a competition open to everyone.

That the work is done especially with the firm demanding the least amount of money, regardless of the quality, creates concerns about the outcome of the design. The local government which enters into process of design with the firm demanding an economical payment designs the public space based on only one architectural idea and work. However, for a quality design, it is required that there are certain specializations and ideas, and that these ideas are discussed. In addition, while designing construction projects, project originator is responsible for

---

4 Municipal Law No. 9469, Law Number: 5393 Adoption Date: 3/7/2005 Official Gazette it was published: Date: 13/7/2005 No: 25874 Code: Array: 5 Volume: 44 (Article 18 & Article 69)[13].

5 *Law on Metropolitan Municipalities No. 8901, Law Number: 5216 Adoption Date: 10/7/2004 Official Gazette it was published: Date: 23/7/2004 No: 25531 Code: Array: 5 Volume: 43 (Article 7/g) [14].*

designing the project in a way that includes all the details and information on the material without any possibility that there can be an uncertainty during the construction. The firm winning the tender has to precisely follow the project. The changes during the construction can only be made with the approval of the project originator[15]. However, some local governments buy the project design and the intellectual property rights with regard to the project from the project originator. This situation is highly undesirable in terms of quality work. In this circumstance, in local governments, the decisions taken during the construction stages can sometimes be wrong as a result of the changes in the authority of the officials at the department responsible for the projects, and the lack of experience. Consequently, structures which are completely different from the designed projects and of low quality are created.

4 The Assessment of the Reflection of Political Decisions on Design Considering the Example of "Süleymanpaşa Multi-Storey Underground Car Park and Square Project"

Tekirdağ Province gained metropolitan municipality status in 2012. Tekirdağ Center became a district named Süleymanpaşa after 2012. After gaining metropolitan municipality status, the city has entered an urban renewal process. City's development process has gained speed; new subcenters have started to be built. Most of the public buildings have started to move to their new buildings constructed in the subcenters developed recently, and the old city center has become a part of a new landscape plan preserving the cultural properties.

Before becoming a metropolitan municipality, Tekirdağ Municipality had announced a tender for the design of the square project. The visual work prepared for the concept project has been shared with the public during the construction stage (Fig. 1). The contract for the construction was put out to tender after the Metropolitan Municipality had been established. The project tendered under the title of 'Süleymanpaşa District Underground Car Park and ÖPA-1[6] Square Design"[7] in 2018 was completed and opened for use in July 2019.

6 ÖPA-1: It is an abbreviation for "Özel Proje Alanı" which means "Special Project Site" in master plans in Turkey.

7 EKAP*, tender registration number: 2018/88762 [16].
It is the institution lodging tenders. EKAP stands for "Electronic Public Procurement Platform".

Fig. 1: Square Project, Concept Project, 3D Visual Work (Photo: Tekirdağ Metropolitan Municipality Archive, 2019)

However, when the completed project and designed project are compared, it is found out that there are many differences which will affect the quality of the public space negatively. It is unknown in what stage and by whose decisions these differences have occurred.

When we examine the completed form of the square design, at first, it is seen that the trees in the concept project are not planted (Fig. 2). There is not a place for the trees to be planted on the pavement in the project the underground car park is built. Because of the fact that roots of trees expand as the tree grows, the problem of the planting area could not be solved by flooring. However, when the older photographs of the project site are examined, it can be seen that there are dense groups of trees (Fig. 3). According to the information received from the Tekirdağ Metropolitan Municipality Website, 35 trees and 1000 shrubs have been planted [17]. After the destruction of the existing trees was criticized by the public, the solution was found in adding plant pots and trees which were not matching the project, and in planting flowers in crates (Fig. 4).

Fig. 2: General View of the Square (Photo: ÜlkerBarış, 2019)

Fig. 3: The City Hall and the Project Site in 1990s(Photography: Anonymous)

When we think of the square as an activity space, the fact that it is treeless is beneficial for that; however, that the broad firm ground has not been done correctly has been disadvantageous. The pavement seen in a grid structure in the concept project has not been constructed during the construction (Fig. 5).

Fig. 4: Flowers Planted in Crates and the Trees in Plant Pots (Photo: Ülker Barış, 2019)

Another important issue is that the exits of the car parks have not been built at the same height and with the same material as in the project. Three different exits from the car park to the elevation of the square have been designed. These exits were named (A), (B) and (C) (Fig. 5). A machine roomless elevator has been used in the project. Because of the condition that the clear height of the top floor must be 4.10 meters for the machine roomless elevators, the heights of the structures providing exits from the car park have been increased. Furthermore, (A) and (B) exits have been designed at the points where the elevation is at its highest level. As a result of these two aspects, the structures' negative effect on the silhouette of the historic governor's building has increased.

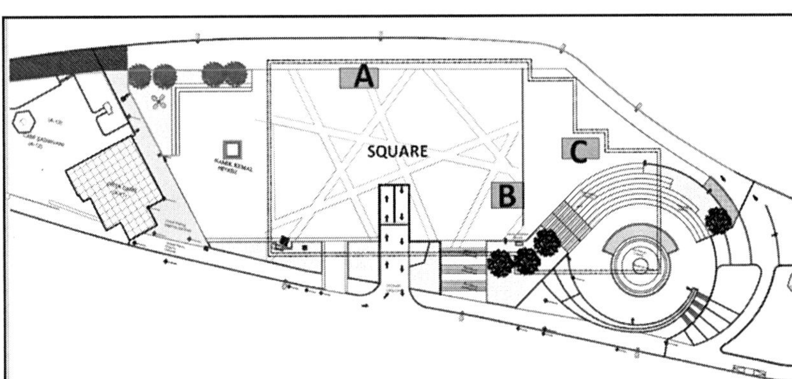

Fig. 5: Square Project Layout and Car Park Exits (Drawing, received from Tekirdağ Metropolitan Municipality and simplified. Illustration, Ülker Barış, 2019)

The car park exits which do not spoil the view of the Historic Governor's Building in the project images and illustrations do stand in front of the historic structure in the completed project (Fig. 6, Fig. 7).

Because the Governor of Tekirdağ demanded that the car park exit (A), spoiling the historic silhouette, was removed as it created "visual pollution," it was removed by the Tekirdağ Metropolitan Municipality in August 2020. The structure ended up being unsuitable because the design which was proper in the design phase was changed during the construction, and the structure was removed as a result of the reaction. The amount of the financial loss occurred due to this wrong decision is unknown.

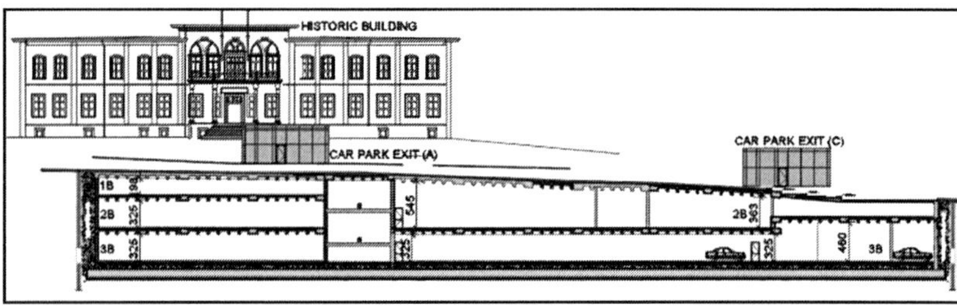

Fig. 6: Floor Heights and the Effect of Car Park Exits on the Silhouette of the Historic Building (Drawing, received from Tekirdağ Metropolitan Municipality and simplified. Illustration, Ülker Barış, 2019)

Fig. 7: The Historic Governor's Building under Restoration and the Car Park Exit (A), spoiling the scene (Photo: Ülker Barış, 2019)

The material used in the car park exit structures is not the same as the one in the concept project (Fig. 8). The wall and the ceiling were made completely of cladding glass (Fig. 8). Facade window was preferred in a dark color. However, the fact that the entire wall and the roof are made of glass causes a greenhouse effect. When the weather is sunny, it is very hot and airless inside the building, which does not fulfill the comfort requirements. A material that would fit the nature and the environment and would not spoil the view was chosen as the siding material in the concept project. However, the facts that the size of the structure was enlarged and that dark-colored glass was chosen as the material have caused a bad view.

Fig. 8: Concept Project and the present elevator exits (Photo: Ülker Barış, 2019)

One of the elements which affect the users the most and guide them is flooring. With its pattern, texture and color, the flooring acts as a carpet which identifies the place for the pedestrian. In addition to being aesthetic, it affects the user with its functionality and comfort. The texture of the flooring guides the pedestrian's movement and activity choices; it determines the direction and prevents them from entering private spaces [10]. When the floor covering of the project is examined, it can be seen that there are some linear patterns on the floor; however, these lines do not have any guiding role (Fig. 5).

Andesite paving stones with the dimension 30cm x 30cm x 4cm have been used for the flooring. They have been used in two different colors in the project. However, the colors chosen and the pattern are insufficient for guiding. Even though the project has been completed recently, andesite paving stones are covered with mold in patches and have deteriorated (Fig. 9). The ends of the stones and the job of combining different materials have been done with poor workmanship (Fig. 11). Storm drains have not been placed according to the symmetry of grout. The stairs built between the elevation difference have an image that seems to be added later because the steps are at different heights, and they do not harmonize with the rest of the project (Fig. 10).

Fig. 9: The deterioration and mold on Andesite Paving Stones (Photo: Ülker Barış, 2019)

Fig. 10: Stairs & Storm Drains (Photo: Ülker Barış, 2019)

Fig. 11: The Details of Combination of Different Floor Covering Materials (Photo: Ülker Barış, 2019)

5 In Lieu of Conclusion

Public spaces are very important for the citizens. Among these public spaces, squares, as we have analyzed, have great importance for the planning of a city, and its history as well as its citizens. While squares are being planned, what should be considered is not only the present but also the future because a square which is built today is going to be a space lasting for generations like many examples we see in Europe, accumulating historical traces.

The study we have done considering Tekirdağ Süleymanpaşa Multi-Storey Underground Car Park and Square Project aims to examine this project and the problems related to such projects. Fast construction process and quick revision decisions lead up to products of low quality. As we have seen in the project, wrong design decisions and wrong choices of materials not only cause the expenses and labor to go for nothing but also force the citizens to live around a design of poor quality for a long time. As an alternative, because the design is of low quality, there will be a renovation process conducted by another local government, and this will cause an excessive waste.

Even though the owners of these places are citizens, the authority in terms of design is local governments. Local governments are responsible for constructing these spaces on behalf of the public. However, the process does not run smoothly in our country because of the work distribution in the administrative bodies and the problems resulting from tender system. Furthermore, transferring the intellectual property rights has negative effects as regards the outcome of projects. For instance, in Germany, the public chooses its structures via contests if possible. Otherwise, projects are tendered and chosen by the tender procedure. In this case, the result of the tender is determined by the scoring system for the offices. While the office's work and portfolio are a significant determinant of its score, the project cost is the least influential criterion. As the project cost is calculated according to the legal cost, there can be minor changes in the tender cost [18].

A certain process model should be prepared for the design and the construction of these spaces in our country as well, and production of designs of poor quality should be prevented. Unfortunately, local governments value the quantity of designs more than the quality. For this reason, local governments should be more considerate in terms of the designs they create on behalf of the citizens and adopt the idea that doing "the best" is better than doing more work. They should establish a system controlling the projects and should not avoid acting on the advice of experienced people and experts. They should be meticulous about the good workmanship which is a problem in the country, and they should realize the importance of creating a perfect product.

References

1. Arendt, H., *İnsanlık Durumu,* translation: Bahadır Sina Şener, Publications of İletişim, 1994, İstanbul.
2. Carr, S., Francis, M., Rivling, L. G., & Stone, A. M., Publicspaces. Cambridge: Cambridge University Press, 1992.
3. Erdönmez, M.E. & Uzgören, G., *Kamusal Açık Alanlarda Mekân Kalitesi ve Kentsel Mekân Aktiviteleri İlişkisi Üzerine Karşılaştırmalı Bir İnceleme,* Megaron YTU of Architecture Faculty, E-Magazine, 2017. 12(1): p.41–56.
4. Gürallar, N., *Kamu - Kamusal Alan - Kamu Yapıları - Kamusal Mekân: Modernite Öncesi ve Sonrası için Bir Terminoloji Tartışması,* Journal of Architecture, vol:350, November - December 2009.
5. Habermas, J., *Kamusallığın Yapısal Dönüşümü,* translation: Tanıl Bora, Mithat Sancar, Publications of İletişim, 1999, İstanbul.
6. Kuban, D., *Kent ve Mimarlık Üzerine İstanbul Yazıları,* Publications of Yapı Endüstri Merkezi, 1998, İstanbul.
7. Marcus, C. C., Francis, C., *People Places 'Design Guide lines for Urban Open Space'.* Van Nostr and Reinhold Company, 1998, New York.
8. Norberg-Schulz, C., *GeniusLoci: Towards a Phenomenology of Architecture,* 1998, New York, Rizzoli.
9. Önal, F., *Kamusal bir alan Olarak Parkları Yeniden Düşünmek,* Journal of Mimarist, 2014. Vol(1): p.27–33.

10. Önder, S. & Aklanoğlu, F., *Kentsel açık mekân olarak meydanların irdelenmesi,* Selçuk University, journal of Agriculture Faculty, 2002. 16 (29): p.96–106.
11. Tekeli, İ., *Önsöz, Katılımcı Demokrasi, Kamusal Alan ve Yerel Yönetim,* Ed.: A.Y. Sarıbay, H.B. Kahraman, E. F. Keyman, İstanbul, Demokrasi Kitaplığı, 1999, p. 8--10.
12. URL1: *[*http://www.tdk.gov.tr/*]* (22.10.2020).
13. URL2:*[*https://www.mevzuat.gov.tr/MevzuatMetin/1.5.5393.pdf*]* (25.11.2020).
14. URL3:[https://www.mevzuat.gov.tr/mevzuat?MevzuatNo=5216&MevzuatTur =1&MevzuatTertip=5*]* (25.11.2020).
15. URL4:*[https://www.resmigazete.gov.tr/eskiler/2005/06/20050602-4.html]* (08.01.2021).
16. URL5:*[*https://ekap.kik.gov.tr/EKAP/Ortak/IhaleArama/index. html*]*(08.01.2021).
17. URL6:*[*http://www.tekirdag.bel.tr/birim_haber_detay/27512*]*(08.01.2021).
18. URL7:*[*https://www.arkitera.com/haber/almanyanin-yarisma-kulturu-masaya-yatirildi/*]* (03.11.2020).
19. URL8:*[*http://www.beylikduzukentbellegi.com/Icerik/Goruntule/469*]* (11.01.2021); Erdönmez, M.E. & Akpınar, A.Ö.,*Kentsel Kamusal Alanlar – Beylikdüzü Örneği,* 2017.

Alper SAĞLIK, Busenur YAMAN, Fatma AKBULUT, Şura YILDIRIM, Tuğçe ÖZDEMİR

Examining the Visual Quality in Environment Choices: The Sample of Coastal Usage in Çanakkale (Between Donanma Tea House and Military Base)

Abstract: Human-beings vitally need recreational areas in cities due to overpopulation and urbanisation. The coastal sides in some cities become the opportunity both for those getting rid of miscellaneous urbanisational matters and building infrustructures and facilities for the possible recreational areas in there. The tourists using coastal sides initially describe these parts with their visual perspectives. These perspectives create various demands and desires on tourists so as to opt for using coastal regions over again. Therefore, visual quality is significant on coastal parts of cities. In this paper, the regions between Donanma Tea House and Military Base in Çanakkale are separated in four zones and each part is not only researched for study but 50 people living in these zones are also surveyed with questionnaire method. Study tried to find that how visual quality effect of coastal areas influences the demands and desires of tourists, why visitors enjoy to spend time on coastal lines and this paper may provide a help for next similar works.

Keywords: Çanakkale, Visual Quality, Urbanisational Landscape, Coastal Line, Coastal Side

1 Introduction

Human-beings have tent to find a place to live and change it according to their needs throughout history. Their these efforts have resulted in both positive and negative development on environment in this period [1]. Along with Industrial Revolution, overpopulation and industrial/technological improvements are increased in urbans; therefore, unplanned urbanization and the growing beyond control occur in cities. These elements both physically and mentally affects residentials in negative ways [2].

Nowadays, cities are in a change because of overpopulation and constantly developing technology. Hence, It is fact that environment had better be altered in order to improve the welfare of individuals [3]. The conditions between people and environment are vital to have a correct change.

The facilities keeping the residents and tourists away from such negative elements as congestion and communal problems in urban are important. Recreational areas and waterfronts can be reasonable samples for these facilities. Especially, the coastal sides where recreational facilities are built can be better opportunity so that tourists have leisure and cities are visually developed [4]. Coastal sides not only visually but also financially have contributions for cities. Some infrastructures including transportation, tourism, trade and cultural events on coastal sides provide economical inputs for regions. Using coastal sides giving maximum benefits is also important to bridge the links between urban and waterfronts since transportation opportunities, recreational facilities, values and visual impacts on coastal sides are likely to shine the charisma of cities.

Human-beings vitally need recreational areas in cities due to the problems hailing from overpopulation and urbanisation. The coastal sides in some cities become the opportunity both for those getting rid of miscellaneous urbanisational matters and building infrustructures and facilities for the possible recreational areas in there. The tourists using coastal sides initially describe these parts with their visual perceptions since visual perceptions are likely to create various demands and desires on tourists so as to opt for using coastal regions over again. Therefore, visual quality is significant on the coastal sides of cities. The landscape works on coastal sides in order to develope visual quality are possible to create new demands and desires and they may inspire the other studies.

1.1 Environment and Choice

Environment is the total of all the factors which constantly interact with each other and affect natural, physical and social elements interacting with each other in a system containing artificial habitats and creatures [5]. The environment changed constantly and having a dynamic structure is formed according to daily needs and external reasons; thus, it is understood that the description of environment containing all natural, social and artificial habitats is to detect, understand and interpret a dynamic system involving various elements [6]. There are plenty of values like culture, nature, aesthetics, history etc. which highlight physical environment [7]. Environment has a complicated structure related with natural and physical elements, climatic and geographical reasons, cultural, communal, social and financial features partly or completely [8].

Human-beings have changed and regulated the environment due to such reasons as protecting themselves from natural factors, sheltering and consumption. In order to cope with sheltering needs, they have built closed places and they have used solid in order to manage their consumption and trade for many years and as a result, urbans and towns are built. Residents usually demand to alter their habitat; hence, this attempt leads to either positive or negative results on environment. That is why, the choices are important because both environment and choice mostly trigger to come up physical and natural environment and the perception toward environment [6].

Choice is a result hailing from the pleasure feeling. On one hand, the choices of individuals have been associated with their needs throughout history [9]. On the other hand, the picked places are those related with the needs of people and providing maximum benefits for them [10]. As a result, physical elements and visual quality play an important role on the stages of choosing a habitat.

1.2 Visual Quality

Vision is the combination of emotional and logical detection in mind when people use their eyesight [11]. Visual perception is the knowledge learnt from around via main senses [12] and human-beings use their senses so that they correctly detect environment. Among them, the most frequently one people use is visual perception. Therefore, visual perception commonly manipulates the decisions individuals make. For instance, visual perception has an impact on a location, a recreational area on that location, the duration people spend in that area and those using that recreational area [13].

In order to have better results from the studies related with visual quality, multidisciplinary works should be understood well. Initially, the description, meaning and purpose of landscape branch should be specified so as to detect visual quality [6].

Landscape is the picture which is visually produced with topographical, natural and cultural values in a certain location. Another description emphasizes that landscape is a combination in which past, present, culture and nature are linked with each other. Also, landscape everlastingly has provided some opportunities for human-beings to meet history, geography and climate for a long time [6, 14].

Acar (2003) claims that the description of landscaping in European Landscape Convention is that landscape is a field where people detect its features and the interactions between human and nature. Landscape architect is the person who aesthetically bridges links between nature and human-beings and builds the places in the scope of the demands and desires of individuals abide by these links [6].

Acar (2003) also believes that landscape architects both build the places providing active or passive, functional and aesthetical recreation opportunities and design areas meeting the demands and needs of individuals [6].

Landscape has naturally, culturally and compositionally been changed in progress of time. Landscaping had better be sustainable since next generations are expected to have places in existing or better conditions [6, 14].

Porteus (1996) thinks that Aristotle mathematically evaluated beauty. According to Aristotle, the description of beauty is the harmony in a certain proportion and tune. Therefore, human-beings mostly regard the imagines they could not perceive as ugly. Being beautiful is to be perceived by individuals. Shortly, the elements as big as or as small as people cannot perceive can not be joint in the category of beauty [10, 6].

Environment is a factor formed with various elements and these elements play a vital role on environment at minimum or maximum level. The interaction among the elements, the features of them and the relations between them with main factors have a huge influence on environment. Moreover, the correlations of the elements creating a habitat are unique in only that environment [6].

Yürekli (1977) believes that visual perception contains two elements which are those observing and observed; that is why, two different psychological perceptions have an impact on visual perception and they need to be evaluated separately. However, due to the fact that those observed is commonly environment, the scarcity of some approaches occurs. The reason why this scarcity is happened is that human-beings and environment cannot be independent from each other. Such elements as environment, environment and human, visual quality and cultural relations are usually manipulated with both external and internal factors [6].

Daniel and Vinning (1983) claim that environment should initially be described with the description of "quality" instead of the description of "visual quality". There are two meaning of quality in both linguistic and landscape branches [16].

In linguistic discipline, quality is the combination of the features creating an element, its characteristics, the factors creating quality and the level of its perfectness. According to some researchers, the description of visual quality is akin to the meaning of beauty. In landscape field, quality is the combination of the interaction between human, nature, aesthetical and personal expectation [6].

Perfectness is another factor creating quality. Quality is not explained with only vision but function and benefit features of quality are also the other ways to explain it. The level of the enjoy individuals have towards the objects and views around them can also specify quality. That is, human-beings personally evaluate environment with visual quality in negative or positive way. Environmental psychology and visual quality are linked with each other. The psychology is associated with natural and man-made resources, such personal conditions as the socio-cultural and literacy levels of those using the sources and their personal expectations and some environmental situations including climate and weather [17]. Kalın (2004) believes that due to the fact that visual quality involves in socio-cultural, environmental, ecological and psychological factors, environment cannot easily be analyzed [6].

1.3 The Usage of Coastal Side

Coastal sides have such functions as trade, leisure, recreation, hospitality, social and cultural activities. Coastal sides also provide miscellaneous opportunities like social interaction among people at offshores and beaches [18, 19].

The facilities located on coastal sides should be designed according to the demands and desires of majority. These facilities should also be synchronized with natural structure and they need to be planned well [20, 21].

During physical planning, communal needs and environmental facts have to be balanced and the culture of coastal side and cultural designs should be matched with each other [22]. The ideas of the tourists who commonly spend time on coastal sides should be cared in planning stage since such visitors may realize main needs on these places. That is, the demands and desires of coastal side tourists, destinations and activities have to be synchronized well. Furthermore, the experts of coastal side tourism should play an important role in planning stage and the choices of coastal side tourists need to inform about necessary parts of study. All of these methods are likely to enlighten the dark sides of study wanted to research. Thus, the data handling from study is possible to inspire further works. Consequently, the project associated with this paper may give chance for better open-air recreational areas, meet the demands and desires of tourists well and increase the number of possible recreational facilities [21].

During planning stage about coastal sides, the desires and demands of the coastal side tourists who know destination well, spend time commonly on the destination have to be picked as participant for study. In this way, the experts of coastal side tourism should assess the findings from researches so as to have better results. According to the results of assessment, potential recreational areas on coastal side can aesthetically be designed and the designs of the indoor facilities can be managed well [21].

1.4 Coastal Law

According to Coastal Law (2020);

Article 1- This law is ruled to protect natural and cultural features of coastal sides, sea, natural and man-made lakes and river banks and specify the rights and the open use of public on these areas.

Article 2- This law involves in sea, natural and man-made lakes, river banks and the regulations on coastal sides and the rights of public uses on these areas.

According to Coastal Law (2020);

Shoreline: It is the field which is at least 100 meters from soil to shore edge line.

Coastal Line: It is the line which is made with calm sea, natural and man-made lakes and rivers on the area where water and land meet.

Shore Edge Line: It is the natural line which is usually rocky, sandy, reedy, stony or muddy part of the area which is between the coastal line, from sea, natural/man-made lakes and rivers, and soil.

Coast: It is the area between coastal line and shore edge line.

Dock: It is the coastal facility which is built in amount of sufficient water level in order to protects water vehicles against the negative effects of wind and sea, have storage, repair, support, management unites and technical and social infrastructures and where ships board passengers, load and empty mass goods and are knotted.

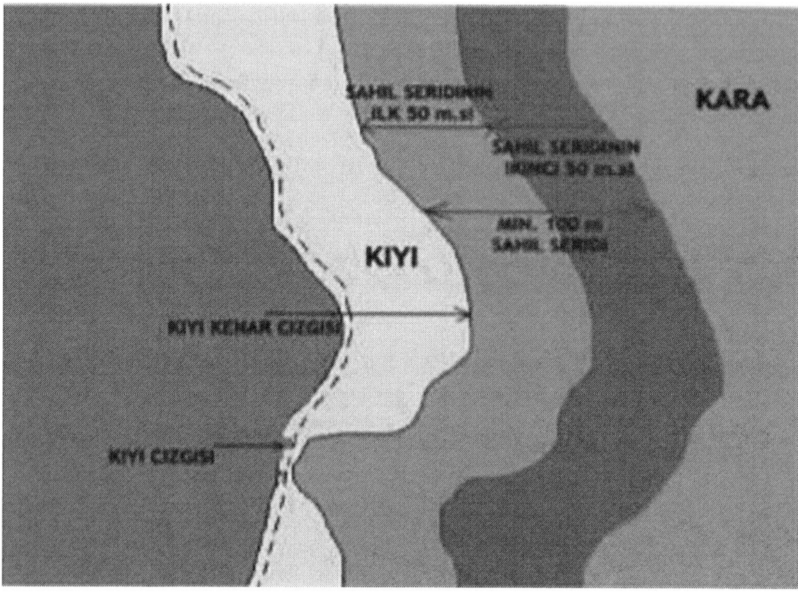

Fig. 1: The Figure of the Description of Coastal Law [26]

1.5 Çanakkale the Coastal City

Çanakkale is a Turkish City which is settled on Gallipoli and Biga Peninsulas in East Thrace, where is Southeast part of Balkans in East Europe [23].

Since 3000 B.C (Early Bronze Age), Çanakkale has been a settlement location and it used to be called as Dardanelles and Hellespontos. Çanakkale Strait does not link only Minor Asia and Europe but it also bridges Black Sea and Mediterranean Sea. Therefore, there are plenty of historical and cultural heritages in Çanakkale. During history, a number of civilizations were founded because of martial and economic reasons. Therefore, this region was either conquered everlastingly or became the target of migrations many times and this increased cultural interaction and the number of looting attempts in history. Hence, multicultural structure has outstandingly been strengthened in this period. Çanakkale Strait divided the city in two piece. One of them is in Europe side which is called as Kilitbahir and the other part is in Asia Minor side which is called as Çanakkale [23].

First settlements in Çanakkale used to be located on the coastal side of city in past. However, the rate of urbanization in Çanakkale is increased at the inside of city which is far away from coast nowadays [24] (Fig. 2).

Fig. 2: The Coastal Side of Çanakkale [25]

The coordinates of Çanakkale is 25° 40′-27°30′ East Longitudes and 39°27′-40°45′ North Latitudes. That is, Çanakkale approximately occupies 9,933 km^2 of land. Moreover, the largest part of Çanakkale is located on the South side of Marmara Sea and rest of them is located in the small part of North Aegean State [23].

The borders of Çanakkale is limited with Avlaka Cape, which is at Far West side of Turkey, and Baba Cape, which is at Far West side of Minor Asia. According to Çanakkale İl Kültür ve Turizm Müdürlüğü (Culture and Tourism Headship of Çanakkale), there are 671 km of coastal line length, 12 counties, 23 municipalities, 81 neighborhoods and 576 villages in Çanakkale. According to Türkiye İstatistik Kurumu (Stats Institution of Turkey) in February 4th 2020, the population of Çanakkale is 542.157 [25].

Even though several climate types can be felt in same duration in region because of the location of Çanakkale, Mediterranean climate is more dominant in city. In Çanakkale, the coldest temperature is measured as -4,2 ℃ in February, the hottest temperature is measured as 35,8 ℃ in August, the annual average of temperature is 14.7 ℃ and the annual average of humidity is 72,6 %. Additionally, the number of windy days is much more in Çanakkale than the other cities around Çanakkale [23].

Much as Northern winds are active in Çanakkale, South wind (Kıble), Southwest wind (Lodos) and Northeast wind (Poyraz) are much more active in there. In Çanakkale, the annual average amount of rainfall is between 854.9 m^3 and 662.8 m^3. Although the amount of rainfall in summer is quite low, the highest amounts of rainfall belong to the days in December, January and February. Furthermore, the maximum number of snowy days in there is only eight [23].

2 Material and Methods

2.1 Material

Research is approximately managed on 1,2 km part of coastal side in Çanakkale (Fig. 3). This specific area is called as Old Knurls (Fig. 4), which is between navy and military bases. On Old Knurls, there are such places as Şakir'in Yeri, Donanma Tea House, Custom Dock, Morabbin Park and Dock Square and these places are commonly preferred by tourists so as to do their recreational activities. Moreover, there are some officer's lodge for educators, military and police department members, hotels, restaurants and the places of entertainment and rest along coastal side. Additionally, there are also some popular places including Watchtower, Central Park, Yalı Street and Republic Square.

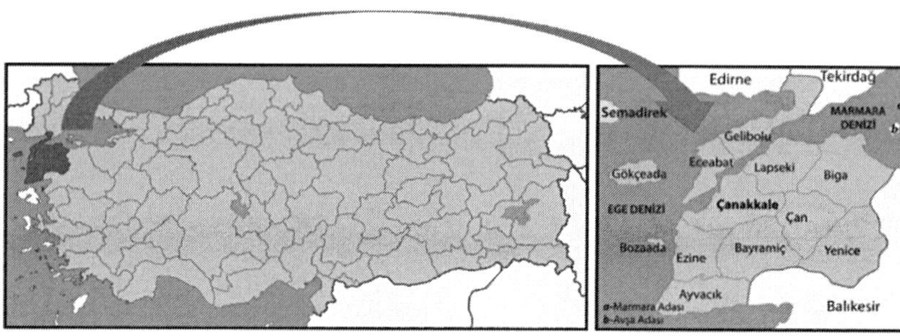

Fig. 3: The Location of Çanakkale on Map [25]

Fig. 4: The Coastal Side of Çanakkale (Map was marked on Google Map in order to show the area where study is managed)

2.2 Method

Firstly, both domestic and international studies related with this paper are reviewed at the literature review stage; hence, the data containing the knowledge about the tourist motivations expressing the coastal sides and visual quality of Çanakkale. Secondly, the method stages in this study are explained with "The Diagram of Method Stages". Thirdly, Old Knurls are divided in four zones. Respectively, the first zone is the area between Donanma Tea House and Dock, the second zone is the area between Dock and Troy Square, the third zone is Troy Square and the forth zone is the area between Troy Square and Military Base. In this part, the pictures emphasizing the coastal sides and their visual quality parameters of Çanakkale are taken from various parts of city in November and December of 2020 so that the pictures can be used in this paper. Lastly, the S.W.O.T. analysis of the coastal side of Çanakkale is made.

In this study, the thoughts of local public about the tourism, coastal sides of Çanakkale and the factors playing important roles on these two elements and the influences of the relations between city and community on the landscape architecture of Çanakkale are expressed and their views are professionally analyzed with the perspectives of external, internal and landscape architectures so that the image of city is described in this study. Furthermore, the relation between local public and city is paraphrased and the various ways of commenting such terms as urban, environment and visual quality are observed.

This paper is likely to be akin to some studies like "Kent Parklarının Görsel Peyzaj Algısının Peyzaj Mimarlığı Öğrencileri Tarafından Değerlendirilmesi" (Surat, 2017) and "Çevre Tercih ve Değerlendirmesinde Görsel Kalitenin Belirlenmesi ve Geliştirilmesi" (Kalın, 2004), since similar choices, environmental characteristics, methods and questionnaire papers are used in this study.

2.2.2 Preparing Questionnaire Questions

In this paper, questionnaire method is opted for collecting data. Participants filled the questionnaire papers which are divided in three chapters and 40 pictures taken from four different parts of Çanakkale are used in the papers. These pictures are shown in this study as Figure 5, 6, 7 and 8 below.

Fig. 5: The Pictures Shown on Questionnaire Papers about the First Zone (between Donanma Tea House and Dock)

Fig. 6: The Pictures Shown on Questionnaire Papers about the Second Zone (between Dock and Troy Square)

Fig. 7: The Pictures Shown on Questionnaire Papers about the Third Zone (Troy Square)

Fig. 8: The Pictures Shown on Questionnaire Papers about the Forth Zone (between Troy Square and Military Base)

In the first chapter of questionnaire paper, the questions produced with Likert Scale containing four degrees [27,28,29] were asked to participants so that their perspectives, hedonic and perception levels about urban view and visual quality were learnt. The admiration criterions of participants were asked and they were required to give score to the questions as 1, 2, 3 or 4 (Tab. 1).

Tab. 1: The Grading Range Used in Research

Score	Admiration Conditions
4	Really Liked
3	Liked
2	Disliked
1	Not Really Liked

In the second chapter of paper, the assessments of environmental choices, its criterions and the characteristics of place about existing conditions of Çanakkale and its coastal side were determined (Tab. 2). In order to specify the level of visual quality of landscapes on coastal side, the questions produced with Likert Scale containing four degrees were asked to participants and they were required to grade the questions as 1, 2, 3 or 4 (Tab. 3). In this chapter, the studies of these authors famed about their aesthetical and visual landscape works are used: Arriaza et al.2004; Önder and Polat 2004; Acar and Kurdoğlu 2005; Kaplan et al., 2006; Müderrisoğlu and Eroğlu 2006; Bulut and Yılmaz 2007; Kıroğlu 2007; Çakcı and Çelem, 2009; Bulubut et al., 2010; Irmak and Yılmaz 2010; Özhancı and Yılmaz 2011; Akten and Çelik, 2013; Aytaş and Uzun, 2015; Karaşah and Var, 2016 [27].

Tab. 2: The Choice Standards and the Characteristics of Place Used in This Paper

Choice Standards	Characteristics of Place
Meaningfulness	With Identity-Without Identity
Durableness	Continuous-Non Continuous
Mysteriousness	Exiting-Boring
Complexness	Simple-Complicated
Legibleness	Well kept-Neglected
Consistency	Tidy-Untidy

Tab. 3: The Grading Range Used for the Characteristics of Place

Score	**With Identity-Without Identity**	**Continuous-Non Continuous**	**Exiting-Boring**	**Simple-Complicated**	**Well Kept-Neglected**	**Tidy-Untidy**
4	Multi Identity	Too Continuous	Too Exiting	Too Simple	Quite Well-Kept	Too Tidy
3	With Identity	Continuous	Exiting	Simple	Well-Kept	Tidy
2	Without Identity	Non-Continuous	Boring	Complicated	Neglected	Untidy
1	No Identity	Too Non-Continuous	Too Boring	Too Complicated	Too Neglected	Too Untidy

Roughly, 50 participants filled the papers. The pictures having same resolution degree were digitally shared and questionnaire papers were digitally filled. The collected data were turned into numeric data with arithmetic mean. Then, the results of analysis are shown in various tables and graphical displays.

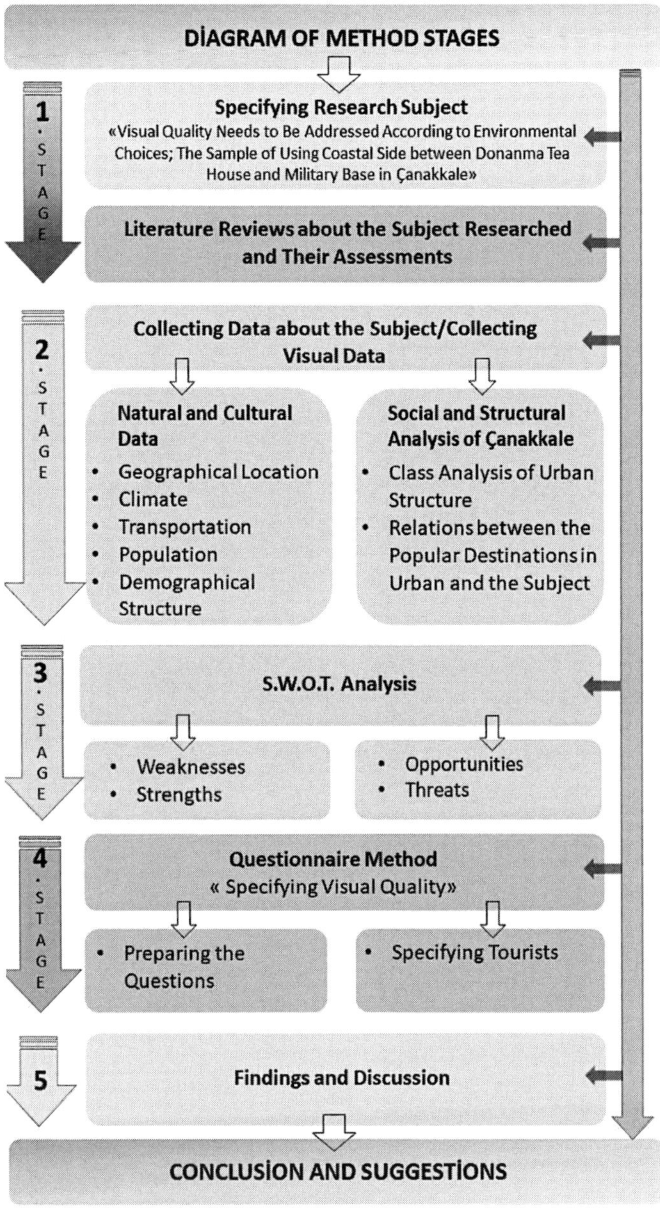

Fig. 9: Diagram of Method Stages

3 Findings and Discussion

3.1 Examining the Landscapes on Coastal Side in Çanakkale

Easy accessibility initially needs to be provided for both tourists and local community so as to visit coasts, urban and touristic destinations in designing period (bicycle, pedestrian, walking ways etc.). Coasts are the locations which give biological and psychological supports to sea creatures and create cultural and natural values for tourists [44].

In order to specify the visual quality in a place, the combination of staffs, plants and interior design plays quite important role. Therefore, all landscape elements in the place should be taken into account for a better visual quality analysis [6].

The data collected in the stages of planning, design and determining changes are likely to help designers to increase visual quality in landscape places [45].

Çanakkale has always been such a versatile city containing natural, cultural, historical and touristic values that a lot of tourists has visited here for a long time. Furthermore, There are not only many recreational opportunities in Çanakkale for both local public and tourists but redesigning project for Old Knurls in Çanakkale is also possible to provide social, cultural and recreational places.

Old Knurls in Çanakkale is both separated in various zones and examined in details so as to have better results from study (Fig. 10).

Fig. 10: Examining Zones on Old Knurls in Çanakkale

3.1.1 First Zone (between Donanma Tea House and Dock)

Donanma Tea House region which is the first step of study is free recreation side of the first zone. There are many deficiencies about landscapes on the coast side which is parallel to Old Balıkhane Street. In this side, there are insufficient number of sitting unites. Even though there are modern streetlights on the coastal side, the sitting unites and streetlights are together. This side is both narrow and has hard soil to sit. Therefore, there is not any plant habitat. However, if some green bushes which are durable against sea salt are planted beside streetlights, path and footway, hard soil may be softened. Moreover, there are one way road and bicycle way which is parallel of sea line; nonetheless, this bicycle way is on footway and this may be dangerous for both pedestrians and riders in rush hours (Fig. 11).

Fig. 11: The Images of the First Zone (Original)

3.1.2 Second Zone (between Dock and Troy Square)

The natural and man-made structure of Dock Square is compatible with other places in the second zone. This side is pretty close to city center, the paving stones in here are natural granite stones, there are plenty of sitting unites under trees, modern streetlights and traffic signs. Even so, there are some problems; nevertheless, if the amount of plantation is increased, sitting unites are softened to sit and new recreational places meeting the demands and desires of people, matters can be coped with soon. Hence, the duration of leisure cannot be only increased but urban can also have new socialization areas (Fig. 12).

It is realized that plantation is lacked in this zone just as every zones in this paper lack do. In order to deal with this scarcity, the bushes having large tops, staying always green, being durable against sea salt and strong wind should be preferred. Thus, the amount of green areas is increased.

Fig. 12. The Images of Dock Square (Original)

Due to the first place seen by those using ferry, Dock Square is quite vital for the charisma of city because the historical and cultural values and the elements advertising city on this side are likely to create positive thoughts about city in the mind of tourists. Therefore, the first interactive war museum which is called as "Çanakkale'nin Evlatları" (Sons of Çanakkale) was built on Dock Square.

From Dock Square through Troy Square, there are many restaurants and cafes on the right wing of coastal side and such tree families as *Planatus orientalis* and *Populus alba* on the across of cafes. Additionally, there are sitting unites under each of these trees. However, because of the fact that the structure of these sitting unites bothers those require to sit in long term, many tourists frequently avoid to spend time on them (Fig. 13).

Fig. 13: The Sitting Unites under Trees (Original)

3.1.3 Third Zone (Troy Square)

Troy Square and Morabbin Park in the third zone play an important role in analysis, since there are such elements advertising city as the Troy Horse used in Troy the Movie, the model of Ancient Troy Town in pyramid form and shadow clock. Therefore, this part is so popular that number of domestic and international tourists do not only visit here but they also accept this side as meeting point (Fig. 14).

The color of the bicycle way from Donanma Tea House through Dock Square and Morabbin Park and which is the parallel of Kayserili Ahmet Paşa avenue separates road and pedestrian way (Fig. 14).

Fig. 14: The Images of Troy Square (Original)

There is also a basketball court, surrounded with wooden sitting stands of which height is one meter from ground, nearby the Troy Horse.

3.1.4 Forth Zone (between Troy Square and Military Base)

There are square, resting, sitting and free recreational places in the fourth zone. One of them is Piri Reis Square. In this place, the World Map of Piri Reis is figured with paving stones on ground and there is one of his busts. Furthermore, there are the sitting unites under trees and independent ones, both tall streetlights and small ones surrounded with bushes and trash baskets on the corner of both bicycle and pedestrian ways (Fig. 15).

Fig. 15: The Images of the Forth Zone (Original)

Although this zone is preferred by those having different interests, the number of recreational places is insufficient. For instance, some elements including playing grounds, chess tables, the bushes being durable against sea salt and strong winds in the green areas and the stands which ideally strengthen the relations between individuals and water activities may not increase only the popularity of destination but also strengthen the aesthetic view of city.

The concept of the places on coastal side is not differ from each other. Much as ground designs of some places are different, the materials and their colors used in them are similar.

Moreover, the number of both natural and man-made structures is not sufficient. Especially, sun rays affect these elements to increase the amount of the accumulation of heat in summer season on Old Knurls; that is why, people become unrested. Similarly, due to the scarcity of the objects creating shade on such recreational places as resting, dining and sitting, individuals mostly do not opt for spending time at outside in daytime of summer season.

The number, quality and position of sitting unites on coastal side are also insufficient. That is, these elements are either far away from shore or uncomfortable. Therefore, some people usually bring their foldable chairs in order to sit on the closer part of coastal side at the duration when spending time at outside is comfortable.

3.2 S.W.O.T. Analysis of Old Knurls in Çanakkale

Old Knurl in Çanakkale is analyzed in some categories involving Strengths (Tab. 4), Weaknesses (Tab. 5), Opportunities (Tab. 6) and Threats (Tab. 7).

Tab. 4: Strengths

Strengths
Geographical location of urban
Closeness of urban
Easy access
Preferred by those from all age groups
Closeness of public transportation
Cultural and historical places
Dining and resting facilities
Bicycle and pedestrian ways
Sitting unites
Recreational places
Activity areas (including cycling, hooking, dining, sports, skateboard etc.)
The view of strait
Water transport
Streetlights
Well plantation
Closeness of public institutions
The potentials of shore activities
Marine
Nature
Protected areas nearby Old Knurls

Tab. 5: Weaknesses

Weaknesses
Noise pollution hailing from traffic
The scarcity of shade equipment
Insufficient number of sitting unites
Uncomfortable sitting unites
Insufficient number of security
Too loud airplays at facilities
Unseparated pedestrian and bicycle ways on some parts of coastal side
Scarcity of green areas
Far coastal locations of sitting unites
Insufficient number of playgrounds
Crowded tourist groups
Weak entertainments of cultural and historical values
Low pedestrian circulation at facilities

Tab. 6. Opportunities

Fırsatlar
Existing tourism potential of region
Closeness of public institutions
Various recreational areas for those having miscellaneous hobbies
Closeness of urban
Infrastructures of recreational activities
Closeness of both public and water transportation
High tourism potential of region
Coastal activity accustomedness of local community
High water sport tourism potential
Usage potential of coastal side
Potential of advertising unfamed historical and cultural values

Tab. 7: Threats

Threats
Insufficient number of parking lot
Much more amount of hard soil than soft soil
Sliding ground in bad weather (because of using materials)
Accumulation of heat in daytime of sunny days
No barriers between pedestrian way and road
Overpopulation
Soil and water pollution
Unplanned urbanization
Scarcity of city planning
Insufficient works of local governments
Low number of bicycle way

3.3 Assessments Graphics of Questionnaire Analysis about the Visual Quality of Old Knurl in Çanakkale

3.3.1 First Zone (between Donanma Tea House and Dock)

The assessment results of the first zone are that 44,7 % is “not liked”, 23,4 % is “not really liked”, 21,5 % is “liked” and 10,4 % is “really liked”.

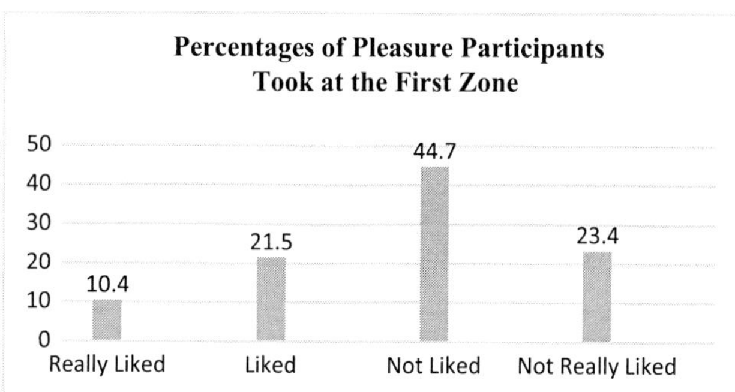

Tab. 8: Percentages of Pleasure Participants Took at the First Zone

3.3.2 Second Zone (between Dock and Troy Square)

2. Bölge mekanı beğeni durumu sonucunda yaklaşık %47,6 oranında "beğenmedim" sonucuna ulaşılmıştır. Bundan sonra yaklaşık %26 ile "beğendim", %17, 7 ile "hiç beğenmedim", %8,4 ile çok beğendim sonuçları takip etmiştir.

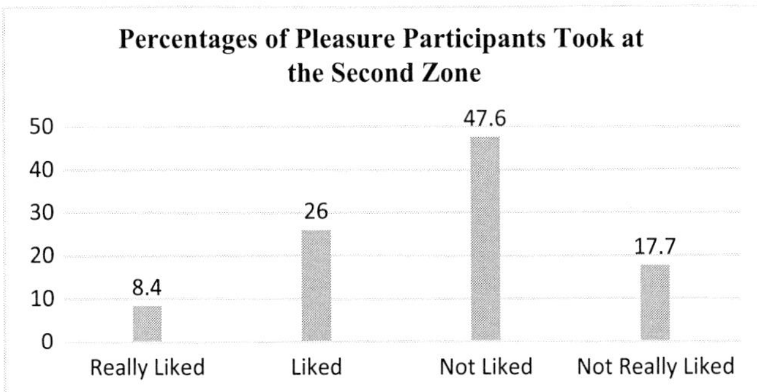

Tab. 9: Percentages of Pleasure Participants Took at the Second Zone

3.3.3 Third Zone (Troy Square)

The assessment results of the third zone show that 41,6 % belong to those saying "Liked", 29,8 % "Not Liked", 18,7 % "Really Liked" and 9,9 % belong to the participants who answered "Not Really Liked".

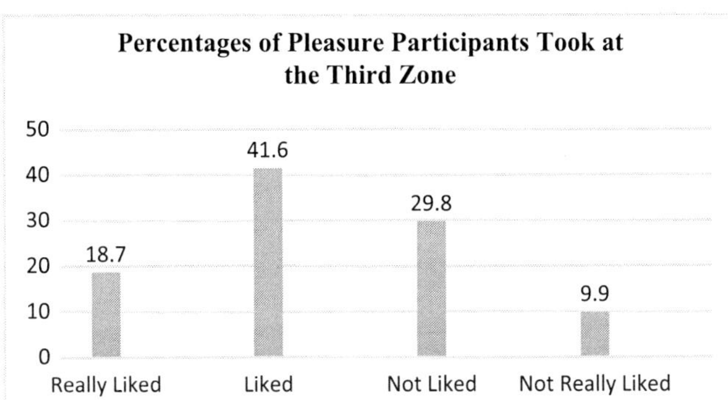

Tab. 10: Percentages of Pleasure Participants Took at the Third Zone

3.3.4 Forth Zone (between Troy Square and Military Base)

The ratios of the assessment results of the forth zone are respectively 49,7 % (Not Liked), 33,7 % (Liked), 9,5 % (Really Liked) and 7,1 % (Not Really Liked).

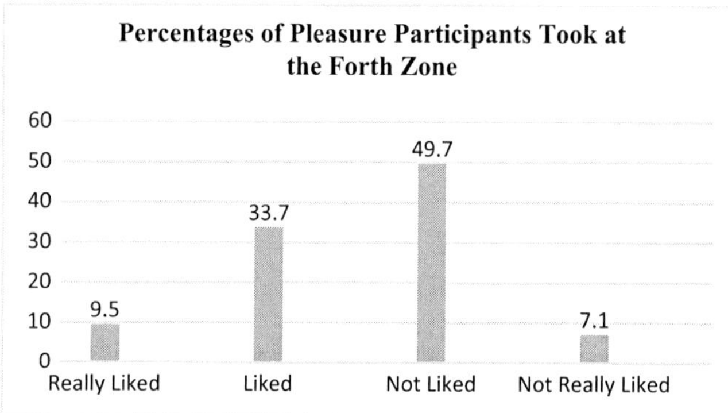

Tab. 11: Percentages of Pleasure Participants Took at the Forth Zone

3.3.5 Tables of the Arithmetic Averages of the Scores Participants Gave for Old Knurls on the Questionnaire Papers

The tables of the arithmetic averages of both the pleasure scores (Tab. 12) and visual quality parameter scores (Tab. 13) participants gave for Old Knurls are given below.

Tab. 12: Arithmetic Averages of the Pleasure Scores Participants gave for Old Knurls

1. First Zone (between Donanma Tea House and Dock)	2,15
2. Second Zone (between Dock and Troy Square)	2,24
3. Third Zone (Troy Square)	2,66
4. Forth Zone (between Troy Square and Military Base)	2,42

Tab. 13. Arithmetic Averages of the Visual Quality Parameter Scores Participants gave for Old Knurls

	First Zone	**Second Zone**	**Third Zone**	**Forth Zone**
With Identity-Without Identity	2,09	2,22	2,69	2,41
Continuous-Non Continuous	2,19	2,36	2,47	2,50
Exiting-Boring	2,03	2,04	2,58	2,34
Simple-Complicated	2,33	2,52	2,55	2,65
Well Kept-Neglected	2,29	2,45	2,72	2,60
Tidy-Untidy	2,17	2,32	2,54	2,62
Average	2,18	2,31	2,59	2,52

5 Conclusion and Suggestions

This study is managed in order to determine the combination of visual landscape perceptions and landscape structure on coastal side in Çanakkale. Visual perception may be differ from personal perception. Therefore, various results from those having miscellaneous personalities are found. This paper illustrates that pleasure and visual quality parameters are related with each other. Moreover, visual landscape and landscape structure are also associated with each other [46]. Additionally, care plays an important role on visual landscape and landscape structure, as well and the more landscape structure looks well-kept, tidy and natural, the more area is attractive.

The averages of the highest scores given for pleasure are respectively 2,66 (the third zone), 2,42 (the forth zone), 2,24 (the second zone) and 2,15 (the first zone). Furthermore, the averages of the highest scores given for visual quality parameters are respectively 2,59 (the third zone), 2,52 (the forth zone), 2,31 (the second zone) and 2,18 (the first zone). It is certain that visual quality parameters positively influence pleasures and the visual quality level of area manipulates the pleasures of tourists.

The results of research demonstrate that the "Meaningfulness" choice standard (Tab. 2) which is directly proportionate to pleasure condition value and reaches too close pleasure condition values of all choice standards becomes the characteristic of the place of "With Identity-Without Identity". Moreover, the results show that the scores given below average in all categories imply designs to be without identity, non-continuous, untidy and boring. Consequently, it is important that the places which are with identity, continuous, tidy, attractive and have landscape elements should be paid attention during planning and design stages so that the areas being an eyesore are built over again.

The high scores given for visual quality parameters imply places to provide tidy and continuous access; that is why, such places should initially be preferred. Additionally, when man-made or natural landscape elements are well-kept in areas, places do not only provide positive impact for urban ecologically, but they also become more attractive. Furthermore, the places which are exiting and with identity bring much more tourists and visitors to destination.

The assessments in the visual quality works of landscape are based on the thoughts of individuals since it is vital that personal choices should meet their demands and desires [27]. Therefore, design and planning stages of landscape had better be managed according to their needs and visual quality parameters.

5 References

1. Özgeriş, M., Karahan, F., Rekreasyonel Tesislerde Görsel Kalite Değerlendirmesi Üzerine Bir Araştırma: Tortum Ve Uzundere (Erzurum) Örneği, 2015.
2. Akten M. Isparta İlindeki Bazı Rekreasyon Alanlarının Mevcut Potansiyellerinin Belirlenmesi. Süleyman Demirel Üniversitesi Orman Fakültesi Dergisi, 2003. Seri: A (2), 115–132
3. Ak,M.,K. Akçakoca Kıyı Bandı Örneğinde Görsel Kalitenin Belirlenmesi Ve Değerlendirilmesi Üzerine Bir Araştırma, 2010. (Doctoral dissertation, Ankara Üniversitesi Fen Bilimleri Enstitüsü Peyzaj Mimarlığı Anabilim Dalı).
4. Önen, M., Kentsel kıyı mekanı olarak akarsuların rekreasyonel kullanım potansiyelinin irdelenmesi: Eskişehir Porsuk Çayı ve İstanbul Kurbağalıdere Örneği. Doctoral dissertation, Fen Bilimleri Enstitüsü, 2007.
5. Yücel, M., Aslanboğa, İ., Korkut, A., Peyzaj Mimarlığı Terimleri Sözlüğü. 2008. TMMOB Peyzaj Mimarları Odası Yayın No: 2008/4.
6. Kalın, A. Çevre Tercih Değerlendirilmesinde Görsel Kalitenin Belirlenmesi ve Geliştirilmesi: Trabzon Sahil Bandı Örneği, Doctoral Dissertation, Karadeniz Teknik Üniversitesi Fen Bilimleri Enstitüsü, 2004, Trabzon.
7. Erdoğan, E. Çevre ve kent estetiği. Bartın Orman Fakültesi Dergisi, 2006. 8(9), 68–77.
8. İzgi, U., Mimarlıkta Süreç, Kavramlar-İlişkiler, Yapı Endüstri Merkezi Yayınları, 1999, İstanbul. ISBN. 975-7438-88- X.
9. Kaplan, A., Hepcan, Ç.C., Ege üniversitesi kampüsü 'Sevgi Yolu'nun görsel (etki) değerlendirme çalışması, Ege Üniversitesi Ziraat Fakültesi Dergisi, 2004. 41 (1):159–167.
10. Çakcı, I., Peyzaj Planlama çalışmalarında görsel peyzaj değerlendirmesine yönelik bir yöntem araştırması, Doktora Tezi, Ankara Üniversitesi Fen Bilimleri Enstitüsü Peyzaj Mimarlığı Anabilim Dalı, 2007, Ankara. 9–49

11. Elinç, H. Görsel Kalite Değerlendirmesi Yöntemi Ile Antalya Ili Alanya Ilçesindeki Abdurrahman Alaettinoğlu Ve Alanya Belediye Başkanları Kent Parklarının Irdelenmesi, Doctoral Dissertation, Selçuk Üniversitesi Fen Bilimleri Enstitüsü, 2011.
12. Göler, S. Biçim, Renk, Malzeme, Doku ve Işığın Mekân Algısına Etkisi. Mimar Sinan G.S.Ü. Fen Bilimleri Enstitüsü, Yüksek Lisans Tezi, 2009, İstanbul.
13. Tüfekçioğlu K., H., Tarihsel Çevrede Görsel Peyzaj Kalite Değerlendirmesi İstanbul Yedikule Örneği. İstanbul Teknik Üniversitesi Fen Bilimleri Enstitüsü Peyzaj Mimarlığı Anabilim Dalı, Yüksek Lisans Tezi, 2008, İstanbul. 136 s.
14. Lavagna, Monıca, Life Cycle Assessment in edilizia, 2008.
15. Porteous, J. D., Environmental Aesthetics: Ideas, Politics and Planning, Routledge, London and New York, 1996.
16. Daniel, T. C. ve Vinning, J., Methodological Issues in the Assessment of Landscape Quality, In I. Altman abd J. F. Wohlwill (Eds.), Human Behavior and Environment: Volume 6. Behavior and the Natural Environment, Plenum Press, New York, 1983.
17. Acar, C., Dağ ekosistemlerinde görsel kalite ve görsel kaynak yönetimi-ekolojik ve görsel indikatörler. Kaçkar Dağları Milli Parkı Çevre Eğitimi, 2003, Ayder-Çamlıhemşin- Rize
18. İncedayı, D.O., Kültür Politikası ve Kıyılar. Mimarist Dergisi, l6 (19), 38–44 s., 2006.
19. Hall C.M. ve Page, S.J., The Geography of Tourism and Recreation. Routledge, New York, 2006.
20. Taneri, E., Kıyı Kullanımı. Kıyılar Kolokyumu, 7. Türkiye Dünya Şehircilik Günü Daimi Komitesi, KTÜ, 1983, Trabzon.
21. Özdemir Işık, B., Demirel, Ö., Kıyı Kulllanıcılarının Kıyı Alanı Rekreasyon Planlama Kararları Üzerindeki Etkisi: Trabzon Kıyı Bandı Örneği, İnönü Üniversitesi Sanat Ve Tasarım Dergisi, 2014. 4 (10), 1–11 s.
22. Koçan, N. ve Ateş, O., Effect Of Structuring The Visual Aspects Of Coastal Landscape: BartinInkumu Example Reviews. Gümüşhane University Journal of Scienceand Technology Institute, 2011. 1(1), 54- 56.
23. URL 1: https://canakkale.ktb.gov.tr/
24. Alkan, Y. Çanakkale Kentsel Yapısına Eleştirel Bir Yaklaşım: Butik Kent. Mehmet Akif Ersoy Üniversitesi Fen Bilimleri Enstitüsü Dergisi, 2019. 10(2): p:150–158
25. URL 2: https://tr.wikipedia.org/wiki/%C3%87anakkale
26. Gedikli, B., Kıyıyı Yaşamak, Kıyıyı Planlamak, 7. Kıyı Mühendisliği Sempozyumu,2011, Ankara. 23- 33s.

27. Surat, H., Kent Parklarının Görsel Peyzaj Algısının Peyzaj Mimarlığı Öğrencileri Tarafından Değerlendirilmesi, Bartın Orman Fakültesi Dergisi, 2017. 19(1): 70–80s.
28. Akten M, Yılmaz O, Gül A. Determination of Recreational Land Use Factors For Land Use Planning: Case Of Isparta Plain. Turkish Journal of Forestry, 2009. 2: 119–133.
29. Yeşil, M., Yılmaz, H., Tozanlı Havzası Tokat-Almus ilçesi ekolojik temelli kırsal peyzaj planlaması. Akademik Ziraat Dergisi, 2013. 2:2 63–74.
30. Arriaza M, Canas-Ortega JF, Canas-Madueno JA, Ruiz- Aviles P. Assessing The Visual Quality Of Rural Landscapes. Lanscape and Urban Planning, 2004. 69:115- 125.
31. Önder, S., Polat, A.T., Konya İli Karapınar İlçesi'nin Ekoturizm Yönünden Görsel Kalite Değerlendirmesi ve Swot Analizi. Selçuk Tarım Bilimleri Dergisi, 2004. 18.33: 80–86.
32. Acar C., Kurdoglu B. C. Kaçkar Dağları Milli Parkında Görsel Kalite Değerlendirmesi. Korunan Doğal Alanlar Sempozyumu, SDÜ, 2005, Isparta, p. 219–226.
33. Kaplan A, Taşkın T, Önenç A., Assessing the Visual Quality of Rural and Urban-fringed Landscapes surrounding Livestock Farms. Biosystems Engineering, 2006. 95 (3), 437–448.
34. Müderrisoğlu H., Eroğlu, E., Bazı İbreli Ağaçların Kar Yükü Altında Görsel Algılanmasındaki Farklılıklar, Süleyman Demirel Üniversitesi Orman Fakültesi Dergisi, 2006, Isparta. Seri: A, Sayı: 1, 136–146.
35. Bulut Z., Yılmaz H. Determination of landscape beauties through visual quality assessment method: a case study for Kemaliye (Erzincan/Turkey), 2007. Environmental Monitoring and Assessment Volume 141, Numbers 1–3, 121–129.
36. Kıroğlu, E., Erzurum Kenti ve Yakın Çevresindeki Bazı Rekreasyon Alanlarının Görsel Peyzaj Kalitesi Yönünden Değerlendirilmesi. Atatürk Üniversitesi Fen Bilimleri Enstitüsü Basılmamış Yüksek Lisans Tezi, 2007, Erzurum.
37. Çakcı I., Çelem H. Kent Parklarında Görsel Peyzaj Algısının Değerlendirilmesi. Ankara Üniversitesi Ziraat Fakültesi, Tarım Bilimleri Dergisi, 2009. 15(1):88–95.
38. Bulut Z., Karahan F., Sezen I. Determining visual beauties of natural waterscapes: A case study for Tortum Valley (Erzurum/Turkey), 2010. Scientific Research and Essay Vol. 5 (2), 170–182.
39. Irmak M.A., Yılmaz. H. Farklı Peyzaj Karakter Alanlarına Göre Doğal ve Kültürel Kaynak Değerlerinin Görsel Analizi: Erzurum Örneği. Gaziosmanpaşa Üniversitesi Ziraat Fakültesi Dergisi. 2010. 27(2), 45–55

40. Özhancı, E., Yılmaz, H., Rekreasyon Alanlarının Görsel Peyzaj Kalitesi Yönünden Değerlendirilmesi; Erzurum Örneği. Iğdır Üniversitesi Fen Bilimleri Enstitüsü Dergisi, 2011. 1(2): 67–76.
41. Akten M, Çelık M. Evaluation of visual landscape perception for Incilipinar and Adalet Park cases. Journal of Food, Agriculture & Environment, 2013. 11(2), 1532–1538.
42. Aytaş İ., Uzun S. Düzce kent merkezindeki yaya alanlarının görsel peyzaj kalitesinin belirlenmesi. Journal of the Faculty of Forestry Istanbul University 2015, 65(1): 11.
43. Karaşah B., Var, M., Botanik Bahçelerinde Ziyaretçi Tercihlerinin Belirlenmesi 'Nezahat Gökyiğit Botanik Bahçesi Örneği'. Kastamonu University Journal of Forestry Faculty, 2016. 16(1).
44. Şimşek D. S., Korkut A. B. Kıyı Şeridi Rekreasyon Potansiyelinin Belirlenmesinde Bir Yöntem Uygulaması: Tekirdağ Merkez İlçe Örneği. Tekirdağ Ziraat Fakültesi Dergisi, 2009, 6(3).
45. Asur, F., Alphan, H. Görsel Peyzaj Kalite Değerlendirmesi ve Alan Kullanım Planlamasına Olan Etkileri, YYÜ TAR BİL DERG (YYU J AGR SCI), 2008. 28(1): 117–125.
46. Bernasconi C Strager MP, Maskey V, Hasanmyer M (2009). Assessing public preferences for design and environmental attributes of an urban automated transportation system. Landscape and Urban Planning 90 155–167.

E. Seda ARSLAN

Vulnerability of Ecosystem Services under the Impacts of Climate Change: A Literature Review

Abstract: Due to climate change (CC) effects, ecosystems and their services are become more and more vulnerable. In recent years, scientists have attention to this subject for providing protection and management strategies against future climate conditions. Today, it is obviously seen that, changing climate conditions would significantly change the distribution, composition and benefits of the ecosystem services (ES) in the future. The aim of this study is to identify the relation between CC and ES scrutinizing of the articles focused on the vulnerability of ecosystem services against CC impacts. Forty seven articles are reviewed in terms of their titles and topics. In this study articles are reviewed via answering research questions related to their tools and techniques for determining the effects of CC on their focusing ES types. Results showed that there is widely concerning about the impacts of CC on ES especially related to the water and forest ecosystems. The overall picture from the result of related studies is analyzing the impacts of CC from different points of view like ecological and economical on the provisioning of ES. Studies clearly indicated the vulnerability of the ES under the future climate scenarios.

Keywords: Ecosystem services, climate change, future climate conditions

Introduction

Changing climate is today's reality. International Panel of Climate Change (IPCC) has been reported that global temperature is increasing day by day [1]. As a result of CC landscape are also changing and ecosystems are precisely under pressure [2]. Changing climate is threaten not only landscape and ecosystems sustainability but also effected unfavorably their services and values [3, 4].

ES are means of the benefits that humans derive from ecosystems [5]. These benefits represent the provision (fresh water, food and timber, etc.) regulation (climate regulation, pest regulation and pollination, etc.), supporting (nutrient cycling, soil formation, purification of air and water, etc.) and cultural services (recreation and tourism, cultural heritage and aesthetic values, etc.) provided by directly or indirectly from ecosystems and crucially important for human well-being [5–8].

The availability of ES depends on so many factors and components in the natural environment such as biodiversity, climate and soil [9–11]. For instance, soil is a key factor for conducting the earth's cycling systems, such as carbon and nitrogen. Changes of the carbon and nitrogen level can cause the alteration of the

Earth's temperature in long term and can be explained as CC. As a result, it directly causes ES and biodiversity loss [12–14]. Biodiversity plays an important role for providing the ES [15]. Because of this reason changing biodiversity patterns means the changing ES supply[16]. Today as a result of CC, species are shifted to different landscapes or have to adapt different environmental conditions. Several scientists have reported distribution shifts of the species according to the impact of CC under the future climate scenarios [17–28]. It is obvious that ecosystems, their services and biodiversity are inherently dependent on climate [29]. Due to global warming effects, ecosystems are altered in terms of the capability of provisioning ES and effect their sustainability [30]. CC is improving its effects on ES, so they become more and more vulnerable. Scientists are examined CC impacts in terms of their observed and protected impacts on biodiversity and ES [31–36]. So it is important to understand the link between CC and changing ecosystems.

The aim of this study is to identify the relation between CC and ES scrutinizing of the articles focused on the vulnerability of ES against CC impacts. The articles are reviewed in the base of the research questions listed in the following:

1. What is the research question or hypothesis of articles in relation to the impacts of CC? How they describe this relation in terms of the vulnerability of ES?
2. What type of the ES are they focused on?
3. Which techniques or tools are they using to determine the effects of CC on ES? Are they qualitative or quantitative?

Material and Method

Forty seven (47) articles from Web of Science Core Collection are the main materials of this research [2, 9, 30, 37–80].

Search results are collected from ISI Web of Science database between the years 1975–2020 for the terms '*climate change impacts*' and '*ecosystem services*' occurring in researches 'titles' and 'topics'. The research is conducted in December 2020 and analyzed the result according to field (Web of Science Categories), number and distribution of research countries and types of the impact in terms of the ecosystem service that they focused.

Results and Discussion

The first scientific study identified the impacts of CC on ES was published in 2011 and in the last couple of years it is getting increase in terms of the publishing rate in accordance with the web of science searching records.

Most articles focused on the environmental sciences-studies and ecology in terms of their research area according to the Web of Science Categories (Fig. 1).

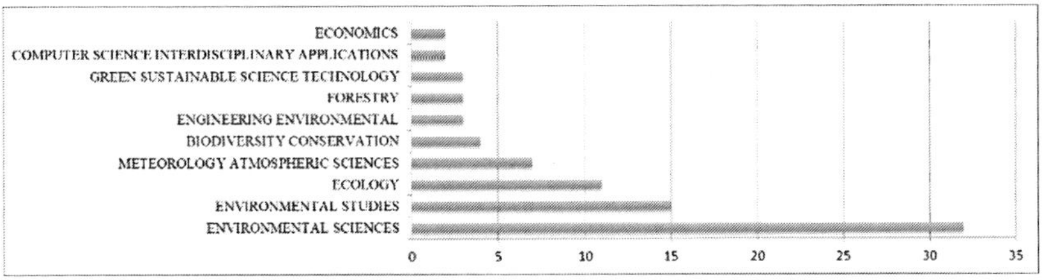

Fig. 1 Web of science categories of the published articles

The geographical distribution represented by the articles the most papers are published in China (11 articles), USA (9) and United Kingdom (4). Fig. 2 shows the number of studies in a world map.

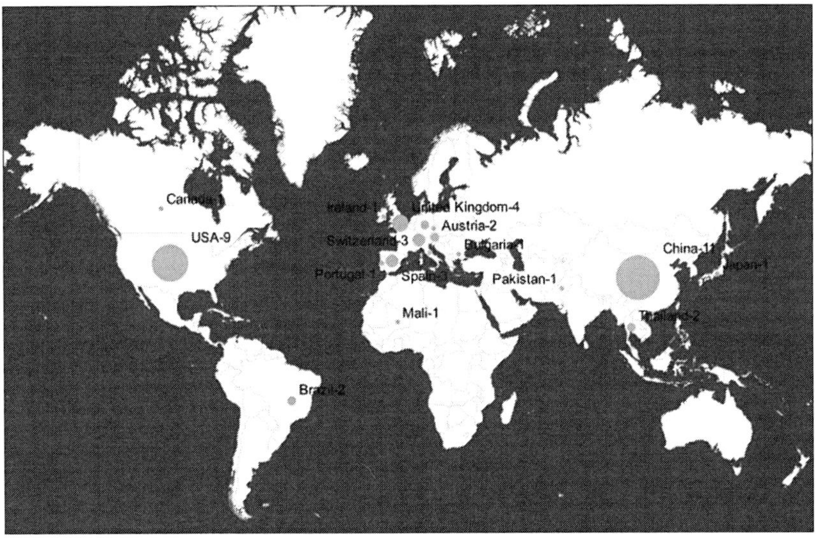

Fig. 2 Number of the studies from Web of Science database

When it comes to types of the CC impacts in terms of the ES that they focused, they are several point of views related to the topic and title. The articles are reviewed for answering the research questions. All articles have been classified according to the types of the ecosystems services that they focused on.

Following tables (Tab. 1, Tab. 2, Tab. 3, Tab. 4, Tab. 5, Tab. 6, Tab. 7, Tab. 8 and Tab. 9) represented the summary of these papers in terms of the year, research questions (RQ) listed in the introduction part and types of the ecosystem focused on their study.

Tab. 1 Summary of the papers published in 2020

Author	**RQ1-RQ2-RQ3**	**Ecosystem Type**
Cheng & Li	The authors emphasized the importance of providing a reasonable management model for ecological health of the rivers to become sustainable ES production	Water ecosystems
Cobacho et al.	Shellfish-based management strategies model were developed for providing ES against CC effects	Water-related ecosystems
Dai et al.	The authors suggested soil conservation and ecological restoration process for protecting ES provision using mapping techniques	Terrestrial ecosystems
J. Li et al.	The authors highlighted the dynamic changes of providing ES and interpreted the relation between ES provision and CC effects.	Water ecosystems
Lundholm et al.	The authors proposed a forest management decision support system for providing ES in forest ecosystems under the future climate conditions	Forest ecosystems
Wang et al.	The authors emphasized the relation between water yield and productivity of ES. They brought an attention to CC effects on ES production for management of natural resources	Water ecosystems
Zarrineh et al.	The authors clearly emphasized the effects of CC on provisioning ES. An integrated model were created to analyze CC impacts on ES production	Water ecosystems
X. Li et al.	The authors pointed out the necessity for understanding the impacts of CC on ES and provided knowledge regarding the consequences of CC on water infrastructures.	Water ecosystems

Tab. 2 Summary of the papers published in 2019

Author	**RQ1-RQ2-RQ3**	**Ecosystem Type**
Bai et al.	The authors quantified water related ecosystem services and assessed the effects of land use and CC water related ES by using InVEST model and environmental scenarios	Terrestrial and Water Ecosystems
Cristal et al.	The authors created a decision support tool to assess CC impacts on multiple ES for forest management	Forest ecosystems
Ferreira et al.	The authors were examined the potential of forests to buffer negative effects of CC on water related ecosystems services	Water ecosystems
Lyu et al.	The authors were modelled hydrological ES to estimate the effects of future climate conditions	Water ecosystems
Pan & Choi	The authors analyzed the impacts not only CC but also land use/land cover on hydrologic ES. They created a framework for hydrologic ES under the future climate scenarios	Water ecosystems
Peng et al.	The Integrated Valuation of Ecosystem Services and Tradeoffs (InVEST) model was applied to quantify hydrologic ES to identify density changes of hydrologic ES under the CC.	Water ecosystems
Shahzad et al.	Vulnerability of provisioning forest ES are estimated under the future CC effects. Study result represented that; multiple ES provided by forest ecosystems can be declined as result of CC.	Forest ecosystems
Underwood et al.	The authors analyzed the alteration in different types of ES between current and future climate conditions. They used five general circulation models under RCP 8.5 CC scenario.	Terrestrial ecosystems

Tab. 3 Summary of the papers published in 2018

Author	RQ1-RQ2-RQ3	Ecosystem Type
Costa et al.	The authors predicted the distribution of bat species which have several functions for producing multiple ES under future climate conditions	Forest ecosystems
Liu et al.	The authors explored the impacts of CC on four related to the agricultural ecosystems. They created a land-use scenario using representative concentration pathways (RCPs) to analyze and responses of ES against CC.	Agricultural ecosystems
Mahmoud & Gan	ES in coastal regions were analyzed using Earth observation data. CC effects were detected with relation to land use projection. It is clearly indicated that, land use transformation can increase the greenhouse gases and cause the CC and directly affect the provision of ES.	Coastal ecosystems
Monprapussorn	The author focused on how to project future ES of coastal ecosystems using future climate conditions and land use data. The author brought attention to the importance of the engagement of stakeholders to the adaptation process of CC.	Coastal ecosystems
Tang et al.	The authors conducted a model to clarify ES provided by agricultural ecosystems under the future climate conditions. The authors developed three different scenarios to analyze the response of ES to the global warming	Agricultural ecosystems

Tab. 4 Summary of the papers published in 2017

Author	**RQ1-RQ2-RQ3**	**Ecosystem Type**
Boone et al.	An ecosystem model were integrated to determine sensitivity of rangelands under the CC effects	Rangeland ecosystems
Bugmann et al.	The impacts of CC and forest management on ES were assessed using 3 different scenarios	Forest ecosystems
Fan et al.	The authors developed a framework via analyzing the spatial distributions of six water related ES under current and future climate conditions.	Water ecosystems
Hao et al.	The authors showed that; alterations of climate and landscape pattern are the result of providing capacity losses of ES in grassland ecosystems	Grassland ecosystems
Sanogo et al.	The authors emphasized the vulnerability of ES in agricultural ecosystems under the future climate conditions	Agricultural landscapes
Schirpke et al.	Resilience potential of ES were assessed under the future CC impacts in grassland ecosystems	Grassland ecosystems
Zlatanov et al.	The authors emphasized the vulnerability of forest ES expected as a result of CC using climate sensitive models	Forest ecosystems
Trisurat et al.	The authors quantified water yield and determined potential impacts of climate change on rice cultivation	Agricultural and water ecosystems

Tab. 5 Summary of the papers published in 2016

Author	**RQ1-RQ2-RQ3**	**Ecosystem Type**
Grimm et al.	Impacts of CC on ES were identified and the research findings showed the threaten on water supplies, species distribution and phenology	Terrestrial and water related ecosystems
Sample et al.	The authors created a methodology for river basin management evaluating twelve different ES against CC impacts	Water ecosystems

Tab. 6 Summary of the papers published in 2015

Author	**RQ1-RQ2-RQ3**	**Ecosystem Type**
Dunford et al.	ES adaptation strategies were created within the CC scenarios for water related ES provision	Water ecosystems
Kirchner et al.	The authors developed an integrated framework to quantify indicators for ES and economic development in agricultural landscapes under the future climate conditions	Agricultural ecosystems
Lane et al.	The authors analyzed potential physical and economic impacts of CC on freshwater fisheries and coral reefs in terms of providing ES	Water-related ecosystems
Shi et al.	This study was conducted to assess the ES conservation area under the change of land use associated with future climate conditions	Water ecosystems

Tab. 7 Summary of the papers published in 2014

Author	**RQ1-RQ2-RQ3**	**Ecosystem Type**
Fezzi et al.	The authors represented a prediction model of the contribution of the ecosystem to the provisioning ES generated by agriculture ecosystems	Agricultural ecosystems
Sara et al.	The authors focused on the identification of ES provided by mountain ecosystems and they represented a conservation approaches for the sustainability of ES under the future climate conditions	Terrestrial ecosystems
Rocca et al.	A broad-scale assessment tool for ES were conducted in terms of characterization the effects of CC on water resources, air quality and biodiversity.	Forest ecosystems and woodlands

Tab. 8 Summary of the papers published in 2013

Author	**RQ1-RQ2-RQ3**	**Ecosystem Type**
Bangash et al.	The authors evaluated the impacts of CC on water related ES productivity and erosion control services.	Water ecosystems
Briner et al.	The authors analyzed the changes in ES driven by CC effects. They quantified different kind of impacts in forest and agricultural ecosystems	Forest and agricultural ecosystems
Gosling	The author reviewed the impacts of future climate changes on ES in the large scale and discussed the vulnerability.	Water and forest ecosystems
Lorencova et al.	The authors provided a large scale analysis of the impacts of CC on multiple ES in agricultural landscapes	Agricultural landscapes
Marques et al.	The impacts of CC on water provision were analyzed under the low flow regime as a consequences of CC	Water ecosystems
Nelson et al.	The authors suggested a model to evaluate national and adaptation strategies for ES under the CC scenarios	Water ecosystems

Tab. 9 Summary of the papers published in 2012

Author	**RQ1-RQ2-RQ3**	**Ecosystem Type**
Civantos et al.	The authors provided an assessment of potential impacts of CC on the biodiversity and their associated ES. They modelled the future distribution of the species that provided ES	Forest ecosystems
Heubes et al.	Non-timber forest product provision were analyzed using niche modelling under the future climate conditions and land use change	Forest ecosystems
Shaw et al.	The authors created a framework to provide better understanding the effects of CC on ecosystem service production and its consequences	Terrestrial ecosystems

Results showed that there is widely concerning about the impacts of CC on ES especially related to the water and forest ecosystems. Studies are generally focused on creating developing strategies to quantify the impacts of CC, evaluating the impacts of provision of the services or mapping the future adaptation of the services. The main aim of the studies is analyzing the impacts of CC from different points of view like ecological and economical on the provisioning ES. Studies clearly indicated the vulnerability of the ES under the future climate scenarios.

Several types of ecosystems are shifting in time and space as a result of CC and other related factors like land use land cover change and biological drivers in time and space [52]. Due to all of these factors, ecosystems are losing their capacity of producing services what is crucial for the human well-being. According to the results; ES will be diminish because of the changing climate and strategies and management plans should be established [43].

Conclusion

This study focus on the articles published international journals and collected from ISI Web of Science database between the years 1975–2020 for the terms *'climate change impacts'* and *'ecosystem services'* occurring in researches 'titles' and 'topics' with the purpose of bringing attention to the vulnerability of ecosystem services under the future climate conditions. Results showed that; there is a growing concern about the CC effects on provisioning multiple ES. On the other hand, articles are presented new tools and techniques for providing different types of ecosystems and also strategies for protecting ES supply under the future climate scenarios.

References

1. IPCC, *The Intergovernmental Panel on Climate Change*, R.K. Pachauri and L.A. Meyer, Editors. 2014: Synthesis Report. p. 50.
2. Li, X., et al., *Ecosystem Services under Climate Change Impact Water Infrastructure in a Highly Forested Basin.* Water, 2020. **12**(10): p. 2825.
3. Dawson, T.P., et al., *Beyond Predictions: Biodiversity Conservation in a Changing Climate.* Science, 2011. **332**(6025): p. 53–58.
4. Williams, J.J. and T. Newbold, *Local climatic changes affect biodiversity responses to land use: A review.* Diversity and Distributions, 2020. **26**(1): p. 76–92.
5. MEA, M.E.A., *Ecosystems and human well-being.* Vol. 5. 2005: Island press United States of America.
6. Bolund, P. and S. Hunhammar, *Ecosystem services in urban areas.* Ecological economics, 1999. **29**(2): p. 293–301.
7. Daily, G., *What are ecosystem services.* Global environmental challenges for the twenty-first century: Resources, consumption and sustainable solutions, 2003: p. 227–231.

8. Daily, G.C., *Introduction: what are ecosystem services.* Nature's services: Societal dependence on natural ecosystems, 1997. **1**(1).
9. Civantos, E., et al., *Potential Impacts of Climate Change on Ecosystem Services in Europe: The Case of Pest Control by Vertebrates.* Bioscience, 2012. **62**(7): p. 658–666.
10. Díaz, S., et al., *Biodiversity loss threatens human well-being.* PLoS Biol, 2006. **4**(8): p. e277.
11. Sandifer, P.A., A.E. Sutton-Grier, and B.P. Ward, *Exploring connections among nature, biodiversity, ecosystem services, and human health and well-being: Opportunities to enhance health and biodiversity conservation.* Ecosystem services, 2015. **12**: p. 1–15.
12. Haddad, N.M., et al., *Habitat fragmentation and its lasting impact on Earth's ecosystems.* Science advances, 2015. **1**(2): p. e1500052.
13. Pereira, H.M., et al., *Scenarios for global biodiversity in the 21st century.* Science, 2010. **330**(6010): p. 1496–1501.
14. Uzun, A. and Ö.K. Örücü, *Adenocarpus complicatus (L.) Gay türünün iklim değişkenlerine bağlı günümüz ve gelecekteki yayılış alanlarının tahmini.* Türkiye Ormancılık Dergisi, 2020. **21**(4): p. 498–508.
15. Ridder, B., *Questioning the ecosystem services argument for biodiversity conservation.* Biodiversity and Conservation, 2008. **17**(4): p. 781–790.
16. Chapin Iii, F.S., et al., *Consequences of changing biodiversity.* Nature, 2000. **405**(6783): p. 234–242.
17. Arslan, E.S., et al., *Distribution of rose hip (Rosa canina L.) under current and future climate conditions.* Regional Environmental Change, 2020. **20**(3): p. 1–13.
18. Du, Z., et al., *Potential geographical distribution and habitat shift of the genus Ammopiptanthus in China under current and future climate change based on the MaxEnt model.* Journal of Arid Environments, 2021. **184**: p. 104328.
19. Kaky, E. and F. Gilbert, *Assessment of the extinction risks of medicinal plants in Egypt under climate change by integrating species distribution models and IUCN Red List criteria.* Journal of Arid Environments, 2019. **170**: p. 103988.
20. Manish, K., et al., *Modelling the impacts of future climate change on plant communities in the Himalaya: a case study from Eastern Himalaya, India.* Modeling Earth Systems and Environment, 2016. **2**(2): p. 1–12.
21. Naudiyal, N., et al., *Potential distribution of Abies, Picea, and Juniperus species in the sub-alpine forest of Minjiang headwater region under current and future climate scenarios and its implications on ecosystem services supply.* Ecological Indicators, 2021. **121**: p. 107131.
22. Pecl, G.T., et al., *Biodiversity redistribution under climate change: Impacts on ecosystems and human well-being.* Science, 2017. **355**(6332).

23. Rupprecht, F., J. Oldeland, and M. Finckh, *Modelling potential distribution of the threatened tree species Juniperus oxycedrus: how to evaluate the predictions of different modelling approaches?* Journal of Vegetation Science, 2011. **22**(4): p. 647–659.
24. Sarıkaya, A. and Ö. Örücü, *Prediction of potential and future distribution areas of Anatolian Chesnut (Castanea sativa Mill.) by using maximum entropy (Maxent) modeling depending on climate change in Turkey.* International Journal of Ecosystems and Ecology Science 2019. **9**(4): p. 699–708.
25. Arslan, E.S., et al., *Kokulu Ardıç'ın (Juniperus foetidissima Willd.) Günümüz ve Gelecekteki Potansiyel Yayılışının Makine Öğrenmesi ile Modellenmesi.* Avrupa Bilim ve Teknoloji Dergisi, 2021(22): p. 1–12.
26. Çoban, H.O., Ö.K. Örücü, and E.S. Arslan, *MaxEnt modeling for predicting the current and future potential geographical distribution of Quercus libani Olivier.* Sustainability, 2020. **12**(7): p. 2671.
27. Şen, İ., O. Sarıkaya, and Ö.K. Örücü, *Current and Future Potential Distribution Areas of Carphoborus minimus (Fabricius, 1798) in Turkey.* Folia Biologica (Kraków), 2020. **68**(4): p. 141–148.
28. Akyol, A. and Ö.K. Örücü, *Investigation and evaluation of stone pine (Pinus pinea L.) current and future potential distribution under climate change in Turkey.* Cerne, 2019. **25**(4): p. 415–423.
29. Groffman, P., et al., *Climate change, ecosystems, biodiversity and ecosystem services.* 2014.
30. Nelson, E.J., et al., *Climate change's impact on key ecosystem services and the human well-being they support in the US.* Frontiers in Ecology and the Environment, 2013. **11**(9): p. 483–493.
31. Hallett, C.S., et al., *Observed and predicted impacts of climate change on the estuaries of south-western Australia, a Mediterranean climate region.* Regional Environmental Change, 2018. **18**(5): p. 1357–1373.
32. Johnston, A., et al., *Observed and predicted effects of climate change on species abundance in protected areas.* Nature Climate Change, 2013. **3**(12): p. 1055–1061.
33. Palomo, I., *Climate Change Impacts on Ecosystem Services in High Mountain Areas: A Literature Review.* Mountain Research and Development, 2017. **37**(2): p. 179–187.
34. Tompkins, E.L., et al., *Observed adaptation to climate change: UK evidence of transition to a well-adapting society.* Global environmental change, 2010. **20**(4): p. 627–635.
35. Weatherdon, L.V., et al., *Observed and projected impacts of climate change on marine fisheries, aquaculture, coastal tourism, and human health: an update.* Frontiers in Marine Science, 2016. **3**: p. 48.

36. Zografou, K., et al., *Signals of climate change in butterfly communities in a Mediterranean protected area.* PLoS One, 2014. **9**(1): p. e87245.
37. Bai, Y., T.O. Ochuodho, and J. Yang, *Impact of land use and climate change on water-related ecosystem services in Kentucky, USA.* Ecological Indicators, 2019. **102**: p. 51–64.
38. Bangash, R.F., et al., *Ecosystem services in Mediterranean river basin: Climate change impact on water provisioning and erosion control.* Science of the Total Environment, 2013. **458**: p. 246–255.
39. Boone, R.B., et al., *Climate change impacts on selected global rangeland ecosystem services.* Global Change Biology, 2018. **24**(3): p. 1382–1393.
40. Briner, S., C. Elkin, and R. Huber, *Evaluating the relative impact of climate and economic changes on forest and agricultural ecosystem services in mountain regions.* Journal of Environmental Management, 2013. **129**: p. 414–422.
41. Bugmann, H., et al., *Impacts of business-as-usual management on ecosystem services in European mountain ranges under climate change.* Regional Environmental Change, 2017. **17**(1): p. 3–16.
42. Cheng, B. and H. Li, *Impact of climate change and human activities on economic values produced by ecosystem service functions of rivers in water shortage area of Northwest China.* Environmental Science and Pollution Research, 2020. **27**(21): p. 26570–26578.
43. Cobacho, S.P., et al., *Impacts of shellfish reef management on the provision of ecosystem services resulting from climate change in the Dutch Wadden Sea.* Marine Policy, 2020. **119**: p. 104058.
44. Costa, W.F., et al., *Bat diversity in Carajas National Forest (Eastern Amazon) and potential impacts on ecosystem services under climate change.* Biological Conservation, 2018. **218**: p. 200–210.
45. Cristal, I., et al., *A Decision Support Tool for Assessing the Impact of Climate Change on Multiple Ecosystem Services.* Forests, 2019. **10**(5): p. 440.
46. Dai, E., et al., *Quantitative Assessment of the Relative Impacts of Land Use and Climate Change on the Key Ecosystem Services in the Hengduan Mountain Region, China.* Sustainability, 2020. **12**(10): p. 4100.
47. Dunford, R.W., et al., *Ecosystem service provision in a changing Europe: adapting to the impacts of combined climate and socio-economic change.* Landscape Ecology, 2015. **30**(3): p. 443–461.
48. Fan, M., H. Shibata, and L. Chen, *Assessing high impacts of climate change: spatial characteristics and relationships of hydrological ecosystem services in northern Japan (Teshio River watershed).* Mitigation and Adaptation Strategies for Global Change, 2018. **23**(4): p. 525–552.
49. Ferreira, P., et al., *Can forests buffer negative impacts of land-use and climate changes on water ecosystem services? The case of a Brazilian megalopolis.* Science of the Total Environment, 2019. **685**: p. 248–258.

50. Fezzi, C., et al., *Valuing Provisioning Ecosystem Services in Agriculture: The Impact of Climate Change on Food Production in the United Kingdom.* Environmental & Resource Economics, 2014. **57**(2): p. 197–214.
51. Gosling, S.N., *The likelihood and potential impact of future change in the large-scale climate-earth system on ecosystem services.* Environmental Science & Policy, 2013. **27**: p. S15-S31.
52. Grimm, N.B., et al., *Climate change impacts on ecosystems and ecosystem services in the United States: process and prospects for sustained assessment.* Climatic Change, 2016. **135**(1): p. 97–109.
53. Hao, R., et al., *Impacts of changes in climate and landscape pattern on ecosystem services.* Science of the Total Environment, 2017. **579**: p. 718–728.
54. Heubes, J., et al., *Impact of Future Climate and Land Use Change on Non-timber Forest Product Provision in Benin, West Africa: Linking Niche-based Modeling with Ecosystem Service Values.* Economic Botany, 2012. **66**(4): p. 383–397.
55. Kirchner, M., et al., *Ecosystem services and economic development in Austrian agricultural landscapes - The impact of policy and climate change scenarios on trade-offs and synergies.* Ecological Economics, 2015. **109**: p. 161–174.
56. Lane, D., et al., *Climate change impacts on freshwater fish, coral reefs, and related ecosystem services in the United States.* Climatic Change, 2015. **131**(1): p. 143–157.
57. Li, J., H. Chen, and C. Zhang, *Impacts of climate change on key soil ecosystem services and interactions in Central Asia.* Ecological Indicators, 2020. **116**.
58. Liu, Y., J. Bi, and J. Lv, *Future Impacts of Climate Change and Land Use on Multiple Ecosystem Services in a Rapidly Urbanizing Agricultural Basin, China.* Sustainability, 2018. **10**(12): p. 4575.
59. Lorencova, E., et al., *Past and future impacts of land use and climate change on agricultural ecosystem services in the Czech Republic.* Land Use Policy, 2013. **33**: p. 183–194.
60. Lundholm, A., et al., *Evaluating the Impact of Future Global Climate Change and Bioeconomy Scenarios on Ecosystem Services Using a Strategic Forest Management Decision Support System.* Frontiers in Ecology and Evolution, 2020. **8**: p. 200.
61. Lyu, R., et al., *The impact of urbanization and climate change on ecosystem services: A case study of the city belt along the Yellow River in Ningxia, China.* Computers Environment and Urban Systems, 2019. **77**: p. 101351.
62. Mahmoud, S.H. and T.Y. Gan, *Impact of anthropogenic climate change and human activities on environment and ecosystem services in arid regions.* Science of the Total Environment, 2018. **633**: p. 1329–1344.

63. Marques, M., et al., *The impact of climate change on water provision under a low flow regime: A case study of the ecosystems services in the Francoli river basin.* Journal of Hazardous Materials, 2013. **263**: p. 224–232.

64. Monprapussorn, S., *Impact of climate and land use change on ecosystem services: A case study of Samutsakorn province, Thailand.* Ecological Informatics, 2018. **47**: p. 45–49.

65. Pan, F. and W. Choi, *Impacts of Climate Change and Urban Expansion on Hydrologic Ecosystem Services in the Milwaukee River Basin.* Climate, 2019. **7**(4): p. 59.

66. Peng, L.-C., et al., *Climate Change Impact on Spatiotemporal Hotspots of Hydrologic Ecosystem Services: A Case Study of Chinan Catchment, Taiwan.* Water, 2019. **11**(4): p. 867.

67. Rocca, M.E., et al., *Climate change impacts on fire regimes and key ecosystem services in Rocky Mountain forests.* Forest Ecology and Management, 2014. **327**: p. 290–305.

68. Sample, J.E., I. Baber, and R. Badger, *A spatially distributed risk screening tool to assess climate and land use change impacts on water-related ecosystem services.* Environmental Modelling & Software, 2016. **83**: p. 12–26.

69. Sanogo, K., et al., *Farmers' perceptions of climate change impacts on ecosystem services delivery of parklands in southern Mali.* Agroforestry Systems, 2017. **91**(2): p. 345–361.

70. Sara, G., et al., *The impact of climate change on mediterranean intertidal communities: losses in coastal ecosystem integrity and services.* Regional Environmental Change, 2014. **14**: p. S5-S17.

71. Schirpke, U., et al., *Future impacts of changing land-use and climate on ecosystem services of mountain grassland and their resilience.* Ecosystem Services, 2017. **26**: p. 79–94.

72. Shahzad, L., et al., *Assessing the Impacts of Changing Climate on Forest Ecosystem Services and Livelihood of Balakot Mountainous Communities.* Pakistan Journal of Botany, 2019. **51**(4): p. 1405–1414.

73. Shaw, M.R., et al., *The impact of climate change on California's ecosystem services.* Climatic Change, 2011. **109**: p. 465–484.

74. Shi, C., et al., *Land Use Zoning for Conserving Ecosystem Services under the Impact of Climate Change: A Case Study in the Middle Reaches of the Heihe River Basin.* Advances in Meteorology, 2015. **2015**: p. 496942.

75. Tang, Z., et al., *Impacts of Land-Use and Climate Change on Ecosystem Service in Eastern Tibetan Plateau, China.* Sustainability, 2018. **10**(2): p. 467.

76. Trisurat, Y., et al., *Basin-wide impacts of climate change on ecosystem services in the Lower Mekong Basin.* Ecological Research, 2017. **33**(1): p. 73–86.

77. Underwood, E.C., et al., *The impacts of climate change on ecosystem services in southern California.* Ecosystem Services, 2019. **39**: p. 101008.
78. Wang, L.-J., et al., *Simulating the Impact of Future Climate Change and Ecological Restoration on Trade-Offs and Synergies of Ecosystem Services in Two Ecological Shelters and Three Belts in China.* International Journal of Environmental Research and Public Health, 2020. **17**(21): p. 7849.
79. Zarrineh, N., K.C. Abbaspour, and A. Holzkaemper, *Integrated assessment of climate change impacts on multiple ecosystem services in Western Switzerland.* Science of the Total Environment, 2020. **708**: p. 135212.
80. Zlatanov, T., et al., *Impact of climate change on vulnerability of forests and ecosystem service supply in Western Rhodopes Mountains.* Regional Environmental Change, 2017. **17**(1): p. 79–91.

Ayşe DURUKAN KOPUZ

The Rise of Nationalist Tendencies in the 1940s; Second National Architectural Movement

"The architectural style must bear the character of the nation and the revolution." Sedad Hakkı Eldem, 1939

Abstract: This paper tries to explain the basic characteristics of Second National Architectural Movement which took place between 1940 and 1950's of Turkey in Republican period, upon eight distinctive buildings in Ankara and İstanbul. The planning and façade typologies were evaluated at the end of the paper. The architectural designs for new state purposes, as well as their reception in the highly new developing context of Turkish Republic, requires the study to emphasize the principles as well as the professional perspective and background of the architect of the contemporary architectural context. Meanwhile, the public buildings produced during this period, their design reflecting the transition to the new government and a dual structure consisting of modern and traditional texture can be observed.

Keywords: *Second National Architectural Movement, Sedad Hakkı Eldem, Nationalism, Emin Onat, Turkish Architecture.*

1 Introduction

During the late 1930's, within the discussions on architectural theory, the tendency towards a symmetrical and monumental architectural style started to become more observable in Turkish architecture. All over the world, international style and modernism which were dominant theories in 1930's, have been abandoned due to the objectives of nationalist regimes especially in European countries such as Germany and Italy. Soon after the solutions/reflections of this international tendency towards nationalism have been spread to Turkey also (1). At the beginning of 1940's the new Architecture was dominated with monumental and classical formalism based on a kind of Neo-classicism imported from the Central Europe.

The mentioned period, which we can call a national architectural trend, consisting of superficial ornaments added to the large-mass buildings, reflecting the state authority of the period, again taking as an example from previous of the Turkish building culture, continued until to 1950's. Sedat Hakkı Eldem and Emin Onat, who are considered to be the most important masters of Turkish architecture between 1940 and 1950, together with Paul Bonatz, who also came from Germany, could not get rid of the Western-based mass understanding in the application of regional and national characteristics that emerged as a reaction to the international forms of the period. Sedat Hakkı Eldem, advocated that "the architectural style cannot be imported from abroad, that every country should be local". These words became the theory of mentioned period named as "Second National Architectural Movement" between 1940 and 1950 years in Turkey.

Substantially, this paper is part of an elective course named as "Republican Period in Turkey" (2), which aims to reveal the architectural typology of buildings of early Republican Period, justified of not creating any theoretical attempt within the field of architecture. A part of the course is the possibility of different planning for understanding the second national architectural movement of architecture and different ways of classifying them, which may in turn enhance the development of other forms of the next period within the field of architecture in Turkey.

So, the aim of this research is to highlight the architectural reading of Second National Architectural Movement that was dominated in Turkey between 1940's to 50's. The research will focus on the projects of especially public buildings not only between these years but also before and after these years, because of understanding the typology changes better.

However, Second National Architectural movement corresponds to a very interesting period in terms of Turkish Architecture, the studies or typological arrangement on public buildings are quite limited such as First National Architectural Movement (3). This research is trying to make a contribution to this necessity, especially within the eight distinctive buildings in Ankara and İstanbul in Turkey.

2 The Evolution of Second National Architectural Movement

The Second National Architecture Movement, which developed in an effort to establish a connection with traditional architecture, evolved as an extension of the rational-functional architecture movement.

In 1940, one of the most famous architect of this period Sedad Hakkı Eldem, has told the concept of this movement with an article on "Architect" magazine with the title of "toward domestic architecture". He stated his thoughts on what needs to be done, if architecture is changing depending on how the current regime in Europe, how to develop depending upon the regime of national architectural trends in Turkey and began to consider what should be done for it. It was decided that the following criteria should be considered first;

- Thankness to the land
- Compliance with climatic conditions
- Being able to connect with traditional architecture
- Creating domestic materials and workmanship (4)
- Suitable for people of hometown.

However, mentioned period shows some differences from the functional architectural characteristics in some ways. The basic features of nationalist period are; rhythmic-looking colonnades with two or three stories high on the facades, prismatic masses with horizontal flat roofs or hidden roofs and rectangular window arrangements. Especially the use of cemented plaster or artificial stone, which is generally called edelputz plaster was on the buildings with strong gray color. It can be summarized as the use of new types of industrial materials such as chrome iron and straight line items together with plain geometric forms such as square, rectangle, circle or semicircle (5).

Besides these characteristics, we can say that the concepts of buildings after 1939 have been so monumental and heavy. The facade was placed as a stone covering in the form of cut stone masonry, by lining the rows of old flat bricks in between. Windows and doors have flat lintels and a flattening belt is placed on them. The top of the building is covered with a roof and finished with wide eaves, and the interior architecture is flat brick and tile covering. The halls with bay windows were built in the form of Bosphorus mansions and the plan was shaped like a clover leaf, concrete ceilings and eaves were covered with wood. Some places are even covered with vaults and domes. When all these elements are examined, the existence of a formalist approach is seen (6).

Another feature of this period is, especially educational university buildings were began to be built such as, İstanbul University Faculty of Art and Science and Ankara University Science Faculty. Whereas in 30's new industrial complexes were emerged through modern architecture (7). Within the scope of this study, public buildings; İstanbul University Observatory, Yalova Thermal Hotel, General Directorate of TCDD State Railways, Istanbul University Faculty of Arts and Science, Ankara University Science Faculty, Istanbul Courthouse, Istanbul Radio House and Atatürk Library were examined with their plan and facade features and their typologies were revealed.

Istanbul University Observatory, 1934–1936, Beyazıt

During the beginning of 1930's new types of buildings were decided to be built upon modernization movements all over the country. One of them is the observatory building which is a rarely seen institution structure in Istanbul University.

Within 1933 University Reform in Turkey, it was decided to build an observatory in the Faculty of Science to start astronomy research and to allow astronomy students to make observations in İstanbul University. The plans for the observatory building were drawn by Architect Arif Hikmet Holtay (1896–1968), a faculty member of the Fine Arts Academy, in 1934 (8).

By designing this building, he made a contribution to "new architecture" applications to become one of the most original examples and has been among the architectural icons of the early Republic of Turkey. It is a rare example not only for the period of the structure in Turkey, but also the rational attitude that architecture has been kept in the forefront.

Anyhow, Holtay's functionality of an application's architecture reflects the characteristics of period. This important project designed by Holtay for Istanbul University was served to the Director of the Institute of Delegation and Kandilli Observatory at the beginning of the 1933–1934 academic year. By the request of Erwin Finlay-Freundlich, it was decided to build in the garden of the University Central Building for the Institute of Astronomy in its present location (9).

So, considering Freundlich's wishes, the foundation of the building was laid in December 1935. Contractor Load. Eng. Ekrem Hakkı Ayverdi completed the construction in six months. Holtay designed a symmetrical building with a floor area of 20 by 10 meters. The entrance to the building is provided by the copper-domed cylindrical observation tower that houses the telescope (10).

The new building, built on approximately 1300 m^2 floor area, is reinforced concrete; consists of ground floor, basement, and first normal floor. In the new building, which consists of four blocks, there are 4 lecture halls for 72 and 144 people, planetarium, observatory, laboratories, library, seminar room, canteen, lecturer rooms, technical workshop, and shelter (11).

The spaces opening to the middle hall on the ground floor, which is reached by steps from the entrance, consist of three rooms reserved for the director and staff of the institute and a room reserved for seminars and lessons. In addition, a library serving employees in the building has been planned (12).

On the basement floor, spaces for other necessary uses are designed in the same plan organization. As a precaution against vibrations, the separate base of the observation tower on which the telescope stands were also accessible from this floor where necessary interventions were made. Access to other floors is provided from the ladder bucket, which is integrated into the rectangular mass from the symmetry of the entrance and the observation tower where the telescope is located. While the stairway tower provides a transparent view with its wide horizontal window arrangement, it has given movement to the geometric mass of the building with the geometric joints on the façade (13).

With this mass organization, the building also references Bauhaus-style structures, one of the popular examples of modern architecture in Europe, with an externally readable organization of different geometric masses. In this structure, which shows the rationalization brought to the construction art by the Bauhaus movement, which Holtay had the opportunity to follow closely in Germany, with a plain and simple mass order. The building, which has a functional plan, made the plan scheme readable with its mass organization. It has been an important example reflecting the architectural elements of the period, based on the structural principles of the Bauhaus with its horizontal window bands on the facade, corner-rotating windows, horizontal joints and terrace roof cover (14). As a result, rectangular mass with a circle tower represents the contemporary face of its time.

Fig. 1. İstanbul University Observatory building perspective facades in 1936 (15)

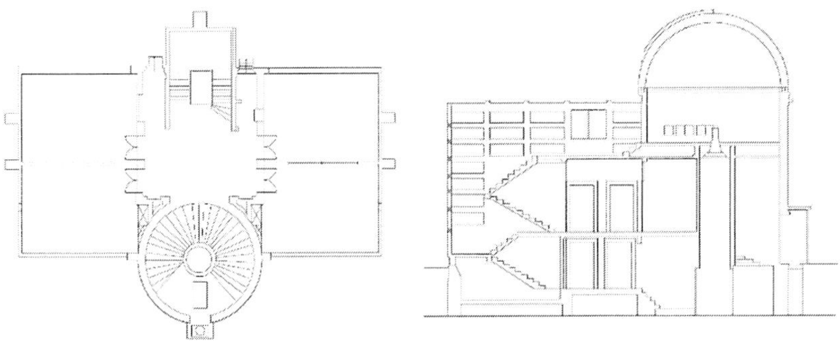

Fig. 2. The basement plan and section of Observatory (Redrawn by Sena Durgut).

Thermal Hotel, 1934–1937, Yalova

Sedad Hakkı Eldem's Thermal Hotel is a project that received the first prize of a competition in which 21 projects participated and Ernst Egli was also a jury member. Eldem describes the Yalova Thermal Hotel as "the first product of the National Architecture movement". While the vertical element repetitions and modular accents observed in the mass point to the opposite, the roof details, porches and interior features surrounded by lattices reflect the characteristics of the Turkish House and traditional architecture. The building is extremely critical in terms of carrying these stylistic features to a large scale and adapting them to a different function (16).

T shaped building by reinforced concrete carcass system, with basement, suspended floor, first and second floors were placed through a valley with trees and a stream flowing through it. About architectural concerns, the building is divided into two parts from entrance axis: hot spring/spa and hotel. Floor entered from one end of T's short arm hot spring, the other from the foyer to the long arm of the hotel is the living hall, then the restaurant. The hot spring/spa is located in this section on the ground floor and the basement. The restaurant, with sliding doors merges with the terrace in front of the building. On the mezzanine, staff bedrooms and second class bed rooms are available (17). The vertical structural system of the bedroom floors, eaves, balconies were covered with marble plaster with the rhythmic repetition of its elements. On the ground floor, the front of the entrance is partially evacuated and an entrance terrace with a porch under the bedroom floors has been created. In the words of Eldem came out of a hand (...)" and "(...) furniture, door balls, lamps, carpets, fabrics and curtains were specially made according to the architect's pattern." (18).

For the first time in this public building, apart from small-scale buildings, Turkish house and traditional civil architecture references were seen on the architecture of this scenery. Unfortunately, the hotel was demolished in 1983.

Fig. 3. Yalova thermal hotel in 1934 (19)

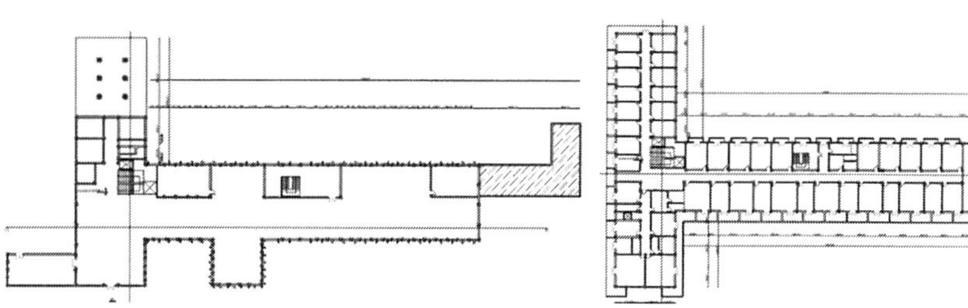

Fig. 4. Ground and first floor plan of thermal hotel (Redrawn by Kasım Gül).

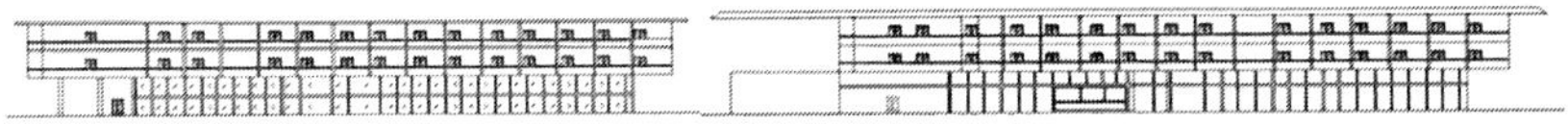

Fig. 5. Front and back facades of building (Redrawn by Kasım Gül).

General Directorate of (TCDD) State Railways, 1939–1941, Ankara

The General Directorate of State Railways is located in the northwest of Ankara Station building, on Talat Pasa Boulevard. Through the northwest of the building is the Ankara Metropolitan Municipality Headquarters, the Turkish Aeronautical Association Museum to the north, the Ankara Arena Sports Hall to the northeast, and the Ankara Station to the southeast (20).

The building was designed by Architect Bedri Uçar from the Ministry of Public Works Building Zoning Directorate Project Office and tendered to "Haymil" Construction Company for 1 million liras. The construction of the building started in 1939, but after a while, the II. world war started and the construction took place with great difficulties. Meanwhile the company successfully built the building and delivered it in 1941. About this building, German architect Prof. Bonatz after touring Ankara, told like this: "I think this is the most beautiful building in Ankara" (21).

It is striking that there is a difference between the architectural design and application of the building. The first project, which was created by adding wings to the right and left sides of the four blocks that form a large open area in the middle of four floors on the basement, was applied differently due to the change of the first place where the building was planned to be built and the war broke out. The two wings on the right and left were not implemented due to the increase in material prices due to the war and the transition to a narrower area, as the architect stated (22).

The entrance of the building is reached by wide garden stairs from the main road. From a magnificent and logical entrance with a marble vestibule, you can enter the big hall from there. Although this hall is actually considered one and a half times deeper, the program has sacrificed the depth of the hall since a conference hall was requested for the central courtyard and the actual depth of the plot was not available. The hall floor and surrounding walls are covered with colorful Hereke and Bilecik marbles. The top of the hall is covered with a stained glass outlined by the architect, detailed and applied by decorator Mazhar Nazım Resmor. There are public manager and proxy offices on the first floor. Other floors are allocated to rooms such as offices and saloons belonging to various services of the administration (23).

The floors of the building, which is built as reinforced concrete as a carrier system, are hollow blocks and the facade is stone-coated. The basement of the two side blocks was built as a shelter. Six of these parts were naturally built as a raft general, and the foundations of the other blocks were made as continuous pillars. Fill body walls are partially covered with stone and partially artificial stone from the outside. The floors are made of parquet, linoleum or mosaic depending on the size of the rooms (24).

Fig. 6. The first design of entrance facade of General Directorate of State Railways with two wings (25).

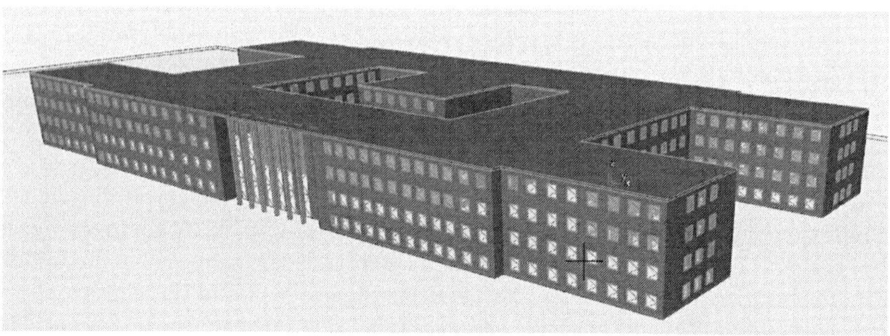

Fig. 7. The render of first architectural design of the building.

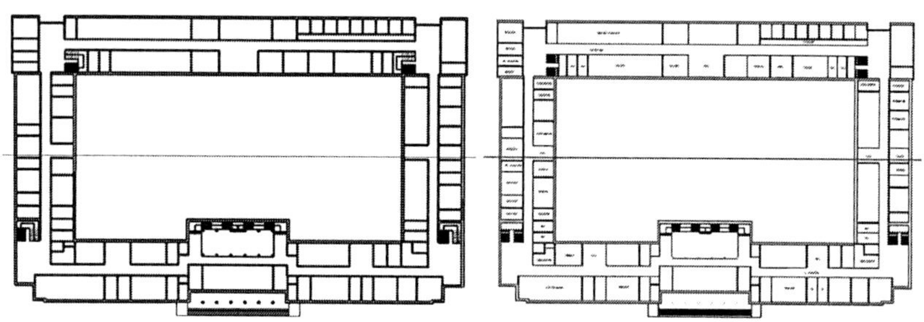

Fig. 8. Basement and first floor plan of General Directorate of State Railways (Redrawn by Adilhan Gül).

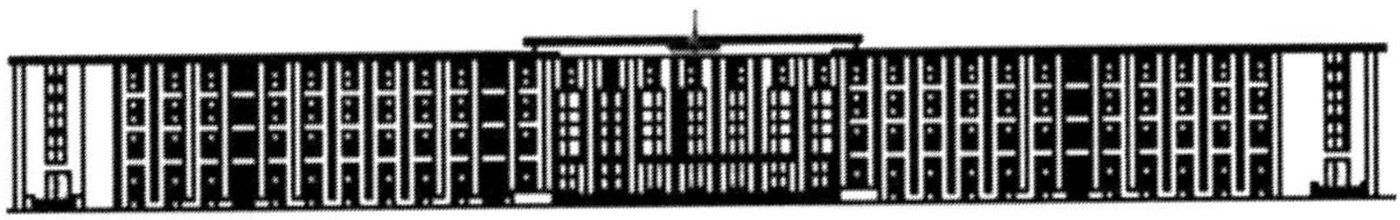

Fig. 9. The entrance facade of the GDRS building (Redrawn by Adilhan Gül).

Istanbul University Faculty of Science and Literature, 1942–1949, Beyazıt

The Science and Literature faculty was built on the land of Zeynep Hanım Mansion, which was burned in 1942. The mansion was built in 1864 by Zeynep Hanım, the daughter of Kavalalı Mehmet Ali Pasha and the wife of Grand Vizier Yusuf Kamil Pasha. The science and literature faculty building was designed by architects Sedat Hakkı Eldem (1908–1988) and Emin Onat (1908–1961) on this land in 1952 and operates within the building complex (26).

When the architecture of building is examined, it is seen that it is influenced by the traditional education institutions as madrasahs. The central courtyard with two iwans (eyvan) and two wings of a porch is the part where it can be seen most clearly in the design. The portico, wide canopies and corner protrusions, which are reflected on the facade also attract attention. This situation is not only in the outer lines, but inside, it is supported by the details of traditional covering systems such as the mirrored vault applications on the upper floor ceilings of the stairwells, the barrel vaulted space above the honor hall behind the northern wing of the portico. On the inside, against the traditional lines of the exterior, the reflection of the educational institutions is seen, except for some applications on the covers. Wide corridors are connected with halls that provide the distribution of wide stairways. In these, the inner courtyards with porticoes are kept alive with four corners. In addition to the rooms and seminar libraries allocated to academic staff and research groups, there are nine large classrooms and small classrooms handled as lecture halls. Today, in the face of the expanding teaching staff and the increasing number of students, there is a naturally congested situation in the rooms and private teaching institutions. In the building, which is often criticized that there is too much volume loss in the interior, the corridors and halls that cause this loss are only enough for the intense student traffic and accumulation today. The congestion that occurs as an area provides the opportunity to breathe in the width of the volume provided by high ceilings. It is possible to understand that the large gaps that are the subject of criticism are very useful today during the busy hours of the student (27).

A university building inevitably encompasses the task of representing the regime, especially if it is the most foreseen university in the country. At least he was tasked with this building, then two great architects, Sedad Hakkı Eldem and Emin Onat, who left their mark on the architecture of the 1940s to be described as

National Architecture, must have approached the design problem in this way. It is certain that Sedad Hakkı Eldem wanted to nationalize (or localize) the building. However, if one of the most important representative structures of the period is in question, it is inevitable that the building has huge proportions and emerges as an indicator of the regime. It can be said that the building is, in a sense, a gigantic Turkish house interpretation. Window proportions, jambs, long eaves, high arches, alternate walls and other similar elements were used in this project. In addition, giant refined crown door models and long vaults borrowed from Ottoman monumental architecture are found. Although the historicist orientation in the project is quite noticeable, it can easily be said that the plan scheme, that is, the way of using the building, is functional (28).

The process that can be symbolized with the Faculty of Science and Literature has brought many innovations in higher education life. This educational institution has set an example for many countries in terms of both its contribution to education and its structural features.

Fig. 10. Perspective drawings of Faculty building by Eldem (29)

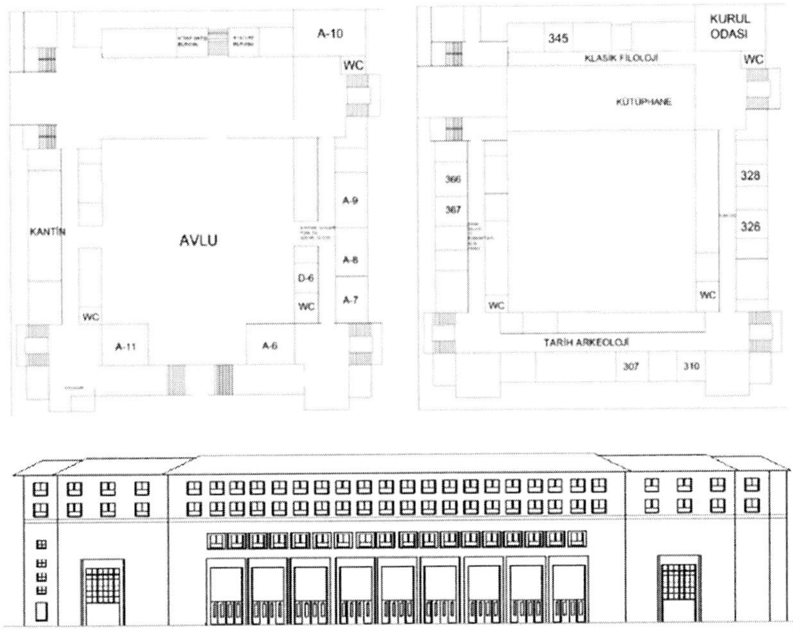

Fig. 11. Plans and facade of Faculty (Redrawn by Abdullah Çelik).

Ankara University Science Faculty, 1943–1945, Beşevler, Ankara

The Faculty of Science is designed within two rectangular planned masses on both sides of the campus entrance, porches, lightening arches, stone walls, stalactite and porcupine eaves and brick decorations It is composed of monumental architecture of nationalism with its alternating masonry, cornices and solid geometry.

The faculty consists of two separate buildings, which have a courtyard system, come together on a symmetrical axis system. There is a pair of monumental porticoes as entrance doors in parts of the buildings opening to the main courtyard, and their ceilings are with ribbed cassette system. These porches have corner chamfers with muqarnas on their feet. Again, there are lightening arches on the columns in the same space, and there are brick decorations in these arches. Each building consists of corridors that form two courtyards, and there are classrooms and rooms at the edges of these corridors. The octagonal auditorium building separates the two courtyards. In this plan scheme, there are references to the Turkish madrasa plan scheme. This building is staged within itself and its ceilings are wooden lattices. A similar application regarding ceilings is found in the main entrances of the buildings, but various perspective studies and wooden models were made before the building was built for these spaces (30).

The reinforced concrete mass of the two-storey building is covered with cut stones and bricks, and on the facades, lightening arches inspired by the first era Ottoman architecture. These are; brick decorations, corner bevels with stalactite and hedgehog eaves (31). The same weight is felt with the use of alternating stones and bricks in the interior. Cornices are used on ceiling and wall joints. During this period, economic and political reasons were effective in the use of stone in buildings, and the state encouraged the implementation of these buildings, which were carried out in a strict monumental style (32).

The presence of stone was heavily felt in this building and thus we see a strong monumental architecture on this building.

Fig. 12. Ankara Science Faculty (33)

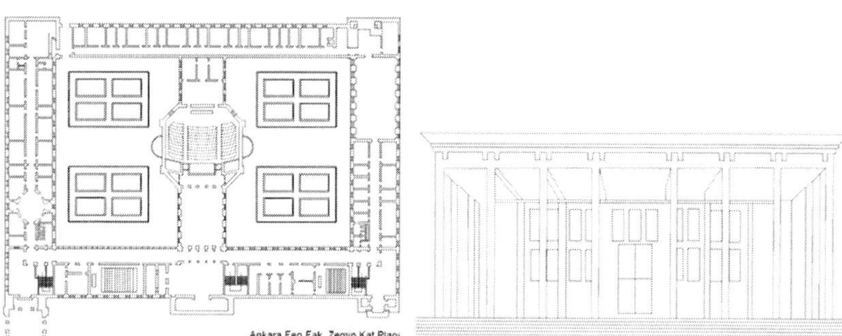

Fig. 13. Floor plan and perspective of Ankara Science Faculty (Redrawn by İlayda Turhan).

Fig. 14. Façade of Ankara Science Faculty (Redrawn by İlayda Turhan).

Istanbul Radio House, 1945–1949, Harbiye

In 1939, a project competition was opened by the Press and Broadcasting General Directorate to meet the studio needs of Istanbul Radio. A total of 74 projects participate in the architectural project competition held for Istanbul Radio House between Turkish engineers and architects. Doğan Erginbaş works with his colleagues Ömer Güney and İsmail Utkular for this competition and the project they prepared is chosen as the winner (34).

The project of Erginbaş and his friends, who put forward a classical style architectural project, is exhibited in Ankara Community Center and Istanbul. The construction of the Istanbul Radio House, whose foundations were laid on November 11, 1945 and recorded as one of the symbols of the early Republican era architecture, was completed within 4 years and was opened by the President İsmet İnönü on November 19, 1949. Turkey's first date to the Radio House building, four-story building is being constructed; There are studios, record and film reception units and program service offices on the ground floor and the first floor, while on the other floors there are main command and control offices and technical and administrative service offices. Istanbul Radio House is a structure where radio programs are held in its studios and continues its service until today (35).

Fig. 15. Perspective drawing of Radio House (36)

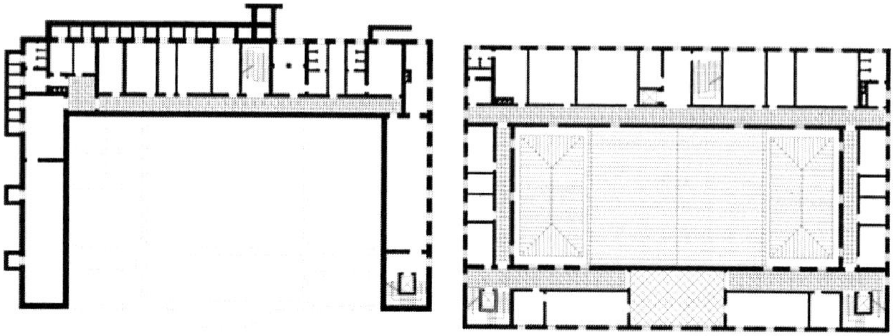

Fig. 16. Basement and 3rd Floor Plan of Radio House (Redrawn by İrem Önder)

Fig. 17. Front and side facades of Radio House (Redrawn by İrem Önder)

Istanbul Courthouse, 1951–1955, Sultanahmet

It is located behind the Ibrahim Pasha Palace in Sultanahmet Square and used as the Istanbul Courthouse. Turkish architects Sedat Hakkı Eldem and Ord. Prof. Emin Halid Onat's project has won the competition held in 1949, as first place (37).

The buildings' construction started in 1951, was put into service in 1955. Built on an area of 4,000 m^2, the building has 6 floors, 2 of which are below ground level. In fact, this building consists of the first block of the 2-block building envisaged in the project, which is considered as the "court block", the second block was not built. Following the death of Emin Halid Onat, Sedat Hakkı Eldem developed a new project for the construction of the 2nd block, taking into account the archaeological findings in the vicinity, partially covering them in 1961. However, this project could not be implemented, although it also received the approval of the Council of Monuments (38).

In the text of the jury proposals of the competition, descriptions indicating that the building will again contain a serious representation mechanism are used: The building should be an immortal monument of the Republic era that will live throughout history. Besides, it has to be "magnificent", "great", "harmonious", "monumental" and "majestic". It is important to underline that Paul Bonatz is also on the jury of this competition (39). The jury report was announced on 24.08.1949. Sedad Hakkı Eldem, winner of the second competition Emin Onat's design also follows these suggestions. Therefore, Tanyeli's commented on Eldem – Onat project for its structure, "The building is almost all historical ending national trends in studies as a rational, functional modernist product the perspectives of the project, though considered Second National of the 1940's with its monumental qualities they are not far outside the architectural line. "(40).

In the design, units are placed around a created inner street. Repeated blocks along an axis create a strong sense of order. Therefore, the structure is dominated by modularity. In this context, although the building is considered as a functional modernist product, components such as long eaves, vertical window proportions and high porches, which Sedad Hakkı Eldem cannot give up, indicate that the building is actually a historicist revival (41).

Eldem - Onat design, positioned behind the Ibrahim Pasha Palace a hearing set around a kind of inner street with blocks and courtyards between them. At the end of the long arm, the heavy penalty halls and of the two larger blocks containing the prosecution offices consists of it. These two blocks working together, one Sultanahmet Square and the other with Divanyolu. It has two main entrances opening to the Avenue. These two different entrances at the elevations are five stories high with a central hall and the flamboyant stair setup connects. On Divanyolu entrance, Eldem's previously encountered elements of architectural language eaves, reinforced concrete carcass left without plaster and vertical window proportions. The three-story high porch was located in design. The long arm behind Ibrahim Pasha Palace was built only. In the following years, Eldem changed structure in the project for completion but none of his suggestions realised (42).

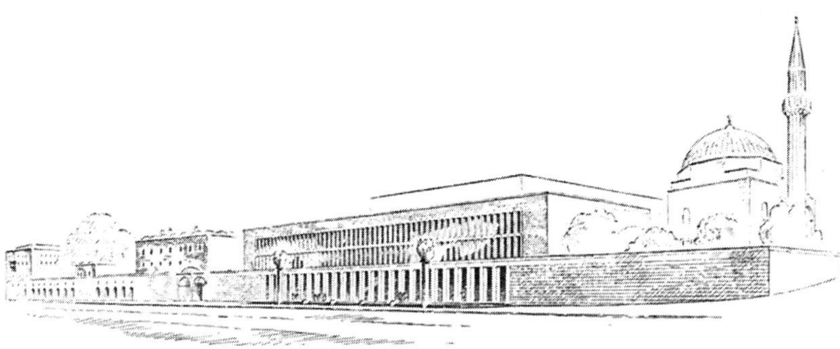

Fig. 18. Perspective from İstanbul Courthouse Building first Project (43)

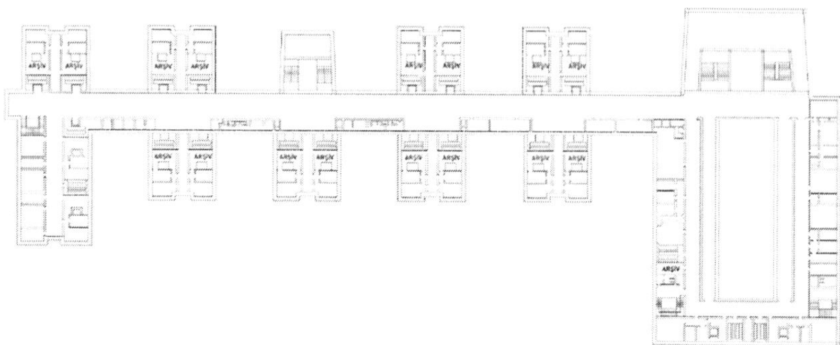

Fig. 19. Plan of İstanbul Courthouse Building (Redrawn by Aslı Yılmaz).

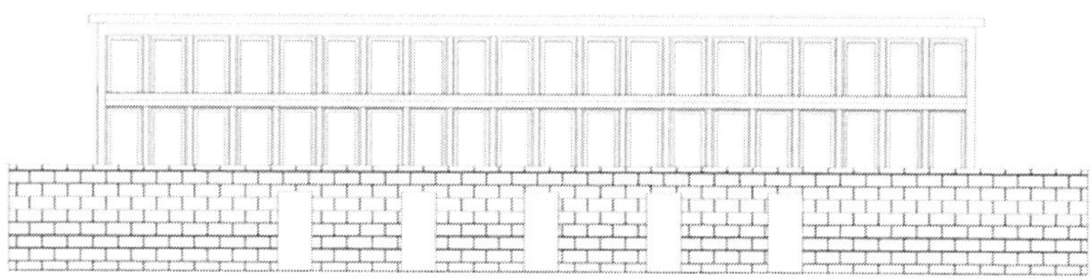

Fig. 20. Facade of İstanbul Courthouse Building (Redrawn by Aslı Yılmaz).

Atatürk Library, 1973–1975, Taksim- Istanbul

The building was designed by Architect Sedat Hakkı Eldem as a cultural complex consisting of a library, museum and exhibition halls, but only the library was implemented. The Atatürk Library that was built in Taksim, overlooking the Bosphorus, has a usage area of 5,000 square meters. Reflecting Eldem's effort to synthesize traditional Turkish architecture and modernism, it is a three-storey reinforced concrete building made of hexagonal elements (44).

The bookcase is created from hexagonal elements to accommodate 60 degree angles in the Sheraton Hotel behind it. Although these are in an axial order in the central building, they are freely grouped in the buildings on the right and left, thus facilitating the transition to the garden parts. Established on a centralized system, there are vertical connections in the middle of the composition, namely freight elevators and control points. There are private halls on the mezzanine and the general reading hall on the upper floor. This place is made up of six domed units surrounded by a central core. Thus, it was necessary to make the domes of the building lighted over the depth of the building. The lead-covered domes do not remind us of some applications of our classical architecture as a group (45).

The main mass of the structure is defined by smaller hexagonal modules surrounding a large hexagon defining the center. The hexagonal module in the center is connected to the warehouse of approximately 600,000 books in the basement with the elevator system, and this system is surrounded by a service counter on two reading floors. The seven-domed pyramid cover, also larger and higher in the center, covered with lead and with a luminous lantern on top, completes the mass. Five hexagonal modules surrounding the central module containing the service counter on the upper floor complete the general reading and index rooms. The last hexagon, which is half floor down, contains the staircase in the center that connects with the lower floor. The ground floor includes a private reading room and book maintenance/repair services. Outside of the main mass, and in the middle of three other terrace-roofed hexagonal modules adjoining the hexagon, the main entrance includes the conference hall and exhibition halls on both sides. Two high chimneys on either side of this trio complete the composition (46). It is noteworthy that the hexagonal plan scheme, which emerged as a theme used by Eldem in the Turkish restaurant for Hilton, is used in this library also.

Fig. 21. Site plan and sea façade of library by Eldem (47).

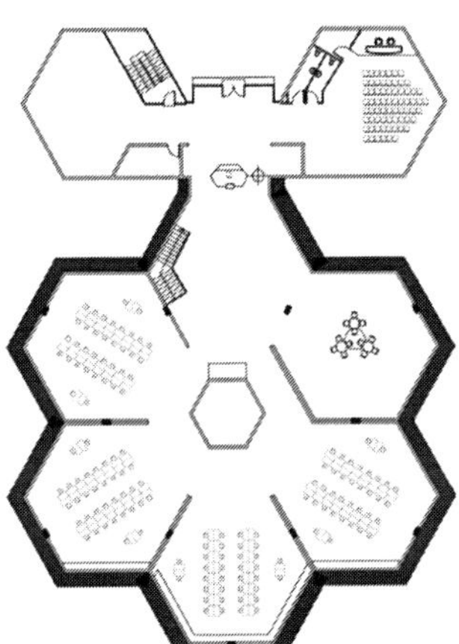

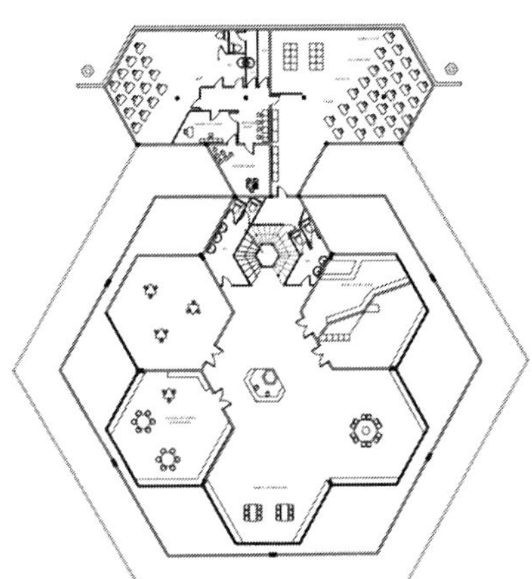

Fig. 22. Plans of library by Eldem (Redrawn by Ahmet Fuat Duman).

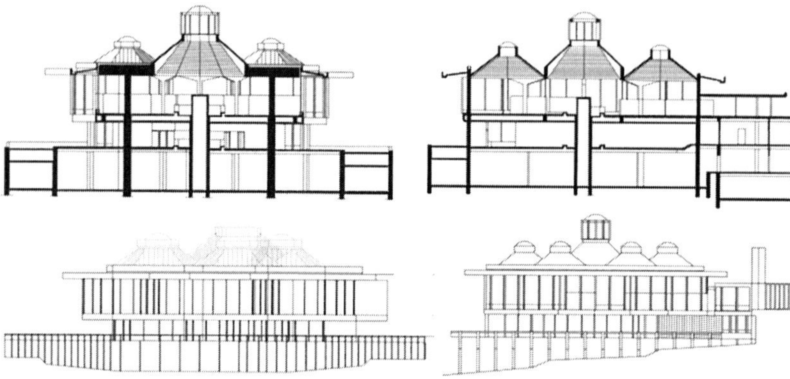

Fig. 23. Sections and facades of library (Redrawn by Ahmet Fuat Duman).

Conclusion

Covering the years 1940–50, there are some common characteristics of the buildings belonging to Second National Architecture Period. It is seen that a nationalist approach is dominant in the proportions and details of architectural elements such as windows and eaves in buildings. When the plan characteristics of the public buildings are examined, it is seen that they have a rectangular or square plan scheme. The most distinctive feature of the buildings is that they are symmetrical and monumental. Architecture in this period has a historicist aspect: In some ways, it evokes the exaggerated monumentality western architecture, and some details such as the eaves are related to traditional Ottoman residential architecture. When evaluated in terms of construction technology; it can be said that the highest technology standard of those years was reached.

However, in large-scale structures such as Ankara faculty of science and Istanbul faculty of science and literature, it has given importance to the need for function, material and carrier system, and traditional elements are used only in formal details.

The same as, when Sedat Hakkı Eldem was designing the Ankara Science Faculty with Emin Onat, he did not find a solution different from the monumental state buildings of the 1930's in terms of mass understanding. The building, which is planned symmetrically on a two-way axle system, has been subjected to a superficial formation, contrary to the structural features of the main reinforced concrete mass, reflecting all the National Architectural features of the period, facades are covered with cut stone, superficial lightening arches, brick ornaments, stalactite corner chamfers, and hedgehogs. The eaves were used on the building surface as the form features inspired by the early Ottoman Architecture. Thus, the new understanding of National Architecture did not go beyond an ornamentation or shaping effort that was free from the waxed surface decoration of the First Architecture period (48).

As a result, the common features of the structures examined in this study are (Tab. 1); the language of form, stone cladding used on the façade, entrances highlighted by colonnaded façade openings, window proportions and façade decoration elements referencing classical elements have been simple and rational examples of the approach. The entrance axes of the state buildings are explained with monumental-scale stairs, rich entrance halls, courtyards, squares and entrance colonnades were used as in architectural forms.

Tab. 1. Republican Period Architecture Public Buildings Plan Analysis

İstanbul University Observatory building	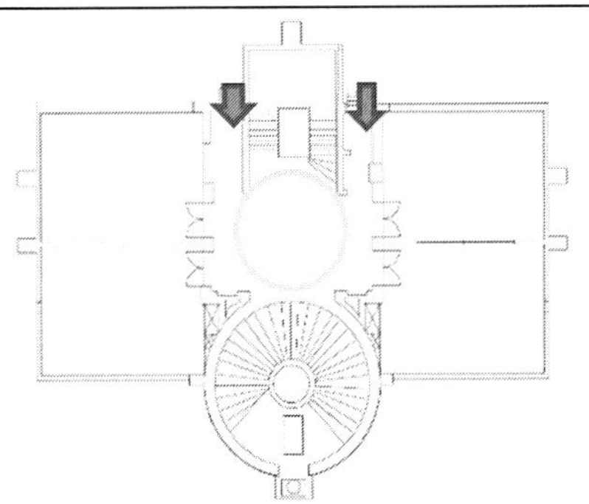	-rectagular plan with a circle observation tower at the axes, -horizontal window bands and corner rotating windows
Yalova Thermal Hotel	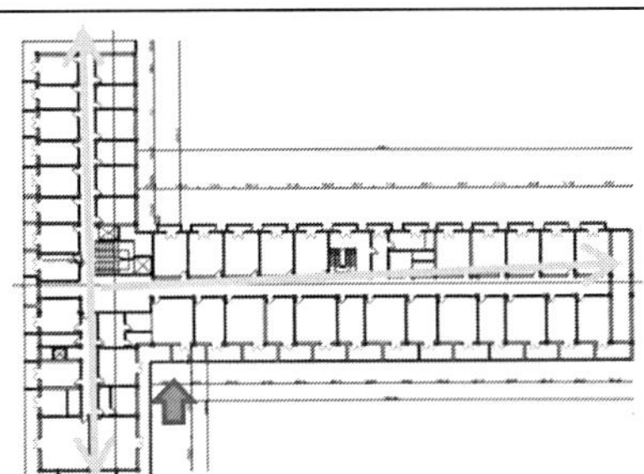	-horizantal **T** shaped plan. -porches surrounded by lattices, vertical element repetitions and modular accents

(continued)

Tab. 1. Continued

General Directorate of TCDD State Railways	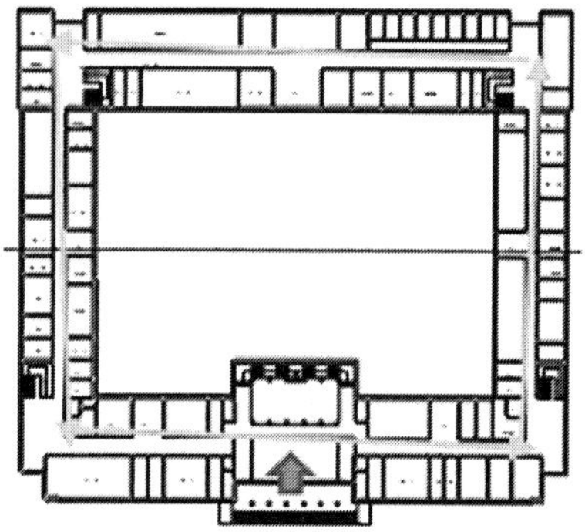	-Square plan with a courtyard, entrance axes with monumental-scale staircases, -facades are covered with cutstone
Istanbul University Faculty Science and Literature	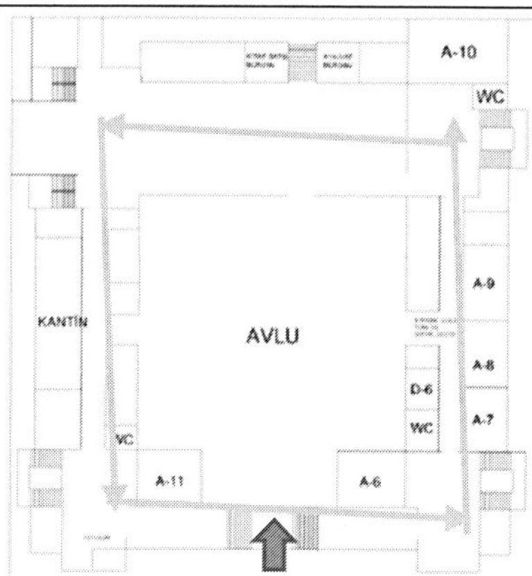	-Square plan with a courtyard, rich entrance halls, huge entrance colonnades -Facade is with porches, wide canopies and corner overhangs
Ankara University Science Faculty	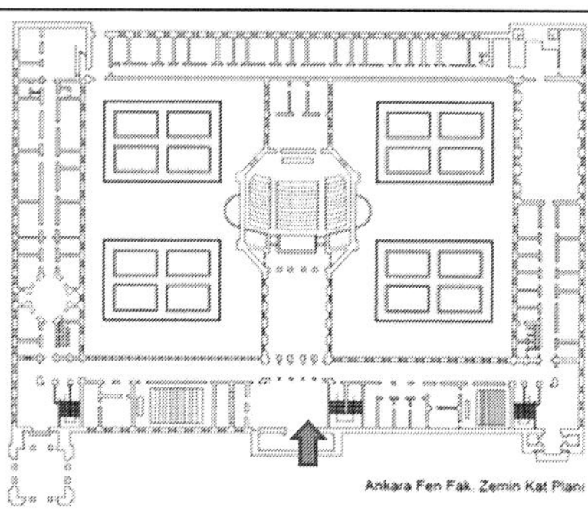	-Symetrical two rectangular planned masses, -stone walls, stalactite eaves, brick decorations on facade

Tab. 1. Continued

Istanbul Radio House	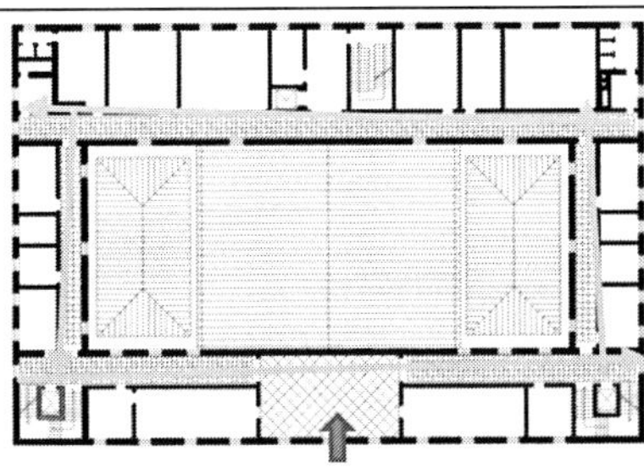	-Symetrical rectangular, entrance axes with monumental-scale vertical lines, facades are covered with cutstone
Istanbul Courthouse	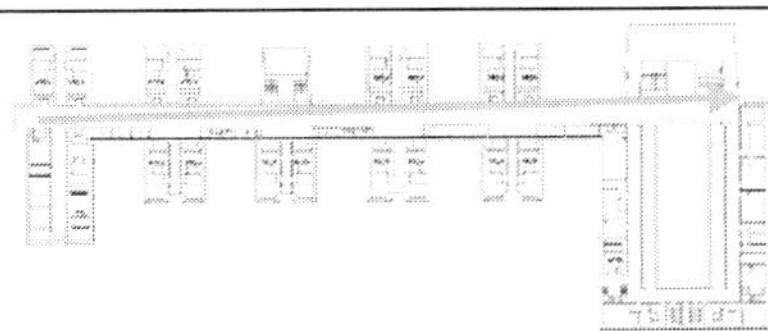	- Repeated blocks along an axis, the structure is dominated by modularity. the three-story high porch, long eaves and vertical window proportions
Atatürk Library	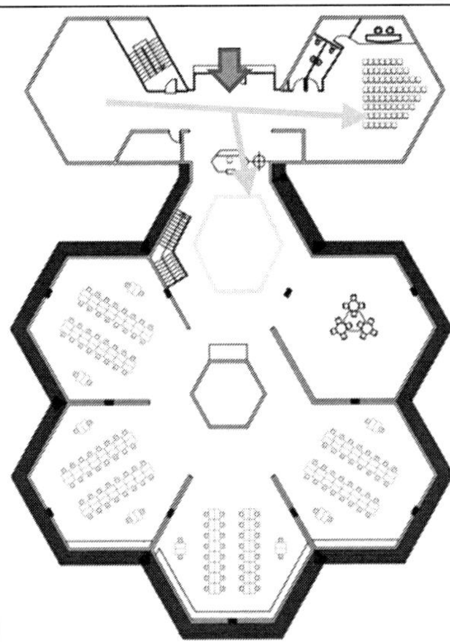	- hexagonal module, -the main mass, and in the middle of three other terrace-roofed hexagonal modules adjoining the hexagon, -two high chimneys on either side of this trio complete the composition

Acknowledgement

Thanks for the architectural redrawings of buildings to the students of "Republican Period in Turkey" course; Adilhan Gül, Ahmet Fuat Duman, Aslı Yılmaz, Sena Durgut, İrem Önder, İlayda Turhan, Kasım Gül and Abdullah Çelik.

References

1. Özdel, İ, Turkish Architectural Periodicals during the Republican Period, 1923–1980, Cankaya University Journal of Humanities and Social Sciences, No. 7 (2), 2010: 517–550
2. Durukan Kopuz, A. "Republican period in Turkey" elective course in Tekirdağ Namık Kemal University, Architectural Department, 2020.
3. Durukan Kopuz, A., An Example Building from the First National Architecture Period in Adana; Adana Train Station, Design+Theory Journal 12 (22), 2016: p. 21–33

4–5. Alsaç, Ü., The Evolution of Architectural Thought during the period of the Republic of Turkey, Karadeniz Technical University Press, Trabzon, 1976: 12–25

6. Sözen, M., Tapan, M., Turkish Architecture of 50 years, Culture broadcasting of İş Bank, 122 Queen Press, İstanbul, 1973:255
7. Durukan Kopuz, A. "Alpullu Sugar Factory and Workers' Houses" Metu Journal of the Faculty of Architecture 35.2, 2018: 29–54.
8. Arkitekt, İstanbul University Observatory, No: 1936–04 (64) 1936: 97–102

9–10. Küreğibüyük, Z. İ., Arif Hikmet Holtay in the Context of the Republican Architecture, Unpublished Master Thesis, İstanbul Technical University, 2011.

11. Bayhan, B., First Refresh of Istanbul University in Turkey (https://www.arkitera.com/haber/istanbul-universitesi-turkiyenin-ilklerini-yeniledi), 2013, (accessed 01.10.2020)

12–13–14. Küreğibüyük, Z. İ., Arif Hikmet Holtay in the Context of the Republican Architecture, Unpublished Master Thesis, İstanbul Technical University, 2011.

15. Arkitekt, İstanbul University Observatory, No: 1936–04 (64) 1936: 97–102
16. Can, A., Telling Sedat Hakkı Eldem I: 1930–1950, SALT Research and Programs.
17. Arkitekt, Yalova Thermal Hotel, No: 1938–03 (87), 1938: 67–81
18. Tanju, B., A Catalog trial: Sedad Hakkı Eldem, In Bülent Tanju and Uğur Tanyeli (Eds.), Sedad Hakkı Eldem II: Retrospektif Istanbul: Osmanlı Bankası, 2009: 29–285.

19. Arkitekt, Yalova Thermal Hotel, No: 1938–03 (87), 1938: 67–81

20–21–22. General Directorate of State Railways, survey and restoration project, (http://anitsal.com/proje/506/%E2%80%A2_tcdd_genel_mudurlugu_binasi_rolove_ve_restitusyon_projeleri) (accessed 01.10.2020)

23–24–25. Arkitekt, Public Administration building, (D.D.Y Umumî İdare Binası), No: 1941/42-11-12 (131–132) 1941: 241–247

26. Günergun, F., Kadıoğlu, S., 2006, Notes on the Settlement History of Istanbul University, Osmanlı Bilimi Araştırmaları VIII/1, 2006:154.

27. Buildings, Education Facilities. (https://www.imo.org.tr/resimler/ekutuphane/pdf/16633_34_10.pdf) (accessed 05.10.2020)

28. Can, A., Telling Sedat Hakkı Eldem I: 1930–1950, SALT Research and Programs.

29. Tanju, B., A Catalog trial: Sedad Hakkı Eldem, In Bülent Tanju and Uğur Tanyeli (Eds.), Sedad Hakkı Eldem II: Retrospektif Istanbul: Osmanlı Bankası, 2009: 29–285.

30. Ergüler, M., Analysis of Sedad Hakkı eldem Buildings, İTU Msc Thesis, 1996:104–114

31. Yavuz, Y., Architectural Form Concern in Republican Period Ankara, Architecture, 121–122 Ankara, 1973: 31

32. Bozdoğan, S., Modernism and Nation Building: Turkish Architectural Culture in the Early Republic, University of Washington Press, 2001:34

33. https://archives.saltresearch.org/handle/123456789/79586?mode=simple (accessed 05.10.2020)

34. Arkitekt, İstanbul Radio House Project Competition, No: 1945-07-08 (163–164) 1945: 143–157

35. https://tr.wikipedia.org/wiki/%C3%96mer_G%C3%BCney (accessed 05.10.2020)

36. Arkitekt, İstanbul Radio House Project Competition, No: 1945-07-08 (163–164) 1945: 143–157

37. Arkitekt, Istanbul Justice Building Project Competition, No: 1949-07-10 (211-212-213-214) 1949: 179–194

38. http://www.kalinti-istanbul.com/item/eski-adliye-binasi/ (accessed 25.10.2020)

39. Arkitekt, Istanbul Justice Building Project Competition, No: 1949-07-10 (211-212-213-214) 1949: 179–194

40. Tanyeli, U; Sedad Hakkı Eldem, Boyut Print Group,İstanbul, 2001: 77–82

41-. Arkitekt, Istanbul Justice Building Project Competition, No: 1949-07-10 (211-212-213-214) 1949: 179–194

42. Tanju, B., A Catalog trial: Sedad Hakkı Eldem, In Bülent Tanju and Uğur Tanyeli (Eds.), Sedad Hakkı Eldem II: Retrospektif Istanbul: Osmanlı Bankası, 2009: 29–285.
43. Arkitekt, Istanbul Justice Building Project Competition, No: 1949-07-10 (211-212-213-214) 1949: 179–194
44–45. Arkitekt, Atatürk Library, No. 374, 1979: 43–46.
46. Tanju, B., A Catalog trial: Sedad Hakkı Eldem, In Bülent Tanju and Uğur Tanyeli (Eds.), Sedad Hakkı Eldem II: Retrospektif Istanbul: Osmanlı Bankası, 2009: 29–285.
47. Arkitekt, Atatürk Library, No. 374, 1979: 43–46.
48. Yavuz, Y., Architectural Form Concern in Republican Period Ankara, Architecture, 121–122 Ankara, 1973: 31

Aybike Ayfer KARADAĞ

An Example of Ecology-Based Analysis in Landscape: Landscape Sensitivity Analysis

Abstract: Despite all its wealth and power, landscape is the most sensitive victim of the changing face of life. This is an evidence of man's attitude towards the world in which he lives. While this evidence carries traces from the past, it also leaves clues to the future. At the same time, it guides the human being with all the details of the landscape existence. However, today it is not easy for human beings to see these details, perhaps to read the secrets of life, to understand what they read, and to pass on what they understand to their lives. This study arises from this concern. Landscape sensitivity analysis, as one of the methods that will contribute to the understanding/reading of the landscape, is explained in the study.

Keywords: Landscape, sensitivity, vulnerability, adaptive capability, resilience, sustainable approach.

1 Introduction

Today's world is living a period in which human beings are transformed into a consumer society, the demand for production has increased dramatically, technology has come to terms with nature, and environmental problems have reached irreversible proportions. This period is shaped by population growth, industrialization and technological development.

Today, people are planning the world with the desire to live in cities. However, in 1800, only 2 % of the world's population lived in cities, and cities were not at all suitable for a good life [1]. However, with the changing world, this number has changed rapidly in the direction of increase. Especially in the 1950s after the world wars, when the population started to increase, 30 % of the world population started to live in cities [2]. Nowdays, urbanization has been one of the most important global trends. Cities now host 55 % of the world's population [3]. This inconceivable trend leads to the production of scenarios in which approximately 68 % of the world [2], in other words 7 out of 10 people in the world, will live in cities by 2050 [4].

Today, this unprecedented rate of urban sprawl brings with it global environmental problems and natural disasters. Urban sprawl, especially in forests, wetlands and agricultural areas, causes the loss of habitats, degradation and fragmentation of landscapes. Cities that tend to be consumer, require large natural resources and generate increasing amounts of waste cause an increase in air, water and soil pollution [5]. In addition cities are major contributors to climate change.

According to UN Habitat, cities consume 78 % of the world's energy and produce more than 60 % of greenhouse gas emissions [6]. Especially climate change has also increased the frequency of some natural disasters (hurricanes, droughts, floods, landslides and earthquakes) in the last 20 years. While this situation caused more loss of life and property day by day, it also brought about a traumatic life. In addition, every year, approximately 26 million people live in extreme poverty with natural disasters, ie $ 1.90 or less per day [7, 8]. To summarize the current state of the world; population is increasing rapidly and increasing population is concentrated in cities. Cities are spreading rapidly and often unplanned, with the power they get from technology and industry to carry this population. Cities turn into one of the main sources of environmental problems and natural disasters. Nowadays, sustainable living approaches have come to the fore as way to solve these problems and struggle [9–11].

Sustainable approaches; they are perspectives that support developments that enable future generations to meet their needs by protecting their own needs. These approaches are scientific solutions that touch every aspect of life, such as planning, design, management, etc. Sustainable approaches basically ensure the ecological order in the world and the vital union of human beings and establish the balance of their mutual relations. It takes its strength from scientific data [12, 13].

Landscape sensitivity analysis, which is the subject of this study, is one of the sustainable spatial planning approaches. Sustainable spatial planning is a planning approach that produces spatial solutions on the basis of ecologica balance, defines human life limits, observes natural processes, etc. These approaches generally have ecology-based analyzes and evaluations that are rational, reliable, defensible, etc. [14–16]. In this study, landscape sensitivity and analysis methods are evaluated on the basis of the landscape perception and the formation of the landscape.

2 Ecology Based Analysis in Landscape Planning

It is necessary to understand the landscape to understand ecological analysis in landscape planning. Understanding landscape starts with understanding its conceptual development. In this context, first the concept of landscape is explained as well as landscape planning. Afterwards, the subject of ecological analyses regarding the landscape is discussed.

2.1 Landscape Concept and Landscape Planning

Landscape is an expression that we use in the concrete and abstract dimensions of our lives from the earliest periods of the past to the present. The fact that the landscape is the subject of scientific research has created a necessity to reveal the development process of this concept. In this context, when the relevant literature is examined, a rich history regarding the landscape resurfaces.

The first traces of the conceptual development of landscape coincides with the beginning of the Middle Ages as stated in Braune and Ebbinghaus (1969) and

SchuÈtzeichel (1969). The first references to the term land-scape have been in Germanic languages. *The oldest written evidence was found in old High German manuscripts.* "Lantscaf" or "lantscaft'' statements were encountered in the Tatian manuscript Codex 56, written around the year 830 A.D. The term "Lantscaf" as given in Hard (1976) was used to translate the Latin term "regio" which means area, territory, region. In the medieval period, this term was synonymous with "regio" and "territorry", one of the themes still in use today. According to Geerts and Heestermans (1992) and Vriesand Tollenaere (1991), this term was known as "Lantschap" in the Middle Dutch period from the 13th century onwards. There it defined an administrative entity, a particular area, or was used as a synonym for land or the native country of a person. Old English manuscripts, i.e. the Anglo-Saxon Genesis, used the term "Landscipe", again as a synonym for "region". As given in Onions (1969) and Simpsons and Weiner (1989), when Dutch landscape paintings came to England at the end of the 16th century, "lands-cape" was used to refer to a work of art, a painted scene. In these pictures landscape was no longer just the backdrop for the figures as it was during the Middle Ages; it became subject on their own right. In 15th and 16th centuries landscape painting originated as an independent genre and became popular in the 17th and 18th centuries. Early stages of the conceptual use of landscape revealed that it was viewed as region and perception. As stated by Humboldt (1849), Humboldt, Goethe, Ritter, Steffens and others introduced the term "landscape" to scientific circles in the transition from the 17th century to the 18th century. Humboldt defined the term as the physical reality of a landscape; for him, landscape was the sum of the properties of a region [17]. In 19th century, landscape, for Ritter, was "a particular form of a particular land"; for Troll, "it is part of the earth's surface and forms the spatial unit of a particular character due to external relations and cooperation between phenomena"; for Humboldt it was "a complex of properties of a particular region." These definitions take a conceptual geographical approach. The first nature-related definition of landscape was published in Poland in 1912. Wodziczko's definition was that landscape represents the entire nature within a limited fragment of Earth's surface, whose state of equilibrium is a result of ongoing self-regulating processes [18].

At the beginning of the 20th century Vidal de la blache and others revived "landscape as a scientific concept. It was common at that time to look at the summation of the properties of a region and mentally combine it [17]. As stated by Sauer (1963), Olwig (1996) and Muir (1999), Sauer introduced this term to American geography by using the ideas of German Romanticism and natural philosophy. According to Sauer, landscape was a combination of different forms, both physical and cultural. Since Hartshorne landscape is preferable alternative for region and space concept, he found territorial concept confusing. He also suggested it to be removed from geographic research. However, Sauer's vision resulted later in the first important symposium on Man's Role in the Changing Face of the Earth [17, 19]. As stated by Antroop (2006), this situation has led geographers to lose their interests on the landscape. Geographers have reduced their interest in landscaping in regional synthesis and visual/aesthetic research. Landscape synthesis was used by very few researchers during this period [20].

Today, as stated by Zonneveld (1995), landscape is defined as the combination of ecosystems composed of abiotic/physical (climate, micro-climate, geology, geomorphology, hydrology, soil, topography) and biotic (vegetation, fauna, biodiversity) components. In addition, the other component of this ecosystem, cultural component that exists with humans is added to the definition (residential areas, agricultural areas, transportation, historical and archaeological sites, etc.). In this context, landscape is defined as the complex network of relationships formed by human-ecosystem interaction [20]. With an ecology-based approach, landscape has been defined as a heterogeneous land area composed of ecosystems that are similarly repeating and interacting with each other [21–24]. Also, for an ecologist, landscape is the vegetation and associated faunal populations draped over the geomorphology that give it most of its color and texture. The brochure of the International Association of Landscape Ecology's 8th annual meeting in the UK Territory in 1999 states that "landscape consists of mosaics of different habitats arranged in complex patterns at different scales" [25]. In European Landscape Convention signed in 2000 (ELC) landscape is defined as "the area formed as perceived by people as a result of its characteristics, interaction and action with human and/or natural factors" [26]. Landscape has also been defined as "the visible interaction of abiotic, biotic and human processes developing on the earth's surface over time" [27]. It is a mental/factual element that includes not only abiotic (physical) elements and natural processes, but also the information we perceive with our senses [17].

Landscape is a concept used by various disciplines. The approach of other science fields to the concept is different from each other. For example, the concept of landscape in prehistoric archeology is closely linked to the term cultural landscape. In Classical Archeology, landscape is perceived as the backdrop or environment whose characteristics would shape the mentality of its inhabitants and thus, the expression of the artist. Historians' approach to landscape involves the analysis of complex interactions between human factors and natural factors and the study of environmental perception of people in the past. Geographers define natural landscape as a "region" that represents, as much as possible, a unit according to its climate, vegetation, modeling of the surface, geological structure and soil. Although it occurs frequently in landscape geology, it is not clearly defined. Most likely, geologists have a similar landscape definition as physical geographers; and, they use it to describe the geomorphological features encountered in a particular environment. Geologists use landscapes as analogues for past environments and relate processes observed in the modern landscape (e.g. erosion or sedimentation) to past processes based on realistic approaches [28]. As a result, the concept of landscape has scientific uses to express human and nature interaction.

As seen, the concept of landscape has been used in every period of history and in various fields of science. While these uses reveal the landscape perception in life, they also express many features in relation. However, in order to understand the analysis and necessity of the landscape, it is necessary to go beyond these definitions and to know its features.

The most prominent feature of the landscape is that it is a spatial element that we can physically see and on/in which we perform various activities. Spatial features of the landscape are geosphere and biosphere while its perceptional feature is noo-sphere (human-sphere). Geosphere and biosphere reflect its natural features; noo-sphere reflects its cultural features. In addition, landscape with its dynamic structure gains character over time and changes with the effect of natural (biotic, abiotic) and cultural components [17, 20]. In short, landscape is a space that gains all its features in its existence over a period of time.

Landscape actually contains land uses (housing, transport, agriculture, recreation) and natural areas, or a combination of these. [29]. The landscape creates an environment for ecosystems and especially for humans. In this context, it refers to the whole, not to the parts of the whole [17]. The natural and cultural components of the landscape provide ecosystem services. In addition, natural assets can also be called "natural capital", which are elements of nature that produce value or benefit to humans [30].

Landscape has an important public interest in the cultural, ecological, environmental and social fields and constitutes a resource conducive to economic activity and whose protection, management and planning can contribute to job creation. It contributes to the formation of local cultures; in this context, it is an essential component of the natural and cultural heritage and contributes to human well-being and consolidation. Landscape is an important part of the quality of life for people in every location: in urban areas and in the countryside, in degraded areas as well as in areas of high quality, in areas recognized as being of outstanding beauty as well as everyday areas [26]. In this context, they contribute to what is fundamental to the existence of societies i.e. a sense of place and social well-being [30]. Thaitakoo (1998) and Vos and Zonneveld (1993) combined contribution of landscape to people within four groups [31].

1. Production function: providing capacity to produce such as mineral materials, agriculture production, hydraulic energy, etc.
2. Carrying function (supporting function): providing space and capacity for human construction and transportation.
3. Regulation function: providing self-regulate sustainable environment based on the resilience of ecological systems by various cybernetic loops keeping the environment suitable for society.
4. Information function: providing information regarding the various potentials and limitations and linking up with aesthetic, scientific, and historical interest of humankind.

Landscape sustainability is important thanks to all these values it provides to human life today. In the European Landscape convention developed for this purpose, it is stated that "landscape is one of the key elements of individual and social welfare and hence, its protection, management and planning is a right and responsibility for everyone". Therefore, the Convention emphasized the importance of landscape planning to develop strong forward action to develop, restore or create landscapes [26]. Landscape is the interface between social and environmental processes. Human communities, on the other hand, face many social, economic, political and environmental problems and opportunities. Problems and opportunities in particular lead to certain planning problems. At this point, landscape planning deals with issues concerning the mutual relationship between human and nature. Thus, landscape plans present a local scale development strategy; provides flexible guidelines for policy makers, land managers and land users on how to protect, restore or develop an area [29]. Landscape planning is similar to land use planning in that it is related to the macro environment. However, landscape planning focuses on resources and systems in planning, and management decisions [32].

Landscape planning observes the principles of sustainability. In this context, The Federal Agency of Nature Conservation has stated that "landscape planning supports sustainable development by creating planning preconditions that will enable future generations to live in an ecologically unspoiled environment" [33]. This statement is correct; because landscape planning contributes significantly to the sustainable conservation and long-term development of landscape, nature and natural resources at all levels (local, county or entire regional state) [34].

Landscape planning aims to prevent environmental degradation or to minimize its effects. In this context, landscape planning questions the suitability of land for certain uses, and appropriate areas are defined. In this process, contribution is made to the integration of landscape planning and ecological principles into spatial planning [35]. At this point, it is appropriate to mention the concept of ecological planning. Ecological planning produces procedures for primarily studying the biophysical and sociocultural systems of a place to reveal where specific land uses can best be applied [29]. In other words, ecological planning enables the use of biophysical and sociocultural knowledge to suggest opportunities and constraints in decision making about the landscape. In this process, ecological planning offers a specialization in landscape planning that focuses on spatial planning to achieve clear goals [36].

Today, increasing environmental concerns aim for sparing use and sustainability of natural resources, elimination of pressures on resources and appropriate planning [37]. In this context, ecological planning is accepted as the basis for sustainable development [29]. At the same time, it is increasingly recognized around the world that more sustainable approaches are needed for landscape planning and management. New tools are needed to effectively apply sustainable principles to planning and management. The spatial dimension of sustainability combines processes and relationships between different land uses, ecosystems and biotopes at different scales and over time. Therefore, ecological knowledge is essential when planning sustainability [38]. In addition, contemporary planning procedures also include formal procedures such as sustainability assessment, (strategic) environmental impact assessment, comparative assessment of alternative options, etc. These procedures require methods that address environmental protection issues in clear and legally defensible manner [39]. Because today, systems that are not based on transparent and provable foundations create insecurity and lose their validity over time.

2.2 Ecology Based Analysis

The relationship between landscape and ecology, as stated by Tansley (1935), is based on the long-term relationships between geography and ecology, which can be traced back to the origins of ecology. In this context, the theoretical foundations and methods of ecology have been applied to geographical objects, especially landscapes. In addition, the theoretical foundations and methods of geography has been applied to solve human ecology and the analysis of human-nature interaction problems. These relationships between geography and ecology have led to the development of a number of approaches such as Landscape Ecology, Geoecology, Ecological Geography (Ecological Geography), Ecological Geomorphology, Ecological Landscape studies, etc. As Sochava (1978) first stated on this matter, landscape ecology has developed in German Geography with the works by Troll (1939), Schmithussen (1942), Neef (1961) and Haase (1964) [40]. Landscape ecology presented approaches supported by ecology and landscape sciences, in which landscape functions are analyzed and the structural change of landscape is examined [41]. Landscape ecology, according to Grodzynskyi (1993), is an independent field located between ecology and landscape studies; but for Dierke-Worterbuch (1997), it is a professional field among geography, regional planning and environmental protection studies. Landscape studies based on ecology, which developed within the framework of geography between 1950–1970, spread to many countries of Europe, America and Asia in the 1980s [40]. Today, these studies are carried out wherever science reaches.

Ecological studies of landscapes as a discipline relies on a logical theoretical concept which includes an object and themes of investigation and system of specialized terms and notions; and which possesses a clear goal and concrete tasks. Landscape analysis, focused on solving problems in the human environment, can be considered an ecological analysis of the landscape. Ecological analysis of landscapes is a methodological concept that investigates the state, ecological conditions and problems of the human environment in order to protect, reproduce and improve the complex ecological features of the landscape. It is based on theoretical principles of Landscape Studies and use of scientific principles of Landscape Ecology, Geosystems Theory, Ecological Geography, Geoecology and other fields [40].

The content of the ecological analysis of the landscape is explained consistently with the concepts, characterizing the state of the natural environment of human activities, such as "natural or geographical environment, natural and ecological potential of landscape complexes, landscape ecological conditions (eco-conditions), ecological factors (eco-factors), ecological situations (eco-situations), ecological problems (eco-problems), ecological state (eco -state) of landscape complexes etc." In a sense, the analysis aims to reveal these concepts. For this reason, it is very important to first understand these concepts. These concepts are explained below [40].

- As given in Okhrana (1982), natural environment is the natural environments where people live and move, natural bodies and natural conditions. The natural environment consists of landscape complexes at different levels that can self-organize and develop. According to Hrynevetskyi and Schevchenko (1991), natural conditions that affect people's life, health, work and other activities are called eco-conditions. In this context, the eco conditions of the landscape are natural and anthropogenic characteristics of the landscape complexes on which people's living activities depend. Certain ecological features of landscape systems are eco-factors of the landscape. They can also be characterized by landscape eco-parameters [40].
- Positive or negative ecological consequences for humans (ecological consequences), Isachenko (1995) states, is linked to the eco-factors of the landscape. The consequences of the impact of ecological factors on certain regions may or may not coincide with the boundaries of landscape complexes. In case of non-overlap, certain habitats, areas and regions are formed with a specific eco-state called eco-anomalies. Eco-conditions and eco-problems, which remain stable during a certain period of time, form landscape eco-situations. Landscape eco-conditions, eco-problems and eco-situations reflect the eco-state of landscape complexes, i.e., the state of environment on the whole [40].
- The emergence of eco-problems of the landscape is linked to the changing eco-conditions under the influence of destructive rhythms of economic activities, and natural and geographical processes. It is the adverse nature changes that accompany the damage to the structure and functioning of landscape complexes and consequently lead to negative ecological, social, economic and other consequences [40].
- Eco-conditions and ecological problems that remain constant over a period of time create landscape eco-states. Landscape eco-conditions, eco-problems and ecological situations reflect eco-state of landscape complexes, that is, the general state of the environment. Eco-state of a landscape complex is interpreted in terms of human ecological needs. The evaluation of eco-conditions of landscape complexes under the influence of natural and anthropogenic factors is important in determining the suitability of landscape to human activities [40].
- The problem of optimization, improvement of eco-status of landscape complexes, i.e. ecological optimization, is an important issue. Here it is important to justify ecological norms (eco-norms) differentiated with respect to landscape complexes at various levels. Landscape eco-normalization is the normalization of anthropogenic loads on landscapes [40].

There are many ecology-based landscape analyses that reveal the ecological conditions/characteristics, ecological problems, situations, anomalies of the landscape. Various analyses are introduced on each passing day. The examples below belong to the most used landscape analyses [40].

Landscape character analyses enable us to evaluate the elements that make a landscape unique. Especially, the elements such as geology, topography, soil, vegetation, land uses, and settlements provide the landscape character. According to Swanwick (2002), landscape character is "a pattern formed by single, different and ongoing elements in a landscape". With these features, a landscape is distinguished from another one. Landscape character type is landscapes that differ from each other but are homogeneous in character. Landscape characters with similar characteristics can be found in different parts of the world. Understanding that an area is different/similar to another one contributes to better planning for the future by taking environmental and socio-economic factors of that area into consideration. In this context, landscape character analysis aims to reveal the nature of landscape [41, 42]. Landscape character analysis provides interpretation and evaluation of eco-conditions and complex structure of landscape and transforms it into spatial data.

Biodiversity analyses of the landscape have been proposed for species rather than genetics or ecosystem. In this context, important flora and fauna areas are graded according to the quality of protection based on expert opinions and this information is transformed into spatial data [43]. These analyses are also quite important in terms of revealing the nature and character of the landscape.

Biotope analyses ensure that the habitats created by the natural and cultural landscape for species are identified and transformed into spatial data. Thus, decisions can be made for the conservation of biological diversity [44]. At the same time, this analysis enables the natural and cultural characteristics of the landscape and the changes in them to be evaluated. There are many different methods in biotope analysis or mapping.

Water process analysis is one of the landscape function analysis and can also be expressed as the water function of the landscape. The main purpose of this analysis is to transform (map) the process of infiltration of precipitation from vegetation cover, soil, rock layers by the effect of gravity and slope into spatial data. Within the scope of the analysis, vegetation cover, permeability of soil and rock layers are evaluated holistically. The results allow grading areas with high and low permeability in terms of infiltration. The field usage decisions data obtained can be used in sensitivity, vulnerability, suitable analyses, etc. [43, 45–51]. Water process analyses are very important in terms of revealing the sensitivity situation in the landscape.

Erosion process analysis is one of the landscape function analyses and is also referred to as potential erosion analysis. The main purpose of this analysis is to transform erosion process of the soil, which is important for healthy sustainability of the natural and cultural processes of the landscape, into spatial data. Within the scope of the analysis, various methods and many data are analyzed to reveal the erosion risk. In this context, canopy cover, rock erosion, slope, precipitation, etc. data are evaluated in terms of erosion risk [43, 45, 46]. This analysis also reveals the problems/anomalies in the landscape.

Landscape ecology based analyses are carried out to reveal and map fragmentation(s) in the structure of landscape. Landscape structure analyses can be carried out in the level of landscape, patch class and patch. By the help of these analyses, the structure, function and change of landscape can be figured out. In the research, analyses have been made in the level of patch class. The data obtained is also very important in terms of temporal and spatial change. In this context, analysis based on landscape ecology reveals change analysis [43, 45, 46]. These analyses provide the expression of ecological problems in the landscape.

Bioclimatic comfort analysis is a holistic evaluation of the parameters of the climate (temperature, humidity, wind, etc.), one of the natural components of landscape, based on their positive effects on human life. Within the scope of this analysis, climate data are evaluated spatially [52]. Bioclimatic comfort analyses reveal the ecological state of the landscape and reveal suitable areas.

Ecological risk analysis is the evaluation of probability of adverse ecological impacts occurring as a result of exposure to one or more stress factors, and the probability of damage. Ecological risk analysis should follow a flexible process to organize and analyze data, forecasts, uncertainties while estimating the likelihood of adverse ecological impacts. An ecological risk analysis structured in this way offers decision-makers the opportunity to evaluate existing scientific knowledge along with other factors (social, legal, political or economic, etc.) that should be considered in making environmental decisions. Environmental processes (events) that can be evaluated on a physical and socio-economic basis also change the character of the ecosystem of which they are a part. For an effective ecological risk analysis, the effects of the processes that lead to the said negative change are evaluated within the scope of their observability and measurability; and thus, determining them for the current area is the primary priority. Within the scope of these analyses, environmental processes that carry ecological risks are determined quantitatively and spatially and converted into spatial data [53].

Suitability analysis is a tool used to determine the most suitable locations for future land uses. The most important step in suitability analysis is to be able to describe the interactions between human action and natural processes for spatial planning. Compliance techniques enable environmental managers and planners to analyze the interactions between location, development actions, and environmental elements. Analysts can then map these interactions in a variety of ways. For example, it may indicate (1) which land uses will have the least negative impact on environmental processes, (2) qualitative estimates of the environmental impacts of projected developments and (3) the most and least favorable locations [54]. The data obtained as a result of the suitability analysis provide practical data that can be used to determine policies and make decisions for many different land uses (agricultural land, garbage area, camp and picnic area, cemetery, settlement, etc.) [55–59].

Onc of the important analyzes that are effective in the decision-making process regarding the landscape, in global changes and its sustainable goal is vulnerability analysis. Vulnerability is generally called the deterioration/change of sensitive landscapes by being subjected to intense stressor and its failure to absorb and adapt to this situation. Vulnerability has three primary components: exposure to the stressor, sensitivity to a range of stressor variability, and resilience following exposure (Fig. 2.1) [35, 60–66]. Vulnerability, as G. F White (1974) defined it, is the degree to which a system, subsystem, or system component is likely to experience harm due to exposure to a hazard, either a perturbation or stress/stressor. This definition and the concept it addresses are not new. Vulnerability analysis can be done at any spatial or temporal scale appropriate to the problem in question [67].

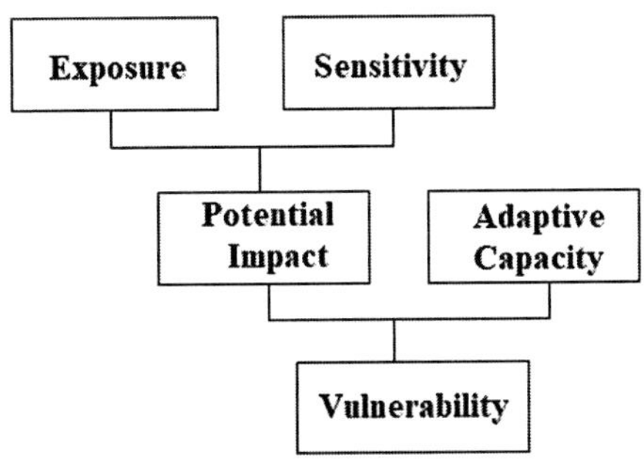

Fig. 2.1 Vulnerability process [68].

Vulnerability analysis enables to evaluate the landscape on the basis of its natural structure, processes and human effects; It is an approach that integrates many different processes and analyzes. In this context, the following concepts are important to perceive vulnerability analysis.

Landscape capacity/sustainability or adaptation capacity is the capacity to resist external influences/pressures that preserve the characteristics of a system, to manage risks and impacts, to respond to challenges, to adapt by developing effective approaches [62, 67, 69]. In other words, landscape capacity is about how much a landscape can absorb or accommodate development without a fundamental change in its character [70]. The ability or capacity to accommodate landscape change or development (defined in terms of type and scale) is based on the following relationship: *"landscape capacity (to accommodate specific type and scale of change) = overall landscape sensitivity + landscape value"* [71]. In this context, landscapes with high sensitivity on their character or structure are landscapes with low adaptation capacity. Stressor and duration of exposure to this factor are qualities that can affect the adaptation process.

Resilience is defined as the ability of a system to recover, rearrange and develop in the face of external stresses and disturbances [63]. Resilience can also be expressed as the capacity to adapt [65]. In this context, landscapes with high sensitivity also have a high resilience.

Vulnerability refers to the fate of landscapes with low capacity of response and resilience in the process of exposure, based on the state of vulnerability. In this context, Thomas (2001) stated that unstable systems behave erratically, while stable systems resist change until the threshold values of system parameters are exceeded [72]. In Fig. 2.2, interactions related to the vulnerability process are described. Also the comparative analysis of the concepts puts in evidence important similarities and differences, and in some cases contradictions, between the concepts as specified, or utilized, in different fields of inquiry. The comparison also shows that there is no generally accepted meaning for these concepts. The lack of general agreement on the concepts when considered one by one becomes more visible when they are taken together (Fig. 2.3) [73].

The basic philosophy of the landscape analysis given above is based on the idea that the landscape has a complex ecological system. The sensitivity evaluated in this study is the process that is in the main character of the landscape. Understanding the sensitivity process will contribute to spatial planning aimed at expressing the ecological structure of the landscape, which consists of complex relationships. As Ahern (2005) said, regardless of the method chosen or adapted, the issues of “spatial and temporal process, transferability and unpredictability of

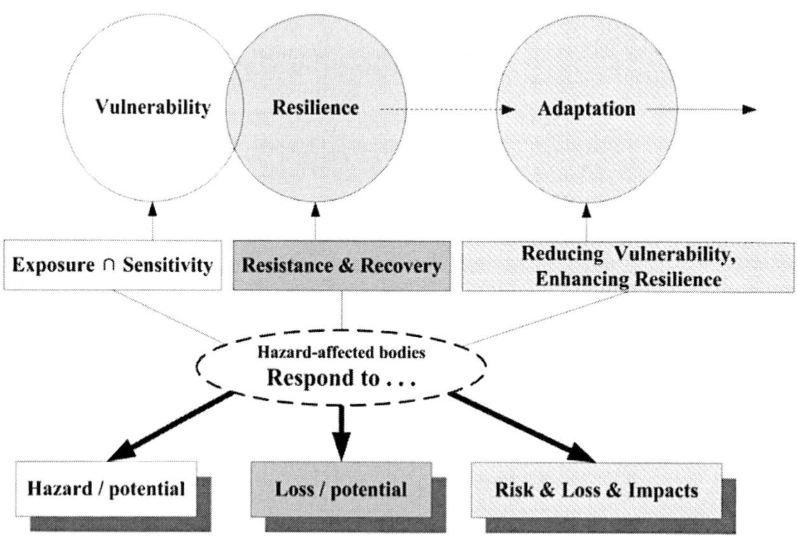

Fig. 2.2. The relationship between vulnerability, resilience and adaptation [74].

human input" always constitute uncertainty. Yet, strategies from science are useful to help reduce uncertainty. Planners work in the "real world" where there is an imperative to act. The world does not stop or wait as planners try to collect data to reduce uncertainty [36]. For this reason, it is necessary to model the analysis methods well and in this context, the ecological structure of the landscape should be examined well.

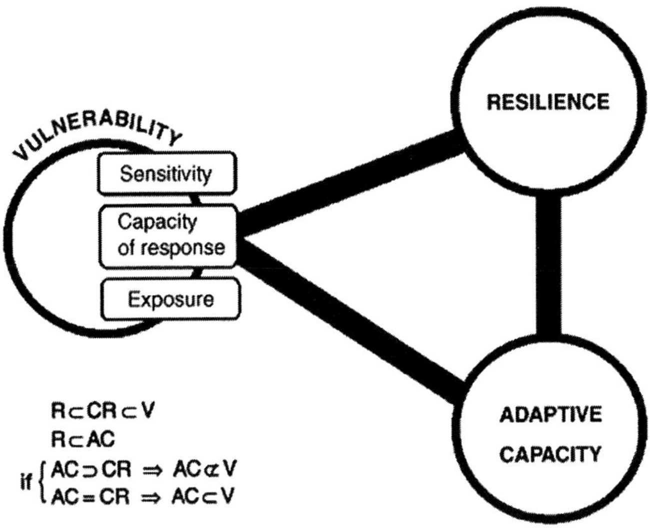

Fig. 2.3. The conceptual relations among vulnerability, resilience, sensitivity and adaptive capacity.The signs represent relationships between sets: C ¼ ''subset of''; g ¼ ''not a subset of''; R, V, AC, and CR stand for resilience, vulnerability, adaptive capacity, and capacity of response, respectively [73].

3 Landscape Sensitivity Analysis

Landscape sensitivity is one of the words which was present to describe the sensitive features, red lines and borders of the landscape. For, everything that exists needs a security area to survive. In this section, the concept of landscape sensitivity, sensitivity analysis and the criteria that are important in revealing the sensitivity, as well as the measurement methods that will make the sensitivity concrete, are explained.

3.1 Landscape Sensitivity Concept

Landscape sensitivity depends on the process of landscape formation, development and change. Landscape includes not only entities whose growth, decay, flow and transformation processes are maintained (rocks, soils, plants, animals, etc.), but also dynamic systems that contain energy and material stores [72]. In this context, landscape, consisting of many superimposed environments, is a semiotic interface between resources and organisms. This interface is shaped by ecological processes [75–77]. These processes are related to the physical (landform, superficial deposits, rock) and the biological systems (plant, animal and human) components, interactions and existence [78].

Landscape is the product of a long evolutionary history, reflecting the processes and interactions of geomorphology, topography, geology and climate as well as their changes over time [79]. The color and texture of landscape, shaped by its geomorphological history, mainly comes from the vegetation. This is largely a reflection of the soil, climate and their associated history [25]. The composition of landscape is particularly relevant to the Quaternary history. Quaternary history, natural processes and especially the occurrence of geomorphological processes have evidence of plant communities responding to both climatic changes and improvements. In addition, this evidence has information about the development of natural environment before and after human influence [79]. Therefore, understanding the relationship between landscape, nature and humans is key to defining the co-evolution between natural and human processes [80]. In addition, landscape has tips for the past and the future, with the influence of external forces changing and triggering its instability [81]. For this reason, landscape has always been a matter of curiosity for the future of humanity and the world.

The concept of landscape sensitivity was first used in 1979 to help thinking about geomorphological systems, particularly in understanding episodes of erosion and sedimentation [72]. On the issue, Milles et al. (2001) stated that landscape sensitivity comes from geomorphology [25]. Usher (2001) stated that "landscape sensitivity requires considering natural systems that determine landscape and its structure, and then the processes that change these natural systems, by taking time and space into consideration." This reveals that the sensitivity is based on the science of geomorphology [78].

Landscape is the product of geomorphological, ecological, and similar formation and development processes and they have complicated functioning. Therefore, as Usher (2001) stated, landscape sensitivity is not an easy term to define [78].

The concept of landscape sensitivity was first described by Thomas and Allison in 1993 as "the potential for change and the ability/response of the physical system to resist change in response to external influences of natural and anthropogenic origin" [82–83]. Thomas also used landscape sensitivity to express the response of landscape systems to deterioration in different temporal and spatial scales [72]. On the basis of the first definitions, various definitions that emphasize the change in landscape have been proposed. Some of them are given here. For example, landscape sensitivity has been defined as the ratio of the change in a system to the change in a landscape component, and hence, the greater the ratio, the greater the sensitivity [78]. In another definition, landscape sensitivity is interpreted as the vulnerability of the landscape character to change [70]. With a similar approach, landscape sensitivity is expressed as the conditional instability of landscape system and its rapid response to external influences [81]. In another similar definition, landscape sensitivity is "the capacity of natural components of landscape to change their characteristics and character under external influences; In other words, it is a type of response of the ecosystem to external factors/pressures and corresponds to the intensity of the response, the scale of the changes and their consequences". Sensitivity is understood as the capacity of natural ingredients to change their properties and character under external influences [69]. The definitions show that the landscape sensitivity is related to landscape's tendency to change, and its capacity to absorb the effects of deterioration [84]. In fact, definitions are made depending on the stressor exposure, and landscape's potential to be harmed [63]. In other words, the measure of the landscape's resilience to change is the extent to which it can accommodate change, without adverse effects on the character of the landscape.

3.2 Sensitivity Criteria

3.2 Sensitivity Analysis and Sensitivity Criteria

Generally, there was a need to reveal the sensitivity of landscape for three reasons. These are (i) determining what environmental conditions require how much change in the landscape; (ii) revealing which parts of the landscape are affected and how much; (iii) disclosing how long the transition of landscape features will take before the quasi-equilibrium state is reached [82]. Landscape sensitivity analyses developed in this context have been defined as the examination of how ecological variables respond to changes in external conditions [85]. Brunsden stated that the answer to the question of where sensitive landscapes are "should be sought in the resistance of landscapes and their answers to the forces of change" [72].

Usher (2001) emphasized about determining sensitivity that it is best to start thinking about landscape and the main natural systems that determine its structure and then the processes that change these natural systems. In addition, he also stated that both the changes and their magnitude have spatial and temporal components and that these components are an integral part of precision [78].

They identify four central key principles for the understanding and appropriate use of Landscape Sensitivity Assessment. They are [83].

- the emphasis placed on landscape character.
- the division between the process of characterization and the making of judgements to inform decisions.
- the roles for both objectivity and subjectivity in the process.

- the potential for application at different scales.

Landscape sensitivity criteria are the basis for answering the landscape sensitivity problem and developing resource management concepts. Therefore, the selection of criteria (indicators/parameters) is a very important issue. Criteria are the main cause, in other words, the source of sensitivity [83]. Expressing sensitivity, Thomas and Allison (1993) stated that "natural and human-induced external factors are the driving force in landscape changes and developments" [72]. Thiemann et al., (2018) expressed that the sensitivity criterion should contain the problems [83].

When the studies focusing on landscape sensitivity are examined (ecological sensitivity, environmental sensitivity, landscape capacity, vulnerability, resilience), it is seen that various natural and human-originated features are evaluated as criteria. Two approaches have attracted attention in determining the criteria. These are (i) selecting criteria that are considered problematic in the landscape that follows a reductionist attitude and favors temporal-spatial scale; (ii) selecting criteria for landscape/landscape character components that may cause sensitivity.

The first approach in selecting the criteria takes the holistic approach of temporal and spatial scales into consideration. Time and space are important factors affecting sensitivity [72]. In studies developed with this philosophy, erosion and sedimentation criteria were evaluated as temporal scale; habitat fragmentation and landscape pattern as criteria in spatial scale [72, 84]. In addition, within the framework of this approach, landscape related problems such as soil erosion, surface runoff, flood, landslide, etc. criteria were also evaluated [49, 83, 86]. On the other hand, this approach is a reductionist approach. For, while evaluating erosion, surface runoff, et al., the sub-reason (sensitive landscape component) that may cause these processes is also evaluated.

Landscape pattern is an important criterion in terms of temporal and spatial evaluations [25, 81, 84, 87, 88]. Landscape pattern is a concrete reflection of the spatial heterogeneity of the landscape [89] and it is the result of various ecological processes that affect different spatial and temporal scales [90]. Landscape fragmentation, on the other hand, increases landscape sensitivity [81].However, habitat loss and fragmentations are the most compelling triggers of global biodiversity crisis. For, landscape connectivity is critical for species survival at many scales. Habitat fragmentation reduces disintegration, increases mortality and decreases genetic diversity [88]. Also, the sensitivity of vegetation types in a landscape to change is impossible to understand without understanding the dynamics between and within the patches [25]. For this reason, landscape fragmentation is increasingly recognized as an important environmental indicator (sensitivity criterion) in sustainable land use and biodiversity areas [87].

Another approach used in criteria selection in landscape sensitivity analysis is the evaluation of landscape/landscape character components. In this context, aspect, erosion, elevation, distance to a water body, river network density, groundwater level, land capability, soil texture, land type, landscape type, biological diversity, road-effect zone, land use, conservation status, etc. were each evaluated as criterion [69, 86, 91, 92]. This issue also stated that in some studies that stress that the perceptual/visual features of landscape should be taken into account along with the physical character of landscape in sensitivity analysis [70, 93].

In some of the studies on sensitivity analysis, these two approaches are not differentiated. Many important landscape sensitivity components such as topography, aspect, elevation, slope, drainage density, ground water salinity, drinkable water resources, soil texture, land capability, annual rain fall, temperature, topographic wetness index, plan curvature, normalized difference vegetation index, distance to road, water infiltration, erosion, flood, earthquake, landslide, liquefaction, proximity to settlement, etc. are evaluated together [93,94,96].

No matter which approach is chosen, an issue needs to be considered. This issue is the recognition that, in the process of selecting and evaluating the criteria, any one or more landscape components may be sensitive to change in certain situations. It is also that a change in a component can cause instability in any part of the system since landscape is an internally complex system. Sub-units in this system can exhibit simultaneous but opposite behavior [72, 78]. Thus, sensitivity criteria should not be determined without understanding the processes that are effective in the formation, development and change of landscape on which the work is performed.

3.3 Analysis Methods

Analyzing the sensitivity of landscape and revealing the results is quite important in terms of understanding landscape, determining future trends, planning and managing. Analysis processes are based on modeling and mapping. Models are produced in order to contribute to prediction development and interpretation by imitating the nature of the interaction between human movement and natural processes [29].

Regardless of the scope and methodology adopted in landscape sensitivity analysis, the following principles should be adhered to [30]:

- A landscape sensitivity assessment should be as simple, clear, concise, proportionate and transparent as possible;
- Results should be easily understood by all;
- Complex approaches are not desired;
- There should be no complex risk modifications and/or documents rejected by decision makers;
- For the sake of consistency, appropriate definitions related to good practice should be used.

Landscape sensitivity analyses show a deductive approach. This process is based on concepts built on some theoretical basis. In this context, the first step involves understanding the phenomenon being studied and the main processes involved. In other words, in this step, the existence of landscape in which sensitivity will be revealed is examined. The second step is the reductionist step and involves defining the main processes to be included in the study and how they are related. The third step involves selecting criteria and assigning values and weights [72]. Sensitivity works. The most important issue to be considered in the evaluation process is to understand the system and to be able to produce data with a reductionist approach and avoiding excessive data [72, 84].

Spatial data, in other words, different methods are used in the process of generating maps in the evaluation of criteria. These methods include determining sensitivity criteria based on expert opinions; grouping criteria scoring based on sensitivity; evaluation of criteria with various methods (index accumulation method (IAM), geographic information systems (GIS) analysis, remote sensing methods; various indexes, holistic evaluation formulas; and mapping of the results. In these processes, analytical hierarchy process, Delphi technique, etc. can be utilized. Below are some examples of these methods.

Zhang Mingwu et al (2001), compared two methods used in sensitivity studies. The first method used in the study is the ecological unit method. In this method, the study area is first divided into ecological units. Criteria for sensitivity (elevation, water protection zone, biodiversity, land use, road protection zone) were identified and the weight factor which expresses the effect of each criterion on sensitivity was determined by analytical hierarchy process (AHP). Sensitivity indices related to subclasses of the criteria (sensitivity indexes) were determined by the Delphi technique. In the light of these values obtained, the index accumulation method was used to evaluate the ecological sensitivity of each ecological unit. As a result, the higher the evaluation values of the ecological unit, the higher the ecological sensitivity is accepted, and the results are mapped by GIS analysis. The second method is the GIS method. In this method, criteria have been set according to the ecological characteristics of the area (vegetation type, soil erosion, runoff depth). Regarding the subclasses of these criteria, sensitivity grades were grouped as *"High, medium, low, no"*. In addition, weighting scores for each criterion were determined by analytical hierarchy model. Finally, the data from the GIS spatial analysis module were evaluated holistically based on weight score. Within the scope of the study, it was observed that the results obtained in both methods were basically the same. However, the first method can be used in case of comparative shortage of existing data. The second method can be used in case of the abundant Remote Sensing (RS) and geographic information systems (GIS) data. The analytical results of GIS method are meticulous and objective [86].

In their study, Whang and Bian (2010) determined 8 criteria for ecological sensitivity (protection areas of drinking water source, groundwater salinity, soil texture, topography, vegetation cover, soil erosion intensity, and slope). Sensitivity levels related to criteria subclasses were grouped and scored as such (2: insensitive, 4: slightly sensitive, 6: moderately sensitive, 8: severely sensitive, 10: extremely sensitive). In addition, score weightings for each criterion were determined. An improved multiple index comprehensive evaluation method is adopted to analyze the ecological sensitivity of land use. The following formula is used. Where N is the score of the comprehensive evaluation of the eco-sensitivity, wi and ni are weight and score of the No. i index, n_7 and n_8 show that the 7th and 8th index of the eco-sensitivity analysis, which are based on single-factor calibration method [94]. Aretona et al., (2015), focusing on land cover characteristics and land use (vegetation cover, nature protection area, forests, agricultural land and its variety, wetlands, etc.) grouped the sensitivity of these areas in 5 groups (very high, high, medium, low, very low) and mapped it with GIS. He also included the data into ecological vulnerability analyses [63].

In their landscape sensitivity analysis, Karadağ ve Şenik (2019) have determined the sensitivity criteria first by considering ecological characteristics of the study area and adopting a reductionist approach. The sensitivity of criteria subclasses was scored between 1–5 points (5: very high, 4: high, 3: medium, 2: low, 1: very low) by experts. All scores were entered into the database of sensitivity criteria using ArcGIS. All data were then analyzed with ArcGIS (union analysis). Total sensitivity in the data obtained was calculated with the following formula: "*Landscape sensitivity total score = Erosion sensitivity score + Landslide sensitivity score + Water infiltration sensitivity score + Habitat sensitivity score.* The scores obtained were divided into 5 groups according to the sensitivity scores. In the study, it was stated that the lack of GIS-based data on the ecological sensitivity of the area would provide a deficiency in the definition of total sensitivity [49].

Chi et al. (2019) evaluated the landscape pattern in their study to determine the ecological sensitivity of landscape in temporal and spatial scale. The landscape pattern indicates the spatial characteristics of landscape composition, configuration, and function, and it represents the landscape ecological condition. Two new indices, namely, landscape ecological condition index (LECI) and landscape ecological sensitivity index (LESI), were proposed using remote sensing and geographic information system methods in the study. In the study, landscape patterns belonging to 4 different years at different heights were evaluated using satellite photographs. The researchers stated that the data required for the model could be easily obtained with remote sensing and geographic information system methods, and the calculation processes of the indices were clear, simple and repeatable [84].

In his work to determine vulnerability, Hussain et al., (2018) first determined the criteria (annual rain fall, degree of slope, temperature, land use land cover, topographic wetness index, aspect, drainage density, plan curvature, normalized vegetation index, elevation, distance to road, proximity to settlement). Subsequently, the subclasses of each criterion were scored between1–5. Weighting values of all criteria were determined by analytical hierarchy process. The data were then subjected to weighted overlay analysis with ArcGIS and the results were mapped [96].

In some studies conducted to determine landscape sensitivity, evaluations were made according to landscape character and visual sensitivity. In this context, each area defined according to the landscape character analysis was divided into smaller sub-areas (areas with a radius of approximately 1 km). The sensitivity of each sub-area was evaluated according to 6 criteria (physical and natural character, settlement form and edge, settlement setting, visual character, perceptual qualities, historic character). In this context, a judgment was made as to what 'key sensitivities' were applicable to the sub-area. Then, for each sub-domain, evaluations were made based on 5 sensitivity scores (very high, high, medium, low, and very low) [97–100]. Using a similar method, Perry (2017) defined the characteristics of the sensitivity degrees primarily according to landscape character and visual sensitivity. Then, he highlighted overall landscape sensitivity by evaluating these two sensitivities as a whole. The holistic evaluation created an evaluation matrix (Tab. 3.1) [71].

Tab. 3.1. Overall landscape sensitivity matris [71].

Landscape character sensitivity	*High*	**Moderate**	**High**	**High**
	Moderate	**Low**	**Moderate**	**High**
	Low	**Low**	**Low**	**Moderate**
		Moderate	*Moderate*	*Moderate*
		Visual Sensitivity		

4 Conclusion

Humankind, that is exists in nature as a part of the ecological system, with the power of his mind and the ambition in his creation, he has turned into a being that conflict with nature. This threatening attitude, which started with the industrial revolution, has upended the ecological balance today. Over time, these ecological imbalances have turned into environmental problems and natural disasters, and have turned into situations that threaten people. This situation has led to the formation of a life philosophy in harmony with nature. Today, many approaches originating from this philosophy have been produced in many different fields. Ecology-based approaches/attitudes in spatial planning that organize the spread of people around the world is one of them. Landscape sensitivity analysis is one of the evaluations developed within the scope of these approaches.

Landscape sensitivity analysis provides an assessment at the landscape scale. Evaluation is an analysis model developed on the basis of the ecological system and natural processes that make up the landscape. Time and spatial evaluations can be made with the data evaluated within the scope of the model. The results of the analysis are an approach that reveals the boundaries of human in the landscape and, from another perspective, the red lines of nature. In this context, the areas that people should protect in the landscape and the areas that can be used in ecological balance are defined. This definition is produced on the basis of scientific, objective, rational, transparent, interpretable, defensible, developable, etc. data.

In landscape sensitivity analysis, the setting of the model, in other words, the choice of criteria is very important. The formation of the landscape and the quality (scale, production technology) of the data to be analyzed are the main guides in the selection of criteria. In this context, a complicated model should not be used. The process should be analyzed with as much reduced data as possible and a simple method.

The data obtained within the scope of landscape sensitivity analysis are quantitative (area) and visual (map) data. These data are the basic data of landscape planning, landscape character analysis, optimal area selection, vulnerability analysis, risk analysis evaluations. At the same time, these data guide the planning and management processes such as landscape, space, basin, protected area, etc. It provides reliable and transparent strategies for landscape-spatial-nature policies and sectoral decisions.

As a result, landscape sensitivity analysis offers an ecological approach that provides spatial evaluations where the human mind is used correctly and ambition can be controlled on the basis of ecological sensitivities. In a way, it is one of the secrets of sustainable life.

References

1. Boyle Torrey, B. Urbanization: An Environmental Force to be Reckoned With, 2004. Access link: www://richmondvale.org/en/blog/6-major-effects-of-urbanization-that-will-make-you-think (date of access: 30.01.2021).
2. United Nations, *World Urbanization Prospects The 2018 Revision,* 2019. https://population.un.org/wup/Publications/Files/WUP2018-Report.pdf (date of access: 30.01.2021).
3. Clos, J.A. *New Urban Agenda for the 21st Century: The Role of Urbanisation in Sustainable Development,* 2016. OECD Regional Outlook 2016 Productive Regions for Inclusive Societies. Access link: https://www.oecd-ilibrary.org/sites/9789264260245-9-en/index.html?itemId=/content/component/9789264260245-9-en (date of access: 30.01.2021).
4. World Bank, *Urban Development,* 2020. https://www.worldbank.org/en/topic/urbandevelopment/overview
5. Anonymous *6 Major Effects of Urbanization That Will Make you Think!* 2017. Access link: https://richmondvale.org/en/blog/6-major-effects-of-urbanization-that-will-make-you-think (date of access: 30.01.2021).
6. United Nations, Cities and Pollution, 2020 https://www.un.org/en/climatechange/climate-solutions/cities-pollution
7. Banholzer, S., Kossin, J., Donner, S. *The Impact of Climate Change on Natural Disasters,* 2014. Reducing Disaster: Early Warning Systems For Climate Change (pp.21–49) Edit: A. Singh, Z. Zommers, Springer.

8. World Bank, *Nature-based Solutions: a Cost-effective Approach for Disaster Risk and Water Resource Management, 2019.* Access link: https://www.worldbank.org/en/topic/disasterriskmanagement/brief/nature-based-solutions-cost-effective-approach-for-disaster-risk-and-water-resource-management (date of access: 30.01.2021).
9. Cengiz, A. E., Çavuş, C. Z. and Koç, T. *The Relationship Between the Irrigated Agricultural Areas and the Urbanization in Çanakkale and Kepez Settlements.* 2014, Turkish Journal of Geographical Sciences, 12(1): p. 69–88.
10. Açmaz Özden, M., Cengiz, A. E. (2015). *The Value of Urban Green Infrastructure for Sustainability of Urban Ecosystems.* Tourism Environment and Sustainability, Ed: C. Avcıkurt, M. Dinu, N. Hacıoğlu, R. Efe, A.Soykan, 2015, St. Kliment Ohridski University Press, Sofya, p.507–535.
11. Cengiz, A. E. *Ecological Approaches in making Urban Land Use Decisions Science.* Ecology and Engineering Research in the Globalizing World, Ed: R. Efe, I. Chrıstov, E. STRAUSS, A. A. Gad, I. 2018, St. Kliment Ohridski University Press, Sofia, p: 40–57
12. Cengiz A. E., Pekin Timur, U., Karadağ, A. A. and Demiroğlu, D., A Contribution to Sustainable Urban Development: Urban Agricultural Lands. Urban and Urbanization, Ed: R. Efe, Recep, T.T., Onay, I. Sharuho, E.,Atasoy, 2014, St. Kliment Ohridski University Press, p. 85–104.
13. Cengiz A. E, Karadağ A. A, Demiroğlu D., *Importance of Landscape Planning for Sustainable Spaces.* International Congress of the New Approaches and Technologies for Sustainable Development. Suleyman Demirel University, 2017, September, Isparta, Turkey, p. 1, 4–16.
14. Demiroğlu, D., Pekin Timur, U., Karadağ, A. A. and Cengiz A. E., *Ecology-Based Contemporary Urbanism Approaches, Kitap adı: Urban and Urbanization,* Urban and Urbanization, Ed: R. Efe, Recep, T.T., Onay, I. Sharuho, E.,Atasoy, 2014, St. Kliment Ohridski University Press, p. 579–600.
15. Demiroğlu, D. and Karadağ, A. A. *Ecosystem Services Approach to Spatial Planning in Turkey.* I. Internatıonal Urban Studies Congress-Problems in Present-Day City, 2015, April Eskişehir, Turkey, p. 153–168.
16. Karadağ, A. A., Demiroğlu, D. and Cengiz A. E. *Importance of Ecological Analysis in Spatial Planning.* International Congress on Agriculture and Forestry Research (AGRIFOR), 2019, April, Marmaris, Turkey, p. 835–843.
17. Tress, B. and Tress, G., *Capitalising on Multiplicity: A Transdisciplinary Systems Approach to Landscape Research.* Landscape and Urban Planning, 2001. 57(1): p. 143–157.
18. Litwin, U., Bacior, S. and Piech, I., *The Methodology of Valorising and Assessing Landscape.* Journal of Ecological Engineering, 2012. 18(1): 210–230.

19. Antrop, M., *A Brief History of Landscape Research.* The Routledge Companion to Landscape Studies, Routledge, ed: P. Howard, I. Thompson and E. Waterton, 2018, Routledge Press, p.1–15.
20. Çetinkaya, G. and Uzun, O., *Landscape Planning.* 2014, Birsen Press, İstanbul, p:1–40.
21. Forman, R. T. and Godron, M., *Landscape Ecology.* 1986, John Wiley & Sons Press, New York, p.4–28.
22. Dunning, J. B., Danielson, B. J. and Pulliam, H. R., *Eco-Logical Processes that Affect Populations in Complex Landscapes.* – Oikos, 1992. 65(1): p. 169–175.
23. Turner, M. G., Gardner, R. H. and O'Neill, R. V., *Landscape Ecology in Theory and Practice.* 2001, Springer Press.
24. Boström, C., Pittman, S. J., Simenstad, C. and Kneib, R. T., *Seascape Ecology of Coastal Biogenic Habitats: Advances, Gaps, and Challenge.* Marine Ecology Progress Series, 2011. 427: p. 191–217.
25. Miles, J., Cummins, R. P., French, D. D., Fardner, S. and Orr, J. L,. *Landscape Sensitivity: An Ecological View.* Catena, 2001. 42: p. 125–141.
26. ELC, *European Landscape Convention.* 2000. Access link: https://rm.coe.int/1680080621 (Date of access: 30.01.2021).
27. Pungetti, G., *Islands, Culture, Landscape and Seascape.* Journal of Marine and Island Cultures, 2012. 1: p. 51–54.
28. Förster, F., Großmann, R., Iwe, K., Kinkel, H., Larsen, A., Lungershausen, U, Matarese, C., Meurer, P., Nelle, O., Robin, V. and Teichmann, M., *What is Landscape? Towards a Common Concept within an Interdisciplinary Research Environment.* Journal of Ancient Study, 2012. Special Volume (3): p. 169–179.
29. Steiner, F., *The Living Landscape: An Ecological Approach to Landscape Planning.* 2008, Island Press, London, pp:491.
30. Tudor, C., *An Approach to Landscape Sensitivity Assessment – to İnform Spatial Planning and Land Management.* 2019, Natural England, pp: 1–47 (1–9).
31. Noor, A.M., *Charting the Territory of Landscape.* 2019. Access link: https://uitmshahalam.academia.edu/AnuarNoor (date of access: 30.01.2021).
32. Cengiz, C., *Ecological Landscape Planning, with a Focus on the Coastal Zone,* Landscape Planning, 2012, ed: M. Ozyavuz. 2012, In Tech Press, p. 55–84.
33. Pietsch, M. and Richter, K., *Analyzing Structure and Functions - Can Landscape Metrics Improve The Landscape Planning Process?* Specifics: Discussing Landscape Architecture, Ed: C. Soerensen and K. Liedtke, 2014, p. 350–355.
34. Pietsch, M., *GIS in Landscape Planning.* Landscape Planning, Ed: M. Ozyavuz, 2012, In tech Press, p.55–84.
35. Golobič, M. and Žaucer, L.B., *Landscape Planning and Vulnerability Assessment in The Mediterranean.* Regional Activity Centre for the Priority

Actions Programme, 2010. Access link: www.pap-thecoastcentre.org/pdfs/Landscape%20Vulnerability.pdf (date of access: 19.02.2019).

36. Ahern, J., *Theories, Methods and Strategies for Sustainable Landscape Planning.* Landscape Research to Landscape Planning: Aspects of Integration, Education and Application, Ed: B. Tress, G. Tress, G. Fry, P. Opdam, 2005, Springer Press, p.119–131.
37. Cengiz, A. E. and Gönüz, A., *Urban Land Uses with Respect to Ecology: A Case Study for Çanakkale City Centre.* Journal of Agricultural Faculty of Atatürk University, 2011. 42 (1): p. 79–89.
38. Botequilha-Leitão, A. and Ahern, J., *Appling Landscape Ecological Concepts and Metrics in Sustainable Landscape Planning.* Landscape and Urban Planning, 2002. 59: p. 65–93
39. Ndubisi, F. and Steiner, F.R., *Ecological Planning: A Historical and Comparative Synthesis.* 2002, The Johns Hopkins University Press, p, 281.
40. Melenyk, A., *Ecologıcal Analysıs of Landscapes.* Methodology of Landscape Research. Ed: V. Andreychouk, 2008. Commission of Cultural Landscape of Polish Geographical Society, p.151–169.
41. Uzun, O., *Identification of Turkey Landscape, Landscape Character Types and Areas,* International Geography Symposium on the 30th Anniversary of TUCAUM, 2018, October, Ankara, Turkey, p. 974–990.
42. Cengiz, A.E., Karadağ, A.A. and Demiroğlu, D., *Contribution of Landscape Character Analysis to Sustainable Urban Planning: Gaziantep Case/Turkey.* 7. International Workshop on "Environmental Engineering and Sustainable Development", 2019, June, Alba Iulia, Romania, p.1–19.
43. Uzun, O., İlke, E. F., Çetinkaya, G., Erduran, F. and Açıksöz, S., *Landscape Planning. 2012,* Lazer Ofset Matbaa Tes. San. ve Tic. Ltd. Şt., Ankara, Turkey, p. 189.
44. Güngöroğlu, C., Musaoğlu, N., Türkkan, M., Yöntem, O., Yılmaztürk, A. and Çayıt, G., *Classification and Mapping of Biotope Types Through GIS-based Remote Sensing Techniques –The Example of the Köprülü Kanyon National Park,* (2008). South-West Anatolia Forest Research Institute, Antalya, Turkey, p:223.
45. Uzun, O., and Gültekin, P., *Process Analysis in Landscape Planning, The Example of Sakarya/Kocaeli, Turkey.* Scientific Research and Essays, 2011. 6(2): p. 313–331.
46. Karadağ, A. A. and Yıldız K., Evaluation of Landscape Functions in the Sample of Hendek District. Düzce University Faculty of Forestry Journal of Forestry, 2013. 9(1), 77–96.
47. Karadağ, A.A., Cengiz A.E., Demiroğlu D. *An Approach to The Development of Land Use Decision on the Protection of Water Resources.* International

Conference on Natural Science and Engineering (ICNASE'16), 2016, March, Kilis, Turkey, p. 3149–3161. (in Turkish)

48. Karadağ, A. A., *A Research on Determination of Vulnerable Landscapes in Terms of Groundwater: Duzce Case, Turkey*, 2019. Fresenius Environmental Bulletin, 28(4A): p. 3231–3241.
49. Karadağ, A.A. and Şenik B., *Landscape Senitivity Analysis As An Ecological Key: The Case of Duzce Province.* 2019, Applied Ecology and Environmental Research, 17(6), p. 14277–14296.
50. *Karadağ, A. A. and Cengiz A. E,. Investigation of Water Process Analysis Method*, New Approaches to Spatial Planning and Design - Planning, Design, Applications, Ed: M. Özyavuz, 2019, Peter Lang Pres, p. 639.
51. Çoban, A., Demiroğlu, D. and Karadağ A. A., An *Ecological Based Approach for Determining of Suitable Landfill: The Case of Kilis/Turkey. 2020,* Journal of Environmental Biology, 41(2), p. 445–452.
52. Gümüş, A. E., *Analysis of Bioclimatic Comfort of Ankara Province*, 2012, SDU Faculty of Forestry Journal, 13: p. 48–56.
53. Erdoğan, M. A., *Developing an Ecological Risk Analysis Method for Büyük Menderes Basin*, 2012. Thesis (PhD), İstanbul Technical University, Institute of Science and Technology, Istanbul, pp:216.
54. Collins, M. G., Rushman, M.J. and Steiner, F., *Land-Use Suitability Analysis in the United States: Historical Development and Promising Technological Achievement.* 2001, Environmental Management, 28(5): p. 611–621.
55. Karadağ, A. A., Gültekin, P. and Kayapınar, E., *A Research on the Determination of Cemeteries Site Selection: The Case of Akçakoca (Düzce).* 2018, Düzce University Faculty of Forestry Journal of Forestry, 14(2): 17–33 .
56. Karadağ A. A., Başaran N. and YILMAZ KAYA M., *Importance of Determination of Suitable Agricultural Land Use in Combating Global Warming: The Case of Duzce, Turkey.* 2019, Journal of Environmental Protection and Ecology (JEPE), 21(5): p. 1757–1768.
57. Karadağ, A.A., Cengiz, A. E., *Demiroğlu, D., Determination of Sensitive Landscapes for Landfill: Sivas Case/Turkey.* 2019, Journal of Environmental Protection and Ecology (JEPE), 20(3): p. 1234–1242.
58. Karakuş, C. B., Demiroğlu D. and Karadağ A.A., *Evaluation of Land Use/ land Cover Change (LULCC) and Meteorological Data in Terms of Sustainable Agriculture in Kilis/Turkey.* 2020, Journal of Environmental Protection and Ecology (JEPE), 21(4): p. 1186–1193.
59. Gültekin P., Kaya S., Uzun S. (2018), *Assessment of Natural Landscape Elements for Camp and Picnic Areas Site Selection: Example of Duzce Topuk Plateau*, Düzce University Journal of Science & Technology, 6: p. 162–175.

60. Adger, W. N., *Vulnerability.* 2006, Global Environmental Change, 16(3): p. 268–281.
61. Birkmann, J. and Wisner, B., *Measuring the Un-Measurable: The Challenge of Vulnerability.* 2006. UNU Institute for Environment and Human Security, p. 64. Access link: www.d-nb.info/1029694141/34 (Date of access: 31.01.2021).
62. Damiano, H. R., Umberger, M., Monitz, G., Friedman, J., Hariri, M., Tehranifar, R., Grassi, C. and Lindsey, E., *NYC's Risk Landscape: A Guide to Hazard Mitigation. – The NYC Department of City Planning & The NYC Mayor's Office of Recovery and Resiliency.* 2014. Access link: https://www1.nyc.gov/assets/em/downloads/pdf/hazard_mitigation/nycs_risk_landscape_a_guide_to_hazard_mitigation_final.pdf (Date of access: 26.03.2019).
63. Aretano, R., Semeraro, T., Petrosillo, I., De Marco, A., Pasimeni, M. R. and Zurlini, G. *Mapping Ecological Vulnerability to Fire for Effective Conservation Management of Natural Protected Areas.* 2015, Ecological Modelling, 295: p. 163–175.
64. Campbell, C., Conger, S., Gould, B., Haegeli, P., Jamieson, B. and Statham, G. *Technical Aspects of Snow Avalanche Risk Management.* 2016, Canadian Avalanche Association, Revelstoke, BC, Canada. Access link: https://cdn.ymaws.com/www.avalancheassociation.ca/resource/resmgr/standards_docs/tasarm_english.pdf (Date of access: 31.01.2021).
65. Vaillant, N. M., Kolden, C. A. and Smith, A. M. S., *Assessing Landscape Vulnerability to Wildfire in The USA.* 2010, Current Forestry Reports 23: p. 201–213.
66. Menezes, J. A., Confalonieri, U., Madureira, A. P., Duval, I. B. and Santos, R. B. D., Mapping human Vulnerability to Climate Change in the Brazilian Amazon: The Construction of a Municipal Vulnerability Index. 2018, PLoS One 13(2): p. 1–30.
67. Turner, B. L., Kasperson, R. E., Matsone, P. A., McCarthyf, J. J., Corell, R. W., Christensene, L. and Schillerb, A. *A Framework for Vulnerability Analysis in Sustainability* Science. 2003. Proceedings of the National Academy of Sciences of the United States of America, 100(14): p. 8074–8079.
68. Marshall, N. A., Marshall, P. A., Tamelander, J., Obura, D, Malleret-King, D. and Cinner, J. E., *A Framework for Social Adaptation to Climate Change.* 2010, International Union for Conservation of Nature. Access link: https://portals.iucn.org/library/sites/library/files/documents/2010-022.pdf (Date of access: 31.01.2021).
69. Kesoretskikh, I., Zotov, S. and Drobiz, M., Assessing Spatial and Temporal Changes in the Landscape Vulnerability in the Kaliningrad Region as an Element of Sustainable Spatial Planning. 2015, Baltic Region, 4(26): p. 122–136.

70. Anderson, C. and Grant, A., *Landscape Capacity Study for Wind Turbine Development in East Lothian.* 2005, Final Report, East Lothian Council, p. 6–7. Access link: www.eastlothian.gov.uk (Access date: 20.10.2020)
71. Perry, M., *Landscape Sensitivity and* Capacity *Assessment.* 2017, Amec Foster Wheeler Environment & Infrastructure UK Limited, p. 4–10.
72. Thomas, M. F., *Landscape Sensitivity in Time and Space - an Introduction.* 2001, Catena 42: 83–98.
73. Gallopin, G. C., *Linkages between vulnerability, resilience, and adaptive capacity.* 2006. Global Environmental Change, 16: p. 293–303.
74. Zhou, H., Wang, X. and Wang, j. *A Way to Sustainability: Perspective of Resilience and Adaptation to Disaster,* 2016. Sustainability, 8(8): p. 737.
75. Farina, A., *Landscapes and their Ecological Components. 2001.* Access link: www3.udg.edu/cgpt/Almo%20Farina/Biosphere.PDF (Date of access: 19.02.2019).
76. Farina, A. The Landscape as a Semiotic İnterface Between Organisms and Resources. 2008, Biosemiotics, 1(1): p. 75–83.
77. Farina, A. and Pieretti, N. *The Soundscape Ecology: A New Frontier of Landscape Research and its Application to Islands and Coastal System.* 2012, Journal of Marine and Island Cultures 1(1): p. 21–26.
78. Usher, M. B., *Landscape Sensitivity: from Theory to Practice.* 2001, Catena 42: 375–383.
79. Gordon, J. E. and Sutherland, D. G., *The Quaternary in Scotland. Quaternary of Scotland,* Ed: J. E. Gordon, D. G. Sutherland, 1993, Springer, Dordrecht p. 11–47.
80. Mazzino, F. and Burlando, P., *Cultural Landscapes: Negotiation Between Global and Local.* 2010, A-Z ITU Journal of the Faculty of Architecture, 7(2): p. 14–20.
81. Mas, J.F., Gao, Y. and Pacheco, J.A.N., *Sensitivity of Landscape Pattern Metrics to Classification Approaches.* 2010, Forest Ecology and Management, 259: p. 1215–1224.
82. Knox, J. C., *Agricultural Influence on Landscape Sensitivity in the Upper Mississippi River Valley.* 200, Catena. 42: p. 193–224.
83. Thiemann, S., Schubert, H. and Schütt, B., *Integrated Watershed Management.* 2018. Postgraduate Program on Sustainable Development and Social Inequalities in the Andean Region, Latin American Institute, Free University of Berlin, p. 66–94. Access link: https://www.programa-trandes.net/Ressources/Manuales/Manual_Thiemann_Schubert_Schuett.pdf (Access date: 20.10.2020)
84. Chia, Y., Zhang, Z., Gao, J., Xie, Z., Zhao, M. And Wanga, E., *Evaluating Landscape Ecological Sensitivity of An Estuarine Island Based on Landscape*

Pattern Across Temporal and Spatial Scales. 2019, Ecological Indicators, 101: p. 221–237.

85. Barabás, G., Pásztor, L., Meszéna, G. and Ostling, A., *Sensitivity Analysis of Coexistence in Ecological Communities: Theory and Application.* 2014. Ecology Letters, 17: p. 1479–1494.
86. Zhang, M., Haijiang, J., Desuo, C. and Chunbo, J., *The comparative study on the ecological sensitivity analysis in Huixian karst wetland, China.* 2010, Procedia Environmental Sciences, 2: p. 386–398.
87. Jochen A. G., Jaeger, J., Schwarz-von Raumer, H. G., Esswein, H., Müller M and Schmidt-Lüttmann, M., *Time Series of Landscape Fragmentation Caused by Transportation Infrastructure and Urban Development: a Case Study from BadenWürttemberg, Germany.* 2007, Ecology and Society, 12(1): p. 22
88. Cushman, S. A., Landguth, E., L. and Flather, C., H., *Climate Change and Connectivity: Assessing Landscape and Species Vulnerability.* 2010. Access link: www.fs.fed.us/rm/pubs_other/rmrs_2010_cushman_s014.pdf (Date of access: 19.02.2021).
89. Penghua, Q., Songjun, X., Genzong, X., Benan, T., Hua, B. and Longshi, Y. *Analysis of The Ecological Vulnerability of the Western Hainan Island Based on its Landscape Pattern and Ecosystem Sensitivity.* 2007, Acta Ecologica Sinica, 27(4): p. 1257–1264.
90. Bolliger, J., *Simulating Complex Landscapes with a Generic Model: Sensitivity to Qualitative and Quantitative Classifications,* 2005. Ecological Complexity 2: p. 131–149.
91. Bede-Fazekas, A., Czúcz, B. and Somodi, I., *Vulnerability of Natural Landscapes to Climate Change - A Case Study of Hungary.* 2017, Quarterly Journal of the Hungarian Meteorological Service 121(4): 393–414.
92. Özhancı, E. and Yılmaz, H. *Sensitivity Analysis in Landscape Ecological Planning; the Sample of Bayburt.* 2018, Journal of Agricultural Faculty of Bursa Uludağ University, 32(2): p. 77–98.
93. LUC, *Hertfordshire Landscape and Visual Sensitivity Study of Potential Mineral Sites.* 2018, Glascow, Scotland, P. 4. Access link: www.hertfordshire.gov.uk/ (Date of access: 30.01.2021).
94. Wang, X. and Bian, Z., *The Implications of Ecological Sensitivity on Exploitation of Unutilized Land: A Case Study in Ji'Nan City, China.* 2011, Procedia Environmental Sciences 10: p. 275–281.
95. Aydın, H., Tezer,A., Turkay, Z., Yılmaz Kaya, M., Kutay Karaçor, E., Güler, İ., Uzun, O., Okay, N., Terzi, F., Köylü, P., Satılmış, E., Satılmış, E., Kara, D., *Resilience Through Participatory Planning for the Integrated Ecological Risks in Düzce.* 2018, Resilience Journal, 2(2): p. 105–121.
96. Hussain, A., Singh, G. and Rawat, G., S. *Landscape Vulnerability Assessment Using Remote Sensing and GIS Tools in the Indian Part of Kailash Sacred*

Landscape. 2018, The International Archives of the Photogrammetry, Remote Sensing and Spatial Information Sciences, XLII(5): p. 20–23.

97. LUC, *Landscape Sensitivity Assessment of Additional Settlements in Tunbridge Wells.* 2017, Report, Tunbridge Wells Borough Council, p. 22. Access link: www.tunbridgewells.gov.uk/__data/assets/pdf_file/0007/343834/Landscape_Sensitivity_Assessment_Report_-_Main_Report.pdf https://uitmshahalam.academia.edu/AnuarNoor (Date of access: 30.01.2021).
98. LUC, *Selby District Landscape Sensitivity Study.* 2019, Report, Selby District Council, p. 34, London. Access link: www.selby.gov.uk/sites/default/files/Selby%20LSS%20Report%20Final.pdfhttps://uitmshahalam.academia.edu/AnuarNoor (Date of access: 30.01.2021).
99. LUC, *Black Country Landscape Sensitivity Assessment.* 2019, Report, The Black Country Councils, p. 51, Access link: www.blackcountryplan.dudley.gov.uk/media/13883/black-country-lsa-front-end-report-final-lr_redacted.pdfhttps://uitmshahalam.academia.edu/AnuarNoor (Date of access: 30.01.2021).
100. JACOBS, *Maidstone Landscape Capacity Study: Sensitivity Assessment.* 2015, Jacobs U.K. Limited,. p:1–3. Access link: maidstone.gov.uk/__data/assets/pdf_file/0007/90475/Maidstone-Landscape-Capacity-Study-Sensitivity-Assessment-January-2015.pdf (Date of access: 30.01.2021).

Demet A. DİNÇAY, Çağıl Y. TOKER, Elif B. ÖKSÜZ, Sena SEMİZOĞLU, Özge CORDAN

A Transdisciplinary Student Design Workshop for an Inclusive In-door Navigation App

Abstract: Designing accessible environments is often found challenging by architects and interior designers. Common design solutions may not be suitable for every occasion. Individual design solutions are required for a better experience for people with special needs. Accordingly, transforming a built environment for people with physical disabilities has become a transdisciplinary issue for designers. The experience of physical space towards smart devices has significantly changed the definition of accessibility in today's world. Hence, this study issued the accessibility of space by reviving the 'transdisciplinarity' in design education. In this study, the authors seek to develop a transdisciplinary approach to accessibility by gathering interior and game designers for an indoor navigation app for mobile devices. By highlighting the differences in designers' approaches to accessibility in space, the authors explain why accessibility is about to become a transdisciplinary issue for future designers.

Keywords: Inclusive design, transdisciplinary workshop, navigation app design

1 Introduction

Technology is the center of today's world, leading us to a more livable environment. Our moves, choices, negative or positive comments are being kept in the data collection system to present us with a better way of living. Many companies work to improve mobility in cities and buildings while the awareness for an inclusive environment increases. People are more conscious about their possibilities, and they take individual action to create inclusive spaces if necessary. Transforming a built environment for people with physical disabilities has become a transdisciplinary issue for designers. According to the World Health Organization (WHO), there are currently more than 2 billion disabled people in the world, which is 37.5 % of the world population [1]. That means 37.5 % of the world population is dealing with difficulties to endure their daily lives; and the greater percentage of it prefer to isolate themselves from the outside world.

Design is one of the keys to prevent their isolation from society; and designers are expected to provide accessible environments for everyone. Since the '80s, architects and designers have been working on universal design principles. Unfortunately, it is still not a common topic in the design world. Therefore, it is essential to embed and widen the awareness of universal design knowledge and provide efficient education for future designers to practice innovative ways. However, designing accessible environments is often found challenging by architects and interior designers. Architects and interior designers mostly focus on designing an accessible environment for all although common design solutions may not be suitable for every occasion. Some individuals may need different design solutions depending on their disabilities or habits. At this point, transforming a built environment for people with physical disabilities has become a transdisciplinary issue for designers. While technology may help to customize design solutions for different individuals; design thinking may help to develop innovative and creative technologies. In other words, when architects and designers collaborate with game designers or system architects, their knowledge in different areas will help to develop versatile solutions for inclusion. In this context, this study issued the accessibility of space by reviving the 'transdisciplinarity' in design education.

In this study, the authors discuss the outcomes of their workshop series, the voluntary work of the Project PUDCAD's members from Istanbul Technical University that gathers interior and game designers for an indoor navigation app design for mobile devices by developing a transdisciplinary approach.

2 About the Project PUDCAD

By partnering with Universities and Non-governmental Organizations from Turkey and European countries, Istanbul Technical University coordinated The Project PUDCAD (Practicing Universal Design Principles in Design Education through a CAD-based Game) [2], between 2017–2020 under the Erasmus+ KA203 Programme. The project was proposed to support and strengthen universal design education in international networks. Regarding the European Universal Design Standards and the current state of inclusive design education among international design networks, this project offered a design game on a CAD-Based platform that aims to learn and practice their universal design knowledge skills through an empathetic approach. It provided an innovative way to learn and practice universal design principles for undergraduate students, triggered the awareness of accessibility, and let future designers and architects develop accessible and innovative design ideas.

The Project PUDCAD involved four different outputs, including an educational game, which helps students to learn about basic and advanced universal design principles and trains them with an entertaining context. The academic partners of this project were deliberately selected regarding their broad experiences, based on their proficiencies in universal design, education, and research. Also, the non-profit organization partners were reasonably selected to create an effective multidisciplinary research platform. The partner institutions and organizations of the PUDCAD Project were Istanbul Technical University, Department of Interior Architecture; Technische Hochschule Ostwestfalen-Lippe, University of Applied Sciences and Art, School of Interior Architecture; LAB Institute of Design and Fine Arts, University of Florence, Department of Design; University of Politecnico di

Milano, Department of Architecture, Bahcesehir University, Department of Game Design; The Association for Well-being of Children with Cerebral Palsy and the Occupational Therapy Association in Turkey.

3 Creating an Inclusive Environments with Technology

Currently, three different related topics can be found in the literature about inclusion; Accessibility, universal design, and inclusive design [3,4,5]. Accessibility is usually the main target for the other two; however, it gives the user limited possibilities when considered alone. On the other hand, universal design is a broader term that aims to include as many people as possible. While it is true that it provides accessible environments for the majority, it focuses mostly on an average user's needs in his environment. That means, solutions for the masses may not be working for some individuals. Kat Holmes (2018) argues that "when an object is designed for all, that doesn't mean everybody would be able to use it." Holmes also suggests that focusing on the users who are left out; only then can we be aware of diverse needs and create various design solutions to make everyone's participation possible [6]. Hence, the inclusive design tries to fill this gap and employs other disciplines to provide flexible and adaptive solutions. But it is only possible when designers become more cautious about user experience, especially for excluded ones, and skilled at designing with these technological advances. Novice designers need to learn how a physical space can provide versatile solutions for different individuals and how technology can be adapted. Educators in design fields let their students to know more about inclusion. Therefore, creating learning experiences towards accessibility guidelines with a transdisciplinary approach would be a good start for future designers' education.

In this context, the design should be a process that solves complex problems by understanding them thoroughly, looking from different angles, integrating diverse knowledge. Designers should empathize and know their users, their socio-cultural context and, needs, besides technology, and involve other disciplines' perspectives to broaden their ideation skills for a successful design thinking process. To achieve this, they should start collaboration and dialogue between different practices and partners [7]. Focusing on the issue, the authors extend the boundaries of the design thinking process and bring new perspectives to the process by following a *transdisciplinary* approach.

Jean Piaget first used the term in 1970 at a seminar on interdisciplinarity, teaching, and research problems in universities, organized by the Organization for Economic Co-operation and Development (OECD) and the French Ministry of Education. He discussed disciplines shouldn't be integrated individually but interact dynamically within a total system. Later Nicolescu suggested that transdisciplinarity is a strategy, evolved from the earlier research fields of multidisciplinarity and interdisciplinarity, and it involves what is between disciplines, across disciplines, and beyond disciplines [8]. (Fig. 1)

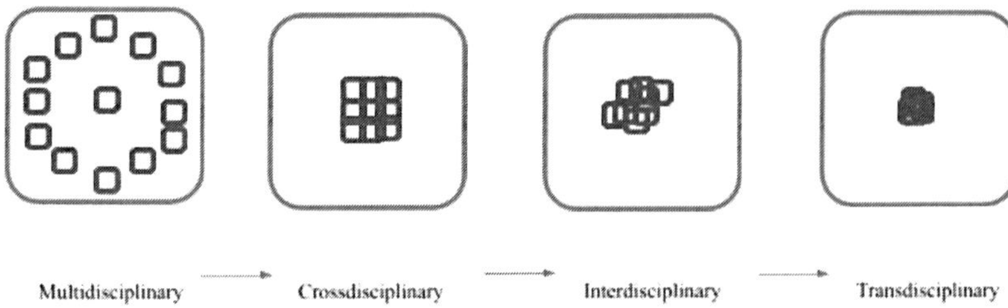

Fig. 1. Differences in interaction between disciplines. (Nicolescu, 2002)

Transdisciplinarity holds great potential to make contributions to a sustainable change in inclusive design education. It invites all disciplines to fully integrate during the design process, leading to holistic thinking. Transdisciplinarity, will create a platform for all disciplines to question their way of thinking, understand each other's domains, methods, and values to complete each other [9]. From that time on, the way transdisciplinary being implemented in design education is continuously evolving as the term itself. Even though it sounds like a new term in design education, we can see the integration of disciplines in the past. Bauhaus school is a great example that brought technology, science, art, architecture, and design together to achieve innovation [10]. Therefore, it is not a surprise that transdisciplinary approaches are also highly recognized in education and research. Since technological advances pursue designers to adopt new skills from other disciplines, transdisciplinary design education approaches became inevitable. In 1972, Jantsch, following Piaget's works, pointed out that transdisciplinarity would help coordinate the education/innovation system as a dynamic multi-level, multi- goal system [11]. In other words, transdisciplinarity provides an integrated education/innovation model in which all disciplines are connected without a hierarchical following and evolving continuously to generate new knowledge. Similarly, Go-beil and Nicolescu (1997) suggested implementing the methodology by setting up several transdisciplinary research workshops in a few pilot universities at the International Congress in Switzerland, 1997 [12]. Later on, in 1997 and 2013, in several international forums such as Transdisciplinarity, stimulating synergies, integrating knowledge, France, 1998; World Conference: Higher Education in the Twenty-First Century: Vision and Action", Paris, 1998 and "ARISE 2" (Advancing Research in Science and Engineering), U.S.A, 2013, transdisciplinarity was stated to be a meta discipline that should be integrated to universities' curriculum [13].

Regarding the above discussion, the authors used transdisciplinarity as a model for their workshop; and created a design task that requires a transdisciplinary approach to understand inclusion of space. The authors aimed at designing a learning experience for the students from different disciplines and widening their perspectives for universal accessibility at different scales.

4 PUDCAD Supporting Workshop; Designing an Inclusive Indoor Navigation App

Within the contribution of six design instructors from interior architecture and game design departments, the authors organized a workshop to be conducted in two sessions. Regarding their background in design education, these students were divided into two main groups for each session[14]. While the first session was conducted with the participation of 35 interior architecture students from Istanbul Technical University, the second session was conducted with 25 game design students from Bahcesehir University.

Aims and Task Description

Since the intention of the authors' voluntary work was to help students to interpret accessibility in design at different scales from different perspectives, the authors invented a design theme considering the user experience in digital and physical spaces. For that, they issued the navigability in complex buildings and the use of indoor navigation applications by the people with diverse needs. By highlighting the differences in designers' approaches to accessibility in space, the authors aimed to explain why accessibility is about to become a transdisciplinary issue for future designers. To reveal the impacts of the transdisciplinary approach to accessibility for novice designers' understanding of inclusive design, the authors organized and instructed a workshop in two sessions for two different design groups from game design and interior architecture departments.

Methodology

To support the transdisciplinary approach, the instructors used one of the design thinking methods, simplifying the design task by addressing five critical questions and commonly used by various domains. By directing the questions of *"who, what, where, when, and why,"* or the 5Ws in short [15], the instructors wanted the students to acknowledge the main objectives of the design problem. They wanted their students to identify the real users of their design, reveal their design intentions, and present effective design solutions. Hence, while *'what'* was used to define the navigation app as the output of the workshop, *'who' (or to whom)* was used to describe the end-users of the app. Additionally, while the answers to *'why'* intends to emphasize accessibility issue towards different disciplines, *'where'* corresponded to Taşkışla Campus Building of ITU, the workshop venue. Followingly, the instructors also determined their task evaluation criteria for the same Ws. Therefore, as design outputs, the expectation was limited to the graphic interface design of the app and its demonstration for five screenshots. For a design output, all student groups were expected to present and promote their app designs in 5 screenshots, with 150 word-length descriptive information (Fig. 2). But the students were allowed to show some of the perceivable features of the mobile devices (cam, mic, sound, vibration) as visuals in their design presentations.

For students to create the paper prototype of the navigation app and complete the design problem, the instructors wanted the students use paper prototyping and basic digital tools such as Adobe XD, Adobe Photoshop. These tools are used for creating interface layouts since building a mobile application requires technical knowledge about programming and coding, and not all the participants are familiar with these [16]. In this regard, along with the use of analog design tools during the ideation and paper prototyping processes, the students were also allowed to use various visualization software programs for their design presentations.

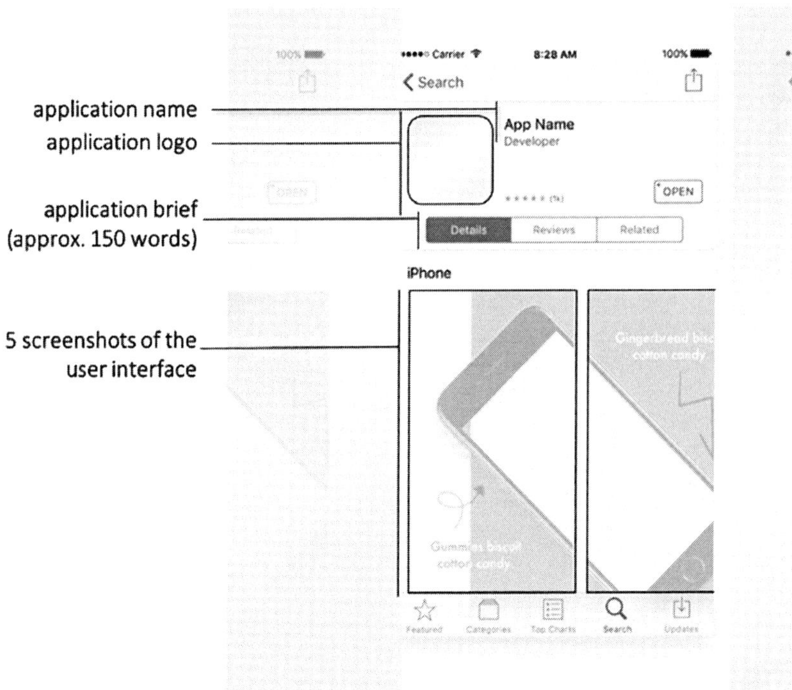

Fig. 2. Requested Visuals and Descriptions for the Students' Designs. (© ITU PUDCAD Team)

Participants:
For the workshop, the instructors looked for the participants, who have developed their spatial design and representation skills by practicing either in physical spaces or digital spaces. In this regard, the participants were selected among sophomore design students from interior architecture and game design departments.

Process:
The workshop was conducted in the same format for each session; and each session was planned to be completed in a half-day. In both sessions, the instructors started

with short lectures on the principles of universal design and the implications of accessibility guidelines on mobile applications. Later, they introduced the workshop theme by staging a discussion on the wayfinding in complex buildings and the role of in-door navigation (assistive) technologies for people with diverse needs. After the discussion session, the students were divided into groups of five to work on their design tasks. Fig. 3 shows the steps of the workshop process.

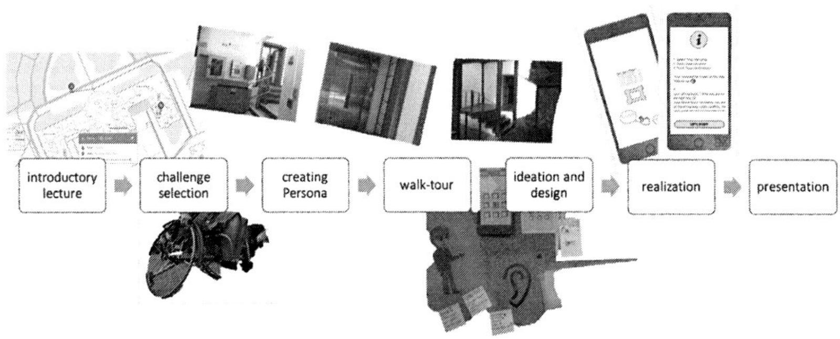

Fig. 3. The workshop format. (© ITU PUDCAD Team)

Inspired by the Microsoft's Inclusive Design Kit [17], the instructors assigned seven different user profiles with certain physical impairments/diverse needs, such as a person on a wheelchair, a person with a hearing problem, or a person carrying a heavy object for their design task etc. In the second part, each group picked one end-user profile to work on their design tasks. The groups started an online research to learn more about the accessibility guidelines for their end-user profiles and address solutions for their diverse needs with their design proposals.

In the third part, the groups had taken a walk-tour in the campus building. By using the floor plans of the Taşkışla Campus building, each group drew a route for their end-user profiles to use in their application design. To determine 'the navigability of space' [18], the students were asked to empathize with their end user profiles and become the navigators inside the building. Hence, the instructors asked students to draw the shortest path for their user profiles by

- discovering or infer their present location in the building,
- finding a route to the destination, and
- accumulating their way-finding experience in the space.

During walk tours, the groups were encouraged to take notes on the accessibility installations they had noticed in the building. They were asked to address the spatial features of the building and evaluate the use of these accessibility installations in their design proposals. They were allowed to take photos and videos use in their design presentations. After the walk-tours, the groups were gathered in the studio to design with the guidance of the workshop instructors (Fig. 4). At the end of both sessions, the groups presented their work.

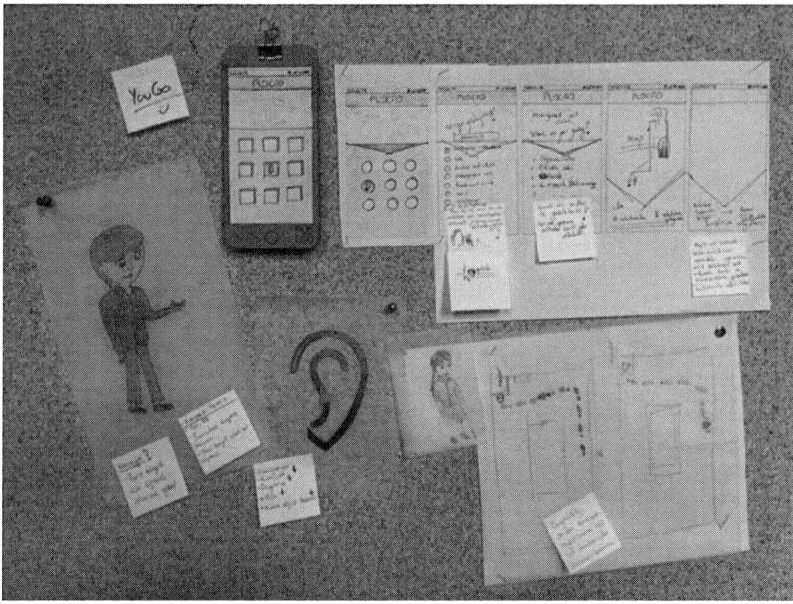

Fig. 4. A photograph of the design board from one of the groups. (© ITU PUDCAD Team)

4.1 Outcomes of the Workshop Sessions

4.1.1 ITU Session with Interior Architecture Students

Fig. 5. (a, b, c, d, e, f, g). ITU session, opening screens for the apps belong each group. (© ITU PUDCAD Team)

ITU-G1 (a) app usage steps:
Opening screen with logo and information button > Information screen > Persona selection with suitable icons > Destination selection (from – to), along with a "You're here" map > Map with multiple layers showing different stories.

ITU-G1 app features:
The app has a simple coloured scheme with white and yellow. A special logo is designed referring to the Taşkışla building. Architectural plan drawings are used along with exploded perspectives to show different stories of the building.

ITU-G2 (b) app usage steps:
Welcome screen with language selection > destination selection page with icons representing main places along with a "need help" live chat section > a chat screen with a student/representor > map page including buttons for each floor and "you are here" mark > route showing map with search section

ITU-G2 app features:
In this app simplistic features such as icons are used. The colour scheme is red and white. Simplified and coloured architectural drawings are used in the app making it easier to read user's whereabouts and the room numbers around.

ITU-G3 (c) app usage steps:
Opening screen with Taşkışla plan doodle, search bar and location icons > Route map > Vibration and audio description navigating the user for turns, steps and other building related obstacles > When on elevator the phone vibrates the floors > Destination reached screen

ITU-G3 app features:
Group 3 dealt with visual impairment. Along with visual maps and conventional navigation interface it uses smart mobiles' accessibility tools such as voice recognition, audio description and vibration. Colour scheme consists of mostly white with coloured icons. The map base is a coloured architectural drawing.

ITU-G4 (d) app usage steps:
Selecting the app icon from ITU Mobile app > Opening screen with language selection and app logo > Choosing between voice command and touchscreen > Destination selection screen with icons representing different space categories > Possible destinations are shown on an exploded isometric map to choose > the route for the destination and estimated steps, duration and distance is shown.

ITU-G4 app features:
The app is designed as a plug-in application that comes with ITU Mobile app, but its design differs from it. The colour scheme is black and grey along with purple highlights. Isometric architectural plan drawings are used. It gives the user to choose the preferred way of interaction in between voice recognition and typical touchscreen features. Though in some other apps voice recognition icon can be seen, this group mentioned and used way of interaction as a design element.

ITU-G5 (e) app usage steps:
Impairment selection screen > Destination search screen with ready to go selections and a search bar > Possible destinations screen regarding the user's whereabouts > Route screen with estimated time and steps > route screens with amplified sound feedback that gives the user the opportunity to listen the close places (such as playing bird pipe sounds when walking inside near the middle courtyard)

ITU-G5 app features:
This app features coloured architectural drawings as a part of its navigation. It offers sign language videos along with some amplified sound feedbacks about surrounding places. Colour scheme is white other than the route map screens.

ITU-G6 (f) app usage steps:
Opening screen with colour scheme selection > "Ulaş/Keşfet" swiping screen for two different modes of the app, first for conventional navigation and the second for exploration interface > Navigation screen with pinned locations and routes > Explore screen with photos to reveal where the user is by asking whether if they see the landmark in the photo

ITU-G6 app features:
This app gamifies the process by adding the explore feature. This feature allows users to look out for the landmarks in the building such as the Venus and Sinan statues. It has several colour schemes to allow users to pick the most suitable and accessibility. It also uses direct feedback from users to locate their whereabouts by showing photos and asking them if they see it or not.

ITU-G7 (g) app usage steps:
Opening screen with user selection and log-in section > Campus selection > Destination selection with icons representing different categories such as canteens, WCs etc. > Route selection with alternatives > Navigation map with route and estimated time and start button

ITU-G7 app features:
The app is planned as a universal navigation app for ITU campuses, therefore it begins with campus selection. It has a white and pastel coloured scheme. Coloured architectural drawings are used as the base of the map. Icons are used for representing spaces, also the ramps, stairs and escalators are represented with related icons. It offers alternative routes regarding to the circulation icons.

Fig. 5 shows the opening screens that created by each group in ITU section.

4.1.2 BAU Session with Game Design Students

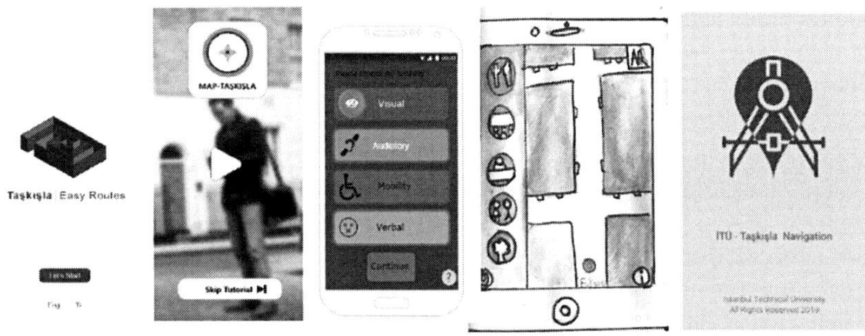

Fig. 6. (a, b, c, d, e) BAU session, opening screens for the apps (© ITU PUDCAD Team)

Bau-G1 (a) app usage steps:
"Let's start" page with language options -> Sign in page with several connection options such as 'connect with ITU Portal, Facebook, Google' along with e-mail option -> user type selection (student, guest, guide) -> Selection of impairment type (touch, speak, hear, see) -> Choosing a start location -> Choosing a destination -> A map screen showing 1the direction in line, 2step by step explanation of the route, 3floor level information, 4small icons to indicate building attributes related to the movement such as steps and elevators, 5a "get help" button -> a map screen for the upper floor to show the changes in the app between different floors, same features -> "You have arrived!" screen
"Get help!" button -> A screen with "qr code, video map, call security, connect to guide" options.

Bau-G1 app features:
In this app students opted for simple design with basic text and blue monochrome colouring. They designed a logo indicating the maze-like nature of Taşkışla and used a simplified version of the building plan just with contouring. The app contains a few common and easy to understand icons such as check mark, question mark, step and elevator icons. Also, the app contains a video map option.

Bau-G2 (b) app usage steps:
Opening screen with tutorial video, logo and 'skip tutorial' button -> Full screen video tutorial -> Map screen with 1help, 2elevation, 3search, 4foodcourt, 5WC, 6elevator, 7code reader buttons -> Map screen showing the exact location with 'You're Here' indicator and code reader feature -> Search screen with keyboard
This group's persona has a hearing impairment.

Bau-G2 app features:
In this app, students limited the text use to bare minimum while using different colours and shapes to identify different wings of the building. The app is based on the camera and code reading function of mobile devices. 3D voxel mapping is offered to use to capture the exact location, also 360o photos are embedded in the app. The group used architectural plan drawings of the building.

<u>*Bau-G3 (c)*</u> app usage steps:
Loading screen with logo -> 5 step app introduction giving information about the capabilities of the app including 1finding classrooms, 2finding instructors according to their weekly plan, 3finding common spaces (library, WC, food court), 4contacting staff, 5handsfree option with audio assistance and vibration feedback -> Selection of impairment type (visual, auditory, mobility, verbal) -> Map screen showing the exact location with 'You're Here' indicator and 'Where to go' option icons such as classroom, WC, deanery, etc. -> A screen for instructor search -> Requesting help screen with several possible problem options around the building such as a wheelchair ramp that doesn't work, in which user can reach out for the authorised person.

Bau-G3 app features:
This group of students used a mix of coloured graphics, text and icons. The map is a coloured version of architectural plan drawings. Impairment related icons are shown on the map, such as the wheelchair ramps. Students created a simple app introduction showing the features.
<u>*Bau-G4 (d)*</u> app usage steps:

Opening screen with destination icons, options, information and AR button -> choosing destination -> start screen -> Map screen with a line indicator for the route -> as an option AR mode can be used, in which the camera is activated -> AR mode screen with a real-time visual which is embedded with direction indicators such as arrows, location specific icons and route alternatives with their duration.

Bau-G4 app features:
In this app students opted for a diminished version of the building map and included augmented reality (AR) features. The non-AR navigation interface is designed with a simple colour scheme consisting blue and grayscale. For the AR interface the group used big, simple and visible navigation and information icons.

<u>*Bau-G5 (e)*</u> app usage steps:
Opening screen with logo and credits -> selection of impairment type -> confirming impairment type -> Map screen showing the exact location with 'You're Here' indicator, search and menu buttons, and a slider that contains possible destination points -> Choosing a destination -> Map screen showing the destination along with specific directions such as '6m right' etc.

Bau-G5 app features:
This group designed a simple colour schemed app with pale pink and navy. They designed a logo combining navigation and mapping elements. Architectural plan drawings are graphically altered and used as the base for navigation. The interface consists of big, easy-to-understand icons and buttons along with pop-up menus.

Fig. 6 shows the opening screens that created by each group in BAU section.

4.2 Results and Discussions about the Workshop

Both sessions' outcomes were evaluated and compared under the 5W1H. All groups managed to address the diverse needs of the user profiles they selected and offer possible accessibility solutions with their interface design and completed their design tasks. Hence, despite the differences in the forms of their presentations the groups from both departments have reflected their knowledge of accessibility guidelines and gained new insights for the accessibility concept in inclusive design. However, when the design solutions of both designer groups were compared, highly unexpected result was noticed.

Participants were asked to present their app in 5 steps using paper prototyping tools. While interior architects used inactive and conservative presentation techniques, game design students used interactive presentation along with traditional techniques. In interior architects' presentations, user interface steps can be seen together with explanations below or near. The participants explained their design ideas along with visuals in which they mostly used colored architectural plan drawings as the base for navigation. On the contrary, game design students explained their ideas with visual slides referring to their process and aims, and later went on with the app itself in steps or as an animation. They used real-time images with architectural drawings. It is obvious that methodological differences come through professions' approaches. On the one hand, interior architects handle design issues in 3-dimensions, they design to answer the current needs of the user, and as the designed medium is a physical asset, every aspect should be solved in a detailed manner. On the contrary, game designers' design medium is in the digital domain and handled in 4-dimensions including time. Therefore it defines a process or a journey. It is possible to say that game designers' solutions contain less space-organization related details since they usually work in a digital domain.

Another reason that can explain the difference between the outcomes is the way of handling the design process by two distantly related professions. While the solutions of interior architecture designers were mostly fixated with user experience in digital space, the game design students were quite the opposite (Fig. 7–8).

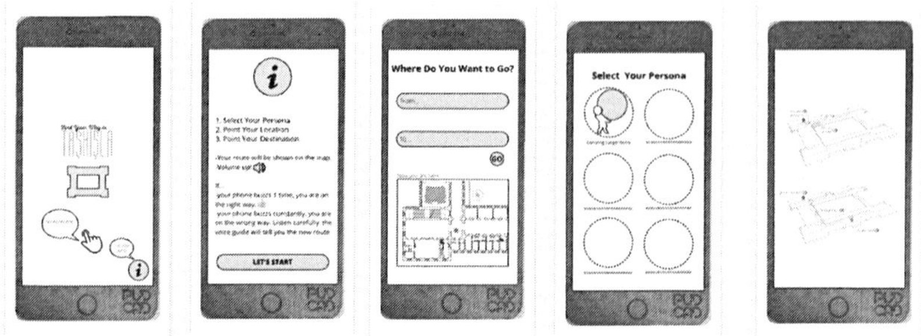

Fig. 7. The screenshots of one of the interior architecture groups' design (© ITU PUDCAD Team)

Fig. 8. The screenshots of one of the game design groups' design. (© ITU PUDCAD Team)

It was noticed that almost every group of interior architecture focused considerably on the perceptual experiences of their users and provided design solutions based on the user's physical interaction with mobile devices. None of them offered to use any AR or MR technologies that require interaction with a physical space. As opposed to that, the game design students were less concerned about the perceptual experience of the users and more concerned about the user's interaction with the physical space. They prefer to use architectural elements of the building to alert their user.

It is possible to state that each student group focused on user needs from their own professions' perspective. While interior architecture students dealt with aspects of user experience in the physical domain, game design students dealt with user experience aspects in the digital domain. Outcomes of the workshop highlighted some issues which can be related to the differences in participants' backgrounds and professions' design approaches.

It can be seen from the workshop outcomes that, game design students and interior architects have different approaches for defining user needs. Interior architecture students showed more empathy towards the disabled, also gathered more information around the building about the tricky spots, while game design students offered pragmatic solutions such as adding a 'call an authorized person' button.

When examining the outcomes of the workshop, the familiarity of students with the interaction design tools and processes should be considered. While game design students regularly deal with interaction issues on interfaces, it is relatively a new field for interior architecture students. Interior architecture students tried to transfer their design practice to this field. Therefore, it is possible to claim that the differences and variety in interface feature such as the locations and functions of buttons, AR/VR technologies use etc., between two sessions are caused by this unfamiliarity. Discussion of the outcomes indicate that the reason of this differences between outcomes can be explained by designing processes of these disciplines, and show that each profession can handle universal design issues differently and may offer different ways for the design solutions.

5 Conclusion

Recent developments in technology caused designers to adopt new skills from other disciplines, and a transdisciplinary approach became inevitable in both design practice and education. This workshop basically aimed to reveal and analyze the approaches of designer groups from two different disciplines for the same mobile navigation application as a design problem. The outcomes of the workshop showed each discipline has a particular way of handling the design problem. The secondary purpose was to enable design students to experience teamwork in an unfamiliar design area. Design problems which define with different point of views can solve in a better way. Transdisciplinarity is one of the most productive working collaborations, since it allows multiple ways to contribute various perspectives from different professions.

One side result of the workshop was – considering students as designers of the future – to raise some empathy for the disability and restriction of movements and gain awareness about Universal Design via this workshop. The empathy part of the workshop helped game design students gain awareness about accessibility and UD via role-playing. Similarly, interior architecture students learned more about intangible accessibility issues in the digital domain.

With the recent developments in technology, transdisciplinarity theme is enhanced, and it also became possible to reach out to a broader audience. Though these advancements are widely accepted and used, when it comes to accessibility – which should be a priority – these advancements are not used well enough. This study points out the diversity that transdisciplinarity brings into accessibility issues. All the workshop results support the importance of making collaborations that positively affect professions' and public's awareness of Universal Design issues, as each profession contributes from their perspective and as all the research and applications in the field bring overall wellness.

References

1. *World Report on Disability.* World Health Organization, 2011. Retrieved September 4, 2020, from https://www.who.int/disabilities/world_report/2011/report.pdf
2. "THE PUDCAD PROJECT." *PUDCAD PROJECT*, 2020, Retrieved September 4, 2020, from www.pudcadgame.com/tutorial-project/.
3. Imrie, Rob. "Universalism, universal design and equitable access to the built environment." *Disability and rehabilitation* 34.10 (2012): 873–882.
4. Mace, R. *What is universal design,* The Center for Universal Design at North Carolina State University. 2004,1997 Retrieved November, vol. 19.
5. Robert F. Erlandson - *Universal and Accessible Design for Products, Services, and Processes.* 2007.CRC Press
6. Holmes, Kat . *Mismatch: How Inclusion Shapes Design.*2018.Research gate January 2019,doi: 10.23860/MGDR-2019-04-02-09
7. Moreno, L,A and Villalb, E. *"Transdisciplinary Design: Tamed complexity through new collaboration."* Strategic Design Research Journal .2018 .11, no.1 p: 42–50.
8. Nicolescu,B. *Multidisciplinarity, Interdisciplinarity, Indisciplinarity, and Transdisciplinarity: Similarities and Differences.*2014.RCCPerspectives,(2), p:19–26.Retrieved September 4, 2020, from http://www.jstor.org/stable/26241230
9. Leblanc T. *Transdisciplinary Design Approach.* In: Kotzé P., Wong W., Jorge J., Dix A., Silva P.A. (eds) *Creativity and HCI: From Experience to Design in Education. IFIP* – International Federation for Information Processing, 2009 vol 289. Springer, Boston, MA. https://doi.org/10.1007/978-0-387-89022-7_8
10. Correia, RM, Alexandra P, and L. A. Soares. "Architectural design digital change: interactivity policy." *Architectural design digital change: interactivity policy* (2018): 798–804.
11. Jantsch, E. *Inter- and Transdisciplinary University: A Systems Approach to Education and Innovation.* High. Educ. 1972, 1, p: 7–37
12. Gobeil, M., & Nicolescu, B. *Le projet CIRET-UNESCO Évolution transdisciplinaire de l'université [The CI-RET-UNESCO project Transdisciplinary evolution of the university*].1997 http://ciret-transdisciplinarity.org/projet_ciret_unesco.php#fr

13. Mokiy,V *International Standard of Transdisciplinary Education and Transdisciplinary Competence. Informing Science: The International Journal of an Emerging Transdiscipline*, 2019. 22, p:73–90. https://doi.org/10.28945/4480
14. Cordan, Ö., Yurdakul Toker, Ç., Öksüz, E. B., Dinçay, D. & Semizoğlu, S., *Designing an inclusive navigation app for Taşkışla Building,* Design and Ergonomics Conference, May 2019, Florence
15. Roberts J.C., Headleand C.J., Ritsos P.D. (2017) Introduction: Think, Prep, Sketch. In: Five Design-Sheets: Creative Design and Sketching for Computing and Visualisation. Springer, Cham. https://doi.org/10.1007/978-3-319-55627-7_1
16. Ballantyne, M., Jha, A., Jacobsen, A., Hawker, J. S., & El-Glaly, Y. N. (2018, November). Study of accessibility guidelines of mobile applications. In *Proceedings of the 17th international conference on mobile and ubiquitous multimedia* (pp. 305–315).
17. https://www.microsoft.com/design/inclusive/ (Accessed June 23, 2020)
18. Foltz, M. A. (1998). *Designing navigable information spaces* (Master's thesis, Massachusetts Institute of Technology, Dept. of Electrical Engineering and Computer Science).

Reva ŞERMET, Nehar BÜYÜKBAYRAKTAR, Fürüzan ASLAN

Nature-Based Approach in Ensuring Urban Sustainability: Green Infrastructure Strategy

1 Introduction

The anthropocentric environment conception arguing that the natural resources are indefinite and the raison d'être of the entire living beings in nature is to serve the human being has caused devastating and sometimes irrevocable destruction on the environment. The countries, upon the Industrial Revolution, have been focusing on manufacturing and consuming more for the purpose of achieving quantitative growth and thus, resulted in an incremental pressure on natural resources.

Disproportionate rate of self-renewal of natural resources with the rate of said consumption results in rapid increasing of the world's population and the advancing technology leads to exposure of resources to the risk of extinction and causes damages sustained by the species along with their habitat. These highly detrimental effects of the conventional development/improvement approach focused on quantitative growth have been discussed on international platforms and has been replaced by a sustainable development approach taking into account the problems such as poverty, equality and environment. The objective is the minimization of the damage and developing a further protective conception and mindset against the requirements and needs of the future generations.

Despite the inexistence of a space targeted by the sustainability phenomenon, more than half of the world's population living in urban areas with increasing numbers day by day, and the urban areas being the locations exposed to the highest ecocide to the greatest extent have revealed the phenomenon of sustainable urbanization. Various solutions are developed by the authorities for the purpose of accomplishing the phenomenon of sustainable urbanization, which is a new phenomenon. Ecosystem services and green infrastructure strategy are the primary goals of the sustainable urbanization endeavors, seeking to achieve the main goals of harmonization of the urban ecosystem with the built environment, reducing the ecological footprint of the urban areas, increasing inhabitability and quality of life, ensuring social justice as well as minimizing the requirement for the natural resources. Integration of the green infrastructure endeavors to the urban planning echelons substantially contributes to the sustainable urbanization movement.

2 Sustainable Development

Various opinions are available regarding the revival of the sustainability concept. Dating back to the Middle Ages and even Ancient Greece, the idea of sustainability, was included for the first time in the document of World Charter for Nature adopted by the International Union for Conservation of Nature in 1982 in the 19th century. It has been highlighted in the document that it is required to manage the environmental system, organisms, terrestrial, marine and atmospheric resources benefited by the humans in a manner ensuring the optimum sustainability and the integrity of the environmental systems not to be jeopardized (Yazar, 2006). Defined by Meadowcroft (1997) as "Perpetuating an existence of an actual being", the concept of sustainability is applied in accordance with different objectives in different fields such as economy, politics, sociology, environment and architecture (Low, 2005). The definition of the concept of sustainable development covering these entire fields recognized on international scale can be found in the report named "Our Common Future" issued by the World Commission on Environment and Development. Based on the report, the sustainable development is defined as "*development that meets the needs of the present generation without compromising the ability of future generations to meet their needs*" (UN WCED,1991).

Sustainable development refers to the renouncement of the economic development conception of the period aiming merely at quantitative growth to be replaced by a qualitative development model in where the sustainability of environmental and social values is ensured along with the economic development. The concept focuses on preventing environmental problems arising due to developments in the economy and the technology.

The concept of sustainable development is required to be addressed in 3 dimensions such as economic, environmental and societal. Despite the fact that these three dimensions interact with each other, required to be addressed together. The intersections of these dimensions define the sustainable development.

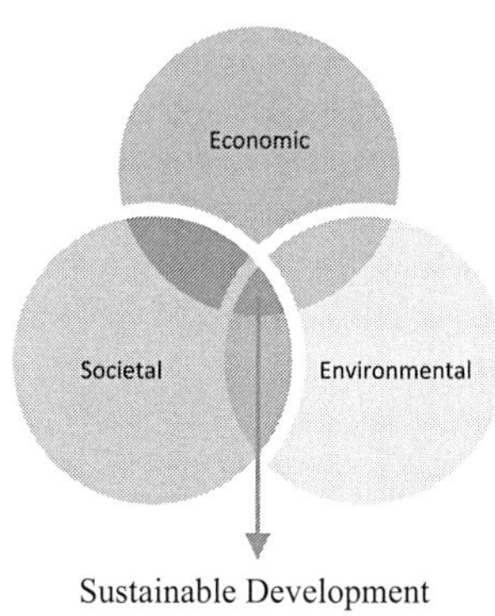

The Conception of Sustainable Development has been adopted, developed and formed through the endeavors performed by the United Nations. The said endeavors are stated in Tab. 2.1 in chronological order.

Tab. 2.1 Development of the Conception of Sustainable Development in Chronological Order

Year	Endeavor
1972	Conference on the Human Environment
1980	World Conservation Strategy
1982	World Charter for Nature
1987	Our Common Future Report
1992	Conference on Environment and Development
1993	Commission on Sustainable Development
2000	Millennium Summit
2002	Johannesburg Summit
2012	Rio+20 Summit
2015	Brussels Summit
2015	Antalya Global Development Goals
2016	Delhi World Sustainable Development Summit

It is stated in the Our Common Future (Brundtland) Report in where the concept of sustainable development is defined for the first time in scientific literature that the sustainable development cannot be limited solely with the ecology and the urban and cultural sustainability are of utmost importance as well as the ecological sustainability.

3 Sustainable Urban Development

High rate of literacy, advantages such as developed healthcare service, facilitation of the accessibility to the social services, cultural and political involvement mechanisms has caused increased migration to urban areas (World Urbanization Prospects, 2014).

"Urbanization", defined as "migration of population to rural areas to urban areas", "growth of cities in terms of both population and area" or "increasing urban population ratio within the total population" (Keleş, 2001; Akçakaya,2016), has driven the cities towards insufficiency.

Today, the great majority of the world's population lives in cities and this rate has been increasing with each passing day. For the first time in 2007, the urban population has exceeded the rural population, and subsequently the rate has demonstrated an increase and the urban population has reached to 54 % in 2014. Accordingly, by the year 2050, the urbanization rate is expected to increase to 66 %, and two-thirds of the world population, corresponding to roughly 6.5 billion people, are expected to start living in cities (World Urbanization Prospects, 2014).

Cities are structures having influence on the local and global environment and ecosystem through various ways. For instance, cities both form the locality and have the ability to change it. Second of all, the expansive food, water and energy consumption in cities result in carbon emissions and wastes to a greatest extent. Moreover, numerous subsystems within the cities are based on various local and global ecological functions and activities.

Rapid and unplanned urbanization, developing based on conditions such as inability to form sufficient infrastructure in cities and failure to ensure fair sharing of urban opportunities, practicing upon the ever-advancing technology by ignoring the environmental awareness and the scarcity of the resources started to impose substantial threat to the sustainable development. (World Urbanization Prospects, 2014).

Sustainable urban development, sidling with the sustainable development is a concept shaped in line with its principle, policy and actions as spatial reflections.

The quest of the concept of sustainable development on cities originates from the fact that the major cities are both the leading consumers of natural resources and the main producers of pollution and waste (Karakurt Tosun,2019). While the cities, having numerous positive features in terms of development, grow rapidly both horizontally and vertically with the increasing population and technological developments, it is a regarded and considered as an undeniable fact that it harms the natural resources within its perimeter.

Urban sustainability can be interpreted within the frame of fulfilment of the urban needs without jeopardizing the sustainability of local and urban resources (Alberti, 1996).

Goals identified for the implementation of sustainable development on the urban level can be complied as; increasing the quality of life, presence of an alternative in development, combating poverty, providing solution for the problems of employment and nutrition, healthy living, supporting and developing the biological diversity base, technological restructuring, controlling population growth, utilizing renewable energy resources, seeking for clean and potable water and elimination of the risks. In this context, the solution herein is required to be produced by taking into account certain aspects such as the direct transformation of the physical structure and the habitat, utilization of renewable and nonrenewable natural resources, gas emissions, wastes, human health and welfare (Alberti, 1996).

Endeavors have been performed by many international organizations, particularly, the United Nations, European Union and the European Council towards the practicability of the sustainable development within the urban areas. Such endeavors comprising the foundations of the conception of the sustainable urbanization are stated on Tab. 3.1 along with their organizers in a chronological order.

Tab. 3.1 Development of the Conception of the Sustainable Urban Development in Chronological Order

Year	Endeavor	Organizer
1972	Conference on the Human Environment	UN
1976	Habitat I	UN
1983	European Regional/Spatial Planning Charter	CoE
1985	European Charter of Local Self-Government	CoE
1987	Our Common Future Report	UN
1990	Sustainable Cities Programme	UN
1992	European Urban Charter I	CoE
1992	Conference on Environment and Development	UN
1994	European Conference on Sustainable Cities and Towns	EU
1996	Habitat II	UN
1997	Urban Audit Programme	EU
1999	European Spatial Development Perspective	EU
2000	Lille Action Programme	EU
2000	Guiding Principles for Sustainable Spatial Development of the European Continent	CoE
2001 2007	Sustainable Cities Programme	UN HABITAT - UNEP
2002 2012	World Urban Forums	UN
2004	Urban Regulation	EU
2005	Bristol Accord	EU
2006	Thematic Strategy on the Urban Environment	CoE
2007	Leipzig Declaration	EU
2008	European Urban Charter II	CoE
2009	Action Plan on Urban Mobility	EU
2010	Europe 2020 Strategy	EU
2016	Habitat III	UN

The entire regulations have contributed substantially to the emergence of the mutually complementary principles in terms of sustainable urban development.

The sustainable urban principles according to the UN Habitat are stated in Fig. 3.1.

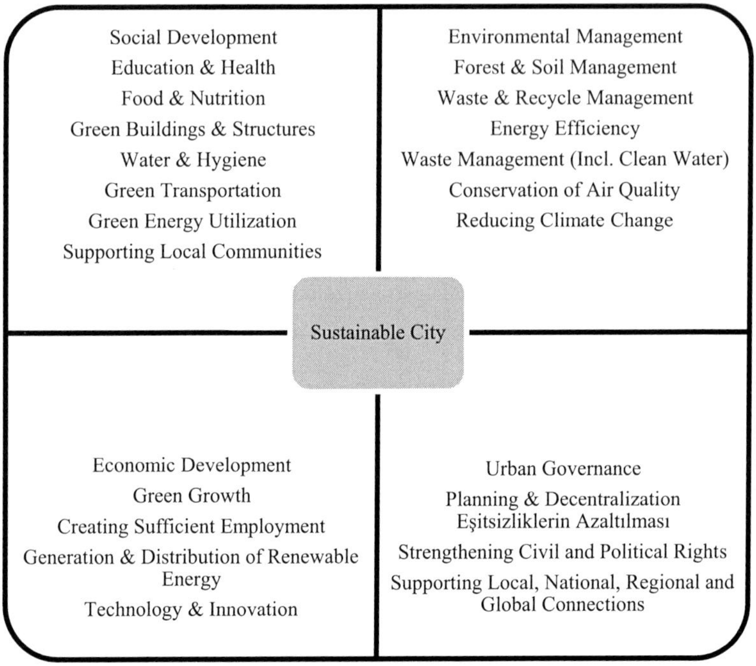

Fig. 3.1 Sustainable Urban Principles

One of the endeavors performed for the implementation these principles is the one for the green infrastructure.

4 Green Infrastructure Approach

Issues such as the insufficiency of urban gray infrastructure systems in resolution of the actual urban problems, negativities introduced by urbanization accelerating along with the population increase, unhealthy environments in terms of society caused by environmental pollution, disasters caused by climate change result in major problems in terms of cities. The green infrastructure system, introducing a nature-based solution to these issues, has become a substantial ecological tool in ensuring the sustainability of cities in recent years.

Green infrastructure is a concept frequently discussed within around the last twenty years and with the policies developed upon. Developed based on the idea of ecological network, the concept suggests a nature-based solution for the issues introduced by urbanization. The green infrastructure is defined in the Green Infrastructure Declaration (EU Commission, 2013) issued by the European Commission as the network consisting of natural or semi-natural interconnected areas, strategically planncd, designed and managed for the purpose of providing ecosystem benefits by conserving and preserving the ecosystem services. Benedict and McMahon (2006) propounding and developing the concept have addressed the green infrastructure on the basis of an ecological tool ensuring the social, economic and environmental health and defined as a network of interconnected natural areas and other open areas preserving natural ecosystem values and functions, ensuring clean air and water and providing various benefits and advantages to humans and wildlife.

Urban green infrastructure consists of areas and corridors connecting the areas. Factors of the green infrastructure system with the ability to operate and function at different scales from country to parcel level, may vary based on scale differences (Tab. 4.1). The green infrastructure on urban scale is consisting of "areas" such as urban forests, urban parks, neighborhood parks and green roofs, etc. and "corridors" consisting of linear lines such as planted avenues, motorways, railroads, energy transmission lines and stream corridors ensuring the interconnectivity of such areas. Areas are connected to each other by means of corridors and ensure such mobility, increase species diversity and accordingly support the formation of enriched areas in urban spaces in terms of biodiversity.

Tab. 4.1 Components of green infrastructure based on different scales (Güneş, 2017).

Neighborhood Scale	**City Scale**	**City-Region, Regional and National Scale**
- Street trees, fences - Green roofs and green walls - Pocket parks, Private gardens - City plazas, Sports fields - Town & Village green areas Pedestrian and bicycle paths - Cemeteries and church Cemetery - Public spaces - Pools and streams - Small woodlands, ditches - Playgrounds, school gardens - Local nature resources, wastelands and abandoned land, etc.	- Urban/Region parks - Channels, streams - Forest parks - Country parks - Municipality plazas - Main recreational areas Lakes, rivers and basins - Agricultural lands - Forest - Mining Sites - Landfills etc.	- Regional parks - Rivers and basins, channels - Strategic long distance routes - Forests, woodlands & motorway and railway networks Designed green belt and strategic clearances - Agricultural lands - National parks - National, regional and local landscape - Common grounds - Open and rural areas, etc.

Multifunctional infrastructure systems in the ecosystems dedicated to providing abiotic, biotic and cultural functions (Table 4.2) provide numerous benefits and advantages to the city and urban societies with these functions. In terms of ecosystem services offered in urban spaces, they provide certain benefits and advantages such as minimizing the effect of urban heat island, filtering and draining rainwater, enhancing the biodiversity, ensuring the mobility of species, regulating the climate, reducing air and water pollution, etc. and in terms of people living in the city, such benefits and advantages include creating spaces offering recreational opportunities, creating aesthetic areas with high visual quality and healthy and habitable environments in terms of public health and increasing the quality of life, etc. Moreover, the benefits and advantages in terms of combating the climate change and adaptation to the climate change include reducing the impact of disasters such as floods and storms, enhancing the resistance of ecosystems and reduce the impact of urban heat island. Furthermore, "ecosystem services offered by each component comprising the green infrastructure system vary depending on the ecological characteristics of these areas (location, size, spatial distribution, structure, density of the flora, etc.)" (Coşkun Hepcan, 2019).

Tab. 4.2 Functions of Green Infrastructure Systems (Ahern, 2007; Shakouri, 2016)

Abiotic	**Biotic**	**Cultural**
• Interaction between ground and underground waters • Soil development process • Maintenance of the hydrological regime • Nutrient cycle buffering • Entrapment of carbon and greenhouse gases • Climate buffering and modification	• Habitat for common species • Habitat for special species • Motion path for species and corridors • Biomass production • Provision of genetic reserve • Supporting flora and fauna interactions	• Direct experiencing of natural ecosystems • Physical recreation • Interpreting and experiencing cultural history • Provision of feeling of loneliness and inspiration • Provision of social interactions internal opportunities • Stimulating abstract/artistic expressions • Environmental education

Green infrastructure refers to a system performing multiple functions within the same spatial system, operating and functioning in integration with gray infrastructure, and consisting of interconnected green areas. It is proven in various research and studies that interconnected green areas yield higher level of performance. Therefore, ensuring the connectivity is among the fundamental principles in the green infrastructure approach. Upon establishment of the connectivity, the quality of the ecosystem services offered to city is increased and thus, the sustainability of the ecological processes is ensured accordingly. Ensuring an integrated operability of the green infrastructure with the gray infrastructure consisting of factors such as bridges, transportation networks, sewage systems and energy lines is the second fundamental principle of green infrastructure planning.

"Such an integration shall further contribute to the development of social infrastructures such as wastewater systems, healthcare facilities and the development of gray infrastructure such as environmental-friendly transportation, accessibility and mobility strategies." (Pauleit et al., 2018; Pauleit, 2019). The third fundamental principal is the multifunctional green infrastructure system. This principle is associated with the green infrastructure system offering multi-services. For instance; while the air pollution is minimized, urban heat island effect can be reduced, and rainwater can be drained. The final principle of the green infrastructure planning is to ensure the involvement of the relevant stakeholders. The entire relevant actors are required to be included and involved within the planning processes for green infrastructure planning to be able to respond to the requirements and needs of the entire stakeholders (Gülgün Aslan and Yazıcı, 2016).

The principles related to the green infrastructure planning and planning method are tried to be identified through various projects performed in the EU. Benedict and McMahon developed the concept and content of the green infrastructure set forth the following provisions related to the principles of the green infrastructure planning in an article published in 2006;

- Connectivity is essential.
- Content – scope is of utmost importance.
- Green infrastructure is required to be based on a scientifically constructed sound land use planning theory and practice.
- The green infrastructure does and should serve as a frame for conservation and development.
- The green infrastructure is required to be planned and maintained prior to development.
- The green infrastructure is a critical public investment required to be financed in advance.
- The green infrastructure provides benefits and advantages to the nature and humans.
- The green infrastructure is susceptible towards the requirements ad needs of the land owners and other stakeholders.
- The green infrastructure requires connectivity with the activities within the society (Parlak and Atik, 2020).

In consideration of the historical development of the concept, it is acknowledged that the green network systems utilized for land use and water management in the United States have started to be included within the agenda of the EU since the late 20th century. Becoming prevalent primarily in the United Kingdom and then within other European countries, the system was suggested through the European

Strategy for Biodiversity for the restoration and recovery of the habitat areas distorted and disrupted and subsequently, the scope has been expanded by being included within the covenants such as Water Directive, Nitrate Directive, Flood Directive, Seventh Action Plan for Environment (Gülçin, 2018).

The EU Strategy for Green Infrastructure has been established in 2013, setting out the goals of collecting data and information on green infrastructure, managing green infrastructure projects, and ensuring the integration of green infrastructure with other applicable EU policies. Projects such as Urban GreenUp funded by the EU, aiming to diminish and reduce the effects of climate change, improve the water management through nature-based solutions and accordingly ensure the urban sustainability; Green Surge, aiming for the climate change adaptation and to produce solutions for the major urban issues related to the human health and welfare; Grow Green, aiming to establish healthy and habitable cities resistant to climate and water, have been addressed through the green infrastructure approach.

As regards to the projects and research conducted related to urban green infrastructure planning, it is revealed that certain aspects rise to the prominence such as combating the climate change and/or ensuring the adaptation of the cities to the climate change, rainwater management, restoration and recovery of the distorted landscape, determination of potential areas for the green infrastructure and ensuring the involvement of the general public within the planning process.

Conclusion

Cities bear the qualification of being the spaces in where the development and advancement of the humanity become concrete. Economic, technological, social and cultural developments experienced in cities have resulted in a rapid and unplanned urbanization with the increasing population and during this period, urban ecology has been ignored and cities have started to become unhealthy habitats.

Issues such as the social and economic injustice as well as the environmental pollution, polarization and alienation pose threat to the urban sustainability. Seeking for a solution against this threat has become inevitable for the entire humanity.

The majority of the world's population living in the cities in our today's time has resulted in a sustainable urbanization conception. This approach is based on the assessment of the factors ensuring the economic and social development conjointly with the approaches towards the conservation of the nature.

One of the concrete practices supporting the approach is the green infrastructure strategy. The green infrastructure strategy augments the welfare in economic and social terms while ensuring the urban sustainability on local and regional scale by preserving the natural resources (European Commission, 2007; Parlak and Atik, 2020). When evaluated in this context, the strategy introduces nature-based solutions to the issues resulted from urbanization, provides benefits and advantages in terms of urban sustainability with ecosystem services, and thus offering a highly effective method in order to achieve the sustainable urbanization goals.

In consideration of the benefits provided by the urban green infrastructure through the perspective of sustainable urbanization, it is acknowledged to contribute to the following provisions;

- Environmental sustainability by offering ecosystem services (reducing air pollution, regulating the climate, etc.), increasing and maintaining the quality of biodiversity, collection, retention and ensuring the use of the rainwater, preventing the natural disasters, ensuring the continuation of ecological processes in a sound and robust manner;
- Societal sustainability by ensuring the public involvement within the process of planning, supporting the local identity, providing recreational opportunities, contributing for the establishment of healthy communities by creating healthy environments, increasing the quality of life, establishing habitable and healthy environments;
- Economic sustainability by generating employment opportunities, preserving and developing the natural resources.

The planning process is managed and guided by equally assessing the ecological, economic and socio-cultural data within the planning of the green infrastructure addressing to the city through a holistic perspective in multidimensional manner instead of single dimension. Therefore, it is required to preserve the landscape connectivity, establishing connectivity among the current/potential components of the green infrastructure and ensuring the integration of the green infrastructure with the gray infrastructure for the purpose of ensuring the sustainability of the urban spaces and the processes accordingly.

The management plans are required to be established in terms of establishing more habitable environments by supporting the urban economy, ensuring the societal welfare, ensuring integration of the green infrastructure systems with the urban infrastructure systems to decrease the environmental problems and including it within the processes of spatial planning and design and increasing the compatibility with the future spatial, economic, politic and sociological changes for ensuring the sustainability of the applied green infrastructure plans.

References

Ahern, J. (2007). Green infrastructure for cities: The spatial dimension. In: Novotny, V & Brown, P (Eds.) (2007) Cities for the Future Towards Integrated Sustainable Water and Landscape management. IWA Publishing, London. Pg. 265–283

Akçakaya, O. (2016). "Function of Local Governments in the Context of Achieving and Measuring of Urban Sustainability". Ardahan University Journal of Faculty of Economics and Administrative Sciences. Number 4, Autumn 2016, ISSN: 2148 - 7154.

Alberti, M. (1996), "Measuring Urban Sustainability", Environ Impact Assess Rev., C. 16.

Benedict M. ve McMahon, E. (2006), Green Infrastructure-Linking Landscapes and Communities, Island Press, USA.

Coşkun Hepcan, Ç. (2019). Kentlerde iklim değişikliği ile mücadele için yeşil altyapi çözümleri. İklim Değişikliği Eğitim Modülleri Serisi 12.

European Commission, (2017). Urban GreenInfrastructure: Connecting People and Nature forSustainableCities. GreenSurge

Gülçin, D. (2018). Yeşil altyapi bağlaminda açik/yeşil alan sistemlerinin uygulama olanaklarinin araştirilmasi: aşaği büyük menderes havzasi örneği, Doktora Tezi, Çukurova Üniversitesi.

Gülgün Aslan B. ve Yazıcı, K (2016), Yeşil Altyapı Sistemlerinde Mevcut Uygulamalar, Ziraat Mühendisliği Dergisi, Sayı:363

Güneş, M. (2017), yeşil altyapi kapsaminda yeşil ağ plani ve kent kimliği etkileşiminin irdelenmesi: ankara cumhuriyet dönemi sınırı örneği, Doktora Tezi, Ankara Üniversitesi.

Karakurt tosun, E. (2019). "Sürdürülebilir Kentleşme Kent Modelleri Üzerine Bir İnceleme".Bursa, Turkey.

Keleş, R. (2001). ''Sustainable Urbanization and Its Policy Implications the Case of Turkey'', (Ed. CAMARDA, D. ve L. GRASSINI) Interdependency Between Agriculture and Urbanization: Conflicts on Sustainable Use of Soil and Water, Bari: CIHEAM, Options Méditerranéennes: Série A. Séminaires Méditerranéens; N.44, 119–135.

Low, N. (2005). The green city: sustainable homes, sustainable suburbs. UNSW Press.

Meadowcroft, J. (1997). 'Planning for Sustainable Development: Insights from the Literatures of Political Science', European Journal of Political Research 31.

Parlak E ve Atik, M. (2020). Dünyadan ve ülkemizden mavi – yeşil altyapi uygulamalari, Peyzaj - Eğitim, Bilim, Kültür ve Sanat Dergisi 2/2. 86–100.

Pauleit, S. (2019). Urban green infrastructure for better living adapting cities & planning practice, Türkiye Peyzajları III. Ulusal Konferansı Yeşil Altyapı Bildiriler Kitabı, syf:27–35. Antalya.

Pauleit, S., Hansen, R., Rall, E.L., Zölch, T., Andersson, E., Luz, A., Santos, A., Szaraz, L., Tosics, I., &Vierikko, K (2017). Urban Landscapes and Green Infrastructure. In. Shugart H. (Ed.) Oxford Research Encyclopedia of Environmental Science. Online Publication Date: Jun 2017, https://doi.org/10.1093/acrefore/9780199389414.013.23

Project. Seventh Framework Programme.

Shakouri, N. (2016). Kentlerde yağmursuyu yönetimi kapsaminda yeşil altyapi peyzaj planlama ve tasarim yaklaşimi: sakarya-hendek örneği, Doktora Tezi, Ankara Üniversitesi.

UN WCED (1987). Report of World Commission on Environment and Development. "Our Common Future" United Nations.

UN WCED (1991). Report of World Commission on Environment and Development. "Our Common Future" United Nations. Ortak Geleceğimiz, Belkıs Çırakçı (çev.), Türkiye Çevre Sorunları Vakfı Yayını, Ankara, Turkey.

World Urbanization Prospects (2014). "The 2014 Revision", U.N. Department of Economic and Social Affairs. New York, U.S.A.

Yazar, K.(2006). "A Proposal of An Urban Planning Method for the Medium Sized Cities Within the Framework of Sustainable Urban Development" Ankara University, Ankara,Turkey.

Feran AŞUR and Elif AKPINAR KULEKÇİ

Courtyard Gardens Regulation Principles and Plant Selection in the Historical Process

Abstract: Culture, lifestyle, traditions, religious beliefs, protection, sense of privacy, climate and history play an important role in shaping the historic courtyard buildings and gardens. Courtyard gardens have been an element of these cultural activities. These spaces also create a cool environment during the hot summer seasons. Wind, has been an important factor affecting building design since ancient times. The effects of wind on buildings and reduction of energy consumption in terms of is important to demonstrate this effect through experiments. In this context, determining the location, orientation, building height, courtyard area, used materials and courtyard type of the courtyards of the buildings can create a new data for the designer and the implementer. The attention paid to the reconstructing of historical building designs in order to carry historical values into the future is the mainstay of this study. In this study the given information about the general characteristics the courtyard gardens built in different periods. In addition, within the content of the study, general information about too the preferred plant species for each period in the design of courtyard gardens is included. Thus, the reflections of past designs on modern designs of today were considered and evaluated from the point of view of landscape architecture.

Keywords: Courtyard, garden, plant selection in courtyards, courtyard and landscape design.

1 Introduction

Gathering is a common need in almost all civilizations that have existed in the historical process. The way people share together for various purposes has led societies to create spaces for these needs. The places that took their place in history as the agoras of the Ancient Greek Period, the forums of the Old Roman Period and the piazzas of the medieval cities are areas designed for similar functions. In Çatalhöyük, which is regarded as the first regular settlement example in history, Mellaart's [1] drawings in 1970show the first examples of open spaces reserved for common use. James Mellaart [1] mentions that the Çatalhöyük houses are built together, the entrance to the houses is provided from the roofs and there are narrow passages leading to the courtyards between the houses, and that these courtyards are places that provide air and light and are also used as cult areas (Fig. 1) [2, 3]. This settlement belonging to the Neolithic period has a history of about 9,000 years and Çatalhöyük, which is located at a point that dominates the Konya Plain in Çumra, is included in UNESCO World heritage sites [4].

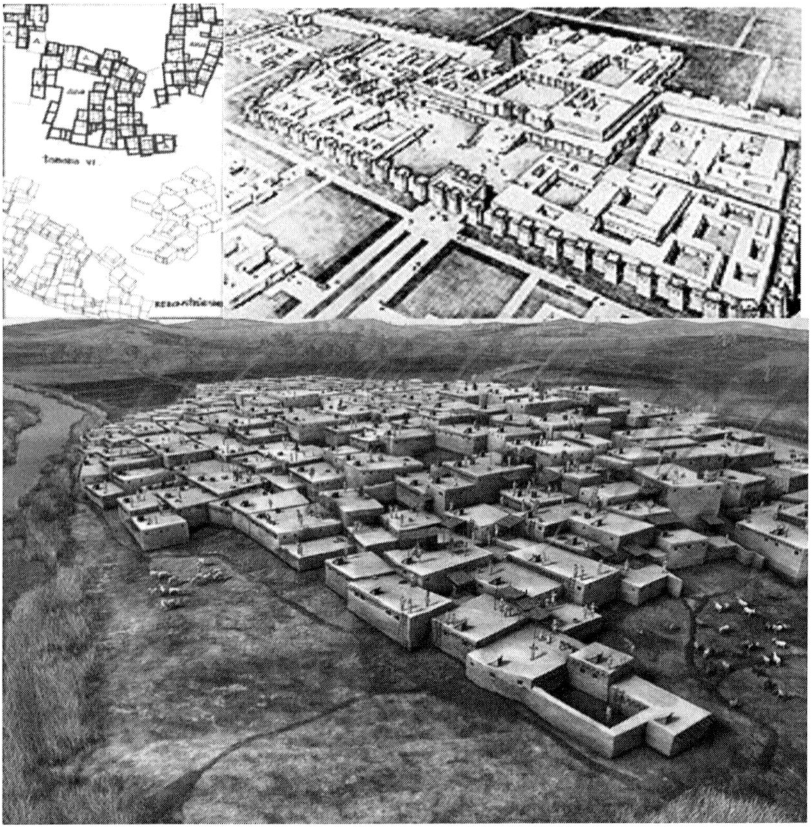

Fig. 1. Views from Çatalhöyük campus [3, 5].

Human beings, who are in constant contact with the environment they live in throughout the historical process, have shaped and changed nature in line with their own desires since the day they existed. This shaping first developed in the form of protection and struggle and later became a continuous and organic form with the discovery of the need for shelter. This situation is; it has caused people to attribute new meanings to the outdoor. With the transition to social life, these open spaces have gained more importance and this situation has led to the formation of different spatial constructions. The courtyard defined as a covered or open area in the middle of the buildings; It has developed as a space that has certain boundaries over time and is planned and organized according to certain purposes by people [6].

The attention paid to reconstructing historical building designs in order to carry historical values into the future is the main basis of this study. In this study, information is given about the general features of the courtyard gardens built in different periods. In addition, within the scope of the study, general information about the preferred plant species for each period in the design of courtyard gardens is included. Thus, the reflections of past designs on today's modern designs have been handled from a landscape architecture perspective and evaluated.

2 The Historical Development Process of Courtyards

In the period from the first appearance of the courtyards to the present day, the landscape planning of the courtyards shaped by the influence of Christianity, which began to spread in the medieval period with the first courtyards, has changed with Islamic culture. In particular, the issue of privacy manifests itself from courtyards created by the influence of Islamic culture. In addition, Far Eastern culture also has different scale courtyards, and it is possible to find different courtyard gardens in the early and medieval periods [7, 8]. (Iron 2019; Lightning 2015).

Examples of the first courtyard (Mesopotamian, Egyptian, Iranian, Greek and Roman courtyards), no matter what civilizations they belong to, are important for landscape design, the arrangement is functional, suitable for ecological data, has high aesthetic values, and the use of resources is economic [9].

In ancient times, courtyard structures were first encountered in southern Tunisia, and then encountered in North Africa, West Africa, South Africa and Morocco [10]. Although elliptical forms were adopted as courtyards in the period 8000 -5500 BC, rectangular forms began to be used over time [11]. The use of the courtyard during this period is for protection purposes. At that time, entering houses through the roof is also one of the most important reasons for this. The fact that the hearth, which was used in the Neolithic period, was located in the courtyard indicates that actions such as eating and drinking took place [12].

Courtyards were built in rectangular or L-shaped form to gain volume, in the Chalcolithic period 5500–3000 BC [13]. During this period, the courtyard was created for defense purposes. The surroundings of the buildings are built with stones like castles and all the structures in the courtyard open to a courtyard or courtyards called life [1]. In the period of 3000–2000 BC, the circular form was preferred as the courtyard form and although important developments were seen in terms of economy, it did not reflect on the buildings as a class difference [14]. In the years 2000–1200 BC, eaves were built in the spaces around the courtyard to collect heat in winters and to prevent harmful rays in summers [15]. Courtyard, it is seen in the Roman and Greek civilizations in the Classical period between 480 - 330 BC. In the buildings with courtyards, they have made eaves that prevent both privacy and harmful radiation to the buildings around the courtyard [15]. In the period of 476 AD the courtyard was used to provide privacy, above the rooms below ground level used as cellars were also used as courtyards [16].

The four-column scheme in the square, which we come across in Seljuk architecture and especially in Anatolian closed courtyard examples, is a continuation of the Asian architectural tradition. In the Seljuk architecture, it is seen that the courtyards were built considering the privacy and climate conditions [8].

Today, the courtyard is a system preferred to provide protection and privacy, which is usually located in the middle of the buildings and sometimes on the front, back and sides [17]. Many western studies have been made to dissemination the use of courtyards in modern architecture.

The scope of this study, there are structures with courtyards built in different periods throughout history and their general characteristics (Tab. 1).

Tab. 1. Courtyard Buildings Constructed in Different Periods in the Historical Process and Their General Features

The Period of the Courtyard Garden	**General features**	**Views of the Courtyard Structures of the Period**
Mesopotamian Courtyard Structures	Mesopotamia, located between the Tigris and Euphrates rivers, has been a home to rich civilizations throughout its history due to the soil quality and fertile vegetation. Sumerian, Akkadian and Assyrian civilizations are known as the civilizations established in this region that enabled the region to develop in agriculture, science and technical sense. Small courtyards created by Mesopotamian civilizations in the middle of residential buildings are the oldest known examples of the courtyard dating back to 3000 BC. Mesopotamian dwellings are a type of residence with a courtyard surrounded by deaf walls on four sides to protect from the negative effects of the climate [18, 19, 20, 21].	 Figure 2.A view of the Mesopotamian palace structure and courtyard [21].
Ancient Iranian Courtyard Structures	The Iranians interacted with the Egyptian and Mesopotamian civilizations. As a result of this interaction, the courtyard system separated from each other by walls in a formal garden layout surrounded by high walls thus became the characteristic feature of the first Iranian Gardens. (Figure 3). The garden is a living space where water is an indispensable element for Iranians due to the climate and soil conditions of the region. The fact that water and fountains are placed on the most important axis of the courtyard reveals how valuable water is for Iranians. With the influence of Egyptian and Mesopotamian cultures, trees were planted in a formal order to provide the desired amount of shade, thus facilitating circulation in the courtyard [9, 22, 23, 24].	 Figure 3. The Courtyard Garden of the Old Persian Period Architecture [25].

(*continued*)

Tab. 1. Continued

The Period of the Courtyard Garden	General features	Views of the Courtyard Structures of the Period
Ancient Egyptian Period Courtyard Structures	According to Egyptian beliefs, the living space gained importance after death. For this reason, they saw the houses they lived in as shelters. They did not pay attention to the permanent construction materials. On the other hand, they carefully built the temples and tombs with a courtyard plan. (Figure 4). In civilization of Egyptian, the garden has been an indispensable complement of living spaces. The green space was accepted as the food of the soul, and gardens were considered as a place where people would want to be in after death [26, 27]. The courtyard, surrounded by walls in the Islamic period, is generally built to regulate the air temperature. By balancing the temperature of the cold air leaking into the rooms in the evenings, the cold air remains stranded in the courtyard like a humidification tank until the late hours of the next morning. [28]. Also the courtyard provides privacy.	 Figure 4. A view of the ancient Egyptian temple and courtyard structure [27].
Ancient Greek Period Courtyard Structures	The outdoor space was very important for the ancient Greeks, in terms of both sports activities and public communication. As a result of cultural interactions, the origins of the gardens created by arranging rooms around a central courtyard are based on the Mesopotamian courtyard understanding [19]. Ancient Greek gardens have been shaped with ecological conditions in line with the social life and religious beliefs of the people. Care has been taken to create courtyards that meet the need to be outdoors in introverted residential life. The courtyard has become a part of the house in the form of a peristyle (portico with columns around the courtyard) where the daily life of the household passes [29]. Aesthetic concerns gained importance as a result of the eastern expeditions in the 4th century BC in the Greek gardens, where economic utilization was at the forefront in the first examples. The Greeks, who thought that the gods were related to natural phenomena, dedicated their gardens to the Gods. However, these gardens dedicated to the gods were replaced by ornamental gardens in the Hellenistic civilization, with a fondness for luxury in the 1st century BC [30]. In the pictures where living spaces are depicted in the Greeks, the courtyard, which has a small area, is covered with mosaic or stone and is not arranged as a garden. By decorating the walls with baskets in which plants are planted, the courtyard has become a garden. This basket or flower pot, in which the Greeks grow plants to dry in the sun, is called Adonis gardens. Reminding the "pronaos" of Greek temples, it is in the form of a forecourt/porch, and there are two wooden circular poles between its walls [31, 32, 33, 34,35, 36, 37], (Figure 5).	 Figure 5. Ancient Greek architectural building and courtyard ruins [36].

The Period of the Courtyard Garden	General features	Views of the Courtyard Structures of the Period
Roman Period Courtyard Structures	Roman civilization has shaped the cultural accumulations of various civilizations according to their worldview requirements. In the identity of the heir of the Greek civilization; They devoted gardens to the Gods, however, they did not neglect to create gardens for pleasure [38]. In Roman architecture, villas with inner courtyards in the city are of great importance. According to the Romans, the courtyard is an inner garden, which is the center of family life, in which there are service areas, the pool called the impluvium (rain pool) is placed to collect the rain water. Apart from the residences, there are also courtyards of the baths, which are health complexes in Roman architecture where social communication activities take place. Independently designed fountains are fountains built at the focal point of a large square or in a courtyard. It was built in the Roman period, resembling the "U" shape of the narrow and long houses on three sides of the courtyard [39]. Also, 8–7 BC. It is stated that there are megaron-type houses in many Anatolian cities dated to the 1st century. In the courtyards, an entrance is made through the outer door or directly to the courtyard. Generally, in the north of the courtyard, there is a living space, namely oikos (room, residence), which can also be two-storey [39]. There are storage areas, workshops or shops on the other side of the courtyard [40, 41, 42, 43], (Figure 6).	 Figure 6. A view of the Roman courtyard structure [43].
Courtyard Structures Constructed in the Shape of a Monastery and Castle in Europe	Courtyard Buildings Constructed in the Shape of a Monastery and a Castle in Europe the churches built to meet the needs of people to worship and the monasteries that are the center of religious teachings are structures with courtyards in the middle. These courtyards, which are located in religious buildings where worship are held and where travel and resting needs are met, are called "cloister" [30]. In the courtyard arrangements of the monasteries, which are of great importance as a source of life in medieval architecture, care was taken to be self-sufficient by serving all physical and spiritual needs (Figure 7) During this period, the soil was cultivated. In addition to vegetables and fruits, medicinal plants were grown to cure diseases. Warehouses for irrigation of plants, fountains or wells to meet drinking needs and ponds where fish are grown for food have become indispensable water elements of the courtyard. [17, 44, 45]. The courtyards, which feed the monasteries as a source of life, have become indispensable places in the middle of the buildings, where more thinking, walking and resting activities are carried out, aesthetic and functional concerns are prioritized, and vegetative design understanding is realized [32, 44, 17, 19, 46].	 Figure 7. A view from the Courtyard structure Built in the Shape of a Monastery and Castle in Europe [46].

(continued)

Tab. 1. Continued

The Period of the Courtyard Garden	**General features**	**Views of the Courtyard Structures of the Period**
Byzantine Courtyard Structures	Byzantine period archaeological excavations and surveys are very important in terms of uncovering and defining Byzantine settlements, whether urban or rural. As a matter of fact, rectangular planned buildings consisting of two, three or four rooms opening to a courtyard constitute the typical rural living unit of the Byzantine period [47]. In particular, ground surveys can provide information about the organization and resources of an entire region, as they can scan much larger areas than an excavation area can cover. Byzantine civilization synthesized the eastern and western garden cultures with the unique values. This means that Byzantine courtyards should be considered separately from Medieval European courtyards. (Figure 8). Byzantines, under the influence of eastern culture, began to give a religious meaning to the tree. This approach has been effective on garden designs. Although Byzantine architecture derives its source from villa architecture in Roman culture, it was also influenced by the use of courtyards in eastern culture [45]. Byzantine palaces consist of complexes with multiple courtyards. Byzantines gave importance to vegetative design in courtyards. Care has also been taken to use non-stagnant waters. The water gushing from animal figures lined up around pools covered with precious metals such as gold and silver is an indicator of eastern culture. The courtyards are adorned with marble columns and statues due to the Roman culture. [44, 46, 47, 48].	 Figure 8. A view of the Byzantine courtyard structure [48].

The Period of the Courtyard Garden	General features	Views of the Courtyard Structures of the Period
Courtyard Structures in Ancient Chinese Architecture	The fact that their cities are surrounded by four walls has enabled the formation of courtyards in Chinese architecture. According to Chinese belief, every part of nature such as mountains, rivers, seas and sky symbolizes the transformation of the soul into matter. For this reason, it is aimed to create a place where people can find and develop the spiritual beauties of nature by keeping them equated with nature parts in garden art [49]. Chinese culture dates back to ancient times like the civilizations of Egypt and Mesopotamia, the widespread use of wood in architecture prevented us from having an idea about the first garden examples. Privacy is very important in Chinese culture. For him, the buildings are located in a large garden surrounded by walls. Unlike palaces and temples reaching the sky with their magnificent architectures, Chinese family residences consist of multi-angle courtyard complexes placed on a single-storey and slightly higher platform. The courtyards, which are the family's private living spaces, are separated from the surroundings by high walls covered with stone or plaster. in addition to the shadow play of leaves and branches falling on the windows and walls and creating a background, the view that opens in the form of a moon allows the garden to watch from a different perspective. With the influence of Chinese culture, more pebbles, mosaics or stones were used instead of grass for the floor covering. It is believed that the rocks are the skeleton of the earth and the rivers located in large garden parts are also blood vessels [50]. In Chinese culture, water is expressed as a source of life, which is proof that water is an indispensable design element in Chinese gardens. [8, 49, 51], (Figure 9).	 Figure 9. A view from the ancient Chinese architecture courtyard structure [51].

(continued)

Tab. 1. Continued

The Period of the Courtyard Garden	**General features**	**Views of the Courtyard Structures of the Period**
Courtyard Structures in Ancient Japanese Architecture	In Japan, the garden has been an indicator of the people's love and longing for nature. Its origin is BC III. Although dating back to the 19th century, the idea of landscaping has changed under the influence of Chinese Buddhist missionaries (Figure 10). Thanks to the information reached today, the water, plants and rocks used in Chinese gardens, whose properties we know, have been indispensable design elements of Japanese gardens. [47]. In Japanese home architecture, care was taken to ensure that the garden, which is considered as an extension from interior spaces to nature, is compatible with the building. Bridges integrated with the diversity of the water elements in the garden, bird nests placed in different parts of the garden for the birds to accommodate, tea houses where tea ceremonies are held have been the structures that decorate the garden. The distinctive features of Japanese gardens are as follows; It consists of rock gardens that are organized as small examples of wild nature, dry gardens symbolizing water by arranging sand, stones and gravel in different ways, and water gardens that can be large enough to create islands in the middle. In Japanese culture, privacy is provided with regulations that realize the idea of living in nature [52, 53, 54, 55, 47, 56].	 Figure 10. A view of the old Japanese architecture courtyard structure [56].
Old Spanish Period Courtyard Structures	As a result of Islam reaching Spain via North Africa; In Spain, where the summers are hot and dry, the need for shade and coolness and the need to provide privacy with a certain introverted plan have become the most prominent features of the garden. In addition to the influence of the climate and social structure, the fact that it was settled on the ruins of Roman civilization was also effective in the formation of Spanish gardens (Figure 11). As in the Roman atriums, the gardens in Spain are formed from courtyard complexes surrounded by high and closed deaf walls, with passages between them. Under the influence of Islam, the Spanish courtyards have become a kind of open living room of the building, where the ideal of heaven on earth is tried to be realized, the coolness and music of the flowing water can relax and relax [57, 45]. Courtyards, thin columns and horseshoe-shaped arches, which are generally seen as outdoor spaces that take shelter in the shadow of the building, and the decoration of structures with geometrical stone and tile reliefs are indicators of Islamic decorative art. According to the traditional Islamic perception from the history of Islam until today, it is known that the lands of Spain, Portugal and Southern France (Iberian Peninsula or Iberia) still carry the Andalusian Islamic culture. [58, 57, 59, 60, 47, 61, 62].	 Figure 11. A view of the old Spanish courtyard structure [62].

The Period of the Courtyard Garden	General features	Views of the Courtyard Structures of the Period
Courtyard Structures in Ancient Indian Culture	The most important examples of garden architecture shaped by the influence of Buddhism are the monasteries called vihara, which means the place where time is pleasant. In Indian culture, the courtyard is a sacred place where the pool, which is considered to be sacred, is located, surrounded by chambers of dervishes and where rituals are performed. The fact that the front of the rooms opening directly to the courtyard is surrounded by porches was preferred in order to ease the circulation in the courtyard [63, 64, 19], (Figure 12). Hittite Tabet texts also mention many temples. These temples are often called "the house of God" in the text. The abundance of water in the region provided great opportunities in the use of water as a design element in the arrangement of the courtyard surrounded by high walls. The sacred building called the baradari, which is like a mausoleum in the center of the courtyard, has given the courtyard a sacred identity with the influence of Islamic culture. It is thought that the inner courtyard and columned hall in the center of the Great Temple may be the venues used for these festivals, and the Taj Mahal, a common product of Turkish-Indian architecture, is a monumental tomb located in a large courtyard. A monumental door enters the garden where the Taç Mahal is located, and it is thought to be a veil that covers a woman's face and needs to be gently opened [64, 65, 66, 19, 9, 67].	 Figure 12. A view of the courtyard structure in ancient Indian culture [67].

3 Most Used Plants in Courtyards Created in Different Cultures

In the historical development process of the courtyards; Different cultures have been formed depending on the living conditions required to the differences in the geographical locations of the societies. In the light of these different cultural values, the variety of plants used in living spaces has increased depending on the climatic conditions and soil characteristics of the geographic region where the societies live. The most used plants in courtyard landscaping in Egyptian, Greek, Roman, Medieval European, Far East and Islamic cultures are classified below [9, 68, 69].

In Egyptian Culture;

- *Origanum vulgare* (thyme), *Rosmarinus officinalis* (rosemary), *Anethum graveolens* (dill), *Nymphaea* spp. (Water lilies), *Mandragora autumnalis* (mandrake), *Centaurea turcica* (cornflowers), *Iris* spp. (irises), *Lilium* spp. (lilies), *Chamaerops humilis* (Palm trees), *Ficus kariensis* (ficus), *Robinia pseudoacacia* (acacias), *Platanus* spp. (plane trees) were used.

In Greek Culture;

- *Pyrus communis* (pear), *Malus communis* (apple), *Ficus carica* (ficus), *Punica granatum* (pomegranate), *Olea euopea* (olive tree), *Cissus antartica* (vine) were used.

In Roman Culture;

- *Platanus orientalis* (oriental plane tree), *Acer pseudoplatanus* (sycamore leaf Maple), Cupressus sempervirens (cypress), *Laurus nobilis* (bay), *Citrus limonum* (lemon), *Ficus carica* (ficus), *Pyrus communis* (pear), *Juglans regia* (walnut), *Olea europea* (olive), *Myrtus communis* (myrtle), *Buxus* spp. (boxwoods), *Nerium oleander* (oleander), *Punica granatum* (pomegranate), *Rosmarinus oficinalis* (rosemary).

In the Monastery and Castle Courtyards in Europe;

- Palm trees, *Lilium candidum* (white lily), *Rosa* spp. (rose), *Iris florentina* (irises), *Buxus* spp. (boxwood), *Taxus baccata* (yew), *Laurus nobilis* (laurel), *Myrtus communis* (myrtle), *Citrus limonum* (lemon), *Citsus sinensis* (orange), *Malus communis* (apple), *Punica granatum* (pomegranate), *Pyrus communis* (pear), *Syringa* spp. (lilacs), hyacinths, *Tulipa* spp. (tulip), *Jasminum* spp. (jasmine), *Hibiscus* spp. (marshmallow flower) is used.

In Far Eastern Culture;

- *Magnolia* spp. (magnolias), *Prunus domestica* (plum), *Punica granatum* (pomegranate), *Prunus persica* (peach), *Citrus sinensis* (orange), *Pinus* spp. (pines),

Cedrus spp. (cedars), *Juniperus* spp. (junipers), *Salix* spp. (willows), *Musa flexete* (banana), *Bambusa* (bamboos), *Chrysanthemum* spp. (chrysanthemums), *Wisteria sinensis* (purple panicle), *Rosa* spp. (hugging roses), *Nymphaea* spp. (water lilies) are used.

In Islamic Culture;

- *Platanus* spp. (plane trees), *Cupressus* spp. (cypress), *Populus* spp. (poplars), *Ulmus* spp. (elm), *Aesculus* spp. (ash), *Quercus* spp. (oaks), *Salix* spp. (willows), *Prunus domestica* (plum), *Prunus armeniaca* (apricot), *Prunus persica* (peach), *Prunus dulcis* (almond), *Prunus avium* (cherry), *Malus communis* (apple), *Pyrus communis* (pear), *Punica granatum* (pomegranate), *Ficus carica* (ficus), *Morus* spp. (berries), *Citrus sinensis* (orange), *Citrus limonum* (lemon), *Rosa* spp. (rose), *Narcissus* spp. (narcissus), *Dianthus spp.* (carnations), *Tulipa spp.* (tulips), *Hibiscus* spp. (marshmallows), hyacinths, lily of the valley, marigold flowers, gillyflower, *Jasminum* spp. (jasmine), *Viola* spp. (violets), *Nymphaea* spp. (water lilies) are used.

The courtyards created in Iran under the influence of Islamic culture are suitable for Iranian ecology;

- *Populus* spp. (poplars), *Acer pseudoplatanus* (sycamore-leaved Maple), *Ulmus* spp. (elm), *Salix* spp. (willows), Aesculus spp. (ash), gum trees were used.

Cypresses are used because they symbolize death, almonds because they symbolize a life full of hope, pomegranate symbolizes abundance, dates symbolize life.

The prevalence of different climate and ecological conditions in Spain distinguishes Spanish courtyards from other Islamic courtyards in plant use. In Spain;

- *Laurus nobilis* (laurel), *Myrtus communis* (myrtle), *Citrus sinensis* (orange), *Taxus baccata* (badger), *Cupressus* spp. (cypress), *Cedrus* spp. (cedars) and palm trees, *Diospyros* spp. (dates), *Punica granatum* (pomegranate) and *Jasminum* spp. (jasmine), *Iris* spp. (irises), daffodils, *Rosa* spp. (roses) are Used.

4 Conclusion and Recommendations

The accumulation that various civilizations have designed and produced for centuries and left to future generations after them is of great value to be able to understand and make inferences from the lens of the past. The courtyard gardens, where all these savings come to life, are the most important indicators of the superior creativity of societies with their fascinating looks, different shapes and forms. These unique values, cultural and architectural heritages are the most important traces left by the past to the future. Therefore, preserving these works and transferring them to the future also means transferring the identity of societies to the future. When the prehistoric courtyard gardens are examined, it is seen that tourism, industry, migration, urbanization and uncontrolled population movements have undergone many changes and developments in a rapid wear process as a result of unplanned construction. However, these places with unique cultural values and heritage values should be evaluated within the scope of interdisciplinary integrated conservation strategies and kept alive in a balance of protection and use.

As can be seen in this study, if the courtyard architecture covering different periods in the historical process is evaluated, it is seen that the form and spatial organizations of the courtyard gardens reflect certain periods. Theoretically, the basic needs played a decisive role in the shaping of the courtyards: Storage, open air, meeting areas, creating resting places. The courtyard structures with the building masses and high walls formed in line with the needs of people coming together to improve the living conditions of the people are living spaces that are sheltered from the noise of the street, where privacy and climatic comfort conditions are provided, and created by the influence of cultural, social and psychological values. Visual energy, which means a clear perception of the space and easy selection of design elements, is the most important factor in the arrangement of courtyards as prestige spaces. It reveals the difference of historical building designs by taking the examples of the buildings that have witnessed historical periods to the present day, taking the buildings under protection and evaluating them as a whole with their courtyards and performing their restorations. The courtyards, which meet the need for sheltered outdoor living, create privacy conditions thanks to their limited space, and provide natural lighting and ventilation of the building in which they are located, should be considered as urban green areas within the building.

References

1. Mellaart, J. Excavations at Hacılar, The BritishInstitute of Archaeology at Ankara, 1970. Edinburgh: 392–395.
2. Erdem, İ. "Şanlıurfa'da Yapı Avlularinin Doğal Havalandirmasinin İncelenmesi". Hasan Kalyoncu Üniversitesi, 2018.Yüksek Lisans Tezi.
3. URL-1,2021. http://212.174.161.8/kudeb_dokuman/catalhoyuk.pdf (Accessed: November 2020).
4. Demirel, B. N. G.and Pilehvarian, N. K. Osmanlı Sultan Camilerinde 'Avlu'. Journal of Near Archıtecture 2018. **1(2):**11–33.
5. URL-2,2021. https://arkeofili.com/10-maddede-neolitik-yerlesim-catalhoyuku-anlamak/ (Accessed: November 2020).
6. Bozkurt, S. G., and Altınçekiç, H. Anadolu da Geleneksel Konut ve Avluların Özellikleri ile Tarihsel Gelişiminin Safranbolu Evleri Örneğinde İrdelenmesi. *İstanbul Üniversitesi Orman Fakültesi Dergisi*, 2013. ***63*(1),** 69–91.
7. Demir, H. Anadolu Selçuklu Dönemi Külliye Düzenlemesinde Cami ve Medrese'de Ortak Avlu Kullanımı. Hacettepe Üniversitesi Türkiyat Araştırmaları *(HÜTAD)*, 2019. **(30),** 143–166.

8. Şimşek, O. Büyük Selçuklu Mimarisinde Avlu Kimliği. FSM İlmî Araştırmalar İnsan ve Toplum Bilimleri Dergisi, 2015. **5.** 455–479.
9. Hindistan, A. (2006). Avluların Peyzaj Tasarım Kriterler Yönünden Değerlendirilmesi. İstanbul Teknik Üniversitesi, Fen Bilimleri Enstitüsü, Peyzaj Mimarlığı Bölümü, Yüksek Lisans Tezi, s.134.
10. Awad, O. An Investıgatıon of the Effects of Ventilative Cooling in the Design of Low Energy Buildings, Department of Dokuz Eylül University Graduate School of Natural and Applied Sciences,: 2015. **10–14.**
11. Acar, E. "Anadoluda Tarih Öncesi Çağlardan Tunç Çağı Sonuna Kadar Konut ve Yerleşme" Türkiye Ekonomik ve Toplumsal Tarih Vakfı, İstanbul, 1996. **380–394.**
12. Esin, U. "Akeramik Neolitik Evrede Aşıklı Höyük" XI. Türk Tarihi Kongresinden Ayrı Basım, Türk Tarih Kurumu, 1994. Ankara: 30–33.
13. Aydın, Ö. "Anadolu Mimarlığında Avlu -Neolitik Dönemden Tunç Çağı Sonu" Gazi Üniversitesi Yayınlanmamış Yüksek Lisans Tezi-2000. Sayı**, 93665:** 26–28.
14. Al-Mumin, A. A." Suitability of Sunken Courtyards in the Desert Climate of Kuwait" Energy and Buildings 2001. Skin: 33 (2): 103–111.
15. Heinrichs, C. The Courtyard Housing Form as Traditional Dwelling. 1989. The Courtyard as Dwelling: 5–7.
16. Schoenauer, N. Seeman, S. The Court-Garden House. 1962. McGill-Queen's Press MQUP.
17. Bozkurt, G. Antik Priene Kenti Konut Mimarisinde Açık ve Kapalı Mekan İlişkilerinin Özellikleri, Yüksek Lisans Tezi, 2002. Ankara.
18. Naiboğlu, N. Kentleşmenin Kökeni Mezopotamya'da İlk Kentler. *MSGSÜ Sosyal Bilimler Dergisi*, 2019. **3(20)**:211–224.
19. Özköse, A. "Avluların Tarihsel Süreç İçindeki Evrimi", Doktora Tezi, Gazi Üniversitesi Fen Bilimleri Enstitüsü, 1995. Ankara.
20. Zettler, R. L. ve Horne, L. Treasures from the royal tombs of Ur. Philadelphia: University of Pennsylvania, 1998. Museum of Archaeology and Anthropology.
21. URL-3,2021. https://acikders.ankara.edu.tr/pluginfile.php/11800/mod_resource/content/1/3.%20Hafta%20Mezopotamya %20Uygarl %C4% B1%C4%9F% C4%B1%20Bah%C3%A7e%20Sanat.pdf (Accessed: November 2020).
22. Khooshroo, S. *İran'ın Geleneksel Mimarisi Sürdürülebilir Mimaride Bir İmge* Doctoral dissertation, 2017. İstanbul Aydın Üniversitesi Fen Bilimleri Enstitüsü.
23. İlter, İ. "Türk-İran Mimarlık Yapıları Üzerine Bir Deneme", Vakıflar Dergisi, XVI, Ankara 1982. pp.131–134.

24. Çınar, H. S., ve Erdoğan, R. Geleneksel bahçe ve mimari üslubun ortaya çıkışında doğal faktörlerin rolü: İran örneği. *Bartın Orman Fakültesi Dergisi*, 2019. ***21*(3),** 685–699.
25. URL-4,2021. https://www.google.com.tr/search?q=avlulu+iran+bah%C3%A7eleri+&t2ahUKEwjj5r7U28buAhUwAm (Accessed: January 2021).
26. Elinç, Z. K., Kaya, L. G., and Elinç, H. Efes Antik Kenti Yamaç Ev-1'in Peristil Bahçesi Restitüsyonu ile Antik Dönem Bahçe Mimarisi'nin İncelenmesi. International Journal of Social and Economic Sciences, 2013. ***3*(2),** 99–110.
27. URL-5,2021.https://www.turna.com/blog/antik-misir-tapinaklari (Accessed: November 2020).
28. Tharwat, O. El Kyam Fy El Emara El Islameya, Shrouk Yayınları, 1994. Kahire.
29. Arseven, C. E., "Ev" ve "Peristil", M.E.B. Devlet Kitapları Sanat Ansiklopedisi, 1950. Cilt-2, s: 1607, Milli Eğitim Basımevi, İstanbul.
30. Kayakent, T. Tarih İçinde Bahçe Olgusu ve Eski Türk Bahçelerinin Günümüz Bahçelerine Dönüşüm Süreci, *Yüksek Lisans Tezi*, 1999. İTÜ Fen Bilimleri Enstitüsü, İstanbul.
31. Kortanoğlu, R. E.and Kortanoğlu, M. S. Kıta Yunanistan Mimarisinde Birden Fazla Girişi ve Mekânı Olan Anıtsal Yapılar: Tarihöncesi ve Tarihsel Uzamlar, Kütleler ve Çevirileri. 2020. Cedrus, *8*, 117–134.
32. Soyöz, U. Mimari Kimlik ve Aidiyet: Bergama ve Kos Asklepios Kutsal Alanların Roma Dönemi Yapılaşması. Mekanlar/Zamanlar/İnsanlar: Kimlik Aidiyet ve Mimarlik Tarihi, 2029. **s.17.**
33. Serin, U. Karya'daki Geç Antik ve Bizans Dönemi Yapı ve Yerleşimleri Üzerine Bazı Gözlemler. *METU Journal of the Faculty of Architecture*, 2016. ***30*(1).**
34. Görkay, K. T. D. And Morkoç, S. Y. Antik Dönem Yunan evleri Doctoral dissertation, 2008. Ankara Üniversitesi Sosyal Bilimler Enstitüsü Arkeoloji Anabilim Dalı.
35. Spillecke, M. C. "Antik Yunan Bahçeleri", Bahçelerin ve Parkların Tarihi, Derleyen:Hans Sarkowicz, Çeviren Ersel Kayaoğlu, Dost Kitabevi Yaşam ve Kültür Serisi, 2003. Ankara.
36. URL-6,2021. https://chiubaka.com/travel/greece-day-6-the-oracle-at-delphi/ (Accessed: January 2021).
37. Kortanoğlu, R. E. and Heidegger, M. Yunan Tapınağı Olarak İsimlendirilmiş Yapının Temel Doğasına Yüklenmiş Anlamlar. 2015. In *Colloquium Anatolicum14* (pp. 83–101).
38. Turani, A. Dünya Sanat Tarihi, Evrim Matbaacılık Ltd., 1992. Remzi Kitabevi, İstanbul.

39. Akurgal, Ekrem (1999). Anadolu'nun Kültür Tarihi. Ankara: Tübitak Popüler Bilim Kitapları
40. Nevett, L. C. House and Society in the Ancient World. 1999. Cambridge University.
41. Aslan, N. Roma Dönemi Konut Mimarisinde Avlu Kullanımı ve Kültürel Kimlik İlişkisi: Ephesos, Priene, Aphrodisias, Kremna, Sia, 2019. Zeugma Örnekleri Işığında.
42. Farrington, A. The Roman Baths of Lycia. An Architectural Study. Likya'daki Roma Hamamları. Mimari Araştırmalar, British Institute of Archaeology at Ankara, 1995. London.
43. URL-7,2021. http://factsanddetails.com/world/cat56/sub369/item2072.html (Accessed: November 2020).
44. Barbet, A. Zeugma II. Peintures Murales Romaines, 2005. Varia Anatolica 17. Paris: Institut Français d'Etudes Anatoliennes Georges Dumézil.
45. Aydın, R. F. "Tarihi Türk Bahçeleri İstanbul Örneği", 1993. *Yüksek Lisans Tezi*, İTÜ Fen Bilimleri Enstitüsü, İstanbul.
46. URL-8,2021. http://www.zamekceskykrumlov.eu/description-of-objects-within-the-castle/ (Accessed: December 2020).
47. Akdoğan, G.,Bahçe ve Peyzaj Sanatı Tarihi**,** Ankara Üniversitesi Ziraat Fakültesi Yayınları 536, Ders Kitabı 1974. no: 309, Ankara.
48. URL-9,2021.https://www.pallasweb.com/deesis/tekfur-saray-blanchernai-byzantium.html (Accessed: January 2021).
49. Shimizu, K. Bibliography on Saljuq Studies, 1979.Tokyo, Institute for the Study of Languages and cultures os Asia and Afrika.
50. Zorlu Bekiroğlu, D. "Tarihsel Süreç İçerisinde Su Öğesinin Peyzaj Planlamada Kullanımı**",** Yüksek Lisans Tezi, 1992. Yıldız Teknik Üniversitesi Fen Bilimleri Enstitüsü, İstanbul.
51. URL-10,2021. https://www.shutterstock.com/search/classical+chinese+architecture (Accessed: December 2020).
52. Matsushita, S. Comparative Study Of The Structure Of Traditional Timber Housing In Turkey And Japan. Orta Doğu Teknik Üniversitesi, Fen Bilimleri Enstitüsü, 2004. Yüksek Lisans Tezi, Ankara.
53. Güngör, S. *Türk Evinin Geleneksel Japon Evi ile Yapısal Açıdan Karşılaştırılması.* 2019. Fatih Sultan Mehmet Vakıf Üniversitesi, Lisansüstü Eğitim Enstitüsü, Yüksek Lisans Tezi.
54. Demirkol, H. Jean Nouvel'in Çok Katmanlı Mimarisi Üzerinden Bir Kapalı Bahçe Değerlendirmesi. *Art-e Sanat Dergisi*, 2010. **3(5),** 1–16.
55. Nishi, K. ve Hozumi, K. What is Japanese Architecture: A Survey of Traditional Japanese Architecture. 1996. Tokyo: Kodansha International Ltd.

56. URL-11,2021. https://www.pexels.com/photo/view-of-building-exterior-327483/ (Accessed: December 2020).
57. Ayrancı, I. Uluslararası Sosyal Bilimler Sempoyumu. 18–24 Eylül 2017, İspanya. *Kırıkkale İslami İlimler Fakültesi Dergisi*, 2017. ***2*(4),** 131–134.
58. Şeyban, L. İspanya'da Endülüs Mirası Üzerine Bir Değerlendirme. *Akademik Tarih ve Düşünce Dergisi*, 2020. ***7*(4),** 2177–2196.
59. Bloom, J. B. Sheila, Islamic Arts, London, Phaidon Press Limited, 1997.
60. Stierlin, H. Islam: Early Architecture from Baghdad to Cordoba, 1996. **Vol.1.**
61. Erdoğan, R., Oktay, H. E. and Özçatalbaş, O. Türk İslam Bahçe Sanatında Avlular. 2016. s.17.
62. URL-12,2021. https://www.pexels.com/photo/view-of-building-exterior-327483/ (Accessed: January 2021).
63. Cezar, M. Anadolu öncesi Türklerde Şehir ve Mimarlık. Türkiye İş Bankası Kültür Yayınları. İstanbul, 1977. s.53.
64. Salihoğlu, E. F. and Demirarslan, D. "Hitit Uygarlığında Büyük Tapınak Mimarisi ve Etkileri". *Hitit Üniversitesi Sosyal Bilimler Enstitüsü Dergisi*, 2018. ***11*(3),** 1761–1779.
65. Bryce, T. Thé Kingdom of the Hittites, 2005. New York: Oxford University Press, Inc.
66. Güterbock, H. G. Yazlikaya: Apropos yeni bir yorum. Yakın Doğu Araştırmaları Dergisi, 1975. **34 (4),** 273–277.
67. URL-13,2021. https://www.dnaindia.com/analysis/column-understanding-the-wisdom-of-indian-residential-architecture-2610238 (Accessed: December 2020).
68. Aşur, F., Yazici, K. "Bitkisel Tasarım Çerçevesinde Bitki Kullanım Kültürü; İran Bahçe Örneği" Gaziosmanpaşa Bilimsel Araştirma Dergisi, 2018.**7(1),** 34–42.
69. Gheshlagh Sofla, N., Akpınar Külekçi E., Yılmaz, H., Current Academic Studies in Architectural Sciences -2018. Chapter: "Secret Gardens in Anatolian Seljuk Architecture: Double Minaret Madrasah as an Example" Gece published, Editor: Prof. Dr. Hasan Babacan, Printing & Bindingı:1, p. 354, ISBN:978-605-288-221-4.

Banu GÖKMEN ERDOĞAN

An Evaluation of Contemporary Additions to Refunctioned Architectural Heritage

Abstract: Some buildings used in the past lose their functions over time due to the changing lifestyle and standards; developing production systems, and increasing technological opportunities in the historical process. In this study, it is aimed to comparatively evaluate the contemporary additions to the re-functioned architectural heritage in terms of restoration discipline.

Preserving the buildings that have lost their functions without being used is economically difficult as well as difficulties in conservation practice due to insufficient financial resources and manpower in terms of regular control and maintenance. Giving new functions to buildings that have lost their original functions is also an accepted practice in the regulations in terms of preserving the architectural heritage. The architectural heritage, which continues to be used with a new function, will continue to be a component of cultural identity by maintaining its presence in the social memory. But, there are two important problems encountered in giving new functions to the architectural heritage. The first problem is that the given function's incompatibility with the original form of the heritage. Due to the new function a lot of unreasonable changes are required, so that the purpose of transferring the historical structure to the future in a healthy way is completely abandoned. The second problem is that the additions made by a design approach ignoring the restoration charters to the historical building are surpasses the historical building. In both cases, the aim of transferring the architectural heritage to the future has not been achieved.

In the study, at first, refunctioning; contemporary additions; items related to the use of modern techniques and materials were examined in internationally valid regulations. Afterwards, national and international examples of refunctioned architectural heritage were investigated. The architectural heritage buildings which had contemporary additions, and restorated by contemporary techniques and materials were selected as examples. As a result of the study, contemporary addition made at different levels were evaluated in terms of the preservation of architectural heritage with 5 criteria generated from charters.

Keywords: Contemporary approaches to architectural heritage, re-functioning, adaptive reuse, contemporary additions.

1 Introduction

The historical buildings, which transfer the culture and identity of the period they were built to today, show the change of the built environment and society in which they live, with their features such as materials, construction techniques and space organizations. With these features, historical buildings provide the necessary infrastructure for societies to survive and pass on their cultures to future generations. [1]. However, in the historical process, some structures used in the past lose their original functions over time, due to the reasons such as changing lifestyle and standards, developing production systems, increasing technological opportunities, etc. As an example of losing function that occurs due to changes in lifestyle and standards, the multi-room mansions that served the culture of living in the same house in the past being replaced by today's modern residences can be shown. There are also functional losses due to the opportunities provided by technological developments. For example, by the reason of the fact that water networks did not reach the houses in the past, hammams, which were of great importance for people have lost their functionality to a large extent today. Similarly, developing production technologies have caused many industrial heritage to become idle.

The greatest risk that loss of function poses on the architectural heritage is "abandonment". The building, which was abandoned and stayed in an idle state for a long time, starting deterioration and in some cases has the risk of collapsing completely. It is economically difficult to preserve buildings that have lost their functions without being used, and it creates difficulties in conservation practice due to insufficient financial resources and manpower for regular control and maintenance. However, if the building is re-functioned, the user, who will be responsible for the maintenance of the building, will also contribute to the regular repairs of the building by identifying new structural problems that may arise as well as performing regular maintenance. In addition, the architectural heritage used by giving a new function will continue to be a component of cultural identity by maintaining its existing place in the memory of the society. "Giving new functions proper for buildings whose original functions are discontinued", which is also accepted in the restoration regulations, makes it easier to keep them alive and to be transferred to the future in a healthy way. In recent years, "Adaptive reuse" application has emerged as a new approach in the transformation of heritage structures that have lost their function. According to Bullen and Love (2011) the most successful adaptive reuse projects are those that respect and retain a building's heritage significance as well as add a contemporary layer that provides value for the future. When a building can no longer function with its original use, adaption is the only way that a building's fabric heritage significance can be preserved and maintained [2].

Regardless of the name, "new functioning", "function change" or "adaptive reuse" for architectural heritage, two important problems are arising in practice.

The first problem is that the function given is incompatible with the original architecture and features of the building. So many changes are made for the new function that is incompatible for the historical structure that the purpose of "transferring the historical structure to the future in a healthy way" disappears. After the changes were completed the originality of the building has been lost to a large extent or completely. The second problem is that contemporary additions to the historical building surpass the historical building. The main issue here is that the contemporary design is very dominant and incompatible with the architectural heritage. Some of these types of designs can be considered architecturally successful when considered independent of the architectural heritage, but they are unsuccessful applications in terms of preservation when designed in addition to the heritage structure. Therefore, the new function should be decided after extensive evaluations, taking into account the universal value and uniqueness of the architectural heritage.

The conditions of giving new functions to the architectural heritage, making contemporary additions and using modern techniques and materials are clearly stated in the relevant regulations of the restoration charters. The contemporary additions that are planned to be made in the historical building or texture should be of the least interfering with the historical texture and structures, readable from the outside and not concealing. In addition, contemporary additions should be compatible with the architectural heritage and should not dominate the historical structure in terms of size, proportion, height, material and color [3]. Before moving on to examples, the relevant principles of the internationally accepted Carta Del Restauro (1931) and the Venice Charter (1964) are given briefly below.

In Carta Del Restauro, it is stated that new functions can be given to the monuments that have survived to the present day, not far from their original function and in a way that the necessary adaptations in the building will not cause significant damage. In the Venice Charter on the subject, it has been accepted that continuing to use the monuments will facilitate preservation; however, it has been emphasized that the changes required by the new function can be made provided that no changes will be made to the existing plans and decorations [4]. (Tab. 1)

It has been pointed out that for additions to the architectural heritage, in both charters that the additions should definitely be distinguishable from the original structure. However, it was emphasized that the least possible additions should be made and new additions should be lean and harmonious with the whole. It is also stated that the material to be used in the addition should not be the same as the original material. [5]. (Tab. 1)

On the use of modern techniques and materials, the issue emphasized in both charters is that the scientific validity of modern techniques and materials to be used due to the insufficiency of original traditional techniques of the building [5]. (Tab. 1)

Tab. 1. Carta Del Restauro and Venice Charter

	Carta Del Restauro (1931)	Venice Charter (1964)
New function	**Article 4:** It is acceptable to give usages for the existing, that is, standing monuments, only not far from its original function	**Article 5:** Preservation of monuments can be facilitated by their use. Changes required by the new function are allowed as long as no changes are made to the plans or decorations for the new function.
Additions	**Article 7:** New additions to consolidate, integrate partially or completely or re-use should be kept to a minimum and should reflect a simple and structural order. **Article 8:** Additions must be precise and clear. It should be made with materials other than the original material.	**Article 9:** Any addition should be distinguished from architectural composition and bear the hallmark of its era. **Article 12:** The missing parts should be compatible with the whole while being completed. It should be distinguishable from the original. **Article 13:** Additions are only permissible if the original condition of the structure will not be compromised.
Contemporary Techniques and material	**Article 9:** When old construction techniques are insufficient to strengthen the structural system or to integrate the structure, it may be appropriate to use modern techniques.	**Article 10:** In cases where traditional techniques are insufficient, reinforcement can be made with modern techniques validated by scientific data and experiments.

2 Contemporary Additions to the Architectural Heritage

According to the ICOMOS (2013), the value of architectural heritage is not only in its appearance. Its value is also in the integrity of all its components as a unique product of the specific building technology of its time. Therefore, removing of the inner structures maintaining only the façades does not fit the conservation criteria [6]. This means that the main purpose have to be preserving the architectural heritage with its all components when giving it a new function. "Preserving by using" can be considered as one of the approaches in sustainable conservation studies. The most important point in this process is that the building can be kept alive, not the given function. [7].

New spaces or building elements are added to historical buildings for different purposes. These reasons can be listed as follows:

- The need for new space required by the new function;
- Providing disabled access;
- Compliance with disaster regulations such as fire and earthquakes;
- User comfort; providing indoor air quality
- Linking existing historical buildings
- Integrating the missing parts of the historical building that have not reached today,
- Consolidation/Strengthening, etc.

Contemporary materials of the era are preferred when making attachments. These materials provide great opportunities to the designer and make his designs more free [8]. Besides, the additions reflecting the technology of the era can be distinguished from the original historical structure in accordance with the regulations.

3 Examples of Contemporary Additions

In this section, new functioned 6 architectural heritages with different levels of intervention were examined under two separate headings, namely: "Foreign Examples" and "Domestic Examples".

3.1 Foreign Examples

Almost all over the world, there are many historical buildings that lost their first built purpose, and re-functioned for any purposes. In this section re-functioned 3 architectural heritage examples (a church from Netherlands, a post Office from Canada and a church from Germany) are presented:

3.1.1 Dominican Church, Maastricht, Netherlands

The Dominican Church is a Gothic church building in the center of the Dutch city of Maastricht. The church was built in the 13th century as a monastic church. At the end of the 18th century the monastery was dissolved and the church fulfilled various functions [9]. The church has been used for secular uses for over more than a century but despite its historic and architectural significance, most of the functions; such as warehouse, flower exhibition, a place for carnival parties etc. [10]; did not respect the monumental value of the building. The function, which was given as a bookstore in 2007, was made without damaging the originality of the building. Because the church is an impressive monument in its own right, it needed to be kept as an open space as much as possible but the available floor area was only half of the required commercial area. Therefore, a volume of two floors high was built in the church. In the meantime, no changes were made to the exterior of the church [11]. (Fig. 1)

Fig. 1. Selexyz Dominican Bookstore Inside Views (top), section drawing (bottom left), façade (bottom right)[26]

3.1.2 Former Galt Post Office, Cambridge, Canada

The masonry structured building by Canadian architect Thomas Fuller was completed in 1887 as a government building and after a while it was turned into a post office [12]. The building listed in 1983 as a representative one of small urban post offices by Thomas Fuller. Fuller was Chief Architect of the Department of Public Works from 1881 to 1896. The buildings heritage value is defined by Parks Canada Directory of Federal Heritage Designations as below:

> *"The high standard of design imposed by Fuller is evident in the former Galt Post Office. Typical of his designs, the building features a round-arched entrance portico, symmetrical twin façades and a clock tower, in this case uniquely adapted to be visible from both the street and the Grand River. The steeply pitched mansard roof, as well as the use of contrasting colors and textures and the accomplished stonework, come together to create a striking blend of the Gothic, Second Empire and Romanesque styles."* [13]

Until 2012, Cambridge's masonry post office, listed as a National Historic Site was in serious disrepair. A project was developed which involved wrapping a 9,000-square-foot transparent pavilion around two sides of the old structure, while sensitively rehabilitating the original building [14] (Figs. 2 and 3).

Fig. 2. Facade of the Building (Left: 1907, Right: 2018) [27]

Fig. 3. Back Facade View from Grand River (2018) [28]

3.1.3 Saint Kolumba Church, Cologne, Germany

The Church of St. Kolumba was the oldest and richest parish church of late Gothic in Cologne [15]. The Church of St. Kolumba was destroyed by bombs during World War II (Fig. 4). In the city center of Cologne, it was decided to design a new museum in the area where the ruins of the St. Kolumba and Böhm Chapel, which was destroyed by bombing in the past, in order to preserve and evaluate the historical heritage, and Peter Zumthor won the competition opened in 1997 [16].

Fig. 4. The Ruins after World War II [15]

A museum has now been built on the historic grounds, at 12 m above ground level. Realization of this ambitious building was only possible by the use of micro piles bored into the sandy ground all the way through the over 600-year-old walls of the destroyed church [18].

It has been stated that it is aimed to make distinguishable the addition and the historical parts of the building by the new design. Thus, the museum was built on the Gothic wall ruins of St. Kolumba Church [16]. (Fig. 5) On the upper two levels of the museum Zumthor's modernist architectural language was seen [17].

Fig. 5. Kolumba Museum [29]

3.2 Domestic Examples

In this section, 3 domestic examples with different intervention levels are given, which are used today with new functions because of the original function does not continue. The examples are a gin factory, a warehouse and a cistern, respectively.

3.2.1 Tarsus Ginning Factory, Tarsus, Turkey

The Tarsus Ginning Factory, considered to have been built by the British in the second half of the 19th century, is one of the first industrial buildings of its time [19]. This wooden-roofed stone structure was built originally to be used for the separation and storage of cotton fibers from the core [20]. In the north of the factory located in the city center, there are more than twenty examples of traditional civil architecture as well as important monumental structures such as St. Paul's Church, Ulucami and Kırkkaşık Bedesteni. The archaeological site, located in the southwest of the area and known as Gözlükule, forms the important "cultural landscape" of Tarsus with its historical values, topographic location and vegetation. [19].

The building has gone under reconstruction by Boğaziçi University with the permission and supervision of the Adana Regional Cultural Heritage Board, and refunctioned as a research center for the archaeological excavations, conservation laboratory, and storage facility for study collection, library and exhibition space. In 2019, Boğaziçi University Tarsus-Gözlükule Archeology and Research Center was awarded Europa Nostra Cultural Heritage Conservation Prize [20] (Fig. 6).

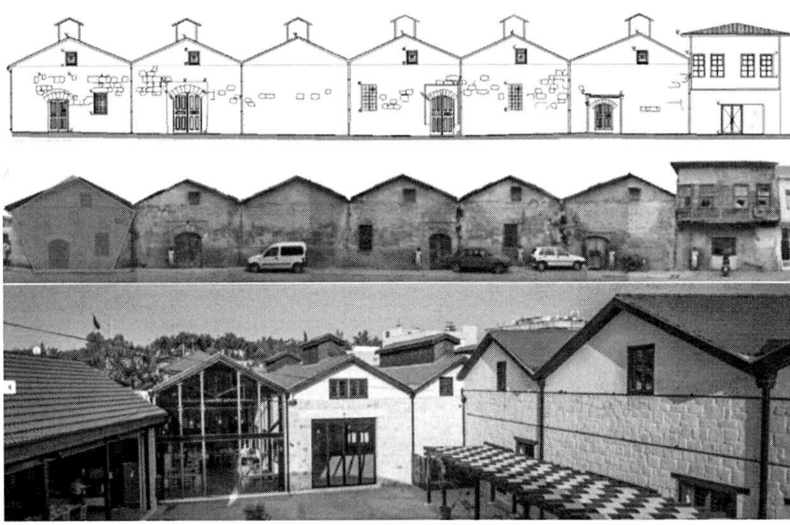

Fig. 6. Boğaziçi University Tarsus Gözlükule Excavations Research Center [20]

3.2.2 Tantavi Warehouse, Konya, Turkey

With the arrival of the Istanbul-Baghdad Railway project in Konya in 1896, it was necessary to construct warehouse buildings around the station where exported or imported goods were stored. Built by Hafız Ragıp Efendi, who studied in the city of "Tanta" in Egypt, today the building is known as the "Tantavi Warehouse". The building was built in 1903, according to the inscription on the wooden decoration in the middle of the overhanging eaves of the gable roof, on the façade facing north of the building, which does not exist today [7] (Fig. 7).

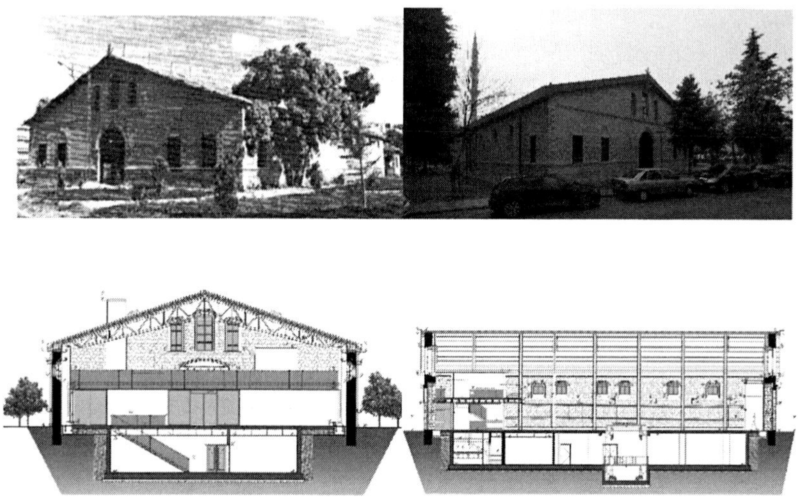

Fig. 7. Tantavi Warehouse before Restoration (top left), after Restoration (top right), Restoration Project Drawings (bottom) [22]

After being used for a long time with its original function, the building, which was in ruins for many years, was listed as architectural heritage in 1990. In 1993, the Provincial Directorate of Culture made an application to the Preservation Board to transform the building into a cultural center, Meram Municipality into a library, and the neighborhood mukhtar into a health center. The Preservation Board decided to repair the building and use it for cultural purposes [21]. The building was reopened in 2019 as a culture and art center where events such as symposiums, conferences, fashion shows, and exhibitions will be held.

The original exterior walls of the building were preserved during the interventions. A second steel structure was built inside the building and the roof was placed on this structure. Additional spaces (artist rooms, etc.) and service units (kitchen, wet areas, technical units) required for cultural and artistic activities to be carried out have been resolved in the basement floor added to the building [22].

3.2.3 Cistern, Istanbul, Turkey

The cistern located under the Antique Hotel in Beyazıt was unearthed during the construction of the hotel in 1984. The cistern, 20x15 m in dimensions, has a rectangular plan and contains 13 columns [23]. As a result of the investigations made by historians and experts, it was determined that the ruins belong to the lower floor (cistern) of a large building belonging to the Late Roman and Early Byzantine periods.

The hotel was built on the cistern and today it is called Antique Hotel. The cistern is used as a meeting and invitation room in the basement of the hotel [24] (Fig. 8).

Fig. 8. Cistern excavation (top) Antique Hotel (left bottom) and the Cistern under the Hotel now [24]

4 Evaluation

Historical buildings lose their original functions over time due to various reasons. Unused structures remain idle, and be damaged or collapsed over time. For this reason, it is a frequently preferred method to give a new function that do not spoil the uniqueness of the heritage. The main purpose, which has not to be unforgotten here is to transfer the architectural heritage to the future in a healthy way. However, it is seen that this purpose has been ignored in some examples in both past and today, and especially the architectural heritage remaining in precious areas in the city center over time is perceived only as a building stock. For this reason, sometimes profit-oriented functions are given in the functionalization of the architectural heritage and irreversible damages can be given to the building through function change applications.

In re-functioning, another rule accepted in international regulations is reversibility. This is usually to prepare a project that envisages that new additions are made without damaging the old structure (this may require an independent structure), and that items such as stairs, entrances, porches, eaves, and partitions can be removed. Because in this situation the problem is to leave a special aesthetic and cultural message to the future as an accumulation of civilization. The architectural heritage preserved by giving a new function because there is no economic opportunity, the structure should be ready for new investigations and evaluations in the future, and should be able to return to its current state in other conditions in the future [25].

Within the scope of this study, 6 architectural heritage structures were evaluated via 5 criteria adapted from the relevant articles of the preservation charters. (Tab. 2) These criteria are as follows:

- Criterion 1: The addition has not been required much intervention to the architectural heritage.
- Criterion 2: The addition can be understood from the outside.
- Criterion 3: The addition did not close the architectural heritage.
- Criterion 4: The addition is in harmony with the historical building in terms of size, ratio, height, material and color.
- Criterion 5: The addition is reversible.

Tab. 2 Criteria Table. C: Abbreviation of "criterion"

Original Name	**Recent Name**	**C1**		**C2**		**C3**		**C4**		**C5**	
		Y	**N**	**Y**	**N**	**Y**	**N**	**Y**	**N**	**Y**	**N**
Dominican Church	Slexy Bookstore										
The Former Galt Post Office	Idea Excahnge Office										
Saint Kolumba Church	Kolumba Museum										
Tarsus Ginning Factory	Gözlükule Excavation Observatory										
Tantavi Warehouse	Tantavi Cultural Center										
Cistern	Antique Hotel										

When the examples were evaluated according to the criteria, it was determined that; Dominican Church, Tantavi Warehouse and Tarsus Ginning Factory fulfilled all the relevant conditions in the charters in the change of function while The Former Galt Post Office has met 4 of the criteria. Although, Kolumba Museum was an awarded project by its design, it was seen that by new function it only met two of the criteria in the charters. Lastly, in the cistern, even though it is stated on the Hotel's website that the remains of the cistern have been preserved, it was determined that none of the 5 criteria according to the regulations are provided.

5 Conclusion

New functions that will not damage to the original state can be given to the structures that will be difficult to preserve when not in use. The rules of giving new functions are clearly expressed in the preservation charters. As seen in Tab. 2, if the very basic 5 criteria have been provided, both the architectural heritage can be preserved and the building can be used with the new function.

The purpose of re-functioning must be defined correctly. The main purpose of re-functioning historical buildings is to transfer the architectural heritage to future generations in a healthy way. Otherwise, there will be the risk of irreversible loss of heritage identity and even loss of the architectural heritage completely whose function has been changed. In such cases the architectural heritage that is supposed to be preserved will not go beyond being a decor to the addition.

As a conclusion, additions to the historical building must comply with the principles specified in the restoration charters for whatever purpose. For this reason, rigorous research and comprehensive evaluations are required before design and implementation of additions. Also, irreversible additions should be avoided.

References

1. Taştan, H. and Manisa, K. (2019). *"Tarihi Yapılarda Kullanım Dönüşümünün Sosyal Boyutu; Bağlarbaşı Elektrik Fabrikası Ve Tramvay Deposu Örneği"*, Omer Halisdemir University Journal of Engineering Sciences, Volume 8, Issue 2, (2019), 1251–1264
2. Bullen, P. A. and Love, E. D. (2011). *"Adaptive Reuse of Heritage Buildings"* Structural Survey Vol. 29 No. 5, 2011 pp. 411–421.
3. Çelebi Karakök, M., E. and Gökarslan, A. B. (2017). *"Tarihi Dokuda "Çağdaş Ek" Kavramının Atölye Ortamında Deneyimlenmesi: Mass' Workshop 2015"*, Tasarım+Kuram. DOI: 10.23835/tasarimkuram.413581
4. Ahunbay, Z. (1999). *"Tarihi Çevre Koruma ve Restorasyon"*, YEM Yayın, İstanbul. ISBN: 975-7438-38-3
5. ICOMOS, 2003. http://www.icomos.org.tr/Dosyalar/ICOMOSTR_tr0033791001536913477.pdf

6. ICOMOS, 2013. http://www.icomos.org.tr/Dosyalar/ICOMOSTR_en0033912001536913477.pdf
7. Yalaz, E.T. & Yaldız, E. (2020). "*Yeniden Kullanım Sonrası Yapısal Müdahalelerin Değerlendirilmesi, Tantavi Ambarı Örneği*", Artium, 8 (2), 105–117
8. Gökmen Erdoğan, B. (2015). *"Korunması Gerekli Taşınmaz Kültür Mirasında Cam Malzeme Kullanımı Üzerine Bir Araştırma"*, 5. Tarihi Eserlerin Güçlendirilmesi ve Geleceğe Güvenle Devredilmesi Sempozyumu, İMO, 1–2 Ekim, s:333–346, Erzurum. http://www.imo.org.tr/resimler/ekutuphane/pdf/17519_04_49.pdf
9. Dominicanenkerk (Maastricht) - Wikipedia (Accessed 03.01.2021).
10. http://crossroadsmag.eu/2008/03/between-two-selexyz-dominicanen-as-church-and-bookstore/ (Accessed 03.01.2021).
11. Plevoets, B.& Van Cleempoel, K., (2009). "*Retail-reuse of historic buildings: developing guidelines for designer and conservators"* https://www.researchgate.net/publication/242419497_Retail-reuse_of_historic_buildings_developing_guidelines_for_designer_and_conservators/citations
12. https://www.dezeen.com/2019/11/15/idea-exhange-old-post-office-rdha-canada/ (Accessed 04.01.2021)
13. Historic Sites and Monuments Board of Canada. https://www.pc.gc.ca/apps/dfhd/page_nhs_eng.aspx?id=374&i=50766
14. https://nationaltrustcanada.ca/nt-awards/cambridge-old-post-office-and-idea-exchange
15. McCune Bruhn, H. (2007). "*The Parish Monstrance of St. Kolumba: Community Pride and Eucharistic Devotion in Cologne around 1400"*, Athanor, 25, p: 17–27.
16. Öktem Erkartal, P and Ökem, H. S. (2015). *"Mimari Tasarımda Dokunma Olgusu ve Dokunsal Haritalamaya İlişkin Bir Alan Çalışması"*, Megaron, 2015;10(1):92–111, İstanbul.
17. Moffatt, L. (2009). *"Laura Moffatt reports on Pritzker Prize winner Peter Zumthor, the Kolumba Art Museum and Gerhard Richter's window in Cologne Cathedral"* Art and Christianity, (58), 8+.
18. Dietz, K. and Schürmann, A. (2006). "*Foundation Improvement of Historic Buildings by Micro Piles, Museum Island, Berlin and St. Kolumba, Cologne"*, http://www.ismicropiles.org/ISM/Proceedings/2006/
19. XXI. (2018). *"Katmanlar Arasında"*, V: 167, p:46, İstanbul. https://xxi.com.tr/i/katmanlar-arasinda
20. http://www.tarsus.boun.edu.tr/ Research Centers: Tarsus-Gözlükule Excavations Research Center (Accessed 11.01.2021)
21. Fırat, N. (1996). *"1900–1930 Yılları Arasında Konya'da Türk Mimarisi"* (Yayımlanmamış Doktora Tezi). Ankara Üniversitesi,

22. https://www.arkiv.com.tr/proje/tantavi-ambari-restorasyonu/11374 (Accesses 11.01.2021).
23. Altuğ, K. (2013). "İstanbul'da Bizans Dönemi Sarnıçlarının Mimari Özellikleri Ve Kentin Tarihsel Topografyasındaki Dağılımı" (Yayınlanmamış Doktora Tezi), İstanbul Teknik Üniversitesi, İstanbul.
24. http://www.antikhotel.com/tr/pages/menu/antik-cisterna/tarhce-ve-bulgular (Accessed 06.01.2021).
25. Kuban, D., 2000. "*Tarihi Çevre Korumanın Mimarlık Boyutu Kuram ve Uygulama*", YEM Yayınları, İstanbul.

Figure References

26. Figure 1: Selexyz Dominicanen Bookstore https://archello.com/project/bookstore-selexyz-dominicanen
27. Figure 2: https://ideaexchange.org/oldpostoffice/idea/thomas-fuller-architect-extraordinaire
28. Figure 3: https://www.archdaily.com/918530/idea-exchange-old-post-office-community-center-rdha
29. Figure 5: https://www.architectureanddesign.com.au/features/product-in-focus/bricks#

Gildis TACHIR and Hatice KIRAN ÇAKIR

Instantly Pandemic to Converted Metropolitan City Plan

Abstract: There is a linear relationship between the pandemic and the city. The transmission rate of the pandemic is higher in areas where the population is dense, that is, in cities. The pandemic is transmitted from human to human through contact. In this study, the covid-19 pandemic and city's relationship were examined, and it was understood that it was crowded at the basis of the pandemic. Population density is mostly experienced in crowded cities. The crowded city is a metropolitan city with a population exceeding 500,000 people. It is envisaged that the isolation of the crowd can be provided by converted planning of the cities. In this sense, it is emphasized that the converting conditions of cities during the pandemic are evaluated by urban building areas and capacity ratio calculation method, and the necessity of each city to be converted plan in the city vision in the case of pandemic.

Keywords: Population density-crowd, Metropolitan City, Pandemic, COVID-19.

1 Introduction

More than half of the world's population lives in cities [1]. Urban planners should ensure that the impact of population density is taken into notice in the city's vision planning. The mean of population density is commonly used to refer to the number of people per land area (square kilometre). It is the ratio of the population to the land area [2]. In 1998, the world's population density was almost 40 persons per sq. km. The U. S. population density was only 28 persons per sq. km., a level substantially below the world average. The population density in metropolitan city of China at 130 people per sq. km. is more than three times the world average, even though vast arid areas in the western part of China are sparsely populated. India is one of the more densely populated countries of the world. Its density of 307 people per square km. puts it at almost eight times the average world population density [3]. The world's population rate of 60 % in 2030 is expected legislature in urban areas [1].

The world population has reached 7 million. An important part of this number consists of people living in developing countries such as Africa and Asia. It is thought that 342 million people will be added to the world population by 2030 only from China and 80 million more in Indonesia [4]. This situation shows us that the world population is getting more density and more crowded.

Along with the industrialization and modern period, the vast majority of the growing world population lives in cities. In Japan, 1/4 of its total population lives in the Tokyo Metro, and from this point of view, urbanization is over 80 % in Japan. This rate is 89 % in England, and same 89 % in Netherlands and 78 % in Spain. Also the ratio of the urban population in Turkey is 66.7 % [5]. The fact that a large proportion of the country's population lives in cities makes the growth of urban areas and building masses inevitable. For example, while the area covered by the Istanbul metropolitan area was 334 km2 in 1950, this number reached 2819 km^2 in 2000. It is understood that the city has increased almost 8.5 times with its spatial growth, expansion and integrated settlements [6]. In parallel with this, when the population growth of Istanbul is analysed, it is seen that the population was 1.268.771 in 1955 [7]. It was determined that this number was 15.067.724 in 2018 [8]. It is seen that the population density of the city has increased 15 times. Increasing population causes cities to become crowded. Increasing crowded population constitutes physical and social problems. Some of these are natural disasters, (earthquake, flood, erosion, hurricane, etc.) geographical climate changes, pandemic and infectious diseases, (Plague, Tuff, Cholera, Malaria, HIV/AIDS and recently seen COVID-19 etc.) Its effects on the mobility, crowd and density of the population are more destructive [9]. One of the biggest consequences of living in masses with the growth of urban areas is that it becomes difficult to control the rate of spread of diseases that are classified as global pandemic. The most recently seen COVID-19 pandemic [10] is a disease that is transmitted through human contact and intimacy. For this reason, crowded cities, areas with contact and rapprochement are the environments where the transmission in the pandemic increases more easily.

Metropolitan cities are spaces where coexistence is more affected by poverty, pollution and diseases. Outbreaks will bring greater destructions in cities where population density/crowd is increasing. Cities as dynamic, feeding each other and living organisms are affected by the changes created by global problems [11]. In reducing and controlling this effect, it is envisaged that the planning of converting urban spaces, which can be directed to the application in unexpected situations in which the size ratio balance of the population and building massive is evaluated, is included in the city's vision planning by taking part in the disaster and risk management.

Besides this, pandemic disease is spreading faster in cities where the population is dense [12], [11]. Indeed, converted to city plan is a key factor to preventing pandemic spreading. There is a clearly defined relationship between population density and pandemic spread rate [13].

It is clear that the rate of spread of the pandemic can be reduced to a minimum rate by controlling the city density [14]. From a realistic point of view, pandemic is a big problem which will never really disappear [15], but it can be controlled by converted city planning that have evolved for the urban vision of city planners.

2 Literature Review

What is Pandemic?

Pandemic is for the most part disease outbreaks that become widespread as a result of the spread of human to human infection [11], [16], [17]. There have been many significant disease outbreaks and pandemics recorded in history, including Spanish Flu, Hong Kong Flu, SARS, H7N9, Ebola and Sika [18], [19], [20] and the last seen COVID-19 pandemic. The 2019-nCoV common disease continues to spread worldwide. COVID-19 widespread disease began in Wuhan, China and spread globally, affecting almost the whole world. While the world's population is affected by about 382,000 viruses, the number of people who died is 16,555 - these numbers are increasing day by day [21]. Countries take decisions and impose sanctions in order to decrease and control the speed of the epidemic. The measures taken are implemented on the state, regional and urban scale. It is understood that the urban scale is an important step in the implementation of measures that spread from micro scale to macro scale. It is anticipated that a prepared converting plan should be planned against the pandemic of the city.

Disease, 13 May 2020 by the date, at least 213 countries and more than 4,395,332 COVID-19 cases have been reported in the region, which has resulted in more than 50 thousand deaths and 1 million infected (Tab. 1), [22].

Tab. 1. A country, where COVID-19 is overtaking, has infected number of China, 13 May 2020.

	Country	**Cases**	**Deaths**	**Region**
1	United States	1.418.504	84,187	North America
2	Spain	271,095	27,104	Europe
3	Russia	242,271	2,212	Europe
4	United Kingdom	229,705	33,186	Europe
5	Italy	222,104	31,106	Europe
6	Brazil	180,737	12,635	South America
7	France	178,225	26,991	Europe
8	Germany	173,824	7,792	Europe
9	Turkey	143,114	3,952	Europe, Asia
10	Iran	112,725	6,783	Asia
11	China	82,926	4,633	Asia

As time progresses, it is seen that the numbers increase exponentially, 28 July 2020, 958,703 deaths and 30,905,162 infected Tab. 2, [22].

Tab. 2. A country, where COVID-19 is overtaking, has infected number of China, 28 July 2020.

	Country	Cases	Deaths	Region
1	United States	7,406,729	210,797	North America
2	India	6,229,474	97,541	Asia
3	Brazil	4,780,317	143,010	South America
4	Russia	1,176,286	20,722	Europe
5	Colombia	824,042	25,828	South America
6	Peru	811,768	32,396	South America
7	Spain	758,172	31,614	Europe
8	Mexico	738,163	77,163	North America
9	Argentina	736,609	16,519	South America
10	South Africa	672,572	16,667	Africa
11	France	550,690	31,893	Europe
12	Chile	461,300	12,725	South America
13	Iran	453,637	25,986	Asia
14	United Kingdom	446,156	42,072	Europe
15	Bangladesh	362,043	5,219	Asia
16	Iraq	358,290	9,122	Asia
17	Saudi Arabia	334,187	4,739	Asia
18	Turkey	317,272	8,130	Asia
19	Italy	313,011	35,875	Europe
20	Pakistan	312,263	6,479	Asia
21	Philippines	311,694	5,504	Asia
22	Germany	290,466	9,556	Europe
23	Indonesia	287,008	10,740	Asia
24	Israel	239,806	1,547	Asia
25	Ukraine	208,959	4,129	Europe
26	Canada	156,961	9,291	North America
27	Ecuador	135,749	11,312	South America
28	Bolivia	134,641	7,931	South America
29	Romania	127,572	4,825	Europe
30	Qatar	125,760	214	Asia

Tab. 2. Continued

	Country	Cases	Deaths	Region
31	Morocco	121,183	2,152	Africa
32	Netherlands	117,551	6,393	Europe
33	Belgium	117,115	10,001	Europe
34	Dominican Repub.	111,900	2,101	North America
35	Panama	111,853	2,364	North America
36	Kazakhstan	107,908	1,725	Asia
37	Kuwait	104,568	607	Asia
38	Egypt	103,079	5,914	Africa
39	Oman	98,585	935	Asia
40	Sweden	92,466	5,890	Europe
41	Poland	91,514	2,513	Europe
42	Guatemala	90,968	3,238	North America
43	China	85,403	4,634	Asia

It is seen that on May 13, 2020, Chine fell to 11th place according to Tab. 1 and Tab. 2 fell to 43rd place. Codiv-19 is advancing rapidly and continues to influence of many cities. When this table is evaluated, it is seen that the population density is dominant in metropolitan cities, more than 500.000 people live. In this sense, it is predicted that the cities should be prepared the converting city plan for the pandemic in unexpected situations.

3 City and Pandemic Relationships

The city is the place where the human population lives intensely [23]. The city is defined primarily by its “size”, that is, by its population and in another definition for the city is its “density” that is, by its population in physical area. Size varies with the “intensity” of use by people and building (low density 1 person, high density 10.000 people per km^2) [23], [24], [25]. The density of the population is an important factor in interactions to people to each another [26]. These places for cities are commercial, communication-information, educational-scientific, cultural-entertainment, sport-recreation, social-health protection, social-political buildings. From this point, it is necessary to ensure the isolation of the crowd in these places, because there is a linear relationship between the pandemic and the cities. Increasing the crowd increases the pandemic transmission rate and the isolation of the crowd in these places prevents pandemic spread.

The city is an area where people interact continuously with intensive use. Therefore, metropolitan cities of the population density are more than 500.000 is the center; the transmission is experienced the most in case of pandemic disease. The form of the city determines the relationship between population density and building function. This relationship has been the most important factor in forming the city throughout history. The massive buildings that this is urban spaces where people share their social activities; shopping centers, religious building, educational buildings, public buildings, accommodation buildings, etc. [27].

It is known that while the cities defend themselves against the military force in the history, the city itself is covered with walls, but today it cannot be achieved with physical elements (the city walls surrounding the city). However, the fact that pandemic, which is one of the threat variables, affects the cities with high population density/crowd, shows the necessity of planning the city for self-defense. It is expected that the control spaces of the city will be converted during the pandemic. It is clear here that the concept of the defense space of the city has changed into the concept of control space in the historical process.

When pandemic disease (such as COVID 19) is seen, a race against time is needed to win the fight against the epidemic virus. The fight against coronavirus spread is not just an emergency that can be overcome. It should be involved in the pandemic control process with active cooperation of different professions and disciplines (medical industry, transportation, logistics, state administration, city district planning, architecture, product manufacturing, technology manufacturers, etc.) [14]. The pandemic disease, which has a linear relationship with the city, is also understood in the urban planners and researches are underway to evaluate the urban vision of the metropolitan city planning that was converted during the pandemic.

The metropolitan city is affected as centers with high population density during the global pandemic. Considering that there is a linear relationship between population density and pandemic propagation rate [28], population density must be separated and so that reduced to the infected rate. It is envisaged to provide control of the source (human) by dividing the effect of the virus on the population into three groups [14];

1. Virus source control: preventing the spread of the virus source by isolating the typical common instantly infected person in a closed area (hospital or converted unit hospitals).
2. Virus spread control: in order to prevent the continuation of this spread against the population or population with restrictive measures, people in the quarantine period (e.g. 14 days COVID 19 etc.) indoors (large mass building of the city; shopping centers, accommodation, dormitory and educational building) to prevent the spread of the virus by isolating it.
3. Virus following Control: to eliminate the spread of the virus by isolating the people who are not infected with the virus source in closed spaces (residences or house).

Therefore, it is true that, the spread of the pandemic can be prevented by reducing human density. However, the fact that pandemic disease, which is one of the threat variables, affects the cities with high population density indicates a planning requirement for the city to control itself. Thus, the returned to planning of the city should be prepared during the pandemic

4 Converted Metropolitan City Plan

Urban Functions and Its Building

An urban function expresses a relationship of the social needs and the urban form. In this context, the urban is shaped massive building and dwellings. Considering this configuration, it is observed that 1) population density and 2) urban building size is considered in ideal urban approaches [29], [30], [31], [32]. The urban characteristics are the size of the urban populations, [33], [34], [35]. The city is considered in five groups according to its demographic structure [23];

- Small settlement: settlements with fewer than 10,000 people
- Large settlement: settlements with a population between 10,000 and 50,000
- Small City: settlements with a population between 50.000 and 100.000,
- Big city: settlements with a population between 100.000 and 500.000
- Metropolitan city: settlements with more than 500,000 people.

Population and city size have directed relations with each another. It is situations population's density is automatically increase, while the city size is decrease and city population is increase. Urban buildings size is depended of building functions. The following functions belong to the urban functions [36], developed), (Fig. 1):

1) *Commercial activities*: retail trade, handicraft-and-service, catering and tourism, accommodation building, financial and business services and some of the manufacturing activities.
2) *Communication-information*; the basic components of this function are the perceptive, information and traffic. The perceptive role of the center comprises a unique empirical experience of an observer, his individual perception, conception and understanding of the center are, physical structures and spatial relationships. All the information about the city is concentrated in the center. The traffic connections of the center with the city and the wider gravity area ensure an easy access to a widest possible range of users and visitors.
3) *Educational-scientific*; the buildings of the educational-scientific branch are the schools and institutions of all the levels of education (primary, secondary, high, dormitory), as well as of scientific and research activities.
4) *Cultural-entertainment*; the function of the culture in a city is most overtly expressed in its center, through the numerous buildings where people may satisfy their demands: archives, libraries, cinemas, community cultural centers,

galleries, concert halls, religious building museums, theaters, exhibition pavilions.

5) *Sport-recreation*; If they are in the center itself, the sport arenas are most often indoors. Open, green and recreation areas in the center are usually used for the mass, and seldom for the top sports, and most often are they intended for recreation, strolling, socializing and entertainment.
6) *Social-health protection*; this function is performed by the institutions which are a part of the specialized centers: most frequently, those are the pharmacies and some ambulance or polyclinic services.
7) *Social-political*; it consists of the various activities of the social and political organs and associations, youth, amateur and professional organizations and associations.

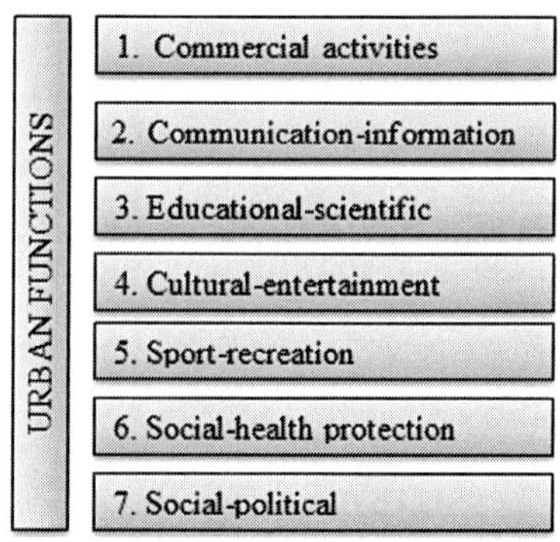

Fig. 1. Urban functions and its building.

As seem that, urban functions buildings are massive building, and are the places that the population uses dense. The potential of the city to have massive building should be evaluated in the city's transformation planning during the pandemic.

5 Findings

Converted Spaces

The most important feature of pandemic (COVID 19, etc.) disease is that it is transmitted very quickly by passing it from person to person. The more important is to provide social isolation. Considering this situation, the transformation of the city should be planned in the setup of categorizing the affected patients and directing them to returned places. In the categorizing process should be fallowing to step to step. With this approach (Fig. 2);

1. Step: The condition of the contaminant source should be determined.
2. Step: Contact status of the source should be determined
3. Step: Determined urban functions:
    - If it is external contact, it should be directed to building functions with commercial activities and educational- scientific
    - If internal contact, it should be separated as intensive care and intubated patient.
4. Step: Determined source to urban space,
    - Under quarantine: Dormitory and accommodations
    - Intensive care and intubated patient: Hospitals
    - Under surveillance: Shopping center, exhibition center, sport center and stadium

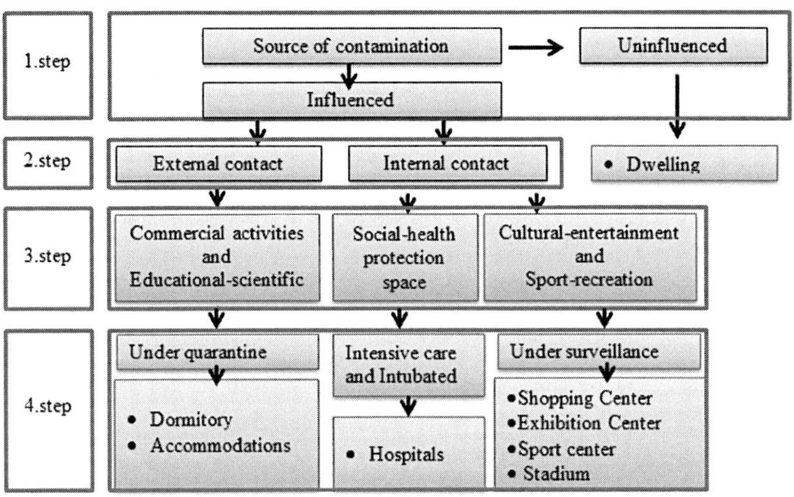

Fig. 2. Urban functions and source (patient person) relationship

In this context, in the implementation of these steps, urban functions will be change, while steps 3 and 4 are planning in the urban plan. In these situations, urban buildings will be change (Fig. 3);

- Accommodations and dormitories are converted to quarantine space, while in this situatıon is need to unique space.
- Traffic and roads are returned to information's areas, while this is under police control

- Shopping center, exhibition center, sport center, and stadium are returned to unit's hospitals, because these buildings cover large spaces.
- Hospitals are not converted and, because the infrastructure of the hospitals, emergency interventions are ready.
- Public buildings are converted to management crisis, while management is required from a central location.

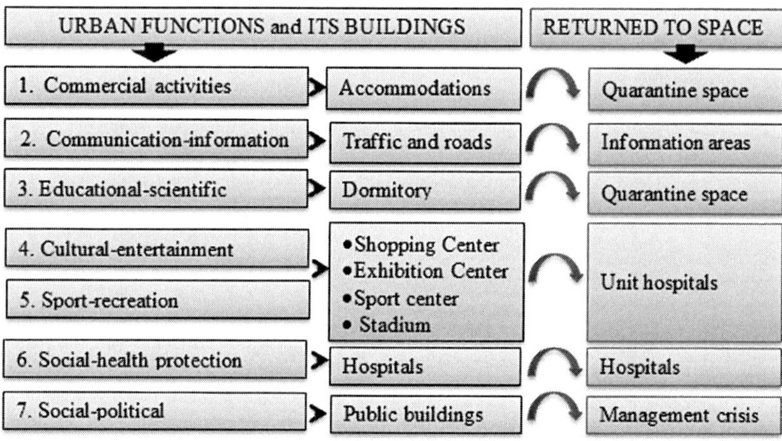

Fig. 3. Urban functions and its buildings to converted spaces

In cases where the number of infected people in cities and the increase of the source of the virus is insufficient, it is expected that the large mass building on large areas of the city will converted into emergency hospitals-the virus source control place of the virus source- another mass building that will undergo a quarantine period - the virus spread control place - and be converted into evaluation.

In means where the number of infected people in cities and the source of the virus increases, the health buildings are insufficient, to the emergency hospitals of the city's large mass built on large areas or quarantine will spend the period in space or as transformed/returned evaluation is expected. Covidien-19 common diseases of the US-seen New York City, the world's largest exhibition area of 170,000 m^2 Jahvist Centres 250 beds, a total of 1,000 people, to 4 emergency federal hospital being converted by one week or 10 day [37]. Iran, which is another town common, not be controlled, located on the large area building which 45,000 m^2 shopping centre's was converted to 250-bed capacity control spaces, targeted is the 3,000-bed capacity [38]. In another dramatic example, with the insufficient morgue services in Spain, Marcia building, the biggest skating- rink of Madrid, has been converted into a morgue space [38].

When the health building standards are examined, the rooms with more than one bed (such as ward type intensive care or emergency observation) should have at least 7 square meters of free floor space per bed or wheeled bed. The distance between the beds or stretchers will be at least 110 cm. The corridors must be at least 200 cm wide so that the stretcher or patient bed can travel a minimum of a foot next to a staff member [39], (Fig. 4).

These calculations show that, during the pandemic period, the plans of the massive structures of Metropolitan cities to serve emergency situations should take their place in the city strategies. In the city's vision and strategies, the studies about the converted plans of the cities should be evaluated and decisions should be taken for their implementation in emergency situations.

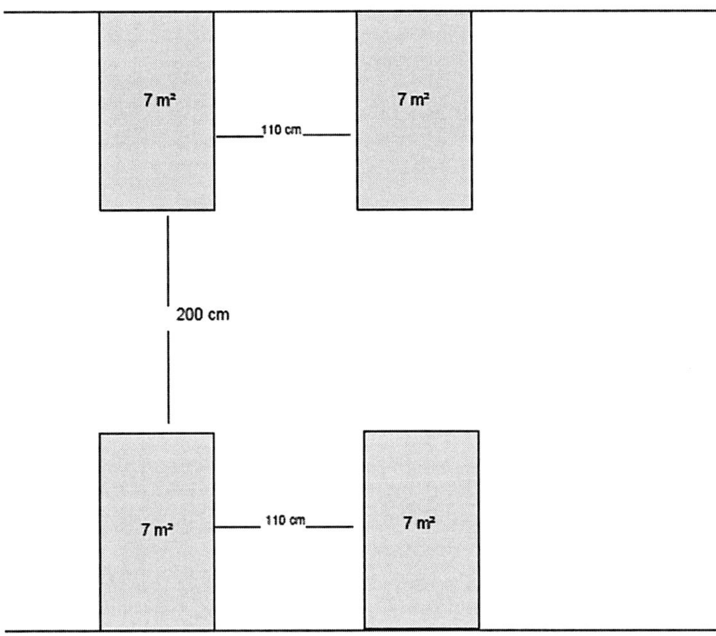

Fig. 4. Bed plan in intensive care or emergency observation

6 Results

For Calculations

When a general evaluation is made by calculating over China, where the COVID-19 virus is first seen; the number of patients affected in China 83.448

(81.171 infected and 3,277 died), [10]. This puts the country's population (1.386 billion), 0.16 % of group consists of ration. Metropolitan (500.000 inhabitants) a city considered; 83.448 x 500,000/1,386 billion= 30. 060 people will be affected is assumed. Outbreak instantly infected directing the converted control spaces population density was being expected to be isolated. Converted control space is evaluated planning spaces; 7 m^2 square spaces are required for a bed [39]. When the planning of the converted control areas is evaluated, it is understood that 170.000 m^2 area has a capacity of 1000 people, for 30.060 people; It was determined that 170,000 m^2 x 30,060 people/1000 people = 5,110,200 m^2 control areas are needed. When a general evaluation is made within the scope of this calculation;

- 1000 for person ⟶ 170.000 m^2
- 10.000 for person ⟶ 1.700.000 m^2
- 100.000 for person ⟶ 17.000.000 m^2 space is required.

In this sense, it should be ensured that the planning of the converted control areas by making the ratio calculation of the population and the mass density of the mass and the evaluation of the risk management situation.

7 Conclusions

The city is a dense area where human interaction is high. Pandemic (COVID 19, etc.) is a disease that is transmitted by interaction and spreads very rapidly. From this point of view, the common factor that the city and the pandemic is interaction. In this context, there is a need for a city planning to minimize the human interaction instantly pandemic in the city.

Presently, each city must prepare itself for a pandemic. The importance of this is to prepare a converted urban planning. It is clear to manage the source in the urban plan, so that it is to separate the source (human) by categorizing the source at the time of a pandemic. Directing the source to converted spaces will reduce the interaction, making it easier to struggle, the pandemic.

Converted to space of city is planned according to the state of the source. The city's accommodation and dormitory buildings should be prepared to source with spend in a quarantine process, shopping centres, exhibition hall, sports centres and stadiums should be prepared to be slightly infected and hospitals should be prepared to patients with severe conditions for a converted planning.

As a result, the converted to planning of each city should be prepared in case of a pandemic in city visions and strategy. This plan will facilitate the struggle of the city with the pandemic and will ensure the control of the spread of the source.

References

1. Moir E., Moonen T., Clark G. *What Are Future Cities? Origins, Meanings and Uses,* Compiled by The Business of Cities for the Foresight Future of Cities Project and the Future Cities Catapult, Government Office for science, 2014, United Kingdom.
2. Madulu Ndalahwa F., *Population Distribution and Density in Tanzania: Experıences from 2002 Population and Housing Census*, 2002, Institute of Resource Assessment University of Dar as Salaam.
3. Long J.F., Rain D.R., and Ratcliffe M. R., *Population Density vs. Urban Population: Comparative GIS Studies in China, India, and the United States,* Population Applications of Spatial Analysis Systems (SIS) at the IUSSP Conference in Salvador,2001, Brazil.
4. United Nations (UN), 2019, http://www.un.org/en/ga/president/65/issues/sustdev.shtml
5. Başol K., *Doğal Kaynaklar Ekonomisi Ders Notları, Aklıselim Ofset Tesisleri*, 1991, İzmir.
6. Karakuyu, M., The *Physical Growth of Istanbul, A Case Study*, Unpublished Master of Arts in Geography Thesis Fatih Üniversitesi, 2001, İstanbul.
7. Toprak Z., *Nüfus Dünden Bugüne İstanbul Ansiklopedisi*, 1994, vol. 6, s.108–111
8. http://www.tuik.gov.tr/UstMenu.do?metod=kategorist, TÜİK, 2018.
9. Dönmezer S., *Toplumbilim, Beta Basım Yayın*, 1994, İstanbul.
10. Chinese Centers for Disease Control and Prevention. Novel coronavirus, 2020,www.chinacdc.cn/jkzt/crb/zl/szkb_11803/jszl_2275/202001/t20200121_211326.html
11. Ercoşkun, Ö. Y., *A Paradigm Shift Toward Urban Resilience, Green and Ecological Technologies for Urban Planning: Creating Smart Cities* (Ed.) Ö.Y. Ercoşkun, Hershey, PA: Information Science Publishing, 2012, p:1–16.
12. Laurence J. C. Ma and Chusheng L., *Development of Towns in China: A Case Study of Guangdong Province*, Population and Development Review, 1993, 19, 3: 583–606.
13. Li R., Richmond P. and Roehner B.M., *Effect of population density on epidemics*, Physical A: Statistical Mechanics and its Applications, Volume 510, 15 November 2018, Pages 713–724, DOİ: 10.1016/j.physa.2018.07.025.
14. Chunwen X., Xilian L., Chuck Y. and Shi-Jie C., *The 2019-nCoV Epidemic Control Strategies and Future Challenges of Building Healthy Smart Cities*, Indoor and Built Environment, 2020, 0(0) 1–6, The Author(s) 2020, Article reuse guidelines: sagepub.com/journal spermissions. DOI:10.1177/1420326X20910408journals. sagepub.com/home/ibe.

15. Chowell G., Bettencourt L.M.A., Johnson N., Alonso W.J., Viboud C., *The 1918–1919 influenza pandemic in England and Wales: spatial patterns in transmissibility and mortality impact*, Proceedings of the Royal Society B., 2008
16. Qiu W., Rutherford S., Mao A., Chu C., *The Pandemic and its Impacts, Health, Culture and Society*, 2017, Vol 9–10 (2016–2017), ISSN 2161-6590 (online), DOI 10.5195/hcs.2017.221, http://hcs.pitt.edu.
17. Merriam Webster Sözlüğü, 11 Eylül 2020 date accepted http://www.merriamwebster.com/dictionary/boutique.adresinden.
18. WHO, *Comparative Analysis of National Pandemic Influenza Preparedness Plans*.pdf, 2011.
19. Rewar S., Mirdha D., and Rewar P., *Treatment and Prevention of Pandemic H1N1 Influenza,* Annals of Global Health, 2015, 81(5), 645–653. DOI: http://dx.doi.org/10.1016/j.aogh.2015.08.014
20. Maurice J.,*Cost of protection against pandemics is small*, The Lancet, 2016, 387(10016), e12.
21. National Health Commission of the People's Republic of China. March 23 2020: Daily briefing on novel coronavirus cases in China, http://en.nhc.gov.cn/DailyBriefing/2020-03/23/ c.html
22. https://www.worldometers.info/coronavirus/#countries
23. The European Environment (EEA), *Technical Report No 25: Environmental Indicators: Typology and Overview,* State and Outlook, European Environment Agency, 2005, Copenhagen.
24. Aksu, L., *Dünya'da ve Türkiye'de Nüfus Analizleri*, Istanbul Journal of Sociological Studies 2011, 0(25): 219–311, https://dergipark.org.tr/en/download/article-file/101080.
25. Tachir G., *Integrated Accommodation Approach in The Relation of Coastal Cities and Ecological Tourism: Alexandroupoli and Dadia*, Unpublished Doctoral Thesis, Trakya University Institute of Natural Sciences, 2019, Edirne.
26. Spreiregen P.D., *Urban Design, the Architecture of Towns and Cities,* 1965, Corpus ID: 128699719, Geography.
27. Lynch, K., *A Theory of good city form. Cambridge*, Massachusetts, 1981, London: The MIT Press.
28. Centres for Disease Control and Prevention (CDC), *Novel Coronavirus (2019-nCoV) Situation Summary*, 2019, www.cdc.gov/coronavirus/2019-nCoV/summary. html#risk-assessment
29. Geen E., Lowe J. and Walker K. (eds), *Man and the Modern City: Ten Essays,* 1966, P.4, Pittsburgh: University of Pittsburgh Press
30. Glasgow City Council, *A Fifty Year Vision for the Future: Future*, Glasgow 2011 – 2061: Summary of Full Consultation Report, 2011.

31. Hall P. and Pfeiffer U., *Urban Future 21,* E & FN Spon, 2000, London.
32. World Bank, (2012). *Urban Development Strategy for Vietnam Cities System to 2050.*
33. Lynch, K., *The Image of the City.* 1960, Massachusetts Institute of Technology Cambridge, Massachusetts, London
34. Karaman, A., *Kentsel Tasarım: Kuramlar,* İlkeler, Roller, Mimarist, 2008, Vol. 29, 35–42.
35. Coccossis, H. ve Constantoglou, M. E., *Land Use and Water Management in a Sustainable Network Society,* 45th Congress of The European Regional Science Association. Vrije Üniversitesi, 2005, Amsterdam.
36. Mitković P., Dinić M., *City Center Organızatıon and Its Influence on The Cıty Structure,* Facta Universities Series: Architecture and Civil Engineering, 2004, Vol. 3, No 1, 2004, Pp. 41 – 56, DOI: 10.2298/FUACE0401041M.
37. Liz Y., Jahvist Center May Be Converted to Medical Facility, Reporter, New York Business Journal Mar 21, 2020, 1:11pm EDT.
38. CNN Turk, Home Newsletter, March 22, 2020.
39. Turkish Design for Health Structures Minimum Design Standards 2010 guide, (2010). T.C. Sağlık Bakanlığı İnşaat ve Onarım Daire Başkanlığı. https://sbu.saglik.gov.tr/Ekutuphane/Yayin/414

Seyhan SEYHAN and Öner DEMİREL

Intra-Forest Recreation Areas as Protected Area Status in the World/Their Legal Status and Current Status

Abstract: Protected areas are evaluated under different conservation status in accordance with the resource values that the area has, and these areas are included as different conservation status in the relevant laws and regulations of different countries of the world and Turkey. There are also protected statuses reserved for recreational and tourist uses within these protected statuses. These statuses are located under different names in the world and in Turkey. In Turkey, one of the areas where recreational and tourist uses are allowed within the Protected Area status under the National Parks Law No. 2873 is Intra-Forest Recreation Areas.

In this study, our country's protected area status from someone who Intra-Forest Recreation (promenade) in the first of the field in the world and Turkey are dealt with legal and managerial aspects of the historical development and in the sequel describes the methodology and planning stages for the planning of these areas.

Keywords: Forest, Recreation, Intra-Forest Recreation, National Park, Natural Park, Natural Monument

1 Introduction

Among the natural resources, especially forests have become international common assets that extend beyond national borders with their importance and functions. Therefore, international agreements are made for the protection of forest areas with rich biodiversity, and each country participates in these international agreements and ensures the protection of living assets and important ecological systems in their own countries.

From the past to the present, forested areas are areas that people tend to utilize to meet many needs such as resting and entertainment. This situation has not changed today, and due to many reasons such as technological developments, excessive population growth in cities, industrialization and increase in living levels, all individuals working or not working in cities have turned to forest areas and rural areas in order to enjoy their remaining free time, rest and have fun. [1]. For individuals living in a region or region, the first areas that come to mind in terms of the use of land and water and the need for various types of outdoor recreation are forest areas. These areas are the most important recreation areas where individuals can perform and apply activities that require large areas outside the city, apart from fragmented and limited green areas, and which require the existence of natural character areas in the city [2, 3].

Forest areas that can offer natural resources collectively are the most suitable recreational living environments. The recreation type that occurs in these living environments both in terms of the characteristics and benefits it provides is defined as "Forest Recreation". [4]. In the 1920s, the desire to be in natural areas initiated the tendency of individuals to natural areas, but later on, this situation changed and directed the target group towards forest areas. "Intra-Forest Recreation", which emerged for the first time in the USA since the beginning of the 20th century, has an important place in the life of human beings with its renewal, revitalization, taste of the beauties of life, physical and perceptual features among the activities related to outdoor and rural recreation [5, 6].

2 Functions and Benefits of Forest

One of the most important natural resources in the world is forests. 19. y.y. As a result of the industrial revolution that lasted until its second half, there were changes in the structure of the demand for natural resources and especially for forests. The ecological and socio-cultural characteristics of forests such as hydrological, recreation, soil protection, and carbon storage have come to the fore. As a result of the studies carried out by the General Directorate of Forestry and Regional Directorates in line with the ecosystem infrastructure and participatory approaches, three basic functions of forests have been determined. The percentage ratio of these functions to the general forest areas of the country is shown in Fig. 1, and the distribution according to the area covered (ha) is shown in Tab. 1. [7].

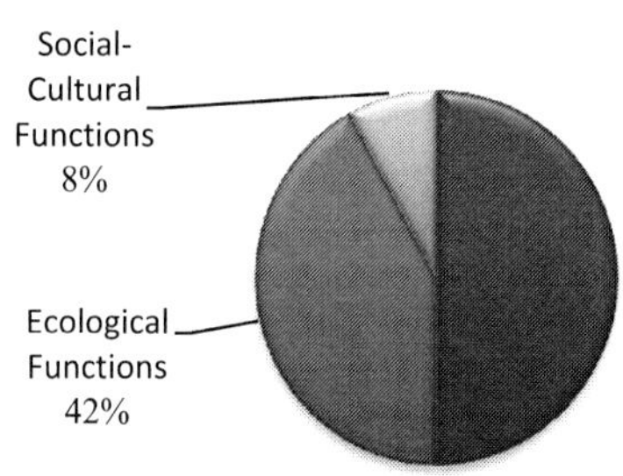

Fig. 1. General forest area ratios of basic functions

Tab. 1. Distribution of forest areas by basic functions [7]

Basic Functions	Total Size (Ha)
Economic Functions	11.243.094
Ecological Functions	9.287.847
Social-Cultural Functions	1.811.994

Considering the functions and benefits of forests, changes have occurred over time in people's perspectives and demands regarding forests. When people's demands for traditional wood resources are fulfilled, the recreation function of forests, which includes health and wellness functions, recreational activities and landscape value, is increasingly valued [8].

3 Historical Development of Intra-Forest Recreation Areas

Nature conservation activities on forest lands started after the 1850s. In this direction, there are various laws/regulations/decree/implementation communiqué enacted on various dates regarding the protection, management and use of forests and areas of protection value from past to present.

In the "Land Law", the first regulation on forestry in Turkey and published in 1858, land ownership was mainly addressed and aimed at solving legal problems related to land. 104 of the act.in the article, forests were considered" mountains specific to everyone's benefit (cibal-i mubaha)", and thus a process began in which there was no control over the cutting of forests, and forests were considered common property of everyone [9].

Other laws, regulations, decrees, circulars, communiques and constitutions related to this issue in Turkey are given below.

- 18 January 1870 Forest Regulation
- No. 504 entered into force on April 22, 1924 Fenner Fulfills a wide range of existing Forest Law on the Administration and Operation Procedure them in Turkey
- Land Hunting Law No. 3167 dated 5 May 1937
- Land Hunting Law No. 4915 dated 11 July 2003
- Forest Law No. 3116 dated February 18, 1937
- Forest Law No. 6831 dated 31 August 1956
- Fisheries Law No. 1380 dated April 4, 1971
- National Parks Law No. 2873 dated 9 August 1983
- Environment Law No. 2872 of 9 August 1983
- Law No. 2863 on the Protection of Cultural and Natural Properties dated 21 July 1983
- National Parks Regulation No. 19309 dated December 12, 1986
- Recreational Areas Regulation No. 26305 dated September 30, 2006
- The Law on the Amendment and Adoption of the Decree Law on the Organization and Duties of the General Directorate of Forestry dated 31 October 1985 was rearranged and the Decree Law No. 645 on the Organization and Duties of the Ministry of Forestry and Water Affairs on 29 June 2011.
- Recreational Areas Regulation No. 28578 dated March 5, 2013
- Circular on Procedures and Principles of Separating Forests Belonging to the State as a "Forest Recreation Area" dated 10 May 2013

- Regulation on the Amendment of the Regulation on Recreational Areas No 29222 dated December 31, 2014
- Recreational Areas Implementation Communiqué No. 296
- Recreational Areas Implementation Communiqué No. 300 dated 14 May 2015 and numbered 992685
- 1924 Constitution
- 1961 Constitution
- 1982 Constitution

All these legislations have created certain protected area statuses by supporting the nature conservation idea. These protected area statuses are also shown in Tab. 2.

Turkey has a total of 1632 units in terms of national and international protected area under protection. Of these, the number of National Parks that allow recreation use is 44, the number of Natural Parks is 247, and the number of Natural Monuments is 116 [10].

Tab. 2. protected area status determined in the regulations in Turkey

Conservation Forests **Gene** **Conservation Areas** **Seed Stands** **Seed Gardens** **Research Forests**	"Forest Law" dated 31 August 1956 and numbered 6831
Fisheries Production (Production) Fields	"Fisheries Law" No. 1380, dated April 4, 1971
Special Environmental Protection Area	"Environment Law" dated 9 August 1983 and numbered 2872; Decree no. 383 dated 13 November 1989
Cultural Heritage **Natural Beings** **Sites**	"Law on Protection of Cultural and Natural Properties" dated 21 July 1983 and numbered 2863
Intra-Forest Recreation Areas	"National Parks Regulation" dated December 12, 1986 and numbered 19309 "Recreational Areas Regulation" numbered 26305 dated 30 September 2006 "Regulation on Planning to be Made in Protected Areas" No. 28242 dated March 23, 2012 "Promenade Places Regulation" numbered 28578 dated 5 March 2013 "Regulation on the Amendment of the Regulation on Recreational Areas" numbered 29222 dated December 31, 2014 "Recreational Areas Implementation Communiqué" No. 300 "Recreational Areas Implementation Communiqué" No. 296

Tab. 2. Continued

Wildlife Protection Areas Wildlife Development Areas	"Land Hunting Law" No. 4915 dated 11 July 2003 "Land Hunting Law" No. 3167 dated 13 May 1937
National Park nature Park Natural Monument Nature Protection Area	"National Parks Law" No. 2873 dated 9 August 1983

4 Recreation Dimension of The Forest

Intra-forest recreation areas are one of the first natural areas that come to mind in meeting the "recreation usage demand" that has increased intensely in recent years. It was thought that it would be convenient to have a protected area status such as Conservation Buildings or National Conservation Status, after the studies carried out to ensure that these areas can be protected to meet the recreation demand for many years without deterioration. Within the framework of this system Protection Status, it is aimed to provide services to meet the recreation use demanded by the public, to create working environments for scientific studies and research, together with the protection of natural, cultural and historical beauties in the natural area [1].

Individual and society with healthy environments created by forest areas; By affecting their physical, psychological and physical health positively, they heal themselves, renew themselves, rest, experience nature and gain energy [11]. The recreational and touristic activities carried out in this natural climate offered by the forests have a very positive contribution to the psycho-pedagogical development of children, as well as their knowledge of nature and creating a nature awareness. Especially frequent weekend walks, sportive activities in the forest and nature walks have very positive effects on the mental health of people.

Forest recreation is evaluated under two headings as intensive recreational use and less intense or dispersed recreational use when considered in terms of usage. Intensive recreational use is the application of activities such as events, picnics and certain types of walks in areas that are centralized and created for their use. Activities in less intense or scattered recreational use consist of activities carried out in large areas such as long walks, fishing and land hunting, climbing and primitive camping [2]. Forest areas where all uses are combined and applied are generally defined as "intra- forest recreation areas "[12].

The "Regulation on Recreation Areas" dated March 5, 2013, "Recreation Areas in the Forest" in general, "to meet the various recreation, entertainment and sports needs of the society, to contribute to the beauty of the country and to allow touristic activities, reserved for use with the necessary structures, facilities and equipment, they are areas subject to a forest regime that meet daily or overnight needs and have recreational and aesthetic resource values [13].

The most distinctive feature of these areas is that intra-forest recreation areas find a presence area within a natural or artificial forest ecosystem and attract a high rate of visitors regardless of the activity and activity groups thanks to this ecosystem [14]. In addition, intra-forest recreational areas, which play an important role in the relationship between man and nature, contain many activities that will adapt to this. These activities such as picnicking, camping, hiking, excursions, relaxing, improving self-anxiety, watching the scenery, climbing, fishing, swimming, exploring nature and wildlife; keeps individuals healthy and prosperous with physical, mental and social functions [15, 16].

5 Legal Status and Legislation, Organization and Management of Intra-Forest Recreation Areas in the United States of America

In the United States, National Forests and Grasslands are managed by the Forest Service, the Agency of the United States Department of Agriculture [17]. The National Forest System was created by the 1891 "Land Revision Act" signed under Benjamin Harrison. In the United States, 2.3 billion acres of land across the country meets the potential use for recreational purposes. Forest recreation activities constitute the most important characteristic features of these recreation areas. The development of the economy of the United States, especially after the Second World War, and the desire of people to move away from the cities not only increased the importance of forest areas as recreation areas, but also directed people to these areas [6].

6 Planning Approaches in the United States of America

In the United States, management plans are developed by the National Park Service on areas within the National Park System. It provides basically guidance on how the legal responsibility for the conservation of resources should be and how it should be carried out, while providing support for appropriate visitor use, recreation and recreation opportunity for future generations of sites. These plans are prepared by interdisciplinary teams. The team includes regional planners, landscape architects, environmental design experts, natural and cultural resource experts, environmental management experts, interpretation experts, inspectors and staff working in the field of nature conservation and experts in other fields. These plans also have many different types of plans such as general management plans, visitor use management plans, commercial services plans, implementation plans, development plans, wild and natural river plans, wildlife plans. [18]. According to the United States Department of Agricultural Forest Service Strategic Plan (FY 2015–2020), management plans such as urban forest plans and landscape planning help to eliminate the negative effects of new and existing developments on open space. It works to sustain the benefits of tourism, recreation, natural character and open spaces and the benefits of ecological support systems by supporting planning activities on a local scale [19]. The planning approach (Fig. 2) developed by the National Park Service includes the management, timing and mission-appropriate information about decision-making for the future of each national park system. While planning, methods and tools to solve problems in a way to minimize conflicts and contradictions, it provides useful solutions that can be part of a strategy by ensuring that the resource values of the areas are transferred and protected to future generations without harming them [18].

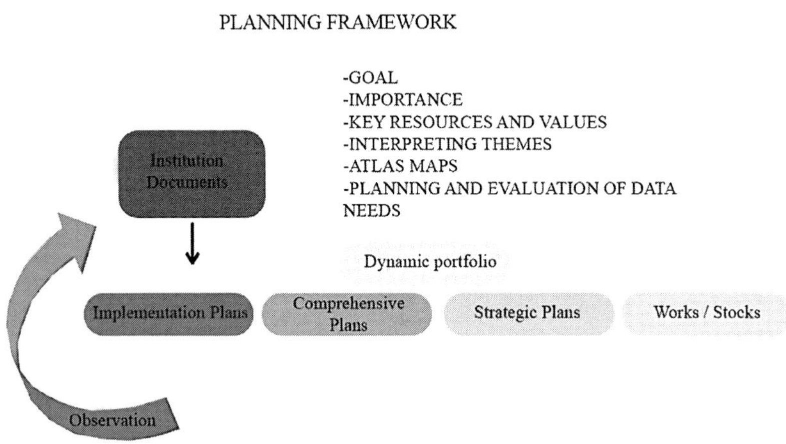

Fig. 2. Planning framework developed for protected area statuses by the national park service [18]

Planning studies in the United States of America are carried out in 12 stages. These steps are shown in Fig. 3. [20]

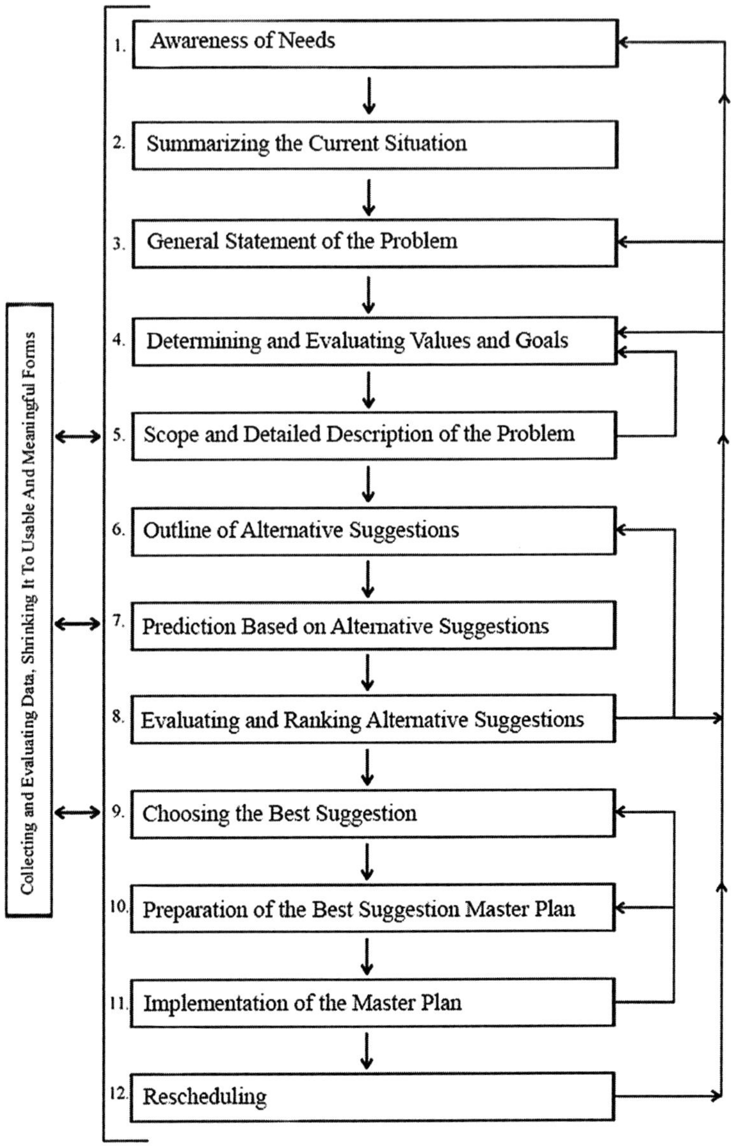

Fig. 3. Steps of the planning process developed by the National Park Service for protected area statuses in the United States [20]

7 Protected Area Planning Practice in Turkey

Long Term Development Plan and Management Plan

Planning principles for national parks and other conservation areas in Turkey are regulated and 9 August 1983 dated 2873 National Parks Act. According to the 4th article of this law, "*the development plan covering the establishment, development and operation in order to realize the conservation and utilization purposes, taking into account the characteristics and qualities of the places designated as national parks, is prepared by the Ministry of Forestry and Water Affairs, with the positive opinions of the relevant ministries and, when necessary, actual contributions and put into effect*" with this provision explained the long-term development plan and the relevant institution. In addition, in this law, the necessary projects for the places designated as other nature protection areas (nature park, nature monument and nature protection area) are prepared and put into effect by the Ministry of Forestry and Water Affairs, taking the opinion of the Ministry of Culture and Tourism. [21]. According to the relevant 4th article of this law, General Directorate of Nature Conservation and National Parks prepared the "Long Term Development Plan and Management Plan General Technical Specification". According to this specification; The long-term development plan is *"the basis for all kinds of settlement and other usage decisions in the envisaged conservation and development zones and for the development areas of settlement units, and the implementation projects and implementation programs of all scales, determining the decisions and provisions related to these, and the scientific report as a whole and 1/25,000 or a lower scale physical plan as required depending on the size of the area"*[22].

The end of the 1960, 1969, and for certain areas of conservation property in Turkey at the beginning of 1970 at the invitation of Turkish experts, came to our country in U.S. Support of experts from the National Parks Department and also in the U.S. With the help of the International Development Agency (USAID), "National Park Long Term Development Plans" were prepared under the leadership of experts from the United States of America and Canada [9]. These plans were completed in 1971. These plans, which will be used as a guide for conservation measures, public utilization of nature conservation areas, identification, development, general management and large-scale plan feature in national park services, are not conclusions but recommendation plans that present developments and suggestions. Planning studies in Turkey entered a period of about 20 years pauses during the period from 1971 up to the 1990s. Following the enactment of the National Parks Law No. 2873 in 1983, taking into account the plans made in 1971 and using the planning technique, the Commander-in-Chief Historical National Park Long Term Development Plan was prepared by the General Directorate of National Parks and Hunting-Wildlife in 1991 [23].

After the Yozgat Çamlığı was declared the first national park in 1958, the "Regulation on the Separation, Administration and Management of National Parks" was issued in 1959, numbered 6885. By the publication of the "Regulation on the Separation, Planning, Implementation and Management of National Parks" numbered 14456 on 8 February 1973, the regulation numbered 6885 was abolished. With the regulation dated 1973, the concept of planning was included in the legal texts for the first time. Later, this regulation was also repealed.

For all protected areas in Turkey in the 1990s accelerated the preparation of the Long-Term Development Plan, it has opened tenders to private sector companies and universities for 33 areas. While plans were made in some areas determined within the protected areas, some were not made [18]. Long-term development plans of 16 areas were prepared between 1971–1997. A long-term development plan of 34 national parks has been prepared in Turkey by March 2015. 6 of them are known to be under construction. In Turkey, these plans are prepared by the companies operating on this issue in accordance with the technical specifications prepared by the National Parks Department. The planning process takes approximately one year. There are three basic stages in this process. Within these stages, Long-Term Development Plan plans are being made [24]. These Are 1/Analysis Stage, 2/Synthesis Stage, 3/ Planning Stage.

Management plan "*with the status of nature protection areas and natural monuments areas of protection of the continuity of the provision of the value of the resource, management and planning principles for the promotion; and projected work for the scientific and educational activities in the field are accustomed for all types of application programs or application to be made in determining the provisions for forming decisions and development plans based on the report a whole, 1/25.000 or more as needed, depending on the size of the area sub-scale physical plan*" is defined as [22]. Management planning of nature conservation areas in Turkey Long Term Development Plan and Environmental Plan prepared by conventional planning approaches such as forest management plan. The Ministry of Forestry and Water Affairs, the Ministry of Environment and Urbanization of the Republic of Turkey and the Ministry of Culture and Tourism of the Republic of Turkey are authorized ministries in this regard. Management plans provide guidance on how to demonstrate the management, use and development of nature reserves. It also sets out principles and goals for managing the future within the framework of protection [26]. Considering the basic principles of management planning of nature conservation areas, it is discussed under six titles [27]. These;

1. To protect resource values in a way that both meets the needs of today and can be passed on to future generations,
2. To implement scientific data effectively with various uses and investments,
3. To create the institutional structure that will ensure harmony and order between sectors,
4. To adopt a multi-disciplinary approach,
5. To minimize or eliminate the negative effects of humans and other living things on resource values,
6. To provide opportunities for community participation and awareness in planning and decision-making.

Management plans are a tool for determining actions and goals to be implemented in a specific time period. In this direction, IUCN has put forward management plans as a guide for nature protection areas. In real terms, there are 13 basic steps in the management planned process. These are shown in Fig. 4 below [28].

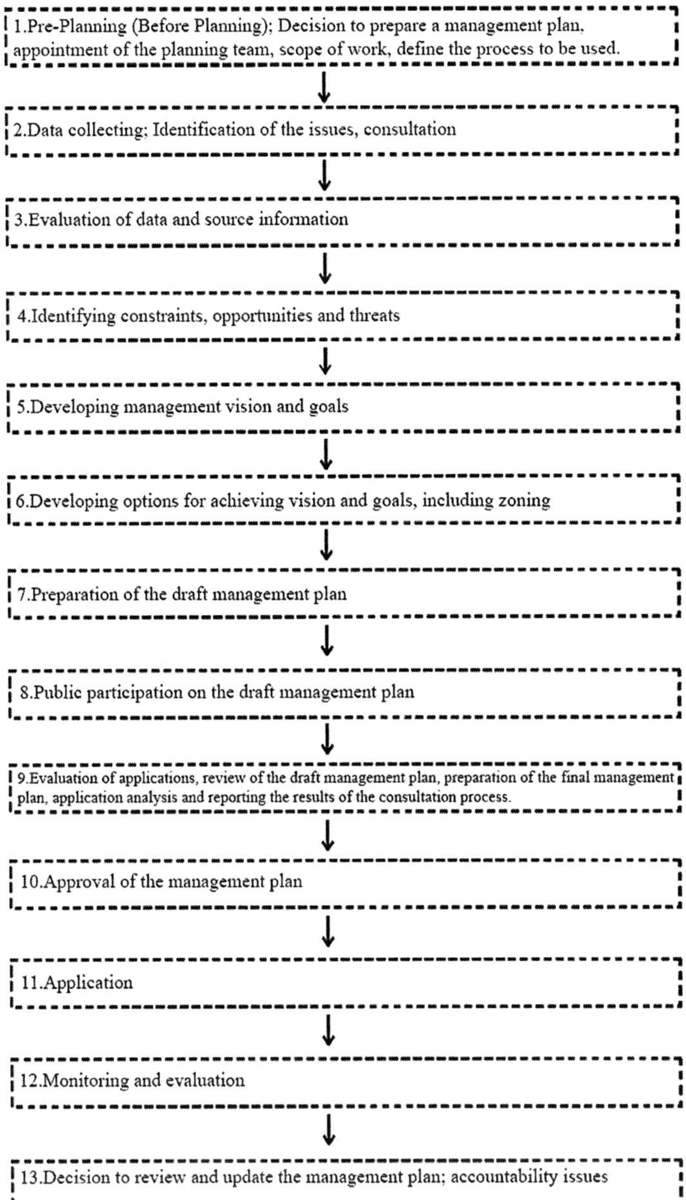

Fig. 4. Planning steps of IUCN's management plan in protected areas [28]

This management plan is a continuous process consisting of three basic elements [28, 29]. This process is shown in Fig. 5.

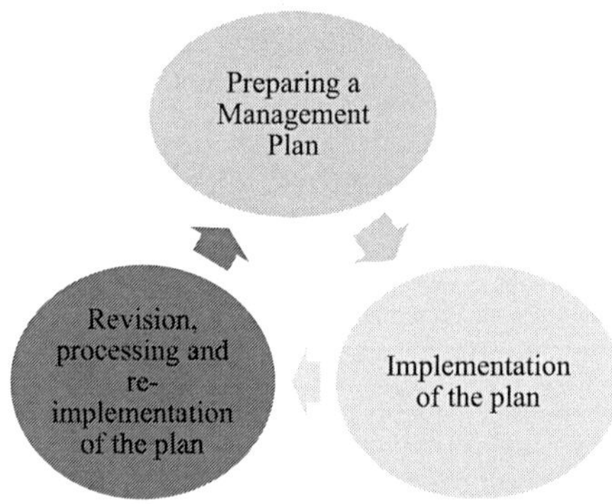

Fig. 5. Three basic elements in the planning process of management plans [28, 29].

Turkey protected area system is one of the causes of the problems encountered in the absence of an updated management plan for the management of conservation areas because of false and incorrect use of natural resources [30]. In today's world, developments such as industrialization and globalization have led to the emergence of many new ideas in the world and our country is taking steps to adapt to these ideas. In this direction, some administrative and legal regulations have taken place in legal texts as a part of our domestic law [26]. In line with this development in Turkey for the successful planning and management of a nature conservation area in the international standards recommended by the IUCN 13 basic steps identified above. These steps IUCN both by the abolished Environment and the Ministry of Forestry in Turkey "Biodiversity and Natural Resources Management Project (BNRMP)" the scope of which is prepared "Turkey Protected Areas Management Planning Guide" is steps consistent with defined stages in the publication [24].

8 Current Situation of Protected Areas Status in Turkey

Protected areas in Turkey, the Ministry of Agriculture and Forestry Nature Conservation and National Parks is the responsibility of the General Directorate there are 13 protected area status. According to data obtained from 2019 to 1958 a total of 1632 pieces of protected areas in the current situation in Turkey today. These areas constitute an area of 3,537,398 hectares of Turkey [10].

Nature Conservation and National Parks General Directorate, consists of 15 Regional Directorate in Turkey. These regional directorates reveal numerical and verbal data regarding regional and urbanly protected areas status. In Fig. 6, the region names of these regional directorates and the cities they cover are shown.

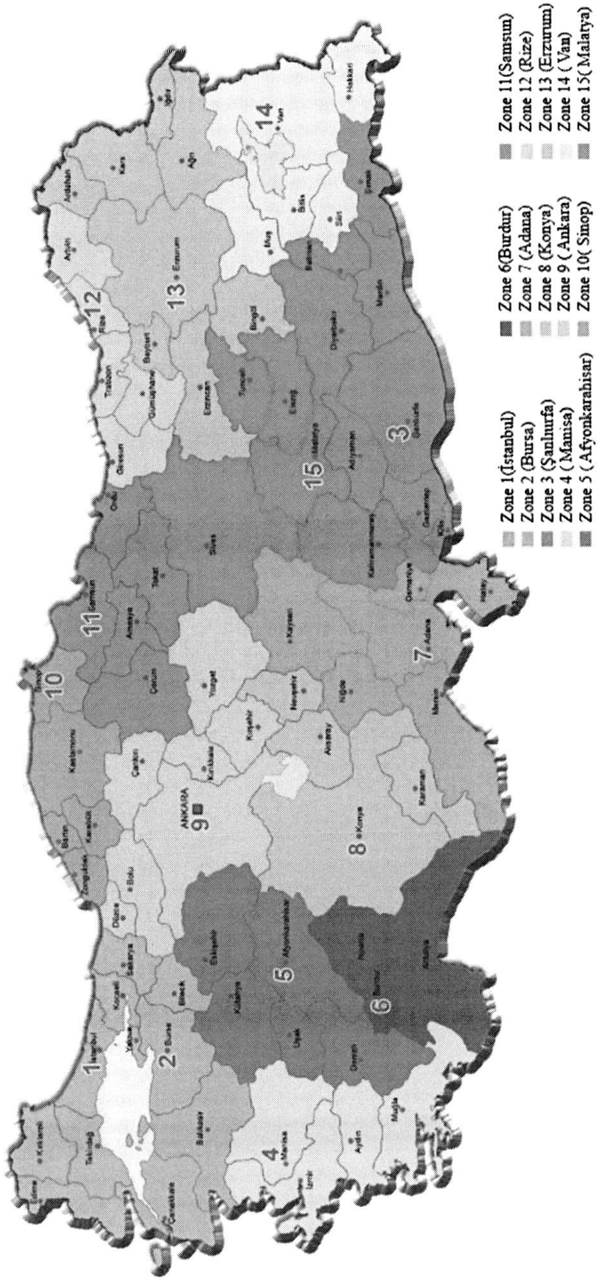

Fig. 6. The regional distribution of protected areas in Turkey [10].

As in the world of nature protection areas in Turkey, recreational and protected under different legislation according to tourist use. Especially with the "National Parks Law" dated 9 August 1983 and the "National Parks Regulation" enacted on December 12, 1986, there are four different characteristics: nature protection areas, National Parks, Natural Parks, Natural Monument and Intra-Forest Recreation Areas in terms of recreational and touristic uses. It is reserved for conservation status. In line with the definitions made within the scope of the National Parks Regulation dated 1986, especially nature parks and recreation areas in the forest are the most appropriate protection statuses in terms of recreation and touristic use of the society.

Considering the protected area statuses within the scope of this study, especially the protection statuses that allow recreational and touristic uses were examined. Accordingly, it has been determined that "National Parks, Natural Parks, Natural Monument and Forest Recreation Areas" allow recreational and touristic uses distribution of the current state of the art is shown in Fig. 7 in Turkey.

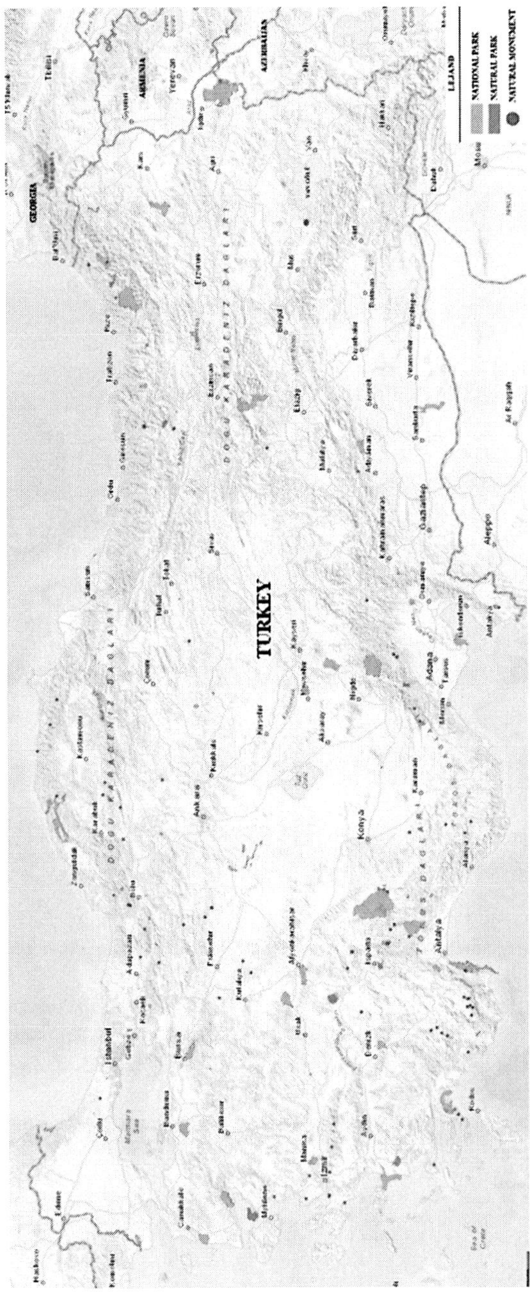

Fig. 7. Current general distribution of national parks, natural parks and natural monument areas in Turkey [10]

Information on the changing number of areas and the areas covered (ha) for each protection status from 1958 to 2019 are discussed in detail in the figures below. Of these;

National Parks: There are currently 44 units available national parks in Turkey. These areas cover a total area of 868,428 hectares across the country.

Nature Parks: Today, the number of nature parks has reached 247 and covers a total area of 106,836 hectares.

Natural Monuments: Natural parks, whose number has reached 247 today, cover a total area of 9,389 hectares.

Intra-Forest Recreation Areas: another conservation status that allows recreational and tourist uses within protected areas in Turkey (Fig. 8). Currently, 1521 "Intra-Forest Recreation Areas" constitute 27207.36 hectares of Turkey's Area [31].

Fig. 8. 125. Yıl Atatürk Fountain Picnic and Recreation Area [32]

9 Recreational Protected Area Studies in the United States of America

In this section, firstly, protection statuses that allow recreational use among the protected area statuses created in the United States of America, which has long-term experience in protected areas, are examined.

The development and change of the nature conservation movement in the United States started with the protection of the wildlife, wild lands and other natural resources of the United States between 1890 and 1920 [33]. Generally, nature protection is done with the conscious and efficient use of the land or natural resources. Following the emergence of this situation, the thought of nature conservation in the United States was seen as the most inspiring idea for humanity and the establishment of national parks in this direction is based on the declaring of the Yellowstone Region of the USA's Wyoming state as a national park in 1872 [34]. In 1876, Appalachian Mountain Club, one of the first private conservation organizations in the United States, was established to protect and maintain wild areas in nature [35].

Theodore Roosevelt, George Bird Grinnell, and other top athletes of the day founded the Boone and Crockett Club in 1887 to address the protection issues that arose at the time. This club is the first true conservation organization in the United States of America. Later, in 1891, as a result of the travels made by the US President Theodore Roosevelt to the Yellowstone region and surrounding areas, the issue of protection gained momentum with the establishment of the "Yellowstone Timberland Reserve". Later, the largest part of this reserve was changed to the Shoshone National Forest, and this forest is the oldest national forest in the United States [36].

From the date Yellowstone was declared the first National park in 1872 until 1916, no work was carried out on the park for forty years. During this time, areas such as the country's parks, reserves and monuments were inspected by the Ministry of Agriculture and the interior at different times [35].

In the United States of America, the first attempts to preserve natural beauties started in the second half of the 19th century and the beginning of the 20th century [34]. In 1916, the National Park Service was established under the Ministry of Interior for the management of national parks. This establishment started its work in Yellowstone National Park (1872), Yosemite National Park (1890), Grand Canyon National Monument (1908), which were previously taken under protection [37, 38]. In 1890, Sequoia National Park and General Grant National Park were also created. In the middle of the twentieth century, conservation efforts continued with the establishment and implementation of federal legislation to protect wildlife, natural resources and wildlife [39]. In addition, today National Parks; They are public resources that should be preserved as heritage for the conservation of recreational values, educational resources, beautiful and unique landscapes, endangered natural communities and species [34].

The laws regarding the protection, management and use of nature conservation areas given under the title of "National Park Timeline" in the United States are given below [40]. These;

- 1 March 1872 Yellowstone National Park Act
- 8 June 1906 Antiquities Act
- 25 August 1916 National Park System Organic Act
- 10 June 1933 The Reconstruction Act
- 21 August 1935 Historic Sites Protection Act
- 3 September 1964 The Wildlife Act
- 1965 Land and Water Conservation Fund Act
- 15 October 1966 National Historic Preservation Act
- 2 October 1968 National Wild and Natural Rivers Act
- 2 October 1968 National Track System Act
- 2 October 1968 Volunteers in the 1969 Parking Act
- 1 January 1970 National Environmental Policy Act

- 18 August 1970 General Authorities Act
- 28 December 1973 The Threatened Species Act
- 27 March 1978 Redwood National Park Development Act
- 10 November 1978 National Parks and Recreation Act
- 31 October 1979 Archeological Resources Protection Act
- 2 December 1980 Alaska National Lands Protection Act
- 16 November 1990 Tombs of the Native United States

Protection and Return Act

- 13 November 1998 Omnibus National Park Management Act

National nature reserves for the federal government of the United States are featured areas designated as important and worthy of protection. These areas have been regulated by the laws above, and their statuses have been set forth under ten main headings. These statuses also made some distinctions within themselves. When these are examined, the National Park System was established in 1872 and the National Park, National Monument, National Conservation, National Historic Site, National War, National Cemetery, National Recreation Area, National Park Road, National Beach, National Trail, National Lakeshore, Affiliated Areas, National The river has been divided into subtitles such as Other Areas and defined these areas within this system. Also other protected areas; National Forests and Grasslands (1891), National Wildlife Refuge System (1903), National Monuments (1906), National Wildlife Conservation System (1964), National Wild and Natural Rivers System (1968). Described the National Marine Protection System and the National Estuary Research Reserves (1972), the National Marine Protected Areas and the National Landscape Protection System in 2000. In addition, the National Landscape Protection System; National Monument, National Reserve, Wildlife Area, Wildlife Study Area, National Wild and Natural River, National Natural Trail, National Historical Trail, Cooperative Management and Conservation Area, Forest Reserve and Extraordinary National Areas are included in the system by defining the titles. Considering all these and looking generally, there are 10 nature conservation area statuses determined in the United States of America [41]. These;

1. **National Park System:** It has been redefined with the "General Authorities Law" enacted on 18 August 1970. This law emphasized that changes should be made to the old legislation in which the National Park System was defined in order to re-evaluate the areas in the category of recreational activities. He declared that protected areas and park types defined under different names will be part of a single system [40]. Other nature protection statuses in this system are given under headings. The statuses that allow the use of recreation, especially under the protection status, are discussed together with their explanations.

    - National Park,
    - National Monument,

- National Reserve,
- National Historic Site,
- National War,
- National Cemetery,
- National Recreation Area: These are areas of extraordinary beauty designated by the United States Congress for the use by visitors of various outdoor activities. In 1961, President John F. Kennedy gave a speech emphasizing the need to acquire federal lands for recreational purposes and described the effort to establish national recreation areas. This movement bore its first fruit, and the Boulder Dam Recreation Area, renamed the Mead Lake National Recreation Area, was declared the "National Recreation Area". Originally, national recreation areas were established through an agreement between the United States Bureau of Reclamation and the National Park Service. In 1963, the President's Recreation Advisory Committee published an executive body policy that lays down criteria for the creation of national recreation areas. All national recreation areas to be created by the United States Congress were also added to the policy (Fig. 9). The National Recreation Areas are managed by the Bureau of Land Management or by the US Forest Service as the National Park Service that Manages the National Recreation Areas [42]. Established to protect the recreational, natural and scientific values of individuals and future generations, they define recreational activities as the center of water and land-based activities associated with the natural environment. In 1980, the first and only National Recreation Area Bureau of land management was established by law in Alaska [43].

Fig. 9. Glen Canyon national recreation area [44]

- National Park Road, National Beach,
- National Trail, National Lakeshore,
- Connected Areas,
- National River,
- Other Areas

2. **National Forests and Grasslands:** National forests are managed by the Forest Service, the Agency of the United States Department of Agriculture [45]. On March 30, 1891, the first national forest was established by the Ministry of Internal Affairs as the "Yellowstone Park and Land Reserve". With the "Organic Law" of 1897, it was aimed to create forest reserves, including protecting the forest, providing water resources and providing timber. With the "Forest Protection Act" of 1891, the President of the United States was given the power to separate forest reserves in the public domain. The newly created "United States Forest Service" for forest reserves with the Transfer Act of 1905 became part of the United States Department of Agriculture [36, 46].

The top priority of the forest service is to protect and improve the health, diversity and productivity of the country's forests and meadows to meet the needs of present and future generations [47]. National forests and grasslands offer the world's largest outdoor recreation opportunities combining a variety of activities with nature. They are areas that include many activities such as hiking, cycling and horse riding, picnicking, camping, hunting, fishing, wildlife and scenery, and exploring historical places [48]. The National Forest System was created by the "Land Revision Act" of 1891, signed under the chairmanship of Benjamin Harrison. National forests (forest reserves) are the great open areas of the United States (Fig. 10). It covers an area larger than the state of Texas of 193 million acres (78 million hectares). National forests also provide opportunities for recreational activities in open spaces and natural environments. These areas have become more important for people living in cities. They are areas that include a wide variety of activities for people such as backpacking in wild life areas with no roads, walking around on difficult roads, fishing and watching the scenery [49].

Fig. 10. Deschutes national forest [50]

3. **National Wildlife Refuge System:** This system is managed by the United States Fish and Wildlife Service, protecting multi-use conservation and recreation areas. The first wild animal sanctuary of the United States was designated as "Pelican Island National Wildlife Refuge" by President Theodore Roosevelt

in 1903. Since then, the National Wildlife Refuge System has grown to include more than 560 shelters (Fig. 11). With the "National Wildlife Refuge System Improvement Act" enacted in 1997, he summarized six wildlife-dependent recreation areas suitable for shelters as hunting, fishing, wildlife observation, wildlife photography, environmental education and environmental interpretation. The United States Fish and Wildlife Service is also working to take action to meet conservation challenges after the inclusion of increased demands for urban attack, habitat fragmentation, degradation of water quality, climate change and energy enhancement [41].

Fig. 11. Wyoming national deer sanctuary [51]

4. **National Monuments**
5. **National Wildlife Protection System**
6. **National System of Wild and Natural Rivers:** It was first established by the United States Congress in 1968. It is jointly managed by four federal agencies: the Bureau of Land Management, the National Park Service, the US Fish and Wildlife Service, and the US Forest Service. Together, these institutions are working to ensure the protection of rivers selected under the "The Wild and Scenic Rivers Act" (Fig. 12). Rivers to be protected under this law; exceptional natural, recreational, geological, fish and wildlife, historical, cultural or other similar criteria are selected according to[41].

Fig. 12. Obed wild and nature river [52].

7. **National Marine Protection System**
8. **National Golden Horn Research Reserves**
9. **National Marine Protected Areas**
10. **National Landscape Protection System:** The National Landscape Protection System, managed by the Bureau of Land Management's, was created in 2000 to protect important landscapes and the protected area has a size of approximately 36 million acres. These areas are protected by multiple definitions, with an overall goal to preserve exclusive and unique features, provide opportunities for hunting, wildlife viewing, fishing, historical research, scientific research, and a wide variety of traditional uses [41] (Fig. 13).

Fig. 13. Headwater forest reserve [53]

11. **Other Protected Areas**
    - National Reserve,
    - National Reserve,
    - Wildlife Area,
    - Wildlife Study Area,
    - National Wild and Natural River,
    - National Natural Trail,
    - National Historical Trail,
    - Cooperative Management and Protection Area,
    - Forest Reserve,
    - Extraordinary National Areas.

All these nature reserves in the United States are managed by a number of institutions and organizations consisting of different federal, state, tribal and local level authorities. There are 34,075 nature protection areas in total [54]. There are a total of 418 areas in the National Park system, among the areas designated as protected areas of the United States. The number of National Recreation Areas in this system that allows recreation use is 18 [55]. Within the National Wild and Natural Rivers System, 209 rivers are protected and allowed to be used for recreational purposes [56]. In the National Wildlife Refuge System, which is another nature protection status, there are approximately 600 shelters and wetlands [19]. In the National Landscape Protection System, the total number of protected areas is 874 [57,58]. There are 154 National Forests and 20 Grazing Lands under the status of National Forests and Grasslands, which allow the recreational use of forests in particular [59].

In the United States, 2.3 billion acres of land across the country meets the potential use for recreational purposes. Forest recreation activities are one of the most important of these recreation areas. The development of the economy of the United States especially after the Second World War and the people's desire to move away from the cities not only increased the importance of forest areas as recreation areas, but also directed people to these areas. However, increasing pressures and problems have emerged on forest recreation as a result of increased use. After discussing this situation, the true value of forest recreation at all levels was realized in 1960. Recreation was considered an important part of forest resources, not a poor sister of other forest products [6]. The "Annual Report of The Forester" between 1891 and 1919, issued before 1960, mentions the increasing importance of recreational activities in national forests, describing the work that has already been done to attract visitors and ensure their comfort and pleasure [60].

Acknowledgement

This publication has been produced from a master's thesis titled "Legal, Managerial and Planning Approaches in Forest Recreation Areas" prepared by Research Assistant Seyhan Seyhan under the supervision of Prof. Dr. Öner Demirel.

References

1. Özdönmez, M., İstanbullu, T., Akesen, A., ve Ekizoğlu, A., *Ormancılık Politikası*, 1996, İstanbul Üniversitesi Yayın No: 3968, Orman Fakültesi Yayın No: 435, İstanbul.
2. Akesen, A., *Türkiye'de Ulusal Parkların Açıkhava Rekreasyonu Yönünden Nitelikleri ve Sorunları*, 1978, İ.Ü. Orman Fakültesi Yayını No: 262, İstanbul.
3. Pamay, B., *Park-Bahçe ve Peyzaj Mimarisi Kitabı*, 1979, İstanbul Üniversitesi, Orman Fakültesi, İ.Ü. Yayın No: 2486, O.F. Yayın No: 264. İstanbul
4. Pehlivanoğlu, M., *Belgrad Ormanının Rekreasyon Potansiyeli ve Planlama İlkelerinin Tespiti*, 1987, Doktora Tezi, İ.T.Ü., Fen Bilimleri Enstitüsü, İstanbul.
5. Pehlivanoğlu, M.T., *Rekreasyon ve Ormaniçi Rekrasyon Planlaması*, İstanbul Üniversitesi Orman Fakültesi Dergisi, 1976, Seri B: **26, 2**, s.142–170
6. Douglass, W.R., *Forest Recreation*. Pergamon Press, 1982, Third Edition. New York, Oxford, Toronto, Sydney, Paris, Frankfurt.

7. OGM, *T.C.Orman ve Su İşleri Bakanlığı, Orman Genel Müdürlüğü,* https://www.ogm.gov.tr/ekutuphane/Yayinlar/T%C3%BCrkiye%20 Orman20Varl%C4%B1%C4%9F%C4%B1-2016-2017.pdf, 2015.
8. Gao, Y., Zhang, T., Zhang, W., Meng, H., & Zhang, Z., *Research On Visual Behavior Characteristics and Cognitive Evaluation Of Different Types Of Forest Landscape Spaces.* Urban Forestry & Urban Greening, 2020, **54,** 126788.
9. Yücel, M., ve Babuş D., Doğa Korumanın Tarihçesi ve Türkiye'deki Gelişmeler. Doğu Akdeniz Ormancılık Araştırma Müdürlüğü, Doa Dergisi (Journal Of Doa), 2005, **Sayı: 11**, s.151 - 175.
10. DKMPGM, *Doğa Koruma ve Milli Parklar Genel Müdürlüğü. Doğa Koruma Statüleri Resmi İstatistikleri. http://www.milliparklar.gov. tr/resmiistatistikler yeni,* 2019
11. Eriksson, L., & Nordlund, A., *How Is Setting Preference Related To Intention To Engage In Forest Recreation Activities?.* Urban Forestry & Urban Greening, 2013, **12(4),** 481–489.
12. Gülez, S., *Park– Bahçe ve Peyzaj Mimarisi,* 1989, Karadeniz Teknik Üniversitesi Orman Fakültesi Ders Teksirleri**, 29, 2,** 213–214, Trabzon.
13. T.C.Resmi Gazete, *Mesire Yerleri Yönetmeliği.* (28578), 05.03.2013.
14. Demirel, Ö., *Çoruh Havzası(Yusufeli Kesimi) Doğal Ve Kültürel Kaynak Değerlerinin Turizm ve Rekreasyon Potansiyeli Açısından Değerlendirilmesi Üzerine Bir Araştırma,* 1997, Doktora Tezi, Karadeniz Teknik Üniversitesi, Fen Bilimleri Enstitüsü, Trabzon.
15. Gülez, S., *Orman İçi Rekreasyon Planlaması,* Karadeniz Üniversitesi Basımevi, Karadeniz Teknik Üniversitesi Dergisi, 1983, **6**,2, Trabzon.
16. Ramanpong, J., Yu, C. P., Chiang, P. N., & Tsai, M. J., *Risk Management In Suburban Forest Recreation Areas: A Retrospective Analysis Of Illness Cases.* Urban Forestry & Urban Greening, 2020, 126710.
17. *https://www.fs.usda.gov/Internet/FSE_DOCUMENTS/stelprdb5346156.pdf,* 14 Kasım 2018.
18. *https://parkplanning.nps.gov/NPSSystemPlan.cfm.* 07 Şubat 2019
19. Anonymous, *Annual Performance Report FY2015 National Wildlife Refuge System, U.S. Fish & Wildlife Service. https://www.fws.gov/refuges/about/ RefugeReports/pdfs/RAPP _Rpt_2015_final.pdf,* 15 Aralık 2015.35.
20. Uleck, R.B., *The Challenge Of Recreation Planning: Methodology And Factors To Consider.* In In: Larson, E. vH., ed. The Forest Recreation Symposium. State University of New York College of Forestry; 1971 October 12–14: US Department of Agriculture, Forest Service, Northeastern Forest Experiment Station. 200–211.
21. T.C.Resmi Gazete, *Milli Parklar Kanunu. (2873),* 09.08.1983

22. Cırık, U., *Milli Parklar ve Uzun Devreli Gelişme Planları,* Planlama.TMMOB Şehir Plancıları Odası Yayını, 2007, **1, 39**, 45–50.

23. Arda, S.S., *Türkiye'de Doğa Koruma Alanı Uygulamaları ve Avrupa Birliği Mevzuatı ile Karşılaştırılması,* 2003, Yüksek Lisans Tezi, Ankara Üniversitesi. Sosyal Bilimler Enstitüsü. Ankara.

24. Anonymous, *Korunan Alan Planlaması ve Yönetimi, Biyolojik Çeşitlilik ve Doğal Kaynak Yönetimi Projesi Deneyimi,* 2007, T.C. Çevre ve Orman Bakanlığı, Doğa Koruma ve Milli Parklar Genel Müdürlüğü, Biyolojik Çeşitlilik ve Doğal Kaynak Yönetimi Projesi, TŞOF Trafik Matbaacılık A.Ş. Ankara.

25. T.C.Tarım ve Orman Bakanlığı, *Orman Genel Müdürlüğü, Odun Dışı Ürün ve Hizmetler Dairesi Başkanlığı Ormancılık İstatistikleri.https://www.ogm.gov.tr/ekutuphaneSayfalar/Istatistikler.As px;* 25 Eylül 2019

26. Düzgüneş, E., *Milli Park Planlamalarında Ziyaretçi Yönetimine İlişkin Yeni Bir Model Önerisi: Altındere Vadisi Milli Parkı (Maçka/Trabzon) Örneği,* 2015, Doktora Tezi, Karadeniz Teknik Üniversitesi, Fen Bilimleri Enstitüsü, Trabzon.

27. Demirayak, F., *Türkiye'de Korunan Alanlar için Yeni Bir Yaklaşım Ortaklaşa Yönetim,* 2006, Doktora Tezi, Ankara Üniversitesi, Sosyal Bilimler Enstitüsü, Ankara.

28. Thomas, L., and Middleton, J., *Guidelines for Management Planning of Protected Areas,* 2003,World Commission on Protected Areas (WCPA) Best Practice Protected Area Guidelines Series No. 10, IUCN Gland, Switzerland and Cambridge, UK. ix + 79pp.Series Editor: Adrian Phillips).

29. Anonymous, *Türkiye Korunan Alanlar Yönetim Planlaması Rehberi, T.C. Çevre ve Orman Bakanlığı,* 2006, Doğa Koruma Ve Milli Parklar Genel Müdürlüğü, Biyolojik Çeşitlilik ve Doğal Kaynak Yönetimi Projesi, Ankara

30. Thomas, L., *Türkiye Korunan Alan Yönetiminde IUCN Kategori Sistemi,* 2006, Çevre ve Orman Bakanlığı, Doğa Koruma ve Milli Parklar Genel Müdürlüğü, Biyolojik Çeşitlilik ve Doğal Kaynak Yönetimi Projesi, Ankara

31. T.C.Tarım ve Orman Bakanlığı, *Orman Genel Müdürlüğü, Odun Dışı Ürün ve Hizmetler Dairesi Başkanlığı Ormancılık İstatistikleri.https://www.ogm.gov.tr/ekutuphaneSayfalar/Istatistikler.As px;* 25 Eylül 2019.

32. *https://karadeniz.gov.tr/125-yil-ataturk-cesmesi-piknik-ve-me sire-alani/.*

33. http://www.loc.gov/teachers/classroommaterials/connections/conservation/histor-y.htm. 05 Kasım 2018.

34. Çelik, C., *Doğal Alanları Koruma: Amerika Birleşik Devletleri Birleşik Devletlerinde Milli Parkların Oluşum Süreci ve İdaresi,* 2015, Türkiye Cumhuriyeti İçişleri Bakanlığı, Türk İdare Dergisi, 481.

35. https://www.loc.gov/collections/national-parks-maps/articles-andessays/brief-history-of-the-national-parks/. 05 Kasım 2018.

36. Gorte, R.W., and Cody, B.A., *The Forest Service And Bureau Of Land Management:History And Analysis Of Merger Proposals. Congressional Research Service Report For Report For Congress 95–117 ENR,* 1995, Washington, DC: Committee for the National Institute for the Environment.

37. Demirel, Ö., *Doğa Koruma ve Milli Parklar,* Karadeniz Teknik Üniversitesi Rektörlüğü, Orman Fakültesi, 2005, Genel Yayın: 219, Fakülte Yayın No:37.

38. Foresta, R.A., *America's National Park And Their Keepers. Resources For The Future Library Collection Forests, Lands, And Recreatıon,*Volume 5. Rff Press Resources For The Future, 2011, New York, London. Forestry and Urban Greening, **4–3–4,93–103.**

40. https://www.nps.gov/parkhistory/hisnps/npshistory/timeline_annotated.htm. 7 Kasım 2018

41. *https://scipol.org/track/science-module-protected-areas-united-states,* 7 Kasım 2018

42. Molloy, J., *Mount Rogers National Recreation Area Guidebook, A Complete Resource for Outdoor Enthusiasts,* 2008, Second Edition. The University of Tennessee Press. Knoxville.

43. *https://web.archive.org/web/20060205114010/http://www.blm.gov:80/nlcs/rec-reation/index.html.* 05 Kasım 2018.

44. *https://www.nps.gov/media/photo/gallery.htm?id=CEF13E09-1DD8-B71B-0B 5CCB2620A82047,* 05 Kasım 2018.

45. *https://www.fs.usda.gov/Internet/FSE_DOCUMENTS/stelprdb5346156.pdf,* 14 Kasım 2018

46. *https://www.webcitation.org/6EaRGckIT?url=http://www.foresthistory.org/A-SPNET/places/National%20Forests%20of%20the%20U.S.pdf,* 22 Kasım 2018.

47. *https://www.fs.fed.us/managing-land/national-forests-grasslands,* 14 Kasım 2018.

48. *https://www.fs.fed.us/recreation/.* 14 Kasım 2018.

49. *https://www.fs.usda.gov/Internet/FSE_DOCUMENTS/stelprdb5346156.pdf.* 14 Kasım 2018.

50. *https://www.flickr.com/photos/forestservicenw/sets/72157662485451432/,* 15 Kasım 2018.

51. *https://www.flickr.com/photos/usfwsmtnprairie/sets/72157629666768844/with/7171363174/,* 15 Kasım 2018

52. *https://www.nps.gov/obed/planyourvisit/2017-photo-contestwinners.ht m,*06 Kasım 2018.

53. *https://www.flickr.com/photos/blmcalifornia/sets/72157653908108571/*.06 Kasım 2018.

54. UNEP-WCMC, *Protected Area Profile for United States of America from the World Database of Protected Areas*, 2018, November. Available at: www.protectedplanet.net

55. https://www.nps.gov/aboutus/national-park-system.htm#collapseCollapsible1540823584188, 07 Kasım 2018.

56. *https://www.rivers.gov/national-system.php*, 06 Kasım 2018.

57. *https://www.blm.gov/programs/national-conservation-lands/about/maps-data-and-resources*, 06 Kasım 2018.

58. *https://www.blm.gov/sites/blm.gov/files/NLCS%20Summary%20Table_Q1 _2019.pdf*, 06 Kasım 2018.

59. Anonymous1a, *Land Areas of the National Forest System, United States Department of Agriculture Forest Service WO, Lands FS-383*.https://www.fs.fed.us/land/staff/lar/LAR2017/LAR_Book_FY2017.pdf. 30 Eylül 2017.

60. Tweed, W.C., *A History Of Outdoor Recreation Development In National Forests 1891–1942. Formerly Historian*, 1989, History Section United States Department of Agriculture Forest Service Washington, D.C.

Sultan Sevinç KURT KONAKOĞLU

Tourism Corridor Planning Approaches in Urban Areas

Abstract: Corridors are linear areas and are intended to meet the needs of urban people in urban areas. Corridors have the characteristics of linearity, multifunctionality, connectivity, sustainability, offering alternative transportation means and contributing to landscape planning. Corridors are named according to their function with different names such as water corridor, wildlife corridor, recreation corridor, tourism corridor. Tourism corridor is the development and use of routes with abiotic, biotic and cultural landscape components based on certain themes for tourism purposes. Tourism corridors with historical and cultural values, local food theme, thermal resource values, transporting to religious areas and enabling agricultural activities enable local and foreign tourists to perform various touristic activities. At the same time, tourism corridors have important open-air classrooms as local and foreign tourists can learn about natural animal species, natural and endemic plant species, ecosystem and ecological processes in these areas. Tourism corridors should be handled with an integrated approach in line with their tourism potentials. Within the scope of the study, the importance and necessity of planning tourism corridors with different themes in urban areas will be emphasized.

Keywords: Corridor concept, tourism corridor planning, urban areas.

1 Introduction

Following the Industrial Revolution, the development of technology has led to a worldwide increase in urban populations, rapid urbanization movements lacking in infrastructure, growing consumption of natural resources, and uncontrolled use of resources, along with an expansion of tourism activities. Because of the deteriorated and fragmented state of the landscape texture, studies have been advanced in order to protect natural resources and to ensure their sustainability in parallel with these negative developments.

The landscape, together with natural and cultural components, forms the basis of the concept of tourism. Throughout history, interaction has always existed between tourism and the natural and cultural spaces we live in, as people have always had the desire to visit different natural and cultural areas. Landscape and tourism are closely related because natural and cultural sites constitute the main appeal of tourism. Consequently, in order to ensure the continuity of tourism, natural and cultural components should be protected and developed.

Since tourism activities put pressure on natural and cultural landscape components, they lead to the deterioration of the landscape texture over time. The landscape texture, which defines the spatial relationship between ecosystems, is considered as a mosaic, composed of patch, corridor, and matrix. According to Forman (1995), the patch refers to the non-linear areas that are homogeneous within themselves and the corridor refers to the use of linear areas that appear different from their surroundings in terms of content and physical structure, whereas the matrix refers to the main skeleton of the landscape [1].

By considering the concept of the corridor that divides the landscape texture or connects the landscape with different features, this study establishes its relationship with tourism and emphasizes the importance and necessity of planning tourism corridors with different themes in urban areas.

2 The Corridor Concept

The corridor is an extending linear segment having an aquatic or terrestrial character that connects two different landscape patches [2,3]. In other words, they are linear elements that, by addressing the landscape processes, provide opportunities for an ecological inter-connection of habitats, for a social inter-connection of cultures, and for usage as recreational areas [4].

Corridors can be riverbeds or vegetation of a natural character in a linear form. They can also be of a man-made character in the form of a canal, railway, road, or linear park. The main function of this type of corridor is greater in urban areas, where it is aimed at meeting human needs such as transportation, recreation, and infrastructure. The width, shape, location, structure, and functions of the corridor are the main criteria that determine its effectiveness. The effectiveness of the corridor also depends on the nature of the habitats with which it connects [1,5,6].

Corridor structures take different characteristics from the landscape patches surrounding them. A creek or streambed can be given as an example of a natural corridor. The unique plants contained in a streambed demonstrate a feature of the streambed corridor that is distinctive among landscape patches. Similarly, built structures like highways, railways, and canals have the characteristics of artificial corridors. The variety of plant species seen along the roadside shows a distinctive feature of artificial corridors [3].

Although corridors can have linear, streambed, or strip characteristics, they also have structural characteristics such as curved, narrow, wide, long and narrow, or twisted forms [7]. According to the studies of Noss (1987), Forman (1995), and Low Choy (2009), the functions of corridors include [1,8,9]:

- Helping to preserve biological diversity by preserving landscape values.
- Preparing an environment by forming a suitable habitat matrix that will enable rare and threatened species to re-colonize,
- Minimizing the fragmentation seen in the landscape and increasing the connection between similar areas,
- Providing genetic variation between isolated habitats,

- Providing alternative passages for species that migrate between different seasonal habitats,
- Connecting various shelter and hunting areas with each other,
- Improving water resources management for control of floods and sedimentation and for maintenance of reservoir capacity, clean water, a sustainable fish population, and fisheries,
- Ensuring sustainability in the landscape through recreational activities in rural green areas including game management, wildlife conservation/enjoyment of nature, hiking, cycling, boating, and skiing.
- Providing recreational, visual, and climatic benefits by creating a green belt surrounding the city.

Corridors are aimed at meeting the needs of urban people in urban areas. Corridors have the features of linearity, multifunctionality, connection, and sustainability, in addition to offering alternative means of transportation and contributing to landscape planning. Corridors are categorized, depending on their function, as water corridors, wildlife corridors, recreation corridors, and tourism corridors.

3 The Tourism Corridor Concept

A tourism corridor is defined as a route with natural and cultural features appealing to tourists and based on certain themes [10]. It is also known as a route that connects natural and cultural sites that are attractive to tourists [11]. In other words, it is a tourism approach, developed in various ways, that offers tourists the opportunity to visit areas having natural appeal along a themed route [12]. Interactions between domestic and foreign tourists and the places they visit may be suggested by local inhabitants.

Tourism corridors connect historical sites, natural areas, waterside paths, nature trails, waterways, and footpaths, and contribute to the planning of regional touristic sites in line with themes, ensuring that these areas are visited by tourists [13].

The concept of a tourism corridor was associated with transportation and began with the 'Orient Express', a train service from Paris to Istanbul established in 1883. Later, the concept was adopted by urban planning, development agencies, and the tourism and public sectors and then was expanded in the 1990s. With the concept of the tourism corridor, the most direct and shortest routes are used to connect two or more urban areas [14]. Since the tourism corridor may refer to a route defined by a theme spanning several countries and even continents, economic cooperation among the countries through which the corridor passes is necessary. Therefore, corridors provide economic and commercial contributions to the countries in which they are found [12,15].

The natural-historical-cultural attractions of tourism corridors include local food, thermal springs, religious sites, agricultural events, and so on, and different themes are created in line with their features. Thus, they provide domestic and foreign tourists with the opportunity to participate in various touristic activities [16].

At the same time, tourism corridors also serve as important open-air classrooms where domestic and foreign tourists can learn about the local wildlife species, natural and endemic plant species, ecosystem and ecological processes in these areas [17].

Tourism corridors also take the form of natural and cultural heritage corridors, according to their inherent characteristics. Natural tourism corridors are based on natural attractions such as scenic landscapes, plants and wildlife, and biological diversity. Natural tourism corridors can be combined with elements such as adventure, wildlife, topography, and education. Mozambique's Limpopo National Park, South Africa's Kruger National Park and the Great Limpopo Transfrontier Park are examples of nature-based tourism corridors [12]. Cultural tourism corridors are based on historical, cultural, spiritual, and economic bonds that are a part of the 'collective memory' of societies. They can also be combined with various aspects such as religion, art, cultural traditions, gastronomy, architectural elements, and historical buildings. Local food can serve as a theme in cultural tourism corridors, e.g., the Italian 'Route of the Buffalo Mozzarella' [18], the 'Alsace Wine Route' [19], Edinburgh's 'Royal Mile' [20], the 'Cypriot Monuments of the UNESCO Religious Route' for visiting religious sites [21], the 'Marian Route' of religious pilgrimage through Spain, France, and Andorra [22], the 'European Cultural Route' [23], the 'Santiago de Compostela Route', the 'Transromanica' route of artistic and architectural heritage, the Italian 'Montecatini Thermal Circuit' for thermal springs [24], the 'Minho Thermal Route' [25], the 'European Route of Historic Thermal Towns' [23], the 'Windsor-Québec City' corridor highlighting local agriculture [26], the 'Beira Agricultural Growth Corridor' [27], the 'German Fairy Tale Route' invoking the cultural heritage of fairy tales and legends [28], and finally, the 'Destination Napoleon', the 'Mozart Ways', and the 'Phoenicians' Route' can be given as examples offering the opportunity to learn about the lives of historical figures and peoples [12].

Tourism corridors should be planned by taking an effective and holistic approach in line with their tourism potential.

4 Tourism Corridor Planning Approaches in Urban Areas

Planning is defined as the process of preparing systematic action programs and making decisions in order to achieve the desired goals for the future. In other words, it is systematically putting theoretical knowledge into action. By incorporating the past, present, and future, planning is a kind of forecasting to predetermine certain events that are supposed to happen in the future. In planning, the idea is to reach the target with the shortest time, the least expense, the lowest environmental impact, the least conflict, and the highest number of predictions. Planning is a process that has no beginning and no end, can adapt to changes, and carries on by demonstrating changes [29,30,31,32,33].

Since tourism is closely related to transportation, communication, accommodation, and the provision of urban services, it can be controlled through planning and converted into a plan decision. Tourism planning is a disciplined arrangement that indicates the periodic goals to be achieved in the tourism sector, the tools that can be used to reach these goals, the possibilities, the work to be done, and the results of these efforts [34]. The output of the tourism planning process is the tourism plan. In cities, the tourism plan is a long-term and comprehensive plan formed by stakeholders related to tourism (local people, tourism organizations, public institutions, tourism sector employees, etc.).

Today, different types of tourism planning have emerged in order to produce alternative solutions based on population growth, development of technology, and updated information. İçöz et al. (2002) [32], referring to the studies conducted by Coşkun (2004) [35], stated that tourism planning should be carried out in the form of periodic tourism plans (short, medium, and long-term), organizational tourism plans (integrated and sectoral tourism planning), and physical tourism plans (international, national, and local tourism plans and regional tourism planning).

Every stage of tourism planning is important for success in tourism development and management and for tourist satisfaction. In today's rapidly changing conditions, planning has become mandatory for every tourism region; however, some tourism regions develop without conscious planning. Environmental, social, and economic problems are encountered in places where tourism develops in an unplanned manner [36]. Therefore, for the satisfaction of the local people, the tourists, and the economy of the region, tourism corridors should be managed using tourism planning in line with their tourism potential. In planning studies, including tourism planning, the cultural structure of the society and the urban identity formed accordingly should be taken into account. In tourism planning conducted in cities, resource values should be properly analyzed, tourism should be well planned, natural-cultural-historical-architectural sites should be protected, and tourism should be developed systematically and effectively.

The tourism corridors considered within tourism planning may exhibit a historical or cultural quality, a local food theme, or cultural heritage elements such as fairy tales and legends, or may provide opportunities for domestic and foreign tourists to visit thermal springs or religious sites, to participate in agricultural events, or to enjoy various touristic activities [16]. Tourism corridors provide routes that include various attractions covering nature-culture-history-architecture and sports as opposed to regions that offer only a single event or attraction to tourists [17].

In planning the tourism corridor, first, tourism infrastructures should be improved in line with the natural (landscapes, plants and wildlife, biological diversity) and cultural (religious heritage, art, cultural traditions, gastronomy, architectural elements, historical buildings, etc.) resources of the cities. In this context, coordination should be established between hotels, tour operators, transportation providers (buses, trains, airlines, railways, and cruise ships), and tourism companies. Since the absence of directional signage on the corridors adversely affects the development of tourism corridor planning, the problem of directional signs along the corridors should be addressed. Well-designed signposts can attract the attention of tourists and increase the appeal for touristic areas. Tourism corridors help tourists find their way to places of interest and connect touristic areas having different features. When tourism corridors are planned, waste of time, money, and effort is avoided.

5 Conclusion

Corridors provide the connection between areas having similar or different functions within the landscape texture. As the necessary link for the correct functioning of ecosystems, the corridor indicates the richness of the landscape and is frequently used in planning studies. Since nature and culture are the basic values of tourism, tourism and planning are closely related in terms of preserving the natural and cultural heritage and transferring it to future generations. Because tourism planning is a matter of supply and demand, the supply of natural and cultural resources and the potential demand for them should be evaluated in tourism planning. The impact of tourism includes many economic, social, environmental, and cultural effects. Tourism planning can be used as a sustainable development tool to eliminate the negative consequences of tourism in cities.

An examination of today's tourism planning approaches shows that in many countries and regions, tourism planning is carried out in order to ensure a controlled development of tourism whereby new accommodation units are opened, transportation is provided to these areas, certain themes are given to the areas with existing natural and cultural values, tourists are brought to visit these areas, tourism products are diversified, and tourism development campaigns are carried out.

Tourism corridors are also one of the current approaches considered within the scope of tourism planning. Tourism corridors connect areas with various tourism attractions for domestic and foreign tourists in line with certain themes. Tourism corridors may have historical and cultural values, local food themes, or thermal springs, and may lead to religious sites, agricultural events, or various touristic activities (experiencing local food, visiting religious shrines/hot springs/baths, hiking). They provide the opportunity to take nature photographs, watch birds/wildlife, visit monumental trees, observe local wine production on site, participate in festivals/celebrations/art events, etc. At the same time, tourism corridors serve the important function of open-air classrooms, where domestic and foreign tourists can learn about wildlife species, natural and endemic plant species, and the ecosystem and ecological processes in these areas. Tourism corridors should be managed using a holistic planning approach in line with their natural and cultural values. The existing corridor attractions in cities should be improved and brought into urban tourism.

Natural and cultural resources in tourism corridors are foreseen as being considered using a conservation-utilization balance and planned along linear axes instead of on a point scale. With the tourism corridor planning approach created in urban areas, tourism types can be diversified and carried out in all four seasons.

Moreover, sources of local culture and handicrafts can be supported by being connected to the restaurants, cafes, and accommodation units, thus promoting these attractions. When tourism corridors are planned in cities, tourists will be directed to these corridors, and they will be encouraged to visit not only the part of the corridor where they are, but also other parts. In this way, the duration of the tourists' stay in the city will increase and they will contribute positively to the city's economy.

References

1. Forman, R.T.T., *Land-Use Planning From Global to Local Challenge, A Dowden and Culver Book*, 1995, Environmental Resource Management Series, Chapman & Hall: New York.
2. Forman, R.T.T., and Godron, M., *Landscape Ecology*. 1986, John Wiley and Sons Ltd: New York.
3. Odum, E. P., and Barrett, G.W., *Turkish Translation of Fundamentals of Ecology*, 2008, Foreword.
4. Dramstad, W.E., Olson, J.D., Forman, R.T.T., *Landscape Ecology Principles in Landscape Architecture and Land – Use Planning*. 1996, Graduate School of Design, Island Press, Harvard University.
5. Cook, E.A., *Ecological Networks in Urban Landscapes*. 2000, PhD Dissertation, Wageningen University, The Netherlands.
6. Deniz, B., Küçükerbaş, E.V. and Eşbah Tunçay, H., *Peyzaj Ekolojisine Genel Bakış*. Adnan Menderes Üniversitesi Ziraat Fakültesi Dergisi, 2006, **3**(2): p. 5–18.
7. Hess, G.R. and Fischer R.A., *Communieating clearly about eonservation corridors*. Landscape and Urban Planning, 2001, 55: p. 195–208.
8. Noss, R.F., *Corridors In Real Landscape: A Reply to Simberloff and Cox*. Conservation Biology, 1987, **1**(2): p. 159–164.
9. Low Choy, D.C., *A joined-up Regional Landscape: A Roadmap to connect the parts. Breaking the Barriers: Engineering Solutions to Ecological Problems*. 2009, Urban Research Program, Griffith University.
10. https://www.avrupa.info.tr/tr/node/6865. Accessed at 19 August 2018.
11. https://www.igi-global.com/dictionary/tourist-corridor/79500. Accessed at 20 October 2020.
12. https://sbb.gov.tr/wp-content/uploads/2019/11/Developing-Multi-Destination-Tourism-Corridors-October_2019.pdf. Accessed at 20 October 2020.
13. https://tompkinscountyny.gov/planvct/tool/tourismcorridorplanning. Accessed at 21 October 2020.

14. https://tr.wikipedia.org/wiki/%C5%9Eark_Ekspresi. Accessed at 22 October 2020.
15. https://www.academia.edu/14264572/Basic_Concept_on_Tourism_Planning_and_Development, Accessed at 20 October 2020.
16. Kurt Konakoğlu, S.S., *Katılımcı Turizm Planlaması Yaklaşımı ile Amasya Kent Örneğinde Mekânsal ve Senaryo Analizi.* 2018, Doktora Tezi, Karadeniz Teknik Üniversitesi Fen Bilimleri Enstitüsü, Trabzon.
17. Kültür ve Turizm Bakanlığı, *Türkiye Turizm Stratejisi 2023.* 2007, Kültür ve Turizm Bakanlığı Yayınları: Ankara.
18. http://www.italia.it/en/travel-ideas/gastronomy/mozzarella-roads-the-story-of-mozzarella-di-bufala-dop.html. Accessed at 11 December 2018.
19. https://www.alsace-wine-route.com/. Accessed at 11 December 2018.
20. http://www.royal-mile.com/. Accessed at 11 December 2018.
21. https://www.cyprushighlights.com/en/2010/09/02/the-cypriot-monuments-of-unesco-religious-route/. Accessed at 11 December 2018.
22. https://www.spain.info/en/reportajes/la_ruta_mariana.html. Accessed at 11 December 2018.
23. http://www.anothertravelguide.com/alternate_routes/europe/italy/italy/anothertravelguide.com/european_cultural_and_religious_routes/. Accessed at 11 December 2018.
24. https://www.viamichelin.com/web/Routes/Route-Montecatini_Terme-51016-Pistoia-Italy-to-Porretta_Terme-_-Bologna-Italy. Accessed at 11 December 2018.
25. http://turismodeourense.gal/en/agrupacion/ruta-termal-del-mino-termas-naturaleza/. Accessed at 11 December 2018.
26. https://en.wikivoyage.org/wiki/Windsor-Quebec_corridor. Accessed at 11 December 2018.
27. http://beiracorridor.org/. Accessed at 11 December 2018.
28. https://www.deutsche-maerchenstrasse.com/en/. Accessed at 11 December 2018.
29. Bettleheim, C., *Studies in the Theory of Planning.* 1967, Asia Publishing House: Bombay.
30. Friedman, J., *Planning in the Public Domain: From Knowledge to Action,* 1987, Princeton University: New Jersey.
31. Şahbaz, R.P., Hukuki, Dini ve Ahlaki Kuralların Turizm Planlamasına Etkisi. Dokuz Eylül Üniversitesi Sosyal Bilimler Enstitüsü Dergisi, 2002, **2**: p. 233–260.
32. İçöz, O., Var, T. and İlhan, İ., *Turizm Planlaması ve Politikası.* 2002, Turhan Kitabevi: Ankara.

33. Ersoy, M., *Kentsel Planlama Kuramları.* 2008, İmge Kitabevi: Ankara.
34. Olalı, H., *Turizm Politikası ve Planlaması.* 1990, Yön Ajans: İstanbul.
35. Coşkun, İ.O., *Turizm Politikası ve Planlaması İçin En Uygun Talep Öngörü Yönteminin Belirlenmesi.* 2004, Yüksek Lisans Tezi, Anadolu Üniversitesi Sosyal Bilimler Enstitüsü, Eskişehir.
36. WTO, *National and Regional Tourism Planning: Methodologies and Case Studies.* 1994, A World Tourism Organization Publication: U.S.A.

Sema MUMCU and Duygu AKYOL

Dealing with the Problems of Urban Life in Residential Landscape Design Studio

1 Introduction

Cities are the developed world's economic, cultural, and political power centers [1]. Therefore, with the current trends in urban population growth, two out of every three people born during the next 30 years is likely to live in a city [2]. Nowadays, however, cities are frequently defined as unsustainable centers that use environmental resources, which has an impact on climate change[1] and are the reason for multiple problems that adversely affect urban life. These problems include overcrowded downtowns and attendant urban densification and sprawl. As a result of these rapid changes, increasing identity loss, irregular and chaotic perception of urbanscapes, traffic problems, increasing loss of green spaces, grey and unfriendly looking environments, higher rates of crime, social inequity, noise, air, and water pollution, waste materials, and unsatisfactory sanitary conditions are seen. Moser (2009:353)[3] stated that in order to understand people's personal well-being and quality of life (QOL) in urban spaces, their relationship to their own living environment presents a crucial issue. It is well known that lack of environmental quality is perceived as an important threat to people's well-being, and nowadays, urban environments are unfortunately becoming more and more threatening to residents' daily lives [4]. The problems noted above are repeatedly mentioned by city dwellers as a threat to their quality of life [4].

QOL is a term of particular interest to both environmental designers and behavioral researchers. Places may be designed to enhance people's QOL; therefore, examining the relationships between the characteristics of the places in which urban life takes place and perceived QOL of the residents is an important issue [2]. However, designers' approaches and adoptions were generally neglected in QOL studies. The results of environmental psychology studies have been found to be sufficient for guiding designs, but studies that bring designers' decisions and research findings together and compare them are not common. The authors of this study have observed that landscape design and designers do not play an important part in quality of urban life (QOUL) studies.

Landscapes can provide restorative, recreational, and esthetic benefits to people, promote both place attachment and identity, and also contribute positively to human perception toward landscapes for stress relief, neighbourhood satisfaction, housing premiums, and physical, social, and educational benefits [5]. Mohd Shahli et al. (2014:311)[6] stated that a pleasing landscape can bring mental and physical benefits to people. Factors such as planned housing and living environment, housing type and quality, presence of open and green areas were defined as components of QOL (Akyol et al., 2019:7991)[7] therefore, concern about the importance of landscape design in supporting QOL is needed, especially in the context of housing development. Landscape design can be used as a tool to make a community more sustainable and contribute to a balanced environment [8]. Therefore, the need to study the effectiveness of landscape design in order to attain the sustainable housing development was emphasized [6]. In their study, Mohd Hussain et al. (2014:319) [9] determined that the majority of the respondents agreed that landscape design can influence their decisions when buying or renting a house and can increase property values. Zhang (2010:83)[10] indicates how home buyers in the 1990s in China began to pay greater attention to the landscape and environment of the community; both the ratio of green space and the quality and design became the most important criterion for assessing the desirability of an estate. These statements reveal the increasing importance of the relationship between housing estates and their landscapes. Fleischer (2007:287)[11] described specific ideas about living environment and life-style as developed by the new urban middle class and stated that they aspire to have green space, better air quality, and spaciousness, among other physical characteristics, but also privacy and exclusivity in their new places of residence. The researchers showed that residential compounds have become the basis for identity and lifestyle formation and are crucial to the process of social differentiation [11]. Similarly, Zhang (2010:79)[10] stated that residential environments are salient sites for articulating cultural distinctions, social entitlements, and life orientations. These reflect the fact that QOUL is related to different expectations which surpasses mere physical quality. Gated communities, which have been increasingly gaining popularity as a result of these expectations, are being severely criticized. These sites are being designed freely without concern for their relationships to other parts of the city. Furthermore, they create disharmony with their surroundings, disrupting the integrity of the city, and disconnect the landscape [12]. In this sense, the question of designing a good environment and lifestyle presents an issue of social importance that must be considered together by different disciplines. In brief, housing estates are multilayered fields of application that consist of different aspects such as sustainability, personal well-being, QOL, and social structure. How are these subjects reflected in landscape design? While the growth of enclosed high-income neighbourhoods has mostly been studied in terms of architectural design, representation, residential structure, and behavior, little attention has been paid to the landscape architecture component of this issue. Awareness of landscape architecture students about the problems of urbanization and urban life, how they reflect it in their design approaches, their level of knowledge about current issues, and attributes of housing estate design are all important research subjects. Thus, with appropriate interventions to landscape design curricula, deficiencies can be corrected, and the students can be supported with innovative and creative approaches.

How Are the Problems of Urban Life and Gated Communities Related?

Since efforts relating residential environmental design to escape from problems of urban life are becoming widespread in the field of environmental design, knowing what these problems are and how they can be solved is important. Cities are being increasingly recognized as both a problem and a solution to sustainable development worldwide [1]. Many studies in the field of planning and design accept that the problems of urban life are related to various subjects such as QOL or QOUL, neighbour satisfaction, and sustainability. The problems of urban life and spaces that are frequently defined in these studies can be environmental, socio-economical, and/or psychological.

Environmental problems include noise, pollution (poor air quality, underdeveloped drinking water supply infrastructure,) lack of adequate infrastructure (water, energy, sewerage system, and drainage), traffic (increased road traffic), and transportation (transportation designed only for cars, poorly designed transportation and/or lack of satisfactory transportation systems), security problems, struggling for services, waste disposal or solid waste management, dependence on imported foods from remote areas, tremendous amounts of energy consumption, lack of adequate housing, lost cultural and historical heritage, damage to surrounding ecosystems, insufficient neighbourhood facilities, unaesthetic, disorderly, and ugly built environments-decreasing visual quality, and identity loss [13; 14; 15; 1; 3; 4; 16; 17]. Socio-economic problems can be listed as demographic problems, social inequality, urban poor, social exclusion of certain population groups or segregation, socio-economic deprivation and inequities, various social problems stemming from poor housing conditions, higher rates of crime, neighbourhood degradation, inequities in healthcare, well-being and healthcare accessibility, crowding, and lack of civilized manners due to overcrowding [14; 1; 18; 19]. Psychological problems mostly depend on other problem groups. City dwellers are reputed to be more exposed to environmental stressors such as constraining daily life in a big metropolis, concern about environmental deterioration, fear of crime, and chaotic, unsavory and disorienting daily life [14;4]. All of the above-mentioned problems affect the inhabitants' health [1], and these adverse conditions in an urban setting create stressful situations or experiences for some of the people living there [2]. Since lack of environmental quality is perceived as an important threat to people's well-being, their relationship to their own living environment is crucial for understanding their personal well-being and QOL. For Moser (2009:353)[3], this was an important issue for sustainable development.

In Turkey, many studies about neighbourhood satisfaction and/or QOUL associate the problems of urban life with the new trends in housing estates, especially those of gated communities. The reason behind the popularity of gated communities is seen as the search for solutions to the problem of urban space. The same situation has also been expressed in some other countries. For example, in case of Budapest[20], the main factor in the demand for gated communities is the eagerness of the affluent upper middle classes to escape from environmentally unfavorable and overcrowded areas of downtown. Similarly, for Hook and Vrdoljak (2002:196)[21], the rationale for building and living in gated a community rests on the assumption that unregulated and uncontrolled space is dangerous space. According to these authors, an increasing fear of crime is a reflection of a fear about the future, which is fed by rapid demographic, economic, and social change. Gated communities offer a haven for peace and quiet, order, and pleasure which stand exactly in the middle of a chaotic and hostile world [22]. For gated communities in Turkey, the reasons for urban residents' preferences of luxury housing estates at the periphery of cities are explained by the desire to escape the stress and anxiety caused by overcrowded downtowns, traffic, and air pollution[12]. Pérouse and Danış (2005:100) [16] point to advertisements that picture urban centers as places that have become chaotic and evil and must be avoided and persistently speak of identity loss and nostalgia for neighbourhoods. How these advertisements use pollution, degradation, chaos, and traffic to express the problems of urban life was also determined [16]. Similarly, while discussing the housing trends in Istanbul, Geniş (2007:785)[14] explained how the city has lost its identity and become chaotic and irregular. New urbanism justified by gated communities in Istanbul views the city as a place of decay, chaos, and unsavory encounters, and instead proposes 'alternative urban living' [14]. Görgülü and Kaymaz Koca (2007:3)[12] emphasize advertisements about new lifestyles that have been proposed as the long-awaited solution against the problems of urban life. In these advertisements houses are depicted as being in close touch with nature, far away from unwanted problems and stress, filled with peace and safety but at the same time close to the city center. Several authors have shown that these statements are commonly used in the housing market. Statements of assurances in advertisements such as 'an idealized exclusive life' or 'a prestigious new life' as opposed to urban life were also mentioned [12; 16]. In her study, Güzey (2014:95)[23] determined that the factors in moving to a gated community were the condition of the physical environment, security, children's comfort and security, and prestige. Similar trends in other countries that have been mentioned by researchers reflect the desire of the middle classes to cultivate a secure and distinctive lifestyle [10].

As a result of these factors, since the 1980's, gated communities have become widespread in Turkey as they have in other parts of the world. Their role in the daily urban life is important; although the quality of the physical environment they offer is of high quality, they are being severely criticized as adversely affecting QOL because of isolation, individualization, alienation, and social and cultural disconnectedness that they have caused [24]. This movement ignores the history and patterns of cities, cannot stand the sight of 'the others' in the city, and is being fed by the eagerness of higher income classes for establishing an identity of 'exclusivity and prestige' [16].

Landscape Design and Gated Communities

The increasing popularity of gated communities in the search for solutions for the problems of urban life has already been noted. In addition to this, gated communities are being used by people as a social practice to distinguish themselves from others and for portraying themselves as modern citizens[11]. Depending on this tendency, developers compete to attract potential home buyers and emphasize the 'exclusive and prestigious' image in order to market their product [16]. According to some researchers, in some cases, eagerness for prestige seems to surpass even both fear of crime and the urge for self-segregation [20]. In this way, housing has become one of the prime markers of socio-economic status; where and how one lives are indicators of one's financial well-being. According to Gadecki and Smigiel (2009:201)[25], security and nature are the most important features of these pledged lifestyles. In order to produce a new kind of place, evoking a vision of 'paradise' helped to highlight these features [25]. This determination reveals the role of landscape design in creating lifestyles promised by new housing estates. Therefore, in this section, the roles of landscape design in searching for solutions to solve urban problems and introducing gated communities' lifestyle images and how components of landscape design are being linked to promised lifestyles will be discussed.

According to Park and Kim (2017:80)[5], one aim of community design and planning is to encourage the construction of high quality landscapes because human beings are a vital element of landscape who are the legitimate recipient of landscape benefits. Gadecki and Smigiel (2009:205)[25] defined landscape as a key element in understanding gated housing estates' phenomenon. According to these authors, landscaping is more than a form of designing surroundings or arranging space and places: it is a way of arranging and projecting social and cultural values. Under the conditions of a consumer-oriented society, affiliation and access to a particular landscape constitutes one of the most serious indicators of identity, especially with regards to social class[25]. Thus, in the case of gated communities, the landscape works both as an exclusion and an inclusion factor since it is appropriated and controlled by one group and made unavailable to others [25]. Spacious, clean, and well-protected open spaces, featuring different architectural styles and made of better construction materials that includes well-kept lawns, flowers, and trees present a particularly attractive environment reflective of residents' higher socio-economic status [10]. Both access to and usage of 'the world of nature' plays a significant role in this spatial manifestation. Access to nature is firmly controlled by the fence, and the so-called 'natural environment' is always professionally improved and maintained by gardeners [25].

In this context, Zhang (2010:84) determined that gated communities can be classified into three groups according to the lifestyles they offered: 1.) modern living; 2.) vernacular modern; and 3.) eco-living. A new term, 'special zones of eco-living', was coined to describe a new type of residential space with a close, harmonious relationship to the environment in which humans are allowed to return to nature. The envisioned 'natural world' is a human-made environment that features plants, trees, ponds, and fresh air, and/or proximity to forests, mountains, lakes, and rivers. The key selling point of this type of housing is 'being close to nature' or 'living amid the natural world' [10]. The seamless merging of nature and humans and the ultimate return to nature and truth ideal is reflected in the popular concept of eco-living. Also, in Turkey developers of gated communities pay attention to their products' closeness to nature since it increases the value of the houses [16]. Along with its aesthetic dimension, the emphasis on nature has recently been much more associated with the promotion of sustainable development. Sustainability practice in housing development has resulted in the development of various green strategies, especially for improving environmental performance [26]. The term 'green strategy' is defined as the approach that developers have adopted to improve and gain sustainable competitive advantages by contributing to environment protection, ecological responsiveness, and social responsibility. Researchers have shown that assuming social responsibility by adopting green strategies, housing developers can establish a good image. It was also revealed from the discussions with professionals involved in the case studies that the green image can result in higher housing prices [26].

Another role of landscape design is to strengthen social ties as opposed to segregation. Mahmoudi Farahani and Lozanovska (2014:224)[27] state that the created environment is able to influence the feeling of sense of community either directly or indirectly through increasing the chance of interactions among residents. According to Aiello et al. (2010:273)[28], neighbourhood attachment mainly depends on social relationships and an affective bond with the neighbourhood based on the presence of stimulating, interesting places and the absence of gloomy ones. According to these studies, green areas represent the main source of pleasant places in the residential environment [28]. In the case of gated communities landscape components help to bring together the residents who share 'same world views and values' and move away from the others in city. Developers reflect this as cultural exclusivity and reinforce it through culture and art activities such as festivals during mid-season months [16].

2 Research Problem and Hypotheses

How landscape architecture students perceive urban life in the process of a housing estate landscape design, how they reflect this to their design decisions, and what kind of environments and lifestyles they are predicted to design are research questions of this study. These subjects mainly deal with studies of neighbourhood satisfaction and QUOL. Studies on neighbourhood satisfaction promotes better understanding of the key satisfaction or dissatisfaction sources among residents, factors that influence their satisfaction levels, and how they are most likely to react to dissatisfiying housing conditions, all of which are important in informing housing policy and planning intervention [29]. However, in this study, unlike in previous ones, the subject was discussed from the student designers' point of view.

Moser (2009:351) stated that congruity between an individual and the environment can be achieved through a positive relationship between the objective environmental qualities accompanied by satisfied expressions concerning this environment. An integrated approach that gives a more adequate picture of people's relations to their environment deals with the relationship between the objective features of the environment and individuals' subjective judgments about it [3]. Aiello et al. (2010:266) shared the same point of view and explained two different components of residential environment quality as the objective characteristics of the neighbourhood and the subjective perceptions of these. Similarly, Marans (2012:10) recognized that QOL has both an objective and a subjective component and required an understanding of both components and relationships between them. Along with the affective reactions to everyday places and the activities performed in the urban settings, the objective attributes of the residential environment and the subjective evaluation of its quality were all defined as aspects of person–environment relationships[28]. In this study the 'objective conditions and subjective outcomes' approach was adopted. However objective and subjective attributes are determined through design approaches and environmental attributes of student designers. Following the terms of QOL in this paper, it was aimed to discuss how student designers understand and define the objective and subjective components in the landscape design process for creating a housing estate.

3 Materials and Method

The study was conducted in Karadeniz Technical University, Landscape Architecture Department. In context of Spatial Behavior course students were asked to prepare homework about their environmental design studio project of that term. Spatial Behavior is a compulsory course in the 7th semester of the curriculum and aims to teach students basic concepts in landscape design, such as spatial organization, human-environment interaction and congruency, and types of spatial behavior. During the 7th term, students also complete Environmental Design Project (EDP) VI course that is the last of the design studios that are accompanied by a tutor within the scope of the curriculum. The analyzed students' projects in this study belonged to the EDP-VI course. In EDP-VI, students are expected to develop a landscape design for a housing estate. The projects included in this study were all conducted with the same studio approach, in a class of 6 groups with different tutors and 9–11 students in each group. While giving the design subject and the site, the tutors did not limit the conceptual approaches to a specific title. The students determined their conceptual approaches based on various findings and as a result of their research and analysis within the scope of the design site and subject. In this latest project, students are expected to develop an advanced, creative, and powerful conceptual approach to the design problem. In this study, the materials (e.g., CAD drawings, sections, etc.) submitted in the project class final were not used; instead, the poster presentations about their projects were used, which contained written and visual narratives prepared by the students as part of their homework. The students were requested to explain their approaches in conceptual, formal and pragmatic dimensions and informed that the score from homework will comprise 50 % of their final note thus encouraged to pay strict attention. They were informed that the presentation will be used for a research about their design approaches but detailed information about study's aim was not told in order to prevent any influence on their representation. By doing so the number of students who considered and discussed urbanization problems without prejudice was determined. The students were not restricted with regards to their written and visual materials that were related to their projects. They were free to prepare their presentations as they wished, if in line with the titles given. Thirty-two senior students' poster presentations that explained their problem definitions and landscape design approaches of a gated community were analyzed. The student projects involved the same site, which was located in Trabzon, Turkey. Trabzon began to experience rapid urbanization process especially since 2000s. The location where the students' projects belong to is a new and rapid developing housing area of the city in recent years that mainly addresses the middle-high income level. While the area was dominated by rural character previously, it is currently surrounded with problems such as densification of concrete high-rise buildings, loss of green space, traffic density.

Content analysis method was adopted in the study. Content analysis is a technique for systematically describing the form and content of a written or spoken material. This technique is described as suitable for any kind of material including publications, recorded interviews, and reports and so on. Content refers to the specific topic or themes in the material and quantification (expressing data in numbers) of these forms the basis of a content analysis [30]. The explanations in presentations were evaluated and categorized in terms of their problem definition and 'objective conditions-subjective outcomes' components of proposed landscape design. Nineteen of 32 student poster presentations that framed design approaches depending on urban life or residential environment problems were analyzed in depth. The others who used concepts without relating it to any definition of a design problem or did not make any clear problem statements were excluded. The explanations were evaluated and categorized by a group of three researchers including the researchers of this study. All three researchers are landscape architects and experienced lecturers in landscape design. Landscape architects evaluated the sentences of problem statements and design approaches; decided and coded in which category each belonged to. Each concept was included into the most voted category. Analyses included two phases: 1) in the first phase, 270 sentences in total were categorized depending on whether the sentence explained the design problem-approach definition or 'objective conditions-subjective outcomes' of suggested landscape design and 2) in the second phase, these two groups were categorized into sub-groups in order to determine the factors that played role in their definitions. In the first phase, any sentence that included adverse effect of urban life or designers' suggestion as a solution was grouped as design approach, and any sentence that included a type of a space (objective condition) or the outcome of the experience of any space (subjective outcome) was grouped as 'objective conditions-subjective outcomes'. In the second phase, the main groups were subdivided depending on their similar contents; in this way, the factors that played a role in guiding students' decision were determined and understanding students' awareness about urban life problems and housing estate landscape designs became possible.

Definition of Design Area

The website for a specific estate publicized the project as: "We aim to create a living environment in which you'll live in pleasure" [31]. The area is located on the periphery of the city center, being 20 km far away. The project area was 20.000 m^2 total and included eight high-rise buildings with 128 condominiums (Fig. 1). Fourteen square meters (14.000 m^2) of the area was planned as green space. The estate developers declared in their website that the project brought a new perspective to the area with its unique landscape, social facilities, safety services, and social class. The landscape design included a beach, swimming pool, children's swimming pool, children's playground, tennis, basketball and volleyball courts, cafe, water shows, seating and viewing terraces, hobby gardens, walking routes, and video surveillance system [31]. The area was given to students as not designed, but they were free to make use of features listed in advertisements.

Fig. 1. The design area under construction and its surroundings [32].

4 Results

Problem Definitions

In total, 36 sentences were categorized as design approach definitions. Detailed analyses of these definitions revealed the urban problems to which students paid attention, about which problems their level of awareness was the highest, and which ones were generally neglected in the design process. Analyses of the definitions showed that in addition to problems, the students also should have paid attention to advantages of the design area. Most frequently, urbanization problems were referred to as decreasing green areas, unsustainability, the chaos of urban life and spaces, densification of buildings-concrete built environment that addressed physical dimensions, and also psychological and social urbanization problems, the stress of urban life, weakening of neighbour bonds, over crowdedness, and monotony of urban life. Students explanations to these problems could be exemplified as: "The residents of this estate need to engage in different kinds of individual and social activities in order to remove the stress of the day." "The part of the city where this estate is located is becoming increasingly concreted recently." "Humans caught in the chaos of the city began alienate from others." "The human-nature relationship has disappeared and human-technology relationship has replaced it." "As our lifestyles increasingly become unsustainable, I wanted to design this residence as a solution to this problem." It was determined that students showed a tendency toward discussing different problems that corresponded to physical, psychological, and/or social dimensions. As opposed to solutions to urbanization problems, students generally tended to suggest solutions through sustainability terms. Sustainability terms were discussed in different ways. Some of the students paid attention to the traditional life, which they defined as a relationship with rural life and stated that "Trabzon in its current state is a city living very close to nature and partly rural life is still being sustained." Some students considered nested urban areas and natural areas and suggested sustainability of this as: "The local mountains, streams, greens were taken into consideration while determining design decisions." Other students focused on the sustainability of QOL, which can be exemplified as: "My aim is to improve and sustain the quality of life." Other than these ideas, there were also some other ideas that students adopted and discussed with sustainability as one of the main responsibilities of a landscape architecture: "In landscape architecture, it important to design residential landscapes in terms of sustainability principles." Depending on these findings, general sustainability, including sustainability of natural resources, traditional lifestyles, and QOL, were approaches that were used (Tab. 1).

Tab. 1. The classification of students' definitions of design approaches

Design Approaches	
Threats	**Advantages and Aims**
. Stressful urban life . Building densification . Chaotic urban spaces and life . Unsustainable way of life . Air pollution . Disrupted human-nature relationship . Decreasing green areas . Weakened neighbour bonds . Monotony of urban life . Crowdedness	. Closeness to city center . Being surrounded by different land use types . Sustainability Sustainability in general Sustainability of natural resources Sustainability of traditional lifestyles Sustainability of QOL

'Objective Conditions-Subjective Outcomes' Components of Landscape Designs

The 'objective conditions-subjective outcomes' components approach that is widely accepted in QOL studies was adopted in this study and used as a theoretical framework in order to analyze students' design suggestions. Thus, understanding how students relate the problems and approaches that they defined with landscape design can became possible. The second phase analyses of objective conditions revealed five groups, while analyses of subjective outcomes revealed three groups. In total, 234 analyzed sentences used to determine the groups and how these are related to each other are shown in the model of QOL components (Fig. 2).

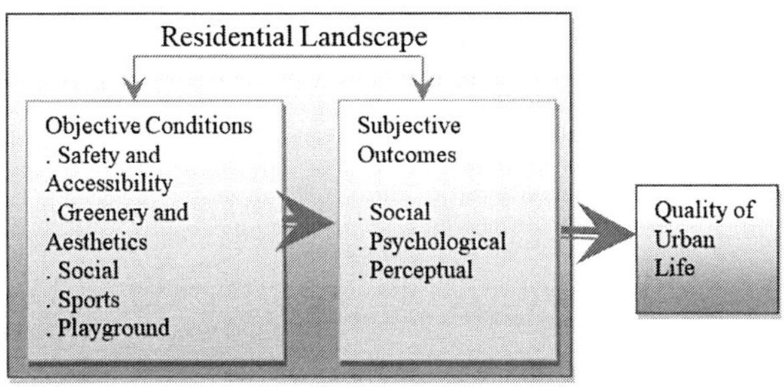

Fig. 2. The determined components of residential landscape in terms of QOUL

Objective Conditions

The space types and spatial attributes that were addressed in students' definitions were grouped as objective conditions. Accordingly, 184 definitions were appointed to this group, which referred to 32 space types. After classification of these answers depending on the functional similarities, five groups were determined: 1) safety and accessibility; 2) greenery and aesthetics; 3) social; 4) sports; and 5) playground. Students mainly paid attention to social interaction, greenery, and safety groups (Fig. 3). Students emphasized social life in residential open spaces by suggesting spaces for social activities such as cafes, seating and resting equipment, open theaters, and beaches (Tab. 2).

One important factor about gated communities that was revealed was the emphasis on open space physical quality and its relationship to the offered lifestyles. A similar factor revealed in this study was the greenery and aesthetics group. In this group, water elements and different garden types such as fragrance or rock gardens or orchards were suggested (Tab. 2). However, students offered these in the context of the sustainability approach.

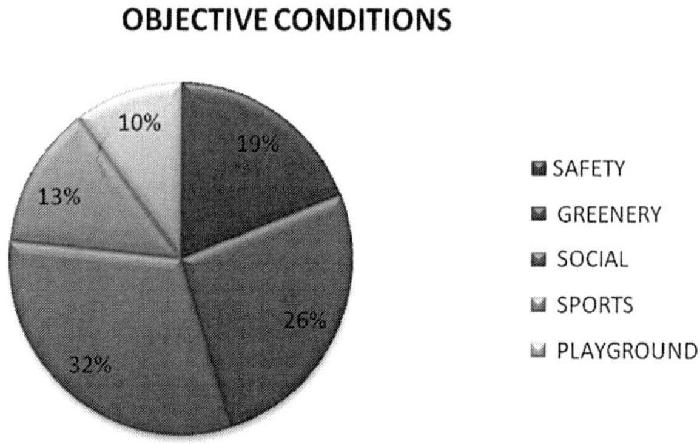

Fig. 3. The classification and distribution of 'objective conditions' in students' definitions

Tab. 2. 'Objective conditions' components of suggested landscape designs

Objective Conditions				
Safety and Accessibility	**Greenery and Aesthetics**	**Social**	**Sports**	**Play Ground**
. Gate/controlled entrance . Security and information unit . Surveillance system, alarm system, fire alarm . Lighting . Pedestrian system . Traffic road . Square	. Water show . Fountain . Color garden . Fragrance garden . Rock garden . Orchard . Green elevations . Hobby garden . Green edges	. Cafe . Seating and resting equipments . Open theater . Beach . Picnic area . Viewing platforms . Barbecue . Semi-closed, and open areas . Sunbathing	. Swimming pool . Courts . Basketball . Volleyball . Tennis . Golf . Fitness . Jogging Track	. Children Play Ground

The safety and accessibility group dealt mainly with safety precautions. The space types or spatial attributes included in this group were gate-controlled entrance, security units, and different types of security systems such as video surveillance, alarm systems, fire alarms, and lighting. Also, pedestrian routes and traffic roads were included in this group since they are generally accepted as contributing to the feeling of safety in environmental design studies. The factors in this group reflected the main characteristics of gated communities. Gated communities are accepted as related to the need for living in a safe and enclosed environment as opposed to the unsafe city; students' design approaches showed similar perceptions of gated communities (Tab. 2).

Since sports are accepted as contributing to an individual's well-being and QOL, the suggested objective conditions included various types of sport facilities. Other than these four objective conditions factors, children's playground was the last type of space that was suggested by all students.

Subjective Outcomes

The experiences and feelings included in students' explanations were suggested as the aimed outcomes for residents' interaction with outdoor spaces and were grouped as subjective outcomes. Fifty sentences were included in this group, and in their detailed analyses, the sentences referring to social interactions were grouped as social, sentences referring to psychological health and well-being were grouped as psychological, and finally, the sentences referring to perception of various spatial attributes were grouped as perceptual. Students mainly reflected psychological outcomes in their design, while social and perceptual outcomes followed this (Fig. 4).

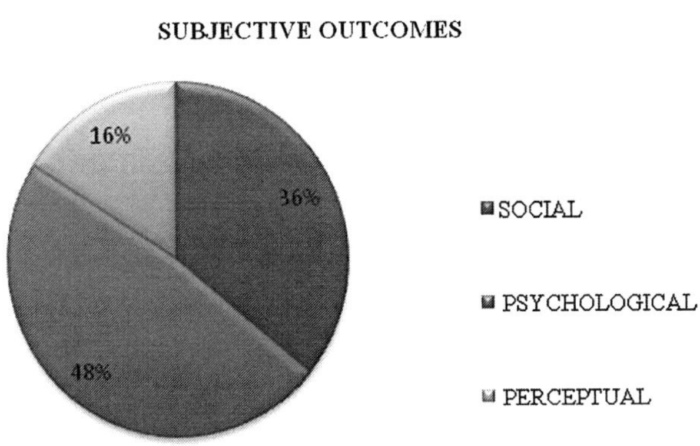

Fig. 4. The classification and distribution of subjective outcomes in students' definitions

The most cited psychological outcomes such as safety, peace, relaxing, well-being, tranquility, privacy, and happiness referred to various positive feelings related to psychological well-being (Tab. 3). "Outdoor spaces where residents will feel themselves in safe will be built." "I aimed to move away residents such an environment (chaotic city) and market them feel in peace." "However, as landscape architects, it's our duty to help people to move away from such environments, and guide them to more comfortable ones." These subjective outcomes mentioned by students are similar to ones used in gated housing estate advertisements in order to appeal people.

Social outcomes focused on relationships between residents and included togetherness, sharing, communication, socialization, liveliness, and interactions. When dealing together with the social spaces that were the most cited factor in the objective conditions group, students' emphasis on the social function of residential landscapes becomes apparent. "People will spend time together and be in touch." "In these spaces residents will come together and share the life." "We must spend much more time with each other, make visit, and engage in activities together."

Tab. 3. 'Subjective outcomes' components of suggested landscape designs

Subjective Outcomes		
Social	**Psychological**	**Perceptual**
. Togetherness . Sharing . Communication . Socialization . Liveliness . Interaction . Coalescence . Utility	. Safety . Peace . Relaxing . Well-being . Tranquility . Privacy . Happiness . Comfortable . Pleasure . Escape . Freedom . Harmony with nature	. Naturalness . Beauty . Colorful . Distinctness

Perceptual outcomes refer to effect of various spatial attributes (Tab. 3). "By bringing nature into the site, a sense of naturalness was created." "The fruit trees around buildings will create the sense of beauty with their seasonal changes." "In design process changing the area into a colorful place was aimed."

5 Discussion

Students' emphasis on sustainability in design solution approaches was regarded positively. Students' approaches to sustainability in various perspectives reflected their high levels of awareness about the term. However, relating these different sustainability approaches with landscape design was found to be poor. Students offered varied types of gardens and green areas but never discussed how these would connect to the ecosystem of or green areas in the city. Zhang et al. (2011:158) stated that along with its aesthetic dimension, the emphasis on nature is much more associated with the promotion of sustainable development. In this context, students' emphasis on nature and greenery conforms to this statement. However, sustainability practices in housing development have resulted in the development of various green strategies, especially for improving environmental performance. The term 'green strategy' is defined as the approach that developers adopted to improve and gain a sustainable competitive advantage by contributing to environmental protection, ecological responsiveness, and social responsibility [26]. Students can learn about these green strategies in landscape design.

The factors determined under the objective conditions component exhibit high levels of similarity with the ones defined in the QOUL. Bonenberg (2015:1721) [33] defined building neighbourly bonds, integration of residents, improvement of safety in the area, improvement of the natural structure, and landscape quality of the housing estate, providing sports and recreational facilities, playgrounds for children, and inclusion of green corridors as social and environmental goals for QOL in housing estates. The group for objective conditions determined in this study corresponded to these definitions. Thus, it can be concluded that students' decisions about spatial components in terms of improving QOL can be evaluated as successful.

The subjective outcomes component defined by students in this study also exhibited great similarities with the previous studies. According to Raposo (2006:55), gated communities offer a haven of peace and quiet and order and pleasure, which stand exactly in the middle of a chaotic and hostile world. Similarly, landscape architecture students generally cited chaos of the city and offered 'Safety, peace, relaxing, tranquility, pleasure etc.' as their designs' psychological outcomes. Görgülü and Kaymaz Koca (2007:3) explained the reasons for urban residents' preferences for luxury housing estates in Turkey as the stress and anxiety caused by overcrowded downtowns. The same relationship was expressed by the students in this study.

The most important study outcome concerned the common criticism of gated communities as fostering social segregation. Students' suggestions about lifestyles never referred to this aspect of gated communities. In Turkey statements of assurance in advertisements generally emphasize 'an idealized exclusive life' or 'a prestigious new life' as opposed to an urban life [12; 16]. Depending on this tendency, developers compete to attract potential home buyers and emphasize a 'being exclusive and prestigious' image in order to market their product [16]. However, we determined that students' problem definitions, design approaches, and/or QOL components' definitions never included such emphasis. Students in this study showed a contrasting tendency by relating open-green areas with socialization and sustainability. On the contrary, spacious, clean, and well protected open spaces that feature different architectural styles made of better construction materials that include well-kept lawns, flowers, and trees present a particularly attractive environment that is reflective of residents' higher socio-economic status [10].

Students emphasis on social attributes through problem definitions of weakening neighbour bonds, and 'objective conditions-subjective outcomes' components indicated that students accepted social interactions and cohesion as important components of QOUL. As Mahmoudi Farahani and Lozanovska (2014:224) described, the physical environment is able to influence the feeling of sense of community either directly or indirectly through increasing the chance of interactions among residents. According to Aiello et al. (2010:273), neighbourhood attachment mainly depends on social relationships and an affective bond with the neighbourhood based on the presence of stimulating places and absence of gloomy ones. Yılmaz et al. (2017)[34] emphasized how green areas positively contribute to social well-being along with physical and mental well-being. Green areas represent the main source of pleasant places in the residential environment (Aiello et al., 2010:273). By making outdoor residential spaces more attractive, natural elements (trees and grass) may draw residents to these spaces. Coley et al. (1997:486) showed that trees especially are an important variable in creating sociopetal outdoor spaces-spaces that attract people to them. Thus, by increasing face-to-face contact among neighbours, and enhancing residents' social functioning, green spaces contribute to the community's social cohesion [35]. Also, Park and Kim (2017:89) revealed that landscape composition, especially tree covers in a subdivision, had an impact on housing prices.

6 Conclusion

Analyzing landscape architecture students' problems with definitions of and design approaches to housing estate landscape design allowed understanding of their awareness levels and deficiencies about urbanization process and urban life problems. Additionally, how students related problem solutions to landscape design, and the spatial attributes to which they paid attention or neglected were also determined. Students discussed various urban life problems that corresponded to different dimensions such as physical, psychological, and social. However, when the complexity of the problems in urban areas was taken into consideration, limitations in students' definitions and the need for being informed about these subjects become clear. For example, students avoided relating landscape design with infrastructure (water, energy, sewerage system, and drainage) and never discussed how to contribute positively to issues such as waste disposal, solid waste management, energy consuming, or saving in their landscape designs. Anyway, since the solution to such problems requires multidisciplinary teamwork, and the students develop their designs individually in a limited time period, such deficiencies can be regarded as normal. However, the need for information and courses in landscape architecture curricula about urbanization processes and the problems they create, how the solution of these are technically related with landscape design, and the duties of landscape architects when they are a part of a multidisciplinary team must be voiced.

The analyses of landscape architecture students' design approaches to a gated community landscape design indicated that they paid attention to urban life problems such as chaos and stress of the city, decreasing green areas, unsustainability, and weakening neighbour bonds. These analyses revealed the need for curriculum interventions in order to enhance their knowledge about urbanization processes, its problems, and approaches to solutions. Sustainability was a common solution approach among students; their various perspectives to this term such as sustainability of natural resources, traditional life style, or personal development were accepted as reflection of their awareness. However, they can be informed in detail about green strategies, thus strengthening the relationship between sustainability term and landscape design can be achieved. Another important finding of the study revealed the congruence between defined 'objective conditions-subjective outcomes' components, was especially emphasized in socialization by students and other studies. This study showed students' tendency about using landscape design as a tool for achieving sustainability and socialization as a solution to urbanization problems.

References

1. Krstinic Nizic, M., and Baresa, S. (2013). The role of cities in protecting environmental quality. *UTMS Journal of Economics*, *4*(3), 325–339.
2. Marans, R.W. (2012). Quality of Urban Life Studies: An Overview and Implications for Environment-Behaviour Research. *Procedia-Social and Behavioral Sciences*, *35*, 9–22.
3. Moser, G. (2009). Quality of life and sustainability: Toward person–environment congruity. *Journal of Environmental Psychology*, *29*, 351–357.
4. Moser, G., and Robin, M. (2006). Environmental annoyances: an urban-specific threat to quality of life? *Revue européenne de psychologie appliquée*, *56*, 35–41.
5. Park, Y., and Kim, H.W. (2017). The cross-level impact of landscape patterns on housing premiums in micro-neighborhoods. *Urban Forestry & Urban Greening*, *24*, 80–91.
6. Mohd Shahli, F., Mohd Hussain, M. R., Tukiman, I., Zaidin, N. (2014). The Importance Aspects of Landscape Design on Housing Development in Urban Areas. *APCBEE Procedia*, *10*, 311–315.
7. Akyol, D., Özkan, D.G., and Dedeoğlu Özkan, S. (2019). Measuring The Quality of Urban Life in The Trabzon Province. *Fresenius Environmental Bulletin*, *28*, 7991–8000.
8. Gause, J. et al. (2007). *Developing Sustainable Planned Communities.* Urban Land Institute, Washington D.C.

9. Mohd Hussain, M.R., Tukiman, I., Zen, I. Hj., and Mohd Shahli, F. (2014). The Impact of Landscape Design on House Prices and Values in Residential Development in Urban Areas. *APCBEE Procedia, 10,* 316–320.
10. Zhang, L. (2010). *In Search of Paradise: Middle-Class Living in a Chinese Metropolis.* Cornell University Press, Ithaca, NY.
11. Fleischer, F. (2007). "To Choose a House Means to Choose a Lifestyle" The Consumption of Housing and Class-Structuration in Urban China. *City & Society, 19* (2), 287–311.
12. Görgülü, T., and Kaymaz Koca, S. (2007). Türkiye'de Barınma Biçimlerinde Yaşanan Değişimler: Son Dönemde Yapılan Tüketim Odaklı Konutlar. *Mimarlık, 337,* 1–4.
13. Alpak, E.M., Özkan, D.G., Mumcu, S., and Özbilen, A. (2016). Complexity Historicity and Coherence Preference and Quality of the Changes in the Urban Scene. *Anthropologist, 24,* 262–277.
14. Geniş, Ş. (2007). Producing Elite Localities: The Rise of Gated Communities in Istanbul. *Urban Studies, 44*(4), 771–798.
15. Keleş, R. (2012). The Quality of Life and the Environment. *Procedia-Social and Behavioral Sciences, 35,* 23–32.
16. Pérouse, J.F., and Daniş, A.D. (2005). Zenginliğin mekânda yeni yansımaları: İstanbul'da güvenlikli siteler. *Toplum ve Bilim, 104,* 92–123.
17. Reith, A., and Orova, M. (2015). Do green neighbourhood ratings cover sustainability? *Ecological Indicators, 48,* 660–672.
18. Tanülkü, B. (2013). Gated Communities: Ideal Packages or Proeessual Spaces of Conflict? *Housing Studies, 28*(7), 937–959.
19. Van Kamp, I., Leidelmeijer, K., Marsman, G., and De Hollander, A. (2003). Urban environmental quality and human well-being Towards a conceptual framework and demarcation of concepts; a literature study. *Landscape and Urban Planning, 65,* 5–18.
20. Cséfalvay, Z. (2011). Gated Communities for Security or Prestige? A Public Choice Approach and the Case of Budapest. *International Journal of Urban and Regional Research, 35*(4), 735–52.
21. Hook, D., and Vrdoljak, M. (2002). Gated communities, heterotopia and a ''rights'' of privilege: a 'heterotopology' of the South African security-park. *Geoforum, 33,* 195–219.
22. Raposo, R. (2006). Gated communities, commodification and aestheticization: The case of the Lisbon metropolitan area. *GeoJournal, 66,* 43–56.
23. Güzey, Ö. (2014). Neoliberal urbanism restructuring the city of Ankara: Gated communities as a new life style in a suburban settlement. *Cities, 36,* 93–106.

24. Gültekin, A.T. (2015). Konut Kullanıcılarının Kentsel ve Sosyal Yaşam Kalitesi Farkındalığı: Ankara/Çayyolu-Prof. Dr. Ahmet Taner Kışlalı Mahallesi Örneği. *Ankara Araştırmaları Dergisi, 3*(2),165–182
25. Gadecki, J., and Smigiel, C. (2009). A Paradise Behind Gates and Walls. Gated Communities in Eastern Europe and The Promise of Happiness. In A. Bartetzky, M. Schalenberg (Eds.), *Urban Planning and The Pursuit of Happiness European Variations on a Universal Theme* (18th—21 St Centuries), Jovis, Berlin, pp.198–217.
26. Zhang, X., Shen, L., and Wu, Y. (2011). Green strategy for gaining competitive advantage in housing development: A China study. *Journal of Cleaner Production, 19*, 157–167.
27. Mahmoudi Farahani, L., and Lozanovska, M. (2014). A Framework for Exploring the Sense of Community and Social Life in Residential Environments. *Archnet-IJAR, 8*(3), 223–237.
28. Aiello, A., Ardone, R.G., and Scopelliti, M. (2010). Neighbourhood planning improvement: Physical attributes, cognitive and affective evaluation and activities in two neighbourhoods in Rome. *Evaluation and Program Planning, 33*, 264–275.
29. Ibem, E.O., and Aduwo, E.B. (2013). Assessment of residential satisfaction in public housing in Ogun State, Nigeria. *Habitat International, 40*, 163–175.
30. Sommer, R., and Sommer, B. (2002). *A practical Guide to Behavioral Research; Tools and Techniques.* New York, Oxford University Press.
31. URL-1 http://www.ornaparkresidence.com/ozellikler/ Access date: 20 February 2018
32. URL-2 http://www.ornaparkresidence.com/galeri/ Access date: 20 February 2018
33. Bonenberg, W. (2015). Public space in the residential areas: the method of social-spatial analysis. *Procedia Manufacturing, 3*, 1720–1727.
34. Yılmaz, S., Düzenli, T., and Dinçer, D. (2017) Evaluation of factors related to well-being effects of urban green spaces on users. *Fresenius Environmental Bulletin, 26*(12), 7789–7800.
35. Coley, R.L., Sullivan, W.C., and Kuo, F.E. (1997). Where Does Community Grow? The Social Context Created by Nature in Urban Public Housing. *Environment and Behavior, 29*, 468–494.

Alev P. GÜRBEY and Gülden KARABUDAK

Forest Bathing as an Eco-Wellness Tourism Activity

Abstract: Human survival is directly tied to our relationship with the natural environment. During the last century, research has been increasingly drawn toward understanding the human-nature relationship and has revealed the many ways humans are linked with the natural environment. Natural outdoor areas and natural features such as forests, parks, trees, and gardens provide opportunities to enhance public health and wellbeing. Today, forest bathing is considered as a form of eco wellness tourism and a popular health promotion approach by being one of the ecotherapy activities by including seven eco wellness factors (Physical Access, Sensory Access, Connection, Protection, Preservation, Spirituality, and Community Connectedness). In this study's scope, a representative online survey was done with 'Certified Forest Bathing Guides and Practitioners' from different countries to reveal the guided forest therapy tours' framework.

Keywords: human-nature relationship, nature-based concepts, eco wellness, ecotherapy, tourism, forest bathing guides.

1 Introduction

As the world population increases, anthropogenic impact on nature has increased in line with rising settlement development, industry and production and food and natural resource supply. People who work extensively to meet these needs are often socially isolated from the concept of being a part of nature and in some cases may experience definable physical and mental health disorders as a result. Interaction with nature is critically important to human wellbeing and development but sadly has become compromised and diminished in modern times [1]. Opting for healthy human and ecological communities is a decision we can make that will require us to find new ways to live and to run our economies. Placing the human economy above the natural world's wellbeing creates a destructive unsustainable relationship [2].

In the late nineteenth-century, the concept of environmental protection was born with the slogan "drive back to nature" and catalyzed a reaction to the growing industrialization and urbanization era [3,4]. This trend has led to the emergence of some new concepts such as sustainability more recently.

Human survival is directly tied to our relationship with the natural environment [5]. During the last century, research has been increasingly drawn toward understanding the human-nature relationship and has revealed the many ways humans are linked with the natural environment. Some examples of these include human preference for scenes dominated by natural elements, the sustainability of natural resources, and the health benefits of engaging with nature [6]. Natural outdoor areas and natural features such as forests, parks, trees and gardens are known to provide opportunities to enhance public health and wellbeing [7].

Since the development of the concept of nature as a therapy in the 1990s, several studies using various methodologies have been conducted to examine the forested environments' effect on health promotion and wellbeing [8]. As one of the largest nature components, forests represent an enormous source of tree and plant material with known or potential medicinal or nutritional value. Besides, forests might have an essential but indirect part in the local economy, such as a background to ecotourism [9]. Being a multi-sensual environment, forests perfectly fit an operational scope for a wellness service product for ecotourism because of its invigorating effects [10].

Forest therapy tourism which can be classified under ecotourism or forest-based wellbeing tourism provides healing opportunities for people to restore mental and physical wellbeing through nature [11,12].

2 Nature-Based Concepts in the Human-Nature Relationship

Nature is a separate variable with all its components the other hand, human-being is entirely a different variable both take place in the world. Human-nature variables interact in different conditions every day at different rates. In line with these interaction possibilities, various concepts have emerged and will continue to emerge. Living in the days of COVID-19 showed that the emergence of an unexpected and challenging virus might affect the human-nature relationship. Gürbey (2020) mentioned that COVID-19, as a kind of social shock, causes physical and psychological harm to people in the short, medium and long term [13]. There are likely to be many other far-reaching and varied impacts of our mismanagement of the natural environment in the future too. We wanted to touch upon the concepts that we think are closely related to our research under these current conditions and how their impacts on the individual could be managed sustainably.

Psychoterratic

Emotions are primary and primordial forces that motivate us to action. The word "emotion" is derived from the Latin "ēmovēre", to disturb, at the root of which is "movēre", to move. Emotional experiences tied to the state of the earth have long been discussed, depicted, praised, and lamented in art, poetry, and literature. However, few clear concepts have been set to describe mental distress and physical illness as subject of environmental degradation. Australian sustainability professor Glenn Albrecht has created two diagnostic categories: psychoterratic and somaterratic health and illness. These make the connection between the state of the earth (terra) and mental (psyche) and bodily (somatic) health [14].

Psychoterratic disease arises from a negative relationship to our home environment, be it at local, regional or global scales. The negative relationship involves a loss of identity, loss of an endemic sense of place and a decline in wellbeing. Conversely, an enduring and positive relationship to a loving home environment delivers the benefits of a robust endemic sense of place and wellbeing. Negative relationships with the environment lead to negative emotions, and positive relationships also create negative emotions. Somaterratic illness arises from the direct toxic pollution of the environment (arsenic in water) or harmful changes to life's biophysical conditions, such as heat stress from anthropogenic global warming [15].

Humans are now by far the most powerful change agent on the planet. Their impacts are fundamentally transforming the face of the physical landscape of the earth and altering natural patterns and rhythms, including the climate. Accordingly, the main concepts defined here are physical and mental health syndromes related to the earth due to environmental and climate change in this anthropogenic period when environmental pressures increase [16].

The concept we focus on here is "psychoterratic": mental health problems related to the world are caused by environmental change that is negatively felt and perceived. On the one hand, human beings have been trying to cope with the intense stress of modern life, and on the other hand, they have become more and more distant from nature every day. Disconnection from nature can adversely affect human mental health. It is noted that the essential medicine for this ailment is to reconnect with nature and take forest walks at regular intervals. There are plenty of ways to taking nature as medicine: contact with the soil, a visit to the park, a quiet forest walk and more. To put it in a single sentence, we can describe the trauma experienced due to staying away from nature using the term psychoterratica. So, the more often we connect with nature, the greater our chances of maintaining our physical and mental health.

Albrecht (2017) has set the "psychoterratic typology" that lists the emotions that occur due to the positive and negative relationships established with the environment as in Tab. 1;

Tab. 1: Albreht's Psychoterratic Typology [17]

The Psychoterratic Typology			
Negative Emotions	Origin	Positive Emotions	Origin
Biophobia	Kellert & Wilson 1995	Biophilia	Wilson 1984
Ecoagnosy	Albrecht 2017	Ecoliteracy	Orr 1991
Ecoamnesia	Kahn 1999	Sumbiophilia	Albrecht 2018
Ecoanxiety	Leff 1990	Eutierria	Albrecht 2010
Ecocide	Galston 1970s	Terraliben	Albrecht 2018
Eco-Necrophilia	Fromm 1965	Biophilia	Fromm 1965
Ecoparalysis	Rees 2007	Soliphilia	Albrecht 2009
Ecophobia	Sobel 1995	Ecophilia	Sobel 1995
Global Dread	Albrecht 2003	Soliphilia	Albrecht 2009
Meteoranxiety	Albrecht 2014	-	
Nature-Deficit Disorder	Louv 2005	Eutierria	Albrecht 2010
Nostalgia	Hoffer 1688	Endemophilia	Albrecht 2010
Solastalgia	Albrecht 2003	Topophilia	Tuan 1974
Terrafurie	Albrecht 2017	-	
Tierracide	Albrecht 2017	Ecoliben	Albrecht 2017
Tierratrauma	Albrecht 2013	Eutierria	Albrecht 2010
Topoaversion	Albrecht 2013	Topophilia	Tuan 1974
Toponesia	Heneghan 2013	-	

Biophilia

Many researchers have studied social, psychological, and physiological benefits of both active and passive interactions with natural environments for years [18]. According to Souter-Brown (2015), contact with nature is an effective "upstream health promotion" tool for whole populations [19]. Understanding how natural systems affect human physical and mental wellbeing can be explained by the concept of biophilia underlying much of this examination of humans and nature [1]. Human beings' innate tendency to prefer natural places and sounds, and to be curious about, attracted to, or at least attentive of, other living creatures can also be explained within the framework of this concept [20].

In the case of biophilia, we can either choose to engage our inherent tendencies to affiliate with nature or separate from and impoverish our connections to the natural world. Unfortunately, modern society has erected many obstacles to the beneficial experience of nature. Most problematic is an increasing disconnect from the natural world, often viewed as merely a resource to be exploited or a nice but not necessary recreational amenity [21].

Solastalgia

Solastalgia is a new concept developed to provide further meaning and clarity to environmentally-induced distress instead of nostalgia. When separated from a loved home, solastalgia, the melancholia or homesickness experienced by individuals, is the distress produced by environmental change impacting people. At the same time, they are directly connected to their home environment. Solastalgia is the pain or sickness caused by the loss or lack of solace and the sense of isolation connected to the present state of one's home and territory [14].

Muller (2020) explains the term as "the environment's chaotic transformation is damaging many of our favourite places—and causing a shared homesickness [22].

Solastalgia manifests in the loss of the sense of belonging to a particular place and a feeling of desolation about its disappearance. It is an intense desire for the place where one is a resident to be returned to a state that gives solace. The diagnosis of solastalgia is based on recognizing the distress within an individual or a community about the loss of endemic sense of place and the loss of a sense of control of its destiny. The positive prescription for solastalgia is personal and community involvement in the protection, restoration and rehabilitation of their place bioregion/country and the return of an endemic sense of place in both individuals and communities [23].

Unfortunately, most of the world's communities suffer from this ailment. The most important remedy to get rid of this ailment is the rehabilitation of damaged forests that have been impacted by such things as the construction of resource-extraction sites, new modern buildings and intensive technology needs. Additionally, the restoration of natural coastal areas plundered for tourism, protection of thousands of extinct animal and plant species and engagement in community education help to reduce solastalgia in individuals. The cure for the pain and the healing of desolation's intense feeling is also in nature.

Ecoagnosy

Ecoagnosy is the ignorance of ecology (agnosy is a synonym for ignorance). We experience less and know less about ecology in nature with each generation. Hence we cannot forget that which we have never known. At this point, maybe another term is needed to capture the essence of echoagnosia. Ecoamnasy can, perhaps, be reversed within some people of a certain age with a re-awakening of childhood memories of a once verdant and rich natural world, however, 'ecoagnosy' or ignorance of ecology leads to nothing but a path to extinction [24, 25].

Therefore, our children are suffering from ecoagnosy, a socially induced form of ecoretrogression. The disturbing idea is that the current generation is less ecologically literate, attuned, aware and emotional than previous generations. Consequently, they are unable to respond to the enormous risks posed by ecosystem distress and climate change. The most dangerous facet of ecoagnosy is that those who suffer from it often have no idea that it afflicts them [24, 25].

Topoaversion

Topoaversion is the feeling that you do not wish to return to a place that you once loved and enjoyed when you know that it has been irrevocably changed for the worse. It is not topophobia where you can fear a place while entering it; topoaversion is a strong enough feeling to keep you from ever returning to visit the place that was once beloved. The concept has its origins in topos (place) and aversion (to turn away) As the pace of development quickens, topoaversion is likely to increase in many people as a felt emotional response to the changes that take place. It is somewhat ironic, as unique places on earth become major tourism and ecotourism destinations, that management of the impacts of increased visitation demands changes to the way people and their needs are met [25].

Eutierria

Eutierria is a good and positive feeling of oneness with the earth and its life forces where the boundaries between self and the rest of nature are obliterated. A deep sense of peace and connectedness pervades consciousness It arises when "the human-nature relationship is spontaneous and mutually enriching (symbiotic)". The prefix "eu" comes from the Greek word for "good"; the root "tierra" means Earth [25, 26].

It is one of the essential terms collected and invented by Glenn Albrecht that may help reorient modernity's approach to the rest of nature.

Eutierria refers to secular experiences but echoes the "oceanic" feeling identified in various world religious traditions. When it occurs, your perception of the boundaries between yourself and all else—the thoughts and feelings setting you off from the rest of the cosmos—seem to evaporate. The distinction between you and nature (or in the religious versions nature and God) breaks down. You become one with the universe. A reassuring sense of harmony and connection with the world infuses your consciousness. It is an experience that matches up with the knowledge of your dependence on and connection to the world. By placing the individual at the centre of concern rather than the whole community of life, modernity can create loneliness. Seeking out eutierria can help heal the divide [27].

Tierratrauma

When people possess a deep connection to the earth, they can experience deep emotional trauma when powerful forces directly impact that connection. Tierratrauma from (Tierra, earth and trauma) describes the moment when a person experiences a sudden negative environmental change [25]. Your favourite tree is being cut down, fire is destroying your local area, a bulldozer is demolishing your beloved streetscape, or you are witness to an oil spill that smothers all life on "your" beach. This is not medically defined post-traumatic stress disorder (PTSD), this is not solastalgia derived from chronic change, this is acute earth-based existential trauma. We needed a concept fort the commonly-felt, gut-wrenching mental anguish. Tierratrauma can be contrasted to eutierria [28].

Both solastalgia and tierratrauma are sure to proliferate as climate-related disasters increase with climate change. Fire, drought, storms, and floods increase in intensity and frequency; this trend will likely continue for the foreseeable future. Because of the temporal displacement of the impacts of the pollution humans are emitting today, future generations are guaranteed to experience a greater degree of solastalgia and tierratrauma in their lives then we are in the present day [25].

These words are essential to have because they help describe the relationship we, as humans, have with other people, animals, and the earth as a whole. At the same time, we must remain aware of our separateness from each of these so they can live out their lives in their wholeness while we do with ours. It is a state of being separate but one, individuals and a community, independent but together [29].

3 Ecotherapy and Eco Wellness

Ecotherapy

Ecotherapy is a subset of ecopsychology. Ecopsychology is the study of the relationship between the human psyche (body-mind-soul-spirit) and the rest of nature. Ecotherapy (also called applied or clinical ecopsychology) focuses on healing that dangerously dysfunctional relationship. Ecotherapy is an umbrella term for gathering techniques and practices that lead to circles of mutual healing between the human mind and the natural world from which it evolved. As an umbrella term for nature-based physical and psychological healing methods, ecotherapy represents a new form of psychotherapy that acknowledges nature's vital role and addresses the human-nature relationship [30]. Ecotherapy is evidenced by numerous approaches such as green exercise, green views, horticultural therapy, wilderness therapy, body therapy through movement, art therapy and animal-assisted therapy. Sometimes, ecotherapy can be just taking more traditional talk therapy outside into a garden, public space, forest or beach. Ecotherapy often incorporates elements of mindfulness practices [31].

Ecotherapy differs from psychotherapy in its focus on transforming our relationship to the natural world and its reliance on non-risky and non-intrusive interventions. A certificate in ecotherapy is not a license to do psychotherapy. Nevertheless, ecotherapy techniques are being taught to practising psychotherapists, whose concentration on mending relationships and inner conflicts benefit from placement in the broader ecological context in which all human activity unfolds [32].

The types of interventions used are many in ecotherapy. Activities can be carried out individually or with the guidance of a therapist. Some of the more common ecotherapy activities can be listed as horticultural therapy, forest bathing, physical exercise in a natural environment, and conservation activities [33, 34].

As mentioned above, forest bathing is one of the activities of ecotherapy. By activating the senses, the forest bath walk, which allows participants to communicate with their environment and themselves, can calm the nervous system, stop fighting or soften rumination and mood disorders. Recent studies have proven that forest bathing reduces stress hormone production, improves feelings of happiness, free up creativity, lower heart rate and blood pressure, boosts the immune system, and accelerates recovery from illness. All these pieces of evidence substantially support the ecotherapy approach as an adjunct treatment element.

Ecowellness

In 2012, Reese and Myers introduced the construct of EcoWellness to address the lack of literature linking nature and holistic wellness models and outcomes [35]. The EcoWellness construct initially included three dimensions, exposure to natural environments (Access to nature) contributes to the development of one's sense of self (environmental identity) and both spiritual wellbeing and community connectedness (transcendence). Although the three components of EcoWellness have been described individually, they are related. The concept EcoWellness was later broadened to include the following seven factors after additional research [36].

1. ***Physical Access*** is the ability to physically interact in or with what the individual considers to be nature.
2. ***Sensory Access*** is the ability to touch, smell, see or hear nature in the absence of physical nature contact.
3. ***The connection*** comprises pleasant cognitions and emotions elicited by one's relationship with nature.
4. ***Protection*** (also termed nature self-efficacy) is feeling effective when navigating natural settings and having a sense for what might contribute to one's survival and enjoyment when in or with nature.
5. ***Preservation*** (also referred to as environmental agency) acts on behalf of or advocating for the natural world, including recycling or supporting an environmental cause.
6. ***Spirituality*** is feeling connected with one's conception of a higher power and/or life-guiding principles elicited through one's connection with nature. Finally,
7. ***Community Connectedness*** is feeling connected with others in nature.

4 Practice of Forest Bathing within the Eco Wellness Tourism

The Agency of Agriculture, Forestry and Fisheries of Japan (MAFF) introduced Forest Bathing, approximately 40 years ago, to allow the Japanese to heal through nature. Forest therapy has become an emerging therapeutic approach used by psychologists, ecopsychologists, biologists, physicians, and healthcare workers globally. They draw on interdisciplinary research from a range of disciplines that

Fig. 1. Forest Bathing Sessions from Turkey (From Gülden Karabudak)

Fig. 2. Forest Bathing Sessions from Turkey (From Gülden Karabudak)

describe this form of use for forests and experiences to heighten the understanding, importance, and benefits of contiguous natural areas and engage in such passive forms of nature-based tourism and recreation. Forest Bathing has also become a unique brand which links forest visits to health and wellness-oriented ecotourism (Figure 1, 2)[12].

The definition of wellness dates back to the 1960s. The term was developed by combining the words "wellbeing" and "fitness". According to Smith and Puczko (2014), wellbeing is related to happiness and satisfaction, whereas wellness deals with all life dimensions. Wellness can be defined as "the bridge between medical and well-being"[37]. Balance and spirituality are two core dimensions of wellness [38]. It is crucial in wellness and wellness tourism that a deliberate and proactive approach is taken to improving the different domains of one's life to bring them into balance [37].

The term "wellness" in the tourism market was expended in holistic and spiritual ways by different researchers. One of the most comprehensive definitions of wellness tourism is made by Sheldon and Bushell (2009) as a holistic travel model that integrates physical health, beauty, longevity, and spiritual awareness and a connection with community, nature, or the divine mystery [38, 39].

Medina-Muñoz & Medina-Muñoz(2013) mentioned that wellness tourism had become a fashion label for tourist products for some reasons which include concern about physical, mental and social wellbeing and the necessity to cope with stress, and the need to get away from the routine of everyday life [40, 41].

Nowadays, the sectors of wellness, medical and health tourism are growing significantly, at the same time changing rapidly and profoundly, thus creating space for new and innovative types of products and destinations [42].

In forest-based wellness tourism, the forests and the forest environment are the core resources of the product. The forest-based wellness tourism product can include utilizing natural resources of the forests, physical and relaxation/stress relief activities in the forest, and educational activities [43].

Forest therapy is considered as a form of wellness tourism and a popular approach for health promotion. Three different therapy programs are applied to benefit from the therapeutic effects of the forest. These programs can be categorized under guided, self-guided or strolling alone. The influence of experiential types of forest therapy program on physiological responses and mood states may differ. The guided forest therapy program's effect on mood states improvement mostly and strolling alone increases parasympathetic nerve activity, suggesting a higher power of being in a resting state [44].

Eco wellness provides a practical lens through which counsellors can determine the relevance of implementing nature-based counselling interventions with the diverse populations they serve [45]. Therefore to evaluate guided forest bathing programs under eco wellness tourism will be more appropriate.

Careful assessment of how three dimensions of eco wellness interact with and affect clients' holistic wellbeing can provide practitioners with many powerful eco wellness counselling strategies [45].

Today, there are some societies and institutions which offers forest bathing or therapy guides training. These are; Forest Therapy Society (FTS), The Association of Nature and Forest Therapy (ANFT), Global Institute of Forest Therapy (GIFT), International Nature, Forest Therapy Alliance (INFTA) and The Forest Therapy Institute (FTI).

Forest Therapy Society (FTS) specializing in dealing with forest therapy programs, was established in 2007. The society was created to support practical forest therapy activities. Using the "Forest Therapy Base" as a domain, they also provide training certified forest therapist" and "forest therapy guides [46].

The Association of Nature and Forest Therapy (ANFT) was founded in 2012 in the USA. They provide training for forest therapy guides [47].

Global Institute of Forest Therapy (GIFT) was established as the Canadian Chapter of the Association of Nature and Forest Therapy in 2015, in Canada. In 2018, they became an independent organization supporting healthcare practitioners and other trainees to become certified guides in forest therapy ventures [48].

International Nature, Forest Therapy Alliance (INFTA) was founded in 2017 in Australia to support members of the association in their endeavours as certified forest therapy guides [49].

European and International versions of Forest Therapy Institute (FTI) were established in 2019 to support and train forest bathing guides and forest therapy practitioners [50].

5 Case Study with Forest Bathing Guides and Practitioners

In this study's scope, a representative online survey was done with sixty-seven (67) 'Certified Forest Bathing Guides and Practitioners' from different countries. Since we could not find any precise data about the number of Forest Bathing Guides and Practitioners worldwide, a survey was applied to guides and practitioners whom we could contact. We reached the contact information of guides' via institutions' websites (FTS, ANFT, GIFT, INFTA and FTI), social networks (Facebook, Instagram) and digital platforms for tourism (Airbnb, Booking, Trip Advisor). Therefore, this research is a preliminary study and will be studied for further expansion in the future.

According to the information obtained from the online surveys conducted with the 67 participants; 16,4 % of the participants are from Germany, 13,4 % are from Portugal, 13,4 % are from Spain, 11,9 % of the participants are from Canada, 10,4 % are from the USA, 6 % are from Italy, 4,5 % are from England, 3 % are from Holland, and 3 % are from Ireland. The other participants are from Chile, Denmark, France, Hungary, Lithuania, New Zealand, Romania, Scotland, Singapore, Sweden, Taiwan and Turkey with a 1,5 per cent each (Tab. 2).

The Forest Bathing guides' main occupation is mostly in health care, the social sector, teaching and education sectors by 62,6 %. This is followed by employment

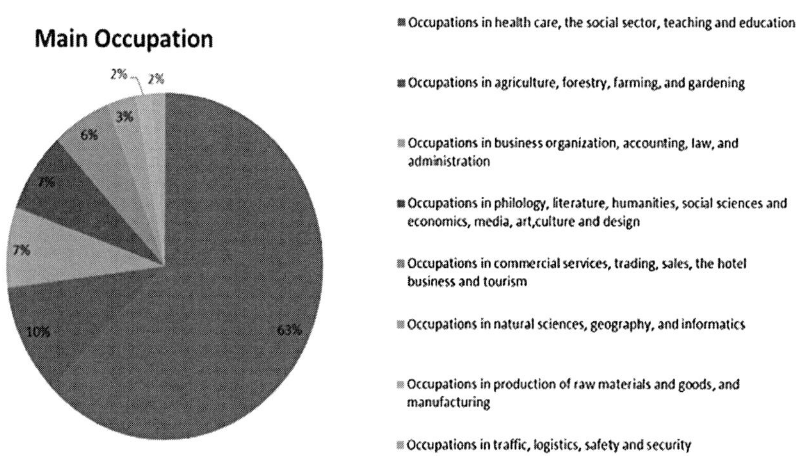

Fig. 3: The main occupation of the forest bathing guides

Tab. 2: Locations' of the forest bathing guides

Country	Region/City	Number	Percentage (%)
Germany	Bad König	1	16,4
	Bommental	1	
	Cologne	1	
	Düsseldorf	2	
	Michelstadt	1	
	Neuwied	1	
	Oberstdorf	1	
	Ratingen	1	
	Stuttgart	1	
	Weyarn	1	
Portugal	Almada	1	13,4
	Lisbon	4	
	Oeiras	1	
	Seia	1	
	Tomar	1	
	unknown	1	
Spain	Arruazu	1	13,4
	Bilbao	2	
	Burgos	1	
	Madrid	1	
	Sevilla	1	
	Terrassa	1	
	Vic	1	
	Vigo	1	
Canada	Georgina	1	11,9
	Kingston	1	
	Kitchener and Tobermory	1	
	Montreal	1	
	Orillia	1	
	Toronto	1	
	Winnipeg	2	
USA	Copiague-NY	1	10,4
	Denver-CO	1	
	Los Angeles- CA	1	
	New York-NY	1	
	Portland-MI	1	
	Seattle-WA	1	
	Woodbine-MD	1	
		44	65,5

Tab. 2 Continued

Country	Region/City	Number	Percentage (%)
Italy	Emilia Romagna region Visso Udine Biella	1 1 1 1	6
England	Leicester London Winchester	1 1 1	4,5
Holland	Amsterdam unknown	1 1	3
Ireland	Dublin	2	3
Chile	Valdivia	1	1,5
Denmark	North Sealand region	1	1,5
France	Bordeaux	1	1,5
Hungary	Piliscsaba	1	1,5
Lithuania	Vilnius	1	1,5
New Zealand	Auckland	1	1,5
Romania	Cluj Napoca	1	1,5
Scotland	Stromness	1	1,5
Singapore	-	1	1,5
Sweden	Stockholm (Lidingö)	1	1,5
Taiwan	Taipei	1	1,5
Turkey	İzmir	1	1,5
		23	34,5
TOTAL		67	100

in agriculture, forestry, farming, and gardening (10,4 %), business organization, accounting, law, and administration (7,5 %), philology, literature, humanities, social sciences and economics, media, art, culture and design (7,5 %), commercial services, trading, sales, the hotel business and tourism (6 %), natural sciences, geography, and informatics (3 %), production of raw materials and goods, and manufacturing (1,5 %) and traffic, logistics, safety and security (1,5 %) (Figure 3).

Most of the participants are working as a forest bathing guide for 1–3 years (49,2 %), 41,8 % of guides for less than one year and 9 % for 4–6 years.

During a year, 28,4 % of forest therapy guides are providing 1–5 sessions. 22,4 % of them providing 6–10 sessions per year, followed by 11–15 sessions (19,4 %), 16–20 sessions (13,4 %) and more than 21 sessions with a 16,4 % rate. Most of the guides mentioned that the number of sessions is less than usual during the COVID-19 pandemic.

The forest bathing sessions are provided in private forest (26,9 %), state forest (25,4 %), urban parks (22,4 %), national parks (13,4 %), farmlands (1,5 %) and national scenic areas (1.5 %). 8,9 % of guides mentioned that the places' status includes different places (besides, rainforests, arboretums, meadows, school grounds and online).

According to the information received about the vegetation type of the areas, forest bathing sessions are mostly in mixed forest (76,1 %). 8,9 % of sessions are in a broadleaved forest, 6 % in coniferous, 7,5 % in other types and 1,5 % have all types of vegetation mentioned above. 67,2 % of these areas have a medium canopy density, 20,9 % have high and %11,9 have low canopy densities.

Dominant tree species in the areas of forest bathing were mostly stated as *Pinus* spp. (43 %), *Quercus* spp. (27 %) and *Fagus* spp. (31 %). Other following species are *Acer* spp., *Abies* spp., *Betula* spp., *Cedrus* spp., *Picea* spp., *Castanea* spp., *Populus* spp., *Larix* spp., *Platanus* spp. and *Drimys winteri*. 7,3 % of the participants mentioned that their session areas have mixed and diverse vegetation and hard to define the dominant species.

Water resources in these areas stated as a stream (35,8 %), river (16,5 %), lake (10,4 %) and waterfall (3 %). 14,9 % of the areas have all water resources, 10,4 % have other types such as the sea, pond, swamp etc. and 9 % have none.

Forest bathing trail lengths also differ; 62,7 % of the guides stated that the minimum trail length is about 0–1 km, 32,8 % is 1,1–3 km, 3 % is 3,1–5 km, and 1,5 % stated that it depends on the area, weather conditions and participants. Maximum trail lengths were stated mostly 1,1–3km (58,2 %) this followed by 3,1–5 km (23,9 %), more than 5 km (11,9 %), 0–1 km (4,5 %) and 1,5 % stated that it depends on the area, weather conditions and participants.

Before COVID-19 pandemic the maximum participant numbers in forest bathing sessions stated mostly as 9–15 persons/session (55,2 %), this followed by 3–8 persons (23,9 %), 16–22 persons (11,9 %), 23–30 persons (6 %) and more than 31 persons (3 %).

According to the complete survey information, forest bathing sessions can be done in different natural areas but mainly in forests with different status. Biodiversity is an essential component by offering differents dimensions to the participants. The dominant tree species, water resources, length of the trails and other conditions may differ according to the area. The session place and the trail's length can also depend on the number and ages' of the participants and the weather conditions. In some cases, the session duration can be more effective than the length of the therapy trail. Some of the guides stated that sessions with a minimum of two hours are most effective on parasympathetic nervous system activity, burnout prevention and the reduction of stress markers. The first forest sessions were described as fascinating by the participants. Participants were also felt relaxed and grateful and appreciated the calming effect of nature on their minds and bodies combined with a deceleration. COVID-19 measures restrict the frequency and participant numbers.

6 Conclusion

With its massive scale and hugely centralized economy, western industrial society can hardly be seen with any particular landscape or ecosystem. Moreover, our culture's relation to the biosphere can in no way be considered a reciprocal or balanced one with thousands of acres of primary forest disappearing rapidly, and hundreds of species of flora and fauna becoming extinct each month., In this light, we can hardly be surprised by the amount of epidemic illness in our culture, from increasingly severe immune dysfunctions and cancers to widespread psychological distress and depression [51].

Green spaces such as forests, woodlands, urban parks, botanical gardens, community gardens, farmlands, and meadows offer unique and creative opportunities for restoring and improving the physical and spiritual, emotional, neurological and psychological aspects of human health and wellbeing [52].

As an emerging field of research surrounding human interactions with nature, Forest Bathing continues to provide substantial evidence of the natural environments' benefits on human health [53]. Park et al. (2010) stated that forest bathing studies results would support developing a research field dedicated to forest medicine, generating new preventive medicine strategies. The field experiments result also provides a platform for interested enterprises, universities, and local governments to promote the effective use of forest resources in stress management, health promotion, rehabilitation, and disease prevention [54].

Experiencing the diversity of forest environments in a multi-sensory way is therapeutic. The pandemic process has shown us that people can experience different traumas in such social shocks. At this point, Forest Bathing is one of the best opportunities to use the therapeutic effects of forests.

Forest bathing is among the Ecotherapy activities, including seven eco wellness factors (Physical Access, Sensory Access, Connection, Protection, Preservation, Spirituality, and Community Connectedness).

Today, there are many forest bathing guides and practitioners that are providing forest bathing sessions worldwide. The sessions can be done through natural environments.

Biodiversity is an essential component by offering different dimensions to the participants. Therefore, participating in forest bathing sessions in different regions and countries may provide various experiences and have different therapeutic effects. In this context, it is possible to travel domestically or internationally to multiple destinations to take a forest bathing session as an eco wellness tourism activity. Although the pandemic process has affected forest bathing session in this light, it could become one of the most sought-after treatment methods post-COVID.

In Japan, forest therapy trails are promoted in municipal websites' tourism/culture sections for many cities or towns. They offer forest bathing tours to visitors with accommodation, hot springs and food-beverage facilities.

Most of the forest therapy and forest bathing societies and institutions continue to have guide training programmes. The more guides mean more forest bathing opportunities and this make guided-forest bathing more accessible. The visitors can find forest bathing guides through societies and institutional websites, social networks and digital platforms for tourism.

Although it is an activity that can be done on your own, a forest bathing session with a guide allows you to benefit more effectively from the forests, because guides will help enable you to find your answer with the right orientation at the right moments.

Acknowledgement

We would like to thank all of the certified Forest Bathing/Therapy Guides whose helped contribute data to this study by answering our survey. May you forever shine your light on the path through the forests.

References

1. Kellert, S. R. *Building for life: designing and understanding the human-nature connection.* 2005, Island Press, 250 p., ISBN 1-55963-721-8
2. Brown, P.G. and Garver, G. *Right Relationship: Building a Whole Earth Economy.* 2008, Berrett-Koehler Publishers, ISBN 9781576757628, San Francisco.
3. Bramwell, A. *Ecology in the 20th Century: A History.* 1989. Yale University Press, New Haven, CT. 292 pages. ISBN: 0-300-04521-2.
4. Dobson, A. *Ecologism.* 1993, Contemporary political ideologies, In Eatwell R.- Wright And., (p. 216–238), London: Printer Publishers.
5. Schultz, P. W. *Inclusion with nature: The psychology of human-nature relations,* 2002. in Schmuck, P. and P. W. Schultz (eds), Psychology of sustainable development, Massachusetts: Kluwer Academic Publishers, pp. 61–78.
6. Seymour V. *The Human-Nature Relationship and Its Impact on Health: A Critical Review.* 2016. Frontiers in public health, **4**, 260. https://doi.org/10.3389/fpubh.2016.00260
7. Nilsson, K., Sanster, M. and Konijnendijk, C.C. *Forests, Trees and Human Health and Well-being: Introduction.* 2011. pp. 1–19, Nilsson, K., Sangster, M., Gallis, C., Hartig, T., de Vries, S., Seeland, K., & Schipperijn, J. (Eds.). Forests, trees and human health, Springer Science + Business Media. https://doi.org/10.1007/978-90-481-9806-1
8. Oh, B., Lee, K. J., Zaslawski, C., Yeung, A., Rosenthal, D., Larkey, L., & Back, M., *Health and wellbeing benefits of spending time in forests: systematic review.* 2017. Environmental health and preventive medicine, **22**(1), 71. https://doi.org/10.1186/s12199-017-0677-9.

9. Sanesi, G., Gallis, C. and Kasperidus, H.D. *Urban Forests and Their Ecosystem Services in Relation to Human Health*, 2011. Nilsson, K., Sangster, M., Gallis, C., Hartig, T., de Vries, S., Seeland, K., & Schipperijn, J. (Eds.). Forests, trees and human health, Springer Science + Business Media, pp.23–40. https://doi.org/10.1007/978-90-481-9806-1.
10. Johanna, I. and Sami, N. *Forest- and Well-Being Services for Tourists – Case Nuuksio.* 2011. Master's Thesis, Laurea University of Applied Sciences, Finland.
11. Komppula R. and Konu H. *Designing forest-based wellbeing tourism services for Japanese customers- A case study from Finland.* 2017. Nina K. Prebensen, Joseph S.Chen and Muzaffer S. Uysal (Eds.). Co-Creation in Tourist Experiences. Routledge, London and New York. pp. 50–63.
12. Gürbey, A.P., Runlovski, M.N., Işık, I. *Forest Therapy (Shinrin-Ryoho) and Forest Bathing (Shinrin-Yoku): New Function Proposals For Protected Areas.* 2020. Advances in Scientific Research: Engineering and Architecture, St. Kliment Ohridski University Press, Sofia. 237–256, ISBN: 978-954-07-5047-7.
13. Gürbey, A.P. *New Trends in Ecotourism: Forest Bathing/Forest Therapy in the World and Turkey.* 2020. 1st International Forestry and Nature Tourism Congress 2020 (INFONT 2020), 25–27 November 2020, Kastamonu, Turkey.
14. Albrecht, G., Sartore, G.M., Connor, L., Higginbotham, N., Freeman, S., Kelly, B., Stain, H., Tonna, A. and Pollard, G. *Solastalgia: the distress caused by environmental change.* 2007. Australasian psychiatry, **15** (sup1), pp.S95-S98, doi: 10.1080/10398560701701288.
15. Albrecht, G. *Psychoterratic and Somaterratic health and dis-ease.* 2018. Psychoterratica. https://glennaalbrecht.com/2018/07/20/psychoterratic-and-somaterratic-health-and-dis-ease/
16. Albrecht, G. (2011) *Chronic environmental change: Emerging' psychoterratic' syndromes.* 2011. In: Weissbecker, I., (ed.) Climate Change and Human Well-Being: Global Challenges and Opportunities. Springer, pp. 43–56.
17. Albrecht, G. *The Psychoterratic Typology*, 2017a. Psychoterratica. https://glennaalbrecht.com/2017/10/15/the-psychoterratic-typology/
18. Ulrich, R.S. *Biophilia, biophobia, and natural landscapes*, 1993. in S.R. Kellertand E.O. Wilson (eds.), The Biophilia Hypothesis, Island Press, Washington DC, pp. 73–137
19. Souter-Brown, G. *Landscape and Urban Design for Health and Well-Being: Using Healing, Sensory, Therapeutic Gardens.* 2015. Routledge, ISBN 978-0-415-84351-5.
20. Mencagli M. and Nieri M. *The Secret Therapy of Trees: Harness the Healing Energy of Forest Bathing and Natural Landscapes.* 2019. Rodale, 212 p., ISBN 9781984824141, New York.

21. Kellert, S. and Calabrese, E. *The Practice of Biophilic Design*. 2015. www.biophilic-design.com
22. Muller, P. *As climate change alters beloved landscapes, we feel the loss*. 2020. National Geographic, April 2020. https://www.nationalgeographic.com/magazine/2020/04/climate-change-alters-beloved-landscapes-we-experience-solastalgia-feature/
23. Higginbotham, N. and Freeman, S. *Do Open Cuts Ever Heal? Relating Human Distress and Environmental Change*, 2003. Joint Medical and Health Sciences Newsletter 118, Faculty of Health, The University of Newcastle, ISSN 1442-8040, Australia. https://glennaalbrecht.files.wordpress.com/2019/08/open-cuts-solastalgia-2003.pdf
24. Albrecht, G. *Ecoagnosy*. 2017b. Psychoterratica. https://glennaalbrecht.wordpress.com/2017/07/27/ecoagnosy/.
25. Albrecht, G. *Earth Emotions: New Words for a New World*. 2019. Cornell Univesity Press, ISBN 9781501715242 pp.75–76.
26. Albrecht, G. Eutierria. 2018. Psychoterratica. https://glennaalbrecht.com/2018/02/20/eutierria/
27. Worthy, K. *Eutierria: Becoming One with Nature*, 2016. Psychology Today. https://www.psychologytoday.com/intl/blog/the-green-mind/201607/eutierria-becoming-one-nature)
28. Albrecht, G. *Solastalgia and the new Mourning, Mourning Nature: Hope at the Heart of Ecological Loss and Grief*. 2017. ed. Ashlee Cunsolo and Karen Landman, McGill-Queen's University Press, ISBN:978-0-7735-4936-4, Canada.
29. Eutierria. 2020. http://www.eutierria.org/our-story.html
30. Buzzell L., and Chalquist C. *Ecotherapy: Healing with Nature in mind*. 2009. San Francisco: [Berkeley, Calif.]: Sierra Club Books.
31. Summers, J. K., Vivian, D. N. *Ecotherapy – A forgotten ecosystem service: A review*. 2018. Frontiers in Psychology. https://doi.org/10.3389/fpsyg.2018.01389.
32. Ecotherapy News. 2012. International Association for Ecotherapy. http://www.ecotherapyheals.com/summerissuefinal_Layout%201.pdf
33. Chalquist, C. *A look at the ecotherapy research evidence*. 2009. Ecopsychology **1**(2):p. 64–74, Mary Ann Lıebert, ınc.
34. GoodTherapy. *Ecotherapy/Nature Therapy*. 2020. https://www.goodtherapy.org/learn-about-therapy/types/econature-therapy
35. Reese, R. F. *EcoWellness: Construction and validation of the Reese ecowellness inventory*. 2013. (Order No. 3568902). Available from ProQuest Dissertations & Theses Global; Publicly Available Content Database. (1425268211). Retrieved from https://search.proquest.com/dissertations-theses/ecowellness-construction-validation-reese/docview/1425268211/se-2?accountid=11637

36. Reese, R. F., Myers, J. E., Lewis, T. F., & Willse, J. T. *Construction and initial validation of the Reese ecowellness inventory.* 2015. International Journal for the Advancement of Counselling, **37**(2), 124–142. doi:10.1007/s10447-014-9232-1.

37. Smith, M. and Puczko, L. *Future Trends and Predictions, Health, Tourism and Hospitality: Spas, Wellness and Medical Travel,* 2014. Routledge Pub., p. 203–228, ISBN: 978–0–203–08377–2

38. Kucukusta, D., & Heung, V. C. S. *The problems of developing wellness tourism in China: From supply perspective.* 2012. Journal of China Tourism Research, 8, 146–158.

39. Sheldon, P., & Bushell, R. *Introduction to wellness and tourism.* 2009. In R. Bushell & P. Sheldon (Eds.), Wellness and tourism: Mind, body, spirit, place (pp. 3–18). New York,NY: Cognizant Communication Books.

40. Medina-Muñoz, D. R., & Medina-Muñoz, R. D. *Critical issues in health and wellness tourism: An exploratory study of visitors to wellness centres on Gran Canaria.* 2013. Current Issues in Tourism, **16**, 415–435. doi: 10.1080/13683500.2012.748719

41. Hartwell, H., Fyall, A., Willis, C., Page, S., Ladkin, A. & Hemingway, A. *Progress in tourism and destination wellbeing research.* 2018. Current Issues in Tourism, **21**(16), p. 1830–1892, DOI: 10.1080/13683500.2016.1223609

42. Ferrari S. and Gilli M. *Protected natural areas as innovative health tourism destinations,* 2017. The Routledge Handbook of Health Tourism, p. 419–429, Routledge- Taylor& Francis, USA, ISBN 978-1-138-90983-0.

43. Konu, H. *Developing forest-based wellbeing tourism products by using virtual product testing.* 2015. Anatolia – An International Journal of Tourism and Hospitality Research, **26**(1), p.99–102., http://dx.doi.org/10.1080/13032917.2014.921633

44. Yu, C-P., Chen, H-T. and Chao P-H. *Investigating Restorative Benefits of Wellness Tourism: A Study of Forest Therapy Trip and the Role of Forest Therapy Guide.* 2020. 2020 TTRA International Conference, 16–18 June 2020, Canada.

45. Reese, R. F., & Myers, J. E. *EcoWellness: The missing factor in holistic wellness models.* 2012. Journal of Counseling & Development, **90**, 400–406. doi:10.1002/j.1556-6676.2012.00050.x

46. Ohe, Y.; Ikei, H.; Song, C.; Miyazaki, Y. *Evaluating the relaxation effects of emerging forest-therapy tourism:A multidisciplinary approach.* 2017. Tour. Manag., **62**, 322–334

47. ANFT, 2021. The Association of Nature and Forest Therapy https://www.natureandforesttherapy.org/about/association

48. GIFT, 2021. Global Institute of Forest Therapy. https://www.giftoftheforest.com/

49. Kotte, D. *International Nature and Forest Therapy Alliance (INFTA).* 2019. International Approaches in Forest Therapy, International Handbook of Forest Therapy, Cambridge Scholars Publishing.
50. FTI, 2021. Forest Therapy Institute. https://foresttherapyinstitute.com/who-we-are/
51. Abram, D. *The spell of the sensuous: perception and language in a more-than-human world.* 1996. Vintage Books-Random House, Inc., New York, eISBN: 978-0-307-83055-5, 203p.
52. Townsend, M. and Weerasuriya, R. *Beyond Blue to Green:The Benefits of Contact with Nature for Mental Health and Well-being Project.* 2010. Deakin University of Australia, ISBN 978-0-9581971-6-8 .
53. Terrapin. *The Economics of Biophilia Why Designing With Nature In Mind Makes Financial Sense.* 2012. Terrapin Bright Green. https://www.terrapinbrightgreen.com/report/economics-of-biophilia/
54. Park B.J., Tsunetsugu Y., Kasetani T., Kagawa T., Miyazaki Y. *The physiological effects of Shinrin- Yoku (taking in the forest atmosphere or Forest Bathing): Evidence from field experiments in 24 forests across Japan.* 2010. Environ. Health Prev. Med.; **15**:p. 18–26.

Demet Ülkü GÜLPINAR SEKBAN

Evaluation of the Urban Circulation System in Terms of Pedestrian Use; Kaşüstü/ Trabzon Example

1 Introduction

City; It meets the needs of many individuals with a wide variety of identities and characters, such as accommodation, recreation, entertainment, working and socialization, which is formed as a result of many factors that constantly improve and change their structure and shape for certain reasons, including neighborhood units, infrastructure systems, transportation networks, structures, It is a residential unit with structural systems, spatial textures and green areas. (Dinç, Hamurcu, & Eren, 2018; Gülpınar Sekban, Bekar, & Acar, 2019; Keleş, 1980; Lynch, 2010). The main factor in the formation of the city concept is to ensure that the user meets her daily needs with minimum time and minimum effort, as well as providing the user with maximum time and various opportunities for recreational activities such as resting and having fun (Arü, 1965). For this reason, the city, which forms various layers and spaces according to many sociocultural reasons, has an environmental system that can be defined as a result of occupancy and gaps (Kürkçüoğlu & Ocakçı, 2015). Perceivable or legibility is very important for these environments that make up the city with its size, width, variety and complexity (Kürkçüoğlu & Ocakçı, 2015; Lynch, 2010). Because as cities become places that have lost their environmental system due to irregular construction and uncontrolled use, users have difficulty in perceiving the scale, texture and borders within the city (Kavi, 2003). This prevents the formation of a sense of belonging by creating a feeling of loneliness and strangeness in the user. However, environments that are perceived or readable by the users positively affect the user satisfaction and strengthen the image of the city. The formation of the city image depends on many concrete or abstract factors and components. According to Lynch (2010), there are five basic urban elements that constitute concrete components that affect the readability of a city. The main five urban elements are roads (Lynch, 2010). Roads, ie urban circulation systems; They are the transportation networks that define the identity of the city and establish a connection between the places that meet the needs of the user (such as accommodation, rest, eating and drinking). These networks are in direct relation with the daily activities of the user (Dinç et al., 2018). Accordingly, in order to ensure the continuity of the relationship, urban circulation systems encourage the joining of many connections and systematically include these connections. These connections consist of urban circulation systems such as vehicle roads, pedestrian zones and pedestrian roads actively used by the city user.

1.1 Pedestrian Routes in the Circulation System in the Urban

Users are on the move in order to continue their daily activities, and as a result of this mobility, transportation (Surat & Yaman, 2015), and the formation of infrastructure and structures required for transportation, transportation networks, that is, urban circulation systems, are formed. Urban circulation systems; It covers public spaces that meet the access needs that are at the center of the user's various needs, such as streets, vehicle roads, pedestrian areas or pedestrian roads (Gültekin, 2007) which are the main points of urban life (Akyol & Gülpınar Sekban, 2019). These areas are considered as the smallest units that meet the recreational needs of the city, as well as social spaces where they socialize by chatting (Bakan & Konuk, 1987; Moughtin, 1999). Users can evaluate the spaces they can perceive and read. According to the scope and result of this evaluation, they decide whether that space is of habitable quality or not. Therefore, the most effective urban element of the city used in the perception of a city's identity; are the ways.

Roads establish the relationship between layers and user activities by connecting the surrounding urban fabric. However, the only urban element in which the user intensively absorbs the city image during use is the pedestrian roads that he actively uses; It serves to meet the sociological, physiological and economic needs of the user (Shirvani, 1985). It is accepted as the first sign of the image of the city and the symbol of the city, creating the beginning and end of the user's relationship with the city. Pedestrian roads planned and designed according to the user's needs, environmental variables and urban identity; strengthens the image of the city in a healthy and balanced way. Besides, it provides benefits in traffic management, economic recovery, environmental improvement and social cohesion in the city (Brambilla & Longo, 1977). For this reason, it is necessary to use many criteria and concepts together, meticulously and in balance in the planning of pedestrian roads in the urban circulation system (Rapoport, 1991).

Many factors are at the forefront in determining the user's use of pedestrian roads in the city. These factors are very important for the establishment of pedestrian roads in a healthy and orderly manner. At the beginning of the factors affecting usage; environmental variables, suitability to climatic conditions, safety, accessibility, consideration of the topography factor, having a position of reaching the target at a certain distance, cultural factors, socialization, having aesthetic, physical and perceptual characters (Rapoport, 1991). While these factors affect the small-scale movements and orientations of the user on foot, as well as the city's morphology and space sequences, the preferences and movements of the user are shaped and directed. Therefore, in the planning of the pedestrian road and its system, the analysis of urban elements and the examination of urban morphology together with environmental factors constitute the main topics. Therefore, there are many studies in the literature investigating the effects of environmental factors and urban morphology on pedestrian movements.

The purpose of this study is; It is the examination of how pedestrian movements are shaped according to physical and psychological factors in an urban environment where intense structuring begins. Within the scope of the study, it is very important to determine the factors affecting the use by analyzing the use of pedestrian roads. Accordingly, it is aimed to create an ideal pedestrian road system with environmental comfort according to pedestrian movements.

2 Material and Method

2.1 The Reason to Choose the Work Area

Within the scope of the study, Kaşüstü Neighborhood of Yomra district of Trabzon province was considered as the study area (Fig. 1).

Fig. 1. The location of the work area

The reason why the study area was chosen as Kaşüstü District is that the area has started to be constructed rapidly since 2010. It is seen that housing and workplaces are densely located in the study area, but the infrastructure and circulation network has not been fully formed. Pedestrian and vehicle circulation networks are not defined in the study area, and it is observed that there are deficiencies in the defined places.

2.2 Method

Observational and experimental methods were used within the scope of the study. In the first stage of the study, space syntax analysis was performed using the depthmapX program in order to analyze the relations of spatial components with each other. By using DepthmapX program, "connectivity", "visibility" and "integration" analyzes were made. After the spatial syntax analysis, people counted in 3 different regions twice a week using the observation method. The counting study was carried out between 08:00 in the morning and 20:00 in the evening in order to distinguish the intensity of movement during the day for a total of two days, one on a weekday and one on a weekend. In the majority of the studies reviewed, the observation time was accepted as 4–5 minutes (Kubat, Rab, Güney, Özer, & Kaya, 2012). Therefore, the counts in the study were limited to 5 minutes. As a result of these counts, pedestrian density analysis was performed using the ArcGIS program. Thus, the roads and usage intensities preferred by pedestrians were compared with the data obtained from the spatial sequence analysis. "Do the morphological features have an effect on pedestrian movements?" The answer to the question was sought. Finally, a survey was conducted with 296 people within the scope of user satisfaction. The survey questions were shaped according to the general characteristics (connected, readable, comfortable, convenient, enjoyable, safe, safe, universal, accessible) of a walkable environment included in the report called pedestrian areas planning and design guide (NZ Transport Agency, 2009; Özer, 2014).

3 Findings

3.1 Findings of the Spatial Syntax

The spatial syntax method argues that the movements of the pedestrians can be affected by the effects caused by the morphological structure of the city and that these effects can direct the movements of the pedestrians. In the method, axis maps are mostly used to examine the morphological structures of cities. (Hiller, 2002; Özer, 2014). Axis maps consist of the longest and shortest linear lines representing the morphological structure of the city and define these lines with accessibility(Fig. 2).

Fig. 2. Visibilty map of the study area

Analyzing the relationship of the lines obtained in the map with all other lines, the relations of each line with other lines are converted into numerical values. Among these transformations, mostly "connectivity" and "integration" values are used to make sense of the morphological structure of the city. The connectivity value defines the nodes directly connected to each node (Fig. 3).

Fig. 3. Connectivity map of the study area

Evaluates the connection between truth and truth. Integration evaluates the degree to which a node is integrated or separated into a system as a whole (general) or partially (locally) (Fig. 4).

Fig. 4. Integration map of the study area

So it measures how accessible each node is from other nodes. How accessible a node is shows how much potential that node has as a "target point" (Özer, 2014).

3.2 Findings about Pedestrian Movements

Observations were carried out two days a week, on weekdays (Tuesday) and on weekends (Saturday). On both days observed, the weather is clear and without precipitation. Counts in the study area were made simultaneously at the observation points (Tab. 1).

Tab. 1. The total number of persons counted as a result of the observations

Weekday Observation			Weekend Observation		
1. Observation Point	*2. Observation Point*	*3. Observation Point*	*1. Observation Point*	*2. Observation Point*	*3. Observation Point*
325	262	181	174	349	233
Total:768			**Total:756**		

As a result of the observations made during the week within the scope of the study, a total of 768 people were counted at three observation points. In these observations, it was determined that the highest number of people was at the 1st observation point between 08.00-08.05 (Fig. 5). While density was detected at the 1st observation point between 12.30-12.35/15.30-15.35 hours, it was determined that the density was at the second observation point as a result of the observation made between 17.00-17.05/19.55-20.00 (Tab. 2).

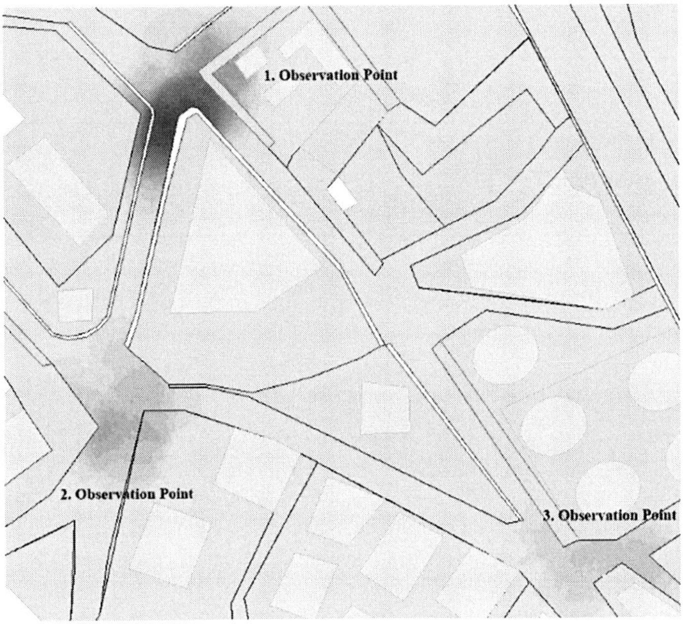

Fig. 5. Density map of the morning observation result

Tab. 2. The change of the number of people counted in the weekday count according to hours

Weekday Observation					
	08.00 08.05	**12.30 12.35**	**15.30 15.35**	**17.00 17.05**	**19.55 20.00**
1.Observation Point	107	69	45	51	53
2.Observation Point	50	49	29	55	79
3.Observation Point	36	43	17	48	37

As a result of the observations made at the weekend, a total of 756 people were counted at three observation points. In these observations, it was determined that the highest number of people was at the 2nd observation point between 08.00-08.05. The density was detected at the 2nd observation point between 12.30-12.35/15.30-15.35/17.00-17.05/19.55-20.00 hours (Tab. 3).

Tab. 3. The change of the number of people counted in the weekend count according to hours

Weekend Observation					
	08.00 08.05	**12.30 12.35**	**15.30 15.35**	**17.00 17.05**	**19.55 20.00**
1.Observation Point	13	39	35	38	49
2.Observation Point	22	79	59	71	118
3.Observation Point	19	64	58	42	50

The results of the observations were combined in the GIS database and density maps were created according to the density of the observation points (Fig. 5).

3.3 Findings of Pedestrian Satisfaction

A total of 296 people participated in the survey prepared to question pedestrian satisfaction. The survey questions were shaped according to the general characteristics (connected, readable, comfortable, convenient, enjoyable, safe, safe, universal, accessible) of a walkable environment included in the report called pedestrian areas planning and design guide. In this context, a total of 12 questions were asked to the participants (Özer, 2014) (Tab. 4).

As a result of the survey study, it was observed that the existing road system did not satisfy pedestrian satisfaction structurally and qualitatively. When the answers given were examined, it was seen that the pedestrian roads in the study area, pedestrian areas in the planning and design guide only support the accessible, safe, enjoyable and readable definitions of the general characteristics of a walkable environment. In addition to these, important definitions such as connected, comfortable, convenient, safe and universal were found to be lacking.

Tab. 4. The questions asked in the survey study and the answers of the participants

Does the road network allow pedestrians to easily reach the point they want to reach?		
Yes	40	13.5 %
No	242	81.75 %
No idea	14	4.75 %
Are pedestrian paths linked to other pedestrian paths, surrounding roads and common spaces?		
Yes	37	12.5 %
No	218	73.6 %
No idea	41	13.9 %
Can pedestrians intuitively understand where to reach?		
Yes	185	62.5 %
No	56	18.9 %
No idea	55	18.6 %
Are there informative signs on pedestrian and vehicle roads, such as traffic signs or place names?		
Yes	59	19.9 %
No	234	79 %
No idea	3	1.1 %
Do you find the current physical condition of the pedestrian roads sufficient?		
Yes	19	6.4 %
No	209	70.6 %
No idea	68	23 %
Are the width and slope of the pedestrian paths within acceptable levels?		
Yes	19	6.4 %
No	272	91.9 %
No idea	5	1.7 %
Are there any disturbances or density on the roads that may prevent walking?		
Yes	239	80.7 %
No	52	17.6 %
No idea	5	1.7 %

(continued)

Tab. 4. Continued

Do pedestrian areas have features such as quirkiness, calmness and cleanliness that allow social interaction?		
Yes	119	40.2 %
No	108	36.4 %
No idea	69	23.4 %
Is vehicle traffic dangerous when walking and crossing the pedestrian path?		
Yes	129	43.5 %
No	123	41.5 %
No idea	44	15 %
Does the walking place promote asocial and criminal psychology?		
Yes	32	10.8 %
No	208	70.3 %
No idea	56	18.9 %
Are there popular routes within easily accessible distances?		
Yes	267	90.2 %
No	29	9.8 %
No idea	0	0 %
Do pedestrian paths have universal accessibility features (suitable for the visually impaired, wheelchairs, strollers, etc.)?		
Yes	27	9.1 %
No	217	73.3 %
No idea	52	17.6 %

4 Conclusion

Roads are very important for the urban identity, defining the city and taking a place in the user's memory are among the main components of the city (Bekar & Gülpınar Sekban, 2018; Güneroğlu & Bekar, 2017). In addition to contributing to the urban identity, roads are concepts that affect the life quality of city residents and enable the city to be defined as livable (Bekar, Gülpınar Sekban, & Acar, 2018). Roads are very effective in shaping the spaces and the preferences of the user. The data obtained in this study supports the result.

The data obtained as a result of the spatial syntax analysis and the data obtained as a result of the observations made to measure the user density overlap with each other. As a result of the spatial syntax analysis, when the places that are thought to have accessibility and usage density were observed, it was observed that the density increased in these areas. Buddha showed that he can benefit from spatial syntax analysis while constructing the urban road system and creating the quantities of the circulation network. Keeping the pedestrian transportation axes in the foreground in terms of quality and quantity should be a priority in circulation axes with high spatial value. In particular, these axles are expected to be safe, enjoyable, readable, comfortable, convenient, safe and have qualities and quantities suitable for universal definitions. It has been concluded that Kaşüstü, which has started to be built recently and has many residential and business areas, is quite lacking in terms of circulation network. Especially pedestrian roads that do not comply with the definitions of connection, comfort, convenience, safety and universality negatively affect the urban experience of the user. The continuation of pedestrian networks with any connection prevents the user from crying the city and harms the user's health. However, city roads and pedestrian circulation axes, which have been proved to be very important in the city layer, should serve as a pioneer in shaping the city.

References

1. Akyol, D., & Gülpınar Sekban, D. Ü. (2019). İzmir – Konak İlçesi İçinde Alsancak ve Konak Meydanını Kapsayan Bir Mekânsal Tanımlamada Kentsel Tasarım Strateji Önerisi Geliştirilmesi. *The Journal of International Social Research, 12*, 521–532.
2. Arü, K. A. (1965). *Yayalar Taşıtlar- Şehir Dokusundaki Yeri- Ulaştırma Düzenleri.* İstanbul: İstanbul Teknik Üniversitesi Mimarlık Fakültesi Yayını.
3. Bakan, K., & Konuk, G. (1987). *Türkiye'de Kentsel Dış Mekanların Düzenlenmesi.* İstanbul: Tübitak Yapı Araştırma Enstitüsü
4. Bekar, M., & Gülpınar Sekban, D. Ü. (2018). Kentlerin Zihinde Bıraktığı İzler Trabzon Örneği. *The Journal of International Social Research*, 331–339.
5. Bekar, M., Gülpınar Sekban, D. Ü., & Acar, C. (2018). Planting Along Area Sides of Highway, Trabzon Coastal Case, Turkey. *Journal of Balkan Ecology, 21*(4), 443–463.
6. Brambilla, R., & Longo, G. (1977). *For pedestrians only: planning, design, and management of traffic-free zones.* New York: Whitney Library of Design.
7. Dinç, S., Hamurcu, M., & Eren, T. (2018). Kentsel Ulaşım İçin Alternatif Tramvay Araçlarının Çok Kriterli Seçimi. *Gazi Mühendislik Bilimleri Dergisi, 4*(2), 124–135.
8. Gülpınar Sekban, D. Ü., Bekar, M., & Acar, C. (2019). Evaluation of Sustainability Potential According to Cittaslow Criteria in Turkey/Trabzon. *Fresenius Environmental Bulletin, 28*(7), 5435–5446.
9. Gültekin, B. (2007). *Kent İçi Yolların, Yaya Kullanımına Yönelik Değerlendirilmesinde Çözümlemeli Bir Yaklaşım: Adana Örneği.* (Master Thesis), Çukurova University, Adana.

10. Güneroğlu, N., & Bekar, M. (2017). Dönüşüm ve kimlik kavramı: Trabzon Örneği. *Kastamonu Üniversitesi Orman Fakültesi Dergisi, 4*, 580–593.
11. Hiller, B. (2002). A theory of the city as object. *Urban Design International, 9*, 31–45.
12. Kavi, F. (2003). *İstanbul'daki Yaya Yolu Düzenlemelerinin Bitkisel Tasarım Açısından Değerlendirilmesiİstan* (Master), İstanbul Technical University, İstanbul.
13. Keleş, R. (1980). *Kentbilim Terimleri Sözlüğü*. Ankara: Türk Dil Kurumu Yayınları.
14. Kubat, A. S., Rab, S., Güney, Y. İ., Özer, Ö., & Kaya, S. (2012). *Application of space syntax in developing. A regeneration framework for Sharjah's heritage area*. Paper presented at the Eighth International Space Syntax Symposium, Santiago de Chile.
15. Kürkçüoğlu, E., & Ocakçı, M. (2015). Kentsel Dokuda Mekânsal Yönelme Üzerine Bir Algı-Davranış Çalışması: Kadıköy Çarşı Bölgesi. *Megaron, 10*(3), 365–388.
16. Lynch, K. (2010). *Kent İmgesi* (Vol. 1). İstanbul: Türkiye İş Bankası Kültür Yayınları.
17. Moughtin, C. (1999). *Street and Square* (Vol. 3). Oxford: Architectural Press.
18. NZ Transport Agency. (2009). *Pedestrian planning and design guide*. New Zealand: NZ Transport Agency.
19. Özer, Ö. (2014). *Kentsel mekanda yaya hareketleri: morfoloji ve çevresel algının etkisi*. (Doctoral Thesis), İstanbul Technical University, İstanbul.
20. Rapoport, A. (1991). Pedestrian Street Use: Culture and Perception. In A. V. Moudon (Ed.), *Public Streets For Public Use* (pp. 80–91). New York: Columbia University Press.
21. Shirvani, H. (1985). *The Urban Design Process*. New York: Van Nostrana Reinhold.
22. Surat, H., & Yaman, Y. K. (2015). Kent İçi Yolların Yaya Kullanımına Yönelik Değerlendirilmesinde Kamu Tercihi. *Kastamonu Üniversitesi Orman Fakültesi Dergisi, 15*(1), 58–72.

A. Esra CENGİZ and Zeynep Burke ÖZDEMİR

Agricultural Landscape Planning on the Basis of Conservation and Sustainable Use of Landscapes

Abstract: Along with the increase of the effect of global climate change, the pandemic life we are experiencing right now has brought with it a series of difficult and complex problems to be solved today. The protection and sustainable use of agricultural landscapes where food is manufactured is one of the most compelling ones. Agricultural landscapes have extreme importance on ecology-economy balance in urban and rural areas and sustainable food supply. In this context, this study explains the necessity of sustainable planning of agricultural landscapes and landscape planning works to be done in this context based on the definitions of natural and cultural landscapes.

Keywords: Agricultural landscape, agricultural landscape planning, conservation, natural-cultural landscape, sustainable use.

1 Introduction

None of the disasters (natural disasters, epidemics, melting of glaciers, global warming etc.) that human beings are experiencing today are coincidental. Due to the destructions manufactured by man, the ecological balance is gradually deteriorating, and nature protects itself in different ways. The pandemic and visibility of climate change are just two of the ways that nature protects itself. Restoration and protection of nature is one of the issues that should be emphasized especially during these hard times.

Due to human activities and natural disasters, the land use/land cover is changing rapidly [1]. Currently, land use/land cover change is one of the fundamental problems of sustainable development. This human-induced change has a devastating effect on the landscape, and this is how the greatest global change occurs with climate today [2, 3, 4, 5, 6].

Although the effect of human on nature continues to increase from the past to the present, the reasons and solutions for the deterioration of the ecosystem has faced a lot of discussion on the world agenda especially after 1970s. The current state of today's technology, the increase in population and similar problems as such have started to manifest themselves with events such as the depletion of natural resources, deterioration in their quality, decrease in biological diversity, and global warming. Humans must face the consequences as reflected through soil on which they live [7, 8, 9].

Soil is a lifeless natural resource that is as essential and indispensable as air and water for the survival of all biological forms [10, 11]. In other words, soil is a natural resource composed of a mixture of inorganic and organic materials, on which people live and pursue their agricultural activities. The soil as an essential element for human life should be used in a balanced and efficient manner [12, 13, 11]. However, due to people's unconscious dangerous activities, it now faces gradual extinction [14, 15].

The continuous increase in the population has brought about the expansion of residential areas. This situation has increased the pressure and interventions on natural and cultural resources. The areas that were used as agriculture, forest and pasture in the past have started to turn into residential areas today [16, 17]. Undoubtedly, agricultural areas are the most affected by these pressures [18].

In the long run, as well as creating ecological and economic consequences, the loss of productive soil resources leads countries/cities to become foreign-dependent by reducing their self-sufficiency potential [18]. In fact, considering Turkey's agricultural land potential, it is observed that only one-third is suitable for agriculture [19] and class I, II and III fertile agricultural lands are mostly used by different sectors [20]. The remaining two-thirds are not suitable for agriculture and therefore, it is extremely important to plan their usage wisely [19]. Urban agricultural areas, where urban pressure is felt most intensely and where land occupations are most brutal, are places with extreme sustainability problems [21, 22]. Establishment or spread of cities on fertile agricultural lands causes unnecessary loss of land [23]. According to Kayıkçı (2005), the importance of agricultural lands has gradually increased since they are considered natural resources with no alternative with a potential as nutritional requirement for present and future generations and as strategic resource that can be used as a 'hunger weapon' [11]. As the agricultural areas tend to shrink, the need for agricultural products is constantly increasing [17]. Due to the high risk inherent in extraordinary periods, it is not acceptable for any country to be externally dependent in terms of agriculture [24]. The current pandemic is also an extraordinary situation, and it is vital that all countries become self-sufficient in terms of agricultural production within the context of approaching food crisis.

For all these reasons, as Bayar (2018) stated, a country with fertile agricultural lands needs to protect and properly evaluate existing agricultural areas to ensure food security, to improve its economy, and to maintain ecological balance [25].

In fact, today all land use is a debatable issue between two contradictory dimensions such as ecological protection and economic life. While ecological protection in rural landscape defends the necessity of protecting natural resources such as water, air and soil, flora and fauna, it may conflict with economic life. This contradiction can only be resolved by establishing a protection-use balance [26]. Ensuring the balance between protection and use on natural and cultural landscapes is possible by correctly determining the potential of the natural and cultural landscape and then using it sustainably in accordance with its potential.

Ecologically based planning approaches are the most effective tools to protect, develop and carry nature into the future in the planning process [27, 28]. As an ecologically based planning approach, landscape planning comes to the fore as the most appropriate and systematic way to correctly determine, protect and use the potential of natural-cultural landscape.

In the pandemic life today, the protection and use of agricultural landscapes along with access to safe food has become increasingly questionable. In this context, this study emphasizes the importance of sustainable use and planning of agricultural landscapes as important cultural components and main source of food production. For this purpose, first, agricultural landscapes as subcomponents of the cultural landscape are explained in detail by giving their definitions. Then, the necessity of sustainable planning of agricultural landscapes and planning studies to be done in this context are explained.

2 Concepts of Landscape and Its Components

In its broadest sense landscape can be expressed as all the things that surround us wherever we are, as well as the appearance of natural and cultural entities [29]. Land morphology, climate, natural vegetation, wildlife and various types of structural elements, agriculture, settlement etc. constitute the physical elements of the landscape [30]. However, landscape refers not only to the combination of all these elements individually, but also to a whole that develops and changes as a result of mutual interaction between human and nature [31]. Forman and Godron (1986) and Forman (1995) emphasized this point in their studies and stated that landscape is a mosaic/heterogeneous piece of earth as mixture of land uses or ecosystems that are repeated and interacted with similar forms over large areas of kilometers [32, 33]. In the European Landscape Convention (2000), which was accepted in 2003 in our country, the same point was mentioned again, and the concept of landscape was defined as "*a field whose character is the result of the action and interaction of natural and/or human element as perceived by humans*" [34].

Landscapes are dynamic and therefore, always changing. The change of landscapes can be caused by natural processes or by human effects [35]. In this context, landscape is divided into two main groups as "natural landscape" and "cultural landscape".

Natural landscapes as areas that preserve their natural pattern that humans never change [36] can be defined as a synthesis of natural life with objects and features that exist spontaneously in nature [29]. Physical factors such as land morphology, climate, natural vegetation, and wildlife create the natural landscape [37]. Increase in the diversity and intensity of human activities from past to present has led natural landscapes to change the ecosystem processes and dynamics into human dominant ecosystems [38]. The cultural landscape emerging as a result of the intense interaction between nature and human is a landscape form that is formed by human groups shaping, changing and processing natural landscape elements as a result of their own activities [39, 37]. In this context, natural elements such as land, soil structure and vegetation are not only a part of the cultural landscape, but they also form its skeleton [40]. As a concept that is becoming increasingly important today, cultural landscape is important for understanding and evaluating the value of landscapes [41]. Cultural landscapes began to be studied in academic circles, especially in geography and other social sciences in the 1900s [42]. In Western geography and other social science literature, cultural landscape is also called the structured or built environment [37].

Cultural landscape that gained an international recognition in 1992 with the work of the International Union for Conservation of Nature (IUCN) [42] is defined by IUCN as follows; *"Cultural landscapes are cultural properties and represent the combined works of nature and of man" designated in Article 1 of the Convention. They are illustrative of the evolution of human society and settlement over time, under the influence of the physical constraints and/or opportunities presented by their natural environment and of successive social, economic and cultural forces, both external and internal."* [43]. The harmony of the elements created by nature and man in these areas; their historical, aesthetic, ethnological and anthropological value; their ability to represent the dominant natural elements; land use patterns; and the textures of traditional life in the region are among other qualifications sought [44]. Cultural landscapes, which have become a new category combining nature and human artifacts in the World Heritage List, generally show a certain relationship between people and nature and can reflect sustainable land use techniques by promoting strong links between culture and sustainability [45]. Conservation of cultural landscapes can contribute to modern techniques of sustainable land use and preserve or enhance natural resources in the landscape. The continued existence of traditional land use patterns supports biodiversity in many parts of the world. Therefore, preserving traditional cultural landscapes also helps preserve biodiversity [43]. Based on all these definitions, cultural landscape, according to Sauer (1945), is the shaping of natural landscape by a cultural group. Here, culture means natural environment and cultural landscape is the product of this interaction [42].

Today, almost every landscape we are in is a cultural landscape, since there is almost no natural area that has not been affected by cultural processes in the human-nature interaction that has been going on for thousands of years [46]. Cultural landscape is at the interface between nature and culture, tangible and intangible heritage, biological and cultural diversity. They represent a closely knit network of relationships, the essence of culture and the identity of people. Cultural landscapes, in the wider context of ecosystem, is the focal point of protected areas and is a symbol of the growing recognition of local communities and natural-cultural heritage, the fundamental links between humanity and its natural environment [47].

As a result, the mutual interaction between nature and human gives different qualities to the landscape, making it possible to define it as natural and cultural, urban and rural. Agricultural landscapes that represent the transition processes between natural and cultural landscapes reveal how natural values turn into cultural values and change [42].

3 Agricultural Landscape as a Sub-Component of the Cultural Landscape

Due to the mutual interactions between human and natural environment, cultural landscape includes the diversity around the region and accordingly, "agriculture" is one of the most important cultural landscape values [48]. Agricultural fields have been areas where people started to change nature and gave the signs of the first cultural landscape. Therefore, "agricultural areas" constitute the first examples of cultural landscape [39]. According to Goudie and Viles (1997), human activities have a great impact on the development of agricultural areas, which constitute an important part of the cultural landscape [49].

Forman and Gordron (1986) describe agricultural landscapes as mosaic whose size, shape and order changes with human rule [32]. According to Akdoğan (1975), agricultural landscape is a landscape that emerges as a result of productivity-based efforts and serves the production purpose of people [36]. Natural elements such as land structure, soil and topographic features, water and climate form the agricultural landscape character [36, 30]. On the other hand, the customs and cultures of people reveal a diverse range of agricultural landscapes varying from country to country and/or from one place to another [30].

The main elements that make up the agricultural landscape are agricultural settlements and various agricultural areas (Fig. 1) [36]. However, the structure of agricultural landscape consists of elements such as opening fields in the natural landscape, field parceling, terracing, irrigation channels, roads, open drainage ditches, scattered and single agricultural holdings, village settlements, village squares, village schools, plateaus, livestock farms and pens, natural meadows and pastures, various agricultural buildings, orchards, green covers of agricultural plants [30]. In addition to these, the processing and consumption of the products produced, and the traditional lifestyles experienced are also reflected on the land. All these features add a visual character to the region. The sustainability of this visual landscape as a resource value is directly proportional to the preservation of agricultural landscape character of the region [50]. Throughout history, agricultural landscapes are indispensable areas for the sustainable development both cities and rural areas with their vital, economic, social and ecological characteristics [32, 51, 52, 53].

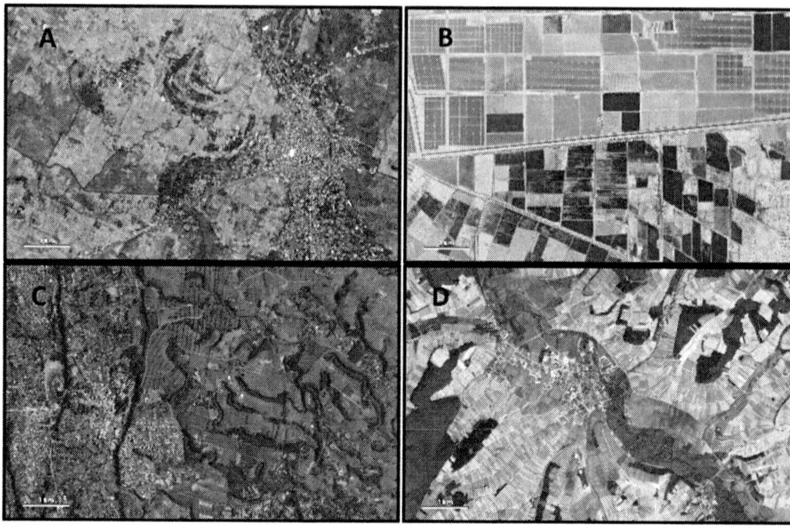

Fig. 1. Examples from various agricultural landscapes (A: rice paddies, B: solar farms, C: tiny subsistence farms, D: vineyards [54]

Agricultural landscapes, one of the most dominant landscapes in the terrestrial area, contribute substantially to the conservation of global biodiversity while providing basic food products [55]. In this context, it should not be forgotten that the concept of sustainability is not only valid for the effective use of natural resources and that sustainability in living spaces is possible by considering the cultural structure. Planning works from any field should examine natural and cultural values together, and decisions should be made accordingly [56]. As stated, the protection and development of agricultural landscapes, which are among the landscape types that contribute to sustainable development with their socio-economic, cultural and ecological aspects, are very important for the sustainable use of agriculture [49]. In this context, a holistic approach to agricultural landscapes at the ecosystem level and adoption of a planning approach that will ensure conservation-utilization balance will contribute to ensuring ecological and economic sustainability on a global scale [57].

In addition to all these, agricultural landscapes today constitute the focal point and solution of the food crisis prediction to be experienced in the near future. Productive, preserved agricultural areas are primarily needed for food production. Therefore, agricultural landscapes of this quality should be determined and used only for agricultural purposes. Agricultural landscapes should be given priority in area use in both urban and rural planning works.

4 Sustainable Use and Planning of Agricultural Landscapes

The 1987 Brundtland Report proposes that the balance between environmental and economic development should be established and it should be sustainable in the face of increasingly severe environmental problems [58]. Sustainable development as used today in the report is defined as *"development that meets today's needs without jeopardizing the ability to meet the needs of future generations"* [23]. Besides, sustainable development also targets environmental, social, cultural and economic sustainability [59].

In fact, the concept of sustainability is based on the idea that natural resources are scarce. Due to the increasing world population and new demands emerging in accordance, the pressure on natural resources rapidly increases [60]. Natural resources in underdeveloped and developing countries of the world are under constant pressure due to reasons such as wrong and unplanned land use, high population growth, soil erosion in sensitive ecosystems, versatile demands for scarce resources, poor rural population, insufficiency or lack of institutional support, etc. [61]. As urban areas grow, the resulting resource requirement and pollution put an increasing pressure on cultivated areas and pristine rural areas [62]. However, in the face of all these pressures, the number of natural resources is fixed and counted scarce. The land and its basic elements constitute one of the scarce natural resources such as forest, water and underground resources [60].

Planning policies pursued in both urban and rural areas constantly put the economy and ecology against each other, and this results in the destruction of natural and cultural environment [63]. Agricultural landscapes as the most damaged cultural areas are rapidly transforming into new residential areas [8] or different areas of use contrary to the expectations. While the natural and cultural potential is not taken into account in agricultural landscape planning, the use of this potential contrary to its purpose and potential will bring along irreparable losses in the future.

For this reason, it has become a necessity to eliminate the pressures on resources and to plan accordingly for the sparing use and sustainability of natural-cultural resources [27, 64, 65, 66, 28, 67]. In this context, it is emphasized that natural and cultural landscape elements are destroyed by human activities knowingly or unknowingly, and this problem can only be eliminated by appropriate land use plans [68]. In the context of sustainable use of the natural and cultural landscape, physical planning approaches with an ecological perspective have a priority in determining the appropriate and optimal land uses for both urban and rural areas. Landscape planning approach is one and the most important of them.

Landscape planning concept handled in physical planning means protecting the quality and structure of an area within certain boundaries; minimizing the pressures from the environment; and ensuring healthy and sustainable landscape use [69]. Landscape planning also aims to protect rare, scarce and unique resources, prevent dangers, protect limited resources with controlled use, and identify areas suitable for development activities [67].

The characteristics of landscape (dynamic and flexible, continuous, authentic, border, hierarchy, etc.) bear the basic principles of sustainability originally developed for the protection of natural and cultural resources. This situation enabled landscape planning to be developed within the framework of sustainability principles. Landscape planning supports this process with analysis methods that evaluate natural processes and cultural structure in a holistic way. The results obtained from the analysis create data for planning at an advanced scale [59]. Existing and

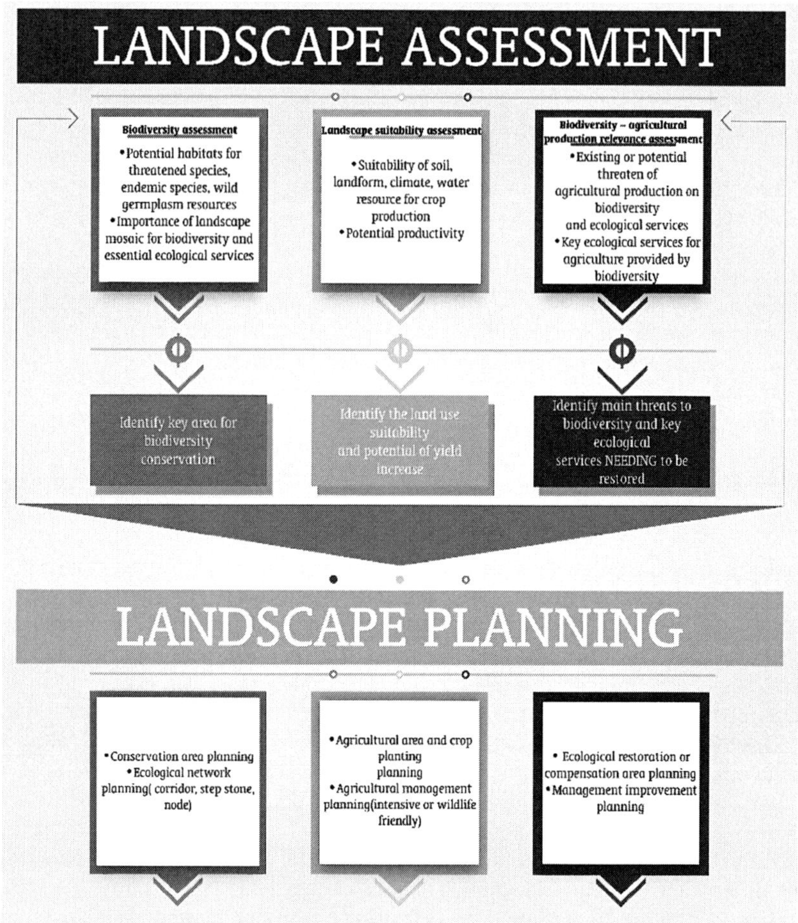

Fig. 2. Process and contents of landscape assessment and landscape planning for reconciling biodiversity conservation and food production in agricultural landscapes [adapted from 55]

potential uses must be taken into consideration, suitability analyzes, and land use maps must be made for this potential in the conservation, sustainable development and reconciling food production of agricultural landscapes, which are among the most important sub-components of the cultural landscape. The most important components of landscape planning are natural-cultural inventory, suitability analysis and land use (optimal land use) maps. In addition to all these, today in land use decisions regarding landscape planning, areas with sensitive landscape are also included in the assessment and in this context, landscape sensitivity as an important sub-component lays a guiding role in landscape planning (Fig. 2).

Preventing problems arising from land use, determining the most appropriate land use type and ensuring sustainability in land use is only possible by considering the user requirements and the existing (natural-cultural) characteristics of the land for various uses. Effective results can be obtained with suitability analyzes and land use maps in the use of land for specific purposes [70, 26, 71, 72, 73, 74, 75]. Area suitability analysis is a study conducted to determine the suitability level of a piece of land for specified uses [70, 72, 74, 75]. In the suitability analysis or in determining the appropriate areas for the selected area use, existing natural and cultural landscape potential is determined, and the area is evaluated according to the usage type and purpose. Regions reserved for certain purposes in urban and

Tab. 1: Natural and cultural factors used in determining suitable agricultural areas and selected references

Natural-Cultural Factors	**Selected References**
Slope	[74, 76, 78, 79, 80, 81, 82, 83, 84]
Aspect	[74, 76, 80, 81, 82]
Altitude groups	[74, 76, 82, 84]
Proximity to rivers	[78, 80, 82, 83, 84]
Large soil groups	[83, 84]
Land capability classes	[74, 76, 78, 79, 80, 82, 83, 84]
Limiting soil features	[74, 76, 78, 79, 80, 81, 82, 84]
Potential erosion risk	[74, 76, 78, 79, 80, 81, 82, 83]
Soil depth	[78, 76, 79, 80, 81, 82]
Soil infiltration	[78, 80, 81, 83]
Geology/Rock permeability	[82, 83, 84]
Vegetation	[76, 80]
Climatic features	[76, 78, 80, 82, 84]
Landslide susceptibility	[83]
Proximity to city centers or roads	[74, 80, 81, 82, 83, 84]

rural areas and called as zoning reflect the land use [76]. Area usage maps are needed especially in determining misuse, in monitoring the changes in land use and in the early stages of land use planning works [77].

For example, natural-cultural data such as slope, aspect, altitude groups, proximity to rivers, large soil groups, land capability classes, limiting soil features, potential erosion risk, soil depth, soil infiltration, geology/rock permeability, vegetation, climatic features, landslide susceptibility, proximity to city centers or roads can be used in the suitability analysis of potential agricultural areas (Tab. 1) [78, 79, 80, 74, 81, 76, 82, 83, 84].

In determining suitable agricultural areas, the mentioned natural-cultural landscape data are evaluated in a holistic manner within the framework of the selected multi-criteria analysis method, and can be mapped on the basis of sustainable living and ecological principles [83]. Maps belonging to area suitability analysis basically allow the creation and evaluation of land use plans. Various landscape evaluation methods can be used in the creation and evaluation of land use plans. Tassinari et al. (2013) have developed a multi-criteria analytical model aimed at characterizing rural areas through the integrated use of various parameters and indicators [85]. The methodological framework developed is summarized in the scheme of Fig. 3.

Maps belonging to area suitability analysis basically allow the creation and evaluation of land use plans. Developed by Mc Harg (1969) and also known as the "Mc Harg Method" aims to determine ecologically the most suitable method for land uses (settlement, agriculture, forest, industry, open-green area, recreation-tourism, etc.) and to ensure the sustainability of natural resources [27, 86, 87]. Priorities are given according to ecological importance in determining potential land use types. For this purpose, firstly, potentials and some basic issues are put forward [27, 87]. This technique involves overlapping analysis maps obtained from inventory studies to define the areas that provide opportunities for selected land uses and then impose restrictions [88]. All landscape evaluation methods developed by different landscape planners at different times are based on the overlay analysis technique developed by McHarg.

In the management and analysis of the land used for different purposes (forest, tourism, natural life and agriculture, etc.) in the world, considering natural and cultural factors in a fully integrated manner has gained importance to distinguish homogeneous ecological units [89] and has been the subject of various studies in recent years [90, 91, 92].

They emphasized that the management and utilization of agricultural resources play a key role in areas such as agriculture and forestry, but there is a need for multi-criteria decision support systems and Geographical Information Systems (GIS) in order to use and manage these resources in most appropriate ways [93, 90, 94, 92, 95, 96, 71, 97, 98]. For this purpose, a wide variety of GIS software (ArcGIS, Global Mapper, MapInfo, MapSource, ER Mapper, QGIS, etc.) are frequently used in the registration and query stages [88].

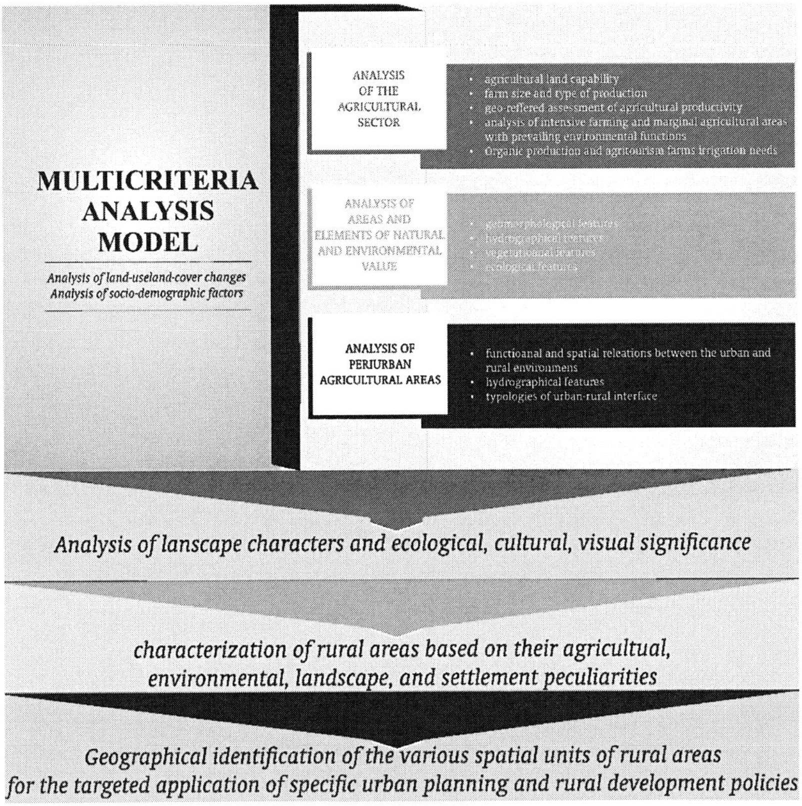

Fig. 3. The methodological framework of the multi-criteria analysis model developed for land-use planning of rural areas [adapted from 85].

Landscape sensitivity maps, on the other hand, are maps that reveal a holistic assessment of events such as erosion, landslides, surface water permeability, earthquakes, etc. [99, 100]. Landscape sensitivity maps are the areas which are called as the function of landscape, which need to be treated more sensitively in land use decisions, and which should be protected most of the time [101]. Sensitivity maps ensure that optimal areas are determined by overlapping the appropriate areas in the process of making area usage/selection decisions. Taking sensitive landscapes into consideration is very important for sustainable planning to reduce the pressure while determining the areas such as agriculture, settlement, industry, etc. areas [102]. Therefore, landscape sensitivity maps are important denominators in spatial ecology-based decision-making processes. The use of spatial planning in the process of land use decisions (agriculture, settlement, industry, etc.) is now a necessity at this place where natural and cultural landscapes are today. The use and preparation of these maps should be a legal obligation [99].

Sustainable planning should be supported by powerful accurate data production and practical, updatable databases that allow analysis. In this context, it is very important to determine the agricultural landscape uses, to keep up with their current situation and to reveal agricultural policies and economic data of the countries in a healthier manner. For this reason, it is essential to map agricultural landscapes and to update data on these areas [103]. Maps of suitable agricultural areas, landscape sensitivity and optimal agricultural areas obtained as a result of landscape planning analysis should be evaluated during spatial planning process in order to protect agricultural areas. Otherwise, occupation or pressure on agricultural landscapes cannot be prevented [83].

In addition, monitoring is also a very important issue along with the determination of agricultural areas. In particular, global warming increases agricultural areas and the pressure to be created on these areas. For this reason, the changes in suitable agricultural lands should be continuously analyzed based on meteorological data. The change in suitable agricultural land maps created should be evaluated temporally [104]. At the same time, revealing agricultural landscape change will guide the decisions to be taken regarding the integrity and management of rural landscape [49].

5 Conclusion

Determining agricultural landscape potential and using it according to its potential is one of the most important issues today in Turkey where large part of the population and national income depends on agriculture [105, 106]. Despite all the negative conditions of agricultural sector in Turkey, it has a great potential in the development process and contributes to the national economy in various ways [25]. For this reason, protection and development of agricultural landscapes is extremely important for the sustainability of agriculture [49].

Although agriculture is facing enormous challenges in the 21st century, sustainable planning approaches to agriculture are now needed to feed the growing human population [107]. It is extremely important to protect nature and the land on which we earn income, to make maximum use of its current potential, and to transfer it to future generations. This will only be possible by using the existing lands based on a plan and ensuring continuous maintenance and control [76].

Mechanization in agriculture that started after the Industrial Revolution and the decrease in agricultural value of the soil has rapidly increased the disintegration of agricultural structure from past to present. In this process, the agricultural sector, like every other sector, has had its fair share of technological developments. Modern farming methods applied to get more products from the soil in the short term and to feed the increasing population have damaged the ecological structure of the soil in the long run and caused a decrease in agricultural yield. In the current century, food production and access to clean food together with the increasing food crisis rhetoric is now a necessity rather than a need in both developed and developing countries. Deterioration of the quality of agricultural landscapes and change in its usage purposes is now a global problem. The dramatic effects caused by this problem are felt for the first time in such a great extent and all over the world. It has now become necessary to determine the current structure of the agricultural landscape and to use it in accordance with its potential.

In this century where global climate change has increased dramatic effects on the entire ecosystem along with the current pandemic, planning approaches for sustainable planning of the agricultural landscape offer a scientific solution today.

Landscape planning approach that provides a holistic evaluation of natural and cultural data based on ecological principles is a very important planning approach in creating sustainable land uses in both urban and rural areas [59]. Landscape planning approaches that contribute to the creation of sustainable land uses today is considered one of the versatile and effective decision-making tools in the protection, development and sustainability of agricultural landscapes.

References

1. Tağıl Ş., *Peyzaj Patern Metrikleriyle Balıkesir Ovası ve Yakınında Habitat Parçalılığında ve Kalitesinde Meydana Gelen Değişim (1975–2000).* Ekoloji Dergisi, 2006. **15** (60): p. 24–36.
2. Thomas W.L., *Man's Role in Changing the Face of the Earth.* 1956, Chicago: University of Chicago Press.
3. Turner II, B.L., Clark, W.C., Kates, R.W., Richards, J.F., Mathews, J.T. and Meyer, W.B., *Global and Regional Changes in the Biosphere over the Past 300 Years.* The Earth as Transformed by Human Action. Ed: Turner II, B.L., Clark, W.C., Kates, R.W., Richards, J.F., Mathews, J.T. and Meyer, W.B., 1990, New York: Cambridge University Press (Environment and Society Graduate School of Geography and George Perkins Marsh Institute Press).
4. Meyer, W.B. and Turner II, B. L., *Changes in Land Use and Land Cover.* A Global Perspective. Ed: Meyer, W.B. and Turner II, B. L., 1994, New York: Cambridge University Press.
5. Chase, T.N., Pielke, R.A., Kittel, T.G.F., Nemani, R.R. and Running, S.W., *Simulated Impacts of Historical Land Cover Changes on Global Climate in Northern Winter.* Climate Dynamics, 1999. **16:** p. 93–105.
6. Houghton, R.A., Hackler, J.L. and Lawrence, K.T. *The U.S. Carbon Budget: Contribution from Land-Use Change.* Science, 1999. **285** (5427): p. 574–578.
7. Sılaydın, M.B., *Şehir Planlamanın Paradigmal Sorgulanması ve Ekolojik Dengenin Korunması Bağlamında Yeni Bir Süreç Önerisi (Kuramsal Bir Deneme).* 2006, PhD Thesis, Dokuz Eylül University, İzmir-Turkey.

8. Açıksöz S., Topay M. and Yılmaz B., *Arazinin Yanlış Kullanımından Kaynaklanan Sorunlar: Bartın Kenti Örneği.* Süleyman Demirel Üniversitesi Orman Fakültesi Dergisi, 2008. **A** (1): p. 155–167.
9. Tümertekin, E. and Özgüç, N. *Ekonomik Coğrafya Küreselleşme ve Kalkınma.* 2009, İstanbul: Çantay Kitabevi.
10. Keleş, R. and Hamamcı, C., *Çevre Politikası.* 2005, Ankara: İmge Kitabevi.
11. Sezgin, D., *Kentsel Saçaklanmanın Verimli Tarım Topraklarının Amaç Dışı Kullanımına Etkisi: Ankara Örneği.* 2010, Master Thesis, Gazi University, Ankara-Turkey.
12. Goudie, A., *The Human Impact on the Natural Environment.* 2006, Oxford, UK: Blackwell Publishers.
13. Özçağlar, A., Özgür, E.M., Somuncu, M., Bayar, R., Yılmaz M., Yüceşahin, M., Yavan, N., Akpınar, N. and Karadeniz, N., *Çamlıhemşin İlçesinde Doğal ve Beşeri Kaynak Tespitine Bağlı Olarak Geliştirilen Arazi Kullanım Kararları.* Coğrafi Bilimler Dergisi, 2006. **4** (1): p. 1–27.
14. Zhan, X., Sohlberg, R., Townshend, J., Dimiceli, C., Carroll, M., Eastman, J., Hansen M.C. and DeFries R.S., *Detection of Land Cover Changes Using MODIS 250 m Data.* Remote Sensing of Environment, 2002. **83** (1–2): p. 336–350.
15. Sandal, E.K. and Gürbüz, M., *Mersin Şehrinin Mekansal Gelişimi ve Çevresindeki Tarım Alanlarının Amaç Dışı Kullanımı.* Coğrafi Bilimler Dergisi, 2003. **1** (1): p. 117–130.
16. Aksoy, M.A.C., *Türkiye'de Tarım Topraklarının Amaç Dışı Kullanılması.* 1997, Master Thesis. Ankara University, Ankara-Turkey.
17. Taş, B., *Tarım Alanlarının Planlaması Sürecinde SWOT Analizi Kullanımına Bir Örnek: Sandıklı İlçesi.* Coğrafi Bilimler Dergisi, 2011. **9** (2): p. 189–207.
18. Akseki, H. and Meşhur, M.Ç., *Kentsel Yayılma Sonucu Yapılaşmaya Açılan Verimli Tarım Alanları: Konya Kenti Deneyimleri.* MEGARON, 2013. **8** (3): p. 165–174.
19. Şenol, S., *Tarım Topraklarının Kullanımı ve Korunması.* Küresel Krizin Eşiğinde Tarım Sempoyumu, 11–15 Ocak 2012, TMMOB Ziraat Mühendisleri Odası, Ankara-Turkey, p. 100.
20. Ünal, M. and Başkaya, H.S. *Kocaeli İli Topraklarının Amaç Dışı Kullanımı.* Ekoloji ve Çevre Dergisi, 2000. **9** (36): p. 9–11.
21. Cengiz A.E., Pekin Timur, U., Karadağ, A.A. and Demiroğlu, D., A Contribution to Sustainable Urban Development: Urban Agricultural Lands. Urban and Urbanization, Ed: R. Efe, T.T., Onay, I. Sharuho, E. Atasoy, 2014, Sofia: St. Kliment Ohridski University Press, p. 85–104.
22. Demiroğlu, D., Pekin Timur, U., Karadağ, A.A. and Cengiz A.E., Ecology-Based Contemporary Urbanism Approaches, Kitap adı: Urban and

Urbanization, Urban and Urbanization, Ed: R. Efe, T.T., Onay, I. Sharuho, E. Atasoy, 2014, Sofia: St. Kliment Ohridski University Press, p. 579–600.

23. WCED (World Commission on Environment and Development), *Our Common Future.* 1987, Oxford: Oxford University Press. Access link: https://sustainabledevelopment.un.org/content/documents/5987our-common-future.pdf (Date of access: 20.11.2020).
24. Doğan, A., *Ekonomik Gelişme Sürecinde Tarımın Katkısı: Türkiye Örneği.* Sosyal ve Ekonomik Araştırmalar Dergisi, 2009. **9** (17): p. 365–392.
25. Bayar, R., *Arazi Kullanımı Açısından Türkiye'de Tarım Alanlarının Değişimi.* Coğrafi Bilimler Dergisi, 2018. 1**6** (2): p. 187- 200.
26. Van Lier, H.N., *The Role of Land Use Planning in Sustainable Rural Systems.* Landscape and Urban Planning, 1998. **41**, p. 83–91.
27. Mc Harg, I.L., *Processes as Values.* Desing With Nature. 1992, New York: The American Museum of Natural History, pp. 279.
28. Steiner, F., *The Living Landscape: An Ecological Approach to Landscape Planning.* 2008, London: Island Press, pp. 491.
29. Gökalp, D. D. and Yazgan, M. E. *Kırsal Peyzaj Planlamada Agroturizm ve Agriturizm.* KMÜ Sosyal ve Ekonomik Araştırmalar Dergisi, 2013. **15** (24): p. 25–29.
30. Gül A., *Peyzaj-İnsan İlişkisi ve Peyzaj Mimarlığı.* Süleyman Demirel Üniversitesi Orman Fakültesi Dergisi, 2000. **A** (1): p. 97–114.
31. Deniz, B., Küçükerbaş, E.V. and Eşbah Tunçay, H., *Peyzaj Ekolojisine Genel Bakış.* ADÜ Ziraat Fakültesi Dergisi, 2006. **3** (2): p. 5–18.
32. Forman, R.T.T. and Godron, M., *Landscape Ecology.* 1986, New York: John Wiley.
33. Forman, R.T.T., *Land Mosaics: The Ecology of Landscapes and Regions.* 1995, New York: Cambridge University Press.
34. Official Gazette, *Avrupa Peyzaj Sözleşmesi'nin Onaylanmasının Uygun Bulunduğuna Dair Kanun (Madde 1-Tanımlar).* 17 Haziran 2003 tarih ve 24141 sayılı Resmi Gazete, 2003. Access link: https://www.resmigazete.gov.tr/eskiler/2003/07/20030727.htm (Date of access: 09.01.2020).
35. Benliay, A. and Yıldırım, E., *Peyzaj Planlama Çalışmalarında Peyzaj Metriklerinin Kullanımı.* Türk Bilimsel Derlemeler Dergisi, 2013. **1**: p. 7–11.
36. Açıksöz, S. and Tanrıvermiş, E., *Planlamada Kırsal Peyzajın Önemi, Sorunları ve Öneriler.* 2000, Kırsal Çevre Yıllığı, Ankara: Kırsal Çevre ve Ormancılık Sorunları Araştırma Derneği Yayını, pp. 44–65.
37. Arı, Y., *Amerikan Kültürel Coğrafyasında Peyzaj Kavramı.* Doğu Coğrafya Dergisi, 2005. **10** (13): p. 311–339.

38. Alberti, M., *Advances in Urban Ecology-Integrating Humans and Ecological Processes in Urban Ecosystems.* 2008, New York: Springer Science Business Media.
39. Ürgenç, S.İ., *Kırsal Peyzaj.* 2000, İstanbul: Yıldız Teknik Üniversitesi Yayınları.
40. Korgavuş, B., *Sosyo-Ekonomik ve Doğal Çevre Faktörlerinin Kültürel Peyzaja Etkileri: Rize Merkez İlçesi Örneği.* 2012, PhD Thesis, İstanbul Technical University, İstanbul-Turkey.
41. Antrop, M., *Rural – Urban Conflictsand Opportunities.* The New Dimensions of the European Landscapes, Ed: R.H.G. Jongmna, 2004, Dordrecht: Springer, p. 83–91.
42. Atik, M., *Antalya Akdeniz Örnekleri ile Kırsal Kültürel Peyzaj Karakterleri.* Peyzaj Araştırmaları ve Uygulamaları Dergisi, 2019. **1**: p. 18–25.
43. UNESCO, *Operational Guidelines for the Implementation of the World Heritage Convention (Cultural Landscapes).* 2008, UNESCO World Heritage Centre, Paris. Access link: http://whc.unesco.org/archive/opguide08-en.pdf#annex3 (Date of access: 30.01.2021).
44. IUCN (International Union for Conservation of Nature), *IUCN World Heritage Outlook 2014 – A Conservation Assessment of All Natural World Heritage Site*s. 2014, IUCN International Union for Conservation of Nature Gland, Switzerland. Access link: https://portals.iucn.org/library/sites/library/files/documents/2014-039.pdf (Date of access: 30.01.2021).
45. Rössler, M. and Chih-Hung Lin, R., *Cultural Landscape in World Heritage Conservation and Cultural Landscape Conservation Challenges in Asia.* Built Heritage, 2018. **2**: p. 3–26.
46. Baylan, E. *Kültürel Peyzaj ve Yere Bağlılık: Van Kenti Örneği.* Yüzüncü Yıl Üniversitesi Tarım Bilimleri Dergisi, 2019. **29** (1): p. 10–23.
47. Rössler, M., *World Heritage Cultural Landscapes: A UNESCO Flagship Programme.* Landscape Research, 2006. **31** (4): p. 333–353.
48. Erduran Nemutlu, F., *Kültürel Peyzaj Kaynaklarının Kırsal Alan Kullanımında Önemi: Çanakkale Örneği.* ÇOMÜ Ziraat Fakültesi Dergisi, 2017. **5** (1): p. 87–97.
49. Gülçin, D., *Landscape Metrics for Assessment of Agricultural Land Use Change: the Case of Aydın Province.* International Congress on Agriculture and Forestry Research (AGRIFOR), 8–10 April 2019, Marmaris-Turkey, p. 242–251.
50. Erduran Nemutlu, F. and Doğan, Z., *Kültürel Miras Olarak Tarımsal Peyzajın Önemi: Karabiga, Çanakkale Örneği.* ÇOMÜ Ziraat Fakültesi Dergisi, 2018. **6** (Özel Sayı): p. 161–168.

51. Akyol, M., *Kentsel Tarımın Tarihi Gelişimi ve Kentsel Tarım Tasarım Kriterlerinin Belirlenmesi.* 2011, Master Thesis, İstanbul Technical University, İstanbul-Turkey.

52. Cengiz, S., Görmüş, S. and Ateşoğlu, A., *Uzaktan Algılama Aracılığıyla Tarımsal Peyzaj Karakterizasyonu.* 5. Uzaktan Algılama-CBS Sempozyumu, 14–17 Ekim 2014, İstanbul-Turkey. Access link: https://www.kongresistemi.com/root/dosyalar/uzalcbs2014/210.pdf (Date of access: 18.12.2020).

53. Çolak, H.E. and Memişoğlu, T., *Trabzon İlinde Tarımsal Arazi Kullanımındaki Zamansal Değişimin CBS ile Belirlenmesi.* Afyon Kocatepe Üniversitesi Fen ve Mühendislik Bilimleri Dergisi, 2018. **18:** p. 946–958.

54. Planet Stories, *Weird & Wonderful Agricultural Landscapes.* 2016. Access link: https://medium.com/planet-stories/weird-wonderful-agricultural-landscapes-9db0ff336b98 (Date of access: 20.01.2021).

55. Liu, Y., Zhao, G. and Yu, Z., *Developing Integrated Methods for Biological Conservation and Sustainable Production in Agricultural Landscapes.* Biocultural Landscapes, Ed: S.-K. Hong et al., 2014, Springer Science+Business Media Dordrecht, p. 45–67.

56. Cengiz Gökçe, G. and Açıksöz, S., *Visual Landscape Analysis in Relation of Cultural Landscape Components with Tourism: The Case of Beydili Village of Nallıhan.* I. Eurasia International Tourism Congress: Current Issues, Trends, and Indicators (EITOC-2015), 28–30 May 2015, Konya-Turkey, p. 409–419.

57. Yılmaz, K.T., *Sürdürülebilir Kentleşme ve Peyzaj Mimarlığının Katılımı, Kentleşme Şurası 2009 Süreci.* Peyzaj Mimarlığı IV. Kongresi, 21–24 Ekim 2010, Selçuk, İzmir-Turkey, p. 11–20.

58. Çöteli, M., *Kültürel Peyzajın Korunması ve Sürdürülebilir Kentsel Gelişmenin Yaratılması Yönünde Bir Politika Önerisi.* Sosyal Bilimler Enstitüsü Dergisi, 2012. **32** (1): p. 71–110.

59. Cengiz, A.E., Karadağ, A.A. and Demiroğlu, D., *Importance of Landscape Planning for Sustainable Spaces.* International Congress of the New Approaches and Technologies for Sustainable Development, 21–24 September 2017, Isparta-Turkey, **1** (1): p. 4–16.

60. Atış, E., Nurlu, E., Miran, B. and Kenanoğlu, Z., *Tarımsal Arazinin Sürdürülebilir Kullanımını Etkileyen Ekonomik ve Ekolojik Faktörler Üzerinde Bir Araştırma.* 2004, E.Ü. Bilimsel Araştırma Projeleri Komisyonu, E.Ü. Araştırma Fonu, 1997 ZRF-014 no'lu Proje, Ege Üniversitesi Ziraat Fakültesi Yayın Alt Komisyonu, ISBN: 975-94133-1-0, Bornova-İzmir.

61. Yılmaz, E., *Bir Arazi Kullanım Planlaması Modeli: Cehennemdere Vadisi Örneği.* 2005, Çevre ve Orman Bakanlığı Yayınları, No: 253, ISSN: 1300–7912, Ankara-Turkey.

62. Erdem, Ü., *Çevre Bilimi-Sürdürülebilir Dünya.* Ed: Prof. Dr. Ümit Erdem, 2000, Ege Üniversitesi Çevre Sorunları Uygulama ve Araştırma Merkezi Yayınları, No: 1, ISBN: 975-483-465-2, Bornova, İzmir-Turkey.
63. Tunçer, M. and Ercoşkun, Y. Ö., *Kentsel Rant Baskısıyla Şekillenen Antalya İçin Ekolojik Planlama.* Ekolojik Mimarlık ve Planlama Sempozyumu, 27–28 Nisan 2007, Antalya-Turkey, p. 102–109.
64. Dramstad, W.E., Olson, J.D. and Formon, R.T.T., *Landscape Ecology Principles in Landscape Architecture and Land-Use Planning.* 1996, Washington: Island Press.
65. Marsh, W.M., *Landscape Planning Environmental Applications.* 1997, New York: John Wiley&Sons.
66. Makhzoumi, J. and Pungetti, G., *Ecological Landscape Desing and Planning.* 1999, London: Taylor ve Francis.
67. Ahern, J., *Greenways as Strategic Landscape Planning: Theory and Application.* 2002, Wageningen University, The Netherlands.
68. Hobbs, R., *Future Landscapes and The Future of Landscape Ecology.* Landscape and Urban Planning, 1999. **37** (2–4): p. 181–200.
69. Yeşil, M. and Yılmaz, H., *Tozanlı Havzası Tokat-Almus İlçesi Ekolojik Temelli Kırsal Peyzaj Planlaması.* Akademik Ziraat Dergisi, 2013. **2** (2): p. 63–74.
70. Rossiter, D.G., *A Theoretical Framework for Land Evaluation.* Geoderma, 1996. **72** (3–4): p. 165–190.
71. Malczewski, J., *GIS-Based Land-Use Suitability Analysis: A Critical Overview.* Progress in Planning, 2004. **62**: p. 3–65.
72. Groot, R., *Function-Analysis and Valuation as a Tool to Assess Land Use Conflicts in Planning for Sustainable, Multi-Functional Landscapes.* Landscape and Urban Planning, 2006. **75** (3–4): p. 175–186.
73. Abdul Mohit, A. and Mahmut Ali, M., Integrating GIS and AHP for Land Suitability Analysis for Urban Development in a Secondary City of Bangladesh. Jurnal Alam Bina, 2006. **8** (1): p. 1–20.
74. Akbulak, C., *Analitik Hiyerarşi Süreci ve Coğrafi Bilgi Sistemleri ile Yukarı Kara Menderes Havzası'nın Arazi Kullanımı Uygunluk Analizi.* Uluslararası İnsan Bilimleri Dergisi, 2010. **7** (2): p. 557–576.
75. Bagheri, M., Sulaiman, W.N.A. and Vaghefi, N., *Land Use Suitability Analysis Using Multi Criteria Decision Analysis Method for Coastal Management and Planning: A Case Study of Malaysia.* Journal of Environmental Science and Technology, 2012. **5** (5): p. 364–372.
76. Demir, M., Demircioğlu Yıldız, N., Bulut, Y., Yılmaz, S. and Özer, S., *Alan Kullanım Planlamasında Potansiyel Tarım Alanlarının Ölçütlerinin Coğrafi Bilgi Sistemleri (CBS) Yöntemi İle Belirlenmesi (İspir Örneği).* Iğdır Üniversitesi Fen Bilimleri Enstitüsü Dergisi, 2011. **1** (3): p. 77–86.

77. Şenol, S., *Spot Uydu Verileriyle Göksu Deltası Arazi Kullanım Haritasının Hazırlanması.* II. Uzaktan Algılama ve Türkiye'deki Uygulamaları Semineri, 1994, Harita Genel Komutanlığı Hacettepe Üniversitesi-UKAM, Bursa-Turkey.
78. Akten, M., *Isparta Ovası'nın Optimal Alan Kullanım Planlaması Üzerine Bir Araştırma.* 2008, PhD Thesis, Süleyman Demirel University, Isparta-Turkey.
79. Karakuş, C.B., Demiroğlu, D. and Sarıtaş, Z., *CBS Kullanılarak Sivas Kent Merkezi ve Yakın Çevresinin Doğal Potansiyelinin Saptanması ve Uygun Tarım Alanlarının Belirlenmesi.* 2. Uzaktan Algılama ve Coğrafi Bilgi Sistemleri Sempozyumu, 13–14 Ekim 2008, Kayseri-Turkey, **1** (1): p. 238–255.
80. Zengin, M., *Ardahan-Kura Nehri ve Yakın Çevresi Alan Kullanımlarının Belirlenmesi ve Optimal Alan Kullanım Önerileri.* 2007, PhD Thesis, Atatürk University, Erzurum-Turkey.
81. Cengiz, T., Akbulak, C., Özcan, H. and Baytekin, H., *Gökçeada'da Optimal Arazi Kullanımının Belirlenmesi.* Tarım Bilimleri Dergisi, 2013. **19:** p. 148–162.
82. Demiroğlu, D. and Memlük, Y., *Sivas Kentsel Gelişim Alanının Kentin Peyzaj Özelliklerine Göre Değerlendirilmesi.* Journal of Tekirdag Agriculture Faculty, 2015. **12** (3): p. 68- 81.
83. Karadağ, A.A., Başaran, N., Yılmaz Kaya, M., *Importance of Determination of Suitable Agricultural Land Use in Combating Global Warming: The Case of Düzce, Turkey.* Journal of Environmental Protection and Ecology (JEPE), 2019. **21** (5): p. 1757–1768.
84. Erbeşler Ayaşlıgil, T., *Optimal Peyzaj Uygunluk Analizi Yöntemi: Anamur İlçesi Örneği.* MEGARON, 2020. **15** (2): p. 332–342.
85. Tassinari, P., Torreggiani, D. and Benni, S., Dealing with Agriculture, Environment and Landscape in Spatial Planning: A Discussion about the Italian Case Study. Land Use Policy, 2013. **30**: p. 739–747.
86. Tozar, T., *Doğal Kaynakların Sürdürülebilirliği için Geliştirilen Ekolojik Planlama Yöntemleri.* 2006, Master Thesis, Yıldız Technical University, İstanbul-Turkey.
87. Herrington, S., *The Nature of Ian McHarg's Science.* Landscape Journal, 2010. **29** (1): p. 1–20.
88. Cengiz, A.E., *Ekolojik Açıdan Kentsel Alan Kullanımları: Çanakkale Kent Merkezi Örneği.* 2011, PhD Thesis, Çanakkale Onsekiz Mart University, Çanakkale-Turkey.
89. Erdal, Ü., Sökmen, Ö., Ölgen, M. K. 2017 Orta Gediz Havzasında (Salihli-Turgutlu) Tarımsal Amaçlı Bölgelendirme Yöntemi İle Organik Tarım Alanlarının Belirlenmesi. 1. GAP Organik Tarım Kongresi 17 - 20 Kasım 2009, Şanlıurfa-Turkey. Access link: https://orgprints.org/32396/1/GAP-%C3%BClfet%20erdal.pdf (Date of access: 18.12.2020).

90. Aspinal, R.J., Miller, D.R. and Birnie, R.V., *Geographical Information Systems for Rural Land Use Planning*. Applied Geography, 1993. **13** (1): p. 54–66.
91. Fisher, G., Makowski, M., Antoine, J., 1996, *Multiple Criteria Land Use Analysis*. IIASA Working Paper, WP -96-006, International Institute for Applied Systems Analysis, Laxenburg-Austria.
92. Host, G.E., Polzer, P.L., Mladenoff, D.J., White, M.A. and Crow, T.R., *A Quantitative Approach to Developing Regional Ecosystem Classifications*. Ecological Applications, 1996. **6** (2): p. 608–618.
93. Hopkins, L.D., *Methods of Generating Land Suitability Maps: A Comparative Evaluation*. Journal of American Institute of Planners, 1977. **43** (4): p. 386–400.
94. FAO (The Food and Agriculture Organization), *Agro-Ecological Zonning: Guidelines*, 1996. FAO Soils Bulletin 73, Rome-Italy.
95. Malczewski, J., *A GIS-Based Approach to Multiple Criteria Group Decision-Making*. Geographical Information Systems, 1996. **10** (8): p. 955–971.
96. Collins, M.G., Steiner, F.R. and Rushman, M.J., *Land-Use Suitability Analysis in the United States: Historical Development and Promising Technological Achievements*. Environmental Management, 2001. **28** (5): p. 611–621.
97. Malczewski, J., *GIS-Based Multicriteria Decision Analysis: A Survey of the Literature*. International Journal of Geographical Information Science 2006. **20** (7): p. 703–726.
98. Cengiz, T. And Akbulak, C., *Application of Analytical Hierarchy Process and Geographic Information Systems in Land-Use Suitability Evaluation: A Case Study of Dümrek Village (Çanakkale, Turkey)*. International Journal of Sustainable Development & World Ecology, 2009. **16** (4): p. 286–294.
99. Karadağ, A. A. and Şenik, B., *Landscape Senitivity Analysis as an Ecological Key: The Case of Düzce Province*. Applied Ecology and Environmental Research, 2019. **17** (6): p. 14277–14296.
100. Karadağ A.A., Demiroğlu, D. and Cengiz, A.E. (2019). *Importance of Ecological Analysis in Spatial Planning*. International Congress on Agriculture and Forestry Research (AGRIFOR), 8–10 April 2019, Marmaris-Turkey, p. 835–843.
101. Karadağ, A.A. and Yıldız, K., *Peyzaj Fonksiyonlarının Hendek İlçesi Örneğinde Değerlendirilmesi*. Düzce Üniversitesi Ormancılık Dergisi, 2013. **9** (1): p. 77–96.
102. Karadağ, A.A., Gültekin, P. and Kayapınar, E., *Mezarlık Yer Seçim Kriterlerinin Belirlenmesi Üzerine Bir Araştırma: Akçakoca (Düzce)*. Düzce Üniversitesi Ormancılık Dergisi, 2018. **14** (2): p. 17–33.
103. Başaran, N., Yılmaz Kaya, M., Karadağ, A.A. and Çiftçi, T., *Determination of Agricultural Landscape Areas in Düzce City*. International Symposium on

Urbanization and Environmental Problems: Transition/Transformation/ Authenticity, 28–30 June 2018, Eskişehir, Turkey, **1** (4): p. 403–411.

104. Karakuş, C.B., Demiroğlu, D. and Karadağ, A.A., *Evaluation of Land Use/ land Cover Change (LULCC) and Meteorological Data in Terms of Sustainable Agriculture in Kilis/Turkey.* Journal of Environmental Protection and Ecology (JEPE), 2020. **21** (4): p. 1186–1193.
105. Kiper, T. and Arslan, M., *Safranbolu-Yörükköyü Tarımsal Turizm Potansiyelinin Kırsal Kalkınma Açısından Değerlendirilmesi.* Süleyman Demirel Üniversitesi Orman Fakültesi Dergisi, 2007. **A** (2): p. 145–158.
106. Karadeniz, N., Özdemir, G., Toksöz, M. B. and Durmaz, D., *Ankara İli Nallıhan İlçesi Sarıyar Kasabası Kırsal Peyzaj Karakteristiği ve Kalkınma Potansiyelinin Belirlenmesi.* 2013, Bilimsel Araştırma Projesi Sonuç Raporu, Ankara Üniversitesi Bilimsel Araştırma Projeleri, Ankara-Turkey.
107. Landis, D. A., *Designing Agricultural Landscapes for Biodiversity-Based Ecosystem Services.* Basic and Applied Ecology, 2017. **18:** p. 1–12.

Müberra PULATKAN Dilek BEYAZLI Zeliha AKTÜRK

Participatory Approach in Design

Abstract: User-centered design is an approach dominated by innovative design tools and interaction between the actors that are the requirement of a rational communication as an alternative to the traditional design process. Due to the advances in user-centered approaches and interdisciplinary studies, new content and methods were introduced to design approaches and applications. A design process where the users have a say in spatial design includes all stakeholders, especially the users with an interactive and compromising approach, in addition to the relationship between the designer and the final product. The present study aimed to discuss the participatory design experiences that contributed to domestic and international literature based on the targeted level of participation, implemented methods and techniques, and most importantly, the role of the user.

Keywords: Participatory approach, user participation, community design, urban park design.

1 Introduction

Design is a complex phenomenon that includes various dimensions and disciplines. A multidimensional design process could guide individuals of different disciplines and different levels to collaborate with the designer and the user [1]. Thus, the concept of user participation has been considered as part of the design process [2]. The concept of user participation, which has a social value since it combines various disciplines, is included in the design process and post-process user experiences [3]. Participatory design is an approach associated with the power of change adopted for the public in the development and administration of the environment. It derives its strength from the ability to go beyond traditional professional limitations and cultures. Social participation is based on the principle that it is better for citizens to participate and be active in the development and administration of the environment than treating the citizens as passive consumers in the development and administration of the environment [4].

Although it has been considered that the idea of participation was based on the public forums in Ancient Greece, it is known that the participatory approach was first discussed in studies conducted by a group of Dutch scientists in early 1950s. In the early 1960s, the discussion on the role of the users became prevalent in professional circles. However, in the second half of the 60s, due to a violent and increasingly politicized discussion, the term "participation" started to be employed [2].

The topic of "participation in design" was initially discussed at Manchester Conference in 1971. In the conference, designers expressed their views on user participation in design process [5]. The first methods and propositions were developed in this conference, and then scientific participatory approaches were popularized globally by Open House and SAR publications [3].

According to Sanoff (2008), participation could be described as a transformative power employed in the development and administration of environments for humans. It was based on the fact that individuals develop common solutions for specific problems, and these solutions are created through development of a vision, strategic planning, and deliberative democracy [6].

Allowing the user, the individual who experiences the space, to have a say in the spatial design is the main aim of participatory approaches [3]. Participation aims to include the users in the planning and design of their physical environment and to improve self-esteem among the people [7]. If the public is involved in the decision-making and becomes a part of the development, they own the program more, leading to better public decisions [4].

In short, participatory planning aims to take relevant and influential individuals into account in decision-making, planning and implementation processes to ensure continuous urban environment quality. This process focuses on encouragement and support, aims to improve individual and their life quality [8].

Sanoff (2000) used the concept of "Community Design", and defined participation as the direct involvement of the society in decision-making processes. He employed the concept of community design to include the concepts of community planning, community architecture, social architecture, community development and community participation, and stated that all these concepts emphasized the inclusion of the habitat of local people in the physical and social development [4].

Afrassiabi (1985) categorized urban renewal participants in three main groups [9]:

- Designers/Planners
- Local government agencies
- Local residents

Participatory design approach allows the designer to interact with individuals during the design process. Thus, as it enables individuals to learn about their needs and manage the process accordingly, it prevents the development of a gap between the design and individuals after the design process is complete and people start living with the design. It allows the creation of a design that was adopted by individuals and experience no difficulties with the function of the design in their lives [10]. Designs that neglect user desires and requirements have to change when they are a part of the user's life. After all, it is them who would live with the built environment [11]. Thus, the more people involved in the decision-making process and are part of the development, the more they own the program, leading to better public decisions [4].

Lang (1987) stated that to understand the successful harmony between the user and the space or vice versa, it is necessary to first interpret the relationship between the designer and the user, proposing a method that does not determine the way people live but based on how they live and feel [12]. Lang recommended that a user-centered design approach should be prioritized [13].

2 Advantages and Disadvantages of Participation

Twigg (2001) itemized the advantages of participation as follows [14]:

- Thanks to participation, individuals could express their actual needs and priorities. Problems could be identified accurately, and adequate measures could be implemented in design.
- Participation provides the opportunity for preemptive interventions to future problems based on individual directions and ideas.
- A versatile discussion environment could be developed with a participatory approach. This environment helps easily solve the problems.
- The process of participation, work and success leads to a strong society. Thus, it improves local organizations, cooperation, harmony, and critical analysis capacity, and strengthens the sense of community against possible difficulties.
- Participatory incentives are sustainable.
- It is the right of individuals to ensure social participation in planning and implementation, and to express their views about the environment that affect their lives. Thus, participation contributes significantly to the development of a democratic society.
- It could be less costly when compared to other approaches since it is sustainable for an extended period of time, due to the fact that the ideas are tested before acceptance.
- Solutions proposed by local residents instead of those proposed by external agencies improve sustainability.
- Since professionals are in close relationship with local residents, solutions are easier to find.

However, despite the advantages of participatory planning, it also has disadvantages and these are listed below [8]:

- Possible conflicts between priorities, interests, values and functions,
- Inconsistencies between the degree of participation and the project,
- Inadequate financial resources for active participation,
- Participation acting as an obstacle to productivity due to the extent of the project duration,
- Red tape that prevents productive results,
- Conflicts between various interests and demands that challenge problem solutions.

Based on the above-mentioned disadvantages, active participation could be possible by balancing all advantages and disadvantages [8].

Sanoff (2000) categorized participation methods into three main groups [4]:

1. Awareness Methods
2. Indirect Methods
3. Group Interaction Methods

Awareness Methods: Awareness methods can be described as preliminary development methods that aim to raise awareness and desire for participation through information before the actual participation.
Indirect Methods: Indirect methods could include questionnaires and surveys conducted to obtain the views of a certain group, to determine associated behavior and to collect data.
Group Interaction Methods: Group interaction methods include focus groups and workshops (charettes) that could include various number of individuals. Focus groups include selected individuals who represent the rest of the community due to common demographics or issues/problems and aim to discuss specific issues and administered by a facilitator.

Apart from these categories, there are methods that employ the defined technology. Various programs that allow video conferencing based on communication technologies could offer a virtual environment where ideas and documents could be shared using personal computers. In the internet environment, virtual platforms where participants can comment on a map and enter their suggestions on a space are among the methods that allow participation [15].

3 Participatory Approach Design Experiences

In this section, two game designs conducted with children, previous applications in Turkey, and projects conducted in other countries in different scales and with different methods are presented.

3.1 Game without Borders Park Project – Turkey

The Game without Borders Park Project was introduced in 2013 in the Nilüfer district in Bursa province on 4700 square meters [16]. It was coordinated by Nilüfer City Council and included several stakeholders and organizations. In the two-semester program, the first semester included theoretical education and the second semester included mainly practical education. In the first semester, seminars that included theoretical instruction were organized about the elements of built environment. The second semester included three successive stages: Basic design, analysis and preliminary design, and playground design in a design studio. The seventeen-week design training included successive steps [17] (Fig. 1).

Fig. 1: Game Without Borders Project design process [18]

The Game without Borders Project was a collective production platform where children and young adults participated in the local administration of the city, had the opportunity to participate in urban space design decisions in their residential area, and supported the democratic life culture and participatory urban identity. One of the goals of the project was the design of a playground design by the participants where disabled and typical children could play together after the built environment training [19].

3.2 Design with Children Game – Turkey

The project was conducted within the scope of a thesis at KTU Faculty of Architecture, City and Regional Planning Department. The main aim of the thesis was to create a real workshop experience with the method of design games to experience the participatory design approach that included the users and the designer, where participants were the children, and to develop a user-centered design model. It aimed to raise knowledge and awareness about the spaces where children spent time. Furthermore, it aimed to investigate spatial aesthetic, technical and functional issues, to allow children to create their designs by developing their skills, and to encourage children to explore their habitat [18].

The study area was the Karayemiş Park located in Trabzon Konaklar District. A workshop was organized in the field. The association between the design process and the design game made adoption of the process easier and quicker by the participants during the workshop. The workshop was conducted with 12 9–10 years old primary school students. The process was conducted in conjunction with various workshops. The workshop was conducted in three stages: Preparation (pre-design) stage, Planning & Design stage, Decision stage (to select between the designed alternatives) (Fig. 2) [18].

Fig. 2: *Design with Children Game design process* [18]

At the end of the process, the most useful, different and most liked sections of all the products were combined based on the consensus of the groups, and a playground design was finalized with children.

3.3 Rize Park Project – Turkey

Rize Municipal Park, an urban park located in Rize province in Turkey, was redesigned in 2011 based on a participatory process and implemented on an area of 3050 m2 in 2015. The project was the first project implemented in Rize with a participatory design approach. The project was developed by a design team (urban planning, architecture and landscape architecture) from Karadeniz Technical University at the request of Rize Municipal Administration and implemented by the municipality [20].

In the first phase, a meeting was organized with the park occupants. Participants were provided with visual presentations, physical data about the region, and information about urban park functions and activities. In the meeting, the demands and expectations of the participants and users were discussed. Their preferences in design elements and park activities were determined. Furthermore, they were allowed to mark their preferences on the current map of the park. Thus, their participation in the process included visual material [20] (Fig. 3).

Fig. 3: Community meetings and a voyage to my dreams activity

In the second phase, the demands of other urban residents and park occupants who did not participate in the meeting were also collected with a survey to ensure their participation. At the end of the process, the project was developed by the design team and presented to the contractor (Fig. 4). Urban residents were involved in the design process and took an active role in the design of their environment. After the construction phase, a second survey was conducted with the park occupants. In this project, it was observed that the participatory design positively affected user satisfaction, and the design was adopted by the occupants. The participatory design process played a key role in ensuring user satisfaction [20].

Fig. 4: *Rize park project* [21]

3.4 Pocket Park Design in Informal Settlements in Cairo – Egypt

In 2017, a study was conducted to transform problem areas and deserted streets into pocket parks to improve the quality of life in a high density urban neighborhood in Cairo. The aim of the study was to design pocket parks that meets the needs and behavior of the occupants. The project criteria included the discussion of existing location, accessibility, spatial design, activities, environmental issues, landscape elements and maintenance issues [22].

Local residents were asked to assess current problems in the area and indicate their priorities. Participants also provided suggestions about the design components of the new park and the type of activities the park should offer. The survey conducted with the participants revealed that the priority of the participants included safety, lighting and regular maintenance in these deserted areas, which were used as solid waste dumps. Furthermore, sitting and resting areas, children's playgrounds, tree shade, edible garden, bicycle tracks were among the top recommendations. Participants especially requested areas where their children could play. In addition to participant suggestions, designer observations demonstrated that the pocket parks should preferably include trees and lawns, since there were no green spaces in these areas (Fig. 5) [22].

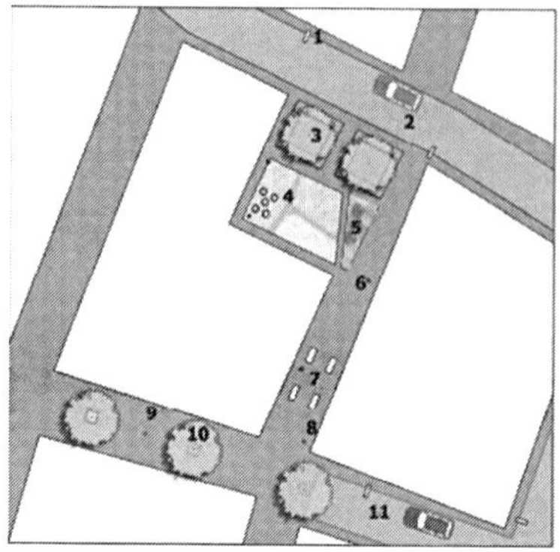

Fig. 5: Cairo Pocket Park Design [22]

3.5 Frontone Park – Italy

Historical park Frontone, located in Perugia, Italy, was the oldest park in the city, and renovated with a participatory method. The municipality, historical and cultural institutions, universities and certain associations contributed to a team of 17 individuals who administered the participatory process. In the user community, 108 people between the ages of 24–82 were involved in the process. Participants were included in the process via a card game named after the city (Perugine) [23].

In the first stage, a website that included documents about the park was developed. The first step was "Get to know the park". The second step was to create a deck where each card represented a proposed design element for the park's renovation and helped the participant to decide. In the 3rd and 4th stages, the participants bid with the cards they selected. Experts also analyzed the selected cards for their compliance with historical values and sustainability of the urban green network. In step 5, the card game began. They selected the cards that represented restoration alternatives for the park. In step 6, experts developed a design alternative for each user hand of cards, and participants were asked to vote for the "winning hand". The design alternative corresponding to the winning hand became the final project for the renovation of the park (Fig. 6) [23].

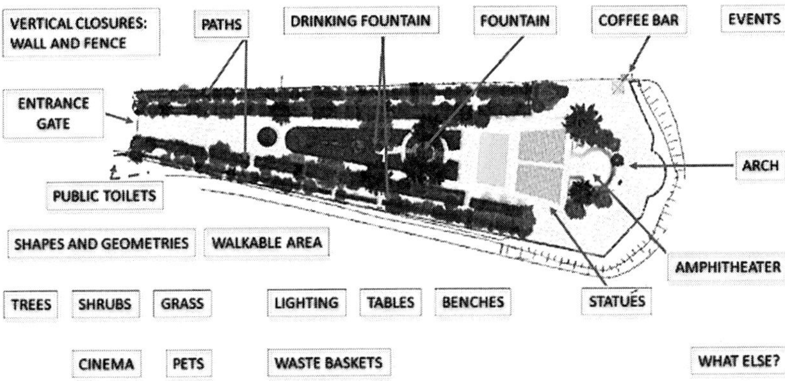

Fig. 6: The interactive map for Frontone Park analysis [23]

3.6 Boston Wharf District Park – USA

The most significant section of the Boston Wharf District Park Project was The Rose Fitzgerald Kennedy Green Axis, which aimed to use the green area acquired with the completion of the Big Dig project, which has been under construction for 20 years in the city and entailed the transfer of the elevated highway to the underground [15]. Boston Rose F. Kennedy Green Road is a 2.5 km long urban green road on an area of about 61 thousand square meters [24]. This project, which provided a connection between Boston urban center and the port, was constructed through an intensive participation process (Fig. 7).

Fig. 7. Wharf District Park Project [25]

In 2002, the Rose Fitzgerald Kennedy Green Axis project was announced to the public and a competition was organized for the design of three associated parks. After four months, EDAW and Copley Wolff Design Team won the competition for the design of Wharf District Park. The company developed an engagement process that involved the public, artists, designers, state and local governments to draft the master plan for downtown Boston. In the planned process, the aim was to ensure participation through group meetings and one-on-one meetings with stakeholders within a few months of the completion of the design. Thus, the local government, central government, non-governmental organizations, local historians, design experts, students and the public were included in the process [26]. A total of 133 public meetings were organized within a year. In public meetings, the concept plan was introduced to the public and photographs of similar areas were shown to raise awareness about the size of the project after the field analysis. Then, the design process was initiated. The final design was submitted and accepted in the autumn of 2004 [15].

3.7 Saint Etienne Place au Changement Project – France

Place au Changement Project is located in Saint-Étienne, France, on a 700 square meter plot at the northeast of the urban center, in an area which is at the intersection of several neighborhoods intersect, including non-zoned and empty lots. This lot was used as a gas station until 2000, when the gas station was demolished to develop a real estate project in the lot, while the land was used as a parking [15] (Fig. 8). It is a square development conducted on a parcel that was evacuated to construct a new building but remained empty due to economic problems, and the implementation followed a year-long design phase in 2011, when a series of workshops were conducted [10].

Fig. 8. Pre-development state of the lot [27]

A competition was organized in 2011 to create solutions for the problems induced by the vacant and dysfunctional project area. The Collectif Etc. team, who won the competition, proposed a model that aimed to design a public area, instead of a final project [15].

The local residents were informed through calls and signs posted in various parts of the neighborhood to build a square park for the residents and to ensure their participation in the development. Based on the public interest and preferences, daily workshops were organized on three themes to ensure participation. These workshops included carpentry activities, gardening activities and graphic design activities [15] (Fig. 9).

Fig. 9: Carpentry and gardening workshops [27]

In the carpentry workshop, which was the first workshop, the pre-determined urban furniture were produced by the skilled volunteering members of the community. The second workshop, the gardening workshop, was organized to design and plan the green spaces in the square. In the graphic design workshop, the elements of the imaginary building were painted in real dimension on the façade of the building facing the square, especially with the participation of children [15].

After the project was completed in 2011, the area was used by the public for various activities. In 2013, certain changes were implemented in the park, also with the participation of the public. In the central section, sets were constructed and flowered, the mural and a few urban furniture were colored and new urban furniture were added [10] (Fig. 10).

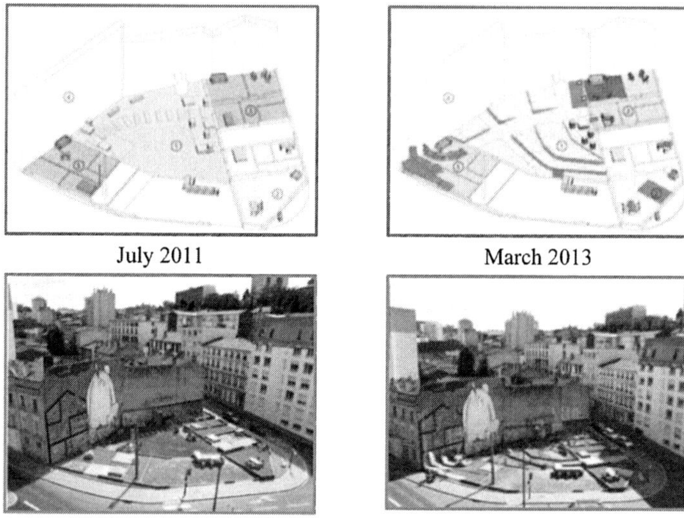

Fig. 10: The reorganization of Saint Etienne park in 2011 and 2013 [28]

3.8 Leinefelde ZukunftsWerkStadt Project – Germany

This project was implemented in Germany in 2007 with the participatory design approach. In the 90s, the Municipality of Leinefelde adopted an innovative, integrated, participatory approach to reverse depopulation, poor economy and the large number of vacant and mismanaged homes. In the project, group interaction method, open-ended methods, workshops, public meetings were employed. Furthermore, surveys, one-on-one interviews and workshops were also conducted. Project stakeholders included urban planners, designers, architects, occupants, landscape architects, civil engineers, and institutions such as the municipality of Leinefelde and local housing communities. The participatory approach was adopted by stakeholders and target groups, and active participation was achieved in the project. Furthermore, active participation in local development programs played a key role in establishing deep ties in the society [10] (Fig. 11).

Fig. 11. *Leinefelde ZukunftsWerkStadt project* [29, 30]

4 Conclusion

When the participants have the opportunity to have a say in decisions about their environment and take part in the design and planning processes, they could adopt these spaces better and take the responsibility to protect these spaces. Thus, ensuring the participation of the community in projects conducted with a participatory design approach would increase the occupancy in these spaces. Playing an active role in the design of the environment is quite important for the occupants to adopt urban design and life. Sanoff (2000) suggested that the benefits of participatory work included the improvement of social capital and promotion of a sense of community, and participatory research is an attitude associated with the transformative power employed in the development and operation of human environment [4].

It is very important for the public to produce a urban rhetoric about their environment, and to play a role in participation-oriented projects. The spatial requirements and design decisions of the participants determine the quality and sustainability of the built environment. The public is directly or indirectly effective on the planning and design of the built environment.

The contribution of the individuals, who are a part of the society, to the design of the built environment should be ensured to strengthen the sense of belonging to their environment, and their awareness about their voice in decisions about their common habitat should be raised. As their knowledge on the built environment develops, the participants would be more concerned about and courageous to take responsibility for their environment.

References

1. Petric, J., Maver T., Jonti G. and Ucelli G., 2002. *Virtual Reality in the Service of User Participation in Architecture*, CIB w78 Conference, Aarus School of Architecture, Aarus, p.12–14.
2. Habraken, J.N., 1985. *The Appearance of the Form*, Awater Press, Cambridge.
3. Baba, E. C., 2010. *Tasarım Sürecinde Katılımcı Yaklaşımlara Dair Bir Değerlendirme İstanbul Levent Bölgesinde Alan Çalışması*, Tasarım-Kuram Dergisi, 9–10, p. 21–53.
4. Sanoff, H., 2000. *Community Participation Methods in Design and Planning*, Wiley, New York.
5. Baheshti, M.R., 1985. *Introduction, Proceedings of the International Design Participation*, Conference–Design Coalition Team, TU/e, Eindhoven, 6.
6. Sanoff, H., 2008. *Multiple Views of Participatory Design*, Archnet-IJAR, Volume 2- Issue, p. 57–69.
7. Özaloğlu, S., ve Kennedy, N. F., 2009. *Yönetimin kalitesi kentlerin yaşanabilirliğinde gizli: Katılımcı mimarlık: Nereye kadar?*, Mimarlık Dergisi, 346.
8. Gülersoy, N. Z., Özsoy, A., Tezer, A., Genli Yiğiter, R. ve Günay, Z., 2012. *Mevcut Kentsel Dokuda Çevresel Kalitenin İyileştirilmesi, Stratejik Kalite Planlaması Modeli*, İstanbul Teknik Üniversitesi, İstanbul, 11–12, 63.
9. Afrassiabi, A. H., 1985. *Design Participation in The Context of Urban Renewal, Design Coalition Team, Volume One*, Proceedings of The International Design Participation Conference.
10. Bulut, H. B., 2016. *Mimarlıkta Katılımcı Tasarım Yaklaşımlarının Analizi Ve Küçükçekmece Örneğinde İrdeleme*, Yüksek Lisans Tezi, Yıldız Teknik Üniversitesi, Fen Bilimleri Enstitüsü, İstanbul.
11. Faga, B., 2006. *Designing Public Consensus: The Civic Theater of Community Participation for Architects, Landscape Architects, Planners, and Urban Designers*, John Wiley and Sons, United States.
12. Lang. J., 1987. *Creating Architectural Theory, The Role Of The Behavioral Sciences in Environmental Design.* Van Nostrand Reinhold Company, New York.
13. Artu, M., 1985. *Modern Mimarlık Hareketinin Uygulama Yapan Mimarların Tasarımlarına Etkisi*, Mimarlık Dergisi, 85, 5–6.
14. Twigg, J., 2001. *Guidance notes on participation and accountability.* Londra, UK.
15. Akyol, İ., 2014. *Açık Alan Tasarımında Katılım Yöntemlerinin Etkinlik Düzeyi Değerlendirilmesi*, Yüksek Lisans Tezi, İstanbul Teknik Üniversitesi, Fen Bilimleri Enstitüsü, İstanbul.

16. https://www.arkitera.com/haber/niluferde-oyun-engel-tanima z-parkinin-temelleri-atildi/. 07.04.2020

17. Arın, S., 2015. *Çocuklara Yönelik Yapılı Çevre Eğitimi: Bursa için Katılımcı Bir Model*, Doktora Tezi, İstanbul Teknik Üniversitesi, Fen Bilimleri Enstitüsü, İstanbul.

18. Aktürk, Z., 2019. *Katılımcı Tasarım Deneyimi: Çocuklarla Tasarım Oyunu.* Yüksek Lisans Tezi, KTÜ Fen Bilimleri Enstitüsü, Trabzon.

19. Arın, S. and Özsoy, F. A. (2015) . *Children Participation in Urban Space Design.* İdeal Kent, 182–201.

20. Turan, S. Ö., Pulatkan, M., Beyazlı, D., & Özen, B. S. 2016. *User evaluation of the urban park design implementation with participatory approach process.* Procedia-Social and Behavioral Sciences, 216, 306–315.

21. Turan, S. Ö., Pulatkan M. and Beyazlı, D., 2011. *Rize Belediye Parkı Tasarımı*, Turkiye

22. Abd El Aziz, N. A. (2017). *Pocket park design in informal settlements in Cairo city, Egypt.* Landscape Architecture Regional Planning, 2(1), 51–60.

23. Menconi, M. E., Tasso, S., Santinelli, M., & Grohmann, D. (2020). *A card game to renew urban parks: Face-to-face and online approach for the inclusive involvement of local community.* Evaluation and Program Planning, 79, 101741.

24. Kurtaslan Öztürk, B., 2014. *Kentsel Yeşil Yollar ve Sürdürülebilirlik: Boston Rose Fitzgerald Kennedy Yeşil Yolu Örneği*, Ulusal Hakemli Tasarım ve Mimarlık Dergisi, 2, 1.

25. https://external-preview.redd.it/AHKQrKRTd9678IsHN1aDYvVLQZ1F lF6ZDtNHD_AGozM.jpg?auto=webp&s=fbb4de9599a6ebca6c9e8a795d21151 18b483260. 21.10.2017

26. Graves, H., 2004. *Wharf Dıstrıct Parks On The Rose Kennedy Greenway*, Boston.

27. http://www.publicspace.org/en/works/g027-place-au-changement. 07.04.2020.

28. http://www.collectifetc.com/realisation/place-au-changement -chantier-ouvert/ 07.04.2020.

29. The Building and Social Housing Foundation web sitesi, www.bshf.org/ worldhabitat-awards/winners-and-finalists/zukunftswerk stadt-leinefelde/ 19.02.2019

30. https://www.world-habitat.org/world-habitat-awards/winners-and-finalists/ zukunftswerkstadt-leinefelde/#award-content 19.02.2019

Demet DEMİROĞLU

Resistant Cities to Climate Change and Planning Interaction

Abstract: Today, climate change is regarded as one of the biggest environmental problems facing the world. The implementation of climate change mitigation/adaptation actions and strategies in cities, which are the largest production and consumption centers where more than half of the global population lives, has a key role in creating change-resistant cities. At this point, cities that are suitable/integratedly planned in various areas such as land use, green infrastructure, transportation, building, disaster and waste management will become more resistant to the negative effects of change as well as better adapting to change. In this study, which focuses on the importance of spatial planning in creating climate resistant cities; first of all, the two-way interaction between climate change and cities has been summarized; then, within the framework of the resilient city concept definitions, a framework regarding the interaction between mitigation and adaptation areas for climate change-resistant cities and the spatial plans of these areas is presented.

Keywords: Climate change, urban resilence, spatial planning.

1 Introduction

As evidenced by many international scientific and reliable researches, together with the Intergovernmental Panel on Climate Change (IPCC) studies, the climate is changing in the globe. In this context, global warming and climate change as one of the most important problems of the current century [1, 2]; It is considered as a cause of increasing natural disasters such as dust and sandstorms, droughts and floods [3, 4]. Studies on this issue have shown that the warming experienced since the middle of the 20th century is mainly due to the increase in greenhouse gas concentrations due to emissions from human activities and agrees upon that this increase is largely due to fossil fuels and land use changes [5,6,7,8]. Cities that contain a significant part of the world's population and economic activities are considered the most important areas responsible for climate change, as they are the places where these activities are most intense [5,8,9].

According to the IPCC 5th Assessment Report, in and around urban areas; Problems such as drought caused by climate change, increase in precipitation, floods and overflows and associated erosion, difficulties in water supply and air pollution tend to increase [5]. While cities are responsible for the disasters and other dangers that occur as a result of environmental destruction caused by the activities carried out in them; On the other hand, they are the biggest victims of these disasters, shocks and stresses. It has been observed that there has been a significant increase in the number of globally recorded disasters starting from the 2000's. It is seen that the characteristics of the disasters that took place in the last 10–20 years have changed, and have produced results that affect not only a certain region, but also larger areas in the economic, social and environmental aspects within the domino effect. In particular, the number, frequency and severity of hydrological disasters (floods and overflows, etc.) and climatic disasters (hurricanes, extreme temperatures, drought and fires, etc.) are increasing. Cities are highly dependent on ecosystem services. These disasters cause serious environmental and economic damages in cities, loss of lives and significant damages in socio-cultural areas [10].

The capacity of cities' preparedness, response, and adaptation to the problems, dangers, disasters and risks they face determines their resilience, which is called urban resilience [11]. When considered in the context of climate change, the urban resilience approach requires a planning that can overcome the problems created by climate change and develop strategies to support adaptation to climate change [12, 13, 14].

At this point, it is extremely important to consider the reality of climate change in city plans and to make long-term city plans that anticipate climate risks in order to make urban life more sustainable and cities more resistant to disasters caused by climate change. Because city plans directly affect the shaping of energy consumption patterns that arise in different areas of need (e.g. housing, transportation, production, etc.) in the living environment they produce. Considering that urban energy consumption patterns are directly related to greenhouse gas emissions, it is seen that urban planning has a serious role both in combating climate change and adaptation to climate change. Related studies; highlights the important relationship between land use, urban form, urban infrastructure and energy consumption and greenhouse gas emissions [7,15,16,17,18,19]. An urban ecosystem in which nature is too far away from the city is doomed to self-exhaustion in the long run. In ideally defined urban ecosystems, urban areas must be a set of systematic processes that are self-sufficient and also feed natural processes [20].

The formation of cities involves the production of different spaces that vary between the urban scale and the architectural scale. Each scale produces different patterns of energy use in itself, while it contributes to different sizes in terms of greenhouse gases released into the atmosphere depending on the amount of fossil fuels used during energy production. Urban planning includes the analysis of many sub-issues such as the main-form of the city, the layout of the city parts, neighborhood textures, street/street patterns, building typologies within a holistic system. When considered in the context of climate change, this analysis should be designed in a way that minimizes carbon emission and therefore creating the lowest level of fossil fuel dependent energy need [7,17,18,19,21].

In this study, which focuses on the importance of planning in creating climate resistant cities, first of all, the two-way interaction between climate change and cities (both the contributions of cities to climate change and the risks caused by climate change for cities) is summarized, and then, within the framework of the definitions of the resilient city concept, for cities resistant to climate change. areas of reduction and adaptation are specified. In the last part of the study, a framework is presented on how to integrate these mitigation/adaptation areas into spatial plans.

2 Cities and Climate Change

There is a two-way interaction between climate change and cities. Cities are settlements that both affect and are affected by climate change. For this reason, it is useful to give this section as two subtitles.

2.1 The Effects of Cities on Climate Change

Nowadays, it has been proven that 95 % of climate change is caused by human activities [5] and climate change is recognized as one of the biggest environmental problems that world faces. The United Nations Framework Convention On Climate Change, which entered into force in 1994 defines Climate change as: "*a change of climate which is attributed directly or indirectly to human activity that alters the composition of the global atmosphere and which is in addition to natural climate variability observed over comparable time periods*" [22]. Climate change in IPCC usage refers to a change in the state of the climate that can be identified (e.g. using statistical tests) by changes in the mean and/or the variability of its properties, and that persists for an extended period, typically decades or longer [23].

In the urban era we are in, cities hosting 55 % of the total population [24] constitute 80 % of the global gross income. Considering that 65 million people migrate to cities every year, this situation not only makes cities the engine of the global economy, but also makes them important actors of climate change. Today, it has been proven that cities are responsible for 71–76 % of global carbon emissions, 75 % of natural resource consumption and 60–75 % of primary energy consumption. Global carbon footprint rates are particularly concentrated in high-income cities [19,24]. Cities, which are multi-sector formations, play an active role in the formation of greenhouse gas emissions resulting from the consumption of non-renewable energy sources. Greenhouse gas emissions of cities are mainly caused by urban activities such as industry, trade, construction, transportation, and building and use of buildings. The building sector is among the sectors that create the highest greenhouse gas emissions. Greenhouse gas emission is generated by the energy consumed in buildings to provide thermal comfort through heating and cooling systems. Similarly, systems, infrastructure and networks used in the transportation sector are among the main sectors in emission generation. Like these, different sectors that form the basis of urbanization also contribute to climate change to different degrees [6,15,17].

In the 5th Assessment Report of the IPCC [5] which is accepted as the most up-to-date reference in this regard, it is stated that the greenhouse gas emission, which increased with the industrial revolution, increased by 78 % between 1970 and 2011and the greenhouse gas emissions between 2000 and 2010 are the highest values ever measured. Parallel to this increase, the report emphasizes that the change in land and ocean temperatures in the globe started after 1880 and the temperature increased by an average of 0.85 °C (0.65 to 1.06 °C) between 1880 and 2012. Also, it is stated that the total increase between the average temperature in the 50-year period covering the years 1850–1900 and the average temperature in the 9-year period covering 2003–2012 is 0.85 °C. According to more recent data from the World Meteorological Organization (WMO), the global average temperature for 2016 is 0.83 ℃ above the average of the 1961–1990 reference period. In the related report, it was stated that this average varies according to latitude degrees and that in the same year, the annual average temperatures on the high latitudes of Russia, Alaska, Canada and Norway coasts were at least 3 ° C above the 1961–1990 average [25]. There is a warning that if effective measures are not taken to reduce greenhouse gas emissions and if the existing policies/practices are satisfied, the increase will exceed 1.1 °C by 2100 and even the temperature increase will be 1.5 °C between 2030 and 2052 [5, 26].

Observations and projections for 100 cities examined within the scope of Urban Climate Change Research Network (UCCRN) Second Assessment Report on Climate Change and Cities (ARC 3.2) are as follows [27]:

- The temperatures in the cities around the world are increasing due to both climate change and the heat island effect in the cities. When the annual average temperature increase trends in the cities are examined within the scope of ARC3.2, it was observed that the average annual temperature in 39 cities increased by an average of 0.12 to 0.45 ° C for each 10-year period between 1961 and 2010.
- It is estimated that the annual average temperatures in the 100 cities examined within the scope of the report will increase between 0.7 and 1.5 ° C in 2020s, 1.4 and 3 ° C in 2050s, and 1.7 and 4.9 ° C in 2080s.
- It is predicted that there will be excessive fluctuations in the annual average amount of precipitation in these 100 cities, in this context, the amount of precipitation will be increased in the 2020s (-7 %, + 10 %), 2050s (-9 %, + 15 %) and 2080s (-11 %)., + 21 %).
- 52 coastal cities were studied with ARC3.2. and It was estimated that the sea level rise will be between 4 and 9 cm in the 2020's, between 15 and 60 cm in the 2050's and between 22 and 124 cm in the 2080's in those cities.

Scientific estimations for the future predict that cities will be increasingly affected by climate change. In addition, according to estimates, the global urban population will increase from approximately 4 billion today to 6.5 billion in 2050, and urban infrastructure will inevitably grow with this population. Subsequently, the water, energy, sewerage, transportation, communication and service infrastructures of cities will be directly affected by climate change and this effect will expose the city and its inhabitants to some economic, social, health and environmental risks [28].

2.2 Risks Created by Climate Change for Cities

While cities are at the center of activities that contribute to climate change; On the other hand, they are the places most affected by climate change. Due to climate change, temperature fluctuations, urban heat island formations are observed in cities, and floods and overflows, drought, disease and deaths increase due to the increase in extreme weather events. Extreme weather events also affect urban infrastructures, threatening urban services such as water treatment and drainage, transport, energy, drinking water and sewage. Settlements in coastal areas are exposed to flooding as a result of increasing sea levels. More than 2/3 of the world's largest cities are vulnerable to rising sea levels as a result of climate change. Living conditions, which are already difficult in cities due to climate refugees, are getting more difficult and urban poverty is increasing. The number of urban poor people, most of whom are victim of natural disasters triggered by climate change, is expected to reach 325 million by 2030. This situation will increase the struggle for scarce resources. According to the risk analysis carried out in 616 major metropolis today, it is known that 1.7 billion people (i.e. approximately 25 % of the world population) and approximately half of the global GDP are under risk of flooding and overflow [5,15,26,29,30].

The destructive and negative effects of climate change on cities are intertwined and each triggers the other. It is very important to accurately and comprehensively define the current and future impacts of climate change, the risks and threats that cities may face in the decisions to be taken and in the policies and strategies to be determined. In this way, it will be possible to determine effective adaptation measures in line with the nature of risks and threats faced by cities [7,31,32]. Planning for Climate Change: A strategic, values-based approach for urban planners report [7] emphasizes that the severity of the effects that cities experience/may experience due to climate change vary according to the geographical location of the city, its specific geographical characteristics, its sensitivity to climate change hazards, and the rate of vulnerable population (such as vulnerable groups - poor, women, youth, elderly people).

The carbon footprints of cities also differ. Carbon emission rates generally vary according to the size and planning of the cities, the energy sources used and the life styles of the population. Although buildings, transportation and energy consumption practices have the highest share in urban-based carbon emissions, industrial production also contributes significantly to urban-based greenhouse gas emissions in cities where industrialization is concentrated, depending on their economic structures. Accordingly, cities with larger populations, higher income levels and higher consumption cause more carbon emissions into the atmosphere [19].

Other risks created by climate change for cities, such as temperature increase, increase in precipitation, rise in sea level, increase in extreme weather events (such as storms, hurricanes), are given in Tab. 1 and the risks arising from climate change and the potential primary and secondary effects of these risks are summarized in the related table [7]. When the table is examined, it is seen that these destructive effects have a wide range in the field of natural environment, public health, social life and economic activities.

3 Resilience and Urban Resilience

In recent years, the popularity of the concept of "resilience" has increased in both academic and policy fields. [33,34,35]. The concept specifically provides insights into climate change [36,37,38], complex socio-ecological systems and their sustainable management [11].

So what exactly is meant by the term 'urban resilience'? The etymological origins of resistance originate from the Latin word "resilio" which means "to return" [39]. Although the concept has a long history in engineering, psychology, and disaster literature [40], the academic paper by ecologist Holling (1973) on the resilience of ecological systems is often cited as the origin of modern resistance theory [39,34].

Because it is argued that the concept of resistance, and especially ecological resilience, is more suitable for complex systems that are in a continuous flow state, and therefore must adapt to change and uncertainty. Meerow *et.al* (2016) found that cities are definitely complex and dynamic systems, and most definitions of urban resilience are more compatible with ecological resilience [11].

The term resilience was introduced in scientific research, initially associated with ecological systems by Holling (1973): "*Resilience determines the persistence of relationships within a system and is a measure of the ability of these systems to absorb changes of state variables, driving variables, and parameters, and still persist*". In this definition, resilience is the property of the system and persistence or probability of extinction is the result [41].

The concept of urban resilience is essentially a transformational structure rather than a fixed feature, and involves the change and development of the urban system in the face of changing conditions [10]. The origins of the term "resilience" refer to strength and resilience, but in new practices in ecology, socio-ecological systems, disaster management, and urban sustainability, it is understood that resilience requires flexibility, learning, and change [12, 41].

Therefore, in the context of urban resilience, it is a matter of reaching and developing its new normal state rather than returning to its previous state after the shock or stress experienced. As a resilient city is exposed to impact, it will move from a fixed and stable state to a process of adaptation and change in the process [11]. Undoubtedly, the realization of these stages is directly related to the capacity of the urban system to absorb/overcome the impact, adaptation and transformation. There are different opportunities such as economic, social resources, infrastructure, skills, innovation and social networks that will enable such a development in cities [20,42].

Tab. 1. Climate change impacts

CLIMATE HAZARD	POTENTIAL PRIMARY IMPACTS	POTENTIAL SECONDARY IMPACTS
Increased Temperatures	• Groundwater depletion • Water shortages • Drought • Degraded air quality (smog)	• Water shortages • Distress migration to cities/towns due to droughts in rural areas • Reduced food supply and higher food prices • Potential energy price increases (e.g. from reduced hydroelectricity generation in places where it exists) • Exaggerated urban heat island effect • Increased energy demands for cooling • Need for higher or additional wastewater treatment • Population health impacts (e.g. increased mortality during heat waves, decreased access to food/nutrition)
Increased Precipitation	• Increased flooding • Increased risk of landslides or mudslides on hazardous slopes	• Reduced food supply and higher food prices • Property damage (homes and businesses) • Disruption of livelihoods and city/town economies • Damage to infrastructure not designed to the standards necessary to withstand the occurrences being experienced • Distress migration to cities due to floods in rural areas • Displacement and population movement from informal settlements built on steep slope hazard lands, etc. • Increased vector borne diseases (malaria, dengue,encephalitis) and water borne diseases (acute diarrhoea, cholera, dysentery)
Sea Level Rise	• Coastal flooding • Salt water intrusion into • groundwater supplies in coastal areas • Increased storm surge • Hazard • Coastal erosion	• Displacement and population movement from coastal areas • Property damage (homes and businesses) • Damage to infrastructure not designed to the standards necessary to withstand the occurrences being experienced • Disruption of livelihoods and city/town economies • Reduced food supply and higher food prices • Population health impacts (e.g. injuries, increased mortality and illness) • Loss of productive/residential land due to erosion
Increased Extreme Weather Episodes (storms, cyclones, hurricanes)	• More intense flooding • Higher risk of landslides/ mudslides on hazardous slopes • Intense and disastrous wind speeds	• Property damage (homes and businesses) • Damage to infrastructure not designed to standards of occurrences being experienced • Population health impacts (e.g. injuries, increased mortality, distress) • Disruption of livelihoods and city/town economies • Reduced food supply and higher food prices

Resilience has also been an important concept in current debates on climate change and adaptation. IPCC defines resilience as *'the ability of a social or ecological system to absorb disturbances while retaining the same basic structure and ways of functioning, the capacity of self-organization, and the capacity to adapt to stress and change.'* [23]. This definition is broadly consistent with definitions from ecological sciences [43, 12].

To better understand how the term is defined and used across disciplines and fields of study, Meerow *et al.* (2016) [11] reviewed forty years of academic literature on urban resilience since 1973. The article proposes a new definition of urban resilience that is flexible enough to be further adopted by a range of disciplines and experts to advance science and practice. "*Urban resilience refers to the ability of an urban system-and all itsconstituent socio-ecological and socio-technical networks acrosstemporal and spatial scales-to maintain or rapidly return to desiredfunctions in the face of a disturbance, to adapt to change, and to quickly transform systems that limit current or future adaptivecapacity".* In this definition, urban resilience is dynamic and offers multiple pathways to resilience (e.g. permanence, transition, and transformation). Acknowledges the importance of temporal scale and advocates general adaptability rather than specific adaptability [11].

In relation to the concept of resilience, which has become increasingly important in the cities and climate change literature, common terms such as "climate proof", "climate resistant" and "resilient city" emphasize the idea that cities, urban systems and urban decision makers should be able to quickly return from climate-related shocks and stresses[36]. The need to discuss urban climate resilience lies in the threat to cities' continued well-being. Developing cities are closely related to the development, progress and growth of human society, and climate change is considered a disturbance that jeopardizes continued progress. Although cities generally have a natural capacity for survival and some believe that cities already have great resilience, they are not immune to the enormous gravity of external changes, an inertia that interferes with the process of systemic change [44].

There is a wide range of definitions of urban climate resistance in the scientific literature [11]. . Meanwhile the concept is defined as "*A disaster resilient city can be understood as a city that has managed . . . to: (a) reduce or avoid current and future hazards; (b) reduce current and future susceptibility to hazards; (c) establish functioning mechanisms and structures for disaster response; and (d) establish functioning mechanisms and structures for disaster recovery"* by Wamsler *et al.* (2013) [45]; Leichenko (2011) [36] defines it as "*The ability of a city or urban system to withstand a wide array of shocks and stresses";* Lu and Stead (2013))[46] defines it as:*"the ability of a city to absorb disturbance while maintaining its functions and structures"* and Thornbush et al. (2013) [47] use the definition "*a general quality of the city's social, economic, and natural systems to be sufficiently future-proof".* Although the definitions of urban climate resistance differ from each other, the common point of the definitions is that the definition of resistance in terms of the general capacity to cope with climate effects and damages [11,34,41].

4 Mitigation and Adaptation Areas for Cities Resilient to Climate Change and the Relationship of These Areas with Planning

International concerns are growing about how to address the impacts of climate change for urban areas, especially in developing countries where cities are growing rapidly and a large proportion of the urban population is poor or vulnerable to climate-related damage. Relevant literature points to the need for measures, including infrastructure and capacity investments, to adapt to projected climate impacts [7,15,48,49].

When the studies including the measures taken to create cities resistant to climate change are examined; It is seen that two concepts called reduction (mitigation) and adaptation come to the fore. The concept of mitigation defined as, "*Activities that help to reduce the rate or magnitude of climate change by reducing humangenerated greenhouse gas emissions or land use practices that contribute to climate change, such as deforestation*" [7]. United Nations Framework Convention on Climate Change (UNFCC) defines adaptation as "*the adjustments in ecological, social, or economic systems in response to actual or expected climatic stimuli and their effects or impacts. It refers to changes in processes, practices and structures to moderate potential damages or to benefit from opportunities associated with climate change*". Adaptation does not mean that the negative impacts will be avoided, only that they will be less severe than if no planning had occurred [7].

Mitigation is the general name of practices aimed at reducing/zeroing greenhouse gases that cause climate change and protecting/increasing carbon sinks on the planet. Giving up fossil fuel consumption is key to reducing/stopping greenhouse gases. On the other hand, adaptation defines the practices that aim to ensure that society and communities live with the least damage against the current negative effects of climate change. Mitigation and adaptation should be addressed in a coordinated and complementary way in their solutions. The goal of mitigation policies is to reduce current and future greenhouse gas emissions, especially from energy generation, land use, waste management, industry and transportation sectors.. The main purpose of adaptation policies is to be resistant to the effects such as sea level rise, floods, droughts, storms and heat islands, and to prepare cities for adverse effects. The most important reason why adaptation and mitigation policies are handled separately from each other is that, historically, mitigation policies seek a solution to a global issue, while adaptation policies are seen as policies implemented at a local scale [7,17].

Cities, which are important in reducing the effects of climate change and adapting to changes, are also the information and innovation centers of economic, political and cultural activities. The development and implementation of spatial plans based on climate change adaptation and mitigation, taking into account the conditions of cities such as economy, culture and geography, with their assets and capacities, will play an important role in minimizing these effects [7,14,46,50].

UN planning for climate change report [7] emphasizes that the links between sustainable urban planning and adaptation to climate change are strong and clear, and good planning practices - due to their nature and also climate smart planning practices - should be implemented in cities. At this point, the planners' responsibilities are as follows:

- To minimize the risk occurring in or near areas with flood, slope or coastal danger and to improve the land development activities,
- Improving infrastructure for rainwater management, solid and liquid waste management, access to safe drinking water and transport of goods and people,
- Protecting towns and cities and their surrounding ecosystems and environmentally sensitive areas,
- Improving disaster risk reduction (especially weather and climate related events), including improving response capacities for disasters,
- To support local economic development in order to reduce poverty and improve the quality of life.

Cities are both perpetrators and victims of climate change are also one of the most important solution areas. Cities have a great potential and opportunities to come up with the necessary solutions to stop and reduce the devastating effects of climate change. While spatial plans are produced under the planning and coordination of local governments, the main areas where mitigation and adaptation measures required are; ***Land use/urban form, transportation, water resources management, disaster management, urban green infrastructure, waste management and buildings*** [7,15,16,17,18,19,50].

a Land Use and Urban Form

Planners are directly involved in shaping and controlling physical land use and development patterns (urban form). Paying attention to physical land use and urban form is an important way for planners to manage and adapt to the impacts of climate change.

The planners have a wide range of responsibilities for making the most appropriate land use plans within the protection-use balance suitable for the natural (water resources, geology, soil, topography, climate, etc.) and cultural landscape characteristics (such as socio-economic structure, demographic structure) of the city and its environment [51,52,53,54,55,56,57,58]. Urban plans, which should be the main determinant of urban development, should be designed in a way that protects natural resources and covers the future of the city for long periods. In this context, it is imperative to adopt the principle of coordination rather than authority confusion, especially in the legislation, plan preparation and implementation processes for zoning plans. Because the zoning plans, which are responsible for solving the basic shelter needs of the people, should take the most appropriate urban growth process as a basis, while protecting natural resources as a principle [59]. In all urban planning based on protecting and maintaining the ecological structure, the natural and cultural resource inventory of the area must first be determined and then a planning must be made according to the existing potential [60].

Planners play an important role in the protection and development of biodiversity, environmentally sensitive areas and ecosystems. For example, planners can help minimize zone use changes in environmentally sensitive areas, especially in important coastal habitats such as estuaries, wetlands and mangrove forests, or prohibit city development towards these areas. Keeping the urbanization outside of these areas can help protecting the city against river floods, erosion and sea storm surges [7]. In addition to the environmental benefits of these actions (e.g. habitat protection), other climate benefits can be realized. For example, open-green spaces in and around the city, which are important components of green infrastructure, can help cool the air and provide shade to help limit urban heat island impacts. At the same time, they act as carbon sinks to reduce carbon emissions and makes important contributions, especially in urban water resources management [61,62,63,64,65].

In addition to being involved in land use determination, planners can help shape settlement patterns to reduce and minimize exposure to climate-compromised terrain (e.g. steep and uneven slopes, flood zones, coastal areas subject to sea level rise and storm surges) [7].

Urban form is an important component of a city's adaptive capacity and resilience, and how sensitive people, places, institutions and industries are to climate change risks. From a mitigation perspective, urban form is also the main driver of a city's greenhouse gas emissions. From this point of view, urban plans that encourage more compact development models may affect many sub-systems of the city, such as the degree to which the city benefits from local climatic conditions (e.g. prevailing wind direction, solar radiation, etc.), urban transport systems network and infrastructure systems; thus may help to reduce the energy consumption and associated greenhouse gas emissions. In the recent years, within the scope of the studies aimed at preventing and adapting to climate change, it is seen that the compact city form has been increasingly emphasized. In this framework, compact city form and urban development have recently become important components of urban climate policy. In the compact city, unlike the fringed city, it has been proven that the energy used for heating and cooling, the energy used in urban transportation and infrastructure costs are reduced [7,15,16,17,18,19,66,67,68,69,70].

With the compact design of cities or the prevention of urban fringes, transportation distances are shortened and time and energy spent for transportation can be saved. If the density of an average neighborhood is doubled, the vehicle use of each household decreases by 20–40 %. Public transport investment costs are also reduced, as more passengers will be transported over shorter distances [19,71]. Moreover, there is an opportunity to save money and energy spent on infrastructure in compact cities. For example, in fringed cities, it is known that 70 % of the water network cost is caused by pipes and 30 % of the energy consumed in urban infrastructure is spent for pumping clean water and collecting waste water [72].

Compact cities also provide effective protection of natural areas around the city, such as agriculture and forest areas, which will act as biological sinks and reduce the effect of urban heat islands [15,67,68,70].

In addition to the main-form of the city, the building layout, which affects the formation of neighborhood textures in the sub-parts of the city, also plays a role in the formation of energy consumption patterns due to use. For example, in cold and windy climate zones, the amount of energy consumed for heating in buildings that were constructed in an adjacent and seperate manners will differ from each other. In this regard, it is important to analyze the prevailing climatic conditions well and to determine the appropriate structure layout for the local area. Another point is that the formation of urban form and building layout should be considered in a holistic way. On a purely building scale, parcel-based solutions cannot be an effective strategy for reducing the carbon emissions that the city's final form will produce. For example, the adjacent construction layout plays an active role in reducing the energy consumed for heating purposes in residential buildings, while continuous canyons present in this layout can lead to the formation of unwanted wind corridors. Similarly, seperately constructed multi-store building blocks, which are seen in fringing urban forms, trigger vehicle-dependent life models, especially in hot and arid climates, since they are not effective in creating urban shadows. This is the main reason why the ongoing zoning practices produce cities that are insensitive to climate effects [17,21,69].

b Transportation

The transportation sector is one of the key sectors for climate change due to its high energy consumption. Two-thirds of the world's oil consumption belongs to the transportation sector. The transportation sector accounts for 27 % of global carbon dioxide emissions based on energy use. By 2050, a total of two billion more vehicles are expected to be added to the existing automobile stock worldwide. It is estimated that this will increase the transport-related emissions by 120–230 %. Much of this increase will be the result of urban population growth and land cover expansion at unprecedented rates. As more and more people move to cities, urban transport systems will play an increasingly critical role in driving responsible climate action and reducing global emissions [73,74]. According to the calculations, in order to keep the temperature increase at 1.5 ° C, it is necessary to reduce the carbon dioxide emissions from transportation by 2050 to 70–80 % below the 2015 level [15].

Urban transportation is of particular importance within the transportation sector. 60 % of the total distance traveled around the world takes place in cities. This places great responsibility on cities in the transformation that should be experienced in the transportation sector [73,74,75].

Planners can help reduce the greenhouse gas emissions by trying to reduce vehicle miles traveled and urban congestion through strategies such as compact, high-density, mixed-use development. In addition, with the use of environmentally friendly, clean and sustainable methods in urban transportation planning, transportation becomes accessible to everyone, equitable and fair. For example, urban transportation, which can also be done by non-motorized methods, is carried out by using motor vehicles in many cities, and this increases the energy consumption and pollutant emissions caused by transportation. Whereas, by promoting modes of transport such as walking and cycling, energy consumption from part of daily urban journeys can be reduced to zero. [7,32,76,77,78]. Creutzig (2016) [79] has detected that effective urban planning can reduce green house gas emissions from urban transport between 20 % and 50 %.

c Water Resources Management

Climate change affects water resources in terms of both water quality and water quantity. The water-related impacts of climate change in many cities around the world show its effects as exposure to extreme drought or excessive rainfall. This threat posed by climate change on water resources reveals a serious need for water resources management in terms of the sustainability of the social and economic activities of the cities. The rational and sustainable use of water resources is an issue that can be achieved with the integration of spatial and inter-sectoral planning and decision-making processes [80, 81,82].

For this reason, strategies and actions related to water resources in climate change adaptation plans should be handled in an integrated manner with water resources strategies, which take place in different places in the hierarchy of spatial plans. Ground waters, artificial or natural reservoirs, urban areas and agricultural areas require different water management strategies [83]. Water management adaptation strategies should be formulated by considering the local risks of climate change. Lack of water or excessive water exposure in urban areas can be eliminated by different adaptation strategies [17]. Planners, for example, are critical to improving a city's rainwater management infrastructure. This, in turn, a key determinant about a resilient city's ability to manage and adapt to rainfall and flood events (ie more frequent and heavy rainfall) associated with climate change. While some developments are inevitable in climate change susceptible areas, planners can help organize and control building forms (e.g. keeping habitats above flood risk levels) and can make designs to incorporate stormwater management features into these areas (e.g. seepage areas, permeable surfaces, water retention areas). [64,65,84]. Urban planners also play a role in determining the location and scale of river banks and other "hard" flood management systems [7]. Developing effective adaptation strategies depends on well estimating what the local geography will encounter in the context of climate change It is important that these predictions are based on scientific models, projections and calculations [17,85].

d Disaster Management

In cities, it is necessary to identify possible risk areas, disaster types arising from climate change and to take measures. Different regions or neighborhoods of cities can have different microclimates and can be affected by climate change in different ways. While extreme temperature increases due to urban heat island formation are more common in some regions, other regions may be exposed to excessive rainfall. For this reason, the collection of smaller scale (neighborhood, district and street level) and high resolution climate data is needed to better understand how cities are affected by climate change and to develop meaningful adaptation strategies [17]

In order to understand how the urban climate system changes in different subregions within the city in terms of temperature, precipitation and wind formation and to analyze effectively how these changes pose risks, the participation of all actors of the city and communication between the actors is important. Risks should not be approached only as technical issues that should be determined by professionals, but areas where different actors and user groups of the city may be affected by climate change should be determined. Participatory methods in risk assessment enable a better understanding of micro-climate points, especially in cities, to identify key vulnerabilities in a realistic way, and to develop specific adaptation strategies for these vulnerabilities [17, 85, 86].

Planners are playing an increasingly large role in disaster preparedness, such as providing adequate community shelters and schools, regulations on backup energy sources, and Public Alert-Information Systems along with access to emergency services.The other responsible people in these areas are technical supporters, policy makers, project and policy enforcers [7].

e Urban Green Infrastructure

One of the most important tools to reduce the effects of climate change as well as increase urban resilience and sustainability is the development of urban green infrastructure. Planning natural and semi-natural areas such as green roads, parks, rain gardens, urban forests, urban agricultural areas, green roofs and green walls in the form of an integrated network of urban green areas has many benefits for the city and city dwellers in terms of ecological, social and economic aspects [87]. For example, improvement in air quality [88]; water management [89, 90]; reduced water pollution [91]; carbon capture and storage [92]; reduced energy use in buildings [93]; reduced heat island effect [91]; ecological benefits [94]; social benefits-social sustainabilty (inclusion, cohesion, more sustainable behaviour patterns) [95,96,97]; human health and well-being [98].

In this context, among the urban policy options to be implemented for combating and adapting to climate change, the protection and development of green infrastructure components and the natural environment of the city have an important place. Protecting and increasing open-green areas (including agriculture and forest areas around the city) and establishing continuous open green space connections should be among the priority targets of the city administrations, both in the city and its immediate vicinity [53,58,61],

While designing the open-green areas system, it is necessary to consider the balance of open-green areas with soft (green) and hard ground that can meet the needs of local climatic conditions. The pattern of medium and small-scale green areas distributed evenly in the sub-parts of the city can function as a breathing space that provides thermal comfort in hot and arid climate zones, as well as sinks that absorb water in rainy and temperate zones. An open area system, where present wide open areas integrated together is one of the strategies preferred to get maximum benefit from solar radiation in cold climate zones [17]. The use of green space systems as an effective mitigation strategy is directly related to the selection of landscape elements to be used in the designed green areas system. Details such as carbon-absorbing plant species, planting suitable for local climatic conditions [99,100,101], urban density-green area ratio on hectare basis [61] are important for the effective realization of green strategies.

It should also be kept in mind that open-green areas should be designed and implemented with sustainable and innovative drainage and infrastructure systems. In this context, the main blue-green infrastructure tools that prevent rainwater from turning into wastewater (Green roofs, rainwater harvesting, rain gardens, shallow rain transmission channels, green parking lots, permeable surfaces/sidewalks, green streets, passages and highways, etc.) creates an effective result in water resources management [81,102]. In addition, it should be taken into consideration by the city administrations that the use of green roof and green facade applications, especially in high-density urban textures where open space supply is limited, have become increasingly common in recent years and that such applications have positive results [93,103].

f Waste management

The main source of methane (CH4) emissions, the second most important greenhouse gas after carbon dioxide, is landfills. There is a continuous increase in the amount and content of solid waste with the increase in the population, the growth of cities and the change in consumption patterns worldwide. The total amount of waste produced worldwide is calculated as 1.3 billion tons annually and it is estimated that this amount will increase to 2.2 billion tons by 2025. The largest share in solid waste production belongs to cities. 3–4 % of global human-induced greenhouse gas emission comes from urban waste. For these reasons, making solid waste management in cities environmentally friendly and sustainable holds an important place in combatting the climate change [19,104]. At this point, the most important area of responsibility for the planners in solid waste management is the location of solid waste storage zones in the cities. Zones that are not planned in accordance with the natural and cultural structure of the city and not taking landscape sensitivity areas into account will cause water, soil and air pollution [50,105,106,107].

Making waste management climate-friendly firstly means the reducing the level of methane gas released into the atmosphere. In addition, it is possible to reduce indirect and direct carbon emissions through practices such as collecting and transporting wastes under a sustainable waste management, recycling, compost, and energy recovery. While the rate of reduction in emissions with sustainable solid waste management is 10–15 % in a city, the rate of reduction in emissions has been determined as 20–25 % with sustainable solid waste management + waste reduction studies [108].

An effective solid waste management system has been implemented in Yokohama, Japan's second largest city, since 2003. With the implementation of the system, although the population of the city increased by 170,000 people between 2001–2008, the amount of solid waste was reduced from 465 kg per person to 260 kg per person [69]. In the same period, a total 840,000 tons of carbon dioxide equivalent emission reduction was achieved [72].

g Structures/Buildings

The building sector has an important share in global energy consumption. According to 2017 figures, 36 % of global energy consumption is caused by the construction and use of buildings. 39 % of global energy-related greenhouse gas emissions arise from the construction, use and production of construction materials. In buildings, activities such as heating, cooling, lighting, cooking and interior equipment are the main sources of energy consumption [109].

Local land use plans and zoning laws can be used to regulate the location, type and scale of development in climate-susceptible areas. Numerous location and design features can be encouraged and/or required to make areas and buildings more resistant to climate effects. For example, habitats may need to be placed at a suitable height in order to keep them above flood hazard levels. Tree planting and other sun shading may be required to reduce urban heat island effects and improve the safety and well-being of the building occupants. Buildings can be designed to withstand other potential climate change impacts, such as increased high winds associated with storms. From a mitigation perspective, planners can encourage more environmentally friendly building design to reduce energy and water consumption [7,21,110]. Enerji tüketimini ve karbon ayak izini azaltma amacıyla planlanan çevre dostu binalarda alınan önlemler; aktif ve pasif tasarım stratejileri ve uygulamaları ile desteklenmelidir [110,69].

The spatial plan types, whose details are explained in Chapter 4 and that include mitigation and adaptation measures that fall under the responsibility of planners, and the objectives of the plans and their possible effects on potential climate change are shown in the Tab. 2.

5 Conclusion

It is a scientifically proven fact that the greenhouse gas emissions that cause climate change, one of the most important problems of our day, emerge as a result of the activities carried out by people in cities. At this point, cities, which are the major production and consumption centers where more than half of the world's population live, have critical roles in the mitigation and adaptation of climate change.

Tab. 2. Plans and Climate Change [7]

TYPE OF PLAN	GENERAL PURPOSE POTENTIAL	CLIMATE CHANGE FEATURES
City Plan & Physical Land Use Plan	• Identifies areas (zones) for different types of development (i.e. housing, commercial, industrial, etc.) • Identifies development hazard areas (steep slopes, flood plains, etc.) • Provides long-term policy direction on land use and development, transport and overall community development	• Highlights development "hot spots" or "no development areas" where climate change impacts are likely to be most severe • Sets policy direction on "climate friendly" or "climate resilient" infrastructure and servicing (i.e. stormwater management) • Formally recognizes climate change and highlights related impacts (i.e. builds public awareness, political support) • Land capability, suitability and the feasibility of different development alternatives are analysed to determine appropriate spatial relationships that form the basis of generalized future land use maps
Storm Water Management Plan	• Improves storm water management, including drainage and infrastructure	• Identifies climate change-related storm water or flooding hazard areas • Directs new infrastructure to "safer" areas not as exposed to climate change impacts (i.e. can attract or pull development to serviced areas) • Accelerates investment in existing coastal and river flood defence programmes to protect existing development in flood prone areas • Identifies options to increase permeability of paved areas in drought prone and flood affected areas.
Transport Plan	• Improves road, pedestrian, transit and bicycle connections and infrastructure	• Identifies and improves "weak links" in transport networks that are threatened by climate change impacts (e.g. bridges threatened by storm surges, roads subject to flooding) • Identifies and designates emergency transportation networks (e.g. roads and transport links that could be used during a climate change impact event to transport people, supplies and any required relief supplies) • Prioritizes transportation network improvements that improve transportation connections for climate vulnerable groups • Supports climate change mitigation through reduced traffic congestion, prioritizing non-motorized transportation
Disaster Risk	• Reduction Plan • Improves disaster response preparednes • Improves early warning capacities • Identifies "hot spots" (i.e. areas and groups vulnerable to disasters)	• Identifies climate change disaster risks, likelihoods and adaptive capacity • Supports, facilitates and expedites infrastructure and planning improvements to reduce climate change-related disaster impacts • Ensures provision of adequate community shelters and schools, access for emergency services, introduces regulations on back up sources of energy and water supply
Sewer / Liquid Waste Management Plan	• Improves waste water/sewer management, including and infrastructure	• Identifies development or construction guidelines for "climate proof" facilities (i.e. facilities that are located and built to withstand and function during climate change impact events) • Identifies and prioritizes high risk areas where new facilities are most needed to reduce climate change impacts amongst vulnerable groups • Identifies options to reduce or reuse wastewater (grey water) for urban agriculture and horticulture
Water Management Plan	• Improves water supply, management and distribution • Improves water conservation	• Identifies climate change-related risks to municipal water supply, treatment and distribution and adaptive measures to counter them • Identifies water conservation and water demand strategies and tools to better manage and adapt to future potential water shortages

Scientific-based sustainable urban planning and innovative urban design studies are very important in order to strengthen the resilience to climate change in cities and to better manage the risks in the cities. The principles set out in the areas of climate change mitigation and adaptation in the headings (land use/urban form and urban structure order, transportation, water resources management, disaster management, urban green infrastructure, waste management, buildings) in this study in terms of creating more climate resistant cities must be adapted to the plans.

In fact, climate change adaptation planning is not an easy process and specific steps may require technical support and assistance from climate experts and other climate change stakeholders [7]. When considered in the context of urban planning, there is a need for scientific tools that will enable the interpretation, modeling and mapping of climatological data in a way that can input into the spatial plans language. This modeling and maps are essential in terms of integrating the climatological data required for all spatial plans from upper to lower scale into plan decisions. These tools also create the need for technical personnel from professional groups specialized in climate change issues in municipalities. Among the technological strategies related to the climate change, there is the development of digital applications that will facilitate information sharing and produce information with multi-actor processes [17]. At this point, in the fight against climate change, the arrangement of energy, transportation, water use, waste management systems in accordance with the understanding of smart cities enables the realization of appropriate policies and practices quickly, together with real-time monitoring of needs [19, 111].

It is also important to note that although the urban planners are certainly key players, they should not be the only actors involved in climate change planning. Developing relevant decision support systems for improved science-policy integration and organizational learning to assist city authorities in their efforts to mainstream climate change adaptation to urban planning practices is a valuable step at this point. In this context, it is necessary to integrate scientific knowledge with local policy decision-making and to "translate" scientific results into policy recommendations. Training and dissemination of research results through workshops, conferences and networks on this subject, organizing seminars on specific aspects of climate change adaptation and risk reduction with the science-policy interface as the main subject, creating opportunities for information exchange for researchers, enforcers and policy makers, and to involve enforcers in research projects from the very beginning and to include local governments more in universities [45].

It should not be forgotten that the most successful city plans and strategies will be developed with community contribution and ultimately community support. Changing ongoing consumption patterns or adaptation to changing conditions will be strengthened with the construction of an effective information and education system. Outreach activities that planners typically undertake in all types of planning and policy development projects, as well as community meetings, focus groups, polls, newsletters and advertisements; Informing stakeholder working groups and the public about their links to the policy or program area on climate change provides important opportunities for climate change impacts and risks [7].

References

1. Houghton, J., *Global warming.* The complete briefing, 4th edn. Cambridge University Press, 2009, Cambridge.
2. Adger, W. N., Barnett, J., Brown, K., Marshall, N. and O'brien, K., *Cultural dimensions of climate change impacts and adaptation.* Nature climate change, 2013, 3(2), p. 112–117.
3. Denton, F., Wilbanks, T., Burton, I., Chandani, A., Gao, Q., Lemos, M. C., ... and van Ypersele, J. P. *Climate-resilient pathways: adaptation, mitigation, and sustainable development.* In: Field CB (ed) Climate change 2014: impacts, adaptation, and vulnerability. Part A: global and sectoral aspects. Contribution of working group II to the fifth assessment report of the intergovernmental panel on climate change. Cambridge University Press, Cambridge, pp 1101–1131
4. Field, C.B, Barros, V.R, Mach, K. et al. *Climate change impacts, adaptation, and vulnerability,* 2014, In: Field CB (ed) Part A: global and sectoral aspects. Contribution of working group II to the fifth assessment report of the intergovernmental panel on climate change. Cambridge University Press, Cambridge, p 1132
5. IPCC, *Climate Change 2014: Synthesis Report. Contribution of Working Groups I, II and III to the Fifth Assessment Report of the Intergovernmental Panel on Climate Change,* 2014, [Core Writing Team, R.K. Pachauri and L.A. Meyer (eds.)]. IPCC, Geneva, Switzerland, 151 pp.
6. Corfee-Morlot, Jan, Lamia Kamal-Chaoui, Michael G. Donovan, Ian Cochran, Alexis Robert and Pierre-Jonathan Teasdale, *"Cities, Climate Change and Multilevel Governance",* 2009, OECD Environmental Working Papers N 14, 2009, OECD publishing.
7. United Nations (UN)-Habitat, *Planning for Climate Change: A strategic, values-based approach for urban planners.* 2014, United Nations Human Settlements Programme (UN-Habitat),
8. UN-Habitat, *Urbanization and Development Emerging Futures, World Cities Report,* 2016, United Nations Human Settlements Programme.
9. Olsson, L., H., Barbosa, S. Bhadwal, A., Cowie, K., Delusca, D., Flores-Renteria, K., Hermans, E., Jobbagy, W. Kurz, D. Li, D.J. Sonwa, L. Stringer, *Land Degradation. In: Climate Change and Land: an IPCC special report on climate change, desertification, land degradation, sustainable land management, food security, and greenhouse gas fluxes in terrestrial ecosystems,* 2019 [P.R. Shukla, J. Skea, E. Calvo Buendia, V. Masson-Delmotte, H.-O. Pörtner, D. C. Roberts, P. Zhai, R. Slade, S. Connors, R. van Diemen, M. Ferrat, E. Haughey,

S. Luz, S. Neogi, M. Pathak, J. Petzold, J. Portugal Pereira, P. Vyas, E. Huntley, K. Kissick, M. Belkacemi, J. Malley, (eds.)]. In press.

10. Galderisi, A., *Urban resilience: A framework for empowering cities in face of heterogeneous risk factors.* Z magazine, 2014, **11**(1), p.36–58.
11. Meerow, S., Newell, J. P. and Stults, M., *Defining urban resilience: A review.* Landscape and urban planning, 2016, **147**, p.38–49.
12. Tyler, S., and Moench, M., *A framework for urban climate resilience.* Climate and development, 2012, **4**(4), p.311–326.
13. Wamsler, C., Brink, E. and Rivera, C., *Planning for climate change in urban areas: from theory to practice.* Journal of Cleaner Production, 2013, **50**, p.68–81.
14. Davoudi, S., Crawford, J., Mehmood, A. (Eds.), *Planning for Climate Change: Strategies for Mitigation and Adaptation for Spatial Planners.* 2010, Earthscan, London.
15. IPCC, *Global Warming of 1.5°C. An IPCC Special Report on the impacts of global warming of 1.5°C above pre-industrial levels and related global greenhouse gas emission pathways, in the context of strengthening the global response to the threat of climate change, sustainable development, and efforts to eradicate poverty.*2018, [Masson-Delmotte, V., P. Zhai, H.-O. Pörtner, D. Roberts, J. Skea, P.R. Shukla, A. Pirani, W. Moufouma-Okia, C. Péan, R. Pidcock, S. Connors, J.B.R. Matthews, Y. Chen, X. Zhou, M.I. Gomis, E. Lonnoy, T. Maycock, M. Tignor, and T. Waterfield (eds.)]. In Press.
16. Çolakoğlu, E., *İklim Değişikliği, Sürdürülebilir Kentler ve Kentsel Planlama Etkileşimi.* İklim Değişikliği Alanında Ortak Çabaların Desteklenmesi Projesi (iklimİN), 2019, T.C. Çevre ve Şehircilik Bakanlığı Çevre Yönetimi Genel Müdürlüğü, Ankara.
17. Peker, E., and Aydın, C. İ., *Değişen İklimde Kentler: Yerel yönetimler için azaltım ve uyum politikaları.* 2019, İPM-Mercator Politika Notu.
18. Tuğaç, Ç., *Türkiye'de kentsel iklim değişikliği için eko-kompakt kentler.* Ankara Üniversitesi Ernst Reuter İskân Ve Şehircilik Uygulama ve Araştırma Merkezi Yayınları: 23, 2019, Ankara.
19. Uncu, B. A., *İklim İçin Kentler. Yerel Yönetimlerde İklim eylem Planı,* Dijital Düşler Basım San. Ve Tic. A.Ş., 2019, İstanbul.
20. Açmaz Özden, M., Cengiz, A. E. (2015). *The Value of Urban Green Infrastructure for Sustainability of Urban Ecosystems.* In: Tourism Environment and Sustainability, Cevdet AVCIKURT, Mihaela DINU, Necdet HACIOĞLU, Recep EFE, Abdullah SOYKAN (Eds.), St. Kliment Ohridski University Press, Sofya, ISBN 978-954-07-4021-8, p.507–535.
21. Çelebi, G., Gültekin, B.A., Harputlugil, G., Bedir, M., and Tereci, A.,. *Yapı Çevre İlişkileri.* TMMOB Mimarlar Odası Sürekli Mesleki Gelişim Merkezi, 2008, ISBN: ISBN/ISSN: 978-9944-89-645-0

22. UN, *United Nations Framework Convention On Climate Change.* 1992
23. IPCC, *Climate Change 2007: Synthesis Report. Contribution of Working Groups I, II and III to the Fourth Assessment Report of the Intergovernmental Panel on Climate Change* [Core Writing Team, Pachauri, R.K and Reisinger, A.(eds.)]. IPCC, 2007, Geneva, Switzerland, 104 pp.
24. UN, *Department of Economic and Social Affairs, Population Division,* 2019, World Urbanization Prospects: The 2018 Revision (ST/ESA/SER.A/420). New York: United Nations.
25. World Meteorological Organization(WMO), *Statement on the State of the Global Climate in 2017.* Geneva 2, Switzerland.
26. Bazaz, A., Bertoldi, P., Buckeridge, M., Cartwright, A., Coninck, H., Engelbrecht, F., ..., Waisman, H., *C40 Summary for urban policy makers-what the IPCC special report on global warming of 1,5C Means for Cities.* 2018, https://www.c40.org/researches/summary-for-urban-policymakers-what-the-ipcc-special-report-on-global-warming-of-1-5-c-means-for-cities.
27. Rosenzweig, C., W. Solecki, P. Romero-Lankao, S. Mehrotra, S. Dhakal, T. Bowman, and S. Ali Ibrahim, *ARC3.2 Summary for City Leaders. Urban Climate Change Research Network.* 2015, Columbia University. New York.
28. WBGU – German Advisory Council on Global Change, *Humanity on the move: Unlocking the transformative power of cities. Berlin*: WBGU. 2016
29. World Bank, *Guide to climate change adaptation in cities.*2011, https://openknowledge.worldbank.org/bitstream/handle/10986/27396/653590WP0v200B0Urban0Handbook0Final.pdf?sequence=1&isAllowed=y
30. UN-Habitat,. *Local governments' pocket guide to resilience.* 2015, https://urbanresiliencehub.org/wp-content/uploads/2018/07/Local-Governments-Pocket-Guide-to-Resilience.pdf
31. Abeygunawardena, P., Vyas, Y., Knill, P., Foy, T., Harrold, M., Steele, P., ... Sperling, F., *Poverty and climate change: reducing the vulnerability of the poor through adaptation.* 2003, Washington, DC: World Bank
32. Aşıcı, A. A., *İklim için Yeşil Ekonomi Politikaları. İstanbul Politikaları Merkezi,* 82, 2017.
33. Brown, K., *Global environmental change I: A social turn for resilience?.* Progress in human geography, 2014, **38**(1), p.107–117.
34. Meerow, S., and Newell, J. P., *Resilience and complexity: A bibliometric review and prospects for industrial ecology.* Journal of Industrial Ecology, 2015, **19**(2), p.236–251.
35. OECD, **Resilient cities**, 2016, preliminary version.
36. Leichenko, R., *Climate change and urban resilience.* Current Opinion in Environmental Sustainability, 2011, **3**(3), p. 164–168

37. Pierce, J. C., Budd, W. W., and Lovrich, N. P., *Resilience and sustainability in US urban areas.* Environmental Politics, 2011, **20**(4), p.566–584.
38. Zimmerman, R., and Faris, C. *Climate change mitigation and adaptation in North American cities.* Current Opinion in Environmental Sustainability, 2011, **3**(3), p.181–187.
39. Klein, R. J. T., Nicholls, R. J., and Thomalla, F., *Resilience to natural hazards:How useful is this concept?* Environmental Hazards, 2003, **5**(1), 35–45.
40. Matyas, D. and Pelling, M., *Positioning resilience for 2015: The role ofresistance, incremental adjustment and transformation in disaster risk management policy.* Disasters, 2014, **39**(SI), S1–S18.
41. Ribeiro, P. J. G., and Gonçalves, L. A. P. J., *Urban resilience: A conceptual framework.* Sustainable Cities and Society, 2019, **50**, 101625.
42. World Bank-IFC, *Climate Investment opportunities in cities: an IFC analysis.* 2018, https://www.ifc.org/wps/wcm/connect/875afb8f-de49-460e-a66a-dd2664452840/201811-CIOC-IFC-Analysis.pdf?MOD=AJPERES&CVID=mthPzYg
43. Carpenter, S., Walker, B., Anderies, J.M., and Abel, N., *From metaphor to measurement: Resilience of what to what?* Ecosystems, 2001, **4**, p.765–781.
44. Kim, D., and Lim, U., *Urban resilience in climate change adaptation: A conceptual framework.* Sustainability, 2016, **8**(4), p.405.
45. Wamsler, C., Brink, E.; Rivera, C., *Planning for climate change in urban areas: From theory to practice.* Journal of Cleaner Production, 2013, **50**, p.68–81.
46. Lu, P.; and Stead, D., *Understanding the notion of resilience in spatial planning: A case study of Rotterdam, The Netherlands.* Cities, 2013, **35**, p.200–212.
47. Thornbush,M.; Golubchikov, O., Bouzarovski, S., *Sustainable cities targeted by combined mitigation-adaptation efforts for future-proofing.* Sustainable Cities and Society, 2013, **9**, p.1–9.
48. Balk, D., Montgomery, M.R., McGranahan, G., Kim, D., Mara, V., Todd, M., . . . Dorelian, A., *Mapping urban settlements and the risks of climate change in Africa, Asia and Climate and Development South America.* In J.M. Guzma˜n, G. Martine, G. McGranahan, D. Schensul, & C. Tacoli (Eds.), Population dynamics and climate change, 2009, (pp. 80–103)., New York: UNFPA.
49. Wilbanks, T., Lankao, P.R., Bao, M., Berkhout, F., Cairncross, S., Ceron, J.-P., and Kapshe, M., *Industry, settlement and society.* In M. Parry, O. Canziani, J. Palutikof, P. van der Linden, & C. Hanson (Eds.), Climate change: Impacts, adaptation and vulnerability. Contribution of working group II to the fourth assessment report of the intergovernmental panel on climate change, 2007, (pp. 357–390). Cambridge, UK: Cambridge University Press.

50. UN-Habitat, *Guiding Principles for Climate City Planning Action.* 2016, https://e-lib.iclei.org/wp-content/uploads/2016/02/Guiding-Principles-for-City-Climate-Action-Planning.pdf
51. Karakuş, C. B.,,Cerıt, O., Demiroğlu, D., *Sivas Kenti ve Yakın Çevresinin Mevcut Arazi Kullanım Yapısı ile Alan Kullanım Potansiyellerinin Karsılastırılması.* Cumhuriyet Üniversitesi Yer Bilimleri Dergisi, 2012, **29**(2), p. 72–87.
52. Demiroğlu, D., Pekin T., U., Karadağ A., A., Cengiz, A., E., *Ecology-Based Contemporary Urbanism Approaches.* In: Urban and Urbanization, 2014, St.Kliment Ohridski Universty Press. Sofia
53. Demiroğlu, D., and Karadağ A., A., *Ecosystem Services Approach To Spatial Planning In Turkey.* Internatıonal Urban Studies Congress-Problems In Present-Day Cıty, 2015, **1**(1), p. 252–270.
54. Karadağ, A., A.,Cengiz A., E. and Demiroğlu, D., *An Approach to the Development of Land Use Decision on the Protection of Water Resources.* International conference on natural science and engineering,,2016, **1**(1), p. 3149–3161.
55. Cengiz A., E., Karadağ, A., A. and Demiroğlu, D., *Importance of Landscape Planning for Sustainable Spaces.* International Congress of the New Approaches and Technologies for Sustainable Development, 2017, **1**, 4–15.
56. Karadağ, A. A., Demiroğlu, D. and Cengiz A., E., *Importance of Ecological Analysis in Spatial Planning.* International Congress on Agriculture and Forestry Research (AGRIFOR), 2019, p.835–843.
57. Cengiz, A. E, *Ecological Approaches in making Urban Land Use Decisions.* In: Science, Ecology and Engineering Research in the Globalizing World, St. Kliment Ohridski University Press, Sofia, 2018, p.40–57.
58. Cengiz A. E., and Demiroğlu, D., *Integrating Ecosystem Services into Spatial Planing.* In: New Approaches to Spatial Planning and Design - Planning, Design, Applications, 2019, Peter lang.
59. Cengiz, A. E., Çavuş, C. Z. and Koç, T., Çanakkale ve Kepez Yerleşmelerinde Sulu Tarım Alanları Kentleşme İlişkisi. Coğrafi Bilimler Dergisi, 2014, **12** (1), p. 69–88.
60. Cengiz, A. E. and Gönüz, A. Ekolojik Açıdan Kentsel Alan Kullanımları: Çanakkale Kent Merkezi Örneği. Atatürk Üniversitesi Ziraat Fakültesi Dergisi, 2011, **42** (1), p.79–89.
61. Demiroglu, D., Yucekaya, M., Coban, A., and Gokce, D., Investigation of Green Infrastructure for Sustainable Cities: the Kilis Case. Journal of Environmental Protection and Ecology, 2014, **15**(3), 1199–1207.
62. Demiroğlu, D., Karadağ, A. A., Cengiz A. E. and Günaydın, A., S., *Su Duyarlı Kent Planlama Ilkelerinin Gelistirilmesi:Kilis Kent Örnegi.* TMMOB Peyzaj

Mimarları Odası 6. Peyzaj Mimarlıgı Kongresi(Uluslarası Katılımlı), 2016, **1**(1), 513–526.

63. Demiroğlu, D., Karadağ, A. A., Cengiz A. E. and Günaydın, A., S., *Evaluation of Sustainable Water Using at Open and Green Spaces Kilis 7 Aralık University Main Campus.* International conference on natural science and engineering, 2016, **1**(1), p. 3171–3183.
64. Karadağ, A. A. and Demiroğlu, D. *Contribution of Green Infrastructure Systems on Water Management.* In: Environmental Sustainability and Landscape Management, 2016, St.Kliment Ohridski Universty Press. Sofia
65. Demiroğlu, D., Karadağ, A. A., Cengiz A. E. and Çoban, A., *Evaluation of Contribution of Green Infrastructure System to Urban Water Management.* 2. International Conference on Agriculture, Forest, Food Sciences and Technologies (ICAFOF'18), 2018, p. 422–431.
66. Golany, G. S., *Urban design morphology and thermal performance.* Atmospheric Environment, 1996, **30**(3), p. 455–465.
67. Williams, K., Joynt, J. L., and Hopkins, D., *Adapting to climate change in the compact city: the suburban challenge.* Built Environment, **2010,** 36(1), 105–115
68. Balaban, O., *Climate change and cities: a review on the impacts and policy responses.* METU Journal of the Faculty of Architecture, 2012, **29**(1), 21–44.
69. Balaban, O., and De Oliveira, J. A. P., *Sustainable buildings for healthier cities: assessing the co-benefits of green buildings in Japan.* Journal of Cleaner Production, 2017, **163**, 68–78.
70. Bibri, S. E., Krogstie, J., and Kärrholm, M., *Compact City planning and development: emerging practices and strategies for achieving the goals of sustainable development.* Development in the Built Environment. 2020, Compact City Planning and Development: Emerging Practices and Strategies for Achieving the Goals of Sustainable Development (lu.se)
71. Dodman, D., *Urban form, greenhouse gas emissions and climate vulnerability.* In Population dynamics and climate change, 2009, p. 64–79, London: UNFPA-IIED.
72. Suzuki, H., Dastur, A., Moffatt, S., Yabuki, N. and Maruyama, H., *Eco2 Cities: Ecological Cities as Economic Cities.* 2010, Washington, DC. The World Bank.
73. SLoCaT, *Transport and Climate Change Global Status Report 2018.* https://www.changing-transport.org/wp-content/uploads/TCC-GSR2018-May9-web.pdf
74. C40 Cities Climate Leadership Group, *Good Practice Guide: Low Emission Vehicles.* 7_C40_GPG_LEV.original.pdf (c40-production-images.s3.amazonaws.com)

75. Van Audenhove, F. J., Korniichuk, O., Dauby, L., and Pourbaix, J., *The future of urban mobility 2.0: imperatives to shape extended mobility ecosystems of tomorrow.* 2014.
76. Demiroğlu, D., Günaydın A.S.,Yücekaya, M., Çoban, A., *Building Sustainable Transportation Model in University Campuses: Kilis 7 Aralık University.* 2nd International Conference on Agriculture, Forest, Food Sciences and Technologies (ICAFOF), 2018, **1**(1), p.705–715.
77. Bamwesigye, D., and Hlavackova, P., *Analysis of sustainable transport for smart cities.* Sustainability, 2019, **11**(7), p.21–40.
78. Zhao, X., Ke, Y., Zuo, J., Xiong, W. and Wu, P., *Evaluation of sustainable transport research in 2000–2019.* Journal of Cleaner Production, 2020, **256**, 120404.
79. Creutzig, F., *Evolving narratives of low-carbon futures in transportation.* Transport reviews, 2016, **36**(3), 341–360.
80. Bariş, M. E., and Karadag, A. A., *Water resources management issues in Turkey and recommendations.* Journal of Applied Sciences, 2007, **7**(24), 3900–3908.
81. Aksungur, N., and Firidin, Ş., *Su kaynaklarının kullanımı ve sürdürülebilirlik.* SUMAE Yunus Araştırma Bülteni, 2008, **8**(2), p.1–3.
82. Karadağ, A. A., *Türkiye'deki su kaynaklari yönetimine ilişkin sorunlar ve çözüm önerileri.* TMMOB 2. Su Politikaları Kongresi. 2008.
83. Salerno, F., *Adaptation Strategies for Water Resources: Criteria for Research,"*, Water, 2017, **9**, 805, p.1–9
84. Demiroğlu, D., Çoban, A., Cengiz A. E., *Sustainable Use of Water in Urban Design.* Hacettepe J. Biol. & Chem, 2016, **44**(2), p.191–201
85. Mechler, R., and Bouwer, L. M., *Understanding trends and projections of disaster losses and climate change: is vulnerability the missing link?.* Climatic Change, 2015, **133**(1), 23–35.
86. O'Brien, G., O'keefe, P., Rose, J., and Wisner, B., *Climate change and disaster management.* Disasters, 2006, **30**(1), 64–80.
87. De la Sota, C., Ruffato-Ferreira, V. J., Ruiz-García, L., and Alvarez, S., *Urban green infrastructure as a strategy of climate change mitigation. A case study in northern Spain.* Urban Forestry & Urban Greening, 2019, **40**, p.145–151.
88. Setälä, H., Viippola, V., Rantalainen, A. L., Pennanen, A., and Yli-Pelkonen, V., Does urban vegetation mitigate air pollution in northern conditions?. Environmental pollution, 2013, **183**, p. 104–112.
89. Gill, S. E., Handley, J. F., Ennos, A. R., and Pauleit, S., *Adapting cities for climate change: the role of the green infrastructure.* Built environment, 2007, **33**(1), p.115–133.

90. Demiroğlu, D., Karadağ, A. A., Cengiz A. E. and Çoban, A., *Evaluation of Contribution of Green Infrastructure System to Urban Water Management.* 2. International Conference on Agriculture, Forest, Food Sciences and Technologies (ICAFOF'18), 2018, p. 422–431.

91. Manning, W. J., *Plants in urban ecosystems: Essential role of urban forests in urban metabolism and succession toward sustainability.* The International Journal of Sustainable Development & World Ecology, 2008, **15**(4), p. 362–370.

92. Dubbeling, M., Campbell, M. C., Hoekstra, F. and van Veenhuizen, R., *Building resilient cities.* Urban Agriculture Magazine, 2009, **22**, 3–11.

93. Jaffal, I., Ouldboukhitine, S. And Belarbi, R., *A comprehensive study of the impact of green roofs on building energy performance.* Renewable Energy, 2012, **43**, p. 157–164

94. Costanza, R., d'Arge, R., De Groot, R., Farber, S., Grasso, M., Hannon, B., ... and Van Den Belt, M., *The value of the world's ecosystem services and natural capital. nature,*1997, **387**(6630), 253–260.

95. Sharifi, A. and Murayama, A., *A critical review of seven selected neighborhood sustainability assessment tools.* Environmental Impact Assessment Review, 2013, **38**, 73–87

96. Demiroğlu, D., Cengiz A. E. and Karadağ, A. A., *The Role of OpenGreen Spaces in the Ensuring of Social Sustainability.* International Congress of the New Approaches and Technologies for Sustainable Development, **2017**, 1, p. 212–222.

97. Demiroğlu, D., Çoban, A., Karadağ, A. A. and Cengiz A. E., *Migration-Asylum Seekers and Open-Green Spaces: Kilis Case/Turkey.* Journal of Environmental Protection and Ecology, 2019, **20**(3), p.1282–1291

98. Nordh, H., Hartig, T., Hagerhall, C.M. and Fry, G., *Components of small urban parks that predict the possibility for restoration.* Urban forestry & urban greening, 2009, **8**, p.225–235

99. Onur, B. E. and Demiroğlu, D., Kentsel sürdürülebilir mekânlar: Ekolojik parklar. İstanbul Üniversitesi Orman Fakültesi Dergisi, 2016, **66**(1), p.340–355.

100. Güvenç, İ., & Demiroğlu, D. (2016, November). *The Evaluation of Main Campus Area of Kilis 7 Aralık University in Terms of "Xeriscape" Approach.* In 3rd International Symposium on Environment and Morality (ISEM2016), **1**(1), p.414–425.

101. Demiroğlu, D., Yücekaya, M., Günaydın, A. S., and Tasçıoğlu, S., *Ecological approach to urban parks: The case of urban parks in Kilis, Turkey.* Feb-Fresenius Environmental Bulletin, **2017**, 7142.

102. EPA, 2021. www.epa.gov/green-infrastructure/what-green-infrastructure

103. Demiroğlu, D.,,Karakan, E., Ercan, B. and Çoban A., *A multidisciplinary overview of Roof Gardens.* 2nd International Conference on Civil Enviromental Engineering, 2017, **1**(1), p.1317–1325.

104. Marquez, R.,*"Waste to Resources: An incredible opportunity to reduce GHG emissions and transform communities"*,C40 Cities, 2017, https://www.c40.org/blog_posts/waste-to-resources-an-incredible-opportunity-to-reduce-ghg-emissions-and-transform-communities#_ftn5

105. Karadağ, A., A.,Cengiz A., E. and Demiroğlu, D., *Determination of Sensitive Landscapes for Lanfill: Sivas Case/Turkey.* Journal of Environmental Protection and Ecology, 2019, **20**(3), p.1234–1242

106. Çoban A., Demiroğlu, D. and Karadağ, A., A., *An ecological based approach for determining suitable landfill: The case of Kilis/Turkey.* Journal of Envıronmental Biology, 2020, **13**(41), p.445–452.

107. Karakuş, C. B., Demiroğlu, D., Çoban, A. and Ulutaş, A., *Evaluation of GIS-based multi-criteria decision-making methods for sanitary landfill site selection: the case of Sivas city, Turkey.* Journal of Material Cycles and Waste Management, 2020, **22**(1), 254–272.

108. UNEP-ISWA. The International Solid Waste Association - Global Waste Management Outlook: GWMO_summary_web.pdf (iswa.org)

109. UN, *Global Alliance for Buildings and Construction Global Status Report 2018.* Towards a zero-emission, efficient and resilient buildings and construction sector. https://wedocs.unep.org/bitstream/handle/20.500.11822/27140/Global_Status_2018.pdf?sequence=1&isAllowed=y

110. Bradshaw, V., *The building environment: Active and passive control systems.* 2010, John Wiley & Sons.

111. Demiroğlu, D. And Karadağ, A., *Sustainable and Integrated Framework for Local Governments: Smart City,* In: New Approaches to Spatial Planning and Design - Planning, Design, Applications, 2019, Peter Lang.

Ahmet BENLİAY and Arzu ALTUNTAŞ

The Evaluation of the Colors Used in Layouts in Project Competitions: Examples from Turkey

Abstract: In terms of project competitions, it is of great importance that a presentation is effective, interesting and understandable. Color is one of the most important factors in terms of project presentation technique. It not only makes the presentation more understandable, but also makes it more effective and engaging. In this study, the color information used in the plot which received the first prize in the competition held in Turkey in 20 years set by software running on the Google API-based Vision and interpreted using Artificial Neural Networks. The revealed color values are classified and discussed on the basis of year, scope, scale and document type criteria. Estimates for the titles in each criterion with the help of artificial neural networks were created and discussed. As a result, it has been determined that green, gray and brown colors are used more; sheets with low saturation value and high brightness value are preferred.

Keywords: Project competitions, color, artificial neural networks, Google Vision API.

1 Introduction

"Competition", which means trying to be superior to others, can seem like situations where ambition and selfishness are manifested. However, its main purpose is to find the beter [1]. In fact, competitions are a system that is used especially in the design of public structures and spaces due to reasons such as providing opportunities for different project production methods, providing equal opportunities among professionals, and being open to the public [2].

Competitions have a long history. Especially in Ancient Greece, important buildings and monuments were built as a result of competitions. The competitions have reached their present form with the Industrial Revolution. Thus, they have reached the status of appealing to a wider audience in terms of both the teams participating in the competition and the social communities that the competition concerns [3]. In our country, it is the period when the tradition of the Republican competition started to form. In this period, public buildings in Ankara and the reconstruction works of newly established urban settlements in Anatolia were organized with the help of competitions [4]. Today, the competitions have diversified with the change and shaping of living conditions and needs.

2 Competitions and Competition Types

In the most basic sense, racing means that living things struggle with natural conditions, make an effort to survive, or observe, take risks. In the dictionary of Turkish Language Association [5], it means "trying to be superior to others, competing". Today, competitions that participate individually or as a team are organized on various platforms and used as a tool to measure the knowledge, ability and competition levels of participants. Creativity is prioritized in design competitions and defined that "Competition projects are a good exercise and application environment for the creation of original designs where ideas are developed to create better quality spaces, physical and social problems are solved together, considering the user comfort and needs [6, 7].

In design competitions, there are competition regulations that explain the procedures and rules that the participants have to comply with and provide information about the competition process. In the Regulation on Architecture, Landscape Architecture, Engineering, Urban Design Projects, City and Regional Planning and Fine Art Works Competitions within the scope of the Public Procurement Law No. 4734 in our country [8], the competition defined: "the organization that the idea, design, project, plan or work required for the realization of a subject related to architecture, landscape architecture, engineering, urban design projects, city and regional planning and fine art works will be selected by a jury that is open to multiple participation and will be formed according to the rules of this regulation".

Design competitions are organized for specific purposes, and the participants of the competition prepare and present all the outputs that the jury expects from them within the framework of the subject of the competition. Projects prepared within the rules specified in the specifications are first preselection. Finalists who pass this stage are rewarded with awards specified by unanimity or majority of votes as a result of another voting by the jury.

It is possible to classify the competitions in various categories [1]. In terms of scale, competitions are divided into regional, national or international levels.

- Regional Competitions: These are competitions that are generally held on the basis of provinces.
- National Competitions: These are the competitions opened in accordance with the TMMOB Chamber of Architects Competitions Regulation.
- International Competitions: These types of competitions are competitions held according to the regulations determined by the UIA (International Union of Architects) [9]. In the UNESCO Conference in 1958, certain standards were set for architecture and urban planning competitions and the UIA was asked to inspect them. UNESCO standards help project owners with the explanations and application examples they contain [8].

According to their scope, the competitions can be held as open (free), limited, invited or pre-selected.

- Open (Free) Competitions: These are competitions with no restrictions, as well as only basic rules (not being among those determining and appointing the jury members and reporters, not being among the jury members and rapporteurs and their first-degree relatives, partners, assistants and employees etc.) aiming to provide an open and fair competition to all participants.
- Limited Competitions: These are competitions organized by the organization with certain restrictions (such as age, education level).
- Invited Design Competitions: These are the competitions where the employer determines who will participate in the competition. Invitations to these competitions are opened for reasons such as accessing the work that institutions want, requiring confidentiality or requiring realization in a short time, requiring special expertise, etc.
- Pre-Selection Competitions: These are competitions that are made upon the request of the jury or the employer for some prerequisites for the competition, such as identity file, portfolio, letter of intent.
- It is possible to divide the competitions into two according to the opening purpose:
- Project Competitions: These are competitions held to build a project. Thanks to these competitions, both the project and the team that will realize this project are determined.
- Idea Competitions: These are contests that are not opened for construction purposes, but rather are organized for the purpose of researching a field and producing data.

Competitions are divided into one-stage (gradual), two-stage (gradual) or multi-stage (gradual) according to the way the jury evaluates the projects after the project delivery.

In addition, it is possible to classify design competitions according to professional disciplines. According to this;

- Architectural design competitions
- Urban design competitions
- City and regional planning design competitions
- Landscape architecture design competitions
- Interior architecture design competitions
- Engineering design competitions
- Fine arts design competitions
- Industrial design competitions are organized.

2.1 Evaluation Criteria of Competitions

Studies are completed only when measurable results are reached. During the evaluation and conclusion phase of the competitions, the jury makes open voting within itself, and in its reports, the evaluation processes, the reasoned decisions in which the positive and negative aspects of the eliminated designs are stated, and the statements regarding the process are included in a regular manner. The jury can make evaluations at different levels such as pre-selection criteria and final selection criteria in order of priority in order to bring designs that are more suitable for the evaluation criteria. The jury checks the conformity of the designs to the specification at different stages. These evaluation criteria can be shaped within the framework of various concepts such as aesthetic values, ergonomics, schematic design, safety, maintenance, repair and operation, recycling, economy, commercial potential, manufacturability, technical competence, environmental awareness, considering the existing infrastructure, suitability for the field, etc [7]. One of the first criteria that the jury pays attention to in the evaluation is undoubtedly the layout and schematic compliance. In this regard, the colors preferred in the preparation of the sheets come to the fore.

2.2 Color

According to [5], the sensation created by the light reflected by the objects in the eye is called color. In the literature, there are studies investigating the semantic effect of color and some determinations on this subject. As a result of color studies conducted by many researchers, it has been concluded that colors have positive and negative effects on people [10]. Red activates the power, joy and happiness and creates a dynamic and lively effect in the place where it is used. On the negative side, red reveals fear, excessive anger, uncontrolled passion, and increases blood pressure. Orange energizes, stimulates creativity and ambition, and is effective in creating a warm atmosphere. However, the dominant use of orange color is a cause of neurosis and restlessness. Yellow, a joyful color, reveals wisdom, intuitive understanding, and creates a striking effect in the environment in which it is used. The golden yellow represents the wealth and fame in the place where it is used. On the negative side, yellow, very bright and remarkable, it stimulates the nervous system too much and can cause discomfort. Green, the color of nature, is a color used to create a peaceful and calm environment that relieves physical and mental fatigue. However, when green is used predominantly, the representation of calmness can evoke laziness, while dark green can create a boring, unsettling environment. Blue is known as the color of calmness, hope and harmony. With this feature, it is preferred in environments where a quiet and calm environment is desired. Negatively, blue can create a sad and depressing effect, and the space may feel dull, static, boring. Dark blue, located between purple and blue, combines logic, intuition and discipline with creativity, while viewed from a negative perspective, it is a color that brings stagnation and mental fatigue [11, 12]. It is known that white and black is achromatic. White is the symbol of purity and cleanliness, white light is the symbol of power. But under white light, everything looks its own color, all the flaws of the place are visible. Black represents the absence of light, signifies the end of life, reduces or sometimes destroys the boundaries in space [13].

The aim in competitions is to reach the best. For this, the best projects are selected and awarded by the competition jury. In this study, it is aimed to reveal whether the color preferences used in the layouts of the projects that received the first prize in the competitions are effective in receiving awards.

3 Material and Method

The main material of the study is the projects that received the first prize in the national competitions in Turkey which are in between 2001 and 2020. In this 20-year period, 80 national competitions under the titles of Architecture, Urban design and Landscape Achitecture were examined. A total of 200 visuals belonging to the first prize projects were evaluated.

The GPA 1.0 software which has been previously created by the authors, was updated for the study. With its new name, GPA 2.1 software evaluates images by using Google Vision API. Software can determine the dominant color, the number of colors used and the total Hue, Saturation and Brightness (HSB) color values in the images by this new update.

Artificial neural network models are powerful data mining tools that allow the use of small numbers of data with different characteristics. Neural Designer software, where artificial neural network models can be created easily, can create usable algorithms with its easy interface. Neural Designer software was used to interpret the obtained data and create prediction models.

The study is based on the determination and interpretation of the colors used in the project documents that won the first place in national competitions. Firstly, the visuals have been categorized by Year of Competition, Scope of Competition (Architecture, Urban design, Landscape architecture), Project Scale (Building, Square, Neighborhood, City, Region) and Document Type (Preliminary Project, Final, Detail, Perspective).

These images have been evaluated by GPA 2.1 software to determine dominant color, number of color and total HSB values. While the GPA 2.1 software defines the colors in the evaluation of the colors in the visuals, the page borders are cut off and the white color is excluded.

The generated data was imported into the Neural Designer software to train artificial neural networks model. By using the trained algorithm, dominant color, number of colors used and total color value estimations were created according to the competition subject, competition year, project scale and document type.

In the last stage of the study, the data determined by the prediction model were presented, examined and discussed.

4 Results

The competition projects that received the first prize examined within the scope of the study vary within themselves. It differs not only in terms of the use of color, but also in terms of the layouts within the presentations. These differences are generally seen as changes in the scope of the competition. The document type in the presentations is one of the elements that change the use of color. The distribution of the visuals evaluated within the scope of the year of competition, scope of competition, project scale and document type is given in Tab. 1.

Tab. 1. Distribution of 200 Visuals by Criteria

Visuals by Year									
2001	1	2002	6	2003	3	2004	9	2005	17
2006	4	2007	1	2008	7	2009	20	2010	11
2011	16	2012	14	2013	10	2014	12	2015	10
2016	15	2017	18	2018	12	2019	9	2020	5

Visuals by Scope		
Architecture	Urban Design	Landscape Achitecture
54	67	79

Visuals by Scale				
Building	Square	Neighborhood	City	Region
70	20	18	31	61

Visuals by Document Type				
Preliminary	Analysis	Final	Detail	Perspective
36	21	48	20	75

The dominant color of the images, the ratio of this color, the number of colors used in the presentation (the software is calibrated to determine those with more than 10 % value) and the combination value of the colors in the presentation sheet were determined by the GPA 2.1 software. This information, which is classified according to Visual codes, was combined with the tag information and transferred to the Neural Designer software. After the learning phase, the Neural Designer software was asked to create a forecast according to each criterion. The color information, ratios, amount of color used and total color information values, which are estimated according to the year criteria created by the software, are given in Tab. 2.

Tab. 2. Color Estimation for Year Criteria

	Dominant Color				Information		Total Color Value			
Year	**Color**	**H**	**S**	**B**	**Ratio**	**Count**	**Color**	**H**	**S**	**B**
2001		102	5	7	% 38	3		97	5	62
2002		107	41	56	% 39	4		106	36	53
2003		38	26	64	% 44	4		42	23	70
2004		135	80	66	% 31	5		94	28	62
2005		347	16	72	% 39	4		19	23	65
2006		43	24	55	% 46	4		41	29	61
2007		0	0	78	% 63	4		0	0	68
2008		331	14	76	% 35	5		69	12	72
2009		120	25	72	% 37	5		107	21	68
2010		52	22	69	% 38	4		77	16	66
2011		46	36	86	% 40	5		43	25	78
2012		64	45	58	% 42	5		64	12	66

Artificial neural networks prediction model shows that the dominant colors and total color values in year Criteria are generally green, gray and brown colors varieties. The average ratio in dominant color in visuals was determined for Saturation as 31 and Brighness value as 63. The average ratio of dominant color usage in visuals by years was determined as % 41. The average ratio in total color value in visuals was determined for Saturation as 16 and Brighness value as 68.

The color information, ratios, amount of color used and total color information values, which are estimated according to the Scope criteria created by the software, are given in Tab. 3.

Tab. 3. Color Estimation for Scope Criteria

	Dominant Color				**Information**		**Total Color Value**			
Scope	**Color**	**H**	**S**	**B**	**Ratio**	**Count**	**Color**	**H**	**S**	**B**
Architecture		20 1	1 1	5 8	% 31	5		16 2	3	7 7
Urban Design		44	3 5	5 6	% 37	4		43	2 0	6 3
Landscape Architecture		12 8	2 4	6 8	% 41	4		10 9	1 5	6 6

The average ratio in dominant color in Scope Criteria was determined for Saturation as 23 and Brighness value as 60. The average ratio of dominant color usage in visuals for Scope was determined as % 36. The average ratio in total color value in visuals was determined for Saturation as 13 and Brighness value as 69.

The color information, ratios, amount of color used and total color information values, which are estimated according to the Scale criteria created by the software, are given in Tab. 4.

Tab. 4. Color Estimation for Scale Criteria

	Dominant Color				Information		Total Color Value			
Scale	**Color**	**H**	**S**	**B**	**Ratio**	**Count**	**Color**	**H**	**S**	**B**
Building		200	14	55	% 30	5		160	7	72
Square		61	44	45	% 37	4		50	11	60
Neighborhood		92	60	63	% 41	4		93	12	74
City		64	45	58	% 42	5		69	12	66
Region		118	47	56	% 30	5		120	13	65

The average ratio in dominant color in Scale Criteria was determined for Saturation as 42 and Brighness value as 55. The average ratio of dominant color usage in visuals for Scale was determined as % 36. The average ratio in total color value in visuals was determined for Saturation as 11 and Brighness value as 67.

The color information, ratios, amount of color used and total color information values, which are estimated according to the Type criteria created by the software, are given in Tab. 5.

Tab. 5. Color Estimation for Document Type

	Dominant Color				Information		Total Color Value			
Type	**Color**	**H**	**S**	**B**	**Ratio**	**Count**	**Color**	**H**	**S**	**B**
Preliminary		84	10	85	% 38	5		91	11	75
Analysis		59	29	72	% 37	4		52	26	69
Final		71	18	70	% 40	4		51	16	68
Detail		109	23	73	% 41	4		104	19	68
Perspective		117	35	65	% 33	5		114	29	61

The average ratio in dominant color in Type Criteria was determined for Saturation as 23 and Brighness value as 73. The average ratio of dominant color usage in visuals for Type was determined as % 36. The average ratio in total color value in visuals was determined for Saturation as 20 and Brighness value as 68.

It has been determined that there are significant differences in the use of color in terms of scope, scale and document type criteria. The main reason for this is that each layout is different in terms of features. It would be unreasonable to expect the layout of an architectural structure to be prepared on the same color foundations and the layout of landscape architecture project. Estimates show that the model is functional, especially for the demonstration of these features.

The most important result that emerged in the study is that the saturation values of the colors used are low. Also it has been determined that the Brightness (illuminance) values are high. Another important result of the study is the number of colors used. In particular, competitors used a dominant color on their visuals between 31 % and 48 % in total and the amount of color that exceeds 10 % on the images did not exceed 5.

5 Conclusion

Design is a cognitive action to solve a problem. One of the best ways to get solution suggestions for a problem is through competitions. Competitions are one of the best ways to get solutions and suggestions for a problem. Competition is one of the best ways to put forward an idea or a thought for a designer. Also the advantage of choosing the right idea from a pool where many design ideas are collected is of great importance.

The main reason for this method to be preferred is that artificial neural network models can make statistical evaluation with less data. The estimates created by the artificial neural networks algorithm are the evaluation of the 200 project images with first prize competition awards. In order to create this algorithm, the feature of learn in artificial neural network was used. It should be noted that the results are only estimates and are not the average values of the actual competition documents. These values are evaluated only on the basis of criteria. Regardless of the number of images found in each criterion, the algorithm re-evaluates and creates new predictions. For this reason, the average values emerging on the basis of the criteria differ.

Along with the fact that the information contained in the documents participating in the competition is very important, but the presentation technique used is as important as the information they contain. The use of color is the most important component of a presentation. These results show that the images that do not use too many colors, have low saturation value and high brightness value are successful. Although the algorithm predicts different colors in the year criteria, the colors used are generally determined as hues of green, gray and brown color variants.

The project competitions can soon change forms and can include virtual reality, augmented reality, technics and possibilities provided by modern technology. In the future, studies can be carried out that combine different algorithms that includes other factors that affect the results. However, it should not be forgotten that due to the nature of the design concept, the use of color may differ for each designer and each design idea.

References

1. Sezgin, N. *Türkiye'de Belediye Binalarının Elde Edilmesinde Mimari Tasarım Yarışmalarının Rolü.* Yüksek Lisans Tezi. Yıldız Teknik Üniversitesi, Fen Bilimleri Enstitüsü, Mimarlık Anabilim Dalı, Mimari Tasarım Programı, İstanbul, 2016, 104 s.
2. Meltem, İ.A. *1930–2010 Yılları Arasında Bir Proje Elde Etme Yöntemi Olarak Türkiye'deki Mimari Tasarım Yarışmalarının İrdelenmesi.* Yüksek Lisans Tezi. Yıldız Teknik Üniversitesi, Fen Bilimleri Enstitüsü, Mimarlık Anabilim Dalı, Mimari Tasarım Programı, İstanbul, 2010, 104 s.
3. Çimen D. Urban *Design Competitions as Discursive Practice in Turkey: 1980–2009,* PhD Thesis, Middle East Technical University, September 2010, Ankara
4. Erten, S., Çimen, D. (2011). *Türkiye'de Kentsel Tasarım Yarışmalarının Proje Olarak Uygulanma(ma) Süreçleri.* Planlama Dergisi, Sayı: 51, 2011, 41–49 s.
5. TDK (Türk Dil Kurumu), 2020. https://sozluk.gov.tr/
6. Ayataç, H. & Ketboğa, S. *Alternatif bir üretim biçimi olarak kentsel tasarım proje yarışma süreçlerinin değerlendirilmesi.* Yarışmalar ve Mimarlık Sempozyumu 2016 Kitabı, 2016, (ss.10–22). http://www.yarismaylayap.com/yarismalar-ve-mimarliksempozyumu-2016-2/
7. Aktürk, S. *Mimarlık Eğitimi ve Meslek Pratiği İlişkisinde Tasarım Stüdyoları ve Yarışmalar Üzerinden Başarım Tabanlı Bir Bütünleşik Ölçme – Değerlendirme Modeli.* Yüksek Lisans Tezi. İstanbul Teknik Üniversitesi Fen Bilimleri Enstitüsü, Mimarlık Anabilim Dalı, Proje ve Yapım Yönetimi Programı, İstanbul, 2019, 125 s.
8. KİK (Kamu İhale Kurumu) Yarışmalar Yönetmeliği. *4734 sayılı Kamu İhale Kanunu, Mimarlık, Peyzaj Mimarlığı, Mühendislik, Kentsel Tasarım Projeleri, Şehir ve Bölge Planlama ve Güzel Sanat Eserleri Yarışmaları Yönetmeliği.* T. C. Resmi Gazete, 24973, 24 Aralık 2002.
9. Toprak, Z. *Dünden Bugüne İstanbul Ansiklopedisi (DBİA), "Altıncı Daire-i Belediye",* 01:220-224, Tarih Vakfı Yayınları, 1993. İstanbul.
10. Meerwein G., Rodeck B., Mahnke F.H. *Color-Communication in Architectural Space.* Birkhäuser Basel, 2007, ISBN: 978-3-7643-8286-5, 152 pp.

11. Banks A., Fraser T. *The Complete Guide to Colour/The Ultimate Book for the Colour Conscious,* 2004, Ilex, Lewes, East Sussex, UK.
12. Sharma, R. *Renklerle Terapi,* Nokta Kitap, 2007.
13. Manav, B. *Renk-Anlam-Mekan İlişkisi.* The Turkish Online Journal of Design, Art and Communation, 2015, TOSDAC, 5(3): 22–27.

Alper SAĞLIK and Çağrı SAVAŞ

The Place of Landscape Design and Applications in Psychological Treatment

Abstract: It is an attractive area that can appeal to users from all strata with its rich plant diversity, reinforcement and structural elements, landscapes and activity opportunities it offers. In a landscape study to be done, the designer should first know the user group who will use the space, create aesthetic spaces that will allow them to have a good time by taking into account the special needs of them. Today, depending on the changes in living conditions, mental crises such as stress, anxiety, and depression increase in people. One of the factors that will support improvement in the treatment process of these ailments is the specially designed light green areas. These areas have taken names such as healing, therapy, meditation and health garden today. When designing these areas, landscape architects work with people from different disciplines such as doctors, architects, psychologists, and know the patients' needs and requests, and shape their designs. In this study, topics such as the effects of landscaping areas designed for people undergoing psychological treatment on their recovery status during the treatment process, criteria of a good therapy garden, history of healing gardens were examined and a study investigating the effect of gardening therapy on patients was investigated.

Keywords: Landscape design, therapy garden, healing garden, horticultural therapy.

1 Introduction

The landscape is defined in the European Landscape Convention as "an area that its features are formed as perceived by people, as a result of the interaction and action of human and/or natural factors". Landscape architects, on the other hand, are a discipline that designs spaces by establishing a connection between nature and people in line with aesthetic concerns, taking into account the needs and desires of people.

Landscape architects create spaces that meet the needs and demands of their users, and that offer them functional, active or passive activity opportunities in accordance with aesthetic measures. The science of psychology also examines the issues like how the users perceive these designed spaces, how and why they react to which space (sitting, resting area, plant growing area, ornamental pool, viewing area, sculpture, etc.).

The science of psychology is a discipline that studies people, human behavior and their causes. Environmental psychology as a newly developed branch of psychology; examines the effects of environmental factors (such as space size, heat, light, color, texture, smell) on people's behavior, emotions and thoughts [1].

In a landscape area that will be designed for people receiving psychological treatment, patient profiles should be determined very well. Designers should be informed about the health conditions of the patients such as sentiment levels and stimulants, and the plants, structural elements, colors, textures, scents that will be preferred in the design study should be specially selected to support the comfort and treatment of the patients.

In the research, it has been proven that green areas have positive effects in terms of improving and controlling the increasing levels of stress and anxiety in people in today's conditions. It is psychologically beneficial for people to see the green areas as a view, even without making active contact directly with them [2].

With the increase in urbanization, the increasing concretion in cities, population in the urban, and excessive time spent in closed areas cause an increase in the demand for green areas in humans [3].

As a result of the studies, it has been stated that spending time in nature reduces stress and relaxes people and also accelerates the healing processes of patients. Considering this situation, aesthetic and functional healing gardens, which can affect people physically, spiritually and mentally, can appeal to different patients and age groups, using herbal and structural materials, have started to be implemented in health institutions such as hospitals, rehabilitation centers and nursing homes [4].

Healing gardens are places that are specially designed and applied to make their users feel better physically and psychologically [3].

In order for a landscape area to become a healing garden, it should contribute to the psychological treatment of its users and support them in their recovery process in a positive way [5].

People with psychological disorders differ from mentally and emotionally normal people to perceive, feel and interpret the places they live in. Therefore, designs should be applied considering the needs of people in the spatial studies to be done [6].

Since ancient times, people have turned to nature in all kinds of physical and psychological ailments. They have accepted the healing aspect of nature as a source of healing for physical, spiritual and mental ailments for humans. On the one hand, nature has improved people's quality of life and, on the other hand, offered them treatment methods. The gardens of the churches in the Middle Age, the gardens of the hospitals in the Ottoman Empire are specially designed places that played an important role in the healing process of the mentally ill patients who were treated there. Today, these places can be considered as hospitals, rehabilitation centers, health centers, nursing homes gardens.

This study is about examining the positive and negative effects of designs made with living, non-living materials on the treatment processes of people receiving psychological treatment.

2 Material and Method

2.1 Material

In order to examine the place of landscape applications, which constitute the main material of the study, in psychological treatment and to reach scientific data on this subject; defining concepts such as landscape, psychology, landscape designs, determining the effects of open spaces on people, determining the positive psychological effects of nature on people, determining the designs made in this direction and determining the sub-branches, determining the psychological effects of the designs on the users and determining the design criteria of these areas, and the literature research with the keywords determined from foreign sources (healing gardens, health garden, horticulturel threapy, psychiatric disability, environmental therapy etc.) has been done; and many books, theses, articles, journals and visual images related to the subject were examined, and findings were obtained.

2.2 Method

The method of the study consists of data collection, analysis and evaluation stages.

During the data collection phase, in line with the purpose and scope of the study; the findings obtained through literature research were examined to provide data on issues such as the definition of healing gardens designed especially for psychological treated and supported users, the historical process from ancient times to the present, the types of these gardens and what kind of treatment method was followed, and the rehabilitation center and healing gardens whose designs can be taken as an example.

The scientific data and visual images obtained during the analysis phase were divided into subtitles and each title was evaluated and analyzed within itself.

In the evaluation phase, with the findings obtained, the descriptions of open space activity spaces designed especially considering the special needs of patients receiving psychological treatment were made and their history was analyzed with examples, then the classification of these open spaces was made, and detailed investigations were made in subheadings about the design criteria of these spaces and gardens and their effects on patients. Again, one of the findings obtained in the literature reviews, the horticultural therapy method, was examined and evaluated in a study conducted on psychological patients. In addition, the place and importance of the vegetative and structural designs of the outdoor spaces of the health institutions (hospital, rehabilitation center, nursing home, health center, etc.) providing psychological treatment in the treatment processes of the patients were examined, the herbal and structural designs of the healing gardens of the 3 selected health centers, it was examined in line with the design criteria of healing gardens.

Findings and Discussion

Many methods have been used from past to present in the treatment of people with psychological disorders. One of the oldest of these is the outdoor spaces that support the treatment, which is now known as the therapy garden, healing garden, and health garden. These spaces are specially designed for patients (physically disabled, mentally handicapped, alzheimer's, mental diseases, etc.) and the most suitable environments are offered for their use according to their requirements.

Healing (Therapy, Health) Gardens

Close access to the natural landscape or garden can increase people's ability to cope with stress and thus support their psychological well-being [7]. It is known that nature increases psychological/emotional and positive emotions (calmness, stillness) and decreases negative emotions (anger, anxiety, fear) [8].

Healing gardens are generally specially designed places in the gardens of health institutions that provide psychological treatment. These gardens are designed considering the needs of the patients, and it is tried to improve the well-being of the patients with selected herbal and structural materials [9].

Fig. 1. Therapy Garden Samples [10]

The people who will use the garden should be provided with an environment that will help them feel calm and serene, and that they can hear relaxing sounds such as bird sounds and water sound instead of being isolated from the outside noise [7].

The plant species selected for healing gardens should be carefully selected according to seasonal changes (defoliation periods, blooming periods, color changes...). Plants, birds and butterflies should be preferred from species that will allow fauna, which draws attention both visually and audibly, in the garden [9].

When designing therapy gardens, it is important for the patient group and age range to be used by the garden, for the functionality of the garden and the safety and comfort of the user. It is very important for the designer to consider the needs and comfort of the user rather than aesthetic concern.

People's access to gardens or a natural landscape can strengthen their ability to cope with stress and anxiety levels and thus support healing processes [7].

Interviews has been made with users in 4 hospital gardens in San Francisco in the USA. 90 % of the interviewees have said that the time they spent in the hospital garden had a positive effect on their mood changes. The users have been asked how they felt after spending time in the garden and what caused them to feel better (Tab.-1) [7].

Tab. 1. Users' Opinions about the Garden [7]

How do you feel after spending time in the garden?	What will help you spend better time in the garden?
■ More comfortable, calmer: 79 % ■ Refreshed, stronger: 25 % ■ Better thinking/coping: 22 % ■ I feel better, more positive: 19 % ■ Spiritual peace: 6 % ■ No mood change: 5 %	■ Trees, plants, nature: 59 % ■ Smells, sounds, fresh air: 58 % ■ Places to spend time alone or with community: 50 % ■ Landscapes, textures: 26 % ■ Practical features, benches etc.: 17 % ■ I don't know: 8 %

The History of Healing (Therapy, Health) Gardens

It was accepted that gardens were supportive of therapies in psychological treatments at certain periods in history but were not taken into consideration from time to time. In the Middle Ages in Europe, gardens were designed to support the treatment of patients, especially in monastery gardens (Fig. 2), and the healing properties of the plants grown in these gardens and the calming properties of the gardens were used in the treatment of patients [8]. At the beginning of the 19th century, the hospital gardens have been specially designed by taking into consideration the comfort of the patients and the windows of the rooms facing the garden and the people who receive psychological treatment can easily reach.

Fig. 2. Examples of Medieval Period Monastery Gardens [11,12]

In 20th century, the changing socio-cultural structure, economic conditions and developments in science in the world have caused the importance of these gardens in treatment to be ignored. Today, as a result of the studies carried out, considering the supportive aspect of the healing gardens in treatment, gardens have been started to be designed according to the needs of the patients and used in treatment [4].

In the Seyahatname of Evliya Celebi, it gives information about culture, art, political, social life and health conditions in the 17th century Ottoman period. In the light of this information, we are able to learn about the health institutions of that period, Darussifa (hospital), mental hospital, and the treatment methods there.

While talking about the architecture of the Darussifa in Edirne, which he mentions in his Seyahatname, he mentions that there are eight arches and winter rooms under each of these arches. Each of these mentioned rooms also has two windows. He states that while one window has a view of hyacinth (garden), the other is facing the fountain. Especially in the spring season, he mentions that both visual and fragrant plants such as gillyflower, basil, redbud, rose, daffodil, hyacinth are given to the patients for healing and that patients spend time in this garden at certain times of the day (Evliya Celebi; Seyahatname - III, 1999:261) (Fig. 3).

Fig. 3. Edirne Sultan II. Bayezid Hospital [13,14]

The Design Features of Healing (Therapy, Health) Gardens

When designing a space, the use of design elements (color, texture, form, size) together in a balanced way affects the user's positive or negative perception of the space. The colors and odors of the plants used in the place, and the textures and dimensions of structural elements guide the users mentally and affect their psychological state. In particular, the colors preferred in design have a strong emotional impact on people. For example, with the colors preferred in a place, the place can be defined as warm, friendly, safe, peaceful, calm or disturbing, cold [15].

When designing therapy gardens, the user should be given the opportunity to choose activities in the area so that the user can spend a comfortable and quality time. They should be able to make the decision to socialize by spending all kinds of active and passive activities such as sitting, resting, playing, socializing, watching the scenery either individually or with other people (Fig. 4). For this, the designer should design both activity venues held in groups and places where individual can spend time and choose suitable equipment. For example, if a four or five-person arbor area is used in the design, two-person or single-person benches should also be preferred in the area. When the user is offered a choice, the user feels special and feels safe and relaxed in the area.

Fig. 4. Legacy Emanuel Medical Center Gardens [16]

There are 10 elements identified for healing gardens as places that help reduce stress levels [7]:

1. To create practical areas for physical movement and exercise.
2. To give users the opportunity to make a selection within the area.
   With the garden design to be made, the user can be in a sunny or shady area; he/she must make the choice himself/herself to sit alone or with others and spend time. This reduces the feeling of being controlled in the user, thus the level of stress and anxiety is reduced.
3. To socialize people bringing them together.
4. To establish a direct relationship with plants.
   The stress levels should be reduced by enabling the users to come into direct contact with the plants (touching the plants, tasting, planting, harvesting, etc.) and by making designs that affect their senses (sniffing, watching).
5. Direction signs in the garden.
6. A design where people of all ages and from all strata (patients, healthcare personnel, companions, disabled) can spend time comfortably in the garden, and provide suitable conditions for children, disabled people and the elderly.
7. Similarity.
   With this criterion, which is especially important for Alzheimer's patients, similar to familiar environments, designs should be made with the plants and furniture they are familiar.
8. Silence.

9. Comfort.
10. In touch with art.

Investigation of the Psychological Effects of Design Elements

There are two kinds of attention in humans. The first of these is directed attention. It can be explained as a voluntary orientation of a person to an area, an object or a person while the second one is involuntary attention which means focusing on one direction, event or situation. This focus may be the glint on a still water surface or the sudden horn sound [17]. The point where the users' attention is desired to be drawn in the design area can be achieved with design elements.

Color: Colors are grouped as warm, cool and neutral according to their psychological effects on people. Warm colors (orange, yellow, red) are active colors that the perception by the buildings and attracting attention rate are higher. They create feelings of excitement, vitality, and activeness. In excessive use, it may cause feelings such as fatigue, violence, nervousness and lack of concentration. Cool colors (blue, green, purple) are passive colors. It gives a feeling of confidence, calmness, stillness and relaxation. The excessive use of cold colors, on the other hand, may cause feelings of gloom, stagnation, excessive emotionality and demoralization [18].

In this direction, the use of warm colors should be used in a balanced way with cold colors in the selection of plants that will be preferred in outdoor designs to be used by people receiving psychological treatment, and the preferences that will have a negative effect on the mental state of the patients should be avoided (Fig. 5).

The color changes (blooming flowers, yellowing or greening leaves) in the plants in the garden depending on the changing seasons also help users gain time awareness [9].

Fig. 5. Use of Color in Therapy Gardens [10,19,20]

Texture: In the planting design made with coarse or fine textured plants, the space can be perceived as narrow and cramped or wide, spacious and bright.

Form: With the forms of the living and non-living elements used in the space, the desired emotions of the people who will use the space can be awakened. While natural form plants evoke a sense of comfort, plants with geometric forms (conical, pyramid, column) evoke a serious, formal ambient feeling that combines the point of view [18].

Examining the Healing Gardens of Some Health Institutions

Today, healthcare institutions have started to transform their existing open spaces into healing gardens open to the use of patients, healthcare professionals, visitors, attendants and the public. The main reasons for this are first of all, these gardens support the healing process of the patients, the working staff can relax and work more motivated in these places in terms of spiritual and psychological aspects, and thus the increased hospital costs are reduced thanks to the patients whose recovery process is shortened.

Scottsdale Healthcare Thompson Peak Hospital - Healing Garden (USA/ Arizona)

The hospital is a health institution that provides treatment for both physical and some psychological disorders (addicts). The hospital, which has an inner courtyard visible to most patient rooms; has transformed the courtyard into a healing garden and created a place for both patients and healthcare professionals. In the planting work carried out in the garden, which allows users to spend time individually and together, xerophyte landscape design was preferred. The sound generated by the water flowing from the rectangular mini ornamental pool designed in the garden creates a relaxing and calming effect in the environment. Thanks to the birdhouses hanging on the trees in the garden, it was aimed to attract birds to the area, and thus, the stress and anxiety levels of the people using the garden were aimed to be reduced by the sound of the water and the bird sounds (Fig. 6).

a b

Fig. 6. View from the Courtyard of the Hospital [21,22]

Els Center of Excellence - Sensory Arts Garden (USA/Florida)

A sensory garden for children and adults is designed in this center, which is a school and rehabilitation center for individuals with autism disorder. Using the sensory aspect of nature, a therapeutic and educational garden has been designed in the 13.000m^2 garden to support the movement, the activity, and the smell, taste, touch, hearing, vision of the users. Organic lines were preferred in the garden designed by a landscape architect and a therapist. Considering the psychological and physical needs of individuals with autism, all structural and vegetal elements have been specially designed, and spaces have been created. By surrounding of the garden was limited with a vegetative fence, a green texture was created, and a safer and aesthetically higher quality appearance was provided. Water wall and water cannon are designed for users in the garden. With the sound of flowing water, a relaxing, anxiety-relieving and relaxing environment is provided, and patients' stress levels are reduced. In addition, this water wall and water ball arise the curiosity of the children and contribute to their sensory development by touching the water.

Suitable areas are designed in the garden for individual or group use. Here, users can spend time using ergonomic garden furniture specially designed for them, suitable for their needs. The planting design made in the garden has been selected from specially selected plants to appeal to the senses. For the sense of smell; Rosmanirus officinalis, Slavia officinalis, for the sense of taste; Aloysia citriodora, Plectranthus amboinicus, for the sense of sight; Philodendron x 'Prince of Orange', Phlebodium aureum 'Blue Star', for the sense of touch; Plants such as Asplenium antiquum 'Leslie', Selaginella kraussiana 'aurea' were preferred in the garden.

In addition, Adonidia merrillii, Bursera simaruba, Wodyetia bifurcata plants have been among the preferred plant species due to their suitability to the climate of the region and their visual emphasis effect, space-defining and aesthetic appearance (Fig. 7).

Fig. 7. Els Center of Excellence - Sensory Arts Garden Images [23]

UM Rehabilitation and Orthopedics Institute Healing Garden (USA/ Maryland)

The purpose of this healing garden, which is located in the garden of the university hospital, is to positively affect the mental state of the patients, which are worn out during the treatment process, to facilitate the transition of patients from treatment to normal life, and to create a place where the hospital staff can spend their free time. The patient profile of the hospital is mostly composed of people who have had serious illnesses and accidents and receive physical and psychological treatment. With the healing garden, it was aimed to awaken the different senses and abilities of the patients, to make them feel good physically and mentally in the open area, to spend quality time, and thus to alleviate their mental states such as depression, stress and anxiety.

In the garden, there are seating elements for the comfortable use of the disabled, a mini ornamental pool with fish, a wooden bridge, a sensory garden area where aromatic plants are used, materials that allow users to sowing and planting in the garden, and shaded spaces specially designed to spend group or individual time are placed. The designer designed the road axle, ramps, and stairs with different materials and textures to teach maneuvering in the garden so that patients who are new to wheelchairs, crutches and prostheses can use these tools more comfortably in normal life.

For patients who want to plant in the garden, there are plant beds that are raised in a way that every patient can reach. In addition, plants with sensory effects such as Azelia sp., Aloysia sp., Ocimum sp. were also preferred in the garden (Fig. 8).

Fig. 8. UM Rehabilitation and Orthopedics Institute Healing Garden Images [24–25]

Horticultural Therapy

As a sub-branch of healing gardens, horticultural therapy is a horticultural activity that its main material is plants and is practiced in the presence of supervisors, in order to support the treatment of people who are especially physically-mentally and psychologically disturbed and who are treated in hospital and to support the well-being of patients (Fig.-9). People who participate actively in open space activities stated that they feel better psychologically [26]. Scientists have done some research to find answers to questions such as how this treatment method has an effect on people with psychological disorders, whether it causes an advantage, disadvantage or a neutral situation in their treatment.

Fig. 9. Horticultural Garden Samples [27–29]

A horticultural therapy was designed by the New Life Psychiatric Rehabilitation Association (NLPRA) in Hong Kong, and a study was conducted to examine the effects of this therapy on patients with the voluntary participation of 82 psychiatric patients. The objectives of the study are to reduce the stress and anxiety levels of patients with therapy, to increase their participation in activities such as gardening and to increase their social interactions. Before starting the horticultural treatment, the participants were asked to take some tests and score their stress, anxiety, and depression. In the treatment, first of all, basic information about gardening was given to the participants, and then practical sessions such as planting, maintenance, plant identification, harvesting of grown products, and a small garden design were applied. At the end of the 8-session application, in which all participants actively participated in the program, it was determined that the horticultural therapy provided mental relaxation in the participants, but there was no significant change in stress and anxiety variables. Also, there was no significant increase in socialization among the participants during their horicultural sessions [30].

It is necessary to admit that healing gardens do not have therapeutic properties. These special gardens only reduce the stress level, facilitate the recovery of the body to a more balanced state, and support the treatment process [7].

Conclusion

Healing gardens with names such as therapy gardens, health gardens, and meditation gardens have been included in the health systems in the world due to their positive contributions to the healing processes and have been the subject of many studies in the literature. However, these gardens, which do not receive sufficient attention in our country, should be given more importance, and especially in the hospital gardens, these special gardens designed by a professional team for both patients and healthcare personnel and other user groups should be included.

A well-designed hospital garden not only provides a beautiful green space but can also reduce the stress levels of patients spending time there and help improve treatment outcomes [8].

Patients receiving treatment in hospitals should be allowed to use the hospital garden, unless they have a condition that would prevent them from going outdoors. Light green areas are very effective in reducing the negative emotions of patients. In addition, studies have proven that patients who use hospital gardens have a shortened recovery period.

Especially for a therapy garden to be built in hospital gardens, the designer must not only consider the needs of the patients, but also the needs of different user groups such as hospital staff, attendants, visitors, children, disabled and elderly people, so that people spend time in a more comfortable environment in the garden.

The therapy gardens must be places that are reachable by everyone and are designed by the designer in different areas such as hospitals, nursing homes, rehabilitation centers, children's dormitories, prisons and in a part of public green areas, according to different age groups, people with different diseases (physical, mental, etc.) [31].

The designer should put his/her aesthetic concerns in the second plan and primarily design safe and functional spaces that appeal to the needs of the patients according to their psychological conditions, calming, relaxing and supporting their treatment.

While designing in therapy gardens, organic, soft lines should be studied more. Sharp, clear lines should be avoided, and a natural environment should be provided. Spatial analysis should be very simple and easily perceivable by users. In vegetation design, instead of the species that can attract thorny, poisonous, harmful fauna to the area, a natural area should be created with species that are suitable for the climate of the region, where a wide variety of natural species are used, and seasonal changes are taken into account.

Including the water element in the designs of the gardens can make the patients feel calmer and more relaxed. In addition, as materials such as fences and wire mesh used to limit the garden may cause a feeling of imprisonment in patients, vegetal border elements should be preferred instead of these materials [6].

Gardening activities (horticultural therapy) also help patients to establish social interaction. This socialization environment contributes positively to the recovery of the individual during the treatment process [32].

Horticultural therapy did not cause any positive change in the improvement of psychological problems such as stress, anxiety and anxiety as a result of research and studies; it just contributed to patients' instant mental relaxation, socialization and sharing, task awareness and supported their well-being.

As a result, in order to support the treatment process and to provide the best, most comfortable and safe environment, the garden that appeals to everyone and that is functional and aesthetic must be designed according to the user with the common study of different disciplines such as doctors, architects, landscape architects, psychologists.

References

1. Cüceloğlu, D., *İnsan ve Davranışı.* 2012, Remzi Kitabevi, Istanbul.
2. Ulrich, R.S., *Human Responses to Vegetation and Landscapes.* Landscape Urban Plan, 1986. **13**: p. 29–44.
3. Aksu, Ö. V., and Demirel, Ö., *Hastane Bahçelerinde Peyzaj Tasarımları: Trabzon Kenti Örneği,* Kastamonu University Journal of Forestry Faculty, 2012. **12**(2): p. 236–250.
4. Keçecioğlu, P., *Ruh Sağlığı Kurumlarında İyileştirme Bahçelerinin İrdelenmesi ve Peyzaj Tasarım İlkelerinin Belirlenmesi.* 2014, Master Thesis, Istanbul Technical University Graduate School of Natural and Applied Sciences Open Archive.
5. Sakıcı, Ç., and Var, M., *Ruh ve Sinir Hastalıkları Hastane Bahçelerinin (Açık Alan Terapi Üniteleri) Düzenlenmesi Kriterleri.* Kastamonu University Journal of Forestry Faculty, 2014. **14**(1): p. 101–112.
6. Poyraz, A., *Akıl Hastanelerinde Hasta-Mekan Etkileşimi.* 2015, Master Thesis, Hacettepe University Libraries Open Access Unit.
7. Marcus, C.C., *Healing Gardens in Hospitals.* IDRP Interdisciplinary Design and Research e-Journal, 2007. I(I).
8. Ulrich, R.S., Health Benefits of Gardens in Hospitals. 2002, Plants for People International Exhibition Floriade.
9. Uslu, A., and Shakouri, N., *Zihinsel ve Fiziksel Engelliler İçin Hortikültürel Terapi.* Kastamonu University Journal of Forestry Faculty, 2012. **12**(1): p. 134–143.
10. Hospital Gardens that Heal, Healthy Savvy and Wise. [ONLINE] Available at: https://healthysavvyandwise.com/hospital-gardens-that-heal/. [Accessed 1 December 2020].
11. Clever H the Mag. 2015. Healing from the Monastic Garden. [ONLINE] Available at: https://cleverhthemag.com/2015/09/23/healing-from-the-monastic-garden/. [Accessed 1 December 2020].
12. 31 Daily. 2018. MEDICINAL HERB GARDENS SERIES. [ONLINE] Available at: https://www.31daily.com/medicinal-herb-gardens/. [Accessed 2 December 2020].
13. Arkeolojik Haber. 2018. Edirne Darüşşifası'nın ziyaretçi sayısı bayramda arttı. [ONLINE] Available at: https://www.arkeolojikhaber.com/haber-edirne-darussifasinin-ziyaretci-sayisi-bayramda-artti-13504/. [Accessed 2 December 2020].

14. Kültür Portalı. 2014. İkinci Bayezid Külliyesi Sağlık Müzesi - Edirne. [ONLINE] Available at: https://www.kulturportali.gov.tr/turkiye/edirne/gezilecekyer/iibeyazid-kulliyesi-saglik-muzesi. [Accessed 1 December 2020].
15. Özçalık, M. and Eskisarılı, Ş., *Peyzaj Mimarlığında Renk Olgusu*, International Journal of Uyghur Studies, 2019. **14**: p. 60–74.
16. Oregon Live. 2019. Healing in a garden: Horticultural therapy feels good. [ONLINE] Available at: https://www.oregonlive.com/hg/2015/09/healing_garden_at_home_horticu.html. [Accessed 2 December 2020].
17. Grahn, P., Pálsdóttir, A.M., Ottoson, J., Jonsdottir, I.J., *Longer Nature-Based Rehabilitation May Contribute to a Faster Return to Work in Patients with Reactions to Severe Stress and/or Depression.* International Journal of Environmental Research and Public Health, 2017. **14**(11): p. 1310.
18. Özer, B., *İnsan Psikolojisi ve Peyzaj Tasarımı.* 2005, Master Thesis, Ankara University Institute of Science and Technology.
19. Ontario Sores. The Building and Grounds. [ONLINE] Available at: https://www.ontarioshores.ca/cms/One.aspx?portalId=169&pageId=28886. [Accessed 2 December 2020].
20. The Field. 2018. Therapeutic Gardening for an Adult Inpatient Psychiatric Unit. [ONLINE] Available at: https://thefield.asla.org/2018/04/17/therapeutic-gardening-for-an-adult-inpatient-psychiatric-unit/#more-8070. [Accessed 2 December 2020].
21. Gensler. Scottsdale Healthcare Thompson Peak Hospital Healing Garden. [ONLINE] Available at: https://www.gensler.com/projects/scottsdale-healthcare-thompson-peak-hospital-healing-garden?k=healing&l=search&s=date. [Accessed 3 December 2020].
22. AzCentral. 2014. West Valley hospitals use healing gardens to reduce patient stress. [ONLINE] Available at: https://www.azcentral.com/story/news/local/southwest-valley/2014/04/17/west-valley-hospitals-healing-gardens/7830061/. [Accessed 3 December 2020].
23. Dirt Works. Sensory Arts Garden Els Center of Excellence. [ONLINE] Available at: http://dirtworks.us/portfolio/sensory-arts-garden-els-center-excellence/. [Accessed 4 December 2020].
24. Nature Sacred. The Healing Garden at UMMS Rehabilitation & Orthopedic Institute. [ONLINE] Available at: https://naturesacred.org/sacred_place/umms-rehabilitation-orthopedic-institute-2/. [Accessed 4 December 2020].
25. Nature Sacred. 2014. Healing Therapeutic Garden at Kernan Hospital. [ONLINE] Available at: https://naturesacred.org/case_study/healing-therapeutic-garden-at-kernan-hospital/. [Accessed 4 December 2020].
26. Eryılmaz, G., and Unur, K., *Açık Alan Rekreasyon Etkinliklerine Katılan Bireylerin Pozitif ve Negatif Duygu Durumlarının Karşılaştırılması.* Gastroia: Journal of Gastronomy and Travel Research, 2019. **3**(4): p. 562–578.

27. Horticultural Therapy Institute. 2020. HTI Restructures Curriculum During Pandemic. [ONLINE] Available at: https://htinstitute.org/community/blog/page/5/. [Accessed 5 December 2020].
28. Horticultural Therapy Institute. 2020. HT Programs for Children with Autism. [ONLINE] Available at: https://htinstitute.org/community/blog/page/8/. [Accessed 5 December 2020].
29. Helmsley Walled Garden. Social and Therapeutic Horticulture. [ONLINE] Available at: https://www.helmsleywalledgarden.org.uk/horticultural-therapy/social-and-therapeutic-horticulture/34. [Accessed 5 December 2020].
30. Siu, A.M.H., Kam, M., Mok., I., *Horticultural Therapy Program for People with Mental Illnes: A Mixed Method Evaluation.* International Journal of Environmental Research and Public Health, 2020. **17**: p. 711.
31. Arslan, M., and Ekren, E., *Yaşlı Kişilerin Sağlığı ve Etkinlikleri İçin Terapi Bahçeleri.* Journal of Ahi Evran University Social Sciences Institute, 2017. **3**(2): p. 361–373.
32. Seixas, M., Williamson, D., Barker, G., Vickerstaff, R., *Horticultural Terapy in a Psychiatric in- Patient Setting.* BJPsych International, 2017. **14**(4): p. 87–89.

Meltem GÜNEŞ TİGEN and Murat ÖZYAVUZ

Universal Design in Playgrounds

Abstract: Children develop their cognitive, physical and social skills through play. Every child has the right to play. Designing open spaces where all children with or without disabilities in all spheres of life can play together under equal conditions forms the basis of the universal design (UD) concept. UD is a concept that goes beyond the accessibility issues of the built environment. According to the understanding of UD, all built environments should be accessible to everyone regardless of age, gender, socio-cultural factors, economic status and disability. In attempts to correct conceptual confusion, the present study first examined such design approaches as "universal design", "design for all", "inclusive design", "barrier-free design", "accessible design", which are frequently encountered in the literature and are used interchangeably as well as playing a major role in planning and design of more accessible cities and having similar or different meanings. Subsequently, we addressed the UD criteria, which are critically important for building an accessible playground for everyone, serving people without any type of discrimination and enhancing their development. To this end, this study aimed to raise awareness in our country by examining sample playgrounds designed in accordance with UD criteria in the world. These sample playgrounds reveals the importance of the universally designed playground thanks to its success in providing a more useful playground for children with disabilities. This can serve as an inspiration to build similar playgrounds in our country. This study highlighted the necessity of widespread playgrounds designed for everyone, and where all children with or without disabilities, regardless of their age, gender and social status, can not only play in accordance with standards and regulations but also can have equal rights to play. As a result, it is possible to build children's playgrounds in accordance with the principles of UD, which adopts an inclusive/socialization mission for the future of our children. Thus, environments with play value that provide benefits not only for children with disabilities but for every child regardless of ability can be created. UD should be considered as a design approach that supports and encourages new developments in children's playgrounds, which is a significant public place at the same time.

Keywords: Universal design, Inclusive design, Design for all, Accessible design, Barrier-free design, accessibility, playgrounds.

1 Introduction

With the increase in the number of people with disabilities after the Second World War, the interest towards disabled people increased and this interest caused an increase in the laws regarding both the elderly and the disabled. The increase of these laws has also encouraged "Disability-Specific Design" in the design industry [1]. Accessible and usable products, services and environments have been created to prevent discrimination [1]. These studies were approved by the decision of the United Nations General Assembly on 13 December 2006 and were accelerated with the "The Convention on the Rights of Persons with Disabilities (CORD)", which entered into force on 3 May 2008 [2]. With the increasing awareness of the elderly and disabled people, the design of appropriate environments that will meet their needs in order to integrate them into society has attracted the attention of city planners and designers [1]. With these developments, the design approach has started to change towards designing accessible spaces where they can access services equally, independently and safely, not only for individuals with functional limitations, but also for all individuals in the society (elderly, temporary or permanent disability, children, pregnant, excessively short, excessively long, the extremely thin, overweight etc.) [3, 4], regardless of their social, cultural and economic status.

Accordingly, design approaches aimed at removing barriers for the disabled people have emerged in the design of products, services and environments [1]. One of the biggest factors in the development of Universal design, one of these approaches, is the design approach that takes users' needs into account from the beginning of the design process. Recently, the term user-centered design has been used to describe the design that defines and addresses the user's needs, abilities and limitations [1]. Universal design has started to spread in various fields including architecture, urban design and landscape architecture.

Although there are some variations in terms used to describe different planning and design processes in urban design such as non-discriminatory and non-stigmatizing "universal design, inclusive design and design for all", Universal design remains the most common [5, 6]. In the first phase of the study, the conceptual confusion was tried to be eliminated by making the definitions of concepts that are frequently encountered in the literature and sometimes used interchangeably as well as playing a major role in planning and design of more accessible cities and having similar or different meanings.

According to the UN's World Population Estimates Report, 27 percent of the world's population consists of children under the age of 15 with different abilities and different socio-economic status [4]. "Address Based Population Registration System" shows that the child population is 27.5 percent in Turkey by the end of 2019 [7]. The calculations of the World Health Organization, 12 percent of Turkey's population is disabled [8] and the number of disabled children in Turkey is estimated to be 3 million [9]. According to these data (2019), approximately 13 percent of our children in Turkey are disabled. As stated in many sources, children develop their cognitive, physical and social skills through play. Every child has the right to play and this right was secured for the first time by the World Convention on the Rights of the Child on 20 November 1989, together with many rights specific to the child [10].

Play is regarded as an activity that not only helps to develop social, cognitive and physical skills, but also allows them to develop their self-awareness, socialize, explore their surroundings and gain experience [11]. As stated in many sources, playing games is the primary need of the child and is an important part of the child's development.

Designing playgrounds for every child with or without disabilities where they can play outside together and under equal conditions by integrating with nature, is based on the concept of "universal design".

In the continuation of the study, firstly, the features of the universally designed playgrounds were mentioned. After that, some examples of successful playgrounds designed with the guidance of Universal design criteria and giving disabled children the chance to participate fully in social and physical games were selected and examined in order to attract attention to this issue in our country.

2 Design Concepts Based on Accessibility

The concepts that are frequently encountered in the literature and sometimes used interchangeably, are non-discriminatory and effective in planning and design of more accessible of cities, with similar or different meanings are explained in Tab. 1.

In the emergence of all these concepts, the accessibility of elderly and disabled individuals to services and places was taken as a basis [12]. While "barrier-free design" and "accessible design" approaches are a design concept that especially takes into account the needs of individuals with disabilities, they differ from the Universal design approach.

One of the great features of universal design is when the purposeful design is very subtle. For example, while the accessible curved mirror above the sink in the bathroom is "Accessible Design", the full-length mirror next to the Washbasin is the "Universal design" approach [13].

It is also known as "Inclusive design", "Design for all" in Europe and "Universal design" in the USA (Design Council, 2008). These terms have the same underlying moral values, but there are subtle differences in their perspectives and practices [12, 14]. Inclusive design differs from Universal design, which is a separate design style, with the design principles offered by Mace [12, 15].

Tab. 1. Design concepts based on accessibility

Concept	Originating from
Barrier-free design 1950s [16, 17] Accessible design 1970s [18]	1961, USA: American National Standard ANSI A117.1 became the first Barrier-Free Design standard [16]. 1963, England: Architect Selwyn Goldsmith published "Design for the Disabled", an architectural access guide [18]. 1990, USA: These standards were legalized and demonstrated that the Americans with Disabilities Act (ADA) of 1990 clearly prohibits discrimination based on disability [16].
	Definition
	Barrier-free design and accessible design is a design process in which the needs of disabled people are taken into account. It is used to describe the effort to remove physical barriers from the "built environment" for people with disabilities [19].
Universal design 1983 [18]	**Originating from**
	Architecture in the USA in the early 1990s by architect Ronald L. Mace [20, 21]
	Definition
	As explained by Mace (1985), "Universal design can be defined as the design of Products and environments that can be used by all people to the greatest extent possible, without the need for adaptation or special design" [15].
Inclusive design 1994 [18, 22]	**Originating from**
	The term 'Inclusive Design' was first used by Roger Coleman in 1994 in UK [18, 22].
	Definition
	The British Standards Institution (2005) defines it as "the design of generic products and services that can be accessed and used by as many people as reasonably possible, without the need for special adaptation or special design" [23]. Inclusive design originated with product design [23] is defined in the Architecture and Built Environment Commission (CABE) as the process of designing, building, managing and placing places and spaces that enable them to work not only for some groups but for as many people as possible [24].
Design for all	**Originating from**
	In Europe
	Definition
	The European Institute for Design and Disability defined it as "Design for all is design for human diversity, social inclusion and equality" [25]. The European Institute for Design and Disability (EIDD) changed its name to "Design for All of Europe" in 2006 [16].

The origin of Universal design, which proposes a standardized design that can be used by every user [24], are based on Barrier-free design and accessible design approaches [16]. Design for all is the European Union term which is equivalent to Universal design, and they are used interchangeably in recent studies. [16]. The implementation of design for all uses the analysis of human needs and desires consciously and requires the participation of end users at every stage of the design process [16].

Inclusive design, Design for all and Universal design are based on an approach that focuses on increasing the widest possible range of use.

Concept of Universal design covers different scales from micro scale such as product design to macro scale such as interior design, landscape and urban planning which deals with the design and planning of the architectural and physical environment [26]. In this context, it is critical to design children's playgrounds, which is one of the working areas of the landscape architecture profession and which is extremely important for the healthy development of children, in accordance with the Universal design approach, where all children can play equally regardless of their abilities.

From 1994 to 1997, the Universal Design Center at North Carolina State University led by Ronald Mace developed a series of Universal design guidelines with architects, product designers, engineers and environmental design researchers [20, 27]. The Universal Design Principles in this guide and examples of playgrounds related to these principles are shown in Tab. 2.

Tab. 2. Universal Design Principles and examples of playgrounds built based on these principles [20].

Principle 1: Equitable Use "The design is useful and marketable to people with diverse abilities. Everyone with different ages, abilities and socio-cultural characteristics can access playgrounds".	**Figure 1.** Unwin Park, City of Surrey, B.C. Canada [28]. A park designed for all. 
Principle 2: Flexibility in Use "The design accommodates a wide range of İndividual preferences and abilities".	**Figure 2.** Sandbox design available for all children, such as the father using a wheelchair playing with his children in the sandbox [29]. 

Tab. 2. Continued

Principle 3: Simple and Intuitive Use "Use of the design is easy to understand, regardless of the user's experience, knowledge, language skills, or current concentration level".	**Figure 3.** Sensory panels to explore on the playground [30]. The child can sense how the wheels will move. 
Principle 4: Perceptible Information "The design communicates necessary information effectively to the user, regardless of ambient conditions or the user's sensory abilities".	**Figure 4.** Frimley Park Lodge, Surrey, England [31]. Different colored groundfloor have made the park more perceptible for children. 

(*continued*)

Tab. 2. Continued

Principle 5: Tolerance for Error "The design minimizes hazards and the adverse consequences of accidental or unintended actions".	**Figure 5.** A child sitting on an elevated platform entrance, leaving her wheelchair on the floor [13]. 
Principle 6: Low Physical Effort "The design can be used efficiently and comfortably and with a minimum of fatigue".	**Figure 6.** Harper's Playground, Portland, USA [32]. Using the accessible ramp road, which offers equal opportunities for all children, users can access playground equipment from various parts of the park with low physical effort. 

Tab. 2. Continued

Principle 7: Size and Space for Approach and Use "Appropriate size and space is provided for approach, reach, manipulation, and use regardless of user's body size, posture, or mobility".	**Figure 7.** Boy in a wheelchair using play equipment with other friends [33]. 

It is emphasized that these seven principles are "applicable to evaluate existing designs, to guide the design process and to educate both designers and consumers about the features of more useful products and environments" [20].

3 Features of universally designed playgrounds

A universally designed playground should take into account the mobility, sensory and cognitive abilities users need to interact with the playground [34]. When addressing mobility flooring, ramps and the height of objects are also important [34].

"Universally-designed playgrounds feature" published by the American Society of Landscape Architects (ASLA) in his work "Universal Design: Playgrounds" [35] and in the project named "Universally Designed Playgrounds: An Outdoor Play Place for All" presented by Emmett (2012) [34], it is stated that the universally designed playgrounds not only support mobility but also address the sensory needs of many children who have difficulties in processing or enjoying their tactile, visual, auditory and movement sensations. These two sources are summarized and the features that should be in Universally Designed Playgrounds are given in Tab. 3.

Tab. 3 Common features of universally designed playgrounds [34, 35]

Multiple forms of play/creative play space Multi forms of play environments, diverse and more creative, active and social play space in many playgrounds are critical for the social and motor development of children. Adding everyday items such as car and bike tires, cardboard boxes, pieces of fabric, and pipes to playgrounds can be enriched the play.
Range of risk Some risky games are beneficial for children, but there should be levels of physical difficulty of appropriate difficulty and height that they can be comfortable with for different children and parents..
Ease of access The use of ramps with handrails going to and on the game equipment is also important for wheelchair users. All sides of the children's playground should be easily and safely accessible to every child.
Accessible equipment Play equipment, unlike standard playground equipment, should be easily adaptable to meet the needs of all children, not only those with physical disabilities, but also those with cognitive disabilities. In addition, auxiliary equipment that will allow the use of these equipment should be used.
Safe water and sand When water and sand are designed safely, it provides an environment for all children, especially autistic, blind and deaf children, to use their tactile systems.
Range of sensory engagement Many autistic children need different intensities of sensory interaction in order to be stimulated, sometimes larger, more interesting areas throughout the playground, sometimes small quiet places to play. There should also be different stimuli for different sensory engagement. For example, placing a panel with bells or drums to support auditory stimulation will help to encourage participating and playing instruments in a fun and motivating way.
Materials Materials of tactile and sensory interest should be balanced together with materials that are easy to navigate for children with mobility impairments. Universal design principles provide another option, which can consist of a flat rubber surface that provides safety for children but allows participation of children with disabilities.
Interpretive signage Interpretive signages should point out places where different types of sensory games are present, in a way that everyone can see and understand.
Enclosure Safely play spaces can be created with the effective use of vegetative or structural containment elements.

Tab. 3 Continued

Ample seating In order for the parents accompanying the children to be able to observe them from all sides of the playground, the seating units should be spread all over the playground and should be designed in sufficient width by considering the possibility of their special situations.
Clear line of sight to restrooms Toilets should be placed in a way that allows easy access to all children, including those with cognitive disorders.
Landform design Landform design is used to enhance sensory play and promote unstructured play.
Non-toxic, non-thorny plants Plants provide excellent sources of sensory stimulation for all children. However, care should be taken to use herbs that are safe for children that will not cause allergen and toxic effects.

The "flexibility in use", which is actually meant to be explained in these concepts, is compatible with the universal design principle and can be applied to any playground. It can help promote positive play in disabled and non-disabled children at a very low cost [34].

4 Playgrounds Designed According to Universal Design Principles

Clement Park, USA

Clement Park is a regional park in Littleton, Colo., serving people with different abilities and offering them all kinds of recreational opportunities. In the playground built in accordance with the universal design criteria as well as The American Society for Testing and Materials (ASTM) standards, there are playground equipment placed for children of all ages (2–5 and 5–12 years old) [36].

It offers many entertainment opportunities with its athletics, picnic and special playgrounds. The music-themed playground includes different sensory stimuli (such as sight, touch, hearing and smell), such as a trombone-like playground structure and various musical instruments, and different stimuli to support fine motor development [36].

In addition, there are activities at the appropriate level of difficulty and height where disabled people can do physical activity by taking risks from time to time in the game (Fig. 8).

Moving around and accessing game equipment in the playground is provided by ramps, the most suitable solution for everyone. As the floor covering, soft materials such as ready-made grass, plastic-based flooring, Engineered wood fiber are used both to reduce impact and allow wheelchair users to move around comfortably. Besides, wheelchair swings and movable bridges in the playground allow children with disabilities to have fun. The slides were placed on the elevated hills by designing the landform, thus creating more interesting spaces for children (Fig. 8).

Park is a successful example designed in accordance with universal design principles, accessible to all and providing equal usage rights for all.

Fig. 8. Clement Park- multiple forms of play, and slides placed according to the landform [36].

Ballum Bumps Regional Playscape, Australia

Ballum Bumps Regional Playscape (Frankston, Australia, Playce) is a regional playground designed for everyone and awarded as best playground by the 2018 Natural Landscape Architecture Award [37].

This playground serves all age groups from toddlers to adults. It attracts the attention of toddlers and babies with the wooden play space on the east side a wide ramp, climbing equipment and slide network. There are picnic places and fountains in the natural shade of the trees surrounding the park [38]. The park, which has both a strong and unpretentious design, also serves as many recreational resources including track, skate park, rock climbing and ball sports [39] (Fig. 9).

Fig. 9. Ballum Bumps Regional Playscape, Frankston, Australia, Playce [38, 39].

Another playground in the park is the adventure play space, created with aging equipment for older children and has a suitable natural wood structure [38]. Additionally, the playground has many small hills that allow children to play with play equipment as well as create their own play form using the landform [35] (Fig. 9).

The playground is a good example that is universally designed with creative and diverse playgrounds, with activities suitable for everyone from 7 to 70 with disabilities.

Magical Bridge Playground in Palo Alto, CA

The Magical Bridge Foundation (MBF), one of the first unique design solutions in the country, opened in 2015, incorporating all seven Universal Design principles, has become the pioneer of innovative playgrounds and playground equipment for all individuals [40, 41].

The playground gives children of different abilities and ages the opportunity to play independently and cooperatively. Children can walk among trees and have unobstructed access to other activities in the area. The playground has a wide variety of areas for different types of activities to meet the various developmental needs of children, such as turning, swinging, sliding, climbing and exploring natüre [41].

Play zones are designed to be scaffolded with various supports. Activities and equipment in each region are designed to encourage the development of similar sensory-motor and cognitive skills, and the regions are labeled with large font and Braille, ensuring that they are perceptible to everyone. In addition, the design of activities and equipment fosters interaction between individuals with different abilities in similar activities [40].

The park area also includes retreats areas for children with autism, sensory stimuli such as a music zone and whimsical artwork, and a two-story wooden playhouse suitable for wheelchair Access [40, 41].

Magical Bridge Playground, which increases its users day by day, is a park that has managed to benefit everyone in the community, regardless of ability, by going beyond the minimum requirements of the Americans with Disabilities Act (ADA) standards. It provides a fun, dynamic and inclusive game opportunity for the increasing autistic population, the visually and hearing impaired, physically limited, medically fragile and even the aging community.

While today's typical park designs overlook and exclude so many people, the successful results in these example parks, which are fun, aesthetic and interesting at the same time, demonstrate the importance and necessity of adopting universal design principles that appeal to everyone in current and future playgrounds.

Fig. 10. Magical Bridge Playground multiple forms of play [40, 41].

5 Conclusion

The number of children with disabilities is unfortunately increasing day by day [8]. As a result, the importance of universally designed playgrounds also increases. Traditional playgrounds are not suitable for children with physical or mental disabilities. The "disabled park" or "barrier-free park" approaches offered as a solution for this stigmatize disabled people, thereby marginalizing them. In addition to that, since they cannot find playmates, many disabled children face isolation and play alone in sites designated for them [42]. However, all disabled and non-disabled children should play equally and interact with each other.

While barrier free or accessible design focuses on the needs of people with disabilities, UD covers a broad spectrum of human abilities. A universal approach aims to go beyond minimum standards to meet the needs of the greatest number of people [13]. Accordingly, design approaches that tend to assume only disabled children have been replaced by the UD approach, which adopts the notion of "design for everyone". UD is a concept that goes beyond the accessibility issues of the built environment. It can be defined as a concept that is effective in bringing together people with different physical, mental or psychological abilities and from different social, cultural and economic conditions [27].

Given the sampled parks incorporating UD principles, no additional special equipment were used and they were designed in a way that were accessible to everyone in various and creative ways. As shown in the sampled parks, universally designed playgrounds promote access and equal use for everyone. Such playgrounds also reveals the importance of the universally designed playground thanks to its success in providing a more useful playground for children with disabilities. This can serve as an inspiration to build similar playgrounds in our country.

UD is vital not only for those with physical disabilities, but also for children with cognitive disabilities and non-disabled children [34]. The target audience should be all children – not just for children who experience disability. If we adopt a human-centered design approach while applying the UD Principles, it is possible to create environments with play value so that it benefits every child regardless of disability and ability [13, 34].

Not only children but also their accompanying parents use children's playgrounds. These parents may also have various limitations such as being elderly, temporarily or permanently disabled, pregnant, extremely short, long, thin, and obese. When parents allow their children the freedom to play by their own when appropriate, they also have the opportunity to be together with their children in the playground. Thus, the UD approach is also significant for the use of parents.

Given the children's playgrounds in our country, they either appeal to those children who are healthy or they are named as "disabled park "or" non-disabled park". In other words, the vast majority of the playgrounds adopt an approach that stigmatize children with disabilities. These approaches are far from the UD approach. While every universal design product is also an accessible design product, not every accessible design product is a universal design product [43].

Therefore, in addition to playgrounds designed in accordance with standards and regulations, our country also needs playgrounds designed for everyone and where all children, regardless of their disability, age, gender and social status, can have equal rights to play in terms of healthy child development. With this in mind, this study attempts to underline the importance and the necessity of widespread playgrounds built with the understanding of UD.

Consequently, creating accessible, barrier-free and inclusive children's playgrounds that will foster children's development can be ensured through the implementation of the principles of UD, which sustain an inclusive/socialization mission.

References

1. National Disability Authority, *History of Universal Design,* http://universaldesign.ie/What-is-Universal-Design/History-of-UD/.
2. Resmi Gazete, *Milletlerarası Sözleşme,* 2009. https://www.resmigazete.gov.tr/eskiler/2009/07/20090714-1.htm
3. Uslu A., Güneş M. *Engelsiz Kentler-"Herkes Için Erişilebilir Kentler",* 2017. International Journal of Landscape Architecture Research, pp. 30–36.
4. United Nations, *World Population Prospects 2019,* https://population.un.org/wpp/
5. Hitch, D., Larkin, H., Watchorn, V., Ang, S. *Community mobility in the context of universal design: Inter-professional collaboration and education.* Aust. Occup. Ther. J. 2012, 59, 375–383.
6. Larkin, H., Hitch, D., Watchorn, V. and Ang. S. *Working with Policy and Regulatory Factors to Implement Universal Design in the Built Environment: The Australian Experience.* ISSN 1660-4601. Int. J. Environ. Res. Public Health 2015, 12, 8157–8171; doi:10.3390/ijerph120708157.
7. Türkiye İstatistik Kurumu, *İstatistiklerle Çocuk, 2019.* https://tuikweb.tuik.gov.tr/PreHaberBultenleri.do?id=33733
8. World Health Organization, *World Report on Disability 2011.* ISBN 978 92 4 068521 5 (PDF). https://www.who.int/disabilities/world_report/2011/report.pdf.
9. Sağlıkon, "turkiye-nufusunun-yuzde-12si-engelli", 02-12-2019. https://www.saglikon.com/2019/12/turkiye-nufusunun-yuzde-12si-engelli.html
10. Unicef, *Çocuk Haklarına Dair Sözleşme.* https://www.unicef.org/turkey/çocuk-haklarına-dair-sözleşme
11. Stagnetti, K. *Understanding play: The implications for play assessment.* 2004. Australian Occupational Therapy Journal, 51, 3–12.
12. Clarkson, P. J., Coleman, R. *History of Inclusive Design in the UK.* 2015. Applied Ergonomics, Volume 46, Part B, January 2015, Pages 235–247.
13. National Center on Acccessibility, *Designing For Inclusive Play: Applying The Principles Of Universal Design To The Playground,* 2007. https://ncaonline.org/designing-for-inclusive-play-applying-the-principles-of-universal-design-to-the-playground/
14. Centre for Universal Design Australia, *Inclusive Design: Manufacturing, Design and Retail Views,* May 2016, www.rica.org.uk, http://universaldesignaustralia.net.au/wp-content/uploads/2016/11/Inclusive-Design-manufacturing-design-and-retail-experts-views-May-2016.pdf
15. Mace, R.L., Hardie, G.J., Place, J.P., Accessible Environments: Toward Universal Design, 1991. In: Preiser, W.E., Vischer, J.C., White, E.T. (Eds.),

Design Intervention: Toward a More Humane Architecture. Van Nostrand Reinhold, New York, USA.

16. Persson, H., Ahman, H., Arvei, A., Jan Gulliksen, Y. Universal Design, Inclusive Design, Accessible Design, Design For All: Different Concepts—one goal? On the concept of accessibility—historical, methodological and philosophical aspects. Springer-Verlag Berlin Heidelberg 2014, Universal Access in the Information Society 14(4). DOI: 10.1007/s10209-014-0358-z
17. The Center For Universal Design, North Carolina State University, *A Brief History of Universal* Design. The Universal Design File, https://projects.ncsu.edu/ncsu/design/cud/pubs_p/docs/udffile/chap_1.pdf
18. Maisel, J. L., Steinfeld, E., Basnak, M., Smith, K. and Tauke, M. B. *Inclusive Design Implementation and Evaluation.* WDD18_Architecture_Chapter_Compliation_Routledge, Series Editor: Ryan E. Smith, First published 2018 by Routledge.
19. Media Services - Lane Community College, Universal Design Project 2020, *Universal Design vs. Accessibility and Barrier-free Design.*https://media.lanecc.edu/users/howardc/PTA103/103UniDesign2/103UniDesign25.html
20. The Center for Universal Design. North Carolina State University, *The Principles of Universal Design*, 1997. Version 2.0 - 4/1/97. Raleigh, NC: NC State University, Author. https://projects.ncsu.edu/ncsu/design/cud/about_ud/udprinciplestext.htm
21. Steinfeld, E., Maisel, J., Levine, D. *Universal Design: Creating Inclusive Environments*, 2012. John Wiley & Sons Ltd.: London, UK.
22. Coleman, R. *The Case for Inclusive Design – an Overview*, 1994. 12th Triennial Congress. International Ergonomics Association and the Human. P John Clarkson, Roger Coleman Factors Association of Canada, Toronto, Canada.
23. University of Cambridge, Inclusive Design Toolkit, *What is inclusive design?* Engineering Design Department. http://www.inclusivedesigntoolkit.com/whatis/whatis.html
24. Tanuwidjaja, G., Rieger, J. & Franz, J. *Understanding The Differences Between Universal Design and Inclusive Design Implementation: The Case of an Indonesian Public Library*, 2020. In 4th Australian Universal Design Conference, 2021-05-17 - 2021-05-18, Melbourne, Australia.
25. The European Institute for Design and Disability (EIDD), *The EIDD Stockholm Declaration 2004.* Adopted on 9 May 2004, at the Annual General Meeting of the European Institute for Design and Disability in Stockholm.
26. Dostoğlu, N., Şahin, E., Taneli, Y., Tasarıma Kapsayıcı Yaklaşım: Herkes Için Tasarım, 2009. Mimarlık Dergisi, 347, 23–27.
27. Universal Design Handbook, 2011. Wolfgang F. E. Preiser Editor-in- Chief Korydon H. Smith Senior Editor, Second Edition, Architectural Record. ISBN: 978-0-07-162922-5

28. British Columbia, *New universally accessible playground opens in Surrey, B.C.* Unwin Park, City of Surrey, B.C. Canada https://www.cbc.ca/news/canada/british-columbia/surrey-new-accessible-playground-1.5320087
29. Making Healthy Places, makinghealthyplaces.com, photo: CDC/Richard Duncan, Senior project manager, North Carolina State University, Center for Universal design.
30. Child in the City, *Designing a play space for everyone*, 2018, https://www.childinthecity.org/2018/04/09/designing-a-play-space-for-everyone/?gdpr=accept
31. Surrey Life, *The Children's Playground at Frimley Parkodge*, https://www.getsurrey.co.uk/whats-on/family-kids-news/gallery/childrens-playground-frimley-park-lodge-16617710
32. Barber, M. *Why Cities Need Accessible Playgrounds*, 2018. https://archive.curbed.com/2018/7/20/17582148/accessible-playgrounds-design-ada-standards-inclusive.
33. Inclusive Playgrounds, *Inclusive Play Design Process*, https://www.inclusiveplaygrounds.net/
34. Emmett, M. *Universally Designed Playgrounds: An Outdoor Play Place for All*, May 10, 2012, submitted by Emmett, M. the degree of Master of Occupational Therapy from the University of Puget Sound.
35. American Society of Landscape Architects, Universal Design: Playgrounds. https://www.asla.org/universalplaygrounds.asp
36. Landscape Structure, *Clement Park*, https://www.playlsi.com/en/commercial-playground-equipment/playgrounds/clement-park/
37. Landscape Australia, ballam Bumps, https://landscapeaustralia.com/articles/2018-AILA-National-Awards-Ballam-Bumps/
38. Ballam Park, https://tothotornot.com/ballam-park/2021
39. Landscape Architecture Projects, Ballum Bumps Regional Play Space.https://www.landscapearchitectureprojects.com/projects/2018/ballumbumps
40. Ofiesh, N. and Poller, L. *A Playground For The Entire Community: The Design of Magical Bridge Playground* Universal. https:playworld.com and https:magicalbridge.org.
41. American Society of Landscape Architects, *Magical Bridge Universal Playground*. https://asla-ncc.org/portfolio-items/magical-bridge-universal-playground/
42. Tamm, M. and Skar, L. How I Play: Roles and Relations in the Play Situations of Children with Restricted Mobility, 2000. Scandinavian Journal of Occupational Therapy, 7, 174–182.
43. Tandoğan, O. Evrensel Tasarım Kavramı ve Kentsel Peyzaj İle İlgili Örnekler Üzerinden Değerlendirilmesi, 2017. Artium Dergisi, Cilt 5, Sayı 2, 51–66, 2017

Timur KAPROL

Experiencing the Mixed-Use from Structure to City Scale in the Architectural Design Studio within Architectural Design and Planning

Abstract: Within the context of the Architectural Design Studio 5 of the Department of Architecture in Kırklareli University, architect candidates were demanded to design an "Exhibition Center" in the city center of Kırklareli, during the Spring Semester of 2019–2020 Academic Year, in the scope of grounding the theoretical studies in the Architectural Design Studios on scientific research and developments, conveying the obtained data to the architect candidates so that they can use these in the design, i.e., approaching the experience based on scientific data as a method. Withing the context of the continuation of the city wall, students were demanded to make a design based on service to the public who continue their lives in residential and commercial buildings, on empty plots which are in the building blocks that are situated in the city center. The architect candidates were expected to include social content functions such as a mini movie theatre, library, book café, and an art gallery in buildings that can house cultural activities. The pedestrian axes of the buildings were aimed to be in connection with the present social and green spaces. The theoretical information that will direct the design was given to the architect candidates by the City Planning Specialist Asst. Prof. Dr. Ayşe ALTUN through a seminar as part of the course. The topic of the seminar was "mixed-use". After the necessary theoretical infrastructure study, the social structure of the area was defined through an analysis of design areas and continued with possible designs. In terms of the synthesis provided by the analysis, spatial organizations were preferred for building design. The first part of the article defines the needs of mixed-use on an urban and architectural scale, and the second part elaborates on the theory of mixed-use and compact use: the purpose of analysis on city planning on a world scale has been to obtain findings on both an urban and architectural scale that will serve the end product through scientific studies and practices on the subject. The third part exemplifies three out of sixty studies carried out in the architecture studio practice. All three examples of this study highlight functions that visualize public use, as well as the use of the pedestrian axes nearby in the building. By assembling the obtained scientific and social data in the area analysis, the positive sides created by the mixed-use in the design were given in the fourth part that includes the conclusion and evaluation. The study aims to convey to the architect candidates that recreational spots can be created in the city by creating socio-cultural spaces in the studio practices in which insights on architectural scale in an urban context are given gradually.

Keywords: Mixed-use, Exhibition Center, Pedestrian Walk, Building for Social Use

1 Introduction

At the beginning of 21st century, the wish for a more functional, habitable, and sustainable planning of the urban terrain in city planning, as well as the smart city projects, have taken place in designers' ideas. While describing the modern city, a mixed-use tendency has been dominant contrary to a division of uses. Developments and alterations that take place in cities necessitates a regional and local approach instead of centralization. It can be seen that every city develops as a result of both economic and socio-cultural infrastructure in terms of natural, constructed, and social environmental data, and these conditions are reflected in the structuring of the cities. This concept has brough about a "mixed-use" in cities and applications have begun in this context.

The development of mixed-use still maintains its ambiguity. Schemas that belong to certain mixed-use suggestions introduce advantages in terms of design and management. Mixed-use spaces can even be seen in traditional spaces. For instance, Kaleiçi neighborhood of Edirne houses a historical covered bazaar, market places, Selatin Mosques, and right next to these, there are residential areas. Spatial organization that is the advantage of living conditions of the era it was built in shows itself on the building scale in urban construction. There is an unwritten, yet humane and technical-in-its-own-right approach. While a closed city plan was dominant in the citadel organization beforehand, it can be seen that both in Bursa and Edirne, an open city organization was applied during the Ottoman Era. In Bursa and Edirne, which are both seen as cities to be maintained for future by UNESCO, the demand of the political power was taken into consideration for the city planning, and upon looking at the commerce, it can be seen that the livelihood of the city has been prominent in the current urban organization. However, during the 20th century that housed certain urban organization in these two historical Ottoman capitals; the expansion of residential areas show significance both in the third dimension and on a planning scale. Not only these two cities but also other cities in Turkey such as Ankara or Gaziantep which have zoning plans designed by German urban designer and architect Jansen (1), have the same expansion for residential and commercial areas. In fact, the nighttime solitude of commercial areas had become inevitable.

Nonetheless, could it be possible for these areas to be lively around the clock? As an answer to this question, this study aims to resolve issues whether mixed-use can solve these problems, how urban planning and urban design approach urban development theoretically and end urban solitude, and whether it can hinder the relative aging in the city. Maintaining the present mixed-use areas has the potential to allow for new developments. Important indicators in restructuring are obtained in the modernizing or newly-developing cities. Mixed-use may not be the only solution to an issue. It should not erase the urban pattern and the life construct which are priorities to culture. Upon analyzing the sustainable life tendency of mixed-use in the repopulation and restructuring of the city, it is found out that making an analysis by examining the historical pattern and reaching a synthesis makes way for habitable and sustainable urban formations. The research article titled "Mixed-use Development: Ambiguous Concept, Simplistic and Wishful Thinking?" investigates the three following questions (2). In order to shed light on certain subjects and context of conditions in England:

(1) What is the development of mixed-use?
(2) How does the development of mixed-use promote urban quality?
(3) Is it possible to develop mixed-use?

As there are no certain answers to these questions, it is necessary to go further than a desktop study on paper. The variety of academic and professional papers about urban design and sustainable development, life styles and changing business patterns, the development process and their successors point out to the city planners that they particularly have to focus on the following subjects when developing mixed-use.

2 Reflections of Mixed-Use in the World as opposed to Compact Use in Urban Planning

In a research carried out by the Royal Chartered Surveyors Institute in England in January, 1996 and financed by land administration authorities, there has been three case studies about the development of mixed-use in situations that include contradictory situations.

As an answer to the question "What is the development of mixed-use?", the research defines mixed-use as a multidimensional concept with an ambiguous development, and at the same time a dimension of the inner texture of the settlement. The texture of a settlement is an important determiner of its character and quality, and its basic features are the nucleus, density, and permeability created out of the organization of roads, and streets and roads. The 'compact city' preferred by the European Council in 1994 has had a quite different inner texture from the 'typical' modern city. Components of the texture of a settlement defines the collection way of people, activities, land uses, buildings, and spaces. While historical cities generally have a close grain, modern cities are criticized because of their sharp functional grains (3).

The entire structuring of mixed-use has been examined on four mediums. The first one is in terms of a district or neighborhoods; the second in terms of streets and other public spaces; the third is on the entirety of buildings or street blocks; and finally, in terms of individual buildings. The reason for this is that the variety of activities are not only the activities in buildings and blocks; also, it is necessary to acknowledge the public use of the street itself. In the study, which also defends the necessity of examining the formations outside a certain focus point, concludes that many mix-use reasons dwell on the experience of public life (4) (Tab. 1).

Tab. 1. Comparison of the purposes of compact planning and mixed-use development in city planning.

The Aim of Compact Development:	***The Aim of Mixed-use of Land:***
*The **compact development**, which takes the advantage of filling and redeveloping and providing central development, makes buildings, recreational areas, streets, roads, and public spaces to reduce touring possibilities, as well as reducing land consumption levels because of car dependency, bringing about energy use and air pollution. Compact development promotes a full use. Urban services such as water lines, sewer system, streets and emergency services necessitate the need for forming facilities to benefit from the existing public facilities and to reduce the need for new facilities (5).*	*Developing different but competent land uses takes into consideration the transportation such as walking, bicycle, and public transport roads in order to ease alternative working modes and reduce the time spent outdoors. **Developing mixed-use** is suitable for residential areas. Neighborhood centered downtowns, transit triangulation points, main streets, and some public commercial centers allow for smart commercial purposes and developing mixed-use (6).*

There are four kinds of locations where mixed-use is and can be used in the formation of the city. The first of these is commercial or social grain or grains in cities, city centers, or towns; the second one is in-city areas such as abandoned, empty, or areas that need renovation as a result of constructed relative aging; the third is suburbs or the upstate; and the fourth is the green spaces which the land planning policy allows. Lastly, the first of the three necessities for the maintenance and promotion of mixed-use is protecting the features of the present mixed-use; the second is a gradual revitalization and reuse as a result of the increasing need for development, and a restructuring of the existing parts of the city including a transformation; and the third is a comprehensive development or a redevelopment and a formation of the wide spaces. A certain situation of mixed-use development is the result of one or more of these. Variables, problems, and opportunities necessitate different mixed-use situations (7).

Going in and out of the city, the activities these constitute, and land use bring about differences in terms of use. Encouraging social activities in all this advancement develops a contribution to life in cities and a sense of belonging in people (see Student Projects). The temporal dimension of land use is important in mixed-use and social activities are determinant in this use. Mixed-use is generally evaluated as containing different uses that invade separate parts. The function that a building, block, or area assumes provides people to come and go on different time schedules and with different purposes; however, it sometimes causes lesser and regular use, despite not dense, for users in activities concerning the facilities. *The necessity of use in every density requires adding functions on various capacities, a successful sharing of land in all areas, and a careful method.* Enrichening the social life in a community is a must-be for social sharing and collective memory.

Nowadays, the lifespan of the property and rent cycle, high costs, the need for new commercial buildings, and finance institutions make it possible to invade old settlements for rent. Nonetheless, this first models slowly develops; assignment and fragmentation take place in subletting processes, and a richer combination of population and local life style may develop. Alterations may be needed in terms of the needs for use and a building, a block, or an area may be adapted. Opportunities which are renewed to comply with the new conditions can be reached with mixed-use. For this reason, in order to produce an 'enthusiastic variety' in the city, four 'indispensable' conditions may clearly be formed in an assertive way in the streets and districts of the city. City life and economic observations were given as follows in Jacobs' analysis and conclusion (8).

1. New functions in multiple components should be proposed in the district or entirety of the city. Service for multiple uses should be considered in this proposal. Buildings that contribute to the city, buildings that are open to the use of people around the clock, the possibility of night use the roof or the exterior shell of a day-time shopping center, serving as a recital area during the evening hours, opening the green spaces planned for the roof to the use of urban people, a recreational organization of the surface floor of a building which was formed by taking a closed parking garage underground, and multi-purpose common facilities for different needs present the urban people open spaces and brings enthusiasm to the city life.
2. Structuring should be on humane grounds
3. , should not suffocate the individual, and accessible streets and round-the-corner opportunities, and buildings and areas that would be triangulation points should be designed.
4. Spatial designs that will address users from all ages and bring them together in the urban texture should be considered.
5. Either for accommodation, commerce, health or for education, the concentration of people should always be adequate.

According to Grayson and Young (1994), the issue concerning how Urban Quality is promoted in the mixed-use proposal holds an important place in the mixed-use development argument: nowadays, the fact that 'urban quality' concept has inevitably changed to a great extent. There have been some attempts to direct and evaluate the urban design to develop a disciplined set of norms in the 1980s (9), (10). Concerns underlying similar reasons, visual, functional, and the inevitable environmental features of areas and urban experience, extensive thoughts on the upper scale to overlap and mingle; in ways effecting the different mixed-use and on different levels; with proposals that contradict instead of strong evidences, and as a result of promoting views that support mixed-use, *areas that have a mixture of uses as well as various buildings, various architectures, districts, and smaller-scale environments are also encouraged to be experienced.* Still, obtaining these formations can also be achieved in an enthusiastic way. Because the design of some mixed-use developments has failed to present visual quality either through isolation or details related to it. Mega-structure buildings have been constructed in the USA, England, and Europe: a suitable design was sometimes adopted in city plannings where these buildings stood. Promoting the "LIFE BETWEEN BUILDINGS" idea was dominant in mixed-use (11). Additionally, the necessity for moderately compact planning was also defended, but 'Defenders of Mixed-use' showed a tendency to ignore Jacobs' analysis. The primary understandable pragmatic reasons have been an increase of population density in cities that produce more with economic concerns. While a rapid crowding made a part of the society rich, it also paved the way for some parts to be discontent.

When we investigate the answers to the question "Is it possible to develop mixed-use?", it can be observed that pre-industrial cities have been a model from many aspects, and the environment has fewer demands in terms of security, economic prudence, and land use.

Planners have to design the spatial organizations the city for new developments and to realize designs that will receive the support of the society. In order to realize these expectations, solutions to answer all spatial needs in terms of urban concentration should be included in city planning. Processes of city structuring are as follows; there is a need for citizens who were educated in three main qualities.

1. Investors with adequate economic power, experienced in production and urban planning
2. Public authorities that are experienced in terms of city
3. 'Volunteer' organizations, non-governmental organizations, volunteer groups, and individuals (12). The human power that can work in cooperation will work together and in unity just as the treadmills of a machine.

The principles which define the 'smart' ways of forming a community that is beneficial to all citizens, at the same time, represent the smart use of both financial and natural resources. The correct way of managing public facilities and buildings should be done according to these principles. These principles were described in a handbook prepared by *Turkish Customs Directorate (TGM) Program* in 1997 (13).

1. A full use of urban services
2. Mixed-use
3. Transportation choices
4. Detailed and human-focused design

Above are the local aims that support the land use and transportation policies of the state for smart development. In addition to the articles that cover aforementioned principles, an additional resource is economic development and the betterment of tax base. The proposals are as follows (14):

1. Repopulating the residential areas, main streets, and centers of the neighborhood.
2. Creating the necessary residential buildings close to businesses and services; proposing urban areas that include businesses close to where people live.
3. Transportation choices and interconnections.
4. Building complexes that are within a walking distance and transit support opportunities whenever it is possible.
5. Suburban accommodations and decreasing the traffic congestion.
6. An alternative and effective use of the present urban services and facilities.
7. Expansion of new facilities.
8. Energy conservation by decreasing the dependency for cars.
9. Public cost conservation on development plans.

Business "places" such as traditional city centers and well-planned centers offer us preferences: choices on how we get there, what we buy, where we work, what we eat, and how we reach entertainment and what kind of entertainment it is. Moreover, spaces such as car parkings should be isolated, so the city can contribute to the human life rather than vehicles (they should be located underground or on the top floors of buildings, away from visual communication) (see student projects).

Shopping centers, the locations of commercial development after the Second World War, have been distant from houses and businesses. Such location decisions have brought about residential areas that permanently depend on a vehicle. In case transportation to such places are limited or non-existent, and if there is not a local avenue and thus needs are not met, the situation will raise discontent against the area among people. Apart from this, traffic congestion and limitations in state highways and other major arteries, and when young, old, disabled, and economically disadvantaged individuals cannot roam without a vehicle will again cause discontent for the individual.

The priority of State Highways, which were built to supply the transport between cities and regions, was to supply the connection between cities; however, the fact that car traffic bringing extra load to the intercity way that is left between the suburbs, new neighborhoods, and urban spaces created as the city grew caused these ways to have a congestion in the traffic. It can be observed that heavy-tonnage vehicles, which are traditionally seen in the interstate traffic and connected to commercial centers, are proceeding with smaller cars (thus, nowadays the city of Kırklareli needs a bypass for heavy-tonnage vehicles).

The fact that vehicles with different heaviness and maneuvering ability are together creates a "friction" effect on the traffic flow on two different categories. Traffic congestion and a rise in security issues, and reduced driving speed have caused a need for facilities that have the proper need for use compatible with new developments, parking lots and driveway designs, landscape design, and signs. The fact that the current roads and transportation opportunities are not enough causes a need for new highways, high-speed trains, new airports, and commercial and touristic ports that will serve marine transportation. All of these necessitate a new investment and an expectation for an answer for the public need creates an important expanse amount in the budget of the government ruling the state. They have required new investments to build new residential areas and car roads.

Urban development gradually brough about some social issues and discontent in people about their living environment such as traffic congestion, a corrupted air and water quality, pejoration in social values, negative influences of the pejorating environmental conditions on human psychology. Citizens expect elected authorities to review land use plans without effecting the commercial development, and create alternative transportation by reducing the dependency on car. Therefore, local planning policies and regulations need to contemplate and work more to promote 'Smart Development'.

The aim of forming pedestrian access is to create car and pedestrian circulation systems and provide pedestrians with secure, functional, appealing, and comfortable access. Every journey in the city should be as to start and end as a pedestrian walking. Accessibility in pedestrian walks and avenues provides ease in transportation; also, the use of bicycle lanes and transit transportation vehicles should be encouraged. Pedestrian arteries, in which dangers is minimized, gives a feeling of comfort and the freedom to choose walking to the people, should provide opportunities for humane sensory experience in physical and visual comfort. If a street or a way is physically uncomfortable and unappealing, people will avoid it or choose to travel with a car. Pedestrian areas which do not include activities such as Suitable Recreation that create pedestrian areas will be uncomfortable. While providing physical comfort it is important to take into consideration the heat, the dominant wind direction, rain, snow in high altitudes, sun, and shade control for people (15).

Including pedestrian walks, creating and developing open spaces for the use of pedestrians, building structures, plazas, and parks that enhance cultural life, supporting social interaction between people in recreation areas that bring people together should be of primary importance to boost social interaction and shape urban identity. Thus, acquiring the share in social life should be stimulated and social life should be given the opportunity to extend. Authorities should create public spaces in which people will obtain such an opportunity, reorganize the aging areas, and create urban spaces that will host social activities such as festivals when they are available. For instance, people should be encouraged to socialize through artistic activities such as talks and recitals therefore creating a space for individual identity to form. Buildings that contribute to the city by encouraging public use should be designed; however, it is imperative to provide the security in these places to provide the security of the experiences obtained. Only new urban areas can present a secure, appealing, comfortable, and sharing space under these conditions.

3 Architectural Design Proposals of Architect Candidates that Include the Principles of Mixed-Use on an Urban and Architectural Scale

In terms of the need for compact approach and mixed-use, student projects focused on 'Architectural Design Proposals' that suggest the use of mixed-use in central Kırklareli withing the context of Architectural Design 5 class of Kırklareli University, Department of Architecture in Architecture Faculty of the Spring Term of the 2019–2020 Academic Year, led by Assoc. Prof. Dr. Timur Kaprol and managed by Architect Mümtaz Akın M.A., Res. As. Architect Pelin Doğan M.A. The class produced 45 projects during the term study and the projects took into consideration the potential areas that can contribute to city life or areas that are used as parking lots in the center.

In the architectural designs, the aim was to produce designs that will include a mixed-use to promote comprehensive use of public services, supports pedestrian walks, socially-oriented and human-centered design, repopulate the residential areas, main streets, and neighborhoods, provide urban areas that include businesses close to where people live, urban spaces and buildings for social use in the walking distance, take the parking lots underground and give way to green areas and pedestrian areas on the surface, and make use of natural resources for land possibilities.

Designers in their projects were expected to revive an Exhibition Centers in and near the densely populated residential areas in the city center. Architect candidates were firstly asked to determine the necessary functions that need to be maintained as they are seen as positive in social life as well as the deficiencies experienced on a neighborhood scale within their study area. After the analysis, each architect candidate determined the area and the urban empty spots in this area. Building designs were made that include spatial organizations that will create pleasant memories, take exhibition to its focus, and cheer social life on this parcel or group of parcels. By designing parking garages in the underground parts of the buildings, pedestrians were given a more comfortable area in the street. Buildings that will contribute to the city were investigated in the proposed designs. By opening rooftops to social use, outdoor film screenings, cabarets, sky observation activities were promoted. Buildings included functions such as cinema, chamber theatre, chamber music, art gallery, café, book café, restaurants as well as functions such as a local authority office, health cabin, vet cabin etc. Additionally, pedestrian axes that will encourage public use and integrate to the present public space were also included. This studio study aims to encourage the architect candidate to design the way of life and learn about all the levels of settlement from the topmost scale to the lowest, produce spatial organizations that will make the individual to feel as part of it, cheer the public life. As a result, the architect candidate acquired the ability to design culturally functional buildings that include residential, commercial, and public buildings as well as social activities, which are also within walking distance, more habitable, and fresh. Below, three out of forty-five projects obtained in the studio study were given as examples as they included the pedestrian axis in the design as a common function. 'Exhibition Center' examples that belong to architect candidates Ahmet Furkan Tırpan, Arif Hakan Kıratlı, and Ebru Çelik are given below (Fig. 1).

3.1.Architect candidate Ahmet Furkan TIRPAN proposed a connection from the upper elevation to the business center opposite the parcel the design stands on. The design has an open stage, fresco experience area, shopping spaces that are art-oriented, exhibition areas, closed concert and conference venues, dance venues, underground parking garages, a green rooftop and rooftop gardens (16).

Fig. 1. Design area of architect candidate Ahmet Faruk TIRPAN and the model and sketch study for mass research (2020) (17).

In a neighborhood where high-rise buildings reside there are residential buildings, shopping places, trade offices, business centers, restaurants, municipality building, banks, and the historical grain. The fact that green spaces are almost non-existent is replaced by buildings for social use where local people will relax and get fresh air that Fig. 2 shows (Fig. 3).

Fig. 2. Architect candidate Ahmet Faruk TIRPAN's search for green texture, rooftop garden and access to the topmost elevation (18).

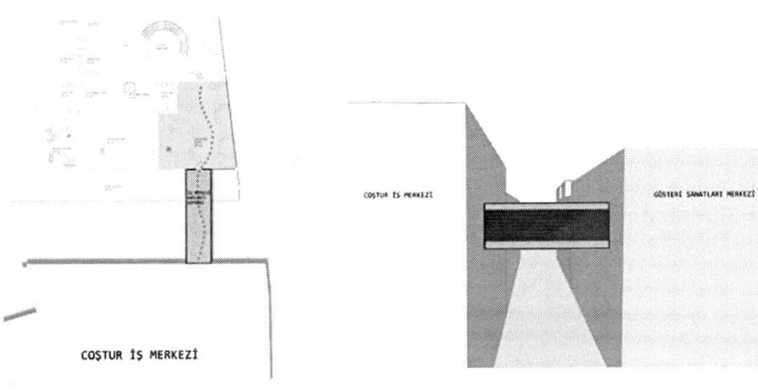

Fig. 3. The outlook of architect candidate Ahmet Faruk TIRPAN's street silhouette on Gaziler Avenue (19).

While developing the design by taking into consideration the 'building block elevation' and 'city wall' provided in the building block, the design paid attention to using green walls, storied forest, and rooftop gardens as there was not green areas near the buildings (Fig. 4).

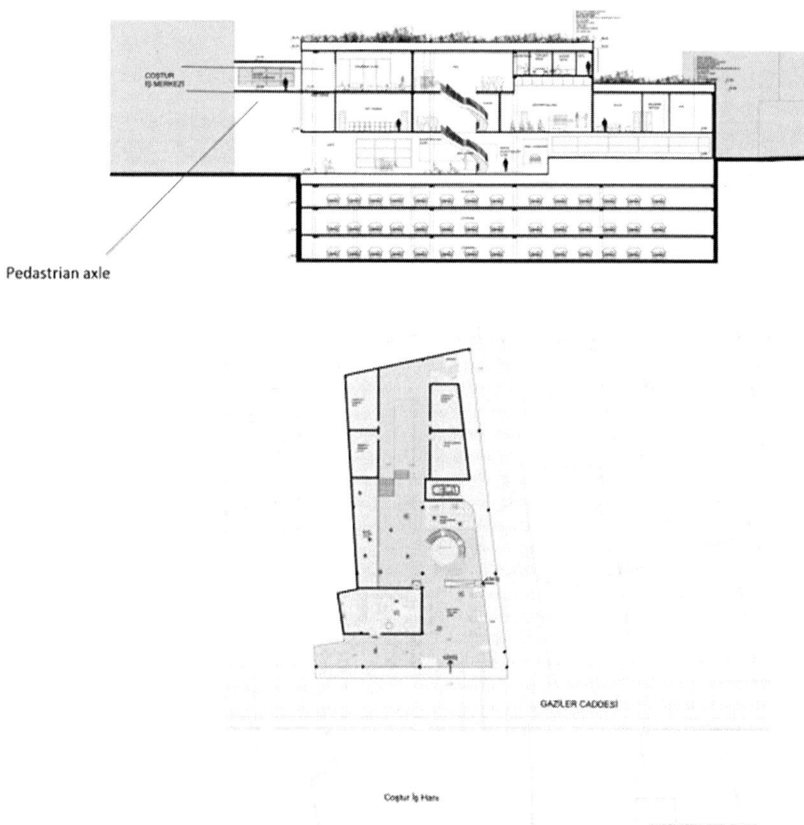

Fig. 4. The ground floor plan and longitudinal section of the Exhibition Center the architect candidate Ahmet Faruk TIRPAN designed (20).

The ground floor of the building on the corner houses a studio, café, an art scene that will serve as a gathering area, and an art shop. The project includes a three-storey underground parking garage with a lift. The connection made with the Coştur Business Center in the section provides people working in the center an opportunity to use the area to get fresh air and relax during the lunch hour. The section also shows the planned three-storey underground parking garage. With this approach, it was aimed to diminish the constraint caused by the parking lots on pedestrians and the traffic in the street (Figs. 5 and 6).

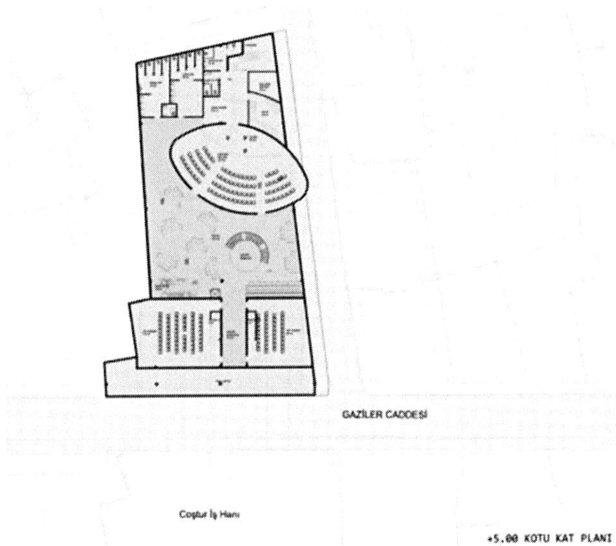

Fig. 5. Floor plan showing the collective exhibition areas on the first floor of the design that belongs to architect candidate Ahmet Faruk TIRPAN (21).

Fig. 6. Visuals of the final version of the shopping street plan and the appearance of the area that includes the music scene on the upper elevation of the shopping street on the ground floor level, designed by the architect candidate Ahmet Faruk TIRPAN (22).

The fresco wall surrounding the music stage, exhibition areas, studio areas, and the café enables the ground floor of the building to make a diffusion with its environment and invites the passers-by on the south façade to visit the interior spaces and its functions (Fig. 7).

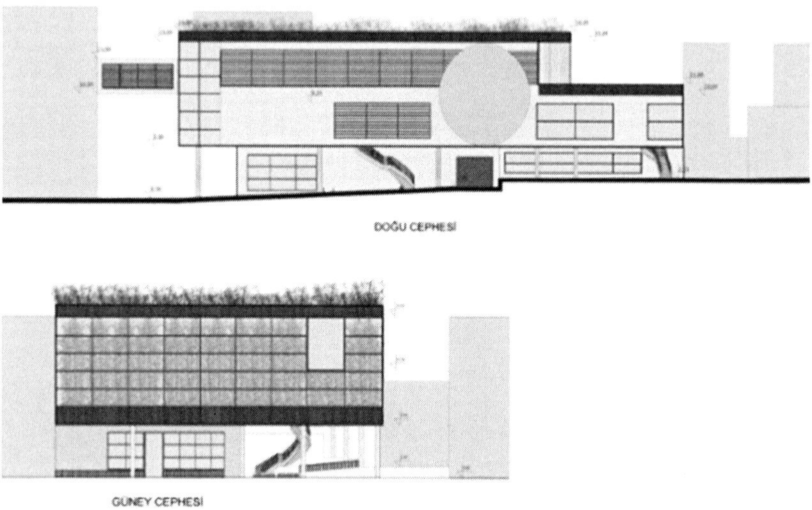

Fig. 7. The drawing of the east façade which connects to the topmost elevation of the business center and the south façade which hosts a dense residential area of the Exhibitions Center designed by the architect candidate Ahmet Faruk TIRPAN (23).

A connection corridor was planned on the topmost elevation of the building on Gaziler Street. This business center opposite the building has a ramp that enables light-tonnage vehicles can go up and this ramp almost functions as an inner street. As this building functions as a connection to the street behind it, this function was also evaluated as a pedestrian walk in the design. Moreover, the closed exhibition center was thought to be seen from the façade. With this façade the perception of the interior space functions has a more inviting organization for the local people.

The building was aimed to see the green area of the business center on the opposite as it is in search of a green wall on the façade looking at the street. The use of recreational space opportunity was given to the business center with a suspension corridor on the topmost elevation.

3.2.Architect candidate Arif Hakan KIRATLI has described his design as "aiming to create an important aorta by concentrically synthesizing multiple functions" in his definitive report for his design on the area next to Atatürk Park overlooking the city center and next to the Kırklareli Governor's Office. He aimed to revive the decreasing social activities in the city and highlight the urban economy and touristic development. The building, which aims to make a connection with the exhibition center close to it, offers integrating functions with the exhibition center. Another aim has been to encourage attending, being curious, and observing the activities in the local authority office, Karakaş Neighborhood office that serves the neighborhood, by including it in the building. The building, with 300–350 people capacity, is designed on an area of 1200 m². The area where the building will supposedly exist is now serving as a parking lot. An underground parking garage was also included in the central parcel, which will serve both the local people of the neighborhood and the visitors. The design also includes a pedestrian axis (the conjunction point of Karakaş Street and Atatürk Park) which meets an urban transition point at Atatürk Park which is one of the important urban circulation axes just as the Fevzi Çakmak and 100. Yıl Avenues. Thus, the building, which will be an inseparable and indispensable part of the city, was planned to be a living space compliant with urban unity and sustainability (24). (Figs. 8 and 9)

Fig. 8. Modeling of Artist Building Proposal" by Architect Candidate Arif Hakan Kıratlı in the province square (25)

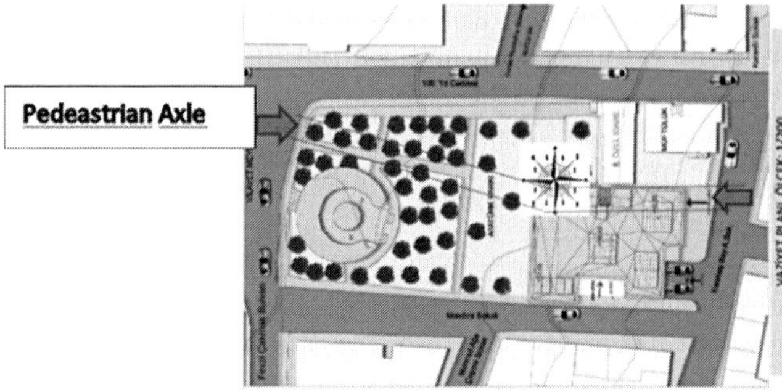

Fig. 9. Layout Plan showing the pedestrian axes and their relation to the environment, and the 'Artists' Building' in the central square of Kırklareli of the architect candidate Arif Hakan KIRATLI (26).

The building, which has entrances from Karakaş Bey Street and Mandıra Street, connects Atatürk Park and Karakaş Bey Street with a traffic-free pedestrian road and a portico transition through the building block next to the Provincial Special Administration Building and Office of Mufti. The ground floor of the building houses an art gallery that can be seen from Karakaş Bey Street and a restaurant space on the façade looking at Atatürk Street. The art gallery in the building with a wide entrance hall will welcome visitors inside and illuminate both streets when the lights are on during the night. Exhibition and gathering areas on the first floor will enable the city people to reach artistic satisfaction, and the spaces on the topmost floor will be public aid oriented (local authority office, gathering and social aid offices) (Figs. 10 and 11).

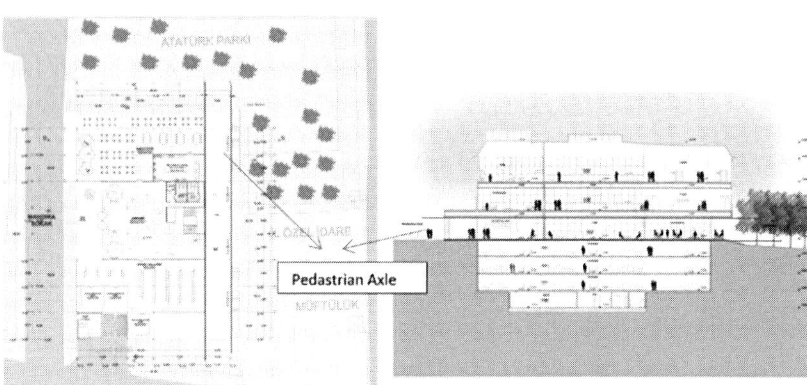

Fig. 10. Ground Floor Plan and Section Drawing showing the 'Portico Pedestrian Axis' connecting Karakaş Bey Street and Atatürk Park, designed by the architect candidate Arif Hakan KIRATLI (27).

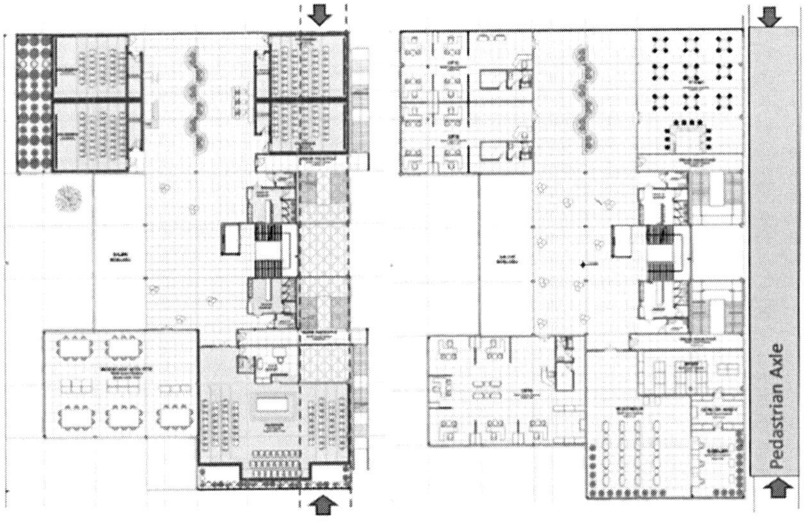

Fig. 11. Architectural drawings of plans showing the first floor and the second floor that has social orientation that include Exhibition Halls, designed by the architect candidate Arif Hakan KIRATLI (28).

3.3. The design of architect candidate Ebru Çelik is on the land right next to the city grain, neighbor to both the residential areas and to the historical grain. It serves as a bridge between the area where residential buildings reside and the business area in the neighborhood it is planned. It is a firm ground today serving as a parking lot. That is why, this negative use is evaluated so as to close the urban split and the area is refunctioned. The ground elevation of the structure, which is taken underground, is thought to be a pedestrian and green area. The elevation differences of the area are functioned differently as to house activities that include artistic and cultural events, a separate area to carry the commercial grain, and a pedestrian axis that connects residential areas. The surface elevation is arranged to have a green area so it will provide local people comfort. It is designed to have pedestrian areas both on the surface area and a secure closed pedestrian space underground. Thus, the area, which was used as an open market and also housed 'Honey Festival' in 1912, leaves its parking lot function to an artistic and commercial center (29). (Figs. 12 and 13)

Fig. 12. Visuals of the green area and layout plan showing the pedestrian use and green area proposal, in architect candidate Ebru ÇELİK's redesign of the area (30).

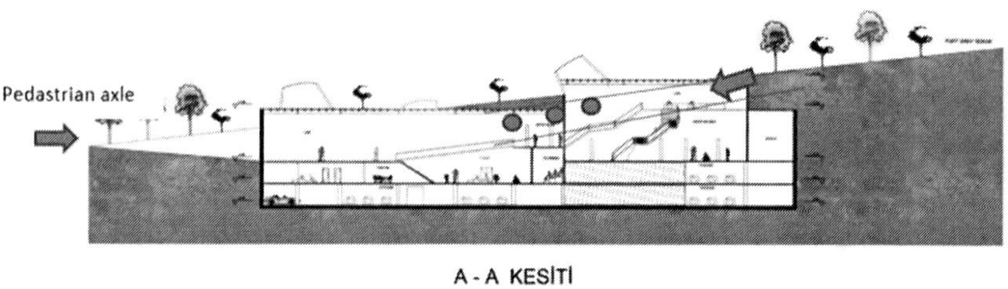

Fig. 13. Land topography section of the pedestrian axis and expression of elevation gain prepared by the architect candidate Ebru ÇELİK (31).

The social-use building, which can be entered to the ground floor from ... Avenue, includes an open exhibition area on a collapsed surface right below the entrance axis. There are closed exhibition halls on the upper elevations of the back plate of this space with a gathering and exhibition function. Additionally, the opportunity to see the activities and exhibitions in the lower elevation from the upper pedestrian axis has also been taken into consideration. It can be seen that artistic events are concentrating in the second section where elevation is rising. The transition from the elevating part of the area to the residential area that has a higher ground has been shown on floor plans. In this rising part, helical stairs are also planned on the axis to reach the studios and exhibition hall which are on the lower elevation (see Fig. 13). The section plan also includes a ceiling organization that will give sunlight and fresh air opportunities to the underground functions through natural light shells which function as chimneys opening up to the surface (32) (Fig. 14).

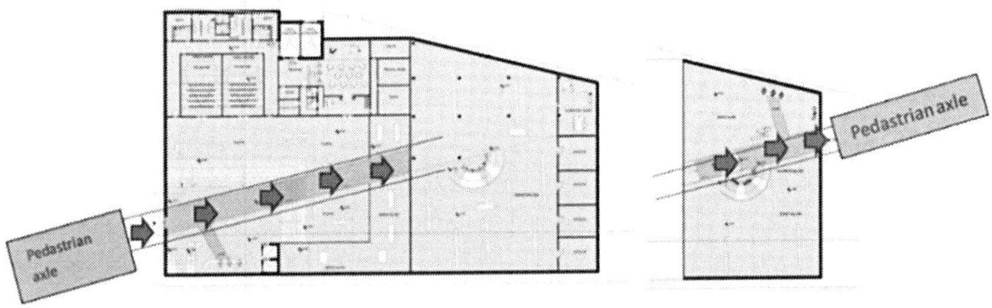

Fig. 14. Drawings of elevated floor plans that show the spatial organization where the pedestrian area is under the ground (33).

4 Conclusion and Evaluation

New functions have been suggested in the mixed-use designs. The daily complete and active use in the building which contributes to the city and is open to use around the clock has been as follows: for instance, the rooftop and outer shell of the building can be used during the night; it can serve as a concert venue during the evening; the green spaces on the roof can be open to public use; parking areas in the redesigned land can be taken underground and the area can become a pedestrian area; recreative events can be organized in the ground elevation of the roof of the building; human-oriented buildings for the prospected multi-functional building plans; defined and accessible streets that do not suffocate people; and buildings and areas that can be used as transition points. Designs that include spatial organizations which appeal to users of all ages and bring them together in the urban texture are reached. These mixed-use designs also encourage the development of urban quality (34).By making urban and architectural plans that are dense and efficient so as not to threaten human health, we reached designs that will unify the present pedestrian axis with spatial organizations that will house urban services. They have created user-oriented functions and spatial organizations to the city people that include design criteria for everyone. The designs which integrate with green areas gave the opportunity to social share and gathering in places close to the neighborhood center. Another interesting feature in the new designs has been presenting people living in the residential areas close to business and service centers the socio-cultural opportunities. These group of buildings in the walking distance make it possible for the city people to use them. The proposals suggesting the inclusion of new buildings to the city also give the opportunity of pedestrian privileged spaces in the city instead of a city where car-use is dominant. The unoccupied areas in the urban texture, buildings with a social use that will answer new needs, exhibition centers that enable social share, and values that pedestrian axis the city obtains have been the main subjects of the architectural projects.

Acknowledge

1. Grayson, L. & Young, K. (1994) Quality of Life in Cities: An Overview and Guide to the Literature, London, British Library
2. Bennett, M. (1978) Mixed Use Development in the Inner City, Estates Gazette, 245, 18 March, pp. 920–923
3. What is Smart City.pdf (accessed 01.10.2020)
4. Altun,Özyetgin, A., (2020) ''Kentsel Tasarımda Karma Kullanım'' Kırklareli Üniversitesi, Mimarlık Fakültesi, 2019–2020 Bahar yy Mimari Proje 5 dersi kapsamında, Şehir ve Bölge Planlama Bölümü Dr. Öğretim Üyesi Ayşe Altun tarafından, Mimari Proje 5 öğrencilerine verilen seminer, Kırklareli.

Bibliography

1. Durukan Kopuz, A. (2019) Planning Experience of Hermann Jansen in Gaziantep City during 1930's, New Approaches to Spatial Planning and Design,, Edited by Murat Özyavuz, Peter Lang, (2019):79–90.
2. Rowley,A. (1996) Mixed-use Development: Ambiguous Concept, Simplistic Analysis and Wishful Thinking?, Planning Practice & Research, 11:1,p.85, https://doi.org/10.1080/02697459650036477, (accessed 01.10.2020)
3. Rowley,A. (1996) Mixed-use Development: Ambiguous Concept, Simplistic Analysis and Wishful Thinking?, Planning Practice & Research, 11:1,p.86, https://doi.org/10.1080/02697459650036477, (accessed 01.10.2020)
4. Rowley,A. (1996) Mixed-use Development: Ambiguous Concept, Simplistic Analysis and Wishful Thinking?, Planning Practice & Research, 11:1,p.87, https://doi.org/10.1080/02697459650036477, (accessed 01.10.2020)
5. Commercial_MixedUse_CodeHandbook.pdf (1997) www.lcd.state.or.us/tgm (accessed 01.10.2020)
6. Rowley,A. (1996) Mixed-use Development: Ambiguous Concept, Simplistic Analysis and Wishful Thinking?, Planning Practice & Research, 11:1,p.87, https://doi.org/10.1080/02697459650036477, (accessed 01.10.2020)
7. Ambrose, P. (1994) Urban Process and Power, London, Routledge.
8. Rowley,A. (1996) Mixed-use Development: Ambiguous concept, simplistic analysis and wishful thinking?, Planning Practice & Research, 11:1, https://doi.org/10.1080/02697459650036477, erişim tarihi 01.10.2020
9. Bentley, I., Alcock, A. et al. (1985) Responsive Environments, (London, Architectural Press).
10. Bentley, I. (1990) Urban design 3, ecological urban design, Architects Journal (24 October), pp. 69- 71.

11. Gehl, J. (1987) Life Between Buildings: using public space (New York, Van Nostrand Reinhold).
12. Ambrose, P. (1994) Urban Process and Power (London, Routledge).
13.–14. Commercial_MixedUse_CodeHandbook.pdf (1997) www.lcd.state.or.us/tgm (accessed 01.10.2020) P.(5–6)
15. Commercial_MixedUse_CodeHandbook.pdf (1997) www.lcd.state.or.us/tgm (accessed 01.10.2020) p.7
16–23. Tırpan A. F., (2020) Kırklareli Üniversitesi, Mimarlık Fakültesi, Mimari Proje 5 dersi kapsamanda hazırlanan, Gösteri Binası Tasarımı
24–28. Kıratlı A. H., (2020) Kırklareli Üniversitesi, Mimarlık Fakültesi, Mimari Proje 5 dersi kapsamanda hazırlanan, Sanat Binası Tasarımı
29–33. Çelik E., Kırklareli Üniversitesi, Mimarlık Fakültesi, Mimari Proje 5 dersi kapsamanda hazırlanan, Gösteri Binası Tasarımı
34. İnceoğlu, M., Aytuğ, A., (2009). Kentsel Mekanda Kalite Kavramı, Megaron Dergisi, Vol:4, No:3, İstanbul.

Oğuz ATEŞ

The Relationship of Bioclimatic Comfort and Landscape Planning

Abstract: Bioclimatic comfort is the climate condition in which a person feels healthy and dynamic and the harmony he provides to his environment with the least energy.

In the literature, all of the comfort assessment criteria are based on the thermal physiological balance of the human body. Heat balance is an important component of outdoor comfort. It affects the thermal feeling of outdoor comfort. However, this approach has proven to be absolutely insufficient to define external comfort criteria. Psychological adjustment is an equally important variable. Criteria that affect people's psychological adjustment should be integrated into any outdoor comfort assessment method.

In this study, the relationship between bioclimatic comfort analysis and landscape planning studies was examined. In addition, the necessity of adding quantitative data to the analysis of bioclimatic comfort, which is qualitatively evaluated in almost all sources, has been revealed.

Keywords: Bioclimatic Comfort, Landscape Planning, Climate.

1 Introduction

As in all countries of the world in recent years in Turkey it has entered a rapid process of urbanization. With this process, technological developments, increase in income level, ease of transportation, etc. elements have accelerated and therefore ecological degradation has occurred in urban settlements [1].

It is a well-known fact that climatic conditions are very important on human life and activities. Researches aimed at determining the optimum climatic conditions for human life in planning studies are quite new. Values related to climate elements were determined and graded, especially in line with researches on human comfort. These values are very important to determine the deviation from the optimal values and to determine the current state of the climatic conditions of the environment [2,3,4]

One of the main indicators of socio-economic and physical development is planned urbanization. At the same time, the more diverse and sensitive the criteria or data that are the basis of planning, the higher the success achieved in planning [5].

In planned urbanization; climate emerges as one of the important variables in determining the regions suitable for human health [5]. Because climate is an important factor that affects people positively or negatively [6, 1]. Climate is an important factor that ensures livability of the geographical environment by integrating socio-cultural processes and natural processes. Therefore, it is primarily effective in the formation of both natural and socio-cultural environment [1].

The purpose of bioclimatic planning is to determine the effect of climatic data on planning and to change or improve the climate in a positive way by considering the comfort of living things. Climate, which plays a decisive role in the lives of living things, constitutes a large part of nature's effects on humans [1].

2 Bioclimatic Comfort

In the literature, all of the comfort assessment criteria are based on the thermal physiological balance of the human body. However, this approach has proven insufficient to define the external comfort criteria [7,8]. Psychological adjustment is an equally important evaluation criterion [9]. Consequently, criteria that affect people's psychological adjustment should be integrated into any outdoor comfort assessment method [10].

In the context of bioclimatic comfort, adaptation includes the steps people can take to match their thermal needs with environmental conditions. Human thermal adaptation includes three different processes: behavioural adaptation (or physical adaptation), physiological adaptation, and psychological adaptation [11,12,13,10]

Behavioural adaptation brings together all the conscious or unconscious actions aimed at changing a person's bodily thermal balance. This type of adaptation can be done in three different ways: personal adjustment (change of clothing, activity, posture, food, drinking, localization), environmental adaptation (changing the environment itself, e.g., opening an umbrella), and cultural adjustment (dress code, community-based activities, for example siesta time). The effects of physical adaptations can often be evaluated quantitatively [10].

Physiological adaptation is a physiological response of the body. It tends to reduce stress related to an uncomfortable situation. In this case it is called habituation to the physiological climate. This type of mechanism is essential in extreme climatic conditions. However, its effect is negligible for the purpose of this study [10].

Psychological adaptation is a change in the perception of sensory information. This kind of adaptation can be attributed to the concept of "habit" found in psychophysics. This concept corresponds to a decrease in the feeling of density when repeatedly exposed to the same environmental stress. Thermal perceptions are directly dependent on past experience and the expectations of field users. Psychological adjustment is very difficult to measure [10].

Many authors affirm the importance of subjective perspective for thermal comfort assessment. However, they focus only on quantitative methods (tables, graphs or formulas). Recently, some authors have also studied specifically psychological comfort, but they too have abandoned quantitative assessment. The combination of quantitative and qualitative parameters allows to improve the thermal comfort possibilities of public spaces[10].

2.1 Quantitative Bioclimatic Comfort Criteria

Bioclimatic comfort is the climate condition in which a person feels healthy and dynamic and the harmony he provides to his environment with the least amount of energy. Bioclimatic factors have 6 important components. Important climate components for bioclimatic comfort can be listed as relative humidity, temperature, wind and radiation [14].

In order to provide bioclimatic comfort, relative humidity, temperature, wind and radiation conditions should be determined and evaluated. The state of bioclimatic comfort depends on the number of hot days, air pollution, oxygen content in the atmosphere and illness due to weather events. It can be determined by factors such as. As seen in Fig. 1, according to Olgyay (1973), the ideal bioclimatic comfort temperature value varies between 21 and 27.5 ℃ and 30–65 % relative humidity. For example, in which the middle latitudes Turkey, sensed temperature value, the relative humidity and wind are considered as suitable for bioclimatic comfort from 17 to 24.9 ℃ [15,16,17,18,19,20,21].

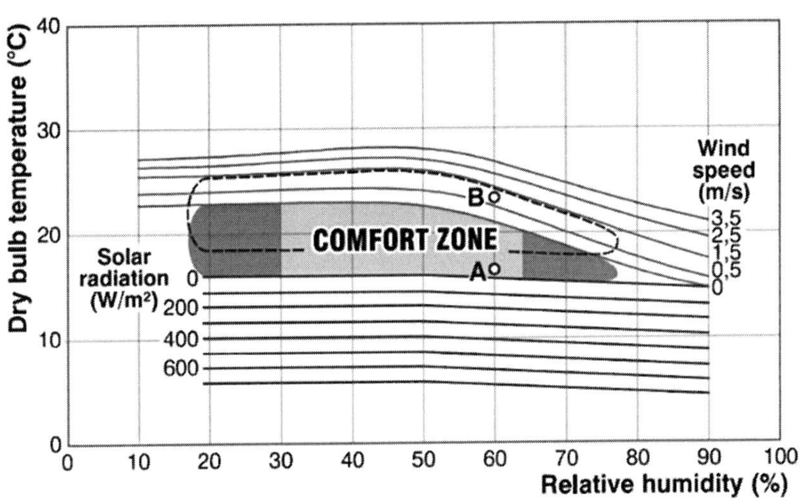

Fig. 1. Olgyay's chart for moderate European climates [10].

Altunkasa (1990) revealed that when all other variables are normal, a temperature of 21–27 ℃ and a relative humidity of 30–65 % provide a comfortable environment. As seen in Fig. 2; It states that temperature or radiation energy or shadow, wind and specific humidity are required to provide bioclimatic comfort at values above or below 21–27 ℃ and 30–65 % relative humidity. It is of great importance for the landscape arrangements to serve the purpose of determining the values above or below the comfort environment and considering them in the planning studies [22, 23].

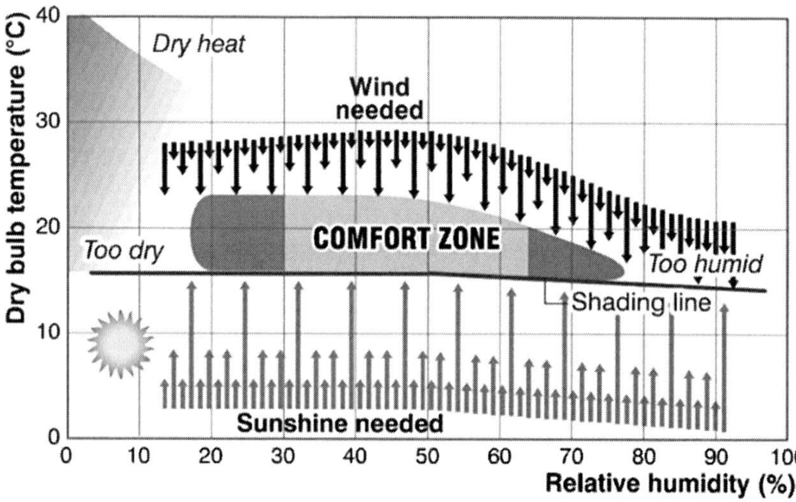

Fig. 2. Interpretation of Olgyay's Bioclimatic Chart [10].

One of the most common indices used to determine the suitability of bioclimatic comfort is the Physiological Equivalent Temperature (PET) index. Physiologically Equivalent Temperature (PET) is the index derived from the human heat energy balance. The Physiological Equivalent index is an easy to understand index. Because the numerical values obtained from the FES index are the temperature unit used today in centigrade (℃) [24].

2.2 Qualitative Bioclimatic Comfort Criteria

Oseland (1994) found in his study that thermal comfort depends on context. He found that the same group of people felt different levels of comfort in three different contexts. This contexts are a climatic room, their work place and their home (becoming increasingly tolerant). They accepted a lower temperature of comfort in their house than in the climatic room [25]. It is revealed that people become even more tolerant in outdoor conditions [10].

The perception of potential control is one of the most essential elements of thermal comfort. Paciuk (1990) affirms that unpleasant stimuli are less irritating when the subjects perceive that they can control them [26].

However, the possibility of moving to another thermal environment also creates the perception that we can control our thermal balance. In a study conducted among people sitting in a physiologically uncomfortable square, they feel more comfortable if they are free to leave as soon as they wish instead of waiting for someone. [9, 10] When people want to go somewhere, they can move whenever they want. This way, they can control their thermal environment [10].

Squares that offer different regions with shade and sun are used longer than those that offer only one type of zone [9, 10].

Urban residents may prefer to sit in the sun or in the shade of a building, refreshed by a breeze or protected from the wind, in contact with the sky or under the roof. Urban Open and Green Spaces are always designed to provide these options. It is obvious that the spaces that offer different atmospheres together create more comfort for the users. Some regions in the city will be more preferred in certain seasons or at certain times of the day. The aim is to create a rich and diverse urban environment [10].

A study conducted in Norway reveals that the minimum comfort temperature is 11 ° C in the autumn months and 9 ° C in the spring. Therefore, people's comfort satisfaction varies according to their expectations and is guided by past experiences. Generally, the temperature before Spring is colder than before Autumn [27, 10].

Recent memories and expectations of pedestrians are effective in two ways. First, they play a role in physiological balance by affecting what a person wears. Second, they prepare it psychologically, affecting the level of satisfaction [10].

Vegetation and climatic elements are two types of natural elements that easily fit into urban spaces. The presence of sunlight or air movement can be very attractive to people who are very used to artificial environments. The most basic feature of natural spaces is the variability in time and space [10].

Many studies show that the interaction of natural elements with humans is quite positive. For example, according to Reiter and De Herde (2001), natural light adapts better to the human eye than artificial light [28]. According to Ulrich (1984), patients who can see a natural view from hospital rooms recover faster than those who look at the white wall [29]. A study by Fjeld (1998) in Norway found that employees working in an office decorated with beautiful plants reduced fatigue, headache and concentration problems by 23 % [30]. These studies prove that vegetation is an indispensable element for human comfort [10].

Vacationers' outdoor activities are generally places with a high level of disturbance. For example, tanning under the extreme sun at the seaside, skiing, mountain climbing, deep diving, etc. How can we reconcile these activities with our definition of thermal comfort? When we encounter these extreme environmental conditions in our fixed and artificial residential areas, we feel disrupted in comfort [11]. Human beings really need naturalness and especially environmental stimulation [10].

3 The Relationship of Bioclimatic Comfort and Landscape Planning

The increasing urban texture with each passing day, the heavy traffic and the increase of hard floors are one of the most important factors that cause a decrease in bioclimatic comfort. Hard floors such as cement, asphalt, brick, etc. that cover the city surface create significant increases in bioclimatic comfort parameters. This equipment cannot absorb and reflect heat, light, radiation and noise, so it can increase these rates. In addition, they release the energy they absorb in the daytime with long thermal wave radiations and at night with heat waves. The open green areas sandwiched between the urban texture and the plants in them contribute to the regulation of the climate. These areas not only improve the visual landscape quality, but also absorb light, wind force and heat, reduce thermal radiation and act as a green screen to reduce noise, making significant contributions to urban ecosystems [31].

Landscape planning studies and climate data should be combined while selecting living spaces and recreation areas. Planning of climatic data and design must go hand in hand. Analyzes to be made in large-scale city planning and design allow us to pay attention to the problems of users. The analyzes to be made include aesthetic, ecological and meteorological functions as well as physical and emotional sensitivity while creating a comfort zone. The areas that will emerge with the synthesis of meteorological, ecological, functional and aesthetic aspects should create economically viable plans. As a result of these syntheses, different comfort zones can be found. All parameters of meteorological values play a very important role in guiding landscape management, landscape planning, landscape design and urban planning. Meteorological parameters are extremely necessary in the development of new residential areas and selection of disposable requirements. People can feel more comfortable in the perceived temperature and therefore bioclimatic comfort map locations can be used to select new settlements. Thus, bioclimatic comfort maps can make areas suitable for residential use more pristine. When done correctly, this planning uses a number of factors derived from the analysis of the meteorological parameter map to aid landscape design. For example, if an open area with a few productive days per year can be determined through the determination of meteorological parameters, it can be used as the area for the design of a swimming pool [32].

Modern life is increasingly moving us away from the natural environment. Considering that urbanization will grow faster every day, access to the natural environment will become even more critical. Accordingly, we should increase the natural elements in our cities, design areas that offer similar features to those provided in the natural world, prepare climate projections and integrate them into landscape planning studies.

4 Conclusion

Nowadays, it is clear that long-standing development models only create a large number of environmental and social damages that emerge gradually. These problems, which seriously limit the potential of future generations, also reduce our quality of life.

It is time to feel responsible for the more harmonious development of our natural and cultural heritage. In an increasingly urbanized society, one of the main goals of sustainable development should be the quality of the urban environment. Therefore, it is necessary to examine all aspects of urban and architectural design that affect the comfort of people, energy management and the quality of residential areas.

One of the problems that have been tried to be solved since the beginning of humanity is how to control the environment and create the best conditions for human activities. However, modern cities have been strongly affected by economic constraints. The environmental quality of the spaces between buildings is seldom taken into account. Now is the time to update the analysis criteria for urban planning studies.

The sustainable development of cities mainly depends on the high environmental qualities of open and green space systems. These areas are needed to enrich the life experience in the city center. Microclimate is one of the most important stages of this process. In the past, people knew that climate adaptation was an important architectural goal. Huntington (1951) explains in his book "Principles of Human Geography" that climate is related to racial heritage and cultural development. Climate is one of the three main factors that determine a civilization [33, 10].

Since the planning and design to be made should be carried out in order to adapt to the environment by consuming less energy, the aim should be primarily providing bioclimatic comfort [23].

It is of great importance for the landscape arrangements to serve the purpose of determining the values above or below the comfort environment and taking them into consideration in the planning studies.

Landscape planning for determining bioclimatic comfort states can be improved by applying the criteria and ideas discussed in this study. Planning and design using the wrong conditions for bioclimatic comfort can create extremely unfavourable conditions because natural vegetation will be subjected to intense hot or cold stress in areas where natural conditions are not sufficient to provide bioclimatic comfort.

Despite the complexity of the issues mentioned, some design and planning criteria may be taken into account that will affect bioclimatic comfort in urban areas. These are;

- Changing the microclimate by regulating one of the physical parameters (radiation, wind, etc.),
- To encourage activities compatible with comfort activities,
- Changing the perception of comfort with artificial materials (eg umbrellas),
- considering several climate variables in the same location (shade and sun, wind and shelter, etc.),
- Creating transition gaps between different regions,
- Increase naturalness through vegetation or environmental stimulation.

It is not entirely correct to express the outdoor bioclimatic comfort only with qualitative criteria. It needs to be defined as a combination of qualitative and quantitative criteria. It is necessary to evaluate the thermal balance of people quantitatively and qualitatively the criteria that affect their psychological adaptability.

References

1. Çetin, M., Topay, M., Kaya, L. G., Yılmaz, B. *Biyoiklimsel Konforun Peyzaj Planlama Sürecindeki Etkinliği: Kütahya Örneği.* Süleyman Demirel Üniversitesi Orman Fakültesi Dergisi, 2010. **A**(1): p. 83–95.
2. Evans J.M., *The Comfort Triangles: A New Tool for Bioclimatic Design*, Doctoral Thesis, 2007, Delft University of Technology, Delft.
3. Güçlü, Y., *Alanya-Samandağ Kıyı Kuşağında Konforlu Olma Süresi ve Deniz Turizmi Mevsiminin İklim Koşullarına Göre Belirlenmesi.* Türk Coğrafya Dergisi, 2008. **50**, p. 1–20.
4. Dönmez, Y., Özyavuz, M., Çabuk, S., Çorbaci, Ö. L., Determination of Bioclimatic Comfort Zones by Geographic Information Systems: Karabük Province, Turkey. J. Int. Environmental Application & Science, 2018. **13**(1): p. 41–49
5. Altunkasa, F. ve Gültekin, E., *Şanlıurfa'da İklimle Dengeli Kentsel Alan Kullanım İlkeleri ve Yeşil Alan Sistemlerinin Belirlenmesi.* Çukurova Üniversitesi Ziraat Fakültesi Dergisi, 1991. **6**: p. 23–38.
6. Çınar, İ., *Fiziksel Planlamada Biyoiklimsel Veriler Kullanarak Biyokonforun Oluşturulması Üzerine Fethiye Merkezi Yerleşimi Üzerinde Araştırmalar.* Yüksek Lisans Tezi, 1999, Ege Üniversitesi Fen Bilimleri Enstitüsü.
7. Nikolopoulou, M., Baker, N., Steemers, K., *Thermal comfort in outdoor urban spaces: understanding the human parameter.* Solar Energy, 2001. **70** (3): p. 227–235.
8. Höppe, P., Different aspects of assessing indoor and outdoor thermal comfort. Energy and Buildings, 2002. **34**(6), p. 661–665.
9. Nikolopoulou, M. and Steemers, K. *Thermal comfort and psychological adaptation as a guide for designing urban spaces.* In Proceedings of PLEA 2000. Cambridge.

10. Reiter, S. and De Herde, A., *Quantitative and qualitative criteria for comfortable urban public spaces.* Research in Building Physics, 2003 Belgium. p. 1001–1011.
11. Baker, N., *We are all outdoor animals.* Proceedings of PLEA 2000. London.
12. de Dear, R., Brager, G., Cooper, D., *Developing an Adaptive Model of Thermal Comfort and Preference, Final Report.* 1997. ASHRAE RP- 884. Ashrae.
13. Nikolopoulou, M. and Steemers, K., *Thermal comfort and psychological adaptation as a guide for designing urban spaces.* Energy and Buildings, 2003. **35** (1): p. 95–101.
14. Topay, M. and Yılmaz, B., *Biyoklimatik Konfora Sahip Alanların Belirlenmesinde CBS den Yararlanma olanakları Muğla İli Örneği.* 3. Coğrafi Bilgi Sistemleri Bilişim Günleri, 2004. İstanbul, Turkey, p. 1–12.
15. Olgyay, V., *Design with climate: bioclimatic approach to architectural regionalism.* Princeton University Press. 1973.
16. Topay, M., Cinar, I., *Determining of bioclimatic comfort structure of Suleyman Demirel University East Campuses situated in the back of Mediterranean region.* Geophysical Research Abstracts, 10. 2008.
17. Attia, S., & De Herde, A., *Bioclimatic architecture: design strategies in Egypt.* Sustainable Energy Technologies. 2009.
18. Bezlova, D., Doncheva-Boneva, M., *Protected areas at the black sea cost as natural resources for development of ecotourism.* Journal of Environmental Protection and Ecology, 2011. **12**(3), p. 1179–1185.
19. Topay, M., *Importance of thermal comfort in the sustainable landscape planning.* Journal of Environmental Protection and Ecology, 2012. **13**(3), p. 1480–1487.
20. Tsiourlis, G., Konstantinidis, P., Xofis, P., *An ecological assessment method: application to the vegetation units of the Lagadas country (Greece).* Journal of Environmental Protection and Ecology, 2012. **13**(3), p. 1560–1569.
21. Çetin, M., *Determining the bioclimatic comfort in Kastamonu City.* Environmental Monitoring and Assessment, 2015. **187**(10):640.
22. Altunkasa M.F., *Determination of climate-balanced urban green space planning principles in Adana and the example of multi-purpose development of a green field.* Institutional Faculty of Agriculture, 1990. **5**: p. 9–54.
23. Çetin, M., *Peyzaj Planlamada Biyoklimatik Konfor Alanların Belirlenmesi: Cide Kıyı Şeridi Örneği.* Turkish Journal of Agriculture - Food Science and Technology, 2016. **4**(9):800.
24. Bulgan, E., Yılmaz, S., *Farklı Kent Dokularının Yaz Aylarında Biyoklimatik Konfora Etkisi: Erzurum Örneği.* Iğdır Univ. J. Inst. Sci. & Tech., 2017. 7(4): p. 235–242.

25. Oseland, N.A., *A within-groups comparison of predicted and reported thermal sensation votes in climate chambers, offices and homes.* In Healthy Buildings 1. 1994.
26. Paciuk M., *The role of personal control of the environment in thermal comfort and satisfaction in workplace.* In Coming of age. USA. 1990.
27. Culjat, B., Erskine, R., *Climate-responsive social space: a Scandinavian perspective.* Cities Designed for Winter. Helsinki. 1998.
28. Reiter, S. and De Herde, A., L'éclairage naturel des bâtiments. Belgique: Ministère de la Région Wallonne. 2001.
29. Ulrich, R.S., *View through a Window may influence Recovery from Surgery.* Science 224. 1984.
30. Fjeld, T., *The Effect of Indoor foliage plants on the Health and Discomfort Symptoms among Office Workers.* Indoor Built Environment 7. 1998.
31. Sağlık, A., Kahraman, O., Sağlık, E., Devecioğlu, N. E., Kelkit, A., *Importance of the Bioclimatic Comfort on Landscape Design: Case Study on Canakkale Eski Kordon.* Journal of Environmental Protection and Ecology 17, 2016. **4**, p. 1612–1619.
32. Çetin, M., Adıgüzel, F., Kaya, O. and Sahap, A., *Mapping of bioclimatic comfort for potential planning using GIS in Aydin.* Environment, Development and Sustainability, 2018. **20**, p. 361–375.
33. Huntington, E., *Principles of Human Geography.* 6th edition. New York: John Wiley and Sons. 1951.

Mehmet Kıvanç AK and Tahsin YILMAZ

Evaluation of Urban Roundabout Circulation in the Context of Plant Design - Visual Quality - Safety

Abstract: Dealing with population density is getting harder day by day due to the increasing migration to our cities. The problems brought about by this density affect the inhabitants both psychologically and physically. It is an undeniable fact that urban traffic problems are one of the most significant problems of the cities. The faulty design of roundabouts is an important issue in traffic problems. Especially roundabouts are the parts of the roads with the highest safety risk. In this study, urban roundabout circulations were examined in terms of plant design and safety and supported with case applications. The aim of the study is to evaluate the forms and places of use of the plant and plant groups recommended for the safety of drivers and pedestrians. As a result of the evaluation, suggestions were revealed for the safe use of roundabouts and approaches for drivers and pedestrians.

Keywords: Junction, roundabout, plant design, visual quality, traffic safety.

1 Introduction

In recent years, urbanization, population growth, economic and social changes in the social structure due to the developments in industry and technology have increased the importance of highways. This transformation has required road transport to be carried out in an increasingly comfortable and safe manner. In this way, it has been important to establish the relation of highways with the driver properly and to prevent negative effects on the environment [1].

Highways are sloped or flat land strips in urban and rural areas for the transportation of people from one place to another for purposes such as relocation, transportation, tourism and trade, user and traffic safety established according to predetermined dimensions and standards [2].

Highways are the oldest transportation elements that provide connections between various environments. Apart from this, highways are landscaping applications that allow travelers to pass through a beautiful perspective [1].

Streets, boulevards, crossroads, and medians, which constitute the transportation system in the whole city, are important public spaces that meet the need for active or passive recreation as well as the basic areas of use of people [3].

Planting design, taking into account its functional and aesthetic features in establishing a harmonious relationship between the highway and its surroundings, make a modern highway perfect in terms of traffic and construction technique.

Since the region contains species-rich in plant material, it brings the possibilities of utilization of these plants around the highway to the fore. In this respect, the plant species desired to be used in the vicinity of the road should be determined by studies [4].

The most important effects of road trees in traffic regulation on vehicle users are directing and emphasizing. Besides, it can provide ease of direction determination by preventing the driver from confusion at wide intersections [5].

The main aesthetic and functional contributions of urban green spaces, including highways, can be listed as follows:

- Eliminates monotonous looks by softening sharp lines
- Attracts the attention of the city people by displaying different views seasonally throughout the year, bringing them closer to nature.
- Regulates temperature by absorbing sunlight through its leaves
- Displays aesthetic images to those who travel along the way.
- Regulates the traffic.
- Prevents the accumulation of sand or snow on the roads outside the city.
- Act as a windscreen for the surrounding areas [6].

Apart from connecting the road to the landscape, roundabout plant design has important benefits, especially in protecting the driver from the headlight lights of the opposite traffic. Planting design of roundabouts has benefits in terms of reducing and stopping the speed of runaway vehicles as well as shielding headlight lights [7].

It has been determined that 50 % of the accidents that occur in our country occur at inner-city intersections and 30 % of them occur at intercity intersections [8,9]. According to statistical analysis, intersections, where accidents occur at a high rate, are potentially critical points for accident occurrences, and proper design is of great importance in terms of minimizing pecuniary loss and intangible damages [10].

One of the most common intersection types in the city is roundabouts. The designs of these intersections are important for both pedestrian and vehicle safety. Functional design of roundabouts in terms of plant design can minimize safety risks. In this study, how the safety perceptions of roundabouts change in connection with plant design and visual quality have been revealed.

2 Material

The main material of the study is five simulation images prepared by the writers for evaluation. "Auto-CAD 2018" and "Lumion 10.5 Pro Student" programs used for the preparation of images also were used as material. Another important material of the study is the groups to which the questionnaire was applied. Within the framework of the study, 100 questionnaires were applied to the sample group. The literature on the subject also constitutes the other materials of the study. In obtaining these kind of literature, searches were made under the headings of landscape design, intersection design, intersection safety, visual quality, visual perception and plant design, and publications on the subject of the study were obtained.

3 Method

As stated in the introduction; In this study, a planting design modeling process has been tried to be developed in the context of visual quality and safety in urban roundabouts. In this context, a virtual scene was prepared by means of simulation, and different plant designs were tried to be developed on the same scene. On the other hand, evaluation of plant design; was carried out according to visual quality evaluation criteria. Also; In addition to visual quality criteria, evaluations were made in the context of safety. The subjects were asked to evaluate the simulation images, which constitute the main material of the study, through a questionnaire. In this context; the psychological approach model has been adopted as the method of the research.

According to this method; the beauty of the landscape is a product of the cultural, social and psychological structure of the viewer. This model tries to evaluate the preferences of the people towards the landscape using statistical methods (such as multiple regression analysis, factor analysis). In this method, data are usually obtained with the help of questionnaires. [11].

Preparation of the Questionnaire

Evaluation of simulation images of roundabouts was carried out through a questionnaire.

While creating the survey questions, the following factors were taken into consideration.

- Understandability of the prepared survey
- Homogeneous distribution of users
- Preferring face-to-face interviews during the completion of the questionnaires.

Different plant views were used in the 5 photographs prepared by the writers. Differences in plant height were the most important factor in determining the appearance of plants. The most important reason for this is to determine to what extent the plant height affects the driver and pedestrian safety. In this context; Simulation images were obtained by selecting the ground cover plant group in image number 1. Shrubs were used in image 2, medium-tall trees in image 3, tall trees in image 4, and tall conifer trees in image 5, and a total of 5 simulation images were obtained. Since the scene used in the images is designed the same for each plant group, dispersion of the perception is prevented. This allowed the subjects to evaluate the images more objectively (Tab. 1).

The survey questions prepared consist of 3 groups. These groups are;

- Individual characteristics of the subjects
- Opinions of the subjects about landscaping
- Evaluation of simulation images

Tab. 1. Criterias for evaluating the images of the subjects.

	5 4 3 2 1
Criterias for Evaluating the Simulated Images of the Subjects	Regular -- Irregular
	Wide angle -- Acute angle
	Simple -- Complex
	Natural -- Artificial
	Visually effective ---------------------------- Not visually effective
	Safe -- Unsafe

4 Findings

In this study, in which visual quality was evaluated in the context of driver and pedestrian safety, simulation images were prepared on a sample stage at roundabouts and various data were obtained through a survey study. The results of the survey study are meaningful, and the data on the findings are shared in this section.

The 5 simulation images subject to the study were evaluated by a group of 100 people. The entire group of subjects consists of people driving vehicles with a driving license. The subjects evaluated the images according to a 5-step scale on the Likert Scale. According to this; The individual characteristics of the subjects, simulation images and findings regarding the images are as follows;

Individual Characteristics of Subjects

- The subjects are 38 % female and 62 % male according to gender.
- The subjects, according to their age; 8 % are 18–19, 34 % 30–39, 48 % 40–49 and 10 % over the age of 60.
- 12 % of the subjects graduated from primary school, 16 % from secondary school, 44 % from high school and 28 % from university.
- While 44 % of the subjects provide their transportation with their own vehicles, 56 % of them use public transportation vehicles.
- All subjects have a driving license.

Simulated Photo: Number 1

For this group, plants with ground cover characteristics were selected below eye level. In this planting design with maximum viewing angle, the dominance of the environment is at the top level (Fig. 1 and Tab. 2).

Fig. 1. Simulated photo number 1

Tab. 2. The Subjects' Evaluation Status of Image Number 1

Photo 1	5	4	3	2	1	**Average**
	%	%	%	%	%	
Regularity	19	27	28	13	13	3,26
Wide angle status	43	39	16	2	0	**4,23**
Simplicity	39	30	21	7	3	**3,95**
Naturalness	7	14	28	33	18	2,59
Visually Effectiveness	12	16	28	23	21	2,75
Safe State	34	30	18	13	5	3,75

p=p<0,05

Simulated Photo: Number 2

For this simulation, shrubs were preferred. This group of plants is located above eye level while in the vehicle (Fig. 2 and Tab. 3).

Fig. 2. Simulated photo number 2

Tab. 3. The Subjects' Evaluation Status of Image Number 2

Photo 2	5	4	3	2	1	Average
	%	%	%	%	%	
Regularity	22	25	27	11	15	3,28
Wide angle status	22	25	33	17	3	3,46
Simplicity	13	27	30	22	8	3,15
Naturalness	11	23	24	33	9	2,94
Visually Effectiveness	14	19	32	24	11	3,01
Safe State	22	27	28	15	8	3,4

p=p<0,05

Simulated Photo: Number 3

Medium height trees and boughing above eye level were used for this group (Fig. 3 and Tab. 4).

Fig. 3. Simulated photo number 3

Tab. 4. The Subjects' Evaluation Status of Image Number 3

Photo 3	**5**	**4**	**3**	**2**	**1**	**Average**
	%	%	%	%	%	
Regularity	27	26	26	14	7	**3,52**
Wide angle status	29	32	16	15	8	3,59
Simplicity	19	26	33	16	6	3,36
Naturalness	35	26	22	16	1	3,78
Visually Effectiveness	37	28	20	12	3	3,84
Safe State	36	31	24	9	0	3,94

p=p<0,05

Simulated Photo: Number 4

High trees with a bough above eye level were used for this group (Fig. 4 and Tab. 5).

Fig. 4. Simulated photo number 4

Tab. 5. The Subjects' Evaluation Status of Image Number 4

Photo 4	5	4	3	2	1	Average
	%	%	%	%	%	
Regularity	22	24	26	17	11	3,29
Wide angle status	33	37	16	9	5	3,84
Simplicity	23	31	31	11	4	3,58
Naturalness	40	27	22	11	0	**3,96**
Visually Effectiveness	41	27	22	6	4	**3,95**
Safe State	38	30	25	7	0	**3,99**

p=p<0,05

Simulated Photo: Number 5

Conifer trees were used for this group. Environmental control is minimal for drivers and pedestrians (Fig. 5 and Tab. 6).

Fig. 5. Simulated photo number 5

Tab. 6. The Subjects' Evaluation Status of Image Number 5

Photo 5	5	4	3	2	1	Average
	%	%	%	%	%	
Regularity	18	21	30	17	14	3,12
Wide angle status	0	2	16	39	43	1,77
Simplicity	12	24	29	24	11	3,02
Naturalness	37	22	20	19	2	3,73
Visually Effectiveness	22	28	22	17	11	3,33
Safe State	12	13	27	25	23	2,66

p=p<0,05

The subjects' evaluation of images

The prominent findings according to the survey evaluation results of the 5 images that form the main material of the study are as follows;

- The most regular image is photo number 3 with a score of 3.52
- The least regular image is the photo number 5 with a score of 3.12
- The most wide angle image is the photo number 1 with a score of 4.23
- The least wide angle image is the photo number 5 with a score of 1.77
- The simplest image is the photo number 1 with 3.95 points
- The least simple image is the photo number 5 with a score of 3.02
- The most natural image is the photo number 4 with a score of 3.96
- The least natural image is the photo number 1 with 2.59 points
- The image with the most visual impact is photo number 4 with a score of 3.95
- The image with the least visual impact is the number 1 photo with a score of 2.75
- The safest image is photo number 4 with a score of 3.99
- The least safe image, photo numbered 5 with a score of 2.66 was preferred.

5 Conclusion

Planting design, visual quality and traffic safety analysis have also become an important criterion in the planning and design of the environment. Human comfort has a significant relation with the roundabouts since encountering a lot of in city centers.

According to the data obtained based on the findings, the evaluations can be listed as follows;

- It can be said that there is a significant relationship between plant height and driver and pedestrian safety in urban roundabouts. A safe viewing angle appears to be captured when the height of the plants is below the driver's eye level. In this case, while safe driving increases, visual quality decreases.
- When using conifer trees with the height of the plants used above the eye level of the driver, an unsafe driving is realized with high visual quality but less environmental control.
- When the plants used in the roundabout and its surroundings consist of high and medium-tall trees boughing above the eye level of the driver and pedestrian; Both a safe traffic and a design with high visual quality are presented.

The simulated images prepared in accordance with the research method do not constitute an example for all studies as they serve a virtual space, but it is thought that the quality criteria specified in the method stage can be used for other similar studies. For this reason, plant design should be tested by researchers who examine the issues of visual quality and traffic safety.

References

1. Koçan, N. Gümüşhane-Trabzon Karayolunda Ulaşım ve Peyzaj, Güfbed/Gustı J/ 1(2). 2011. 89–96.
2. Çetin M. Bartın Safranbolu Karayolu Güzergâhı ve Yakın Çevresinin Doğal-Kültürel ve Görsel Peyzaj Değerlerinin İrdelenmesi Üzerine Bir Araştırma, 2005. ZKÜ Fen Bilimleri Enstitüsü, Yüksek Lisans Tezi, 117 s, Zonguldak.
3. Ertin D. G., Meral M. V. & Zülfikar C. Yaya ve Taşıt Trafiği Açısından Bitkisel Tasarım; Edirne Örneği, X. Ulusal Ekoloji ve Çevre Kongresi, 2011. Bildiri Özetleri Kitabı. Çanakkale, 4–7 Ekim 2011, ss. 54–55.
4. Acar, C. Rize Arası Karayolu ve Yakın Çevresinin Doğal, Sosyo-Kültürel ve Görsel Değerlerinin Peyzaj Gelişimindeki Rolü ve Peyzaj Planlama Açısından İncelenmesi, 1993 Karadeniz Teknik Üniversitesi, Fen Bilimleri Enstitüsü, Yüksek Lisans Tezi. Trabzon.
5. Sakıcı, Ç. Ayan, E., Kapucu, Ö., Türkdoğdu, H. Kastamonu Kentiçi Karayolu Bitkilendirmelerinin Kullanıcılar Tarafından Değerlendirilmesi ve Tür Tespiti, Kastamonu Üni., Orman Fakültesi Dergisi, 2014, 14 (2) 237–245.
6. Atay İ. Şehiriçi ağaçların tekniğine uygun bakımı ve budanması. Ormancılık Eğitim ve Kültür Vakfı Yayınları, 1990. No:2, İstanbul. Ss:1–12.
7. Küçük, V. Isparta Kentiçi Yol Ağaçları Yönetim Planı, 2010. Süleyman Demirel Üniversitesi, Fen Bilimleri Enstitüsü, Doktora tezi, 176 s., Isparta.
8. Özinal, Y., Uz, V. Dönel Kavşak Geometrik Elemanlarının Kavşak Güvenliği Üzerine Etkisinin Literatür Işığında Değerlendirilmesi. 2020. Politeknik Dergisi, (), 1–1. DOI: 10.2339/politeknik.630947.
9. EGM, ‘‘Karayolu Trafik Kaza İstatistikleri’’, 2012. Emniyet Genel Müdürlüğü Yayınları, Ankara.
10. Bozkurt, H., Yılmaz, H. Dönel Kavşakların Kavşak Güvenliği Üzerine Etkisinin Analizi. ©2020 Published in 4th International Symposium on Natural Hazards and Disaster Management 24–25 October 2020. (ISHAD 2020 Bursa - Turkey) https://doi.org/10.33793/acperpro.03.02.25.
11. Elinç H. Görsel Kalite Değerlendirmesinde Yönetimi İle Antalya İli Alanya İlçesindeki Abdurrahman ve Alanya Belediye Başkanları Kent Parklarının İrdelenmesi, 2011. Yüksek Lisans Tezi, Selçuk Üniversitesi.

Mehmet İNCEOĞLU and Bircan İNAN

How BIM May Change Architectural Education

Abstract: The software that has entered our lives with the developing computer science in recent years has brought new methodologies and new concepts in many fields. Along with these developing computer sciences, new methodologies have been developed in a wide range of fields such as art, science, engineering, health, social, human, and economic fields. New methodologies have been developed in the field of architecture. With the beginning of computer programs, methodologies have changed with 2D CAD environments and 3D drawing software entering the field of architecture. However, Building Information Modelling (BIM) systems have replaced 2-dimensional or 3-dimensional drawings with computer science that has developed in recent years. Thanks to BIM programs, n-dimensional studies can be carried out to evaluate not only the project design stages of the buildings but also the construction stages and the sustainability and operation of the structure. However, BIM methodology did not exist only in the field of project management and projecting processes and architecture. Architects reveal their designs with the methodology they use. Thanks to the BIM methodology, these designs can create sustainable structures by performing various simulations. At the same time, thanks to BIM, interdisciplinary cooperation has increased, and a joint work can be carried out easily with all project stakeholders. With the simulation systems and interdisciplinary working environments developed thanks to the BIM methodology, many countries make the use of BIM methodology mandatory today. However, making the use of BIM methodology mandatory also ensures that this methodology is performed in certain standards. Many countries in the world are preparing various guidelines and guidelines on the use of BIM methodology today. With the widespread use of BIM methodology, it has led to the formation of certain standards within the BIM methodology training at certain standards. With the specific concepts brought about by BIM methodology education, it has become inevitable that new education understandings have come within architectural education. Maintaining the understanding of education in architecture schools with the understanding of T-type education is a focus for BIM methodology education. At the same time, with the participation of BIM education included in the study in architectural education, the changes to be made in architectural education curriculum have been evaluated. The current status of BIM methodology education on selected architectural schools based on certain criteria in our country and abroad has been discussed. In the schools examined, the classes in which BIM methodology trainings started were examined with various criteria such as the necessity status of the courses.

Keywords: Building Information Modelling (BIM), Architecture Education, Sustainability

BUILDING INFORMATION SYSTEM (BIM)

There are many different definitions in the literature for Building Information Modelling. However, Eastman made one of the most basic and first definitions of BIM as "developing the electronic representation/model of a building that supports all its main activities throughout life". According to this definition, the main activities of the data drawn and obtained in 2-dimensional environments can be interpreted as developing with the help of a model. With the help of the model created together with these processes, indexes can be defined as analyses, simulations and production of data. [1]

BIM has defined that it is not only a modelling methodology, it can be used as a platform that can easily and accurately obtain the necessary analyses, simulations and all other necessary data of the relevant project to be carried out with this methodology, and this new methodology is a new methodology that can be used jointly among all project stakeholders, not only by architects .[2]

Unlike different drawing methodologies or computer-aided design methodologies used by architects and engineers in previous periods, the concept of BIM should be evaluated as a methodology since both design stages, construction stages and all life cycles after construction can be modelled and simulated in a virtual environment.

One of the main objectives of the creation of BIM is to create a common language and a common platform among the project stakeholders. It is aimed to eliminate language conflicts that are present in the different disciplines of the project stakeholders during both the design phase and the manufacturing phase of the relevant project and to gather the data of all project stakeholders in their own disciplines on a common platform. Thanks to this common platform, data flow will be able to be provided between project stakeholders. The data of the different disciplines formed by the project stakeholders will be combined in the layered structure and structure model of BIM and it will be ensured that the data can be easily understood by other project stakeholders. [3]

As it is possible to work with other methodologies, BIM methodology has emerged in new definitions for BIM together with these methodologies. In terms of Geographical Information Systems (GIS), it is seen that the relevant project stakeholders produce numerical data in the design and manufacturing processes of the project in the BIM methodology. It has been observed that many new types of data can be created by integrating these data into GIS systems in comprehensive projects. [4]

Together with the BIM methodology, today's design tools are improving. Thanks to the visual programming languages that BIM methodology also brings with it, it is seen that compelling designs such as parametric design for designers in Computer Aided Design (CAD) environments can be easily made thanks to visual programming languages. BIM methodology shows that a strong design methodology definition can also be made by removing projects from a definition such as analysis and a common language platform methodology .[5]

It should not be considered as a methodology or tool to model using only advanced computational design elements for the BIM methodology. The BIM methodology will also allow designs made using smart design tools together with visual programming languages to exceed project design and the resulting model to be easily shared with other project stakeholders. Azhar, Khalfan, and Maqsood stated that the designs made in the CIS methodology and these designs are not sufficient today. However, it explains that the designs made in the studies conducted within the scope of BIM methodology can be supported with smart tools and sufficient modelling can be done for the project stakeholders. However, these areas are not limited to the new modelling systems that BIM methodology reveals. In his study, Mathews showed us that different and comprehensive models can be made from CIS methodology by using visual programming languages within the scope of BIM methodology and that these models can be created by supporting with smart design tools. Thus, it has been possible to use various design tools such as visual programming languages within the scope of BIM methodology.

Based on all these definitions made for BIM technology, BIM is a technology that

- Helps the development of new design trends by using new design tools in the design processes of projects and pave the way for the realization of today's complex designs,
- Project stakeholders can create a language in which an accurate and reliable data flow can be ensured by using a common language and a common platform in the design and projecting processes of the structure,
- All life cycles can be modelled in the geography where the building is located both physically and functionally before construction and the efficiency of the project can be increased by making the necessary adjustments during the project design phase with the data obtained thanks to these models,
- Works with many methodologies with specific international standards such as GIS and provide various data to project stakeholders with this collaboration.
- Having the technology that can be used not only in the national sense but also as a methodology in the international sense and that the software connected to this methodology can work at all international standards,
- The structure will be digitized and constructed in digital environment,
- It is a methodology in which all fabrication, manufacturing and processes can be programmed during the construction process of the building.

BIM methodology and technology are expanding in the definitions as they develop. With BIM technology, it is aimed not only to increase the productivity in the vital cycle or project design of the structure but also in the construction process. BIM methodology can be defined as a sustainable methodology in which the necessary data related to all the design, construction and post-construction processes of the structure can be created, transferred and in addition, data can be exchanged with all project stakeholders on a common platform and all the processes of the project can be mastered and create a base technology within the future methodologies.

History of BIM

Construction activities have started and developed since the transition to settled life in the history of humanity. Before the product to be made in many activities of humanity from past to present, a prototype was made for that product and production was started in the following stages. This product development technique is still in use. However, this is not the case in the construction sector. This case in construction activities has been replaced by perspective drawings and models on certain scales of the project to be carried out. The use of models instead of the prototype of the project has been a very old methodology used by architects. Although these models used as prototypes were model models in the early periods, computer models have now taken the place of models.

In 1988, the CIFE unit was established by Paul Teicholz in Stanford and many students and researchers within this unit conducted studies on object-based parametric modelling, which was introduced as a new concept. The researches conducted here have formed the foundations of BIM technology. [6]

The Archicad program, developed by Graphisoft, a company founded in Hungary in 1984, was released as the first BIM program with BIM methodology features. With the Archicad program, it is planned to be a drawing tool based on the implementation of all the ideas put forward during the period for BIM. [7]

Thanks to the breakthroughs made in the field of BIM immediately after these studies, the Revit program of the most used Autodesk company of the BIM system was developed and launched in the beginning of the 2000s and the BIM system became widespread.[8]

BIM methodology can introduce a new modelling system by developing all modelling methodologies used so far with today's technology. However, this is not the only difference between the BIM methodology and the previous modelling methodologies. BIM methodology has enabled multidimensional modelling by integrating information flow into this modelling system together with the new modelling system it has put forward. Thanks to these data integrated into the new modelling system, data flow can be provided at every stage of modelling as well as the processing of these data can be ensured through new smart software. Thanks to these data processed, we can be informed about the project design, construction, and post-construction periods in the life cycle of the project.

This smart software developed thanks to the BIM methodology will be the cornerstone of new methodologies to be introduced after the BIM methodology in the future. It can be said that the foundations of the methodologies that can be done completely with artificial intelligence in the future are revealed thanks to BIM methodology.

Scope of BIM

Today, the functions and physical properties of the buildings are constantly changing due to technological developments, increasing human needs, the development of materials used in the buildings and the economic factors that come with increasing competition in the construction sector. When BIM is evaluated in this context, it will be able to respond to these ever-increasing needs. In today's projects aimed at increasing these needs, it has been observed that 2-dimensional or 3-dimensional drawing and modelling methodologies will be insufficient.

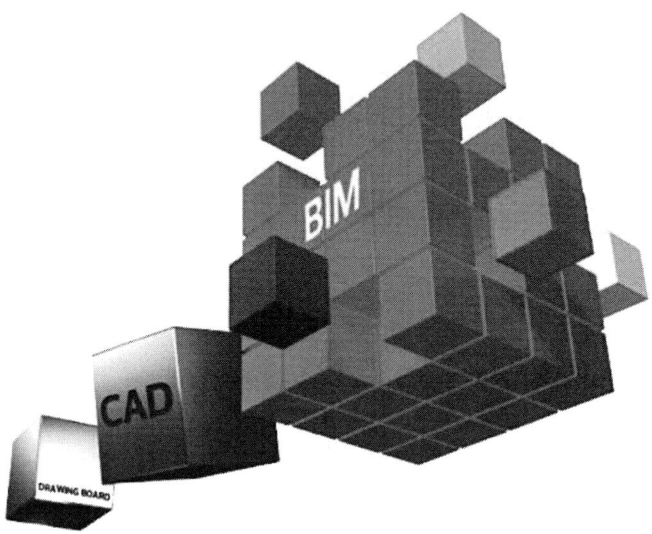

Fig. 1 *Development of BIM methodology*

Thanks to the BIM methodology, it is possible to create a common working language among the project stakeholders and to create the building processes that can be worked at the highest efficiency together with today's building and construction techniques (Fig. 2.3).

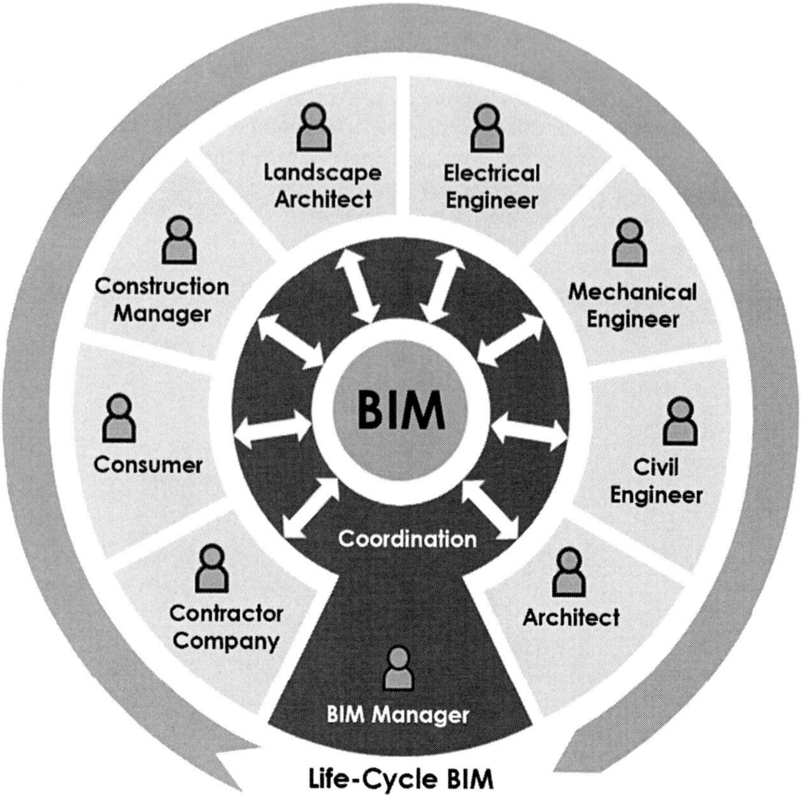

Fig. 2 *Project stakeholders in BIM methodology*

BIM technology has also developed continuously with today's technology and has become an indispensable methodology of construction and projecting processes for stakeholders. BIM methodology and software are used in comprehensive functional and size structures built and built today. Based on the definitions made for the BIM methodology, in fact, each word of the word BIM gives us information about these processes. The word "Structure" can give us information about the type of structure to be built. The word "information" can give us information such as planning, cost, manufacturing, supply, quantity along with all calculations to be required during the design and construction process, and the word "Modelling" can give us information such as the architecture, sustainability and modelling of the structure.[9], [10]

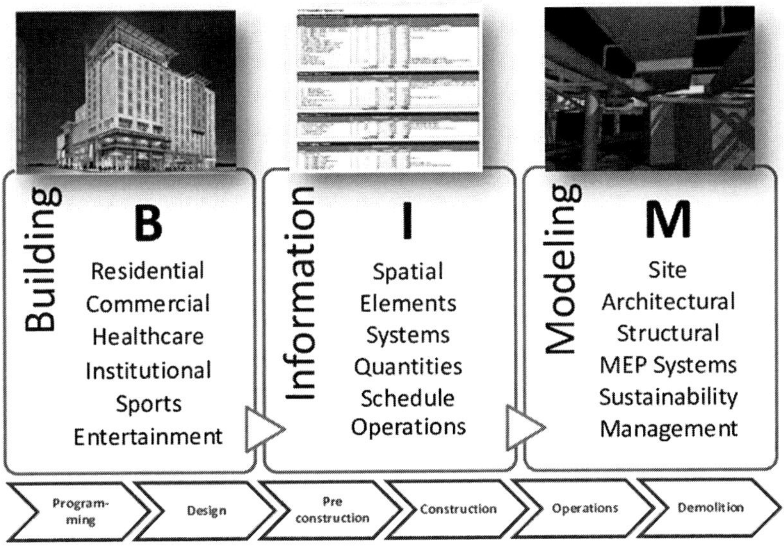

Fig. 3 *Description of the BIM methodology*

Along with the BIM methodology, many concepts used today have changed and new ones have also been added. BIM methodology, together with the changing and newly added concepts, has enabled the conceptual definition of many cases related to these processes that cover all the processes that the structure will undergo before and after construction.

BIM and Sustainability

In recent years, sustainable projects have emerged with the environmental policies adopted by many states around the world. The concept of sustainability, which has emerged as a result of this policy change caused by environmental factors such as the developing climatic changes in recent years, has rapidly become widespread and has led to major changes in the construction sector. As a result of the environmental policies adopted by many states, it is expected from the construction sector in seriously sustainable projects. Sustainable projects will be able to leave minimal damage and waste to the environment and minimal carbon footprint will be observed in their construction. The fact that these structures are both sustainable and economically feasible has brought about changes in building materials and the use of BIM methodology in projects has become almost essential thanks to this material change and integration. Because, thanks to the BIM methodology, the necessary data flow will be provided to ensure sustainability not only in the design phase of the project but also in both construction and post-construction life cycles .[11]

In the literature reviews, the buildings that we encounter as the concept of "Green Building" consume the least amount of energy during the construction phase and in the post-construction operation processes. Emission values of the materials to be used should be kept to a minimum and materials that will adversely affect public health should be avoided [12]. The construction sector uses 40 % of the raw materials produced in the world today. This is an indication of how important the correct use of materials and advances in material technology are to the world [13]. The 6D dimension within the scope of BIM methodology will be able to transfer the environmental data to the project together with the data to be obtained from these materials and ensure that the correct data are transferred to the project stakeholders in material selection.

In the projects made in the periods before BIM methodology, 2-dimensional drawings made in CAD environment were transferred to analysis programs. In these programs, the data obtained by conducting various analyzes such as energy analysis of the project, shading and solar simulations, air flow analysis and climatic reactions were sent to the project stakeholders. However, due to the fact that the drawings and analyzes were in different programs and that all the drawing data could not be transferred accurately from time to time, there were also incorrect data formations. However, thanks to the 6D dimension, these analyzes will be performed in single BIM-based software and the accuracy share in the results of the analyzes will increase, and at the same time, a possible change in the design will be automatically reflected in these analyzes and a faster and more reliable data flow will be offered to the project stakeholders .[14]

Around 40 % of the world's carbon emissions result from the energy consumed by buildings. This energy is spent to ensure the comfort of the user in the building. However, this emission value may decrease if necessary, simulations are made in BIM-based software according to climatic changes and geographical conditions where the building is located while the project is still in the design stage and the necessary data are shared with the project stakeholders. If the designs of the structure and structure elements made by the project stakeholders are arranged according to the analyzes to be made in 6D dimension while they are still at an early stage, a correct project for sustainability will be made in the later processes of the project. [15]

Standardization and Use of BIM Methodology

With the widespread use of BIM methodology and the use of public institutions, the need to standardize the methodology within the framework of certain concepts has arisen. It ensured that the BIM methodology could be carried out with certain standards and qualities for standardization, project design, project management and operations in all life cycles of the project with the use of the BIM methodology. [16]

Standardization process aims to be able to be performed within the framework of certain inputs and outputs throughout the processes of the works to be performed. In the process, the inputs into the system and the outputs resulting from the process should be within certain quality indicators. However, if the inputs and outputs in the transactions are different except in exceptional cases, this standardization is not successful and this standardization needs to be renewed again. [17]

In the construction sector, as in other sectors and industry areas, certain standards and rules are acted on. These standards can not only improve the quality of the projects to be carried out but also save time and increase compliance. Standardization procedures for BIM methodology can be measured by the fact that the data to be entered into the system are at certain standards during the use of the methodology and these inputs entered into this system can be measured by extracting outputs of certain standards and quality.

It can be said that the studies conducted for the standardization of BIM methodology are carried out by organizations in 3 different fields. With these studies, it can be said that 3 different types of standard areas are formed for BIM methodology. These areas are:

- Corporate Standards
- National Standards
- International Standards

These standards aim both to spread the use of BIM methodology within certain standards and to increase the quality in the use of BIM methodology. It can be said that the whole building sector agrees on the implementation of standardization studies conducted in line with these objectives. [18]

In order to ensure the adaptation of the BIM methodology in the sector, certain strategies for the use of BIM need to be developed. In order to implement and develop these strategies to be developed, institutions and organizations from the specific public and private sectors should support them. Today, strategies and guidelines for the use of BIM have been developed by the building SMART organization and Autodesk for the development and dissemination of strategies for the use of BIM, and it is aimed to increase the use of BIM and to use BIM at certain standards.

There are various studies on the rapidly spreading use of BIM methodology today. According to RICS (2014) report, various data on the use of BIM in regions around the world have been shared. According to the published report, the region where the use of BIM methodology is the most used with 71 % usage rate is America, and it is seen that India has the lowest use with 14 % usage rate. As a result of the study, it is revealed that the countries and regions where BIM methodology is used the most are the countries and regions that encourage the use of BIM the most. [16] (Fig. 4).

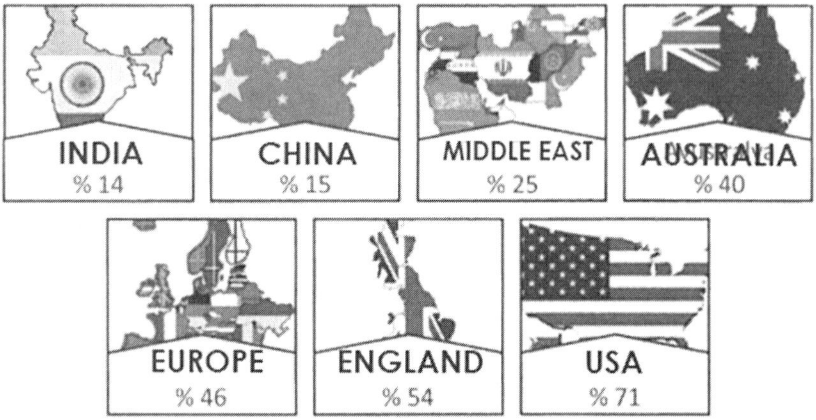

Fig. 4 BIM utilization rates around the world. [16]

Public institutions play the most important role in the widespread use of BIM and in the establishment of certain standards. Public bodies publish a number of guidelines for the widespread use of BIM and its use in the sector to certain standards [19]. With the help of these published guidelines, use of BIM ensures that certain standards of gold can be made. At the same time, the compulsory use of BIM with these guidelines in some countries significantly increases the prevalence of BIM use.

In countries where the use of BIM is mandatory, pilot regions and projects were primarily selected. In the UK, which is considered to be the pioneer of standards on the use of BIM today, it is considered mandatory to use BIM within certain standards and at the same time, its use is used within the framework of certain LOD levels. Via the guidelines and specific strategies developed for the use of BIM in the UK, it has become a leader in the use and standardization of BIM worldwide. Thanks to this path followed by England, the guidelines and standards have set an example for many countries around the world and have been created with certain awareness about the use of BIM worldwide.

In order to disseminate and standardize the use of BIM methodology, architects must first be trained in accordance with BIM methodology. The BIM methodology does not only exist with certain processes and computer programs. BIM methodology has brought along new working principles and thinking styles along with the innovations it has brought today. For this reason, BIM education does not only consist of theoretical knowledge about program use or processes, but also students and professionals who will use BIM need to receive the necessary training about the discipline and order of collaboration with other project stakeholders.

Current State of BIM in Architecture Education

With the development of BIM methodology and the spread of BIM-based software, existing education systems in universities have been discussed. BIM methodology is a multidisciplinary methodology that includes not only architectural discipline but also many disciplines. Therefore, the BIM methodology trainings to be held in universities and institutes should not be continued through a single discipline. However, in this study, the place of BIM methodology in architectural education will be examined.

In the studies conducted in the USA between 2007 and 2011 on the education of BIM methodology, the main learning strategies put forward by universities were examined. These training strategies are;

- BIM-based software courses,
- High-level interactive learning programs
- Interdisciplinary learning programs
- BIM-based project studios

It has been observed that these strategies, which are seen in the examinations, actively affect the education processes. In the studies conducted in these years, it was seen that many universities in the USA focused on the BIM methodology and the scope of their education on BIM was investigated and compiled. When the curricula of these trainings on BIM methodology are examined, it can be seen that most universities have just met the BIM methodology.[20]

When the curricula of the architectural schools examined with this study are examined, it is seen that the curricula in the schools focus on the scope of BIM methodology instead of BIM-based program education. Evaluating the concept of BIM as a methodology and giving lessons in more detail about the fields of activity within the scope of this methodology ensured that the BIM methodology could be understood correctly as well as creating an awareness of BIM methodology on students (Tab. 1).

Tab. 1 *According to the studies conducted in the USA in 2013, BIM education in universities and their scopes* [20]

Routine Name (institution)	**Program Goals**
CAD YBM Advanced Topics (University of Massachuesettes Amherst)	Using the knowledge of advanced topics in Architectural CAD software in design and construction; working with 3D modelling and creating parametric structure model and extracting data; Ability to use CAD-BASED tools to solve technical problems in the planning process; Understanding commonly used BIM software such as Revit, Navisworks, Google SketchUp and AutoCAD
YBM (Stanford University)	Creation, management and application of BIM; processing and tool to create 2D and 3D computer representations of building components and geometries; Organizing and running models to produce architectural views and construction documents, processing and animation and to interface with analysis tools
YBM (Southeast Missouri State University)	Understanding BIM concepts of Building Information modelling; reviewing software and technology available for BIM; using BIM software to build a building model; using BIM to control obstacles and conflicts in a building construction project; exploring construction scheduling and sorting using BIM; exploring cost estimation using BIM; discovering how BIM can help Facility Management
YBM (Southern Polytechnic State University)	Developing the required knowledge and skills to create and modify CPM using Revit Architecture, Revit Structure and Revit MEP. Performing quantity deviations and cost estimates. Identifying constructability problems using CPC
YBM (University of Arkansas)	Basic functions of YBM for residential and commercial construction; Understanding the geometry, spatial relations, geographical information, quantities and properties of building components. Creation of virtual building models that can be used for quantity increases
CPC for Commercial Buildings (Purdue University)	Understanding the use of CCM for commercial construction; Understanding the geometry, spatial relationships, geographical information, quantities and characteristics of building components

Tab. 1 Continued

Routine Name (institution)	**Program Goals**
Building Information Modelling and Integrated Applications (University of Southern California)	Understanding the transition from representation to simulation; familiarity with existing YBM technologies; understanding the new coordination and cooperation tools of design and construction. Understanding the connectivity and continuity of existing and designed CPC information and other important information to the model. Understanding new project delivery systems and technologies for "integrated implementation". An overview of how innovative technologies can be integrated into existing AEC applications
CAD and YBM for Construction Managers (Oklahoma State University)	Interpretation and production of the construction drawing using computer-aided drawing. Theory and use of Building Information Modelling software
Construction Documentation and BIM Practices in Engineering and Construction (University of Maryland)	Understanding the construction documentation methods; ability to read and coordinate the construction drawing prepared by various project consultants. Knowledge of managing and implementing YBM technologies in design and construction
Industrial Applications of Virtual Design and Structure (Stanford University)	Research on the management of Virtual Design and Construction (STI) programs and projects in the building industry. Learning from YBM's industrial practices and its relationship with Integrated by interacting with experts and professionals in real estate, architecture, engineering, construction and technology providers. Project Delivery, Sustainable Design and Construction. Assessing the maturity of STI planning, adoption, technology and performance in practice
Industrial Applications of Virtual Design and Structure (University of Washington)	Understanding competence in program and project management processes, YBM applications, documentation and quality control; expertise in 3D visualization, site planning, problem detection and 4D modelling, web-based project management systems, YBM tools, construction contracts, design and construction procedures. Awareness of designer and contractor rights

1.1 T-Type Professional Training and BIM

The advancement of technologies affects every sector and every discipline as well as learning processes. Especially by supporting the opportunities provided by Web 1.0 and Web 2.0 breakthroughs with information and communication technologies, it has been ensured that all individuals around the world can easily access information. After the web breakthroughs, it has become very easy for each individual to access and share information over time. With these developments in the fields of information and communication, some changes have also been made in the learning processes in universities and institutes. With these changes, learning methods have been developed by experimenting in addition to the learning methods applied in the previous periods, working in the relevant field, cooperating and interacting with technological opportunities [21], [22]. Considering all these new learning methods, it will reveal the importance of interdisciplinary cooperation in learning and developing knowledge, and in addition to this, it shows that specialization in only one discipline will not be sufficient today. Today's universities and institutes aim to educate individuals as individuals at an adequate level for information societies by acting with a multidisciplinary approach. However, the individuals to be raised for these purposes will be sufficient for the information societies.

The importance of interdisciplinary cooperation of people, especially in the fields of architecture and engineering, is increasingly seen today. With the transition to the information society, the need for T-type people is increasing. T-type human understanding and identification was first defined by David Guest [22], [23]. The concept of t-type is an understanding that can enable the person to cooperate in other disciplines in the close relationship depending on the discipline to which he/she is affiliated and support teamwork. While I-type individuals in the industrial society are experts only in their field and their knowledge of other disciplines is not sufficient for a possible teamwork and interdisciplinary working environment, T-type individual understanding will enable them to cooperate with all disciplines around them. With the T-type approach, it will be possible to create more efficient working environments by teamwork of experts in different professions. "Education is to provide access to more information, to produce it and to create an environment to go and learn and meet students." [24] (Fig. 5)

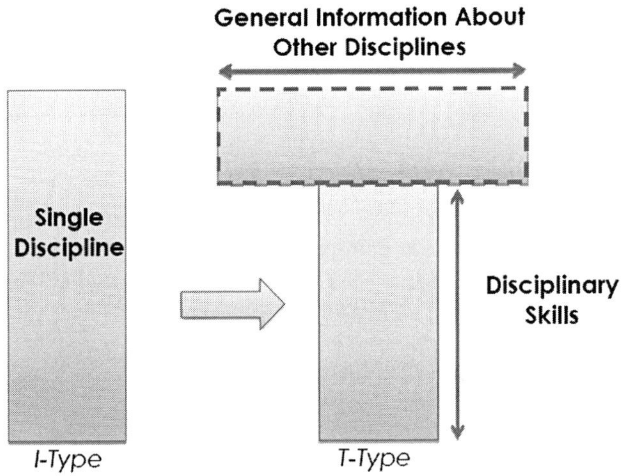

Fig. 5 *I-type and T-type diagram* [22]

T-type understanding is not actually a newly introduced concept. T-type understanding is actually a new concept and methodology derived from I-type understanding. I-type single-branch expertise can be explained as the development of the expertise of the individual in his/her field of expertise interactively with different disciplines. While the discipline in which the individual is an expert is shown vertically, information in different disciplines is shown horizontally. In order to provide education in the T-type methodology, students should be given basic information from different disciplines as a course in addition to the education in the discipline they will be experts in. Otherwise, the individual will not be able to learn basic information from different disciplines. In addition to this, with the concept of "all interdisciplinary empathy" included in the learning processes, the knowledge of the student in his/her own discipline will expand in parallel with the expansion of his/her knowledge in different disciplines.

Students studying together with T-type methodology in the field of architecture will be able to easily communicate and exchange information with individuals of different disciplines and will be able to produce common solutions to possible problems on design together with the teams they are affiliated with. They will also be able to work by meeting the requirements of the interdisciplinary working environment in the BIM methodology. Because programs based on the BIM methodology provide an interdisciplinary working environment only for individuals. These programs cannot work in an interdisciplinary manner. Individuals will do this study among themselves. This is why one of the most important conditions for working in BIM methodology will not be good software or technical equipment, but well-equipped individuals who can work in BIM methodology. The difference of the BIM methodology from other methodologies is not only new software or new technological developments, but also the BIM methodology can be seen as the new working methodology of individuals suitable for new information societies.

1.2 BIM's Place in Architectural Education

Today's universities are still provided with I-type person education. However, this understanding of education is not in question to act together with BIM methodology. This will lead to a significant change in the educational understanding of educational institutions (universities, institutes) that will provide BIM methodology education. This change may lead to the formation of multidisciplinary working environments in the education of BIM methodology by bringing a multidisciplinary education approach.

There are a wide variety of perspectives on BIM methodology education. Ambrose (2007) examines the architectural design effects of BIM methodology in architectural education in variable architectural design processes. In this study, together with the virtual modelling and other aspects of BIM methodology, a critical perspective is presented on how this methodology can be included in architectural education and how it can be applied in architectural design trainings. According to the opinions in the study, this new methodology, which is rapidly spreading around the world, should be accepted in academic circles. It is understood that the old pedagogical and paradigms still used in universities will not be sufficient for future studies. However, before the BIM methodology, it is said that there is a connection between the thinkers produced by the designer on abstract concepts and the building designs. This connection is not established with BIM methodology and it is stated that the designer only combines spatial and functional contexts and makes the necessary designs through software. This shows that BIM methodology should not be used directly in architectural education. By using the BIM methodology, it is stated that the lecturers can integrate the BIM methodology into architectural education by producing a number of educational processes that will enable students to see, think and create contexts in their design processes. Instead of defining BIM as a methodology that is slightly more technological than paper, it has been observed that BIM methodology is a very comprehensive and innovative concept. Ambrose's study examined the effects of new educational processes and learning methods to be developed by lecturers on architectural productions with the use of BIM methodology in architectural education and studio studies.[25]

Apart from some pioneering practices today, architectural education studios focus on the designs and design skills of the students. It can now be seen that these traditional studio design processes prevent students from producing new uniforms using new media tools. In a study by Cheng (2006), this issue was mentioned. In this study conducted, the basics of architectural education are discussed. Cheng saw no difference between "design thinking" and "problem solving". The designer should develop a design idea by thinking in a multi-faceted way about a problem he/she has addressed in the design process. The use of BIM methodology in these processes will prevent the designer from dealing with many unnecessary data and will ensure that this process can be carried out comprehensively.[26]

According to Cheng, with the inclusion of BIM methodology in architectural education, a major renewal process will start in architectural education curricula. With this change, numerical and parametric form productions will start instead of concepts such as aesthetic anxiety that is still ongoing in the world and architectural design processes and products are thought to develop all over the world. These new forms to be made with certain algorithms may turn into architectural forms with various variables such as certain restrictions and materials. In his study, Cheng states that today's architectural education is still dominated by form-centered curricula. However, in the new curricula including BIM, lecturers think that design thoughts will disappear within this methodology. With the new curriculum, it is feared that my importance to the concept of "how to" in educational processes will prevent the concept of "why". In conjunction with these cases, Cheng took a critical approach to moving from a form-centered training in the old curriculum to a new methodology training where the concept of "why" weakened. [26]

The use of BIM methodology in studio studies in architecture education can be described as a great advantage for students. Ambrose argued that BIM methodology could be more useful by being included in studio trainings as well as basic design trainings. Because design education and design processes support a connection between the designer's thoughts and design. However, Ambrose says that this connection between the designer and design will not be very healthy in the BIM environment. At the same time, the trainings on BIM methodology in his/her studies in Cheng and the studio studies constructed with BIM reveal that students can turn to the "how" question instead of the "why" question. Thus, it is estimated that there may be some deficiencies in the educational outcomes for students on basic design education. However, he argues that BIM methodology should exist in today's technology and rapidly in an industry-centered economic development, as well as that BIM methodology can be useful in architectural education in a responsive manner. When these two views are examined, it is understood that it will not be useful for BIM methodology to participate in architectural education in every sense. Because architecture education and its scope are constructed in a versatile way. Since the use of BIM methodology during design lessons given to students and trainings in different directions will cause students to think responsively, the links between design thinking and the abstract concepts that the designer will establish may be weakened. In this case, there will be deficiencies in design aspects due to the inability of students to establish an adequate connection on abstract concepts during their education.

Clayton (2006), on the other hand, discussed the architectural education process since the 1950s and conducted studies on how BIM methodology would be useful in architectural education. The study showed that there were radical changes in architectural education curricula with the spread of the Bauhaus school throughout the world in the 1950s. However, it should be noted that even if these curricula undergo many changes, their foundations were laid in the 1950s. These curricula, which were prepared in the 1950s, were prepared in line with the ideologies of Bauhaus-Beaux Arts schools and it was envisaged that students would be educated based on old methodologies in practical studies such as studio studies. It is understood that these curricula, which are still taught in many architectural schools, include environmental systems and building knowledge courses, which are many old technology products. However, today, it has been seen that world policies and architectural policies have changed along with major changes such as the end of the oil age and global warming. In response to this change, important concepts such as sustainability have been put forward in the field of architecture. However, it has been observed that the methodologies we use in the current case are insufficient for the studies to be carried out together with these concepts.[27]

In his study on architecture education, Clayton thinks that BIM methodology will open a new page in architecture education. BIM methodology education thinks that time losses in education processes will be minimized by enabling students to produce products faster against hand-drawings in past curricula. At the same time, the necessary simulations of the designs to be made with the help of 4D and other dimensions in the BIM methodology can be made to provide the students with the necessary calculations about the environmental systems. With these simulations to be made using BIM-based software, students will be able to learn by getting a more realistic education on a wide variety of subjects such as building elements and materials in their designs and studio processes and by having the chance to try them. Students will be able to make sustainable structure designs using the necessary simulation tools, as well as experiencing design with these concepts not only theoretically but also in the project design processes .[27]

Today, there is still controversy about the trainings to be given on BIM methodology in architectural schools and the new education models to be created. However, the studies revealed that BIM methodology should be included in architectural curricula. At the same time, in parallel with the rapidly increasing use of BIM methodology today, students need to be prepared for this case. However, the fact that BIM methodology is fully integrated into architectural education will lead to some mistakes. The use of a result- and response-oriented methodology in design courses, which is at the core of architectural school curricula, may weaken the links between the abstract concepts that students will develop. The fact that students prefer the "how" question rather than the "why" question may pose big problems for students. Therefore, it would be correct to use BIM methodology only in certain and limited areas in architectural education.

Current State of BIM Education in Architecture Schools

BIM methodology training in architecture schools will not be possible only with the addition of new theoretical courses. Thanks to the interdisciplinary working environments brought about by the BIM methodology and the cooperation based on project stakeholders, students need to receive basic trainings related to different disciplines. In this part of the study, various analyzes have been conducted on BIM education in architecture schools selected from different countries. As a result of these analyses, it will be examined in which classes and frequency BIM education is conducted.

Architectural schools to be studied were selected from different geographies and different continents and information was sought about educational models built on all BIM methodology in the world. At the same time, considering the undergraduate education periods of these schools, the courses related to BIM methodology were evaluated in which periods and in which scope (Fig. 6).

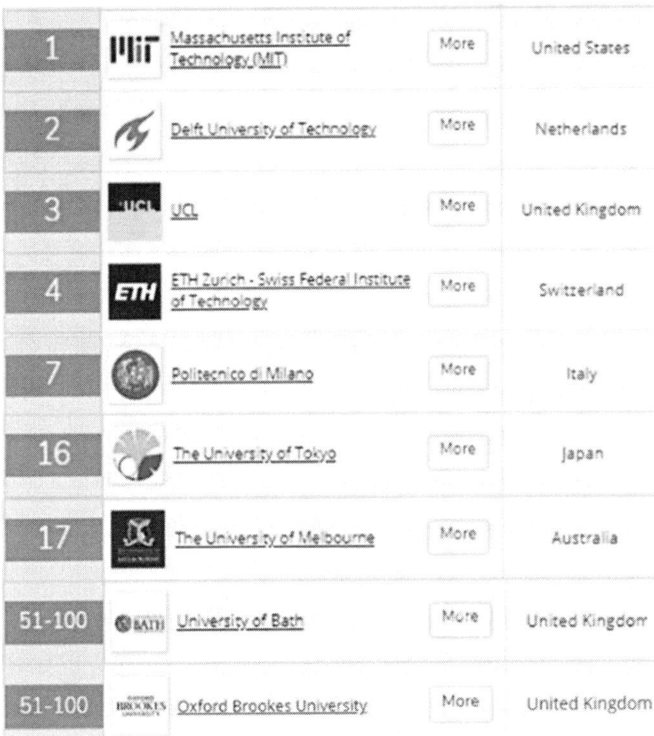

Fig. 6 *The locations of the selected schools in the world ranking* [28]

When the undergraduate education periods of the schools examined abroad are considered, it is observed that they vary between 3–5 years (Tab. 2). This variable case in undergraduate education periods also affects the existing curricula of schools. This variability in the duration of education in the schools examined may create variations on the grade in which the schools started BIM methodology education or on the trainings given according to the classes.

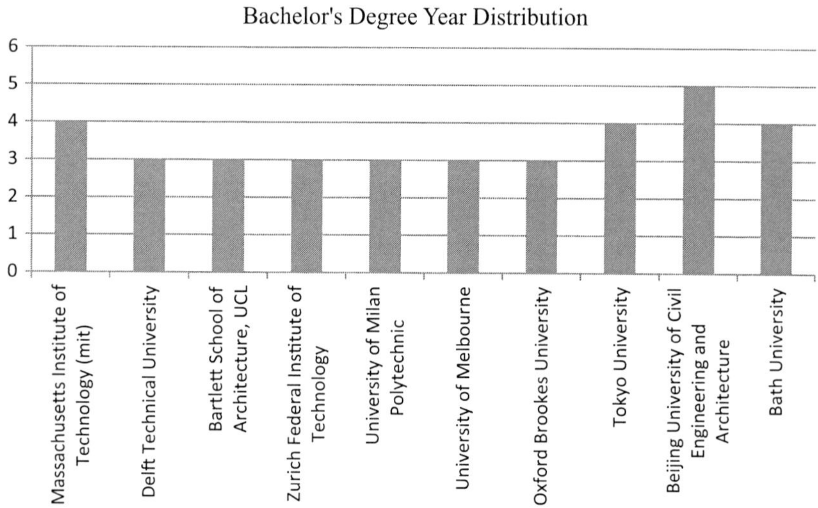

Tab. 2 *Table of the year of undergraduate education of the architectural schools examined.*

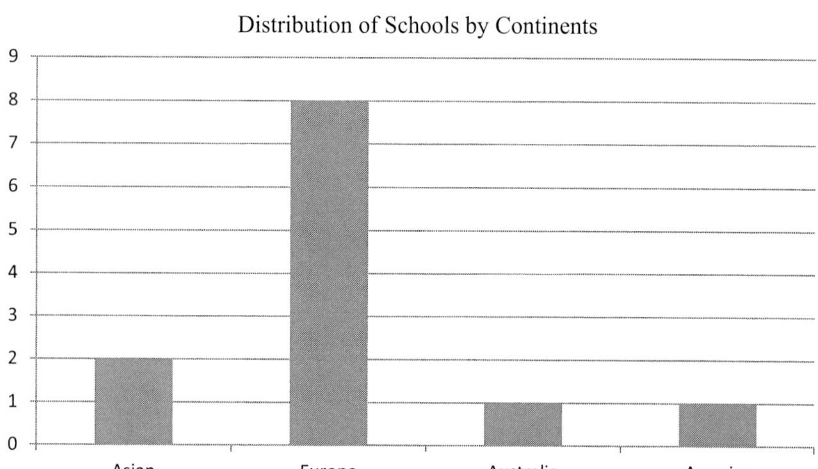

Tab. 3 *Distribution table according to the continents where the architectural schools examined are located.*

They have been accredited by obtaining the necessary approvals from the accreditation institutions of all the architectural schools examined in their countries. In addition, the architectural schools examined were selected from among the successful schools according to the education index rankings made in a way that covers the architectural schools in their own continents and also all over the world.

When the current curricula of the architectural schools examined in Europe, Asia, Australia and the Americas were examined, some similarities were observed on the educational models developed by the schools in connection with the BIM methodology.

When the current curriculum of 1st grade architecture schools examined is examined, it is seen that there are courses covering basic information related to BIM methodology during this period. Most of the architectural schools examined have compulsory or elective courses in 1st grade related to BIM methodology (Tab. 4).

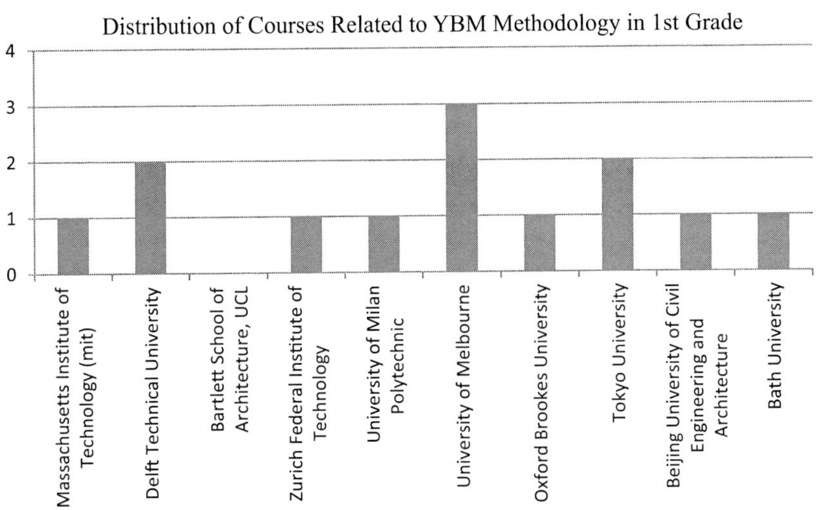

Tab. 4 *Distribution table of courses related to BIM methodology in 1st grade according to the current curricula of the architectural schools examined.*

According to the curricula available in the architectural schools examined, it is seen that there are courses related to BIM methodology, most of which are at least 1 in the 1st year education programs (Tab. 4). The scope of these courses is prepared by considering:

- Theoretical information on BIM methodology
- Basic training on BIM-based software

The courses in this class or the elective courses recommended to be taken in this class serve as a preparatory training for the future trainings and courses on BIM methodology.

In the years after the 1st grade, the courses given depending on the undergraduate education periods of the schools were organized as courses directly or indirectly related to the BIM methodology. In line with the preparatory courses taken in the 1st grade;

- Studio studies using BIM methodology,
- Interdisciplinary studio studies
- Analysis trainings and applications with BIM-based software,
- Basic training sessions on different disciplines in line with the T-type education approach
- Elective courses are held on various topics within the scope of BIM methodology.

Looking at the curricula available in the architectural schools examined, courses related to BIM methodology education were mostly not organized as compulsory courses. Therefore, BIM, which is arranged as an elective course, is available in courses on education. This enables students to work freely with different methodologies in their current curricula (Tab. 5).

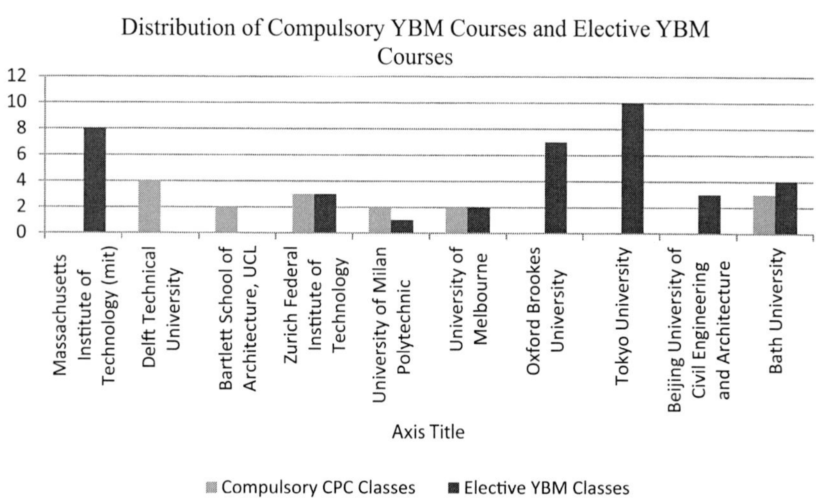

Tab. 5 *Distribution table of compulsory courses and elective courses related to BIM methodology in the current curricula of the architectural schools examined.*

BIM Education Status in Architecture Schools in Turkey

In the curricula of the architectural schools to be examined in Turkey, courses related to BIM methodology will be examined. In the process of determining the architectural schools to be examined in Turkey, the accredited status of the architectural schools to be examined has been taken into account by the "Architectural Accreditation Board (MIAK)" (Fig. 7).

MiAK		2020	
		Accredi-tation	Accreditation Period
	YILDIZ TECHNICAL UNIVERSITY	6 Years	28 June 2017 - 28 June 2023
	EASTERN MEDİTERRANEAN UNIVERSITY	6 Years	9 July 2018 - 9 July 2024
	ATILIM UNIVERSITY	2 Years (Managed)	13 March 2020 - 13 March 2022
	FATIH SULTAN MEHMET VAKIFUNIVERSITY	6 Years (Conditional)	13 March 2020 - 13 March 2026
	BAHCESEHIR UNIVERSITY	3 Years	13 March 2020 - 13 March 2023
	NEAR EAST UNIVERSITY	2 Years (Managed)	13 March 2020 - 13 March 2022
	YEDİTEPE UNIVERSITY	2 Years (Managed)	13 March 2020 - 13 March 2022
	EUROPEAN UNIVERSITY OF LEFKE	3 Years	18 June 2018 - 18 June 2021
	GEBZE TECHNICAL UNIVERSITY	3 Years	9 July 2018 - 9 July 2021
	MIMAR SINAN FINE ARTS UNIVERSITY	6 Years	12 Feb. 2014 - 31 Dec. 2020
	ISTANBUL KÜLTÜR UNIVERSITY	6 Years (Conditional)	22 April 2014 - 31 Dec. 2020
	ESKISEHIR TECHNICAL UNIVERSITY	6 Years (Conditional)	11 June 2014 - 31 Dec. 2020
	CYPRUS INTERNATIONAL UNIVERSITY	6 Years (Conditional)	11 June 2014 - 31 Dec. 2020

Fig. 7 *MIAK accreditation board accredited architecture undergraduate programs* [29]

When evaluated according to the current list, within this section;

- Yıldız Technical University
- Mimar Sinan Fine Arts University
- Eskişehir Technical University

shall be examined.

1.2.1 Yıldız Technical University

Yıldız Technical University was founded in 1911 under the name of Mekteb-i Âlisi in Istanbul. In 1942–1943, an architecture department was opened as Istanbul Technical School (ITO) within Yıldız Technical University, one of the most established universities in Turkey. Today, the Faculty of Architecture, Department of Architecture within the university continues their undergraduate, master and doctorate education. There are 2 undergraduate programs as English and Turkish within the department, but the architecture department with a 4 year undergraduate education period is also accredited by MIAK. [30]

In the curriculum of the Department of Architecture, "Architecture in Computer Environment" course is available with the course code "MIM2081", which is one of the compulsory courses in the first semester of the second year. Within the scope of this course, it is aimed to learn basic information about CAD software, BIM methodology and BIM-based programs, which are computer aided design tools of the students. Together with this course, students will be able to use 3D drawing tools and will be able to obtain basic information about 3D printing technologies with 3D design tools. [31]

One of the compulsory courses in the undergraduate curriculum of the Department of Architecture is "Basic Computer Science" course with the course code "MIM2082" in the 1st semester of the 2nd year. Within the scope of this course, various trainings and practices are designed on 3D modelling programs as well as various programs and techniques about visualization. In addition, there are training sessions on the "Rhinoceros" program that can work with visual programming languages in the course content. [32]

"Integrated Design in Building Information Modelling" course is available with the course code "MIM4510", which is one of the elective courses in the undergraduate curriculum of the Department of Architecture. This course is also conducted in conjunction with the BIM methodology. [33]

The course code "MIM4111", which is one of the elective courses of the undergraduate curriculum in the Department of Architecture, and the basic information about the BIM methodology are given within the scope of the course "Building Information Modelling". In addition to theoretical training on methodology, basic training on BIM-based software is also provided. [34]

Within the scope of "MIM4121" course code and "Computational Design" course, which is one of the elective courses in the undergraduate curriculum of the Department of Architecture, trainings are given about 3D modelling trainings and computational designs related to visual programming languages. [35]

1.2.2 Mimar Sinan Fine Arts University

Mimar Sinan Fine Arts University was founded in 1882 in Istanbul under the name of "Sanayi-i Nefise Mektebi". Mimar Sinan Fine Arts University Mimar Department of Architecture, one of the oldest and long-established architectural education institutions in Turkey, currently provides education in undergraduate, graduate and doctorate fields. Architecture department undergraduate education period is 4 years and accredited by MIAK. [36]

Within the undergraduate curriculum of the Department of Architecture, "Computer-Aided Modelling" course is available with the "Mim239" course code, which can be taken electively in the fall semester of the 2nd grade. 3D modelling trainings are available in connection with computer-aided modelling within the scope of the course. [36]

There are "Mim232" course code, "Computer Aided Drawing and Design" course and "Mim238" course code and "3D Modelling and Architectural Animation" courses in the 2nd grade spring semester of the Architecture Department. Within the scope of these two courses, basic training is provided about BIM-based software. [36]

The basic information on BIM methodology is given within the scope of the "Building Information Modelling" course coded "Mim357", which is one of the elective courses that can be taken in the 3rd grade fall semester of the architecture department. [36]

1.2.3 Eskişehir Technical University

Eskişehir Technical University was established in Eskişehir by leaving Anadolu University in 2018. The Department of Architecture within the Faculty of Architecture and Design within Eskişehir Technical University provides undergraduate, master and doctorate education. Architecture department undergraduate education period is 4 years and accredited by MIAK. [37]

When the undergraduate curriculum of the Department of Architecture is examined, "Computer Applications in Architecture - I" course with "MİM261" course code is available from elective courses. Within the scope of the course, basic training is provided about the "AutoCAD" program, which is one of the 2-dimensional drawing programs. [38]

Basic training sessions are given about 2D and 3D modelling programs within the scope of "Computer Applications in Architecture - II" course with the "MİM262" course code, which is one of the elective courses of the Department of Architecture. At the same time, "Computer Applications in Architecture - III" course with the course code "MİM263" is available as a continuation of this course. Within the scope of this course, 2-dimensional drawing and 3-dimensional modelling programs are provided. ([39] and [40])

When the current curriculum structure of Eskişehir Technical University architecture department is examined, it is understood that there are no direct courses about BIM methodology education. However, in the current curriculum, it is seen that there are courses in accordance with the T-type education understanding in connection with BIM methodology education. With the course code "MİM335" in the 2nd semester of the 3rd grade, general information about mechanical and installation projects is provided within the scope of the "Building Installation" course and basic information about mechanical and installation projects is provided to the students. [41]

The "MİM214" course code in the current curriculum and the "Structure Analysis-I" course provide the static calculations and the basic information in the projects. This course consists of 3 different courses as a series of courses in the curriculum and is first given in the 1st semester of the 2nd grade. Students receive basic trainings about static projects and basic trainings about different but connected disciplines. [42]

It is based on examining the relations between architecture and different but related disciplines within the scope of the "MÜH302" course code and "Interdisciplinary Practices" course, which is one of the elective courses in the curriculum of the architecture department. [43]

Conclusion

In this study, it was emphasized that the concept of BIM should be understood as a methodology. The concept of BIM, which should be evaluated as a methodology, was evaluated in a broad scope throughout the study and it was aimed to explain that it is a comprehensive methodology. At the same time, the BIM methodology should not be limited to architectural discipline only. It was also explained that this methodology can be used by conducting interdisciplinary studies. Together with these studies for BIM methodology, this study has also been conducted on the place of BIM methodology in today's architectural education curricula.

Before examining the training curricula on the BIM methodology, the understanding of t-type education in accordance with the BIM methodology was emphasized. Because before changing the education curricula of schools in accordance with the BIM methodology, it is necessary to change the understanding of education first. The concept of interdisciplinary work is a critical concept for BIM methodology. BIM methodology is based on the concepts of cooperation and communication. For this methodology put forward in the light of these concepts, the understanding of I-type education, which is still ongoing in today's schools, will be insufficient. However, with the trainings to be given by adopting the T-type education approach in today's schools, individuals who are closer to and able to communicate with individuals in different disciplines will be more successful than I-type individuals in the BIM methodology. Therefore, in this thesis study, the understanding of T-type education was examined before the architectural curricula were examined.

In the study conducted, the position of BIM methodology in architectural education was also discussed. Today's current literature reviews have been conducted and many opinions on this subject have been included and at the same time BIM methodology training programs on different schools have been examined.

With this study, BIM education in different universities in many continents such as Europe, Asia and America are considered as a methodology education. At the same time, it is seen that education is provided with an understanding of interdisciplinary cooperation. However, it is seen that the BIM training in our country is carried out only theoretically through BIM software. However, it is seen that there is no comprehensive training by evaluating the concept of BIM as a methodology.

The use of BIM based on BIM methodology, which is rapidly becoming widespread in the world and at the same time based on certain standards, is also seen to be included in the curricula of education in today's architecture schools. However, this is not the case in architectural schools in our country. Despite this methodology, which is planned to be completely passed in the future, architects and engineers in our country should be able to work in this methodology. However, in order to work efficiently with BIM methodology, it is necessary to work together with the concepts brought by this methodology. Today, this working environment can only be in architecture schools. For this reason, architectural schools in our country should also move to BIM methodology education together with the understanding of interdisciplinary work.

References

1. O. Ozener, "Studio education for integrated practice using building information modeling," Texas A&M University, 2009.
2. A. Strafaci, "What does BIM mean for civil engineers?," *CENews.*, 2008. http://216.119.86.79/whitepapers/civil/What Does BIM Mean for Civil Engineers.pdf (accessed Mar. 29, 2020).
3. S. Ofluoğlu, "Yeni Nesil Mimari Yazılımlar," 2009. http://www.sayisalmimar.com/yayin/ybm.pdf (accessed Mar. 29, 2020).
4. J. Przybyla, "The Next Frontier for BIM: Interoperability with GIS," *J. Build. Inf. Model.*, vol. Fall, no. 1, pp. 14–18, 2010.
5. M. Mathews, "BIM collaboration in student architectural technologist learning," *J. Eng. Des. Technol.*, vol. 11, no. 2, pp. 190–206, 2013.
6. V. Quirk, "A Brief History of BIM," 2012. https://www.archdaily.com/302490/a-brief-history-of-bimhttps://www.archdaily.com/302490/a-brief-history-of-bim (accessed Mar. 29, 2020).
7. B. Martens and H. Peter, *ArchiCAD Best Practice: The Virtual Building™ Revealed*, Second. Springer Science + Business Media, 2007.
8. İ. Kıvıcık, "An investigationinto the building information modeling applications in the construction project management," Istanbul Technical University, 2016.
9. S. Azhar, M. Khalfan, and T. Maqsood, "Building information modeling (BIM): Now and beyond," *Australas. J. Constr. Econ. Build.*, vol. 12, no. 4, pp. 15–28, 2012, doi: 10.5130/ajceb.v12i4.3032.
10. T. Akkoyunlu, "Kentsel dönüşüm projeleri için BIM uygulama planı önerisi," İstanbul Teknik Üniversitesi, 2015.

11. B. Ilhan and H. Yaman, "BIM and Sustainable Construction Integration: An IFC-Based Model," *MEGARON*, vol. 03, no. 10, pp. 440–448, 2015.
12. A. Lassen and C. Merschbrock, "Investigating 'Green BIM' in a Norwegian construction project: an institutional theory perspective," *Build. Inf. Model. Des. Constr. Oper.*, vol. 149, pp. 519–530, 2015.
13. R. C. Hill and P. A. Bowen, "Sustainable construction: Principles and a framework for attainment," *Constr. Manag. Econ.*, vol. 15, no. 3, pp. 223–239, 1997.
14. I. Motawa and K. Carter, "Sustainable BIM-based Evaluation of Buildings," *Procedia - Soc. Behav. Sci.*, vol. 74, pp. 419–428, 2013.
15. A. Schlueter and F. Thesseling, "Building information model based energy/exergy performance assessment in early design stages," *Autom. Constr.*, vol. 18, no. 2, pp. 153–163, 2009.
16. B. Tosun Atay, "YBM Standartlarının Karşılaştırılmalı İncelenmesi Ve Ulusal Standardizasyon Çalışmaları Açısından Değerlendirilmesi," İstanbul Teknik Üniversitesi, 2019.
17. P. Lillrank, "The quality of standard, routine and nonroutine processes," *Organ. Stud.*, vol. 24, no. 2, pp. 215–233, 2003, doi: 10.1177/0170840603024002344.
18. M. Hooper, "BIM Anatomy: An investigation into implementation prerequisites," p. 287, 2012, [Online]. Available: http://lup.lub.lu.se/luur/download?func=downloadFile&recordOId=2972126&fileOId=2972151.
19. J. Wong and J. Yang, "Research and application of Building Information Modelling (BIM) in the Architecture, Engineering and Construction (AEC) industry: a review and direction for future research," in *6th International Conference on Innovation in Architecture, Engineering & Construction (AEC)*, 2010, pp. 356–365.
20. N. Lee and D. a Hollar, "Probing BIM Education in Construction Engineering and Management Programs Using Industry Perceptions," in *49th ASC Annual International Conference Proceedings*, 2013, pp. 1–8.
21. S. Barile, M. Saviano, and C. Simone, "Service economy, knowledge, and the need for T-shaped innovators," *World Wide Web*, vol. 18, no. 4, pp. 1177–1197, 2015.
22. G. Alkawı, "Mimarlık Eğitiminde BIM Tabanlı Disiplinlerarası İşbirliği Önerisi," Mimar Sinan Güzel Sanatlar Üniversitesi, 2016.
23. J. Gregory and L. Crispin, *More agile testing: Learning journeys for the whole team.* Crawfordsville: Addison-Wsley, 2015.
24. S. Önür, "IDS for ideas in higher education reform," in *1st International Conference on Improving Construction and Use Through Integrated Design Solutions*, 2009, no. 259, pp. 52–71.

25. M. A. Ambrose, *BIM and integrated practice as provocateurs of design education.* Nanjing: Southeast University Press Publication, 2007.
26. R. Cheng, “Suggestions for an integrative education. In Report on integrated practice,” Minnesota, 2006.
27. M. J. Clayton, “Replacing the 1950 ’ s Curriculum,” in *Proceedings of the 25th Annual Conference of the Association for Computer-Aided Design in Architecture*, 2006, pp. 48–52.
28. https://www.topuniversities.com/university-rankings/university-subject-rankings/2020/architecture (accessed July 14, 2020).
29. https://architecture.mit.edu/overview/graduate-degrees (accessed July 14, 2020).
30. http://www.mmr.yildiz.edu.tr/mmr/1/Tarih%C3%A7e/16 (accessed July 26, 2020).
31. http://bologna.yildiz.edu.tr/index.php?r=course/view&id=8280&aid=38 (accessed July 26, 2020).
32. http://bologna.yildiz.edu.tr/index.php?r=course/view&id=9660&aid=38 (accessed July 26, 2020).
33. http://bologna.yildiz.edu.tr/index.php?r=course/view&id=9174&aid=38 (accessed July 26, 2020).
34. http://bologna.yildiz.edu.tr/index.php?r=course/view&id=9174&aid=38 (accessed July 26, 2020).
35. http://bologna.yildiz.edu.tr/index.php?r=course/view&id=6681&aid=38 (accessed July 26, 2020).
36. https://www.msgsu.edu.tr/faculties/mimarlik-fakultesi/mimarlik-bolumu (accessed July 26, 2020).
37. https://www.eskisehir.edu.tr/akademik/fakulteler/262/mimarlik-ve-tasarim-fakultesi/genel-bilgi (accessed July 26, 2020).
38. https://www.eskisehir.edu.tr/akademik/fakulteler/ders/141447/bilgisayar-destekli-tasarim-i/ders-ogrenme-ciktilari (accessed July 26, 2020).
39. https://www.eskisehir.edu.tr/akademik/fakulteler/ders/141448/bilgisayar-destekli-tasarim-ii/ders-ogrenme-ciktilari (accessed July 26, 2020).
40. https://www.eskisehir.edu.tr/akademik/fakulteler/ders/141449/bilgisayar-destekli-tasarim-iii/ders-ogrenme-ciktilari (accessed July 26, 2020).
41. https://www.eskisehir.edu.tr/akademik/fakulteler/ders/144455/yapi-tesisati/ders-tanitim (accessed July 26, 2020).
42. https://www.eskisehir.edu.tr/akademik/fakulteler/ders/144401/struktur-analizi-i/ders-tanitim (accessed July 26, 2020).
43. https://www.eskisehir.edu.tr/akademik/fakulteler/ders/144462/disiplinlerarasi-uygulamalar/ders-tanitim (accessed July 26, 2020).

Ayşe ÖZYETGİN ALTUN

Six Key Issues for Resilient Community Assessment in the Context of Urban Planning Against Disaster Risk

Abstract: Disaster risk reduction and management is the management, organization and planning business that calculates the probabilities of disasters, aims to prevent damage and/or reduce harm, and aims to restore society, economy, and physical structure to order after the disaster. In the last two decades, especially with the Hyogo Framework Action Plan of 2005, the concept of resilience has brought significant openings to disaster risk management and has been internationally accepted.

The concept of resilience refers to the capacity of systems that they develop in order not to lose their key components against regular, irregular changes and stress factors, shocks, and surprises. The concept includes the meanings of defense, adaptation, and survival with transformation and change. There are two complementary approaches to disaster risk reduction and management that develop with the idea of resilience. The first is to provide disaster risk management for resilient communities and cities. The second is building resilient communities and cities for disaster risk management. The main goal is for societies and cities to sustain their existence by improving their existing social, economic, and spatial values.

However, these values are vulnerable to natural threats, which are uncertain when and how severe they will occur due to some deficiencies, incorrect evaluations, or immaturity. This state of openness or situations that increase the severity of being affected by threats are called vulnerabilities. In addition to normalization of these vulnerabilities, resilience is evaluated according to the institutional, spatial, and technological capacity that can be developed according to the disaster issue.

Various indicators and index studies have been developed to evaluate resilience. These studies can be studied in different details by separating resilience with spatial sampling according to country, region, city, and neighborhood scales. However, these different approaches are evaluated mainly under four basic headings: social dynamics, physical environmental conditions, economic dynamics, and institutional structuring. It is important to make these assessments to improve institutional capacity and to ensure balanced transfer of resources according to deficiencies.

Urban and regional planning are documents that produce solutions in terms of maintaining balance in land use, guiding construction, developing balanced sectoral distribution, and managing infrastructure in this context. They are guidelines on the social-economic-ecological sustainability of corporate structuring and spatial arrangements that take into account justice. Including community in conjunction with participatory processes in the production process of plan decisions; It prevents the plan decisions from being an external pressure factor on the society and provides the return of internalization and sense of ownership of the living environment. This situation is considered an important input in disaster risk reduction and management. In summary, in the process of disaster risk reduction and management, city plans are guidelines describing spatial development, also it is a means of institutional structuring involved in the flow of information and communication that establishes the community-space connection.

In this context, urban planning during disaster risk reducing and management for resilient communities should not only focus on reducing harm before the disaster, but also include the goal of returning life to normal in its safest form after the disaster. It should also include spatial, economic, and ecological resilience and social resilience as a whole. City plans produced with this vision should also be able to make a comprehensive assessment of social resilience. The purpose of this study is to evaluate which issues, how social resilience assessment can be addressed within the scope of urban planning and how it can be integrated into the urban planning process and implementation decisions. Six main issues have been identified for evaluation. How these issues can be included in the urban planning process and planning decisions has been discussed.

Keywords: Social resilience, Natural Disasters, Urban Planning, Resilient Community Indicators, Social and Economic vulnerability indicators.

1 Introduction

The concept of resilience means the capacity of systems to avoid losing key components to regular, irregular change and stress factors, shocks, and surprises. UNISDR (2009) defines the concept of resilience to include the meanings of being able to stand up to, endure, absorb the effect, adapt to the effect, recover as a result of impact, and recover from the negative effects and changes.

Jha et 2013 (2013, pg. 22.) defines social resilience as follows,

> "Social resilience is the capacity of a community or society to cope with, and adapt to, disturbances or changes. It includes the ability of communities and society as a whole to absorb disturbances, self-organize to adjust to current and new stresses, and build and increase capacity for learning and adaptation. A resilient community is able to respond positively to changes or stress and despite them maintain its core functions as a community. Policy makers should strive to create an environment that enables communities to participate and make decisions. This increases collective resilience and facilitates the work of public authorities in both creating and disseminating risk information." [2]

Cooperation with society for resilient community is considered the key input of disaster risk management [3]. In the second priority action defined in the HFA 2005–2015 Framework Action Plan, the presence of knowledge on disaster risks and their transfer to society has been identified as an important input for the construction of resilient communities [4]. At the same time, local knowledge of the society provides important inputs in assessing and monitoring risk in disaster risk assessment management [2].

The phenomenon of social resilience is discussed in two main areas in the literature [5, 6]. In these studies, it is important to understand issues such as social networks, solidarity; ecosystem services and the dependence of individuals, societies on resources, managing them and monitoring systems; culture, memory; production, and transfer of social information; natural disasters, climate change [5,6,7,8]. The second is social resilience studies defined in psychology, health, and social sciences research. In the studies in this context, child, adolescent psychology; family life and health, social solidarity and relationships; disaster management and recovery; rural area; cultural issues come to the forefront [6]. In fact, it is seen that there is a large intersection in the context of the subjects on which the two main areas work. However, the difference between them is that the studies in psychology, health and society give importance to individual outcomes; the socio-ecological system branch focuses on the system and systemic dynamics [5].

Maclean et al. (2014; pg.151–152) adopted both approaches and viewed social resilience as the capacity of societies to be empowered, which defines the possibilities of adapting, transforming, and empowering in the face of economic or environmental problems. In this context, it is stated that social resilience can be developed with six components.

- "Knowledge, skills and learning,
- Community networks,
- People-place connections,
- Community infrastructure,
- Diverse and innovative economy,
- Engaged governance", [6].

The first four components are about society's self-containedness. The existence of social knowledge and memory specific to local conditions is important. Individuals should be able to recognize, care about the natural and built environment they live in and know what kind of dangers exist. It is important for individuals to be able to recognize their social environment due to the great need for local knowledge for the right response at the time of disaster. The main task of individuals in disaster risk management is to know what to do in the event of a disaster, to know what can happen in the reconstruction process afterwards and to be self-organized. This information and equipment can be accessed by individual efforts willingly or created by training programs developed by local governments and non-governmental organizations. Of course, there is also an economic dimension to this process, individuals or institutions should be able to take precautions and measures in advance for risk sharing and transfer. The participation of the community in these systems is carried out in some cases with incentives and in some cases by sanctions. Therefore, it is important to develop the capacity of communities to learn, take precautions and organize systematically for risk reduction. Tran et al. 2009 explains that in order to develop this capacity, the subgroups that make up the social structure can be implemented in three different ways, by themselves, by receiving some external support, and by institutions working only on subjects that require expertise [9]. Institutional capacity has to define the sensitivities of the community during the harm reduction and preparation stages of disaster management and take measures to improve these sensitivities. At the beginning of these measures, sub-social groups are identified and education on disasters comes to the forefront. Accordingly, the city planning dimension directs applications as a specialty and plays a role in communication and information flow between the different sub-groups that make up the social structure.

In the historical process, societies have improved with the emergence of various economic sectors (such as agriculture, trade, industry, service). The issues that stand out in our age are the sectors that prioritize the production and sharing of information such as information, communication and technology. Not only is there no profit from working and producing in these sectors, but investment and money flow to these sectors are also gaining importance. Societies with advanced socio-economic development are moving faster on these issues. Societies with medium, lower and low levels of development are included as the market of this system. However, being able to exist in various economic sectors supports the maintenance of alternative systems in natural disasters. Various creative economies that enable the development of local capacity and support contemporary conditions are considered important as they will give birth to alternatives in the disruptions to today's economic policy.

It is seen that the positions of resilient societies in the flow of information in the risk management process are emphasized with great emphasis on participation issues. The HFA 2005–2015 framework action program highlights social participation in the opening of primary action. It is stated that the participation of the community can be encouraged in disaster risk management by ensuring;

- Adaptation to the featured policies of society,
- Supporting relationship networks,
- Strategic management of voluntary resources,
- Loading duties and responsibilities to individuals,
- Selection of important human resources and officials according to the nomination system [4].

While the development of resilient societies is an important tool in reduction and management disaster; disaster risk reduction and management is an important input in order to ensure resilient societies [10]. In this context, the production process of urban planning decisions and plan decisions plays an important role in the construction of community resilience. In this context, city planning, which will play an effective role, should be able to evaluate spatial, economic and ecological damages as well as social and economic damages and resilience and include them in planning studies. In the meantime, the urban planning process, which can also be a means of communication with the community, can be an important tool if the community is informed about disasters, increases awareness and communicates with the community about the development of individual measures.

2 Natural Disasters, Resilient Society and Urban Planning

The phenomenon of sustainability, which has been accepted since the 1990s, describes socially, economically and environmentally balanced development. In this process, it was understood that urban planning should play an egalitarian, fair and urban ecosystem-protecting role in ensuring the order of communities and settlements [11, 12, 13]. However, the production of urban plans, due to the socio-economic order globally, cannot realize the ideal of producing egalitarian, fair, healthy and safe living spaces for each person of society.

In particular, disaster case results show the ineffectiveness of urban planning in this area. The poor, uneducated and unequipmented population living in unhealthy and unsafe living areas is the population most affected by disaster damage [14, 15]. Shaw et al. (2009 and 2016), Shing (2015) emphasizes the importance of this issue, especially in developing countries [16, 3,1 7]. Urban planning in developing countries follows urban growth rather than leading urban development [2]. Ranke (2016), Pelling (2003), Shaw et al. (2009, 2016) indicates that urban planning plays a big role in increased damage to cities; they use definitions such as non-strategic urban planning, inactive urban planning, poor urban planning, etc. Cities with high social, economic, ecological and spatial vulnerabilities are losing their stability and order in the face of threats that are an element of oppression. When urban planning lags behind urban growth, it loses its ability to be supportive and regulatory to the urban system.

Vulnerability means aspects of systems, societies or assets that increase or facilitate the likelihood of being adversely affected by threats [1]. In other words, the deficiencies of the system when it comes to developing a defense mechanism against the threat are called vulnerability [1]. Vulnerability in the occurrence of disaster risk is called internal factors and has a size directly proportional to the risk phenomenon, hazard and vulnerability dimensions.

Cardona (2007) states that damageability in disaster zones is in depth associated with social processes, and states that society is associated with issues of vulnerability, sensitivity and lack of resilience in the face of various threats [18]. Strong institutional structure and reduced damageability are needed in return for disaster damages is also stated. Similarly, Ranke (2016; pg. 267) states that the following questions should be answered in assessing the vulnerabilities of communities.

- o "Who are the vulnerable groups?
- o What factors make them vulnerable?
- o What are the social and economic outcomes of the disaster exposure?
- o What are the needs of the vulnerable groups to increase their level of resilience?
- o What are the capacities and abilities the vulnerable groups need increase their level of resilience?"

The aim of the community resilience approach is to define, design and realize social, economic and spatial mechanisms that will reduce these negative effects or provide the fastest recovery after the impact. Pelling (2003) states that the idea of resilience defines proactive stance in the face of disaster risk. Proactive stance means developing measures against effects in advance, being prepared for the effect and being able to respond to the effect. Proactive stance against natural disasters can be achieved by improving the capacities of societies and administrations in this direction.

Which responsibilities are assumed by urban planning in this proactive stance? First of all, urban planning should be able to focus on both social, socio-economic and socio-ecological systemic resilience and individual resilience specified in Berkes (2013). Evaluation criteria should be clear according to the scale studied. The characteristics of each country and local unit are different, these characteristics should be able to distinguish. For example, the complexity level and social diversity of a metropolis with a population of ten million and the complexity level of a small-scale city with a population of hundreds of thousand people are very different.

The contribution of city plans in disaster risk management is mainly prominent in three issues. It is considered an important component and a guide for reducing spatial damageability [2,4,19,20,21,23]; it is also considered an important tool that plays a role in communication and information flow both socially and administratively [3, 14, 15, 16, 17, 20, 21, 22, 23].

3 Indicators for Resilient Community Assessment

Different indicators and indices are used in disaster risk management for the assessment of vulnerability, threat and resilience issues and monitoring the system, management and practices. It is indicated that the indicators should be discussed in the details required on national, regional and local scales [15, 18, 24]. In addition, indicators vary according to the type of disaster, the social, economic and environmental characteristics of the settlement.

The production of information on threats and vulnerability within the scope of disaster risk management and the sharing of this information with the at-risk population, decision makers and society are an important element. In the second action of HFA 2005–2015, it is emphasized that indicator systems should be developed at national and local scale in order to accelerate and facilitate the production and sharing of this information. [4].

As an example, it is seen that the choice of indicators varies mainly according to geography and cultural structure. For example, the indicators indicated in the Cutter et al. (2010) study were defined according to the flood and storm cases, which are one of the main threats of the country; while in the case of Turkey, Kundak and Türkoglu (2007), Özceylan (2010), Taşkın (2012), İstabul Metropolitan Municipality (2018) vulnerability indicator studies were defined in related to the earthquake phenomenon [25, 26, 27, 28]. However, in the indicators defined, the experiences of countries related to that phenomenon and the structures they have developed against disaster risk over time (plans, associations, insurances, etc.) are among the indicators of resilience.

Macro scale indices and indicators used provide the opportunity to evaluate in countries, regions and urban areas (Tab. 1). In addition, these indices evaluate past experiences. For example, the system of disaster risk management indicators developed by Cardona (2007) offers an assessment method including disaster economic loss index (DDI), local disaster index (LDI), prevalent vulnerability index (PVI), disaster risk management index (RMI) headings [18]. Yaprak (2020) evaluated disaster exposure and resilience through the province by interpreting the values varying by year with the four-part index presented in Cardona (2007) for Turkey. With this model, a general opinion evaluating social and economic resilience to disasters is developed through the province, it can be interpreted what is missing and the necessary investments can be directed [29].

Microscale indices consist of more detailed indicators for evaluation in local, urban and neighborhood units (Table 1). The indicators used in these studies differ as developed according to local qualities. Indicators of social resilience on a micro scale are considered to include cultural and economic components of society such as employment, poverty, education level, employment of women, proportion of the elderly and disabled, health facility capacity and usage opportunities [16, 3, 19, 18, 24].

Tab. 1: Indicators that increase and reduce resilience.

	Global-Country -Region-Macro	**Local-City-Neighborhood-Mikro (Turkey Example)**
Demographic	Population growth [18, 26]	Population Density [25, 26, 27]
	Population density [18, 26]	
	Dependent population ratio (<5, <12, <15, 65>) [24, 26,30]	Population growth rate [27]
	Immigrant population [24]	Dependent population age and gender [25, 26, 27, 28, 30]
	Birth rate [26]	
	Average household size [26]	Household size [8, 27]
	Female population [26]	
Socio-Cultural	Percentage of population unable to be disabled [24, 26]	Chronic disease condition [28]
	Percentage of population covered by health insurance [24, 30]	Disability [28, 30]
		General Health insurance ownership [27]
	Percentage of population not speaking the local language [24]	Literacy status and level of education [27, 28]
	Place attachment [24]	Being trained in earthquake preparedness [28]
	Voluntary participation in non-governmental organizations [24, 26]	Speaking the local language [28]
	Having a previous disaster experience [24]	Duration of residence in the city [28]
	Percentage of earthquake insurance ownership [26]	Membership in non-governmental organizations [8]
	Percentage of population trained in disaster preparedness [24]	Having solidarity networks [8, 28]
	Social capital [24]	Risk perception and values (such as fatalism, not taking precautions, predisposition not to take responsibility, etc.) [28]

Tab. 1: Continued

	Global-Country -Region-Macro	Local-City-Neighborhood-Mikro (Turkey Example)
Socio-Economic	Poverty index [18]	Guarantee, insurance provided by the profession of owned [8, 28]
	Unemployment rate in the workforce [18,30]	Housing ownership [8, 28, 30]
	GINI coefficient [24]	Secondary housing ownership [28]
	Employment percentage [24]	Total income and indebtedness of the household [28]
	Percentage of female labor force participation [24]	Motor vehicle ownership [28]
	Housing ownership rate [24]	Access status to tools that meet the need for communication (internet, telephone, television, etc.) [28]
	Low national income [30]	Basic white goods ownership [28]
Access to Urban Services	Number of hospital beds per 100,000 [24, 26]	Access to a health institution [28]
	Number of doctors per 10,000 people [24, 26]	Access to open, semi-open space
	Emergency shelter alternatives [24, 26]	Access to the main transport link
	Urban growth rate [18]	Number of medical facilities in the neighborhood [25]
		Hospital bed capacity in the neighborhood [25]

4 Resilient Community Assessment within the Scope of Urban Planning

Physical, economic, and institutional vulnerabilities can react differently to the type of natural threat, vary, but regardless of the severity of the threat and the vulnerabilities of the physical environment, the vulnerabilities of social dynamics are open to all kinds of threats. Therefore, place-independent social building sensitivities should be integrated into planning studies [21, 23, 30]. In this context, urban planning should cover social resilience with a holistic and synergistic approach as a discipline that attaches importance to understanding both the functioning of the urban system and the well-being of individuals. When indicators evaluating resilience examined six main topics are revealed on how urban planning can include community resilience.

Population density is calculated as the ratio of the number of people living in an area to the residential area. The smaller the area and the more of the living population, the more density there is. Therefore, it is the most basic indicator used to make an overall assessment of people who will be exposed to disaster risk regardless of the type of threat. As population density increases, disaster risk size increases [15, 18, 24, 27]. Population density is used on all scales in the assessment of social resilience, but the methods of calculation vary according to the scale in which it is used. In urban planning, plan decisions are produced by making population size projections based on the population density account calculated through how much area the population size will need. Within the framework of resilient community assessment, population density should be planned as low in areas with high risk of disasters. Regions with high-risk degrees according to the sensitivities of the natural environment should be identified first. Integrated risk maps of the threat type and physical environment should be created, and population density decisions should be made inverse proportion to the degree of risk. At this stage, it should be remembered that the risk reduction decisions of the physical space and the decisions of population density should coincide. What this value will be can of course be determined by the character specific to the location.

The dependent population describes the population whose main topic is low self-rescue ability at the time of any disaster or after a disaster, or who completely need the support of others. Age, gender, chronic diseases, disability issues are included [24, 28,30]. The youngest age of the dependent population has different assessments as 5, 12 or under 15 years of age [24, 25, 28, 30], and over 65 years of age is also counted as dependent population. The high dependent population increases the likelihood of harm. The proportion of dependent population is a subject that can be contained in the resilience assessment to be made on all scales. However, when looking at urban planning decisions, more meaningful solutions can be produced if they are contained on a micro scale. Different needs may need to be met according to the diversity character of the dependent population, and plan decisions may not cover all of this diversity, especially in large cities. However, in order to meet these needs, a general solution can be produced if open areas and medical support units are located within walking and access distances on a neighborhood scale.

Access distances are defined by different standards in the world. In the case of Turkey, the distance of access to educational units or religious facilities and parks defined for 15 years and younger is defined as 500 meters. Therefore, capacity increase can be achieved if the dependent population is identified on a micro scale and solutions such as open space, medical support area and so on can be developed within a distance of up to 500 meters by containing dependent population in the resilience assessment to be developed for urban planning in order to ensure that the dependent population is safer at the time of disaster and after the disaster.

Awareness is examined in order to define the attitudes and behaviors of society in the face of disasters. The level of resilience is considered to have increased as equality and quality at the level of education increased [24, 27, 28]. The community's knowledge of local types of threats, the knowledge of preparatory training increases the capacity for resilience [24]. Balamir (1999) states that the fatalist society approach, which did not take responsibility and precautions, negatively affected resilience [31]. Similarly, in the socio-economic damageability study developed for the 2018 Istanbul Metropolis, Istanbul Metropolitan Municipality evaluated how society perceives risk, from which communication sources they received information, how much responsibility they took for disasters, or which stakeholders they considered responsible. Social groups that do not take responsibility and precautions and do not receive information from scientific sources have been evaluated as more vulnerable groups [28]. Insuring, especially those who have been included in compulsory earthquake insurance in Turkey, is also considered as conscious behavior [26]. Urban planning cannot produce plan decisions in order to increase social awareness about disasters but carrying out planning through participatory processes can ensure that it plays a role in the development of this awareness [9, 21]. In the planning process, it can be considered an important input for local governments to carry out this process in frequent dialogue with the community.

In this context, the social capital indicator should also be addressed. As social capital increases, resilience increases [24]. Indicators collected under the social capital dimension include their relationships with individuals and their large neighbor units and communities [24]. Adger (2004) states that social capital is very difficult to measure, but at the beginning of the development of adaptation capacity against disasters is the strong capacity to participate in social life [32]. It states that the resilience of individuals increases as their relationship with society becomes stronger. The indicators collected under the social capital dimension are summarized under three main headings as social vulnerability, spatial belonging, citizenship participation. It is possible to achieve gains by carrying out participatory and transparent processes and urban planning activities and by adopting this situation socially and administratively.

Communication and social solidarity themes aim to evaluate local social communication and relations. Poor communication and relationship decrease resilience. The population formed by migration reduces resilience [24]. Because this population may not have developed local relationship networks and place attachment [24]. In addition, the lack of local language increases the likelihood of damage due to communication restrictions during and after the disaster [24].

Issues such as social solidarity, family, relatives, membership in associations, membership in political parties also increase resilience [28]. Micro-evaluation of these issues will give more meaningful results. Urban planning can contribute to the social cohesion of the migrating population with participatory processes. In terms of social solidarity, it is important to accept that the type of solidarity developed with local relations is not disturbed by planning studies.

Socio-economic themes measure resilience by evaluate employment, poverty income status, insurance, housing ownership, business size, women's participation in the workforce and dependence variables on singular sector employment [24]. Being employed is among the variables because it provides social and income security. The fact that the economy is based on various sectors increases resilience [24]. However, since dependence on household economy ecosystem units increases individuals' dependence on the ground, if there is no diversity, it reduces resilience. For these reasons, in the context of social resilience assessment, the city plans should be able to produce decisions within the framework of the development of various economic sectors by including socio-economic resilience.

As access to urban infrastructures and equipment, increases the capacity of the community to access transportation, education, communication and health facilities, the capacity to become a disaster-resistant society increases. Economic access is as important as spatial access. In order to have economic access, egalitarian and balanced distribution must be ensured. In order to serve various standards of similar quality, such urban services should be distributed in order to serve various standards, rather than choosing places according to the socio-economic profile of settlement. The threshold distances vary depending on the central or sub-unit of the service unit. Countries have different standards for detecting minimum distances to service units. The walking distance in local administrations in Turkey is considered to be 500 meters. The probability break down of transportation systems at the time of disaster requires diversity in transportation connections

5 Conclusion

Cities experience significant social and economic destruction in the context of various natural threats. It is argued that it is necessary to prevent these destructions and to be able to return to life with as little impact as possible. In this context, the theory of resilience sheds light on the approach of disaster risk reduction and management.

In this context, urban planning supports institutional capacity by building the bridge between spatial and social evaluations. However, a socially and institutionally adopted process is needed. City plans should be able to reflect these assessments on their spatial decisions by containing assessments of social sensitivities and resilience, and they should be able to take part in communicating with the community and ensuring the flow of information by ensuring social participation in the process.

Numerous indicator relationships of various scales have been defined for the evaluation of social resilience. In this research, which tried to determine the social structure indicators that urban planning can establish spatial-social structure relationship within the scope of disaster risk assessment, population density, dependent population, consciousness, communication and solidarity, socio-economic structure, urban infrastructure, and access to equipment were identified as the key issues of resilient community evaluation for urban planning.

Resources

1. UNISDR, *Terminology on Disaster Risk Reduction.* 2009.
2. Jha K. A., Minner, T.W., and Stanton-Geddes Z., *Building Urban Resilience Principles, Tools and Practice.* 2013, The World Bank: USA.
3. Shaw, R., Rahman, A., Surjan, S., Parvin, Ara, G., *Urban Disasters and Resilience in Asia.* 2016, Elsevier: USA.
4. UNISDR, *Hyogo Framework.* 2007.
5. Berkes, F, and Ross, H., *Community Resilience: Toward an Integrated Approach.* Society & Natural Resources, 2013. 26:1, 5–20.
6. Maclean, K., Cuthill, M., and Ross, H., *Six Attributes of Social Resilience.* Journal of Environmental Planning and Management, 2014. 57:1, 144–156.
7. Adger, W. N., *Social and ecological Resilience: are they related?* Progress in human Geography, 2000. 24, 37–364.
8. Özyetgin Altun, A., and Tezer, A., A *Preliminary Study on Defining Urban Resilience for Urban Planning: The Case of Sultanbeyli, Istanbul.* Tasarım Kuram, 2019. 15(28): 77–95
9. Tran, P., Kaneko, F., Shaw, R., Lorna P. V., Hidetomi, O., *Urban Disaster Risk Analysis, Action Planning, And Implementation Management.* Urban Risk Reduction: an Asian Perspective ed: R. Shaw, H. Srinivas, and A. Sharma, 2009, Emerald Group Publishing Limited. Volume 1, 13–36.
10. Zhu, W., Zhao, P., and Wang, Y., *Modeling on the Quantitative Resilience Assessment of Urban Emerging Community Based on Systematics.* Systematics in Proceedings of ACM SIGSPATIAL GIS International Conference on Advances in Geographic Information Systems, 2018. https://doi.org/10.1145/3284103.3284121
11. Silberstein, J., and Maser, C., *Land-Use Planning for Sustainable Development.* 2014, Social-Environmental Sustainability Series CRC Press, Taylor & Francis Group: NW.
12. Birkmann, J., Tetzlaff, G., and Zentel, K. *Addressing the Challenge: Recommendations and Quality Criteria for Linking Disaster Risk Reduction and Adaptation to Climate Change.* 2009, DKKV Publication Series 38: Bonn.
13. Tezer, A. and Türkoğlu, H., *Zarar azaltma ve Şehir Planlama.* Afet Zararlarını Azaltmanın Temel İlkeleri ed. M. Kadıoğlu, and E. Özdamar, 2008. T. C. İçişleri Bakanlığı Japonya Uluslararası İş birliği Türkiye Ofisi: Ankara.
14. Pelling, M., *The Vulnerability of Cities: Natural Disasters and Social Resilience. 2003,* Earthscan: London.
15. Ranke, U., *Natural Disaster Risk Management; Geosciences and Social Responsibility.* 2016, Springer International Publishing: Switzerland.

16. Shaw, R., Rahman, A., Surjan, S., Parvin, A. G. *Urban Disasters and Resilience in Asia.* 2016, Elsevier Butterworth-Heinemann: USA.
17. Shing, R. B., *Urban Development Challenges, Risks and Resilience in Asian Mega Cities.* 2015, Springer: Japan.
18. Cardona, O. *Indicators of Disaster Risk and Risk Management.* 2007, Inter-American Development Bank: Washington, D. C.
19. UNISDR, *Sendai Framework.* 2015.
20. Reyes, M. L., *Risk Sensitive Land Use Planning: Towards Reduced Seismic Disaster Vulnerability, The Case of Marikina City, Metro Manila, Philippines.* 2005, Kassel University Press GmbH: Kassel.
21. Özyetgin Altun, A. *Afet Riski Yönetimi Kapsamında Kent Planlama; İstanbul Planları ve Uygulamalar.* 2017, Unpublished Doctoral Thesis, Mimar Sinan Fine Arts University: İstanbul.
22. Goldschalk, *Urban Hazard Mitigation: Creating Resilient Cities.* 2003. Natural Hazards Review, 2003, 4(3):136–143.
23. Balamir, M., *Afetlere İlişkin Planlama Etkinlikleri ve Sakınım Planlaması.* Kentsel Planlama Ansiklopedik Sözlük ed. M. Ersoy, 2016. Ninova: İstanbul.
24. Cutter, S., Emrich, C., and Burton C. G., *Disaster Resilience Indicators for Benchmarking Baseline Conditions.* Journal of homeland security and emergency management. 2010. Volume 7, issue 1 (51).
25. Kundak, S., and Turkoglu, H., *İstanbul'da Deprem Risk Analizi.* İtudergisi/a Architecture, Planning, Design. 2007. 6(2), p. 37–46.
26. Özceylan, D. *Afetler için Sosyal ve Ekonomik Zarar Görebilirlik Endeksi Geliştirilmesi: Türkiye'deki İller Üzerine Bir Uygulama.* 2010. Unpublished Doctoral Thesis, Sakarya University: Sakarya.
27. Taşkın, K., *Mahalle Ölçekli Depremsel Zarar Görebilirlik Endeksi: Sakarya Örneği.* 2012. Unpublished Doctoral Thesis, Unpublished Doctoral Thesis, Sakarya University: Sakarya.
28. Istanbul Metropolitan Municipality, Social Vulnerability Analysis Report, https://depremzemin.ibb.istanbul/calismalarimiz/tamamlanmis-calismalar/istanbul-ili-genelinde-afetler-karsisinda-sosyal-hasar-gorebilirlik-arastirmasi/ access date 03012021
29. Yaprak,Ü., *Afetler için Yaygın Savunmasızlık ve Yerel Afet Endeksleri Hesaplaması: Türkiye İlleri Örneği.* 2020. Yayınlanmamış doktora tezi. Gümüşhane Üniversitesi: Gümüşhane.
30. Balta, M. Ö., Kentsel Risklerin Planlama Temelinde Analizi ve Dirençli Kent Planlama Yaklaşımı. 2013, Unpublished Doctoral Thesis, Gazi University: Ankara.
31. Balamir, M., *Plancının Kasandra Yazgısı Afetler, Kentler, Yasal ve Kurumsal Yapılanma.* 2019. TMMOB Şehir Plancıları Odası: Ankara.

32. Adger, W. N., Brooks, N., Bentham, G., Agnew, M., and Eriksen, S. *New indicators of vulnerability and adaptive capacity.* 2004. Tyndall Centre Working Papers for Climate Change Research. http://www.tyndall.ac.uk/publications/tech_reports/tech_reports.shtml 18.01.2016, İstanbul.

Elif SAĞLIK

Semiotic Reading in Landscape Components

Abstract: Buildings, green spaces and urban water elements make up the fabric of a city. The spaces remaining outside the buildings in the city are called urban open spaces. Design and planning in urban open spaces are made by landscape architects and presented to the use of the city-dweller. The components used in these designs become a part of that city. Thus, it appears as a sign in the open spaces of the city. The parts can be a bench, a tree, a road, a fountain, a cover element or a pattern of a floor covering. These parts are actually the sign itself. These signs used or seen by the city-dweller have a meaning for them. It is possible to use the semiotic reading method in the discovery of this meaning. Within the scope of the method, the reading is made as "the landscape component that emerges with the code consisting of the concepts such as a sign by the landscape architect to the city-dweller, the connotation and denotation meaning of it, and the color, form and texture of it". In this direction, it has been tried to make semiotic reading in landscape components for Pier Square, which is the focal point in Çanakkale city, and its immediate surroundings. In the area, components such as the city square, artistic monuments and objects, boundary element, and sitting unit and floor covering were evaluated. As a result of the reading process made according to the method, the semantic dimension was revealed.

Keywords: Çanakkale, landscape component, semiology, urban landscape design, urban open area.

1 The Concept of Semiology

In order to solve or evaluate a problem or a situation, it is first tried to determine what the smallest part is. Therefore, the need to break the problem or situation into the smallest pieces arises. Starting with the easiest to understand part, the complex, compound and difficult to examine part is reached respectively. The approach of reaching the solution of the problem by dividing it into pieces is a traditional solution method and constitutes the framework of semiology. Semiology takes its place in the dictionary of the Turkish Language Association (TLA) as "Science that examines the structure and functioning of any sign system used for communication purposes, sign science, semiotics" and "The use of signs in language or their application with language" [1].

The term semiology, which was created by taking the word linguistics as an example, was defined at first glance as "the branch of science that studies the signs" or "the scientific study of the signs". However, it is stated that today's field of activity of semiology has gained a different dimension than the sum of the words "sign" and "science" [2].

The basic structure of the concept of semiology is the signs. In order to understand semiology, it is necessary to know what the sign means. Signs are defined by TLA as "the thing, symptom, symbol, indicator that serves to indicate something" and within the scope of linguistics "the language unit consisting of the fusion of meaning and form, signifier and signified, signal" [1]. According to this definition, words, signals, symbols, objects function as signs. This set of signs, which are frequently encountered in daily life, constitutes a meaningful whole. Examining the signs, searching for their meaning, examining the usage patterns, and determining the effect they have on the individual or in the environment are the actions that explain that the signs are a whole system [3]. As it can be understood from the actions in question, the signs do not exist alone. Each sign gains value with a system of signs. Therefore, the concept of system, that is, the system as a whole, is important in semiology [4].

Gaining meaning for a sign as a sign is the result of a complex process of signification that begins with the emergence of human consciousness. This signification has been in the interest of many thinkers in the historical process and has been tried to be explained with a scientific discipline [5]. Indicative science got out of the monopoly of linguistics and literature researchers after the 1950s and started to show its effect in all areas of life. The existence of semiology has been accepted in many fields such as architecture, city planning, industrial design, fashion, mythology, anthropology, and cinema. It is considered a very important scientific method for researchers [3]. It is defined not by an object of science, but by its methodology. In this case, it can be said that semiology does not have a certain object but develops analysis patterns and tools for certain phenomena. It establishes relationships between these phenomena by asking the question "what is (its) meaning" [6]. The most influential book written on semiology after the 1950s is the work "Elements of Semiology" written by Roland Barthes. Barthes stated in this work that the only purpose of the elements is to reveal analytical concepts depending on linguistics [7]. Barthes examined these elements under four main headings stemming from structural linguistics. He revealed this classification based on structural linguistics as binary oppositions:

- Language and speech
- Signified and signifier
- *Syntagm and system*
- *Denotation and connotation* [8]. The evaluation of the system that form the basis of semiology reveals a communication process. In this context, it is possible to create a ranking order in determining what the visible and perceived thing is [4, 9]. The sign reading process chart prepared by Erkman is presented in Fig. 1 [10].

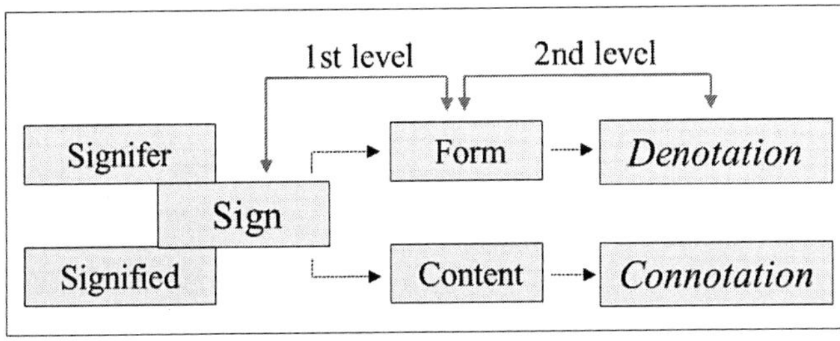

Fig. 1. Sign reading process chart

In semiology, while the signifier consists of forms such as picture, sound and word; the signified is the image itself that first appears in the mind when looking at the signifier. In this context, it is possible to say that the plane formed by the signifiers in the planar setup of the sign is the narrative plane, and the signified plane is the content plane [11]. In the analysis of the form and content plane, signification is made with the exploration of denotation and connotations. If the perceived concept and the object seen in a display are the same; there is a denotation. If the visible object evokes other things; there are connotations [12].

In other words, denotation is the first level of signification and refers to the known. The connotation, on the other hand, highlights the ideological thought and cultural values, and by becoming subjective, it means completely different meanings other than denotation. Thus, a new content is formed at the connotation level [10]. Çalak says that the connotations create alternatives to the way of seeing on this subject, saying: "Narrow streets, masonry, stone or wooden structures are actually separate codes, and we can understand that each of them refers to the land, climate and therefore culture rather than its denotation, and that it gives us many clues." [13].

Apart from linguistic signs, there are many different types of signs such as our behavior, pictures, photographs, posters, traffic signs, billboards, fashion, architectural regulations, movies, theaters. In a life interwoven with a web of signs, we must know the meaning of these signs used by society and must use them correctly in order to communicate [14]. These explanations suggest that in reality everyone is a sign reader. However, being able to read the signs correctly and obtain comprehensive results requires the application of a correct method [11].

As shown in Fig. 2, a method structure consisting of sub-components is used in the process of reading the sign. Within the scope of this method, the communication scheme is used to reveal the fiction between the signifier and the signified. In addition, it should be kept in mind that the signs and ways of expression vary in the readings where the communication scheme will be used. According to Roman Jakobson's communication scheme, it seems that the speaking subject needs to use a context (real world), a code (text, note, number, etc.) and a channel (book, magazine, television, etc.) to send a meaningful message to its receiver. [15].

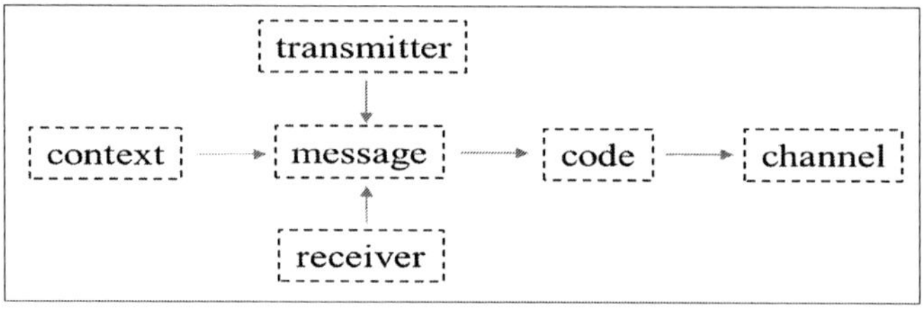

Fig. 2. Communication scheme

In the landscape architecture discipline where the subject of semiotic reading is discussed, living (plants) and non-living materials (floor covering, urban reinforcement, enclosure elements, etc.) are used to create messages. The landscape architect presents his/her message to the city and the city-dweller with various project designs (Fig. 3).

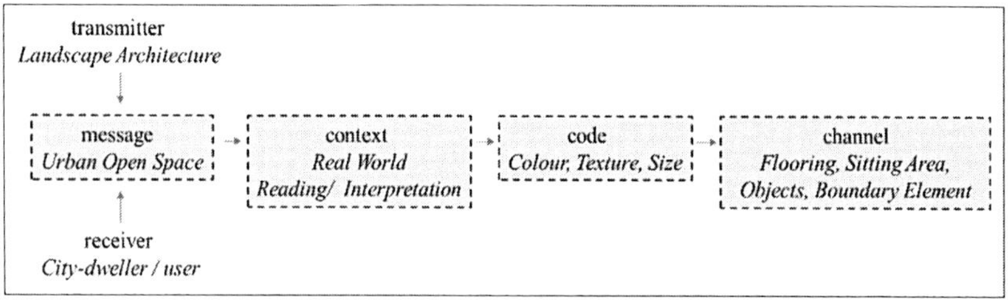

Fig. 3. Semiotic analysis/reading

As can be understood from the definitions made for semiology, there is no typical, bordered, defined field goal. It is possible to benefit from semiology in everything because we can communicate by using signs in all areas of life and everything that happens around us. At this point, it is possible to benefit from the signs that offer culture, space, discipline, opportunities and occupation, and that are a communication tool to understand and make sense of cities showing continuity [16, 17]. Urban life takes place in places that have symbolic/imaginary integrity with neighborhoods, avenues and streets, and where different meanings are presented to the residents. Being a dynamic structure, the city is a study material for many disciplines. In this context, the discipline of landscape architecture, which should be handled in urban scale, takes its place as one of the fields in which semiology is used. When the components that make up the city and come to the forefront in the urban identity are considered in terms of semiology, the meanings of the signs can be observed and expressed through concrete and physical objects. Thus, it is possible to understand the city with the semiotic interpretation of landscape parts in the perspective of landscape architecture.

Within the scope of this study, the reading of the city of Çanakkale will be made using the semiotic method. It is necessary to present information about the city before making a semiotic reading in the landscape architecture perspective.

2 Çanakkale

The city of Çanakkale is a city of unique beauty that has an important place in Turkish and world history. Çanakkale is a geography where blue and green colors dominate the city. It is located in the Marmara region of Turkey (Fig. 4). The city, which also expresses the North Aegean, is located on the western border of our country. It has a coast to the Aegean Sea and the Marmara Sea. Çanakkale city is undertaking a transit task to reach the Aegean Region of Turkey's major cities such as Bursa-Istanbul. This task it assumed unknowingly creates a great source of mobility for Çanakkale.

Canakkale is impassable!...
Stop, wayfarer! This ground you unknowingly come and tread on, is where an epoch sank...

These words are important expressions that reveal the historical value of the city of Çanakkale and have taken place in the collective memory. It is possible to say that Çanakkale is a window to history in the light of these expressions settled in the social memory. Çanakkale city, where countless martyrs were given and countless struggles were experienced, is hosts many historical and touristic values such as Gallipoli Peninsula Historical Site, Martyrdom Monument, Troy Ancient City, Troy Museum, Kilitbahir Castle, Naval Museum, Nusret Mine Ship, Passage Hallio, Assos, Behramkale. In this context, the city heavily welcomes local and foreign tourists who come to visit for history and sea tourism.

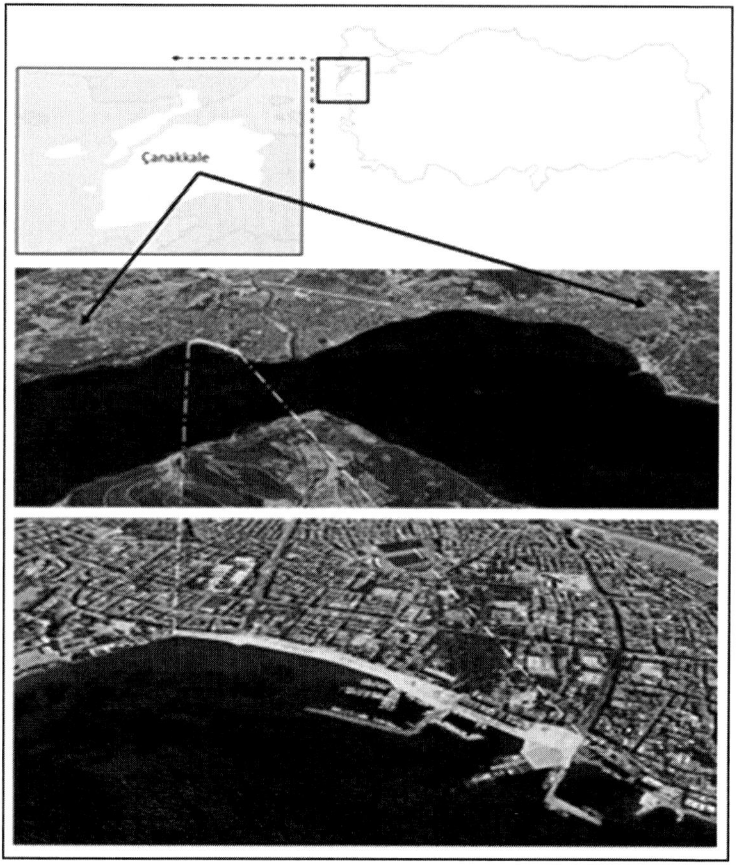

Fig. 4. The location in Çanakkale and Turkey of the field that the semiotic analysis will be done [18]

For the semiotic reading of the city of Çanakkale, the region of the city that has the characteristics of welcoming/meeting/gathering/socializing/resting/viewing balcony/terrace was chosen. It is possible to say that the selected region is the focal point according to these qualifications. It is the first and important point in the city for local and foreign tourists. The selected area is Pier Square and its immediate surroundings, located in the center of the city and revealing the identity of the city.

3 Pier Square and Its Immediate Surroundings

It is the focal point of the city. Access to the square and its immediate surroundings are provided by all kinds of transportation (land, sea, pedestrian, bicycle). In general, the city hosts the meeting/greeting/gathering/socializing/resting/viewing balcony/viewing terrace event. Pier Square is an important open area that welcomes those who come to the city by sea.

The immediate surrounding of Pier Square consists of the promenade, where is called Kordon, extending in east and west directions. Kordon is suitable for recreational activities. It is open to the use of all city dwellers and has a strong inclusion. Pier Square and its immediate surroundings are widely used by the city-dwellers in terms of the peninsula and the strait views.

There are natural and structural elements along the sides of Pier Square and its immediate surroundings. The natural element is the Dardanelles Strait, that is, the Marmara Sea; structural elements are the accommodation establishments, restaurants, and Kayserili Ahmet Paşa Street, which serves as an urban transportation axis. The starting point of the work area where visual reading will be made is Pier Square, and the end point is the Golf Tea Garden.

The length of the field is approximately 1179 meters; Its width varies between 20–55 meters due to its narrow/wide regions. There are various activity areas in Pier Square and its immediate surroundings where have a linear form. While the main activity is the walking; sitting/viewing/resting areas, handicraft products sales area, competition area, Trojan Horse and its immediate surroundings, exhibition areas, fishing and sales areas are the most important urban sub-areas for Pier Square and its immediate surroundings.

Fig. 5. Workspace and its form

4 Urban Landscape Design and Semiology Relation

In urban landscape designs, the design process starts with a main idea and is shaped by a concept. Landscape designs in urban open spaces serve as communication between the city and the city-dweller. At this point, hard and soft landscape elements are signs in a communication process between the landscape architect and the city-dweller. By presenting these signs to the city-dweller, practical benefit is provided, at the same time, the groundwork is prepared for the formation of semantic functions [19].

Every element that constitutes the components of urban landscape designs has semantic value. These elements in the city's identity and collective memory exist directly with their semantic value. Some adjective pairs are also used in semantic reading for landscape design elements within the scope of semiology [20].

5 Pier Square and Semiotic Reading Relating to Its Immediate Surrounding

Landscape components for semiotic reading in the study area were determined as Pier Square, Trojan Horse, Troia Historical National Park Exhibition Area, Trojan Wars Glass Pyramid, Sundial Area, Piri Reis Mosaic Map, sitting and border units and floor covering, as a result of fieldwork and literature review. The semiotic reading process of these components takes place according to the following stages:

- Message - Sign
- Context - Denotation (signifier)
  Connotation (signified)
- Code - Form
- Channel - Landscape component.

Pier Square

Message; Signifier: Open space, square
Context; Denotation: Gathering, meeting, resting, concert, show
Context; Connotation: City hall, density, 24/7 usage, access to the sea, spaciousness
Code: Form, size, vehicle and pedestrian road traffic, ferry pier entrance in open area
Channel: Urban open space, town square, urban focus

Cities are mentioned with certain centers. The most important center within the city of Çanakkale is Pier Square, which is the focal point of the city (Fig. 6). Pier Square is an area that makes its presence felt predominantly with its scale and focal point among the open areas of Çanakkale city.

Fig. 6. Images of Pier Square

Since Pier Square is located in the center of the city, it is a meeting place for citizens. The area derives its name from the ferry pier. The ferry pier is the entrance gate of the city used by people who prefer to reach Çanakkale by sea.

The use of ferries creates a constantly dynamic environment in the area. There are public transportation stops on Kayserili Ahmet Paşa Street, which is the edge of the square. "Pier Square Stop" is an important point for people using public transportation. It is quite easy to be included in Pier Square and its immediate surroundings on foot for the city-dwellers using these stops. In Pier Square, there are sitting and resting places for the citizens. It is possible for the city-dwellers to watch the square in panoramic view from the equipment where they sit while resting, or rest while sitting. Seating units are constructed in various forms. It has been arranged so that individual, double, triple or more people can use it at the same time. At a point where the seating equipment is located, the symbolic element of the city was used. The symbolic element is White Flowering Magnolia tree which is a live landscape element. This element was planted by Mehmet Bey in 1933 when the city was first established. It remains in existence until today. It was registered as a "monumental tree" in 2004 with the decision of the Çanakkale Cultural and Natural Heritage Preservation Board [21]. Granite cube stones and granite parquets have been used on the floor. Spatial change can be perceived with the pattern, texture and color difference made on the floor covering. In the light of these definitions regarding the Pier Square, semantic inferences with the use of adjective pairs in terms of reinforcement, ground and form are as shown in Fig. 7.

airy place	dreary	**new**	old
well-defined	undefined area	**light**	dark
public	private	**regular**	ırregular
aforethought	random	**active**	passive
aesthetically	not aesthetic	**equipped**	simple

Fig. 7. Semantic inferences regarding the Pier Square

In line with semantic inferences; the semantic dimension of the area is revealed as "Pier Square is an active, airy place because it is useful and functional due to its use in all seasons and at all times, it is light and safe enough because it is naturally illuminated, it is aesthetically pleasing and new because it is rich in vegetal materials and landscaping, it is an aforethought place as it is available for spending quality time, it is a well-defined and public place as it is a regular, equipped and accessible to citizens of all ages and groups".

Trojan Horse

Message; Signifier: Trojan Horse
Context; Denotation: A gift to the city from the producer of the movie Troy
Context; Connotation: Power, war, ancient history, recent history
Code: Horse figure, voluminous wood texture
Channel: Artistic monument presented to the city

Trojan Horse is located along the promenade forming the city's coastal line (Fig. 8). The Trojan Horse, which is a symbol of the epic of Ilyada and Odysseia who lived in ancient times, was presented to Çanakkale after being used in the Hollywood-made movie Troy. This artistic monument, which emphasizes the past, is defined among the sign elements of the city of Çanakkale as an extension of the recent historical process.

Fig. 8. Images of the Trojan Horse

In the light of these definitions regarding the Trojan Horse, semantic inferences with the use of adjective pairs in terms of the perception of the city-dweller are as shown in Fig. 9.

new	old	**public**	private
impressive	ordinary	not unique	**unique**
well-defined	undefined area	**regular**	irregular

Fig. 9. Semantic inferences about the Trojan Horse

In line with semantic inferences, the semantic dimension of the urban sign emerges as "The Trojan Horse is an artistic monument that is unique in terms of its value, impressive and magnificent due to its form, has a new appearance because it is regularly maintained, is located in a defined and regular area in a common urban space open to all urban residents because it is at the focal point of the city".

Troy Ancient City Exhibition Area - Sundial - "Troia War" in the Glass Pyramid

Message; Signifier: model of ancient city, plane, pyramid
Context; Denotation: Troia old town history, sundial, Troia War
Context; Connotation: Closeness to history, science, artistic object symbolizing those who lost their lives in the Troia War
Code: Exhibition showcase, planar object, pyramid made of glass
Channel: Object, artistic and scientific work product that presents the elements of urban history to the citizens

The model in which the ancient city of Troia is exhibited is located in the showcase (Fig. 10). The model, which started to be exhibited in Çanakkale coastal region in 2007, reflects the 6th layer of the ancient city. It is positioned in the rear of the Trojan Horse. It is an important exhibition area where people who visit the city find the opportunity to see the ancient city on a small scale. The area where there are also literary sources about the city is an important sub-region in Kordon [22].

The sundial was made by Çanakkale Onsekiz Mart University (COMU) Astrophysics Research Center and Observatory in 2007. A sundial that measures time with the movement of the sun is exhibited on the Çanakkale Kordon (Fig. 10). The sundial positioned on the back of the Trojan Horse is designed to show the day, month and horoscope as well as the clock [23].

The work titled "Troia War" in the glass pyramid was made by Ceramic Artist Erdinç Bakla. Erdinç Bakla donated this work, which he made in 2009, to the city of Çanakkale. It is exhibited in the northern part of the Trojan Horse in the Kordon region (Fig. 10). There are five separate rectangular sections in the work located in a glass pedestal, which is expressed with abstract symbols first time in Turkey.

These parts are the walls surrounding the city of Troia; the spaces between the sections are the city walls; the black marbles surrounding the Troia walls are the Akha soldiers; white triangular marbles depict Troia soldiers [24].

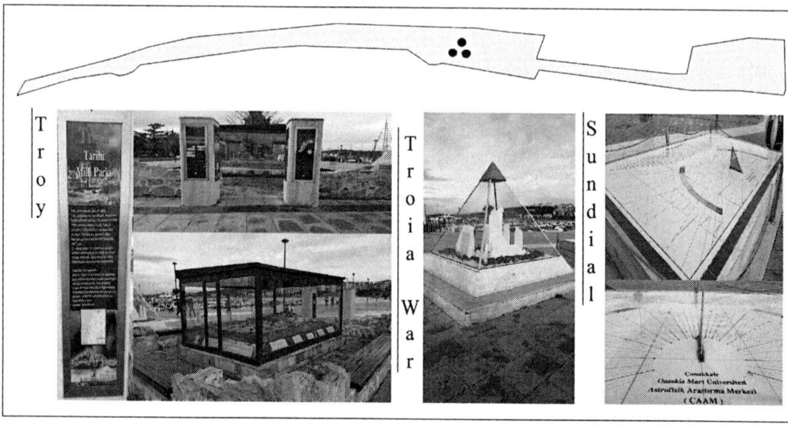

Fig. 10 . Images about the Troy Ancient City Exhibition Area - Sundial - "Troia War" in the Glass Pyramid

The model displaying the ancient city of Troy, the sundial, and the areas where the works named "Troia War" are exhibited in the glass pyramid are located very close to each other. For this reason, they constitute an important focal point for citizens and tourists as the Trojan Horse Square. In the light of these definitions, semantic inferences with the use of adjective pairs regarding the works in terms of the perception of the city-dweller are as shown in Fig. 11.

new	old	**public**	private
impressive	ordinary	not unique	**unique**
well-defined	undefined area	**regular**	ırregular

Fig. 11. Semantic inferences about the "Troia War" in the Glass Pyramid - Troy Ancient City Exhibition Area - Sundial

In line with the semantic inferences, the semantic dimensions of the sundial, the model exhibiting the ancient city of Troy, and glass pyramid titled "Troia War", emerges as "they are typical in regard of their value because of being only in Çanakkale, and they are impressive and inviting because they are focal point, they are the exhibition space presented to the city dwellers in a common space, considered in a defined area in terms of their positions and relations with each other".

Boundary and Sitting Units

Message; Signifier: Railing, seating
Context; Denotation: Border, security, rest, watching
Context; Connotation: City balcony, watching Dardanelles Strait, rest, sunbathing, breathing, view
Code: Railing, brown, wood texture, color, sitting under the tree
Channel: Boundary railing, authentic and under-tree seating that invites to relax

Boundary elements play a role in ensuring the relationship of the city-dweller with the environment in terms of technical and security. Therefore, they are remarkable items in urban open areas. In addition to its functionality, it is expected to be strong in terms of aesthetic and visual impact. The only boundary element used in the Çanakkale promenade area was applied as a combination of wood and steel rope (Fig. 12). The wooden texture was used as an extension of the flooring design on the floor. The area where the railing is located is in a circular form and it is the largest sub-area after the Pier Square.

It is possible to say that the seating units in Pier Square and its immediate surroundings are suitable for resting, chatting, waiting for someone or just spending time. It also fulfills the function of creating social space. In general, the use of wood and concrete is preferred. It is designed as a fixed bench, under-tree seating unit and a sitting pocket. Panoramic view of the Kordon and Dardanelles Strait is possible in all seating units suitable for individual or group use (Fig. 12).

Fig. 12. Images about the boundary and sitting units

Semantic inferences with the use of adjective pairs in terms of urban perception in the light of these definitions regarding limiting and sitting units are as shown in Fig. 13.

well-scaled	under-scaled	**safe**	dangerous
functional	non-functional	**sufficient**	insufficient
private space	**common space**	not unique	**unique**
defined space	undefined area	**new**	old
thought-out	random	**inviting**	unattractive

Fig. 13. Semantic inferences about the boundary and sitting units

In line with semantic inferences, the semantic dimension of urban reinforcements emerges in the form of "boundary elements and sitting units are well-scaled and functional due to their ergonomic features; unique, safe, thought-out and defined space because it creates social space; inviting because it is highly perceptible; common space and sufficient elements because it allows everyone to access".

Piri Reis Bust and Map

Message; Signifier: Piri Reis bust and map
Context; Denotation: Famous captain, Ottoman captain, map
Context; Connotation: Map, geography
Code: Mosaic flooring, colors, figures
Channel: Historical value, thematic mosaic coating on the floor

The full name of Piri Reis from Gallipoli, Piri Muhyiddin b. Hadji Mehmed. He introduced himself on his map as Piri Reis, son of Hadji Mehmed, nephew of Kemal Reis. Kemal Reis, on the other hand, is his uncle, a famous sea veteran/pirate. While working for his work, he referred to the maps of Alexander and Christopher Columbus and completed it in 1513 [25]. The map exhibited in the Çanakkale Kordon area was applied on a 9-meter-diameter ground using mosaic stones in accordance with the original [26]. There is the bust of Piri Reis right next to the map (Fig. 14). In addition, there are three informative frames embedded in the ground.

Fig. 14. Images of Piri Reis Bust and Map

The semantic inferences of this work, made by Piri Reis, with the use of adjective pairs in terms of city-dweller perception are as transferred to the figure in Fig. 15.

well-scaled	under-scaled	**defined space**	undefined area
functional	non-functional	not unique	**unique**

Fig. 15. Semantic inferences regarding the Piri Reis Bust and Map

In line with semantic inferences, the semantic dimension of the Piri Reis map and its bust emerges as "it is unique in terms of the application of the map which is a unique work in terms of geography; it is functional, well-scale and defined area because it makes a difference in floor covering".

Flooring

Message; Signifier: Materials covering the floor
Context; Denotation: Flat and solid ground to walk on
Context; Connotation: Material variety, orientation
Code: Colors, size change in coating material, pattern of materials, rhythm
Channel: Variability in ground coverage along the axle

Floor covering materials or in other words, flooring elements are the first and basic elements of the place. It is a design element on which pedestrians and vehicles move and gather. It is important for these actions to meet the comfort and carrying capacity. In this regard, pattern works come to the fore in the floor coverings in the Pier Square and its immediate surroundings (Fig. 16).

Spaces were created by placing differences on the floor for the examination of the walking, resting and exhibited works used throughout the Kordon.

Fig. 16. Images of flooring

The semantic inferences with the use of adjective pairs in terms of the city-dweller perception of the flooring materials and patterns used in the Pier Square and its immediate surroundings are as shown in Fig. 17.

perceptible	imperceptible	**defined space**	undefined area
not unique	**unique**	**thought-out**	random
Safe	Dangerous	**functional**	non-functional

Fig. 17. Semantic implications for flooring

In line with semantic inferences, the semantic dimension of the floor covering applied in the work area emerges as "they are unique and thought-out as they are made within the design concept; functional, safe and perceptible because they provide spatial divisions; they are the elements that make the sub-spaces defined in the promenade because they are easily perceived".

6 Instead of Conclusion: Semantic Dimension Discovery in City-Dweller Perception

Landscape architecture is a tool for organizing and developing relationships with the environment. Therefore, landscape architecture has a symbolic dimension. It looks for a logical form to fulfill this task. These forms are transferred to the city-dweller through symbols. Communication between the landscape architect and the urban space is actually a communication that takes place through urban design. The landscape architect abstracts signs from their meaning and utility qualities while making formation decisions during design. The values abstracted in this communication process are interpreted by the city-dweller. Semiotic reading is required to reach the interpretation of the city-dweller.

Semiology starts from the view that everything is a meaningful whole. An object, structure, road, bench, statue or a tree becomes a sign for semiology. It is necessary to figure out what the signs mean in a whole and what they mean for people. This is a science method used in analysis. Within the scope of this method, the communication scheme between the sign and the user is created while the semiotic reading of the landscape architecture components is done. In the communication scheme, first of all, what the sign is, the discovery of the denotation and connotation about this sign, and how the codes and channels of the sign found meaning are examined. Thus, the semantic dimension of urban landscape designs is reached.

Within the scope of the study, it is aimed to reach the semantic dimension of Pier Square, which is the focal point of the city of Çanakkale, and its immediate surroundings. In this direction, semiotic reading was used as a method. The stages of semiotic reading were implemented in urban open space. The variables from the floor covering to the artistic monument, from the sitting unit to the city square scale were read. Thus, every landscape component used in urban open areas is actually a sign, and it would be a correct foresight to think that these signs have a meaning in the perception of city-dweller. The study is expected to be a reference for semiotic reading-based studies in urban scale.

References

1. Türk Dil Kurumu Sözlükleri. [ONLINE] Available at: https://sozluk.gov.tr/ [Accessed 16 December 2020].
2. Rifat, M., *Göstergebilimin ABC'si.* 2014, Say Yayınları, İstanbul.
3. Güneş, A., *Göstergebilim Tarihi,* E-Journal Of New World Sciences Academy, 2013. **8**(4): p. 332–348.
4. Erkman, F., *Göstergebilime Giriş.* 2016, Bilge Kültür Sanat Yayınları, İstanbul.
5. Çetindemir, A., *Göstergebilim Nedir.* Yom Sanat, 2005, **22**: p. 5–8.

6. Kıran, A., *Çağdaş Bir Düşünme Biçimi Olarak Göstergebilim,* Journal of Lingustics, 2009, **22**:p. 1–16.
7. Gottdiener, M., *Postmodern Göstergeler.* 2005, İmge Kitabevi Yayınları, Ankara.
8. Bircan, U., *Roland Barthes ve Göstergebilim.* Social Sciences Research Journal, 2015. **13**(26): p. 17- 41.
9. Pellegrino, P., *Semiotics of Architecture.* Semiotic Anthropology, 2006, p. 212–216. [ONLINE] Available at: 3-s2.0-B0080448542013936-main.pdf
10. Hancı, H., The Influence of Semiology on Students of Graphic Design. 2008, Master Thesis, Gazi University Graduate Studies, Drawing-Crafts Teaching.
11. Yaşar, D., Semiology and Architecture İstanbul Kemankeş Avenue as a Text. 2017, Master Thesis. Beykent University Graduate Of Sciences.
12. Teker, U., *Grafik Tasarım ve Reklam.* 2002, Dokuz Eylül Yayıncılık, İzmir.
13. Çalak, I. E., *Göstergebilim Yöntemi ile Kent Okumaları.* Yapı Dergisi, 2007, **311**.
14. Gür, C., Ferdinand de Saussure ve Yapısalcılık. Yom Sanat, 2005, **22**: p. 9–13.
15. Güneş, A., *Bir Eserin Mimarisini Göstergebilimsel Bir Yaklaşımla Okuma ya da Mimari Göstergebilim: Divriği Ulucamii ve Darüşşifası.* Erciyes İletişim Dergisi "akademia" 2013. **3**(2): p. 74–86.
16. Kurak Açıcı, F. and Sönmez, E., Trabzon Kent Merkezinin Göstergebilim Yaklaşımı ile Okunması. Journal of Architecture and Life, 2020. **5**(1): p. 181–192.
17. Foote, K. E. and Azaryahu, M., *Semiotics.* [ONLINE] Available at: 3-s2.0-B9780080449104007410-main.pdf
18. Google Earth Pro
19. Alpak, M., Düzenli, T. and Tarakçı, E., *Seating Furniture Design: Pragmatic, Synthactic And Semantic Functions.* International Furniture Congress, 2018, December. [ONLINE] Available at: (pdf) oturma donatısı tasarımı: pragmatik, sentaktik ve semantik işlevler (researchgate.net)
20. Okuyucu, Ş. E., *Avlunun Göstergebilimsel Yöntemle Analizi; Denizli Erbakır Fen Lisesi Avlusu.* Online Journal of Art and Design, 2018. **6**(3).
21. Pier Square, [ONLINE] Available at: https://www.canakkalematbuat.com/?Syf=22&Mkl=882327 [Accessed 16 December 2020].
22. Troy Ancient City Exhibition Area. [ONLINE] Available at: https://www.canakkaletravel.com/haber/turkiyenin-en-buyuk-maketi-bakimda.html [Accessed 25 December 2020].
23. Sundial. [ONLINE] Available at: https://www.canakkaletravel.com/haber/dunyanin-ilk-takvimli-gunes-saati-canakkalede.html [Accessed 25 December 2020].

24. “Troia War” in the Glass Pyramid. [ONLINE] Available at: https://www.canakkaletravel.com/haber/troia-savasi-cam-piramit-icinde-sergileniyor.html [Accessed 25 December 2020].
25. Piri Reis Bust and Map. [ONLINE] Available at: https://www.atlasdergisi.com/kesfet/bilim-haberleri/piri-reis-ve-haritasi.html [Accessed 25 December 2020].
26. Piri Reis Bust and Map. [ONLINE] Available at: https://www.canakkaletravel.com/haber/kordon-boyuna-mozaik-piri-reis-haritasi-yapiliyor.html [Accessed 25 December 2020].

Özge CORDAN and Talia ÖZCAN AKTAN

Home Environment Appropriation: Syrian Migrants in Sultanbeyli, Istanbul

Abstract: Syrians have been fleeing to Turkey since 2011 due to the ongoing civil war and conflict in Syria. This study focuses on displaced and forcibly migrated Syrians' practices of appropriation in the process of remaking home. The purpose of the study is to define means of appropriation regarding Syrians' cultural, psychological, and spatial needs for their residential interiors during relocation. Thus, Sultanbeyli, a district of Istanbul, was exemplified as a case. The data were obtained from volunteer Syrian families in their home environments through visual methods and interviews. According to the results, appropriation has a crucial role in meeting Syrians' needs for personalization, reflection of identity, belonging to a particular place, and self-expression of both communal and individual spatial and object-based placements. This study intends to generate and disseminate new knowledge for designers, researchers, policymakers, and the public regarding Syrians' appropriation needs to provide more livable and supportive living environments.

Keywords: Appropriation, forced migration, place attachment, remaking home, residential interiors.

1 Introduction

Fleeing from the civil war and conflict in the Syrian Arab Republic, Syrians have been migrating to neighboring countries since 2011. They have been subjected to the largest mass and forced migration since World War II [1]. Turkey applied an "open border policy" [2] to Syrians at the beginning of the crisis and hosted around 3.6 million Syrians [3]. Due to the ongoing civil war in Syria, the number of displaced Syrians has risen in Turkey. In 2014, a temporary protection regulation was issued, and the status of Syrians was defined as "Syrians under temporary protection" [2].

At the beginning of the crisis, the Syrians were housed in temporary accommodation centers. During the process, 93 % of Syrians have dispersed to various provinces and districts in Turkey [4] for reasons such as education, work, accommodation, health, and security. In this regard, most of them relocated to relatively poorer areas of big cities to provide their need for shelter, which is one of the most basic human needs. Istanbul has also been affected by this inner migration, and "suburban neighborhoods in major cities such as Istanbul have become the main sites of settlement for Syrian refugees" [5]. The kinship relations, cultural, religious, and economic aspects have been effective factors for choosing the living environment. Spatial harmonization is also an important issue in coexistence [4] for migrants' life satisfaction and well-being.

Rapoport elaborates on pull and push factors in the selection of inhabitation. According to him, "habitat selection involves both rejecting or leaving undesirable, unsuitable, unsupportive, or inhibiting environments (what are called pushes) and seeking out desired, suitable, and supportive environments (what are called pulls)" [6]. Life satisfaction and quality affect residential attachment [7]. Without an emotional bond to residential attachment, migrants perceive their new residences negatively compared to their previous ones [8].

Residential attachment may help displaced and forced migration populations acquire a sense of belonging in their new physical and social environment over time. Leutloff-Grandits states that "identity and place are essentially linked and that the best way of healing the wounds of war and expulsion is by enabling the return to the prewar homes (or by creating similar territorialized belongings)" [9]. In the forced migration process, the home symbolizes the migrant's presence in relocation and represents the home and homeland lost or left behind. Home is also capable of being both a haven and a place of alienation and discrimination [10].

Appropriation, in this sense, is essential for ensuring place attachment, identity, and sense of belonging to a particular place. This study focuses on appropriation practices of displaced and forcibly migrated Syrians. In this regard, the district of Sultanbeyli was chosen as the case for this study. The field study was carried out with the voluntary participation of Syrian families in their residences in the Fatih neighborhood in Sultanbeyli. The data were collected through visual and interview methods. Thus, multiple means of appropriation were defined based on data analysis.

2 Place Attachment and Appropriation in Forced Migration

Place attachment identifies the relationship between people and places [7, 11]. Fornara et al. describe the distinctive features of place attachment through its affective, emotional, and feeling aspects [7]. It is associated with "attachment to a place as a function of time" [12]. In forced migration, there could be transformations in the sense of belonging and place attachment. Therefore, this situation affects migrants' life satisfaction and well-being.

Place attachment is also associated with longing for places (i.e., home and homeland) in the process of displacement [13]. Therefore, migrants seek an emotional bond that is a part of human nature to attach to their new life and living environment. Esentepe and Günçe posit that "[f]eeling attached to the current environment enables migrants to feel more secure and comfortable in relocation through social ties such as relationship with locals, neighbors and so on" [11]. Place attachment provides a basis for transforming a space into a place, and appropriation of place preserves positive feelings, such as security and safety, in opposition to the fear of getting lost [14].

People make a space their own and differentiate it from others [15] and tend to appropriate their surrounding environments to meet their needs for personalization, identity, self-fulfillment, self-expression, and sense of belonging. Graumann describes the term "appropriation, which literally means making (something) one's own and taking for one's own use, presupposes that it is features of the spatiotemporal environment that arouse, foment, afford, and sustain environment-related intentionality" [16]. According to Lefebvre, the appropriation of space means modifying a natural space to meet the needs of users [17].

Appropriation is a mutual reconciliation between people and places over time. Drawing attention to this reciprocal relationship, Grauman emphasizes that people "change by changing their environments" [16]. In the process of migration, the dynamics of the appropriation of a particular place can also change. In other words, appropriation for migrants turns into a process of existence and proof of identity in relocation. This situation is particularly evident in home environments.

2.1 Appropriation of the Home Environment in the Forced Migration Process

While a house is a physical space [18], a home is more than that [19]. Home refers to the interaction between people and place [20] and has visible and invisible aspects. The home concept can be enriched by its physical, social, and cultural dimensions [21]. Fox defines home as a "physical structure," "territory," "identity," and a "social and cultural unit" [22].

Home can act as a catalyst for migrants' and refugees' adaptation to a particular place and plays an important role in reflecting on their previous lives because they try to reconstruct their previous living environment in the newly settled area. Ralph and Staeheli state that "home is both expand migrants outwards to distant places and embed them in their localities and relationships" [23]. In other words, home for migrants refers to both their previous and present living environments [24–26]. In this sense, home-making is a process related to previous life experiences and has a triggering effect to connect people to a particular place.

Residential attachment to physical and social environments is an important phenomenon for migrants. People, specifically forced migration populations, reflect their memories, experiences, future desires, and wishes in their new living environments. In this regard, refugees' home-making process is an attempt to reproduce familiar cultural values [26] and cultural traces allow people to feel at home, specifically in a lack of familiarity [27]. Kabachnik et al. state that "[h]ome plays an integral role in identity formation, socialization into family and cultural norms and values, is a site of powerful memories and emotions, and is deeply implicated in people's desires and hopes for the future" [28].

In this sense, appropriation has an essential role for migrants as a communication tool, which describes a psychosocial process. People's means of appropriation may vary according to their individual and communal culture, lifestyle, and habits. Appropriation can also be acquired via both space- and object-based placements. The appropriation of space maximizes the sense of belonging, strengthens attachment to a particular place [17], and is based on spatial organization and the use of planes and surfaces. Whether vertical or horizontal, a plane defines a space and controls the physical and visual limits of a space [29]. Additionally, people contact with the surfaces at first [29]. The people taking root in the space through the placement of things (such as furniture and objects), the arrangement of the space, the removal of some elements from the space, and the establishment of control and authority over the space [15] are means of appropriation of a particular place.

Objects, whether furniture, goods, or decorative items, are one of the most important elements for providing identity to a particular place. Objects loaded with meaning establish physical and cultural relationships with their environment [30]. In this regard, object-based appropriation can be ensured through different means of appropriation. Semi-fixed elements are essential for personalization and expressing identity [6], and furniture and decorations carry special meanings, specifically for migrants—leaving them behind equates to leaving behind a part of the story [31]. Little items, such as objects and photographs from migrants' homeland, are also important to personalize their residential interiors [32]. The objects act as reminders [31] and reconstruct a part of migrants' previous life and former identities in the new living environment [33]. Thus, these "transitional objects" [33] as mementos provide familiarity for migrants and meet their needs for identity, personalization, attachment, and belonging in relocation. They also help transform a house into a home.

3 The Study: Materials, Methods, and Procedures

Istanbul hosts the largest immigrant Syrian population in Turkey, and the district of Sultanbeyli, which is an old slum area, hosts the most Syrians in the Anatolian side of Istanbul. Besides the services in terms of shelter, health, education, and employment provided for Syrians by the Sultanbeyli Municipality, the reasons for choosing Sultanbeyli as the case for this study are the spontaneously occurring coexistence of Syrians and the local inhabitants and their similarities regarding religious beliefs, cultural and living habits, and economic status.

According to the 2017 Syrian Coordination Center (SUKOM) database [34], the area between Seyhan and Şalgamlı Streets in the Fatih neighborhood, where the majority of Syrians have settled in this neighborhood, was determined as the field study area (Fig. 1). Due to most of the Syrians living in apartment-type buildings, the study was limited to three-to-five-story apartments in this neighborhood.

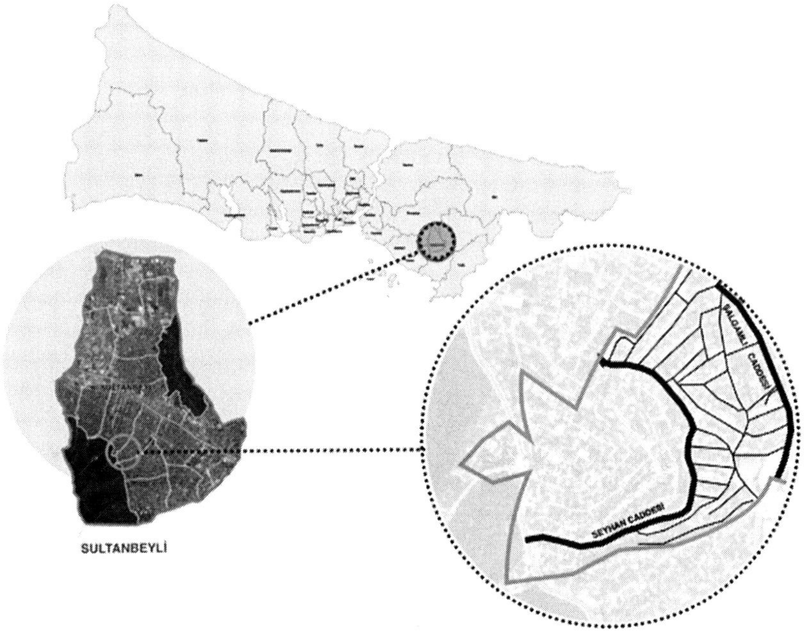

Fig. 1: Field Study Area, Sultanbeyli, Istanbul [3]

Researching home environments prioritizes the family. Home is associated with "family life" [35], so the participant group consists of Syrian families living in the Fatih neighborhood found in the SUKOM database. Those families were contacted by a native translator, and the study was conducted with 45 adults aged 18 and over from 18 Syrian families who participated voluntarily. Most of the participants are from Aleppo and its environs, and they belong to lower- and middle-income groups. Besides the approval from the Ethical Committee of the Social Sciences Area of Istanbul Technical University, the participants were informed about the study, signed the consent form, and were interviewed at their residences. The field study was carried out in two rounds of visits in 2019.

The aim of the multiple forms of data collection using the visual method and the interview technique was to define the different means of appropriation. Sanoff was consulted to obtain visual data [36], and Zeisel was referred for structuring the interview questions [37].

While visual research methods are valuable sources for analyzing the visual and behavioral characteristics of a physical environment, the interview technique is useful for understanding and revealing people's thoughts, beliefs, feelings, desires, privileges, and so on. The data were recorded through photographs, drawings, and annotated notation charts. The interviews were also used to determine spatial and object-based appropriation. The semi-structured and group interviews with the families were carried out via in-person questioning by a native female translator in Arabic and then translated to Turkish. A clinical psychologist also contributed to the field study to communicate better with the participants.

To analyze the visual data, schematic sketches of the floor plan in each house were drawn to indicate the plan type, space organization, and furniture layout. The physical traces, which are based on Zeisel's classifications of physical traces as "by-products of use," "adaptation for use," "display of self," and "public messages" [37], were visualized on floor plans with the help of photographs and written information by considering all planes (walls, floors, and ceilings) and their surfaces. Interviews were analyzed using a content analysis and interpreted by the researchers.

4 Findings

According to the visual data results, Syrians—whether in an extended or nuclear family structure—have hierarchical, introverted, crowded, and large households. Even nuclear families consist of approximately six to eight family members, mostly with children of different ages. The generations (children and parents or elders) live together under the same roof in Syrian families, whose members have different roles depending on their status in the household. They try to sustain their living habits and culture in their new living environment. Therefore, their attempts to remake home are associated with their previous lives in Syria.

The visual data analysis shows that Syrians make changes to their living environment according to their needs. If the entrance hall is large enough, they use this space (called "mamak" in Arabic) as a living room per their cultural habits (Fig. 2a). If it is not, they use the existing living room as a gathering space for household members (Fig. 2b). The living room hosts many activities, such as eating, playing, sleeping, and socializing and so on. The elderly people in the family have their place in this room, mostly sitting on the ground cushions. Syrians also eat on a low table in this room instead of in the kitchen.

They organize interior space according to the privacy thresholds, such as bedrooms according to sexes, and separate the toilet and bathroom, from the other spaces in the home. They also divided the bathroom and toilet if they were in the same space. The bedrooms are used for different activities, such as hosting female guests, playing and studying for children, and sleeping and so on. The use of a floor mattress makes the bedroom flexible for different activities, specifically for large households. The kitchen serves for cooking and storing utensils and other goods. Syrians are also familiar with using terraces, balconies, and front porches—they use those spaces for sitting, socializing, and storage. It is also common to completely cover the ground floor with carpets and/or textiles and cover windows and doors with a film layer to meet the privacy needs of the household.

Fig. 2a: A Syrian family's entrance hall

Fig. 2b: A Syrian family's living room

The use of semi-fixed furniture is also common, such as ground cushions for sleeping/sitting and low tables for eating. Most of the furniture and furnishing items in the Syrians' homes are damaged. Some furniture is also used for storage by ignoring their intended functions, such as a cradle serving as a wardrobe and so on. Syrians also used the walls surfaces for storage, especially if they have available spaces, such as niches and so on, in the residential interior. The wall surfaces are used for conveying public messages, personalization, identification, and self-expression. To do so, they use clocks, calendars, birdcages, religious objects, paintings, pictures, posters, writings, small objects, photographs, toys, personal belongings, the Turkish flag, Turkish politicians' posters, the Quran, and so on (Fig. 3a, 3b). Some electrical appliances and kitchen utensils are treated as if they are vital items regarding their spatial position and lace veils over them.

Fig. 3a: A wall painting by a Syrian child

Fig. 3b: A young Syrian woman's bedroom

According to interviewees' responses, Sultanbeyli is a supportive living environment for living close to relatives who also migrated from Syria and for having good relationships with neighbors and local people. While the younger family members are willing to adapt to the new living environment, the older ones are still attached to their previous home in Syria, as reported by one of the participants: "The home there (*in Syria*) was the real home. This is just like a space we are sheltering..."

Syrian families live as tenants and intervene in their existing interiors at little as possible. They adapt home interiors appropriate to their culture and their needs. Based on the crowded living style, spaces are used to serve more than one function, as one of the participants explained: "In everyday life, this is the living room, but in the evening, it turns into a bedroom, we live here." They use the living room according to their cultural needs and make some additions, such as low tables for eating and ground cushions for sitting.

Older people have tendencies to maintain their living habits. One participant emphasized this by saying, "We are sitting on a ground cushion, even if there is a couch." Syrian families expressed their discomfort that the toilets and bathrooms are in the same place in their residence in Sultanbeyli, saying, "The bathroom and toilet in Syria were separated, here they are together, and it bothers us." Due to connected bathroom and toilet spaces and fittings installed without considering Qibla, some toilets were not used.

Religious symbols are important for people who leave their home and homeland behind due to traumatic events such as war or natural disasters. Therefore, the most valuable thing that Syrian families brought with them is the Holy Quran (Fig. 4a, 4b). A Syrian participant explained: "What could we bring other than the Quran!" A few of them stated that they brought some clothes, blankets, photographs, some kitchen utensils, and other things (Fig. 5a, 5b, 5c, 5d). The most important thing that Syrian families wanted to bring with them were their mementos, especially photographs of their memories. As one of the participants said: "Photographs are all of our history, we would love to bring them." During the interviews, electrical appliances, refrigerators, stoves, dowries, and other items were also reported as other items that Syrian families wanted to bring with them.

Fig. 4 (a, b): Families' Qurans brought from Syria

Fig. 5 (a, b, c, d): Some goods of a Syrian family from their home in Syria

While Syrian families long for their homes in Syria, they also emphasize their property ownership and belongings. The longing for belongings is notable for making life easier and indicating status. One of the participants explained: "Everything there was ours, belongings were our own..." and the other one also declared: "Our living room was very nice, our furniture, our LCD TV." Most of the furniture and furnishing items in Syrians' current home in Sultanbeyli are damaged or secondhand acquired through donations without giving any reference to their preferences and tastes (Fig. 6a, 6b). A Syrian participant explained this situation: "We got some from second-handers, some from help."

Appropriations made by Syrian families to feel attached to their new living environment (home) can be categorized as space- and object-based placements. While spatial appropriation is revealed through space organization and the use of planes and surfaces, object-based appropriation is revealed through furniture and goods, mementos, personal belongings, and decorative items.

Fig. 6 (a, b): Second hand or damaged furniture of a Syrian family

4 Discussion

Displacement or relocation could either weaken or evoke attachment bonds with the new residential environment [7]. While there is a strong link between homeownership and belonging, people also appropriate spaces that are not their own and adapt them to their culture, taste, and preferences. As Esentepe and Günçe stated, "modifications or additions to the housing units are a reflection of the needs of well-being" [11].

The immigrant Syrian families living in Sultanbeyli make modifications to their residential interiors for their cultural needs, lifestyles, and living habits for their life satisfaction. They transform their living spaces through spatial organization and treatment of planes and surfaces, as well as object-based placements reflecting their taste and identity. Thus, they try to remake their current living environments resemble their previous homes in Syria. In other words, their attempt at "remaking home in migration focus[es] on territory, culture, and nation between people and place" [38].

Space organization is the adaptation of space to the needs of the users and the actions to be taken in a space [39]. Spatial appropriation also ensures the personalization of a particular place and strengthens the attachment bonds between the home environment and people. According to the study results, the appropriation of space through spatial organization is based on proposing the entrance hall as a living space and using the living space multi-functionally with the help of equipment that can be easily installed and removed. Arranging spaces according to sex; combining indoor and outdoor spaces, such as balconies and terraces, for different needs of the household; separating the bathroom and toilets for privacy needs and orienting the fittings and fixtures in those areas considering Qibla are the main determinants of spatial organization regarding space appropriation.

Planes and surfaces are used to convey messages, reflect preferences and taste, and respond to functional requirements. Syrian families appropriate wall surfaces by using suitable areas (under stairs, balconies, etc.) to meet the storage needs of crowded family. They also use them for their personalization, identification, and belonging needs through hanging paintings, posters, or photographs and drawing pictures on the walls. They also cover the entire ground floor with carpet for serving different activities performed on the ground (eating, studying, sleeping, playing, etc.) and cover the window and door surfaces for privacy.

The use of furniture, goods, and objects for home decoration gives information about users' appropriation of a space. They play a major role in personalization and other means of expressing individual and group identities [40]. Mementos are also crucial for meeting the belonging needs of migrants in the process of remaking homes [41]. These objects, which are carried, accumulated, and loaded with meaning, enable people to establish a connection between past and present life situations. Their spatial location expresses meaning, and they strengthen communication between humans and objects and between humans and space.

Furniture is semi-fixed element that have aesthetic, functional, and memorable values. On the furniture scale, the most prominent examples of appropriation are seen in Syrians' use of ground cushions for sitting, floor mattresses for sleeping, and low tables for eating. Refrigerators, televisions, washing machines, and other electrical kitchen utensils are crucial goods for ensuring status and making life easier. Mementos are important for Syrian families to create a linkage between their previous and present lives. They brought their Qurans, photographs, some kitchen utensils, traditional games and toys as mementos from Syria. Personal belongings/items are important for appropriation to express themselves and convey messages. Decorative items are also crucial objects for identifying tastes and preferences for the embellishment of residential interiors for Syrians. They can serve both functional and aesthetic needs. While some objects (such as a clock, calendar, birdcage, religious objects, and various other objects), are related to the functional needs of the household, some (such as paintings, posters, small objects, and lace) are related to aesthetic needs of the household.

5 Conclusion

This study focuses on the different means of appropriation in the process of homemaking regarding the Syrian immigrant population living in Sultanbeyli, Turkey. Appropriation carries traces of social relationships and social structures and provides continuity for migrants between past, present, and future linkages. It also encourages the reflection of identity and culture, as well as providing place attachment and a sense of belonging.

The main conclusions of the study show that appropriation has a major role in making a particular place home through spatial and object-based placement. Syrians need to remember their former home in terms of their attachment to a particular place to convey messages to others for their personalization and identity, to keep their memories alive for their belonging, and to meet their cultural needs for life satisfaction. Accordingly, appropriation is particularly notable as a means of communication that reflects culture, identity, belonging, and self-expression for Syrians.

The results obtained from the study will be beneficial for designers, researchers, policymakers, and the public in terms of drawing attention to the importance of the relationship between place attachment and home environment to meet users' appropriation needs for life satisfaction and well-being. The study results also generate and disseminate new knowledge regarding Syrian migrants in Turkey. More specifically, the study has the potential both for adapting existing buildings and newly built residences to create livable and supportive environments. In this context, further studies can focus on the different means of appropriation to highlight the transformations over time, specifically for immigrant populations such as Syrians.

Acknowledgments

This paper was prepared from the research project titled "Determining Housing and Interior Design Principles and Developing Design Proposals Suitable for Modern Life and Cultural Conditions to Facilitate the Adaptation of Syrians Under Temporary Protection" and funded by The Scientific and Technological Research and Council of Turkey (TUBITAK) with the grant number 118K441 in 2018. We would like to thank Assoc. Prof. Dr. Şebnem Ertaş, Assoc. Prof. Dr. Elif Sönmez, and Dr. Çiğdem Kudiaki for their collaboration. We would also like to thank the Syrian families for their participation in this research and the Syrian translators for ensuring communication between the families and the researchers.

References

1. Yenilmez, M.İ., *The Impact of Forced Migration in the Middle East: Syrian and Palestinian Refugees.* Perceptions, 2017, **12**(4): p. 183–201.
2. DIGM, *Directorate General of Migration Management: Temporary protection in Turkey,* https://en.goc.gov.tr/general-principles-of-temporary-protection, last accessed: 31 December 2020.
3. Cordan, Ö., Aktan Özcan, T., *Home-Making Process of Syrians through Sustainable and Flexible Reuse of Existing Building Interiors: The Sultanbeyli Case, Istanbul, Turkey.* In Theory and Practice in Sustainable Planning and Design, ed: Murat Özyavuz, 2020, p. 445–460, Berlin: Peter Lang. ISBN 978–3-631-81792-6
4. Erdoğan, M.M., *Suriyeliler Barometresi: Suriyelilerle Uyum İçinde Yaşamın Çerçevesi SB-2017* (Syrians Barometer: A Framework for Achieving Social Cohesion with Syrians in Turkey SB-2017), 2018, İstanbul: İstanbul Bilgi University Publications.
5. Kılıçaslan, G., *Forced Migration, Citizenship, and Space: The Case of Syrian Kurdish Refugees in İstanbul.* New Perspectives on Turkey, 2016, **54**: p. 77–95.

6. Rapoport, A., *Culture, Architecture and Design.* 2005, Chicago: Locke Publishing.
7. Fornara, F., Lai, A.M., Bonaiuto, M., Pazzaglia, F., *Residential Place Attachment as an Adaptive Strategy for Coping with the Reduction of Spatial Abilities in Old Age.* Frontiers in Psychology, 2019, **10**(856): p. 1–9.
8. Anton, C. E., Lawrence, C., *Home is Where the Heart is: The Effect of Place of Residence on Place Attachment and Community Participation.* Journal of Environmental Psychology, 2014, **40**: p.451–461.
9. Leutloff-Grandits, C., *Anthropological Notebooks.* Book Review, 2009, **15**(2): p.83–84.
10. Taylor, H., *Refugees and the Meaning of Home.* 2015, London: Palgrave Macmillan.
11. Esentepe, B., Günçe, K., *The Role of Memory as a Benefit of Place Attachment on Refugees' Attachment to Place: Case Study on Turkish Cypriot Refugees in Refugee Village, Nicosia, North Cyprus,* LIVENARCH VI - 6th International Congress Livable Environments & Architecture - Replacing Architecture, Karadeniz Technical University, Trabzon, 2019, p.274–290.
12. Tuan, Y., *Space and Place.* 2011, London: University of Minnesota Press.
13. Scannel, L., & Gifford, R., *Defining Place Attachment: A Tripartite Organizing Framework.* Journal of Environmental Psychology, 2010, **30**: p.1–10.
14. İlkay, Y., *(Re) production and Appropriation of Open Public Spaces: Representational Moments For Urban Green in Ankara.* PhD Thesis, The Department of Urban Policy Planning and Local Governments, Social Sciences of Middle East Technical University, 2016, Ankara.
15. Bilgin N., *Fiziksel Mekândan İnsani ya da İnsanlı Mekâna (From Physical Space to Humanistic or Manned Space),* Mimarlık Journal, 1990, **90**(3): p.62–65.
16. Graumann, C.F., *The Phenomenoligical Approach to People-Environment studies.* In Handbook of Environmental Psychology, ed: R. B. Bechtel & A. Churchman, 2002, p.95–113, John Wiley & Sons, Inc. ISBN 0-471-40594-9
17. Lefebvre, H., *The Production of Space.* 1991, London: Blackwell.
18. Lawrence, J.R., *What Makes a House a Home.* Environment and Behaviour, 1987, **19**: p.154–168.
19. Dovey, K., *Home and Homelessness, Human Behavior and Environment: Advances in Theory and Research.* In Home Environments, ed: Irwin Altman & Carol M. Werner,1985, p.33–64, New York: Springer Science+Business Media. ISBN 978-1-4899-2266-3
20. Mallet, S., *Understanding Home: A Critical Review of the Literature.* SAGE Journals, 2004, **52**(1): p.62–89.
21. Moore, J., *Placing Home in Context.* Journal of Environmental Psychology, 2000, **20**: p.207–217.

22. Fox, L., *The Meaning of Home: A Chimerical Concept or a Legal Challenge.* Journal of Law and Society, 2002, **29**(4): p.580–610.
23. Ralph, D., Staeheli L.A., *Home and migration: Mobilities, belongings and identities.* Geography Compass, 2011, **5**(7): p.517–530, http://dro.dur.ac.uk/11542/1/11542.pdf
24. Liu L.S., *A Search for a Place to Call Home: Negotiation of Home, Identity and Sense of Belonging Among New Immigrants from the People's Republic of China (PRC) to New Zealand.* Emotion, Space and Society, 2014, **10**: p.18–26.
25. Boccagni, P., *At Home in the Centre? Spatial Appropriation and Horizons of Home Making in Reception Facilities for Asylum Seekers.* HOMInG Working Paper no.10, 2010.
26. Kissoon P., *Intersections of Displacement: Refugees' Experience of Home and Homelessness.* 2010, Newcastle upon Tyne: Cambridge Scholars.
27. Kreuzer, M., Mühlbacher,H.,Wallpach,S., *Home in the Re-making: Immigrants' Transcultural Experiencing of Home.* Journal of Business Research, 2017, **91**: p.334–341.
28. Kabachnik P., Regulska J., & Mitchneck B., *When and Hhere is Home? The double displacement of Georgian IDPs from Abkhazia.* Journal of Refugee Studies, 2010, **23**(3): p.316–36.
29. Brooker, G.,Stone, S., *What is Interior Design?,* 2010, SA: RotoVision.
30. Brooker, G.,Stone, S., *Re-readings: Interior Architecture and the Design Principles of Remodelling Existing Buildings.* 2004, RIBA Publishing.
31. Ureta, S., *Domesticating Homes: Material Transformation and Decoration Among Low-income Families in Santiago, Chile.* Home Cultures, 2007, **4** (3): p.311–336. doi: 10.2752/174063107X247341.
32. Glusac, T., *Architecture and Belonging: Migration, Re-territorialisation and Self-identity.* Degree of Doctor of Philosophy, School of Built Environment, 2015, Curtin University, Bentley, Australia.
33. Mazumdar, S., Mazumdar, S., *Religion, Immigration, and Home Making in Diaspora: Hindu Space in Southern California.* Journal of Environmental Psychology, 2009, **29** (2): p.256–266. doi: 10.1016/j.jenvp.2008.07.004.
34. SUKOM, *Syrian Coordination Center,* https://multeciler.org.tr/eng/, last accessed: 31 December 2020.
35. Hayward, D.G., *Home as an Environmental and Psychological Concept,* Landscape, 1975, **20**: p.2–9.
36. Sanoff, H., *Visual Research Methods in Design.* 2016, London: Routledge.
37. Zeisel, J., *Inquiry by Design: Environment/Behavior/Neuroscience in Architecture, Interiors, Landscape and Planning.* 2006, New York: Norton Publishing.

38. Jansen, S., Löfving, S., *Struggles for Home: Violence, Hope and the Movement of People.* 2009, Oxford: Berghahn Books.

39. Gür, Ö.Ş., *Mekan Örgütlenmesi.* 1996, Trabzon: Gür Yayıncılık.

40. Altman, I., Rapoport, A., Wohlwill, J.F., *Human Behavior and Environment.* Vol. 2, 1980, New York: Springer Science+Business Media.

41. Kurt, S., *Highly-Skilled Migrants' Adaptation Process through Participatory Urban Interior Design.* Master Thesis, International Masters of Interior Architecture Design Programme, Institute of Social Sciences, Istanbul Technical University, 2018, İstanbul.

Öner DEMİREL· Seyhan SEYHAN, Atila GÜL

Protected Areas

Abstract: Protected areas are land and sea parts that are managed and protected in situ (InSITU) within the framework of national and international laws and agreements and other effective means in order to ensure the continuity and protection of biological diversity, natural and cultural resources at a local, national and universal scale. For this reason, areas with protection status have a natural and cultural heritage value that is universal and concerns all humanity. The protected areas by law in our country constitute approximately 7.24 % of the area of Turkey. The number of areas with protected status is steadily increasing. It seems that there are many laws, decrees, regulations and directives that involve the protection of natural and cultural resource values.

Today, the country-wide nature conservation policy for protected areas is not sustainable, the existence of a wide variety and different legislation, the number of protected areas increasing day by day, the confusion of authority and responsibility between different institutions and organizations, the protected areas are evaluated under different statuses and more than one protection status effective and competent protection and sustainable management of protected areas are not realized as desired due to the changing and diversification of demands and needs for protected areas, problems in planning, management and implementation dimensions. This situation has also led to consequences such as the inability to protect the protected areas in scientific and technical dimensions, the inability to meet the demands and needs of the society for use, the social and economic pressure and problems at the local scale, the insufficiency of financial resources, and the insufficiency of the inspection and control mechanism of the areas.

There is definitely a need for a new legal and organizational regulation within the scope of the current legal and administrative dimension of protected areas in our country. In this context, it is of great benefit to enact a framework law named "Protected Areas Law" and to include all areas with protection status. There is a need to classify protection statuses in accordance with International (IUCN) conservation statuses and to define the powers and responsibilities. An ecological network must be created between protected areas, and it will be of great benefit to ensure an ecological and organic connection between other terrestrial and marine natural areas.

Handling the legal, planning/design, management and implemententta tion processes of protected areas in integrity and with sustainable policies and transferring them to future generations should be accepted as a national responsibility and duty.

Keywords: Protected areas, Development of protected areas, Legal Status and Legislation, Conservation Capacity

1 Introduction

The concept of nature is defined as a very complex set and synthesis of systems that are formed as a result of the interaction and interaction of cyclical processes with each other in integrity as a combination of productive, consumer and differentiating components and internal and external factors, designed in perfect balance and harmony, that can constantly renew and change itself [1, 2]. Rapid population growth, developing technology, industrialization, the emergence of environmental pollution with urbanization, forest fires, unconscious and misuse of forest areas, overgrazing, drying of wetlands, chemical drugs used in agricultural land, excessive and unconscious use of natural resources as a result of the effects, the pressure on nature has increased and increased. As a result of these pressures, there are important changes in nature, especially global warming, the natural balance is damaged and the life of all living things in nature, including humans, is adversely affected day by day [3].

As a result of all these effects on nature, people are exposed to a poor quality life caused by air and water pollution. These negativities cause the concept of nature conservation to remain an up-to-date concept, increase the awareness of nature conservation, and approach with continuous and new measures in social and political areas related to nature conservation.

Unhealthy areas have begun to emerge with the fact that human beings have become effective in these areas with many developments such as industrialization, urbanization, population growth, economy, science and technology, and the fact that this situation has increased in recent years has raised concerns for the future in people. The fact that human beings are in constant interaction with nature from the past to the present cause changes and deterioration in natural areas. With the development of conservation awareness in the world for many years in order to eliminate these deterioration and regressions, a process has been initiated for privileged areas, which have come to the fore with their social, cultural, economic, ecological and recreational features, which offer great richness in terms of biological diversity.

The international mobility in the protection of natural and cultural heritages, starting with global environmental organizations and international civil initiatives, has triggered the national conservation mobility by nurturing the national conservation mobility with various public institutions and organizations with private structuring statuses, and have put forward a will that forms the network of protected areas on the world. In this direction, people have created an awareness of protecting nature in natural areas, aimed at protecting nature, preventing degradation and passing it on to future generations [4].

Nature conservation areas, which have a critical importance for people and nature to attain a healthy environment, are important areas in terms of preserving biological diversity, providing clean air and water, and being a source of culture and livelihood for local people and communities [5].

Protected areas are areas that need to be protected, used and passed on to future generations due to their natural and cultural nature and functional nature. In line with this, one of the most important steps for the revival of these protected areas to be made depending on the characteristics and the characteristics of these areas "protect" the decision to the individual or the environment, special, privileged status for the acquisition of a structure or status of protection/legal initiatives is to launch and bring to safeguard the structure. In this way, it is hoped that the goal of conservation can be achieved more easily and permanently [6].

Protected Areas are the protection of the natural parts that are necessary for the continuity of both nature and human life by protecting the plant and animal existence and living areas of nature within certain criteria [7].

"International Union for Conservation of Nature and Natural Resources (IUCN)" defines these areas as "land and sea parts managed by laws and other effective tools to ensure the continuity and protection of biological diversity, natural and cultural resources". Nature conservation areas are very important areas in terms of protecting biological diversity, functioning natural ecosystems, densely managed landscapes and sea areas, ecological processes and being a shelter for species [8]. At the same time, these areas provide opportunities to meet the vital, social, cultural, economic, scientific, educational, aesthetic and recreational needs of the society, local people and visitors, while protecting the natural and cultural resource values [9]. IUCN and the Council of Europe Nature Protection Committee have determined different conservation statuses by classifying nature conservation areas in order to ensure correct use of the natural resources that the earth has.

Protected areas are subject to protection on ecological, economic, psychological, sociological, scientific, educational and moral (ethical) grounds. The whole of the values owned is the common wealth of past and future generations and is a humanitarian responsibility for the individual and society in terms of protection, survival and sustainability. It is an institutional and universal responsibility for states and countries [2].

2 Conservation Approaches and Development of Protected Areas in the World

Before the formation of nature conservation awareness around the world, some studies have been done on this subject, even if it is not used in its present sense. Some tribes living in the periods before Christ protected some forests by considering them sacred. Measures were taken to protect fruit trees in Rome 2000 years ago [4]. The first conservation concept known in history began with an edict issued by the Emperor Asoka of India in 252 B.C. to protect animals, fish and forests. In 1084, the King of England, William-I, ordered a land survey for conservation purposes. This study is titled "Domesday Book" in the whole of England; It has been prepared in such a way as to create an infrastructure for a realistic planning for the management and development of the country, including forests, fishing, agricultural lands, hunting reserves, and the exchange of fertile soil resources. In 1250, eagles, hawks, hawks and heron birds were requested to be protected in

England. Afforestation of agricultural lands and pastures in Dortmund, Germany in 1343 was subject to certain laws [10, 11]. In today's sense, the foundations of the conscious emergence of the idea of nature conservation, the legal framework of nature conservation studies, the separation of protected areas, taking legal measures and making scientific studies on the subject were laid in the second half of the 19th and 20th centuries [7, 12]. With the industrial revolution, it has caused quite radical changes in the field of conservation, as in other issues, and caused the foundation of today's understanding of protection and the concept of protected area.

As the mobility that started with the conservation approach gave results, the protection mobility in Europe took place before America [13]. For the first time, the Drachenfels Forest around the city of Bonn in Germany was taken under protection in 1829, then a part of the Kubany Virgin Forest in Bohemia in 1838 and the Teufelsmauer area in the Harz region in 1852 [14].

In America, in the 1870s, people who earned their livelihood by hunting in the Yellowstone region worked to preserve the existence of animals and plants in the region and to pass on their natural beauty to future generations. Later, in 1872, "Yellowstone National Park", which is the world's first national park with a size of 8670 km2, was declared in the United States Congress (Fig. 1). The announcement of this national park can be accepted as the official beginning of nature conservation thought by society and countries around the world [15, 7]. Today, Yellowstone National Park covers an area of 8987 kilometers. There are more than 10 thousand hot springs, more than 300 geysers and nearly 300 waterfalls in the park. In particular, grizzly bear, American black bear, bison, elk, antelope, jackal and lynx live in the park [16].

Fig. 1. The current state of Yellowstone National Park [17]

Before the end of the 19th century, the idea of nature protection started to be adopted in other countries and spread all over the world. After America, countries such as Australia, Canada, New Zealand and Mexico declared their first national parks. Australia had the first national park Royal National Park (Fig. 2) in 1879, Canada in 1885 Banff National Park (Fig. 3), New Zealand in 1897 Tongariro National Park (Fig. 4), and Mexico in 1898 El Chico National Park [18,7].

Fig. 2. The current state of the Royal National Park [19]

Fig. 3. The current state of Banff National Park[20]

Fig. 4. The current state of Tongariro National Park [21]

The first example of the idea of a national park in Europe is the Abisko National Park in Sweden, which was declared in 1909 (Fig. 5). The first example of France is Vanoise National Park dated 1963, and the Peak District of England dated 1951. The first national park declared in Africa is the 1925 Albert National Park, now called Virunga [12]. Until the First World War, 11 national parks were declared in Europe and approximately 40 in the world. This number reached 31 in Europe and 300 in the world after the Second World War. In the understanding of national parks in Europe, science and research are carried out by prioritizing the "human-nature" relationship, and visitors are allowed to use them for touristic purposes without harming the nature, while the understanding of the national park in America is to protect large natural areas closed to visitors [7].

Fig. 5. The current state of Abisko National Park [22]

In general, "Historical Turning Points in the Development of Protected Areas" are explained in Tab. 1 below [23] These developments have also led to his work on Turkey's nature conservation as well as all over the world.

Tab. 1. Historical milestones in the development of protected areas [23]

Historical Milestones	**Development Process of Protected Areas**
B.C. 10 000	As agriculture began to transform into the relationship between people and nature, local communities considered certain places "sacred", protecting them from certain human uses. Applied differently in different places over the next millennia, the concept became a practical measure that people found useful in both material and spiritual ways.
B.C. 252	Emperor Ashoka of India established protected areas for mammals, birds, fish, forests, and the earliest recorded areas where a government protected certain resources.
A.D. 684	On the order of the King of Srivijaya, the first "Indonesian Nature Reserve" was established on the island of Sumatra. Sumatra is recognized as the center of world-renowned mega-centers, with large protected areas that include the 25 000 square km Tropical Rainforest, a worldwide heritage.
1079	William Conqueror claimed the "New Forest" (England) as a royal hunting reserve, and the countryside should be protected against illegal harvest; Poaching became a law enforcement issue, but obtaining timber from the forest became necessary for Britain's war effort in the 17–19 centuries. Today, the "New Forest" is still a valuable protected area, and in 2005 it became England's newest National Park.
1865	Yosemite (California) was established by the US Congress as the first of a new national level model of protected areas; Yellowstone (1872) was named a national park for the first time.
1872	"Yellowstone National Park", which is the first national park of the world, has been declared.
1882	Latin America's first "El Chico National Park" was established in Mexico.
1903	The Society for the Conservation of the Wild Fauna of the Empire was established in the UK as the first non-governmental organization devoted to international protection, now known as Fauna and Flora International (FFI). Hundreds of other non-governmental protection organizations now support protected areas around the world.
1925	The first National Park in modern terms was established in Asia. (Angkor Wat, Cambodia).
1926	South Africa's Kruger National Park was established

(*continued*)

Tab. 1. Continued

Historical Milestones	Development Process of Protected Areas
1931	Whaling Treaty/Date of Signature 1934
1934	Argentina's Iguazu National Park was established.
1948	IUCN - World Conservation Union was established (as the International Union for Conservation of Nature) as a tool to promote conservation worldwide, establishment of protected areas has always been seen as an important focus area.
1949	Agreement/Date of Signature on the Establishment of the Mediterranean General Fisheries Council 1967
1950	(Paris Convention) International Convention for the Protection of Birds
1951	Convention on Establishment of Paris/Mediterranean and European Plant Protection Organization/Date of Signature 1965
1959	Washington/Antarctica/commissioned in 1961 and 53 countries ratified the treaty. The treaty establishes Antarctica as a scientific site and ensures the freedom of my scientific research; prohibits military activity on the continent. Since 2004, the Antarctic Treaty System Management Center is located in Buenos Aires, Argentina.
1961	WWF was established as a new international non-governmental organization (as the World Wildlife Fund), specifically to mobilize conservation support from the public, this fund marked the beginning of a growing era for international conservation.
1962	The First World Conference on National Parks in Seattle State, Washington, began a more formal world movement to support protected areas called the "UN List of Protected Areas" and proposed a system of categories. Prior to this, each country kept its own records, so no one knew the extent of the world's protected area system.
1963	Mweka African Wildlife Management College was established in Tanzania. In 2003, more than 4200 Africans graduated from the Mweka.
1963	Moscow/Convention/Signature Date Banning Nuclear Weapon Experiments in the Atmosphere, Space and Under Water 1965
1968	UNESCO Human and Biosphere Program started, biosphere reserves created (529 reserves in 105 countries covering 5 million km2 in 2007)
1968	Paris/European Convention for the Protection of Animals in International Transport/Date of Signature 1971

Tab. 1. Continued

Historical Milestones	**Development Process of Protected Areas**
1971	The Ramsar Agreement accepted 1708 sites covering more than 1.5 million km^2 and 152 contractual parties at the end of 2007.
1971	London, Moscow, Washington/Treaty/Signature Date of the Prohibition of Nuclear Weapons and Other Weapons of Mass Destruction and Installation in Seabeds, Ocean Floor and Below Them 1972
1972	The UN Conference on Environment and Development, Stockholm, Sweden ratified new conventions affecting protected areas, and the Nairobi-based United Nations Environment program (UNEP) was established. The World Heritage Convention, adopted in 1972, 166 natural World Heritage sites and 25 mixed natural and cultural sites (covering more than 1.8 million km2) were accepted.
1972	London, Moscow, Washington/Convention/Date of Signature 1975 on the Prohibition of Development, Production and Storage of Bacteriological (Biological) and Toxic Weapons and their Destruction
1973	Washington (CITES)/Convention on International Trade in Endangered Species of Wild Plants and Animals.
1974	Paris/International Energy Program Treaty/Signature Date 1981
1976	Barcelona Convention for the Protection of the Marine Environment and Coastal Zone of the Mediterranean
1977	A training program for protected area personnel has been established by the Tropical Agricultural Research and Training Center in the small city of Turrialba, Costa Rica. It continues to the present day and has provided trained personnel in most of Central America.
1978	IUCN's protected areas categories have been published; A framework has been established for the global assessment of protected area coverage.
1978	London/Convention on Prevention of Pollution from Ships/Date of Signature 1990
1979	Bern/Europe's Wildlife and Habitat Protection Convention
1979	Genoa/Long Range Transboundary Air Pollution Convention/ Signature Date 1983
1980	The World Conservation Strategy, published by IUCN, WWF and UNEP, spread the concept of sustainable development and the partnership between conservation and development.

(continued)

Tab. 1. Continued

Historical Milestones	Development Process of Protected Areas
1980	Protocol on the Protection of Athens/Mediterranean Against Land-Based Pollutants/Signature Date 1983
1981	The Conservation Monitoring Center, established by IUCN, and the Protected Areas Data Unit, established by the UK's National Parks and Protected Areas Commission, provided the first worldwide database of protected areas.
1982	The 3rd World National Park Congress, held on the Indonesian island of Bali, highlighted the importance of protected areas as a key element in national development plans.
1982	Protocol on Specially Protected Areas in Geneva/Mediterranean/ Signature Date 1986
1985	Granada Convention on the Protection of the European Architectural Heritage
1985	Vienna/Vienna Convention for the Protection of the Ozone Layer
1987	According to the report of the United Nations Sustainable Development Commission, Our Common Future (Brundtland Report) Report requires the land to cover 12 percent of the protected area status and support global action to protect biodiversity.
1989	Basel Convention on the Control of Transboundary Movement and Disposal of Hazardous Wastes.
1989	UN/AEK Declaration on the Conservation of Flora, Fauna and Habitat
1989	European Charter of Environment and Health
1991	The Global Environment Fund, created by the World Bank, UNDP and UNEP, has provided a large funding mechanism among new governments, particularly for protected areas negotiated through the CBD.
1992	Valletta Convention on the Protection of the European Archaeological Heritage.
1992	Bucharest/Convention on the Protection of the Black Sea Against Pollution
1992	The 4th World Congress of National Parks and Protected Areas, held in the capital of Caracas, Venezuela, highlighted the link between protected areas and other segments of society.

Tab. 1. Continued

Historical Milestones	**Development Process of Protected Areas**
1992	The Rio de Janeiro, Brazil World Summit set up Agenda 21 and ratified the Framework Convention on Climate Change, the Convention on Biological Diversity and the UN Forestry Principles, which are highly relevant to protected areas.
1994	Paris/United Nations Convention to Combat Desertification in Countries Subjected to Severe Drought and/or Desertification, Especially in Africa
1995	Protocol on Specially Protected Areas in the Mediterranean/Date of Signature 1996
1998	Aarhus Agreement on Access to Information on Environmental Issues, Public Participation in Decision Making and Appeal to the Judiciary
2000	The UN General Assembly approved the Millennium Development Goals for environmental sustainability with Goal 7.
2000	Florence-Europe Landscape Convention
2002	The World Summit on Sustainable Development, held in Johannesburg, South Africa, called for the loss of biodiversity by 2010 and a comprehensive system of protected areas to be established by 2012.
2003	Held in Durban, South Africa, the 5th World Parks Congress emphasized the importance of protected areas for sustainable development and focused on "benefits beyond borders".
2003	Action Plan for the Conservation of Specially Protected Areas (SPA) and Bird Species Listed in Annex 2 of the Protocol on Biological Diversity in the Mediterranean
2004	The Seventh Meeting of the Conference of the CBD Parties adopted a comprehensive Protected Areas Study Program to support the implementation of the CBD's protection components in place.
2008	Held in Madrid in 2008 and Turkey's contributions as laid III. The World Biosphere Reserves Congress resulted in a document called "Madrid Action Plan." The action plan aims to highlight biosphere reserves as sustainable development tools in the 21st century [4].
2009	Cartagena Biosafety Protocol

(*continued*)

Tab. 1. Continued

Historical Milestones	**Development Process of Protected Areas**
2010	The goal of the Nagoya Protocol, which was adopted at the tenth meeting of the Convention on Biological Diversity in 2010 at the tenth meeting of the Conference of the Parties and entered into force in 2014, is the fair and equal sharing of the benefits arising from the use of genetic resources and thus contributing to the protection and sustainable use of biological diversity.
2012	The 2012 World Conservation Congress, held in Jeju/KORE, aimed to create an environment where all segments of society work together, sharing responsibilities and benefits of conservation, in order to put aside differences and provide tools and mechanisms for proper environmental management. Solutions for the protection of biological diversity were sought against the environmental destruction that the world is facing, and as a result, the 2013–2016 Global Program was approved [4].
2012	In 2012, "Environment and Development Rio + 20 Conference" in Rio De Janeiro, focused on Institutional Themes for Green Economy and Sustainable Development.
2012	In the section titled "Protection and Management of the Natural Resource Base of Economic and Social Development" of the report prepared in 2012–2013, the importance of protecting natural resources for sustainable development was emphasized.
2013	The 1st Asian Parks Congress, held in Sendai, Japan in 2013, was the first successful organization that brought protected area practitioners across the Asian continent into a setting. The congress was deliberately taken to this date and it was aimed to include Asian perspectives in the World National Parks Congress to be held in 2014 [4].
2014	The last of the World National Parks Conferences series, which put forward perspectives on the conservation of biological diversity and the role of protected areas in sustainable development and is vital for conservation policies worldwide, was held in Sydney, Australia in 2014. As a result of the conference, it was reached the outputs that conservation of natural assets and protected areas can provide effective solutions to some of the most important development problems in the world [4].

Tab. 1. Continued

Historical Milestones	Development Process of Protected Areas
2015	In 2015, the "Sustainable Development Summit" was held and the 2030 "Sustainable Development Agenda" was formed at this summit and the member states of the United Nations accepted the "17 Sustainable Development Goals". The Sustainable Development Goals are built on the "Millennium Development Goals" adopted in 2000. While the Millennium Development Declaration covers the period 2000–2015, the 2030 Agenda targets a 15-year period from 2015.
2010–2023	National Climate Change Strategy
2011–2023	Climate Change Adaptation Strategy and Action Plan

3 Development in Turkey

During the Ottoman Empire, the first conscious idea of protection belonged to Fatih Sultan Mehmet and 15th in the middle of the century, in order not to fill some streams with mud, he banned the grazing of animals in stream basins, agriculture and construction, and made studies that would ensure strict soil retention by planting discrete grass on steep slopes. Developments in the Ottoman Empire, especially after the Tanzimat period, related to the environment and non-governmental organizations formed in society, reveal the attitude of the Ottoman Empire towards nature conservation. "The Land Law of 1858, the Forest Law issued in 1870, the Mecelle-i Ahkam-ı Courthouse (Civil Law) completed in 1876 [4]. Studies on the idea of Nature Conservation began in the 1950s in Turkey, which has a rich biodiversity and natural resource values [7].

Some of Nature Conservation in Turkey, Europe and America, such as the protective functions of forest areas (to prevent landslides, such as ensuring the formation of Water Resources) began to highlight or declared as a forest protection for [13]. The first regulation on this issue is No. 504, enacted in 1924, "law on Fenni-Style Management and operation of existing Bilumum forests in Turkey" 8th in the article, cutting in areas protected or to be taken by the state was prohibited, while the expropriation of non-state-owned forests was made mandatory, and this law could not be applied for a long time for some reasons and when the provision of the law was not as necessary [14, 24].

In the Republican era, the first forest law was enacted in 1937, the first forest law, numbered 3116, and the concept of "Conservation Forests" in its 43rd article, which made the idea of nature conservation concrete [14, 25]. In Turkey, the idea of Nature Conservation in the modern sense coincides with the 1940s. During this period, some authors wrote articles on this topic in journals and articles, revealing their thoughts and warning the public about it. For the first time in Turkey, the phrase "National Park" was used by Professor Dr. Selahattin Inal, one of the faculty members of the Faculty of Forestry of Istanbul University, in his work "we and our forestry in the face of Nature Conservation" published in 1949 [26]. In 1950 Belgrad Forest "Storage Forest" be declared, (Fig. 6) for the first time has launched an official nature conservation work in Turkey [12, 27].

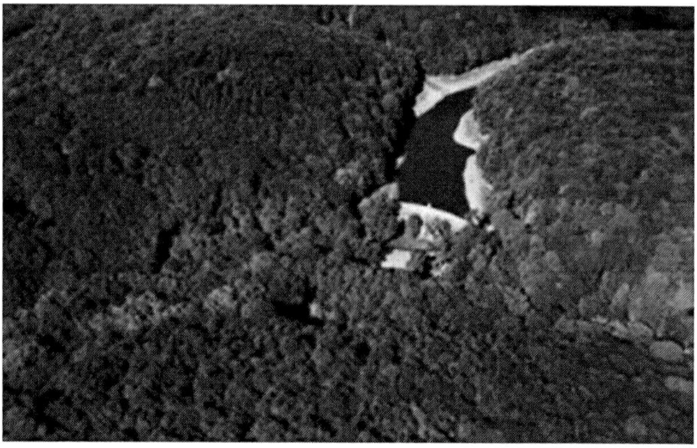

Fig. 6. Istanbul Belgrad Forest[28]

Turkey's first development in the legal sense, nature protection, 31 August 1956 and the 6831 Forest Law and Article 25 of 4. "National Park" of the Turkish legal terms (protection) entry in the legislation. In Article 4 of the Law, forests are divided into "1/Conservation Forests, 2/National Parks and 3/Industrial Forests" in terms of qualifications and character [26,29, 4]. In Article 25 of the Law;

> "*Forest management; Forests that will be deemed necessary due to its location and characteristics and areas that fall under forest regime; It separates and organizes it into national parks, natural parks, nature monuments, nature conservation areas and forest recreation areas in order to allocate it to the benefit of science and science, to preserve the nature, to provide the beauty of the country, to meet the various sports and recreational needs of the society, to allow touristic activities, manages and operates or has it operated when necessary*" [30].

This law was implemented immediately and Yozgat Çamlığı, the first national park of our country, was declared as a national park in 1958 [4]. National Parks Department was established in 1965 in Turkey. The "Regulation on the Separation, Administration and Management of National Parks" dated 1959 and numbered 6885 was abolished by this institution and the "Regulation on the Separation, Planning, Implementation and Management of National Parks" published in the

Official Gazette dated 1973 and numbered 14456 was accepted [31]. In 1976, the "General Directorate of National Parks and Hunting" was established under the Ministry of Forestry and the subject was organized at the desired level [26].

In 1982, this directorate continued as the "Directorate of National Parks Department" under the General Directorate of Forestry. Afterwards, the National Parks Department was attached to the General Directorate of National Parks and Hunting-Wildlife under the Ministry of Forestry. With the implementation of the "National Parks Law" dated 9 August 1983 and numbered 2873 and the "National Parks Regulation" dated December 12, 1986, National Parks were separated from the Forest Law. The Law on National Parks nature conservation areas in Turkey are protected by law. Finally, with the enactment of the "Law on the Organization and Duties of the Ministry of Environment and Forestry" dated 01 May 2003 and numbered 4856, the Ministries of Environment and Forestry were unified. The areas protected by this law are linked to the "General Directorate of Nature Conservation and National Parks" affiliated to this ministry [4]. 10 July 2018 dated and published in the Official Gazette No. 30474 with the presidential decree established the Ministry of Agriculture and Forestry and the decree defines the service units 412 article of the General Directorate of Nature Conservation and National Parks, the ministry continues its activities as a unit.

4 Legal Status and Legislation in the World

The main reasons underlying the emergence of the protected area idea can be grouped under two headings.

1. The nature of the natural and cultural values contained in protected areas and the necessity to **protect the current situation in line with these qualities,**
2. Human uses for protected areas pose a threat and **therefore a restriction should be placed on these uses and change should be prevented.**

Protecting protected natural areas with certain restrictions is not a simple prohibition activity. Behind this effort are **moral, aesthetic, sociological, economic and ecological reasons** that are closely related to our existence. All of the values owned are the common wealth of past and future generations **and one sided disposal should not be made**[2].

The first studies on this subject in the world were carried out at the "Congress on the Conservation of the Flora and Fauna of Africa" in 1933. At the congress; natural areas; It has been examined in two groups as 1 /national parks and 2/Absolute Nature Protected Areas [25]. In the "Nature Protection and Wildlife Protection Convention" made in the Western hemisphere in 1942; Four statuses have been established regarding protected areas: 1/national park, 2/national reserve, 3/nature monument and 4/absolute wildlife reserve [8]. While the "International Union for Protection of Nature (IUPN)" was established in Fontainebleau, France in 1948 with the initiative of UNESCO, an organ of the United Nations, the name of this union, which developed under the roof of the United Nations, was named as "International

Nature Reserve" at the 6th General Assembly meeting in Athens in 1958. And the "Union for Conservation of Natural Resources (IUCN)". This union has become the sole authority that determines the criteria for the systems of nature conservation areas, and six nature conservation area statuses related to this issue have been revealed [32]. The statuses determined by the International Union for Conservation of Nature and Natural Resources (IUCN) are explained below [8].These;

1a. Absolute Nature Reserve: Biological diversity, geological and geomorphological features are strictly protected. People's visit, use and effects are controlled and limited to ensure conservation values.
1b.Wild Habitat: Areas that generally preserve their natural character and influence, largely unaltered or slightly altered areas, where there is no permanent or significant human settlement, managed to protect their natural conditions.
2. National Parks: These are areas that evaluate large natural areas or areas close to nature, ecosystems and characteristics in the ecological process, are compatible with environmental and cultural values, and offer scientific, educational, recreational and visitor opportunities.
3. Nature Monument: It is the protection status with its natural characteristics, biodiversity and habitats that are at the forefront with its features.
4. Habitat/Species Management Area: Areas for the conservation, management and restoration of species or habitats.
5. Land or Marine Landscape Protection Area: It is the protection status that aims to protect and maintain important land and sea landscapes, other values arising from human interactions through traditional management practices.
6. Protected Areas Including the Sustainable Use of Natural Resources: These are protected areas based on the sustainable use of natural resources in cases where protection and sustainable use of natural ecosystems are mutually beneficial.

As Classes of Special Interest of IUCN except these protection statuses; There are Anthropological Reserves/Resource Areas Managed by Natural Biotic Areas/Multiple Uses. As the Classes Used in International Programs; There are Biosphere Reserves and World Heritage Sites.

With the rapid spread of conservation awareness around the world, nature conservation areas in various countries are carried out under the relevant institutions and this idea is supported by the relevant legal statuses.

EUROPARC was established in 1973 under the name of "European Federation of National Parks and Nature Parks" and it is an independent organization that has evolved until today. EUROPARC creates a forum where professional experiences are shared, technical projects and process are brought together. It combines the protected areas (national park, regional park, nature park and biosphere reserve area) in Europe. The purpose of EUROPARC is to protect Europe's wildlife, habitat and landscape values [33].

The Council of Europe has divided protected areas into four classes. As a criterion; scientific value of the field, traditional or contemporary human activities, recreational facilities and accessibility elements are discussed. The first two of these classes, called A, B, C and D, contain strict protection areas, while the other two contain less stringent protection areas [25]

Another regional conservation union was formed as a result of the joint action of Latin American countries. Latin American countries have cooperated on protected areas for many years, and as a result, they organized the "First National Parks and Other Protected Areas Congress" in 1997 with the Santa Marca Declaration. In this congress, it is aimed to bring the current management categories to a common line, to adopt the classification of IUCN and to advance the classification unity to be formed [34].

Another regional union is Asia and the Pacific Region; in 1997, a study titled "Status, Trends and Future Scenarios Conservation of Forests in Protected Areas in the Asia Pacific Region" was published by FAO. This comprehensive study brought together countries in the region to use IUCN categories in protected area management [35].

If the status of legislation in Turkey outside the protected areas have been declared under international agreements are shown in Tab. 2. [25].

Tab. 2. International agreements and contracts within the scope of protected area status was declared in Turkey [25].

Protected Area Status	**International Conventions**
World Heritage List	1982 "Convention on the Protection of the World Cultural and Natural Heritage"
Gene Conservation and Management Areas	1992 "Biological Diversity Convention"
Biogenetic Reserve Areas	1973 "Official Decision No. 2 of the Council of Europe Committee of Ministers" 1973 "Recommendation No. 17 of the Council of Europe Committee of Ministers" 1979 "Convention on the Conservation of Europe's Wildlife and Habitat- Bern"
Biosphere Reserve Areas	1974 "UNESCO's Human and Biosphere Program, 1982 "United Nations Conference on Environment and Development in Rio De Janeiro" 1984 "For the Biosphere Protected Areas at the 8th session of UNESCO Action plan" 1995 UNESCO General Conference 28c/2.4 Directive Seville Strategy 1992 "Convention on Biological Diversity"

(continued)

Tab. 2. Continued

Protected Area Status	International Conventions
Wetlands (Ramsar Areas)	1971 "Convention on Wetlands of International Importance Especially as Waterfowl Habitat-Ramsar Convention"
Sea Turtle Breeding Areas	1979 "Convention on the Conservation of Europe's Wildlife and Habitat-Bern" 1976 "Convention on the Protection of the Mediterranean Against Pollution-Barcelona" 1973 "Convention Concerning the International Trade of Endangered Wild Animal and Plant Species"
Mediterranean Monk Seal Habitats	1979 "Convention on the Conservation of Europe's Wildlife and Habitat-Bern" 1976 "Convention on the Protection of the Mediterranean Against Pollution-Barcelona" 1973 "Convention Concerning the International Trade of Endangered Wild Animal and Plant Species" Land Hunting Law No. 4915 Fisheries Law No. 1380
Turkey's Important Bird Areas	"NATURA 2000 Status" "Bird Protection Regulation of the European Union (79/409/EEC)" "Habitats and Species Conservation Regulation of the European Union (92/43/EEC)"
Turkey's Important Plant Areas	"Endangered Habitat Types "in Bern Convention Annex List II (Final Declaration No: 4 1996)" "Natura 2000 Special Protection Areas (Habitat Regulation Annex List II b and IV b-Annex List I)" 1973 "Convention on International Trade in Endangered Species of Wild Animals and Plants Annex List II"
Turkey's Prime Alanları-Forest Hotspots	WWF identified 100 Forest Areas that include Europe's Biological Wealth in 1994. Of these, 9 of them are in Turkey.
Geological Heritage Sites-Heritage Geographies	In 1991, at the Geological Heritage Preservation Symposium held in Digne, France, a joint declaration was issued by calling all countries to protect the geological heritage.
Blade Tread Areas	No
Ecogeographies	No
Zero Extinction Areas	The definition of sensitive and unique areas in line with the Red Categories of IUCN (one or many species with the most endangered species) is given as "Alliance for Zero Extinction Sites" Zero Point Extinction Areas ".

Tab. 2. Continued

Protected Area Status	International Conventions
Emerald Network Areas	1979 "Convention on the Conservation of Europe's Wildlife and Habitat-Bern"
Natura2000	"NATURA 2000 Status" "Bird Protection Regulation of the European Union (79/409/EEC)" "Habitats and Species Conservation Regulation of the European Union (92/43/EEC)"

Another international status to be evaluated within the scope of the International Protected Areas Network, which has gained a legal dimension as the texts applied to our domestic law within the scope of International Conventions above, is the PAN Parks application.

The Protected Area Network Parks Foundation is a non-profit organization that aims to increase the effectiveness of protected area management and to improve the image of Europe's diverse nature. The Foundation aims to create a protected area network that creates synergies between nature conservation and tourism across Europe. Candidate parks are independently audited whether they meet a set of detailed criteria before obtaining official Pan Parks status. Kure Mountains National Park, Turkey's first PAN Parks has been nominated in 2006 [36].

5 Conservation Capacity in Turkey

7 gene known as one of the most important central Turkey on Earth, is the most important country in terms of geography biodiversity. With its strategic location, Turkey is a key country in terms of the survival of many species. Therefore the responsibility of Turkey in the field of nature conservation is more than any other country. The first law on nature protection period of the Republic of Turkey dated 08.02.1937 date and 3116 "Forest Act". Below are the legal regulations regarding nature conservation.

- 08.02.1937-"Forest Law" numbered 3116, Steps Taken Towards the Protection of Nature
- 05.05.1937- "Land Hunting Law" numbered 3167, regulation of hunting
- 1956- "Forest Law" numbered 6831, the inclusion of the concept of National Park
- 1965, establishment of the National Parks Department, institutionalization
- 1966, the first book on National Parks Planning (Zekai Bayer)
- 23.03.1971-"Fisheries Law" numbered 1380
- 1976, establishment of the General Directorate of National Parks and Hunting
- 16.03.1982-"Tourism Incentive Law" numbered 2634

- 21.7.1983 -2863 "Law on the Protection of Cultural and Natural Heritage"
- 09.8.1983-"Environmental Law" numbered 2872

The first direct legal regulation on nature protection in our country

- It is the National Parks Law No. 2873, dated 9.8.1983.
- 1982, Ministry of Agriculture and Following the merger of the Ministry of Forestry, this General Directorate was closed and organized as the "Department of National Parks" under the General Directorate of Forestry.
- Decree Law No. 383 on the Establishment of the Special Environmental Protection Agency, dated 13.11.1989.
- "Coastal Law" dated 17.04.1990 and numbered 3621
- In 1991, the Institution regained the status of General Directorate.
- Law on the Establishment and Duties of the Ministry of Forestry, dated 21.05.1992 and numbered 3800,
- Land Hunting Law No. 4915, dated 1.7.2003
- 2003, with the "Law No. 4856 on the Organization and Duties of the Ministry of Environment and Forestry", the Ministries of Environment and Forestry were unified, with Article 43 of the Law and the Decree Law No. 443 on the Establishment and Duties of the Ministry of Environment, dated 09.08.1991. Law on the Establishment and Duties of the Ministry of Forestry dated and numbered 3800 has been repealed. With the Law No. 4856, the General Directorate of National Parks and Hunting-Wildlife and the General Directorate of Environmental Protection were evaluated together and restructured as the "General Directorate of Nature Conservation and National Parks".
- Number 644 "Decree Law on the Organization and Duties of the Ministry of Environment and Urbanization" (Official Gazette, 04 July 2011 Date and 27984 Number),
- Number 645 "Decree Law on the Organization and Duties of the Ministry of Forestry and Water Affairs" (Official Gazette, 04 July 2011 Date and 27984 Number) and
- Number 648 "The Decree Law on the Organization and Duties of the Ministry of Environment and Urbanization and the Legislative Decree Amending Certain Laws and Decree Laws" (Official Gazette, August 17, 2011 and Number 28028)
- Regulation No. 6885 on the Separation, Administration and Operation of National Parks (1959)
- Regulation No. 14456 on the Separation, Planning, Implementation and Management of National Parks (1973)
- "National Parks Regulation" No. 19309 (12.12.1986)
- The Regulation on the Identification and Registration of Immovable Cultural and Natural Property Required to be Protected, No. 19660 (10.12.1987)
- Regulation on the Identification and Registration of Immovable Cultural and Natural Property Required to be Protected, No. 20258 (19.8.1989)
- Regulation on the Implementation of the Convention No. 24623 on the International Trade of Endangered Wild Animal and Plant Species (27 December 2001)

- "Regulation on Wildlife Protection and Wildlife Development Areas" numbered 25637 (08.11.2004)
- 25818 "Regulation on the Protection of Wetlands" (17.5.2005)
- Regulation on the Amendment of the Regulation on the Conservation of Wetlands No. 27684 (26.08.2010)
- Regulation on the Procedures and Principles Regarding the Determination, Registration and Approval of Protected Areas No. 28358 (19/7/2012)
- Regulation on Plans to be made in Protected Areas No. 28242 (23/3/2012)
- Regulation on the Identification and Registration of Immovable Cultural Property and Sites Required to be Protected No. 28232 (13.03.2012)
- Regulation No. 28635 on "Natural Assets and Natural Protected Areas and Administration of Places Under the Provision and Possession of the State Located in Special Environmental Protection Areas" (02.05.2013)
- Regulation on the Amendment of the National Parks Regulation No. 28945 (18 March 2014)
- "Regulation on the Conservation of Wetlands" No. 28962 (4/4/2014)
- Regulation on the Amendment of the "Regulation on the Determination and Registration of Immovable Cultural Property and Sites Required to be Protected" No 29231
- Regulation on the Amendment of the Regulation on the Procedures and Principles Regarding the Determination, Registration and Approval of Protected Areas, numbered 30223 (27.10.2017)
- Regulation on the Amendment of the Regulation on the Procedures and Principles Regarding the Determination, Registration and Approval of Protected Areas, numbered 31070 (16.03.2020).

"Nature conservation policies and approach to protected area statuses" that have been carried out for 60 years in our country have been changed and brought legal and administrative changes.

While the Ministry of Forestry was responsible for the protected areas in the 1970s in our country, the Ministry of Agriculture, Forestry and Rural Affairs on 13.12.1983, the Ministry of Environment and Forestry on 08.05.2003 and the Environment and Urbanization established by the Decree Law No. 644 as of 04.07.2011. Ministry of Forestry and Water Affairs established by Decree 645 [29]. With the Presidential Decree on the Presidential Organization published in the Official Gazette dated 10 July 2018 and numbered 30474, the Ministry of Forestry and Water Affairs was abolished and the Ministry of Agriculture and Forestry was established.

Institutions responsible for protected areas today;

- The Republic of Turkey Ministry of Environment and Urban Development (General Directorate of Natural Assets Protection)
- The Republic of Turkey Ministry of Agriculture and Forestry (General Directorate of Nature Protection and National Parks)

- The Republic of Turkey Ministry of Culture and Tourism (Cultural Assets and Museums General Directorate)

For the protection of this wealth, 598 protected areas with a total area of more than 3.2 million hectares, including national parks, were declared (Tab. 3). In addition, other statuses and information about protected areas are given (Tab. 4, 5, 6, 7, 8,).

Tab. 3. Number of Protected Areas of Turkey, size, Legislation and Responsible Organizations

	Name	Number	Area (Hectares)	Legislation to which	Responsible Organization
1	**Nature Protection Area**	31	47.244	2873	TOB
2	**National Park**	45	884.933,7	2873	TOB
3	**Natural Monument**	115	7.143,14	2873	TOB
4	**Nature Park**	250	99.397	2873	TOB
5	**Wildlife Development Area**	84	1.189,308	4915	TOB
6	**Wildlife Protection Area**	1	8.000	4915	TOB
7	**RAMSAR Area**	14	184.487	Ramsar Convention	TOB
8	**Wetland of National Importance**	59	361.606	Regulation on the Protection of Wetlands	TOB
9	**Wetland with Local Importance**	13	1.602	Regulation on the Protection of Wetlands	TOB
10	**Biosphere Reserve**	1	25.258	UNESCO Human and Biosphere Program	TOB
11	**Special Environmental Protection Area**	18	2.582.968	383 KHK	ÇŞB
12	**Natural Site**	2.434	1.991.700	2863	ÇŞB

Tab. 3. Continued

	Name	Number	Area (Hectares)	Legislation to which	Responsible Organization
13	**Natural Being (Memorial Tree)**	8724		2863	ÇŞB
14	**Natural Being (Cave)**	249		2863	ÇŞB
15	**Intra-Forest Recreation areas**	146 (44 A Type) 102 B Type)		December 12, 1986 National Park Regulation	TOB
16	**Urban Forests**	142 (71 province)		Establishment of the Ministry of Forestry and Water Affairs with the Decree No. 645	TOB
17	**Seed Stands**	345	46266		TOB
18	**Gene Forests**	115	16210		TOB
19	**Seed Gardens**	54	1261		TOB
20	**Research Forest**				TOB
21	**Aquaculture Production Areas**	Fisheries Law		Law 1380 of 23 March 1971	TOB

Tab. 4. Data on Natural Sites

Protection Status	Total number	Total Area (Ha)
1st Degree Natural Protected Area	957	1097695
2 nd Degree Natural Protected Area	294	63688
3rd Degree Natural Protected Area	617	140058
Uncertain Degree	83	26207
Sensitive Area to be Protected Definitely	62	58029
Qualified Nature Reserve	247	262181
Sustainable Protection and Controlled Usage Area	294	121087
TOTAL	2554	1768948

The rates where the protected areas within the scope of the "National Parks Law" numbered 2873 of 1983 coincide with the natural site status;

- National Parks: 44.96 % of the 20
- Natural Parks: 45.42 % of 43

Nature Protected Areas: 93.17 % of 10

- Natural Monuments: 4 and 97.74 % of them
- Wildlife Development Area: 40.99 % of 15
- Wetlands: 41.47 % of 60 of them have been given status as natural protected areas.

Tab. 5. The Republic of Turkey Ministry of Culture and Tourism Cultural Assets and Museums are located in areas under the responsibility of the Directorate General

Protected Area Type	**Number**
Other Sites	18.933
World Heritage Site	18
Immovable Cultural Property	108.813
TOTAL	127.757

Tab. 6. Republic of Turkey Ministry of Culture and Tourism Cultural Assets and Museums General Directorate of Protected Areas registered in Turkey General

SITE TYPE	**Number**
Archaeological Site Area	17.958
Urban Site Area	299
Historical Site Area	171
Urban Archaeological Site Area	35
Mixed Sites	95
• Archaeological and Urban Site Area	47
• Archaeological and Historical Site Area	15
• Archaeological-Urban-Historical Site Area	3
• Historical-Urban Site Area	30
Total	18.558
Overlapping Sites (With Natural Site) (as of 17.08.2011)	
Archaeological and Natural Site Area	321
Archaeological-Natural-Historical Site Area	5
Archaeological-Natural-Historical-Urban Site Area	2
Natural and Historical Site Area	13
Natural-Historical-Urban Site Area	4
Urban-Natural-Archaeological Site Area	9
Urban and Natural Site Area	21
Total	375
The Overall Total 18.933	

Tab. 7. UNESCO World Heritage List

UNESCO World Heritage List (2018)
• Registered on the UNESCO World Heritage List worldwide as of 2018 ➢ Cultural 869 pieces ➢ Natural 213 ➢ There are 1121 natural and cultural assets in total, 39 of which are Mixed (Cultural/Natural).
• UNESCO World Heritage List of Turkey; ➢ There are a total of 19 assets, 17 of which are Cultural and 2 are Mixed (Cultural/Natural). On the UNESCO World Temporary Heritage List; • There are 83 assets in total as 2 Mixed (Cultural/Natural), 3 Natural and 78 Cultural.

Tab. 8. Turkey General for the Protection of Immovable Cultural Property (2018 Year-End)

Registered Immovable Cultural Property in Turkey in General	**Immovable Number**
Sivil Mimarlık Örneği	69.104
Religious Buildings	10.147
Cultural Buildings	12.530
Administrative Structures	2.985
Military Structures	1.252
Industrial and Commercial Buildings	4.171
Cemeteries	5.169
Martyrdoms	307
Monuments and Monuments	375
Ruins	2.702
Protected Streets	71
Total	108.813

6 Evaluation

Protected areas are land and sea parts that are managed and protected in situ (InSITU) within the framework of national and international laws and agreements and other effective means in order to ensure the continuity and protection of biological diversity, natural and cultural resources at a local, national and universal scale. For this reason, areas with protection status have a universal nature and a natural and cultural heritage value that concerns all humanity.

Protected area by law in our country, (OGM+ DKMP +ÖÇK + Natural Sites) constitute approximately 7.24 % of the surface area of Turkey. The number of protected areas is gradually increasing. It appears that there are many laws, decrees, regulations and directives involving the protection of natural and cultural resource values. Considering that the legal process for the protected areas in our country has begun to transform the idea of nature conservation into action since 1937, it is seen that there is nature conservation experience and knowledge to date. It can be said that legal regulations and institutional organizations for protection are effective in protecting protected areas, especially by legal means. Especially in the planning and management of existing protected areas in forest areas, the General Directorate of Nature Conservation and National Parks, with its current name, the protection of these areas, long-term development plans, training programs, investment policies, management organization process, together with its services and contributions from the past to the present in nature conservation. It is seen that he has significant experience and knowledge.

However, today, the country-wide nature conservation policy for protected areas is not sustainable, the emergence of various and different legislation, the number of protected areas is increasing, the authority and responsibility confusion between different institutions and organizations, the granting of different status to protected areas or giving more than one protection status, Changing and diversification of demands and needs for protected areas, problems in planning, management and implementation dimensions, etc., adversely affect the effective and competent protection and sustainable management of protected areas. This situation has also led to consequences such as the inability to protect the protected areas in scientific and technical dimensions, the inability to meet the demands and needs of the society, the social and economic pressure and problems at the local scale, the insufficient financial resources, the insufficiency of the inspection and control mechanism of the areas.

Since 2002 has begun working on legislation for nature protection Republic of Turkey Prime Laws and Decisions by the Directorate General 17.05.2012 Date of Turkey submitted to the National Assembly "Nature and Biodiversity Conservation Act Bill" Turkey Grand National Assembly 24 in legislative renovation period and 2. legislative renovation Year, with the number 297 as Turkey has been referred to the Parliament Environmental commission and reports prepared after the evaluations for the bill by the committee, it was submitted to parliament on 18/06/2012. However, these efforts were also inconclusive and this draft was removed from the agenda.

Another case is 1.773.856 ha in the country area, which is 2430. It is the situation related to natural protected areas that cover an area. In accordance with the provisional article 6 of the "Decree on the Organization and Duties of the Ministry of Environment and Urbanization" numbered 644 of 2011, It has been transferred to the Ministry of Environment and Urbanization.

Another problem ball is related to intra-forest recreation areas. In the "National Parks Regulation" (12 December 1986 and 19309 Numbered Official Gazette), the Intra-Forest Recreation Places, which were established as a new protection status in 1986, have been used as «Type A, B, C, and D Their number has reached up to 1459 as "Intra-Forest Recreation Areas", of which D type in-forest recreation areas are considered as "Urban Forests" and their number has reached 142.

With the legal arrangements made on the Intra-Forest Recreation Areas since the date it was first announced, a complete legal and administrative confusion has paved the way for these areas. The legal arrangements made after the first "National Parks Regulation" announced in 1986 are given below;

- "Recreational Areas Regulation" numbered 26305 dated 30 September 2006.
- "Regulation on Plans to be made in Protected Areas" No. 28242 dated March 23, 2012
- "Recreational Areas Regulation" No. 28578 dated March 5, 2013
- "Regulation Amending the Regulation on Recreational Areas" numbered 29222 dated December 31, 2014.
- 296 "Recreation Areas Implementation Communiqué"
- "Recreational Areas Implementation Communiqué" No. 300 dated 05.03.2013.

With the Law on the Amendment of the Tourism Incentive Law No. It is thought that with the transfer of the 8th article of the new law to the Ministry of Culture and Tourism, the practices that will initiate natural destruction and ecological losses have been experienced in these areas.

Currently, the management of these areas, "No. 645" Republic of Turkey Forestry and Water Affairs of the Organization and Duties of the Decree Law "with the General Directorate of Forestry showing operates under" Non-Wood Products and Services Department "under the" Promenade Branch "is below. Article 10 of the "Law on the Amendment of the Tourism Encouragement Law" published in the Official Gazette No. The Ministry is authorized to allocate the facilities that can be used for accommodation purposes among the facilities purchased for the fixed assets of the General Directorate and the recreation areas with accommodation facilities of all types and capacities, and the revenues belong to the General Directorate of Forestry" to authority it is transferred to the Ministry of Culture and Tourism of the Republic of Turkey.

Our country, which became a party in 2003 to the "European Landscape Convention" signed in Florence in 2000, started the studies of the country landscape atlases as a requirement of this contract, and firstly completing the "Yeşilırmak Basin Landscape Atlas", it continues to work on the "Büyükmenderes Landscape Atlas". There is the "Landscape Protection Branch Directorate" under the Sensitive Areas Department of the General Directorate of Nature Conservation and National Parks as the only institution that coordinates these studies and where the word "LANDSCAPE" is mentioned. The scope of influence of this Directorate should be expanded and it should be ensured that nature protection and landscape protection are handled in integrity.

In our country, there is definitely a need for a new legal regulation regarding the legal status of protected areas, that is, natural and cultural areas, under current conditions. In this context, enacting a law named "Protected Areas Law" and it is of great benefit to include all areas with protection status. There is a need to classify protection statuses in accordance with International (IUCN) conservation statuses and to define powers and responsibilities. An ecological network must be created between protected areas and organic connection with other natural areas will be of great benefit.

Handling the legal, planning/design, management and implementation processes of protected areas in integrity and with sustainable policies and transferring them to future generations should be considered as a national task.

In terms of protected areas, more terrestrial areas are primarily considered in our country. In addition, importance should be attached to the creation of protection statuses for our maritime and coastal areas, as well as the wetlands, and to ensure sustainability.

Last Word

Conservation awareness and conservation ethics are a cultural problem and the level of conservation culture is very low in our country. It is the last point that can be reached through the use and implementation of "knowledge". Lack of knowledge, unconsciousness; unconsciousness has reinforced a distorted morality of protection. The discourse and action dimension should be realized in primary, secondary, high school and higher education, starting from kindergarten to create protection awareness and awareness. In addition, a common synergy should be created by giving nature conservation awareness and awareness to NGOs, managers, planners, decision-makers and all stakeholders in all institutions and organizations related to conservation simultaneously and in coordination.

Lack of knowledge and education should be overcome for individuals to have protection ethics on Nature Conservation. It should not be forgotten that the way to have the morality of protection by creating protection awareness passes WITHOUT WANTING TO PROTECT.

References

1. Gül, A., ve Türker, H.B., *Farkındalık ve Bilinçlenme Süreçleri Kapsamında, Doğa Koruma-Etik Ilişkisi*, 2014. SDÜ II. Uluslararası Davraz Kongresi (29–31 Mayıs 2014) Küresel Sorunlar ve Çözüm Arayışları. Bildiriler Kitabı. s.1745–1778. Isparta. ISBN: 978-9944-452- 82-3.
2. Gül, A., *Korunan Doğal Alanların Bütüncül Planlama&Tasarım Süreçleri Eğitimi*, 2018. Sunum Notları. TUBİTAK-2237-A Bilimsel Eğitim Etkinliklerini Destekleme Programı- 27 Ağustos-02 Eylül 2018.

3. Metin, A.E., ve Gül, A., *Doğa Korumada Nudge Teorisinin Uygulanabilirliği*, 2020. II. Uluslararasi 29 Ekim Bilimsel Araştırmalar Sempozyumu. s.144–156. Yayın Tarihi: 30.10.2020. ISBN-978-625-7279-12-3 /ANKARA. https://cb66109b-0e0b-4dac-8f11-8c6697985f7e.filesusr.com/ugd/614b1f_ e1b8d20400294b3dae25b c309b59dac4.pdf
4. Yeşil, M., *Doğa Koruma Yaklaşımlarındaki Değişimlerin Dünyada ve Türkiye'deki Tarihsel Süreci.*Türk Tarım – Gıda Bilim ve Teknoloji Dergisi, 2016. **4**(10): 867–876.
5. Anonim, *Protected Planet Report, How Protected Areas Contribute To Achieving Global Targets For Biodiversity*, 2016. United Nations Environment Programme. pp.1.
6. KIRÇEV, *Türkiye'nin Tabiatı Koruma Alanları*, 2000. Kırsal Çevre ve Ormancılık Sorunları Araştırma Derneği, Ankara.
7. Yücel M., ve Babuş D., *Doğa Korumanın Tarihçesi Ve Türkiye'deki Gelişmeler.* Doğu Akdeniz Ormancılık Araştırma Müdürlüğü, Doa Dergisi (Journal Of Doa), 2005. Sayı: **11**, s.151 - 175.8. Choo, H., Nikrahei, B., Nasar, J.L., and Walther, D. Neural Decoding of Architectural Styles from Scene-Specific Brain Regions. Journal of Vision, 2015. **15**(12): p. 520–520.
8. IUCN, *Guidelines for Applying Protected Area Management Categories. Gland*, 2008. Switzerland: IUCN. x + 86pp. With Stolton, S., P. Shadie and N. Dudley (2013). IUCN WCPA Best Practice Guidance on Recognising Protected Areas and Assigning Management Categories and Governance Types, Best Practice Protected Area Guidelines Series No. 21, Gland, Switzerland: IUCN. xxpp. Ed.Nigel Dudley.
9. Düzgüneş E., *Milli Park Planlamalarında Ziyaretçi Yönetimine İlişkin Yeni Bir Model Önerisi: Altındere Vadisi Milli Parkı (Maçka/Trabzon) Örneği*, 2015 Doktora Tezi, Karadeniz Teknik Üniversitesi, Fen Bilimleri Enstitüsü, Trabzon.
10. Mac Kinnon J., Mac Kinnon K., Child G., Thorrell J., *Managing Protected Areas in the Tropics*, 1986.International Union for Conservation of Nature and Natural Resources and The United Nations Environment Programme, Gland, Switzerland.
11. Yücel M., *Doğa Koruma Alanları ve Planlaması.* Ç.Ü. Ziraat Fakültesi Yayını, 1995. **255**, Adana.
12. Anonim, *T.C.Orman Genel Müdürlüğü, Proje Sonuç Raporu Tabiat Parkı – Mesire Yeri (Orman İçi Dinlenme Yeri) Kavramları Ve Tabiat Parkı Belirleme Kriterleri Üzerine Yeniden Yapılandırma Önerisi: Belgrat Ormanı Örneği*, 2014. Proje Numarası:10.8401/2013–2014, Marmara Ormancılık Araştırma Enstitüsü Müdürlüğü, İstanbul.
13. Erdönmez, C., *Bursa-Keles Kırsal Alanlarındaki Toplumsal Yapının Çevre Sorunları Üzerine Etkisi*, 1997. Doktora Tezi. İÜ Fen Bilimleri Enstitüsü,

14. İnal, S., *Tabiatı Koruma Karşısında Biz ve Ormancılığımız,* 1949. Orman Genel Müdürlüğü Yayını, Özel Sayı 84. Osmanbey Matbaası, İstanbul.
15. Akesen A., *Türkiye'de Ulusal Parkların Açıkhava Rekreasyonu Yönünden Nitelikleri ve Sorunları,* İ.Ü. Orman Fakültesi Yayını, 1978. No: **262**, İstanbul.
16. *https://www.nps.gov/parkhistory/hisnps/npshistory/timeline_annotated.htm.*15.02.2020
17. *Yellowstone Milli Parkı'nın Günümüzdeki Durumu,* https://www.nps.gov/yell/index.htm 19.02.2020
18. Eagles, P., F., J., McCool, S., F. and Haynes, C., D., *Sustainable Tourism in Protec- ted Areas, Guidlines for Planning and Management, World Commission on Protected Areas (WCPA), Best Practice Protected Area Guidliness Series* No: 8, 2002. UNEP,IUCN, WTO Yayınları.
19. *Royal Milli Parkı'nın Günümüzdeki Durumu* https://www.national parks.nsw.gov.au/visit-a-park/parks/royal-national-park20. Steele, F., *The Sense of Place.* 1981, CBI Publishing Company, Inc. 19.02.2020
20. Banff Milli Parkı'nın Günümüzdeki Durumu https://en.wikipedia. org/wiki/Banff_National_Park, 19.02.2020
21. *Tongariro Milli Parkı'nın Günümüzdeki Durumu,* http://www.national park.co.nz/things-to-do.19.02.2020
22. *Abisko Milli Parkı'nın Günümüzdeki Durumu,* http://www.national parksofsweden.se/choose-park---list/abisko-national-park/19.02.2020
23. Chape S., Spalding M., Jenkins M.D., *The World's Protected Areas, Status, Values And Prospects in The 21st Century.The UNEP World Conservation Monitoring Centre,* 2008. University of California Press. Berkeley, USA.
24. Eryılmaz A.Y., Çevre *Korumasının Cumhuriyet Dönemi Ormancılığımızdaki Yeri,* İ.Ü Orman Fakültesi Dergisi, 1990. A(40)**2**, 123–131.
25. Demirel Ö., *Doğa Koruma ve Milli Parklar,* 2005. Karadeniz Teknik Üniversitesi Rektörlüğü, Orman Fakültesi, Genel Yayın:219,Fakülte Yayın No:37.
26. Gülez S., *Park – Bahçe ve Peyzaj Mimarisi* Karadeniz Teknik Üniversitesi Orman Fakültesi Ders Teksirleri, 1989. Serisi:29, Sayı:2, s.213–214, Trabzon.
27. Özdönmez, M., İstanbullu, T., Akesen, A., ve Ekizoğlu, A.,1996. Ormancılık Politikası, İstanbul Üniversitesi Yayın No: 3968, Orman Fakültesi Yayın No: 435, İstanbul.
28. *https://gezievreni.com/belgrad-ormani-istanbul/.20.05.2020*
29. Yücel, M., *Ülkemizde Koruma Mevzuatı, İlgili Kurumlar ve Korunan Alan Statüleri,* 2015. 3.Koruma ve Peyzaj Mimarlığı Sempozyumu, TMMOB Peyzaj Mimarları Odası, 11–13 Aralık 2015, İzmir (Çağrılı Konuşma), İzmir Sanat Merkezi, Kültürpark, İzmir.

30. T.C.Resmi Gazete, *Orman Kanunu,* 1956. Başbakanlık Basımevi
31. Gülez, S., *Orman İçi Rekreasyon Planlaması.* Karadeniz Teknik Üniversitesi Dergisi, 1983. Cilt:6, Sayı:2, Karadeniz Üniversitesi Basımevi,Trabzon
32. Eroğlu S., *Milli Parklar Kanununda Belirlenen Korunan Alan Metodolojisi,* 2014. Denetişim, **13**, 85–90
33. *www.europarc.org.* 11.01.2020
34. Özbay, S., *Doğa Koruma Alanlarında Planlama Çalışmaları ve Ayvalık Adaları Tabiat Parkı Yönetim Planı Önerisi,* 2008. Yüksek Lisans Tezi, İ.T.Ü. Fen Bilimleri Enstitüsü, Peyzaj Mimarlığı Anabilim Dalı, 147 s., İstanbul.
35. FAO, *Status, Trends and Future Scenarios for Forest Conservation including Protected Areas in the Asia-Pacific Region. Food and Agriculture Organization of the United Nations,* 1997. Rome, Italy.
36. Çevre ve Orman Bakanlığı, *Korunan Alan Planlaması ve Yönetimi, Biyolojik Çeşitlilik ve Doğal Kaynak Yönetimi Projesi Deneyimi, Doğa Koruma ve Milli Parklar Genel Müdürlüğü Biyolojik Çeşitlilik ve Doğal Kaynak Yönetimi Projesi,* 2007. TŞOF Trafik Matbaacılık A.Ş., Ankara.

Beste KARAKAYA AYTİN and Murat ÖZYAVUZ

Conservation of Historical Urban Landscapes

Abstract: Today, cities have to struggle with urbanization process due to technological developments and globalization. This adverse condition becomes a threat especially for historical cities in terms of protection. Because historical cities have important cultural heritage areas which include natural, historical and cultural values. Integration of cities with the development of contemporary cities and also ensuring the sustainability of their unique characteristics is only possible by preserving the historical urban landscapes where all these important cultural heritage are displayed. Within the scope of this research, the concept of historical urban landscape has been discussed and how this concept emerged is examined and tangible and intangible values of historical urban landscapes are mentioned. As a result, the importance of conservation historical urban landscapes in historical cities that are gradually losing their unique values is emphasized.

Keywords: Historic urban landscape, tangible historic urban landscape, intangible historic urban landscape

1 Introduction

Cities are dynamic living spaces that cannot be prevented from changing in time. Change is a process in which cities have to be integrated in transferring sustainable development to the future. While this process contributes to the development of the city in positively, it also brings along the problem of losing the cultural characteristics that distinguish cities from another city and give it an identity. This problem has direct cities to an imperative thinking process on how to protect them in urban development.

The concept of protecting cities has changed its content from past to present. First of all, while mentioned about the concept of conservation in terms of buildings, over time, this concept brought up the necessity of protecting the buildings together with their surroundings with a holistic perspective and afterwards, an understanding of conservation emerged on an urban scale. Historical cities are areas where urban protection is a priority. Due to factors such as rapid population growth, industrialization and technological developments all over the world, the natural features of the cities gradually become unhealthy, and historical cities lose their identity and turn into similar areas. Cities lose their vitality and remain inactive places in the conservation works where only traditional features of historical cities are to be preserved.

Historical urban landscapes are the most important areas where the changes of the natural and cultural characteristics of the city in the historical process can be observed. For this reason, it is important to protect the historical urban landscapes, which are the most important indicators that reflect the changes a city has gone through from the past to the present.

2 The Concept of Historic Urban Landscape and Its Occurrence

Historical cities, which are tools that convey national, historical and cultural values, have to adapt to the changing living conditions during the development process, while trying to protect their historical and cultural values and ensure their sustainability. The protection of cities and specially historical cities only begins with the ability to read and define their cultural values and the environments that make a city different and make up its identity. Because cities are the unique environments that are shaped over time with their cultural level, architecture, traditions, lifestyle, social structure and landscape features, and can be defined by all these features.

According to O'Donnell, 2008; The spirit of a historical city is shaped by the tangible elements of that city's landscape. Historical urban landscape features are what express the cultural heritage values and spirit of a city, and reveal its unique landscape identity and character. The tangible and intangible cultural values of the cities are hidden in the historical urban landscape. Historical character and current state of urban landscape; guides the documentation and investigation of the inventory; It also contributes to the planning and evaluation of the future [1].

Prior to the emergence of the concept of historical urban landscape, studies were carried out on the protection and management of historical urban areas. The first of these studies is the "Recommendations on the Conservation of Historical Sites and Their Contemporary Roles", prepared by UNESCO in 1976 in Nairobi. In these decisions, the historical area is defined as "all kinds of buildings, structures and open areas group including archaeological and paleontological areas". Among these "areas", which have a rich variety by nature, especially prehistoric sites, historical cities, neighborhoods, villages, village houses, as a rule, monumental groups of homogeneous structure that should be preserved without any change should be taken into account. In addition, according to the principles included in the recommendations; Each historical or traditional area and its surroundings should be seen as a whole, it should be taken into account that the natural and special balance of the area in question depends on the integrity of its constituent parts, and in this context, the definition includes human activities as well as buildings, spatial features and their surroundings. All these elements that including the activities of people, no matter how they are mediocre, are matter greatly for integrity and should not be disregarded [2].

These recommendations adopted by UNESCO in 1976 in the development of the historical urban landscape concept are important as they are accepted as the first step. The conference "World Heritage and Contemporary Architecture - Historical Urban Landscape Management" held in Vienna in 2005 by the World Heritage

Center, ICOMOS and the Municipality of Vienna was also an important step. In this meeting, the "Vienna Memorandum" was adopted, which promotes an integrated and harmonious relationship between conservation and new urban developments in order to conserve the integrity of the historical landscape. In the memorandum decisions that state the necessity of new constructions to be in harmony with the historical structuring, there are recommendations for detailed studies on how to maintain the historical urban landscape and the development of contemporary architecture together. In the decisions, it was also emphasized that when determining the World Cultural Heritage Sites, the historical urban landscape of the area should be taken into consideration in the nomination and evaluation processes [3]. However, the decisions are insufficient to explain what the historical urban landscape values are and why they should be protected.

The "Recommendation on the Historical Urban Landscape" prepared by UNESCO in 2011, which is intended to be used and adopted as a tool to integrate the policies and practices of the protection of historical environments and the objectives of urban development, is an important step in the protection of cultural heritage. The rapid increase in the world population and the fact that a large part of this population has started to accumulate in cities is the main reason for the adoption of the historical urban landscape concept by UNESCO. As a historical layer formed by the accumulation of cultural heritage and cultures in the UNESCO 2011 Recommendations; It is stated to be a social, cultural and economic value. However, it is emphasized that cities are exposed to pressure and difficulties due to uncontrolled urbanization, demographic changes, mass tourism, climate change, exploitation in the marketing of cultural heritage and similar reasons. In this context, attention has been drawn to the necessity of integrating the protection, management and planning of cultural heritage and identity with local development processes [4]. In addition, in the 2011 Recommendations prepared by UNESCO, which constitutes the beginning of the development of the concept of historical urban landscape, it is emphasized that the conditions have changed since the 1976 Recommendations, the historical cities are now the subject of new urban developments and the historical urban landscape is not protected in historical cities.

According to Article 8 of the 2011 Recommendation prepared by UNESCO, the historical urban landscape has been recognized as "urban areas understood as a result of the cultural and natural values and qualities of the historical plane." Thus, the definition of historical urban landscape was able to cover a wider urban context and geographic space and extended beyond the concept of "historical center" or "kulliye". This comprehensive definition especially includes the topography, geomorphology, hydrology and natural features of the area; historical and contemporary settlements in its surrounding; underground and aboveground infrastructure; open spaces and gardens; land use and organization of the area; It includes perceptions and visual relations and all other elements of the urban structure. In addition, it includes social and cultural practices and values, economic processes and the intangible dimensions of the heritage of diversity and identity [4]. In this context, the historical urban landscape, which is still evolving as a concept, adopts the social, economic, environmental and cultural components of the city for sustainable development by approaching historical cities with a holistic approach [5] (Fig. 1).

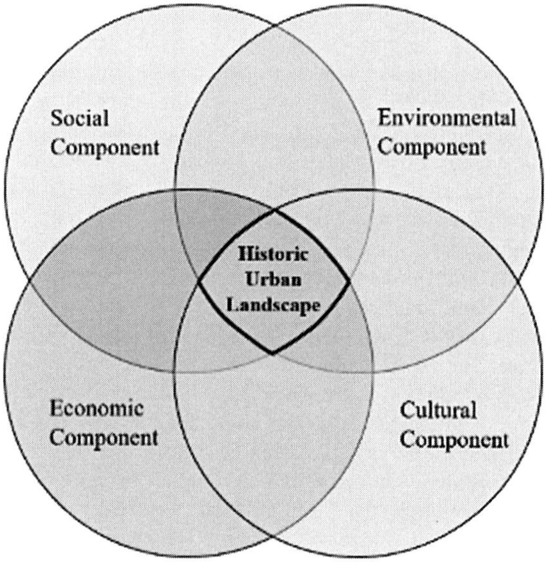

Fig. 1. Historical urban landscape components (modified from Rodwell, 2010)

Historical urban landscape; It provides an understanding of the historical environment by helping to define the complex elements that create a sense of identity and place and constitute the unique values of cities [6]. Historical urban landscapes, one of the important components of cultural heritage; It covers all building groups, structures, open areas, archaeological and paleontological sites in a natural and ecological context, which constitute human settlements in a certain period, that have archaeological, architectural, prehistoric, historical, scientific, aesthetic, socio-cultural and ecological harmony and value [7]. According to Sonkoly (2017), historical urban landscapes should also provide a conceptual framework for the protection of cultural heritage and guide for its management [8].

Historical urban landscapes are compositions that are formed by the natural features of nature, the tangible values created by coming together with the built environment, and the intangible values of the city as a whole with each other and the environment in the historical process.

Historic urban landscapes are envisioned as a means to integrate conservation policies and practices of the built environment, not to replace existing teachings or conservation approaches. In this sense, historical urban landscapes involve the stratification and diversification of vision and methodologies transferred from a tradition with a years of experience [9]. Historical urban landscapes, which are holistic and multidisciplinary at the same time, aim to guide change in historical cities and deal with the comprehensive management of cultural heritages in ever-changing and dynamic environments.

3 Tangible and Intangible Values of the Historical Urban Landscape

Historical urban landscapes are a subcomponent of cultural landscapes and consist of the natural and cultural characteristics of a city. While these features may sometimes be tangible values that include the natural and cultural environment of the city, sometimes they state intangible values that reflect the cultural life.

3.1 Tangible Values of the Historical Urban Landscape

The tangible values of the historical urban landscape reflect the identity and spirit of a city. While determining the historical character of the urban landscape, elements containing cultural heritage are physically observed and documented [10]. According to Can (2009), tangible cultural heritage; It states that they are works that include buildings, historical places, monuments and that should be preserved in order to be passed on to future generations. It also emphasizes that natural cultural heritage is a part of tangible cultural heritage [11]. Tangible values of cultural heritage include monuments, building communities and sites that are featured in terms of history, art or science according to the Convention on the Protection of the World Cultural and Natural Heritage published in the Official Gazette No. 17959 in 1983.

Shamsuddin, Sulaiman, and Amat (2012) stated in their study the importance of protecting tangible and intangible elements that express the historical character of urban areas in order to protect cultural heritage values. Tangible values according to them; urban areas, transportation systems and silhouette [12]. O'Donnell and Turner (2012) tangible values of the historical urban landscape; Natural systems (geomorphology, hydrology, vegetation, topography), spatial organizations, silhouette, transportation systems, buildings and structures are classified as artistic elements in the city [10].

When the tangible values of historical urban landscapes are examined, it is emphasized that they should include both natural and built environment components in the definitions. In this context, when the tangible values of the historical urban landscape are evaluated in the broadest sense according to the definitions, it consists of the components belonging to the natural and cultural environment.

- Natural environmental components

The natural environment is the areas where the natural balance is preserved without human influence, developed according to the ecological conditions of the region. The natural environmental components are geology, geomorphology, topography, soil, water availability, climate, flora and fauna. Wetlands, forest areas and geological formations, the places that spontaneously exist in nature, can be given examples as natural environment. For example, the pond located in the Amsterdam Forest (Amsterdamse Bos) is the component of wetlands (Fig. 2).

Fig. 2. Pond located in the Amsterdam Forest (Original, 2018)

- Cultural environment components

Cultural environments are areas that are shaped by human intervention and have the socio-economic and cultural characteristics of the region. Archeological sites, historical sites, urban sites, buildings, public open and green spaces, silhouette, transportation systems, urban furniture constitute the cultural environment components.

o **Buildings/Structures**

The most important feature that determines the identity of a city is the historical and architectural features of that city. The architectural features of the city are monumental structures and examples of civil architecture. Monumental structures are generally socio-cultural and religious structures that have survived from ancient times. These structures; there may be ancient structures such as gymnasium, temple, stadium, amphitheater, bouleuterion, odeon or palace, basilica, church, bathhouse, library, fountain, cistern, covered bazaar, caravanserai, mosque, Ottoman social complex (Kulliye), synagogue, almshouse, bridge and various public buildings. Fontana Di Trevi, one of the symbols of the Roman city, can be given as an example of these structures (Fig. 3). Civil architecture examples consist of mansions and houses, which are built according to the architectural understanding and materials characteristic to the city, and that reflect the traditional life of the city.

Fig. 3. Fontana Di Trevi (Original, 2016)

o Silhouette

Silhouette are views that enable the perception of the natural and spatial elements that make up a city and the relationship of these elements with each other. When the development of cities in the historical process is examined, it is seen that images with symbolic meanings, especially religious structures, are determinant in the city silhouette. The silhouettes of cities appear as a result of the social, cultural and historical accumulations of the societies that lived in there. Conserving the silhouettes of the symbolic structures that add unique values to the cities is very important in protecting the urban identity. (Fig. 4).

Fig. 4. Prizren city silhouette (Original, 2019)

o Public open and green spaces

They are places that are used by the inhabitants of the city and allow activities such as rest, entertainment and sports for their physical and psychological needs. Public open and green spaces allow people to communicate with each other. As well as social benefits of these areas, they provide ecological contributions to the city with their benefits such as improving air quality and reducing the urban heat island effect. City parks, neighborhood parks, sports fields and squares can be included in this context. Borghese Gardens (Fig. 5) in the city center of Rome can be given as an example.

Fig. 5. Borghese Gardens (Original, 2016)

o Transportation systems

They are public areas that ensure the circulation of pedestrians and vehicles in urban life. These areas, consisting of streets, sidewalks, bicycle paths and pedestrian areas, are places that meet the socio-cultural, recreational and commercial needs of the city dwellers. In addition, they are places that define the character of cities. The number of vehicles is gradually increasing in developing cities. The widening of the roadways creates problems in terms of pedestrian comfort, safety and access. In these situations, pedestrian areas that are closed to vehicle traffic, except for use as emergency and service roads, can meet many needs of the city dweller. It gains importance as areas that can be used safely as well as equipped with versatile functions. In this context, Vaci Utca Street, the most famous pedestrianized street of Budapest, can be given as an example (Fig. 6).

Fig. 6. Vaci Utka Street (Original, 2017)

o **Urban Furniture**

They are the equipment elements that receive the various needs of the citizens in public open and green spaces. Functionality and aesthetics are important in the design of urban furniture. They must be in harmony with the integrity of the place they are in. They should be considered as an element that contributes to the urban identity. The use of urban furniture has a special importance, especially in historical cities. It should be evaluated together with the cultural texture of the city. It should be designed as an element that enriches this texture and complementary to its form, color, texture and material features and identity. Seating elements, lighting elements, information communication signs, sales units, bicycle parks, water elements, waste and recycling units, traffic signs, border elements, maintenance covers of infrastructure facilities, plastic objects, plant crates and floor coverings can be given as examples in this context. A seating unit located in the city center of Valetta, the capital of Malta (Fig. 7) and Frankz Kafka statue (Fig. 8) in Prague can be given as an example.

Fig. 7. Seating Unite. (Original, 2018)

Fig. 8. *Franz Kafka Statue. (Original, 2019)*

3.2 Intangible Values of the Historical Urban Landscape

The scope of cultural heritage conservation is not limited to concepts such as monuments and sites. Traditional performing arts, rituals, festivals, traditional handicrafts transferred from the past to the present are defined as intangible heritage values. The intangible values of the historical urban landscape are a legacy beyond the borders of the regions [13].

Awareness of the role of intangible values in international conservation understanding has been strengthened in 2003 with the adoption of UNESCO's "Convention on the Protection of Intangible Cultural Heritage". According to this convention, intangible cultural heritage is defined as "practices, representations, narratives, knowledge, skills and related tools, materials and cultural places that communities, groups and in some cases individuals define as part of their cultural heritage" [14]. According to O'Donnell and Turner (2012), the intangible values of the historical urban landscape express the interactions and traditions between people and that place in public and private urban landscapes. These landscape-based intangible values are defined as festivals, rituals, traditional music and dance, worship, pilgrimage, traditional cuisine, traditional handicrafts [10]. The fountain and buddha statue used for purification before entering Sensoji, the Buddhist temple in Tokyo's Asakusa District, can be given as an example for places where various traditional religious rituals are performed (Fig. 9,10).

Fig. 9. Buda statue
(Original, 2013)

Fig. 10. Chozuya Fountain
(Original, 2013)

The intangible values of the historical urban landscape which differs from city to city, from region to region, from country to country, are unique to the place. The inventory of tangible and intangible heritage is the basis for understanding the historical character of the urban landscape. It informs the society and individuals about the future planning, management and initiatives.

4 Result

Today, the places where the concept of conservation is mostly discussed are the historical cities that are most damaged or even threatened with extinction as a result of technological developments and rapidly increasing urbanization. According to Oers (2010), as the population living in urban areas increases day by day, the pressures on historical cities are increasing and the protection of historical urban landscapes gains more importance day by day [15].

Rapid urbanization and the transformation of cities have placed the protection of cultural heritage sites at the center of urban planning. Therefore, new approaches have emerged, such as a historical urban landscape for a transformation in conservation, based on the difficulties in this area. The historical urban landscape approach is an understanding of conservation for the management of historical cities under the pressure of new development forms, with the main aim of conserving the importance and qualities of the city [16]. As stated in Article 11 of the definitions section of the Historical Urban Landscapes Recommendation prepared by UNESCO in 2011; While this approach recognizes the dynamic character of urban areas, it aims to protect the quality of the environment in which people live by increasing their productive and sustainable use and improving social and functional diversity. It integrates urban heritage conservation aims with social and economic development aims [4].

Historical urban landscapes, a concept that emerged as a result of the evolution of conservation and urban reorganization approaches, today face the risk of losing their originality and integrity structurally or visually. World cities are faced with this risk, especially in cities where population and economic activities are intense. One of the main problems of these cities is the inability to integrate the natural and cultural characteristics of the city from the past to the present with the new era developments. Historical urban landscapes, which are the most important indicators reflecting the changes a city has gone through from the past to the present, should be protected due to the unique values, which they carry.

* *This research is composed of part of the doctoral dissertation named Edirne City Action Plan Preparation within the scope of Historical Urban Landscapes*

KAYNAKLAR

1. O'Donnell, P., *Fasla, Aıcp Urban Cultural Landscapes & The Spirit of Place.* 2008, 77-Hfcw-13.Doc, Preservation Of The Spirit Of Place, Icomos Quebec.
2. Ahunbay, Z., *Tarihi Çevre Koruma ve Restorasyon.* 2016, Yapı Endüstri Merkezi Yayınları Yayınları, 7. Baskı, Ek 4 Tarihî veya geleneksel alanların korunması ve çağdaş yaşamdaki rolleri konusunda tavsiyeler, 1976 ISBN: 975-7438-38-3, İstanbul.
3. UNESCO, *Fifteenth General Assembly of States Parties To The Convention Concerning The Protection of The World Cultural And Natural Heritage* 2005, Paris, Unesco Headquarters. http://whc.unesco.org/uploads/activities/documents/activity-48-1.pdf
4. UNESCO, *Tarihî Kentsel Peyzaja İlişkin Tavsiye Kararı.* 2011, Http://Www.Une66sco.Org.Tr/Dokumanlar/Somutkulturelmiras/Peyzaj%20tavsiye%20karar%C4%B1.Pdf
5. Rodwell, D., *Historic Urban Landscapes: Concept And Management, Managing Historic Cities.* 2010, ISBN 978-92-3-004175-5, Published in September 2010 by UNESCO World Heritage Centre.
6. UNESCO, *The Hul Guidebook "A Practical Guide to Unesco's Recommendation on The Historic Urban Landscape".* 2016, Https://Historicurbanlandscapes.Wordpress.Com/2016/06/07/Hul-Guidebook/
7. Dinçer, İ., *Kentleri Dönüştürürken Korumayı ve Yenilemeyi Birlikte Düşünmek: "Tarihi Kentsel Peyzaj" Kavramının Sunduğu Olanaklar.* 2013, Iconarp International Journal Of Architecture and Planning Volume 1, Issue 1, Pp:22–40. Issn: 2147-9380

8. Sonkoly, G., *Historical Urban Landscape.* 2017, Budapest, Hungary ISBN 978-3-319-49165-3, DOI 10.1007/978-3-319-49166-0
9. Bandarin, F. ve Oers, R.V. *The Historic Urban Landscape, Managing Heritage in an Urban Century.* 2012, U.K, ISBN 978-0-470-65574-0
10. O'Donnell, P., Turner, M., *The Historic Urban Landscape Recommendation: A New UNESCO Tool For A Sustainable Future.* 2012, Http://Www.Heritagelandscapes.Com/Siteımages/IflaCape%20townHul%20odonnell-Turner%2028july2012(1).Pdf
11. Can, M. *Kültürel Miras ve Müzecilik, Çalışma Raporu.* 2009, Kültür ve Turizm Bakanlığı, http://teftis.kulturturizm.gov.tr/Eklenti/1279,muserrefcanpdf.pdf
12. Shamsuddin, S., Sulaiman, A. B. ve Amat, R.C., *Urban Landscape Factors That Influenced The Character of George Town, Penang UNESCO World Heritage Site.* 2012, Bangkok, Thailand, Procedia Social and Behavioral Sciences, Elsevier
13. UNESCO, *Intangible Cultural Heritage.* 2013, www.unesco.org/culture/ich/doc/src/20435-EN.pdf, [Erişim tarihi: 15.10.2017]
14. UNESCO, *Somut Olmayan Kültürel Mirasın Korunması Sözleşmesi.* 2013, https://ich.unesco.org/doc/src/00009-TR-PDF.pdf [Erişim tarihi: 12.11.2018]
15. Oers, R.V., *Managing Cities And The Historic Urban Landscape İnitiative – An Introduction, Managing Historic Cities.* 2010, ISBN 978-92-3-004175-5, Published in September 2010 by UNESCO World Heritage Centre
16. Martini, V., *The Conservation of Historic Urban Landscapes: An Approach.* 2013, Vol. 1 - Urban Conservation Theories And History Of The Cities, Venice

Ömer Lütfü ÇORBACI and Simten SÜTÜNÇ

Assessment of Amasra in Terms of Visual Landscape Quality With Its Natural, Historical and Cultural Values

In this section, a landscape assessment study in order to develop landscape protection strategies and prepare sectoral guides that would be the basis of landscape character-based landscape planning was exemplified in Amasra district. In the first stage of the study, literature information on the subject of landscape evaluation was presented, and in the second stage, spatial sampling was made.

It seen that at the basis of all planning concepts being created until today is the fact that human beings play an important role in shaping the system that creates the landscape. For this reason, it is very clear that human-scale researches should be conducted in order to understand landscapes. Another issue is that sustainable development involves more than ecological balance. Since human needs, which constitute the social aspect od sustainability, are not sufficiently represented in sustainability research, more efforts should be made to better understand them in the future. Knowing the needs of people and knowing the reasons for these needs is a prerequisite for the design of nature conservation and landscape planning measures that have a chance to be successful in the long run (1). Landscape planning and nature protection measures that conflict with human needs will meet with a counter stance. Even when such measures are compatible with people's needs, training and other actions may be required to encourage planning and adoption of conservation measures. Knowing human needs will help how the interaction between human and landscape will develop. Landscape (Character) Assessment is a tool that contributes to making decisions regarding the two purposes of the sustainable development model, "protection of cultural landscapes with cultural heritage value", "nature protection", and "responsible use of natural resources" (2).

Landscape Assessment

Landscape assessment creates a bridge between the scientific knowledge required to meet the demands of sustainable landscape planning and management and socio-economic problems (3). The degree of interaction between ecological and socio-economic functions depends on the patterns and intensities of land use. Various landscape indicators paly a key role in the assessment of landscape functions for planning purposes (4, 5). Modelling and mapping (6, 7) are the mandatory tools in the indicator-based assessment of landscape functions for sustainable landscape planning (8). An effective landscape management is supported by a landscape assessment. Landscape assessment, if applied soundly and rigorously, should inform both the approach and the decision-making process regarding how landscapes are managed, thereby reducing subsequent costs in the policy making or proposed project.

Material and Method

For this study, which includes the landscape assessment of Amasra district and its immediate surroundings, administrative boundaries were also taken into consideration as well as natural data. Other materials that are of great importance in the application of the developed method consist of data on natural and cultural assets belonging to the district of Amasra and its immediate surroundings where the method is applied, excursions and observations in the field of research, expert person and academician views, interviews with local people, producers, managers and village headmen, previous studies in the field of research by the public and private sectors. In addition, topography, soil, geology, management maps, development plans, data on ground and surface waters, reports, satellite images, and climate data are among the materials used for landscape assessment.

Landscape development strategies and landscape plan based on judgement were created in terms of the continuity of landscapes and in line with the specific purpose on the relative values of different landscapes or different features within themselves. In this process, landscape indicators developed by evaluating the landscape structure, function analysis and landscape change-pressure analysis were brought to the fore. Within the scope of the research, the protection of the landscape and the tourism and recreation sector were taken as the basis. In the assessments, the indicators to be revealed by landscape function analysis and landscape change-pressure analysis were assessed. In the study, three criteria were used to evaluate the indicators in terms of tourism and recreation. These are problem-based, expert opinion-based and broad perspectives that will take into account the entire research area.

All of the goals determined in terms of aesthetics, quality, ecological and cultural structure related to the landscape were assessed in order to establish the general landscape vision, determine the goals and create policies. While determining the vision, all factors affecting the landscape were taken into account. While determining the goals and policies, firstly, separate strategies were created in terms of landscape functions. In this context, general and detailed landscape strategies were produced by interpreting the function-character-pattern in relation. The protection and use value of the landscape was created with the landscape protection-development strategies. In other words, the protection value of the landscape was revelated with the landscape protection and development strategies, and the use value of the landscape was revealed with the sectoral guidelines. In the stage of producing landscape strategies, management strategies were also defined in relation to the protection and use value of the landscape. Landscape protection and development strategies include policies for the restoration, improvement, development and protection of the landscape. Based on the problems and solutions of the tourism and recreation sector in the field of planning, concrete strategies and possible development possibilities of the sectors were presented on the basis of function-character regarding the landscape.

The studies conducted by 9, 10, 11, 12, 13, 14, 15, 16, 17 were examined to reveal the method. Along with these studies, definitions obtained from the project titled TUBITAK-KAMAG 109G074 Landscape Character Analysis and Its Assessment for Tourism/Recreation (Peyzaj-44)[1] carried out by Ankara University Faculty of Agriculture Department of Landscape Architecture between 2010–2012 were also used in realisation of the method to be used during the creation of the landscape plan.

Findings and Discussion

Landscape Vision, Goals and Policies

Based on the landscape inventory and analysis stages, a general landscape vision related to the research area was created. This vision, which will show the desired future situation regarding landscapes, will guide the landscape evaluations and contribute to the assessments as a whole. In line with these definitions, the vision of the research was determined as "the protection and development of natural, cultural, and historical values to meet the needs of future generations and the establishment of sustainable tourism principles in this context". In the next stage, targets were determined for landscape component, landscape component and function. Considering the factors affecting the existing landscape became critically important in setting goals. In this context, maps related to landscape functions were created. These are potential landscape function target maps that include spatial arrangements that will increase the quality of the landscape related to the landscape function in the area, the elimination of pressure reasons, etc. It is aimed to bring the points with low landscape function value due to the pressure caused by the human factor in the area to a high degree of function with improvement goals. In this way, the research area was provided with the status of pure landscape protection area assessment. Tab. 1 shows the general objectives based on the landscape function determined for the research area. These general goals were transformed into detailed goals with function-character based evaluations after general landscape protection and development strategies were produced.

1 Project Coordinator: Prof. Dr. Şükran Şahin
Project Researches and Scholars: Prof. Dr. Halim Perçin, Asst. Prof. Dr. Ekrem Kurum, Asst. Prof. Osman Uzun, Dr. B. Cemil Bilgili, Asst. Prof. Dr. Levent Tezcan, Prof. Dr. İhsan Çiçek, Res. Asst. Volkan Müftüoğlu, Res. Asst. Ömer Lütfü Çorbacı, H. Simten Sütünç, Duygu Doğan, Ersin Ateş, Özlem Koç, Bilge Tarım, Gözde Kurtoğlu.

Tab. 1 General objectives regarding the landscape functions of the research area

Function	Degree of function	Goal/policy
Habitat function	High	Protection of important habitats and biodiversity analysis
	Medium	Ensuring ecological connectivity and creating transition areas
Agro-ecological zone	High	Protection of important agricultural lands
	Medium	Protection of potential agricultural lands
Runoff potential	High	Control of key runoff areas
	Medium	Control of runoff areas
Water permeability	High	Protection of important groundwater recharge areas
	Medium	Protection of permeable areas
Erosion risk analysis	High	Control of key erosion areas
	Medium	Taking soil protection measures
Visual value analysis	High	Protection of important visual resources
	Medium	Protection of visual resources
Landscape elements	High	Protection of important landscape elements
	Medium	Protection of landscape elements

Landscape Strategies: protection – Development – Management

At this stage, strategy development approach based on landscape function value is explained. It is the section where landscape strategies are developed to shed light on spatial planning studies.

- General landscape protection – development – management strategies

General landscape – protection – development – management strategies were created in relation to function-pattern. General strategies were revealed by matching the areas with high landscape value determined as a result of the landscape function analysis with the basic landscape pattern parameters (Tab. 2).

Tab. 2 General landscape protection-development-management strategies

<table>
<tr><th rowspan="2">Landscape function value</th><th colspan="3">Landscape Pattern (1st. Level)</th></tr>
<tr><th>Natural landscape</th><th>Agricultural landscape</th><th>Urban landscape</th></tr>
<tr><td>High
1st. Degree</td><td rowspan="2">Landscape protection area
Strict landscape protection area (SLP)</td><td rowspan="3">Controlled landscape usage area (CLU)</td><td rowspan="3">Controlled landscape usage area (CLU)</td></tr>
<tr><td>2nd. Degree</td></tr>
<tr><td>3rd. Degree</td><td>Partial landscape protection area (PLP)</td></tr>
<tr><td>Medium</td><td>Limited landscape usage area (LLU)</td><td>Strict landscape usage area (SLU)</td><td>Limited landscape usage area (LLU)</td></tr>
<tr><td>Low</td><td>Potential landscape usage area (PLU)</td><td>Potential landscape usage area (PLU)</td><td>Potential landscape usage area (PLU)</td></tr>
</table>

In Fig. 1, the research area landscape protection – development – management strategies are seen for the landscape strategic categories. General landscape protection – development – management strategies should be produced in relation to function-patter. General landscape protection – development – management strategies;

- Landscape protection area,
- Limited landscape usage area,
- Controlled landscape usage area,
- Potential landscape usage area,
- Landscape statuses

Should be defined under five main headings (18).

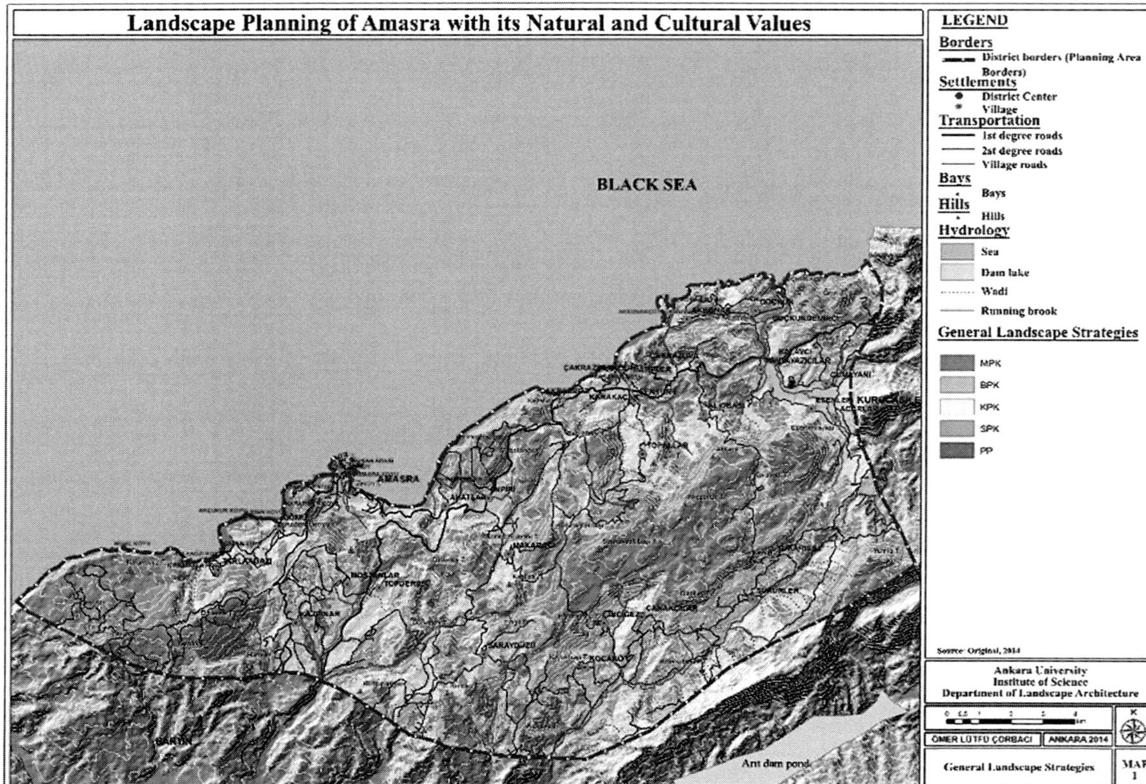

Fig. 1. General landscape strategies.

Landscape protection area: Areas with high landscape function value in natural; landscapes are purely landscape protection areas. In terms of the continuity and improvement of the natural, cultural, and visual landscape, they form the landscapes with the most important conservation value. What is important here is how to identify areas with high landscape value based on landscape function analysis. Its high degree function(s) and key processes should be taken as a basis in determining areas with high landscape value. Key processes in natural, cultural and/or visual aspects should be determined first in developing a function-pattern based strategy for a landscape. The basic landscape functions that make up a landscape are defined as key processes. An area of high value in terms of more than one function is a more important protection area compared to the others. The definition and evaluation of landscape protection areas should be made in three degrees, taking into account the compliance with the current national legislation. In this context, landscape protection areas should be defined in the following three classes in terms of landscape protection-development-management strategies (18).

- **1st Degree Landscape Protection Areas (Pure/Absolute Landscape Protection Area):** They are sensitive areas/landscapes that will definitely be protected. They are areas with a high degree of landscape function in terms of important key processes.
- **2nd Degree Landscape Protection Areas (Pure/Absolute Landscape Protection Area or Partial/Partial Landscape Protection Area):** They are areas of medium value for other functions with a high degree of landscape function in terms of a key processes. If there is a possibility to improve the moderate function depending on the landscape pattern, these landscapes are also delicate landscapes to be protected. If there is no possibility of improvement, it can be named as Partial/Partial Landscape Protection Area as qualified nature protection areas provided that protection measures are taken for the high value landscape function.
- **3rd Degree Landscape Protection Areas (Partial/Partial Landscape Protection Area):** They are qualified nature protection areas. They are landscapes of relatively low value for other functions with a high degree of landscape function in terms of a key process.

Limited landscape use area: Areas with high landscape function value in cultural landscapes (agricultural areas, residential and industrial areas) are sustainable conservation and limited landscape usage areas. Measures are required to ensure the sustainability and operability of landscape functions in these areas. Such a limited development strategy is critically important, especially for key processes. For example, the existence of a landscape that is of high importance in terms of water permeability in an urban development area and city periphery requires limited development in these areas. This situation may be in the form of an environmental arrangement to be carried out based on landscape evaluations or its reflection on zoning plans, revision of the zoning plan or taking an urban transformation decision. If this landscape is an area that has not yet taken place and is not planned yet, it is a place that should be separated as open-green area in the future (18).

In addition to the high function that constitutes the limited landscape use category, another limiting factor is the potential to increase the moderate function value, among other key processes, to a high degree through interventions in the landscape pattern. The use of this factor is limiting if a moderate water function can be raised to a high degree through vegetation improvement or land reclamation efforts, or if the moderate erosion risk turns in to a high risk with land shaping efforts. For example, if planning high volume excavation-filling operations on a slope that has become stable in the long term in urban areas shows the potential to increase the risk of erosion or landslides depending on the soil and rock structure, the intervention should not take place and the static structure of the nature should not be disturbed. Measures are required to ensure the healthy functioning of landscape functions in limited landscape usage areas (18).

Controlled landscape protection area: In all natural and cultural landscapes, areas with moderate landscape function value are limited landscape usage areas. What important here is the possibility that a medium-level landscape in terms of landscape function(s) can be raised to a higher level with improvement measures (planning for water harvesting, creating an ecological corridor to ensure habitat integrity, etc.) or risks (such as erosion, landslides) arising from intervention. In this case, landscape development should be focused on restoring or restoring nature. In other cases, land use or its response should be compatible with the landscape function(s) as a moderate value in key processes is still significant (18).

Potential landscape use area: Here the factor determining the direction of the potential is the possibility of raising a relatively low degree of landscape function(s) in terms of landscape function(s) to a higher degree through improvement measures (planting for water harvesting, creating an ecological corridor for habitat integrity, etc.) or risks (such as erosion, landslide) arising from intervention. In this case, landscape development should be focused on restoring or restoring nature. In other cases, potential landscape development areas are areas best suited to human intervention (18).

Landscape statuses: Especially in sub-regional and local scale Landscape Character Analysis and its Assessment (LCAA) studies, important places for that region should be identified and expressed with the following statuses (18).

- **Important landscapes:** They are unique areas that contribute to character definition with their natural, cultural and visual landscape elements (interesting geomorphological microclimatic structure, etc.). Areas that have an important place in the functioning of ecological and cultural processes in areas where the landscape function is determined to be high should also be evaluated within this scope.

- **Special landscapes:** They are areas that are valued by the locals, beautifully portrayed by people and visited at certain times for recreation. In terms of cultural landscape elements, areas with special values such as local architecture, local handcrafts, local settlement texture, local agricultural culture, local forestry culture should be defined as special landscapes for that region.
- **Symbolic landscapes:** The monumental tree is an area that has an important resource value with its solitary landscape element and it immediate surroundings like a single architectural structure. Areas containing characteristic abstract values are also called symbolic landscapes.

There is a need to develop national criteria for granting status to a landscape for the LCAA process. In this context, national sanctions will be guiding.

- Detailed landscape protection – development – management strategies

It is the stage where landscape strategies are developed to guide spatial planning studies in 1/25.000 and 1/10.000 scale. Detailed landscape strategies cannot be created without determining the upper scale general landscape strategies. Therefore, in the planning process, general and detailed landscape strategies are a whole (18). Detailed strategies for conservation – development is developed depending on the overlap of the landscape function value with general landscape strategies, Landscape Character Areas (LCA) and Landscape Character Types (LCT) and concerning the landscape objectives determined in the previous stages for each landscape function (18). For detailed strategies, the following terminology defined by European Landscape Convention (ELC) should be used:

- **Landscape protection:** Preservation and maintenance of the existing landscape character,
- **Landscape reinstatement:** Reinstatement to improve and strengthen the landscape,
- **Landscape restoration:** Restoration to return to the pre-intervention character,
- **Creating a new landscape:** Creating a new landscape character.

Areas with high landscape value are determined with areas with high function value according to landscape function analysis, and many techniques such as clustering, weight average, etc. can be used in this determination. The important thing is to take into account the value of each of the basic processes (functions) that shape a landscape as well as their value (18). In the development of detailed landscape protection – development – management strategies, studies based on function clusters can be carried out in the following stages (18).

- **Stage 1:** In each cluster, goals are set for each function.
- **Stage 2:** Landscape policies (reinstatement, protection, restoration and creation of a new landscape) are determined for each cluster in line with these objectives.

- **Stage 3:** Detailed landscape strategies are developed for each cluster, taking into account the landscape character type, landscape pattern and landscape elements.
- **Stage 4:** Clusters with similar strategies are grouped.

Detailed landscape strategies developed according to different landscape function degrees for each landscape function are shown in Tab. 3 the database used to produce detailed policies in, Table and the detailed strategy map in Fig. 2.

Tab. 3. Landscape protection-development-management strategies in detail at the research area.

Landscape pattern (1st. Level)		**Degree of landscape function**	
Natural areas (D)	**High**	**Medium**	**Low**
Erosion (ERZ)	ERZ_3D	ERZ_2D	ERZ_1D
Water permeability (SU_GCR)	SU_GCR_3D	SU_GCR_2D	SU_GCR_1D
Habitat function (HBT_FNK)	HBT_3D	HBT _2D	HBT_1D
Runoff potential (YAP)	YAP_3D	YAP_2D	YAP_1D
Agricultural lands (T)	**High**	**Medium**	**Low**
Erosion (ERZ)	ERZ_3T	ERZ_2T	ERZ_1T
Water permeability (SU_GCR)	SU_GCR_3T	SU_GCR_T2	SU_GCR_1T
Habitat function (HBT_FNK)	HBT_3T	HBT_FNK_2T	HBT_FNK_1T
Runoff potential (YAP)	YAP_3T	YAP_2T	YAP_1T
Settlement (Y)	**High**	**Medium**	**Low**
Erosion (ERZ)	ERZ_3Y	ERZ_2Y	ERZ_1Y
Water permeability (SU_GCR)	SU_GCR_3Y	SU_GCR_2Y	SU_GCR_1Y
Habitat function (HBT_FNK)	HBT_FNK_3Y	HBT_FNK_2Y	HBT_FNK_1Y
Runoff potential (YAP)	YAP_3Y	YAP_2Y	YAP_1T
ERZ_3T	It is necessary to take fast and high-level erosion protection measures in agricultural activities.		
ERZ_2T	Erosion protection measures must be taken in agricultural activities.		
SU_GCR_3T	No chemicals in agricultural activities and/or organic farming is required.		
SU_GCR_2T	The use of chemicals in agricultural activities should be minimized, taking into account ground water contamination.		
ERZ_3D	It is necessary to use slope breakers in sloping areas with the improvement and development of the existing lower and upper vegetation cover.		

Tab. 3. Continued

Landscape pattern (1st. Level)	Degree of landscape function
ERZ_2D	In order to protect the existing vegetation, protection/development studies should be carried out for erosion.
SU_GCR-3D	It is necessary to protect important groundwater recharge areas and create planting areas for this purpose.
SU_GCR-2D	Areas that require measures to increase water permeability (water harvest) and protect ground water pollution
HBT_2D	The existing habitat needs to be protect, improved, and interconnected opportunities are created.
ERZ_3Y	Soil protection measures should be developed in urban open and green areas.
SU_GCR_3Y	A rainwater management plan should be prepared and the surface should not be covered with an impermeable material.
SU_GCR_2Y	Rainwater management plan should be prepared and water harvest zones should be established.

Areas with high landscape value are determined with areas with high function value according to landscape function analysis, and many techniques such as clustering, weight average, etc. can be used in this determination. The important thing is to take into account the value of each of the basic processes (functions) that shape a landscape as well as their value separately (18). By evaluating the landscape functions and landscape pattern related to the area according to Tab. 3 27 detailed landscape strategies (protection and improvement of HBT_2D existing habitat and creation of interconnection opportunities) were developed and landscape policies (landscape restoration) were produced accordingly. According to Tab. 4, 5 different general strategies were determined by looking at the landscape functions and landscape patterns related to the area. (MPK/Absolute Landscape Protection Area – while determining the landscape protection strategy, the function values in the area have been taken into consideration and while creating this strategy, it must be a natural area (D) and have at least two high-value functions (such as HBT-3, ERZ-3 etc.) Each landscape character type is determined for Level-2 for Landscape Protection – Development – Management Strategies.

Cluster	Landscape functions				Landscape Character Type (LCT)	Overlay of landscape protection-development-management strategies with landscape pattern	Landscape policies in detail	Landscape development strategies in detail			
	YAP_DGR	HBT_DGR	ERZ_DGR	SU_GCR_DGR	PKT_D2	GNL_STR	POLICY	AYR_YAP	AYR_HBT	AYR_ERZ	AYR_SU_GRC
1	1	3	3	1	NMz-ST-D-YBÖ	MPK	Landscape protection	YAP-1D	HBT-3D	ERZ-3D	SU_GCR-1D
1	1	3	1	3	NMz-ST-D-YBÖ	MPK	Landscape protection	YAP-1D	HBT-3D	ERZ-1D	SU_GCR-3D
2	3	1	2	3	NMz-ST-D-YBÖ	MPK	Landscape protection	YAP-3D	HBT-1D	ERZ-2D	SU_GCR-3D
2	1	2	3	1	NMz-ST-D-YBÖ	BPK	Landscape protection/landscape reinstatement	YAP-1D	HBT-2D	ERZ-3D	SU_GCR-1D
1	3	1	1	3	NMz-ST-T-YBÖ	KPK	Landscape protection	YAP-3D	HBT-1D	ERZ-1D	SU_GCR-3D
3	1	1	2	1	NMz-ST-Y-YBÖ	SPK	Landscape reinstatement	YAP-1D	HBT-1D	ERZ-2D	SU_GCR-1D

Tab. 4. Database used for landscape protection-development-management strategies in detail for the research area

While creating the cluster map; MPK, BPK and KPK were included in the first group as areas with high function degrees, SPK in the second group as areas with medium function value, and PPK in the third group as areas with low function degrees (Fig. 2–3)

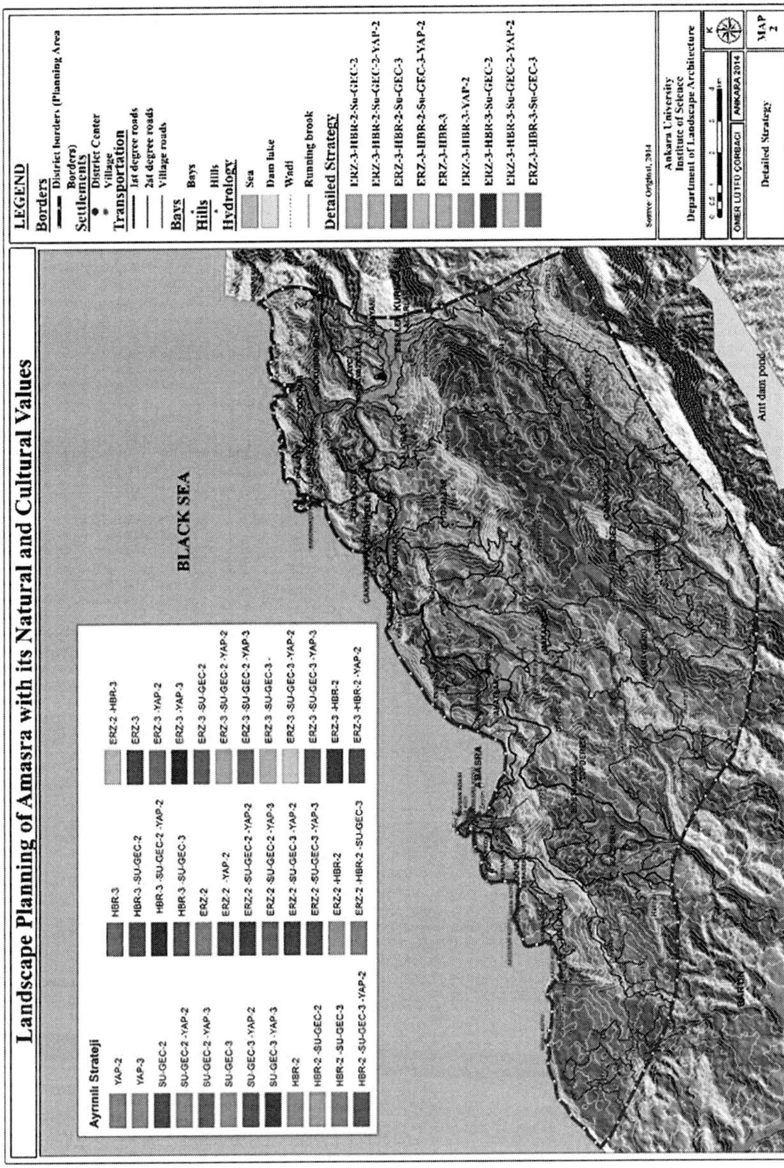

Fig. 2. Detailed landscape strategies.

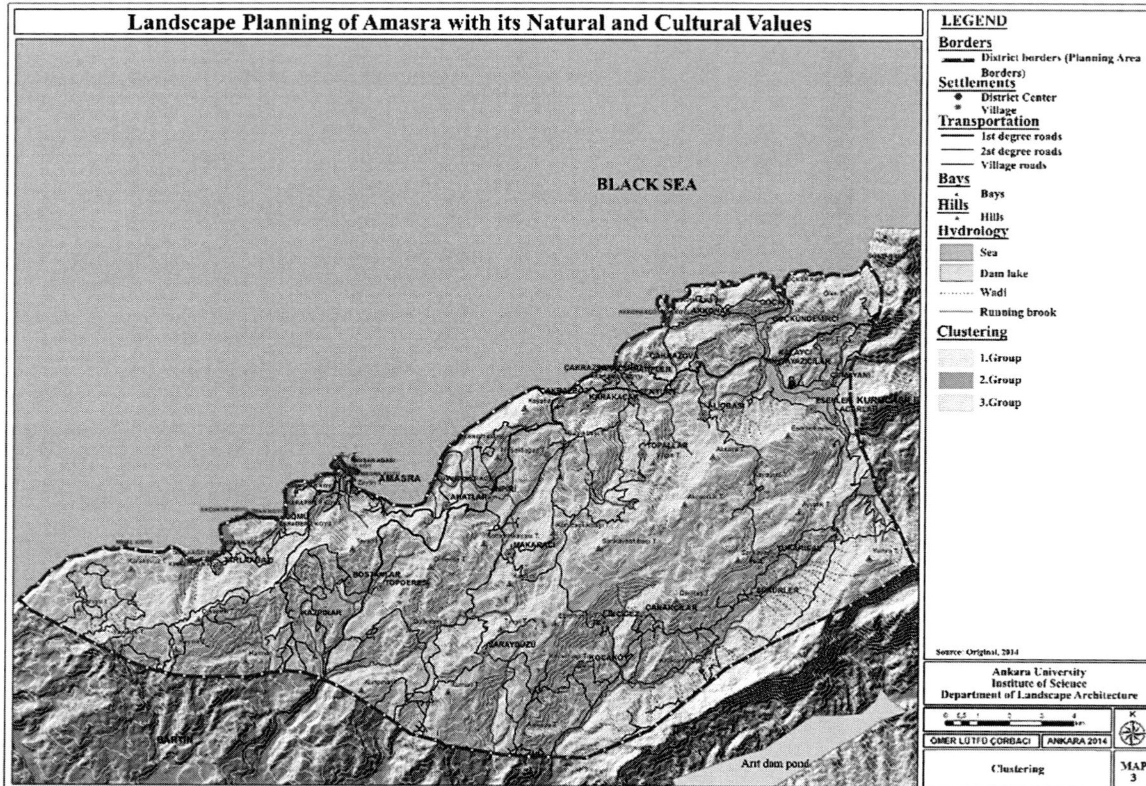

Fig. 3. Clustering map.

Landscape policies:
As a result of the LCAA studies, landscape policies were produced based on character-function. Landscape policies, general and detailed landscape development strategies created based on character-function have been interpreted together with the landscape pattern and developed by taking into account the user requests and goals of the area. In this context, three basic policies can be seen in Fig. 5.

- **Landscape restoration:** These are the works carried out to improve and strengthen the landscape.
- **Landscape protection:** It occurs in the protection and maintenance of the existing landscape due to the high risks or potentials that arise from the analysis of the structure, function and change of the landscape.
- **Restoring the nature:** These are the works carried out to create, support and return the natural landscape to its pre-intervention character.

Sectoral landscape guides:
Another landscape evaluation product that will create a threshold for the National spatial planning process aiming at the rational use of resources are landscape guides. For this purpose, sectoral landscape guides have been presented for the research area, taking into account the landscape survey conducted in the field, its analyses, landscape types and function values, and landscape character areas. Maps of the tourism sector guide (Fig. 5) for the tourism sector and the agro-ecological sector (Fig. 6) maps for the agricultural sector were produced.

Fig. 4. Landscape policies.

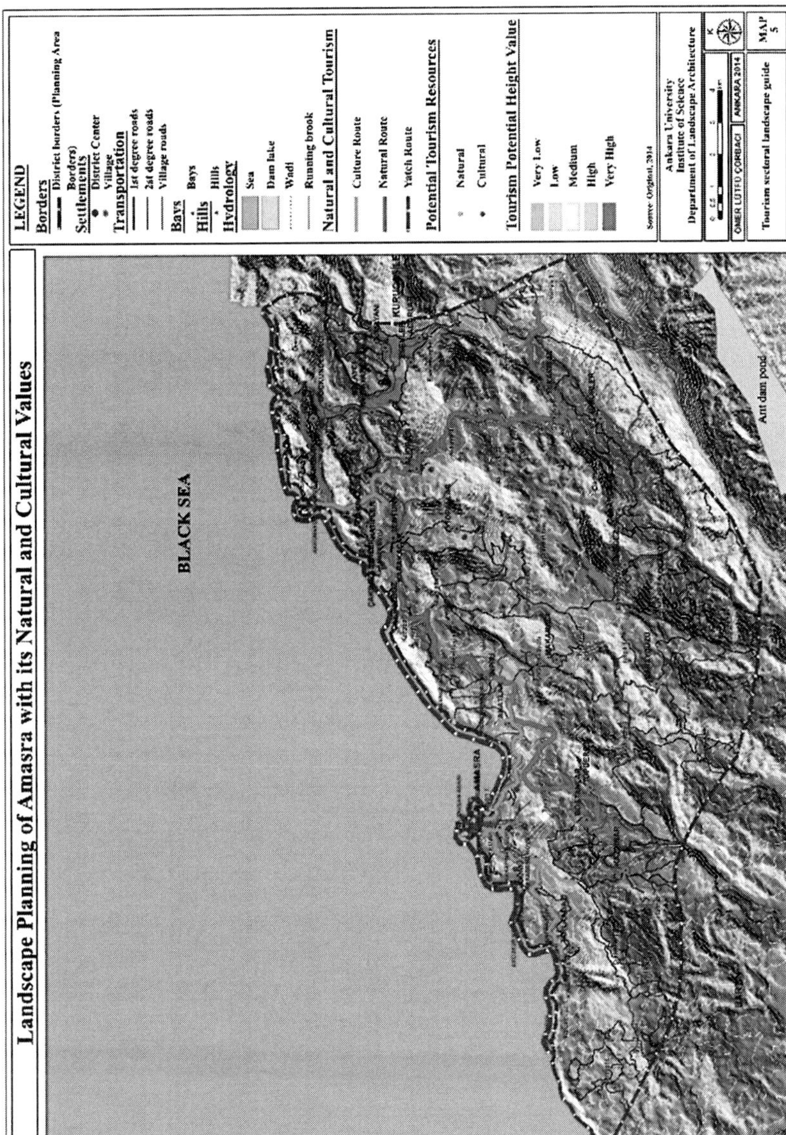

Fig. 5. Sectoral landscape guide for tourism and recreation

Fig. 6. Sectoral landscape guide for agricultural usage

Landscape protection – development – management strategies:
With this research, an exemplary model based on tourism and recreation strategies has been created within landscape planning for the district of Amasra. As a result of the research, considering the balance of conservation-use, landscape policies based on tourism and recreation, including general principles and strategies aiming to take special precautions for the protection, management and planning of landscapes, were produced. Considering the natural and cultural values of the district of Amasra, attention was paid to the suggestions made within the scope of tourism and recreation to be suitable for the environment and culture of the region, to include works that would respect the culture of the environment and the local people rather than meeting the needs of the visitors. Landscape planning studies must be in a protection-use balance. Within the scope of landscape protection – development – management strategies, the entire protection-use balance related to the area has been revealed. Landscape strategies developed should be formed by taking into consideration the higher scale landscape policies and landscape management plans. Such evaluation studies should be done by landscape architects who are specialized in their fields. The results to be obtained here must be collected from the data bank together with the inventory study. In this context, it is necessary to prepare a web-based portal where the information, maps and conservation development and management strategies to be produced will be followed. In this way, planners, investors, decision makers, local people and tourists using the system will be able to easily access information about the area. For landscape planning studies to be more reliable, the evaluation factors created for the usage characteristics must be correct. The natural and cultural evaluation factors for each landscape unit and character type should be determined in a way that will be sensitive to the characteristics of the region by a multi-disciplinary team suitable for the conditions of our country. Within the scope of the strategies created with the inventory and analysis study of the research area, the types of use are evaluated as protection, repair and re-use.

However, different uses have been proposed for different strategies. For example, natural areas with high erosion and high-water permeability must be protected, while agricultural areas with high surface runoff potential have been identified as controlled landscape usage areas.

Strategic landscape planning:
The method suggested within the scope of the research is "Strategic Landscape Planning Method". This method consists of setting and defining goals and developing strategies to achieve these goals. In short, it can be expressed as a process that suggests determining goals for the improvement and development of landscapes for tourism and recreation but with a protection-use balance, and developing strategies to achieve these goals. The objectives within the scope of this research are discussed in terms of sectoral tourism and recreation and agro-ecological zones.

Landscape development strategies:
In the research, "Landscape Development Strategies" were produced by taking into consideration the criteria and indicators to be determined within the framework of sustainable landscape planning in terms of tourism and recreation within the scope of the changing and developing planning approaches and spatial planning. Depending on the space, these strategies, which are obtained by taking into account the ecological and socio-cultural characteristics of the space, consist of a series of interrelated and complementary activities. It is necessary to finalise the site selection for the determined development strategies and to ensure the applicability of the strategies before physical landscape planning and design studies that include detailed solutions at various scales.

Management strategies:
"Management Strategies" have been created in terms of creating an infrastructure for planning and design studies. The recreational design projects to be realised in the next stage should be produced holistically with these landscape strategies and policies. Thus, while ensuring that design decisions are compatible with the landscape goals, the development will be realised with nature and the continuity of the landscape will be guaranteed. Connecting with upper scale general landscape strategies is possible with detailed landscape strategies. Conducting such studies must be included in the planning stages. Considering the important LCAA, types and units determined in the research by local governments in the landscape planning studies to be carried out, will be an appropriate approach for the continuity of the social, economic and natural environment needs of the people living in the region. In this context, the research should be taken into consideration as a base that will shed light on urban and rural landscape planning, offering suggestions for land use planning, restoration and arrangement works that complement each other with nature.

Result

When the authorities and responsibilities regarding landscape planning studies in Turkey are examined, the organisations involved in landscape planning studies consist of the Landscape Conservation Branch Directorate under the Ministry of Agriculture and Forestry, the Ministry of Environment and Urbanisation, the Ministry of Transport and Infrastructure, the Ministry of Energy and Natural Resources, the Ministry of Culture and Tourism, the Ministry of Interior, the Ministry of National Defence and relevant units. Authorised persons from these organisations came together to establish a common data network to collect a combination of landscape planning-related work needs to be done in Turkey. In order to ensure the implementation and continuity of the landscape plans, the legal framework regarding the plans must be fulfilled completely. There is a legal framework that includes the landscape planning issues in Turkey. However, according to the commitments in the ELC, this deficiency should be included in legal regulations. ELC has ensured that landscape planning is handled as a very effective public policy and an international action. Each country that is a party to ELC accepts the protection of natural and cultural values in accordance with the terms of the contract. Landscape plans to be prepared should include a multidisciplinary feature that is compatible with the natural, historical and cultural values, geographic structure, climate, ecology and biological diversity of the area to be studied, and at the same time offers a comprehensive management approach. In this direction, landscape plans should include targets for the future.

References

1. Hunziker M., Hoffmann C. and Wild S. 2001. Die Akzeptanz von Raubtieren, Gründe und Hintergründe – Ergebnisse einer repräsentativen Umfrage in der Schweiz. Forest Snow and Landscape Research 76, 1–2: 301–326.
2. Wascher, D., 2004. Landscape-indicator development: steps towards a European approach. In: Jongman, R.H.G. (ed.) New Dimensions of the European Landscape. Wageningen UR Frontis Series, vol. 4, pp. 237–252.
3. Bastian, O., Lütz, M., 2006. Landscape functions as indicators for the development of local agri-environmental measures. Ecol. Indicators 6 (1), 215–227.
4. Mander, Ü., Müller, F., Wrbka, T., 2005. Functional and structural landscape indicators: upscaling and downscaling problems. Ecol. Indicators 5 (4), 267–272.
5. Fry, G., Tveit, M.S., Ode, Å., Velarde, M.D., 2009. The ecology of visual landscapes: exploring the conceptual common ground of visual and ecological landscape indicators. Ecol. Indicators 9 (5), 933–947.
6. Meyer, B.C., Grabaum, R., 2008. MULBO: Model framework for multicriteria landscape assessment and optimisation. A support system for spatial land use decisions. Landscape Res. 33 (2), 155–179.
7. Willemen, L., Verburg, P.H., Hein, L., van Mensvoort, M.E.F., 2008. Spatial characterization of landscape functions. Landscape Urban Plan. 88 (1), 34–43.
8. Leitão, A.B., Ahern, J., 2002. Applying landscape ecological concepts and metrics in sustainable landscape planning. Landscape Urban Plan. 59 (2), 65–93.
9. Mcharg, I. 1969. Design with Nature. Natural History Press, Garden City, New York 197 pp.
10. Klijn, F. and Udo De Haes, H. A. 1994. A Hierarchical Approach to Ecosystems and its Implications for Ecological Land Classification. Landscape Ecology 9: pp. 89–104, Springer, Netherlands.

11. EPA 1997. An Ecological Assessment of the United States Mid-Atlantic Region: A Landscape Atlas, United States Environmental Protection Agency, EPA/600/R-97/130.
12. Swanwick, C. 2002. Landscape Character Assessment Guidance for England and Scotland. Prepared on behalf of The Countryside Agency and Scottish Natural Heritage. University of Sheffield.
13. Wascher, D. M., Groom, G., Mücher, S., Kindler, A., Blustt, G., Damarad, T., Nieto A. and Delbaere, B. 2005. European Landscape Character Areas-Typologies, Cartography and Indicators for the Assessment of Sustainable Landscapes. Final Project Report as deliverable from the EU's Accompanying Measure project European Landscape.
14. Erol, U. E. 2005. Ekolojik yaklaşımlı peyzaj planlaması (Balabandere vadisi örneği)/An ecological approach to landscape planning (a case study of the Balabandere valley), Doktora Tezi, İstanbul Üniversitesi, Fen Bilimleri Enstitüsü, Peyzaj Mimarlığı Anabilim Dalı, İstanbul.
15. Seguin, J. F. 2007. Peyzaj Atlasları (çeviri: Abdurrahman Güzelkeleş). Avrupa Peyzaj Sözleşmesi'nin Uygulanması Yolunda Türkiye Uluslararası Katılımlı Toplantı Bildiri Kitabı. 17–20 Mayıs 2007, Yayın No: 2008/3, 107 s., Ankara.
16. Chuman, T. and Romportl, D. 2010. Multivariate classification analysis of cultural landscape: An example from the Czech Republic. Landscape and Urban Planning, 98 (3–4), 30, pp. 200–209.
17. Uzun, O., Dilek, F., Çetinkaya, G., Erduran, F. ve Açıksöz, S. 2010. Konya İli, Bozkır- Seydişehir-Ahırlı-Yalıhüyük İlçeleri ve Suğla Gölü Mevkii Peyzaj Yönetimi, Koruma ve Planlama Projesi. 1–2. Ara Rapor. TC Çevre ve Orman Bakanlığı, Doğa Koruma ve Milli Parklar Genel Müdürlüğü Doğa Koruma Dairesi Başkanlığı, Ankara.
18. Şahin, Ş., Perçin, H., Kurum, E., Uzun, O., Bilgili, E., Tezcan, L., Çiçek, İ., Müftüoğlu, V., Çorbacı, Ö.L., Sütünç, S., Doğan, D., Koç, Ö., Ateş, E., Tarım, B. ve Kurdoğlu, G. 2013. PEYZAJ-44: İl Ölçeğinde Peyzaj Karakter Analizi ve Turizm/Rekreasyon Açısından Değerlendirilmesi (PEYZAJ-44). 109G074 Nolu TÜBİTAK KAMAG Projesi, Proje Sonuç Raporu.

İsmet OSMANOĞLU

Cultural Heritage in the Urban Planning Studies of Historical Peninsula in Istanbul between 1930 and 1950

Abstract: This study investigates the approach of the urban planning studies carried out in Istanbul between 1930 and 1950 to the cultural heritage in the Historical Peninsula. The sources of data acquisition of the study are state archives, municipal archives, newspaper archives as well as scientific texts such as books and articles of that period. Istanbul is a very dominant city in the world in terms of the density of its urban heritage. Therefore, in the planning studies of the period, it was tried to be understood whether the cultural heritage in the Historic Peninsula was considered as important data to be protected. The proposals of Lambert-Agache-Ehlgötz made on the scale of the Historical Peninsula, Wagner's planning studies and Prost's plans and reports were examined. The data obtained were evaluated comparatively. With the help of the findings, it was understood whether the historical and cultural heritage was evaluated as a data for planning. It turned out that planning experts did not consider the intense cultural heritage (excluding monuments and archaeological site) in the Historical Peninsula as the main element of planning in the period of 1930–1950. The result should be seen as the determination of an expected situation. However, it is problematic to come to a conclusion without considering the current legal-institutional framework, technical equipment and human resources. It is a fact that it is important to look back from the present and make a due diligence. Nevertheless, it would be unfair to reach a judgment without evaluating each period within its own terms.

Keywords: Istanbul, Historical Peninsula, Urban Planning, Cultural Heritage, Protection, Conservation

Introduction

İstanbul is a city that has a special place in both Turkey and the world. Throughout history, the administrative center has been equipped depending on its specific function. As a result, it has developed depending on the collection - distribution function it has gained. İstanbul's ability to collect and distribute resources as a central city has been the main reason for its growth. The settlement area of the city of Istanbul has very special natural environmental data. This place has developed further with the influence and contribution of the human-made environment. The city has been enriched under the influence of different cultures for centuries. Cultural assets, civil, religious, military and architectural heritage produced by high-level civilizations have intensified in the urban space over time. This historical process has created one of the city spaces that have important natural and cultural assets in the world.

This study investigates the approach of the urban planning studies carried out in Istanbul between 1930 and 1950 to the cultural heritage in the Historical Peninsula. The scope of the study was deemed appropriate to be limited to the period between 1930–50 and the Historical Peninsula. This years is considered the early period of the Republic. The reason for this is that the character of the studies and practices in Istanbul changed as a result of the transition to multi-party life after 1950. The results of post-1950 practices go beyond the limits of this article. Adnan Menderes period has data that can be a very special subject of evaluation.

The legal ground and institutional structure that the analyzed urban planning studies take as reference on cultural heritage should be known. This is because the definition and scope of cultural heritage is an important data for planning. An evaluation made without knowing this will not produce an objective result. In this period, the legal documents in force for the definition and protection of cultural heritage are regulations issued during the Ottoman period (Ibid, 1966). The first of these documents is the Regulations for Historical Monuments (Âsâr-ı Atîka Nizamnamesi) issued in 1906 (BOA,İ..KAN., 4/17; İ..MF.,12/14), and the other is the Regulations for the Preservation of Monuments (Muhafaza-i Abidat Nizamnamesi) which entered into force in 1912 (BOA,İ.MMS.,153/20). These regulations remained in effect for 67 years until The Law of Monuments entered into force in 1973 (BCA,180-9-0-0/47/240; R. Gazete, 1973/14527). The archeological excavation permission given to foreigners in the early period of the Republic was given according to the procedures in this regulation (BCA, 30-18-1-1/16/64).

The sources of data acquisition of the study are state archives, municipal archives, newspaper archives as well as scientific texts such as books and articles of that period. Within the scope of the article, the planning studies aiming to determine the development and direction of the spatial organization of the Historical Peninsula between 1930–1950 were examined. The interesting and voluminous findings that define that period emerged as a result of the investigations. All findings of the research were discussed comparatively. As a result of the discussion, it was understood whether historical and cultural heritage is used as data for planning. The findings obtained are explained in the conclusion section.

The Suggestions by Jacques H. Lambert, Alfred Agache and Hermann Ehlgötz

The studies that can be accepted as the first important planning proposals for Istanbul during the Republic Period were carried out in 1933. During the period when Muhittin Üstündağ was the governor and mayor of Istanbul, studies were initiated to solve the urbanization problems of Istanbul. Both the government and Muhittin Üstündağ consider it appropriate to carry out these studies by foreign experts. For this purpose, it was deemed appropriate to organize an opinion competition among foreign urban planning experts (BCA, 30-10-0-0/81/533/6). It was requested that experts be brought in to make the future development plans of the city of Istanbul and 25 000 lira should be spent for them. This request was submitted to the government by The Ministry of Finance and was approved (BCA, 30-18-1-2/36/32/3).

Burhan Arif, one of the urban planning experts and architects of the period, criticized this competition. He stated that Istanbul has a unique character and doesn't resemble French, German or American cities. According to him, "Istanbul has a unique scale, proportions and distance between buildings. The aesthetic and original form created by these characteristics define the city. However, this originality has begun to disappear". Arif thought that a specialist who didn't live in Istanbul wouldn't be able to know the unique characteristics of the city. He emphasized that high-rise concrete buildings spoil the city's skyline. (Arif, 1933: 154–161).

However, these criticisms will not yield any results and an opinion contest will be organized to which four experts are invited. The urbanists named Jacques H. Lambert, Alfred Agache, Hermann Ehlgötz and Henri Prost were invited to the contest held (Cumhuriyet, 1933a: 2; Arif, 1933: 154). However, Prost didn't participate in this competition. Other three experts came to Istanbul, met with the authorities and made observations in the city. They prepared and submitted reports and sketches containing their views and proposals on urban development and planning issues (Cumhuriyet, 1933a: 2; 1933b: 2; 1933c: 2).

These reports and sketches are the first studies of the Republican Era for planning of Istanbul. The suggestions of the urbanists mostly focused on planning studies such as the arrangement and beautification of the city in line with what modern life brings. The titles of the reports prepared are as follows;

- Rapport de Mission D'Urbanisme İstanbul (Lambert, 1933a, 1933b; BCA, 30-10-0-0, 82/537/1/06.04.1935)
- Büyük İstanbul tanzim ve imar programı (Agache, 1934a; 1934b)
- İstanbul şehrinin umumi planı (Ehlgötz, 1934).

As a result of the evaluation made by the jury members, Ehlgötz won 7 opposing votes and 22 votes in favor. Agache got 7 votes while Lambert couldn't get any votes from any members (Cumhuriyet, 1935a: 5). The scope of the plans includes the Historical Peninsula and its immediate surroundings. The studies and reports prepared were evaluated by the jury and the results were published (Arkitekt, 1935: 61–68).

Ehlgötz didn't open a section under the heading "historical buildings or monuments" in his report (Ehlgötz, 1934). However, he included relations with monuments in every chapter he wrote for the planning of the city. The most important decision of Ehlgötz regarding cultural heritage is that it creates a "cultural zone". It seems that this concept includes all monuments, and the following areas remain in this region; Sarayburnu, Sur-u Sultani, Hagia Sophia, Sultanahmet and Şehit Mehmetpaşa Mosques, the field approximately one hundred meters to the right and left of Divanyolu, Beyazıt and the north of Süleymaniye, the area between Saraçhane and Fatih. According to Ehlgötz; "It is necessary to preserve the old monuments carefully, in this way, the artistic value of the city is ensured, it is necessary to separate these monuments from the main circulation roads and connect them with small promenades, and the old historical road networks must be revived". Ehlgötz proposed a road starting from Sirkeci and continuing along the coast, thus aiming to show the beauty of the walls and the sea. He acknowledged the importance of historical and religious works in Sultanahmet, Beyazıt and Eminönü squares, made a special plan for these regions and added to his report. He proposed that a section be opened under the name of the "Preservation of Monuments Branch" within the organization of the municipality, and recommended that the monuments be protected with rules and regulations.

According to Lambert, who mentioned "revealing the value of existing monuments" in his report (Lambert, 1933a; 1933b); "Although the regulation proposed for the city is prepared for practical reasons, it is also inspired by the idea of uncovering the richness of the city's monuments. The area from Sultanahmet to the sea should be built with terraced and low buildings, and a large water pool should be built to the east of the mosque to reflect the elegant silhouette of this monument. All the historical walls from Yedikule to Ayvansaray should be surrounded by a green field, the vicinity of the castle gates should be arranged well, the surrounding of the historical walls should be arranged as wide grassy squares and gardens, and the remaining areas should be afforested. Since the repair of all the walls will be difficult and expensive, some parts such as Yedikule and the Justinian Palace, which will be determined by archaeologists, should be repaired as historical and military monuments. The road that passes through the high point in the west of the estuary will also reveal buildings such as Fatih, Sultan Selim, Kariye Mosque, Mihrimah Sultan Mosque. The local and ornamental values of the many fountains, tombs and domes on the planned roads route were taken into consideration. These are so many that the most elegant and the best preserved and the most useful ones or only those unique to Istanbul can be chosen to be preserved".

According to Agache, who opened the title of "historical buildings and monuments" in his report (Agache, 1934a; 1934b); "Istanbul's intense architectural and archaeological treasures should be preserved, however, these historical monuments shouldn't be shown with their dry nakedness, As in Rome and Greece, these should be surrounded by beautiful gardens and landscapes. For this reason, archaeological sites historical monuments must be identified, and than the buildings that surround or block the monuments must be removed gradually. The altitude and direction of the new buildings should be determined by regulations in order to keep the harmony of the monuments and to make their facades visible. The monuments should be surrounded by gardens by separating them from modern roads, the old city and the new city should live side by side without interfering and damaging". Agache created a "historical zone" covering 145 historical buildings and sites in his plan, as well as emphasizing cultural tourism through monuments and archaeological sites.

Although zoning and future projections are mentioned in all the reports and sketches, it mainly includes efforts to organize and beautify the city as a result of the urbanism understanding of the period. It is clear that some of Agache and Lambert's views were not sufficiently understood by the commission. In these studies, the issue of cultural heritage has been evaluated mainly through archaeological sites and monuments. Ideas such as preserving monuments and archaeological sites, revealing their beauty by opening their surroundings, supporting them with landscaping, separating from main circulation roads, connecting them with secondary promenades, and paying attention to distance and height in new buildings have been put forward. The idea of surrounding the walls inside and outside with green areas came to the fore. Therefore, the demolition of the buildings surrounding these areas was envisaged in order to reveal the monuments, archaeological sites and historical walls. These buildings that were wanted to be demolished at that time were mostly civil or public historical buildings. The general idea is to protect monuments and archaeological sites and to reveal their beauty.

Lambert's attention to the density and abundance of historical heritage in Istanbul, and the idea of preserving the ones unique to Istanbul by selecting them shows the perspective of the period. He planned green areas around them in order to protect the surroundings of the monumental complexes. On the other hand, Ehlgötz made an important decision regarding cultural heritage by designating the areas where the monuments are located as "cultural zone". The cultural zone includes the Sur-u Sultani, Hagia Sophia, Sultanahmet, the right and left of Divanyolu, the areas between Direklerarası and Fatih. In addition, He recommended that a section be opened under the name of the "Preservation of Monuments Branch" within the organization of the municipality. These recommendations are an important and advanced determination compared to the period. The fact that Agache specifically opens the title of "historical buildings and monuments" in his report is important in terms of showing the importance he attaches to the existence of the historical heritage in the city. The fact that he creates a "historical zone" in his plan is an indication of his sensitivity to the subject.

The Planning Studies of Martin Wagner

The upon failure to obtain a solution that would direct the development of the city, Dr. Engineer Martin Wagner was invited to work in the planning consultancy of Istanbul Municipality (Cumhuriyet, 1935b: 7; BCA, 30-18-1-2/52/14/10). Wagner, who previously worked as an urban planning specialist in the Municipality of Berlin, came to Istanbul in 1935. His views on urban planning are ahead of his time. According to Wagner; an economic city cannot be built without making economic zone plans. He stated that the institutions related to regional planning and country planning, which are still insecure today, will gain more and more importance (Wagner, 1937d: 71–74). He criticizes those who see the city building business only from the point of view of beauty and artistic, and says that they don't understand the age and the main problems (Wagner, 1937e: 57–58).

However, Wagner's report entitled "Istanbul Havalisi'nin Planı" (1936a; 1936b) wasn't found sufficient. As it is understood from the reports and sketches (Wagner, 1936a; 1936b), the problems of the reconstruction and beautification of the city center weren't contented with. It is seen that Istanbul and its surroundings are examined in detail, analyzes based on the data are made, problems such as land use decisions, water, agriculture, food and population, as well as especially transportation and industry issues. He emphasized that no measures should be neglected to enrich the Istanbul hinterland. The fact that Martin Wagner predicts the hinterland of Istanbul to cover the whole region between Thrace and Kocaeli and carries out planning studies in this direction is an indication of his foresight.

He also emphasized the population growth coming to Istanbul and the problems it will bring, land speculation and the loss of agricultural lands in the region. He stated that alongside raw materials, agricultural and industrial production and distribution, trade should also be stimulated. He stated that it would be appropriate to distribute the population in a balanced manner rather than concentrate in a single region. In terms of shaping the architecture of the city he has proposed to continue the cultural and aesthetic traditions in Turkey. He said that it wouldn't be correct to use building styles belonging to European or American cities, and stated that every nation should put the will to develop a building style according to itself (Wagner, 1937c: 276–278).

Their recommendation to make use of used water and garbage dumped into the sea is a result of prescientism (Wagner, 1936a: 305). He accepts that the connection of the two sides of Istanbul (Asia and Europe) is compulsory for freight and passenger transportation. However, he states that this isn't possible now and can be done by future generations (Wagner, 1936b: 334). Wagner recommends that the city's public transportation system (railway, sea, bus, etc.) be planned and managed by the municipality, states that transportation decisions that will lead to land speculation should be avoided (Wagner, 1937a: 143–146). He explains that the land prices are very high in Istanbul compared to the income of the people and this situation prevents the renewal of the city. He develops solutions by saying that the public administration should reduce the land prices with legal measures (Wagner, 1936c: 217–218). However, it is clear that these studies weren't sufficiently understood by the administrators and technical staff of that day (Wagner 1936a; 1936b, 1937a: 143–146; 1937b: 112–114).

Wagner's reports and studies doesn't have a direct section on "cultural heritage, archaeological sites or monuments", and there is no text in the archives that mentions this issue. However, there are details about his approach to this issue in his reports and articles on planning Istanbul.

Wagner is aware that the city itself is directly beautiful, and thinks that it is beneficial to build roads for tourists who want to visit these beauties. He proposes a bridge to the Bosphorus, but wants this bridge not to be built between Sarayburnu and Üsküdar. Otherwise, he states that this is an assassination for Istanbul. The fact that he points to the Rumelihisarı side for the Bosphorus bridge is an indication of his concern for silhouette, on the other hand, he has also envisaged measures to prevent the Golden Horn from filling with mud and sand (Wagner, 1936a: 301–306; 1936b: 333–337).

His thoughts on the civil architectural heritage in Istanbul are parallel to the views of the period. According to him; there is a large population density in places such as Beyoğlu, Taksim, Galata, Küçükpazar, Eminönü, Beyazıd and Kumkapı. This population and building density will make it difficult and expensive to get urban equipment (official buildings, entertainment, health, education, etc.) and infrastructure (road, water, electricity, gas, etc.) to these areas. Most of the residences in these areas cannot meet the needs of modern life. For this reason, old and rundown city areas should be cleaned, and as a result, the people should be distributed to the new residential areas to be opened in Istanbul and the density should be reduced (Wagner, 1937b: 112–114). First of all, he suggested reducing the population and building density in the center of Istanbul, then he envisaged the construction of low-rise houses with gardens suitable for the natural beauty and topographic structure of the city.

In fact, Wagner's ideas about the historical city center are similar to those of other urbanists, for example, he predicts destruction in order to reduce the density of historical neighborhoods and achieve a modern city. However, his desire to reduce the population density of Istanbul and his ideas about making a region and country plan to prevent immigration from Anatolia is an attitude that indirectly affects "cultural heritage" positively.

It is very interesting that the government allowed a contract between Istanbul Municipality and Prost during the period when Wagner was working (BCA, 30-18-1-2; 71/6/20). Wagner has worked in Istanbul until November 1937 (Cumhuriyet, 1937:1), then The Ministry of Public Works assigned as a specialist, but left Turkey in 1938 (Wagner, 1938: 82–86). After leaving Turkey served as a faculty member at the urban planning department in Harvard University, but he has continued his interest in Turkey. He has published articles on the Erzincan earthquake (Wagner, 1940: 12–14) and the road issue of Istanbul (Wagner, 1946: 87–90,94). Wagner has done data-dependent and analytical studies and also included the basic problems of the city and future projections in his studies. Therefore, his planning studies are much more advanced than others. The fact that he wasn't understood sufficiently should be considered a great loss. Probably, there were no cadres who could understand their ideas and put them into physical plan and implementation. After Wagner, Istanbul was able to work with an expert of this level only with the arrival of Piccinato after the second half of the 1950s (Arkitekt, 1956:158; BCA, 30-18-1-2/151/77/2).

The Planning Studies of Henri Prost

H. Prost didn't participate in the idea contest opened in 1933 to plan Istanbul. However, in order to project the Yalova Thermal Spring 1935 seems to have been invited to Turkey. In those years, the planning and organizing of Yalova thermal springs came to the agenda upon the interest of Atatürk and the request of The

Ministry of Economy. Upon this request, with a government decision taken on 03.05.1935, it was decided to transform Yalova thermal springs into a modern complex. It is seen that the person considered to prepare a general plan of the area is H. Prost, known as an urban planning specialist. (BCA, 30-18-1-2; 52/17/1). It is seen that both Wagner and Prost were invited to a meeting organized for the planning of Istanbul in the same year. (Cumhuriyet, 1935c: 2). It is understood that in the days following this meeting, Prost and Istanbul Municipality decided to work together. It is quite interesting that H. Prost, who came for the planning of Yalova thermal springs, made a contract with the Istanbul Municipality in a short time to make the plans of Istanbul. (BCA, 30-18-1-2/71/6/20). This contract amounting to 150 000 francs was made with the permission of The Ministry of Interior. The "Report Explaining the Master Plan of Istanbul" was completed by Prost and his team on 15.10.1937 as a result of two years of work. This report is important in that it is the first official document that explains the basic principles of urban planning in Istanbul (Prost, 1937).

The Ministry of Culture experts also participated in the urban planning studies of Istanbul, the person appointed for this purpose was the Director of Istanbul Museums (BCA, 30-10-0-0/190/302/7). It is an important situation that The Ministry of Culture participates and intervenes in the planning studies in order to give an opinion on historical buildings.

It is understood that Prost was also assigned in the implementation of Istanbul plans with a government decree issued on 07.04.1938 (BCA, 30-18-1-2; 82/25/10). In order to work with Prost, people named Jaubert and M. Sherer, who are French citizens, have been granted work permits by the government (BCA, 30-18-1-2/86/18/5; 30-18-1-2/89/114/15). As a result of the studies carried out under the management of Prost, the 1/5000 scale Master Plan of the Istanbul Side was approved by the municipality in 1939 and sent to The Ministry of Public Works (BCA, 30-10-0-0/81/534/8). The plans were approved by The Ministry of Public Works on 14.06.1939, but they requested that six points in the master development plans be re-studied. In addition, The ministry requested that a program covering three terms, each of which will be five years, be prepared and sent to it in order to implement these plans (BOA, 30-10-0-0/81 // 35/1).

Upon the acceptance of Prost's plan, it was aimed to carry out the investments in Istanbul in line with this plan. For example, with the approval of the plan, the decision to demolish the prison in Sultan Ahmed was canceled because the place where the Courthouse would be built was determined (BCA, 30-10-0-0/81/534/8). In line with the plan, it was envisaged to build a stadium in Dolmabahçe. It was decided to send architect Şinasi Şahingiray and Fazıl Ayson to Italy for 3 months in order to carry out the stadium projects with Italian architect Paolo Vietti Violi (BOA, 0-18-1-2/88/75/19). The Imperial Stables (Istabl-ı Amire) was located where the Dolmabahçe stadium was to be built. These buildings were used by the Ministry of National Defense as animal hospitals and distribution-transit centers. The Ministry of National Defense stated that if another complex is built for them, they may leave this area to the municipality (BOA, 30-10-0-0/81/534/10). In the following process, historical Imperial Stables (Istabl-ı Amire) was demolished and the stadium was built instead.

In the same period, the construction of a hospital in Mecidiyeköy by the Ministry of Health came to the agenda, The Ministry requests these projects to be designed by the French architect Jean Walter. However, both the Prime Ministry and the Ministry of Finance have decided to tender these projects by bargaining method. It is understood that Jean Walter, who is stated to be an expert in hospital projects, also carried out the projects of Ankara Medical Faculty Buildings (BOA, 30-10-0-0/178/226/14). During this period, foreign architects and urbanists produced urban plans and building projects in many settlements in Turkey, It is estimated that the reason for this situation is that there aren't enough experts in the country.

A program including the first five years was prepared by the municipality in 1940 in order to implement the master plans of Istanbul. According to this program, 4 million liras is required for the first 5 years plan applications. However, it has been announced that 2 million lira of this money can be met from the budget of Istanbul Municipality. The Ministry of Finance has been requested to pay the remaining 2 million liras from the state budget. The Minister of Public Works A. Fuad Cebesoy explained that the reconstruction of Istanbul isn't only related to that city but a country issue, and emphasized that government aid is essential for this work (BCA, 30-10-0-0/81//35/1). The first five-year program, which is considered to be implemented, includes 21 topics. However, there are 7 titles within these that directly concern the Historical Peninsula.

These titles include the following topics: Eminönü-Unkapanı, Unkapanı-Şehzadebaşı (Atatürk Boulevard), Koska- Beyazıt, Sirkeci-Yedikule (the expropriation of Teneke District), Aksaray-Topkapi, Sultanahmet Square (arrangement and distribution), Edirnekapı - Eyüp. The issues indirectly affecting the Historical Peninsula are: Azapkapı-Tepebaşı (Atatürk Boulevard), Karaköy Square (BCA, 30-10-0-0 /81/35/1). According to the urban plan of Istanbul, some of the lands belonging to the Treasury and remaining in the green field are requested to be left to the Municipality free of charge (BCA, 30-18-1-2/109/78/13). Subjects included in the five-year program haven't been largely implemented. Some of them were put into practice in the Adnan Menderes period after 1950. However, the practices of the Menderes period caused a great damage on the cultural heritage of Istanbul.

Following these plans, Prost prepared the master plan of 1/5000 scale for Kadıköy, Üsküdar and Çamlıca regions of the Anatolian side. Besides, some of the plans and projects prepared by Prost for many regions of the city are as follows: Eminönü-Uzunçarşı (İB, 1939a) Şişli-Vali Konağı (İB, 1939b), Beşiktaş-Bebek (İB, 1940a) Beyoğlu-Taksim (İB, 1940b), Kadıköy-Fenerbahçe (İB, 1940c), Eminönü-Şehzadebaşı (İB, 1942), Eminönü İnternational Exhibition (İB, 1943), Beşiktaş-Ortaköy-Kuruçeşme (İB, 1949), Beşiktaş - Bebek Square (İB, 1950). He prepared all these application-oriented studies in different scales such as 1/2000, 1/1000, 1/500, and 1/50. It is understood that Prost continued its intense work until 1951.

Apart from his work in Istanbul, it is understood that Prost has been assigned to other positions by the government. It is seen that Prost was assigned for The Bursa Plan (BCA, 30-10-0-0/81/534/5) and a contract was signed with him for Master Plan of Karabük (BCA, 30-18-1-2/ 87/57/2). In addition, he was included in the commission established to determine the monuments and statues to be erected in different districts of the country, and was tasked with examining the wall paintings to be hung on the state buildings (BCA, 30-18-1-2/ 94/16/10). It is understood that Prost had continued its relationship within the framework of friendship until 1959 (BCA, 30-1-0-0/ 7/39/35).

A holistic and analytical approach isn't observed in Prost's studies. He didn't foresee Istanbul's relations with the region and the country and the problems arising from these relations, and only made plans to regulate the development and construction activities. Contrary to Martin Wagner's contemporary planning view, Prost's plans can be viewed as a somewhat advanced version of the 19th century urbanism. Although the plans obtained from Prost's work remained in effect for a long time, they were mostly not implemented. However, these plans remained restrictive decisions on existing construction activities.

Considering Prost's approach to cultural heritage, it is seen that contrary to what is known, it has developed many positive decisions for the Historical Peninsula and the metropolitan area of İstanbul. The Master plan main decisions that have a positive impact on the cultural heritage of Historical Peninsula (Prost, H., 1937) are as follows;

- Avoiding the expansion of effective trade routes, thus preventing the spatial change of commercial areas in different parts of the city,
- Planning new neighborhoods on the Rumeli shores of the Bosphorus and in the development regions of the city in order to reduce the density in the center, and connecting these neighborhoods to the city with a fast transportation network,
- Removal of all warehouses, industrial establishments and coal storages in Kuruçeşme in the Bosphorus, and transferring all these functions to a new location,
- Closing and removing Sarayburnu Port,
- Using Sirkeci railway station only to serve suburban trains,
- Demolition of parasitic buildings in this region after the removal of the Port and Sirkeci railway station,
- Planning an archaeological park between Sarayburnu and Küçük Ayasofya,
- Protection of major historical sites (and monuments),
- The creation of a protection zone both outside and inside the historical city walls and the prohibition of construction in this area.
- Taking special measures about the height and character of the buildings around the monuments, and reflecting these special measures in the development plans of the neighborhoods,
- Transforming the Historical Peninsula into an urban area consisting of low-rise houses with gardens,

- Expropriation of small parcels created as a result of fires, and reorganization of these to build houses for common use.

The main decisions of Prost that have a negative impact on the cultural heritage of the Historic Peninsula are as follows (Prost, H., 1937);

- Extending the Galata port to the Academy of Fine Arts, and providing transportation to this port by trucks and automobiles,
- Theoretically, the decision to demolish the left bank of the Golden Horn until Taksim, and the organizing of this area with a new plan,
- Planning the industry on the shores of the Golden Horn, and allocating the rear part of the Golden Horn for large industrial facilities,
- Building wide industrial roads for trucks on both sides of the Golden Horn,
- Building an international railway station and port in Yenikapı,
- Collecting the warehouses around the port and the railway station (Yenikapı),
- Reorganizing the commercial area in the center, surrounding the outer borders with wide roads and planning parking areas,
- Destruction of historical roads within the walls,
- Planning new roads in the Historical Peninsula and organizing neighborhoods in accordance with these roads,
- Developing wholesale foodstuffs and markets in the Historical Peninsula, and also establishing a wholesale fish market,
- To build all urban facilities (administrative, educational, health, cultural, etc., all official buildings) to the Historic Peninsula,

Most of the plans and reports prepared by Prost have been approved and put into practice. Prost's decisions had positive or negative effects on the cultural heritage of the Historical Peninsula, and these effects have lasted until today.

It is understood that Prost's solution for the preservation of the historical city walls is in line with Lambert's views. The archaeological park proposal corresponds to the concept of "archaeological site" that should be protected, and it is a well thought-out approach. However, the fact that the Prost plans load intensive urban functions on the Historical Peninsula made it difficult to protect the cultural heritage, and caused it to be destroyed and damaged. It is unclear what the effects of the central and local government have on all these plan decisions taken. However, it should be assumed that they are involved in many plan decisions and guide Prost.

Upon Prost's departure from Istanbul, a "revision commission" was established with the decision of the Municipal Council. The commission aims to determine which of Prost's work can be fully or partially implemented. At the end of a work lasting 7 months in 1951, the commission completed its work. In the report they prepared, it is seen that they criticize Prost's work. (IB, 1954a). Subsequently, a "counseling committee" consisting of faculty members was formed to manage the urban planning activities of the Istanbul Municipality.

This commission continued its work until the arrival of Högg in 1956 (Cumhuriyet, 1956: 2; Arkitekt, 1956: 158). A faculty member in the Council of Consultants described traditional civil architecture as "ugly" in the following years (Cumhuriyet, 1964: 6). This attitude towards cultural heritage isn't unique to the relevant faculty member. It is common opinion of the relevant period, and will change in the coming years.

Discussion and Conclusion

Although zoning and future projections are mentioned in the reports and sketches, it mainly includes efforts to organize and beautify the city as a result of the urbanism understanding of the period. In these studies, the issue of cultural heritage has been evaluated mainly through archaeological sites and monuments. Ideas such as preserving monuments and archaeological sites, revealing their beauty by opening their surroundings, supporting them with landscaping, separating from main circulation roads, connecting them with secondary promenades, and paying attention to distance and height in new buildings have been put forward. Since the laws on immovable cultural heritage cover archaeological sites and monuments, it isn't strange that planning studies only take them into account.

The presence of a section called "historical buildings and monuments" in the report prepared by Agache, the creation of a "historical zone" in the plan he made and marking 145 historical monuments show his sensitivity to the subject. Ehlgötz's determination of the area where all the monuments are located as a "cultural zone" is seen as an important decision regarding cultural heritage. His proposal for establishing a "Preservation of Monuments Branch" within the Istanbul Municipality was implemented only after the 1990s. Lambert, whose report has a section on "revealing the value of existing monuments", was inspired by the idea of uncovering the riches of the city's monuments. His idea of creating a protective zone inside and outside the historic walls and making these areas green spaces is a continuation of the same approach. It is understood that Prost's solution for the preservation of the historical city walls is in line with Lambert's views.

Wagner's studies are completely different from the work of Agache, Ehlgötz and Lambert, as well as Prost. It is understood that his suggestions reflect the result of an analysis based on data. He focused on the main problems of the city and the region and future projections. The fact that he includes regional and national data and problems in his works shows that he has an advanced and visionary understanding of planning. It is seen that Wagner's reports and studies don't have a section on "archaeological sites or monuments". His thoughts on the civil architectural heritage in Istanbul are parallel to the views of the period. He predicts destruction in order to reduce the density of historical neighborhoods and achieve a modern city. The fact that he wanted the Bosphorus Bridge to be built in a location that wouldn't affect the silhouette of the Historical Peninsula shows that he is concerned about the historical and natural heritage.

Since previous studies weren't approved and put into practice, Prost's studies should be evaluated differently from this works. Therefore, Prost's decisions directly affected the Historical Peninsula with its positive or negative aspects until today. Prost's studies don't have a holistic and analytical approach. He didn't foresee Istanbul's relations with the region and the country and the problems arising from these relations, and only produced plans that regulate the activities of development and construction. Prost's approach usually includes casual and piecemeal solutions rather than future planning. He focused more on demolishing and evacuating the busy areas of the city center, and planning new roads and buildings in the evacuated areas.

Considering Prost's approach to cultural heritage, it is seen that contrary to what is known, it has developed many positive decisions for the Historical Peninsula and the metropolitan area of İstanbul. Important decisions that have an impact on preventing the destruction of cultural heritage are: protecting major monuments; Establishing an "archaeological park" between Sarayburnu and Küçük Ayasofya; the creation of a green belt where construction is prohibited inside and outside the historic walls; taking special measures regarding the height and character of the buildings around the main monuments dominating the city skyline; not expanding trade routes; establishing protection zones; introducing restrictions for buildings to be built in historical environments. However, the allocation of the Golden Horn coasts for large industry has produced an unfortunate result for this region. In addition, the decisions to install infrastructure and superstructure functions that will attract population, building and traffic density to the Historical Peninsula have created long-term negative effects in terms of cultural heritage.

Since Prost's plans were mostly not implemented for the period 1939–1950, their impact on cultural heritage was limited. However, these plans remained in effect for a long time and became a guide for some of the applications between 1950 and 1960. The evacuation and expansion efforts initiated by A. Menderes in the dense regions of the Historical Peninsula caused damage to the cultural heritage. However, it is seen that this situation didn't satisfy the administrators in the following years. After second half of 1950s it was understood that the city and the hinterland had to be planned as a whole, and it was decided to invite Piccinato.

It is thought that all decisions taken in this period regarding cultural heritage are in parallel with the general attitude in the world. For this reason, it is seen that the conservation problem is limited to monuments and archaeological sites. There is no decision has been found regarding the civil architectural heritage and traditional houses in the contents of plans and reports. However, it is a progressive and predictive approach that urban planners create protected areas under the name of "historical zone", cultural zone ", archaeological park". It is also important that the Ministry of Culture participates and intervenes in the planning studies in order to give an opinion on historical buildings. It is very clear that in the period examined, cultural heritage wasn't considered as the dominant input and an important data for planning. Having such an expectation is only about not understanding the conditions and realities of the period.

Acknowledgments: This article is derived from the author's phd thesis named "A method experiment for the improvement in politics of urban protection in Turkey model: İstanbul Historical Peninsula" (Istanbul Technical University, Institute of Science and Technology, 2004).

References

Agache, A., (1934a). Büyük İstanbul tanzim ve imar programı. İstanbul: Belediye Matbaası,

Agache, A., (1934b). Grand İstanbul un programme d'urbanisation. İstanbul: Imprimerie Selamet,

Arif, B. (1933). İstanbul'un Plânı. Arkitekt, 1933/05 (29). İstanbul: Matbaacılık

ve Neşriyat T.A.Ş.

Arkitekt, (1935). İstanbul Şehir Plânı. Arkitekt, 1935/02 (50). İstanbul: Matbaacılık

ve Neşriyat T.A.Ş.

Arkitekt, (1956). Haberler. Arkitekt. 1956/03 (285). İstanbul: Matbaacılık ve Neşriyat T.A.Ş.

Ehlgötz, H. (1934). İstanbul şehrinin umumi planı. İstanbul: Ahmet Sait Matbaası.

Ibid, (1966). Âsarı Atika (Eski Eserler) Nizamnamesi. Ankara: Milli Eğitim Basımevi.

Lambert, J. H. (1933a). Rapport de Mission D' Urbanisme İstanbul. (BCA, 30-10-0-0, 82/537/1/06.04.1935)

Lambert, J. H. (1933b). A son excellence le Vali d'Istanbul, İstanbul: Milli Neşriyat Yurdu

Prost, H., (1937). İstanbul'un Nazım Planını İzah Eden Rapor. 15.10.1937, İstanbul Belediyesi, Yapı ve İmar İşleri Reisliği, Tasdik: 30.06.1939

Wagner, M. (1936a). İstanbul Havalisinin Plânı. Arkitekt, 1936-10-11 (70–71). İstanbul: Matbaacılık ve Neşriyat T.A.Ş.

Wagner, M. (1936b). İstanbul Havalisinin Plânı. Arkitekt, 1936–12 (72). İstanbul: Matbaacılık ve Neşriyat T.A.Ş.

Wagner, M. (1936c). İstanbul Şehrinin Düzeltilmesi Meseleleri. Arkitekt, 1936–08 (68) İstanbul: Matbaacılık ve Neşriyat T.A.Ş.

Wagner, M. (1937a). İstanbul'un Münakale Tahlili. Arkitekt, 1937-05-06 (77–78) İstanbul: Matbaacılık ve Neşriyat T.A.Ş.

Wagner, M. (1937b). İstanbul Nüfusunun Yayılışı ve Münakale. Arkitekt, 1937–04 (76). İstanbul: Matbaacılık ve Neşriyat T.A.Ş.

Wagner, M. (1937c). İnşa Etmeyen Bir Millet Yaşamıyor Demektir. Arkitekt, 1937-10-11 (82–83). İstanbul: Matbaacılık ve Neşriyat T.A.Ş.

Wagner, M. (1937d). Büyük Şehirler Nasıl Tadil Edilir? Arkitekt, 1937–03 (75) İstanbul: Matbaacılık ve Neşriyat T.A.Ş.

Wagner, M. (1937e). Başlık: Büyük Şehirler Nasıl Tadil Edilir? Arkitekt, 1937–02 (74). İstanbul: Matbaacılık ve Neşriyat T.A.Ş.

Wagner, M. (1938). Türk Şehirleri ve Mevcut Sahalardan İstifade "Ekonomisi". Arkitekt, 1938–03 (87). İstanbul: Matbaacılık ve Neşriyat T.A.Ş.

Wagner, M. (1940). Zelzele Mıntakası için Düşünülmüş Mukavim, Standart Ev Projeleri. Arkitek, 1940-01-02 (109–110) İstanbul: Matbaacılık ve Neşriyat T.A.Ş.

Wagner, M. (1946). İstanbul'un Yol Meselesi. Arkitekt. 1946-03-04 (171–172) İstanbul: Matbaacılık ve Neşriyat T.A.Ş.

The Republic Archives (Presidency of The Republic of Turkey Directorate Of State Archives)

BCA, 30-18-1-1/16/64/8/27.09.1925

BCA, 30-10-0-0/81/533/6/08.02.1933

BCA, 30-18-1-2/36/32/3/30.04.1933

BCA, 30-18-1-2/52/14/10/28.02.1935

BCA, 30-18-1-2;52/17/1/05.03.1935

BCA, 30-10-0-0, 82/537/1/06.04.1935

BCA, 30-18-1-2/71/6/20/27.01.1937

BCA, 30-10-0-0/190/302/7/10.01.1938

BCA, 30-18-1-2/82/25/3/07.04.1938

BCA, 30-10-0-0/81/534/5/03.12.1938

BCA, 30-18-1-2/86/18/5/02.03.1939

BCA, 30-10-0-0/81/534/8/09.06.1939

BCA, 30-18-1-2/ 87/57/2/15.06.1939

BCA,30-10-0-0/81/534/10/28.09.1939

BCA, 30-18-1-2/89/114/15/04.12.1939

BCA, 30-10-0-0/81//35/1/16.01.1940

BCA, 30-10-0-0/178/226/14/12.01.1940

BCA, 30-18-1-2; 94/16/10/01.03.1941

BCA, 30-18-1-2/109/78/13/08.01.1946

BCA, 30-18-1-2/151/77/2/18.02.1959

BCA, 30-1-0-0/ 7/39/35/16.07.1959

The Ottoman Archives (Presidency of The Republic of Turkey Directorate Of State Archives)

BOA,İ..KAN., 4/17/ (1324S-3)H-29-02-1324
BOA, İ..MF.., 12/14/ (1324Ra-1)H-04-03-1324
BOA, İ.MMS., 153/20/(1330 Şaban-11) H-16-08-1330

Cumhuriyet (Newspaper) Archive

Cumhuriyet, (1933a). Cumhuriyet Gazetesi, 15 Nisan 1933, s. 2, İstanbul
Cumhuriyet, (1933b). Cumhuriyet Gazetesi, 06 Haziran 1933, s. 2, İstanbul
Cumhuriyet, (1933c). Cumhuriyet Gazetesi, 19 Temmuz 1933, s.2, İstanbul
Cumhuriyet, (1935a). Cumhuriyet Gazetesi, 27 Şubat 1935, s.5. İstanbul
Cumhuriyet, (1935b). Cumhuriyet Gazetesi, 01 Mayıs 1935, s.7. İstanbul
Cumhuriyet, (1935c). Cumhuriyet Gazetesi, 09 Haziran 1935, s.2. İstanbul
Cumhuriyet, (1937). Cumhuriyet Gazetesi, 20 Kasım 1937, s. 1. İstanbul
Cumhuriyet, (1956). Cumhuriyet Gazetesi, 03 Ağustos 1956, s.2. İstanbul
Cumhuriyet, (1964). Cumhuriyet Gazetesi, 17 Mart 1964:s.6, İstanbul

Archive of the Metropolitan Municipality of İstanbul

İB, (1939a). Eminönü - Uzun Çarşı caddesi ve civarı haritasıdır, 1/500, Prost, Henri İstanbul:İstanbul Belediyesi

İB, (1939b). Şişli - Vali Konağı caddesi - Çocuk bahçesi - Tafsilat resimleri. Plan no: 661, 1/50/Prost, Henri; çiz.; İbrahim Erkoğlu, İstanbul: İstanbul Belediyesi İmar Müdürlüğü.

İB, (1940a). Beşiktaş - Bebek imar palnını gösterir haritadır. = Amenagement du Bord de Mer = Deniz kıyısı Düzenlemesi. Plan No:1428, 1/500, Prost, Henri, İstanbul Belediyesi: İstanbul Belediyesi İmar Bürosu,

İB, (1940b). Beyoğlu - İstanbul şehri Taksim bahçesi Kışla arazisi müstakbel şeklini gösterir haritadır, 1/500, Prost, Henri, İstanbul: İstanbul Belediyesi İmar Müdürlüğü,

İB, (1940c). Kadıköy'ünde Fenerbahçe'de Hatboyu sokağında Aliye Sözel'e ait arazinin ifraz haritasıdır. 1/2000, Prost, Henri; çiz: İbrahim Erkoğlu, İstanbul: İstanbul Belediyesi, İmar Bürosu.

İB, (1942). Eminönü - Şehzadebaşı caddesi ve civarı haritasıdır. Plan No: 2561, 1/ 1000, Prost, Henri, İstanbul: İstanbul Belediyesi

İB, (1943). Eminönü - 1953 Beynelmilel sergi avan projesidir. 1/2000,/Prost, Henri, İstanbul: İstanbul Belediyesi

İB, (1949). Beşiktaş - Ortaköy - Kuruçeşme yoluna ait imar tatbikat planıdır, 1/500, Prost, Henri, İstanbul: İstanbul Belediyesi İmar Müdürlüğü,

İB, (1950). Beşiktaş - Bebek İskele Meydanı ve etrafına ait tanzim planıdır, 1/500, Prost, Henri; haz.: A. Angel; çiz.: İskender, İstanbul: İstanbul Belediyesi İmar Müdürlüğü,

İB, (1954a). Revizyon komisyonu raporu. İstanbul: İstanbul Belediyesi,

Archive of The Official Gazette (Presidency of The Republic of Turkey)

R. Gazete, Tarih: 6/5/1973, Sayı:14527, Kanun No:1710

Makbulenur ONUR and Selver KOÇ ALTUNTAŞ

Assessment of Use in Indoor and Outdoor Subject of Covıd-19 Pandemia Process

1 Introduction

The concept of sustainability increases its importance day by day. In this process, cities lose many values with increasing urbanization (Bekar et. Al., 2018a; Bekar et al 2018b; Gülpınar Sekban et. al., 2018; Gülpınar Sekban et., al. 2019a; Gülpınar Sekban et., al. 2019b; Gülpınar Sekban et al 2019c, Aksoy Işık, 2020, Kayan, Mardinli 2020). City; It is like a puzzle that is considered unfinished when its parts are missing and has to complete the dynamic cycle of urban needs. In this process, mechanisms that are disconnected from each other must constantly complete their missing parts with different decisions (Akyıldız, 2020). Since the world is in a state of motion, the only thing it tries to deal with is not limited to urbanization. Cities that can cope with pandemic and disaster periods should be established.

Covid-19, which emerged in Wuhan, China in 2019, made it necessary to take various measures within the scope of combating the epidemic. In this context; it has caused serious changes in health, economy, education and social life (WHO, 2020). This situation has led to the creation of lists to be made in our country, as it is effective worldwide. This includes the whole area from the isolation of a family in its own living space to collective movements in socializing spaces (Söğüt, 2020). In order to reduce the impact of the epidemic, flexible or rotational working practices have been initiated, and measures have been taken to maintain social distance (Zhao, 2020). These limitations in living spaces have led to gaining experience instead of acquired habits. Experiences include behaviors that are seen as simple in human life such as walking on the street, shopping, eating at the restaurant and using the space equipment according to the needs. This situation has affected the food and beverage industry as well as in every field.

Eating and drinking places arising from a culture and need; (Hattox, 1998), İt offers both closed, semi-open and open spaces, and it is at an important point in our lives with the phenomenon of socialization. When viewed from this angle; Considering the place of eating and drinking places in human life, it is seen that the use of these places is vital with the pandemic process. As of March 2019, various restrictions affecting social life have been implemented. In this context, the curfew, quarantine, new-normalization process imposed in about a year show that consumers' spending on food services has decreased significantly (URL-1,2,2020). In Turkey, approximately 100 thousand small and large businesses, with 2 million employees, has taken a number of measures in the leadership of the Ministry of Health and the Ministry of Culture and Tourism.

The main reason for these shifts in meaning, which were observed in the human-space-object relationship after the pandemic and started to be realized that they were actually working distorted from the beginning, is the belief that the void does not have both an emotional and a physical function. During the pandemic process, what kind of meaning and significance shifts are caused by the verbs of designing, constructing and making prioritizing fullness and fullness over objects (Durukan, Öztürk 2020). COVID-19 has demonstrated the need for new and integrated approaches to urban planning. Urban density, public health, energy use, economic activities, technology use, transportation, urban services such as education and health, open and green space systems and their locations and capacities in the city, and new approaches in architecture and construction It can be predicted that new standards will be developed (Tuğaç 2020).

If the trust problem experienced due to the cause of the pandemic and the importance taken does not provide the physical and psychological factors offered by the food and beverage industry, the user; encounters negative effects while experiencing the space. In order to prevent this, the flexibility of the components and parameters that make the space available has become important recently. New ideas developed in the process, spatial design and strategies reshaped these spaces. In addition, it can be seen that the existing spaces meet the mentioned components and parameters.

Based on all this information; The effects of the current covid-19 pandemic were discussed in every field, and the effects of cafes and restaurants on the public and employment were investigated in our country. It has been stated in the studies that the sectors in question have been made various revisions within the scope of combating the epidemic. Apart from this, as a result of the measures implemented, process recommendations were developed to ensure the protection of employment and the continuity of working life. In the study, space designs where the covid-19 pandemic and food and beverage areas meet on a common ground are discussed. In this period when the adaptation process is very important, drawing attention to functional designs is the original aspect of the study. The study, which is expected to shed light on disaster/pandemic periods, sets the ground for precautions and various applications.

The main purpose of this study is, considering the crisis situations and diseases that may occur worldwide; It was made with the aim of determining the transforming ability of indoor and outdoor spaces that people actively use. It has been discussed whether the eating and drinking areas, which are one of the designed interior spaces, have the ability to adapt and transform with the pandemic. In the study, randomly selected eating and drinking areas with open spaces around the world and ecological features were discussed. As a result of the study, functional designs that can provide resistance to disaster/pandemic periods are exemplified.

2 Method

The method of this study consists of 4 successive stages (Fig. 1).

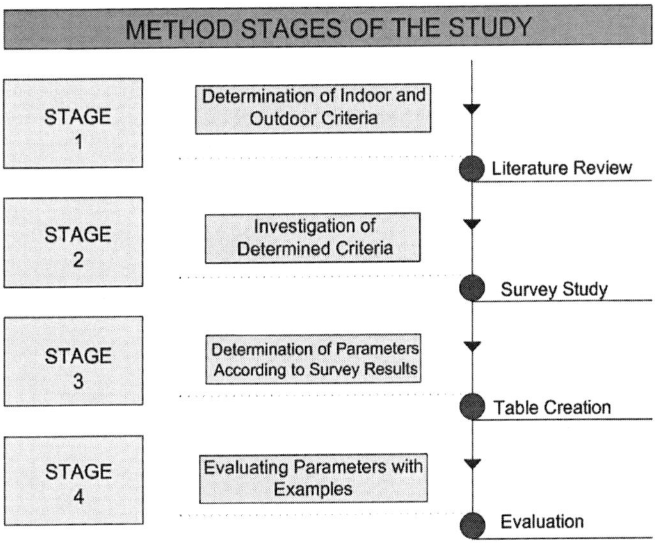

Fig. 1. Study stages

- **In the first stage:** As a result of the first literature study, "Indoor and outdoor criteria" were established. At this point, interior formations provided with openings were taken as basis. Open plan and fluid space concepts and definitions are examined, and these concepts and spaces are addressed by physical and structural environmental factors.
- **In the second stage**: The created criteria were transformed into 20 questions and questioned with a sample group of 100 people. Survey question; "If you want to go to a cafe during a pandemic/disaster period, you would give priority to which parameters".
- **Third stage:** As a result of the survey study, the most preferred 22 parameters were brought into a table.
- **The fourth stage**: this is the last stage of the study. At the end of the survey, the most preferred 22 parameters were analyzed on 5 designed samples

3 Findings

Çalışma kapsamında elde edilen bulgular aşağıda verilmiştir.

3.1 Findings on the First Phase: Setting the Criteria

Within the scope of the literature study, "indoor and outdoor criteria" were established. Rapidly increasing urbanization; It has affected the cities of the world with factors such as population and economy. Therefore, the need for green spaces has increased (Andersson et al., 2015), making a spacious and transparent space setup a necessity. We exist in spatial diversity in indoor and outdoor spaces depending on the shape and conditions of the environment. Conditions perceived in the open area depending on seasonal variables such as snow, fog, daylight, darkness, excessive pressure, and human-created conditions such as natural/artificial light and acoustics in indoor environments diversify our perception of space. (Günal, 2006). According to Lawson; atmosphere is a factor that affects preference. A client will often prefer privacy or sociability, formality or intimacy, excitement or relaxation (Lawson, 1976). Within the scope of this study, the parameters were examined under two main titles as indoor and outdoor criteria. These main headings are divided into subheadings (Tab. 1).

Tab. 1. Main and sub parameters created within the scope of the study

1. Indoor Criteria	**Sub parameters**
	1.1 Physical Criteria
	1.2.Psychological Criteria
2.Open Space Criteria	**Sub parameters**
	2.1 Ecological Criteria
	2.2 Aesthetic criteria

3.2 Findings Regarding the Second Phase: Formation of Survey Questions

The created criteria were converted into 20 questions and questioned with 100 people. Survey question: If you want to go to a cafe during the pandemic/disaster period, which parameters would you prioritize?

3.3 Findings Regarding the Third Phase: Creating Outcome Parameters

As a result of the survey study, the most preferred 22 parameters were brought into a table form (Tab. 2). Participants were asked to give a score from 1 to 5 for the parameters. 1 represents the lowest score, 5 represents the highest score. The highest first 22 parameters and results in the results of the participants are given in

Tab. 1. These parameters; settlement plan suitable for social distance, ventilation, digital menu, contactless payment methods, sensor faucet, spatial air conditioning, disinfectant booths, privacy opportunity, trust effect, spacious space perception, open space existence, plant materials, shelter for fauna, landscape feature, water elements special playgrounds for children, layout plan suitable for social distance, seating units that can be separated when necessary, spacious space perception, open space presence, presence of color effect.

Tab. 2. Points obtained from parameters

No	Parameters	Point
1-	Layout plan that can adapt to social distance (indoor)	+470
2-	Ventilation	+310
3-	Digital menu	+195
4-	Contactless payment methods	+230
5-	Sensor faucet	+110
6-	Sensor light	+125
7-	Spatial air conditioning	+240
8-	Disinfectant cabinets	+120
9-	The possibility of privacy	+405
10-	Trust effect	+450
11-	Spacious space perception	+275
12-	Open space presence	+475
13-	Herbal materials	+472
14-	Shelter opportunity for fauna	+245
15-	Landscape feature	+455
16-	Presence of water elements	+314
17-	Special playgrounds for children	+170
18-	Layout plan that can adapt to social distance (open space)	+410
19-	Separable seating units when necessary	+550
20-	Spacious space plan	+400
21-	Outdoor activity	+440
22-	Presence of color effect	+400

3.4 Findings Regarding the Fourth Stage: Examination of Spaces According to the Criteria

It is the last stage of the study. At the end of the questionnaire, 22 most preferred parameters were examined on 10 designed samples. The 22 parameters created in Tab. 2 were examined on the examples. The existence of that parameter in the analyzed samples was determined with the "x" sign. If it is not and cannot be detected, it is left blank.

In the first part on the catalog; for defining the structure; The name of the building, outdoor and interior photographs are included. In the second part; cafes under review; It has been read over the indoor and outdoor criteria determined in the literature. In this direction, the physical and ecological factors of the indoor and outdoor spaces have been evaluated with their internal parameters. In this context, the table given below has been prepared (Table 4-5-6-7-8).

Eating and drinking places diversified within themselves; It is divided into a wide range as patisserie, restaurant, snack foods. At this point, used as a consumption object in recent years; Cafes, which gain importance due to factors such as their design, corporate identity, the landscape of the space, its environment and its interaction with its surroundings, have been accepted as the universe of the study. As a result of the researches, 10 cafeterias that set an example were examined within the scope of the study.

Tab. 3. Examination of sample no.1 and parameters

Building name	**An'garden Café/Le House**
Building images	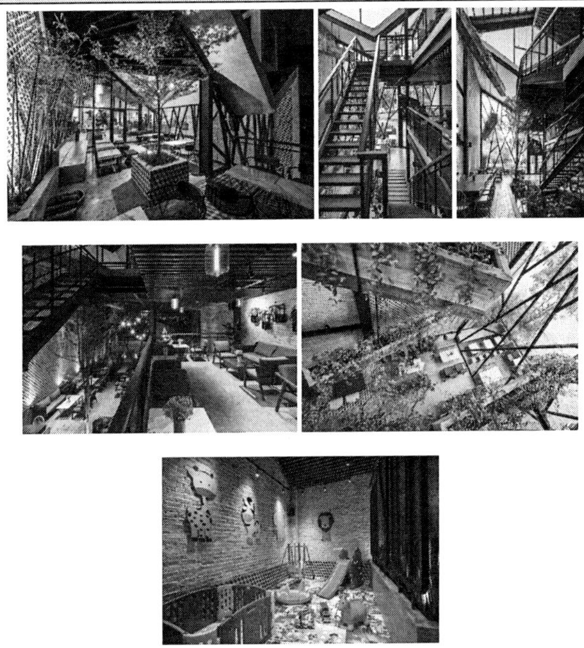 (URL-3,4,5,6,7,8)

(continued)

Tab. 3. Continued

Building name	**An'garden Café/Le House**		
Plan and sections	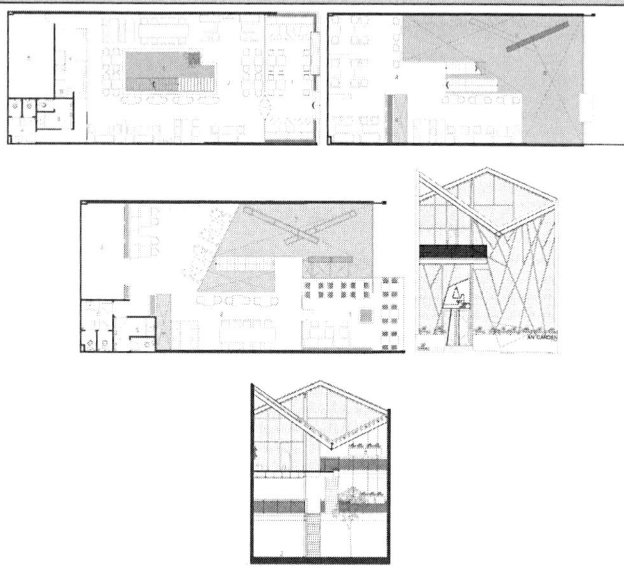 (URL-9,10,11,12)		
Indoor and Outdoor Criteria			
Indoor Criteria		**Open Space Criteria**	
Physical Criteria		**Ecological Criteria**	
Layout plan that adapts to social distance	**X**	Open space presence	**X**
Ventilation	**X**	Herbal materials	**X**
Digital menu		Shelter opportunity for fauna	**X**
Contactless payment methods		Landscape feature	
Sensor faucet		Presence of water elements	
Sensor light		Special playgrounds for children	**X**
Spatial air conditioning	**X**	Layout plan that adapts to social distance	
Disinfectant cabinets		Separable seating units when necessary	
Psychological Criteria		**Aesthetic criteria**	
The possibility of privacy	X	Spacious space plan	X
Trust effect	X	Outdoor activity	X
Spacious space perception	X	Presence of color effect	X

Tab. 4. Examination of sample no.2 and parameters

Building name	**1900 COFFEE/PVA**
Building images	     (URL-13,14,15,16,17)
Plan and sections	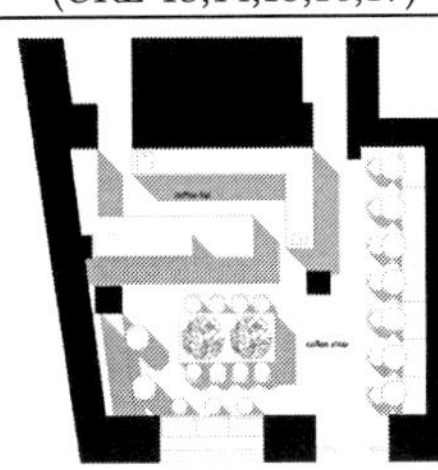 (URL-18)

(continued)

Tab. 4. Continued

Indoor and Outdoor Criteria			
Indoor Criteria		**Open Space Criteria**	
Physical Criteria		**Ecological Criteria**	
Layout plan that adapts to social distance	**X**	Open space presence	
Ventilation	**X**	Herbal materials	**X**
Digital menu		Shelter opportunity for fauna	
Contactless payment methods		Landscape feature	
Sensor faucet		Presence of water elements	
Sensor light		Special playgrounds for children	
Spatial air conditioning	**X**	Layout plan that adapts to social distance	**X**
Disinfectant cabinets		Separable seating units when necessary	**X**
Psychological Criteria		**Aesthetic criteria**	
The possibility of privacy	X	Spacious space plan	
Trust effect	X	Outdoor activity	
Spacious space perception	X	Presence of color effect	X

Tab. 5. Examination of sample no.3 and parameters

Building name	**The 59 Café**
Building images	      (URL-19,20,21,22,23,24)

(continued)

Tab. 5. Continued

Building name	**The 59 Café**		
Plan and sections	(URL-25, 26, 27, 28)		
Indoor and Outdoor Criteria			
Indoor Criteria		**Open Space Criteria**	
Physical Criteria		**Ecological Criteria**	
Layout plan that adapts to social distance	**X**	Open space presence	**X**
Ventilation	**X**	Herbal materials	**X**
Digital menu		Shelter opportunity for fauna	**X**
Contactless payment methods		Landscape feature	**X**
Sensor faucet		Presence of water elements	
Sensor light		Special playgrounds for children	**X**
Spatial air conditioning	**X**	Layout plan that adapts to social distance	**X**
Disinfectant cabinets		Separable seating units when necessary	**X**
Psychological Criteria		**Aesthetic criteria**	
The possibility of privacy	X	Spacious space plan	X
Trust effect	X	Outdoor activity	X
Spacious space perception	X	Presence of color effect	X

Tab. 6. Examination of sample no.4 and parameters

Building name	**Thailand Hi Cafe/balbek bureau**
Building images	     (URL-29,30,31,32,33)
Plan and sections	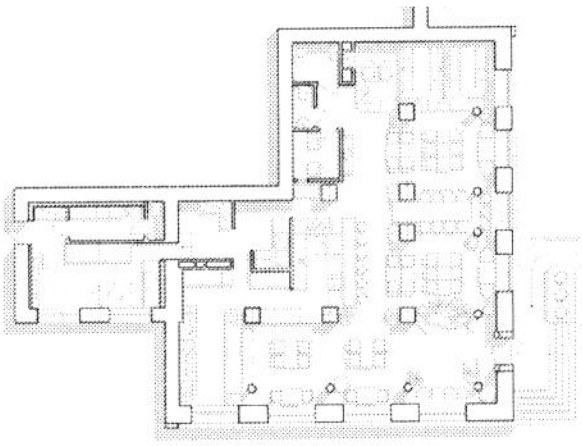 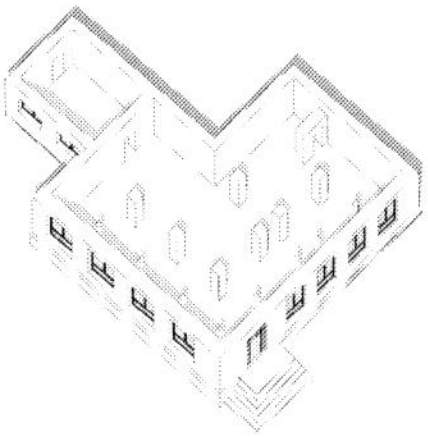 (URL-34,35)

(*continued*)

Tab. 6. Continued

Indoor and Outdoor Criteria			
Indoor Criteria		**Open Space Criteria**	
Physical Criteria		**Ecological Criteria**	
Layout plan that adapts to social distance		Open space presence	
Ventilation	**X**	Herbal materials	**X**
Digital menu		Shelter opportunity for fauna	
Contactless payment methods		Landscape feature	
Sensor faucet		Presence of water elements	
Sensor light		Special playgrounds for children	
Spatial air conditioning	**X**	Layout plan that adapts to social distance	**X**
Disinfectant cabinets		Separable seating units when necessary	**X**
Psychological Criteria		**Aesthetic criteria**	
The possibility of privacy	X	Spacious space plan	X
Trust effect	X	Outdoor activity	
Spacious space perception	X	Presence of color effect	X

Tab. 7. Examination of sample no.5 and parameters

Building name	**Que Dua Restaurant/DDconcept**		
Building images	  (URL-36, 37)		
Plan and sections	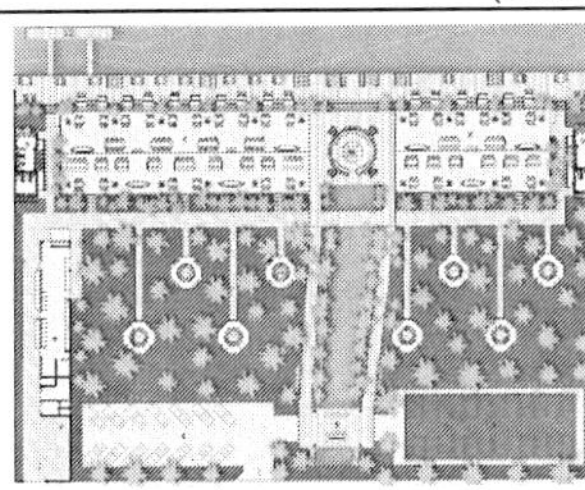 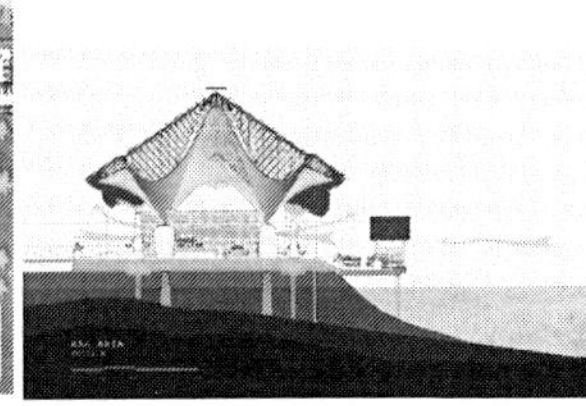 (URL-38,39)		
Indoor and Outdoor Criteria			
Indoor Criteria		**Açık Mekan Kriterleri**	
Physical Criteria		**Ekolojik Kriterler**	
Layout plan that adapts to social distance	**X**	Open space presence	**X**
Ventilation	**X**	Herbal materials	**X**
Digital menu		Shelter opportunity for fauna	**X**
Contactless payment methods		Landscape feature	**X**
Sensor faucet		Presence of water elements	**X**
Sensor light		Special playgrounds for children	
Spatial air conditioning	**X**	Layout plan that adapts to social distance	**X**
Disinfectant cabinets		Separable seating units when necessary	**X**
Psychological Criteria		**Aesthetic criteria**	
The possibility of privacy	X	Spacious space plan	X
Trust effect	X	Outdoor activity	X
Spacious space perception	X	Presence of color effect	X

4 Conclusion

Pandemic period; It brought about sudden events and various unforeseen situations. In the process where changes are made in various branches from the health sector, education, economy to social life, being safe from the environment we live in to every area we spend time will make the time spent physically and psychologically more quality. For this reason, the concept of flexibility in design is of great importance for the continuity of living spaces. In this context, the changes made with epidemics and normalization were discussed in cafes, which are one of the living spaces and have an important place in social life. Physically handled in this process; spatial, air conditioning, lighting and auditory parameters, psychologically addressed; By determining the privacy, behavioral and aesthetic parameters of each cafe subject, some results were obtained. Considering the answers given by the participants, the results obtained are listed as follows;

- Cities should be designed to resist events such as epidemic and disaster periods.
- Decisions should be taken with a comprehensive perspective and in terms of urban sustainability/resilience.
- Design decisions should be made in accordance with the cleaning rules for indoor and outdoor spaces.
- Decisions for indoor and outdoor spaces that can change or adapt to epidemic diseases and similar events should be taken.
- The necessity to consider these risks in design decisions should be increased.
- Orientation should be provided to innovative design decisions.
- Studies on epidemic and disaster periods should be carried out in design-oriented professional disciplines such as architecture/landscape architecture/interior architecture and city-regional planning.
- Functional decisions regarding the relationship between the new world order & design should be made.

As a result, we should not forget that; There are many important criteria that the Covid19 period has had many negative effects on the whole world. It should not be forgotten that this transformation created by the new world order is of key importance for the future.

References

Aksoy Işık, R., (2020). Çevre Sorunsallarından Biri Olan Kentleşme: Mardin Örneği. *International Journal of New Paradigm*, *3*(1), 10–20.

Akyıldız, N. A., (2020). Kentleşme ve Kentsel Gelişim Bağlamında Açık Kamusal Alanların Sürdürülebilir Kentler Açısından Değeri. *Milli Folklor*, *16*(125), 188–201.

Andersson, E., Tengo, M., McPhearson, T., Kremer, P., (2015). Cultural ecosystem services as a gateway for improving urban sustainability. Ecosyst. Serv. 12, 165–168. https://doi.org/10.1016/j.ecoser.2014.08.002.

Bekar, M., Gülpınar Sekban, D. Ü., Acar, C., (2018a). Planting along Area Sides of Highway, Trabzon Coastal Case, Turkcy. *Journal Of Balkan Ecology*, vol.21, 443–463.

Bekar, M., Gülpınar Sekban, D. Ü., (2018b). Kentlerin Zihinde Bıraktığı İzler Trabzon Örneği. *(SAD)-The Journal of International Social Research*, vol.11, 331–339.

Durukan, A., Öztürk, Ö. B., (2020). Küresel salgından iç mimari çıkarımlar: Boşu tasarlamak ve boşluk körlüğünden kurtulmak, *Journal of Social and Humanities Sciences Research*, *7*(56), 1962–1968.

Gülpınar Sekban, D. Ü., Bekar, M., (2018). Evaluation of Gentrification in terms of Landscape Architecture. *Innovative Approaches in Architecture, Planing and Design* (pp.77–82), Ankara: Gece Kitaplığı.

Gülpınar Sekban, D. Ü., Bekar, M., Acar, C., (2019a). Evaluation of Sustainability Potential According to Cittaslow Criteria in Turkey/Trabzon, *Fresenius Environmental Bulletin*, vol.28, 5435–5446.

Gülpınar Sekban, D. Ü., Demirci, Ö., Uzunali, A., (2019b). Tarihi Kentsel Alanların Korunmasında Yayalaştırma Çözümü, Trabzon Arafilboyu Örneği, *The Journal Of Internatıonal Socıal Research*, vol.12, 477–486.

Gülpınar Sekban, D. Ü., Demirci, Ö., Uzunali, A., (2019c). Tarihi Kentsel Alanların Korunmasında Yayalaştırma Çözümü, Trabzon Arafilboyu Örneği. *The Journal Of Internatıonal Socıal Research*, vol.12, 477–486.

Günal, B., (2006). İnsan-Mekân İletişim Modeli Bağlamında Konutta Psiko-Sosyal Kalitenin İrdelenmesi. İstanbul: İstanbul Teknik Üniversitesi, Fen Bilimleri Enstitüsü, İstanbul.

Hattox, R. S., (1998). Kahve ve Kahvehaneler. Çev. Elhüseyni, N. Tarih Vakfı Yurt Yayınları, 142, İstanbul.

Lawson, F., (1976). Hotels, Motels and Condominiums: Design, Planning and Maintenance. Architectural Press, 238, Minnesota.

Kayan, A. Mardinli, İ., (2020). Kentleşme-Sosyal Değişme Ve Ekonomik Gelişme İlişkisinin Değerlendirilmesi. *GÜ İslahiye İİBF Uluslararası E-Dergi*, *4*(4), 262–277.

Söğüt, A., (2020). Covid-19 Pandemisi Sonrası Normalleşme Sürecinin Sürdürülebilirliğe Etkisi. Mimarlık Bilimlerinde Güncel Araştırmalar, 55.

Tuğaç, Ç., (2020). Kentsel Sürdürülebilirlik ve Kentsel Dirençlilik Perspektifinden Tarihteki Pandemiler ve Covid-19 Pandemisi. Yüzüncü Yıl Üniversitesi Sosyal Bilimler Enstitüsü Dergisi, (Salgın Hastalıklar Özel Sayısı), 259–292.

Who.(2020). WHO Director-General's opening remarks at the media briefing on COVID-19 - 11 March 2020, https://www.who.int/dg/speeches/detail/who-director-general-sopening-remarks-at-the-media-briefing-on-covid-19---11-march-2020

URL-1: The Present and Future of Restaurants under the Effect of Covid-19 (Date of Access: 21.12.2020) https://www2.deloitte.com/content/dam/Deloitte/tr/Documents/consumer-business/COVID-19-etkisinde-restoran-sektorunun-bugunu-ve-gelecegi.pdf

URL-2: Covid-19 Restaurant Reports (Date of Access: 21.12.2020) https://www.twentify.com/tr/icgoru-raporlari

URL-3: An'garden Café/Le House Photos (Date of Access: 01.02.2021) https://www.archdaily.com/890221/angarden-cafe-le-house/5a9e491af197cce330000169-angarden-cafe-le-house-photo?next_project=no

URL-4: An'garden Café/Le House Photos (Date of Access: 01.02.2021) https://www.archdaily.com/890221/angarden-cafe-le-house/5a9e458ef197cc5533000122-angarden-cafe-le-house-photo?next_project=no

URL-5: An'garden Café/Le House Photos (Date of Access: 01.02.2021) https://www.archdaily.com/890221/angarden-cafe-le-house/5a9e4734f197cc553300012c-angarden-cafe-le-house-photo?next_project=no

URL-6: An'garden Café/Le House Photos (Date of Access: 01.02.2021) https://www.archdaily.com/890221/angarden-cafe-le-house/5a9e4734f197cc553300012c-angarden-cafe-le-house-photo?next_project=no

URL-7: An'garden Café/Le House Photos (Date of Access: 01.02.2021) https://www.archdaily.com/890221/angarden-cafe-le-house/5a9e49f7f197cc553300013c-angarden-cafe-le-house-photo?next_project=no

URL-8: An'garden Café/Le House Photos (Date of Access: 01.02.2021) https://www.archdaily.com/890221/angarden-cafe-le-house/5a9e4883f197cce330000165-angarden-cafe-le-house-photo?next_project=no

URL-9: An'garden Café/Le House Plan and Section (Date of Access: 01.02.2021) https://www.archdaily.com/953822/1900-coffee-pva/5fe10fd863c017832e000b63-1900-coffee-pva-photo

URL-10: An'garden Café/Le House Plan and Section (Date of Access: 01.02.2021) https://www.archdaily.com/953822/1900-coffee-pva/5fe10fe963c0179859000a04-1900-coffee-pva-photo?next_project=no

URL-11: An'garden Café/Le House Plan and Section (Date of Access: 01.02.2021) https://www.archdaily.com/953822/1900-coffee-pva/5fe1100463c0179859000a06-1900-coffee-pva-photo?next_project=no

URL-12: An'garden Café/Le House Plan and Section (Date of Access: 01.02.2021) https://www.archdaily.com/953822/1900-coffee-pva/5fe10fcd63c0179859000a02-1900-coffee-pva-photo?next_project=no

URL-13: 1900 COFFEE/PVA Photos (Date of Access: 01.02.2021) https://www.archdaily.com/953822/1900-coffee-pva/5fe10fad63c017832e000b60-1900-coffee-pva-photo?next_project=no

URL-14: 1900 COFFEE/PVA Photos (Date of Access: 01.02.2021) https://www.archdaily.com/953822/1900-coffee-pva/5fe10fc063c017832e000b62-1900-coffee-pva-plan?next_project=no

URL-15: 1900 COFFEE/PVA Photos (Date of Access: 01.02.2021) https://www.archdaily.com/955699/the-59-cafe-viet-d-architects/600eb10263c01741d800003e-the-59-cafe-viet-d-architects-photo?next_project=no

URL-16: 1900 COFFEE/PVA Photos (Date of Access: 01.02.2021) https://www.archdaily.com/955699/the-59-cafe-viet-d-architects/600eb1c263c01741d8000044-the-59-cafe-viet-d-architects-photo?next_project=no

URL-17: 1900 COFFEE/PVA Photos (Date of Access: 01.02.2021) https://www.archdaily.com/955699/the-59-cafe-viet-d-architects/600eb06f63c017f5d7000026-the-59-cafe-viet-d-architects-photo?next_project=no

URL-18: 1900 COFFEE/PVA Plans (Date of Access: 01.02.2021) https://www.archdaily.com/955699/the-59-cafe-viet-d-architects/600eb11863c017f5d700002c-the-59-cafe-viet-d-architects-photo?next_project=no

URL-19: The 59 Café Photos (Date of Access: 01.02.2021) https://www.archdaily.com/955699/the-59-cafe-viet-d-architects/600eb18263c017f5d7000030-the-59-cafe-viet-d-architects-photo?next_project=no

URL-20: The 59 Café Photos (Date of Access: 01.02.2021) https://www.archdaily.com/955699/the-59-cafe-viet-d-architects/600eb13c63c01741d8000040-the-59-cafe-viet-d-architects-photo?next_project=no

URL-21: The 59 Café Photos (Date of Access: 01.02.2021) https://www.archdaily.com/955699/the-59-cafe-viet-d-architects/600eb2a763c017f5d7000034-the-59-cafe-viet-d-architects-first-floor-plan?next_project=no

URL-22: The 59 Café Photos (Date of Access: 01.02.2021) https://www.archdaily.com/955699/the-59-cafe-viet-d-architects/600eb2d263c017f5d7000035-the-59-cafe-viet-d-architects-third-floor-plan?next_project=no

URL-23: The 59 Café Photos (Date of Access: 01.02.2021) https://www.archdaily.com/955699/the-59-cafe-viet-d-architects/600eb2d263c017f5d7000035-the-59-cafe-viet-d-architects-third-floor-plan?next_project=no

URL-24: The 59 Café Photos (Date of Access: 01.02.2021) https://www.archdaily.com/955699/the-59-cafe-viet-d-architects/600eb2e963c01741d800004a-the-59-cafe-viet-d-architects-elevation?next_project=no

URL-25: The 59 Café Plan and Section (Date of Access: 01.02.2021) https://www.archdaily.com/890221/angarden-cafe-le-house/5a9e435af197cce330000148-angarden-cafe-le-house-ground-floor-plan?next_project=no

URL-26: The 59 Café Plan and Section (Date of Access: 01.02.2021) https://www.archdaily.com/890221/angarden-cafe-le-house/5a9e4376f197cc5533000115-angarden-cafe-le-house-first-floor-plan?next_project=no

URL-27: The 59 Café Plan and Section (Date of Access: 01.02.2021) https://www.archdaily.com/890221/angarden-cafe-le-house/5a9e4386f197cce330000149-angarden-cafe-le-house-second-floor-plan?next_project=no

URL-28: The 59 Café Plan and Section (Date of Access: 01.02.2021) https://www.archdaily.com/890221/angarden-cafe-le-house/5a9e434cf197cce330000147-angarden-cafe-le-house-front-elevation?next_project=no

URL-29: Thailand Hi Cafe/Balbek Bureau Photos (Date of Access: 01.02.2021) https://www.archdaily.com/890221/angarden-cafe-le-house/5a9e4398f197cc5533000116-angarden-cafe-le-house-section?next_project=no

URL-30: Thailand Hi Cafe/Balbek Bureau Photos (Date of Access: 01.02.2021) https://www.archdaily.com/952712/thailand-hi-cafe-balbek-bureau/5fca756e63c017fead0000dd-thailand-hi-cafe-balbek-bureau-image?next_project=no

URL-31: Thailand Hi Cafe/Balbek Bureau Photos (Date of Access: 01.02.2021) https://www.archdaily.com/952712/thailand-hi-cafe-balbek-bureau/5fca71ef63c01710a8000105-thailand-hi-cafe-balbek-bureau-image?next_project=no

URL-32: Thailand Hi Cafe/Balbek Bureau Photos (Date of Access: 01.02.2021) https://www.archdaily.com/952712/thailand-hi-cafe-balbek-bureau/5fca743063c017fead0000d3-thailand-hi-cafe-balbek-bureau-image?next_project=no

URL-33: Thailand Hi Cafe/Balbek Bureau Photos (Date of Access: 01.02.2021) https://www.archdaily.com/952712/thailand-hi-cafe-balbek-bureau/5fca739b63c01710a8000113-thailand-hi-cafe-balbek-bureau-image?next_project=no

URL-34: Thailand Hi Cafe/Balbek Bureau Plan ve Perspectives (Date of Access: 01.02.2021) https://www.archdaily.com/952712/thailand-hi-cafe-balbek-bureau/5fca73e363c01710a8000115-thailand-hi-cafe-balbek-bureau-image?next_project=no

URL-35: Thailand Hi Cafe/Balbek Bureau Plan ve Perspectives (Date of Access: 01.02.2021) https://www.archdaily.com/952712/thailand-hi-cafe-balbek-bureau/5fca75a663c01710a800012b-thailand-hi-cafe-balbek-bureau-floor-plan?next_project=no

URL-36: Que Dua Restaurant/DDconcept Photos (Date of Access: 01.02.2021) https://www.archdaily.com/955076/que-dua-restaurant-dconcept/6000bd5563c01727af00033a-que-dua-restaurant-ddconcept-photo?next_project=no

URL-37: Que Dua Restaurant/DDconcept Photos (Date of Access: 01.02.2021) https://www.archdaily.com/955076/que-dua-restaurant-dconcept/6000bde263c017e1360003ba-que-dua-restaurant-ddconcept-photo?next_project=no

URL-38: Que Dua Restaurant/DDconcept Plan and Section (Date of Access: 01.02.2021) https://www.archdaily.com/955076/que-dua-restaurant-ddconcept/6000bf5f63c01727af000346-que-dua-restaurant-ddconcept-plan?next_project=no

URL-39: Que Dua Restaurant/DDconcept Plan and Section (Date of Access: 01.02.2021) https://www.archdaily.com/955076/que-dua-restaurant-ddconcept/6000bf7363c017e1360003c3-que-dua-restaurant-ddconcept-bar-area-section?next_project=no

Zhao, Y., (2020). COVID-19 as a catalyst for educational change. Prospects, 1–5. https://doi.org/10.1007/s11125-020-09477-y

Rüya YILMAZ and Derya SERBEST ŞİMŞEK

Ecosystem Services Approach in the Context of Sustainable Environment for Sustainable Development

Abstract: Nowadays, there are some factors such as rapid population growth, industrialization, over urbanization, and a decrease in natural and cultural resources that lead to environmental pollution. Therefore, the Issues of global climate change, depletion of natural resources, degradation of the natural environment, decrease in clean water resources, air pollution, and inefficient soils, are becoming a quite serious international concern. When the natural balance is disturbed as a result of air, water and soil pollution, the balance of the ecosystem is disrupted and causes ecological problems. Ecosystem services, which are the benefits offered to people by the natural environment and healthy ecosystems; It cannot provide the expected benefit due to deteriorating ecosystems.

In this case, the concept of sustainable development should be handled in a holistic way in terms of economic, social and environmental issues. To ensure sustainability; Environmental management system models, consisting of the stages of creating international policies, planning, implementation and management review, taking into account the responsibility of new generations should be established. The sustainable development method occurs with many factors and many unpredictable effects. Therefore, this study evaluates the effects of environmental sustainability index and the applicability of ecosystem services to sustainable development.

Keywords: Sustainable development, environmental sustainability index (ESI), planning, management, sustainable environment.

1 Introduction

Sustainable development concept; Since it covers a wide area with its economic, social and environmental dimensions, the effects of changes are also multifaceted. In addition to its versatility, the dynamic nature of sustainable development makes it difficult to perceive and evaluate the issue. Therefore, sustainable development should be measured with a sufficient number of economic, social and environmental indicators and indices using appropriate methods. Development and development were defined as only economic and technological progress during the 19th and early 20th centuries. In this period, the rapidly increasing environmental problems, environmental pollution and social problems were not given enough attention.

Natural resources support the functioning of the global economy and quality of life. These resources include raw materials such as fuels, minerals and metals, as well as food, soil, water, air, biomass and ecosystems [1]. A well-functioning economy depends on an uninterrupted flow of natural resources and raw materials. In order to meet the needs of the increasing global population, it is possible to protect natural resources, ecosystems and ensure their sustainable management by securing and developing ecosystem services. In this context, the functions and services of the ecosystem should be supported within the framework of the sustainability principle. Therefore, the sustainable use of our existing natural resources with a future-oriented strategy will be an understanding that will also support our economy.

Observing sustainability should be a fundamental principle in development steps towards achieving the goal of a country that has reached a level of economic and social welfare. Within the framework of this principle, a sustainable development approach should be adopted with ecosystem-based approaches that protect the environment and natural resources, use resources efficiently, and based on innovative and technological developments.

Green economy means increasing social welfare and social equality, reducing environmental risks and ecological scarcity. This approach, also referred to as green growth, is the understanding of ensuring economic growth in an environmentally and natural resource friendly manner. In parallel with the understanding of green economy and green growth, by minimizing the pressure on the environment and natural resources, while minimizing the pressure on the environment and natural resources, by creating global financing resources and incentive mechanisms that aim to bring resource consumption to optimum efficiency in production processes, reduce and control the emissions that cause climate change. It is an economical approach that combats poverty and increases ecological efficiency in consumption patterns and products [1]. In the long run, reflecting ecosystem products and services to national accounts will also support the internalization of the green growth model.

Addressing the development route of our country, which has a young population and a high potential economy, with a socially inclusive and environmentally sustainable approach; In terms of environmental quality, it is important to prioritize environmental issues in a balanced way with economic and social issues [1,2].

The use of natural resources and ecosystems with the future in mind is beneficial in the protection of the environment and the sustainable management and consumption of resources.

Being an important source of raw material for the economy, our natural resources are also a part of the environment and natural ecosystem we live in. Any change in this system negatively affects the society negatively. The deterioration of the environment due to all kinds of pollution damages the living groups living in this ecosystem and can cause changes that cannot be controlled at advanced levels. It is important to evaluate and plan the environment not only according to the current climate, social and economic conditions, but also for the protection of the environment.

In addition to protecting the environment, it is economically important to manage resources with a forward-looking perspective. For the purpose of using the existing non-renewable natural resources with maximum efficiency, shaping production and consumption strategies accordingly, optimizing production and consumption processes in terms of resource use and ensuring resource efficiency will ensure that resources are consumed less than the current situation. Considering the reduction in natural resource consumption with a sustainable management style, it will be possible to provide more output with less input. also adds value.

Development in a country; It is possible by ensuring economic, social and environmental sustainability. During this study, it is aimed to evaluate the effects of environmental sustainability index and the applicability of ecosystem services to sustainable development.

Environmental Sustainability Index (ESI), which was established in 2005 at the World Economic Forum by Yale (Environmental Law and Policy Center) and Columbia (Earth Science Center) Universities is a tool that aims to measure the progress of environmental sustainability [3,4]. This index is important in terms of directing sustainability studies. According to the ESI scores, it has revealed which issues should be resolved as a priority for countries to achieve sustainable planning. For this reason, ESI's scoring system for sustainable development is examined. In this study, the preparation purpose, structure, policy and categories, indicators of "International Sustainable Environment Index" developed for the measurement of sustainability and its relation with the environment are emphasized.

2 Sustainable Development

Sustainable development is the organizing principle for meeting human development goals while simultaneously sustaining the ability of natural systems to provide the natural resources and ecosystem services on which the economy and society depend. The desired result is a state of society where living conditions and resources are used to continue to meet human needs without undermining the integrity and stability of the natural system. Sustainable development can be defined as development that meets the needs of the present without compromising the ability of future generations to meet their own needs [5].

The World Commission on Environment and Development in 1987 improved a report named "Our Common Future" which defined as meeting the needs of future generations that clarified the conflict between environment and development. Accordingly, besides the economic dimension of sustainable development, environmental and social issues are considered as other measurements [6,7].

The Brundtland Commission sustainable as "development that meets the needs of the present without compromising the ability of future generations to meet their own needs" [8].

Although the concept of sustainable development is generally evaluated with its environmental dimension, and three different sub-dimensions (economy, environment and society) of this concept are shown in the Fig. 1.

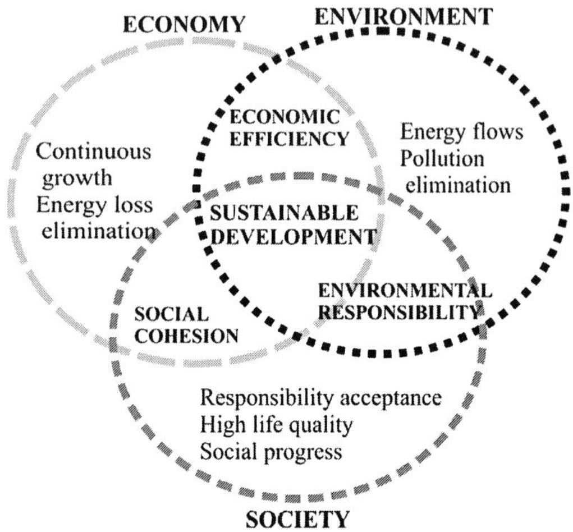

Fig. 1. *Input issues and baseline model of sustainable development* [9]

Sustainability is one the newest degree subjects that attempts to bridge social science with engineering and environmental science with the technology of the future. When we hear the word "sustainability" we tend to think of renewable fuel sources, reducing carbon emissions, protecting environments and a way of keeping the delicate ecosystems of our planet in balance. In short, sustainability looks to protect our natural environment, human and ecological health, while driving innovation and not compromising our way of life.

The 2030 Agenda for Sustainable Development, adopted by all United Nations Member States in 2015, provides a shared blueprint for peace and prosperity for people and the planet, now and into the future. At its heart are the 17 Sustainable Development Goals (SDGs), which are an urgent call for action by all countries-developed and developing-in a global partnership. They recognize that ending poverty and other deprivations must go hand-in-hand with strategies that improve health and education, reduce inequality, and spur economic growth – all while tackling climate change and working to preserve our oceans and forests [7].

3 Environmental Sustainability Index (ESI)

The ESI was created by the Global Leaders of Tomorrow of the World Economic Forum in collaboration with the Yale Center for Environmental Law and Policy (YCELP) and the Center for International Earth Science Information Network (CIESIN) of Columbia University. After an initial pilot study in 2000, the ESI was first published in 2001 and subsequently refined in 2002 and 2005 [3]. The ESI is known as a well established composite sustainability measure that factors in not only ecological but also socio-economic and political foundations of sustainability [10]. The ESI has become a popular indicator in comparative cross-national analyses [11].

The Environmental Sustainability Index (ESI) benchmarks the ability of nations to protect the environment over the next several decades. It does so by integrating 76 data sets–tracking natural resource endowments, past and present pollution levels, environmental management efforts, and the capacity of a society to improve its environmental performance-into 21 indicators of environmental sustainability. These indicators permit comparison across a range of issues that fall into the following five broad categories [3]:

- Environmental Systems (Tab. 1)
- Reducing Environmental Stresses (Tab. 2)
- Reducing Human Vulnerability to Environmental Stresses (Tab. 3)
- Global Stewardship (Tab. 4)
- Societal and Institutional Capacity to Respond to Environmental Challenges (Tab. 5)

Tab. 1. *Environmental Systems category of the Environmental Sustainability Index (ESI)* [3]

Indicator	Variable
Air Quality	Urban population welghted NO_2 concentration
	Urban population welghted SO_2 concentration
	Urban population welghted TSP concentration
	Indoor air pollution from solid fuel use
Biodiversity	Percentage of country's territory in threatened ecoregions
	Threatened bird species as percentage of known breeding bird species in each country
	Threatened mammal species as percentage of known mammal species in each country
	Threatened amphibian species as percentage of known amphibian species in each country
	National Biodiversity Index
Land	Percentage of total land area having very low anthropogenic impact
	Percentage of total land area having very high anthropogenic impact
Water Quality	Dissolved oxygen concentration
	Electrical conductivity
	Phosphorus concentration
	Suspended solids
Water Quantity	Freshwater availability per cabita
	Internal groundwater availability per cabita

Tab. 2. *Reducing Environmental Stresses category of the Environmental Sustainability Index (ESI)* [3]

Indicator	Variable
Reducing Air Pollution	Coal consumption per populated land area
	Anthropogenic NO_X emmisions per populated land area
	Anthropogenic SO_2 emmisions per populated land area
	Anthropogenic VOC emmisions per populated land area
	Vehicles in use per populated land area
Reducing Ecosystem Stress	Annual average forest cover change rate from 1990 to 2000
	Acidification exceedance from anthropogenic sulfur deposition
Reducing Population Pressure	Percentage change in projected population 2004–2050
	Total Fertility Rate
Reducing Waste and Consumption Pressures	Ecological Footprint per capita
	Waste recycling rates
	Generation of hazardous waste
Reducing Water Stress	Industrial organic water pollutant emissions per available freshwater
	Fertilizer consumption per hectare of arable land
	Pesticide consumption per hectare of arable land
	Percentage of country under severe water stress
Natural Resource Management	Productivity overfishing
	Percentage of total forest area that is certified for sustainable management
	World Economic Forum Survey on subsidies
	Salinized area due to irrigation as percentage of total arable land
	Agricultural subsidies

Tab. 3. *Reducing Human Vulnerability to Environmental Stresses category of the Environmental Sustainability Index (ESI)* [3]

Indicator	**Variable**
Environmental Health	Death rate from intestinal infectious diseases
	Child death rate from respiratory diseases
	Childeren under five mortality rate per 1,000 live births
Basic Human Sustenance	Percentage of undernourished in total population
	Percentage of population with Access to improved drinking water source
Reducing Environment Related Natural Disaster Vulnerability	Average number of deaths per million inhabitans from floods, tropical cyclones, and droughts
	Environmental Hazard Exposure Index

Tab. 4. Global Stewardship category of the Environmental Sustainability Index (ESI) [3]

Indicator	**Variable**
Participation in International Collaborative Efforts	Number of memberships in environmental intergovermental organizations
	Contribution to international and bilateral funding of environmental projects and developent aid
	Participation in international environmental agreements
Greenhouse Gas Emissions	Carbon emissions per million US dollars
	Carbon emmissions per capita
Reducing Transboundary Environmental Pressures	SO_2 Exports
	Import of polluting goods and raw materials as percentage of total imports of goods and services

Tab. 5. *Societal and Institutional Capacity to Respond to Environmental Challenges category of the Environmental Sustainability Index (ESI)* [3]

Indicator	Variable
Environmental Governance	Ratio of gasoline price to world average
	Corruption measure
	Govemment effectiveness
	Percentage of total land area under protected status
	World Economic Forum Survey on environmental governance
	Rule of law
	Local Agenda 21 initiatives per million people
	Civil and Political Liberties
	Percentage of variables missing from the Rio to Joburg Dashboard
	IUCN member organizations per million population
	Knowledge creation in environmental science, technology and policy
	Democracy measure
Eco-Efficiency	Energy efficiency
	Hydropower and renewable energy production as a percentage of total energy consumption
Private Sector Resporisiveness	Dow Jones Sustainability Group Index
	Average Innovest EcoValue rating of firms headquartered in a country
	Number of ISO 14001 certified companies per billion dollars GDP
	World Economic Forum Survey on private sector environmental innovation
	Participation in the Responsible Care Program of the Chemical Manufacturer's Association
Science and Technology	Innovation Index
	Digital Access Index
	Female primary education completion rate
	Gross tertiary enrollment rate
	Number of researchers per million inhabitans

4 Ecosystem Services

When the consuming of resources start to threaten economic development, governing by sustainable improvements were implemented.

The policies that were introduced during the 60s and 70s increased the awareness against environmental pollution and the scarcity of natural resources [12,13].

Ecosystems provide a range of services that are of fundamental importance to human well-being, health, livelihoods, and survival [14,15.16]. Interest in ecosystem services in both the research and policy communities has grown rapidly.

The working history of ecosystem services approach reaches back to the 1970s. Ecosystem services is the name for all the products and services provided by the ecosystem. These services, also known as nature's blessings, ensure the continuity of life.

Ecosystem services creates all the demands and needs of people from the available ecosystems. Being used by humans, these services are seen as one-way pragmatism that damage resources without reciprocating any services. This unstable relationship is noticed by the negative changes in the process. Ecosystem services directly affect the relationship between a human's life quality and the ecosystems.

Westman's (1977) "Natural Services" concept was used for the definition of ecosystem service [17].

The current understanding of ecosystem services has been developed in the 1990s. According to Costanza et al., who is known as a milestone of ecosystem services, all the econosystems on Earth, therefore all the nature provides, directly affects more than 1 billion human's lives. [15].

The article titled "Ecosystem Services' Value and Natural Fund" helped place ecological-economical values in literature under the value of ecosystem services [14].

After this, especially in the last 10 years, the importance of this issue increased worldwide and today's ecosystem product and services focused targets were also identified in the international processes.

Humans benefit a great deal from the ecosystems. Such benefits include health provisions, fresh and drinkable water, and food. According to the Millennium Ecosystem Assessment, these benefits are of great importance and should be focused for a sustainable life cycle [15]. In another study by the Economics of Ecosystems and Biodiversity TEEB (2010), more information was provided on the financial importance of these very ecosystems [16]. In conclusion, ecosystem services gained upmost importance at a political level, as shown by the creation of the International science-policy Platform on biodiversity and ecosystem services (IPBES), and the addition of ecosystem services in the 2020 goals set by the 10th Conference of Parties to the Convention on Biological Diversity [18,19].

5 Classification of Ecosystem Services

Ecosystem services within the framework of the previously stated definitions; It can be defined as the benefits offered to humans from ecosystem functions or the direct and indirect contributions of ecosystems for human well-being [15].

In the MEA report, ecosystem services; Resource providing services, organizing services, cultural services and supporting services were classified into 4 main ecological function groups and 30 categories. Classification is shown in Tab. 6.

Information on changes in the area of different ecosystems is needed in order to establish an accounting system for ecosystem conditions and services [21].

Tab. 6. Ecosystem Services Classes [15,20]

Resource provider Services	**Organized Services**	**Supporting Services**	**Cultural Services**
Food	Air quality regulation	Food cycle	Recreation and ecotourism
Biological raw material	Climate regulation	Water cycle	Spiritual and ethical values
Decorative resources	Water flow control	Photosynthesis	Social relations
Genetic resources	Erosion control	Soil formation	Information system
Fresh water	Water treatment and waste control	Primary production	Educational value
Biochemicals and medical Products	Control of epidemics	-	The feeling of place and space
-	Pest control	-	Aesthetic values
-	Pollination	-	Inspiration
-	Natural risk reduction	-	Cultural heritage value
-	-	-	Cultural diversity

6 Conclusion

Sustainable development should form the basis in planning such as ecotourism, landscape reclamation, land use [22,23]. It is necessary to increase the environmental awareness of individuals and the public in order to protect the environment. In many previous studies, they suggested increasing the environmental awareness, environmental attitude and environmental awareness of the public for "Sustainable Environment" [24,25,26].

Our world is confronted with problems such as rapid urbanization and climate change [27]. The ecosystem approach, recognizing that it is an integral part of ecosystems and natural resources in terms of people from different cultural, nature conservation and encourage sustainable use. Therefore; Understanding the functioning and services of ecosystems is essential. Ecosystem services are the contributions or benefits people receive from ecosystems. Studies to evaluate these services have been increasing gradually since the 1970s. Ecosystem services approach, nature conservation, natural resource management and ecological terms to be included in the planning process of ecological values, social and economic assessments are provided.

One of the main challenges of ecosystem management is planning to manage multiple ecosystem services across landscapes. For the regulation of ecosystem services, ecosystem service analysis and ecosystem management should be developed. As a result, we recommend that countries plan ecosystem service by taking EPI scores into consideration in order to ensure sustainable development. Consequently, we recommend that ecosystem services be implemented as part of sustainability and natural resource management. We recommend that ecosystem services practices be implemented as part of sustainability and natural resource management.

References

1. On Birinci Kalkınma Planı (2019–2023), *Çevre ve Doğal Kaynakların Sürdürülebilir Yönetimi Çalışma Grubu Raporu,* 2018, Ankara, 137 s.
2. Malkoç True, E. and Kılıçaslan Ç., *Quality of Urban Environment: Analysis of Sigacik Settlement,* Journal Of Environmental Protection And Ecology, 2019, vol.20, pp.2068–2076.
3. Daniel C.E., Levy M., Srebotnjak T., and de Sherbinin, A. *Environmental Sustainability Index: Benchmarking National Environmental Stewardship.* New Haven: Yale Center for Environmental Law & Policy, 2005.
4. Yale Center for Environmental Law&Policy/Center for International Earth Science Information Network at Columbia University, *Environmental Performance Index Report.* Retrieved 2008-03-18.
5. Yıkmaz, R.F., *Sürdürülebilir Kalkınmanın Ölçülmesi ve Türkiye İçin Yöntem Geliştirilmesi,* T.C. Başbakanlık Devlet Planlama Teşkilatı Müsteşarlığı, Sosyal Sektörler ve Koordinasyon Genel Müdürlüğü, 2011, Ankara, Yayın No:2820.
6. Blewitt, J., *Understanding Sustainable Development,* Rouhtledge, 2014, New York, 394 pages,
7. https://sdgs.un.org/goals
8. WCED, *Brundtland Report,* World Commission on Environment and Development, 1987.

9. Cafuta, M.R. *Open Space Evaluation Methodology and Three Dimensional Evaluation Model as a Base for Sustainable Development Tracking.* Sustainability, 2015. *7*, 13690–13712.

10. Whitford, A. B. and Wong, K. *Political and social foundations for environmental sustainability,* Political Research Quarterly, 2009, 62(1), 190–204.

11. Babcicky, R., *Rethinking the Foundations of Sustainability Measurement: The Limitations of the Environmental Sustainability Index (ESI),* Soc. Indic. Res., 2013, 113:133–157.

12. Gomez-Bagetthun, E., De Groot R., Loma, P.L. and Montes, C., *The history of ecosystem services in economic theory and practice: From early notions to markets and payment schemes,* Ecological economics, 2010, vol.69, pp.1209–1218.

13. Braat, L.C. and De Groot R., *The ecosystem services agenda: bridging the worlds of natural science and economics, conservation and development, and public and private policy,* Ecosystem services, 2012, vol.1, no.1, pp.4–15.

14. Costanza, R., d'Arge, R., de Groot, R., Farber, S., Grasso, M., Hanna, B., Limburg, K., Naeem, S., O'Neill, RV., Paruelo, J., Raskin, RG., *Sutton, P., van den Belt, M. The value of the world's ecosystems services and natural capital,* Nature, 1997, 387: 253–260.

15. Millennium Ecosystem Assessment (MEA), *Ecosystems and Human Well-being: Desertification Synthesis,* World Resources Institute, 2005, Washington, DC.

16. Sukhdev, P., Wittmer, H., Schröter-Schlaack, C., Nesshöver, C., Bishop, J., ten Brink P., Gundimeda, H., Kumar P. and Simmons B., *The Economics of Ecosystems and Biodiversity: Mainstreaming the Economics of Nature: A Synthesis of the Approach, Conclusions and Recommendations of TEEB,* Conclusions and Recommendations of TEEB, Rep. 2010.

17. Westman W.E., *How much are nature's services worth,* Science, 1977, vol.197, no.4307, pp.960–964.

18. Larigauderie, A. and Mooney, H.A., *The Intergovernmental Science-Policy Platform on Biodiversity and Ecosystem Services: moving a step closer to an IPCC-like mechanism for biodiversity,* Current Opinion in Environmental Sustainability, 2010, 2(1):9–14.

19. Mace, M.G., Cramer, W., Díaz, S., Faith, D. P., Larigauderie, A., Le Prestre, P., Palmer, M., Perrings, C., Scholes R.J., Walpole, M., Bruno, A.W., Watson, J.E.M., Mooney, H.A., *Biodiversity targets after 2010,* Current Opinion in Environmental Sustainability, 2010, vol. 2, issues 1–2, pp. 3–8.

20. Reid, W. V., Mooney, H. A., Cropper, A., Capistrano, D., Carpenter, S. R., Chopra, K., Dasgupta, P., Dietz, T., Duraiappah, A. K., Hassan, R., Kasperson, R., Leemans, R., May, R. M., Mc Michael T.A.J., Pingali, P., Samper, C., Scholes, R., Watson, R. T., Zakri, A.H., Shidong, Z., Ash, N. J., Bennett, E., Kumar,

P., Lee, M. J., Raudsepp-Hearne, C., Simons, H., Thonell, J. and Zurek M. B., *Ecosystems and Human Well-being*, European Council, 2003, Washington (DC), USA, Rep.

21. Grunewald, K., Schweppe-Kraft, B., Syrbe, R-U., Meier, S., Krüger, T., Schorcht, M., Walz, U., *Hierarchical classification system of Germany's ecosystems as basis for an ecosystem accounting – methodsand first results,* One Ecosystem (2020) 5: e50648.
22. Yılmaz, O., Mansuroglu, S., Yilmaz, R., *Swot Analysis Of Ecotourism As A Tool For Sustainable Development: A Case Research In North-West Black Sea Coastal Zone Of Turkey,* J Environ Prot Ecol, 2013, Bulgaria, (14) 2, P: 786–798.
23. Yıldırım, T.B., Yılmaz, R., *High Performance Plant Selection for Landscape Reclamation in the Subtropic Climate Zone: A Case Study* Journal of Agronomy, 2005, 4(3): 262–266.
24. Yılmaz, R., Yılmaz, O., *Trakya' da Halkın Çevre Duyarlılığının Saptanması ve Sosyo-Ekonomik Özelliklerin Çevresel Duyarlılık Üzerine Etkileri,* Namık Kemal Üniversitesi Bilimsel Araştırma Projeleri Sonuç Raporu, 2016, Tekirdağ, NKUBAP.00.18.AR.13.03.
25. Yılmaz, R., Yılmaz, O., *Halkın Çevre Bilincinin Saptanması ve Sosyo Ekonomik Özelliklerin Çevresel Bilinç Üzerine Etkileri: Kırklareli Örneğinde,* Ege Üniv. Ziraat Fak. Derg., 2017, 54 (3):301–310.
26. Yılmaz O., *Research on the Environmental Awareness Level and Its Developing Possibilities in Canakkale, Turkey,* Oxidation Communications, 2016, 39(1–2), 723–731.
27. Cities and Climate Change Global Report On Human Settlements, United Nations Human Settlements Programme, 2011. https://unhabitat.org/sites/default/files/download-manager-files/

Shirin IZADPANAH and Kağan GÜNÇE

Designing Enabling Spaces for Pre-Schoolers: Strategies to Boost Children's Control Over Spaces

Abstract: The impact of space design and planning on children's development is well-documented. Enabling spaces for children are defined as spaces that are designed to be age-specific, appealing to children's interests, and encourage them to challenge themselves in a safe and secure environment. The aim of this study is to discuss the design strategies that would help not only designers but planners and educators to create interesting, exciting and challenging learning spaces for pre-schoolers. It is argued that designing and entry for a good start, considering children control and independency in arranging the layouts, respecting children's social interest, including additional spaces and strategies that enhance child-teacher cooperation and designing features that enhance children's imagination and fantasy will contribute to providing spaces that are responsive to children's control and abilities.

Keywords: Pre-schoolers, Visual and Physical Access, Sense of Control, Movement, Interaction

1 Introduction

Enabling spaces for children are spaces that increase a child's control over and interest in that space. The purpose of these spaces is to provide children with a potential environment that enables them to engage in organised activities or play, to socialise, to learn, and to express themselves as an individual [1]. Findings from studies claim that pre-schoolers learn and grow through interaction with the physical world [2, 3] however rapid school construction in response to fulfilling needs for more learning spaces has kept architects busy with standards and in many cases the planning an enabling layout is disregarded [4].

Enabling learning spaces for children was an outcome of child-cantered education which nowadays has gained a considerable amount of notice in early childhood education. This method was introduced as a reaction towards a teacher-cantered structure [5]. In child-centred curricula, activities and stimuli revolve around children's interests and preferences. To develop child-cantered methods, researchers study a variety of children's reactions to different situations in order to understand their interests and interactions [6,7].

Enabling educational spaces for children can be described as environments which allow children to play and learn creatively, provide hands on learning materials, encourage children's various natural activities, nourish children's desires and emotional needs, guide them in developing their skills and support self-directed learning [8].

Based on this definition, enabling spaces are expected to be the constructor of their own knowledge and experience and therefore the design and arrangement of enabling learning environments has to be in parallel with a child's full access to free exploration and engagement [9]. To support children's independent ability, the design and arrangement of learning environments should support children's independence and control, and also provide a range of age-appropriate choices [10].

Preschool settings that employ child-centred approaches attempt to offer learning facilities that are visible and be physically accessible by all children. These enabling learning spaces are suggested to be well organised and contain available and easily accessible materials in order to prevent frustration for a child [11].

Dudek [12] relates children's learning perception to the way they engage with physical spaces. In another study, Dudek [13] claims that children's perception is actually shaped through their movement and sensory experiences in space. In line with a growing number of similar findings, researchers are focusing on improving the design and planning of early childhood educational spaces [14].

It is now known that children acquire their skills and knowledge through a dynamic experience in space [15], however many preschool settings are limited to static spaces. Designing enabling spaces that increase children's control over spaces contribute to children's healthy development since these spaces will be responsive and available to children's attempts at interaction [16].

As authors, we would like draw attention to the need for transforming preschool spaces into physical environments that respond to children's individual needs and abilities to make them feel safe and happy and consequently open to explore and challenge themselves. We have identified this issue during our previous surveys and research on the nature of pre-schoolers' growth and development. This chapter is divided into four parts. The first part focuses on designing a happy start for children. The second part discusses design strategies that will increase children's control over the space and encourage them to self-engage in learning activities. The third part emphasises the importance of considering children's different social interest and competencies in designing preschool layouts. The fourth part discusses the additional strategies that will support children's cooperation in space. Finally, the last part will cover the design requirements of spaces that stimulate children's imagination.

2 Designing a Happy Start

In preschool settings, entrance halls are more than merely a space of transition. The entrance halls of preschool settings can be designed to give insight into the rest of the facility, because impressive halls work as displays which give a message that 'this space is all about children'. Therefore, the amount of space allocated for entrance hall needs reflect the critical role of this place. Studies suggest that over-crowding has a negative influence on young children [17], therefore limited spaces are likely to prevent children from experiencing a happy start.

According to Dudek [18], preschool entrance halls are crucial and emotional spaces for children, because this is where they part with their parents which is sometimes a cause for stress [19]. One of the strategies that can reduce stress in an environment is providing a positive social environment at the entrance. A well-defined sitting area with sufficient comfortable seating is a good contribution to the provision of a social environment in entrances. Considering seating areas in entrance layouts will improve social interaction between children, parents and teachers [20]. Researchers also indicate that quality natural light will improve the mood of both children and adults [21], therefore locating entrance halls in areas that receive enough natural light will improve the level of social quality.

Another strategy in reducing stress at arrival is increasing a sense of security in entrance halls. Creating a visible welcoming desk or reception area is a good attempt to enhance both parents and children's sense of security upon arrival at the setting [22]. The location of this desk is quite important because if the desk is located in a spot with limited views of the entrance or if it is confined to a limited space at the front, the desk will have no contribution to a sense of security and might create a crowded spot that will further increase stress in the environment.

Studies claim that 'distracting events and architectural features that are strategically positioned in or just beyond the threshold can lead to a calming effect' for children [18], therefore strategic distractions in entrance halls is an appropriate contribution to reduce stress and enhance a happy start. Using attractive objects and landscape elements, warm colours and playful textures, and decorative figures of animals or favourite cartoon characters can all be attempts at creating visually appealing entrance halls for children [23,24].

In line with improving visual appearance, hosting interesting activities in entrance halls will take the charm of the space for children to a new level. 'Play' is an activity that is distracting, it is fun, sociable and interesting for children [25,26]. Integrating play with entrance halls can be achieved either by locating an indoor play room near the entrance, or if enough space is available, a playing area can be included in the layout of the entrance hall. The type of play activities on offer must be based on the amount of space. For example, if the available space is inadequate for physically active games, table games such as Lego or other blocks, or even providing a tiny play house in a secluded corner can help integrate play upon arrival.

An exhibition area could be considered in halls with adequate space or an exhibition wall/panel to display children's artwork in smaller spaces will also help improve children's mood upon arrival at the setting. An attractive and well-designed display of children's work will let them know that their efforts and ideas are taken seriously and acknowledged. The height of the displays must consider the eye level of both children and adults [27]. Using design elements such as prominent forms, colours or lighting will emphasise this area upon entry.

Another design strategy that can contribute to a child's happy start is increasing a sense of belonging for children. Providing children with personal space (cubbies, closets, etc.) to leave their personal belongings will increase their sense of belonging in a setting [28]. Having children's personal space at the entrance will provide a place where they can leave their coat, shoes or other belonging when they arrive. It is also essential to consider children's scale, so they can use their lockers/cubbies without assistance.

The final strategy for a good start is defining a practical circulation from the entrance to the rest of the spaces. Studies suggest that locating children's main activity areas close to the entry hall will support a successful transition for children [18, 29]. Movement is how children develop spatial awareness. Defining clear and legible walkways that connect the entrance to other activity areas will help children to navigate with increased awareness [30]. Fig. 1 shows a sample space layout for an entrance hall in a preschool setting.

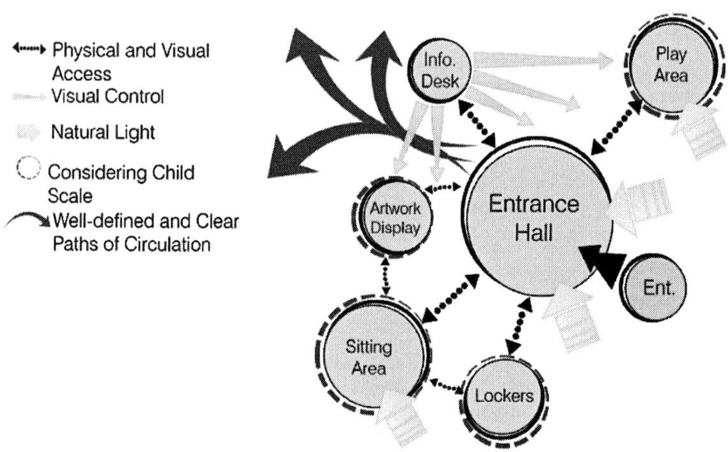

Fig. 1: A sample bubble diagram proposed for planning the entrance hall in preschool settings

3 Designing Spaces that Increase Sense of Control

Most children start to experience life outside their family environment as preschoolers and therefore preschool age is considered as a period during which the child starts to develop self-independence and confidence [31]. Studies have addressed lack of control for pre-schoolers who have been hospitalised or suffer an illness as one of the crucial problems [32, 33], which confirms the importance of a sense of control during preschool development.

To support pre-schoolers' independence and control over the environment, it is necessary to plan spaces that provide visual and physical access to the learning materials and learning stations. One of the design strategies in increasing the possibilities for children to see, reach and interact with learning materials and facilities is improving the lighting quality. While natural light is the best option for early childhood learning environments, providing quality artificial lighting is essential in the absence of large windows or natural light.

One of the problems with the lighting of educational spaces is generally a lack of solutions that allow adjusting or controlling the type and intensity of light for different tasks and activities [18]. Consistent with this shortcoming, proposing adjustable lighting systems is a good attempt in spaces that host a variety of tasks. Tunable lighting (Fig. 2) is an adjustable lighting system that allows users to adjust the colour temperature of a lamp in real time and studies claim that this type of lighting improves performance in a learning environment [34, 35].

Fig. 2: Tunable lighting delivers different light value in a single space based on children and teacher preferences

Lighting can also be used as a strategy to enhance children's visual stimulation and provide attractions. For example, lighting can be used to emphasise specific spots, elements or areas in order to attract children's attention towards those locations (fig. 3).

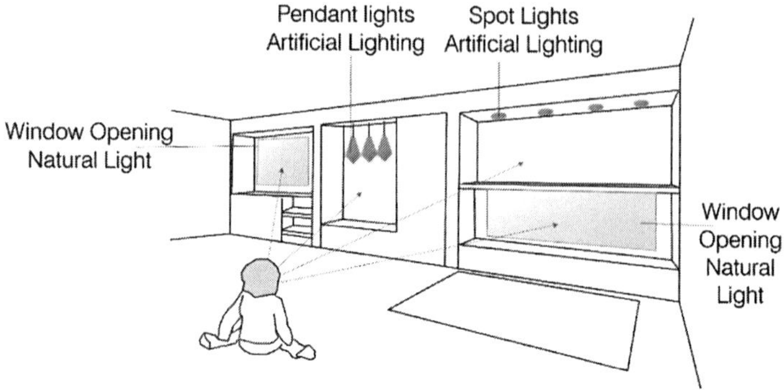

Fig. 3: Emphasising four different spots on one wall by using a combination of natural and artificial lighting

Another strategy that will enhance children's control over the space is providing well-defined circulation paths. Circulation paths within the activity areas or in between the spaces can enhance children's ability to seek materials and opportunities. The first requirement of well-defined circulation is the safe movement of children. A safe circulation path in preschool follows three main steps [36, 37, 38, 39]:

1. Determining the amount of empty space according to furniture volume and number of children
2. Excluding sharp edges or pointed corners from the borders
3. Using flooring materials that are soft and non-slippery

The second requirement for well-defined circulation is logical navigation which is able to lead children to move in-between displays and facilities in comfort and engage with the materials and activities that arouse interest [40]. In a study by Community Plaything (2009), four main design considerations are suggested for improving children's navigation in activity areas [20]. The first suggestion is the optimal positioning of areas which requires finding the best location for each zone/area based on the pattern of activities and interactions. The next suggestion is defining a clear path of movement which according to Weinstein and David (1987, p.75) is "a path that is broad, elongated and easily visible" [23]. The third suggestion is defining clear boundaries. Locating vertical elements (walls, separators, display units, etc.) and using horizontal surfaces (rugs, distinguishing areas by using different flooring materials/colours, hanging surfaces from the ceiling, etc.) will define boundaries for circulation paths and improve legibility. The final suggestion is using low separators that allow children to have visual control over the entire environment. Fig. 4 shows a sample layout for a preschool art station with consideration to the discussed design suggestions.

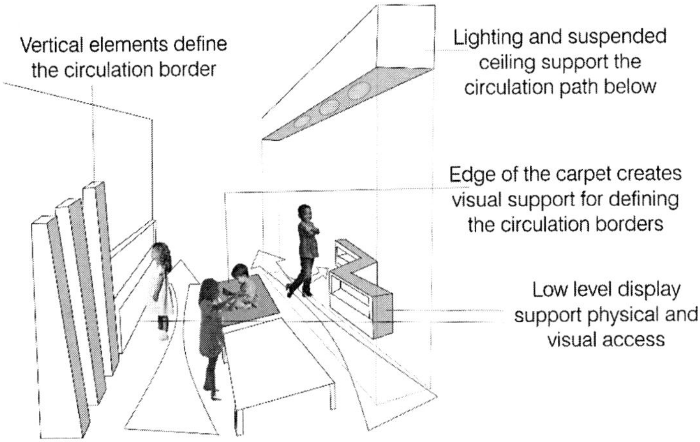

Fig. 4: A sample for defining the circulation path in a classroom art station by using horizontal and vertical elements

The final design strategy that will increase children's control over preschool environments is respecting children's scale and dimensions. In line with this intention, space and all its learning opportunities should be designed and arranged based on children's proportions, abilities and scale [39]. Child-scaled facilities and furniture allow access to activity options without the need for teachers' assistance (Fig. 4).

4 Designing Spaces that Respond to Children's Social Interests

Recent studies argue the importance of focusing on the social and emotional climate of preschools to improve the quality of preschool education [41]. The reason is that pre-schoolers learn the social world by interacting with peers in preschool settings [42].

The design patterns of a space can enhance children's social behaviour [12] and as Weinstein & David (1987, p.162) suggests "Children between 2 and 5 years old, construct a personal identity through their encounters with other people and with the physical environment" [23]. For example, a semi-circular arrangement in contrast to row arrangement increases children's eye contact and therefore, enhances their social and cooperative behaviour [43].

Dramatic play and blocks are two activities that have a significant role in preschoolers' social and linguistic development [44,45,46]. The space layout for both dramatic play and blocks can improve the quality of these activities and as a result enhance socialisation. Providing an appropriate display for play materials is a key strategy in improving the layout for both activities [47]. While costumes, props and toys for pretending are essential in enhancing pre-schoolers' role-playing abilities, providing a variety of blocks (hollow, ramps, boards and etc.) is also essential for improving their block construction skills.

Another strategy to improve the layout of the block and dramatic play area is location. Some preschool settings are able to host each activity in a designated space, however most of the settings offer a multitude of activities as stations, areas or corners in a classroom or in the same space. Studies suggest the benefits of having block and dramatic play areas close to each other to occasionally integrate these activities [48]. It is also suggested to locate the block area away from traffic flow and active circulating paths. Another suggestion is to close off the block play area at least from three sides and define entries and exits in order to prevent arriving and departing children from disturbing the children who are playing [49, 50, 51]. Fig. 5 presents the key design strategies in enabling a space for playing blocks and role play.

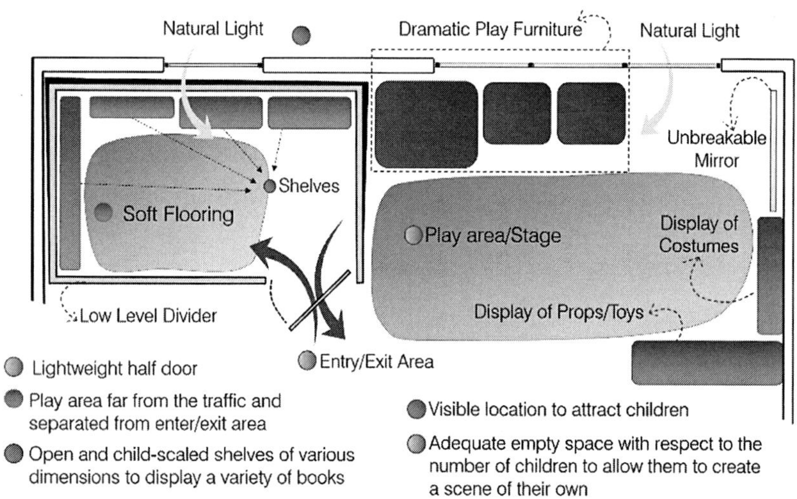

Fig. 5: Sample schematic plan of a block and dramatic play area inside a preschool classroom

In addition to play, activities such as literacy, art, math and science also contribute to the social development of pre-schoolers. The layout of spaces which host the mentioned activities should increase the amount of interaction in order to shape a social climate. In this respect, the physical environment should shape its own community and the layout which supports engagement within this context.

The design strategies listed below will enhance the social climate in preschool classrooms:

- **Size and location of sensory/discovery table in the science centre**: The size of the table should allow children to interact with materials in a group setting and the empty space around the table should be large enough for children to stand or walk around the table.
- **A cosy corner:** Providing an area with soft flooring and pillows will allow children to gather (sit and lie down based on needs) as a group and engage in activities like reading, storytelling or painting.

Providing furniture that allow children to interact in bigger groups.

- **Compact sitting:** Creating a seating layout that increases eye contact between children and teachers.

Like adults, children might not always feel social and can sometimes prefer to engage in smaller groups or even be on their own. Pre-schoolers will be able to create their own social circles if the layout of the classroom or setting provides private spots and different arrangements for different group sizes. Children's private spaces work better when they are in children's size and too small for adult use. It is also a good idea to position these spaces close to large group areas so children can leave the group if and when they want to [52]. These areas need to be visible to a certain degree to provide visual control for children. Fig. 6 shows some sample private areas that can be used in preschool settings based on existing layouts.

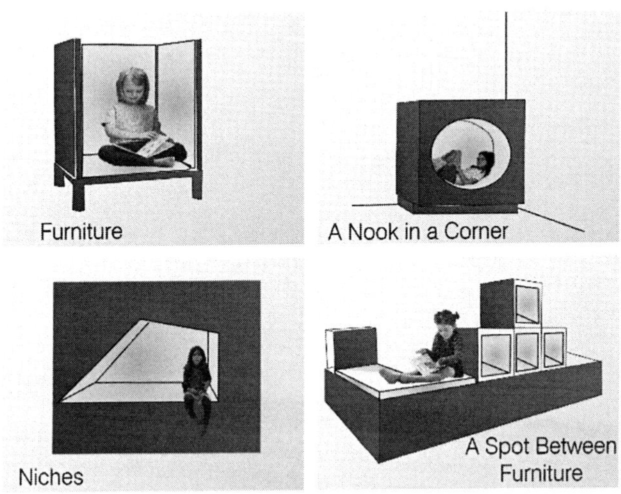

Fig. 6: Example private areas that can be used in creating private areas for pre-schoolers based on existing layout

5 Planning a Layout that Supports Collaboration

In addition to play and learning activities, including activities that engage children in the daily routine of the setting will increase their sense of place, sense of control and confidence and also their collaboration in space. Cooking is one of the activities that has recently become included as a part of early childhood education. This development is in line with further findings suggesting that nutritional behaviour in adulthood is influenced by childhood eating habits [53, 54]. Child care settings play an important role in introducing children to a healthy diet and nutritional choices [55] and have a positive impact on children's willingness to participate in preparing, cooking, serving and cleaning during and after meals and on developing independency [56].

Consistent with the discussions above, children's engagement in cooking and serving activities will contribute to their ability to make healthy and informed nutritional choices and therefore including a space where children can participate in cooking activities seems essential. Preschooler's cooking activities should not be considered as complicated professional attempts. Instead, they should involve simple tasks like tearing vegetables, mixing ingredients, peeling fruit which only require child-sized counter tops or tables surfaces [18].

The layout of the area allocated for children's cooking activities should provide enough space in order to prevent traffic or crowding. Similarly, the circulation area should set forth a logical navigation to lead children to the materials, equipment and areas that they are allowed to use and prevent them from entering or accessing unsafe areas and materials [57]. Kostelnik et al., (2006) suggest storage or display elements that are fully visible and accessible to children, allocating tables/counters with wide surfaces that encourage group interactions and defining a visible area to exhibit and display children's work as strategies that increase their social competence in kitchen [58] (Fig. 7).

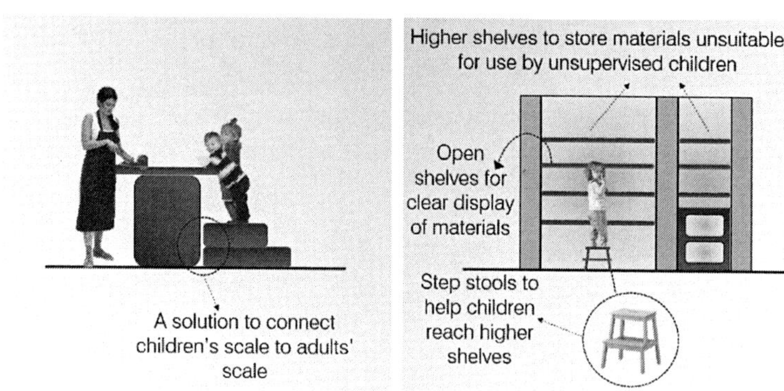

Fig. 7: Design solutions that enhance children's independent interaction in the kitchen

Engaging pre-schoolers in serving during and cleaning after meals is another attempt at increasing collaboration between children. The key strategy in providing an environment that supports children's safety and comfort during meal tasks is establishing a good circulation flow. Some of the strategies to improve the quality of circulation area for meal time tasks are as follows [59]:

- Allocating sufficient space for each child to pass, serve, pour and eat.
- Allocating sufficient space for adults to move easily in-between the tables.
- Adequate space between furniture, so children can move around without interfering with others at the table.
- Positioning tables away from the flow of sanitary needs.

Finally, using lightweight modular furniture in classrooms or common spaces will allow children and teachers to re-arrange the layout to accommodate different activities (Fig. 8). Engaging children in changing the layout of a space will increase collaboration between peers and teachers, while also improving the functionality of the space itself [60, 61].

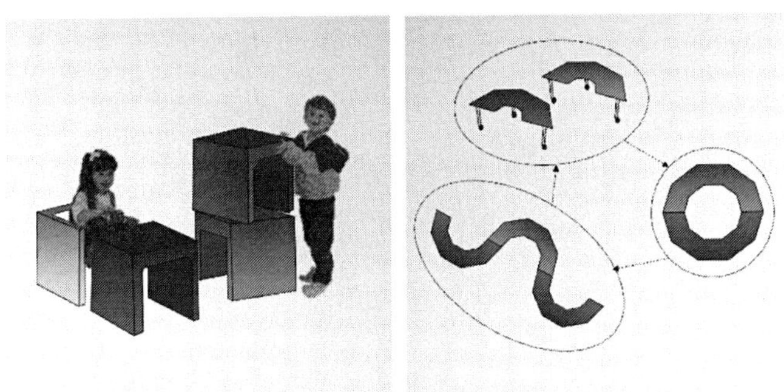

Fig. 8: Modular lightweight furniture will make it easier for children and teachers to rearrange the layout

6 Leaving Space for Imagination

Imagination is an important component of childhood. Preschool age is a stage when children start to shape their own imaginary companions and their imagination and fantasy enhance their social and cognitive development [62]. Although activities like dramatic play, science and art play a significant role in enhancing children's imagination and fantasy, the aura of a space will also stimulate children's imagination.

Studies suggest that in addition to including symbolic elements (tree, boat, train wagon and etc.), abstract forms of furniture will allow the child to shape his/her imagination and support their fantasy and imagination, because a child will perceive them based on his/her imagination [29]. While enhancing children's imagination in a space depends largely on designers' creativity, designers can get ideas directly from children's imaginary characters and build references in the physical world accordingly [12].

Artistic creation is an initial pattern in improving children's imagination. It is also argued that experiencing art is much more important than artistic output in children's art activities [63], so visibly displaying artwork, providing a variety of sitting options to allow children to attain preferred posture, providing good lighting and leaving space to display artwork will improve children's imagination during artistic activities [64, 65, 66].

Outdoor, nature and landscape elements play an important role on enhancing children's eagerness to explore and imagine [67]. However, indoors can never be as stimulating as the outdoors, therefore integrating the indoors and outdoors (Fig. 9), creating natural semi-open spaces between interiors or using indoor plants/green elements [68] can bring a sense of nature inside these settings.

Finally, studies have also noted that versatility and flexibility in a space are features that will enhance children creativity by evoking their imagination and cooperative play [69].

Fig. 9: Direct access from indoor spaces to a central garden provides visual and physical access to the outdoor and green space

7 Conclusion

Preschool age is a stage when children start to develop their sense of control and independence. The layout and characteristics of physical spaces play an important role in enhancing pre-schoolers' sense of control and independence. This chapter was set out to explore the design strategies that respond to children's independence and abilities in preschool settings. The study has sought to emphasise the main features of enabling spaces where children feel interested and in control of their environment.

The general theoretical literature on the subject of pre-schoolers' sense of control and independence point to the issue of considering children's position, movement and interaction in designing spaces for them. This matter has led to the development of various child-friendly educational methodologies. In this study, designing an entrance hall which promises a good start to children, arranging a layout that enhances children's visual and physical access to materials and activities, offering different arrangements for children's different social interest, locating zones and options that encourage collaboration and creating spaces that stimulate children's imagination and fantasy are suggested as key design strategies that will provide children with a sense of control, independence and dynamic experiences.

Finally, this study intends to emphasise the importance of designing a layout that allows free exploration and learning among pre-schoolers by referring to a quote by Dumas (1993, p. 49) that claims "it is important for kids to investigate their world, to touch, feel and manipulate the objects around them. That is how they learn" [70].

Based on our findings, there are gaps in research about specific subjects concerning the design of child-oriented preschool settings. One of these gaps is a lack of adequate studies related to pre-schoolers' patterns of interaction during free play/exploring time. The in-depth investigation of this topic will be very helpful for designers, as it will allow them to consider the findings while designing the layout of spaces.

It would also be helpful to listen to the experiences and perspectives of experienced preschool teachers on the subject of 'increasing pre-schoolers' control over physical space'. As professionals with extensive experience on observing and spending time with pre-schoolers, their suggestions and ideas on these issues will provide key indicators for researchers and designers.

References

1. Kostelny, K., *Starting Up Child Centered Spaces in Emergencies: A Field Manual.* 2008, Christian Children's Fund.
2. Feinberg, S., Kuchner, J. F., & Feldman, S., *Learning environments for young children: rethinking library spaces and services.* 1998, Chicago: American Library Association.

3. Price, D. L., & Gwin, J. F., *Pediatric nursing: an introductory text* (10 ed.). 2007, St Louis, MO: Elsevier Saunders.
4. Hertzberger, H., *Space and learning: Lessons in architecture 3* (English ed.). 2008, Rotterdam: 010.
5. Morgan, H. *Early childhood education: history, theory, and practice.* 2011, U.K.: Rowman & Littlefield.
6. Grotewell, P. R., & Burton, Y. R., *Early childhood education: Issues and developments.* 2008, New York: Nova Science Publishers.
7. MacDonell, C., *Project-based inquiry units for young children first steps to research for grades pre-K-2.* 2007, Columbus, Ohio: Linworth Pub.
8. Gamper, C., *Child-centered education.* 2009, Retrieved from http://www.newlearningculture.com/self-directed-learning.html
9. Niland, A. *The power of musical play: The value of play-based, child-centered curriculum in early childhood music education.* General Music Today, 2009. **23** (1), p. 17–21.
10. Pereira, J. K., & Adcock, S. S., *Child-centered classroom management.* Action in Teacher Education, 2011. **33** (3), p. 254–264.
11. Cochran, N. H., Nordling, W. J., & Cochran, J. L. *Child-centered play therapy: A practical guide to developing therapeutic relationships with children.* 2010, New Jersey: John Wiley & Sons.
12. Dudek, M., *Children's spaces.* 2005, Amsterdam: Elsevier.
13. Dudek, M. *A design manual schools and kindergartens: a design manual.* 2008, Basel: Birkhäuser.
14. Zane, L. *Pedagogy and space: Design inspirations for early childhood classrooms.* 2015, New York: Redleaf Press.
15. Ackermann, K., *Constructing Knowledge and Transforming the World.* A learning zone of one's own: Sharing representations and flow collaborative learning environments ed: M. Tokoro, and L. Steels, 2004, Amsterdam: IOS Press, p. 15–37.
16. Devlin, A.S., Environmental psychology and human well-being effects of built and natural settings. 2020, London: Academic Press, an imprint of Elsevier.
17. Office of the Deputy Prime Minister, *The Impact of Overcrowding on Health & Education: A Review of Evidence and Literature.* 2004, London: Office of the Deputy Prime Minister.
18. Dudek, M., Kindergarten architecture: space for the imagination, 2013, London: Routledge.
19. Essa, D. V. *Introduction to Early Childhood Education*, Student Edition (6th Ed.). 2010, Belmont: Wadsworth.

20. Community Plaything. *Pre-K Space: design for a quality classroom.* 2009, New York: Community Products LLC.
21. Hamlin, T., Grandin, T., Ratey, J., *Autism and the stress effect: a 4-step lifestyle approach to transform your child's health, happiness and vitality*, 2016, London: Jessica Kingsley Publishers.
22. Haywood, J. A. *Enduring connections creating a preschool and children's ministry.* 2007, St. Louis, Mo.: Chalice Press.
23. Weinstein, C. S., & David, T.G. (Ed.). Space for children: the built environment and child development. 1987, New York & London: Plenum Press.
24. Brown, C.R., Interior design for libraries drawing on function appeal, 2002, Chicago: American Library Association.
25. Loebach, J., *The Routledge handbook of designing public spaces for young people processes, practices and policies for youth inclusion*, 2020, New York: Routledge.
26. Hendricks, B.E., *Designing for play*, 2018, Abingdon, Oxon: Routledge, Taylor & Francis Group.
27. Schroeder-Yu, G., *Documentation: Ideas and Applications from the Reggio Emilia Approach*, Teaching Artist Journal, 2008. **6**:2, p. 126–134, DOI: 10.1080/15411790801910735
28. Mayfield, M., *Early childhood education and care in Canada: contexts, dimensions, and issues*, 2001, USA: Prentice Hall.
29. Feinberg, S., & Keller, J. *Designing space for children and teens in libraries and public places.* 2010, Chicago: American Library Association.
30. Leppo, M., & Davis, D., *Movement Opens Pathways to Learning*, Strategies, 2005. **19**(2), p. 11–16, DOI: 10.1080/08924562.2005.10591177
31. Wana, J., *How to choose the best preschool for your child: the ultimate guide to finding, getting into, and preparing for nursery school.* 2010, USA: Sourcebooks.
32. Hockenberry, M.J., & Wilson, D., Wong's Nursing Care of Infants and Children Multimedia Enhanced Version, Place of publication not identified, 2013, Missouri: Elsevier Mosby.
33. Kyle, T., & Carman, S., *Essentials of pediatric nursing*, 2020, Philadelphia: Wolters Kluwer.
34. Sleegers, P.J.C., Moolenaar, N.M., Galetzka, M., Pruyn, A., Sarroukh B.E., & van der Zande, B., *Lighting affects students' concentration positively: Findings from three Dutch studies*, Lighting Research and Technology, 2013. **45**: 159, p. 159–175, DOI: 10.1177/1477153512446099
35. Wessolowski, N., Koenig, H., Schulte-Markwort, M.l., ^ Barkmann, C., *The Effect of Variable Light on the Fidgetiness and Social Behavior of Pupils in School*, Journal of Environmental Psychology, 2014. 39, 101–108. DOI: 10.1016/j.jenvp.2014.05.001

36. Karlen, M., *Space planning basics.* 2009, Hoboken: John Wiley & Sons.
37. Mayesky, M., *Creative activities for young children.* (10th Ed.). 2012, Belmont: Wadsworth Cengage Learning.
38. Gestwicki, C., *Developmentally appropriate practice: curriculum and development in early education* (5th Ed.). 2013, Belmont, Calif.: Wadsworth.
39. Gordon, A. M., & Browne, K. W., *Beginnings & beyond: Foundations in early childhood education.* (9th Ed.). 2013, Belmont: Cengage Learning.
40. Lillard, A. S., *Montessori: The Science Behind the Genius.* 2005, Oxford University Press.
41. Howes, H. C., *Children's Social Development Within the Socialization Context of Child Care and Early Education.* The Wiley-Blackwell Handbook of Childhood Social Development. Ed: Smith, P.K., & Hart, H.C., 2011, Malden, MA: Blackwell Publishing, p. 246–262.
42. Underwood, M. K., & Rosen, L. H., *Social Development: Relationships in Infancy, Childhood, and Adolescence.* 2011, New York, NY: The Guilford Press.
43. Fuhrer, A. M. U., & Hartig, T., *Effects of classroom seating arrangements on children's question-asking.* Learning Environments Research, (2000). **2**, p.249–263.
44. Christie, J. F., & Johnson. E. P., *The Role of Play in Social-Intellectual Development*, Psychology of Education: Social Behaviour and the School Peer Group, Ed: Smith, P.K. and Pellegrini, A.D., 2000, Routledge Falmer, p. 441–468.
45. Power, T. G., Social Play. In The Wiley-Blackwell Handbook of Childhood Social Development. P. Smith & C. Hart (Eds.), 2011, Malden, MA: Blackwell Publishing, p. 455–471.
46. Lytle, D. E. *Play and Educational Theory and Practice.* 2003. London: Praeger.
47. Izadpanah, S., *A Post Occupancy Quality Evaluation Model for Preschool Settings' Interior Spaces (PSIS)*, 2016. PhD Thesis, Eastern Mediterranean University, Gazimağusa.
48. Hereford, N. J., & Schall, J., *Learning through play: A practical guide for teaching young children.* 1991, New York: Scholastic, Early Childhood Division.
49. Riley, M. T., *Environments that Make a Difference.* 1991, Washington: ERIC.
50. Nielsen, D. M., *Block center. In Teaching Young Children, Preschool-K: A Guide to Planning Your Curriculum, Teaching Through Learning Centers, and Just About Everything Else (2nd Ed.).* 2006, Thousand Oaks, CA: Corwin Press.
51. Kearns, K., Birth to big school (3rd Ed.). 2007, New Zealand: Cengage Learning.

52. Assistant Secretary of Defense. *Creating Environments for Preschoolers.* 1982, Washington: Assistant Secretary of Defense, Manpower, Reserve Affairs, and Logistics,
53. Cashdan, E. *A sensitive period for learning about food.* Human Nature. 1994. 5(3), p. 279–291.
54. Skinner, J. D., Carruth, B. R., Bounds, W., Ziegler, P. & Reidy K., Do Food-Related Experiences in the First 2 Years of Life Predict Dietary Variety in School-Aged Children? Journal of Nutrition Education and Behavior, 2002. **34**: 6, p. 310–315., doi:10.1016/s1499-4046(06)60113-9.
55. Briley, M., *Nutrition and the Child-Care Setting. J Am Diet Assoc.* 2011. **111**(9): p. 1298–1300.
56. NJ Department of Education., *Preschool Guidance & Materials: Planning for Meal and Snack Times in Preschool Programs.* 2010, Retrieved from http://www.nj.gov/education/ece/psguide/meal.htm.
57. Salant, K., *Designing a Kid-Friendly Kitchen.* 2011, Retrieved from http://www.katherinesalant.com/columns/column276.html.
58. Kostelnik, M., Whiren, A., Soderman, A., Rupiper, M., & Gregory, K. *Guiding children's social development: Theory to practice (5th Ed.).* 2006. Clifton Park, NY: Thomson Delmar Learning.
59. Fletcher, J., Branen, L., & Price, E., *Building Mealtime Environments and Relationships: An Inventory for Feeding Young Children in Group Settings.* 2005, Moscow: University of Idaho, College of Agricultural and Life Sciences.
60. Puckett, M. B., & Diffily, D. *Teaching young children: An introduction to the early childhood profession.* 2004, Clifton Park: Cengage Learning.
61. Stankovic, D., Milojkovic, A., & Tanic, M., *Physical environment factors and their impact on the cognitive process and social behavior of children in the preschool facilities.* Facta Universitatis - Series: Architecture and Civil Engineering Facta Univ., Arch. Civ. Engl., 2006. **4**:1, p. 51–57.
62. Woolley, J.D., & Gilpin, A.T., *Development of Imagination and Fantasy.* Encyclopedia of infant and early childhood development. eds: M. M. Haith & J. B. Benson, 2020. Amsterdam: Elsevier, Academic Press, p. 430–437.
63. Schwake, S., *Art lab for little kids: 52 playful projects for preschoolers!*, 2013. Beverly: Quarry Books.
64. Libby, W. M. L., *Using art to make art: Creative activities using masterpieces.* 2000. Albany, N.Y.: Delmar.
65. Beaty, J., *Preschool appropriate practices: Environment, curriculum, and development.* (4th ed.). 2014. Belmont: Cengage Learning.
66. Koster, J., *Growing artists: Teaching the arts to young children (6th ed.).* 2015, Stamford: Thomson Delmar Learning.

67. Zamani, Z., *The woods is a more free space for children to be creative; their imagination kind of sparks out there: exploring young children's cognitive play opportunities in natural, manufactured and mixed outdoor preschool zones,* Journal of Adventure Education & Outdoor Learning, 2016. **16**: 2, p. 1–18.
68. Rastegar, H., & Charehjoo, F., *Designing learning Spaces with the Approach of Improving Children's Creativity and Emotional Intelligence.* International Journal of Business and Management Invention, 2017. **6**:2, p. 37–43.
69. Shafaei, M., *Design patterns of kids' spaces, focusing on enhancing their creativity,* Int. J. Architect. Eng. Urban Plan, 2017. **27**(1): p. 19–27.
70. Dumas, Lynne S., *Best Places for Pres*choolers. Working Mother, July 1993, p. 48–51.